MW01055771

Theory and Applications of Digital Speech Processing

First Edition

Lawrence R. Rabiner

Rutgers University and the University of California
at Santa Barbara

Ronald W. Schafer

Hewlett-Packard Laboratories

Upper Saddle River Boston Columbus San Francisco New York
Indianapolis London Toronto Sydney Singapore Tokyo Montreal
Dubai Madrid Hong Kong Mexico City Munich Paris Amsterdam Cape Town

Vice President and Editorial Director, ECS: *Marcia J. Horton*
Senior Editor: *Andrew Gilfillan*
Editorial Assistant: *William Opaluch*
Vice President, Production: *Vince O'Brien*
Marketing Manager: *Tim Galligan*
Marketing Assistant: *Mack Patterson*
Senior Managing Editor: *Scott Disanno*
Production Project Manager: *Clare Romeo*
Senior Operations Specialist: *Alan Fischer*
Operations Specialist: *Lisa McDowell*
Art Director: *Kristine Carney*
Cover Designer: *Suzanne Duda*
Manager, Rights and Permissions: *Zina Arabia*
Manager, Visual Research: *Beth Brenzel*
Image Permission Coordinator: *Debbie Latronica*
Composition: *TexTech International*
Full-Service Project Management: *TexTech International*
Printer/Binder: *Courier Westford*
Typeface: *10/12 TimesTen*

Credits and acknowledgments borrowed from other sources and reproduced, with permission, in this textbook appear on appropriate page within text.

Library of Congress Cataloging in Publication Data

Rabiner, Lawrence R.
Theory and applications of digital speech processing / Lawrence Rabiner, Ronald Schafer.
p. cm.
Includes bibliographical references and index.
ISBN 0-13-603428-4 (alk. paper)
1. Speech processing systems. I. Schafer, Ronald W., 1938- II. Title.
TK7882.S65R32 2010
006.4′54–dc22 2009048536

Prentice Hall
is an imprint of

www.pearsonhighered.com

10 9 8 7 6 5 4 3 2 1

ISBN-13: 978-0-13-603428-5
ISBN-10: 0-13-603428-4

To our children and grandchildren

Sheri, Wendi, Joni
Tess, Nate, Max, Molly, Chloe

and

Bill, Kate, John
Henry, Hannah, Sophia

Contents

Preface

Speech signal processing has been a dynamic and constantly developing field for more than 70 years. The earliest speech processing systems were analog systems. They included, for example, the Voder (voice demonstration recorder) for synthesizing speech by manual controls, developed by Homer Dudley and colleagues at Bell Labs in the 1930s and demonstrated at the 1939 New York World's Fair; the channel vocoder or voice coder, also developed in the 1930s by Homer Dudley at Bell Labs; the sound spectrograph, a system for displaying time-varying speech patterns in time and frequency, developed by Koenig and his colleagues in the 1940s at Bell Labs; and early systems for recognizing spoken words, developed in research labs throughout the world in the 1950s.

Speech processing was the driving force for many of the early developments in the broader field of digital signal processing (DSP) whose roots began to take hold in the 1960s. During this period, pioneering researchers such as Ben Gold and Charlie Rader at MIT Lincoln Labs, and Jim Flanagan, Roger Golden, and Jim Kaiser at Bell Labs began to study methods for the design and application of digital filters for use in simulations of speech processing systems. With the technical disclosure of the fast Fourier transform (FFT) algorithm by Jim Cooley and John Tukey in 1965, and its subsequent widespread application in the areas of fast convolution and spectral analysis, the shackles and limitations of analog technology rapidly fell away and the field of digital speech processing emerged and took on a clear identity.

The authors of the present book (LRR and RWS) worked closely together at Bell Labs during the period from 1968 to 1974, when many fundamental advances in DSP occurred. When RWS left Bell Labs in 1975 for an academic position at Georgia Tech, the field of digital speech processing had developed so much that we decided that it was an appropriate time to write a textbook on methods and systems for digital processing of speech signals. We were confident that the theory of digital speech processing had developed sufficiently (by 1976) that a carefully written textbook would effectively serve as both a textbook for teaching the fundamentals of digital speech processing, and as a reference textbook for practical system design of speech processing systems for the foreseeable future. The resulting textbook, entitled "Digital Processing of Speech Signals" was published by Prentice-Hall Inc. in 1978. In his new position in academia, RWS was able to create one of the first graduate courses in digital speech processing based on this textbook, while LRR continued basic digital speech processing research at Bell Labs. (LRR had a 40-year career with AT&T Bell Labs and AT&T Labs Research after which he also joined academia, jointly teaching at both Rutgers University and the University of California in Santa Barbara, in 2002. RWS had a 30-year career at Georgia Tech and he joined Hewlett Packard Research Labs in 2004.)

The goals of the 1978 textbook were to present the fundamental science of speech together with a range of digital speech processing methods that could be used to create powerful speech signal processing systems. To a large extent, our initial goals were met. The original textbook has served as intended for more than 30 years, and, to our delight, it is still widely used today in teaching undergraduate and graduate courses in digital speech processing. However, as we have learned from our personal experiences in teaching speech processing courses over the past two decades, while its fundamentals remain sound, much of the material in the original volume is greatly out-of-touch with modern speech signal processing systems, and entire areas of current interest are completely missing. The present book is our attempt to correct these weaknesses.

In approaching the daunting task of unifying current theory and practice of digital speech processing, we found that much of the original book remained true and relevant, so we had a good starting point for this new textbook. Furthermore, we learned from both practical experience in research in speech processing and from our teaching experiences that the organization of the material in the 1978 volume, although basically sound, was just not suitable for understanding modern speech processing systems. With these weaknesses in mind, we adopted a new framework for presenting the material in this new book, with two major changes to the original framework. First we embraced the concept of the existence of a hierarchy of knowledge about the subject of digital speech processing. This hierarchy has a base level of fundamental scientific and engineering speech knowledge. The second level of the hierarchy focuses on representations of the speech signal. The original book focused primarily on the bottom two levels but even then some key topics were missing from the presentations. At the third level of the hierarchy are algorithms for manipulating, processing, and extracting information from the speech signal that are based on technology and science from the two lower layers. At the top of the hierarchy (i.e., the fourth level) are the applications of the speech processing algorithms, along with techniques for handling problems in speech communication systems.

We have made every attempt to follow this new hierarchy (called the speech stack in Chapter 1) in presenting the material in this new book. To that end, in Chapters 2–5, we concentrate on the process of building a firm foundation at the bottom layer of the stack, including topics such as the basics of speech production and perception, a review of DSP fundamentals, and discussions of acoustic–phonetics, linguistics, speech perception, and sound propagation in the human vocal tract. In Chapters 6–9 we develop an understanding of how various digital speech (short-time) representations arise from basic signal processing principles (forming the second layer of the speech stack). In Chapter 10 we show how to design speech algorithms that are both reliable and robust for estimating a range of speech parameters of interest (forming the basis for the third layer of the speech stack). Finally, in Chapters 11–14 we show how our knowledge from the lower layers of the speech stack can be used to enable design and implementation of a range of speech applications (forming the fourth layer of the speech stack).

The second major change in structure and presentation of the new book is in the realization that, for maximal impact in teaching, we had to present the material with an equal focus on three areas of learning of new ideas, namely theory, concept, and implementation. Thus for each fundamental concept introduced in this book, the theory is explained in terms of well-understood DSP concepts; similarly the understanding of each new concept is enhanced by providing simple interpretations of the mathematics and by illustrating the basic concepts using carefully-explained examples and associated graphics; and finally, the implementation of new concepts based on understanding of fundamentals is taught by reference to MATLAB® code for specific speech processing operations (often included within the individual chapters), along with extensive and thoroughly documented MATLAB exercises in the (expanded) homework problem section of each chapter. We also provide a course website with all the material needed to solve the various MATLAB exercises, including specialized MATLAB code, access to simple databases, access to a range of speech files, etc. Finally we provide several audio demonstrations of the results of a range of speech processing systems. In this manner, the reader can get a sense of the resulting quality of the processed speech for a range of operations on the speech signal.

More specifically, the organization of this new book is as follows. Chapter 1 provides a broad brush introduction to the area of speech processing, and gives a brief discussion of application areas that are directly related to topics discussed throughout the book. Chapter 2 provides a brief review of DSP with emphasis on a few key concepts that are pervasive in speech processing systems:

1. conversion from the time domain to the frequency domain (via discrete-time Fourier transform methods)
2. understanding the impact of sampling in the frequency domain (i.e., aliasing in the time domain)
3. understanding the impact of sampling (both downsampling and upsampling) in the time domain, and the resulting aliasing or imaging in the frequency domain.

Following the review of the basics of DSP technology, we move on to a discussion of the fundamentals of speech production and perception in Chapters 3 and 4. These chapters, together with Chapters 2 and 5, comprise the bottom layer of the speech stack. We begin with a discussion of the acoustics of speech production. We derive a series of acoustic–phonetic models for the various sounds of speech and show how linguistics and pragmatics interact with the acoustics of speech production to create the speech signal along with its linguistic interpretation. We complete our discussion of the fundamental processes underlying speech communication with an analysis of speech perception processes, beginning with a discussion of how speech is processed in the ear, and ending with a discussion of methods for the transduction of sound to neural signals in the auditory neural pathways leading to the brain. We briefly discuss several possible ways of embedding knowledge of speech perception into an auditory model that can be utilized in speech processing applications. Next, in Chapter 5, we complete our discussion of fundamentals with a discussion of issues of sound propagation in the human vocal tract. We show that uniform lossless tube approximations to the vocal tract have resonance structures elucidating the resonant (formant) frequencies of speech. We show how the transmission properties of a series of concatenated tubes can be represented by an appropriate "terminal-analog" digital system with a specified excitation function and a specified system response corresponding to the differing tube lengths and areas, along with a specified radiation characteristic for transmission of sound at the lips.

We devote the next four chapters of the book to digital speech signal representations (the second layer in the speech stack), with separate chapters on each of the four major representations. We begin, in Chapter 6, with the temporal model of speech production and show how we can estimate basic time-varying properties of the speech model from simple time-domain measurements. In Chapter 7 we show how the concept of short-time Fourier analysis can be applied to speech signals in a simple and consistent manner such that a completely transparent analysis/synthesis system can be realized. We show that there are two interpretations of short-time Fourier analysis/synthesis systems and that both can be used in a wide range of applications, depending on the nature of the information that is required for further processing. In Chapter 8 we describe a homomorphic (cepstrum) representation of speech where we use the property that a convolutional signal (such as speech) can be transformed into a set of additive components. With the understanding that a speech signal can be well represented as the convolution of an excitation signal and a vocal tract system, we see that the speech signal is well suited to such an analysis. Finally, Chapter 9 deals with the theory and practice of linear predictive analysis, which is a representation of speech that models the current speech sample as a linear combination of the previous p speech samples, and finds the coefficients of the best linear predictor (with minimized mean-squared error) that optimally matches the speech signal over a given time duration.

Chapter 10, which represents the third layer in the speech stack, deals with using the signal processing representations and fundamental knowledge of the speech signal, presented in the preceding chapters, as a basis for measuring or estimating properties and attributes of the speech signal. Here we show how the measurements of short-time (log) energy, short-time zero crossing rates, and short-time autocorrelation can be used to estimate basic speech attributes such as whether the signal section under analysis is speech or silence (background signal), whether a segment of speech represents voiced or unvoiced speech, the pitch period (or pitch frequency) for segments of voiced speech signals, the formants (vocal tract resonances) for speech segments, etc. For many of the speech attributes, we show how each of the four speech representations can be used as the basis of an effective and efficient algorithm for estimating the desired attributes. Similarly we show how to estimate formants based on measurements from two of the four speech representations.

Chapters 11–14, representing the top layer in the speech stack (speech applications), deal with several of the major applications of speech and audio signal processing technology. These applications are the payoffs of understanding speech and audio technology, and they represent decades of research on how best to integrate various speech representations and measurements to give the best performance for each speech application. Our goal in discussing speech applications is to give the reader a sense of how such applications are built, and how well they perform at various bit rates and for various task scenarios. In particular, Chapter 11 deals with speech coding systems (both open-loop and closed-loop systems); Chapter 12 deals with audio coding systems based on minimizing the perceptual error of coding using well-understood perceptual masking criteria; Chapter 13 deals with building text-to-speech synthesis systems suitable for use in a speech dialog system; and Chapter 14 deals with speech recognition and natural language understanding systems and their application to a range of task-oriented scenarios. Our goal in these chapters is to provide up-to-date examples but not to be exhaustive in our coverage. Entire textbooks have been written on each of these application areas.

The material in this book can be taught in a one-semester course in speech processing, assuming that students have taken a basic course on DSP. In our own teaching, we emphasize the material in Chapters 3–11, with selected portions of the material in the remaining chapters being taught to give students a sense of the issues in audio coding, speech synthesis, and speech recognition systems. To aid in the teaching process, each chapter contains a set of representative homework problems that are intended to reinforce the ideas discussed in each chapter. Successful completion of a reasonable percentage of these homework problems is essential for a good understanding of the mathematical and theoretical concepts of speech processing, as discussed earlier. However, as the reader will see, much of speech processing is, by its very nature, empirical. Hence we have chosen to include a series of MATLAB exercises in each chapter (either within the text or as part of the set of homework problems) so as to reinforce the student's understanding of the basic concepts of speech processing. We have also provided on the course website (http://www.pearsonhighered/Rabiner.com) which will be updated with new material from time to time—the required speech files, databases, and MATLAB code required to solve the MATLAB exercises, along with a series of demonstrations of a range of speech processing concepts.

ACKNOWLEDGEMENTS

Throughout our careers in speech processing, we have both been extremely fortunate in our affiliations with outstanding research and academic institutions that have provided stimulating research environments and also encouraged the sharing of knowledge. For LRR, these have been Bell Laboratories, AT&T Laboratories, Rutgers University, and the University of California at Santa Barbara, and for RWS, it was Bell Laboratories, Georgia Tech ECE, and Hewlett-Packard Laboratories. Without the support and encouragement of our colleagues and the enlightened managers of these fine institutions, this book would not exist.

Many people have had a significant impact, both directly and indirectly, on the material presented in this book, but our biggest debt of gratitude is to Dr. James L. Flanagan, who served as both supervisor and mentor for both of us at various points in our careers. Jim has provided us with an inspiring model of how to conduct research and how to present the results of research in a clear and meaningful way. His influence on both this book, and our respective careers, has been profound.

Other people with whom we have had the good fortune to collaborate and learn from include our mentors Prof. Alan Oppenheim and Prof. Kenneth Stevens of MIT and our colleagues Tom Barnwell, Mark Clements, Chin Lee, Fred Juang, Jim McClellan and Russ Mersereau, all professors at Georgia Tech. All of these individuals have been both our teachers and our colleagues, and we are grateful to them for their wisdom and their guidance over many years.

Colleagues who have been involved directly with the preparation of this book include Dr. Bishnu Atal, Prof. Victor Zue, Prof. Jim Glass, and Prof. Peter Noll, each of whom provided insight and technical results that strongly impacted a range of the material presented in this book. Other individuals who provided permission to use one or more figures and tables from their publications include Alex Acero, Joe Campbell, Raymond Chen, Eric Cosatto, Rich Cox, Ron Crochiere, Thierry Dutoit, Oded Ghitza, Al Gorin, Hynek Hermansky, Nelson Kiang, Rich Lippman, Dick Lyon, Marion Macchi, John Makhoul, Mehryar Mohri, Joern Ostermann, David Pallett, Roberto Pieraccini, Tom Quatieri, Juergen Schroeter, Stephanie Seneff, Malcolm Slaney, Peter Vary, and Vishu Viswanathan.

We would also like to acknowledge the support provided by Lucent-Alcatel, IEEE, the Acoustical Society of America, and the House-Ear Institute for granting permission to use several figures and tables reprinted from published or archival material.

Also, we wish to acknowledge the individuals at Pearson Prentice Hall who helped make this book come to fruition. These include Andrew Gilfillan, acquisitions editor, Clare Romeo, production editor, and William Opaluch, editorial assistant. I would also like to acknowledge Maheswari PonSaravanan of TexTech International who was the lead individual in the copy-editing and page proof process.

Finally, we thank our spouses Suzanne and Dorothy for their love, patience, and support during the seemingly never ending task of writing of this book.

Lawrence R. Rabiner and Ronald W. Schafer

CHAPTER 1

Introduction to Digital Speech Processing

This book is about subjects that are as old as the study of human language and as new as the latest computer chip. Since before the time of Alexander Graham Bell's revolutionary invention of the telephone, engineers and scientists have studied the processes of speech communication with the goal of creating more efficient and effective systems for human-to-human and (more recently) human-to-machine communication. In the 1960s, digital signal processing (DSP) began to assume a central role in speech communication studies, and today DSP technology enables a myriad of applications of the knowledge that has been gained over decades of research. In the interim, concomitant advances in integrated circuit technology, DSP algorithms, and computer architecture have aligned to create a technological environment with virtually limitless opportunities for innovation in speech processing as well as other fields such as image and video processing, radar and sonar, medical diagnosis systems, and many areas of consumer electronics. It is important to note that DSP and speech processing have evolved hand-in-hand over the past 50 years or more, with speech applications stimulating many advances in DSP theory and algorithm research and those advances finding ready applications in speech communication research and technology. It is reasonable to expect that this symbiotic relationship will continue into the indefinite future.

In order to fully appreciate a technology such as digital speech processing, three levels of understanding must be reached: the theoretical level (theory), the conceptual level (concepts), and the working level (practice). This technology pyramid is depicted in Figure 1.1.[1] In the case of speech technology, the theoretical level is comprised of the acoustic theory of speech production, the basic mathematics of speech signal representations, the derivations of various properties of speech associated with each representation, and the basic signal processing mathematics that relates the speech signal to the real world via sampling, aliasing, filtering, etc. The conceptual level is concerned with how speech processing theory is applied in order to make various speech measurements and to estimate and quantify various attributes of the speech signal. Finally, for a technology to realize its full potential, it is essential to be able to

[1] We use the term "technology pyramid" (rather than "technology triangle") to emphasize the fact that each layer has breadth and depth and supports all higher layers.

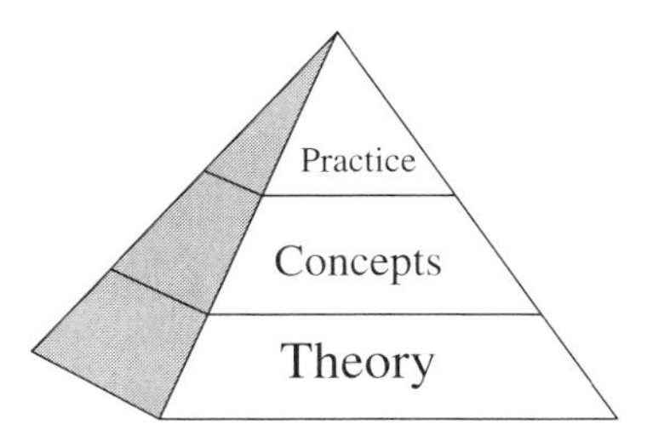

FIGURE 1.1
The technology pyramid—theory, concepts, and practice.

convert theory and conceptual understanding to practice; that is to be able to implement speech processing systems that solve specific application problems. This process involves knowledge of the constraints and goals of the application, engineering trade-offs and judgments, and the ability to produce implementations in working computer code (most often as a program written in MATLAB®, C, or C++) or as specialized code running on real-time signal processing chips [e.g., application specific integrated circuits (ASICs), field programmable gate arrays (FPGAs), or DSP chips].

The continuous improvement of digital implementation technology capabilities, in turn, opens up new application areas that once were considered impossible or impractical, and this is certainly true in digital speech processing. Therefore, our approach in this book is to emphasize the first two levels, but always to keep in mind the ultimate technology payoff at the third (implementation) level of the technology pyramid. The fundamentals and basic concepts of the field of digital speech processing will certainly continue to evolve and expand, but what has been learned in the past 50 years will continue to be the basis for the applications that we will see in the next decades. Therefore, for every topic in speech processing that is covered in this book, we will endeavor to provide as much understanding as possible at the theory and concepts level, and we will provide a set of exercises that enable the reader to gain expertise at the practice level (usually via MATLAB exercises included within the problems at the end of each chapter).

In the remainder of this introductory chapter we begin with an introduction to the speech communication process and the speech signal and conclude with a survey of the important application areas for digital speech processing techniques. The remainder of the text is designed to provide a solid grounding in the fundamentals and to highlight the central role of DSP techniques in modern speech communication research and applications. Our goal is to present a comprehensive overview of digital speech processing that ranges from the basic nature of the speech signal, through a variety of methods of representing speech in digital form, to overviews of applications in voice communication and automatic synthesis and recognition of speech. In the process we hope to provide answers to such questions as:

- What is the nature of the speech signal?
- How do DSP techniques play a role in learning about the speech signal?

- What are the basic digital representations of speech signals, and how are they used in algorithms for speech processing?
- What are the important applications that are enabled by digital speech processing methods?

We begin our study by taking a look at the speech signal and getting a feel of its nature and properties.

1.1 THE SPEECH SIGNAL

The fundamental purpose of speech is human communication; i.e., the transmission of messages between a speaker and a listener. According to Shannon's information theory [364], a message represented as a sequence of discrete symbols can be quantified by its *information content* in bits, where the rate of transmission of information is measured in bits per second (bps). In speech production, as well as in many human-engineered electronic communication systems, the information to be transmitted is encoded in the form of a continuously varying (analog) waveform that can be transmitted, recorded (stored), manipulated, and ultimately decoded by a human listener. The fundamental analog form of the message is an acoustic waveform that we call the *speech signal*. Speech signals, such as the one illustrated in Figure 1.2, can be converted to an electrical waveform by a microphone, further manipulated by both analog and digital signal processing methods, and then converted back to acoustic form by a loudspeaker, a telephone handset, or headphone, as desired. This form of speech processing is, of course, the basis for Bell's telephone invention as well as today's multitude of devices for recording, transmitting, and manipulating speech and audio signals. In Bell's own words [47],

> Watson, if I can get a mechanism which will make a current of electricity vary its intensity as the air varies in density when sound is passing through it, I can telegraph any sound, even the sound of speech.

Although Bell made his great invention without knowing about information theory, the principles of information theory have assumed great importance in the design of sophisticated modern digital communications systems. Therefore, even though our main focus will be mostly on the speech waveform and its representation in the form of parametric models, it is nevertheless useful to begin with a discussion of the information that is encoded in the speech waveform.

Figure 1.3 shows a pictorial representation of the complete process of producing and perceiving speech—from the formulation of a message in the brain of a speaker, to the creation of the speech signal, and finally to the understanding of the message by a listener. In their classic introduction to speech science, Denes and Pinson appropriately referred to this process as the "speech chain" [88]. A more refined block diagram representation of the speech chain is shown in Figure 1.4. The process starts in the upper left as a message represented somehow in the brain of the speaker. The message information can be thought of as having a number of different representations during the process of speech production (the upper path in Figure 1.4). For example

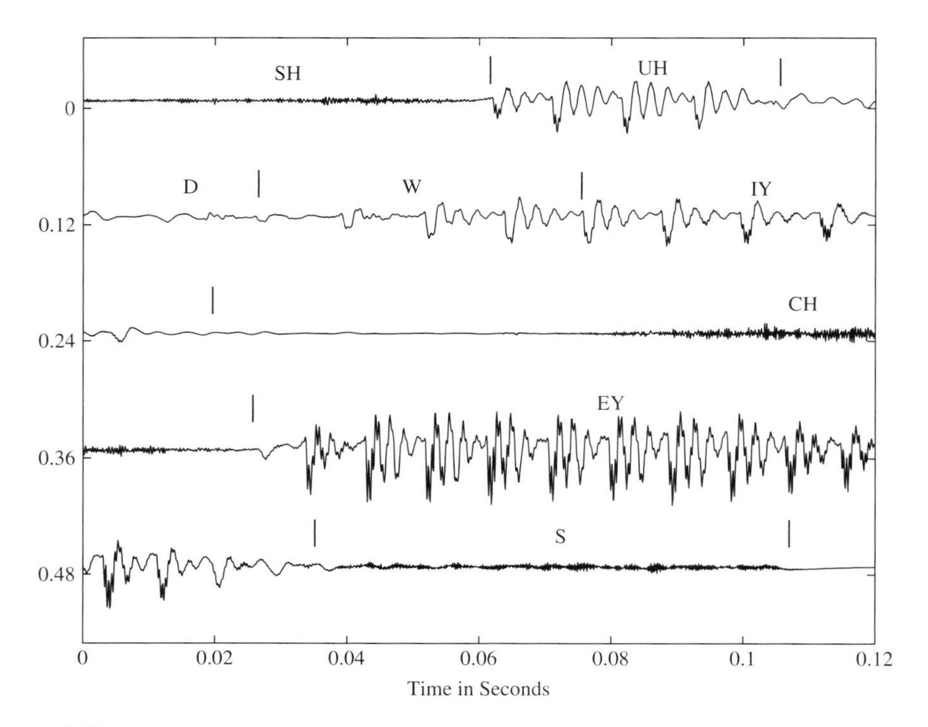

FIGURE 1.2
A speech waveform with phonetic labels for the message represented by the text "should we chase."

the message could be represented initially as English text. In order to "speak" the message, the speaker implicitly converts the text into a symbolic representation of the sequence of sounds corresponding to the spoken version of the text. This step, called the language code generator in Figure 1.4, converts text symbols to phonetic symbols (along with stress and durational information) that describe the basic sounds of a spoken version of the message and the manner (i.e., the speed and emphasis) in which the sounds are intended to be produced. As an example, the segments of the waveform of Figure 1.2 are labeled with phonetic symbols using a computer-keyboard-friendly code called ARPAbet.[2] Thus, the text "should we chase" is represented phonetically (in ARPAbet symbols) as [SH UH D - W IY - CH EY S]. (See Chapter 3 for more discussions of phonetic transcription.) The third step in the speech production process is the conversion to "neuro-muscular controls"; i.e., the set of control signals that direct the neuro-muscular system to move the speech articulators, namely the tongue, lips, teeth, jaw, and velum, in a manner that is consistent with the sounds of the desired spoken

[2]The International Phonetic Association (IPA) provides a set of rules for phonetic transcription using an equivalent set of specialized symbols. The ARPAbet code does not require special fonts and is thus more convenient for computer applications.

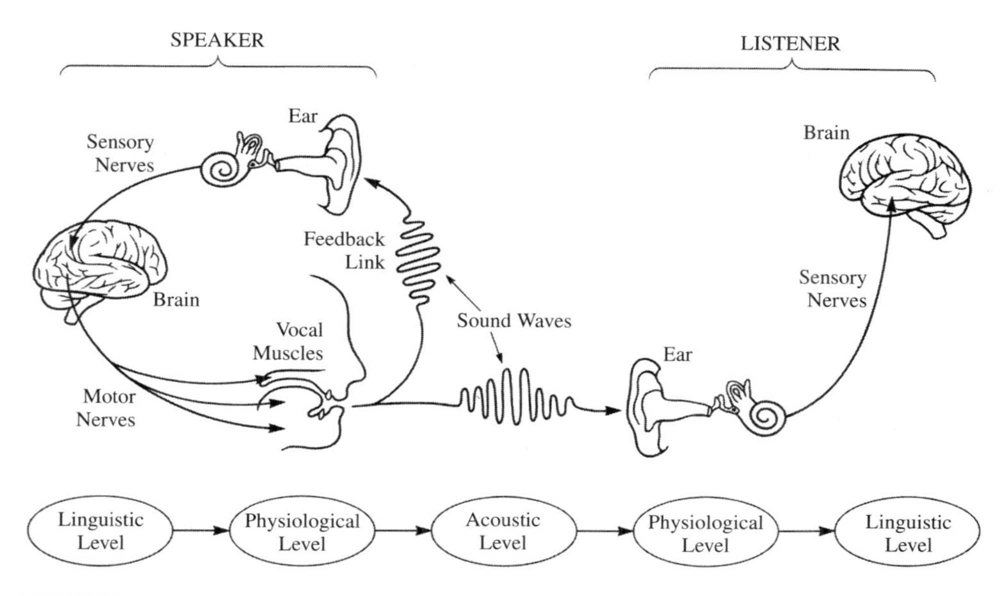

FIGURE 1.3
The speech chain: from message, to speech signal, to understanding. (After Denes and Pinson [88].)

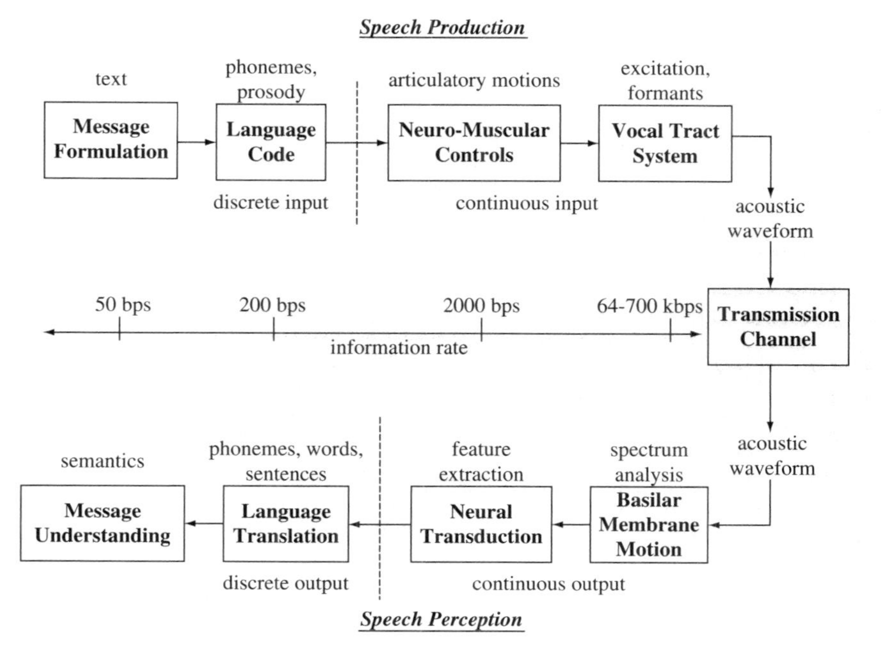

FIGURE 1.4
Block diagram representation of the speech chain.

message and with the desired degree of emphasis. The end result of the neuro-muscular controls step is a set of articulatory motions (continuous control) that cause the vocal tract articulators to move in a prescribed manner in order to create the desired sounds. Finally, the last step in the speech production process is the "vocal tract system," which creates the physical sound sources and appropriate time-varying vocal tract shapes so as to produce an acoustic waveform such as the one shown in Figure 1.2. In this way, the information in the desired message is encoded into the speech signal.

To determine the rate of information flow during speech production, assume that there are about 32 symbols (letters) in the written language. (In English there are 26 letters, but if we include simple punctuation and spaces, we get a count closer to $32 = 2^5$ symbols.) The normal average rate of speaking is about 15 symbols per second. Hence, assuming independent letters as a simple first-order approximation, the base information rate of the text message encoded as speech is about 75 bps (5 bits per symbol times 15 symbols per second). However, the actual rate will vary with speaking rate. For the example of Figure 1.2, the text representation has 15 letters (including spaces) and the corresponding speech utterance has a duration of 0.6 seconds, giving a higher estimate of $15 \times 5/0.6 = 125$ bps. At the second stage of the process, where the text representation is converted into basic sound units called phonemes along with prosody (e.g., pitch and stress) markers, the information rate can easily increase to over 200 bps. The ARBAbet phonetic symbol set used to label the speech sounds in Figure 1.2 contains approximately $64 = 2^6$ symbols, or about 6 bits/phoneme (again a rough approximation assuming independence of phonemes). In Figure 1.2, there are eight phonemes in approximately 0.6 seconds. This leads to an estimate of $8 \times 6/0.6 = 80$ bps. Additional information required to describe prosodic features of the signal (e.g., duration, pitch, loudness) could easily add 100 bps to the total information rate for the text message encoded as a speech signal.

The information representations for the first two stages in the speech chain are discrete, so we can readily estimate the rate of information flow with some simple assumptions. For the next stage in the speech production part of the speech chain, the representation becomes continuous (in the form of neuro-muscular control signals for articulatory motion). If they could be measured, we could estimate the spectral bandwidth of these control signals and appropriately sample and quantize these signals to obtain equivalent digital signals for which the data rate could be estimated. The articulators move relatively slowly compared to the time variation of the resulting acoustic waveform. Estimates of bandwidth and required signal representation accuracy suggest that the total data rate of the sampled articulatory control signals is about 2000 bps [105]. Thus, the original text message is represented by a set of continuously varying signals whose digital representation requires a much higher data rate than the information rate that we estimated for transmission of the message as a discrete textual signal.[3] Finally, as we will see later, the data rate of the digitized speech waveform at the end of the speech production part of the speech chain can

[3]Note that we introduce the term "data rate" for digital representations to distinguish from the inherent information content of the message represented by the speech signal.

be anywhere from 64,000 to more than 700,000 bps. We arrive at such numbers by examining the sampling rate and quantization required to represent the speech signal with a desired perceptual fidelity. For example, "telephone quality" speech processing requires that a bandwidth of 0 to 4 kHz be preserved, implying a sampling rate of 8000 samples/sec. Each sample amplitude can be quantized with 8 bits distributed on a log scale, resulting in a bit rate of 64,000 bps. This representation is highly intelligible (i.e., humans can readily extract the message from it) but to most listeners, it will sound different from the original speech signal uttered by the talker. On the other hand, the speech waveform can be represented with "CD quality" using a sampling rate of 44,100 samples/sec with 16-bit samples, or a data rate of 705,600 bps. In this case, the reproduced acoustic signal will be virtually indistinguishable from the original speech signal.

As we move from a textual representation to the speech waveform representation through the speech chain, the result is an encoding of the message that can be transmitted by acoustic wave propagation and robustly decoded by the hearing mechanism of a listener. The above analysis of data rates shows that as we move from text to a sampled speech waveform, the data rate can increase by a factor of up to 10,000. Part of this extra information represents characteristics of the talker such as emotional state, speech mannerisms, accent, etc., but much of it is due to the inefficiency of simply sampling and finely quantizing analog signals. Thus, motivated by an awareness of the low intrinsic information rate of speech, a central theme of much of digital speech processing is to obtain a digital representation with a lower data rate than that of the sampled waveform.

The complete speech chain consists of a speech production/generation model, of the type discussed above, as well as a speech perception/recognition model, as shown progressing to the left in the bottom half of Figure 1.4. The speech perception model shows the series of processing steps from capturing speech at the ear to understanding the message encoded in the speech signal. The first step is the effective conversion of the acoustic waveform to a spectral representation. This is done within the inner ear by the basilar membrane, which acts as a non-uniform spectrum analyzer by spatially separating the spectral components of the incoming speech signal and thereby analyzing them by what amounts to a non-uniform filter bank. The second step in the speech perception process is a neural transduction of the spectral features into a set of sound features (or distinctive features as they are referred to in the area of linguistics) that can be decoded and processed by the brain. The third step in the process is a conversion of the sound features into the set of phonemes, words, and sentences associated with the incoming message by a language translation process in the human brain. Finally the last step in the speech perception model is the conversion of the phonemes, words, and sentences of the message into an understanding of the meaning of the basic message in order to be able to respond to or take some appropriate action. Our fundamental understanding of the processes in most of the speech perception modules in Figure 1.4 is rudimentary at best, but it is generally agreed that some physical correlate of each of the steps in the speech perception model occurs within the human brain, and thus the entire model is useful for thinking about the processes that occur. The fundamentals of hearing and perception are discussed in Chapter 4.

There is one additional process shown in the diagram of the complete speech chain in Figure 1.4 that we have not discussed—namely the transmission channel between the speech generation and speech perception parts of the model. In its simplest embodiment, as depicted in Figure 1.3, this transmission channel consists of just the acoustic wave connection between a speaker and a listener who are in a common space. It is essential to include this transmission channel in our model for the speech chain since it includes real-world noise and channel distortions that make speech and message understanding more difficult in real-communication environments. More interestingly for our purpose here—this is where the acoustic waveform of speech is converted to digital form and manipulated, stored, or transmitted by a communication system. That is, it is in this domain that we find the applications of digital speech processing.

1.2 THE SPEECH STACK

With the preceding brief introduction to the speech signal and high-level model for human speech communication, we are in a position to fill in more detail in the technology pyramid for digital speech processing. Figure 1.5 shows what we term the "speech stack," which presents a hierarchical view of the essential concepts of digital processing of speech signals. At the bottom layer of the stack is the fundamental science/technology of digital speech processing in the form of DSP theory, acoustics (speech production), linguistics (the sound codes of speech), and perception (of sounds, syllables, words, sentences, and ultimately meaning). These fundamental areas of knowledge form the theoretical basis for the signal processing that is performed to convert a speech signal into a form that is more useful for getting at the

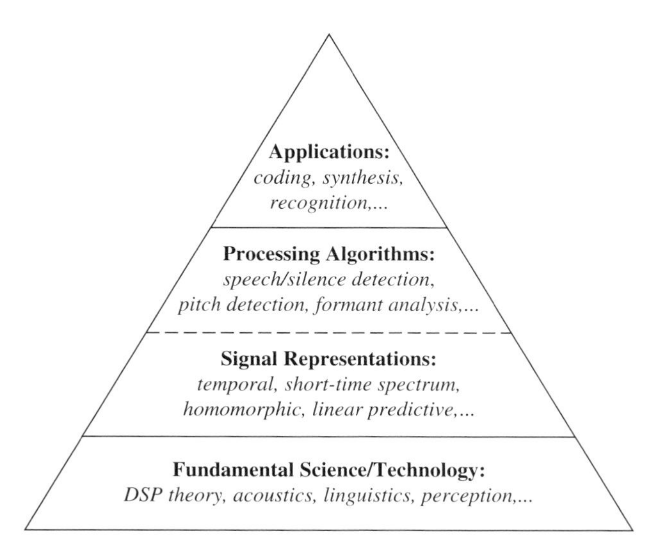

FIGURE 1.5
The speech stack—from fundamentals to applications.

information embedded in it. The fundamentals of DSP and speech science are covered in Chapters 2–5.

The second and third layers of our stack comprise what we have termed the concept layer in the technology pyramid of Figure 1.1. The second layer in the speech stack includes a set of basic representations of the speech signal. These representations include:

- temporal representations (including the speech waveform itself),
- spectral representations (Fourier magnitudes and phases),
- homomorphic representations (the *cepstral* domain), and finally
- model representations such as linear predictive coding (LPC).

It will be shown throughout this book that each of these representations has strengths and weaknesses and, therefore, all are used widely and extensively in modern speech processing systems. Chapters 6–9 are concerned with each of the four basic digital representations of speech signals.

The third layer in the stack is concerned with integrating the various speech representations into algorithms that estimate basic properties of the speech signal. The second and third layers are separated by a dashed line to imply that the boundary is not rigid between the algorithms for computing speech representations and the use of those representations in algorithms for extracting speech features. Examples of speech processing algorithms include methods for determining whether a section of the signal waveform should be classified as:

- speech or silence (background signal),
- voiced speech or unvoiced speech or background signal.

If the section of signal being analyzed is classified as voiced speech, various speech algorithms (known collectively as pitch detection methods) facilitate the determination of the pitch period (or pitch frequency), and a different set of algorithms (known collectively as formant estimation methods) can be used to estimate the set of vocal tract resonances or formants for the speech section of interest. We will see in Chapter 10 that the nature and effectiveness of an algorithm for extracting or estimating a particular speech feature will depend, to a great extent, upon the representation or representations upon which the algorithm is based. Chapter 10 provides numerous illustrations of how the basic representations are employed in speech analysis algorithms.

The fourth and top layer in the speech stack is the set of end-user applications of speech processing. This layer represents the payoff for the technology and consists of a range of applications including speech coding, speech synthesis, speech recognition and understanding, speaker verification and recognition, language translation, speech enhancement systems, speech speed-up and slow-down systems, etc. In Section 1.3 we give an overview of several of these application areas, and in Chapters 11–14 we give detailed descriptions of the three main application areas, namely speech and audio coding (Chapters 11–12), speech synthesis (Chapter 13), and speech recognition and natural language understanding (Chapter 14).

1.3 APPLICATIONS OF DIGITAL SPEECH PROCESSING

The first step in most applications of digital speech processing is to convert the acoustic waveform to a sequence of numbers. This discrete-time representation is the starting point for most applications. From this point, more powerful representations are obtained by digital processing. For the most part, these alternative representations are based on combining DSP operations with knowledge about the workings of the speech chain as depicted in Figure 1.4. As we will see, it is possible to incorporate aspects of both the speech production and speech perception processes into the digital representation and processing. As our discussion unfolds, it will become clear that it is not an oversimplification to assert that digital speech processing is grounded in a set of techniques that have the goal of pushing the data rate of the speech representation to the left (i.e., lowering the data rate) along either the upper or lower path in Figure 1.4.

The remainder of this chapter is devoted to a brief summary of the applications of digital speech processing; i.e., the systems that people interact with daily. Our discussion will emphasize the importance of the chosen digital representation in all application areas.

1.3.1 Speech Coding

Perhaps the most widespread applications of digital speech processing technology occur in the areas of digital transmission and storage of speech signals. In these areas the centrality of the digital representation is obvious, since the goal is to compress the digital waveform representation of speech into a lower bit-rate representation. It is common to refer to this activity as "speech coding" or "speech compression." Our discussion of the information content of speech in Section 1.1 suggests that there is much room for compression. In general, this compression is achieved by combining DSP techniques with fundamental knowledge of the speech production and perception processes.

Figure 1.6 shows a block diagram of a generic speech encoding/decoding (or compression) system. In the upper part of the figure, the A-to-D converter converts the analog speech signal $x_c(t)$ to a sampled waveform representation $x[n]$. The digital signal $x[n]$ is analyzed and coded by digital computation algorithms to produce a new digital signal $y[n]$ that can be transmitted over a digital communication channel or

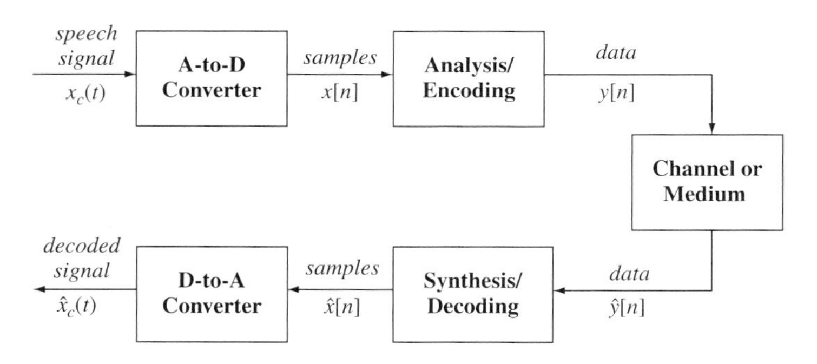

FIGURE 1.6
Speech coding block diagram—encoder and decoder.

stored in a digital storage medium as $\hat{y}[n]$. As we will see, particularly in Chapter 11 for speech and Chapter 12 for audio signals, there are a myriad of ways to do the encoding so as to reduce the data rate over that of the sampled and quantized waveform $x[n]$. Because the digital representation at this point is often not directly related to the sampled speech waveform, $y[n]$ and $\hat{y}[n]$ are appropriately referred to as *data signals* that represent the speech signal. The lower path in Figure 1.6 shows the decoder associated with the speech coder. The received data signal $\hat{y}[n]$ is decoded using the inverse of the analysis processing, giving the sequence of samples $\hat{x}[n]$, which is then converted (using a D-to-A Converter) back to an analog signal $\hat{x}_c(t)$ for human listening. The decoder is often called a *synthesizer* because it must reconstitute the speech waveform from data that often bear no direct relationship to the original waveform samples.

With carefully designed error protection coding of the digital representation, the transmitted ($y[n]$) and received ($\hat{y}[n]$) data can be essentially identical. This is the quintessential feature of digital coding. In theory, perfect transmission of the coded digital representation is possible even under very noisy channel conditions, and in the case of digital storage, it is possible to store a perfect copy of the digital representation in perpetuity if sufficient care is taken to update the storage medium as storage technology advances. This means that the speech signal can be reconstructed to within the accuracy of the original coding for as long as the digital representation is retained. In either case, the goal of the speech coder is to start with samples of the speech signal and reduce (compress) the data rate required to represent the speech signal while maintaining a desired perceptual fidelity. The compressed representation can be more efficiently transmitted or stored, or the bits saved can be devoted to error protection.

Speech coders enable a broad range of applications including narrowband and broadband wired telephony, cellular communications, voice over Internet protocol (VoIP) (which utilizes the Internet as a real-time communications medium), secure voice for privacy and encryption (for national security applications), extremely narrowband communications channels [such as battlefield applications using high frequency (HF) radio], and for storage of speech for telephone answering machines, interactive voice response (IVR) systems, and pre-recorded messages. Speech coders often employ many aspects of both the speech production and speech perception processes, and hence may not be useful for more general audio signals such as music. Coders that are based on incorporating only aspects of sound perception generally do not achieve as much compression as those based on speech production, but they are more general and can be used for all types of audio signals. These coders are widely deployed in MP3 and AAC players and for audio in digital television systems [374].

1.3.2 Text-to-Speech Synthesis

For many years, scientists and engineers have studied the speech production process with the goal of building a system that can start with text and produce speech automatically. In a sense, a text-to-speech synthesizer, such as the one depicted in Figure 1.7, is a digital simulation of the entire upper part of the speech chain diagram. The input to the system is ordinary text such as an email message or an article from a newspaper or magazine. The first block in the text-to-speech synthesis system, labeled linguistic rules, has the job of converting the printed text input into a set of sounds that the

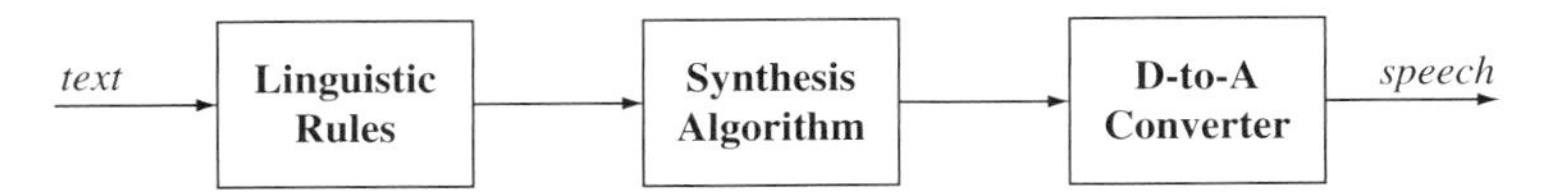

FIGURE 1.7
Text-to-speech synthesis system block diagram.

machine can synthesize. The conversion from text to sounds involves a set of linguistic rules that must determine the appropriate set of sounds (perhaps including things like emphasis, pauses, rates of speaking, etc.) so that the resulting synthetic speech will express the words and intent of the text message in what passes for a natural voice that can be decoded accurately by human speech perception. This is more difficult than simply looking up the words in a pronouncing dictionary because the linguistic rules must determine how to pronounce acronyms, how to pronounce ambiguous words like *read*, *bass*, *object*, how to pronounce abbreviations like St. (street or saint), Dr. (doctor or drive), and how to properly pronounce proper names, specialized terms, etc. Once the proper pronunciation of the text has been determined, the role of the synthesis algorithm is to create the appropriate sound sequence to represent the text message in the form of speech. In essence, the synthesis algorithm must simulate the action of the vocal tract system in creating the sounds of speech. There are many procedures for assembling the speech sounds and compiling them into a proper sentence, but the most promising one today is called "unit selection and concatenation." In this method, the computer stores multiple versions of each of the basic linguistic units of speech (phones, half phones, syllables, etc.), and then decides which sequence of speech units sounds best for the particular text message that is being produced. The basic digital representation is not generally the sampled speech wave. Instead, some sort of compressed representation is normally used to save memory and, more importantly, to allow convenient manipulation of durations and blending of adjacent sounds. Thus, the speech synthesis algorithm would include an appropriate decoder, as discussed in Section 1.3.1, whose output is converted to an analog representation via the D-to-A converter.

Text-to-speech synthesis systems are an essential component of modern human–machine communications systems and are used to do things like read email messages over a telephone, provide voice output from a GPS (global positioning system) in automobiles, provide the voices for talking agents for completion of transactions over the Internet, handle call center help desks and customer care applications, serve as the voice for providing information from handheld devices such as foreign language phrasebooks, dictionaries, crossword puzzle helpers, and as the voice of announcement machines that provide information such as stock quotes, airline schedules, updates on arrivals and departures of flights, etc. Another important application is in reading machines for the blind, where an optical character recognition system provides the text input to a speech synthesis system.

1.3.3 Speech Recognition and Other Pattern Matching Problems

Another large class of digital speech processing applications is concerned with the automatic extraction of information from the speech signal. Most such systems involve

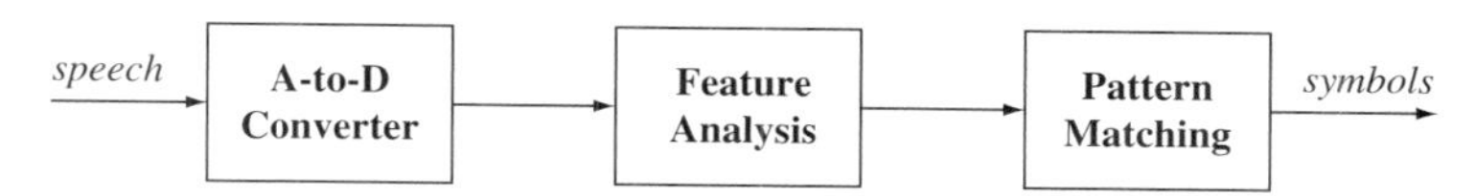

FIGURE 1.8
Block diagram of general pattern matching system for speech signals.

some sort of pattern matching. Figure 1.8 shows a block diagram of a generic approach to pattern matching problems in speech processing. Such problems include the following: speech recognition, where the object is to extract the message from the speech signal; speaker recognition, where the goal is to identify who is speaking; speaker verification, where the goal is to verify a speaker's claimed identity from analysis of their speech signal; word spotting, which involves monitoring a speech signal for the occurrence of specified words or phrases; and automatic indexing of speech recordings based on recognition (or spotting) of spoken keywords.

The first block in the pattern matching system converts the analog speech waveform to digital form using an A-to-D converter. The feature analysis module converts the sampled speech signal to a set of feature vectors. Often, the same analysis techniques that are used in speech coding are also used to derive the feature vectors. The final block in the system, namely the pattern matching block, dynamically time aligns the set of feature vectors representing the speech signal with a concatenated set of stored patterns, and chooses the identity associated with the pattern that is the closest match to the time-aligned set of feature vectors of the speech signal. The symbolic output consists of a set of recognized words, in the case of speech recognition, or the identity of the best matching talker, in the case of speaker recognition, or a decision as to whether to accept or reject the identity claim of a speaker in the case of speaker verification.

Although the block diagram of Figure 1.8 represents a wide range of speech pattern matching problems, the biggest use has been in the area of recognition and understanding of speech in support of human–machine communication by voice. The major areas where such a system finds applications include command and control of computer software, voice dictation to create letters, memos, and other documents, natural language voice dialogues with machines to enable help desks and call centers, and for agent services such as calendar entry and update, address list modification and entry, etc.

Pattern recognition applications often occur in conjunction with other digital speech processing applications. For example, one of the pre-eminent uses of speech technology is in portable communication devices. Speech coding at bit rates on the order of 8 kbps enables normal voice conversations in cell phones. Spoken name speech recognition in cell phones enables voice dialing capability, which can automatically dial the number associated with the recognized name. Names from directories with upwards of several hundred names can readily be recognized and dialed using simple speech recognition technology.

Another major speech application that has long been a dream of speech researchers is *automatic language translation*. The goal of language translation systems is to convert spoken words in one language to spoken words in another language so as to facilitate natural language voice dialogues between people speaking different languages. Language translation technology requires speech synthesis systems that work in both languages, along with speech recognition (and generally natural language understanding) that also works for both languages; hence it is a very difficult task and one for which only limited progress has been made. When such systems exist, it will be possible for people speaking different languages to communicate at data rates on the order of that of printed text reading!

1.3.4 Other Speech Applications

The range of speech communication applications is illustrated in Figure 1.9. As seen in this figure, the techniques of digital speech processing are a key ingredient of a wide range of applications that include the three areas of transmission/storage, speech synthesis, and speech recognition as well as many others such as speaker identification, speech signal quality enhancement, and aids for the hearing or visually impaired.

The block diagram in Figure 1.10 represents any system where time signals such as speech are processed by the techniques of DSP. This figure simply depicts the notion that once the speech signal is sampled, it can be manipulated in virtually limitless ways by DSP techniques. Here again, manipulations and modifications of the speech signal are usually achieved by transforming the speech signal into an alternative representation (that is motivated by our understanding of speech production and speech perception), operating on that representation by further digital computation, and then transforming back to the waveform domain, using a D-to-A converter.

One important application area is *speech enhancement*, where the goal is to remove or suppress noise or echo or reverberation picked up by a microphone along with the desired speech signal. In human-to-human communication, the goal of speech enhancement systems is to make the speech more intelligible and more natural;

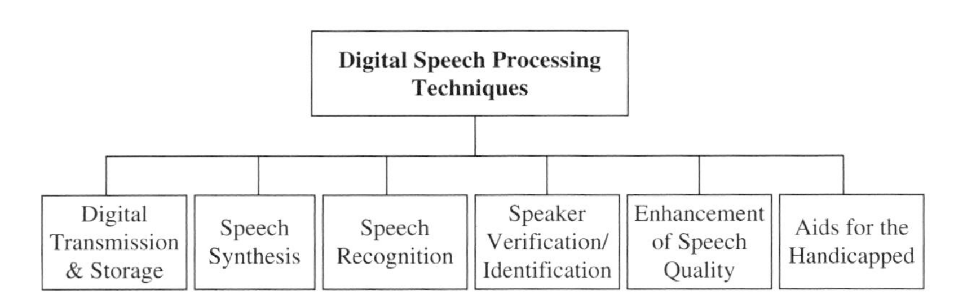

FIGURE 1.9
Range of speech communication applications.

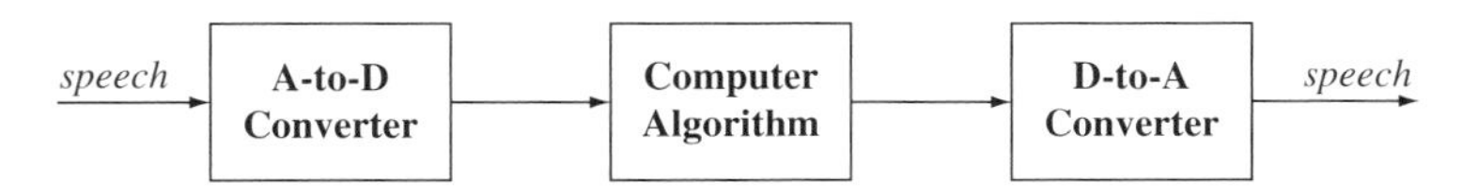

FIGURE 1.10
General block diagram for application of DSP to speech signals.

however, in reality the best that has been achieved so far is less perceptually annoying speech that essentially maintains, but does not improve, the intelligibility of the degraded speech. Success *has* been achieved, however, in making distorted speech signals more useful for further processing as part of a speech coder, synthesizer, or recognizer [212].

Other examples of manipulation of the speech signal include time-scale modification to align voices with video segments, to modify voice qualities, and to speed-up or slow-down pre-recorded speech (e.g., for talking books, rapid review of voice mail messages, or careful scrutinizing of spoken material). Such modifications of the signal are often more easily done on one of the basic digital representations rather than on the sampled waveform itself.

1.4 COMMENT ON THE REFERENCES

The bibliography at the end of this book contains all the references for all the chapters. Many of these references are research papers that established important results in the field of digital speech processing. Also included in the bibliography are a number of important and valuable reference texts that are often referenced as well. Some of these are "classic" texts that hold a special place in the evolution of the field. Others are more recent, and thus they provide knowledge about the latest developments in the field. The following, listed in chronological order by publication date and in categories suggested in this chapter, are texts that we have consulted in our teaching and research. These will be of special interest particularly with regard to application areas where our coverage of topics is less detailed.

General Speech Processing Reference Texts

- G. Fant, *Acoustic Theory of Speech Production,* Mouton, The Hague, 1970.
- J. L. Flanagan, *Speech Analysis, Synthesis and Perception*, 2nd ed., Springer-Verlag, 1972.
- J. D. Markel and A. H. Gray, Jr., *Linear Prediction of Speech*, Springer-Verlag, 1976.
- L. R. Rabiner and R. W. Schafer, *Digital Processing of Speech Signals*, Prentice-Hall Inc., 1978.
- R. W. Schafer and J. D. Markel (eds.), *Speech Analysis*, IEEE Press Selected Reprint Series, 1979.
- D. O'Shaughnessy, *Speech Communication, Human and Machine*, Addison-Wesley, 1987.
- S. Furui and M. M. Sondhi (eds.), *Advances in Speech Signal Processing*, Marcel Dekker Inc., 1991.
- P. B. Denes and E. N. Pinson, *The Speech Chain*, 2nd ed., W. H. Freeman and Co., 1993.
- J. Deller, Jr., J. G. Proakis, and J. Hansen, *Discrete-Time Processing of Speech Signals*, Macmillan Publishing, 1993, Wiley-IEEE Press, Classic Reissue, 1999.
- K. N. Stevens, *Acoustic Phonetics*, MIT Press, 1998.
- B. Gold and N. Morgan, *Speech and Audio Signal Processing*, John Wiley and Sons, 2000.
- S. Furui (ed.), *Digital Speech Processing, Synthesis and Recognition*, 2nd ed., Marcel Dekker Inc., New York, 2001.
- T. F. Quatieri, *Principles of Discrete-Time Speech Processing*, Prentice-Hall Inc., 2002.
- L. Deng and D. O'Shaughnessy, *Speech Processing, A Dynamic and Optimization-Oriented Approach*, Marcel Dekker Inc., 2003.
- J. Benesty, M. M. Sondhi, and Y. Huang (eds.), *Springer Handbook of Speech Processing and Speech Communication*, Springer, 2008.

Speech Coding Reference Texts

- N. S. Jayant and P. Noll, *Digital Coding of Waveforms*, Prentice-Hall Inc., 1984.
- P. E. Papamichalis, *Practical Approaches to Speech Coding*, Prentice-Hall Inc., 1984.
- A. Gersho and R. M. Gray, *Vector Quantization and Signal Compression*, Kluwer Academic Publishers, 1992.
- W. B. Kleijn and K. K. Paliwal, *Speech Coding and Synthesis*, Elsevier, 1995.
- T. P. Barnwell and K. Nayebi, *Speech Coding, A Computer Laboratory Textbook*, John Wiley and Sons, 1996.
- R. Goldberg and L. Riek, *A Practical Handbook of Speech Coders*, CRC Press, 2000.
- W. C. Chu, *Speech Coding Algorithms*, John Wiley and Sons, 2003.
- A. M. Kondoz, *Digital Speech: Coding for Low Bit Rate Communication Systems*, 2nd ed., John Wiley and Sons, 2004.

Speech Synthesis Reference Texts

- J. Allen, S. Hunnicutt, and D. Klatt, *From Text to Speech*, Cambridge University Press, 1987.
- J. Olive, A. Greenwood, and J. Coleman, *Acoustics of American English*, Springer-Verlag, 1993.
- Y. Sagisaka, N. Campbell, and N. Higuchi, *Computing Prosody*, Springer-Verlag, 1996.
- J. VanSanten, R. W. Sproat, J. P. Olive, and J. Hirschberg (eds.), *Progress in Speech Synthesis*, Springer-Verlag, 1996.
- T. Dutoit, *An Introduction to Text-to-Speech Synthesis*, Kluwer Academic Publishers, 1997.
- D. G. Childers, *Speech Processing and Synthesis Toolboxes*, John Wiley and Sons, 1999.
- S. Narayanan and A. Alwan (eds.), *Text to Speech Synthesis: New Paradigms and Advances*, Prentice-Hall Inc., 2004.
- P. Taylor, *Text-to-Speech Synthesis*, Cambridge University Press, 2008.

Speech Recognition and Natural Language Reference Texts

- L. R. Rabiner and B. H. Juang, *Fundamentals of Speech Recognition*, Prentice-Hall Inc., 1993.
- H. A. Bourlard and N. Morgan, *Connectionist Speech Recognition—A Hybrid Approach*, Kluwer Academic Publishers, 1994.
- C. H. Lee, F. K. Soong, and K. K. Paliwal (eds.), *Automatic Speech and Speaker Recognition*, Kluwer Academic Publishers, 1996.
- F. Jelinek, *Statistical Methods for Speech Recognition*, MIT Press, 1998.
- C. D. Manning and H. Schutze, *Foundations of Statistical Natural Language Processing*, MIT Press, 1999.
- X. D. Huang, A. Acero, and H.-W. Hon, *Spoken Language Processing*, Prentice-Hall Inc., 2000.
- S. E. Levinson, *Mathematical Models for Speech Technology*, John Wiley and Sons, 2005.
- D. Jurafsky and J. H. Martin, *Speech and Language Processing*, 2nd ed., Prentice-Hall Inc., 2008.

Speech Enhancement Reference Texts

- P. Vary and R. Martin, *Digital Speech Transmission, Enhancement, Coding and Error Concealment*, John Wiley and Sons, 2006.
- P. Loizou, *Speech Enhancement Theory and Practice*, CRC Press, 2007.

Audio Coding Reference Texts

- H. Kars and K. Brandenburg (eds.), *Applications of Digital Signal Processing to Audio and Acoustics*, Kluwer Academic Publishers, 1998.

- M. Bosi and R. E. Goldberg, *Introduction to Digital Audio Coding and Standards*, Kluwer Academic Publishers, 2003.
- A. Spanias, T. Painter, and V. Atti, *Audio Signal Processing and Coding*, John Wiley and Sons, 2006.

MATLAB Exercises

- T. Dutoit and F. Marques, *Applied Signal Processing, A MATLAB-Based Proof of Concept*, Springer, 2009.

1.5 SUMMARY

In this introductory chapter, we have discussed the speech signal and how it encodes information for human communication. We have given a brief overview of the way in which digital speech processing is being applied today, and we have hinted at some of the possibilities that exist for the future. These and many more examples all rely on the basic principles of digital speech processing, which is explored in the remainder of this text. We make no pretense of exhaustive coverage. The subject is too broad and too deep. Our goal is to provide a comprehensive and up-to-date introduction to the fundamental concepts of this fascinating field. In some areas we will penetrate to significant depth, but in other areas, we will omit details that are application specific. Also, we will not be able to cover all the possible applications of digital speech processing techniques, and those that we do discuss are often the subject of entire books. Instead, our focus is on the fundamentals of digital speech processing and their application to coding, synthesis, and recognition. This means that some of the latest algorithmic innovations and applications will not be discussed—not because they are uninteresting or unimportant, but simply because there are so many fundamental tried-and-true techniques that remain at the core of digital speech processing. We are confident that a thorough mastery of the material of this text will provide a sound basis for contributing many new innovations in the field of digital speech processing.

CHAPTER 2

Review of Fundamentals of Digital Signal Processing

2.1 INTRODUCTION

Since the speech processing schemes and techniques that we discuss in this book are intrinsically discrete-time signal processing systems, it is essential that the reader have a good understanding of the basic techniques of digital signal processing. In this chapter we present a brief review of the concepts in digital signal processing that are most relevant to digital processing of speech signals. This review is intended to serve as a convenient reference for later chapters and to establish the notation that will be used throughout the book. Those readers who are completely unfamiliar with techniques for representation and analysis of discrete-time signals and systems may find it worthwhile to consult a textbook on digital signal processing [245, 270, 305] when this chapter does not provide sufficient detail.

2.2 DISCRETE-TIME SIGNALS AND SYSTEMS

In almost every situation involving information processing or communication, it is natural to begin with a representation of the signal as a continuously varying pattern or waveform. The acoustic wave produced in human speech is most certainly of this nature. It is mathematically convenient to represent such continuously varying patterns as functions of a continuous variable t, which represents time. We use notation of the form $x_a(t)$ to denote continuously varying (or analog) time waveforms. As we will see, it is also possible to represent the speech signal as a sequence of (quantized) numbers; indeed, that is one of the central themes of this book. In general we use notation of the form, $x[n]$, to denote sequences, quantized in both time and amplitude. If, as is the case for sampled speech signals, a sequence can be thought of as a sequence of samples of an analog signal taken periodically with sampling period, T, then we generally find it useful to explicitly indicate this by using the notation $x[n] = x_a(nT)$. For any digital sequence derived from an analog waveform via sampling and quantization, there are two variables that determine the nature of the discrete representation, namely the sampling rate $F_s = 1/T$ and the number of quantization levels, 2^B where B is the number of bits per sample of the representation. Although the sampling rate can be set to any value that satisfies the relation that $F_s = 1/T \geq 2F_N$ (where F_N is the highest frequency present in the continuous-time signal), various "natural" sampling rates for speech have evolved over time, including:[1]

- $F_s = 6.4$ kHz for telephone bandwidth speech ($F_N = 3.2$ kHz);
- $F_s = 8$ kHz for extended telephone bandwidth speech ($F_N = 4$ kHz);
- $F_s = 10$ kHz for oversampled telephone bandwidth speech ($F_N = 5$ kHz);
- $F_s = 16$ kHz for wideband (hi-fi) bandwidth speech ($F_N = 8$ kHz).

The second variable in the digital representation of speech signals is the number of bits per sample for the quantized signal. In Chapter 11, we will study the detailed effects of quantization on the properties of the digitized speech waveform; however, in the present chapter, we assume that the sample values are unquantized.

Figure 2.1 shows an example of a speech signal represented both as an analog signal and as a sequence of samples at a sampling rate of $F_s = 16$ kHz. In subsequent

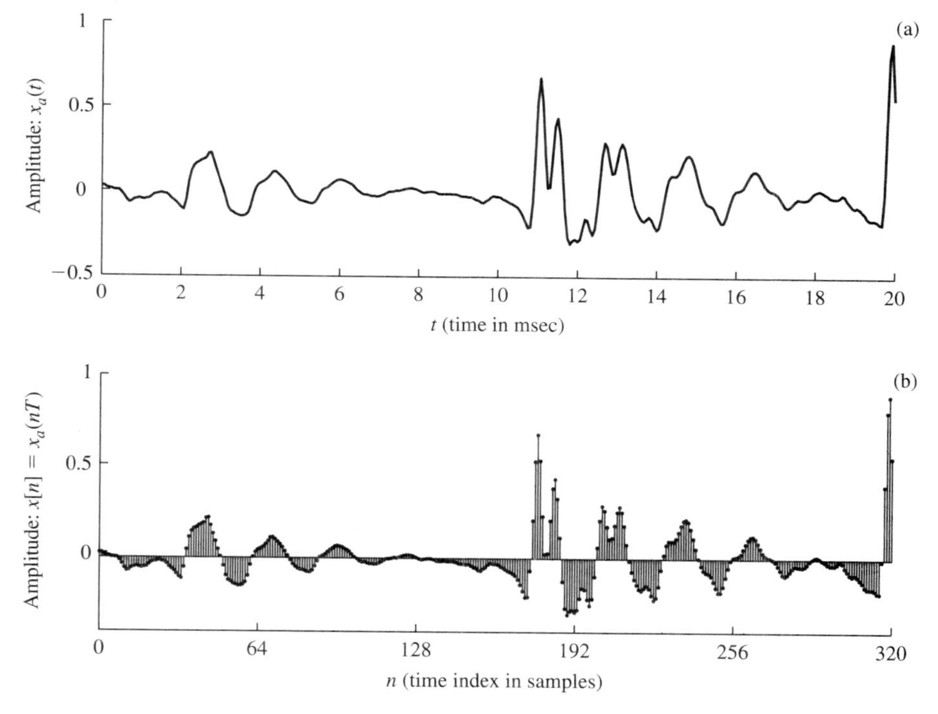

FIGURE 2.1
Plots of a speech waveform: (a) plotted as a continuous-time signal (with MATLAB `plot( )` function); (b) plotted as a sampled signal (with MATLAB `stem( )` function).

[1] As discussed in Section 2.3, we use capital letters to denote analog frequencies as in [270].

figures, convenience in plotting generally dictates the use of the analog representation (i.e., continuous functions) even when the discrete representation is being considered. In such cases, the continuous curve is the envelope of the sequence of samples, with the samples connected by straight lines for plotting (linear interpolation).[2] Figure 2.1 illustrates the use of the MATLAB function `plot` (for the analog waveform at the top of the figure) and `stem` (for the set of discrete samples at the bottom of the figure). It also illustrates an important point about sampled signals in general; when a sequence of samples is plotted as a function of sample index, the time scale is lost. We must know the sampling rate, $F_s = 1/T = 16$ kHz, in order to convert the time duration of the digital waveform (320 samples in this case) to the analog time interval of 20 msec via the sampling period $T = 62.5\ \mu$s.

In our study of digital speech processing systems, we will find a number of special sequences repeatedly arising. Several of these sequences are depicted in Figure 2.2. The unit sample or unit impulse sequence, shown in Figure 2.2a, is defined as

$$\delta[n] = \begin{cases} 1 & n = 0 \\ 0 & \text{otherwise.} \end{cases} \tag{2.1}$$

The unit step sequence, shown in Figure 2.2b, is defined as

$$u[n] = \begin{cases} 1 & n \geq 0 \\ 0 & n < 0. \end{cases} \tag{2.2}$$

An exponential sequence is of the form

$$x[n] = a^n, \tag{2.3}$$

where a can be either real or complex. Figure 2.2c shows an exponential sequence when a is real and $|a| < 1$. If a is complex, i.e., $a = re^{j\omega_0}$, then

$$x[n] = r^n e^{j\omega_0 n} = r^n \cos(\omega_0 n) + jr^n \sin(\omega_0 n). \tag{2.4}$$

Figure 2.2d shows the real part of the complex exponential sequence when $0 < r < 1$ and $\omega_0 = 2\pi/8$. If $r = 1$ and $\omega_0 \neq 0$, $x[n]$ is a complex sinusoid; if $\omega_0 = 0$, $x[n]$ is real; and if $r < 1$ and $\omega_0 \neq 0$, then $x[n]$ is an exponentially decaying oscillatory sequence. We often use the unit step sequence as in

$$x[n] = a^n u[n], \tag{2.5}$$

to represent an exponential sequence that is zero for $n < 0$. Sequences of this type arise especially in the representation of causal linear systems and in modeling the speech waveform.

[2] Reconstruction of a bandlimited continuous-time signal from the samples involves interpolation by filters approximating the ideal lowpass filter [270].

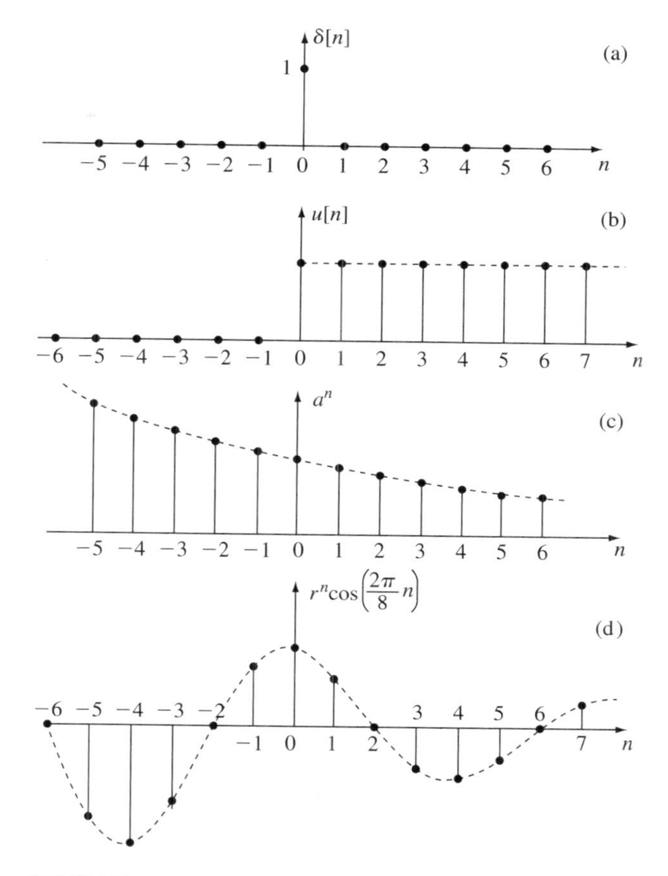

FIGURE 2.2
(a) Unit sample; (b) unit step; (c) real exponential; and (d) damped cosine.

Signal processing involves the transformation of a signal into a form that is, in some sense, more desirable. Thus we are concerned with discrete-time systems, or equivalently, transformations of an input sequence into an output sequence. We shall often find it useful to depict such transformations by block diagrams such as Figure 2.3a. Many speech analysis systems are designed to estimate several time-varying parameters from samples of the speech wave. Such systems therefore have a multiplicity of outputs; i.e., a single input sequence, $x[n]$, representing the speech signal is transformed into a vector of output sequences as depicted in Figure 2.3b, where the broad arrow signifies a vector of output sequences.

The special class of linear shift-invariant systems is especially useful in speech processing. Such systems are completely characterized by their response to a unit sample input. For such systems, the output can be computed from the input, $x[n]$, and

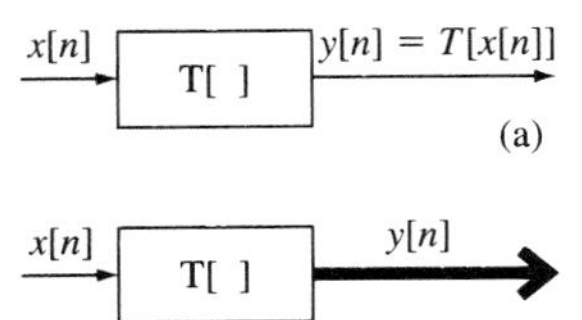

FIGURE 2.3
Block diagram representation of (a) single input/single output system and (b) single input/multiple output system.

the unit sample response, $h[n]$, using the convolution sum expression

$$y[n] = \sum_{k=-\infty}^{\infty} x[k]h[n-k] = x[n] * h[n], \tag{2.6a}$$

where the symbol $*$ stands for the operation of discrete convolution. An equivalent expression, expressing the commutative property of convolution, is

$$y[n] = \sum_{k=-\infty}^{\infty} h[k]x[n-k] = h[n] * x[n]. \tag{2.6b}$$

Linear shift-invariant systems are useful for performing filtering operations on speech signals and, perhaps more importantly, they are useful as models for speech production.

2.3 TRANSFORM REPRESENTATION OF SIGNALS AND SYSTEMS

The analysis and design of linear systems are greatly facilitated by frequency-domain representations of both signals and systems. Thus, it is useful to review Fourier and z-transform representations of discrete-time signals and systems.

2.3.1 The Continuous-Time Fourier Transform

The Fourier transform is a fundamental mathematical concept in both continuous- and discrete-time signal and system theory. In the case of continuous-time signals, we define the continuous-time Fourier transform (CTFT) representation as

$$X_a(j\Omega) = \int_{-\infty}^{\infty} x_a(t)e^{-j\Omega t}dt, \tag{2.7a}$$

where the inverse transform, or Fourier synthesis equation, is

$$x_a(t) = \frac{1}{2\pi}\int_{-\infty}^{\infty} X_a(j\Omega)e^{j\Omega t}d\Omega. \tag{2.7b}$$

Note that we use Ω with units radians/sec to denote continuous-time radian frequency as in [270]. Furthermore, we often use the equivalent relation $\Omega = 2\pi F$, where F represents "cyclic" continuous-time frequency in hertz. It is assumed that the reader has a familiarity with and working knowledge of the properties of continuous-time Fourier representations. Detailed presentations of the CTFT and its use in signal and system analysis can be found in [239, 272].

2.3.2 The z-Transform

The z-transform of a sequence, $x[n]$, is defined as

$$X(z) = \sum_{n=-\infty}^{\infty} x[n]z^{-n}, \tag{2.8a}$$

and the corresponding inverse z-transform is the complex contour integral

$$x[n] = \frac{1}{2\pi j}\oint_C X(z)z^{n-1}dz. \tag{2.8b}$$

It can be seen from Eq. (2.8a) that the direct z-transform, $X(z)$, is an infinite power series in the variable z^{-1}, where the sequence values, $x[n]$, play the role of coefficients in the power series. In general such a power series will converge (add up to a finite value) only for certain values of z. A sufficient condition for convergence is

$$\sum_{n=-\infty}^{\infty} |x[n]|\,|z^{-n}| < \infty. \tag{2.9}$$

The set of values for which the series converges defines a region in the complex z-plane known as the *region of convergence*. In general this region is of the form

$$R_1 < |z| < R_2. \tag{2.10}$$

To show the relationship of the region of convergence to the nature of the sequence, we consider some examples.

Example 2.1 Delayed unit impulse

Let $x[n] = \delta[n - n_0]$. Then by substitution into Eq. (2.8a),

$$X(z) = z^{-n_0}. \tag{2.11}$$

The region of convergence for the delayed unit impulse depends on the value of n_0 and is $|z| > 0$, for $n_0 > 0$, or $|z| < \infty$ for $n_0 < 0$, or for all z for $n_0 = 0$.

Example 2.2 Box pulse

Let

$$x[n] = u[n] - u[n-N] = \begin{cases} 1 & 0 \le n \le N-1 \\ 0 & \text{otherwise.} \end{cases}$$

Then

$$X(z) = \sum_{n=0}^{N-1} (1) z^{-n} = \frac{1 - z^{-N}}{1 - z^{-1}}. \tag{2.12}$$

In this case, $x[n]$ is of finite duration. Therefore $X(z)$ is simply an Nth-order polynomial in the variable z^{-1}, and the region of convergence is everywhere but $z = 0$. All finite-length sequences have a region of convergence that is at least the region $0 < |z| < \infty$.[3]

Example 2.3 Positive-time one-sided exponential

Let $x[n] = a^n u[n]$. Then we have

$$X(z) = \sum_{n=0}^{\infty} a^n z^{-n} = \frac{1}{1 - az^{-1}}, \qquad |a| < |z|. \tag{2.13}$$

In this case the power series is recognized as an infinite geometric series for which a convenient closed form expression exists for the sum. This result is typical of infinite duration sequences that are non-zero for $n > 0$. In the general case, the region of convergence is of the form $|z| > R_1$.

Example 2.4 Negative-time one-sided exponential

Let $x[n] = -b^n u[-n-1]$. Then

$$X(z) = \sum_{n=-\infty}^{-1} -b^n z^{-n} = \frac{1}{1 - bz^{-1}}, \qquad |z| < |b|. \tag{2.14}$$

[3]Note that in this special case, the $(N-1)^{st}$-order polynomial can be expressed as the ratio of an Nth-order polynomial to a first-order polynomial using the general formula for the sum of the first N terms in a geometric series,

$$\sum_{n=0}^{N-1} \alpha^n = \frac{1 - \alpha^N}{1 - \alpha}.$$

This is typical of infinite duration sequences that are non-zero for $n < 0$, where the region of convergence is, in general, $|z| < R_2$. Note that the z-transforms in Eqs. (2.13) and (2.14) have the same functional form. However, the z-transforms are different because they have different regions of convergence. The most general case in which $x[n]$ is non-zero for $-\infty < n < \infty$ can be viewed as a combination of the cases illustrated by Examples 2.3 and 2.4. Thus, for this case, the region of convergence is of the form $R_1 < |z| < R_2$.

The inverse z-transform is given by the contour integral in Eq. (2.8b), where C is a closed contour that encircles the origin of the z-plane and lies inside the region of convergence of $X(z)$. Such complex contour integrals can be evaluated using powerful theorems from the theory of complex variables [46]. For the special case of rational transforms, which are of primary interest for discrete-time linear systems, the integral theorems lead to a partial fraction expansion, which provides a convenient means for finding inverse transforms [270].

There are many theorems and properties of the z-transform representation that are useful in the study of discrete-time systems. A working familiarity with these theorems and properties is essential for a complete understanding of the material in subsequent chapters. A list of important theorems is given in Table 2.1. These theorems can be seen to be similar in form to corresponding theorems for Laplace transforms of continuous-time functions. However, this similarity should not be construed to mean that the z-transform is, in any sense, an approximation to the Laplace transform. The Laplace transform is an *exact* representation of a continuous-time function, and the z-transform is an *exact* representation of a sequence of numbers. The appropriate way to relate the continuous and discrete representations of a signal is through the Fourier transform and the sampling theorem, as will be discussed in Section 2.5.

Operations on sequences may change the region of convergence of the z-transform. For example, in properties 1 and 6, the resulting region of convergence is at

TABLE 2.1 Real sequences and their corresponding z-transforms.

Property	**Sequence**	**z-Transform**
1. Linearity	$ax_1[n] + bx_2[n]$	$aX_1(z) + bX_2(z)$
2. Shift	$x[n + n_0]$	$z^{n_0} X(z)$
3. Exponential weighting	$a^n x[n]$	$X(a^{-1}z)$
4. Linear weighting	$nx[n]$	$-z\dfrac{dX(z)}{dz}$
5. Time reversal	$x[-n]$	$X(z^{-1})$
6. Convolution	$x[n] * h[n]$	$X(z)H(z)$
7. Multiplication of sequences	$x[n]w[n]$	$\dfrac{1}{2\pi j}\oint_C X(v)W(z/v)v^{-1}dv$

least the overlap of the individual regions of convergence of the two z-transforms. The required modifications for the other cases in Table 2.1 are easily determined [270].

2.3.3 The Discrete-Time Fourier Transform

The *discrete-time Fourier transform* (DTFT) representation of a discrete-time signal is given by the equation

$$X(e^{j\omega}) = \sum_{n=-\infty}^{\infty} x[n]e^{-j\omega n}, \tag{2.15a}$$

and the inverse DTFT (synthesis integral) is

$$x[n] = \frac{1}{2\pi}\int_{-\pi}^{\pi} X(e^{j\omega})e^{j\omega n}d\omega. \tag{2.15b}$$

These equations are special cases of Eqs. (2.8a) and (2.8b). Specifically the Fourier representation is obtained by restricting the z-transform to the unit circle of the z-plane; i.e., by setting $z = e^{j\omega}$. As depicted in Figure 2.4, the discrete-time frequency variable, ω, also has the interpretation as angle in the z-plane. A sufficient condition for the existence of a Fourier transform representation can be obtained by setting $|z| = 1$ in Eq. (2.9), thus giving

$$\sum_{n=-\infty}^{\infty} |x[n]| < \infty. \tag{2.16}$$

That is, the region of convergence of $X(z)$ must contain the unit circle of the z-plane. As examples of typical Fourier transforms, we can return to the examples of Section 2.3.2. The Fourier transform is obtained simply by setting $z = e^{j\omega}$ in the given expression. In the first two examples, i.e., the delayed impulse and the box pulse, the

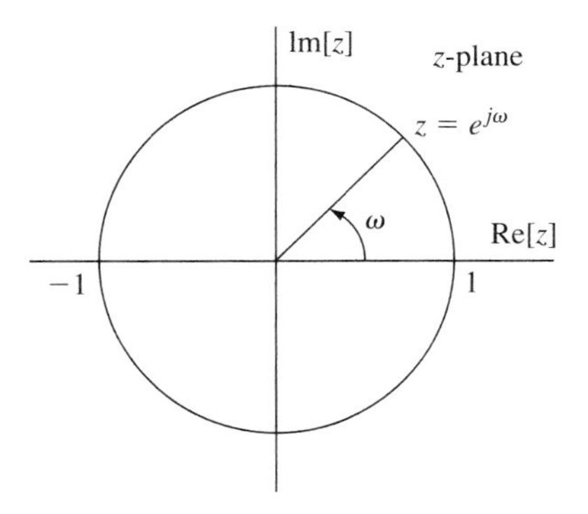

FIGURE 2.4
The unit circle of the z-plane.

Fourier transform is defined since the region of convergence of $X(z)$ includes the unit circle. However, in Examples 2.3 and 2.4, the Fourier transforms exist only if $|a| < 1$ and $|b| > 1$ respectively. These conditions, of course, correspond to decaying sequences for which Eq. (2.16) holds.

An important feature of the DTFT of a sequence is that $X(e^{j\omega})$ is a periodic function of ω, with period 2π. This follows easily by substituting $\omega + 2\pi$ into Eq. (2.15a). Alternatively, since $X(e^{j\omega})$ is the evaluation of $X(z)$ on the unit circle as depicted in Figure 2.4, it follows that $X(e^{j\omega})$ must repeat each time we go completely around the unit circle; i.e., ω has gone through 2π radians.

Some sequences that are very useful models for real signals do not satisfy the condition of Eq. (2.16) for the existence of a DTFT. A specific example is the discrete-time sinusoidal signal discussed in the following example.

Example 2.5 DTFT of a Cosine Signal

The discrete-time cosine signal

$$x[n] = \cos(\omega_0 n), \quad -\infty < n < \infty, \tag{2.17}$$

does not satisfy Eq. (2.16), and therefore the DTFT expression cannot be summed in the normal sense. However if we are willing to include continuous-variable impulse functions (also known as Dirac delta functions) in the DTFT, we can bring discrete-time cosine signals (and any periodic signal) into the DTFT representation. Specifically, the DTFT for the signal of Eq. (2.17) is the periodic continuous-variable impulse train function

$$X(e^{j\omega}) = \sum_{k=-\infty}^{\infty} [\pi\delta(\omega - \omega_0 + 2\pi k) + \pi\delta(\omega + \omega_0 + 2\pi k)]. \tag{2.18}$$

That is, over the frequency range $-\pi < \omega < \pi$, $X(e^{j\omega})$ is comprised of a pair of continuous-variable impulses at $\pm\omega_0$, and that pattern repeats periodically with period 2π. If we use Eq. (2.18) in the inverse DTFT expression of Eq. (2.15b), we obtain

$$\begin{aligned}\frac{1}{2\pi}\int_{-\pi}^{\pi} X(e^{j\omega})e^{j\omega n}d\omega &= \frac{1}{2\pi}\int_{-\pi}^{\pi} [\pi\delta(\omega - \omega_0) + \pi\delta(\omega + \omega_0)]e^{j\omega n}d\omega \\ &= \frac{1}{2}e^{j\omega_0 n} + \frac{1}{2}e^{-j\omega_0 n} \\ &= \cos(\omega_0 n),\end{aligned} \tag{2.19}$$

where the simplification of the expressions on the right uses the "sifting" property of the impulse function [i.e., $e^{j\omega n}\delta(\omega - \omega_0) = e^{j\omega_0 n}\delta(\omega - \omega_0)$], and the property that the area of each impulse is 1. As seen from the result of Eq. (2.19), the formal manipulation results in the desired signal $x[n] = \cos(\omega_0 n)$.

It is especially important to remember that when working with or plotting $X(e^{j\omega})$, the units of normalized frequency, ω, are radians (really dimensionless); i.e., one turn

TABLE 2.2 Real sequences and their corresponding DTFTs.

Property	Sequence	DTFT
1. Linearity	$ax_1[n] + bx_2[n]$	$aX_1(e^{j\omega}) + bX_2(e^{j\omega})$
2. Shift	$x[n + n_0]$	$e^{j\omega n_0} X(e^{j\omega})$
3. Modulation	$x[n]e^{j\omega_0 n}$	$X(e^{j(\omega-\omega_0)})$
4. Linear weighting	$nx[n]$	$j\dfrac{dX(e^{j\omega})}{d\omega}$
5. Time reversal	$x[-n]$	$X(e^{-j\omega}) = X^*(e^{j\omega})$
6. Convolution	$x[n] * h[n]$	$X(e^{j\omega})H(e^{j\omega})$
7. Multiplication of sequences	$x[n]w[n]$	$\dfrac{1}{2\pi}\int_{-\pi}^{\pi} X(e^{j\theta})W(e^{j(\omega-\theta)})d\theta$
8. Parseval's Theorem	$\sum_{n=-\infty}^{\infty} \lvert x[n]\rvert^2 =$	$\dfrac{1}{2\pi}\int_{-\pi}^{\pi} \lvert X(e^{j\omega})\rvert^2 d\omega$

around the unit circle of Figure 2.4 corresponds to ω going from 0 to 2π. If we choose to use normalized cyclic frequency units, f (rather than normalized radian frequency), with $\omega = 2\pi f$, then one turn around the unit circle corresponds to f going from 0 to 1. However both normalized radian frequency units and normalized frequency units do not reflect the analog (physical) frequency associated with the sampling rate of a signal that is the input or output of a discrete-time system. Alternatively we can also consider the use of the unnormalized (or analog) frequency unit, F, where $f = F/T = F \cdot F_s$, which goes from 0 to $F_s = 1/T$ during one turn around the unit circle; i.e., as f goes from 0 to 1. Similarly we can consider the use of the unnormalized (or analog) radian frequency unit Ω, where $\Omega = 2\pi F = 2\pi f \cdot F_s = \Omega \cdot F_s$, which goes from 0 to $2\pi F_s$ during one turn around the unit circle.

By setting $z = e^{j\omega}$ in each of the z-transforms in Table 2.1, we obtain a corresponding set of results for the DTFT, which are given in Table 2.2. Of course, these results are valid only if the DTFTs that are involved do indeed exist. An additional useful result is Parseval's Theorem, which relates the "energy" of a sequence to an integral of the magnitude-squared of the corresponding DTFT.

2.3.4 The Discrete Fourier Transform

A sequence is periodic with period N if

$$\tilde{x}[n] = \tilde{x}[n + N], \quad -\infty < n < \infty. \tag{2.20}$$

As in the case of periodic analog signals, periodic sequences $\tilde{x}[n]$ can be represented by a discrete sum of sinusoids rather than an integral as in Eq. (2.15b). Specifically, $\tilde{x}[n]$ can be represented as a sum of complex exponentials having radian frequencies

$(2\pi k/N)$, with $k = 0, 1, \ldots, N-1$.[4] The Fourier series representation for a periodic sequence with period N is comprised of the "discrete-time Fourier series coefficients"

$$\tilde{X}[k] = \sum_{n=0}^{N-1} \tilde{x}[n] e^{-j\frac{2\pi}{N}kn}, \tag{2.21a}$$

obtained by summing over one period, $0 \le n \le N-1$. These coefficients are used in the corresponding discrete-time Fourier series synthesis expression

$$\tilde{x}[n] = \frac{1}{N} \sum_{k=0}^{N-1} \tilde{X}[k] e^{j\frac{2\pi}{N}kn}, \tag{2.21b}$$

which allows the reconstruction of the periodic sequence from the N discrete-time Fourier series coefficients [270].

This is an *exact* representation of a periodic sequence having period N. However, the great utility of this representation lies in imposing a different interpretation upon Eqs. (2.21a) and (2.21b). Consider a finite-length sequence, $x[n]$, that is zero outside the interval $0 \le n \le N-1$. From this finite-length sequence, we can define what we shall call an "implicit periodic sequence," $\tilde{x}[n]$, as the infinite sequence of shifted replicas of $x[n]$ with period N; i.e.,

$$\tilde{x}[n] = \sum_{r=-\infty}^{\infty} x[n + rN]. \tag{2.22}$$

Since we require $\tilde{x}[n]$ only for $0 \le n \le N-1$ (i.e., only one period) in order to evaluate the Fourier series coefficients in Eq. (2.21a), this implicit periodic sequence can be represented by the discrete-time Fourier series of Eq. (2.21b), which also only requires $\tilde{X}[k]$ for $k = 0, 1, \ldots, N-1$. Thus a finite duration sequence of length N can be exactly represented by the finite computations of Eqs. (2.21a) and (2.21b) if we are careful to evaluate Eq. (2.21b) only in the finite interval $0 \le n \le N-1$. This way of thinking about a finite-length sequence leads to the *discrete Fourier transform* (DFT) representation:

$$X[k] = \sum_{n=0}^{N-1} x[n] e^{-j\frac{2\pi}{N}kn}, \quad k = 0, 1, \ldots, N-1, \tag{2.23a}$$

$$x[n] = \frac{1}{N} \sum_{k=0}^{N-1} X[k] e^{j\frac{2\pi}{N}kn}, \quad n = 0, 1, \ldots, N-1. \tag{2.23b}$$

[4]Signals comprised of a linear combination of a finite or countably infinite set of complex exponentials with different frequencies are said to have a *line spectrum* representation.

Clearly the only difference between Eqs. (2.23a) and (2.23b) and Eqs. (2.21a) and (2.21b) is a slight modification of notation (removing the ˜ symbols, which indicate periodicity) and the explicit restriction to the finite intervals $0 \leq k \leq N-1$ and $0 \leq n \leq N-1$. As the previous discussion shows, it is extremely important to bear in mind when using the DFT representation that all sequences behave as if they were periodic when represented by a DFT representation. That is, the DFT is really a representation of the implicit periodic sequence given in Eq. (2.22). An alternative point of view is that when DFT representations are used, sequence indices must be interpreted modulo N. This follows from the fact that if (and only if) $x[n]$ is of length N, then the implicit periodic sequence for the DFT representation is

$$\begin{aligned}\tilde{x}[n] &= \sum_{r=-\infty}^{\infty} x[n+rN] = x[n \bmod N] \\ &= x[((n))_N].\end{aligned} \tag{2.24a}$$

so that

$$x[n] = \begin{cases} \tilde{x}[n] & 0 \leq n \leq N-1 \\ 0 & \text{otherwise.} \end{cases} \tag{2.24b}$$

The double parenthesis notation used to represent $n \bmod N$ provides a convenient way to express the inherent periodicity of the DFT representation.

2.3.5 Sampling the DTFT

The DFT has another interpretation that is useful in emphasizing the importance of the implicit periodicity. Consider a sequence $x[n]$ whose DTFT is

$$X(e^{j\omega}) = \sum_{n=0}^{L-1} x[n]e^{-j\omega n}, \tag{2.25}$$

where L is the length of the sequence, with L not necessarily equal to N. Generally, we will assume that $L < \infty$, but this assumption is not required for the discussion below. We assume initially that $X(e^{j\omega})$ is known for all ω in $0 \leq \omega < 2\pi$. If we evaluate (sample) $X(e^{j\omega})$ at N equally spaced normalized frequencies $\omega_k = (2\pi k/N)$, for $k = 0, 1, \ldots, N-1$, then we obtain the samples

$$X_N[k] = X(e^{j\frac{2\pi}{N}k}) = \sum_{n=0}^{L-1} x[n]e^{-j\frac{2\pi}{N}kn}, \quad k = 0, 1, \ldots, N-1. \tag{2.26}$$

First observe that if $L = N$, Eq. (2.26) is identical to Eq. (2.23a). That is, in the case of an N-sample finite-length sequence, the samples of the DTFT at frequencies $(2\pi k/N)$ are identical to the DFT values computed using Eq. (2.23a). This is also the case when $L < N$; i.e., when the sequence $x[n]$ is "padded" with zero samples to obtain a sequence of length N. In this case, the zero samples are perfectly correct since $x[n]$ is assumed to be zero outside the interval $0 \leq n \leq L-1$. Thus, if we use the samples $X_N[k]$ in the inverse DFT relation, Eq. (2.23a), to compute

$$x_N[n] = \frac{1}{N}\sum_{k=0}^{N-1} X_N[k]e^{j\frac{2\pi}{N}kn}, \quad n = 0, 1, \ldots, N-1, \tag{2.27}$$

it follows that if $L \leq N$, then $x_N[n] = x[n]$ for $n = 0, 1, \ldots, N-1$.

In the case $L > N$, the result is not so obvious. However, if $X_N[k]$ is defined by sampling the DTFT of a sequence $x[n]$ (finite or not) as in Eq. (2.26), and if the resulting samples are used in the inverse DFT as in Eq. (2.27), then it can be shown [270] that

$$x_N[n] = \sum_{r=-\infty}^{\infty} x[n+rN], \quad n = 0, 1, \ldots, N-1. \tag{2.28}$$

In other words, $x_N[n]$ is equal to the implicit periodic sequence corresponding to the sampled DTFT. This result is referred to as the *time-aliasing* interpretation of the DFT. Clearly, if $x[n]$ is non-zero only in the interval $0 \leq n \leq N-1$ (i.e., $L \leq N$), the shifted copies of $x[n]$ will not overlap, and therefore $x_N[n] = x[n]$ in that interval. If, however, $L > N$, then some of the shifted copies $x[n+rN]$ will overlap into the interval $0 \leq n \leq N-1$, and we say that *time-aliasing distortion* has occurred. This interpretation is particularly useful when using the DFT in computing correlation functions, in implementing filtering operations, and in cepstrum analysis (to be discussed in Chapter 8).

Figure 2.5 shows plots of the time waveform and the resulting line spectrum of a periodic signal (the upper pair of plots) and a finite duration signal (the lower pair of

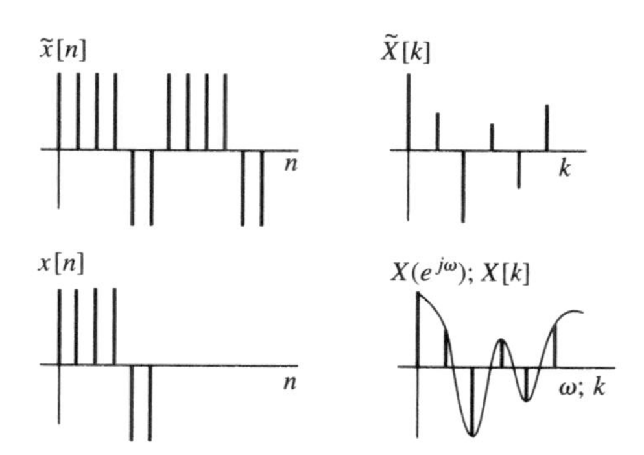

FIGURE 2.5
Plots of the time waveform and the DFT representation of a periodic signal (top set of plots) and a finite duration signal (bottom set of plots). Notice that the line spectrum of the periodic signal and the samples of the DTFT of the finite-duration signal are identical.

plots), which exactly matches one period of the periodic signal. It is clearly seen that the periodic waveform has a line spectrum, $\tilde{X}[k]$, whereas the finite duration signal has a continuous spectrum, $X(e^{j\omega})$. However, when we overlay the line spectrum of the periodic signal onto the continuous spectrum of the finite duration signal, we see that the line spectrum exactly matches the continuous spectrum at the frequencies where the line spectrum is defined, namely at frequencies $\omega_k = 2\pi k/N$, $k = 0, 1, \dots, N-1$. This spectral matching property allows us to treat finite duration signals as though they were periodic in time, so long as we are careful to avoid time aliasing when processing the signal by modifying its DFT.

2.3.6 DFT Properties

The built-in periodicity of the DFT has a significant effect on the properties of the DFT representation. Some of the more important properties are listed in Table 2.3. The most obvious feature is that shifted sequences are shifted modulo N. This leads, for example, to significant differences in discrete convolution.

The DFT representation, with all its peculiarities, is important for a number of reasons:

- The DFT, $X[k]$, can be viewed as a sampled version of the DTFT of a finite-length sequence;
- The DFT has properties very similar (with modifications due to the inherent periodicity) to many of the useful properties of z-transforms and DTFTs;
- The N values of $X[k]$ can be computed very efficiently (with time proportional to $N \log N$) by a set of computational algorithms known collectively as the *fast Fourier transform* (FFT) [64, 245, 270, 305].

TABLE 2.3 Real finite-length sequences and their corresponding DFTs.

Property	Sequence	N-point DFT
1. Linearity	$ax_1[n] + bx_2[n]$	$aX_1[k] + bX_2[k]$
2. Shift	$x[((n - n_0))_N]$	$e^{-j(2\pi k/N)n_0} X[k]$
3. Modulation	$x[n]e^{j(2\pi k_0/N)n}$	$X[((k - k_0))_N]$
4. Time reversal	$x[((-n))_N]$	$X[((-k))_N] = X^*[k]$
5. Convolution	$\sum_{m=0}^{N-1} x[m]h[((n - m))_N]$	$X[k]H[k]$
6. Multiplication of sequences	$x[n]w[n]$	$\frac{1}{N}\sum_{r=0}^{N-1} X[r]W[((k - r))_N]$
7. Parseval's Theorem	$\sum_{n=0}^{N-1} \lvert x[n]\rvert^2 =$	$\frac{1}{N}\sum_{k=0}^{N-1} \lvert X[k]\rvert^2$

The DFT is widely used for computing spectrum estimates, for correlation functions, and for implementing digital filters [142, 270, 380]. We will have frequent occasions to apply DFT representations in speech processing.

2.4 FUNDAMENTALS OF DIGITAL FILTERS

A digital filter is a discrete-time linear shift-invariant system. Recall that for such a system, the input and output are related by the convolution sum expression of Eq. (2.6a) or (2.6b). The corresponding relation between the z-transform of the sequences involved is as given in Table 2.1; i.e.,

$$Y(z) = H(z)X(z). \tag{2.29}$$

The z-transform of the unit sample response, $H(z)$, is called the *system function.* The term *transfer function* is also used interchangeably for system function. The DTFT, $H(e^{j\omega})$, of the unit impulse response, $h[n]$, is called the *frequency response.* $H(e^{j\omega})$ is, in general, a complex function of ω, which can be expressed in terms of real and imaginary parts as

$$H(e^{j\omega}) = H_r(e^{j\omega}) + jH_i(e^{j\omega}), \tag{2.30}$$

or in terms of magnitude and phase angle as

$$H(e^{j\omega}) = |H(e^{j\omega})|e^{j\arg[H(e^{j\omega})]}. \tag{2.31}$$

A *causal* linear shift-invariant system is one for which $h[n] = 0$ for $n < 0$. A *stable* system is one for which every bounded input produces a bounded output. A necessary and sufficient condition for a linear shift-invariant system to be stable is

$$\sum_{n=-\infty}^{\infty} |h[n]| < \infty. \tag{2.32}$$

This condition is identical to Eq. (2.16), and therefore stability is sufficient for the existence of $H(e^{j\omega})$.

In addition to the convolution sum expression of Eqs. (2.6a) and (2.6b), all linear shift-invariant systems of interest for practical implementation as filters have the property that the input and output satisfy a linear difference equation of the form

$$y[n] - \sum_{k=1}^{N} a_k y[n-k] = \sum_{r=0}^{M} b_r x[n-r]. \tag{2.33}$$

By evaluating the z-transform of both sides of this equation, it follows that

$$H(z) = \frac{Y(z)}{X(z)} = \frac{\sum_{r=0}^{M} b_r z^{-r}}{1 - \sum_{k=1}^{N} a_k z^{-k}}. \tag{2.34}$$

A useful observation results from comparing Eq. (2.33) to Eq. (2.34); given a difference equation in the form of Eq. (2.33), we can obtain $H(z)$ directly by simply identifying the coefficients of the delayed inputs in Eq. (2.33) with corresponding powers of z^{-1} in the numerator and coefficients of the delayed output with corresponding powers of z^{-1} in the denominator. Similarly, it is straightforward to write down Eq. (2.33) from an inspection of Eq. (2.34).

The system function, $H(z)$, is, in general, a rational function of z^{-1}. As such, it is characterized by the locations of its poles and zeros in the z-plane. Specifically $H(z)$ can be expressed as

$$H(z) = \frac{A \prod_{r=1}^{M} (1 - c_r z^{-1})}{\prod_{k=1}^{N} (1 - d_k z^{-1})}. \tag{2.35}$$

From our discussion of z-transforms, we recall that a causal discrete-time system with $h[n] = 0$ for $n < 0$ will have a region of convergence of the form $|z| > R_1$. If the system is also stable, then R_1 must be less than unity so that the region of convergence contains the unit circle. Therefore the poles of $H(z)$ must all be inside the unit circle for a stable and causal system. If all the poles *and* all the zeros of $H(z)$ are inside the unit circle, the system is called a *minimum-phase* system [270].

It is convenient to define two classes of linear shift-invariant systems. These are the class of finite-duration impulse response (FIR) systems and the class of infinite duration impulse response (IIR) systems. These classes have distinct properties which we shall summarize below.

2.4.1 FIR Systems

If all the coefficients, a_k, in Eq. (2.33) are zero, the difference equation becomes

$$y[n] = \sum_{r=0}^{M} b_r x[n-r]. \tag{2.36}$$

Comparing Eq. (2.36) to Eq. (2.6b), we observe that

$$h[n] = \begin{cases} b_n & 0 \leq n \leq M \\ 0 & \text{otherwise.} \end{cases} \tag{2.37}$$

FIR systems have a number of important properties. First, we note that $H(z)$ is a polynomial in z^{-1}, and thus $H(z)$ has no non-zero poles, only zeros. Also, FIR systems can have an exactly linear phase. If $h[n]$ satisfies the relation

$$h[n] = \pm h[M - n], \tag{2.38}$$

then $H(e^{j\omega})$ has the form

$$H(e^{j\omega}) = A(e^{j\omega})e^{-j\omega(M/2)}, \tag{2.39}$$

where $A(e^{j\omega})$ is either purely real or imaginary depending upon whether Eq. (2.38) is satisfied with $+$ or $-$ respectively.

The design capability for implementing digital filters that have *exactly* linear phase is often very useful in speech processing applications, especially where precise time alignment is essential. This property of FIR filters also can greatly simplify the problem of designing approximations to ideal frequency-selective digital filters since it is only necessary to be concerned with approximating a desired magnitude response. The penalty that is paid for filters with an exact linear phase response is that a larger impulse response duration is required to adequately approximate sharp cutoff filters.

2.4.2 FIR Filter Design Methods

Based on the properties associated with linear phase FIR filters, there have been developed three well-known design methods for approximating an arbitrary set of specifications with an FIR filter. These three methods are:

- Window approximation [142, 184, 245, 270, 305];
- Frequency sampling approximation [245, 270, 305, 306];
- Optimal (minimax error) approximation [238, 245, 270, 282, 305, 310].

Only the first of these techniques is an analytical design technique, i.e., one based on a closed form set of equations that can be solved directly to obtain the filter coefficients. The second and third design methods are optimization methods that use iterative approaches to obtain the desired filter. Although the window method is simple to apply, the optimal design method is the most widely used design algorithm. This is in part due to a series of intensive investigations into the properties of optimal FIR filters, and in part due to the general availability of a well-documented design program that enables the designer to approximate any desired set of specifications [238, 270, 305].

MATLAB Design of Optimal FIR Lowpass Filter

It is straightforward to use MATLAB to design optimal FIR frequency-selective filters such as lowpass filters, highpass filters, bandpass filters, bandstop filters, and also other types such as differentiators and Hilbert transformers. There are two distinct methods within MATLAB for designing these filters, namely using `fdatool` (which is an interactive design procedure for specifying the FIR filter specifications), or using the function `firpm`.

The following example shows the use of `firpm` to design an optimal lowpass FIR filter, to compute the resulting frequency response, and to plot the log magnitude response. The steps in the procedure are as follows:

- Use the MATLAB function `firpm` to design the FIR filter with a call of the form `B=firpm(N,F,A)` where the resulting filter is a linear phase approximation that has an impulse response length of `N+1` samples, and where `B` is a vector containing the resulting impulse response (the coefficients of the numerator polynomial of the system function). The vector `F` contains the specifications of the ideal frequency response band edges (in pairs and normalized to 1.0 at half the sampling frequency, ω/π rather than $\omega/(2\pi)$ as would usually be the case), and `A` contains the specifications of the ideal amplitude response values in the approximation intervals (also in begin/end pairs).
- Use the MATLAB command `freqz` to evaluate the filter frequency response with a call of the form `[H,W]=freqz(B,1,NF)` where `H` is the complex frequency response of the resulting FIR filter, `W` is the set of radian frequencies at which the frequency response is evaluated (in the range $0 \leq \omega \leq \pi$), `B` is the filter numerator polynomial, namely the impulse response that was computed using `firpm` above, the term `1` is the denominator polynomial for an FIR filter, and `NF` is the number of frequencies at which the frequency response of the filter is evaluated.
- Use the MATLAB command `plot` to plot the log magnitude response of the filter with a call of the form `plot(W/pi, 20log10(abs(H)))`. Note that we plot the quantity $20\log_{10}(|H|)$ as the frequency response log magnitude wherein each factor of 20 dB corresponds to a factor of 10-to-1 in scale of the filter frequency response magnitude. Such a plot enables the designer to rapidly assess the filtering capability of the resulting approximation.

To design a 31-point optimal FIR lowpass filter using `firpm`, we need to set up the specifications as follows:

- `N=30` (which creates a 31-point FIR filter);
- `F=[0 0.4 0.5 1]` [which sets the ideal passband to the range $0 \leq \omega \leq 0.4\pi$ (or in analog frequency terms the edge of the passband is set to $0.2F_s$) and the ideal stopband to the range $0.5\pi \leq \omega \leq \pi$ (or again in analog frequency terms the edge of the stopband is set to $0.25F_s$)];

- `A=[1 1 0 0]` (which sets the ideal amplitude to 1 in the passband and to 0 in the stopband);
- `NF=512` (which sets the number of frequency values at which the frequency response is evaluated to 512).

With the specifications set as above, the following MATLAB program designs and plots the system response characteristics:

```
>> N=30;                          % filter order
>> F=[0 0.4 0.5 1];               % frequency band edges
>> A=[1 1 0 0];                   % gains
>> B=firpm(N,F,A);                % compute impulse response
>> NF=512;                        % number of frequencies to
                                    compute
>> [H,W]=freqz(B,1,NF);           % compute frequency response
>> plot(W/pi,20*log10(abs(H)));   % plot frequency response
```

The impulse and log magnitude (in dB) frequency response of the resulting 31-point optimal FIR lowpass filter are shown in Figure 2.6. It can be seen that the optimal filter is an equiripple approximation in both the passband and stopband.

2.4.3 FIR Filter Implementation

In considering the implementation of digital filters, it is often useful to represent the filter in block diagram form. The difference equation of Eq. (2.36) is depicted in Figure 2.7. Such a diagram, often called a digital filter structure, graphically depicts the operations required to compute each value of the output sequence from values of the input sequence. The basic elements of the diagram depict means for addition, multiplication of sequence values by constants (constants indicated on branches imply multiplication), and storage of past values of the input sequence. Thus the block diagram gives a clear indication of the complexity of the system. When the system has linear phase, further significant simplifications can be incorporated into the implementation. (See Problem 2.11.)

2.4.4 IIR Systems

If the system function of Eq. (2.35) has poles as well as zeros, then the difference equation of Eq. (2.33) can be written as

$$y[n] = \sum_{k=1}^{N} a_k y[n-k] + \sum_{r=0}^{M} b_r x[n-r]. \tag{2.40}$$

This equation is a recurrence formula that can be used sequentially to compute the values of the output sequence from past values of the output and present and past

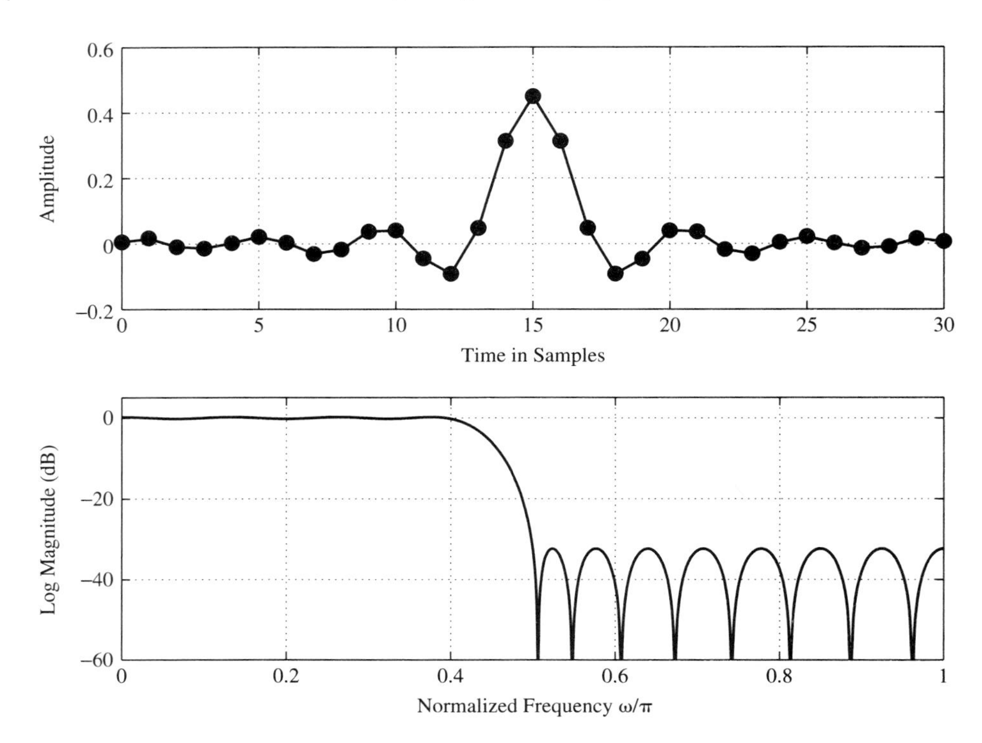

FIGURE 2.6
Plots of the impulse response (top) and log magnitude (in dB) frequency response (bottom) of the $N = 31$ optimal FIR lowpass filter designed using `firpm` in MATLAB.

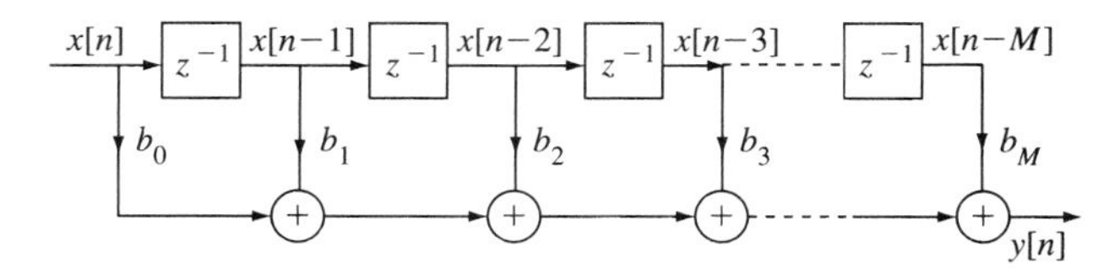

FIGURE 2.7
Digital network for FIR system.

values of the input sequence. If $M \leq N$ in Eq. (2.35), $H(z)$ can be expanded in a partial fraction expansion as in

$$H(z) = A_0 + \sum_{k=1}^{N} \frac{A_k}{1 - d_k z^{-1}}, \tag{2.41}$$

where $A_0 = 0$ when $M < N$. For a causal system, it is easily shown (see Problem 2.16) that

$$h[n] = A_0\delta[n] + \sum_{k=1}^{N} A_k(d_k)^n u[n]. \tag{2.42}$$

Thus, we see that $h[n]$ has infinite duration. However, because of the recurrence formula of Eq. (2.40), it is often possible to implement an IIR filter that approximates a given set of specifications more efficiently (i.e., using fewer computations) than is possible with an FIR system. This is particularly true for sharp cutoff, frequency-selective filters.

2.4.5 IIR Filter Design Methods

A wide variety of design methods are available for IIR filters. Design methods for frequency-selective filters (lowpass, bandpass, etc.) are generally based on transformations of classical analog design procedures that are straightforward to implement. Included in this class are:

- Butterworth approximation (maximally flat amplitude);
- Bessel approximation (maximally flat group delay);
- Chebyshev approximation (equiripple in either passband or stopband);
- Elliptic approximation (equiripple in both passband and stopband).

All the above methods are analytical in nature and have been widely applied to the design of IIR digital filters [245, 270, 305]. In addition a variety of IIR optimization methods have been developed for approximating design specifications that are not easily adapted to one of the above approximation methods [83].

Elliptic Filter Design in MATLAB

MATLAB provides functions for designing each of the types of IIR filters listed above. In this section we show the design of an elliptic lowpass filter with the appropriate calls to MATLAB functions. The first step is the specification of the lowpass filter characteristics, followed by a call to the MATLAB design module `ellip`. For this simple lowpass filter example, we use the following design specifications:

- `N`, the filter order, set to 6 in this example; hence the numerator polynomial, `B`, will have `N+1` coefficients, and the denominator polynomial, `A`, will also have `N+1` coefficients;
- `Rp`, the maximum in-band (passband) approximation error (in dB), set to 0.1 dB in this example;
- `Rs`, the out-of-band (stopband) ripple (in db), set to 40 dB in this example;
- `Wp`, the end of the passband, set to $\omega_p/\pi = 0.45$ in this example.

The sequence of calls to MATLAB functions for designing the elliptic lowpass filter, and for determining its impulse response, its log magnitude frequency response, and the z-plane pole-zero locations, are as follows:

```
>> [B,A]=ellip(6, 0.1, 40, 0.45);   % design elliptic lowpass
                                      filter
>> x=[1,zeros(1,99)];               % generate input impulse
>> h=filter(B,A,x);                 % generate filter impulse
                                      response
>> [H,W]=freqz(B,A,512);            % generate filter
                                      frequency response
>> zplane(B,A);                     % generate pole-zero plot
```

Figure 2.8 shows plots of the impulse response (top left plot—truncated to 100 samples), the log magnitude (in dB) frequency response (bottom left plot), and the pole-zero diagram in the z-plane (plot on right). The equiripple nature of the elliptic filter approximation in the stopband is seen in the log magnitude frequency response plot. Although the very small passband ripple is not visible on this plot, zooming in on the passband in the MATLAB-displayed plot confirms the equiripple behavior there as well. Note that the stopband begins at about $\omega_s/\pi = 0.51$. This value is implicitly determined by specifying `N`, `Rp`, `Rs`, and `Wp`. The MATLAB function `ellipord` determines the system order required to realize prescribed approximation errors and band edges.

2.4.6 Implementations of IIR Systems

There is considerable flexibility in the implementation of IIR systems. The network implied by Eq. (2.40) is depicted in Figure 2.9a, for the case $M = N = 4$. This is often called the direct form I implementation. The generalization to arbitrary M and N is obvious. The difference equation (2.40) can be transformed into many equivalent forms. Particularly useful among these is this set of equations:

$$w[n] = \sum_{k=1}^{N} a_k w[n-k] + x[n], \tag{2.43a}$$

$$y[n] = \sum_{r=0}^{M} b_r w[n-r]. \tag{2.43b}$$

(See Problem 2.17.) This set of equations can be implemented as shown in Figure 2.9b, with a significant saving of memory required to store the delayed sequence values. This implementation is called the direct form II.

Equation (2.35) shows that $H(z)$ can be expressed as a product of poles and zeros. These poles and zeros occur in complex conjugate pairs when the coefficients a_k and b_r are real. By grouping the complex conjugate poles and zeros into complex conjugate

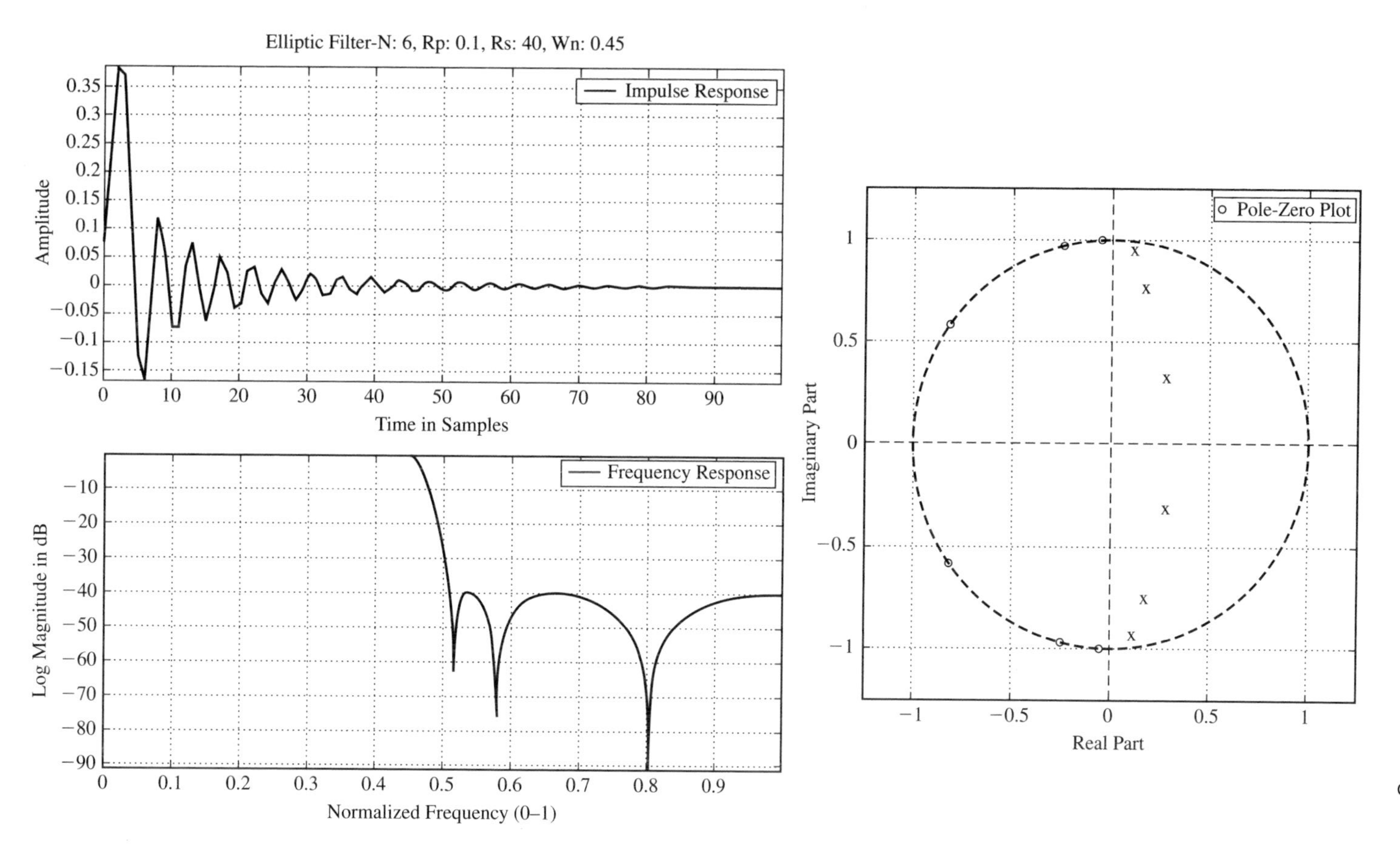

FIGURE 2.8
Elliptic lowpass filter properties: impulse response (top left), log magnitude (in dB) frequency response (bottom left), and pole-zero plot (right side).

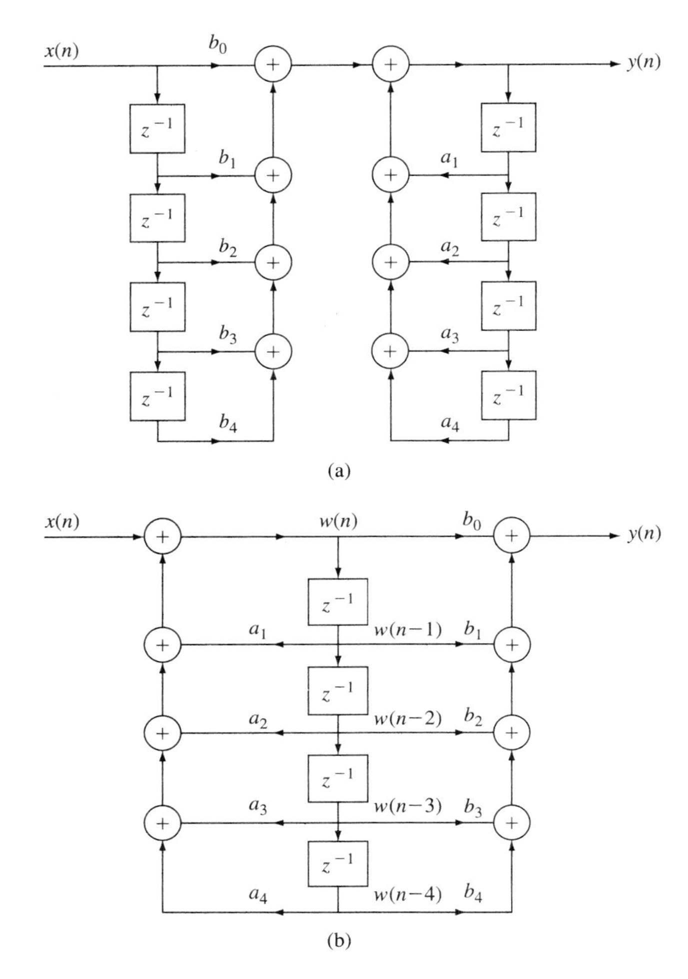

FIGURE 2.9
Direct form IIR structures: (a) direct form I structure; (b) direct form II structure with minimum storage.

pairs, it is possible to express $H(z)$ as a product of elementary second-order system functions, of the form

$$H(z) = A\prod_{k=1}^{K}\left(\frac{1 + b_{1k}z^{-1} + b_{2k}z^{-2}}{1 - a_{1k}z^{-1} - a_{2k}z^{-2}}\right), \tag{2.44}$$

where K is the integer part of $(N+1)/2$. Each second-order system can be implemented as in Figure 2.9 and the systems cascaded to implement $H(z)$. This is depicted in Figure 2.10a for $N = M = 4$. Again, the generalization to higher orders is obvious. The partial fraction expansion of Eq. (2.41) suggests still another approach to

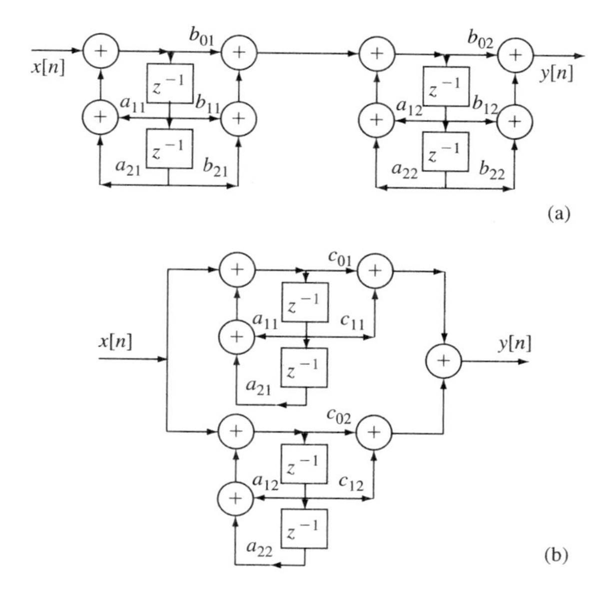

FIGURE 2.10
(a) Cascade form IIR structure; (b) parallel form structure.

implementation. By combining terms involving complex conjugate poles, $H(z)$ can be expressed as

$$H(z) = A_0 + \sum_{k=1}^{K} \frac{c_{0k} + c_{1k}z^{-1}}{1 - a_{1k}z^{-1} - a_{2k}z^{-2}}, \tag{2.45}$$

where $A_0 = 0$ if $M < N$. This suggests a parallel form implementation as depicted in Figure 2.10b for $N = 4$ and $M < N$. (If $M = N$, an additional parallel path with gain A_0 would be added to the diagram.)

All of the implementations discussed are used in speech processing. For linear filtering applications, the cascade form generally exhibits superior performance with respect to round-off noise, coefficient inaccuracies, and stability [245, 270, 305]. All of the above forms have been used in speech synthesis applications, with the direct form being particularly important in synthesis from linear prediction parameters. (See Chapter 9.)

2.4.7 Comments on FIR and IIR Filter Design Methods

The major difference between FIR and IIR filters is that IIR filters cannot have exact linear phase, whereas FIR filters can have this property. In exchange, the IIR filter is often significantly more efficient in realizing sharp cutoff filters than FIR filters [309]. The linear phase property is often very valuable when time delays must be carefully equalized. Linear phase filters are also very attractive in sampling rate changing applications as discussed in the following section.

2.5 SAMPLING

To use digital signal processing methods on an analog signal such as speech, it is necessary to represent the signal as a sequence of numbers. This is commonly done by sampling the analog signal, denoted $x_a(t)$, periodically, to produce the sequence

$$x[n] = x_a(nT), \qquad -\infty < n < \infty, \tag{2.46}$$

where n takes on only integer values.

2.5.1 The Sampling Theorem

The conditions under which the sequence of samples in Eq. (2.46) is a unique representation of the original analog signal are well known and are summarized as follows:

> *The Sampling Theorem*: If a signal $x_a(t)$ has a bandlimited Fourier transform $X_a(j\Omega)$ such that $X_a(j\Omega) = 0$ for $\Omega \geq 2\pi F_N$, then $x_a(t)$ can be uniquely reconstructed from equally spaced samples $x_a(nT)$, $-\infty < n < \infty$, if the sampling rate $F_s = 1/T$ satisfies the condition $F_s \geq 2F_N$.

The above theorem follows from a relationship between the CTFT of $x_a(t)$ and the DTFT of the sample sequence $x[n] = x_a(nT)$. Specifically, if the CTFT $X_a(j\Omega)$ is defined as in Eq. (2.7b) and the DTFT of the sequence $x[n]$ is defined as in Eq. (2.15a), then $X(e^{j\Omega T})$ is related to $X_a(j\Omega)$ by [245, 270, 305]

$$X(e^{j\Omega T}) = \frac{1}{T}\sum_{k=-\infty}^{\infty} X_a\left(j\Omega + j\frac{2\pi}{T}k\right). \tag{2.47}$$

Note that in this relation the DTFT normalized frequency variable ω is expressed explicitly in terms of Ω as

$$\omega = \Omega T. \tag{2.48}$$

This equation is the key link between the continuous-time and discrete-time frequency domains. Observe that Ω has units of radians/sec while T is of course in seconds. Therefore ΩT is in radians as it should be for normalized frequency ω. To see the implications of Eq. (2.47), assume that $X_a(j\Omega)$ is *bandlimited* as shown in Figure 2.11a; i.e., assume that $X_a(j\Omega) = 0$ for $|\Omega| \geq \Omega_N = 2\pi F_N$. The frequency F_N, at and beyond which the continuous-time Fourier spectrum remains zero, is called the *Nyquist frequency*. Now, according to Eq. (2.47), $X(e^{j\Omega T})$ is the sum of an infinite number of replicas of $X_a(j\Omega)$, each centered at an integer multiple of $\Omega_s = 2\pi F_s = 2\pi/T$. Figure 2.11b depicts the case when $F_s = 1/T > 2F_N$ so that the images of the Fourier transform do not overlap into the baseband $|\Omega| < (\Omega_N = 2\pi F_N)$. Clearly, no overlap occurs if $X_a(j\Omega)$ is bandlimited and $F_s = 1/T \geq 2F_N$. When $F_s = 2F_N$ we say that sampling is at the *Nyquist rate*. When $F_s > 2F_N$ as in Figure 2.11b, the signal is said to be *oversampled*. Figure 2.11c, on the other hand, shows the case $F_s < 2F_N$, where the image centered at $2\pi/T$ overlaps into the baseband. This *undersampled* condition,

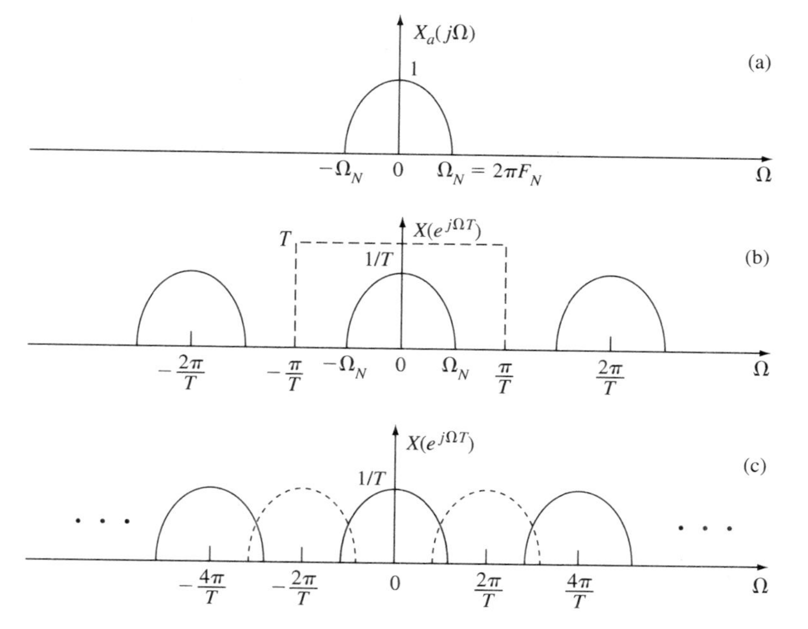

FIGURE 2.11
Frequency-domain illustration of sampling: (a) bandlimited Fourier transform of $x_a(t)$; (b) DTFT of samples $x[n] = x_a(nT)$ when $F_s = 1/T > 2F_N$ (oversampling case); (c) DTFT of samples $x[n] = x_a(nT)$ when $F_s = 1/T < 2F_N$ (undersampling case).

where a high frequency seemingly takes on the identity of a lower frequency, results in *aliasing distortion* or simply *aliasing*. Clearly, aliasing can be avoided only if the CTFT is bandlimited *and* if the sampling frequency $(1/T)$ is equal to at least twice the Nyquist frequency $(F_s = 1/T \geq 2F_N)$.

Under the condition $F_s \geq 2F_N$, it is clear that the Fourier transform of the sequence of samples is proportional to the Fourier transform of the analog signal in the baseband; i.e.,

$$X(e^{j\Omega T}) = \frac{1}{T} X_a(j\Omega), \qquad |\Omega| < \frac{\pi}{T}. \tag{2.49}$$

Therefore, it can be seen from Eq. (2.49) that the CTFT of the original signal can be recovered from the DTFT of the signal samples by (in effect) multiplying $X(e^{j\Omega T})$ by the frequency response function of an ideal lowpass filter having gain T (to compensate for the scale factor $1/T$) and cutoff frequency π/T. The ideal reconstruction filter is shown as the dashed line in Figure 2.11b. The corresponding time-domain relation is the bandlimited interpolation formula [245, 270, 305]

$$x_a(t) = \sum_{k=-\infty}^{\infty} x_a(kT) \left[\frac{\sin\left[\pi(t - kT)/T\right]}{\pi(t - kT)/T} \right]. \tag{2.50}$$

Thus, given samples of a bandlimited analog signal taken at a rate at least twice the Nyquist frequency, it is possible to reconstruct the original analog signal from the samples $x[n] = x_a(nT)$ using Eq. (2.50). Using oversampling and noise shaping techniques, practical digital-to-analog converters can approximate Eq. (2.50) very accurately [270].

2.5.2 Sampling Rates for Speech and Audio Waveforms

We saw in the previous section that in order to preserve the properties of the analog (continuous-time) waveform, we must prevent aliasing distortion by sampling the signal at a rate that is at least twice the highest frequency present in the signal spectrum. For speech and audio signals, this highest frequency is usually defined by the intended application. By filtering the speech or audio signal before sampling, a variety of Nyquist frequencies can be imposed. The following is a brief summary:

Wideband Speech Since most sounds in speech have most of their energy concentrated at frequencies in the band 50–7500 Hz, the intrinsic Nyquist frequency of speech is on the order of 7500–10,000 Hz. Thus a sampling rate of 15,000–20,000 Hz (or higher) is adequate to preserve most of the temporal or spectral detail of the speech signal.

Telephone Bandwidth Speech For telephone communication, the speech spectrum is generally restricted to the band from 300 to 3200 Hz by transmission lines and additional lowpass filters in the communication system. This implies a Nyquist frequency of 3200 Hz, which would allow sampling at a rate as low as $F_s =$ 6400 Hz, but telephone communication applications typically use $F_s = 8000$ Hz (or some integer multiple thereof).

Wideband Audio For audio signals with significant signal energy in the band from 50 to 21,000 Hz, the Nyquist frequency is on the order of 21,000 Hz, and we should sample at a rate of 42,000 Hz (or higher).

The key point is that the Nyquist frequency often depends heavily on the intended application of the sampled signal. Hence, although speech inherently has a bandwidth on the order of 7500 Hz, we often use a lowpass filter to restrict the bandwidth to 3200 Hz for narrowband digital telephony applications so that we can sample the analog signal at much lower rates than would be required for wideband speech. Similarly for audio signals, where we desire to preserve the frequency range for a wide range of audio inputs, we sample at high rates (44.1 kHz, 48 kHz, or even 96 kHz) so as to preserve the sound integrity up to the highest frequencies of hearing, namely 21 kHz.

Sampling is also implicit in many speech processing algorithms that seek to estimate basic parameters of speech production such as pitch period and formant frequencies. In such cases an analog function is not available to be sampled directly, as in the case of sampling the speech waveform itself. However, such parameters change very slowly with time, and thus it is possible to estimate (sample) them at rates commensurate with their inherent communication bandwidth, generally on the order of 100 samples/sec for many speech-derived parameters. Given the samples of a speech parameter, a bandlimited analog function for that parameter theoretically can be constructed using Eq. (2.50), although it would be more likely that we would simply want to change the sampling rate using the techniques to be discussed next.

2.5.3 Changing the Sampling Rate of Sampled Signals

In many examples that we discuss in this book, it is necessary to change the sampling rate of a discrete-time signal. One example occurs when audio is sampled at a 44.1 kHz rate (as is done for CD recordings) and we want to convert the sampling rate to that appropriate for a digital audio tape (DAT) player, namely 48 kHz. A second example occurs when we want to convert wideband speech, sampled at a 16 kHz rate, to narrowband speech (either 8 kHz or 6.67 kHz sampling rate) for telephone transmission or reduced rate storage applications. The processes of sampling rate reduction and sampling rate increase are well understood and are generally called decimation and interpolation.

Figure 2.12a shows a sampled signal $x[n]$ obtained with sampling period T, along with a schematic representation of its discrete Fourier transform, $X(e^{j\Omega T})$. We see that the highest frequency present in the signal is $\Omega = \pi/T$; i.e., the original continuous-time signal was sampled at its Nyquist rate. The process of decimating the signal by a factor of $M = 2$, thereby creating a new sampled signal with sampling period $T' = 2T$, is shown in Figure 2.12b. The decimated signal, $x_d[n]$, is defined at samples $n = 0, 1, 2, \ldots$, corresponding to times $nT' = \{0, T' = 2T, 2T' = 4T, \ldots\}$ and its DFT, $X_d(e^{j\Omega T'})$, has as its highest frequency $\Omega = \pi/T'$, which is half the highest frequency of the original (undecimated) signal. The simple process of throwing away every second sample of the original signal (downsampling) leads to an aliased representation of the

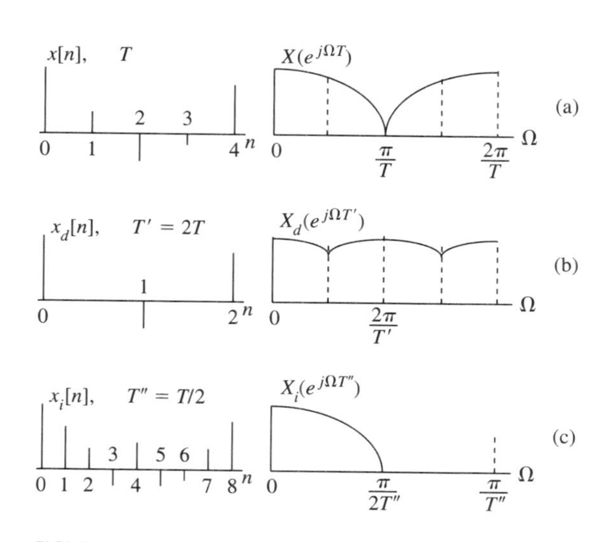

FIGURE 2.12
Illustration of decimation and interpolation processes: (a) original sampled signal; (b) decimated signal ($M = 2$); (c) interpolated signal ($L = 2$).

original signal unless we employ an appropriate lowpass filter prior to the downsampling of the signal (as is assumed in Figure 2.12b) or unless the original sampling rate for $x[n] = x_a(nT)$ was twice the Nyquist rate. We will see later in this section how we can insure that no aliasing occurs in the process of decimating a signal that was not sufficiently oversampled.

Figure 2.12c shows the process of interpolating the signal by a factor of $L = 2$, thereby creating a signal with sampling period $T'' = T/2$. The interpolated signal, $x_i[n]$, is defined at samples $n = 0, 1, 2, \ldots$, corresponding to times $nT'' = \{0, T'' = T/2, 2T'' = T, 3T'' = 3T/2, \ldots\}$. The corresponding DTFT $X_i(e^{j\Omega T''})$ has highest frequency $\Omega = \pi/T = \pi/(2T'')$ in the band $|\Omega| \leq \pi/T''$, since interpolation creates an oversampling condition relative to the sampling rate of the original signal. That is, in the frequency interval between $\Omega = \pi/T$ and $\Omega = \pi/T''$, the DFT of the interpolated signal must be exactly zero since this is the frequency region where no spectral information is available or known. This is not what occurs naturally by simply inserting zero samples between each of the original samples. Again, we will see later how we can ensure that no signal energy appears in the upper frequency band (between $\Omega = \pi/(2T'')$ and $\Omega = \pi/T''$) by using appropriate filtering methods.

2.5.4 Decimation

In discussing both interpolation and decimation, we assume that we have a sequence of samples $x[n] = x_a(nT)$, of an analog signal $x_a(t)$ that is bandlimited such that $X_a(j\Omega) = 0$ for $|\Omega| \geq 2\pi F_N$. Such a bandlimited Fourier transform is depicted in Figure 2.13a. Furthermore, we assume that the signal has been sampled without aliasing distortion with $F_s = 1/T \geq 2F_N$ so that the DTFT of $x[n]$ satisfies the relation

$$X(e^{j\Omega T}) = \frac{1}{T} X_a(j\Omega), \quad |\Omega| < \frac{\pi}{T}. \tag{2.51}$$

As illustrated in Figure 2.13b, this means that

$$X(e^{j\Omega T}) = 0, \quad 2\pi F_N \leq |\Omega| \leq 2\pi(F_s - F_N). \tag{2.52}$$

Since $F_s \geq 2F_N$, there is no overlap (aliasing) between the periodic images of the analog frequency response, and the original analog signal can be recovered exactly as discussed in the previous section.

Now assume that we wish to reduce the sampling rate of the sampled signal of Figure 2.13b by a factor of $M \geq 2$; i.e., we want to compute a new sequence, $x_d[n]$, with sampling rate

$$F_s' = \frac{1}{T'} = \frac{1}{MT} = \frac{F_s}{M}, \tag{2.53}$$

and we wish to derive $x_d[n] = x_a(nT')$ by operations on the original discrete-time signal, $x[n]$, but with no aliasing occurring as a result of the sampling rate reduction operations.

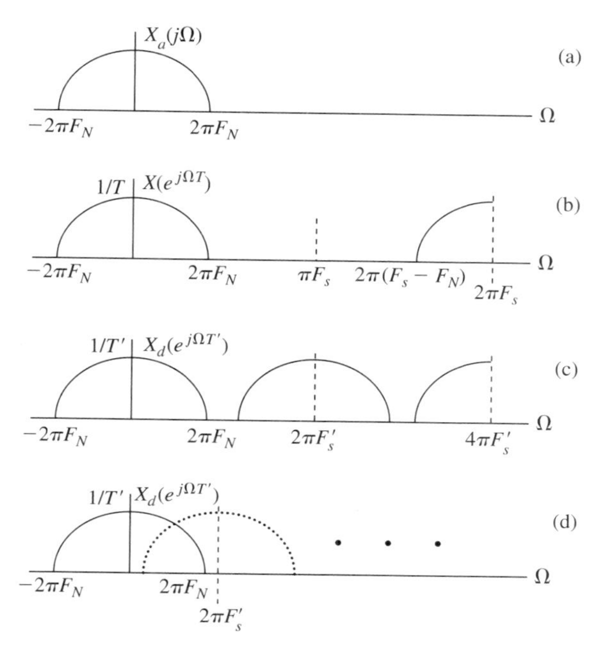

FIGURE 2.13
Illustration of downsampling: (a) bandlimited CTFT; (b) DTFT of sampled signal $x[n]$ where $F_s > 2F_N$; (c) DTFT of $x_d[n] = x[nM]$ after downsampling by $M = 2$; (d) DTFT after downsampling by $M = 4$.

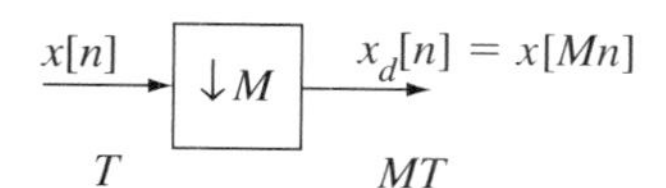

FIGURE 2.14
Block diagram representation of downsampling.

One possible way of obtaining $x_d[n]$ from $x[n]$ is by directly downsampling $x[n] = x_a(nT)$; i.e., by periodically retaining only one out of every M samples of $x[n]$ [ignoring all $(M-1)$ intermediate samples], thereby giving the new discrete-time signal $x_d[n] = x[nM]$. We depict this operation in block diagrams as in Figure 2.14, where the sampling periods of the input and output are indicated explicitly at the bottom of the figure. The resulting downsampled sequence is indeed equal to the desired sequence of samples with sampling period $T' = MT$ since

$$x_d[n] = x[nM] = x_a(nMT) = x_a(nT'). \tag{2.54}$$

However, from our previous discussion of the sampling theorem, we recall that only in the case where the new sampling rate, $F_s' = 1/T'$, satisfies the condition $F_s' \geq 2F_N$ will the samples of $x_d[n]$ be an unaliased representation of the original analog signal; i.e., the original sampling rate, $F_s = 1/T$ (for the signal $x[n]$), must be larger than the minimum (Nyquist) sampling rate for the analog signal, $x_a(t)$, by a factor of M (or greater).

Figure 2.13c illustrates the case of downsampling $x[n]$ by a factor of $M = 2$ when the original sampling rate for $x[n]$ was 4.8 times the frequency, F_N; i.e., $F_s = 1/T = 4.8F_N$ as depicted in Figure 2.13b. In this case we can downsample by a factor of $M = 2$ and still exactly recover the original analog signal with no aliasing since there is no overlap in the periodic images of the analog signal spectrum.

It can be shown that the DTFTs of $x[n]$ and $x_d[n]$ are related by the expression [270, 347]

$$X_d(e^{j\omega}) = \frac{1}{M}\sum_{k=0}^{M-1} X(e^{j(\omega-2\pi k)/M}), \tag{2.55}$$

or, expressing the normalized frequency variable in terms of continuous-time frequencies using the substitution $\omega = \Omega T'$ as is appropriate for the new sampling rate, we obtain

$$X_d(e^{j\Omega T'}) = \frac{1}{M}\sum_{k=0}^{M-1} X(e^{j(\Omega T'-2\pi k)/M}). \tag{2.56}$$

Since we have assumed that the new sampling rate satisfies the criterion $F_s' = 1/(MT) \geq 2F_N$, it follows from Eq. (2.56) that in the new baseband $|\Omega| < \pi/T'$,

$$\begin{aligned} X_d(e^{j\Omega T'}) &= \frac{1}{M}X(e^{j\Omega T'/M}) = \frac{1}{M}X(e^{j\Omega T}) = \frac{1}{M}\frac{1}{T}X_a(j\Omega) \\ &= \frac{1}{T'}X_a(j\Omega), \qquad -\frac{\pi}{T'} < \Omega < \frac{\pi}{T'}, \end{aligned} \tag{2.57}$$

as would have been the case if the original analog signal had been sampled with sampling period T'.

Figure 2.13d illustrates the case of downsampling by a factor of $M = 4$ when the original sampling rate was 4.8 times the frequency F_N as depicted in Figure 2.13b. In this case, when we downsample by $M = 4$, the resulting DTFT is aliased (the periodic images of the analog frequency response overlap) and we *cannot* exactly recover the original analog signal from the aliased discrete-time signal.

In order to properly reduce the sampling rate of a discrete-time signal by a factor of M (i.e., with no aliasing), we must first ensure that the highest frequency present in the discrete-time signal is no greater than the equivalent to $F_s/(2M)$. This can be guaranteed if $x[n]$ is first filtered using an ideal lowpass discrete-time filter having frequency response

$$H_d(e^{j\omega}) = \begin{cases} 1 & |\omega| < \pi/M \\ 0 & \pi/M < |\omega| \leq \pi. \end{cases} \tag{2.58}$$

The cutoff frequency of this filter is $\omega_c = \pi/M$, which corresponds to the continuous-time frequency $F_s/(2M)$; i.e. $2\pi F_s T/(2M) = \pi/M$. We can downsample the resulting lowpass-filtered discrete-time signal by the factor of M without incurring aliasing distortion. These operations are depicted in Figure 2.15. Since $W(e^{j\omega}) = H_d(e^{j\omega})X(e^{j\omega})$, the same analysis that led to Eq. (2.57) results in

$$W_d(e^{j\Omega T'}) = \frac{1}{T'} H_d(e^{j\Omega T}) X_a(j\Omega), \quad -\frac{\pi}{T'} < \Omega < \frac{\pi}{T'}. \tag{2.59}$$

Note that the effective cutoff of $H_d(e^{j\Omega T})$ is the continuous-time frequency $\Omega_c = \omega_c/T = \pi/(MT)$. Thus, in the case when such a lowpass filter is required (to eliminate aliasing), the downsampled signal samples, $w_d[n]$, no longer represent the original analog signal, $x_a(t)$, but rather they represent a new analog signal, $w_a(t)$, which is a lowpass-filtered version of $x_a(t)$.

The combined operations depicted in Figure 2.15 inside the dashed box are commonly called *decimation*. The terminology is not standardized, but we shall make the distinction that decimation includes pre-filtering to avoid aliasing, and downsampling is simply the operation of retaining only every M^{th} sample of a sequence of signal samples.[5]

In summary, the signal processing operations associated with decimation by a factor of M are shown in Figure 2.15. The original signal $x[n]$ at sampling period T is lowpass-filtered by a system with impulse response $h_d[n]$ and frequency response $H_d(e^{j\omega})$, giving the output $w[n]$, still at sampling period T. The decimation block, denoted by the down arrow with the symbol M next to the arrow, retains only one out of every M samples of $w[n]$, giving the reduced rate sequence $w_d[n]$ with corresponding sampling period $T' = MT$.

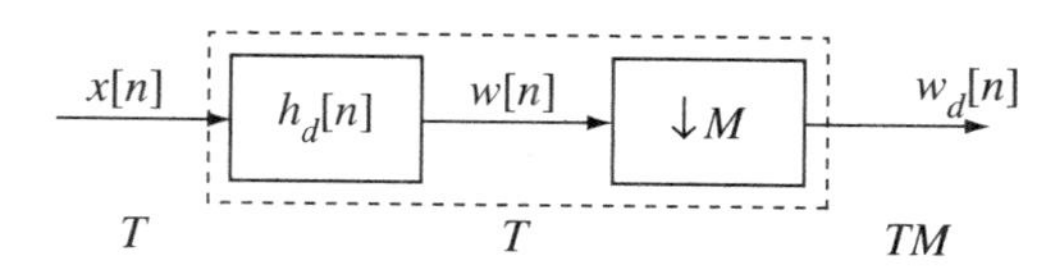

FIGURE 2.15
Signal processing operations for decimation when the input $x[n]$ is not sufficiently oversampled.

[5]Purists object that decimation logically should mean downsampling by a factor of 10, but the term is commonly applied as we have stated.

2.5.5 Interpolation

Now suppose that we have samples of an analog waveform $x[n] = x_a(nT)$ obtained by sampling at the Nyquist rate or higher. If we wish to increase the sampling rate by an integer factor L, we must compute a new sequence corresponding to samples of $x_a(t)$ taken with period $T'' = T/L$; i.e.,

$$x_i[n] = x_a(nT'') = x_a(nT/L). \tag{2.60}$$

Clearly, $x_i[n] = x[n/L]$ for $n = 0, \pm L, \pm 2L, \ldots$, but we must fill in the unknown samples for all other values of n by an interpolation process [270, 347].

One way to look at this problem is to recall that we can reconstruct the signal $x_a(t)$ for all t using Eq. (2.50). In principle, this could be done by reconstructing the original analog signal and then resampling it with the new sampling period T'. While this is not what we want to do in practice, we can derive, however, a relation involving only discrete-time operations by starting with Eq. (2.50). Specifically we can write

$$x_i[n] = x_a(nT'') = \sum_{k=-\infty}^{\infty} x_a(kT)\left[\frac{\sin[\pi(nT'' - kT)/T]}{\pi(nT'' - kT)/T}\right], \tag{2.61}$$

or, substituting $T'' = T/L$ and $x[n] = x_a(nT)$, we can write Eq. (2.61) as

$$x_i[n] = x_a(nT'') = \sum_{k=-\infty}^{\infty} x[k]\left[\frac{\sin[\pi(n-k)/L]}{[\pi(n-k)/L]}\right], \tag{2.62}$$

which gives a direct relationship between the original input samples $x[n] = x_a(nT)$ and the desired output samples $x_i[n] = x_a(nT'')$ when $T'' = T/L$.

To see how Eq. (2.62) can be implemented using discrete-time filtering, consider the upsampled sequence

$$x_u[n] = \begin{cases} x[n/L] & n = 0, \pm L, \pm 2L, \ldots \\ 0 & \text{otherwise,} \end{cases} \tag{2.63}$$

which has the original samples in the correct positions for the new sampling rate, but has zero-valued samples in between. We depict the operation of upsampling by the block diagram of Figure 2.16, where the sampling periods at the input and output are designated at the bottom of the figure.

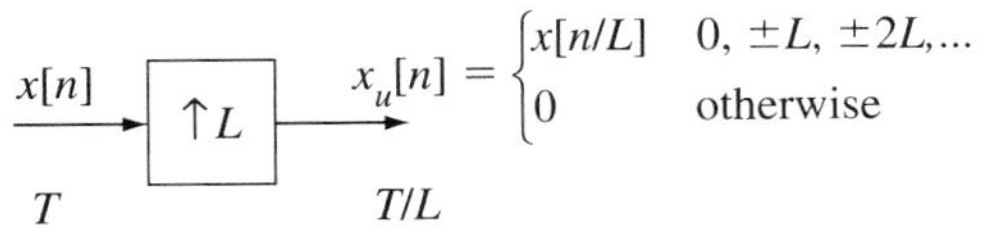

FIGURE 2.16
Block diagram representation of the upsampling operation.

The Fourier transform of $x_u[n]$ is easily shown to be [270, 347]

$$X_u(e^{j\omega}) = X(e^{j\omega L}), \tag{2.64}$$

or equivalently, in terms of Ω with appropriate normalizing for the new sampling period T'',

$$X_u(e^{j\Omega T''}) = X(e^{j\Omega T'' L}) = X(e^{j\Omega T}). \tag{2.65}$$

Thus, $X_u(e^{j\Omega T''})$ is periodic with period $2\pi/T$ (due to the upsampling operation) and $2\pi/T''$ due to its being a DTFT with ω replaced by $\Omega T''$. Equivalently, $X_u(e^{j\omega})$ is periodic with two different periods, namely $2\pi/L$ as well as with the normal period of 2π for a DTFT.

Figure 2.17a shows a plot of $X(e^{j\Omega T})$ (as a function of Ω), and Figure 2.17b shows a plot of $X_u(e^{j\Omega T''})$ (as a function of Ω) showing the double periodicity for the case $L = 2$, i.e., $T'' = T/2$. Figure 2.17c shows the DTFT of the desired signal, showing that if the original sampling had been with period $T'' = T/L$, then for any L,

$$X_i(e^{j\Omega T''}) = \begin{cases} \dfrac{2}{T} X_a(j\Omega) & |\Omega| \leq 2\pi F_N \\ 0 & 2\pi F_N < |\Omega| \leq \pi/T''. \end{cases} \tag{2.66}$$

Now we can obtain Figure 2.17c from Figure 2.17b by multiplying $X_u(e^{j\Omega T''})$ by the frequency response of an ideal lowpass filter with gain L (to restore the amplitude

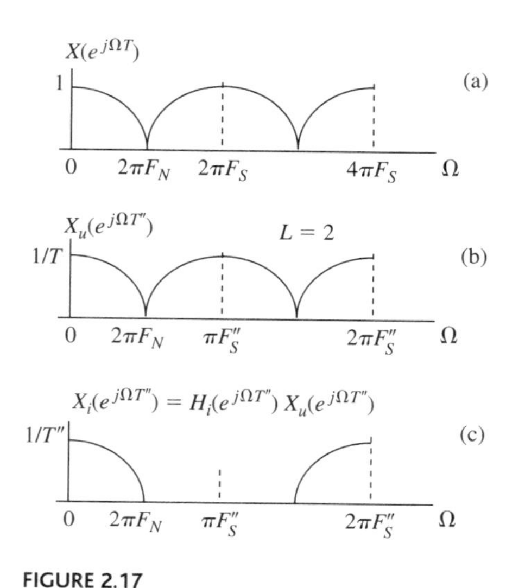

FIGURE 2.17
Illustration of interpolation: (a) DTFT of sampled signal $x[n]$ when $F_s = 2F_N$; (b) DTFT of upsampled (by 2) signal $x_u[n]$; (c) DTFT of filtered output signal $x_i[n]$.

constant to $1/T'$) and cutoff frequency corresponding to the original Nyquist frequency $2\pi F_N = \pi/T$, which corresponds to the normalized discrete-time frequency $\omega_c = \pi/L$. That is, we must ensure that

$$X_i(e^{j\omega}) = \begin{cases} \dfrac{1}{T''}X(e^{j\omega L}) & 0 \leq |\omega| < \pi/L \\ 0 & \pi/L \leq |\omega| \leq \pi. \end{cases} \tag{2.67}$$

This can be achieved with an interpolation filter with frequency response

$$H_i(e^{j\omega}) = \begin{cases} L & |\omega| < \pi/L \\ 0 & \pi/L \leq |\omega| \leq \pi. \end{cases} \tag{2.68}$$

The general interpolation system is depicted in Figure 2.18. The original signal, $x[n]$, at sampling period, T, is first upsampled to give the signal $x_u[n]$ with sampling period $T'' = T/L$. The lowpass filter removes the images of the original spectrum as we have discussed above, resulting in an output signal such that

$$x_i[n] = x_a(nT'') = x_a(nT/L). \tag{2.69}$$

We can demonstrate that the output does satisfy Eq. (2.69) by noting that the upsampled signal can be represented conveniently as

$$x_u[n] = \sum_{k=-\infty}^{\infty} x[n]\delta[n - kL]; \tag{2.70}$$

i.e., we use the shifted impulse sequences to position the original samples with spacing L. Now, the output of the filter is simply the convolution of $x_u[n]$ with the impulse response of the interpolation filter whose ideal frequency response is given by Eq. (2.68). Specifically the corresponding impulse response is

$$h_i[n] = \frac{\sin(\pi n/L)}{(\pi n/L)}. \tag{2.71}$$

Therefore the output of the system of Figure 2.18 is

$$x_i[n] = x_u[n] * h_i[n] = \sum_{k=-\infty}^{\infty} x[k]\left[\frac{\sin[(n-kL)/L]}{[(n-kL)/L]}\right], \tag{2.72}$$

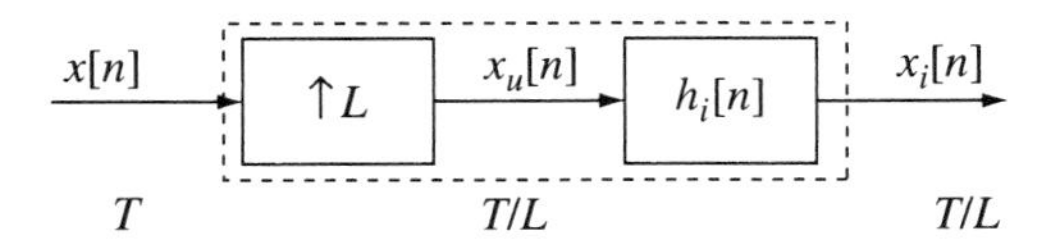

FIGURE 2.18
Representation of interpolation as filtering an upsampled signal.

which is identical to the result in Eq. (2.62). This confirms that if a signal is originally sampled at the Nyquist rate (or above), then we can increase the sampling rate further by the process of interpolation as depicted in Figure 2.18.

2.5.6 Non-Integer Sampling Rate Changes

It is readily seen that samples corresponding to a sampling period $T' = MT/L$ can be obtained by a combination of interpolation by a factor L followed by decimation by a factor M. In general, we cannot reverse the order of the interpolator and the decimator. This is because the lowpass filter of the decimator might reduce the bandwidth of the signal by more than is generally required by the overall system. The system for changing the sampling period by a factor M/L would consist of a cascade connection of the system of Figure 2.18 followed by the system of Figure 2.15. This would result in a cascade combination of the interpolation filter followed by the decimation filter. This combination can be replaced by a single ideal lowpass filter whose gain is L (from the interpolation) and whose cutoff frequency is the minimum of π/L and π/M. This is depicted in Figure 2.19. By a suitable choice of the integers M and L, we can approach arbitrarily close to any desired ratio of sampling rates.

2.5.7 Advantages of FIR Filters

In our discussions above, we have shown that the sampling rate can be changed exactly if we have ideal lowpass filters. In a practical setting, we must substitute practical approximations to the ideal filters in Figures 2.15, 2.18, and 2.19. Thus, an extremely important consideration in the implementation of decimators and interpolators is the choice of the type of lowpass filter approximation. For these systems, a significant savings in computation over alternative filter types can be obtained by using FIR filters in a standard direct form implementation. The savings in computations for FIR filters is due to the fact that for decimators, only one of every M output samples needs to be calculated, while for interpolators, $L-1$ out of every L samples of the input are zero valued, and therefore do not affect the computation. These facts cannot be fully exploited using IIR filters [270, 347].

Assuming that the required filtering is being performed using FIR filters, for large changes in the sampling rate (i.e., large M for decimators, or large L for interpolators), it has been shown that it is more efficient to reduce (or increase) the sampling rate with a series of decimation stages than to make the entire rate reduction with one stage. In this way the sampling rate is reduced gradually, resulting in much less severe filtering requirements on the lowpass filters at each stage. The details of multistage

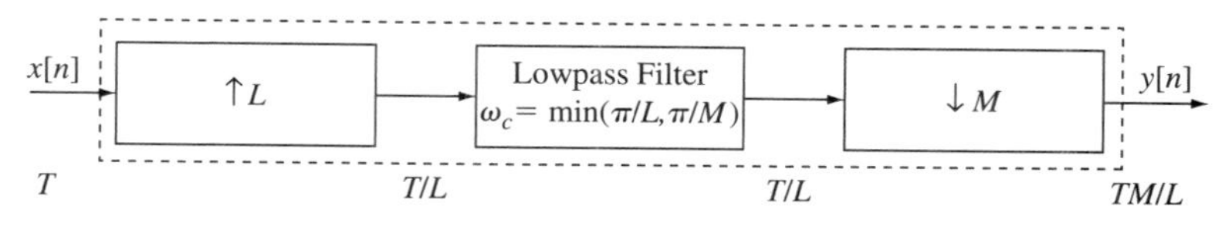

FIGURE 2.19
Block diagram representation of sampling rate increase by a factor of L/M.

implementation of decimation, interpolation, and narrowband filtering are given in Refs. [69, 70, 134, 304].

2.6 SUMMARY

In this chapter we have presented a review of the fundamentals of discrete-time signal processing. The notions of discrete convolution, difference equations, and frequency-domain representations of signals and systems will be used extensively in this book. Also the concepts of sampling of analog signals and digital alteration of the sampling rate as discussed in Section 2.5 are extremely important in all types of digital speech processing systems.

PROBLEMS

2.1. For each of the following systems, $y[n]$ denotes the output and $x[n]$ the input. For each system, determine whether the specified input–output relationship is linear and/or shift-invariant.

(a) $y[n] = 2x[n] + 3$

(b) $y[n] = x[n] \sin\left(\frac{2\pi}{7}n + \frac{\pi}{6}\right)$

(c) $y[n] = (x[n])^3$

(d) $y[n] = \sum_{m=n-N}^{n} x[m]$

2.2. For each of the following questions, justify your response with either a general proof (if you answer yes to the question), or a counter example (if you answer no):

(a) Is the system $y[n] = x[n] + 2x[n+1] + 3$ linear?

(b) Is the system $y[n] = x[n] + 2x[n+1] + 3$ time invariant?

(c) Is the system $y[n] = x[n] + 2x[n+1] + 3$ causal?

2.3. Assume the input to a digital system is of the form:

$$x[n] = a^n \text{ for all } n, \qquad |a| < 1$$

and the output of the system is:

$$y[n] = b^n \text{ for all } n \ (b \neq a), \quad |b| < 1$$

(a) Is the system linear, time-invariant (LTI)?

(b) If the system is LTI, are there more than one LTI systems with the given input–output pair?

(c) Repeat parts (a) and (b) of this problem for the input–output pair:

$$x[n] = a^n u[n], \qquad |a| < 1$$

$$y[n] = b^n u[n], \ (b \neq a), \qquad |b| < 1$$

2.4. Consider the sequence

$$x[n] = \begin{cases} a^n & n \geq n_0 \\ 0 & n < n_0. \end{cases}$$

(a) Determine the z-transform of $x[n]$.

(b) Determine the Fourier transform of $x[n]$. Under what conditions does the Fourier transform exist?

2.5. The input to a linear, time-invariant system is

$$x[n] = \begin{cases} 1 & 0 \leq n \leq N-1 \\ 0 & \text{otherwise.} \end{cases}$$

The impulse response of the system is

$$h[n] = \begin{cases} a^n & n \geq 0, \quad |a| < 1 \\ 0 & n < 0. \end{cases}$$

(a) Using discrete convolution, determine the output, $y[n]$, of the system for all n.

(b) Determine the output using z-transforms.

2.6. Determine the z-transform and the Fourier transform of each of the following sequences. (Each of these are commonly used as "windows" in speech processing systems.)

(1) Exponential window:

$$w_E[n] = \begin{cases} a^n & 0 \leq n \leq N-1, \quad |a| < 1 \\ 0 & \text{otherwise.} \end{cases}$$

(2) Rectangular window:

$$w_R[n] = \begin{cases} 1 & 0 \leq n \leq N-1 \\ 0 & \text{otherwise.} \end{cases}$$

(3) Hamming window:

$$w_H[n] = \begin{cases} 0.54 - 0.46\cos[2\pi n/(N-1)] & 0 \leq n \leq N-1 \\ 0 & \text{otherwise.} \end{cases}$$

Sketch the magnitude of the Fourier transforms in each case. Hint: Obtain a relationship between $W_H(e^{j\omega})$ and $W_R(e^{j\omega})$.

2.7. In the previous problem we examined the properties of the rectangular window,

$$w_R[n] = \begin{cases} 1 & 0 \leq n \leq N-1 \\ 0 & \text{otherwise,} \end{cases}$$

and the Hamming window,

$$w_H[n] = \begin{cases} 0.54 - 0.46\cos[2\pi n/(N-1)] & 0 \le n \le N-1 \\ 0 & \text{otherwise.} \end{cases}$$

Now consider a triangular window of the form

$$w_T[n] = \begin{cases} n+1 & 0 \le n \le (N-1)/2 \\ N-n & (N+1)/2 \le n \le N-1 \\ 0 & \text{otherwise.} \end{cases}$$

(a) Sketch the triangular window for $N = 9$.

(b) Determine the frequency response of the triangular window. (Hint: Think of how you might derive the triangular window from the rectangular window, and then use DTFT properties to derive a simple form for $W_T(e^{j\omega})$.)

(c) Sketch the log magnitude frequency response, $\log|W_T(e^{j\omega})|$ versus ω.

(d) How does $W_T(e^{j\omega})$ compare to $W_R(e^{j\omega})$ and $W_H(e^{j\omega})$ for the same values of N—compare their frequency bandwidths and their peak sidelobe levels.

2.8. The frequency response of an ideal lowpass filter is

$$H(e^{j\omega}) = \begin{cases} 1 & |\omega| < \omega_c \\ 0 & \omega_c < |\omega| \le \pi. \end{cases}$$

($H(e^{j\omega})$ is, of course, periodic with period 2π.)

(a) Determine the impulse response of the ideal lowpass filter.

(b) Sketch the impulse response for $\omega_c = \pi/4$.

The frequency response of an ideal bandpass filter is

$$H(e^{j\omega}) = \begin{cases} 1 & \omega_a < |\omega| < \omega_b \\ 0 & |\omega| < \omega_a \text{ and } \omega_b < |\omega| \le \pi. \end{cases}$$

(c) Determine the impulse response of the ideal bandpass filter.

(d) Sketch the impulse response for $\omega_a = \pi/4$ and $\omega_b = 3\pi/4$.

2.9. The frequency response of an ideal discrete-time differentiator is

$$H(e^{j\omega}) = j\omega e^{-j\omega\tau} \quad -\pi < \omega \le \pi.$$

(This response is repeated with period 2π.) The quantity τ is the delay of the system in samples.

(a) Sketch the magnitude and phase response of this system.

(b) Determine the impulse response, $h[n]$, of this system.

(c) The impulse response of this ideal system can be truncated to a length N samples by a window such as those in Problem 2.6. In so doing the delay is set equal to $\tau = (N-1)/2$ so that the ideal impulse response can be truncated symmetrically. If $\tau = (N-1)/2$ and N is an odd integer, show that the ideal impulse response decreases as $1/n$. Sketch the ideal impulse response for the case $N = 11$.

(d) In the case that N is even, show that $h[n]$ decreases as $1/n^2$. Sketch the ideal impulse response for the case $N = 10$.

2.10. The frequency response of an ideal Hilbert transformer (90° phase shifter) with delay τ is

$$H(e^{j\omega}) = \begin{cases} -je^{-j\omega\tau} & 0 < \omega \leq \pi \\ je^{-j\omega\tau} & -\pi < \omega \leq 0. \end{cases}$$

Determine and sketch the impulse response of this system.

2.11. Consider a linear phase FIR digital filter. The impulse response of such a filter has the property

$$h[n] = \begin{cases} h[N-1-n] & 0 \leq n \leq N-1 \\ 0 & \text{otherwise.} \end{cases}$$

(a) Show that if N is an even integer, the convolution sum expression for the output of such a system can be expressed as

$$y[n] = \sum_{k=0}^{(N-2)/2} h[k](x[n-k] + x[n-N+1+k])$$

and if N is odd

$$y[n] = \sum_{k=0}^{(N-3)/2} h[k](x[n-k] + x[n-N+1+k]) + h[(N-1)/2]x[n-(N-1)/2].$$

Thus, the number of multiplications required to compute each output sample is essentially halved.

(b) Draw the digital filter structures for each of the above equations.

2.12. A stable, linear, shift-invariant system satisfies the linear, constant coefficient, difference equation

$$y[n] = x[n] - \frac{1}{4}x[n-1] + \frac{1}{3}y[n-1].$$

(a) Determine the system function, $H(z)$, and its region of convergence (ROC).

(b) Plot the poles and zeros of $H(z)$ in the z-plane.

(c) Suppose that the input to the system is $x[n] = u[n]$; determine the output signal $y[n]$.

(d) Determine the stable inverse filter impulse response, $h_i[n]$, such that

$$x[n] = x[n] * h[n] * h_i[n].$$

What is the ROC for $H_i(z)$? Hint: You may want to look at the inverse filter in the z-plane, and then solve for $h_i[n]$ from $H_i(z)$.

2.13. Consider the first-order system,

$$y[n] = \alpha y[n-1] + x[n].$$

(a) Find the system function, $H(z)$, for this system.

(b) Find the impulse response of this system.

(c) For what values of α will the system be stable?

(d) Assume that the input is obtained by sampling with period T. Determine the value of α such that

$$h[n] < e^{-1} \text{ for } nT < 2 \text{ msec};$$

i.e., find the value of α that gives a time constant of 2 msec.

2.14. A second-order notch filter has the system function:

$$H(z) = \frac{1 - 2\cos(\theta)z^{-1} + z^{-2}}{1 - 2r\cos(\theta)z^{-1} + r^2 z^{-2}}.$$

(a) Plot the pole/zero diagram for $\theta = 60$ degrees and $r = 0.95$.

(b) Sketch (or use MATLAB to plot it exactly) the log magnitude response, $20\log_{10}|H(e^{j\omega})|$, $0 \le \omega \le \pi$ (or, equivalently, over the region $0 \le f \le 0.5$).

(c) At what frequency, ω_0, does the maximum gain of $|H(e^{j\omega})|$ occur? Does this gain differ greatly from unity?

(d) If the input sequence to the filter is obtained by sampling a speech signal at a rate of $F_s = 8000$ Hz, how should θ be chosen so that the notch occurs at a normalized frequency corresponding analog frequency of 60 Hz?

2.15. Consider a discrete-time system with impulse response

$$h[n] = \left(\frac{1}{2}\right)^n \cos\left(\frac{\pi n}{2}\right) u[n].$$

(a) Determine the system frequency response, $H(e^{j\omega})$, for the system.

(b) Suppose that $x[n] = \cos\left(\frac{\pi n}{2}\right)$. Determine the response of the system, $y[n]$, to the input $x[n]$ using $H(e^{j\omega})$ as found in part (a) above.

2.16. Consider a system function of the form of Eq. (2.35).

(a) Show that if $M < N$, $H(z)$ can be expressed as a partial fraction expansion as in Eq. (2.41), where the coefficient A_m can be found from

$$A_k = H(z)(1 - d_k z^{-1})\Big|_{z=d_k}, \quad k = 1, 2, \ldots, N.$$

(b) Show that the z-transform of the sequence $A_k(d_k)^n u[n]$ is

$$\frac{A_k}{1 - d_k z^{-1}}, \qquad |z| > |d_k|,$$

and thus $h[n]$ is given by Eq. (2.42).

2.17. Consider two linear shift-invariant systems in cascade as shown in Figure P2.17; i.e., the output of the first system is the input to the second.

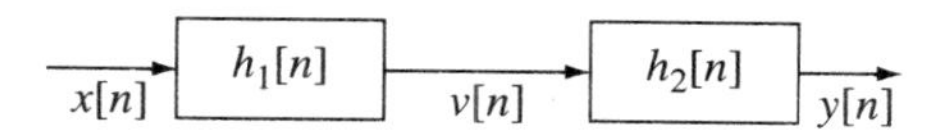

FIGURE P2.17
Cascade of two linear shift-invariant systems.

(a) Show that the impulse response of the overall system is

$$h[n] = h_1[n] * h_2[n].$$

(b) Show that

$$h_1[n] * h_2[n] = h_2[n] * h_1[n],$$

and thus that the overall response does not depend on the order in which the systems are cascaded.

(c) Consider the system function of Eq. (2.34) written as

$$H(z) = \left[\sum_{r=0}^{M} b_r z^{-r}\right]\left[\frac{1}{1 - \sum_{k=1}^{N} a_k z^{-k}}\right]$$
$$= H_1(z) \cdot H_2(z);$$

i.e., as a cascade of two systems. Write the difference equations for the overall system from this point of view.

(d) Now consider the two systems of part (c) in the opposite order; i.e.,

$$H(z) = H_2(z) \cdot H_1(z).$$

Show that the difference equations of Eqs. (2.43a) and (2.43b) result.

2.18. For the difference equation

$$y[n] = 2\cos(bT)y[n-1] - y[n-2],$$

find the two initial conditions $y[-1]$ and $y[-2]$ such that

(a) $y[n] = \cos(bTn), \quad n \geq 0,$

(b) $y[n] = \sin(bTn), \quad n \geq 0.$

2.19. Consider the set of difference equations

$$y_1[n] = Ay_1[n-1] + By_2[n-1] + x[n],$$
$$y_2[n] = Cy_1[n-1] + Dy_2[n-1].$$

(a) Draw the network diagram for this system.

(b) Find the transfer functions,

$$H_1(z) = \frac{Y_1(z)}{X(z)} \quad \text{and} \quad H_2(z) = \frac{Y_2(z)}{X(z)}.$$

(c) For the case $A = D = r\cos\theta$ and $C = -B = r\sin\theta$, determine the impulse responses $h_1[n]$ and $h_2[n]$ that result when the system is excited by $x[n] = \delta[n]$.

2.20. A causal linear shift-invariant system has the system function

$$H(z) = \frac{(1 + 2z^{-1} + z^{-2})(1 + 2z^{-1} + z^{-2})}{(1 + \frac{7}{8}z^{-1} + \frac{5}{16}z^{-2})(1 + \frac{3}{4}z^{-1} + \frac{7}{8}z^{-2})}.$$

(a) Draw a digital network diagram of an implementation of this system in
- Cascade form (using second-order sections)
- Direct form (both I and II)

(b) Is this system stable? Explain.

2.21. Consider the system of Figure P2.21.

(a) Write the difference equations represented by the network.

(b) Determine the system function for the network.

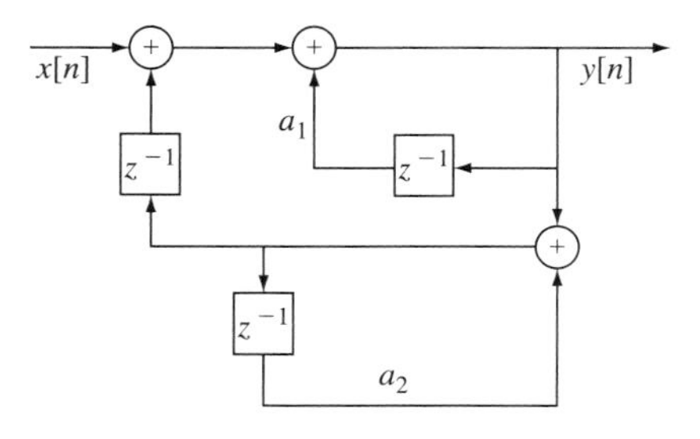

FIGURE P2.21
Linear system block diagram.

2.22. Determine a_1, a_2, and a_3 in terms of b_1 and b_2 so that the two networks of Figure P2.22 have the same transfer function.

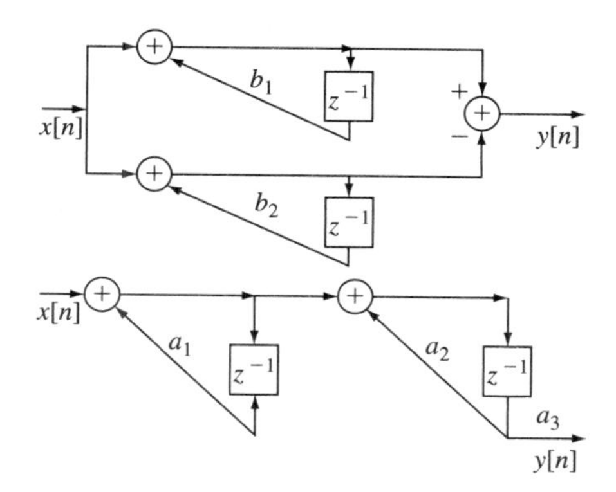

FIGURE P2.22
Two networks that can be made equivalent.

2.23. The system function for a simple resonator is of the form:

$$H(z) = \frac{1 - 2e^{-aT}\cos(bT) + e^{-2aT}}{1 - 2e^{-aT}\cos(bT)z^{-1} + e^{-2aT}z^{-2}}.$$

(a) Determine the poles and zeros of $H(z)$ and plot them in the z-plane.

(b) Determine the impulse response of this system and sketch it for the constants

$$T = 10^{-4},$$
$$b = 1000\pi,$$
$$a = 200\pi.$$

(c) Sketch the log magnitude of the frequency response of this system as a function of analog frequency, Ω.

2.24. Consider the finite-length sequence

$$x[n] = \delta[n] + 0.5\delta[n-5].$$

(a) Determine the z-transform and Fourier transform of $x[n]$.

(b) Determine the N-point DFT of $x[n]$ for $N = 5010$, and 5.

(c) How are the DFT values for $N = 5$ related to those of the DFT for $N = 50$?

(d) What is the relationship between the N-point DFT of $x[n]$ and the DTFT of $x[n]$?

2.25. A speech signal is sampled at a rate of 20,000 samples/sec (20 kHz). A segment of length 1024 samples is selected and the 1024-point DFT is computed.

(a) What is the time duration of the segment of speech?

(b) What is the frequency resolution (spacing in Hz) between the DFT values?

(c) How do your answers to parts (a) and (b) change if we compute the 1024-point DFT of 512 samples of the speech signal. (The 512 samples would be augmented with 512 zero samples before the transform was computed.)

2.26. A *minimum-phase* signal is a causal ($x_{\min}[n] = 0$ for $n < 0$) signal having a z-tranform $X_{\min}(z)$ with all its poles and zeros inside the unit circle. A *maximum-phase* signal has a z-transform with all its poles and zeros outside the unit circle. Show that if $x[n]$ is a minimum-phase signal, then $x[-n]$ is a maximum-phase signal.

2.27. A discrete-time system has system function $H(z) = 1 - az^{-1}$, $|a| < 1$, i.e., $H(z)$ has a real zero at $z = a$.

(a) Show that $H(z)$ can be represented exactly by another discrete-time system system function with an infinite number of poles and no zeros.

(b) What happens to the representation of part (a) when only a finite number of poles are used (i.e., the infinite series is truncated)?

(c) What is the first-order (i.e., single pole) approximation to a real zero at $z = a$? What is the second-order approximation? Where are the poles of the first- and second-order approximations located?

(d) Show that a digital system, $H(z) = 1/(1 - bz^{-1})$, $|b| < 1$, with a single real pole at $z = b$ can be similarly represented by a digital system with an infinite number of zeros.

(e) Use MATLAB to plot the frequency response of the system $H(z) = 1 - az^{-1}$ for $a = 0.9$ along with the frequency responses of the one-pole, two-pole, and 100-pole approximations. Repeat using other values of a.

2.28. Consider methods for designing a system that realizes the system function:

$$y[n] = x[n - D],$$

where D is a non-integer < 1; i.e., you want to design a discrete-time system that delays the input signal by D samples, where D is a fraction less than 1.

(a) Assume $D = 1/2$; how would you realize the transfer function exactly using interpolation and decimation?

(b) Assume we want to approximate the solution of part (a) using a system of the form:

$$y[n] = \alpha x[n] + \beta x[n - 1].$$

What values of α and β would you use?

(c) How would your solution to part (a) change if $D = 1/3$?

(d) How would your solution to part (b) change if $D = 1/3$?

(e) How would your solutions to parts (a) and (b) change if $D = 0.3$ (show two solutions—one with only a single unit delay, and one with a delay of three units).

2.29. An input signal, $x[n]$, defined over the region $-\infty < n < \infty$ is passed through a cubic non-linearity yielding the output signal

$$y[n] = (x[n])^3.$$

(a) If the input to the system is of the form

$$x_1[n] = \cos(\omega_0 n), \qquad -\infty < n < \infty,$$

determine the output DTFT, $Y_1(e^{j\omega})$, and sketch the magnitude of $Y_1(e^{j\omega})$ assuming $3\omega_0 < \pi$.
Hint: Recall the trigonometric relationships

$$\cos^2(x) = \frac{1}{2} + \frac{\cos(2x)}{2},$$

$$\cos^3(x) = \frac{\cos(3x)}{4} + \frac{3\cos(x)}{4}.$$

(b) If the input to the system is the signal

$$x_2[n] = r^n \cos(\omega_0 n)u[n], \quad |r| < 1,$$

sketch the magnitude of $Y_2(e^{j\omega})$ (again assuming $3\omega_0 < \pi$) and assuming a value of $\omega_0 = \Omega_0 T = 2\pi \cdot 500$, $r = 0.9$, and $F_s = 10{,}000$ Hz.

2.30. An analog signal, $x_a(t)$, has spectrum $X_a(j\Omega)$, as shown in Figure P2.30. The analog signal is sampled at rates of $F_s = 10{,}000$, 5000, and 2000 Hz. Sketch the DTFT of the resulting discrete-time signal at each of these three sampling rates.

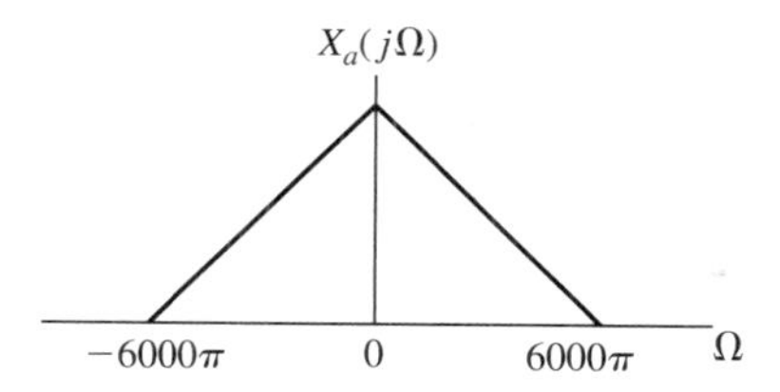

FIGURE P2.30
Spectrum of analog signal.

2.31. An analog signal, $x_a(t) = A\cos(2\pi 200t)$, is sampled at a rate of $F_s = 10{,}000$ Hz, giving the discrete-time signal $x_1[n]$.

(a) Sketch the resulting DTFT, $X_1(e^{j\omega})$.

(b) A second analog signal, $x_b(t) = B\cos(2\pi 201t)$, is also sampled at a rate of $F_s = 10{,}000$ Hz, giving the discrete-time signal $x_2[n]$. Again sketch the resulting DTFT, $X_2(e^{j\omega})$.

(c) Are the discrete-time signals, $x_1[n]$ and $x_2[n]$, periodic signals, and if so, what are their periods?

2.32. (MATLAB Exercise) Write a MATLAB program to do the following:

- Accept an arbitrary speech filename of a speech file in .wav format, i.e., `filename.wav`, and load the file using the MATLAB `wavread` command).
- Listen to the speech file using the MATLAB `sound` command. (Pay attention to the required scaling for the speech file.) Write down the sentence you heard.
- Plot `N` sample values of the speech file (using appropriate MATLAB plot commands), starting at sample `fstart` and ending with sample `fstart+N-1`. Plot `M` samples of speech per line, with up to four lines of plotting per page. The plotting parameters `fstart`, `M`, and `N` should all be specified at run time (i.e., they should be read into the program).

Use the file `s5.wav` that is available on the book website as your speech test file, and use a value of `fstart=2000` and values of `M` and `N` that are equivalent to plotting 100 msec of speech per line and a total of `N=22000` samples of speech being plotted. (It should be clear that the values chosen for `M` and `N` depend on the sampling rate, F_s, of the speech file.)

2.33. (MATLAB Exercise) Write a MATLAB program to plot the speech file as used in Problem 2.32 with the same parameter values for starting sample `fstart`, number of samples per line, `M`, and total number of samples plotted, `N`, but instead of plotting four lines of speech per page, use the supplied MATLAB function `strips_modified.m` to plot all the speech samples in a contiguous format and on a single page.

CHAPTER 3

Fundamentals of Human Speech Production

3.1 INTRODUCTION

In order to apply digital signal processing techniques to speech communication problems, it is essential to understand how humans produce speech signals. Hence, this chapter begins with an overview of sound generation mechanisms in human speech production. We then describe the base set of sounds of any language, namely the phonemes, which define the range and types of sounds that are produced in a given language. We discuss both the acoustic and phonetic properties of American English phonemes, and give examples of their manifestations in both waveforms and spectral representations. We then show how spoken language is manifest in both the textual representation of the spoken material as well as in the waveform and the spectral representation of the resulting speech signal. We show how to (approximately) locate sounds (e.g., phonemes, syllables) in both the acoustic waveform and in the spectral representation of the signal. We end this chapter with a discussion of the articulatory properties of speech sounds, namely the place and manner of articulation, to gain an understanding of how to relate the linguistic and physical properties of speech sounds.

This chapter plays a role similar to that of Chapter 2 in serving as a review of an established area of knowledge. In contrast to the digital signal processing theory in Chapter 2, however, the subject of the present chapter is likely to be less familiar to the engineers and computer scientists who are the target of this text. The references in this chapter to papers and other texts provide much more detail on many of the topics of this and subsequent chapters. Particularly noteworthy are the landmark books by Fant [101], Flanagan [105], and Stevens [376]. Fant's book deals primarily with the acoustics of speech production and contains a great deal of useful data on early vocal system measurements and models. Flanagan's book, which is much broader in scope, contains a wealth of valuable insights into the physical modeling of the speech production process and the way that such models are used in representing and processing speech signals. Stevens' book deals extensively with the interactions between the acoustics and the articulatory configurations of the various sounds of English. These books are indispensable references for serious study of speech communication.

Before discussing the acoustic theory and the resulting mathematical models for speech production (the subject of Chapter 5), it is necessary to understand the basic speech production/generation processes in humans, as well as the speech perception process (discussed in Chapter 4).

3.2 THE PROCESS OF SPEECH PRODUCTION

Speech signals are composed of a sequence of sounds, which, for purposes of analysis and study, are assumed to be physical realizations of a discrete set of symbols. These sounds, and the transitions between them, serve as a symbolic representation of information. The arrangement of these sounds is governed by the rules of language. The study of these rules and their implications in human communication is the domain of *linguistics*, and the study and classification of the sounds of speech is called *phonetics*. A thorough discussion of phonetics and linguistics would take us too far afield. However, in processing speech signals to enhance or extract information, it is helpful to have as much knowledge as possible about the structure of the signal; i.e., about the way in which information is encoded in the signal. Thus, it is worthwhile discussing the main classes of speech sounds before proceeding to Chapter 5 for a detailed discussion of mathematical models of the production of speech signals. Although this chapter contains all that we shall have to say about linguistics and phonetics, this is not meant to minimize their importance—especially in the areas of speech recognition and speech synthesis. We shall have many opportunities to refer to material in this chapter.

3.2.1 The Mechanism of Speech Production

Figure 3.1 shows a mid-sagittal plane X-ray photograph that places in evidence the important features of the human vocal system [108]. The *vocal tract*, outlined by the dotted lines in Figure 3.1, begins at the opening between the vocal cords, or *glottis*, and ends at the lips. The vocal tract thus consists of the *pharynx* (the connection from the esophagus to the mouth) and the mouth or oral cavity. In the average male, the total length of the vocal tract is about 17–17.5 cm. The cross-sectional area of the vocal tract, which is determined by the positions of the tongue, lips, jaw, and velum, varies from zero (complete closure) to about 20 cm^2. The *nasal tract* begins at the velum and ends at the nostrils. The velum is a trapdoor-like mechanism at the back of the mouth cavity. When the velum is lowered, the nasal tract is acoustically coupled to the vocal tract to produce the nasal sounds of speech.

Figure 3.2 shows a more modern way of capturing information about the size and shape of the vocal tract via MR (magnetic resonance) imaging methods [45]. Using MR imaging, a mid-sagittal plane image of the vocal tract, from the glottis to the lips, can be obtained, as shown in Figure 3.2a. Using standard signal analysis methods, the air–tissue boundaries of the vocal tract can be traced, as shown in part (a) of the figure, and parameters describing the size and shape of the vocal tract, such as the lip aperture (LA), the tongue tip constriction degree (TTCD), and the velum aperture (VEL) can be readily estimated, as shown in part (b) of the figure.

A schematic cross-sectional view of the human vocal system is shown in Figure 3.3. The parts of the body involved in speech production include the lungs and chest cavity (as the source of air to excite the vocal tract and as the source of pressure to

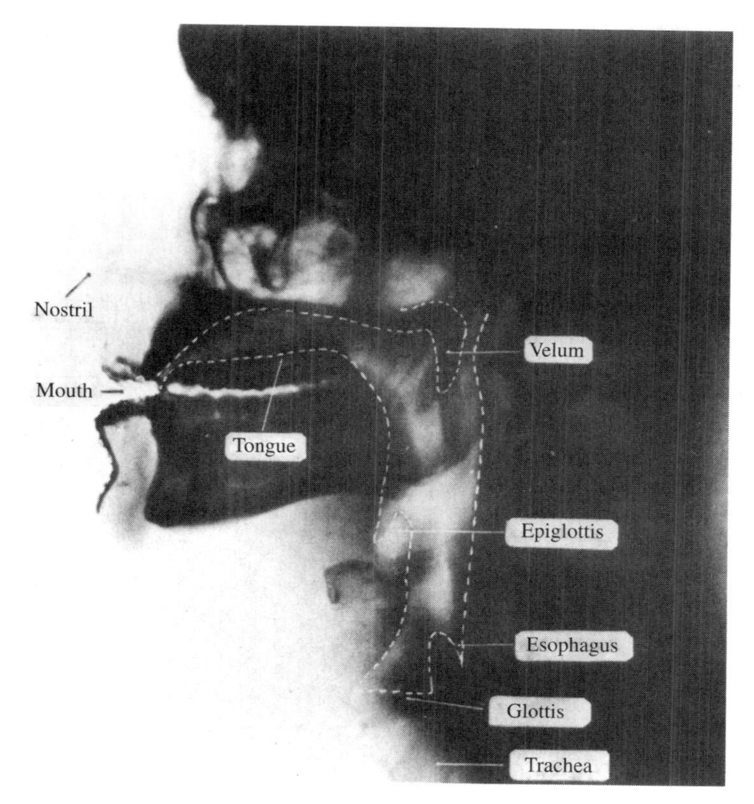

FIGURE 3.1
Sagittal plane X-ray of the human vocal apparatus. (After Flanagan et al. [108]. © [1970] IEEE.)

force the air from the lungs), the trachea or windpipe (which conducts the air from the lungs to the vocal cords and vocal tract), the vocal cords (which vibrate when tensed and excited by air flow), and the vocal tract consisting of the pharynx (the throat cavity), the mouth cavity (including the tongue, lips, jaw, and mouth), and possibly the nasal cavity (depending on the position of the velum).

The speech production mechanism for voiced sounds such as vowels works as follows:

- Air enters the lungs via normal breathing and no speech is produced (generally) on intake;
- As air is expelled from the lungs via the *trachea*, the tensed vocal cords within the *larynx* are made to vibrate by Bernoulli-Law variations of air pressure in the glottal opening;
- Air flow is chopped up by the opening and closing of the glottal orifice into quasi-periodic pulses;

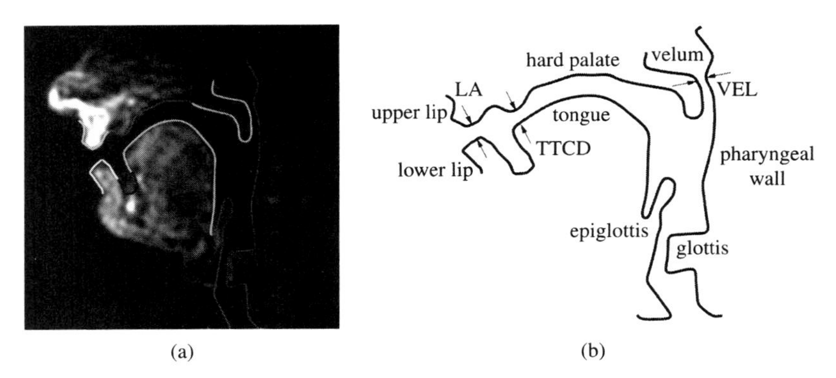

FIGURE 3.2
(a) Example of a typical vocal tract MR (magnetic resonance) image showing contours of interest; (b) a vocal tract schematic showing size and shape parameters. (After Bresch et al. [45]. © [2008] IEEE.)

- These pulses are frequency-shaped when passing through the pharynx (the throat cavity), the mouth cavity, and possibly the nasal cavity. The positions of the various articulators (jaw, tongue, velum, lips, and mouth) determine the sound that is produced.

Figure 3.4 shows two views of the vocal cords, an artist's top view, looking down into the vocal cords, and a schematic longitudinal cross-sectional view, showing the path for air flow from the lungs through the vocal cords and through the vocal tract. The top view shows the two vocal cords (literally membranes) that become tensed via appropriate musculature control of the cartilages, labeled AC for arytenoid cartilage and TC for thyroid cartilage, that surround the vocal cords, labeled VC. When the vocal cords (also called vocal folds) are tensed, they form a relaxation oscillator. Air pressure builds up behind the closed vocal cords until they are eventually blown apart. Air then flows through the orifice that is created and, following Bernoulli's Law, the air pressure drops, causing the vocal cords to return to the closed position. The cycle of building up pressure, blowing apart the vocal cords, and then closing shut is repeated quasi-periodically as air continues to be forced out of the lungs. The rate of opening and closing is controlled mainly by the tension in the vocal cords.

Figure 3.5 shows plots from a simulation of the glottal volume velocity air flow (upper plot) and the resulting sound pressure at the mouth for the first 30 msec of a voiced sound (such as a vowel). The cycle of opening and closing of the vocal cords is clearly seen in the glottal volume velocity flow. Notice that the first 15 msec (or so) of flow represents a period of buildup in the glottal flow, and thus the resulting pressure waveform (at the mouth) also shows a buildup until it begins to look like a quasi-periodic signal. This transient behavior at the onset (and also termination) of voicing is a source of some difficulty in algorithms for deciding exactly when voicing begins and ends, and in estimating parameters of the speech signal during this buildup period.

Pathologies of the larynx sometimes lead to its complete removal, thereby depriving a person of the means to generate natural voiced speech. Detailed knowledge of

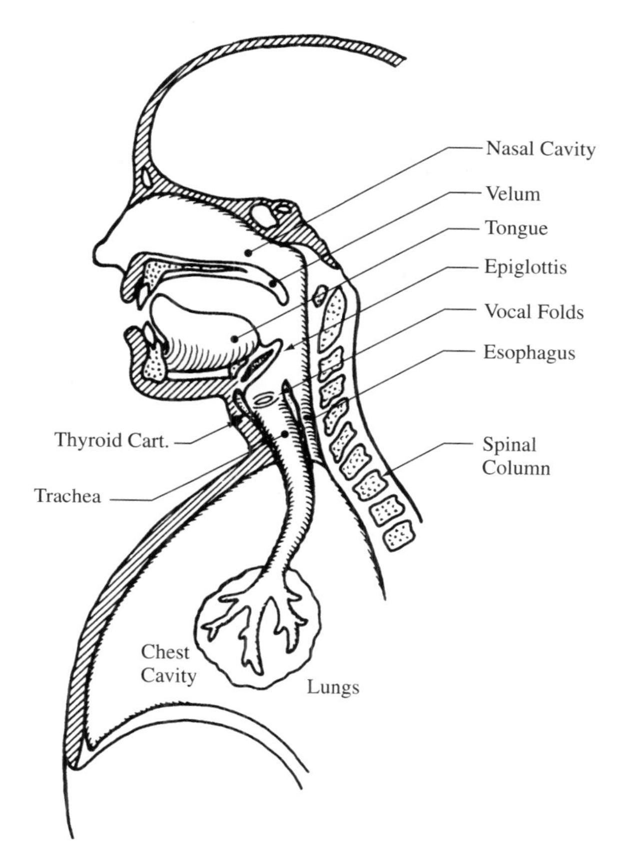

FIGURE 3.3
Schematic view of the human vocal tract. Reprinted with permission of Alcatel-Lucent USA Inc.

the operation of the vocal cords in speech production was the basis for the design of the artificial larynx shown in Figure 3.6 [319]. The artificial larynx was designed and built by AT&T as an aid to patients who had their larynx surgically removed. It is a vibrating diaphragm that can produce a quasi-periodic excitation sound that can be coupled directly into the human vocal tract by holding the artificial larynx tightly against the neck, as shown by the human user in Figure 3.6. The artificial larynx does not cause the vocal cords to open and close (even if they are still intact); however, its vibrations are transmitted through the soft tissue of the neck to the pharynx where the air flow is amplitude modulated. By using the on-off control, along with a "rate of vibration" control, an accomplished human user can create an appropriate excitation signal that essentially mimics the one produced by the vibrating vocal cords, thereby enabling the user to create (albeit somewhat buzzy sounding) speech for communication with other humans.

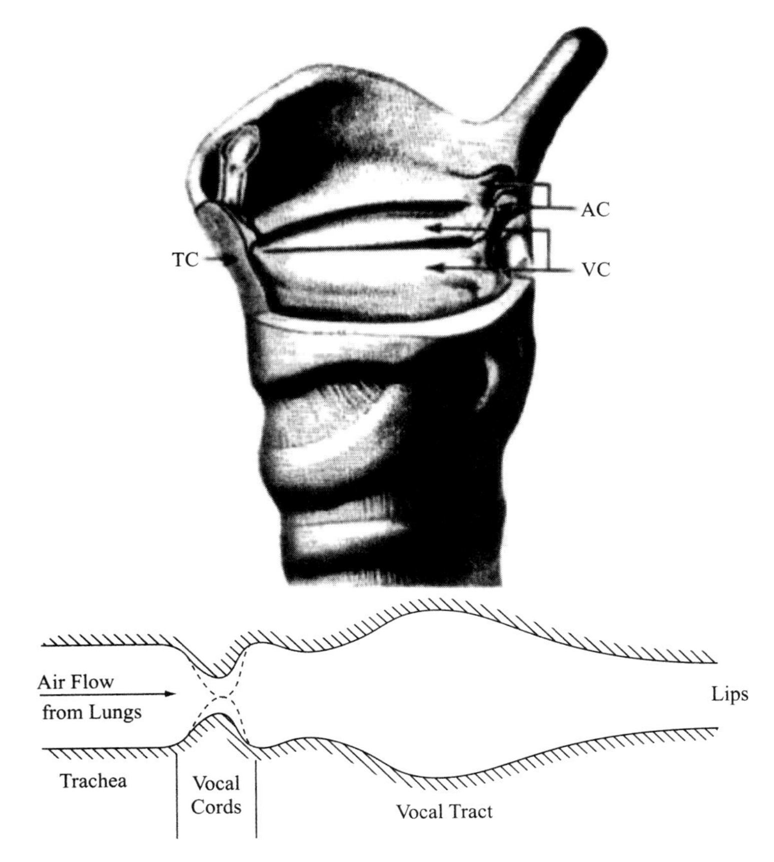

FIGURE 3.4
The vocal cords: (top) an artist's rendering from the top (after Farnsworth [102]); (bottom) a schematic cross-sectional view. Reprinted with permission of Alcatel-Lucent USA Inc.

In studying the speech production process, it is helpful to abstract the important features of the physical system in a manner that leads to a realistic yet tractable mathematical model. Figure 3.7 shows such a schematic diagram [108] of a model that can serve as a basis for thinking about and simulating the vocal system. For completeness, the diagram includes the sub-glottal system composed of the lungs, bronchi and trachea, a mechanical model of the vocal cords, including mechanical components to model the mass and tension of the cords, and a pair of tubes of non-uniform cross-section that model the vocal tract/nasal tract configuration. The sub-glottal system serves as a source of energy for the production of speech. The mechanical model of the vocal cords provides the excitation signal for the vocal tract. The resulting speech signal is simply the acoustic wave that is radiated from this system when air is expelled from the lungs and the resulting flow of air is shaped accordingly by the (time-varying) vocal tract.

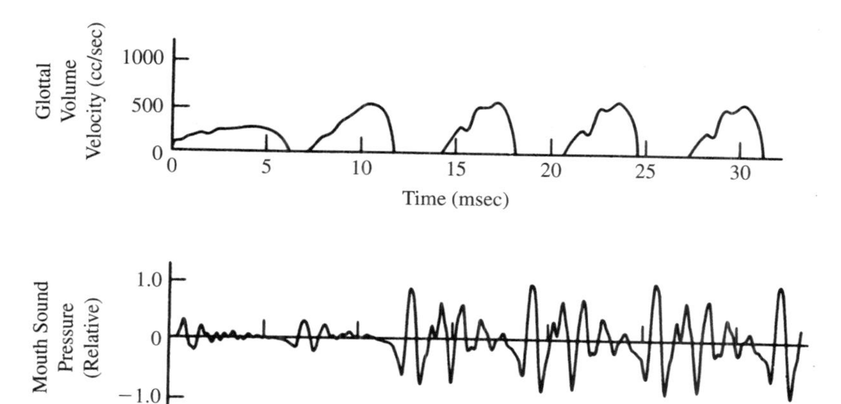

FIGURE 3.5
Plots of simulated glottal volume velocity flow and radiated pressure at the lips at the beginning of voicing.

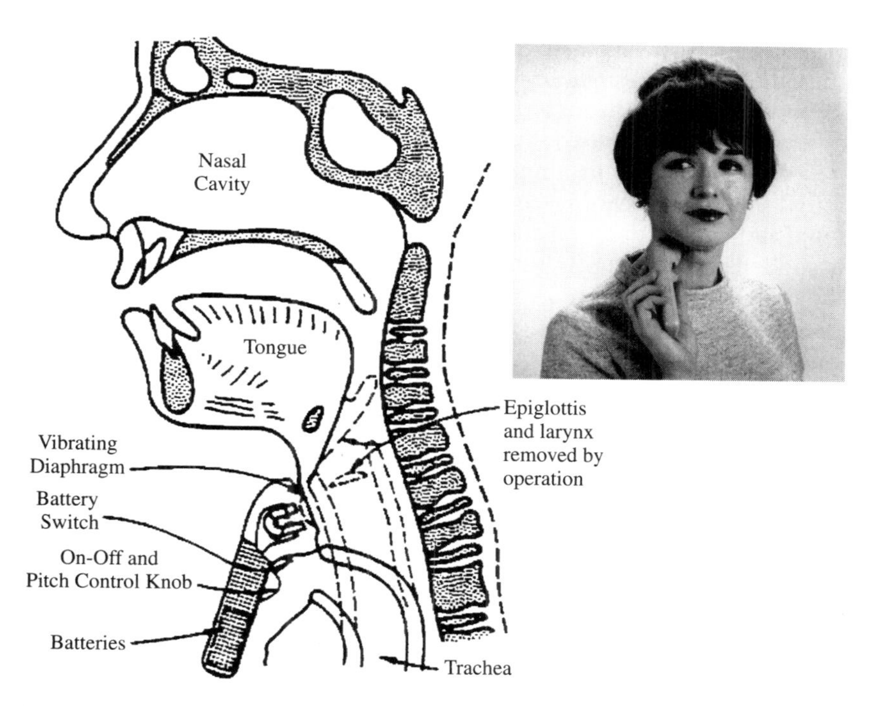

FIGURE 3.6
The artificial larynx, a demonstration of an artificial method for generating vocal cord excitation for speech production. (After Riesz [319]. Reprinted with permission of Alcatel-Lucent USA Inc. © [1930] Acoustical Society of America.)

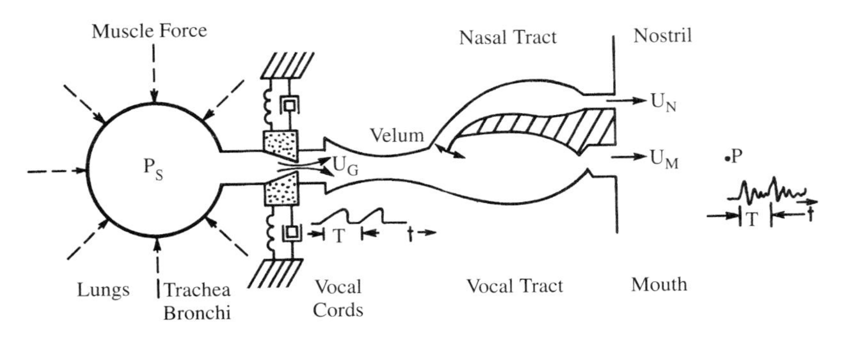

FIGURE 3.7
Schematized diagram of the vocal apparatus. (After Flanagan et al. [108]. © [1970] IEEE.)

In the abstraction of Figure 3.7, the muscle force from the chest muscles pushes air out of the lungs and then through the bronchi and the trachea to the vocal cords. If the vocal cords are tensed (via musculature control as described above), the air flow causes them to vibrate, producing puffs of air at a quasi-periodic rate, which excite the vocal tract (the combination of larynx tube, pharynx cavity, and the mouth) and/or nasal cavity, producing "voiced" or quasi-periodic speech sounds, such as steady state vowel sounds, which radiate from the mouth and/or nose. If the vocal cords are relaxed, then the vocal cord membranes are spread apart and the air flow from the lungs continues unimpeded through the vocal tract until it hits a constriction in the vocal tract, at which point one of two things happens. If the constriction is only partial, the air flow may become turbulent, thereby producing so-called "unvoiced" sounds (such as the initial sound in the word /see/, or the word /shout/). If the constriction is total, pressure builds up behind the total constriction. When the constriction is released, the pressure is suddenly and abruptly released, causing a brief transient sound, such as occurs at the beginning of the words /put/, /take/, or /kick/. Again the sound pressure variations at the mouth and/or nose constitute the speech signal that is produced by the speech generation mechanism.

The vocal tract and nasal tract are shown in Figure 3.7 as tubes of non-uniform cross-sectional area laid out along a straight line. In reality, as is clear from Figure 3.1, the vocal tract bends at almost a right angle between the larynx and pharynx.[1] As sound, generated as discussed above, propagates down these tubes, the frequency spectrum is shaped by the frequency selectivity of the tube. This effect is very similar to the resonance effects observed with organ pipes or wind instruments. In the context of speech production, the resonance frequencies of the vocal tract tube are called *formant frequencies* or simply *formants*. The formant frequencies depend upon the shape and dimensions of the vocal tract; each shape is characterized by a set of formant

[1] As we will show in Chapter 5, the assumption of straight non-uniform tubes leads, nevertheless, to accurate mathematical simulations of sound transmission in the vocal tract.

frequencies. Different sounds are formed by varying the shape of the vocal tract. Thus, the spectral properties of the speech signal vary with time as the vocal tract shape varies.

3.2.2 Speech Properties and the Speech Waveform

The speech signal has the following inherent properties:

- Speech is a sequence of ever-changing sounds.
- The properties of the speech signal waveform are highly dependent on the sounds that are produced in order to encode the content of the implicit message.
- The properties of the speech signal are highly dependent on the context in which the sounds are produced; i.e., the sounds that occur before and after the current sound. This effect is called speech sound *co-articulation*, and it is the result of the vocal control mechanism anticipating following sounds while producing the current sound, thereby modifying the sound properties of the current sound.
- The state of the vocal cords and the positions, shapes, and sizes of the various articulators (lips, teeth, tongue, jaw, velum) all change slowly over time, thereby producing the desired speech sounds.

From the above list of properties, it follows that we should be able to determine some of the physical properties of speech (whether the vocal cords are vibrating or in a lax position, whether the sound is quasi-periodic or noise-like, etc.) by observing and measuring the speech waveform or representations derived from the speech waveform such as the signal spectrum.

One simple example of how we can use knowledge of speech production and the resulting speech waveform to learn the properties of the sounds being produced is given in Figure 3.8, which shows a waveform plot of 500 msec of a speech signal (i.e., each line of waveform represents a time duration of 100 msec). The simplest task we can undertake is to classify regions of the speech waveform as voiced, which we denote as V in the figure, unvoiced, which we denote as U in the figure, and silence (or more appropriately background signal) which we denote as S in the figure. *Voiced* sounds are produced by forcing air through the glottis with the tension of the vocal cords adjusted so that they vibrate in a relaxation oscillation, thereby producing quasi-periodic pulses of air that excite the vocal tract, leading to a quasi-periodic waveform. *Unvoiced* or *fricative* sounds are generated by forming a partial constriction at some point in the vocal tract (usually toward the mouth end), and forcing air through the constriction at a high enough velocity to produce turbulence. This creates a broad-spectrum noise source that excites the vocal tract. Finally *silence* or background sounds are identified by their lack of the characteristics of either voiced or unvoiced sounds and usually occur at the beginning and end of speech utterances, although intervals of silence often occur within speech utterances. The voiced intervals are readily identified as the quasi-periodic waveform regions in Figure 3.8. These regions are labeled with V. Regions of unvoiced speech are somewhat more difficult to identify (because of their confusability with regions of background signal) but some good estimates of these unvoiced speech regions are shown in Figure 3.8 and labeled with U. Finally any remaining regions,

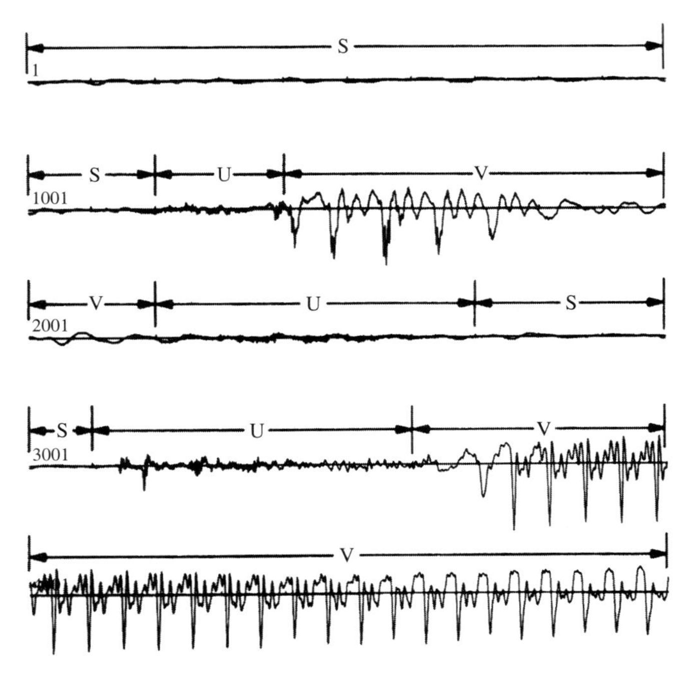

FIGURE 3.8
Example of a speech waveform and its classification into intervals of voicing (V), unvoicing (U), and silence or background signal (S).

almost by default, are recognized as silence (or background signal) regions and labeled with the S label. We give a more formal algorithm for classifying signal regions as voiced, unvoiced, or background in Chapter 10.

If we are given a speech waveform and told what sounds it contains, it seems reasonable to suppose that we could label time intervals with the sounds that they represent. To discuss this in an organized way, it is helpful to define some terminology. We refer to a text representation as the written representation of a string of words of the language. For example, "should we chase" is a string of three words that could occur in English. A phonetic representation for the words gives the sounds that must be made to "pronounce" the words correctly in a given language. The pronunciation of words is given in dictionaries in terms of a set of symbols in some phonetic system. In Section 3.4 we will introduce a phonetic system (called the ARPAbet) for representing the sounds of American English speech. With this system, our example is represented as /SH UH D - W IY - CH EY S/. (We have used "-" to indicate the boundaries of the words in this phonetic transcription.) When a human speaks the words, we call this an "utterance" of the text. In speaking the words, the human attempts to form the sounds represented by the phonetic transcription. The result is an acoustic signal that can be recorded, transmitted, stored, etc. Figure 3.9 shows the waveform for an utterance of the words "should we chase" by a male speaker, sampled at a rate of 10 kHz with 120 msec of speech plotted for each line in the figure. Thus, if the time

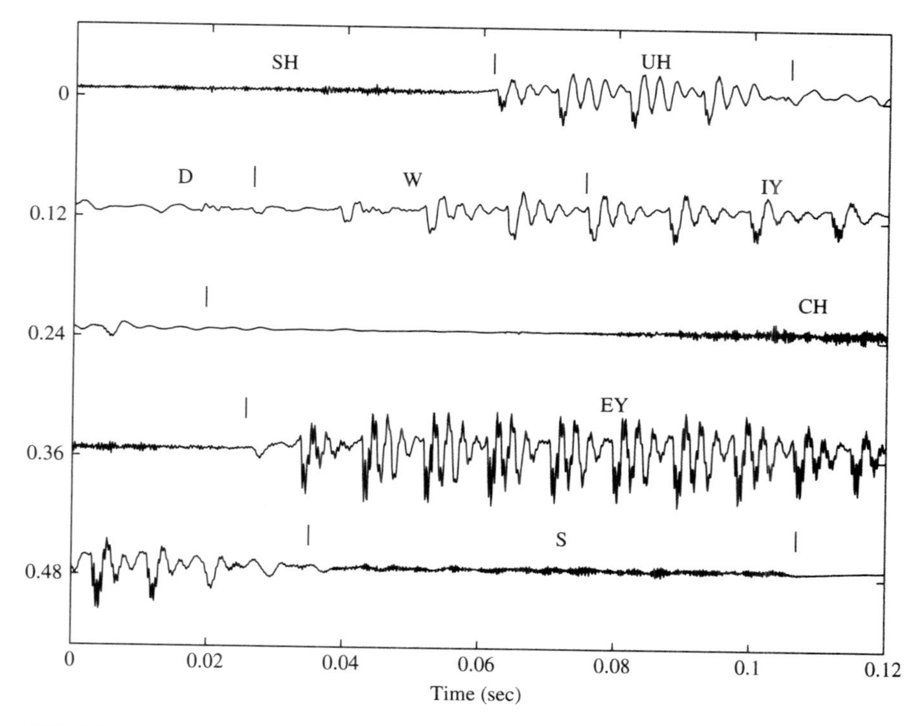

FIGURE 3.9
Waveform of an utterance of "should we chase."

scale in Figure 3.9 is multiplied by 10,000, we obtain a normalized scale in units of samples, with each line comprised of 1200 samples.

Knowing the linguistic (phonetic) transcription of this utterance, we can segment the waveform into the constituent sounds and syllables as shown in Figure 3.9. In general, this segmentation process is not an easy task, especially if we want to obtain a complete phonetic labeling. However, if we say the words "should we chase" slowly, holding our fingers on our throat close to the glottis, it is easy to determine that the utterance begins with an unvoiced interval corresponding to /sh/ (/SH/). This is followed by a longer voiced interval corresponding to the text /ould we/ (/UH D W IY/). This is in turn followed by the unvoiced sound /ch/ (/CH/), then the voiced sound /a/ (/EY/), and finally the unvoiced sound /se/ (/S/). Since no background signal is shown in Figure 3.9, it is relatively easy to pick out the voiced intervals in the waveform, with the remaining intervals being, by default, unvoiced. Table 3.1 provides a summary of the approximate segmentation of the utterance "should we chase" into phonetic units, and the classification of the phonetic units into voiced and unvoiced intervals.

Figure 3.9 shows a suggested phonetic labeling for the entire waveform. As we have already noted, the voiced/unvoiced boundaries are relatively easy to detect. The boundaries for /CH/ are ambiguous due to the fact that /CH/ is the combination of a stop gap (for pressure build up) followed by an explosion. The vowel /EY/ is easy to

TABLE 3.1 Illustration of splitting an utterance of a known text into its phonetic representation along with voiced/unvoiced classification of the phonetic segments along with the approximate locations in the acoustic waveform of the phonetic segments.

Text	Phonetic	Voicing	Samples
/sh/	/SH/	Unvoiced	0 to 600
/ould we/	/UH D W IY/	Voiced	600 to 2600
/ch/	/CH/	Unvoiced	2600 to 3800
/a/	/EY/	Voiced	3800 to 5200
/se/	/S/	Unvoiced	5200 to 6000

spot, as is the beginning of the fricative /S/. However, the boundaries within the long interval corresponding to /UH D W IY/ are more problematic. The criteria we used to create these sample boundaries is crude and will be explained later in this chapter as we describe the basic properties of the various speech sounds. However it should be completely clear, even from such a simple example as the one in Figure 3.9, that it is very difficult to identify and reliably mark the boundaries for sounds, syllables, or even partial syllables. Hence, segmenting a speech waveform into its constituent sounds is a hard task, and it is one that should be avoided if possible. We will encounter these measurement and estimation problems several times throughout this book, so it is useful to see early examples of the difficulties that we will face when we try to design estimation methods for performing these tasks.

Another property of the speech signal that would seem to be readily measured from the speech waveform is the periodicity period (either in samples or in msec) during voiced regions of the signal. Figure 3.10 shows the speech waveform corresponding to a part of the phrase "thieves who rob friends deserve jail." The approximate phonetic labels for the waveform segments corresponding to /TH IY V Z - HH UW - R/ are shown on the plot. The periodicity period is defined as the local period of repetition of the waveform and values of the measured period are shown in Figure 3.10 for the voiced sections of the waveform.[2] (We discuss formal algorithms for performing such a pitch detection task in Chapter 10, but for the time being, we assume that we measure the periodicity period using simple waveform measurements.) The period (also widely referred to as the *pitch period*), as shown by the solid curve above the waveform during the voiced intervals of speech, is seen to vary rather slowly over time. The variation of pitch period over time is often called the speech rhythm since it enables humans to form questions (by making the period fall at the end of a sentence), or make declarative statements, etc.

Table 3.2 shows typical ranges (from minimum to maximum pitch periods), along with the average pitch period for male, female, and child speakers [427]. The pitch

[2]For ease in plotting, the scale of the waveform is adjusted so that it can be shown on the same graph as the pitch period in samples for a sampling rate of 8000 samples/sec.

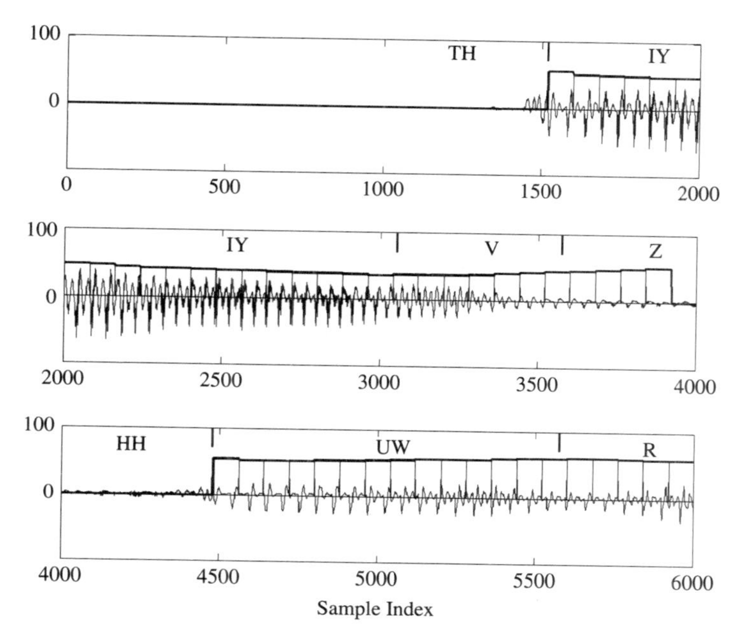

FIGURE 3.10
Pitch period estimates from the speech waveform.

TABLE 3.2 Range of pitch period for male, female, and child speakers, including average, maximum, and minimum periods. (After Zue and Glass, MIT OCW Notes [427].)

	Average (msec)	Maximum (msec)	Minimum (msec)
Male	8	12.5	5
Female	4.4	6.7	2.9
Child	3.3	5	2

period range for a typical male speaker is about 5 msec to about 12.5 msec, corresponding to a pitch frequency range of about 200 Hz down to 80 Hz. For a female speaker the pitch period range is between 2.9 msec and 6.7 msec, corresponding to a pitch frequency range of about 350 Hz down to 150 Hz. The pitch period range for children is even smaller, with pitch periods on the order of 2 msec not being uncommon, corresponding to pitch frequencies of up to 500 Hz.

3.2.3 The Acoustic Theory of Speech Production

Our qualitative discussion of the mechanisms for human speech production led to the schematic model of Figure 3.7. Our subsequent exploration of the characteristics of the speech waveform have provided further understanding of the properties of the speech signal. As we will discuss in much more detail in Chapter 5, the model of Figure 3.7 can

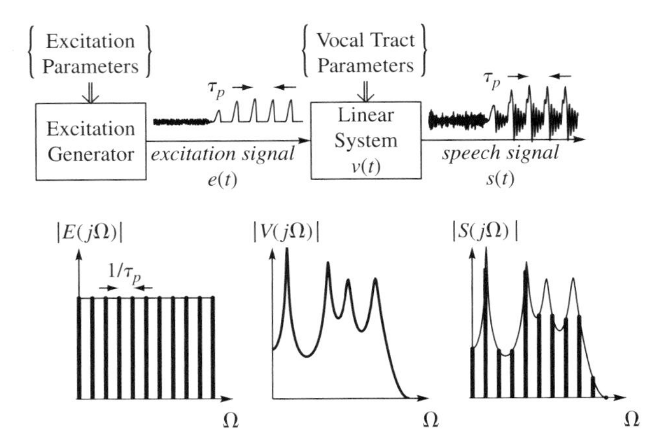

FIGURE 3.11
Linear source/system model of speech production showing temporal and spectral representations of the source, vocal tract, and resulting speech signal.

be analyzed using the mathematics of mechanics, fluid flow, and acoustics. However, for much of our work, an intuitively satisfying and useful approximation is to substitute a source/system model for the detailed physical analysis. In particular, a glance at the waveform of Figure 3.10 suggests that such a waveform could be created by exciting a time-varying linear system with either a locally periodic excitation for voiced speech segments or a random noise excitation for unvoiced segments. Such a simplified model is depicted in Figure 3.11.

If we assume that the excitation signal is $e(t)$, with Fourier transform, $E(j\Omega)$, and the vocal tract impulse response is $v(t)$, with corresponding frequency response, $V(j\Omega)$, then the resulting speech waveform is the convolution of $e(t)$ and $v(t)$, i.e.,

$$s(t) = e(t) * v(t), \tag{3.1}$$

or, in the frequency domain,

$$S(j\Omega) = E(j\Omega) \cdot V(j\Omega). \tag{3.2}$$

Figure 3.11 shows this linear model of speech production along with plots of the temporal and spectral representations of the source, vocal tract, and the resulting speech signal for a voiced section of speech. The excitation signal is represented as a periodic train of pulses with pulse spacing τ_p (modeling the glottal excitation) and with a flat frequency spectrum consisting of isolated spectral lines with spacing $1/\tau_p$. [Strictly speaking, the frequency spectrum of a periodic train of (finite width) pulses is not flat but has a spectral fall-off beginning at a frequency related inversely to the width of the pulses. We will not be concerned with this effect at this time.] The vocal tract impulse response has a continuous Fourier transform that peaks at the resonances (formants) of the particular vocal tract configuration, as seen in the middle part of Figure 3.11. The resulting speech waveform is also periodic with period τ_p, and with a Fourier transform that is the product of the Fourier transforms of the excitation and the vocal tract impulse response; i.e., a line spectrum with frequency spacing of $1/\tau_p$ and an envelope that is determined by the frequency response of the vocal tract, which in turn depends on the shape of the vocal tract tube. This is shown on the right at the bottom of Figure 3.11. The case of unvoiced speech will be much the same except that the random noise excitation will have a continuous flat spectrum.

We will return to the issue of modeling speech signals in Chapter 5. It is sufficient for our purposes now to simply note that the linear source/system model of Figure 3.11 can be applied to represent short segments of the speech waveform. Thus, different segments are characterized by models with the same structure, but different parameters. In this way, the mode of excitation can vary with time, the pitch period can vary with time, and the vocal tract resonance structure can vary with time. In the next section, we show how this point of view leads to the notion of *short-time analysis* of speech.

3.3 SHORT-TIME FOURIER REPRESENTATION OF SPEECH

The time-varying spectral characteristics of a speech signal can be displayed in a graphical representation derived from the speech signal. The short-time Fourier transform of a continuous-time signal $x_c(t)$ is defined as

$$X_c(\hat{t},\Omega) = \int_{-\infty}^{\infty} w_c(\hat{t}-t)x_c(t)e^{-j\Omega t}dt. \tag{3.3}$$

This representation clearly has the form of a continuous-time Fourier transform of the signal $w_c(\hat{t}-t)x_c(t)$. By including the time-localized "analysis window" $w_c(\hat{t}-t)$ in the definition, $X_c(\hat{t},\Omega)$ becomes a function of both Ω, the radian frequency variable for continuous-time signals, and $\hat{t}$, which specifies the analysis time. A two-dimensional function suitable for display as a gray-scale image is obtained by limiting the overall duration of $x_c(t)$, computing $X_c(\hat{t},\Omega)$ for a desired range of values $(\hat{t},\Omega)$, and taking the log magnitude of $X_c(\hat{t},\Omega)$. Such an image is called a *spectrogram* [105, 196, 288].

Speech researchers have relied heavily upon short-time spectrum analysis techniques and the spectrogram since the 1930s. To build a machine to compute and display an approximation to $|X_c(\hat{t},\Omega)|$ in the pre-digital signal processing era required a great deal of ingenuity. One of the earliest systems for computing and displaying the time-dependent Fourier representation of speech was called the *sound spectrograph* or also the *sonograph* [196, 288]. An early version of this machine is depicted in Figure 3.12. Figure 3.13 shows a block diagram representation of the system. A microphone recorded a speech utterance of up to 2 sec duration on a loop of magnetic tape, which was played back repeatedly to measure the spectrum over a range of frequencies between 0 and 5000 Hz. A mechanically controlled and tuned bandpass filter (of appropriate bandwidth) analyzed the speech signal each time around the loop. The average energy in the output of the bandpass filter, at a given time and frequency, was a measure of $|X_c(\hat{t},\Omega)|$. The average energy as a function of time at a fixed setting of the bandpass center frequency was recorded on electrically sensitive teledeltos paper attached to a rotating drum that turned in synchrony with the speech

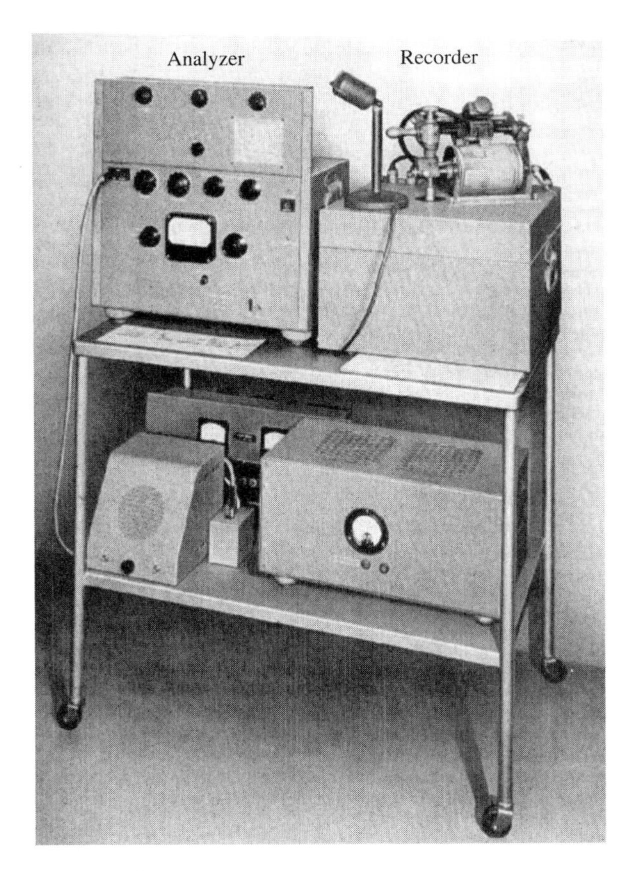

FIGURE 3.12
An early sound spectrograph. (After Potter et al. [288]. Reprinted with permission of Alcatel-Lucent USA Inc.)

signal on the magnetic loop.[3] Each drum rotation corresponded to a new setting of the bandpass filter analysis frequency. This ingenious machine, when properly maintained and adjusted, produced crisp gray-scale spectrogram images; i.e., two-dimensional representations of the short-time spectrum in which the vertical dimension on the paper represented frequency and the horizontal dimension represented time, with the spectrum magnitude represented by the darkness of the marking on the paper. Examples of spectrograms made on such a machine are shown in Figure 3.14.

If the bandpass filter in a spectrograph machine has wide bandwidth (order of 300–900 Hz), the spectrogram has good temporal resolution and poor frequency resolution. This is illustrated by the upper spectrogram in Figure 3.14, which is typical of a *wideband spectrogram*. The resonance frequencies of the vocal tract show up as

[3]The burning of the paper was accompanied by the distinctive smell of ozone!

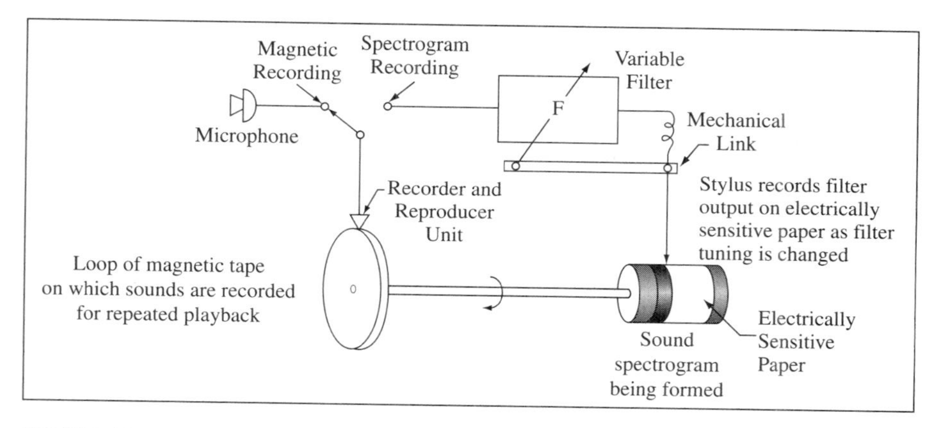

FIGURE 3.13
Diagram of the sound spectrograph [288]. Reprinted with permission of Alcatel-Lucent USA Inc.

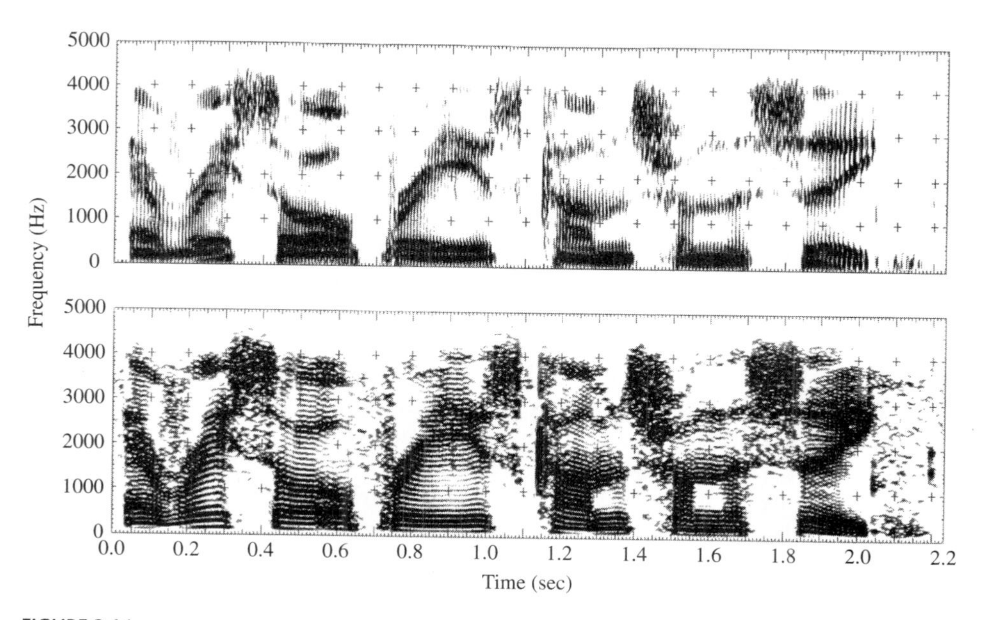

FIGURE 3.14
Wideband and narrowband spectrograms of the sentence "Every salt breeze comes from the sea" made on a sound spectrograph machine.

dark bands in the spectrogram. Voiced regions are characterized by a vertically striated appearance due to the periodicity of the time waveform and its interaction with the short analysis window, while unvoiced intervals are more solidly filled in due to the random nature of the signal in unvoiced intervals. On the other hand, if the bandpass filter has a narrow bandwidth (order of 30–90 Hz), the spectrogram has good frequency resolution and poor time resolution. The lower spectrogram in Figure 3.14 is typical of a *narrowband spectrogram*. In this case, the harmonics of the fundamental frequency are evident in the vertical dimension, leading to a horizontally striated appearance. This is because the window is long enough to include several pitch periods. (Of course the filter bandwidths required for wideband and narrowband spectrograms are highly dependent on the pitch frequency of the speaker; hence a good wideband spectrogram for a male speaker with 100 Hz average pitch might use bandpass filters of 300 Hz bandwidth, whereas for a female speaker of average pitch of 200 Hz, it might use a bandpass filter of 600 Hz bandwidth to get the same apparent frequency resolution.)

With the advent of high-speed computers and image displays, analog spectrograph machines were replaced by flexible software implementations based on a discrete-time formulation of the short-time spectrum. Given a sampled speech signal, $x[n]$, the discrete-time short-time Fourier transform takes the form

$$X_{\hat{n}}(e^{j\omega}) = \sum_{m=-\infty}^{\infty} w[\hat{n} - m]x[m]e^{-j\omega m}, \tag{3.4}$$

where $\hat{n}$ is the analysis time, ω is the normalized frequency for discrete-time signals, and $w[n]$ is a finite-length spectrum analysis window. As we will discuss in more detail in Chapter 7, Eq. (3.4) can be evaluated efficiently at analog frequencies $\Omega_k = (2\pi k/N)F_s$ for $k = 0, 1, \ldots, N/2$ using an N-point fast Fourier transform (FFT) algorithm. The resulting short-time spectrum $|X_{\hat{n}}(e^{j(2\pi k/N)F_s})|$ can be displayed as a function of $\hat{n}$ and k on a computer monitor or printed on a gray-scale or color printer. Figure 3.15 shows a plot of a wideband spectrogram of an utterance of "Oak is strong and also gives shade." This figure is very much like the analog wideband spectrogram in the upper part of Figure 3.14. An ancillary advantage of the digital spectrogram is its greater flexibility. There is no limitation on the length and shape of the analysis window, and the displayed image can be annotated and combined with other plots such as the waveform plot shown at the bottom of the figure. Furthermore, the spectrum magnitude can be represented on a gray scale or some sort of pseudo-color rendering of amplitude can be used. The vertical striations due to the pitch periodicity of voiced sections are clearly seen again in Figure 3.15 as is the noise-like quality during strong unvoiced

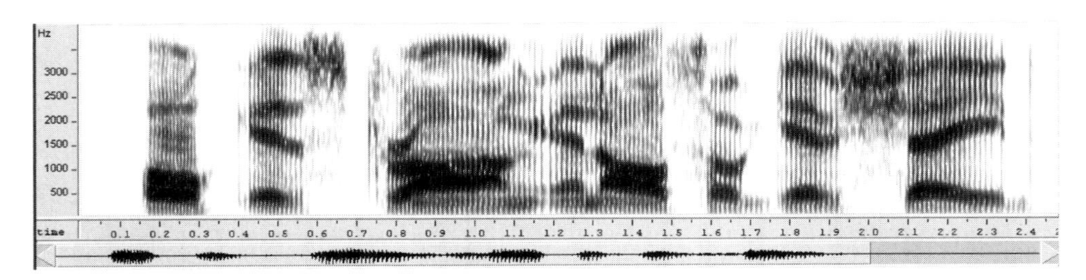

FIGURE 3.15
Wideband spectrogram of the utterance "Oak is strong and also gives shade."

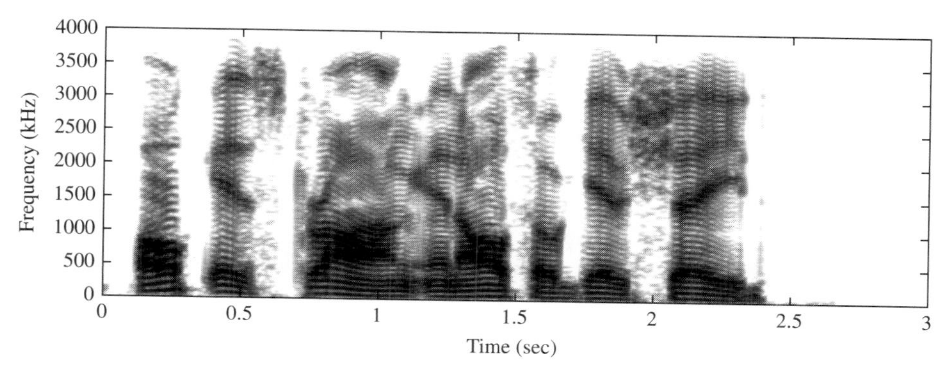

FIGURE 3.16
Narrowband spectrogram of the utterance "Oak is strong and also gives shade."

signal regions. Note that the resonance structure of the speech is strongly manifested during voiced regions where the strong bar-like concentrations of energy that persist over voiced regions of time are the formants or resonant frequencies of the vocal tract configuration that produced the sound shown in the spectrogram. The spectrogram of Figure 3.15 was produced on a standard PC using freely available software called WaveSurfer [413], and was created in a fraction of a second.

Another example of a spectrogram of a speech utterance is shown in Figure 3.16. This is a narrowband digital spectrogram of the utterance "Oak is strong and also gives shade" that was created using MATLAB's built-in spectrogram function. Comparing it with the lower spectrogram in Figure 3.14, we again see horizontal striations due to the pitch harmonics of voiced sections. During regions of unvoiced signal, no such pitch harmonics are seen and the spectrogram again has a noise-like quality. Further, we see that the resonance structure of the speech is not so strongly manifested during voiced regions because of interactions between pitch harmonics and speech resonances.

The sound spectrograph has long been a principal tool in speech research, and its basic principles are still widely used in digital spectrograms. The book *Visible Speech* by Potter, Kopp, and Green [288] was one of the earliest works to recognize the unique aspects of the sound spectrogram for identifying speech sounds and it popularized the use of sound spectrograms for speech research in the early 1950s. This book, although written for the purpose of teaching people literally to "read" spectrograms, provides an excellent introduction to the field of acoustic phonetics. Although virtually no one has been able to learn to read sound spectrographic displays consistently (i.e., convert such displays to coherent and intelligible transcriptions of what was spoken), if one is given the phonetic transcription and both the waveform and its spectrogram, it is much easier to accurately segment the speech signal and assign symbolic labels to the segments. By highlighting the formant frequencies and their motions with time, the short-time spectrum greatly facilitates understanding the relationship between the signal and its mechanism of production [264]. Both the speech waveform and the speech spectrogram are valuable tools for analyzing the speech signal, and we will spend significant time showing how algorithms can be designed to efficiently and reliably estimate the properties of speech from such displays in Chapters 6 and 7. Indeed, it is fair to say that the short-time spectrum concept underlies most of the signal processing techniques that we shall discuss in this book.

3.4 ACOUSTIC PHONETICS

Most languages, including English, can be described in terms of a set of distinctive sounds, or *phonemes*. In particular, for American English, there are somewhere between 39 and 48 phonemes, including vowels, diphthongs, semivowels, and consonants. Table 3.3 gives a standard list of 48 phonemes of American English along with the International Phonetic Alphabet (IPA) representation, the ARPAbet representation [designed by the Advanced Research Projects Agency (ARPA) for transcription on computer keyboards without special fonts], and an example of a word that shows where the phoneme appears. In the rest of this book, we shall use both the IPA and ARPAbet symbols as is convenient. While this may at first seem confusing, this is useful for becoming familiar with both representations. The ARPAbet symbols are generally used in engineering and computer science, but the IPA symbols are most commonly used in linguistics and phonetics publications.

In Table 3.3 the 48 phonemes are divided into five broad classes, namely:

- 14 vowels (from /IY/ to /ER/) and 4 diphthongs (vowel combinations, from /EY/ to /OY/)
- 4 vowel-like (liquid and glide) consonants (from /Y/ to /L/)
- 21 standard consonants (from /M/ to /WH/)
- 4 syllabic sounds (from /EL/ to /DX/)
- 1 glottal stop (/Q/).

The role of phonemes is to provide a link between the orthography[4] of written language and the speech signal that is produced when the sentence corresponding to the orthography is spoken. Thus, for example, the phonetic transcription of the orthography of the name "Larry" is:

Larry → /l æ r i/ (in IPA notation)
or /L AE R IY/ (in ARPAbet notation)

and when the word "Larry" is spoken, we would see clear evidence of the above sequence of phonemes in both the speech waveform and the spectrogram derived from the speech waveform. We use the *phonetic code* as an intermediate representation of language and therefore it is essential to understand the acoustic and articulatory properties of all of the sounds (phonemes) of a language in order to design the best speech

[4]By this term we mean the writing of words with the English alphabet. We have referred to this as the "text" representation.

TABLE 3.3 Condensed list of phonetic symbols for American English.

IPA Phoneme	ARPAbet	Example	IPA Phoneme	ARPAbet	Example
/i/	IY	beet	/ŋ/	NX	sing
/I/	IH	bit	/p/	P	pat
/ɚ/	AXR	butter	/t/	T	ten
/ɛ/, /e/	EH	bet	/k/	K	kit
/æ/	AE	bat	/b/	B	bet
/a/	AA	Bob	/d/	D	debt
/ʌ/	AH	but	/g/	G	get
/ɔ/	AO	bought	/h/	HH	hat
/o/	OW	boat	/f/	F	fat
/ʊ/	UH	book	/θ/	TH	thing
/u/	UW	boot	/s/	S	sat
/ə/	AX	about	/ʃ/, /sh/, /š/	SH	shut
/ɨ/	IX	roses	/v/	V	vat
/ɝ/	ER	bird	/ð/	DH	that
/eʸ/	EY	bait	/z/	Z	zoo
/aʷ/	AW	down	/ʒ/, /zh/, /ž/	ZH	azure
/aʸ/	AY	buy	/tʃ/, /č/	CH	church
/ɔʸ/	OY	boy	/dʒ/, /j/	JH	judge
/y/	Y	you	/ʍ/	WH	which
/w/	W	wit	/l̩/	EL	battle
/r/	R	rent	/m̩/	EM	bottom
/l/	L	let	/n̩/	EN	button
/m/	M	met	/ɾ/	DX	batter
/n/	N	net	/ʔ/	Q	(glottal stop)

processing systems for a given application (especially for speech synthesis and speech recognition applications).

It is also important to understand that phonetic transcriptions from orthographic representations of language are, at best, an idealization of how the speech will be realized in practice. Hence the ideal phonetic realization of the orthography "Did you eat yet" is of the form:

Did you eat yet → /d I d-y u-i t-y ɛ t/
or /D IH D - Y UW - IY T - Y EH T/

but the actual sounds that are produced when speaking this sentence in a perfunctory manner might be of the form:

Dija eat jet → /d I j ə - i t - j ɛ t/
/D IH JH UH - IY T - JH EH T/

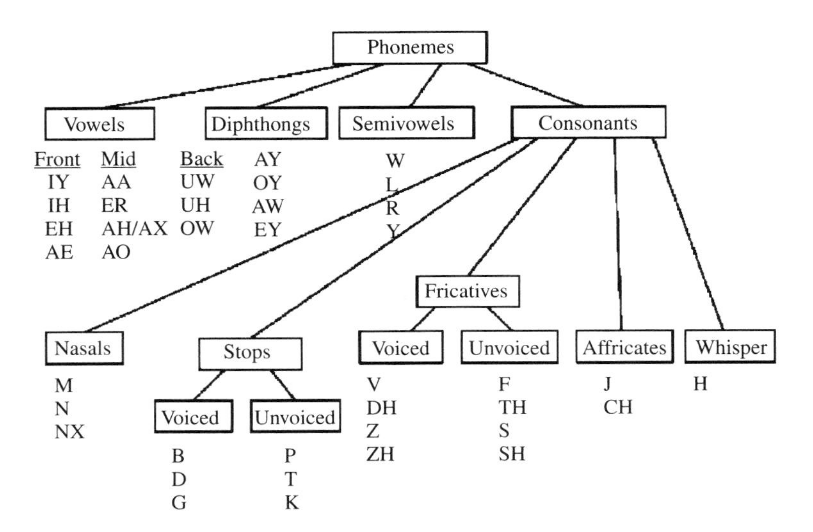

FIGURE 3.17
Phonemes in American English.

which is a highly co-articulated version of the sentence with different phonemes being spoken as a result of the merging of sounds across word boundaries (as well as within words) in natural speech.

There are a variety of ways of studying phonetics and learning about the properties of the phonemes of a given language. For example, linguists often study the distinctive features or articulatory characteristics of the phonemes [60, 166] from the point of view of understanding how the phonemes are produced by humans. Although we will certainly look at the articulatory patterns associated with each of the phonemes of the language, we will be more concerned with developing a good understanding of the acoustic and spectral characteristics of the various phonemes so that we can identify their properties in waveforms and sound spectrograms.

Rather than using the full set of 48 phonemes of Table 3.3, we will work with a reduced set of 39 sounds, as shown in Figure 3.17, which includes

- 11 vowels
- 4 diphthongs
- 4 semivowels
- 3 nasal consonants
- 6 voiced and unvoiced stop consonants
- 8 voiced and unvoiced fricatives
- 2 affricate consonants
- 1 whispered sound.

Here, we have chosen to neglect the syllabic and glottal sounds and some of the highly reduced vowels. The four broad classes of sounds are vowels, diphthongs, semivowels, and consonants. Each of these classes is further broken down into sub-classes, which are related to the manner and place of articulation of the sound within the vocal tract. We discuss these issues later in this chapter.

Each of the phonemes in Figure 3.17 can be classified as either a *continuant*, or a *non-continuant* sound. Continuant sounds are produced by a fixed (non-time-varying) vocal tract configuration excited by the appropriate source. The class of continuant sounds includes the vowels, the fricatives (both unvoiced and voiced), and the nasals. The remaining sounds (diphthongs, semivowels, stops, and affricates) are produced by a changing (time-varying) vocal tract configuration. These are therefore classified as non-continuants.

3.4.1 Vowels

Vowels generally have the longest duration in natural speech, and they are the most well defined of the sounds of the language. They can be held indefinitely, e.g., while singing. Although vowels play a major role in spoken language, they carry very little linguistic information about the orthography of the sentence that is spoken. (There are some languages whose orthography does not include any vowels, e.g., Arabic and Hebrew). As an example, consider the two sentences shown below, the first with the vowel orthography removed, and the second with the consonant orthography removed.

- (*all vowels deleted*) `Th_y n_t_d s_gn_f_c_nt _mpr_v_m_nts _n th_ c_mp_ny's _m_g_, s_p_rv_s__n _nd m_n_g_m_nt.`
- (*all consonants deleted*) `A__i_u_e_ _o_a__ _a_ __a_e_ e__e__ia_ __ __e _a_e, _i__ __e __o_e_ o_ o__u_a_io_a_ e___o_ee_ __i_____ _e__ea_i__.`

Most native speakers of American English would have relatively little difficulty in filling in the vowels in the first sentence and coming up with the transcription, "They noted significant improvements in the company's image, supervision and management." Similarly, it is doubtful that anyone would come up with the sentence that was missing all the consonants. That sentence is, "Attitudes toward pay stayed essentially the same, with the scores of occupational employees slightly decreasing."

Vowels are produced by exciting a fixed vocal tract with quasi-periodic pulses of air caused by vibration of the vocal cords. As we shall see later in Chapter 5, the way in which the cross-sectional area varies along the vocal tract determines the resonant frequencies of the tract (formants) and thus the sound that is produced. The dependence of cross-sectional area upon distance along the tract is called the *area function* of the vocal tract. The area function for a particular vowel is determined primarily by the position of the tongue, but the positions of the jaw, lips, and, to a small extent, the velum also influence the resulting sound. Figure 3.18 shows schematic vocal tract configurations for the vowels /i/, /æ/, /a/, and /u/ (/IY/, /AE/, /AA/, and /UW/ in ARPAbet). In forming the vowel /a/ as in "f<u>a</u>ther," the vocal tract is open at the front and somewhat constricted at the back by the main body of the tongue. In contrast, the vowel /i/ as in "<u>e</u>ve" is formed by raising the tongue toward the palate, thus causing a slight constriction at the front and increasing the opening at the back of the vocal tract.

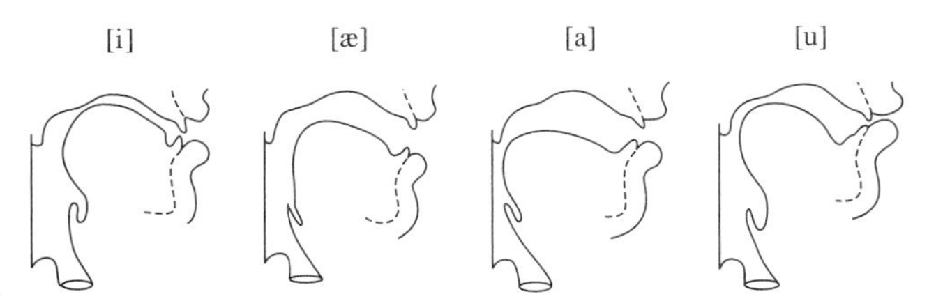

FIGURE 3.18
Schematic vocal tract configurations for the vowels /i/, /æ/, /a/, and /u/ (/IY/, /AE/, /AA/, and /UW/ in ARPAbet). (After Zue and Glass, MIT OCW Notes [427].)

In this manner, each vowel sound can be characterized by the vocal tract configuration (area function) that is used in its production. It is obvious that this is a rather imprecise characterization because of the inherent differences between the vocal tracts of speakers. An alternative representation is in terms of the resonance frequencies of the vocal tract. Again a great deal of variability is to be expected among speakers producing the same vowel. Using an early sound spectrograph such as the one described in Section 3.3, Peterson and Barney [287] measured the formant (resonance) frequencies of vowels that were perceived by listeners to be equivalent. Their results are shown in Figure 3.19, which is a plot of second formant frequency as a function of first formant frequency for several vowels spoken by men and women. The broad ellipses

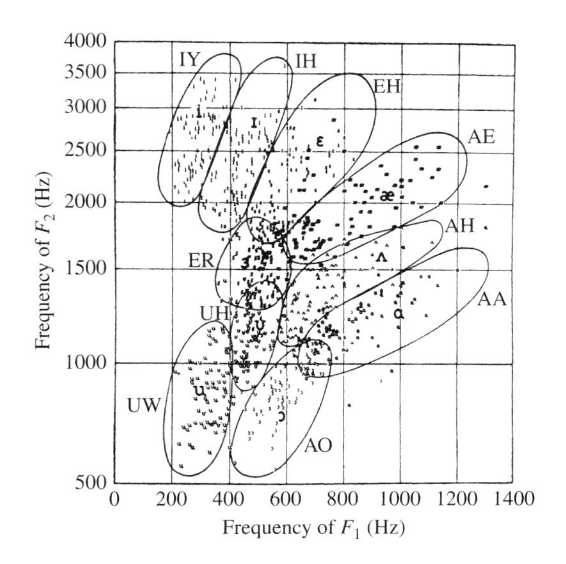

FIGURE 3.19
Plots of second formant frequency versus first formant frequency for vowels by a wide range of speakers. (After Peterson and Barney [287]. © [1952] Acoustical Society of America.)

TABLE 3.4 Average formant frequencies for the vowels of American English. (After Peterson and Barney [287]. © [1952] Acoustical Society of America.)

Average Vowel Formant Frequencies					
ARPAbet Symbol	**IPA Symbol**	**Typical Word**	F_1	F_2	F_3
IY	i	(beet)	270	2290	3010
IH	I	(bit)	390	1990	2550
EH	ε	(bet)	530	1840	2480
AE	æ	(bat)	660	1720	2410
AH	ʌ	(but)	520	1190	2390
AA	a	(hot)	730	1090	2440
AO	ɔ	(bought)	570	840	2410
ER	ɝ	(bird)	490	1350	1690
UH	U	(foot)	440	1020	2240
UW	u	(boot)	300	870	2240

in Figure 3.19 show the approximate range of variation in formant frequencies for each of these vowels. Table 3.4 gives average values of the first three formant frequencies (in Hz) of the vowels for male speakers. Although Figure 3.19 shows that a great deal of variation exists in the vowel formants, the data of Table 3.4 serve as a useful characterization of the vowels.

Figure 3.20 shows a plot of the average second formant frequency versus the first formant frequency for the vowels of Table 3.4. The so-called vowel triangle is readily seen in this figure. At the upper left hand corner of the triangle is the vowel /i/ with a low first formant and a high second formant. This vowel is representative of the class of front vowels, where the vocal tract is constricted by the tongue hump toward the front end. At the lower left hand corner is the vowel /u/ with low first and second formants. This vowel is representative of the class of back vowels, where the vocal tract is most constricted well back in the mouth cavity. The third vertex of the triangle is the vowel /a/ with a high first formant and a low second formant. This vertex represents the mid-vowels. Note that the other vowels fall at intermediate positions close to the triangle connecting /i/, /u/, and /a/. In Chapter 5, we will see from the physics of speech production how vocal tract shape affects the formant frequencies of vowels.

The acoustic waveforms and spectrograms for representative examples of each of the vowels of English are shown in Figure 3.21. The spectrograms clearly show a different pattern of resonances for each vowel. The acoustic waveforms, in addition to showing the periodicity characteristic of voiced sounds, also display the gross spectral properties if a single "period" is considered. For example, the vowel /i/ shows a low frequency damped oscillation upon which is superimposed a relatively strong high frequency oscillation. This is consistent with a low first formant and high second and third formants (see Table 3.4). (Two resonances in proximity tend to boost the spectrum.) In contrast, the vowel /u/ shows relatively little high frequency energy as a consequence of the low first and second formant frequencies. Similar correspondences can be observed for all the vowels in Figure 3.21.

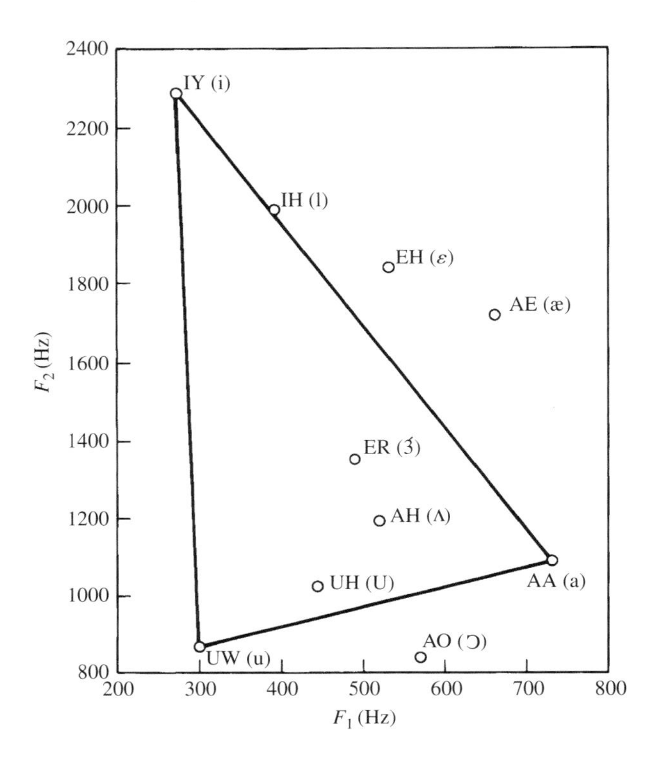

FIGURE 3.20
The vowel triangle.

Figure 3.22 shows a series of spectral plots for the three vowels at the corners of the vowel triangle, namely /IY/ (/i/), /AA/ (/a/), and /UW/ (/u/). The vowels were created synthetically using the three formants from Table 3.4 for each of the vowels to set the pole locations of a set of digital filters. (We will discuss such simulations of the speech production process in detail in Chapter 5.) The left panel of this figure shows the log spectral magnitudes corresponding to the vocal tract frequency response for each of these three vowels. The locations of the formant frequencies are clear for each of these vowels. The middle panel of this figure shows the resulting spectral magnitudes when using a 100 Hz pitch frequency excitation. The individual pitch harmonics, spaced by 100 Hz, are clearly seen for each of the vowel spectra. The resonance structure of the vowel spectra is still well preserved for all three vowels. Finally the right panel of this figure shows the resulting log magnitude spectra for each vowel when excited using a 300 Hz pitch frequency excitation. The harmonics at multiples of 300 Hz are clearly seen in the log magnitude spectra; however, we see that it has become very difficult to reliably determine the resonance structure of each of these vowel sounds with the low rate of sampling of the log spectra due to the 300 Hz periodicity. These spectral plots give us an idea as to the types of problems we will face later on when we try to estimate the resonance structure of the speech signal from spectral or temporal representations of speech.

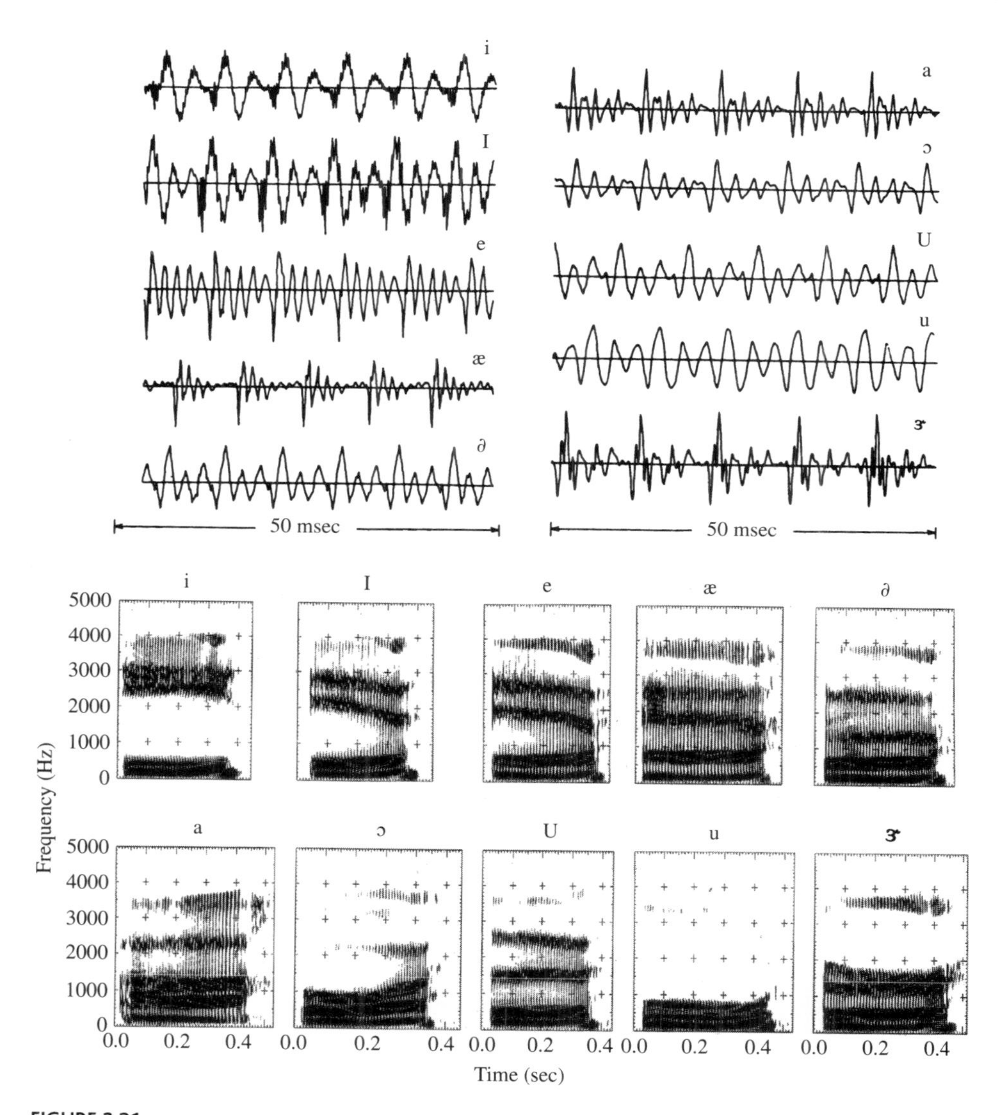

FIGURE 3.21
The acoustic waveforms for several American English vowels and corresponding spectrograms. Note carefully the dark horizontal "voice bar" at a frequency of about 100–200 Hz. This is not a formant frequency. In Chapter 5 we will see that this prominent spectral peak is due to the glottal pulse excitation. The voice bar will appear in many of the spectrograms shown later in this chapter.

3.4.2 Diphthongs

Although there is some ambiguity and disagreement as to what is and what is not a diphthong, a reasonable definition is that a diphthong is a gliding monosyllabic speech item that starts at or near the articulatory position for the initial (first) vowel in the diphthong, and moves to or toward the position for the final (second) vowel in the

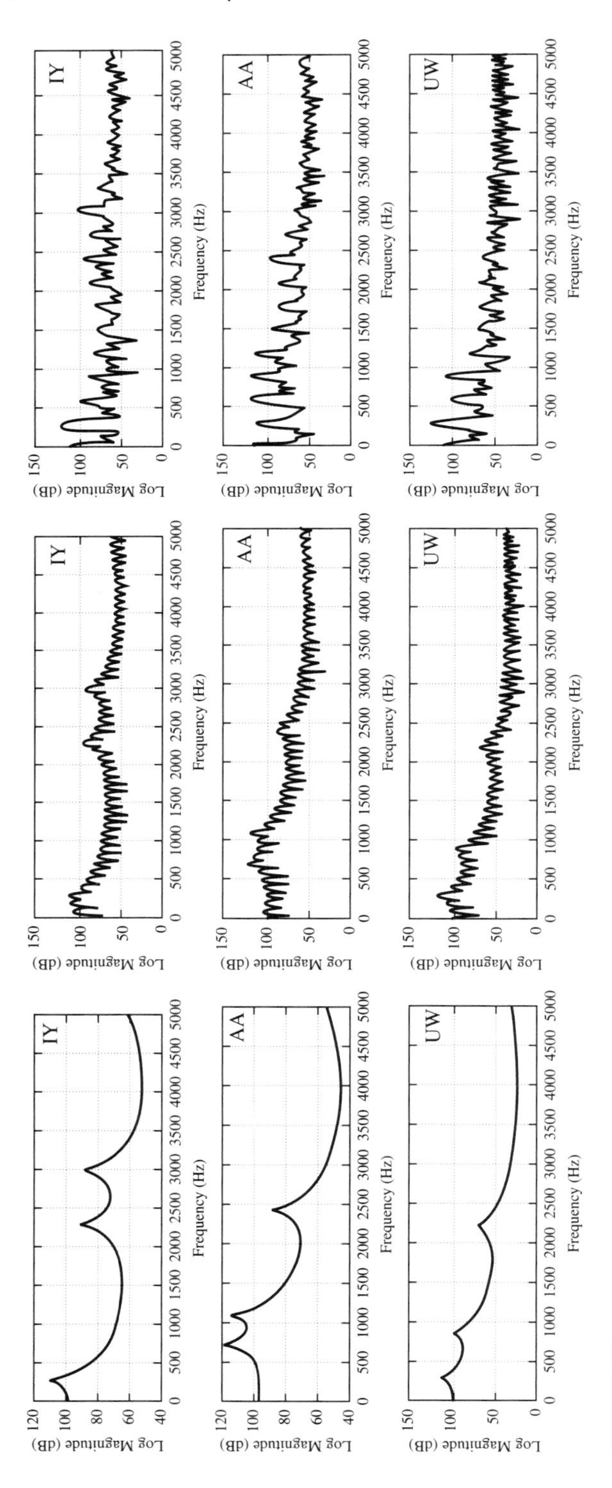

FIGURE 3.22
Canonic vowel spectra for the three vowels /IY/, /AA/, and /UW/ at the corners of the vowel triangle; left pane shows the vocal tract frequency response; middle pane shows the spectra with a 100 Hz pitch frequency excitation; right pane shows the spectra with a 300 Hz pitch frequency excitation.

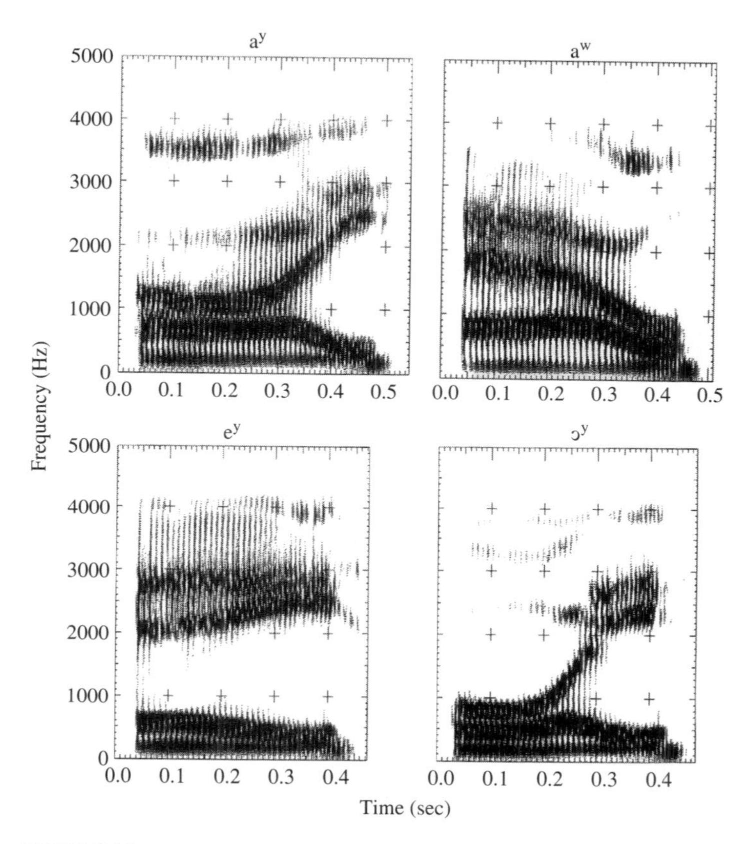

FIGURE 3.23
Spectrograms of the four diphthongs of American English.

diphthong [146]. According to this definition, there are four standard diphthongs in American English, including /e^y/ (/EY/) as in "bay," /a^y/ (/AY/) as in "buy," /a^w/ (/AW/) as in "how," and /$ɔ^y$/ (/OY/) as in "boy."

The diphthongs are produced by varying the vocal tract smoothly between vowel configurations appropriate to the initial and final vowel within the diphthong. To illustrate this point, Figure 3.23 shows spectrogram plots of the four diphthongs of American English. The smooth gliding behavior of the spectral resonances (i.e., the formants) between values appropriate to the initial and final vowels of each diphthong is clearly seen in this figure and characterizes the spectral behavior of the diphthongs.

3.4.3 Distinctive Features of Sounds [60]

All of the sounds of American English, other than the vowels and diphthongs, are characterized by two classes of so-called *distinctive features*, namely the place of articulation (i.e., the point along the vocal tract mechanism of maximal constriction of air flow) and the manner of articulation (i.e., the characteristics of the excitation signal, or of

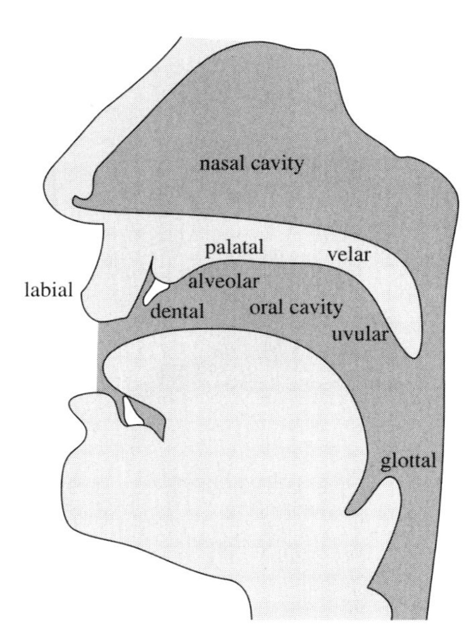

FIGURE 3.24
Places of articulation for the sounds of American English.

the articulator motion that is used to create the sound). This classification of sounds via distinctive features is widely used in linguistics and phonetics. For our purposes, distinctive features provide a good mechanism to express our understanding of the process of sound generation in the human vocal tract.

Figure 3.24 shows the locations of a subset of the places of articulation for the sounds of American English. The places of articulation (and the sounds that are produced at these places of articulation) include the following:

- labial or bilabial (at the lips) where the sounds /p/, /b/, /m/, and /w/ are produced
- labiodental (between the lips and the front of the teeth) where the sounds /f/ and /v/ are produced
- dental (at the teeth) where the sounds /θ/ and /ð/ are produced
- alveolar (the front of the palate) where the sounds /t/, /d/, /s/, /z/, /n/, and /l/ are produced
- palatal (middle of the palate) where the sounds /š/, /ž/, and /r/ are produced
- velar (at the velum) where the sounds /k/, /g/, and /ŋ/ are produced
- pharyngeal (at the end of the pharynx) where the whispered sound /h/ is produced.

The manner of articulation describes either the sound excitation source or the manner of motion of the speech articulators in the production of the static or

Place of Articulation

Manner of Articulation	Labial	Dental	Alveolar	Palatal	Velar
Stop	p b		t d		k g
Fricative	f v	θ ð	s z	š ž	
	Weak (Non-strident)		Strong (Strident)		
Nasal	m		n		ŋ

FIGURE 3.25
Summary of place and manner of articulation for the consonants of American English.

dynamic properties of the sound. The manners of articulation (and the sounds that are produced) include the following:

- glide (smooth motion of the articulators) for the sounds /w/, /l/, /r/, and /y/
- nasal (lowered velum) for the sounds /m/, /n/, and /ŋ/
- stop (totally constricted vocal tract blocking air flow) for the sounds /p/, /t/, /k/, /b/, /d/, and /g/
- fricative (vocal cords not vibrating, with turbulent sound source due to high degree of constriction in vocal tract) for the sounds /f/, /θ/, /s/, /š/, /v/, /θ/, /z/, /ž/, and /h/
- voicing (vocal cords are vibrating throughout the sound) for the sounds /b/, /d/, /g/, /v/, /θ/, /z/, /ž/, /m/, /n/, /ŋ/, /w/, /l/, /r/, and /y/
- mixed source (vocal cords vibrating but turbulence produced at constriction in vocal tract) for sounds /j/ and /č/
- whispered (turbulent air source at the glottis) for the sound /h/.

Figure 3.25 shows a summary of the place and manner of articulation for the consonants of American English using five places of articulation and three manners of articulation. This chart conveniently depicts the breakdown of the consonants of American English into a simple matrix of place and manner options.

3.4.4 Semivowels

The group of sounds consisting of /w/, /l/, /r/, and /y/ (/W/, /L/, /R/, and /Y/) is called the semivowels because of their vowel-like nature. The semivowels /w/ and /y/ are often called *glides* and the semivowels /r/ and /l/ are often called *liquids*. The semivowels are characterized by a constriction in the vocal tract, but one at which no turbulence is created. This is due to the fact that the tongue tip generally forms the constriction for the semivowels, and therefore the constriction does not totally block air flow through the vocal tract. The points of constriction are shown

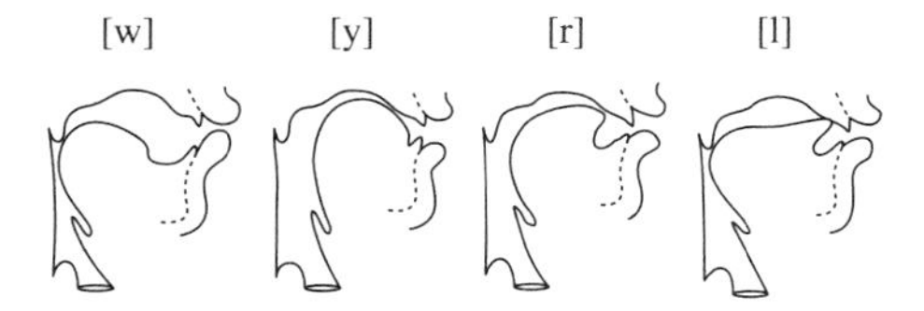

FIGURE 3.26
Articulatory configurations for the semivowels of American English. (After Zue and Glass, MIT OCW Notes [427].)

in Figure 3.26. Instead there is airflow laterally around the sides of the constriction. The semivowels have properties similar to corresponding vowels, but with more pronounced articulations. The vowels most closely corresponding to the four semivowels are the following:

- the semivowel /w/ is closest to the vowel /u/ (as in boot)
- the semivowel /y/ is closest to the vowel /i/ (as in beet)
- the semivowel /r/ is closest to the vowel /ɝ/ (as in bird)
- the semivowel /l/ is closest to the vowel /o/ (as in boat).

Spectrograms of the four American English semivowels, followed by the vowel /i/, are shown in Figure 3.27. The semivowels are generally characterized by a gliding transition in vocal tract area function between adjacent phonemes. Thus the acoustic characteristics of these sounds are strongly influenced by the context in which they occur. This is illustrated in Figure 3.27, where the liquid or glide is at the beginning followed by the vowel /i/. The vowel /i/ has low first formant and high second formant. In the case of /y/, the formants match well, so there is little transitory motion of the second formant. On the other hand, /w/, /l/, and /r/ all have low second formants, so there is significant movement from low to high for the second formant frequency.

The general acoustic properties of semivowels are the following:

- the semivowels /w/ and /l/ are the most confusable pair
- the semivowel /w/ is characterized by a very low first (F_1) and second resonance (F_2), and often with a rapid spectral level decrease for frequencies above the second resonance
- the semivowel /l/ is characterized by a low first (F_1) and second resonance (F_2), and often with a lot of high frequency energy
- the semivowel /y/ is characterized by very low first resonance (F_1) and a very high second resonance (F_2); the semivowel /y/ only occurs before a vowel in the initial position of a syllable
- the semivowel /r/ is characterized by a very low third resonance (F_3) [and hence relatively low first (F_1) and second (F_2) resonances, since they fall below the third resonance].

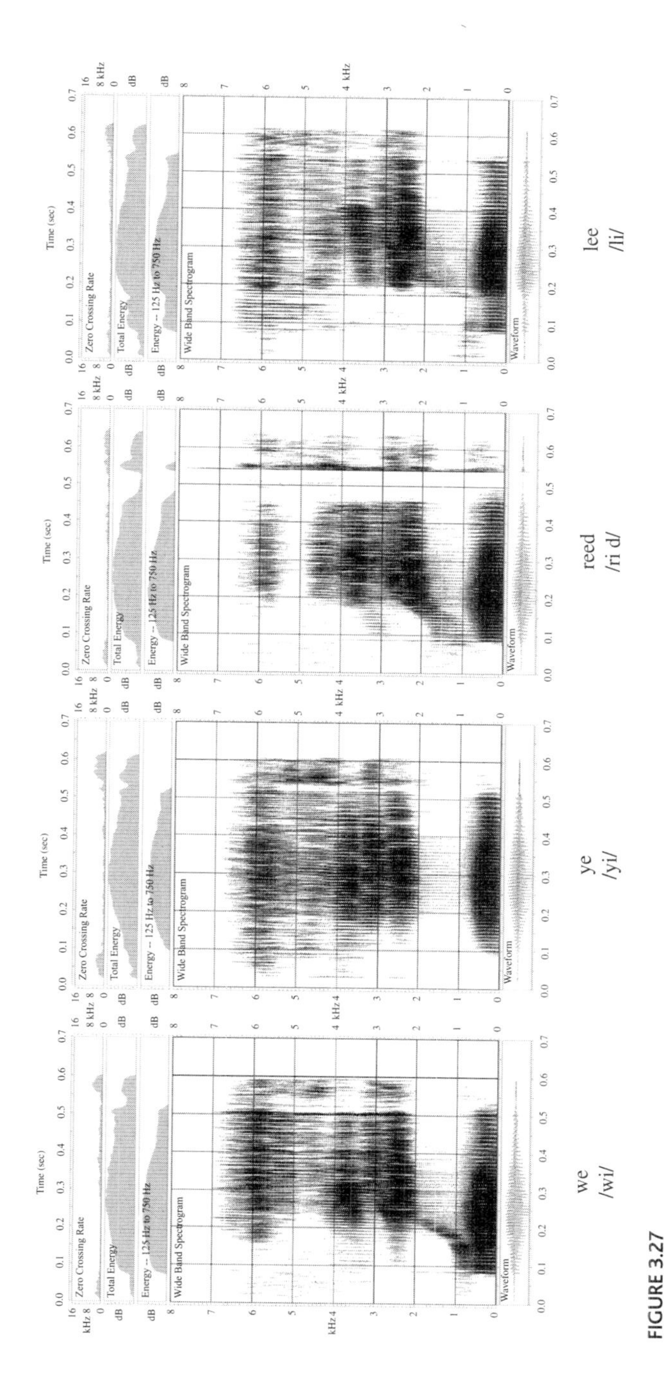

FIGURE 3.27
Spectrograms for the four semivowels of American English followed by the vowel /i/. (After Zue and Glass, MIT OCW Notes [427].)

For our purposes, semivowels are best described as transitional, vowel-like sounds, and hence are similar in nature to the vowels and diphthongs.

3.4.5 Nasals [117]

The nasal consonants /m/, /n/, /ŋ/ (/M/, /N/, and /NX/) are produced with glottal excitation (hence these are voiced sounds) and the vocal tract totally constricted at some point along the oral passageway. The velum is lowered so that air flows through the nasal tract, with sound being radiated at the nostrils. The oral cavity, although constricted toward the front, is still acoustically coupled to the pharynx. Thus, the mouth serves as a resonant cavity that traps acoustic energy at certain natural frequencies. As far as the radiated sound is concerned, these resonant frequencies of the oral cavity appear as anti-resonances, or zeros of sound transmission [105, 117]. Furthermore, nasal consonants and nasalized vowels (i.e., some vowels preceding or following nasal consonants) are characterized by resonances that are spectrally broader, or more highly damped, than those for vowels. The broadening of the nasal resonances occurs because the inner surface of the nasal tract is convoluted, so that the nasal cavity has a relatively large ratio of surface area to cross-sectional area. Therefore, heat conduction and viscous losses are larger than normal.

The three nasal consonants are distinguished by the place along the oral tract at which a total constriction is made, as illustrated in Figure 3.28. For /m/, the constriction is at the lips; for /n/ the constriction is just back of the teeth; for /ŋ/ the constriction is just forward of the velum itself. Figure 3.29 shows typical speech waveforms and spectrograms for the two nasal consonants, /m/ and /n/, in the context of vowel-nasal-vowel (/AH M AA/ and /AH N AA/). It is clear that the waveforms of /m/ and /n/ look very similar (showing sections of a low frequency waveform) and provide very few cues as to which particular nasal was produced. The spectrograms, on the other hand, show two strong features of nasal spectra, namely a significant low frequency first resonance and a well-defined spectral region with virtually no signal energy. The low resonance is characteristic of the long passageway between the glottis (the source for voiced sounds) and the nostrils (where sound is ultimately radiated), and is essentially independent of the nasal sound being produced. The spectral region with virtually no signal energy is related to the oral tract that is totally blocked at the point of constriction of the nasal sound. Depending on the length of the oral tract, spectral zeros are introduced into the sound spectrum, and these are manifest as regions of very low spectral energy [117]. We see from Figure 3.29 that the location of the zero for the /m/ sound is much lower

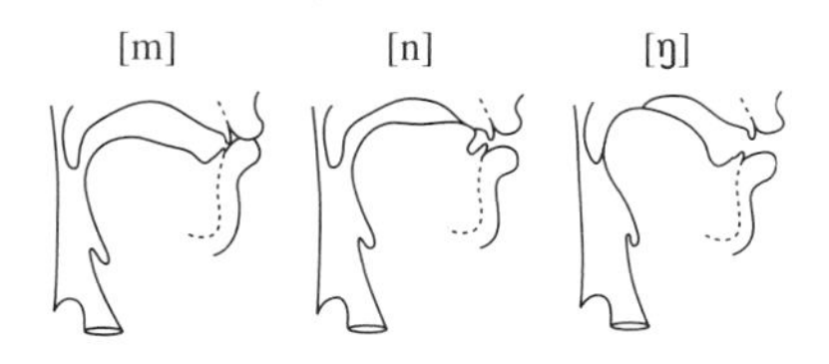

FIGURE 3.28
Articulatory configurations for the nasal consonants. (After Zue and Glass, MIT OCW Notes [427].)

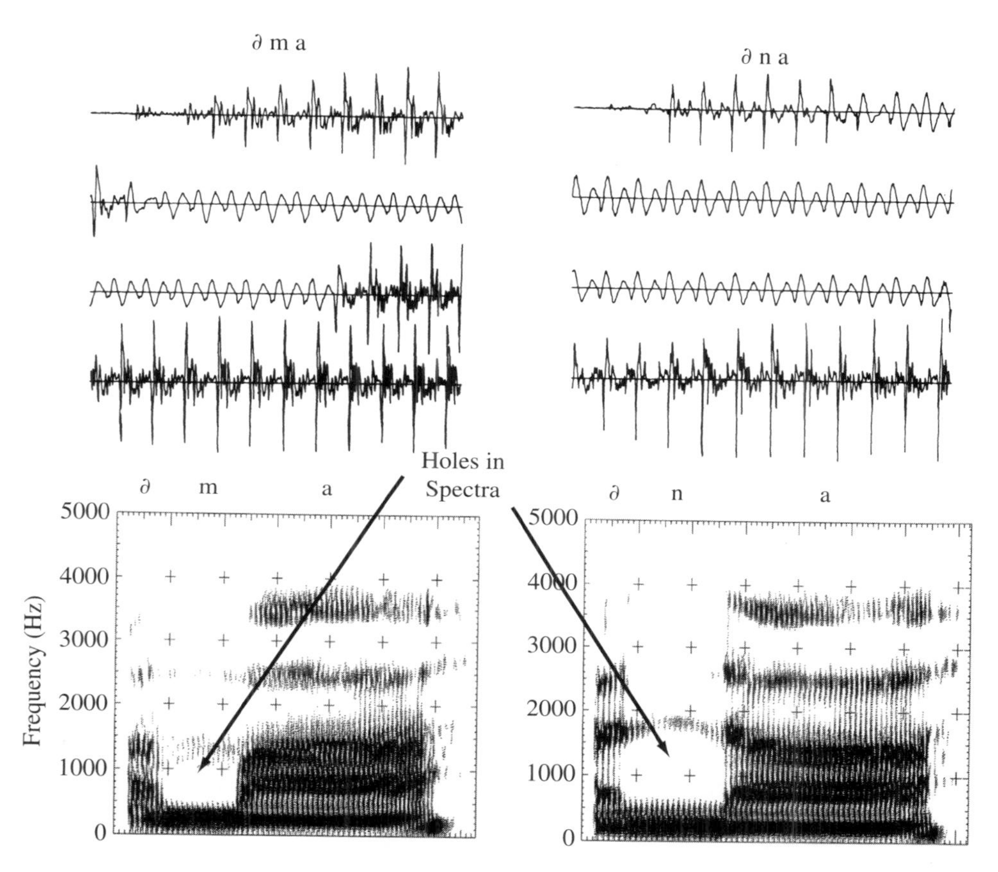

FIGURE 3.29
Acoustic waveforms and spectrograms for utterances /UH M AA/ and /UH N AA/.

in frequency than the location of the zero for the /n/ sound; hence the "hole" in the spectrum for /m/ is somewhat smaller than the hole in the spectrum for /n/, as shown in Figure 3.29.

3.4.6 Unvoiced Fricatives [141]

The unvoiced fricatives /f/, /θ/, /s/, and /š/ (/F/, /TH/, /S/, and /SH/) are produced by exciting the vocal tract by a steady air flow that becomes turbulent in the region of a constriction in the vocal tract. The location of the constriction serves to determine which fricative sound is produced. As shown in Figure 3.30, for the fricative /f/ the constriction is near the lips; for /θ/ it is near the teeth; for /s/ it is near the middle of the oral tract; and for /š/ it is near the back of the oral tract. Thus the system for producing unvoiced fricatives consists of a source of noise at a constriction that separates the vocal tract into two cavities. Sound is radiated from the lips; i.e., from the front cavity. The back cavity serves, as in the case of nasals, to trap energy and thereby introduce anti-resonances (zeros) into the vocal output [105, 141]. Figure 3.31 shows the waveforms and spectrograms of the fricatives /f/, /s/, and /š/ in the context of the

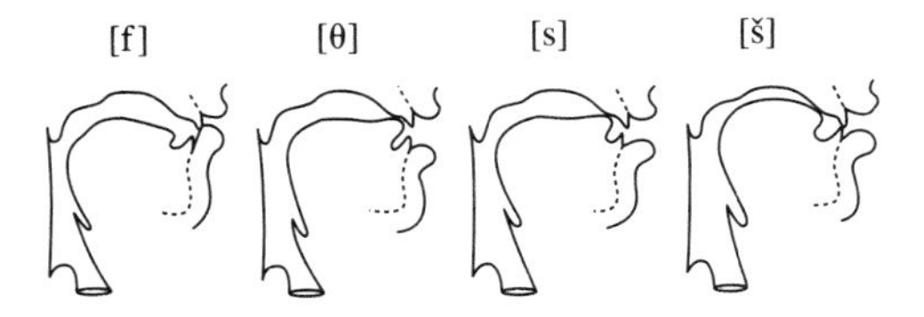

FIGURE 3.30
Articulatory configurations for the unvoiced fricatives. (After Zue and Glass, MIT OCW Notes [427].)

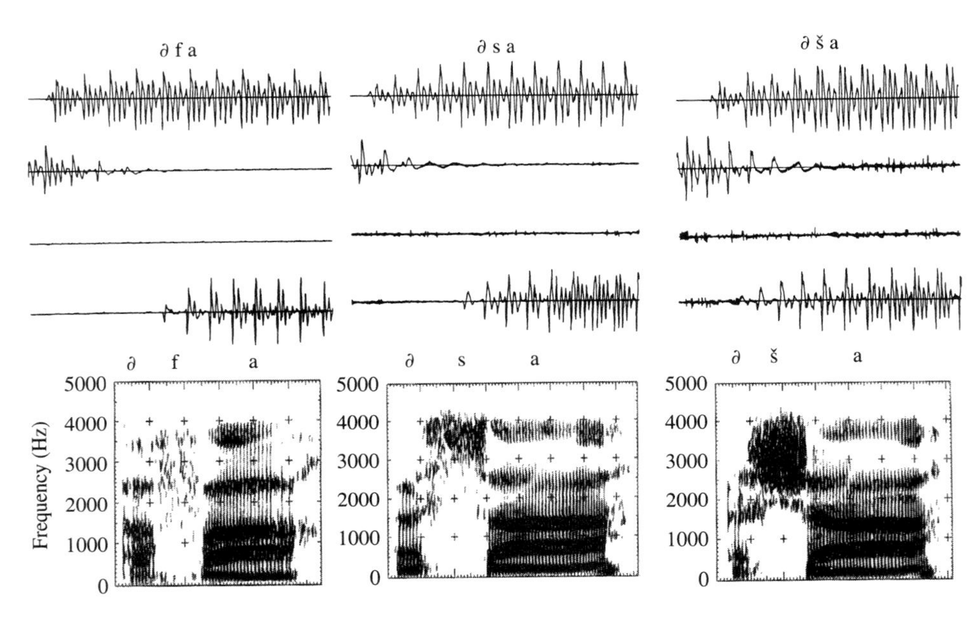

FIGURE 3.31
Acoustic waveforms and spectrograms for utterances /UH F AA/, /UH S AA/, and /UH SH AA/.

preceding vowel (/ə/) and the following vowel (/a/). The noise-like nature of fricative excitation is obvious in the waveform plots, although the signal level for the /f/ waveform is rather low and difficult to see in the figure. The spectrogram plots show that the noise spectrum for the /f/ sound is essentially at or above the 4 kHz range shown in the spectrogram; for the /s/ sound, the lower range of spectral energy is in the 3–4 kHz range; for the /š/ sound, the spectral energy is prominent in the 2–4 kHz range. Figure 3.32 shows another set of spectrograms (this time on a scale that goes to 8 kHz) of the unvoiced fricatives followed by the /i/ sound. Here we clearly see the differences in the spectral concentrations of energy for /f/, /θ/, /s/, and /š/, with /f/ and /θ/ having weak concentrations of energy above 4 kHz, with /s/ having a strong concentration of energy in the 4–7 kHz range, and with /š/ having a strong concentration of energy in the 2–7 kHz range.

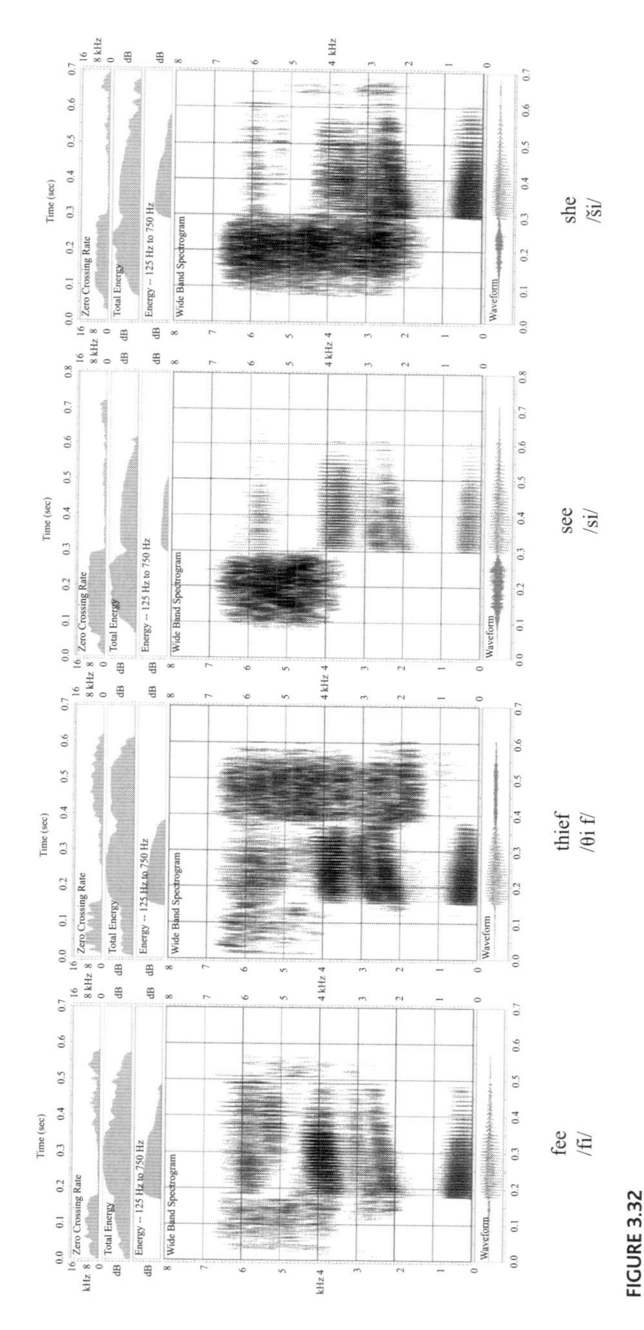

FIGURE 3.32
Spectrograms of the unvoiced fricatives followed by the sound /i/. (After Zue and Glass, MIT OCW Notes [427].)

3.4.7 Voiced Fricatives

The voiced fricatives /v/, /ð/, /z/, and /ž/ (/V/, /DH/, /Z/, and /ZH/) are the counterparts of the unvoiced fricatives /f/, /θ/, /s/, and /š/ (/F/, /TH/, /S/, and /SH/), respectively, in that the place of constriction for each of the corresponding phonemes is essentially the same. However, the voiced fricatives differ markedly from their unvoiced counterparts in that two excitation sources are involved in their production. For voiced fricatives the vocal cords are vibrating, and thus one excitation source is at the glottis. However, since the vocal tract is constricted at some point forward of the glottis, the air flow becomes turbulent in the neighborhood of the constriction. Thus the spectra of voiced fricatives can be expected to display two distinct components. These excitation features are readily observed in Figure 3.33, which shows typical waveforms and spectra for the voiced fricatives /v/ and /ž/. The similarity of the unvoiced fricative /f/ to the voiced fricative /v/ is easily seen by comparing their corresponding spectrograms in Figures 3.31 and 3.33. Likewise, it is instructive to compare the spectrograms of /š/ and /ž/.

An example of the contrast between the spectrograms for the unvoiced fricative /s/ and its voiced counterpart /z/ is shown in Figure 3.34, which shows the two fricatives

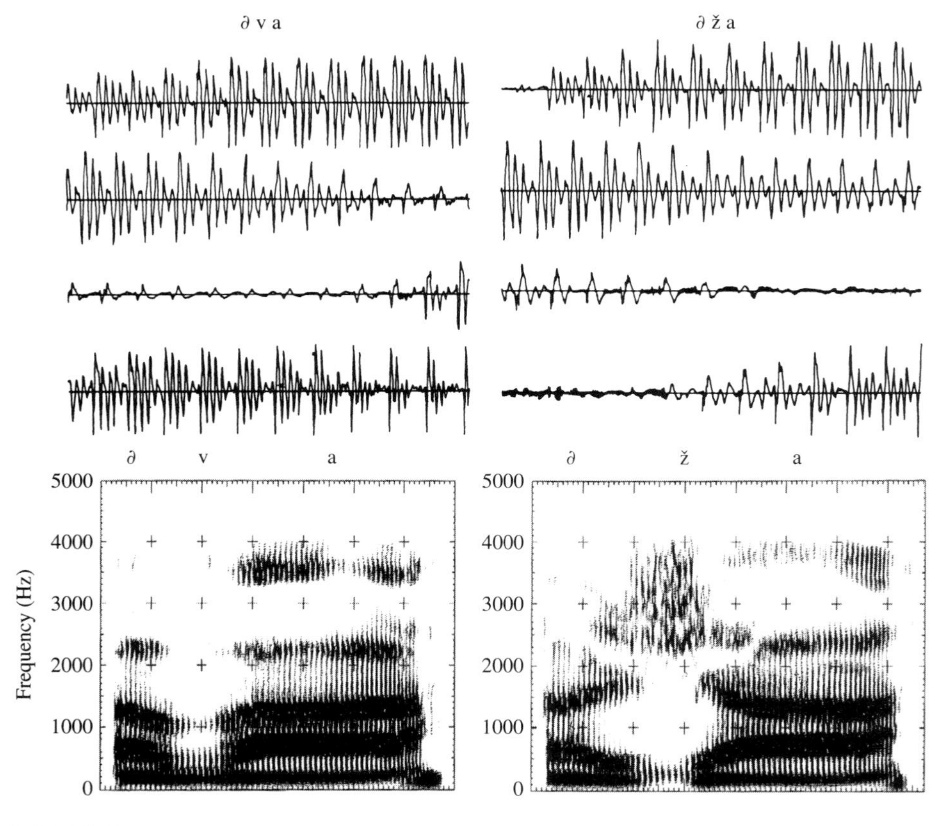

FIGURE 3.33
Acoustic waveforms and spectrograms for utterances /UH V AA/ and /UH ZH AA/.

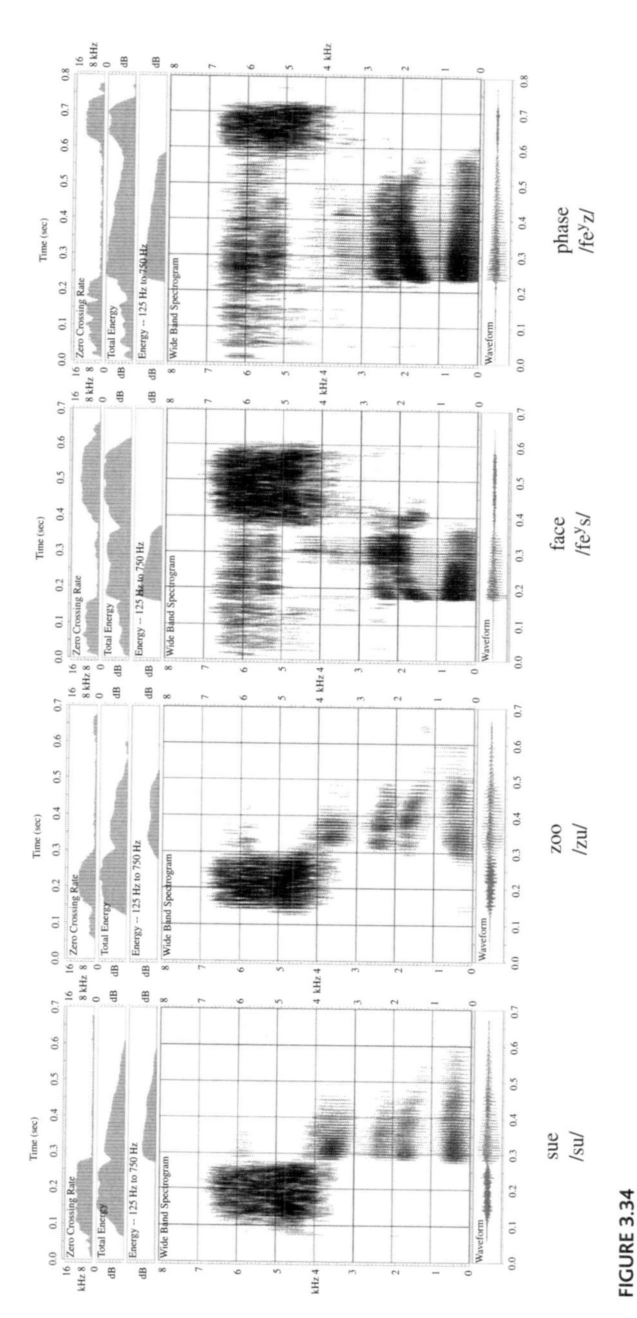

FIGURE 3.34
Spectrogram contrasts for the unvoiced fricative /s/ and its voiced counterpart /z/ in both initial and final positions. (After Zue and Glass, MIT OCW Notes [427].)

in both the initial position (followed by the /u/ sound) and in the final position (preceded by the /e^y/ sound). It can be seen that both /s/ and /z/ have similar regions of energy concentration, but the voiced sound /z/ shows the usual pitch period striation (in time) whereas there is no such pattern for the unvoiced sound /s/, which is produced strictly from a noise-like excitation.

3.4.8 Voiced Stops

The voiced stop consonants /b/, /d/, and /g/ (/B/, /D/, and /G/) are transient, non-continuant sounds that are produced by building up pressure behind a total constriction somewhere in the oral tract and suddenly releasing the pressure. As shown in Figure 3.35, for /b/ the constriction is at the lips; for /d/ the constriction is back of the teeth; and for /g/ it is near the velum. During the period when there is a total constriction in the tract, there is no sound radiated from the lips. However, there is often a small amount of low frequency energy radiated through the walls of the throat (sometimes called a voice bar). This occurs when the vocal cords are able to vibrate even though the vocal tract is closed at some point.

Since the stop sounds are dynamical in nature, their properties are highly influenced by the vowel that follows the stop consonant [85]. As such, the waveforms for stop consonants give little information about the particular stop consonant. Figure 3.36 shows the waveform and spectrogram of the syllable /UH B AA/. The waveform of /b/

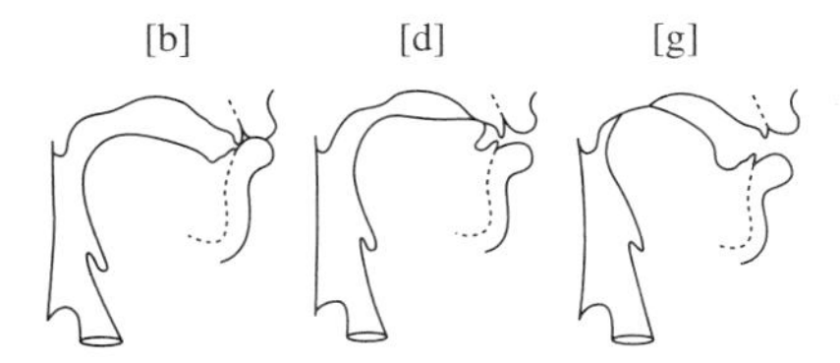

FIGURE 3.35
Articulatory configurations for the voiced stop consonants. (After Zue and Glass, MIT OCW Notes [427].)

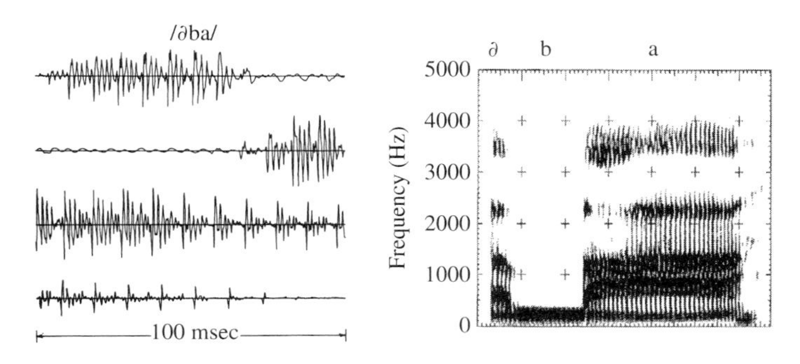

FIGURE 3.36
Acoustic waveforms and spectrograms for utterance /UH B AA/.

shows few distinguishing features except for the voiced excitation and lack of high frequency energy.

3.4.9 Unvoiced Stops

The unvoiced stop consonants /p/, /t/, and /k/ (/P/, /T/, and /K/) are similar to their voiced counterparts /b/, /d/, and /g/ with one major exception. During the period of total closure of the tract, as the pressure builds up, the vocal cords do not vibrate. Thus, following the period of closure, as the air pressure is released, there is a brief interval of frication (due to sudden turbulence of the escaping air) followed by a period of aspiration (steady air flow from the glottis exciting the resonances of the vocal tract) before voiced excitation begins.

Figure 3.37 shows waveforms and spectrograms of the voiceless stop consonants /p/ and /t/. The "stop gap," or time interval during which the pressure is built up, is clearly in evidence. Also, it is readily seen that the duration and frequency content of the frication noise and aspiration varies greatly with the stop consonant.

Further examples of the properties of the unvoiced stop consonants are seen in Figure 3.38, which shows the words "poop" (with initial and final /p/), "toot" (with initial and final /t/), and "kook" (with initial and final /k/). The spectrograms show that the unvoiced stop consonants are typically aspirated with clear regions of noise-like signal at the release of the stop consonant energy. Initial stop consonants do not show the closure interval, but it is readily seen in the final stop consonants of each of the words in Figure 3.38.

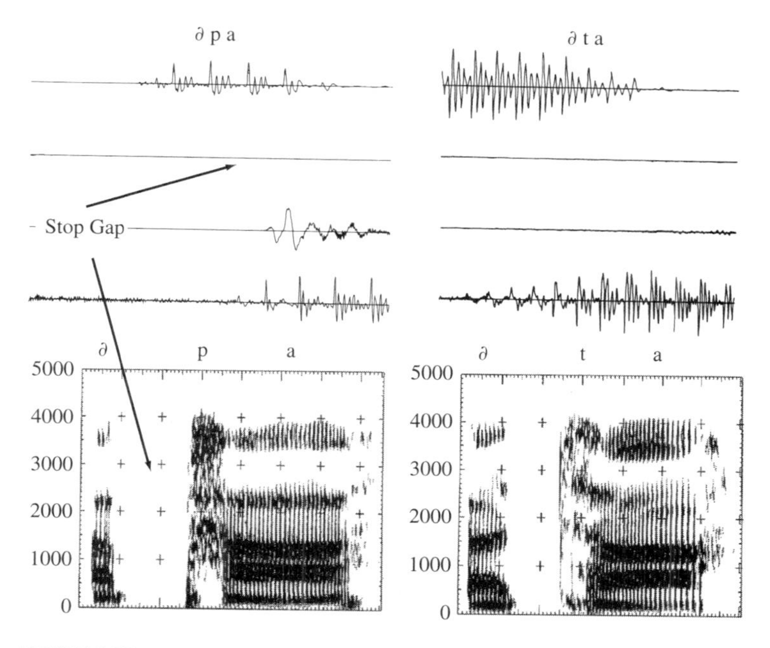

FIGURE 3.37
Acoustic waveforms and spectrograms for utterance /UH P AA/ and /UH T AA/.

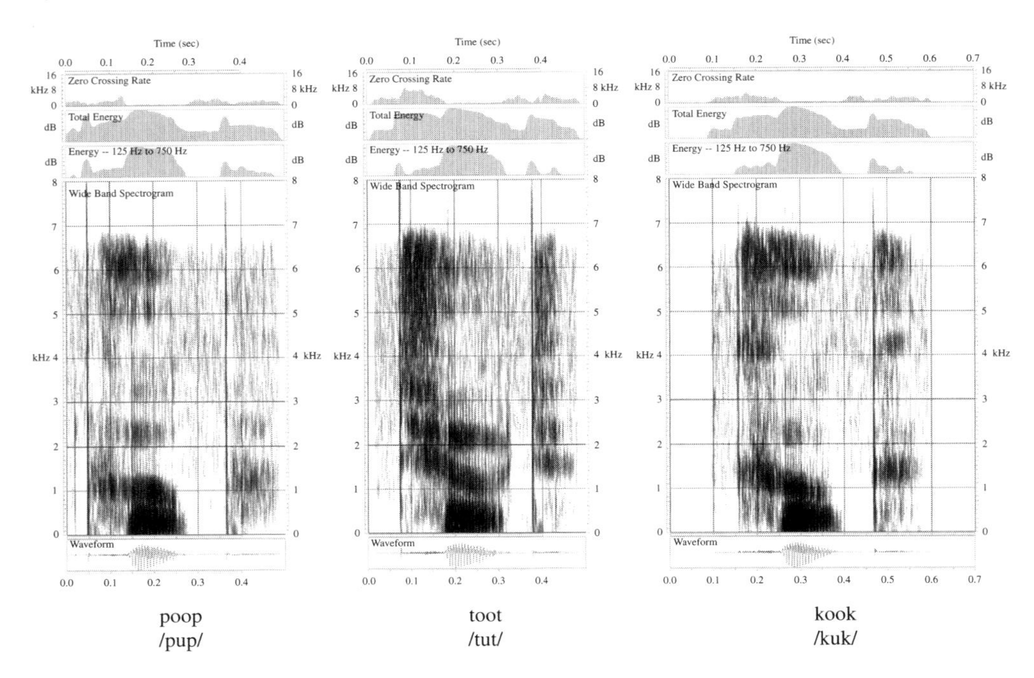

FIGURE 3.38
Spectrograms for the unvoiced stop consonants in both the initial and the final position. (After Zue and Glass, MIT OCW Notes [427].)

3.4.10 Affricates and Whisper

The remaining consonants of American English are the affricates /č/ and /j/ (/CH/ and /JH/), and the aspirated phoneme /h/ (/HH/). The unvoiced affricate /č/ is a dynamical sound that can be modeled as the concatenation of the stop /t/ and the fricative /š/. An example of the waveform is shown in Figure 3.9. The voiced affricate /j/ can be modeled as the concatenation of the stop /d/ and the fricative /ž/. Finally, the phoneme /h/ is produced by exciting the vocal tract by a steady air flow—i.e., without the vocal cords vibrating, but with turbulent flow being produced at the glottis.[5] The characteristics of /h/ are invariably those of the vowel that follows /h/ since the vocal tract assumes the position for the following vowel during the production of /h/.

3.5 DISTINCTIVE FEATURES OF THE PHONEMES OF AMERICAN ENGLISH

Figure 3.39 shows a chart of the distinctive features of the consonants of American English, coded as a set of binary decisions as to place of articulation and manner of articulation [166]. It is believed that the human brain recognizes sounds by doing a

[5]Note that this is also the model of excitation for whispered speech.

	p	k	t	b	d	g	f	thin	s	sh	v	the	z	azure	m	n	ng	l	r	w	h
Place																					
bilabial	+	–	–	+	–	–	–	–	–	–	–	–	–	–	+	–	–	–	–	+	–
labiodental	–	–	–	–	–	–	+	–	–	–	+	–	–	–	–	–	–	–	–	–	–
dental	–	–	–	–	–	–	–	+	–	–	–	+	–	–	–	–	–	–	–	–	–
alveolar	–	–	+	–	+	–	–	–	+	–	–	–	+	–	–	+	–	+	–	–	–
palatal	–	–	–	–	–	–	–	–	–	+	–	–	–	+	–	–	–	–	+	–	–
velar	–	+	–	–	–	+	–	–	–	–	–	–	–	–	–	–	+	–	–	–	–
pharyngeal	–	–	–	–	–	–	–	–	–	–	–	–	–	–	–	–	–	–	–	–	+
Manner																					
glide	–	–	–	–	–	–	–	–	–	–	–	–	–	–	–	–	–	+	+	+	–
nasal	–	–	–	–	–	–	–	–	–	–	–	–	–	–	+	+	+	–	–	–	–
stop	+	+	+	+	+	+	–	–	–	–	–	–	–	–	–	–	–	–	–	–	–
fricative	–	–	–	–	–	–	+	+	+	+	+	+	+	+	–	–	–	–	–	–	–
voicing	–	–	–	+	+	+	–	–	–	–	+	+	+	+	+	+	+	+	+	+	+

FIGURE 3.39
Distinctive features of the non-vowel sounds of American English.

distinctive feature analysis of the coded speech information that gets sent to the brain. Furthermore, the distinctive features of Figure 3.39 have been found to be somewhat insensitive to noise, background signals, and reverberation. Hence, they constitute a robust and reliable set of features that can be used for processes such as speech recognition and speech synthesis. Tables similar to the one in Figure 3.39 represent the distinctive features of the vowel sounds [60, 166, 376]. The book by Stevens [376] is particularly useful for relating the articulatory distinctive features to the properties of the acoustic waveform of speech [60, 376].

3.6 SUMMARY

In this chapter we have discussed the sounds of the American English language and shown how these basic sounds (the phonemes) are used to form larger linguistic units like syllables, words, and sentences. We showed the ideal process of going from an orthographic description of language to a phoneme representation of the sounds within the sentence (the so-called process of phonetic transcription), and we warned about the problems associated with sound co-articulation, both within and across words, so that the resulting sound stream could be very different from the ideal transcription based on dictionary definitions of how individual words would be pronounced in a sentence.

Following this very brief introduction to the linguistic aspects of creating sounds, we discussed the individual sound classes of American English, namely the classes of vowels, diphthongs, semivowels, and consonants, and we discussed their roles in creating sound, the associated articulatory shapes, the resulting speech waveforms and sound spectrograms, and finally the resonance structure of the sounds. We also described how to use the linguistic concept of distinctive features of sounds to classify sounds according to the place and manner of articulation in normal speech.

PROBLEMS

3.1. The waveform plot of Figure P3.1 shows a 500 msec section (100 msec/line) of a speech waveform.

(a) Determine the boundaries of the regions of voiced speech, unvoiced speech, and silence (background signal).

(b) Using hand measurements on the waveform plot, estimate the pitch period on a period-by-period basis and plot the pitch period versus time for this section of speech. (Let the period be indicated as zero during unvoiced and silence intervals.)

3.2. The waveform plot of Figure P3.2 is for the word "cattle." Note that each line of the plot corresponds to 100 msec of the signal.

(a) Indicate the boundaries between the phonemes; i.e., give the times corresponding to the boundaries /K/ /AE/ /T/ /AX/ /L/.

(b) Indicate the point where the fundamental frequency of voiced speech is (i) the highest and (ii) the lowest. What are the approximate pitch frequencies at these points?

(c) Is the speaker most probably a male, female, or a child? How do you know?

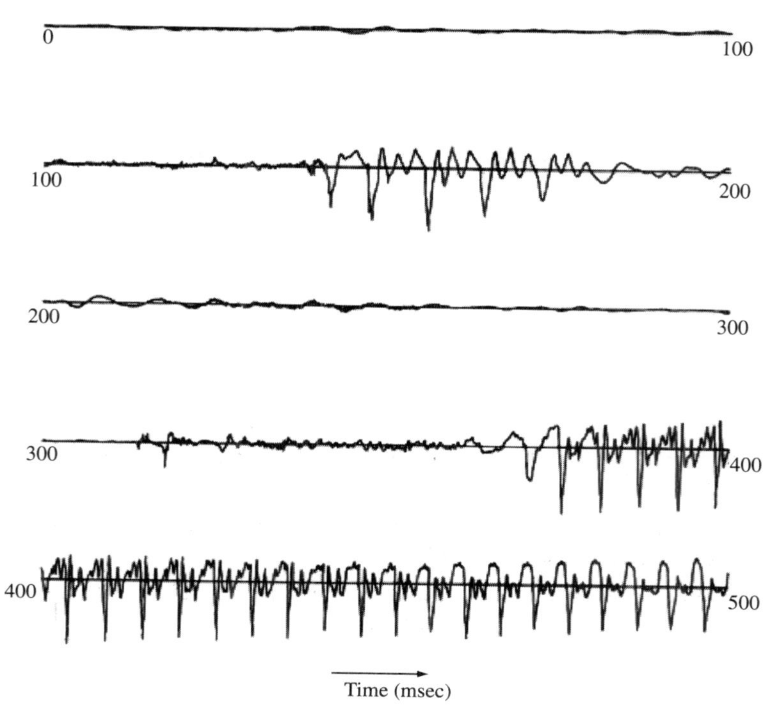

FIGURE P3.1
Time waveform of speech utterance.

3.3. Transcribe into ARPAbet symbols each of the following (note: if there is more than one pronunciation of the word, give ARPAbet pronunciations for the multiple pronunciations):

(a) monosyllabic words: the, of, and, to, a, in, that, is, was, he
(b) bisyllabic words: data, lives, record
(c) trisyllabic words: company, happiness, willingness
(d) sentences: I enjoy the simple life. Good friends are hard to find.

3.4. For the seven waveform plots shown in Figures P3.4a–P3.4g, identify and mark (on the waveform plots) the regions of each of the phonemes in each of the words associated with each waveform plot, namely:

1. "and" [part (a) of the figure]
2. "that" [part (b) of the figure]
3. "was" [part (c) of the figure]
4. "by" [part (d) of the figure]
5. "enjoy" [part (e) of the figure]
6. "company" [part (f) of the figure]
7. "simple" [part (g) of the figure].

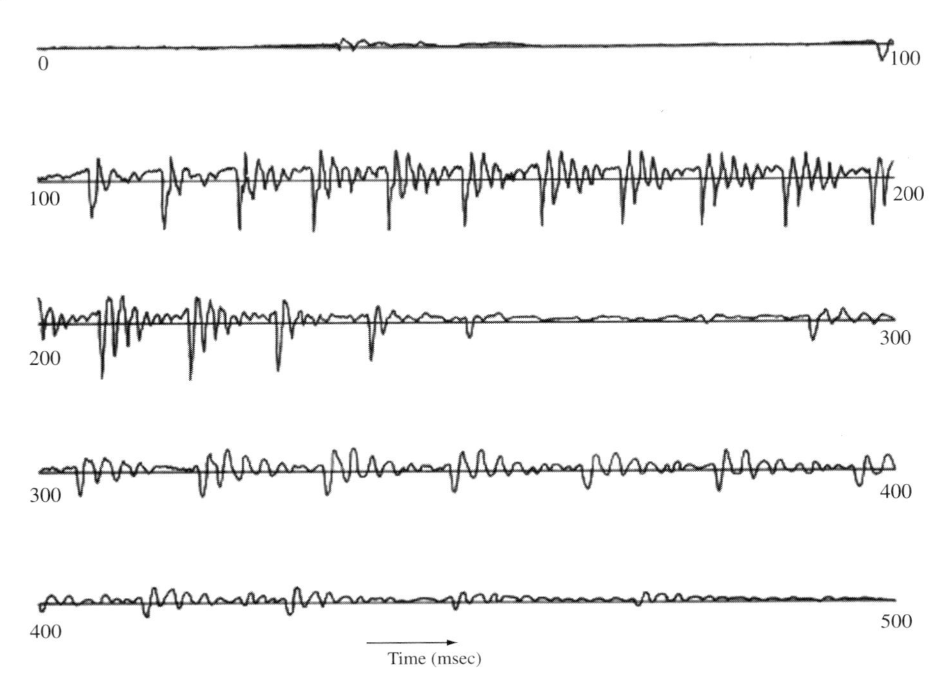

FIGURE P3.2
Time waveform of speech utterance "cattle."

3.5. Shown in Figure P3.5 is a waveform plot, and a wideband and a narrowband spectrogram for the utterance "I enjoy the simple life." Identify the center of the vowel regions in the words "enjoy," "simple," and "life" and measure the pitch period at each of these center points. (Hint: You might want to use the spectrograms to help identify the regions and to measure the fundamental frequency and then to convert to pitch period).

3.6. Segment the waveform in Figure P3.6 into regions of voiced speech (V), regions of unvoiced speech (U), and regions of silence (or background signal) (S). The waveform corresponds to the sentence "Good friends are hard to find."

3.7. Figure P3.7 shows the time waveform of an utterance of the sentence, "Cats and dogs each hate the other." The utterance was sampled at 8000 samples/sec and 2000 samples are plotted on each line, with the samples connected by straight lines to give the appearance of a continuous waveform. The phonetic representation of the text of this sentence in ARPAbet symbols is:

```
K AE T S - AE N D - D AO G Z - IY CH - HH EH T -
DH IY - AH DH ER
```

(Note: The - signifies the end of a *text* word.)

(a) In the signal waveform, the /D/'s in "and" and "dogs" merge. Mark the beginning and end of the merged /D/ on the waveform plot in Figure P3.7 and label it with /D/.

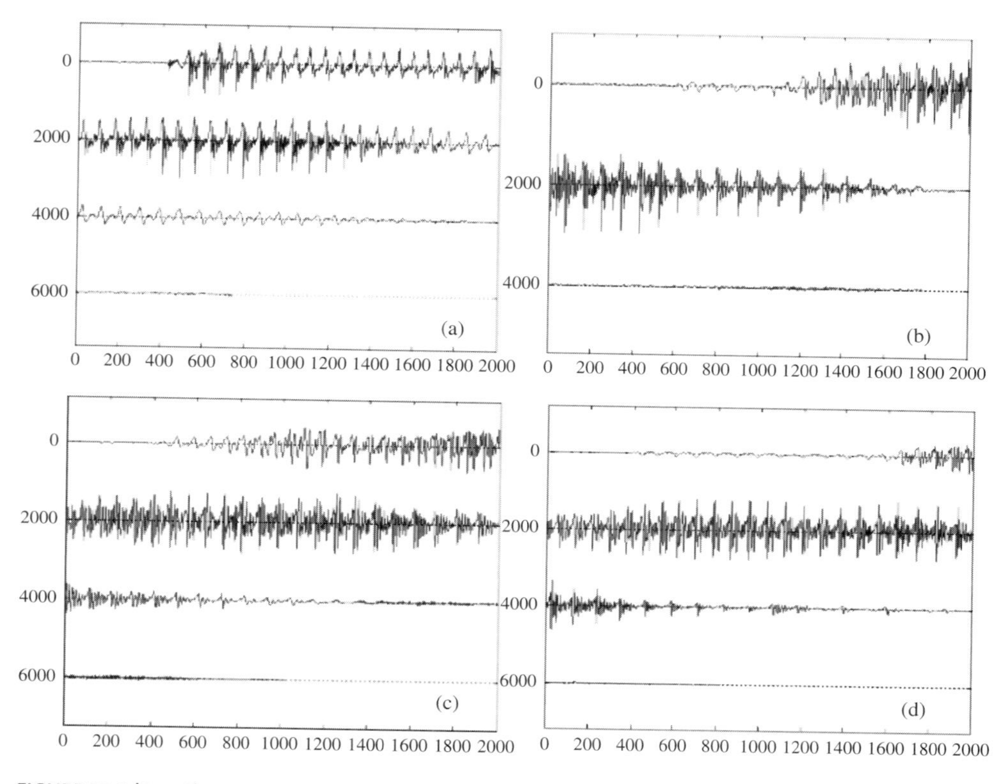

FIGURE P3.4 (Part 1)
Waveforms for words (a) "and," (b) "that," (c) "was," (d) "by."

(b) Mark the beginning and end of the /IY/ of "each" on the waveform plot in Figure P3.7 and label it with /IY/.

(c) Mark the beginning and end of the /CH/ of "each" on the waveform plot in Figure P3.7 and label it with /CH/.

(d) Estimate the *average* fundamental frequency (in Hz) for the voiced segment on the first line of the plot.

(e) Using the approach followed in Section 1.1, estimate the information rate (in bits/sec) of this utterance. Show how you arrived at your answer, and state all assumptions that you made.

3.8. Which of the pair of spectrograms in Figure P3.8 is a wideband spectrogram? Which is a narrowband spectrogram? What distinguishes these two types of spectrograms?

3.9. Figure P3.9 shows two spectrograms of the utterance "Cats and dogs each hate the other." The sampling rate of the utterance is $F_s = 8000$ samples/sec. The phonetic representation of the text of this sentence is:

```
/K AE T S - AE N D - D AO G Z - IY CH - HH EH T -
DH IY - AH DH ER/
```

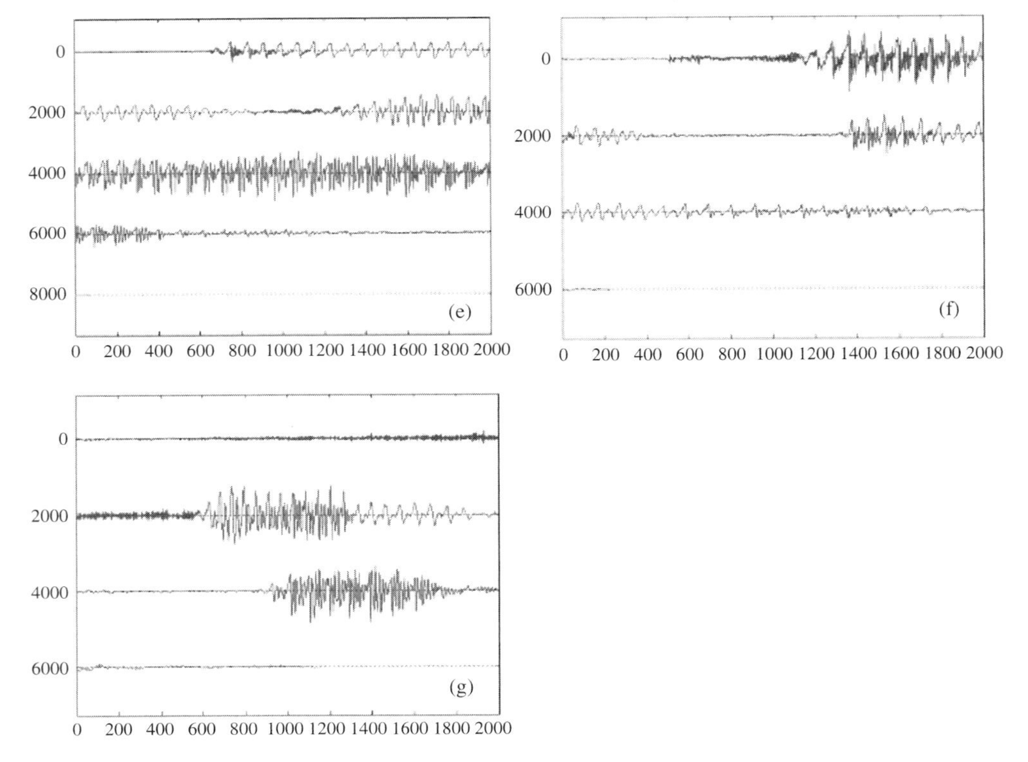

FIGURE P3.4 (Part 2)
Waveforms for words (e) "enjoy," (f) "company," (g) "simple."

(a) Which spectrogram is a wideband spectrogram?

(b) Using the appropriate spectrogram, estimate the fundamental frequency at $t = 0.18$ sec.

(c) Is the fundamental frequency increasing, decreasing, or remaining fairly constant in the time interval between 1.6 and 1.8 sec?

(d) Estimate the first three formant frequencies at time $t = 0.18$ sec.

(e) In this utterance, the /D/'s in "and" and "dogs" merge. Mark the location of the merged /D/ in both spectrograms of Figure P3.9.

3.10. Shown in Figure P3.10 are spectrograms of four isolated words. Which of the following words might be the ones spoken, and which spectrograms match the words you identified:

1.	"that"	8.	"between"
2.	"and"	9.	"very"
3.	"was"	10.	"enjoy"
4.	"by"	11.	"only"
5.	"people"	12.	"other"
6.	"little"	13.	"company"
7.	"simple"	14.	"those"

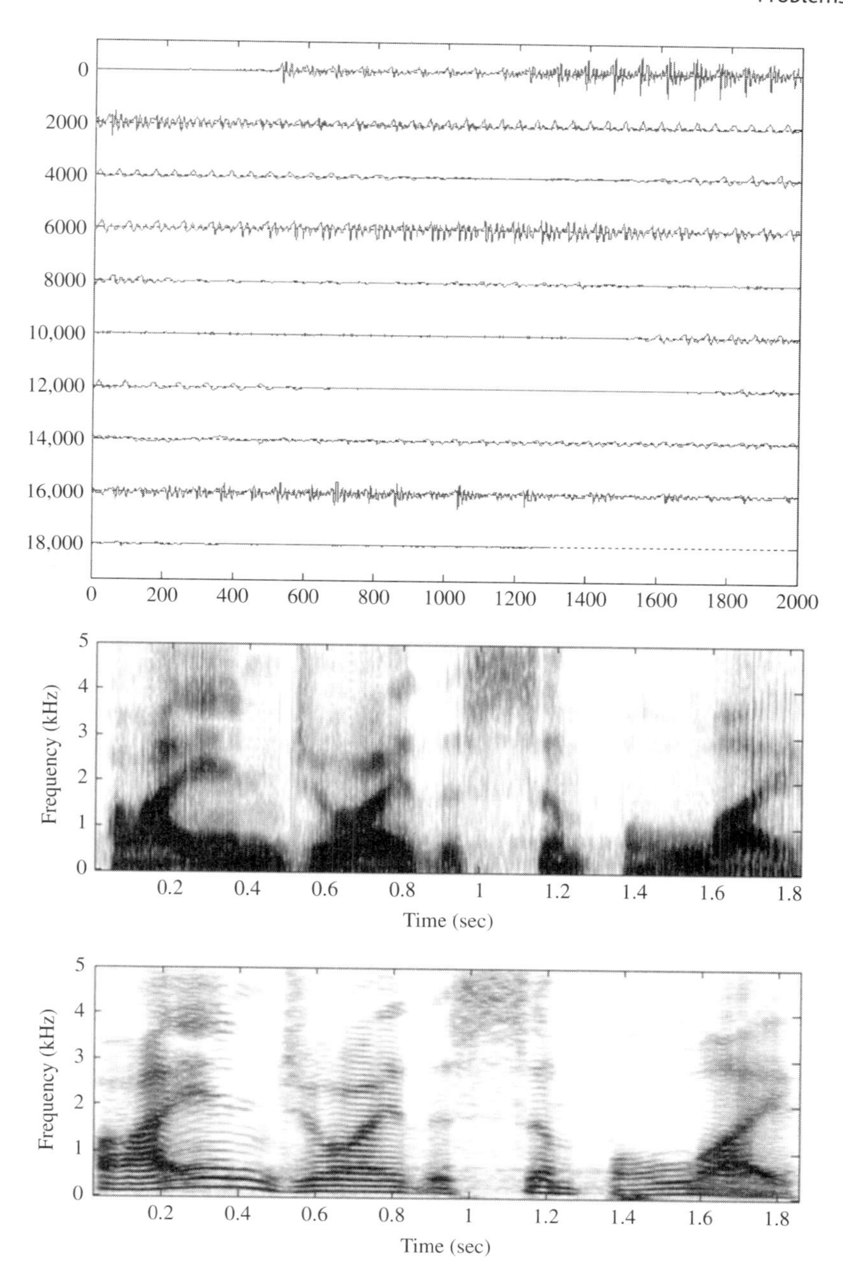

FIGURE P3.5
Waveform and spectrograms for the sentence, "I enjoy the simple life."

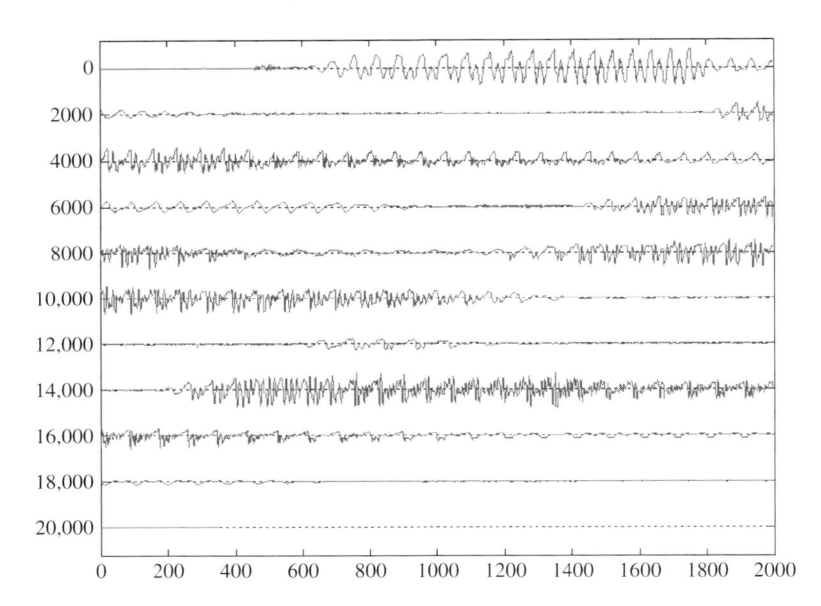

FIGURE P3.6
Waveform for the sentence "Good friends are hard to find."

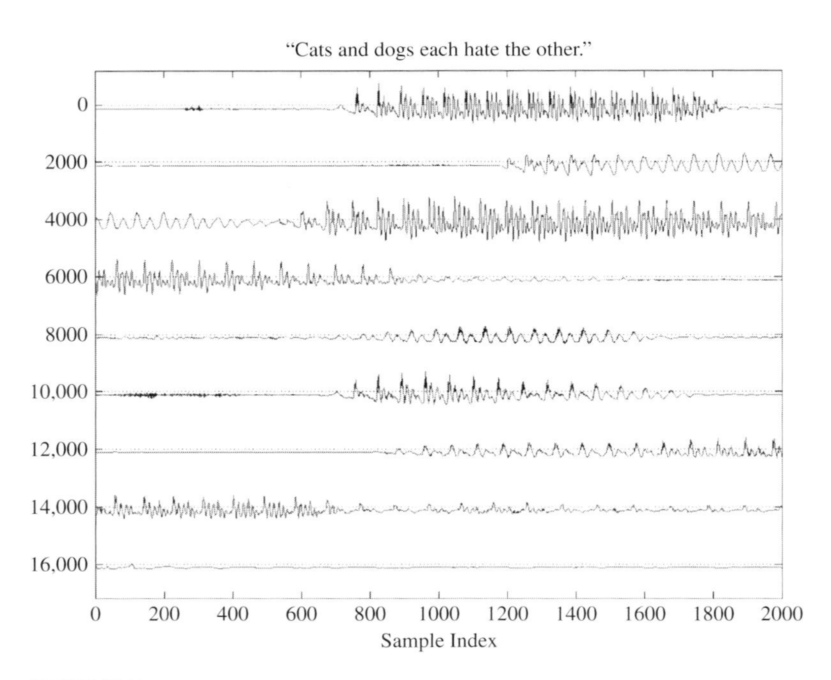

FIGURE P3.7
Waveform of an utterance of "Cats and dogs each hate the other."

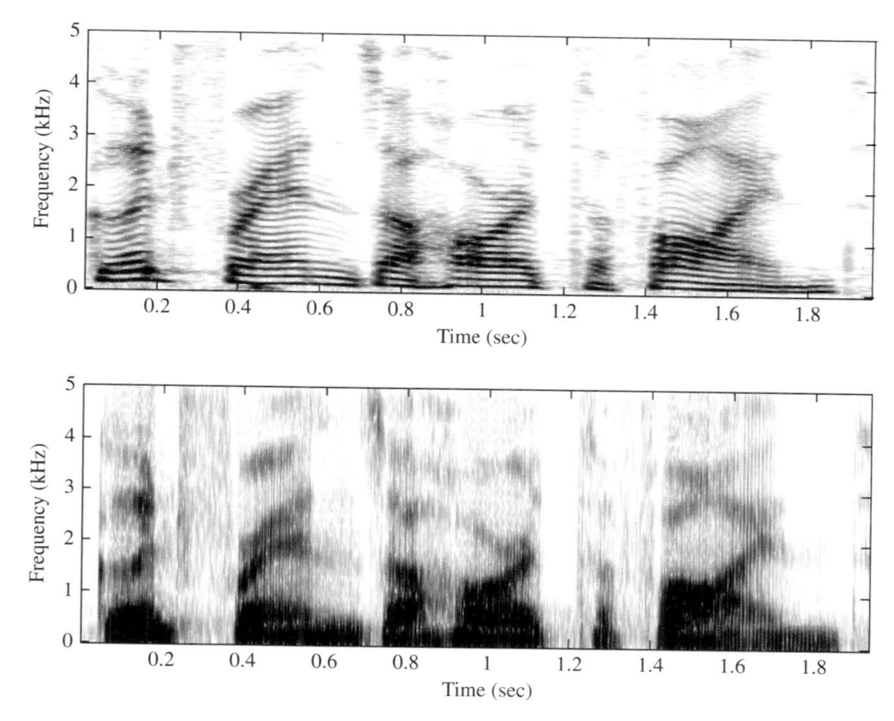

FIGURE P3.8
Spectrograms for Problem 3.8.

3.11. A speech recognition system for voice control of an audio storage system uses the vocabulary words:

1.	"stop"	4.	"record"
2.	"start"	5.	"rewind"
3.	"play"	6.	"pause"

Figure P3.11 shows wideband spectrograms of one version of each of these six words. Using your knowledge of acoustic phonetics, determine which wideband spectrogram corresponds to each word.

3.12. Identify the sounds (from the ARPAbet alphabet) of all the words in the following pair of sentences:

1. She eats some Mexican nuts.

2. Where roads stop providing good driving.

What can you say about the sound sequence at word boundaries for these two sentences?

3.13. Consider the following orthography of two sentences, one without the vowels, and one without the consonants. Try to decode as much of these sentences as you can.

1. T_ g_v_ y_ _ s_m_ _d_ _ _f th_ _m_ _nt _f w_rk r_q_ _r_d _n th_s c_ _rs_.

2. _ea_ i_ _i_ _ _ _a_ i_ a_ _ou_ _ _ _ _o_ _a_ _ _ _ _e _ _a_e i_ _ _i_ _ou_ _e.

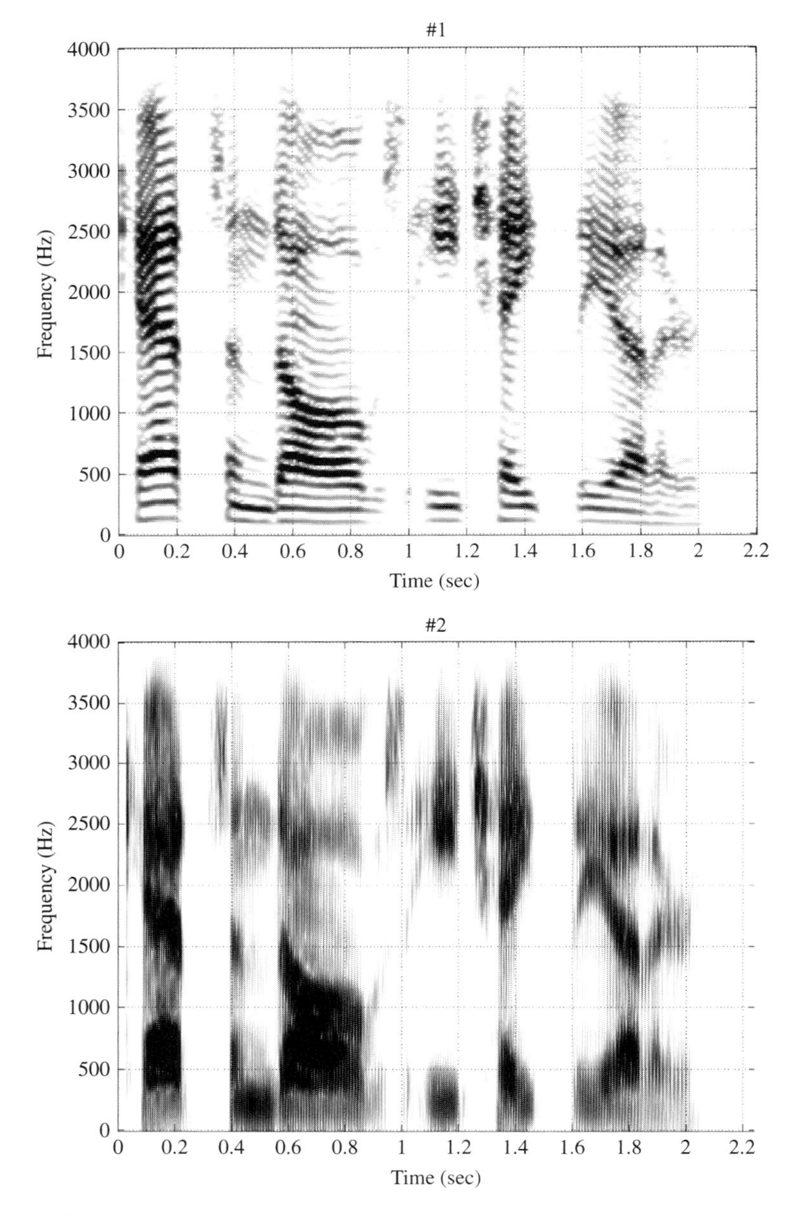

FIGURE P3.9
Spectrograms of utterance of "Cats and dogs each hate the other."

3.14. Figure P3.14 shows the most common 105 words in the Brown corpus of American English words from a corpus of more than a million words. What general statement(s) can you make about the most commonly occurring words in this list? What percentage of these top 105 words have more than one syllable?

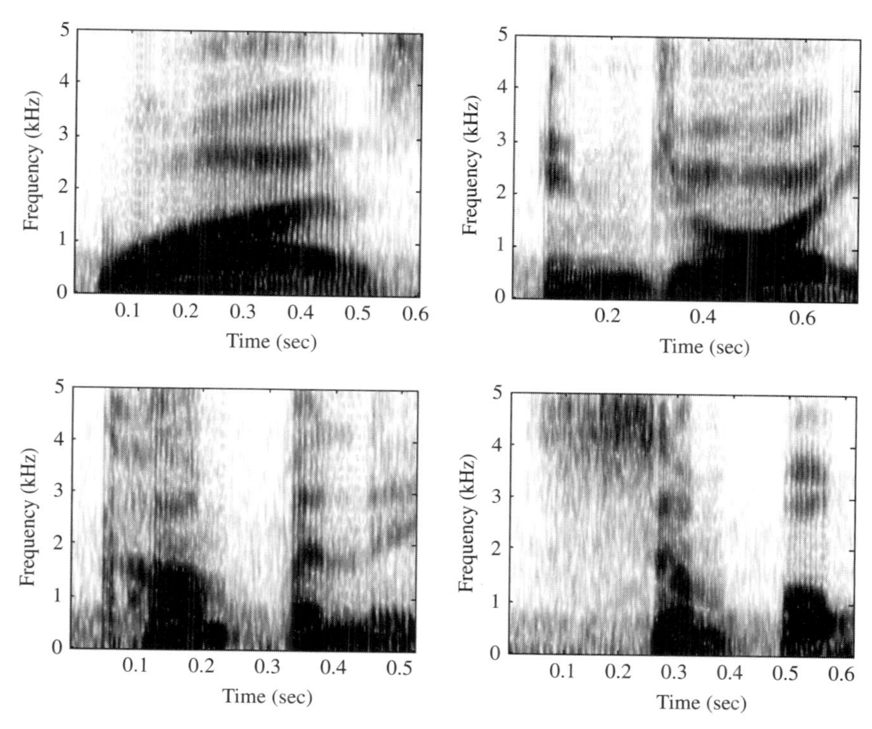

FIGURE P3.10
Spectrograms of four words from a specified list of words.

3.15. Identify the most likely vowels in each of the six spectrograms of CVC (consonant-vowel-consonant) syllables shown in Figure P3.15. What are the frequencies of the first three formants? How do they compare with the table values from the class notes?

3.16. Identify the place of articulation and the manner of articulation for all the sounds in the sentence: "I enjoy the simple life." Show your results as a sequential table of place and manner features, over time (i.e., versus the sounds in the utterance).

3.17. How many consonant pairs can occur within a single word? Name them. What rule do they obey in terms of place and manner of articulation. How many consonant triplets can occur within a single word? Name them. What rule do they obey in terms of place and manner of articulation.

3.18. (MATLAB Exercise) The following MATLAB exercise provides some experience with the process of "segmentation and labeling" of speech. Use the speech file `s5.wav` and its waveform display to do the following:

1. Write out a phonetic transcription of the sentence "Oak is strong and also gives shade" using the ARPAbet symbols.
2. Using the MATLAB functions for plotting and listening to sounds, segment and label the waveform `s5.wav` into regions for each of the sounds in the sentence (including

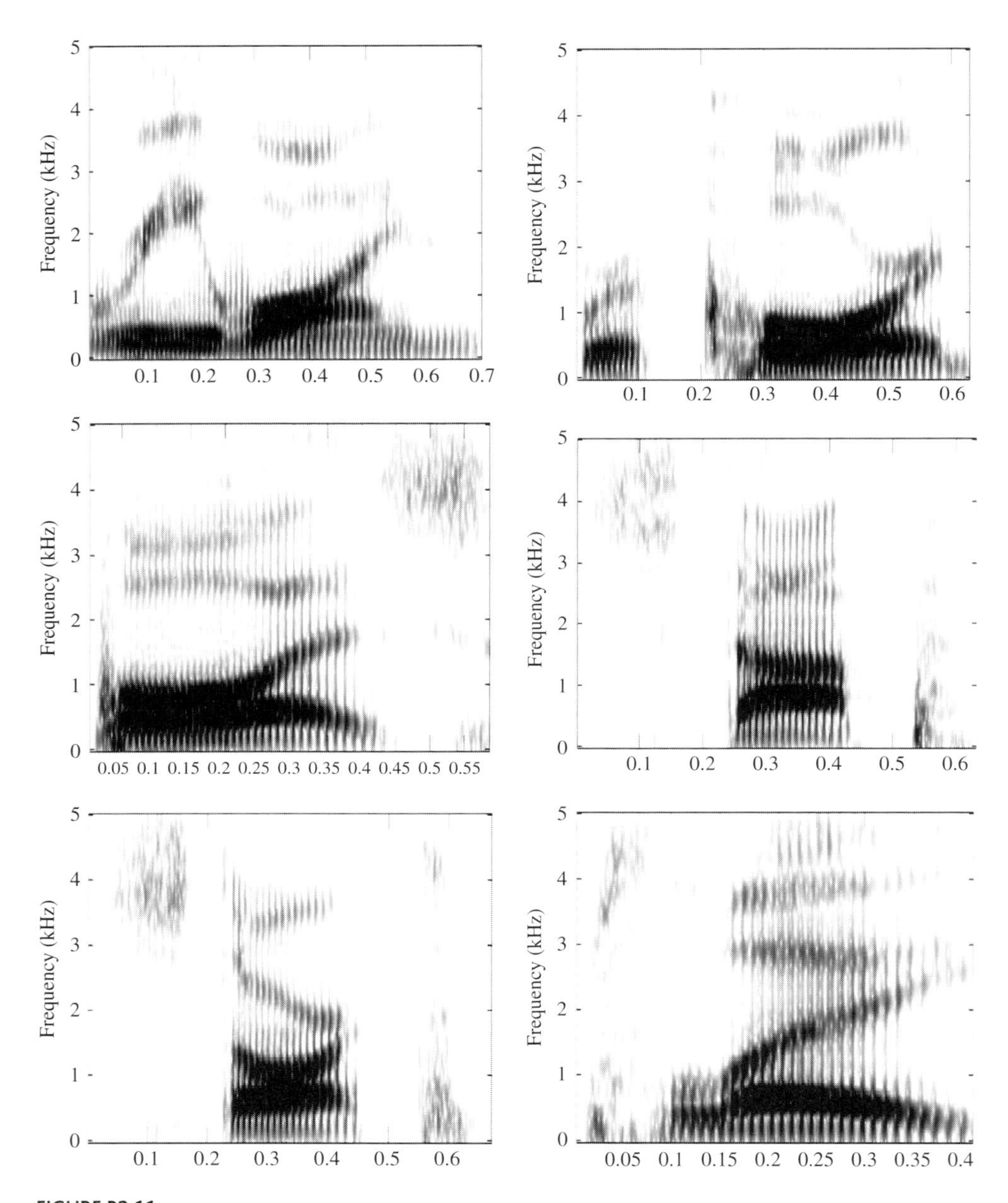

FIGURE P3.11
Spectrograms of one version of each of the control words for an audio storage system.

potential regions of silence at the beginning of the speech, and the end of the speech, and even possibly during the speech).

3. Zero out all the vowel regions of the sentence (replace the speech waveform with zero-valued samples during the vowel regions) and listen to the sentence.

Word	% Frequency	Word	% Frequency	Word	% Frequency
1 The	6.8872	36 all	0.2954	71 then	0.1355
2 of	3.5839	37 She	0.2814	72 do	0.1341
3 and	2.8401	38 there	0.2682	73 first	0.134
4 to	2.5744	39 would	0.2672	74 any	0.1324
5 a	2.2996	40 their	0.2628	75 my	0.1298
6 in	2.101	41 we	0.2611	76 now	0.1293
7 that	1.0428	42 him	0.2578	77 such	0.1283
8 is	0.9943	43 been	0.2434	78 like	0.127
9 was	0.9661	44 has	0.2401	79 our	0.1232
10 He	0.9392	45 when	0.2294	80 over	0.1218
11 for	0.934	46 who	0.2217	81 man	0.1191
12 it	0.8623	47 will	0.2209	82 me	0.1164
13 with	0.7176	48 more	0.2181	83 even	0.1153
14 as	0.7137	49 no	0.2168	84 most	0.1142
15 his	0.6886	50 If	0.2164	85 made	0.1107
16 on	0.6636	51 out	0.2063	86 after	0.1053
17 be	0.6276	52 so	0.1954	87 also	0.1052
18 at	0.5293	53 said	0.193	88 did	0.1028
19 by	0.5224	54 what	0.1878	89 many	0.1014
20 I	0.5099	55 up	0.1865	90 before	0.1
21 this	0.5065	56 its	0.1829	91 must	0.0997
22 had	0.505	57 about	0.1787	92 through	0.0954
23 not	0.4538	58 into	0.1763	93 back	0.0952
24 are	0.4325	59 than	0.1762	94 years	0.0943
25 but	0.4312	60 them	0.1761	95 where	0.0923
26 from	0.4301	61 can	0.1744	96 much	0.0922
27 or	0.4141	62 Only	0.172	97 your	0.0909
28 have	0.388	63 other	0.1675	98 way	0.0895
29 an	0.3689	64 new	0.1609	99 well	0.0883
30 they	0.3562	65 some	0.1592	100 down	0.0881
31 which	0.3505	66 time	0.1576	101 should	0.0874
32 one	0.3245	67 could	0.1574	102 because	0.0869
33 you	0.3234	68 these	0.1548	103 each	0.0863
34 were	0.3232	69 two	0.139	104 just	0.0858
35 her	0.2989	70 may	0.1378	105 those	0.0837

FIGURE P3.14
List of 105 most common words in Brown corpus of American English.

4. Zero out all the consonant regions of the sentence (replace the speech waveform with zero-valued samples during the consonant regions) and listen to the sentence.
5. Determine which of the two modified waveforms (i.e., missing the vowels or missing the consonants) is the most "intelligible." See if someone who had never heard the sentence is able to correctly identify all of the words from either version of the modified waveform.

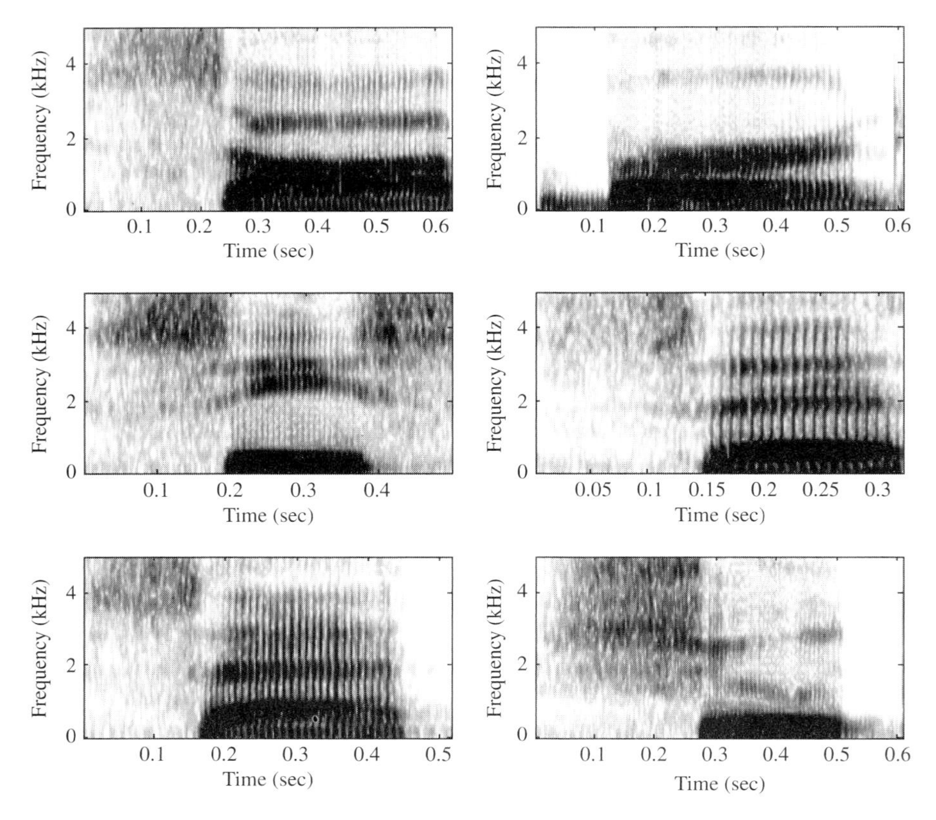

FIGURE P3.15
CVC spectrograms.

3.19. (MATLAB Exercise) Write a MATLAB program to read in a speech file with a sampling rate of $F_s = 16$ kHz and filter it to bandwidths of 5.5, 4, and 3.2 kHz. Listen to each of the resulting files and describe the effect of lowpass filtering on speech intelligibility and quality. (Use the speech file `test_16k.wav` to test your program.)

3.20. (MATLAB Exercise) Write a MATLAB program to change the sampling rate of a speech file from 16 kHz to both 10 kHz and 8 kHz. Listen to the speech at all three sampling rates. Do you hear any significant differences, and if so what are they. (Use the file `test_16k.wav` to test your program.)

3.21. (MATLAB Exercise) Write a MATLAB program to highpass filter a speech file in order to eliminate any potential DC offset and/or 60 Hz hum that may have been recorded with the speech file. The first thing you must do for this exercise is design a digital highpass filter with appropriate cutoff frequencies (edge of stopband and edge of passband) so that the 60 Hz component is attenuated by at least 40 dB. (Hint: You will find that a linear phase FIR filter probably works best for this exercise.) Once the filter has been designed, filter the speech signal and overwrite the original file with the highpass filtered speech.

3.22. (MATLAB Exercise) Write a MATLAB program that enables a sequence of speech files (possibly at different sampling rates) to be played sequentially. Your program should accept input that defines the files to be played and the delay between the end of the current file and the next file to be played. It should also accept input that defines whether a beep sound file delimiter should be played at any point in the playback of the individual files. Test your code by playing out the sequence of files `s1.wav`, `s2.wav`, `s3.wav`, `s4.wav`, `s5.wav`, and `s6.wav` (these are wav files from the TIMIT database that are available on the book website).

CHAPTER 4

Hearing, Auditory Models, and Speech Perception

4.1 INTRODUCTION

In Chapter 3 we discussed how humans produce the acoustic signal that we call speech. Our discussion included the mechanisms of speech generation, the resulting speech properties as reflected in the speech waveform or the speech spectrogram, an introduction to the acoustic theory of speech production, and finally the area of acoustic phonetics (i.e., how various linguistic units are generated and how their acoustic manifestations are reflected in the speech signal).

In this chapter we examine the receiver side of speech communication, namely the processes of speech perception and understanding in humans. We begin by looking at the basic acoustic and physiological mechanisms for speech perception, i.e., the human ear and the ensuing auditory mechanisms that convert sound to neural impulses that are interpreted in the brain. Early on, we will see that understanding of the brain's processing of speech beyond the neural transduction of the cochlea is limited. Hence, rather than speculate about the speech processing mechanisms beyond the auditory cortex of the brain, we instead try to understand characteristics of human perception of a range of signals (including simple tonal and noise sounds), and see what insight this knowledge contributes to our understanding of a complex structured signal such as speech or music. We describe some of the key properties of human sound perception in terms of a range of physical attributes of the sounds and the corresponding psychophysical measures of perception. The key findings from studies of sound and speech perception in humans tell us the following:

- frequency is perceived as pitch on a non-linear frequency scale,
- loudness is perceived on a compressive amplitude scale that rapidly becomes logarithmic above 1000 Hz,
- syllable perception is based on a long-term spectral integration process, and finally
- auditory masking is a key component in sound perception that provides robustness against noise and other interfering signals.

What does speech perception and speech understanding have to do with digital speech processing methods? Why should we study speech perception when our main interest is in how to process speech to better code it, synthesize it, understand it by machine, or process it in ways to increase intelligibility or naturalness? The answer to these questions is naively simple yet profound in impact; by achieving a good understanding of *how humans hear sounds* and *how humans perceive speech*, we are better able to design and implement speech processing systems that are robust and efficient for analyzing and representing speech signals. Throughout this book we will see many examples of speech processing systems that employ knowledge about hearing and speech perception in order to optimize the design or make the resulting system work better across a wide range of environmental conditions.

We begin this chapter with a discussion of the basic physiological model of speech perception, namely the human hearing mechanism. This leads to a discussion of the perceptual correlates of physical quantities such as power and frequency, and then to a discussion of models of the auditory process. We end this chapter with a discussion of human perception of sound and speech in noise, and show how this relates to speech understanding and speech perception in humans.

4.2 THE SPEECH CHAIN

Figure 4.1 shows again the complete process of speech communication between a speaker and a listener. This process, which Denes and Pinson aptly termed the "speech chain" [88], is comprised of the processes of speech production, auditory feedback to the speaker, speech transmission (through air or over an electronic communication system) to the listener, and speech perception and understanding by the listener. A useful point-of-view is that the message to be conveyed by speech goes back and forth among

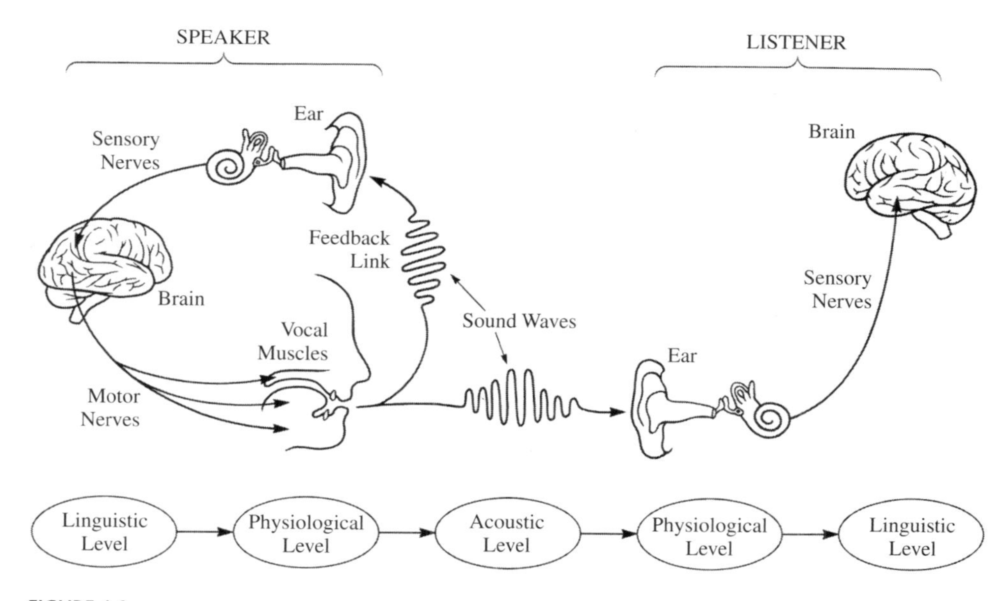

FIGURE 4.1
The speech chain—from production to perception. (After Pinson and Denes [88].)

three distinct levels of representation between the speaker and the listener. They are the *linguistic level* (where the basic sounds of the communication are chosen to express some thought or idea), the *physiological level* (where the vocal tract components produce the sounds associated with the linguistic units of the utterance), the *acoustic level* (where sound is released from the lips and nostrils and transmitted to both the speaker, as feedback, and to the listener), the *physiological level* (where the sound is analyzed by the ear and the auditory nerves), and finally back to the *linguistic level* (where the speech is perceived as a sequence of linguistic units and understood in terms of the ideas being communicated).

In this chapter, our focus is on the listener (or auditory) side of the speech chain. Figure 4.2 shows a block diagram of the physical processes involved in human hearing and speech (sound) perception. The acoustic signal corresponding to the spoken speech utterance is first converted to a neural representation by processing in the ear. The acoustic-to-neural conversion takes place in stages at the outer, middle, and inner ear. These processes are subject to measurement and therefore to mathematical simulation and characterization. The specifics of these processes will be described in later sections of this chapter. The next step, neural transduction, takes place between the output of the inner ear and the neural pathways to the brain, and consists of a statistical process of nerve firings at the hair cells of the inner ear, which are transmitted along the auditory nerve to the brain. Although measurements in nerve fibers have shed significant light on how sound is encoded in auditory nerve signals, much remains to be learned in this stage of the auditory system. Finally the nerve firing signals along the auditory nerve are processed by the brain to create the perceived sound corresponding to the spoken utterance. The processes used in the neural processing step are, as yet, not well understood and we can only speculate as to exactly how the neural information is transformed into sounds, phonemes, syllables, words, and sentences, and how the brain is able to decode the sound sequence into an understanding of the message embedded in the utterance.

Since our ability to observe and measure the mechanisms of speech (and/or sound) perception in the brain is limited, researchers have instead resorted to a "black box" behavioral model of hearing and perception, as illustrated in Figure 4.3. This model assumes that an acoustic signal enters the auditory system causing behavior that we record as psychophysical observations. Psychophysical methods and sound perception experiments are used to determine how the brain processes signals with different loudness levels, different spectral characteristics, and different temporal properties. The characteristics of the physical sound are varied in a systematic manner and the

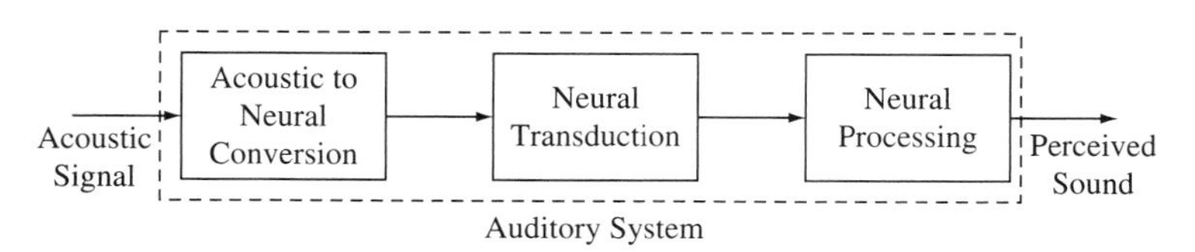

FIGURE 4.2
Block diagram of processes for conversion from acoustic wave to perceived sound.

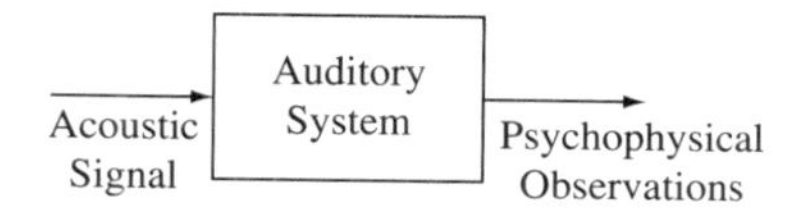

FIGURE 4.3
Black box model of hearing and perception.

TABLE 4.1 Physical attributes and resulting psychophysical observations.

Physical Attribute	Psychophysical Observation
Intensity	Loudness
Frequency	Pitch

psychophysical observations of the human listener are recorded and correlated with the physical attributes of the incoming sound. From such data, we attempt to determine how various attributes of sound (or speech) are processed by the auditory system. For example, Table 4.1 shows two of the most important physical characteristics of sound, i.e., intensity and frequency, along with the most directly related psychophysical counterparts, i.e., the perception of loudness and the perception of pitch. What we will see from a discussion of such psychophysical experiments is that the correspondences between sound intensity and loudness and between frequency and pitch are complicated and far from linear. Attempts to extrapolate from psychophysical measurements to the processes of speech perception and language understanding are highly susceptible to misunderstanding of exactly what is going on in the brain. Nevertheless, such perceptual knowledge, when properly employed to guide speech system design, can be exceedingly valuable. A notable example, which we discuss in detail in Chapter 12, is the use of masking effects in quantization of audio sound signals.

In Section 4.4 we will show how psychophysical observations are related to the physical attributes of sound both for simple sounds (tone-like signals, noise-like signals) and for speech signals. First, however, we will present a brief summary of the anatomy and function of the ear, the sensor that directly enables hearing and ultimately enables speech understanding.

4.3 ANATOMY AND FUNCTION OF THE EAR

A block diagram of the mechanisms in human hearing is shown in Figure 4.4. This figure is a more detailed graphic version of Figure 4.2. The acoustic signal, comprised of pressure variations that propagate through the air, is first processed by the external (outer) and middle ear, where the sound wave is converted to mechanical vibrations along the inner ear or cochlea. The cochlea performs a spatially distributed nonuniform spectral analysis of the sound, producing an auditory nerve representation

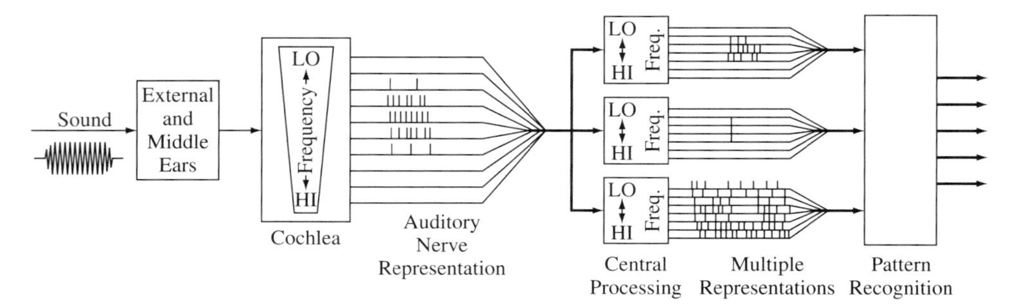

FIGURE 4.4
Block diagram of sound representation in the auditory system. (After Sachs et al. [336].)

in both time and frequency (much like the spectrogram representation discussed in Chapter 3). The first stages up through the cochlea are relatively well understood and can be accurately modeled, but the processing at higher levels in the brain, shown here as a sequence of central processors with multiple representations, followed by some type of pattern recognition, is not well understood. Much remains to be learned about the mechanisms used by the human brain to perceive sound or speech.

Figure 4.5 shows an artist's view of the key parts of the ear. The human ear consists of the following sound processing sections:

1. the outer ear consisting of the pinna and the external canal, which conducts sound to the eardrum at the beginning of the middle ear;
2. the middle ear, which begins at the tympanic membrane, or eardrum, and includes three small connected bones, the malleus (also called the hammer), the incus (also called the anvil), and the stapes (also called the stirrup), which together perform a transduction from acoustic waves to mechanical pressure waves;
3. the inner ear, which consists of the cochlea, the basilar membrane, and the set of neural connections to the brain.

The function of the outer ear is to funnel as much sound energy as possible into the 2 cm long ear canal. Because of the spreading out of sound energy by an inverse square law with distance, it is essential that the receiver of the sound be as large as possible so as to capture as much sound energy as possible. The pinna, by design, is quite large (as compared to the opening of the auditory canal) and, as a result, it increases human hearing sensitivity by a factor of between 2 and 3, as illustrated in Figure 4.6.

The middle ear is a mechanical transducer that converts sound waves that impinge on the tympanic membrane to mechanical vibrations along the inner ear. The mechanism for this transduction is the set of the three smallest bones in the human body, namely the hammer/malleus, the anvil/incus, and the stirrup/stapes. The movements of the tympanic membrane cause these three bones to move in concert. Together they act as a compound lever to amplify the sound vibrations, providing force amplification of anywhere from a factor of 3 to a factor of 15 from the eardrum to the stirrup, thereby enabling humans to hear weak sounds. Muscles around these three tiny bones

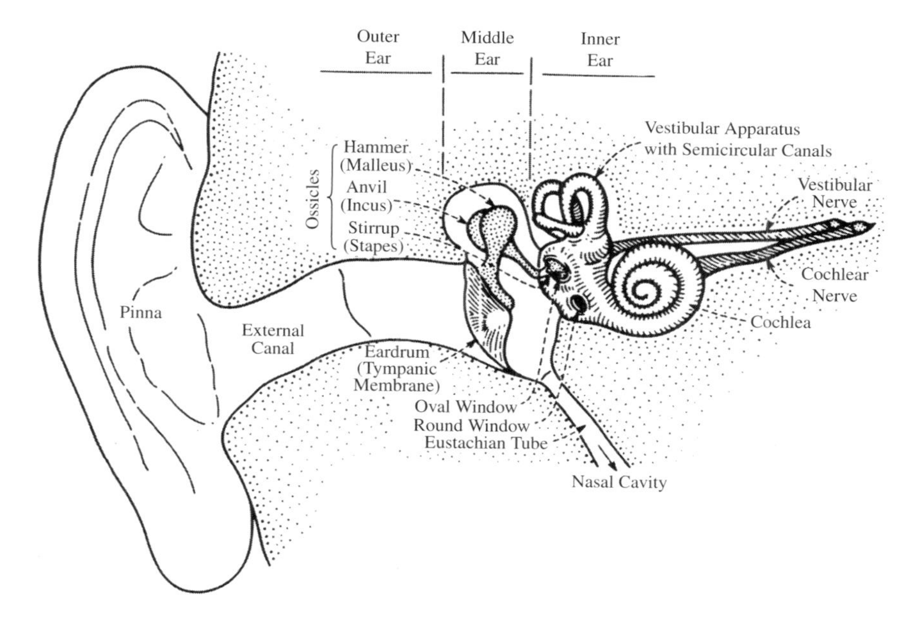

FIGURE 4.5
Schematic view of human ear showing the outer ear, the middle ear, and the inner ear mechanisms for hearing.

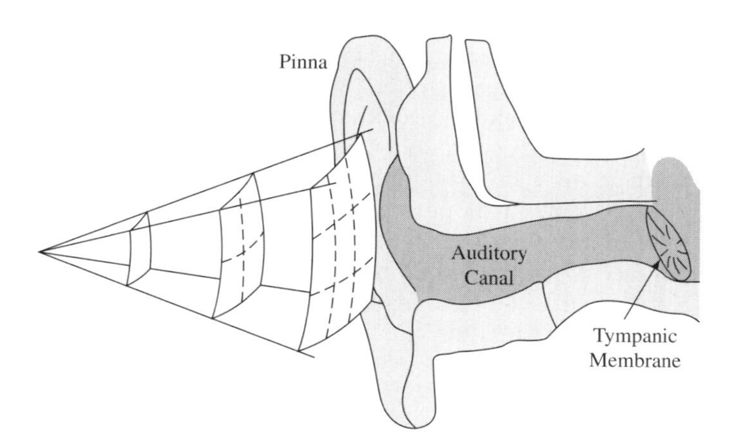

FIGURE 4.6
View of outer ear showing sound amplification obtained by having a relatively large pinna as compared to the opening of the auditory canal [154].

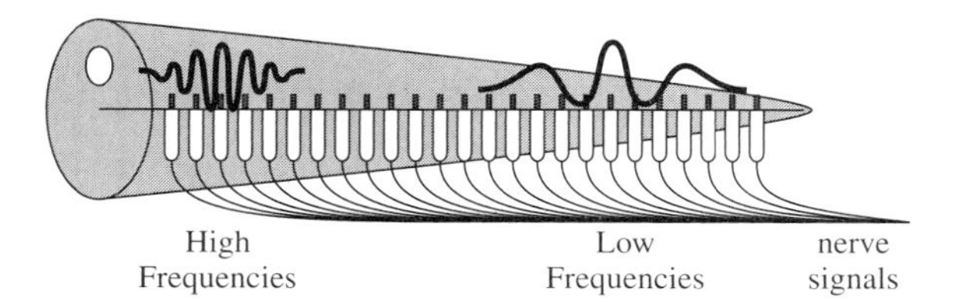

FIGURE 4.7
Illustration of vibration pattern of basilar membrane showing high frequency vibrations at the stapes end and low frequency vibrations at the apical end, with nerve signals emanating all along the basilar membrane [154].

also protect the ear against loud sounds by stiffening and thereby attenuating excessively loud sounds so as to prevent permanent damage to the hearing mechanisms.

The inner ear can be thought of as two organs, namely the semicircular canals, which serve as the body's balance organ, and the cochlea, which serves as the microphone of the auditory system, converting sound pressure signals from the outer ear into electrical impulses that are passed on to the brain via the auditory nerve. The cochlea, a $2\frac{1}{2}$-turn snail-shaped organ of a length of approximately 3 cm, is a fluid-filled chamber that is partitioned longitudinally by the basilar membrane. Mechanical vibrations at the entrance to the cochlea (the stapes end) create standing waves (of the fluid inside the cochlea), causing the basilar membrane to vibrate at frequencies commensurate with the input acoustic wave frequencies, and with largest amplitude occurring at places along the basilar membrane that are "tuned" to these frequencies.

Figure 4.7 illustrates schematically an uncoiled and stretched out basilar membrane from the stapes end (the entrance to the cochlea) to the apical end (the pointed end of the snail-like structure in Figure 4.5). Sound transmitted through the round window into the cochlea is, in effect, spectrally analyzed (on a non-uniform frequency scale) by the basilar membrane, with high frequency vibrations occurring at the stapes end and low frequency vibrations occurring at the apical end. The basilar membrane is lined with more than 30,000 inner hair cells (IHCs) that are set in motion by the mechanical movements along the basilar membrane. The IHCs vibrate at different rates and with different levels and are tuned to different frequencies along the basilar membrane. When the basilar membrane vibration motion is sufficiently large, the IHC vibration evokes an electrical impulse that travels along the auditory nerve to the brain for subsequent analysis.

4.3.1 Basilar Membrane Mechanics

A great deal of study has gone into understanding the mechanics of basilar membrane motion and transduction to electrical impulses that are transmitted to the brain [31, 190, 318, 337]. Some of the important findings about the mechanics of basilar membrane motion are as follows:

- the basilar membrane can be characterized by a set of *frequency responses* at different points along the membrane;

- the basilar membrane can be thought of as a mechanical realization of a non-uniform *bank of filters* (appropriately called cochlear filters);
- the individual filters in the bank of filters are roughly *constant Q* (where Q is the ratio of center frequency to bandwidth of the filter), with logarithmically decreasing bandwidths as we move away from the high frequency (stapes) end of the basilar membrane to the low frequency (apical) end;
- distributed along the basilar membrane are the IHCs, the set of sensors that act as mechanical motion-to-neural activity converters;
- the mechanical motion along the basilar membrane is sensed by local IHCs, causing *firing activity* at nerve fibers that innervate the bottom of each IHC;
- each IHC is connected to about 10 nerve fibers, each of different diameter; thin fibers fire at high motion levels, thick fibers fire at lower motion levels;
- approximately 30,000 nerve fibers link the IHCs to the *auditory nerve*;
- electrical pulses run along the auditory nerve until they ultimately reach higher levels of auditory processing in the brain and are ultimately perceived as *sound*.

The above observations about the workings of the basilar membrane support the notion that the basilar membrane performs a spatio-temporal form of spectral analysis of the incoming acoustic wave signal. The characteristics of this spatio-temporal spectral analysis are as follows:

- different regions of the basilar membrane respond maximally to different frequency components of the input signal; thus a form of frequency tuning occurs along the basilar membrane; we refer to the frequency at which the maximum response occurs along the basilar membrane as the *characteristic frequency*;
- the basilar membrane acts like a bank of non-uniform cochlear filters;
- there is roughly a logarithmic increase in the bandwidths of the individual filters when the filters' center frequency is above about 800 Hz; below 800 Hz, the bandwidth is essentially constant; thus the filter bank behaves like a constant bandwidth set of filters for frequencies below 800 Hz and like a constant Q set of filters for frequencies above 800 Hz.

Figure 4.8 illustrates the characteristics of the basilar membrane filter bank by showing tuning curves of six auditory nerve fibers [190]. A tuning curve shows the level of sound required to elicit a given output level of neural firing activity as a function of frequency. In Figure 4.8, each of the tuning curves is for a nerve fiber at a different location along the basilar membrane, moving from the stapes end to the apical end from top to bottom in the figure. These plots cover the range from about 20 kHz center frequency down to about 1 kHz center frequency. The constant Q nature of the six tuning curves is evident in this figure.

4.3.2 Critical Bands

An idealized version of an "equivalent" basilar membrane filter bank is shown in Figure 4.9. In reality, the bandpass filters are not the ideal filters shown in Figure 4.9,

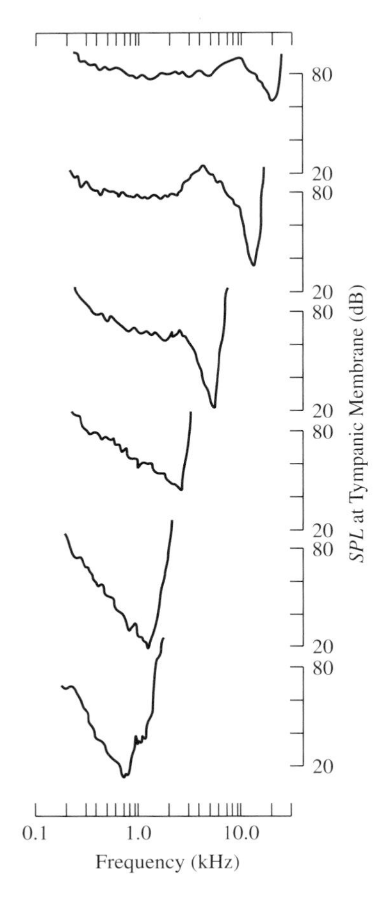

FIGURE 4.8
Tuning curves of six auditory nerve fibers. (After Kiang and Moxon [190]. © [1974] Acoustical Society of America.)

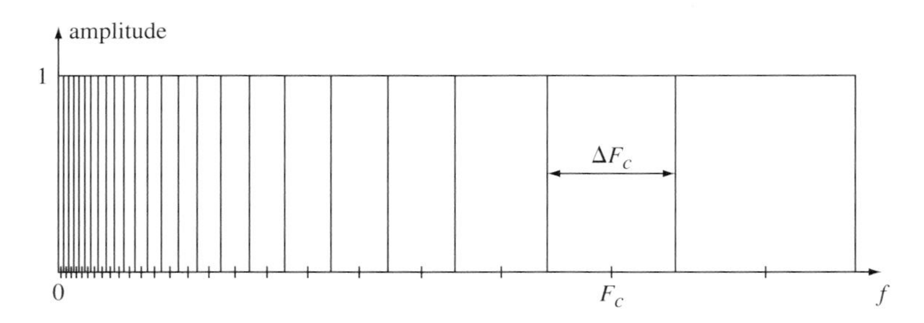

FIGURE 4.9
Schematic representation of bandpass filters according to the critical band theory of hearing.

but instead, the individual filter frequency responses overlap significantly, since points along the basilar membrane cannot vibrate independently of each other. Even so, the concept of bandpass filter analysis in the cochlea is well established, and the critical bandwidths of the basilar membrane filters have been defined and measured using a variety of methods, showing that the effective bandwidths are constant at about 100 Hz for center frequencies below 500 Hz, and with a relative bandwidth of about 20% of the individual filter center frequency above 500 Hz. An equation for the bandwidth as a function of center frequency that fits empirical measurements over the auditory range is as follows:

$$\Delta F_c = 25 + 75[1 + 1.4(F_c/1000)^2]^{0.69}, \tag{4.1}$$

where ΔF_c is the critical bandwidth associated with center frequency F_c [428]. Approximately 25 idealized critical band filters span the frequency range from 0 to 20 kHz. The concept of critical bands is very important in understanding such phenomena as loudness perception, pitch perception, and masking, and it therefore provides motivation for digital representations of the speech signal that are based on a frequency decomposition.

The critical band concept is used in many models for hearing and perception, so to facilitate its consistent use, a new frequency unit, the Bark, was created. The Bark scale is simply the index of the critical band starting with a value $z = 1$ for the top of the first critical band ranging from 0 to 100 Hz. The scale is created by adding one critical band to the next as in Table 4.2. From Table 4.2 we see that the first four critical bands all have a bandwidth of 100 Hz. The bands get progressively wider as the critical band rate increases; however, on the Bark scale the bands are equally spaced.

Considerable success has been achieved in the mathematical modeling of the ear all the way up to the encoding of sound as the neural outputs of the cochlea [124, 143, 214, 360]. The Bark scale is often employed in such models to transform uniform frequency analysis using the fast Fourier transform (FFT) into a Bark scale spectrum. These models are useful because they allow us to use known and well-understood properties of speech/sound perception in speech processing applications. Some examples of these more detailed models are discussed in Section 4.5.

4.4 THE PERCEPTION OF SOUND

Recall that sound is variations of air pressure that are propagated as waves through air or other media. Hearing, in its simplest definition, is the ability in humans and animals to sense sound. The ear is the sensor that enables human hearing of sound. Perception of sound implies making interpretations of the nature of the sensed sound signal. Perception results from the neural processing that mostly occurs after the sound is sensed and its spectrum analyzed by the ear. Sound waves can be characterized by their *amplitude* and *frequency* of variation. These are features that can be represented mathematically and measured using physical devices. Humans, however, sense these features of sound and perceive them as *loudness* and *pitch* respectively, and these perceptual features are not related in a simple way to *amplitude* and *frequency*. For this

TABLE 4.2 Critical band rate z and lower (F_l) and upper (F_u) frequencies of critical bandwidths ΔF_G centered at F_c. (After Zwicker and Fastl [428].)

z	F_l, F_u	F_c	z	ΔF_G	z	F_l, F_u	F_c	z	ΔF_G
Bark	Hz	Hz	Bark	Hz	Bark	Hz	Hz	Bark	Hz
0	0				12	1720			
		50	0.5	100			1850	12.5	280
1	100				13	2000			
		150	1.5	100			2150	13.5	320
2	200				14	2320			
		250	2.5	100			2500	14.5	380
3	300				15	2700			
		350	3.5	100			2900	15.5	450
4	400				16	3150			
		450	4.5	110			3400	16.5	550
5	510				17	3700			
		570	5.5	120			4000	17.5	700
6	630				18	4400			
		700	6.5	140			4800	18.5	900
7	770				19	5300			
		840	7.5	150			5800	19.5	1100
8	920				20	6400			
		1000	8.5	160			7000	20.5	1300
9	1080				21	7700			
		1170	9.5	190			8500	21.5	1800
10	1270				22	9500			
		1370	10.5	210			10,500	22.5	2500
11	1480				23	12,000			
		1600	11.5	240			13,500	23.5	3500
12	1720				24	15,500			
		1850	12.5	280					

reason, the "black box model of speech perception" that was mentioned earlier in this chapter has been used to attempt to understand perceptual phenomena by empirical means. What we know about sound and its perception by humans is the result of countless careful psychophysical experiments that have aimed to answer key questions such as:

- In general terms, what is the "resolving power" of the hearing mechanism?
- What is the sensitivity of human sound perception to variations in the fundamental frequency of sound?
- For speech-like sounds, what is the sensitivity of human sound perception to variations in formant frequencies and bandwidths?
- What is the sensitivity of human sound perception to variations in sound intensity?

Although our direct interest is in answering each question using speech signals, this is not possible or practical since speech is inherently a multi-dimensional signal with a linguistic association. This makes it extremely difficult to measure, with the required precision, any specific parameter or set of parameters, since the linguistic interpretation of the sounds greatly complicates the process. This has led researchers to focus on auditory discrimination capabilities using non-speech stimuli (primarily tones and noise), thereby eliminating all linguistic or contextual interpretations of the sounds. Finally we need to be aware of the differences between *absolute identification* of some property of a sound (e.g., pitch, loudness, spectral resonances) and *discrimination capability* of the sound property. For example, experimentation has shown that humans can detect a frequency difference of 0.1% between a pair of tones, but can only absolutely categorize tones of 5–7 different frequencies. Hence the auditory system of a human is very sensitive to differences, but cannot reliably perceive and categorize them absolutely.

In the remainder of this section we present some of the results of psychophysical experiments that help to establish a perspective on how hearing and perception fit into the overall speech communication chain.

4.4.1 The Intensity of Sound

Sound waves in general are complex patterns of variation of sound pressure. These pressure variations are extremely small compared to the ambient atmospheric pressure. Like other signals, we can represent them as superpositions of sinusoidally varying signals. For this reason, it is common to focus attention on sinusoidal pressure waves that propagate through space. In particular, when considering the effects of sound in hearing, it is conventional to consider a pure tone of frequency 1000 Hz. This frequency is within the range of frequency in which human hearing is most sensitive.

The strength or intensity of sound is a physical quantity that can be measured and quantified. Terms used to describe the strength of sound include acoustic intensity, audible intensity, intensity level, and sound pressure level. The *acoustic intensity* (I) of a sound is defined as the average flow of energy (power) through a unit area, measured in units of watts/m^2. The range of audible intensities is between 10^{-12} watts/m^2 and about 10 watts/m^2, corresponding to the range from the threshold of hearing to the threshold of pain. The intensity at the threshold of hearing (the sound power that can just be detected by a human listener) varies with individuals, but for reference purposes, it is taken to be $I_0 = 10^{-12}$ watts/m^2. The *intensity level*, (IL), of a sound wave is defined relative to I_0 as

$$IL = 10\log_{10}\left(\frac{I}{I_0}\right)\text{ dB}. \tag{4.2}$$

Thus, the IL is the power of a sound source (in dB) relative to the power of a sound source at the threshold of hearing.

For the case of a pure sinusoidal sound wave of amplitude P traveling through space, the intensity (power) is proportional to P^2. The *sound pressure level* (SPL) of a sound wave is defined as

$$SPL = 10\log_{10}\left(\frac{P^2}{P_0^2}\right) = 20\log_{10}\left(\frac{P}{P_0}\right) \text{ dB}, \tag{4.3}$$

where $P_0 = 2 \times 10^{-5}$ newtons/m^2 is the pressure amplitude corresponding to acoustic intensity $I_0 = 10^{-12}$ watts/m^2 at room temperature and atmospheric pressure.[1] Only under these conditions are SPL and IL the same; however, they are generally equated and the simpler term *sound level* is used for either. SPL can be measured with instruments consisting of a microphone, amplifier, and simple circuitry to compute average power [332].

4.4.2 The Range of Human Hearing

The human hearing mechanism has an incredible range between the lowest level sounds that can be heard and the loudest sounds that literally can cause pain and damage to the hearing mechanism. The lowest level sounds that can be heard define what is referred to as the *threshold of hearing,* namely the thermal limit of Brownian motion of air particles in the inner ear. Figure 4.10 shows a photograph of a room known as an *anechoic (without echo) chamber*, which was used to make measurements of the threshold of hearing. This very low echo experimental facility was set up at Bell Telephone Laboratories in the 1920s to enable early research on sound generation, sound perception, and sound pickup to be carried out in an environment where echos and noise were essentially eliminated. The mechanism for absorbing all sound in the walls, ceiling, and floor was the use of sound-absorbing wedges of material that criss-crossed the walls, ceiling, and floor, thereby enabling sound in any direction to be virtually completely absorbed with essentially no reflection (or echo). The room had a very high ceiling (as can be seen by comparing the height of the researcher at the corner of the anechoic chamber with the ceiling dimension) and a very low floor (about 10–20 feet below the wire mesh grid on which people stood and equipment was placed). To illustrate exactly how quiet the anechoic chamber actually was, one of the authors of this book (L. Rabiner) had the following experience: "After entering the anechoic chamber, its massive door was shut. After a minute or so I started hearing a thumping sound, which, of course, turned out to be the beating of my heart." Figure 4.11 illustrates the use of the anechoic chamber for sound experiments. This figure shows two Bell Labs researchers conducting controlled sound propagation tests in the anechoic chamber using a sound reflecting screen and a calibrated microphone.

Based on experiments in the anechoic chamber, researchers were able to measure the SPL at the threshold of hearing for tones, which, by convention, is defined as a level of 0 dB (decibels). Subsequent experimentation has shown that sounds at the threshold of pain have intensities of from 10^{12} to 10^{16} times the sound intensity at the threshold

[1] An alternative name for newtons/m^2 is Pascal (Pa), giving $P_0 = 20\ \mu$Pa.

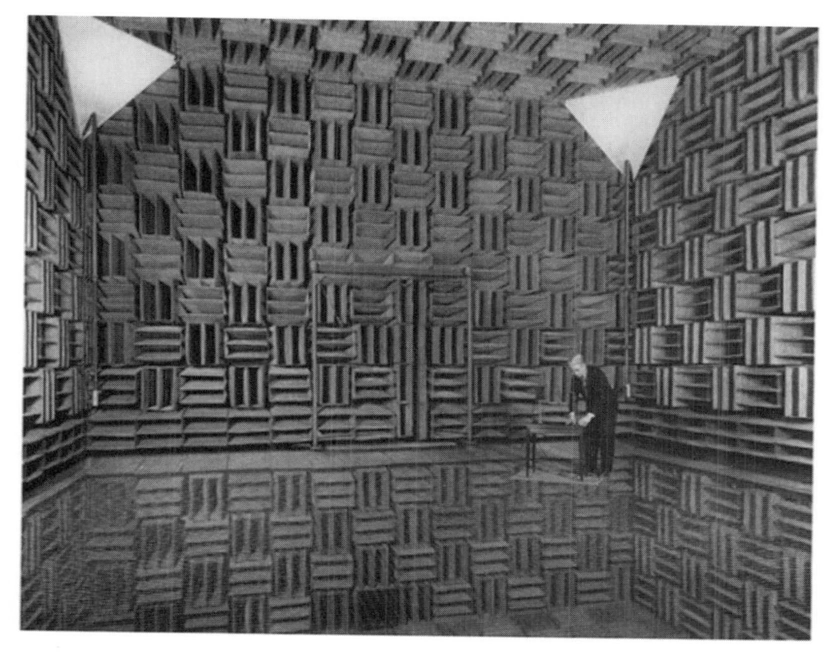

FIGURE 4.10
Anechoic chamber at Lucent Bell Labs, Murray Hill, NJ. Reprinted with permission of Alcatel-Lucent USA Inc.

of hearing, or, on a logarithmic scale, sound pressure levels of 120–160 dB. This range between the threshold of hearing and the threshold of pain represents an incredible range of sound intensities that can be perceived by human listeners. Figure 4.12 shows a cartoon sketch (developed at the House Ear Institute [150]) of the loudness of a range of everyday sounds (in dB) from a faint whisper (30–40 dB or 10^3 to 10^4 greater sound intensities than the threshold of hearing), to a moderately loud conversation (50–70 dB levels), to very loud sounds like blasting firecrackers at a distance of 10 feet (80–100 dB), to extremely loud sounds like a rock concert (110–130 dB) and finally to painfully loud sounds like a jet airplane close up (which produces sound levels of 140–170 dB). Table 4.3 gives a somewhat finer listing of a range of sound sources and the normative *SPLs* in dB. The table lists sound sources at 10 dB *SPL* increments between the threshold of hearing and the threshold of pain.

Figure 4.13 depicts the range of human hearing for pure tones, speech, and music signals. The horizontal axis in this figure is the frequency range (going from 20 Hz to 20 kHz), and the vertical axis shows the *SPL* (in dB) on the left side and sound *IL* (again in dB) on the right side. The bottom curve in Figure 4.13 is a curve of the *threshold of audibility*, namely the acoustic sound pressure level or *IL* of a pure tone (in quiet) that can barely be heard at a particular frequency. For a "standard" tone at a frequency of 1000 Hz, the threshold of audibility is defined to be 0 dB *SPL*

FIGURE 4.11
Sound calibration experiment in the anechoic chamber. (Prof. Manfred Schroeder and James West are the experimenters in this photo.) Reprinted with permission of Alcatel-Lucent USA Inc.

and 0 dB *IL*. It is readily shown that the threshold of audibility changes with frequency (and from person to person), with maximum sensitivity at about 3000–3500 Hz (where the threshold of audibility is about −3 dB), and least sensitivity at 20 Hz (threshold of audibility is about 90 dB) at the low frequency end, and at 20,000 Hz (threshold of audibility is about 70 dB) at the high frequency end. The region of maximum sensitivity (lowest threshold) to signal levels is the range from about 500 Hz (on the low frequency end of the scale) to about 7 kHz (on the high frequency end of the scale) where the threshold of audibility is within a few dB of the threshold at 1000 Hz.

We see from Figure 4.13 that for music and speech, the usual range of signal levels is significantly above the threshold of audibility of tones for the frequency band occupied by these signals, with speech typically being about 30–70 dB above the threshold level, and music being about 20–90 dB above the threshold. Music signals also show broader bandwidths than speech, covering the frequency band from 50 Hz to 10 kHz for most music of broad interest.

Figure 4.13 shows two other curves associated with human hearing, namely the curve of signal level that defines a range of risk to damaging hearing if the range is sustained (curve labeled "contour of damage risk"), and finally a curve defining the threshold of pain. The threshold of damage risk to hearing occurs for sounds at about 90–100 dB *SPL* at 1000 Hz, a level that is only marginally louder than the peak *SPL*

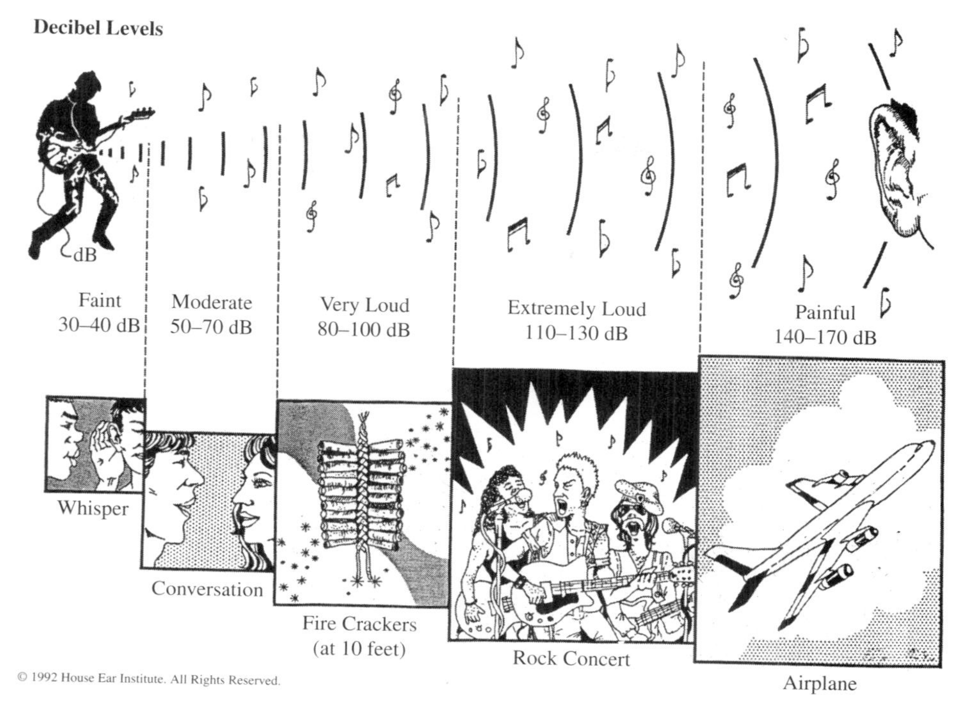

FIGURE 4.12
Sound levels in decibels for a range of sounds from very faint to painful. (After House Ear Institute [150].)

TABLE 4.3 *SPLs* for a range of sound sources.

SPL (dB)	Sound Source	*SPL*(dB)	Sound Source
170	Pain	80	Blow Dryer
160	Jet Engine—close up	70	Noisy Restaurant
150	Artillery Fire	60	Conversation Speech—1 foot
140	Rock Concert	50	Office Background Noise
130	22 Caliber Rifle	40	Quiet Conversation
120	Thunder	30	Whisper
110	Subway Train	20	Rustling Leaves
100	Power Tools	10	Breathing
90	Lawn Mower	0	*Threshold of Hearing*

for music at 1000 Hz, and only varies a small amount over the range from 500 Hz to 5000 Hz. The curve of the threshold of pain is almost flat, occurring at an *SPL* of about 135 dB at 1000 Hz. Damage to hearing can occur rapidly with extended exposure to sounds at these levels, and thus should be avoided at all costs.

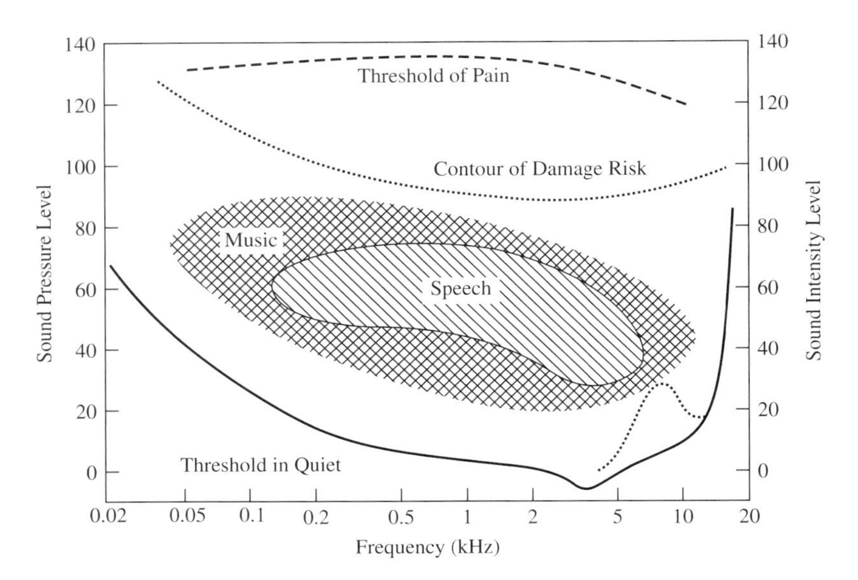

FIGURE 4.13
Range of human hearing for pure tones, speech, and music.

4.4.3 Loudness Level

We recall from our earlier discussion that when we discuss sound perception in humans, we need to distinguish between physical quantities such as IL or SPL, which are objective and measurable, and perceptual quantities such as perceived loudness, which is subjective and can only be determined via psychophysical experiments. In this section we discuss one such perceptual quantity, namely loudness level.

The *loudness level (LL)* of a tone is defined as the IL or SPL of a 1000 Hz tone that sounds (is perceived to be) as loud as the tone of a different frequency that is under investigation. LL is a complex function of the tone frequency and IL, primarily due to the sensitivity of human hearing to both these factors. The unit of LL is the *phon* and, by definition, the LL of a 1000 Hz tone at IL of x dB is x phons.

A set of curves of so-called "equal loudness" contours is given in Figure 4.14 [386].[2] Each curve shows SPL as a function of frequency for a fixed LL (in phons). The bottom curve is for an LL at the threshold of hearing. For a 1000 Hz tone at the threshold of hearing, the LL is approximately 0 phons, corresponding to an SPL of approximately 0 dB. For a tone of 125 Hz at the threshold of hearing, the required SPL is about 20 dB; however, both the 125 Hz tone and the 1000 Hz tone

[2]Earlier sets of equal loudness curves were published by Fletcher and Munson [114] and by Robinson and Dadson [321]. The curves shown in Figure 4.14 are the basis of an international standard (ISO226) for determination of loudness contours.

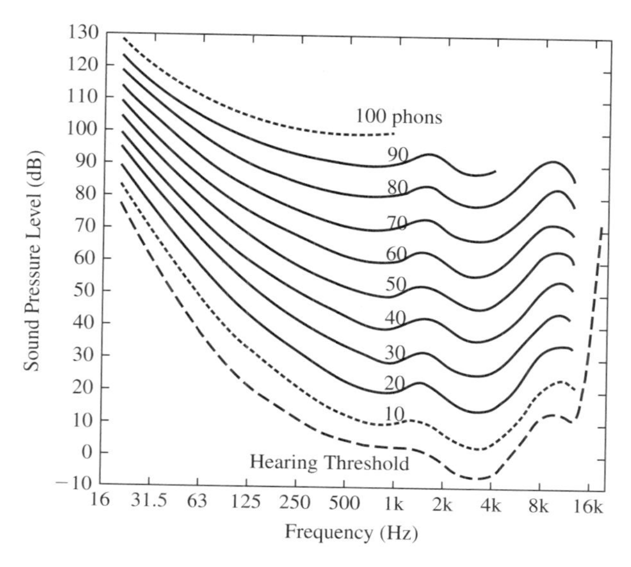

FIGURE 4.14
Equal LL curves for pure tones.

have an LL of 0 phons; i.e., they are perceived to be equally loud. Similarly, for a tone of 8 kHz at the threshold of hearing, the required SPL is about 12 dB. Thus the equal loudness contour at the threshold of hearing spans about a 20 dB range of SPL. The upper contour in Figure 4.14 is for tones with an LL of 100 phons. This contour is relatively flat at an SPL of 100 dB between 250 Hz and 1000 Hz; below 250 Hz the contour rises to close to 130 dB SPL for a 16 Hz tone judged to be equally as loud (100 phons) as a 100 dB SPL tone at 1000 Hz.

The behavior of the equal loudness contours, above 1000 Hz, is interesting as the contours actually rise slightly between 1 and 1.5 kHz, then fall to a minimum value around 3 kHz, and then rise again until 8 kHz. The frequency of maximum sensitivity is about 3000–3500 Hz where a tone judged equally as loud as a 1000 Hz tone has an SPL of about 8 dB less than that of the 1000 Hz tone. This increased sensitivity is due to a resonance effect of the external ear canal. The external canal, being approximately a uniform tube excited by sound from the pinna at one end and terminated by the tympanic membrane at the other, exhibits resonances similar to those to be discussed in Section 5.1.2. In particular, the first resonance, which typically occurs at about 3500 Hz, provides additional emphasis to sounds in this band. Another dip in the equal loudness contours that occurs at about 13 kHz can be attributed to the higher resonance frequencies of the ear canal.

4.4.4 Loudness

A second subjective measure of sound that is of interest is the relative loudness of a pair of sounds. The previous section discussed a measure of equal loudness, LL, where it was shown the two sounds with the same LL (in phons) are perceived as

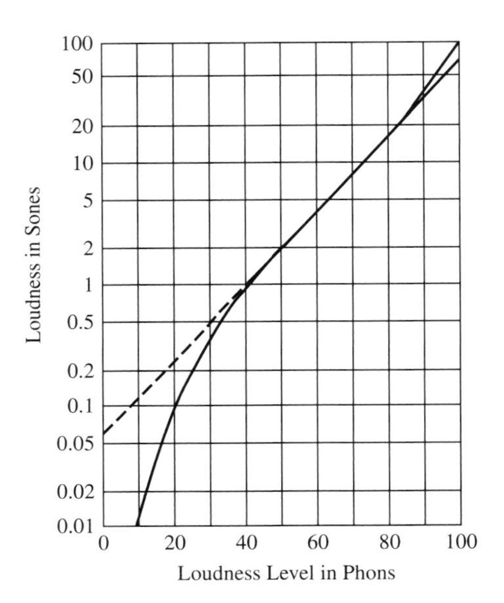

FIGURE 4.15
Loudness (in sones) versus *LL* (in phons) for a 1000 Hz tone. (After Fletcher and Munson [115]. © [1937] Acoustical Society of America.)

being equally loud sounds even though they have physically different *SPLs*. However, the relative perceived loudness of sounds is *not* directly proportional to the *LL* in phons; i.e., a sound with an *LL* of 60 phons is *not* perceived as being twice as loud as a sound with an *LL* of 30 phons. In fact, experiments have shown that for a given frequency, and for *LL*s above 40 phons, an increase in *LL* of approximately 10 phons is perceived as a doubling of loudness. In Figure 4.14, this corresponds to moving from one equal-loudness curve to the one above it.

Specifically, Figure 4.15 shows measurements of loudness, denoted L, in units of sones, as a function of LL, with loudness being defined to have a value of 1 sone at an LL of 40 phons. Above 40 phons, the curve is closely approximated by [332]

$$L = 2^{(LL-40)/10}, \tag{4.4}$$

which expresses the doubling of loudness due to a 10 phon (dB) increase in LL. Taking the logarithm of L in Eq. (4.4) gives

$$\log_{10} L \approx 0.03(LL - 40) = 0.03LL - 1.2, \tag{4.5}$$

which is the straight line approximation plotted in Figure 4.15. For a tone of 1000 Hz, LL (in phons) is, by definition, numerically equal to the IL (in dB), so we can substitute Eq. (4.2) for LL:

$$LL = 10\log_{10}(I/10^{-12}) = 10\log_{10} I + 120. \tag{4.6}$$

Combining Eqs. (4.4) and (4.6), we get

$$\begin{aligned} \log_{10} L &= 0.03(10 \log_{10} I + 120) - 1.2 \\ &= 0.3 \log_{10} I + 2.4, \end{aligned} \tag{4.7}$$

or, equivalently,

$$L = 251 I^{0.3}. \tag{4.8}$$

It can be seen from Eq. (4.8) that the subjective loudness L (in sones) of a 1000 Hz tone varies approximately as the cube root of the intensity I (in watts/m^2).[3] Thus, to double the subjective loudness of a 1000 Hz tone, an approximate eight-fold increase in power is required.

Strictly speaking, the above discussion applies only to pure tones; however, it is possible to use Eq. (4.8) to predict what to expect for more complex sound waves by applying it to signals obtained by dividing the signal spectrum into bands using, for example, octave-band filter banks. A result of such analysis is that sound of a given energy appears louder when the energy is spread across a wide range of frequencies than when concentrated at a single frequency [192].

4.4.5 Pitch

Our ability to perceive sound frequency is, in many ways, as powerful as our ability to perceive ranges of sound loudness. For example, a human can often detect a low level spectral component at high frequencies. (This ability often makes speech coding very difficult, especially in the case of generating very low level spectral tones at half the sampling frequency—the so-called limit cycles of waveform coders.) Similarly a human can often not detect noise or tonal signals in the presence of a strong tonal signal. This property of the interaction of sound and hearing is called *masking*, and we will discuss it in Section 4.4.6.

The perceptual quantity that is related to sound frequency is called *pitch*. (Since frequency is a property of pure tones, most pitch experiments use pure tones.) Just as (subjective) loudness of a tone is influenced by both sound intensity and sound frequency, perceived pitch is similarly influenced by both sound frequency and sound intensity.

In general, pitch (the subjective attribute) is highly correlated with frequency (the physical attribute). The unit of frequency is Hz and the unit of pitch is the *mel* (derived from the word *melody*). For normalization purposes the pitch of a 1000 Hz tone is defined to be 1000 mels. A typical experiment to investigate the relationship of perceived pitch to frequency might proceed as follows: a subject is presented with a pure tone of a frequency F_1 and is asked to adjust the frequency of a second tone until it has half the pitch of the first tone. Using such experiments, we

[3] Assuming an exponent of 1/3 rather than 0.3 leads to slightly different constants in Eq. (4.8) [192]; however the same general conclusions follow.

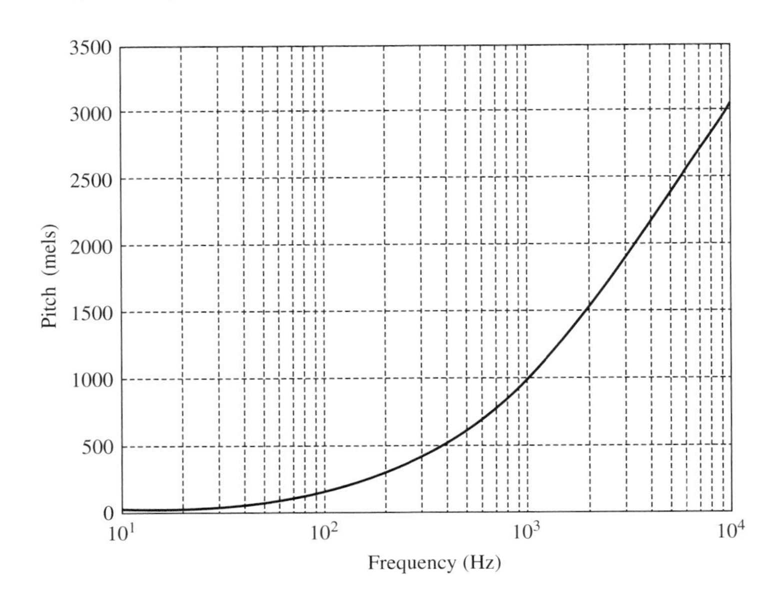

FIGURE 4.16
Chart of the subjective pitch (in mels) versus the actual frequency (in Hz) of a pure tone.

arrive at the curve shown in Figure 4.16, which shows that the relationship between pitch and frequency of a pure tone is not linear. The curve is accurately fit by the equation

$$\text{pitch in mels} = 1127 \log_e (1 + F/700), \tag{4.9}$$

where F is the frequency of the tone [378]. Thus a tone that has a pitch (500 mels) half as large as the pitch of a 1000 Hz tone (1000 mels) has a frequency of about 390 Hz, whereas a tone that has a pitch (2000 mels) twice as large as the pitch of a 1000 Hz tone (1000 mels) has a frequency of about 3429 Hz.

The psychophysical phenomenon of pitch, as quantified by the mel scale, can be related to the concept of critical bands [428]. It turns out that, more or less independently of the center frequency of the band, one critical bandwidth corresponds to about 100 mels on the subjective pitch scale. Thus we calculate that a critical band of width $\Delta F_c = 160$ Hz, centered on $F_c = 1000$ Hz, maps into a band of width 106 mels, and a critical band of width 100 Hz, centered on 350 Hz, maps into a band of width 107 mels. Thus, what we understand about pitch perception reinforces the notion that the auditory system performs a frequency analysis that can be simulated with a bank of bandpass filters whose bandwidths increase as center frequency increases.

One of the characteristics of pitch is that human listeners are extremely sensitive to changes in frequency and can reliably distinguish two tones separated by 3 Hz (or more) if the frequency of the tones is at or below 500 Hz. If the frequency of the tones is at or above 500 Hz, humans can reliably determine that two tones are of different frequencies if they are separated by $0.003F_0$, where F_0 is the frequency of the lower tone. We will refer to this differential sensitivity of humans in Chapter 11 when we discuss coding of speech parameters for transmission and storage.

We have restricted the discussion in this section to the perceived pitch of pure tones. A more interesting and complicated phenomenon is the perceived pitch of complex sounds consisting of combinations of tones and (possibly) noise. The subjective evaluation of pitch for such complex sounds is beyond the scope of this book, but well covered in [428].

4.4.6 Masking Effects—Tones

Masking is the phenomenon whereby some sounds are made less distinct or even inaudible by the presence of other sounds. We experience masking in a variety of circumstances such as noisy environments like factory floors, airport terminals, etc., where we strain to hear announcements or to talk to people. In such cases we say that the desired sound is *masked* by the background sound, and we define the degree of masking as the extent to which the masking sound (the background noise) raises the threshold of audibility of the desired sound (usually as some number of dB).

Masking is a complicated phenomenon and is best studied by restricting both the background sound (called the masker) and the masked sound to tones (or noise). For such cases the masker tone is set to a fixed frequency and the masked sound is increased in level from an inaudible sound to one that is just distinguishable from the masker. Holding the masker tone fixed, we repeat this experiment across a range of masked sound frequencies and intensities and create curves of the threshold shift of the masked tone (in dB) versus the frequency of the masked tone, for a given masker tone frequency. Such a set of plots is shown in Figure 4.17 for both a 400 Hz masker [part (a)] and a 2000 Hz masker [part (b)].

For each plot of Figure 4.17, the masked tone frequency is plotted along the abscissa and the threshold shift (in dB) for different intensities of the masking tone

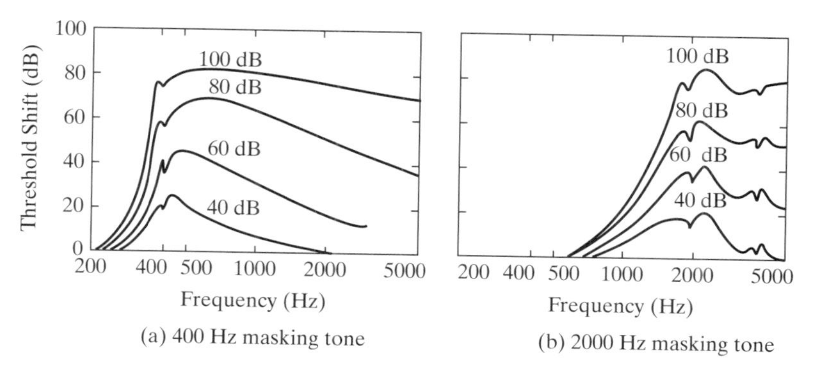

FIGURE 4.17
Masking of tones by tones: (a) 400 Hz masking tone; (b) 2000 Hz masking tone.

is plotted along the ordinate. The curves at the left [part (a)] are for 400 Hz masking tones at *ILs* of 40, 60, 80, and 100 dB, and the curves on the right [part (b)] are for a 2000 Hz masking tone, again at *ILs* of 40, 60, 80, and 100 dB. Several points are immediately obvious from these curves:

- masking is significantly more effective at frequencies above the masking tone than at frequencies below the masking tone;
- masking is significantly more effective at higher *ILs* where the masking curves almost become flat at *ILs* of 80 and 100 dB;
- the masking curves display small notches at frequencies near that of the masking tone; this is due to the creation of beats when the masker and the masked tone are sufficiently close in frequency.

For example, in Figure 4.17 we see that a tone at frequency 1000 Hz and *IL* 40 dB is completely masked by a 400 Hz tone at 80 dB *IL*, but that same tone is well above the threshold of audibility of a 2000 Hz masking tone at 80 dB *IL*.

If we combine the shape of the masking curve (at a given *IL*) for a pure tone masker with the curves of human hearing in Figure 4.13, we get the curve shown in Figure 4.18, which shows the normal hearing threshold curve and its modification to a shifted hearing threshold as a result of a tone masker at a given frequency and *IL*. The tone masker masks all tones whose *IL* falls below the shifted hearing threshold (two masked tone signals are shown in this figure), and only the low frequency tone whose *IL* rises above the shifted hearing threshold is audible. The principle of tone masking and the shifted hearing threshold is an essential component of the audio compression algorithms described in Chapter 12.

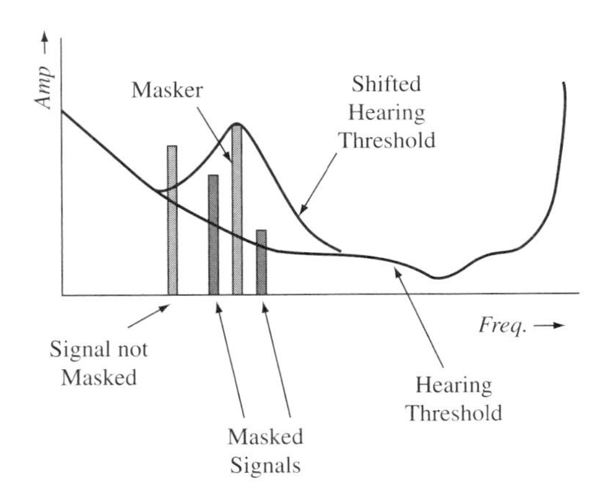

FIGURE 4.18
Auditory masking using a pure tone masker, and showing the shifted hearing threshold that results.

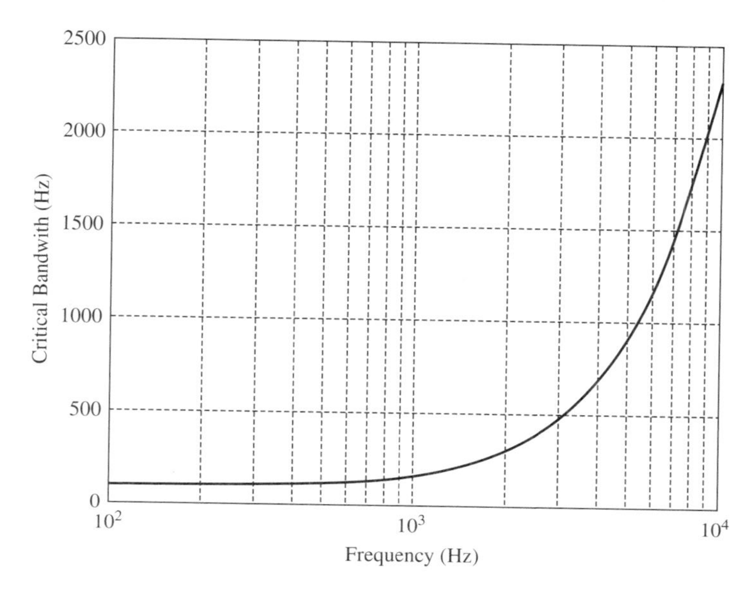

FIGURE 4.19
Critical bandwidth as a function of center frequency of the spectral band.

4.4.7 Masking Effects—Noise

The most common form of masking sound in physical situations is noise. Noise is generated by a variety of sources including motors, generators, fans, etc., and it can also be generated by quantizing the samples of a speech waveform. Generally noise of the above types is broadband; i.e., it has significant spectral level across a broad band of frequencies. To do an analysis of the masking properties of noise, we must break the noise signal into a set of spectral slices of various bandwidths. Hence a key question is what bandwidth should be used for each of the spectral slices. Experimentation has shown that for flat spectrum noise signals, the degree of masking of a tone (at the center frequency of the noise band) by a wideband noise source increases with the bandwidth of the masking noise until it reaches a limit, the critical bandwidth, beyond which any further increase in bandwidth of the masker noise has little or no effect on the degree of masking of the pure tone. This is, in fact, one way to measure the critical bands. Figure 4.19 shows a curve of the critical bandwidths as a function of the center frequency of the spectral band. We recall from our earlier discussion in Section 4.3.2 that the critical bandwidth is about 100 Hz for center frequencies below 500 Hz and then it rises monotonically to a value of about 2200 Hz at a center frequency of 10,000 Hz.

4.4.8 Temporal Masking

Temporal masking of sounds occurs when a transient sound causes sounds that either precede or follow the transient sound to become inaudible. This is illustrated in Figure 4.20, which shows a finite duration masker, along with a region of "backward" or

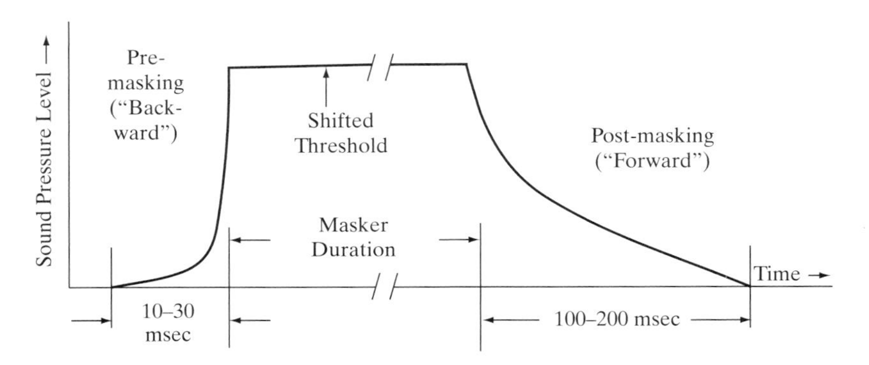

FIGURE 4.20
Illustration of temporal masking showing both backward and forward masking effects. (After Zwicker and Fastl [428].)

pre-masking (approximately 10–30 msec in duration with most of the masking occurring in the interval 20 msec prior to the transient signal), and a region of "forward" or post-masking (approximately 100–200 msec in duration) [428]. The effectiveness of such temporal masking falls almost exponentially before and after the masking sound. Although the pre-masking effect is very short, it is an important component in audio coding since it serves to hide the effect of so-called *pre-echo* that occur due to the quantization noise from block processing of transient sounds in the audio signal. Again we discuss this more fully in Chapter 12.

4.4.9 Masking in Audio Coding

Figure 4.21 shows a plot of the power spectrum (magnitude-squared of the DTFT) for a finite-length segment (called a frame) of an audio signal, along with the predicted masking threshold and the chosen bit assignment for coding this segment of the audio signal. The power spectrum curve shows the typical pattern for a musical instrument (or a set of musical instruments), namely a set of strong, harmonically related resonances and a wide bandwidth spectrum. Using the concepts presented in the previous sections, a curve of the predicted masking threshold is calculated (primarily by summing the masking thresholds for each of the peaks of the power spectrum of the audio signal) and plotted on top of the curve of the audio power spectrum. The spectral regions where the audio power spectrum level exceeds the level of the predicted masking threshold should be coded accurately enough to faithfully preserve the audio spectrum and ensure that the coding error is below the masking threshold. However, whenever the audio power spectrum falls below the predicted masking threshold, *no bits* need to be assigned since the signal in these frequency bands is totally masked by the strong signal in the adjacent bands. Hence the masking threshold provides a very efficient mechanism for deciding how to allocate bits to an audio signal so as to provide the highest quality sound with the fewest number of bits. This process of subjectively coding an audio signal is the basis behind the MP3 and AAC methods of coding audio signals that are used for compressing music and are discussed in Chapter 12.

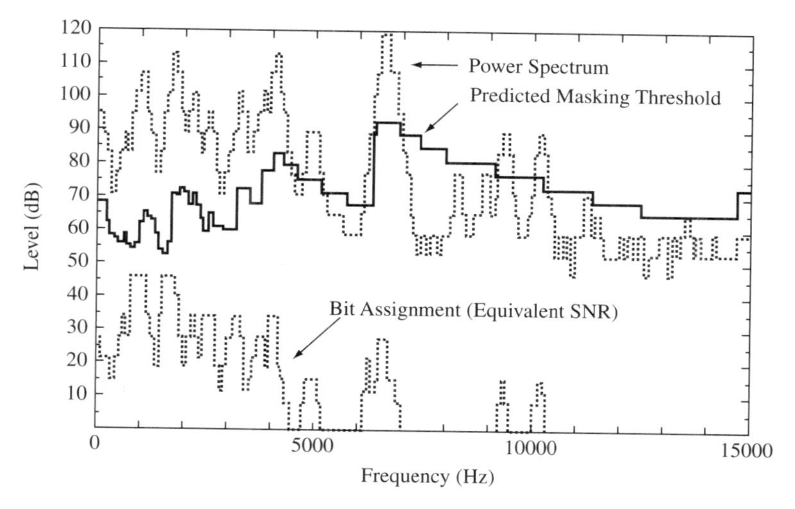

FIGURE 4.21
Plots of the power spectrum of an audio signal, the predicted masking threshold, and the resulting bit assignment (shown as an equivalent signal-to-noise ratio).

4.4.10 Parameter Discrimination—*JND* [105, 112, 377]

An important question that arises in speech processing (especially in pattern recognition systems) is what defines a significant perceptual difference in an objectively measured speech property, such as fundamental frequency, or intensity. To quantify such questions, psychoacousticians have employed the concept of *just noticeable difference* or *JND*, which is defined as the smallest change in a physical parameter such that human listeners can reliably detect the change as measured in standard listening tests. [Another term that is often used to describe the just-discriminable change is the difference limen (*DL*) or the differential threshold.]

Using synthetic vowel sounds generated by a speech synthesizer, the *DL* for the first and second speech resonance frequencies, i.e., the first two formants, was found to be about 3–5% of the formant frequency. Thus for a first formant frequency of 500 Hz, the *JND* in center frequency is about 15–25 Hz. This result was obtained by considering changes in the frequency of only one formant at a time, and finding what level of change was reliably detected by listeners. When *all* formants are changed simultaneously, the measured *JND* was shown to be a complicated function of all formant frequencies and to be very sensitive to how close formants were to each other.

The *JND* in overall intensity of a synthetic vowel has been shown to be about 1.5 dB [105]. Since the first formant is usually the most intense formant in vowel sounds, this *JND* of 1.5 dB can also be considered a rough estimate of the intensity *DL* for the first formant. The intensity *DL* was found to be about 3 dB for the second formant for a near-neutral vowel.

The *DLs* for formant bandwidth were not directly measured, but instead indirect measurements showed that changes on the order of 20–40% in formant bandwidth are just-discriminable [377].

TABLE 4.4 Measured DL/JND for synthetic vowel sounds [105].

Parameter	***DL/JND***
Fundamental Frequency	0.3–0.5%
Formant Frequency	3–5%
Formant Bandwidth	20–40%
Overall Intensity	1.5 dB

Again, using synthetic vowel sounds, the fundamental frequency DL was shown to be about 0.3–0.5% of the fundamental frequency [112]. An interesting observation is that the formant frequency DL is an order of magnitude smaller than the formant bandwidth DL, and the fundamental frequency DL is an order of magnitude smaller than the formant frequency DL. Thus the human sensitivity to pitch is the sharpest, whereas the sensitivity to resonance bandwidth is the least sensitive. Table 4.4 summarizes the measured DL/JND for synthetic vowel sounds.

4.5 AUDITORY MODELS

As we have summarized in Sections 4.3 and 4.4, a significant amount of knowledge has been assembled about the physiology of the ear and about the psychophysical response of the end-to-end auditory system. This has encouraged researchers to attempt to construct detailed auditory models that can predict auditory phenomena, with the goal of obtaining models that can be used in a range of speech applications. Based on the discussion in this chapter, the major perceptual effects that are included in most auditory models are the following:

- spectral analysis on a non-linear frequency scale (usually a mel or Bark scale) which is approximately linear until about 1000 Hz and approximately logarithmic above 1000 Hz;
- spectral amplitude compression (or equivalently, dynamic range compression);
- loudness compression via some type of logarithmic compression process;
- decreased sensitivity at lower (and higher) frequencies based on results from equal loudness contours;
- utilization of temporal features based on long spectral integration intervals (e.g., syllabic rate processing);
- auditory masking whereby loud tones (or noise) mask adjacent signals that are below some threshold and contained within a critical frequency band of the tone (or noise).

Some, or all, of these effects are integrated into virtually all auditory models with the ultimate goal of either gaining insight into the signals being processed by the brain, or enabling improved performance of speech applications (as discussed in

Chapters 11–14). In the remaining parts of this section, we summarize four auditory models that have been created for use in speech processing applications. It will be clear as the discussion unfolds that all of these models bear a strong resemblance to the notion of the spectrogram as introduced in Section 3.3. Spectrogram plots of the short-time Fourier transform use equally spaced analysis frequencies because that is what is most efficient and convenient. Indeed, we will see that most of the auditory models essentially begin with the standard spectrogram and then group frequencies into critical bands to simulate the frequency analysis of the cochlea. Subsequent stages of the models are designed to further simplify the representation to create pattern vectors. Although the models to be discussed rely on details of speech spectrum analysis that are not covered until Chapters 7–9, the models are presented here at a functional level that does not require deep understanding of the signal processing operations.

4.5.1 Perceptual Linear Prediction

One of the most popular (and successful) auditory models is the method of perceptual linear prediction (PLP) as proposed by Hermansky [143].[4] A block diagram of PLP processing is shown in the left panel of Figure 4.22. The specific perceptual effects implemented in PLP are the following:

- critical band spectral analysis using a (non-linear) Bark frequency scale with variable bandwidth trapezoidal shaped filters;
- asymmetric auditory filters with a 25 dB/Bark slope at the high frequency cutoff of each filter and a 10 dB/Bark slope at the low frequency cutoff of each filter;
- use of the equal loudness contour plots to approximate the unequal sensitivity of human hearing to different frequency components of the signal;
- use of the non-linear relationship between sound intensity and perceived loudness by using a cubic root compression of the spectral levels;
- a method of broader-than-critical-band integration of frequency bands based on an autoregressive, all-pole model utilizing a fifth-order analysis.

As shown in Figure 4.22, the first step in the PLP processing chain is a computation of the power spectrum of the speech signal using standard FFT methods (as will be discussed in Chapter 7). The resulting spectral magnitude is shown in the right panel [part (1)] of Figure 4.22. The next step in the processing is to use either a triangular window (for a mel-based frequency axis) or a trapezoidal window (for a Bark-based frequency axis) to integrate the power spectrum within overlapping critical bands, as shown in part (2) of Figure 4.22. Note that the resulting Bark scale is now denoted as a tonality scale rather than a linear frequency scale. The next step in the processing is

[4]A better understanding of this model will be obtained after we discuss the concepts of cepstral analysis of speech (in Chapter 8) and linear predictive coding of speech (in Chapter 9). In this section we emphasize the use of perceptual attributes of speech in each auditory model under discussion.

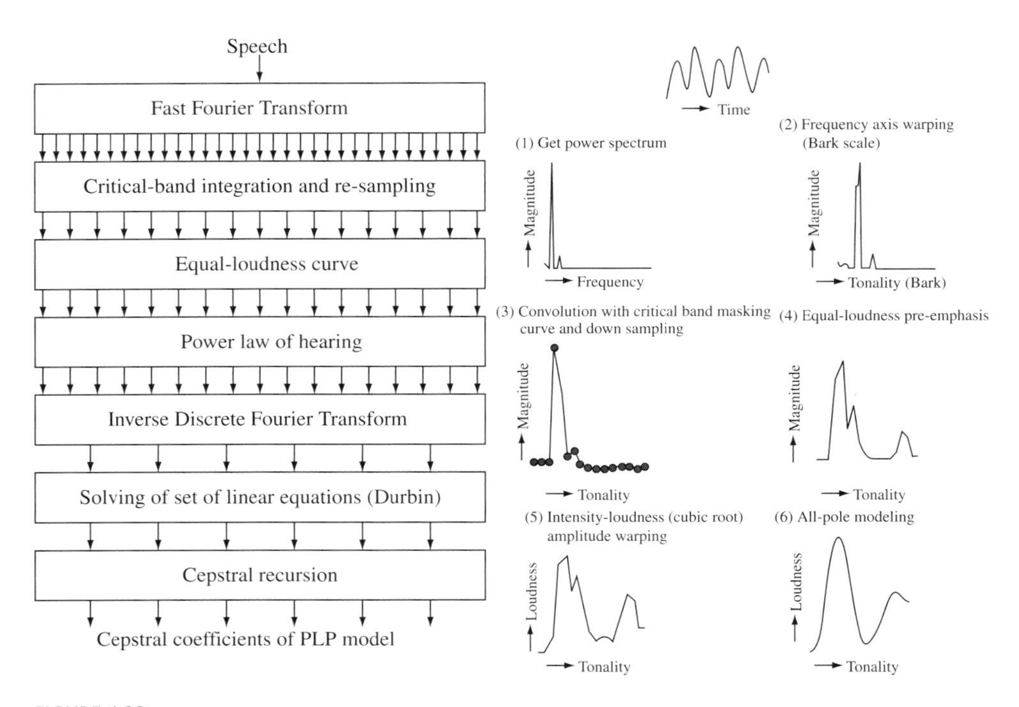

FIGURE 4.22
(Left panel) Block diagram of a PLP analysis system; (right Panel) typical waveforms at various points along the PLP processing chain. (After Hermansky [143].)

pre-emphasis of the spectrum to approximate the unequal sensitivity of human hearing to different frequency components of the signal (this step approximates the equal loudness curves), as illustrated in parts (3) and (4) of Figure 4.22. The fourth step in the processing is compression of the spectral amplitudes with a logarithmic compressor in order to approximate the power law relation between intensity and loudness, as illustrated in part (5) of Figure 4.22. The final processing step performs an inverse FFT (to give a set of cepstral coefficients as discussed in Chapter 8), smoothes the spectrum using a cepstral lifter, and then uses an orthogonal representation to give a set of cepstral coefficients corresponding to the smoothed (frequency warped) spectrum, which is shown in part (6) of Figure 4.22.

The point of PLP is to implement a type of speech spectrum analysis that models that of the human auditory system. The last stage (6) employs the techniques of time-dependent linear prediction and cepstral analysis to derive an efficient parametric representation of the short-time speech spectrum. This representation can be used as the feature vector input to an automatic speech recognition system. Although it may be difficult to understand all the details of the processing of the PLP auditory analysis system at this point in the book, all the signal processing steps will be discussed in detail in later chapters. It should be clear, however, that the PLP processing makes use of many concepts based on human perception of sounds.

4.5.2 Seneff Auditory Model

A completely different type of auditory model was proposed by Seneff [360]. This model attempts to capture the essential features of the response of the cochlea and the attached hair cells in response to speech sound pressure waves. A block diagram of the Seneff model is given in Figure 4.23.

There are three stages of processing in the Seneff auditory model. Stage 1 pre-filters the speech to eliminate very low and very high frequency components (that contribute little or nothing to the model output), and then passes the filtered signal through a 40-channel critical band filter bank where the individual filters are

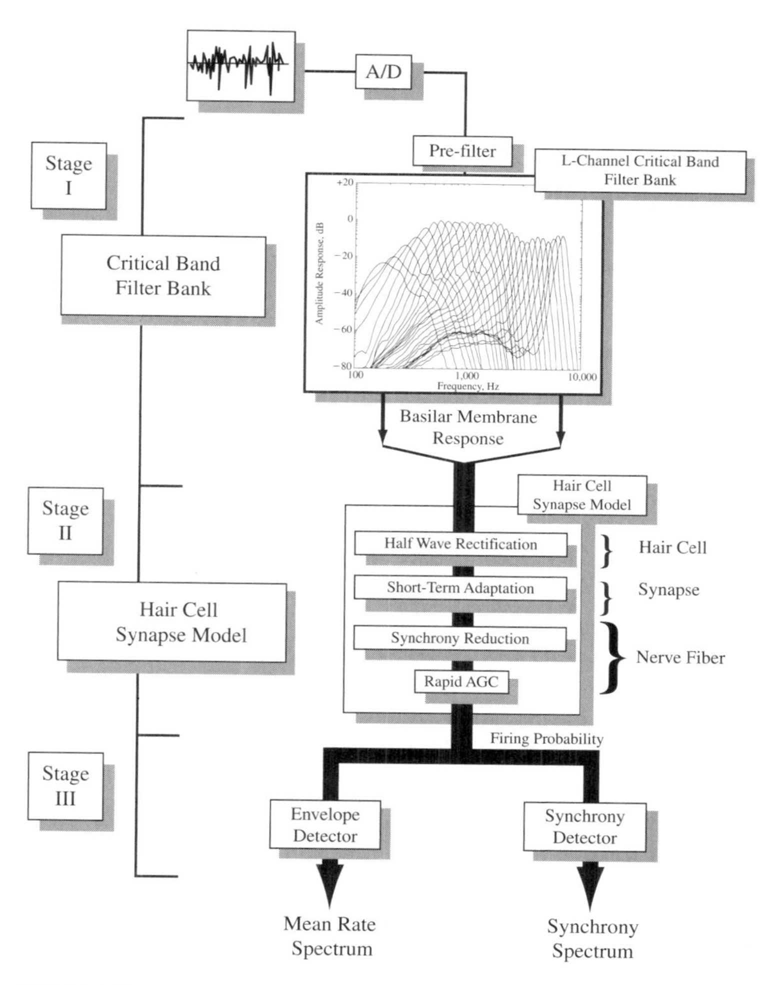

FIGURE 4.23
Block diagram of Seneff auditory model. (After Seneff [360].)

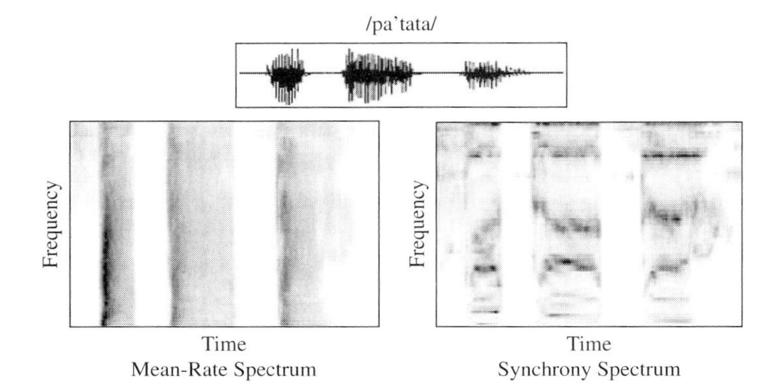

FIGURE 4.24
Examples of mean-rate spectrum and synchrony spectrum for word /pa'tata/. (After Seneff [360].)

non-uniformly distributed according to a Bark frequency scale. This first stage models the basic non-uniform filtering of the cochlea in the inner ear.

Stage II of the Seneff auditory model is a hair cell synapse model of the probabilistic behavior of the combination of IHCs (inner hair cells), synapses, and nerve fibers via the processes of half wave rectification (hair cell firing on positive inputs), short-term adaptation (at the synapses), and synchrony reduction and rapid automatic gain control (AGC) at the nerve fiber. The outputs of the hair cell synapse model are the probabilities of firing, over time, for a set of similar fibers acting as a group.

Stage III of the Seneff auditory model uses the firing probability signals of the second stage as a basis for extracting information relevant to perception, e.g., formant frequencies and enhanced sharpness of onset and offset of speech segments. An envelope detector computes the envelope of the firing probability signals, thereby capturing the rapidly changing dynamic nature of speech, which characterizes (mainly) transient signals. In this manner the envelope detector provides estimates of the mean-rate spectrum, namely the transitions from one phonetic segment to the next, via onsets and offsets in the output representation. A synchrony detector simulates the phase-locking property of nerve fibers, thereby enhancing spectral peaks at formants and enabling the tracking of dynamic spectral changes in the signal.

Figure 4.24 shows examples of the mean-rate spectrum and synchrony spectrum for the word /pa'tata/. The segmentation into well-defined onsets and offsets (for each of the stop consonants in the utterance) is readily seen in the mean-rate spectrum. The speech resonances are clearly seen in the synchrony spectrum.

4.5.3 Lyon's Auditory Model

A somewhat different auditory model, proposed and studied by Lyon [214], is shown in Figure 4.25. The model has a pre-processing stage (simulating the effects of the outer and middle ears as a simple pre-emphasis network), and three full stages of processing for modeling the cochlea as a non-linear filter bank. The first processing stage consists of a bank of 86 cochlea filters implemented at a 16 kHz sampling rate. The filters are

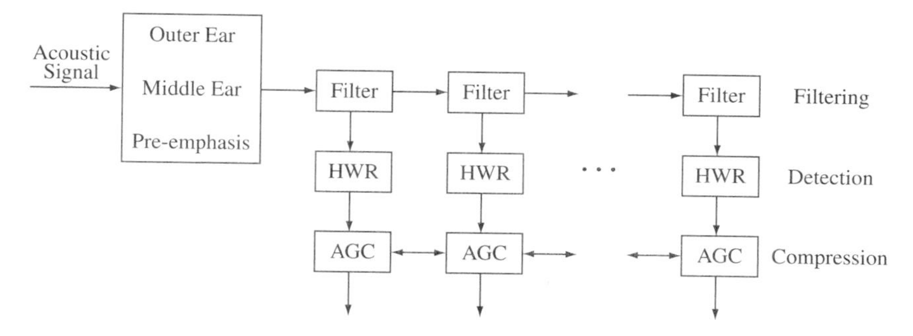

FIGURE 4.25
Block diagram of Lyon's auditory model. (Courtesy M. Slaney, after Lyon [214].)

spaced non-uniformly according to a mel or Bark scale and are highly overlapped in frequency. The second stage uses a half wave rectifier (HWR) non-linearity to convert the basilar membrane signals (from the filter bank) into a representation of the IHC receptor potential or the auditory nerve firing rate.

The third stage of processing consists of a bank of inter-connected AGC circuits that continuously adapt in response to activity levels at the outputs of the HWRs of the second stage in order to compress the wide range of sound levels into a limited dynamic range of basilar membrane motion, IHC receptor potential, and auditory nerve firing rates.

By these processes, the Lyon auditory model converts the speech into a two-dimensional map of neural firing rate versus time and place along the basilar membrane.

In order to gain an insight into the properties of the model signals at the output of the third stage of processing, Lyon proposed the use of two displays, one static and one dynamic. The static display Lyon called a cochleagram in analogy to a spectrogram. The cochleagram is a plot of model intensity as a function of place (warped frequency) and time. An example of a cochleagram for the utterance "Oak is strong and often gives shade" was provided by Malcolm Slaney and is shown in Figure 4.26. The alternative dynamic display proposed by Lyon was based on the idea of computing the short-time autocorrelation of the model output signal (via methods to be discussed in Chapter 6) and displaying the time-varying sequences of autocorrelations as a function of time, in the form of a computer movie. Lyon called this display output a correllogram.

4.5.4 Ensemble Interval Histogram Method

Yet another model of cochlear processing and hair cell transduction was proposed and studied by Ghitza [124]. The model consists of a filter bank that models the frequency selectivity at various points along a simulated basilar membrane, and a non-linear processor for converting the filter bank output to neural firing patterns along a simulated auditory nerve. This model is shown in Figure 4.27 and is called the ensemble interval histogram or EIH model [124].

In the EIH model, the mechanical motion of the basilar membrane is sampled using 165 channels, equally spaced on a log-frequency scale, between 150 and 7000 Hz.

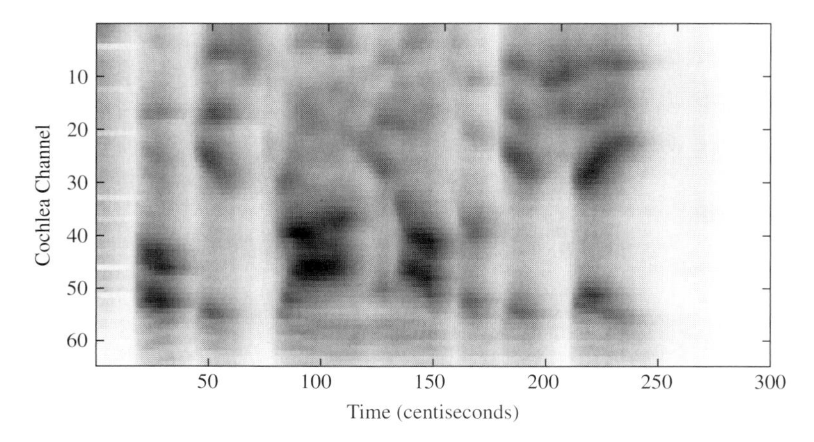

FIGURE 4.26
Cochleagram of utterance "Oak is strong and often gives shade." (Courtesy M. Slaney, after Lyon [214].)

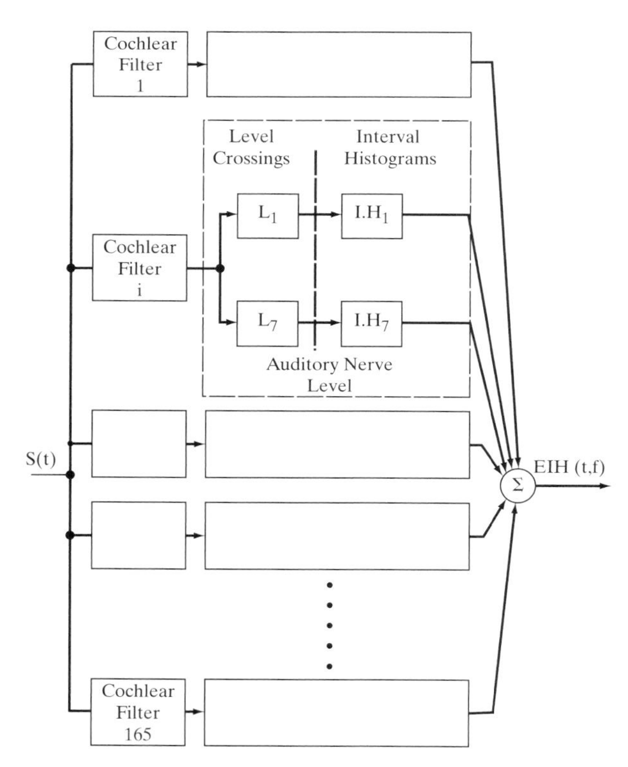

FIGURE 4.27
Block diagram of the EIH model. (After Ghitza [124].)

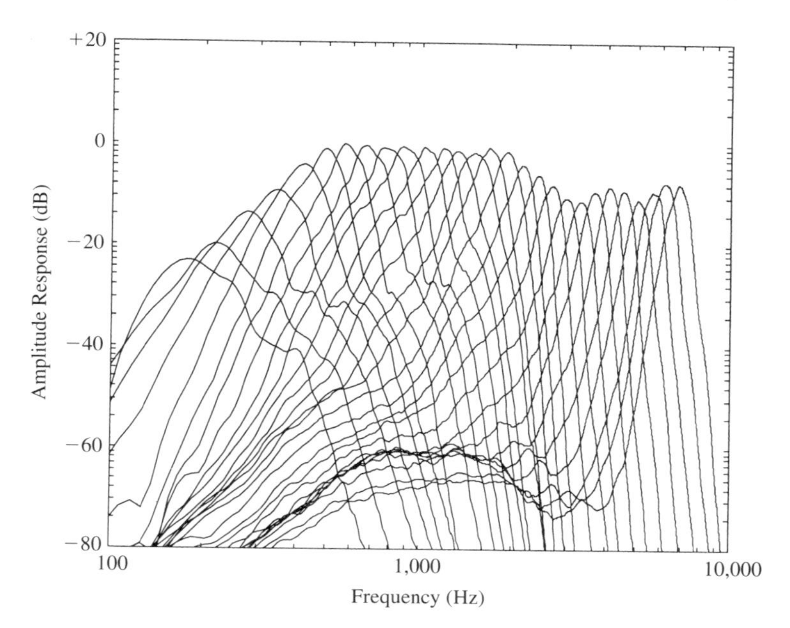

FIGURE 4.28
Frequency response curves of a cat's basilar membrane. (After Ghitza [124].)

The corresponding cochlear filters are based on actual neural tuning curves for cats. The amplitude responses of 28 of these filters (i.e., about 1 in 8 from the EIH model) are shown in Figure 4.28. The phase characteristics of these filters is a minimum phase, and the relative gain, measured at the center frequency of the filter, reflects the corresponding value of the cat's middle ear transfer function.

The processing following the cochlear filter bank is an array of level-crossing detectors that models the motion-to-neural activity transduction of the hair cell mechanisms. The detection levels of each detector are pseudo-randomly distributed (based on measured distributions of level firings), thereby simulating the variability of fiber diameters and their synaptic connections.

The collective output of the level-crossing detectors represents the discharge activity of the auditory nerve fibers. The level-crossing patterns represent the auditory nerve activity, which, in turn, is the input to a second, more central stage of neural processing, which gives the overall EIH. Conceptually, the EIH is a measure of the spatial extent of coherent neural activity across the simulated auditory nerve. Mathematically, it is an estimate of the short-term probability density function of the reciprocal of the intervals between successive firings, measured over the entire simulated auditory nerve in a characteristic frequency-dependent time-frequency zone.

Like the other auditory models described in this section, the EIH model can provide much insight into the spatio-temporal properties of neurally transduced speech signals.

4.5.5 Lessons from Auditory Models

Each of the auditory models discussed in this section has a unique perspective on how to integrate perceptual features of speech into the auditory stream as processed by a model of neural transduction. However, in many ways, the similarities between model features far outweigh the differences, and point to the aspects of speech perception that are essential for high performance speech applications, including the following:

- a non-uniform frequency scale for modeling speech spectra (on either a mel or a Bark scale);
- a logarithmic compression of spectral amplitudes in order to match equal loudness perception;
- some type of AGC to match the masking properties of speech and noise in critical bands;
- both short-time spectral and long-time temporal analysis features in order to characterize a range of speech unit properties, ranging from phoneme-length sounds all the way to syllables.

Throughout this book there are many examples of how knowledge about speech perception is incorporated into the various speech representations (presented in Chapters 6–9), in algorithms for estimating speech parameters (presented in Chapter 10), and in speech applications (presented in Chapters 11–14).

4.6 HUMAN SPEECH PERCEPTION EXPERIMENTS

In the preceding sections of this chapter we discussed some basic physical properties of sound and their corresponding psychophysical attributes that can be measured in sound perception experiments. This discussion contributes to achieving our overarching goal of understanding how knowledge of the processes of speech production and perception relates to the understanding, intelligibility, and naturalness of speech that has been processed, analyzed, synthesized, corrupted by noise and other background disturbances, spectrally distorted by various transmission systems, and finally transmitted over a medium that is possibly fading, highly reverberant, non-linear, or dispersive. We cannot tackle all of these factors at once; in fact it will be difficult to do much more than describe the results of a few simple experiments that attempt to provide some sense of the impact of broadband white noise on the perception of phonemes and words. The insight that we gain from these speech perception, intelligibility, and naturalness experiments should enable us to better understand the constraints associated with the design and implementation of practical speech analysis and synthesis systems.

4.6.1 Sound Perception in Noise

Figure 4.29 shows the measured human-listener confusion matrix for a set of 16 American English consonants embedded in noise at a signal-to-noise ratio (SNR) of +12 dB (i.e., the average signal energy was a factor of 4 greater than the average noise

	p	*t*	*k*	*f*	*θ*	*s*	*ʃ*	*b*	*d*	*g*	*v*	*ð*	*z*	*ʒ*	*m*	*n*
p	240		41	2	1											
t	1	252	1	1						1						
k	18	3	219													
f				225	24			5			2					
θ	9		1	69	185			3				1				
s						232										
ʃ							236									
b				1				242			24	12	1			
d									213	22			1			
g				1					33	203		3				
v								6			171	30			1	
ð				1				1		3	22	208	4			1
z									2	4	1	7	238			
ʒ														244		
m												1			274	1
n																252

FIGURE 4.29

Confusion matrix for SNR=+12 dB for 16 American English consonants. (After Miller and Nicely [243]. © [1955] Acoustical Society of America.)

power) [243]. The consonants were presented in a CV format where C was one of the 16 consonants shown along the rows and columns of Figure 4.29 and V was the vowel /AA/. The row-column entry in the confusion matrix was the number of times that the consonant in a given row was identified as one of the 16 consonants of the columns. Hence an ideal matrix, with perfect human perception, would consist of a diagonal with no off-diagonal entries. The actual confusion matrix of Figure 4.29 is not quite diagonal, but almost all of the off-diagonal entries are contained within a group of sub-matrices. The various sub-matrices include the three unvoiced stop consonants (/p/, /t/, and /k/), the four unvoiced fricatives (/f/, /θ/, /s/, and /ʃ/), the three voiced stop consonants (/b/, /d/, and /g/), the four voiced fricatives (/v/, /ð/, /z/, and /ʒ/), and two of the three nasal consonants (/m/ and /n/). (The third nasal consonant (/ŋ/) was not included since it cannot appear in the CV environment of the test experiment.) What Miller and Nicely concluded from the data of Figure 4.29 was that in a high SNR environment (the +12 dB case in this figure), almost all consonantal confusions are in the *place* of articulation of the sound, but not in the *manner* of articulation of the sound. Thus unvoiced stop consonants were sometimes confused with other unvoiced stop consonants (64 times versus 711 times when the consonant was correctly perceived); they rarely were confused with consonants of a class produced in a different manner (five times). Similar statements can be made about each of the five consonant classes. Thus we conclude that small amounts of noise are sufficient to lead to a low level of confusion as to consonant place of articulation but almost never to an error in the consonant

	p	*t*	*k*	*f*	*θ*	*s*	*ʃ*	*b*	*d*	*g*	*τ*	*δ*	*z*	3	*m*	*n*
p	80	43	64	17	14	6	2	1	1			1				
t	71	84	55	5	9	3	8	1				1	2			
k	66	76	107	12	8	9	4					1				
f	18	12	9	175	48	11	1	7	2	1	2	2				
θ	19	17	16	104	65	32	7	5	4	5	6	4	5			
s	8	5	4	23	39	107	45	4	2	3	1	1	3	2		
ʃ	1	6	3	4	6	29	195		3							
b	1			5	4	4		136	10	9	47	16	6	1		
d							8	5	80	45	11	20	20	26		
g					2			3	63	66	3	19	37	56		
τ				2		2		48	5	5	145	45	12			
δ					6			31	6	17	86	56	21	5		
z					1	1	1	7	20	27	16	28	94	44		
3								1	26	18	3	8	45	129		
m	1							4			4	1	3		17	
n					4			1	5	2		7	1	6	4	

FIGURE 4.30
Confusion matrix for SNR=−6 dB for 16 American English consonants. (After Miller and Nicely [243]. © [1955] Acoustical Society of America.)

manner of articulation. Experiments of this type support the notion that distinctive features play a major role in the higher-level processes of decoding speech utterances.

The obvious next question is what happens to the consonant confusion matrix when the noise level increases, or equivalently when the SNR decreases. The results of the same perception experiment as that shown in Figure 4.29 for an SNR of −6 dB are shown in Figure 4.30. Here we see patterns of consonant confusion that are significantly more complicated than before. In this very low SNR environment, consonant confusions occur in both place of articulation (as before) and manner of articulation (which occurred only infrequently in the high SNR case). Thus for the set of unvoiced stop consonants, there are now 375 confusions (as opposed to 271 correct classifications) within the class of unvoiced stop consonants (versus 64 in the previous experiment), and 105 confusions with other classes of consonants (versus 5 in the previous experiment). This pattern of confusions both within the consonant class and across consonant classes shows that at low SNRs, consonants are perceived to have occurred in more than one class, differing in both place and manner of articulation from the original consonant.

4.6.2 Speech Perception in Noise

As in many other areas, our knowledge of speech perception in noisy environments is rather limited. The problem of speech perception depends on multiple factors including the individual sounds in the utterance (which we believe are perceived based on their linguistic distinctive features) as well as the predictability of the message (the syntactic features of the utterance). To understand the often overwhelming role that syntactic information plays in perceiving speech, consider the utterance beginning with the words "To be or not to be, ..." If this were the soliloquy from Hamlet (as most people would expect it to be), it would take a lot of noise to prevent most listeners from perceiving the following words as the ones that follow the soliloquy exactly, i.e., "that is the question." Similarly, if one correctly perceives the first few words as "Four score and seven ...," again most people would recognize the preamble to Lincoln's Gettysburg Address and would readily fill in the following words as "years ago, our forefathers brought forth ..."

Even semantic information can play a major role in speech perception. This concept can be illustrated in the so-called Shannon game where a listener tries to predict the next word in a spoken sentence, having heard all of the preceding words. Thus for the preamble "he went to the refrigerator and took out a," words like /plum/ and /potato/ are far more likely than words like /book/ and /painting/, although it certainly would be grammatically correct (but semantically unlikely) to have the word /book/ be the next word in the sentence.

A set of experiments was performed by Miller, Heise, and Lichten [242] where they played a series of nonsense monosyllables, words in sentences (with some context), and digits to a group of listeners, at various SNRs from 16 dB to −16 dB. For each such test they recorded and scored the listener's accuracy for each word category and each SNR. Their results are shown in Figure 4.31. The group of 10 digits was perceived the best at each SNR, followed by words in sentences, and finally by the nonsense monosyllables, showing clearly that context is a major contributor to speech perception in noisy environments. Miller et al. [242] found that the SNR at which 50%

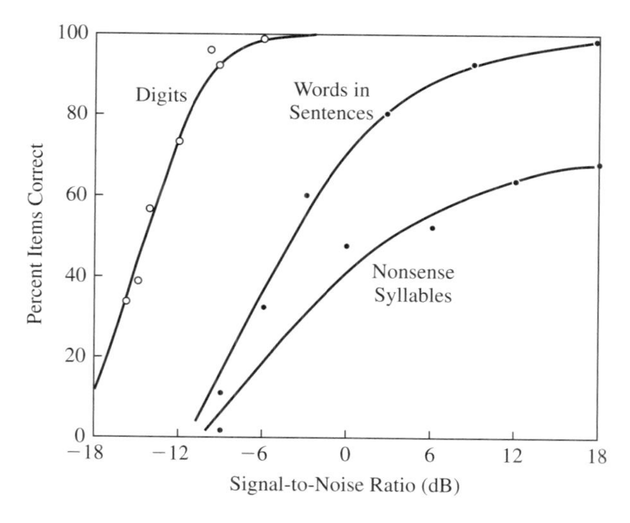

FIGURE 4.31
Intelligibility scores for different types of spoken material as a function of SNR. (After Miller, Heise, and Lichten [242].)

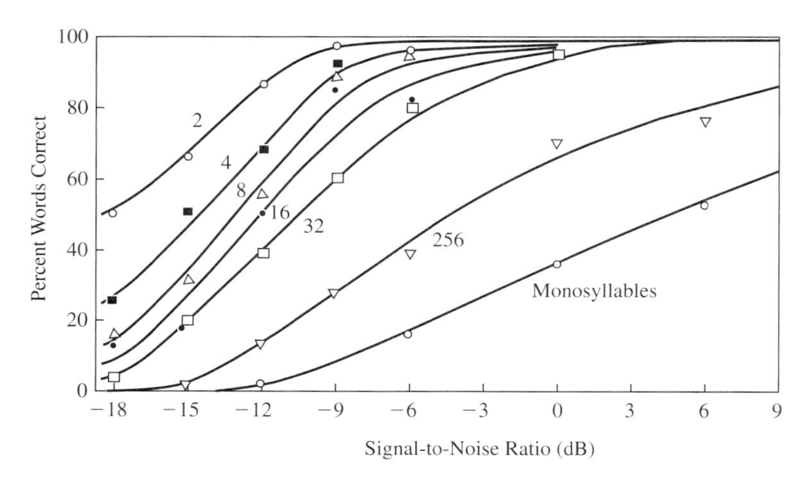

FIGURE 4.32
Effects of vocabulary size upon the intelligibility of monosyllabic words. (After Miller, Heise, and Lichten [242].)

of the perceived input was correct was −14 dB for the digits (choice of 1 out of 10), −4 dB for the words in sentences, and +3 dB for the nonsense syllables.

Finally Figure 4.32 shows a plot of the effects of noise and vocabulary size on the intelligibility of monosyllabic words. The results are shown for varying size vocabularies of monosyllables including sets with 2, 4, 8, 16, 32, 256, and 1024 monosyllables. It can be seen that the context of the small vocabulary tests greatly helps the intelligibility of the monosyllable sets, with the 2-word vocabulary being identified correctly 50% of the time, even at an SNR of −18 dB, whereas the 50% correct point is about −13 dB for 8-word vocabularies, about −9 dB for 32-word vocabularies, −3 dB for 256-word vocabularies, and about 5 dB for the largest set of 1024 monosyllables.

4.7 MEASUREMENT OF SPEECH QUALITY AND INTELLIGIBILITY

In the preceding sections we concentrated on perceptual measures of sound quality and phoneme and word intelligibility in noise in order to understand the effects of auditory processing on basic speech attributes such as frequency, intensity and bandwidth. In digital speech communication systems, the goal is to represent the speech in digital form so that it can be either stored or transmitted to a remote location. One approach is simply to sample the speech signal and represent the individual samples by finite-precision binary numbers. More efficient representations can be derived as discussed in Chapter 11 by additional digital processing of the samples. In deriving such *coded representations*, there are several conflicting goals; we generally wish to achieve the lowest possible bit rate, while minimizing computation and computation delay, and we desire that the analog signal that is reconstructed from the digital representation sound is as much like the original speech signal as possible. When the latter condition is achieved, we say that the *quality* of the representation (and possibly transmission system) is high. Sometimes quality is equated to *naturalness* in the sense that the reconstructed signal sounds like the original speaker and has no unnatural distortions. In other words, speech quality is really an inherent attribute of a speech signal coding, transmission, or processing mechanism.

A standard measure of the quality of a communication system is the SNR, defined as the average power of the input signal divided by the average power of the difference between the original signal and the signal that is reconstructed from the representation used for transmission or storage. The SNR is a single number, whose value is often considered to be proportional to the quality of a communication system, and it has also been applied as a measure of speech quality, i.e., the ability of a digital coder to provide a basis for reconstructing a speech signal that sounds close or identical to the original speech signal. The SNR is a good measure of quality in the case where the distortion introduced by the system can be modeled as an additive noise that is uncorrelated with the signal, and it has been used extensively to design and implement high quality speech coding methods (see the discussion in Chapter 11). However, over time it has been found that the SNR measure is just not good enough as a subjective measure for most speech coding methods. This is especially true for "model-based" speech coders, of the type also discussed in Chapter 11, where the waveform of the speech signal is not inherently preserved, but instead the speech signal is designed to match a speech production model. In such cases the resulting noise, defined as the difference between the coded speech and the original speech, is not an uncorrelated noise source and thus the assumptions behind the calculation of SNR are neither correct nor relevant as a measure of speech quality. Furthermore, the error signal between the coded speech and the original speech signal shows one or more of the following effects:

- clicks or transients in the signal that occur randomly in time;
- error spectra that are frequency dependent (colored) and correlated with the input signal spectrum;
- components of the error signal that are due to reverberation and echo, often from the type of model used to analyze and synthesize the coded signal;
- white noise components due to signal quantization and inherent background noise of the speech recording process;
- delay between the original speech signal and the coded version due to accumulated delay from block coding, speech processing, and transmission; hence the error signal and the original speech signal are often significantly out of alignment, rendering invalid the concept of SNR, which assumes perfect alignment;
- transmission bit errors which cause a type of shot noise in the coded error signal;
- tandem encodings which introduce levels of delay, noise, and other forms of distortion.

4.7.1 Subjective Testing

The SNR is an example of an *objective measure* of speech quality; i.e., a measurement that can be computed given the original signal and the signal reconstructed from a coded digital representation. When simple measures such as SNR fail to adequately predict the quality of a digital representation, subjective judgments of quality are often used instead. A variety of subjective testing procedures have been developed for this purpose. Such procedures involve panels of trained listeners who render judgments on subjective attributes of speech signals having varying degrees of distortion. Examples include the diagnostic acceptability measure (DAM) [409] and diagnostic rhyme test (DRT) [408], which were widely used at one time for evaluating military communication systems and low bit-rate speech coders. More recently, subjective evaluations have focused on the mean opinion score (MOS) [165]. In a typical MOS test, human listeners hear examples of speech processed with various conditions including no coding at all. For each example, the listener is required to give a rating on the following scale:

- Excellent quality–MOS=5 (coded speech is essentially equivalent in quality with original uncoded speech; speech recorded with high SNR);
- Good quality–MOS=4 (coded speech is of very good quality but distinctly below the quality and naturalness of the uncoded speech recording);
- Fair quality–MOS=3 (coded speech is acceptable for communications but is distinctly of lower quality than uncoded original signal);
- Poor quality–MOS=2 (coded speech quality is significantly degraded and barely acceptable for a voice communications system);
- Bad quality–MOS=1 (coded speech quality is unacceptable for communications).

Scores for a given condition or system are averaged over all listener responses to give an MOS score for each different coding condition.

The highest recorded scores in subjective tests using the MOS rating scale on real (not synthetic) speech are about 4.5 for natural wideband speech (bandwidth 50 Hz to 7000 Hz); 4.05 for toll quality telephone speech (essentially undegraded telephony); 3.5–4.0 for communications quality telephone speech (as available and used in cellular communication systems); and 2.0–3.5 for lower quality speech from synthesizers and low bit-rate coders (as might be used for military communications in hostile environments).

To illustrate the MOS scores that are obtained on standardized telephone bandwidth speech coders, Figure 4.33 shows three sets of plots of speech coder MOS scores for a range of coder bit rates from a high of 64 kbps (the G.711 ITU Standard or PCM coder) to a low rate of 2.4 kbps (the MELP coder from NSA) [67]. Three asymptotic curves are shown in this figure, one from 1980, one from 1990, and the third from 2000 (with individual MOS scores of coders). It can be seen that the highest MOS scores are only about 4.1 (for the G.711 coder at 64 kbps) and the lowest MOS scores are in the range 3.1 (the MELP coder at 2.4 kbps). The curve of MOS versus bit rate is fairly flat in the range from 7 kbps to 64 kbps, but falls off at rates below 7 kbps. In Chapter 11, we shall provide a detailed discussion of algorithms for speech coding and the quality levels that are attainable.

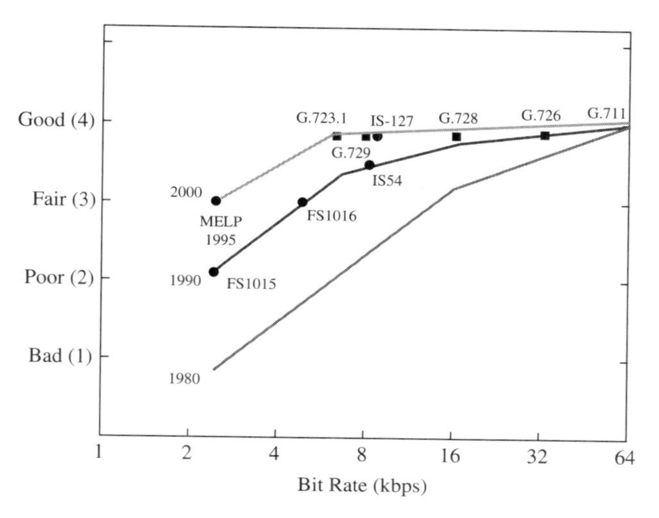

FIGURE 4.33
MOS for a range of speech coders. Labels on points denote specific digital speech coding standards. (After Cox [67].)

4.7.2 Objective Measurement of Speech Quality

A major problem with subjective tests is that they are costly and time consuming to perform, requiring, as they do, large panels of listeners trained to deliver accurate and consistent judgments of speech quality. Objective measures that accurately represent perceived quality are desired since they can be computed immediately and can therefore be used in design and development of systems, and also because they cost little to deploy. This has led to a considerable amount of research on objective measures that are more effective than SNR, but are highly correlated with subjective judgments. An example of early work of this nature was done by Barnwell [23] and colleagues [296], who developed techniques for predicting DAM scores directly by analysis of the coded speech signal. More recently, as the MOS paradigm became widely accepted, researchers turned to the design of objective measures that can extract an MOS-like score by objective comparison of the coded and uncoded speech signals. This work led eventually to the standardization of an objective measurement system called PESQ (perceptual evaluation of speech quality) [320], which is defined in the ITU-T P.862 standard recommendation. Figure 4.34 shows a block diagram of the PESQ measurement system. We do not give a completely detailed description of the method here. However, the block diagram illustrates some of the important features shared by all objective measurements of this type. First note that this system requires the "original" as reference. The first stages are designed to align the test signal to the reference signal in time and amplitude and to estimate and compensate for any spectral shaping or filtering effects, which are often not subjectively important, but can have a major effect in an objective measure. Then the two signals (processed reference and test) are passed through an auditory model of the type discussed in Section 4.5. This model consists essentially of a short-time Fourier transform, as discussed in Chapter 7,

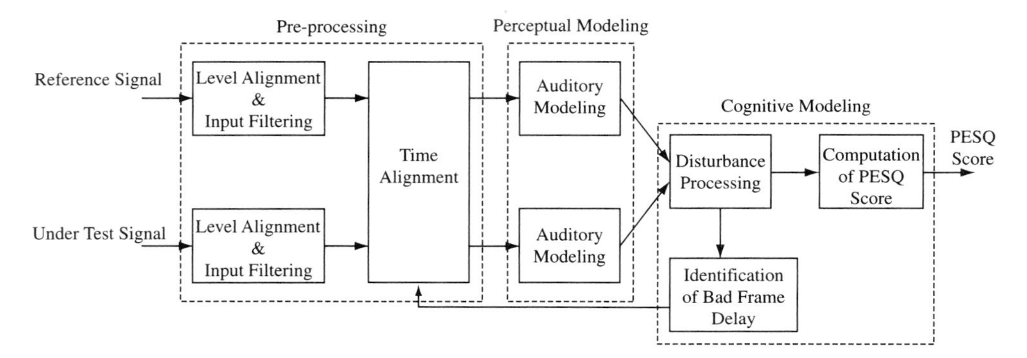

FIGURE 4.34
Perceptual evaluation of speech quality (PESQ).

with frequency warping to a critical band scale and intensity warping to a sone scale. The outputs of the auditory model analysis are compared by computing a "disturbance" between the time–frequency representations. Absolute differences that fall below masking thresholds are ignored. The disturbances are aggregated by averaging over time and frequency. The final result of the "cognitive modeling" stage is a number on the MOS scale. In order to achieve good agreement between objective and subjective measures, several parameters of the cognitive model must be trained by comparison of PESQ outputs to the results of subjective tests (i.e., MOS scores) on the same speech material. The correlation coefficients between subjective test ratings (MOS scores) and corresponding PESQ outputs range from 0.785 to 0.979.

The PESQ measure was designed for evaluation of low bit-rate narrowband coders of the type used in mobile communications and voice-over-IP networks. For wideband audio coders, a similar measure called perceptual evaluation of audio quality (PEAQ) has been recommended as ITU-R BS.1387.

4.8 SUMMARY

In this chapter we have given a brief introduction to the basic knowledge of hearing and perception of sound and speech. We showed that the ear, which is the primary physiological mechanism in hearing, acts as a sound canal, transducer, and spectrum analyzer. The cochlea, in the inner ear, acts like a multi-channel, logarithmically spaced, constant Q filter bank and does a frequency and place analysis of the incoming signal along the basilar membrane and transmits the results of the analysis to the brain via a process of IHC transduction, which are processed by the human brain. We showed that the auditory processing in the ear and the pathways to the brain inherently makes the speech robust to noise and echo, albeit in ways that are not well understood.

Since our understanding of the higher level processing steps in the brain is highly limited, most of this chapter focused on sound and speech perception in humans. We first showed that physical characteristics of sound (frequency, intensity, bandwidth, etc.) are perceived in terms of psychophysical attributes that scale very differently from the physical attributes, e.g., pitch versus frequency, loudness versus intensity. We next showed that sounds like tones or noise are able to mask other sounds (other tones or noise), thereby making it unnecessary to preserve precisely every characteristic of a speech or audio sound, while still preserving very high quality and naturalness in coding the sounds. This is the basis of modern audio coding systems such as MP3 or AAC coders. Finally we examined a few experiments on speech perception and intelligibility in noise and showed how the results pointed to key characteristics of speech that determine the degree of confusion between similar (and often not-so-similar) sounding phonemes or words. We ended the chapter with a brief discussion of a subjective measure of speech quality or naturalness, called the mean opinion score (MOS), and showed how it was used in studies of quality of speech coding methods. We also showed how knowledge of auditory processing could be incorporated into objective measures of speech quality, such as the PESQ method, which can achieve high correlation with well-established subjective measures of speech quality such as MOS scores.

PROBLEMS

4.1. Describe the major function(s) of the outer, middle, and inner ears. How do they achieve this functionality?

4.2. If the input to the ear is the periodic square wave, $s(t)$, shown in Figure P4.1, with period $T = 10$ msec, what frequencies are present at the input to the basilar membrane (assuming all processing at the outer and middle ears is linear). Show the response of the basilar membrane to the periodic square wave at the stapes end and at the apical end.

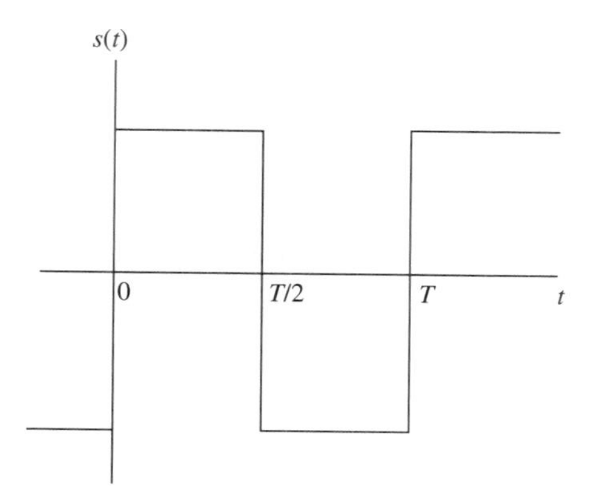

FIGURE P4.1
Periodic square wave input to ear.

4.3. Which of the following pairs of tones is perceived as the louder tone (and by how many phons):

(a) 20 dB IL at 1000 Hz or 20 dB IL at 500 Hz

(b) 40 dB IL at 200 Hz or 30 dB IL at 2000 Hz
(c) 50 dB IL at 100 Hz or 50 dB IL at 1000 Hz.

4.4. What is the perceived pitch (in mels) and what is the critical bandwidth (in Hz) for each of the following tones:

(a) 100 Hz
(b) 200 Hz
(c) 400 Hz
(d) 1000 Hz
(e) 2000 Hz
(f) 4000 Hz
(g) 10,000 Hz.

4.5. At a signal-to-noise (S/N) ratio of 12 dB, which of the following pairs of words are highly likely to be confused in listening tests (and why):

(a) pick-tick
(b) peek-seek
(c) take-bake
(d) king-sing
(e) go-doe
(f) van-than
(g) map-nap
(h) go-no.

Repeat this exercise for the case when the S/N ratio is −6 dB.

4.6. (MATLAB Exercise) Write a MATLAB program to send a sequence of speech files to the A-to-D converter. To enable your program to be widely applicable, follow the steps listed below:

- first create a mat file containing the following information:
  1. `N`=number of speech files to be played sequentially
  2. `files`=list of ASCII speech filenames to be played
  3. `lmax`=maximum number of samples to be played for any of the files
- create a beep signal (a 1 kHz tone of a specified level and duration)
- load the mat file and play out the sequence of files using a beep between each of the files.

4.7. The purpose of this exercise is to enable you to test your own hearing. By searching the Internet, you can find several websites that offer an online system for measuring hearing thresholds. On such website is maintained at the University of New South Wales at URL http://www.phys.unsw.edu.au/jw/hearing.html.

(a) Begin by reading the instructions carefully before listening to any tones. In particular, be careful NOT to listen to the loudest tones first because hearing damage could result. Your task is to determine the lowest level sound that you can hear at a given frequency. You will need a quiet location and a good set of headphones, preferably ones that cover the outer ear completely.

(b) Follow the instructions to measure your threshold of hearing as a function of frequency. The result will not be as accurate as a measurement by a trained audiologist using equipment designed for this purpose, but it will serve to provide some insight into how such measurements are made. What are some of the sources of error in your experiment?

(c) Plot your personal threshold of hearing on the graph in Figure 4.14. How does your measured hearing threshold compare to the baseline hearing threshold shown in Figure 4.14?

CHAPTER 5

Sound Propagation in the Human Vocal Tract

5.1 THE ACOUSTIC THEORY OF SPEECH PRODUCTION

Chapter 3 presented a qualitative description of the sounds of speech and the ways in which they are produced by a human vocal tract. In this chapter we will consider mathematical representations of the process of speech production [101, 105]. Such mathematical representations serve as the basis for the analysis and synthesis of speech and will be used throughout the remainder of this book.

5.1.1 Sound Propagation

Sound is vibration and vibration produces sound. Sound waves are created by vibration and are propagated in air or other media by vibrations of the particles of the media. Thus, the laws of physics are the basis for describing the generation and propagation of sound in the vocal system. In particular, the fundamental laws of conservation of mass, conservation of momentum, and conservation of energy along with the laws of thermodynamics and fluid mechanics, all apply to the compressible, low viscosity fluid (air) that is the medium for sound propagation in speech. Using these physical principles, a set of partial differential equations can be obtained that describe the motion of air in the vocal system [34, 248, 289, 291, 371]. A detailed acoustic theory must consider the effects of the following:

- time and spatial variation of the vocal tract shape (we will consider both spatially constant shapes and spatially varying shapes, but we will not deal with time-varying shapes in this chapter);
- losses due to heat conduction and viscous friction at the vocal tract walls (we will consider both lossless and lossy transmission of sound in the vocal tract);
- softness of the vocal tract walls (sound absorption is another source of energy loss);
- radiation of sound at the lips (radiation of sound at the lips or nostrils is an important source of loss and frequency shaping);

- nasal coupling (the need to model nasal coupling to the vocal tract complicates the simple tube models that we will develop, as it leads to multi-branch solutions);
- excitation of sound in the vocal tract (to properly handle sound excitation sources, we need to fully describe the way the vocal source is coupled to the vocal tract, as well as the source–system interactions that result from the coupling).

The formulation and solution of the equations describing sound generation and propagation is extremely difficult except under very simple assumptions about vocal tract shape and energy losses in the vocal system. A completely detailed acoustic theory incorporating all the above effects is beyond the scope of this chapter, and indeed, such a theory is not generally available. It will be sufficient to survey these factors, providing references to details when available, and qualitative discussions when suitable references are unavailable. Our primary goal in this chapter is to motivate discrete-time system models for the speech signal. To do this, we will simplify the physics as much as possible and see what we can learn about the mechanics of sound propagation in the human vocal tract. We will see that certain types of digital filters are remarkably successful at simulating sound transmission in the human vocal system.

The simplest physical configuration that has a useful interpretation in terms of the speech production process is depicted in Figure 5.1a. Speech corresponds to variations of air flow in such a system. The vocal tract is modeled as a tube of non-uniform, time-varying, cross-section area. For frequencies corresponding to wavelengths that are long compared to the dimensions of the vocal tract (less than about 4000 Hz), it is reasonable to assume plane wave propagation along the axis of the tube. A further simplifying assumption is that there are no losses due to viscosity or thermal conduction, either in the bulk of the fluid or at the walls of the tube. With these assumptions, and the laws of conservation of mass, momentum, and energy, Portnoff [289] showed

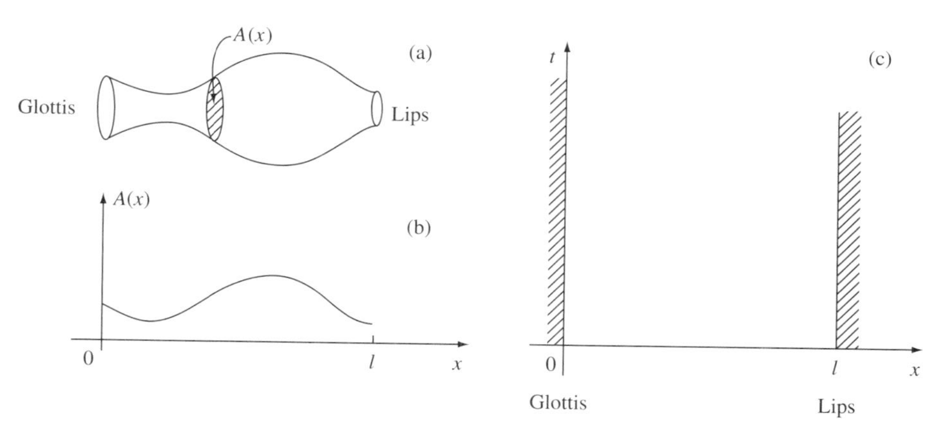

FIGURE 5.1
(a) Schematic vocal tract; (b) corresponding area function; (c) $x-t$ plane for solution of wave equation.

that sound waves in the tube satisfy the following pair of equations:

$$-\frac{\partial p}{\partial x} = \rho \frac{\partial (u/A)}{\partial t}, \tag{5.1a}$$

$$-\frac{\partial u}{\partial x} = \frac{1}{\rho c^2}\frac{\partial (pA)}{\partial t} + \frac{\partial A}{\partial t}, \tag{5.1b}$$

where

$p = p(x,t)$ is the variation in sound pressure in the tube at position x and time t
$u = u(x,t)$ is the variation in volume velocity flow at position x and time t
$\rho =$ the density of air in the tube
$c =$ the velocity of sound (approximately 35,000 cm/sec)
$A = A(x,t) =$ the "area function" of the tube; i.e., the value of cross-sectional area normal to the axis of the tube as a function of distance along the tube and as a function of time

A similar set of equations was derived by Sondhi [371].

Closed form solutions to Eqs. (5.1a) and (5.1b) are not possible except for the simplest configurations. Numerical solutions can be obtained, however. A complete solution of the differential equations requires that pressure and volume velocity be found for values of x and t in the region bounded by the glottis and the lips. To obtain the solution, boundary conditions must be given at each end of the tube. At the lip end, the boundary condition must account for the effects of sound radiation. At the glottis (or possibly some internal point), the boundary condition is imposed by the nature of the excitation.

In addition to the boundary conditions, the vocal tract area function, $A(x,t)$, must be known for all $x, 0 \leq x \leq l$ and for all t. Figure 5.1b shows the area function for the tube in Figure 5.1a, at a particular time, and Figure 5.1c shows the (x,t) plane where Eqs. (5.1a) and (5.1b) must be solved inside the indicated boundaries. For continuant (sustained) sounds, it is reasonable to assume that $A(x,t)$ does not change with time; however, this is not the case for non-continuants. Detailed measurements of $A(x,t)$ are extremely difficult to obtain, even for continuant sounds. One approach to such measurements is through the use of X-ray images. Fant [101] and Perkell [286] provide some data of this form; however, such measurements can only be obtained on a limited scale. Figure 3.1 illustrates the type of X-ray images that can be obtained. More recently, Narayanan and collaborators [45, 251] have used MRI imaging methods to measure time-varying vocal tract shapes. Figure 3.2 shows a single frame of an MRI image sequence. As shown, the boundaries of the vocal tract can be traced by image processing techniques, and the positions of various articulators (e.g., the tongue, the lips, the jaw, the velum) can be tracked. Another approach is to infer the vocal tract shape from acoustic measurements. Sondhi and Gopinath [372] have described an approach that involves exciting the vocal tract by an external source. Atal [8] and Wakita [410, 411] have described methods of estimating $A(x,t)$ directly from the speech signal produced under normal speaking conditions. These methods are based on the technique of linear predictive speech analysis and will be discussed in Chapter 9.

The complete solution of Eqs. (5.1a) and (5.1b) is very complicated [289] even if $A(x,t)$ is accurately determined. Fortunately, it is not necessary to solve the equations under the most general conditions to obtain an insight into the nature of the speech signal. A variety of reasonable approximations and simplifications can be invoked to make the solution feasible.

5.1.2 Example: Uniform Lossless Tube

Useful insight into the acoustics of speech production can be obtained by considering a very simple model in which the vocal tract area function, $A(x,t)$, is assumed constant in both x and t (time invariant with uniform cross-section). This configuration is approximately correct for the neutral vowel /UH/. We shall examine this model first, returning later to examine more realistic models. Figure 5.2a depicts a tube of uniform cross-section being excited by an ideal source of volume velocity flow. This ideal source is represented by a piston that can be caused to move in any desired fashion, independent of pressure variations in the tube. A further assumption is that, at the open end of the tube, there are no variations in air pressure—only in volume velocity. These are obviously gross simplifications that, in fact, are impossible to achieve in practice; however, this is still a useful example because the basic approach of the analysis and the essential features of the resulting solution have much in common with more realistic models. Furthermore, we shall show later in this chapter that more general models can be constructed by concatenation of a series of such uniform tubes.

If $A(x,t)=A$ is a constant, independent of x and t, then the partial differential equations (5.1a) and (5.1b) reduce to the form

$$-\frac{\partial p}{\partial x}=\frac{\rho}{A}\frac{\partial u}{\partial t}, \tag{5.2a}$$

$$-\frac{\partial u}{\partial x}=\frac{A}{\rho c^2}\frac{\partial p}{\partial t}. \tag{5.2b}$$

By differentiating Eq. (5.2a) with respect to t and Eq. (5.2b) with respect to x and eliminating $\partial^2 p/\partial x\partial t$, we obtain the alternative equation,

$$\frac{\partial^2 u}{\partial x^2}=\frac{1}{c^2}\frac{\partial^2 u}{\partial t^2}, \tag{5.3a}$$

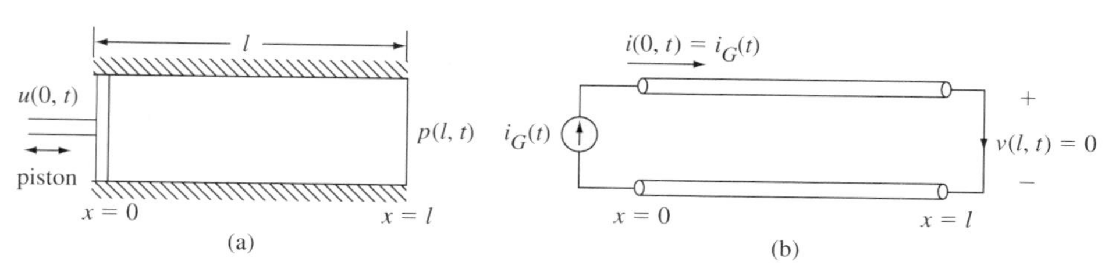

FIGURE 5.2
(a) Uniform lossless tube with ideal terminations; (b) corresponding electrical transmission line analogy.

which is in the classical wave equation form. A similar manipulation leads to

$$\frac{\partial^2 p}{\partial x^2} = \frac{1}{c^2}\frac{\partial^2 p}{\partial t^2}. \tag{5.3b}$$

It can be shown (see Problem 5.2) that the solutions to Eqs. (5.2a) and (5.2b) or Eqs. (5.3a) and (5.3b) have the form

$$u(x,t) = [u^+(t - x/c) - u^-(t + x/c)], \tag{5.4a}$$

$$p(x,t) = \frac{\rho c}{A}[u^+(t - x/c) + u^-(t + x/c)]. \tag{5.4b}$$

In Eqs. (5.4a) and (5.4b), the functions $u^+(t - x/c)$ and $u^-(t + x/c)$ can be interpreted as traveling waves in the positive and negative directions respectively. This is illustrated in Figure 5.3, which shows a positive-going wave at times t_0 and $t_1 > t_0$. It can be seen that the positive-going wave has moved forward by a distance $d = c(t_1 - t_0)$ during the time interval $(t_1 - t_0)$. The relationship between the forward and backward traveling waves is determined by the boundary conditions.

In the theory of electrical transmission lines, the voltage $v(x,t)$ and current $i(x,t)$ on a uniform lossless line satisfy the equations

$$-\frac{\partial v}{\partial x} = L\frac{\partial i}{\partial t}, \tag{5.5a}$$

$$-\frac{\partial i}{\partial x} = C\frac{\partial v}{\partial t}, \tag{5.5b}$$

where L and C are the inductance and capacitance per unit length, respectively. Thus the theory of lossless uniform electric transmission lines [2, 283, 397] applies directly to the uniform acoustic tube if we make the analogies shown in Table 5.1.

Using these analogies, the uniform acoustic tube, terminated as shown in Figure 5.2a, behaves in a way that is identical to that of a lossless uniform electrical

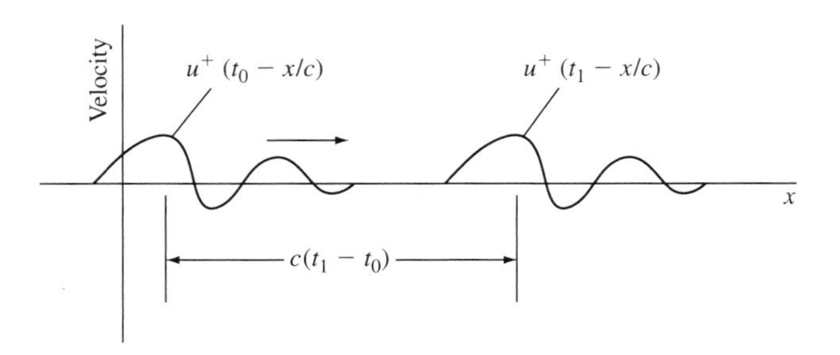

FIGURE 5.3
Illustration of a positive-going wave at times $t = t_0$ and $t = t_1$, showing the wave has traveled a distance $d = c(t_1 - t_0)$ during the time interval $(t_1 - t_0)$.

TABLE 5.1 Analogies between acoustic and electric quantities.

Acoustic Quantity	Analogous Electric Quantity
p: pressure	v: voltage
u: volume velocity	i: current
ρ/A: acoustic inductance	L: inductance
$A/(\rho c^2)$: acoustic capacitance	C: capacitance

transmission line terminated, as depicted in Figure 5.2b, in a short circuit ($v(l,t) = 0$) at one end and excited by a current source ($i(0,t) = i_G(t)$) at the other end.

Frequency-domain representations of linear systems, such as transmission lines and circuits, are exceedingly useful. Similarly, we can obtain similar representations of the lossless uniform tube. The frequency-domain representation of this model is obtained by assuming a boundary condition at $x = 0$ of the form

$$u(0,t) = u_G(t) = U_G(\Omega)e^{j\Omega t}. \tag{5.6}$$

That is, the tube is excited by a complex exponential variation of volume velocity of radian frequency, Ω, and complex amplitude, $U_G(\Omega)$. Since Eqs. (5.2a) and (5.2b) are linear and time invariant, the corresponding solution for $u^+(t - x/c)$ and $u^-(t + x/c)$ must be of the form

$$u^+(t - x/c) = K^+ e^{j\Omega(t-x/c)}, \tag{5.7a}$$

$$u^-(t + x/c) = K^- e^{j\Omega(t+x/c)}. \tag{5.7b}$$

Substituting these equations into Eqs. (5.4a) and (5.4b) and applying the acoustic short-circuit boundary condition

$$p(l,t) = 0 \tag{5.8}$$

at the lip end of the tube and Eq. (5.6) at the glottis end, we can solve for the unknown constants K^+ and K^-, from the following equations:

$$u(0,t) = U_G(\Omega)e^{j\Omega t} = K^+ e^{j\Omega t} - K^- e^{j\Omega t}, \tag{5.9a}$$

$$p(l,t) = 0 = \frac{\rho c}{A}\left[K^+ e^{j\Omega(t-l/c)} + K^- e^{j\Omega(t+l/c)}\right]. \tag{5.9b}$$

The result is

$$K^+ = U_G(\Omega)\frac{e^{2j\Omega l/c}}{1 + e^{j\Omega 2l/c}}, \tag{5.10a}$$

$$K^- = -\frac{U_G(\Omega)}{1 + e^{j\Omega 2l/c}}. \tag{5.10b}$$

Therefore, the sinusoidal steady state solutions for $u(x,t)$ and $p(x,t)$ are

$$u(x,t) = \left[\frac{e^{j\Omega(2l-x)/c} + e^{j\Omega x/c}}{1 + e^{j\Omega 2l/c}}\right] U_G(\Omega)e^{j\Omega t}$$

$$= \frac{\cos[\Omega(l-x)/c]}{\cos[\Omega l/c]} U_G(\Omega)e^{j\Omega t}, \tag{5.11a}$$

$$p(x,t) = \frac{\rho c}{A}\left[\frac{e^{j\Omega(2l-x)/c} - e^{j\Omega x/c}}{1 + e^{j\Omega 2l/c}}\right] U_G(\Omega)e^{j\Omega t}$$

$$= jZ_0 \frac{\sin[\Omega(l-x)/c]}{\cos[\Omega l/c]} U_G(\Omega)e^{j\Omega t}, \tag{5.11b}$$

where

$$Z_0 = \frac{\rho c}{A} \tag{5.12}$$

is, by analogy to the electrical transmission line, called the *characteristic acoustic impedance* of the tube.

An alternative approach avoids a solution for the forward and backward traveling waves by expressing $p(x,t)$ and $u(x,t)$ for a complex exponential excitation directly as[1]

$$p(x,t) = P(x,\Omega)e^{j\Omega t}, \tag{5.13a}$$

$$u(x,t) = U(x,\Omega)e^{j\Omega t}. \tag{5.13b}$$

Substituting these solutions into Eqs. (5.2a) and (5.2b) gives the following ordinary differential equations relating the complex amplitudes:

$$-\frac{dP}{dx} = ZU, \tag{5.14a}$$

$$-\frac{dU}{dx} = YP, \tag{5.14b}$$

where

$$Z = j\Omega\frac{\rho}{A} \tag{5.15}$$

[1] Henceforth, our convention will be to denote time-domain variables with lower case letters (e.g., $u(x,t)$) and their corresponding frequency-domain representations with capital letters (i.e., $U(x,\Omega)$).

can be called the *acoustic impedance* per unit length and

$$Y = j\Omega \frac{A}{\rho c^2} \tag{5.16}$$

is the *acoustic admittance* per unit length. The differential equations (5.14a) and (5.14b) have general solutions of the form

$$P(x, \Omega) = Ae^{\gamma x} + Be^{-\gamma x}, \tag{5.17a}$$

$$U(x, \Omega) = Ce^{\gamma x} + De^{-\gamma x}, \tag{5.17b}$$

$$\gamma = \sqrt{ZY} = j\Omega/c. \tag{5.17c}$$

The unknown coefficients can be found by applying the boundary conditions

$$P(l, \Omega) = 0, \tag{5.18a}$$

$$U(0, \Omega) = U_G(\Omega). \tag{5.18b}$$

The resulting equations for $u(x, t)$ and $p(x, t)$ are, as they should be, the same as Eqs. (5.11a) and (5.11b). Equations (5.11a) and (5.11b) express the relationship between the sinusoidal volume velocity source and the pressure and volume velocity at any point x in the tube. In particular, if we consider the relationship between the volume velocity source (where $x = 0$) and the volume velocity at the lips (where $x = l$), we obtain from Eq. (5.11a),

$$u(l, t) = \frac{1}{\cos(\Omega l/c)} U_G(\Omega) e^{j\Omega t} = U(l, \Omega) e^{j\Omega t} \tag{5.19}$$

where $U(l, \Omega)$ is defined as the complex amplitude of the volume velocity at the lip end of the tube. The ratio

$$V_a(j\Omega) = \frac{U(l, \Omega)}{U_G(\Omega)} = \frac{1}{\cos(\Omega l/c)} \tag{5.20}$$

is the frequency response relating the input and output volume velocities. This function is plotted in Figure 5.4a for $l = 17.5$ cm and $c = 35{,}000$ cm/sec. Replacing $j\Omega$ by s, we obtain the Laplace transform system function

$$V_a(s) = \frac{1}{\cosh(sl/c)} = \frac{2e^{-sl/c}}{1 + e^{-s2l/c}}. \tag{5.21}$$

Writing $V_a(s)$ in the second form in Eq. (5.21) suggests the feedback system interpretation of Figure 5.5 where we see a forward transmission path that is a pure delay (represented by the factor $e^{-sl/c}$) corresponding to the time $\tau = l/c$ seconds required to traverse a distance l at transmission speed c, and a feedback path with a gain of -1 to satisfy the short-circuit boundary condition and a backward transmission path that

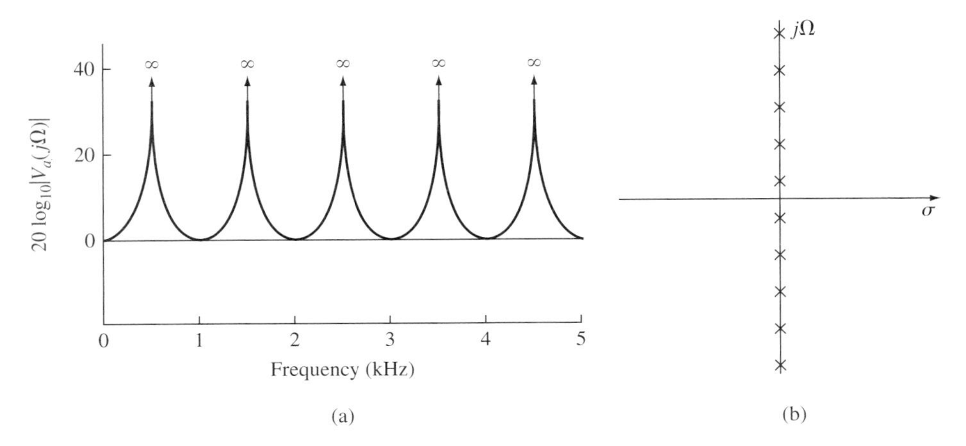

FIGURE 5.4
(a) Log magnitude frequency response; and (b) pole locations for a uniform lossless tube.

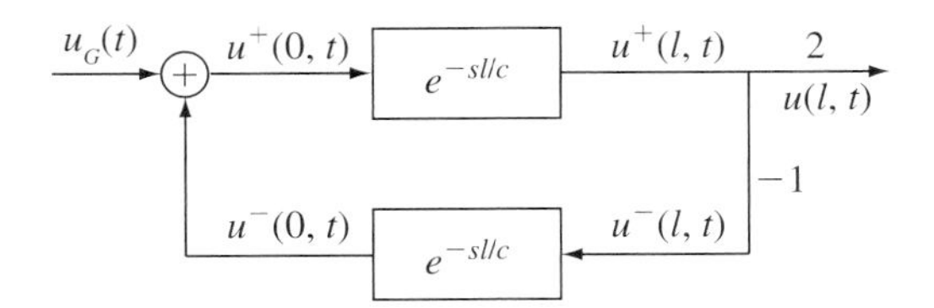

FIGURE 5.5
System interpretation of uniform vocal tract transmission function.

is again a pure delay (with the delay factor $e^{-sl/c}$) and finally a forward path gain of 2. A simple interpretation of the transmission system of Figure 5.5 is as follows: the volume velocity source travels in the forward direction (from the glottis to the lips) with a delay of l/c for each traversal of the vocal tract; the signal is reflected at the lips and travels in the reverse direction, again with a delay of l/c for each backward traversal of the vocal tract; this cycle of reflection at the lips and at the glottis is repeated ad infinitum, leading to the overall transfer function of Eq. (5.21).

The system function $V_a(s)$ has an infinite number of poles equally spaced on the $j\Omega$ axis. The poles are the values of s such that the denominator $(1 + e^{-s2l/c}) = 0$. These values are easily shown to be

$$s_n = \pm j\left[\frac{(2n+1)\pi c}{2l}\right], \quad n = 0, \pm 1, \pm 2, \ldots. \tag{5.22}$$

Thus, for the lossless acoustic tube terminated with an acoustic short circuit, the poles are all on the $j\Omega$ axis of the s-plane as shown in Figure 5.4b. If losses were included, the poles would move off the $j\Omega$ axis into the left half-plane. The poles of the system function of a linear time-invariant system are the natural frequencies (or eigen frequencies) of the system, and they occur at the set of frequencies $F_n = 500(2n+1)$, $n = 0, 1, \ldots$ or $F = 500, 1500, 2500, \ldots$ Hz. The poles also correspond to resonance frequencies of the frequency response of the system. These resonant frequencies are the formant frequencies of the speech signal. As we shall see, similar resonance effects will be observed regardless of the vocal tract shape and also when more realistic energy loss mechanisms are introduced into the model.

At this point it is helpful to recall that the frequency response function allows us to determine the response of the system not only to sinusoids, but to arbitrary inputs through the use of Fourier analysis. Indeed, Eq. (5.20) has the more general interpretation that $V_a(j\Omega)$ is the ratio of the Fourier transform of the volume velocity at the lips (output) to the Fourier transform of the volume velocity at the glottis (input or source). Thus the frequency response is a convenient characterization of the model for the vocal system. Now that we have demonstrated a method for determining the frequency response of acoustic models for speech production by considering the simplest possible model, we can begin to consider more realistic models.

5.1.3 Effects of Losses in the Vocal Tract

The equations of motion for sound propagation in the vocal tract that we have given in Section 5.1.2 were derived under the assumption of no energy loss in the tube. In reality, energy will be lost as a result of viscous friction between the air and the walls of the tube, heat conduction through the walls of the tube, and vibration of the tube walls. To include these effects, we might attempt to return to the basic laws of physics and derive a new set of equations of motion. This is made extremely difficult by the frequency dependence of these losses. A less rigorous approach is to modify the frequency-domain representation of the equations of motion [105, 289]. We will follow this approach in this section.

Losses Due to Soft Walls

First consider the effects of the vibration of the vocal tract wall. The variations of air pressure inside the tract will cause the walls to experience a varying force. If the walls are elastic, the cross-sectional area of the tube will change depending upon the pressure in the tube. Assuming that the walls are "locally reacting" [248, 289], the area $A(x,t)$ will be a function of $p(x,t)$. Since the pressure variations are very small, the resulting variation in cross-sectional area can be treated as a small perturbation of the "nominal" area; i.e., we can assume that

$$A(x,t) = A_0(x,t) + \delta A(x,t), \tag{5.23}$$

where $A_0(x,t)$ is the nominal area and $\delta A(x,t)$ is a small perturbation. This is depicted in Figure 5.6. Because of the mass and elasticity of the vocal tract wall, the relationship between the area perturbation, $\delta A(x,t)$, and the pressure variations, $p(x,t)$, can be modeled by a differential equation of the form

$$m_w \frac{d^2(\delta A)}{dt^2} + b_w \frac{d(\delta A)}{dt} + k_w(\delta A) = p(x,t), \tag{5.24}$$

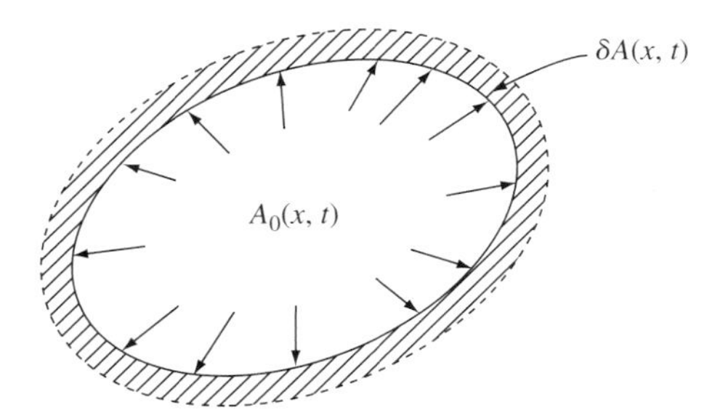

FIGURE 5.6
Illustration of the effects of wall vibration.

where

$m_w(x)$ is the mass/unit length of the vocal tract wall
$b_w(x)$ is the damping/unit length of the vocal tract wall
$k_w(x)$ is the stiffness/unit length of the vocal tract wall.

Substituting Eq. (5.23) into Eqs. (5.1a) and (5.1b) and neglecting second-order terms in the quantities u/A and pA, we obtain the equations

$$-\frac{\partial p}{\partial x} = \rho \frac{\partial(u/A_0)}{\partial t}, \tag{5.25a}$$

$$-\frac{\partial u}{\partial x} = \frac{1}{\rho c^2}\frac{\partial(pA_0)}{\partial t} + \frac{\partial A_0}{\partial t} + \frac{\partial(\delta A)}{\partial t}. \tag{5.25b}$$

Thus, sound propagation in a locally reacting soft-walled tube, such as the vocal tract, is described by the set of equations (5.23), (5.24), (5.25a), and (5.25b). These equations must be solved simultaneously for $p(x,t)$ and $u(x,t)$ given $A_0(x,t)$ and appropriate boundary conditions.

To understand the effect of losses due to soft walls, we can solve for a frequency-domain representation as before. To do so, we consider a soft-walled tube with time-invariant nominal area function $A_0(x)$ as would be an appropriate model for a continuant sound such as a vowel. The excitation is assumed to be an ideal complex volume velocity source; i.e., the boundary condition at the glottis is

$$u(0,t) = U_G(\Omega)e^{j\Omega t}. \tag{5.26}$$

Because the differential equations (5.24), (5.25a), and (5.25b) are all linear and time invariant, the volume velocity and pressure are also of the form

$$p(x,t) = P(x,\Omega)e^{j\Omega t}, \tag{5.27a}$$

$$u(x,t) = U(x,\Omega)e^{j\Omega t}, \tag{5.27b}$$

$$\delta A(x,t) = \delta A(x,\Omega)e^{j\Omega t}. \tag{5.27c}$$

Substituting Eqs. (5.27a), (5.27b), and (5.27c) into Eqs. (5.24), (5.25a), and (5.25b) yields the ordinary differential equations

$$-\frac{dP}{dx} = ZU, \tag{5.28a}$$

$$-\frac{dU}{dx} = YP + Y_w P, \tag{5.28b}$$

where $P = P(x, \Omega)$ and $U = U(x, \Omega)$ and

$$Z(x, \Omega) = j\Omega \frac{\rho}{A_0(x)}, \tag{5.29a}$$

$$Y(x, \Omega) = j\Omega \frac{A_0(x)}{\rho c^2}, \tag{5.29b}$$

$$Y_w(x, \Omega) = \frac{1}{j\Omega m_w(x) + b_w(x) + \dfrac{k_w(x)}{j\Omega}}. \tag{5.29c}$$

Note that Eqs. (5.28a) and (5.28b) are identical to Eqs. (5.14a) and (5.14b) except for the addition of the wall admittance term Y_w and for the fact that the acoustic impedance and admittances are, in this case, functions of x as well as Ω. If we consider a uniform soft-walled tube, then $A_0(x)$ is constant as a function of x, and Eqs. (5.15) and (5.16) are identical to Eqs. (5.29a) and (5.29b).

In order to use Eqs. (5.28a), (5.28b), and (5.29a)–(5.29c) to gain insight into the acoustics of speech production, it is necessary to solve them numerically using either idealized or measured vocal tract area functions. A technique for this solution that was developed by Portnoff [289, 291] is described in Appendix B.

Using estimates obtained from measurements on body tissues [105] ($m_w = 0.4$ gm/cm^2, $b_w = 6500$ dyne-sec/cm^3, and $k_w \approx 0$) for the parameters in Eq. (5.29c), the differential equations (5.28a) and (5.28b) were solved numerically with the boundary condition $p(l, t) = 0$ at the lip end [289, 291]. The ratio

$$V_a(j\Omega) = \frac{U(l, \Omega)}{U_G(\Omega)} \tag{5.30}$$

is plotted on a decibel (dB) scale as a function of Ω in Figure 5.7 for the case of a uniform tube of cross-sectional area 5 cm^2 and length 17.5 cm [289]. The results are similar to Figure 5.4, but different in an important way. It is clear that the resonances are no longer exactly on the $j\Omega$ axis of the s-plane. This is evident since the frequency response no longer is infinite at frequencies 500 Hz, 1500 Hz, 2500 Hz, etc., although the response is peaked in the vicinity of these frequencies. The center frequencies, $F(k)$, and bandwidths,[2] $B(k)$, of the resonances (in Hz) in Figure 5.7 are given in the associated

[2]The bandwidth of a resonance is defined as the frequency interval around a resonance in which the frequency response is greater than $1/\sqrt{(2)}$ (3 dB) times the peak value at the center frequency [41].

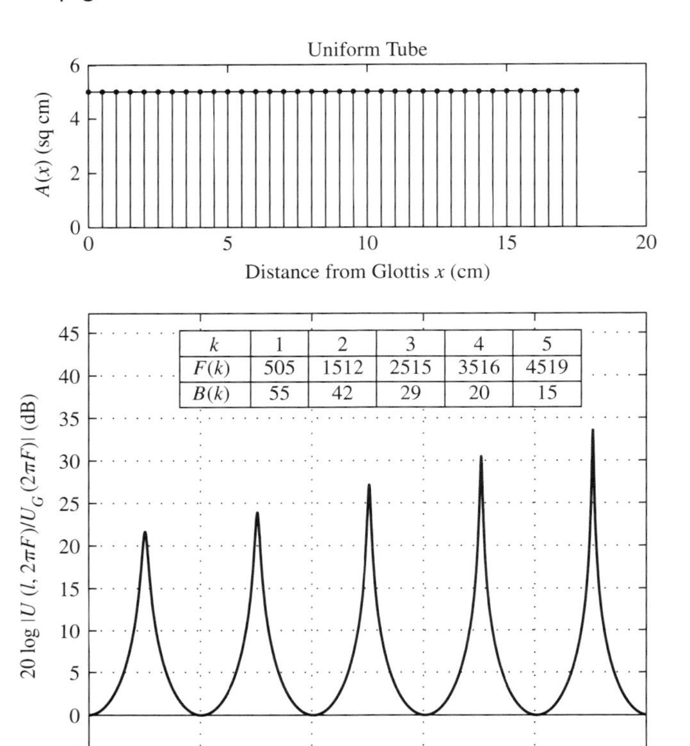

k	1	2	3	4	5
$F(k)$	505	1512	2515	3516	4519
$B(k)$	55	42	29	20	15

FIGURE 5.7
Log magnitude frequency response of uniform tube with yielding walls, no other losses, and terminated in a short circuit ($p(l,t) = 0$). The inserted table gives values (in Hz) for the first five resonance frequencies ($F(k)$) and bandwidths ($B(k)$). (After Portnoff [289].)

table. Several important effects are evident in this example. First, we note that the center frequencies are slightly higher than for the lossless case. Second, the bandwidths of the resonances are no longer zero as in the lossless case, since the peak value is no longer infinite. It can be seen that the effect of yielding walls is most pronounced at low frequencies. This is to be expected since we would expect very little motion of the massive walls at high frequencies. The results of this example are typical of the general effects of vocal tract wall vibration; i.e., the center frequencies are slightly increased and the low frequency resonances are broadened as compared to the rigid wall case.

Effects of Friction and Thermal Conduction

The effects of viscous friction and thermal conduction at the walls are much less pronounced than the effects of wall vibration. Flanagan [105] has considered these losses in detail and has shown that the effect of viscous friction can be accounted for in the frequency-domain representation (Eqs. (5.28a) and (5.28b)) by including a real, frequency-dependent term in the expression for the acoustic impedance, Z; i.e.,

$$Z(x, \Omega) = \frac{S(x)}{[A_0(x)]^2}\sqrt{\Omega\rho\mu/2} + j\Omega\frac{\rho}{A_0(x)}, \tag{5.31a}$$

where $S(x)$ is the circumference of the tube in cm, $\mu = 0.000186$ is the coefficient of friction, and $\rho = 0.00114$ gm/cm^3 is the density of air in the tube. The effects of heat conduction through the vocal tract wall can likewise be accounted for by adding a real frequency-dependent term to the acoustic admittance, $Y(x, \Omega)$; i.e.,

$$Y(x, \Omega) = \frac{S(x)(\eta - 1)}{\rho c^2}\sqrt{\frac{\lambda\Omega}{2c_p\rho}} + j\Omega\frac{A_0(x)}{\rho c^2}, \tag{5.31b}$$

where $c_p = 0.24$ is the specific heat at constant pressure, $\eta = 1.4$ is the ratio of specific heat at constant pressure to that at constant volume, and $\lambda = 0.000055$ is the coefficient of heat conduction [105]. Note that the loss due to friction is proportional to the real part of $Z(x, \Omega)$, and thus to $\Omega^{1/2}$. Likewise, the thermal loss is proportional to the real part of $Y(x, \Omega)$, which is also proportional to $\Omega^{1/2}$. Using the values given by Eqs. (5.31a) and (5.31b) for $Z(x, \Omega)$ and $Y(x, \Omega)$ and the values of $Y_w(x, \Omega)$ given by Eq. (5.29c), Eqs. (5.28a) and (5.28b) were again solved numerically [289]. The resulting log magnitude frequency response for the acoustic short-circuit boundary condition of $p(l, t) = 0$ is shown in Figure 5.8. Again, the center frequencies and bandwidths (in Hz) were determined and are shown in the associated table. Comparing Figure 5.8 with Figure 5.7, we observe that the main effect is to increase the bandwidths of the formants. Since friction and thermal losses increase with $\Omega^{1/2}$, the higher frequency resonances experience a greater broadening than do the lower resonances.

The examples depicted in Figures 5.7 and 5.8 are typical of the general effects of losses in the vocal tract. To summarize, viscous and thermal losses increase with frequency and have their greatest effect in the high frequency resonances, while wall loss is most pronounced at low frequencies. The yielding walls tend to raise the resonant frequencies while the viscous and thermal losses tend to lower them. The net effect for the lower resonances is a slight upward shift as compared to the lossless, rigid-walled model. The effect of friction and thermal loss is small compared to the effects of wall vibration for frequencies below 3–4 kHz. Thus, Eqs. (5.24), (5.25a), and (5.25b) are a good representation of sound transmission in the vocal tract even though we neglect these losses. As we shall see in the next section, the radiation termination at the lips is a much greater source of high frequency loss. This provides further justification for neglecting friction and thermal loss in models or simulations of speech production.

5.1.4 Effects of Radiation at the Lips

So far we have discussed the way that internal losses affect the sound transmission properties of the vocal tract. In our examples we have assumed the boundary condition $p(l, t) = 0$ at the lips. In the electric transmission line analogy, this corresponds to a short circuit. The acoustic counterpart of a short circuit is as difficult to achieve as

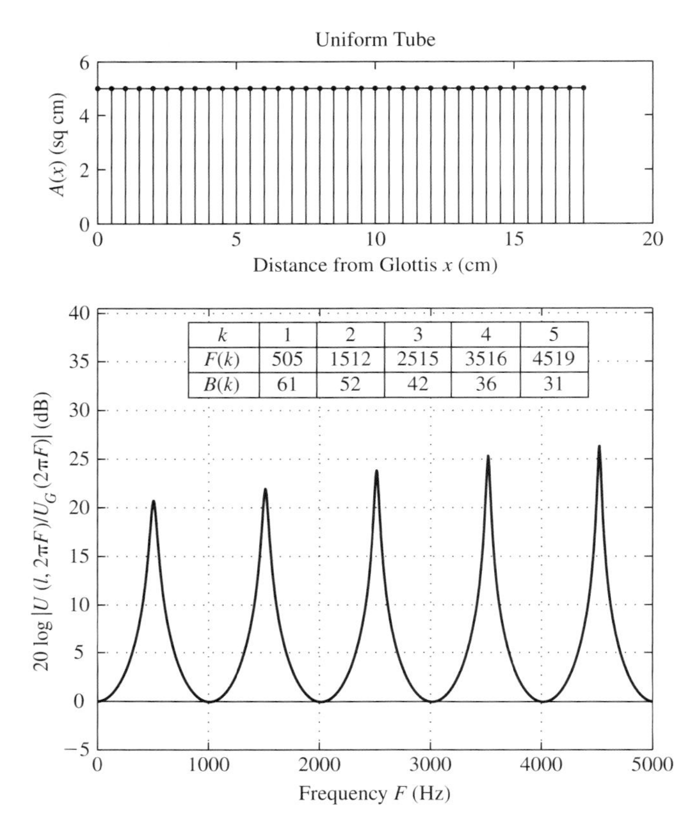

k	1	2	3	4	5
$F(k)$	505	1512	2515	3516	4519
$B(k)$	61	52	42	36	31

FIGURE 5.8
Log magnitude frequency response of uniform tube with yielding walls, friction, and thermal losses, and terminated in an acoustic short circuit ($p(l,t) = 0$). The inserted table gives values (in Hz) for the first five resonance frequencies ($F(k)$) and bandwidths ($B(k)$). (After Portnoff [289].)

an electrical short circuit since it requires a configuration in which volume velocity changes can occur at the end of the vocal tract tube without corresponding pressure changes. In reality, the vocal tract tube terminates with the opening between the lips (or the nostrils in the case of nasal sounds). A more realistic model is as depicted in Figure 5.9a, which shows the lip opening as an orifice in a sphere. In this model, at low frequencies, the opening can be considered a radiating surface, with the radiated sound waves being diffracted by the spherical baffle that represents the head.

The resulting diffraction effects are complicated and difficult to represent; however, for determining the boundary condition at the lips, all that is needed is a relationship between pressure and volume velocity at the radiating surface. Even this is very complicated for the configuration of Figure 5.9a. However, if the radiating surface (lip opening) is small compared to the size of the sphere, a reasonable approximation assumes that the radiating surface is set in a plane baffle of infinite extent as depicted in

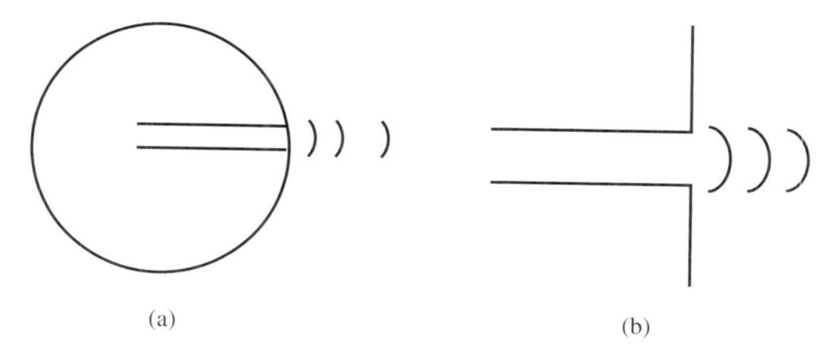

FIGURE 5.9
(a) Radiation from a spherical baffle; (b) radiation from an infinite plane baffle.

Figure 5.9b. In this case, it can be shown [105, 248, 289] that the sinusoidal steady state relation between the complex amplitudes of pressure and volume velocity at the lips is

$$P(l, \Omega) = Z_L(\Omega) \cdot U(l, \Omega), \tag{5.32a}$$

where the "radiation impedance" or "radiation load" at the lips is approximately of the form

$$Z_L(\Omega) = \frac{j\Omega L_r R_r}{R_r + j\Omega L_r}. \tag{5.32b}$$

The electrical analog to this radiation load is a parallel connection of a radiation resistance, R_r, and radiation inductance, L_r. Values of R_r and L_r that provide a good approximation to the infinite plane baffle are [105]

$$R_r = \frac{128}{9\pi^2}, \tag{5.33a}$$

$$L_r = \frac{8a}{3\pi c}, \tag{5.33b}$$

where a is the radius of the opening $[\pi a^2 = A_0(l)]$ and c is the velocity of sound.

The behavior of the radiation load influences the nature of wave propagation in the vocal tract through the boundary condition of Eqs. (5.32a) and (5.32b). Note that it is easily seen from Eq. (5.32b) that at very low frequencies, $Z_L(\Omega) \approx 0$; i.e., at very low frequencies, the radiation impedance approximates the ideal short circuit termination that has been assumed up to this point. Likewise, it is clear from Eq. (5.32b) that for a mid range of frequencies (when $\Omega L_r << R_r$), $Z_L(\Omega) \approx j\Omega L_r$. At higher frequencies ($\Omega L_r >> R_r$), $Z_L(\Omega) \approx R_r$. This is readily seen in Figure 5.10, which shows the real and imaginary parts of $Z_L(\Omega)$ as a function of Ω for typical values of the parameters. The energy dissipated due to radiation is proportional to the real part of the radiation impedance. Thus we can see that for the complete speech production system (vocal tract and radiation), the radiation losses will be most significant at higher frequencies.

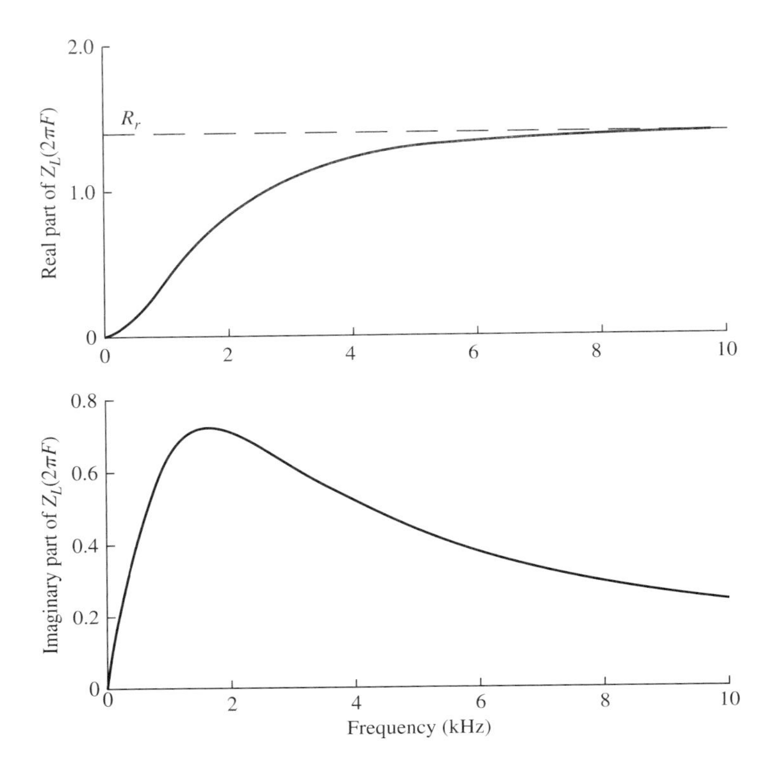

FIGURE 5.10
Real and imaginary parts of the radiation impedance.

To assess the magnitude of this effect, Eqs. (5.28a), (5.28b), (5.29c), (5.32a), and (5.32b) were solved using the technique of Appendix B for the case of a uniform time-invariant tube with yielding walls, friction and thermal losses, and radiation loss corresponding to an infinite plane baffle. Figure 5.11 shows the resulting log magnitude frequency response,

$$V_a(j\Omega) = \frac{U(l,\Omega)}{U_G(\Omega)}, \tag{5.34}$$

for an input $U(0,t) = U_G(\Omega)e^{j\Omega t}$. Comparing Figure 5.11 to Figures 5.7 and 5.8 shows that the major effect is to broaden the resonance frequencies (formant frequencies). As expected, the major effect on the resonance bandwidths occurs at higher frequencies. The first formant bandwidth is primarily determined by the wall loss, while the higher formant bandwidths are primarily determined by radiation loss. The second and third formant bandwidths can be said to be determined by a combination of these two loss mechanisms.

The frequency response shown in Figure 5.11 relates the volume velocity at the lips to the input volume velocity at the glottis. The relationship between pressure at the

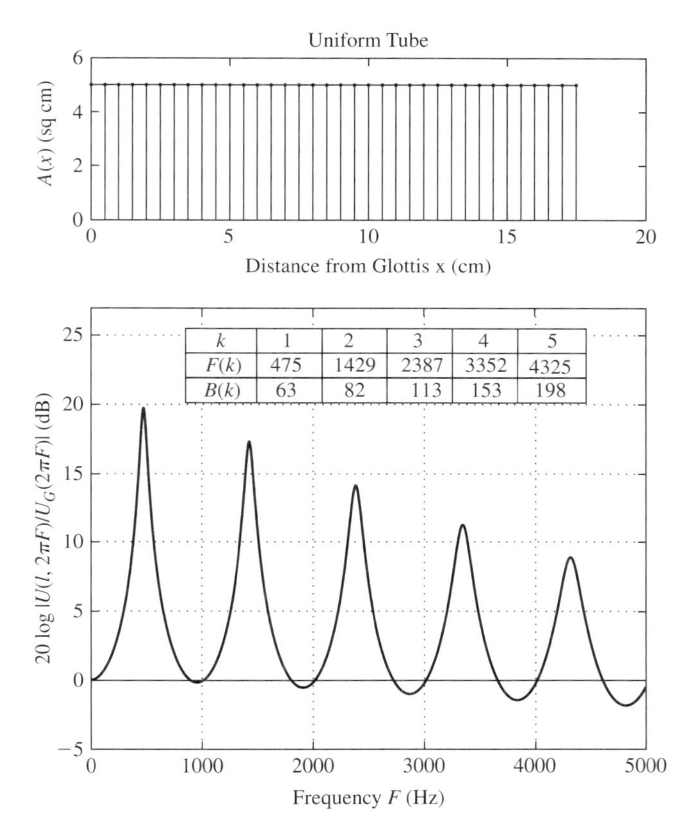

FIGURE 5.11
Log magnitude frequency response of uniform tube with yielding walls, friction and thermal loss, and radiation loss corresponding to an infinite plane baffle. The inserted table gives values (in Hz) for the first five resonance frequencies ($F(k)$) and bandwidths ($B(k)$). (After Portnoff [289].)

lips and volume velocity at the glottis is of interest, especially when a pressure-sensitive microphone is used in converting the acoustic wave to an electrical wave. Since $P(l,\Omega)$ and $U(l,\Omega)$ are related by Eq. (5.32a), the pressure transfer function is simply

$$H_a(\Omega) = \frac{P(l,\Omega)}{U_G(\Omega)} = \frac{P(l,\Omega)}{U(l,\Omega)} \cdot \frac{U(l.\Omega)}{U_G(\Omega)} \tag{5.35a}$$

$$= Z_L(\Omega) \cdot V_a(\Omega). \tag{5.35b}$$

It can be seen from Figure 5.11 that the major effects are an emphasis of high frequencies and the introduction of a zero at $\Omega = 0$. Figure 5.12 shows the log magnitude frequency responses $20\log_{10}|V_a(\Omega)|$ and $20\log_{10}|H_a(\Omega)|$ including wall losses and the radiation loss of an infinite plane baffle. A comparison of Figures 5.12a and 5.12b shows the effects of the zero at $\Omega = 0$ and the high frequency emphasis for the pressure frequency response.

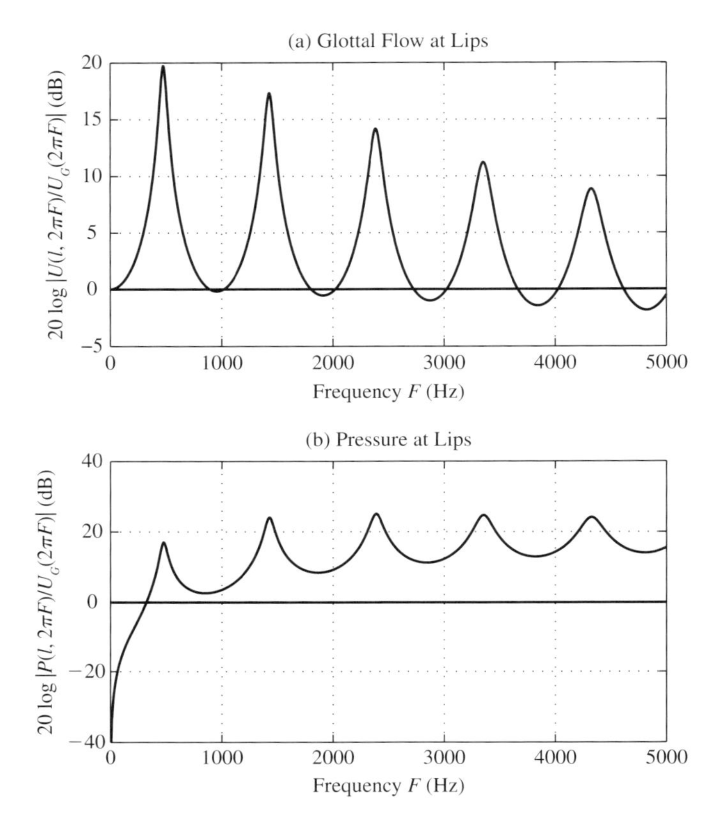

FIGURE 5.12
Log magnitude frequency responses of uniform tube with yielding walls, viscous and thermal losses, and radiation at the lips; (a) volume flow response at lips; (b) pressure response at lips for ideal volume velocity source at the glottis.

5.1.5 Vocal Tract Transfer Functions for Vowels

The equations discussed in Sections 5.1.3 and 5.1.4 comprise a detailed model for sound propagation and radiation in speech production. Using numerical integration techniques, either the time-domain or frequency-domain forms can be solved for a variety of vocal tract response functions. Such solutions provide valuable insight into the nature of the speech production process and the speech signal.

As examples of more realistic area functions, the frequency-domain equations (5.28a), (5.28b), (5.29c), (5.31a), (5.31b), (5.32a), and (5.32b) were used as described in Appendix B to compute frequency response functions for a set of area functions measured by Fant [101]. Figures 5.13–5.16 show the vocal tract area functions and corresponding log magnitude frequency responses $(U(l,\Omega)/U_G(\Omega))$ for the Russian vowels corresponding to /AA/, /EH/, /IY/, and /UW/. These figures illustrate the effects of all the loss mechanisms discussed in Sections 5.1.3 and 5.1.4. The formant

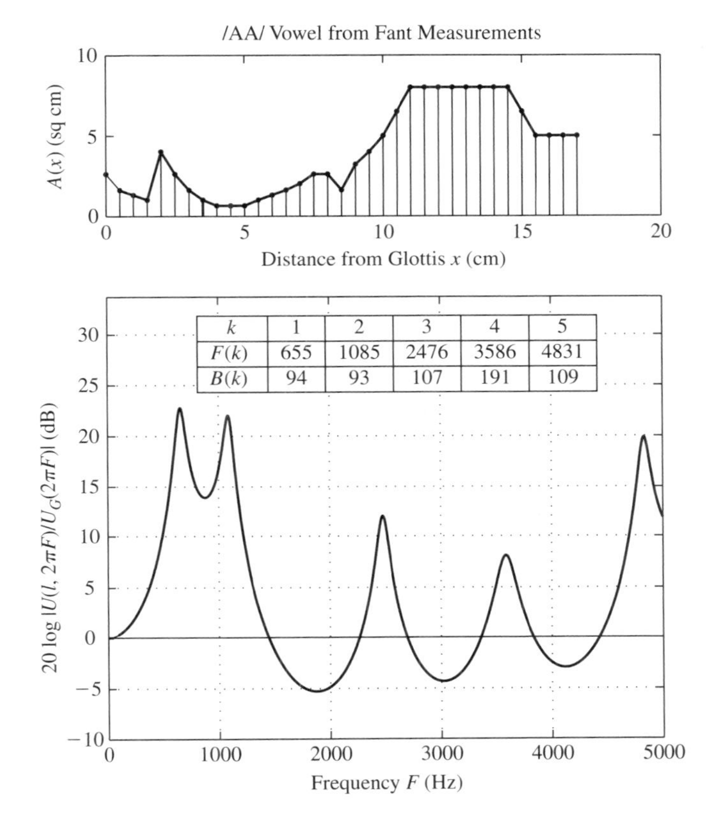

k	1	2	3	4	5
$F(k)$	655	1085	2476	3586	4831
$B(k)$	94	93	107	191	109

FIGURE 5.13
Area function (after Fant [101]) and log magnitude frequency response (after Portnoff [289]) for the Russian vowel /AA/. The inserted table gives values (in Hz) for the first five resonance frequencies ($F(k)$) and bandwidths ($B(k)$).

frequencies and bandwidths compare favorably with measurements on natural vowels of formant frequencies obtained by Peterson and Barney [287] given in Table 3.4 and formant bandwidths by Dunn [94].

In summary, we can conclude from these examples (and those of the previous sections) that:

- the vocal system is characterized by a set of resonances (formants) that depend primarily upon the vocal tract length and spatial variation of the area function, although there is some shift due to losses, as compared to the lossless case;
- the bandwidths of the lowest formant frequencies (first and second) depend primarily upon the vocal tract wall loss;[3]

[3] We will see in Section 5.1.7 that loss associated with the excitation source also affects the lower formants.

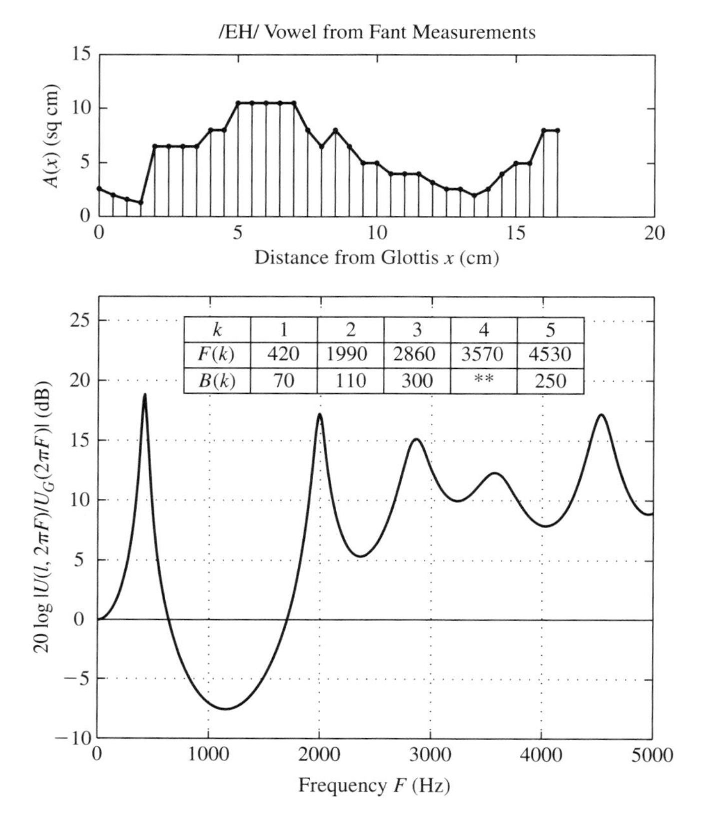

FIGURE 5.14
Area function (after Fant [101]) and log magnitude frequency response (after Portnoff [289]) for the Russian vowel /EH/. The inserted table gives values (in Hz) for the first five resonance frequencies ($F(k)$) and bandwidths ($B(k)$). (**Bandwidth of fourth formant could not be estimated using 3 dB down method.)

- the bandwidths of the higher formant frequencies depend primarily upon the viscous friction and thermal losses in the vocal tract and the radiation loss.

5.1.6 The Effect of Nasal Coupling

In the production of the nasal consonants /M/, /N/, and /NX/, the velum is lowered like a trap-door to couple the nasal tract to the pharynx. Simultaneously a complete closure is formed in the oral tract (e.g., at the lips for /M/). This configuration can be represented as in Figure 5.17a, which shows two branches, one of which is completely closed at one end. At the point of branching, the sound pressure is the same at the input to each tube, while the volume velocity must be continuous at the branching point; i.e., the volume velocity at the output of the pharynx tube must be the sum of the volume velocities at the inputs to the nasal and oral cavities. The corresponding

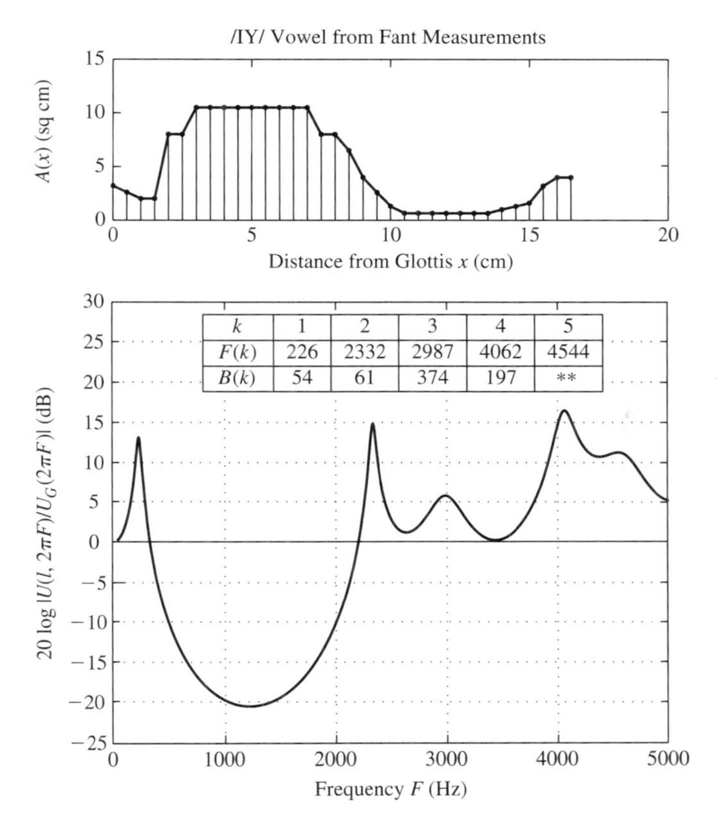

k	1	2	3	4	5
$F(k)$	226	2332	2987	4062	4544
$B(k)$	54	61	374	197	**

FIGURE 5.15
Area function (after Fant [101]) and log magnitude frequency response (after Portnoff [289]) for the Russian vowel /IY/. The inserted table gives values (in Hz) for the first five resonance frequencies ($F(k)$) and bandwidths ($B(k)$). (**Bandwidth of fifth formant could not be estimated using 3 dB down method.)

electrical transmission line analog is shown in Figure 5.17b. Note that continuity of volume velocity at the junction of the three tubes corresponds to Kirchoff's current law at the junction of the electrical transmission line analogy.

For nasal consonants the radiation of sound occurs primarily at the nostrils. Thus, the nasal tube is terminated with a radiation impedance appropriate for the size of the nostril openings. The oral tract, which is completely closed, is terminated by the equivalent of an open electrical circuit; i.e., no flow occurs. Nasalized vowels are produced by the same system, with the oral tract terminated as for vowels. The speech signal would then be the superposition of the nasal and oral outputs.

The mathematical model for this configuration consists of three sets of partial differential equations with boundary conditions being imposed by the form of glottal excitation, terminations of the nasal and oral tracts, and continuity relations at the junction. This leads to a rather complicated set of equations that could, in principle, be

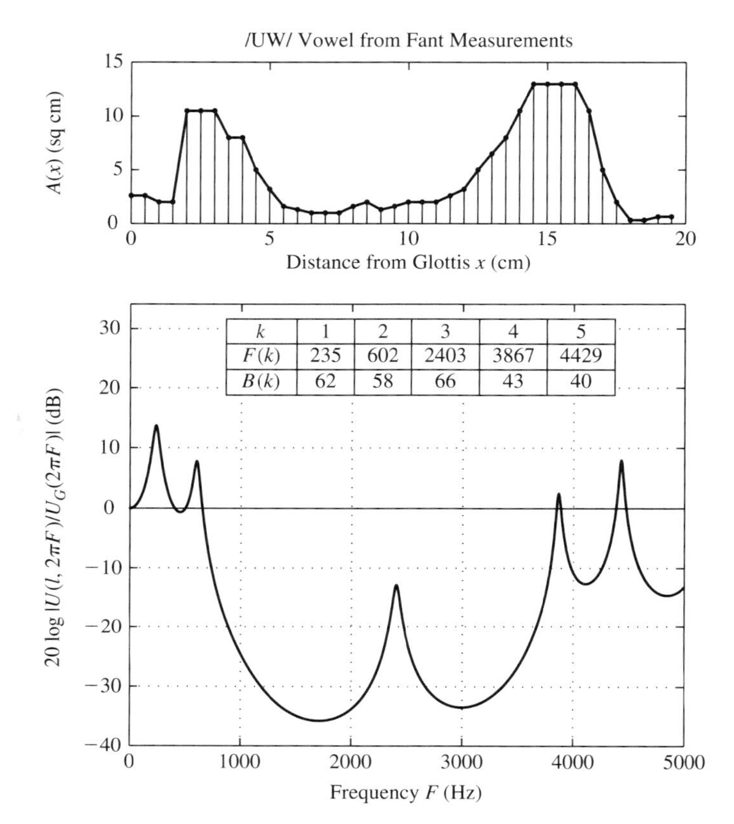

FIGURE 5.16
Area function (after Fant [101]) and log magnitude frequency response (after Portnoff [289]) for the Russian vowel /UW/. The inserted table gives values (in Hz) for the first five resonance frequencies ($F(k)$) and bandwidths ($B(k)$).

solved, given adequate measurements of area functions for all three tubes. However, the transfer function of the complete system would have many features in common with the previous examples. That is, the system would be characterized by a set of resonances or formants that would be dependent upon the shapes and lengths of the three tubes. An important difference results from the fact that the closed oral cavity can trap energy at certain frequencies, preventing those frequencies from appearing in the nasal output. In the electrical transmission line analogy, these are frequencies at which the input impedance of the open circuited line is zero. At these frequencies, the junction is short circuited by the transmission line corresponding to the oral cavity. The result is that for nasal sounds, the vocal system transfer function will be characterized by anti-resonances (zeros) as well as resonances [376]. Flanagan [105] showed in a simple approximate analysis that the nasal model of Figure 5.17 will have a zero at approximately $F_z = c/(4l_m)$, where l_m is the length of the oral cavity from the velum to the point of closure. For an oral cavity of length 7 cm, the zero would be at 1250 Hz.

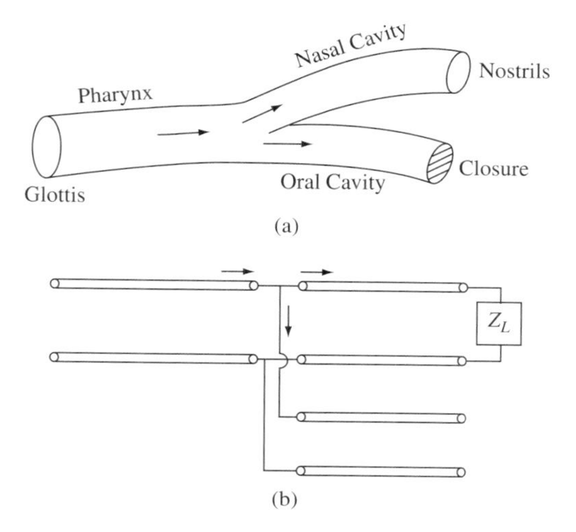

FIGURE 5.17
(a) Tube model for production of nasals;
(b) corresponding electrical analog.

This is consistent with the location of the "holes" in the spectrograms of the nasals /M/ and /N/ shown in Figure 3.29.

It has also been observed [117] that nasal formants have broader bandwidths than non-nasal voiced sounds. This is attributed to the greater viscous friction and thermal loss due to the large surface area of the nasal cavity [376].

5.1.7 Excitation of Sound in the Vocal Tract

The previous sub-sections have described how the laws of physics can be applied to describe the propagation and radiation of sound in speech production. To complete our discussion of acoustic principles, we must now consider the mechanisms whereby sound waves are generated in the vocal system. Recall that in our general overview of speech production, in Section 3.2.1, we identified three major mechanisms of excitation. These are:

1. air flow from the lungs is modulated by the vocal cord vibration, resulting in a quasi-periodic pulse-like excitation;
2. air flow from the lungs becomes turbulent as the air passes through a constriction in the vocal tract, resulting in a noise-like excitation;
3. air flow builds up pressure behind a point of total closure in the vocal tract; the rapid release of this pressure, by removing the constriction, causes a transient excitation.

A detailed model of excitation of sound in the vocal system involves the sub-glottal system (lungs, bronchi, and trachea), the glottis, and the vocal tract. Indeed, a model that is complete in all necessary details is also fully capable of simulating breathing

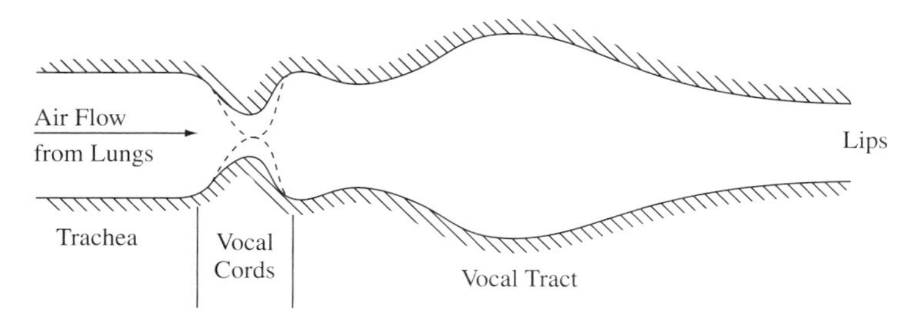

FIGURE 5.18
Schematic representation of the vocal system.

as well as speech production [105]! The first comprehensive effort toward a detailed physical model of sound generation in the vocal system was that of Flanagan [105, 111]. Subsequent research has produced a much refined model that provides a very detailed representation of the process of generation of both voiced and unvoiced speech [107, 110, 111, 155]. This model, which is based upon classical mechanics and fluid mechanics, is beyond the scope of our discussion here. However, a brief qualitative discussion of the basic principles of sound generation will be helpful in pointing the way toward the simple models that are widely used as the basis for speech analysis and synthesis.

The vibration of the vocal cords in voiced speech production can be explained by considering the schematic representation of the vocal system shown in Figure 5.18. The vocal cords constrict the path from the lungs to the vocal tract. As lung pressure is increased, air flows out of the lungs and through the opening between the vocal cords (the glottis). Bernoulli's law states that when a fluid flows through an orifice, the pressure is lower in the constriction than on either side. If the tension in the vocal cords is properly adjusted, the reduced pressure allows the cords to come together, thereby completely constricting air flow. (This is indicated by the dotted lines in Figure 5.18.) As a result, pressure increases behind the vocal cords. Eventually it builds up to a level sufficient to force the vocal cords to open and thus allow air to flow through the glottis again. Again the air pressure in the glottis falls, and the cycle is repeated. Thus, the vocal cords enter a condition of sustained oscillation. The rate at which the glottis opens and closes is controlled by the air pressure in the lungs, the tension and stiffness of the vocal cords, and the area of the glottal opening under rest conditions. These are the control parameters of a detailed model of vocal cord behavior. Such a model must also include the effects of the vocal tract, since pressure variations in the vocal tract influence the pressure variations in the glottis. In terms of the electrical analog, the vocal tract acts as a load on the vocal cord oscillator.

A schematic diagram of the vocal cord model (adapted from [155]) is shown in Figure 5.19a. The vocal cord model consists of a set of non-linear differential equations wherein the vocal cords are represented by mechanically coupled mass elements. The interaction of these differential equations with the partial differential equations describing vocal tract transmission can be represented by a time-varying acoustic resistance and inductance as shown in [155]. These impedance elements are functions of $1/A_G(t)$, where $A_G(t)$ is the area of the glottal opening. For example, when $A_G(t) = 0$

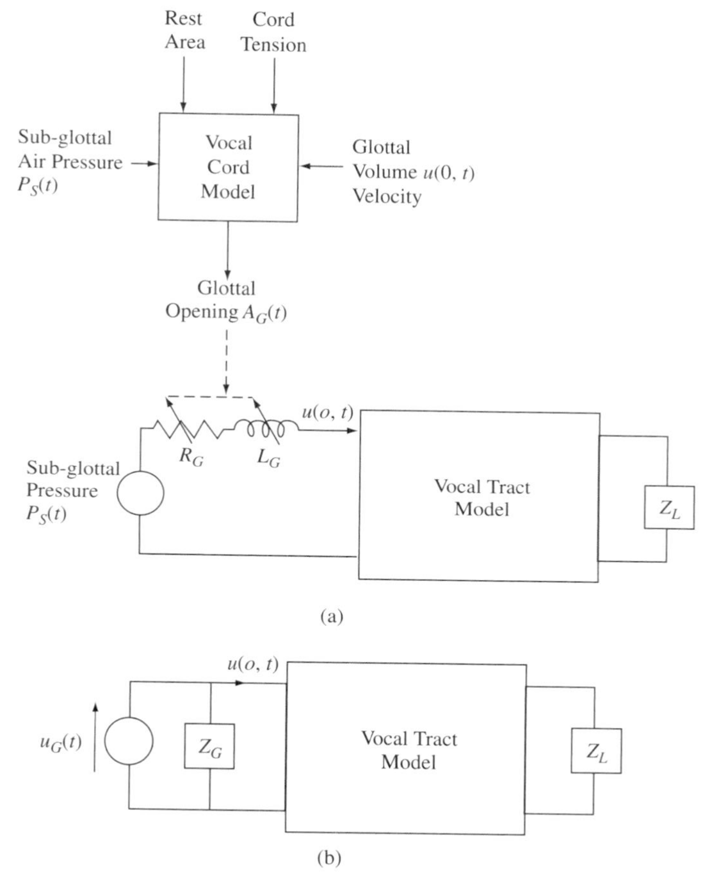

FIGURE 5.19
(a) Diagram of vocal cord model; (b) approximate model for vocal cords.

(glottis closed), the impedance is infinite and the volume velocity is zero. The glottal flow of such a model is shown in Figure 5.20 [155]. The upper waveform is the volume velocity and the lower waveform is the pressure at the lips for a vocal tract configuration appropriate for the vowel /a/. The pulse-like nature of the glottal flow is certainly consistent with our previous discussion and with direct observation through the use of high-speed motion pictures [105]. The damped oscillations of the output are, of course, consistent with our previous discussion of the nature of sound propagation in the vocal tract.

Since glottal area is a function of the flow into the vocal tract, the overall system of Figure 5.19a is non-linear, even though the vocal tract transmission and radiation systems are linear. The coupling between the vocal tract and the glottis is weak, however, and it is common to neglect this interaction. This leads to a separation and linearization of the excitation and transmission system, as depicted in Figure 5.19b.

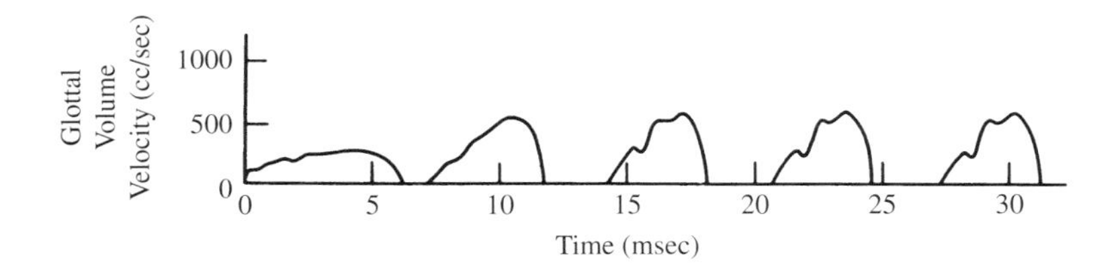

FIGURE 5.20
Glottal volume velocity and sound pressure at the mouth for vowel /a/. (After Ishizaka and Flanagan [155]. Reprinted with permission of Alcatel-Lucent USA Inc.)

In this case $u_G(t)$ is a volume velocity source whose wave shape is of the form of the upper waveform in Figure 5.20. The glottal acoustic impedance, Z_G, is obtained by linearization of the relations between pressure and volume velocity in the glottis [105]. This impedance is of the form

$$Z_G(\Omega) = R_G + j\Omega L_G, \tag{5.36}$$

where R_G and L_G are constants. With this configuration, the ideal frequency-domain boundary condition of $U(0,\Omega) = U_G(\Omega)$ is replaced by

$$U(0,\Omega) = U_G(\Omega) - Y_G(\Omega)P(0,\Omega), \tag{5.37}$$

where $Y_G(\Omega) = 1/Z_G(\Omega)$.

The glottal source impedance has significant effects upon resonance bandwidths for the speech production system. The major effect is a broadening of the lowest resonance. This is because $Z_G(\Omega)$ increases with frequency so that at high frequencies, Z_G appears as an open circuit and all of the glottal source flows into the vocal tract system. Thus, yielding walls and glottal loss control the bandwidths of the lower formants while radiation, friction, and thermal losses control the bandwidths of the higher formants.

Rosenberg's Glottal Pulse Approximation

To illustrate the effect of glottal excitation, we can take advantage of the flexibility of the frequency-domain solution that we have developed in this section. Note that our approach has been to solve Eqs. (5.28a) and (5.28b) at each frequency Ω with a variety of loss conditions. In the examples so far, the complex amplitude of the input $U_G(\Omega)$ was the same for all frequencies. In this case, the complex outputs $U(l,\Omega)$ and $P(l,\Omega)$ are identical to the volume flow and pressure frequency responses of the vocal system. Thus, the plots of vocal tract frequency responses also describe the response of the vocal system to an impulse excitation. However, by choosing $U_G(\Omega)$ differently, our solution can reflect the effect of a glottal pulse excitation. Specifically, Figure 5.21a shows a plot of the equation

$$g_c(t) = \begin{cases} 0.5[1 - \cos(2\pi t/(2T_1))] & 0 \le t \le T_1 \\ \cos(2\pi(t - T_1)/(4T_2)) & T_1 < t \le T_1 + T_2, \end{cases} \tag{5.38}$$

which was developed by Rosenberg in a study of the effect of glottal pulse shape on the quality of synthetic speech [326]. Such pulses, as depicted in Figure 5.21a, are very close approximations to pulse shapes obtained in detailed modeling of the vocal cord system and to pulses derived by inverse filtering [326]. By varying the parameters T_1 and T_2, it is possible to adapt the pulse length to different pitch periods and to model different rates of glottal opening and closing. With this model, we can observe the overall spectral shaping effect of the glottal excitation. Problem 5.1 outlines the derivation

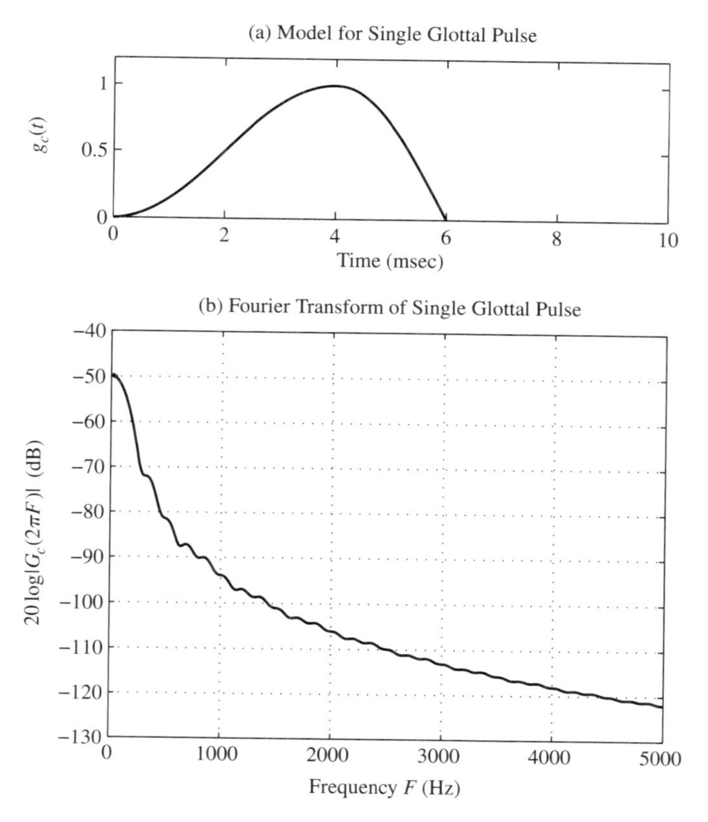

FIGURE 5.21
Rosenberg's glottal pulse model [326]: (a) plot of Eq. (5.38) for $T_1 = 4$ msec and $T_2 = 2$ msec; (b) corresponding log magnitude excitation spectrum $U_G(\Omega)$ when $u_G(t) = g_c(t)$. © [1971] Acoustical Society of America.

of the Fourier transform of the signal $g_c(t)$, and Figure 5.21b shows a plot of the glottal source spectrum (on a dB scale) $U_G(\Omega) = G_c(\Omega)$ [i.e., when $u_G(t) = g_c(t)$] for the condition $T_1 = 4$ msec and $T_2 = 2$ msec. Clearly, the glottal pulse by itself has a lowpass nature. Problem 5.1 also considers the more realistic situation of when the input is a periodically repeated sequence of glottal pulses of the form

$$u_G(t) = \sum_{k=-\infty}^{\infty} g_c(t - kT_0), \tag{5.39}$$

where T_0 is the (pitch) period.

The glottal pulse imposes its lowpass characteristic on the frequency response of the vocal tract. For example, the case of the uniform lossy tube depicted in Figure 5.12 was recomputed by the technique of Appendix B using the glottal spectrum in Figure 5.21b as the input spectrum $U_G(\Omega)$. The corresponding results are shown in Figure 5.22a for the overall volume velocity response and Figure 5.22b for the overall pressure response. A comparison of the corresponding plots in Figures 5.12 and 5.22 shows that the fall-off of the glottal spectrum (additive in dB) lowers the high frequency parts of the spectrum significantly. Also, a comparison of Figures 5.22a and 5.22b shows that the high frequency emphasis of the radiation load is overcome by the glottal spectrum fall-off so that the pressure spectrum also has significantly reduced high frequency content. Also note in Figure 5.22b that the composite spectrum has a very low frequency peak as well as a set of peaks due to the formants. This is due to the combined low frequency differentiator effect of the radiation load and the fact that the peak of the glottal spectrum is at $F = 0$ Hz. Such a low frequency peak is often observed in spectrograms where it is called a *voice bar.*

Unvoiced Excitation

The mechanism of production of voiceless sounds involves the turbulent flow of air. This can occur at a constriction whenever the volume velocity exceeds a certain critical value [105, 107]. Such excitation can be modeled by inserting a randomly time-varying source at the point of constriction. The strength of the source is made dependent (non-linearly) upon the volume velocity in the tube. In this way, frication is automatically inserted when needed [105, 107, 110]. For fricative sounds, the vocal cord parameters are adjusted so that the cords do not vibrate. For voiced fricatives, the vocal cords vibrate and turbulent flow occurs at a constriction whenever the volume velocity pulses exceed the critical value. For plosives, the vocal tract is closed for a period of time while pressure is built up behind the closure, with the vocal cords not vibrating. When the constriction is released, the air rushes out at a high velocity thus causing turbulent flow.

5.1.8 Models Based upon the Acoustic Theory

Section 5.1 has discussed in some detail the important features of the acoustic theory of speech production. The detailed models for sound generation, propagation, and radiation can, in principle, be solved with suitable values of the excitation and vocal tract parameters to compute an output speech waveform. Indeed it can be argued that this

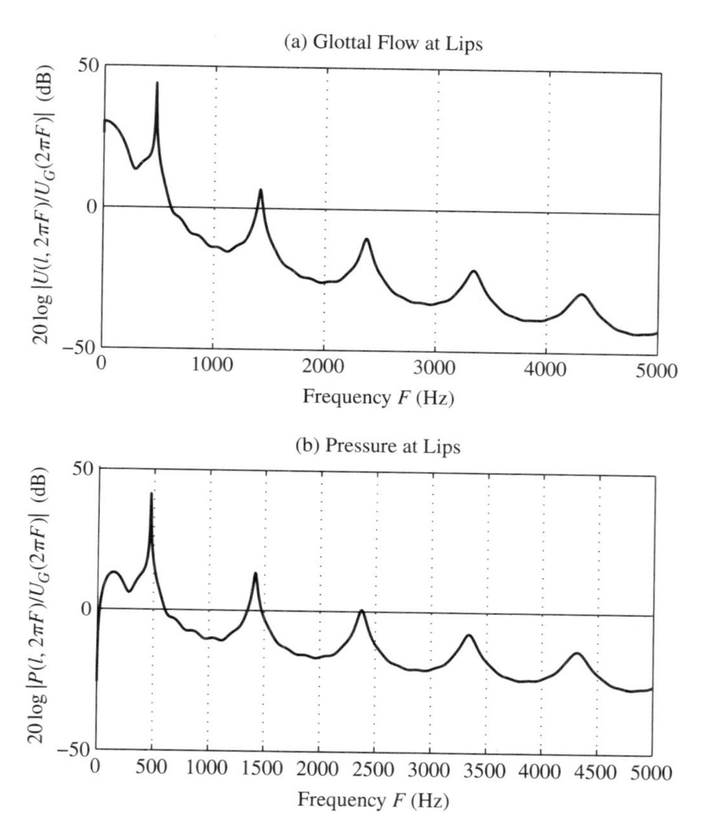

FIGURE 5.22
Effect of glottal pulse shape for uniform tube: (a) plot of log magnitude volume velocity spectrum at the lips for $T_1 = 4$ msec and $T_2 = 2$ msec; (b) corresponding log magnitude pressure spectrum at the lips when $u_G(t) = g_c(t)$.

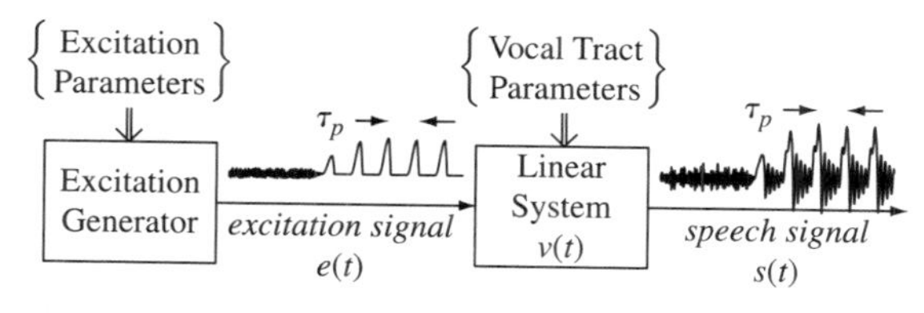

FIGURE 5.23
Source/system model of speech production.

may be the best approach to the synthesis of natural sounding synthetic speech [155]. However, for many purposes, such detail is impractical or unnecessary. In such cases, the acoustic theory that we have discussed in detail is most useful for how it points the way to a simplified approach to modeling speech signals. Figure 5.23 shows a general block diagram that is representative of numerous models that have been used as the basis for speech processing. What all these models have in common is that the excitation features are separated from the vocal tract and radiation features. The vocal tract and radiation effects are accounted for by the time-varying linear system. Its purpose is to model the resonance effects that we have observed and discussed. The excitation generator creates a signal that is either a train of (glottal) pulses, or a randomly varying signal (noise). The parameters of the source and system are chosen so that the resulting output has the desired speech-like properties. When this is done carefully, the model can serve as a useful basis for speech processing. In the remainder of this chapter, we discuss some models of this type.

5.2 LOSSLESS TUBE MODELS

A widely used model for speech production is based upon the assumption that the vocal tract can be represented as a concatenation of variable length, constant cross-sectional area, and lossless acoustic tubes, as depicted in Figure 5.24. The cross-sectional areas $\{A_k\}$, $k = 1, 2, \ldots, N$ of the tubes are chosen so as to approximate the area function, $A(x)$, of the vocal tract. If a large number of tubes, each of short length, is used, we can reasonably expect the resonant frequencies of the concatenated tubes to be close to those of a tube with a continuously varying area function. However, since this approximation neglects the losses due to friction, heat conduction, and wall vibration, we may also reasonably expect the bandwidths of the resonances to differ from those of a detailed model that includes these losses. However, losses can be accounted for at the glottis and lips, and as we shall see here and in Chapter 9, this can be done so as to accurately represent the resonance properties of the speech signal.

More important for our present discussion is the fact that lossless tube models provide a convenient transition between continuous-time models and discrete-time models. Thus we shall consider models of the form of Figure 5.24 in considerable detail.

5.2.1 Wave Propagation in Concatenated Lossless Tubes

Figure 5.25 shows a more detailed concatenated lossless tube model with variable section lengths. Since each tube in Figure 5.25 is assumed to be lossless, sound propagation in each tube is described by Eqs. (5.2a) and (5.2b) with appropriate values of the

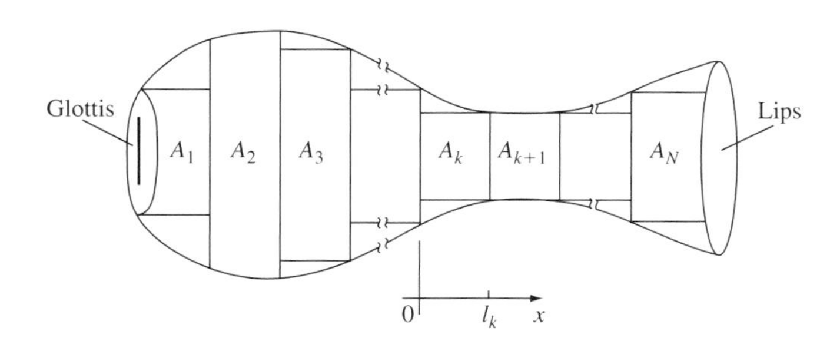

FIGURE 5.24
Concatenation of N lossless acoustic tubes.

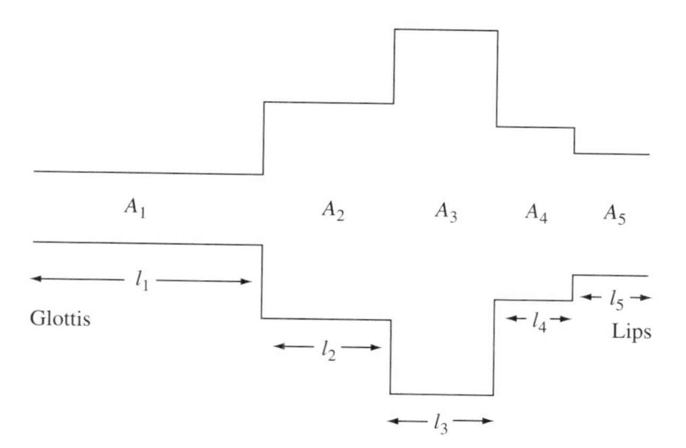

FIGURE 5.25
Concatenation of five lossless acoustic tubes.

cross-sectional area. Thus if we consider the kth tube with cross-sectional area A_k, the pressure and volume velocity in that tube have the form

$$u_k(x,t) = u_k^+(t - x/c) - u_k^-(t + x/c), \tag{5.40a}$$

$$p_k(x,t) = \frac{\rho c}{A_k}\left[u_k^+(t - x/c) + u_k^-(t + x/c)\right], \tag{5.40b}$$

where x is the distance measured from the left-hand end of the kth tube ($0 \leq x \leq l_k$) and $u_k^+(\ )$ and $u_k^-(\ )$ are positive-going and negative-going waves in the kth tube. The relationship between the traveling waves in adjacent tubes can be obtained by applying the physical principle that pressure and volume velocity must be continuous in both time and space everywhere in the system. This provides boundary conditions that can be applied at both ends of each tube.

Consider, in particular, the junction between the kth and $(k+1)^{st}$ tubes as depicted in Figure 5.26. Applying the continuity conditions at the junction gives

$$u_k(l_k,t) = u_{k+1}(0,t), \tag{5.41a}$$

$$p_k(l_k,t) = p_{k+1}(0,t). \tag{5.41b}$$

Substituting Eqs. (5.40a) and (5.40b) into Eqs. (5.41a) and (5.41b) gives

$$u_k^+(t - \tau_k) - u_k^-(t + \tau_k) = u_{k+1}^+(t) - u_{k+1}^-(t), \tag{5.42a}$$

$$\frac{A_{k+1}}{A_k}\left[u_k^+(t - \tau_k) + u_k^-(t + \tau_k)\right] = u_{k+1}^+(t) + u_{k+1}^-(t), \tag{5.42b}$$

where $\tau_k = l_k/c$ is the time for a wave to travel the length of the kth tube. From Figure 5.26 we observe that part of the positive-going wave that reaches the junction is propagated on to the right, while part is reflected back to the left. Likewise

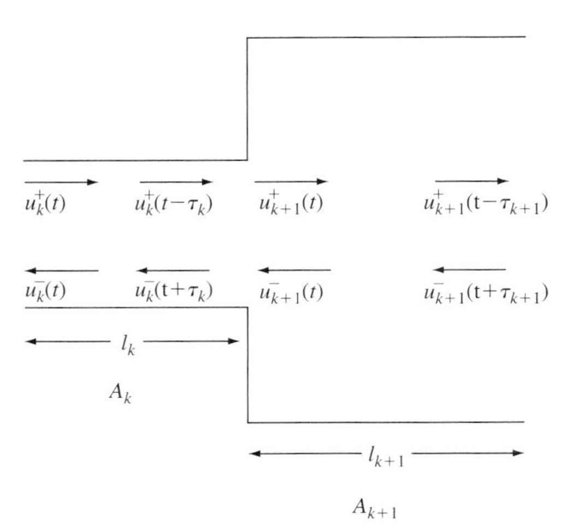

FIGURE 5.26
Illustration of the junction between two lossless tubes.

part of the backward traveling wave is propagated on to the left while part is reflected back to the right. Thus, if we solve for $u_{k+1}^+(t)$ and $u_k^-(t+\tau_k)$ in terms of $u_{k+1}^-(t)$ and $u_k^+(t-\tau_k)$, we will be able to see how the forward and reverse traveling waves propagate across the kth junction. Solving Eq. (5.42a) for $u_k^-(t+\tau_k)$ and substituting the result into Eq. (5.42b) yields

$$u_{k+1}^+(t) = \left[\frac{2A_{k+1}}{A_{k+1}+A_k}\right] u_k^+(t-\tau_k) + \left[\frac{A_{k+1}-A_k}{A_{k+1}+A_k}\right] u_{k+1}^-(t). \tag{5.43a}$$

Subtracting Eq. (5.42a) from Eq. (5.42b) gives

$$u_k^-(t+\tau_k) = -\left[\frac{A_{k+1}-A_k}{A_{k+1}+A_k}\right] u_k^+(t-\tau_k) + \left[\frac{2A_k}{A_{k+1}+A_k}\right] u_{k+1}^-(t). \tag{5.43b}$$

It can be seen from Eq. (5.43a) that the quantity

$$r_k = \frac{A_{k+1}-A_k}{A_{k+1}+A_k} \tag{5.44}$$

is the amount of $u_{k+1}^-(t)$ that is reflected at the junction. Thus, the quantity r_k is called the *reflection coefficient* for the kth junction. It is easily shown that since the areas are all positive (see Problem 5.3),

$$-1 \leq r_k \leq 1. \tag{5.45}$$

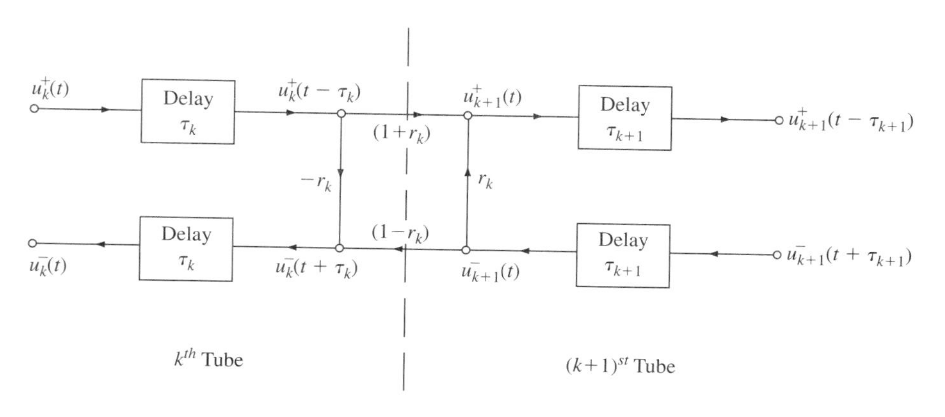

FIGURE 5.27
Signal-flow representation of the junction between two lossless tubes.

Using this definition of r_k, Eqs. (5.43a) and (5.43b) can be expressed as

$$u_{k+1}^+(t) = (1+r_k)u_k^+(t-\tau_k) + r_k u_{k+1}^-(t), \tag{5.46a}$$

$$u_k^-(t+\tau_k) = -r_k u_k^+(t-\tau_k) + (1-r_k)u_{k+1}^-(t). \tag{5.46b}$$

Equations of this form were first used for speech synthesis by Kelly and Lochbaum [188]. It is useful to depict these equations graphically as in Figure 5.27. In this figure, signal flow-graph conventions[4] are used to represent the multiplications and additions of Eqs. (5.46a) and (5.46b). Clearly, each junction of a system such as that depicted in Figure 5.25 can be represented by a system such as Fig 5.27, as long as our interest is only in values of pressure and volume velocity at the input and output of the tubes. This is not restrictive since we are primarily interested only in the relationship between the output of the last tube and the input of the first tube. Thus, a 5-tube model such as depicted in Figure 5.25 would have five sets of forward and backward delays and four junctions, each characterized by a reflection coefficient. To complete the representation of wave propagation in such a system, we must consider boundary conditions at the "lips" and "glottis" of the system.

5.2.2 Boundary Conditions

Let us assume that there are N sections indexed from 1 to N, starting at the glottis. Thus there are N sets of forward and backward waves, $N-1$ junctions (each characterized by a reflection coefficient), and a set of boundary conditions at the lips and glottis.

Boundary Conditions at the Lips

The boundary condition at the lips relates pressure, $p_N(l_N,t)$, and volume velocity, $u_N(l_N,t)$, at the output of the Nth tube to the radiated pressure and volume velocity.

[4]See Ref. [270] for an introduction to the use of signal flow graphs in signal processing.

If we use the frequency-domain relations of Section 5.1.4, we obtain a relation of the form

$$P_N(l_N, \Omega) = Z_L \cdot U_N(l_N, \Omega). \tag{5.47}$$

If we assume, for the moment, that Z_L is real, then we obtain the time-domain relation

$$\frac{\rho c}{A_N}\left[u_N^+(t-\tau_N)+u_N^-(t+\tau_N)\right] = Z_L\left[u_N^+(t-\tau_N)-u_N^-(t+\tau_N)\right]. \tag{5.48}$$

(If Z_L is complex, Eq. (5.48) would be replaced by a differential equation relating $p_N(l_N, t)$ and $u_N(l_N, t)$.) Solving for $u_N^-(t+\tau_N)$ we obtain

$$u_N^-(t+\tau_N) = -r_L u_N^+(t-\tau_N), \tag{5.49}$$

where the reflection coefficient at the lips is

$$r_L = \left[\frac{\rho c/A_N - Z_L}{\rho c/A_N + Z_L}\right]. \tag{5.50}$$

The output volume velocity at the lips is as follows:

$$u_N(l_N, t) = u_N^+(t-\tau_N) - u_N^-(t+\tau_N) \tag{5.51a}$$

$$= (1+r_L)u_N^+(t-\tau_N). \tag{5.51b}$$

The effect of this termination, as represented by Eqs. (5.49), (5.51a), and (5.51b) is depicted in Figure 5.28. Note that if Z_L is complex, it can be shown that Eq. (5.50) remains valid, but, of course, r_L will then be complex also, and it would be necessary to replace Eq. (5.49) by its frequency-domain equivalent. Alternatively, $u_N^-(t+\tau_N)$ and $u_N^+(t-\tau_N)$ could be related by a differential equation. (See Problem 5.5.)

Boundary Conditions at the Glottis

The frequency-domain boundary conditions at the glottis, assuming that the excitation source is linearly separable from the vocal tract, are given in Section 5.1.7. Applying

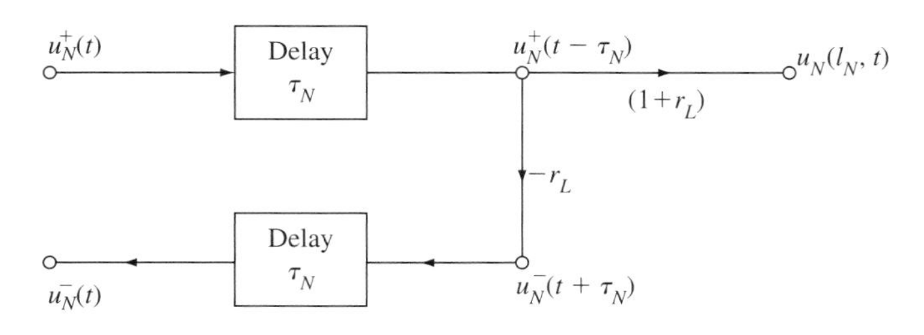

FIGURE 5.28
Termination at lip end of a concatenation of lossless tubes.

this relationship to the pressure and volume velocity at the input to the first tube we get

$$U_1(0, \Omega) = U_G(\Omega) - Y_G(\Omega)P_1(0, \Omega). \tag{5.52}$$

Assuming again that $Y_G = 1/Z_G$ is real and constant,

$$u_1^+(t) - u_1^-(t) = u_G(t) - \frac{\rho c}{A_1}\left[\frac{u_1^+(t) + u_1^-(t)}{Z_G}\right]. \tag{5.53}$$

Solving for $u_1^+(t)$ we obtain (see Problem 5.6)

$$u_1^+(t) = \frac{(1 + r_G)}{2}u_G(t) + r_G u_1^-(t), \tag{5.54}$$

where the glottal reflection coefficient is

$$r_G = \left[\frac{Z_G - \dfrac{\rho c}{A_1}}{Z_G + \dfrac{\rho c}{A_1}}\right]. \tag{5.55}$$

Equation (5.54) can be depicted as in Figure 5.29. As in the case of the radiation termination, if Z_G is complex, then Eq. (5.55) still holds. However, r_G would then be complex and Eq. (5.54) would be replaced by its frequency-domain equivalent or $u_1^+(t)$ would be related to $u_G(t)$ and $u_1^-(t)$ by a differential equation. Normally, since their role is to introduce loss into the system, the impedances Z_G and Z_L can be taken to be real for simplicity without significant effect on the accuracy of the model.

As an example, the complete diagram representing wave propagation in a two-tube model is shown in Figure 5.30. The volume velocity at the lips is defined as $u_L(t) = u_2(l_2, t)$. Writing the equations for this system in the frequency domain, the frequency

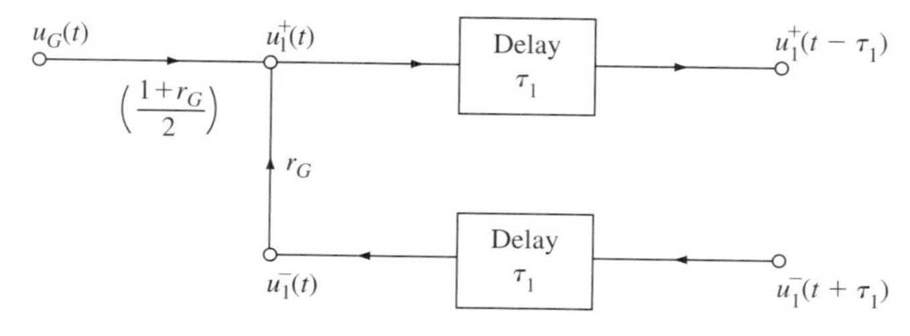

FIGURE 5.29
Termination at glottal end of a concatenation of lossless tubes.

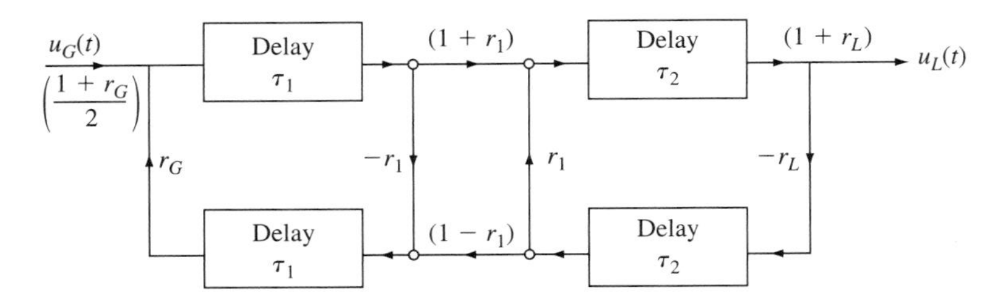

FIGURE 5.30
Complete flow diagram of a two-tube model.

response of the system can be shown to be

$$V_a(\Omega) = \frac{U_L(\Omega)}{U_G(\Omega)} \tag{5.56a}$$

$$= \frac{0.5(1+r_G)(1+r_L)(1+r_1)e^{-j\Omega(\tau_1+\tau_2)}}{1+r_1r_Ge^{-j\Omega 2\tau_1}+r_1r_Le^{-j\Omega 2\tau_2}+r_Lr_Ge^{-j\Omega 2(\tau_1+\tau_2)}}. \tag{5.56b}$$

(See Problem 5.7.) Several features of $V_a(\Omega)$ are worth pointing out. First, note the factor $e^{-j\Omega(\tau_1+\tau_2)}$ in the numerator. This represents simply the total propagation delay in the system from glottis to lips. The Laplace transform system function is found by replacing $j\Omega$ by s in Eq. (5.56b), with the result

$$V_a(s) = \frac{0.5(1+r_G)(1+r_L)(1+r_1)e^{-s(\tau_1+\tau_2)}}{1+r_1r_Ge^{-s2\tau_1}+r_1r_Le^{-s2\tau_2}+r_Lr_Ge^{-s2(\tau_1+\tau_2)}}. \tag{5.57}$$

The poles of $V_a(s)$ are the complex resonance frequencies of the system. We see that there will be an infinite number of poles because of the exponential dependence upon s. Fant [101] and Flanagan [105] show that through proper choice of section lengths and cross-sectional areas, realistic formant frequency distributions can be obtained for vowels. (Also see Problem 5.8.)

As an example, Figure 5.31 shows a series of two-tube models with various values for l_1, l_2, A_1, and A_2 and the locations of the first four resonances from each two-tube model (computed using a value of $c = 35{,}200$ cm/sec). The uniform one-tube model with length $l = 17.6$ cm (shown at the top of the figure) has the classic set of resonances, namely 500, 1500, 2500, and 3500 Hz. The effects of various length and area ratios, for the two-tube model, on the formants of the resulting system are shown in the remaining plots in Figure 5.31, where we see relative movements of each of the formants either above or below the nominal values for the one-tube model.

Figures 5.32–5.34 show examples of the log magnitude frequency response of two-tube model approximations for the vowels /AA/ (Figure 5.32) and /IY/ (Figures 5.33 and 5.34). The two-tube model parameters for the /AA/ vowel are $l_1 = 9$, $A_1 = 1$, $l_2 = 8$, and $A_2 = 7$, where the length values are in centimeters (cm) and the area values are in cm^2. There are two log magnitude frequency response plots shown in Figure 5.32. The more peaked (dashed line) plot is for the boundary conditions,

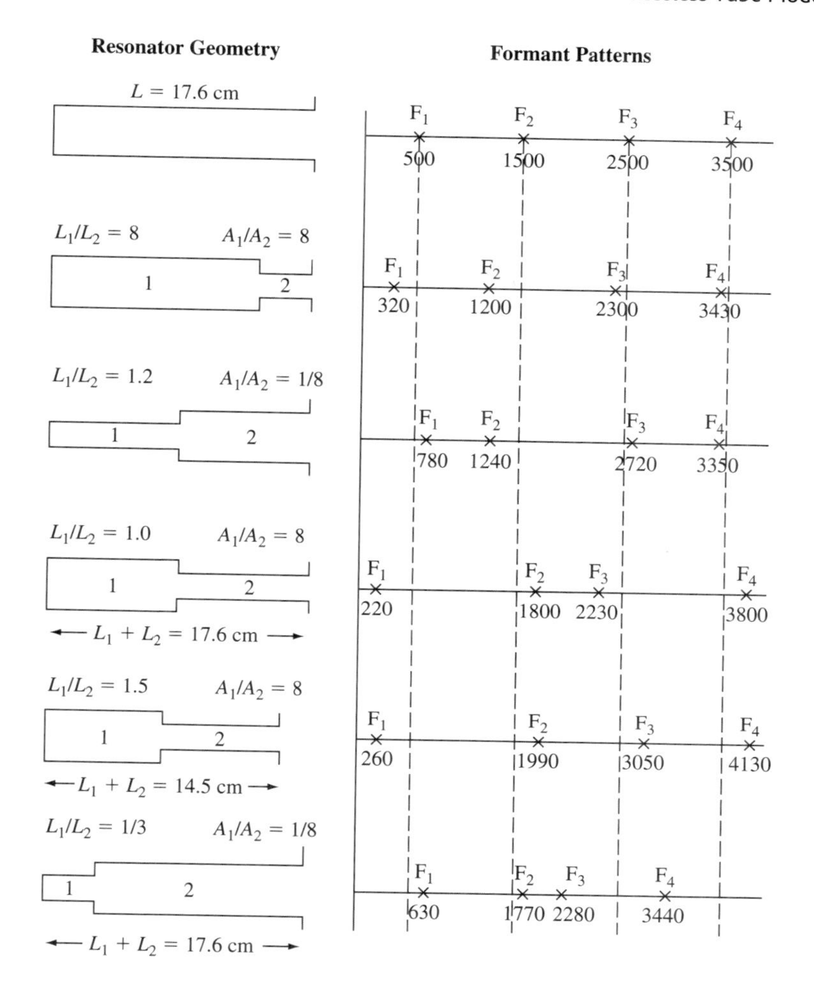

FIGURE 5.31
Examples of various configurations of a two-tube model of the vocal tract, and the resulting first four formant frequencies.

$r_L = 1$, $r_G = 1$, whereas the more damped response is for the boundary conditions $r_L = 1$, $r_G = 0.7$. Clearly the major effect of loss at the glottis is the broadening of the formant bandwidths and the reduction of the peakedness of the resulting frequency response curve.

Figures 5.33 and 5.34 show the log magnitude frequency response plots for two-tube model parameters appropriate to the vowel /IY/, namely $l_1 = 9$, $A_1 = 8$, $l_2 = 6$, $A_2 = 1$, with boundary conditions $r_L = 1$, $r_G = 1$ for the dashed curve in both figures, and with boundary conditions $r_L = 0.7$ for the damped curve in Figure 5.33 and $r_G = 0.7$ for the damped curve in Figure 5.34. The log magnitude frequency responses have the identical formant locations; however the bandwidths differ since $\tau_1 \neq \tau_2$ in Eq. (5.57).

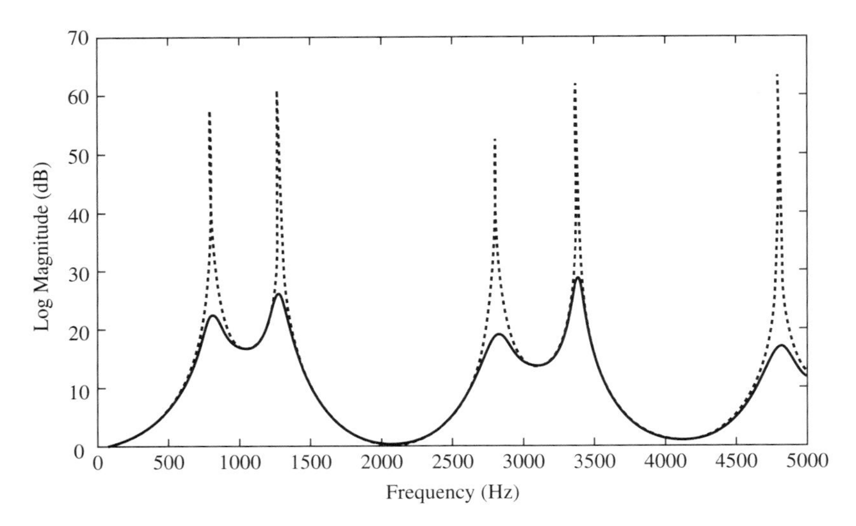

FIGURE 5.32
Log magnitude frequency response of two-tube model with dimensions appropriate for vowel /AA/; $l_1 = 9, A_1 = 1, l_2 = 8, A_2 = 7, r_L = 1, r_G = 1$ for the sharply peaked dashed curve and all parameters the same except $r_G = 0.7$ for the damped curve.

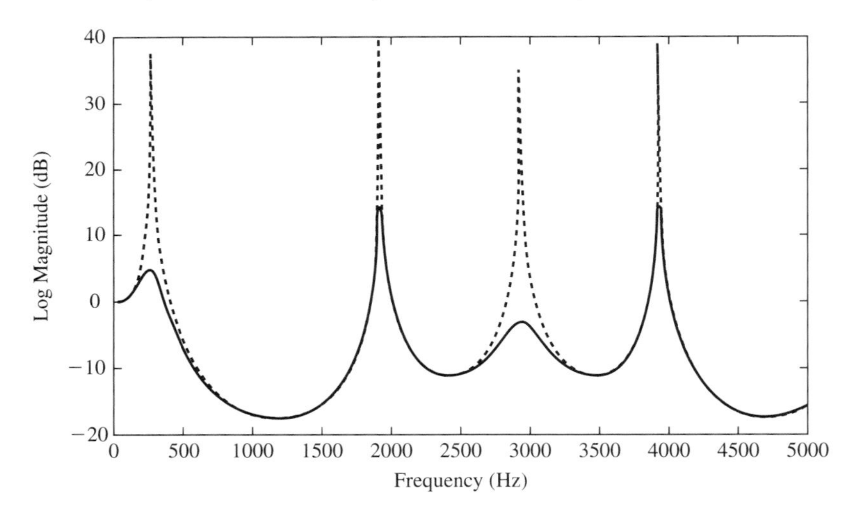

FIGURE 5.33
Log magnitude frequency response of two-tube model with dimensions appropriate for vowel /IY/; $l_1 = 9, A_1 = 8, l_2 = 6, A_2 = 1, r_L = 1, r_G = 1$ for the sharply peaked dashed curve and all parameters the same except $r_L = 0.7$ (loss inserted at lip end) for the damped curve.

5.2.3 Relationship to Digital Filters

The form of $V_a(s)$ for the two-tube model suggests that lossless tube models have many properties in common with digital filters. To see this, let us consider a system composed of N lossless tubes, each of length $\Delta x = l/N$, where l is the overall length of the vocal tract.

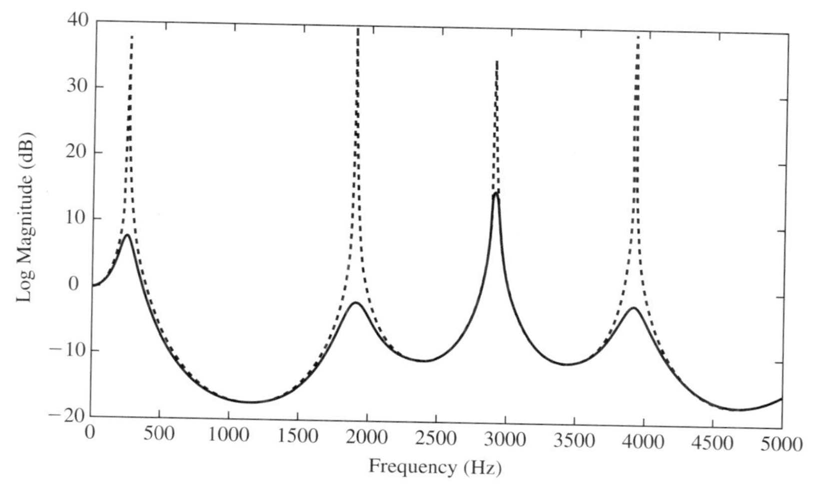

FIGURE 5.34
Log magnitude frequency response of two-tube model with dimensions appropriate for vowel /IY/; $l_1 = 9, A_1 = 8, l_2 = 6, A_2 = 1, r_L = 1, r_G = 1$ for the sharply peaked dashed curve and all parameters the same except $r_G = 0.7$ (loss inserted at glottis end) for the damped curve.

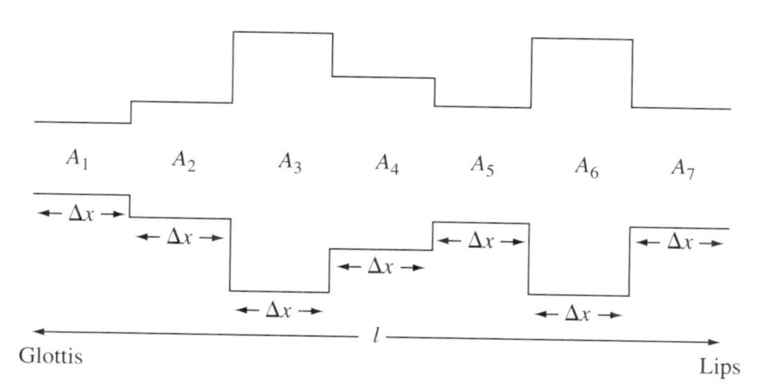

FIGURE 5.35
Concatenation of ($N = 7$) lossless tubes of equal length $\Delta x = l/N$.

Such a system is depicted in Figure 5.35 for $N = 7$. Wave propagation in this system can be represented as in Figure 5.27, with all the delays being equal to $\tau = \Delta x/c$, the time to propagate the length of one tube. It is instructive to begin by considering the response of the system to a unit impulse source, $u_G(t) = \delta(t)$. The impulse propagates down the series of tubes, being partially reflected and partially propagated at the junctions. A detailed consideration of this process shows that the impulse response of the vocal tract (i.e., the volume velocity at the lips due to an impulse at the glottis) is of the form

$$v_a(t) = \alpha_0\delta(t - N\tau) + \sum_{k=1}^{\infty}\alpha_k\delta(t - N\tau - 2k\tau), \tag{5.58}$$

where, in general, each impulse response coefficient α_k depends on the reflection coefficients at all the tube junctions.

Clearly, the soonest that an impulse can reach the output is $N\tau$ seconds. Then successive impulses, due to reflections at the junctions, reach the output at multiples of 2τ seconds later. The quantity 2τ is the time required to propagate both ways in one section. The system function of such a system is of the form

$$V_a(s) = \sum_{k=0}^{\infty} \alpha_k e^{-s(N+2k)\tau} = e^{-sN\tau} \sum_{k=0}^{\infty} \alpha_k e^{-s2\tau k}. \tag{5.59}$$

The factor $e^{-sN\tau}$ corresponds to the delay time required to propagate through all N sections. The quantity

$$\hat{V}_a(s) = \sum_{k=0}^{\infty} \alpha_k e^{-sk2\tau} \tag{5.60}$$

is the system function of a linear system whose impulse response is simply $\hat{v}_a(t) = v_a(t + N\tau)$. This part represents the resonance properties of the system. Figure 5.36a is a block diagram representation of the lossless tube model showing the separation of the system $\hat{v}_a(t)$ from the delay. The frequency response $\hat{V}_a(\Omega)$ is

$$\hat{V}_a(\Omega) = \sum_{k=0}^{\infty} \alpha_k e^{-j\Omega k2\tau}. \tag{5.61}$$

It is easily shown that

$$\hat{V}_a\left(\Omega + \frac{2\pi}{2\tau}\right) = \hat{V}_a(\Omega). \tag{5.62}$$

This is, of course, very reminiscent of the frequency response of a discrete-time system. In fact, if the input to the system (i.e., the excitation) is bandlimited to frequencies

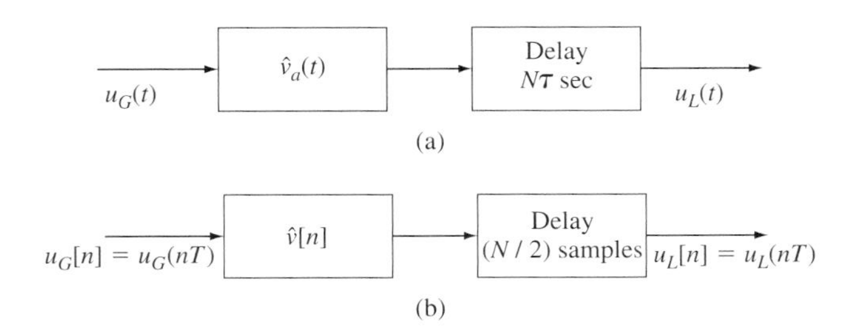

FIGURE 5.36
(a) Block diagram representation of lossless acoustic tube model; (b) equivalent discrete-time system.

below $\pi/(2\tau)$, then we can sample the input with period $T = 2\tau$ and filter the sampled signal with a digital filter whose impulse response is

$$\hat{v}[n] = \begin{cases} \alpha_n & n \geq 0 \\ 0 & n < 0. \end{cases} \tag{5.63}$$

With this choice for $\hat{v}[n]$ and the input bandlimited to frequencies below $\pi/(2\tau)$, the output of the digital filter in Figure 5.36b due to the sampled input signal $u_G[n] = u_G(nT)$ is identical to what would be obtained if we sampled the output of the analog system in Figure 5.36a.

For a sampling period of $T = 2\tau$, the delay of $N\tau$ seconds corresponds to a shift of $N/2$ samples. Thus, the equivalent discrete-time system for bandlimited inputs is shown in Figure 5.36b. Note that if N is even, $N/2$ is an integer and the delay can be implemented by simply shifting the output sequence of the first system. If N is odd, however, an interpolation would be required to obtain samples of the output of Figure 5.36a. This delay would most likely be ignored or avoided in some way (see below) since it is of no consequence in most applications of speech models.

The z-transform of $\hat{v}[n]$ is simply $\hat{V}_a(s)$ with e^{sT} replaced by z; i.e.,

$$\hat{V}(z) = \sum_{k=0}^{\infty} \alpha_k z^{-k}. \tag{5.64}$$

A signal flow graph for the equivalent discrete-time system can be obtained from the flow graph of the analog system in an analogous way. Specifically, each node variable in the analog system is replaced by the corresponding sequence of samples. Also each τ seconds of delay is replaced by a 1/2 sample delay, since $\tau = T/2$. An example is depicted in Figure 5.37b. Note, in particular, that the propagation delay for each section is represented in Figure 5.37b by a transmittance of $z^{-1/2}$.

The 1/2 sample delays in Figure 5.37b imply an interpolation half-way between sample values. Such interpolation is impossible to implement exactly in a fixed rate digital system. A more desirable configuration can be obtained by observing that the structure of Figure 5.37b has the form of a ladder, with the delay elements only in the upper and lower paths. Signals propagate to the right in the upper path and to the left in the lower path. We can see that the delay around any closed path in Figure 5.37b will be preserved if the delays in the lower branches are literally moved up to the corresponding branches directly above. The overall delay from input to output will then be incorrect, but this is of minor significance in practice and theoretically can be compensated by the insertion of the correct amount of advance (in general $z^{N/2}$).[5] Figure 5.37c shows how this is done for the three-tube example. The advantage of this form is that difference equations can be written for this system and these difference equations can be used iteratively to compute samples of the output from samples of the input.

[5]Note that we could also move all the delay to the lower branches. In this case, the delay through the system could be corrected by inserting a *delay* of $N/2$ samples.

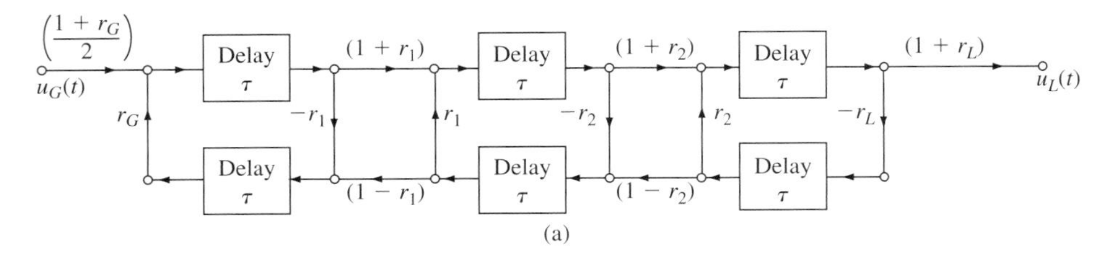

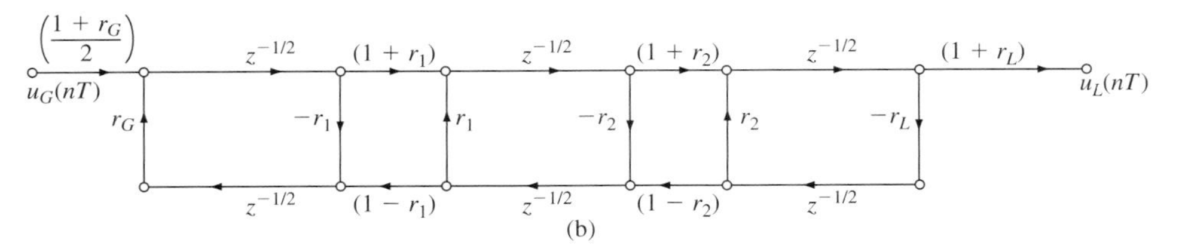

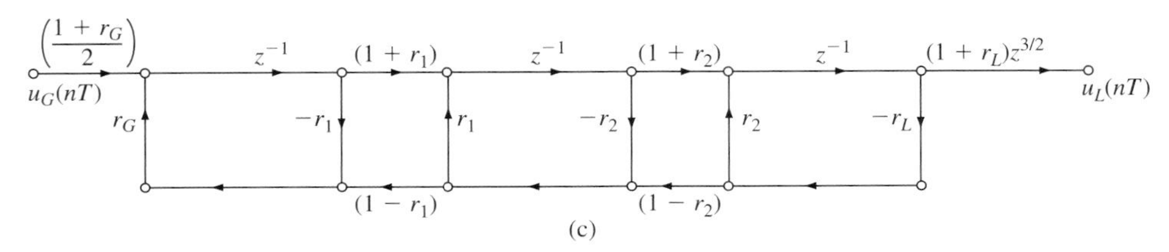

FIGURE 5.37
(a) Signal flow graph for lossless tube model of the vocal tract; (b) equivalent discrete-time system; (c) equivalent discrete-time system using only whole delays in ladder part.

Digital networks such as Figure 5.37c can be used to compute samples of a synthetic speech signal from samples of an appropriate excitation signal [188, 270]. In such applications, the structure of the network representation determines the complexity of the operations required to compute each output sample. Each branch whose transmittance is not unity requires a multiplication. We see that each junction requires four multiplications and two additions to implement each tube section. Generalizing from Figure 5.37c, we see that $4N$ multiplications and $2N$ additions are required to implement an N-tube model. Since multiplications often are the most time-consuming operation, it is of interest to consider other structures (literally, other organizations of the computation) that may require fewer multiplications. These can easily be derived by considering a typical junction as depicted in Figure 5.38a. The difference equations represented by this diagram are

$$u^+[n] = (1+r)w^+[n] + ru^-[n], \tag{5.65a}$$

$$w^-[n] = -rw^+[n] + (1-r)u^-[n]. \tag{5.65b}$$

These equations can be written as

$$u^+[n] = w^+[n] + rw^+[n] + ru^-[n], \tag{5.66a}$$

$$w^-[n] = -rw^+[n] - ru^-[n] + u^-[n]. \tag{5.66b}$$

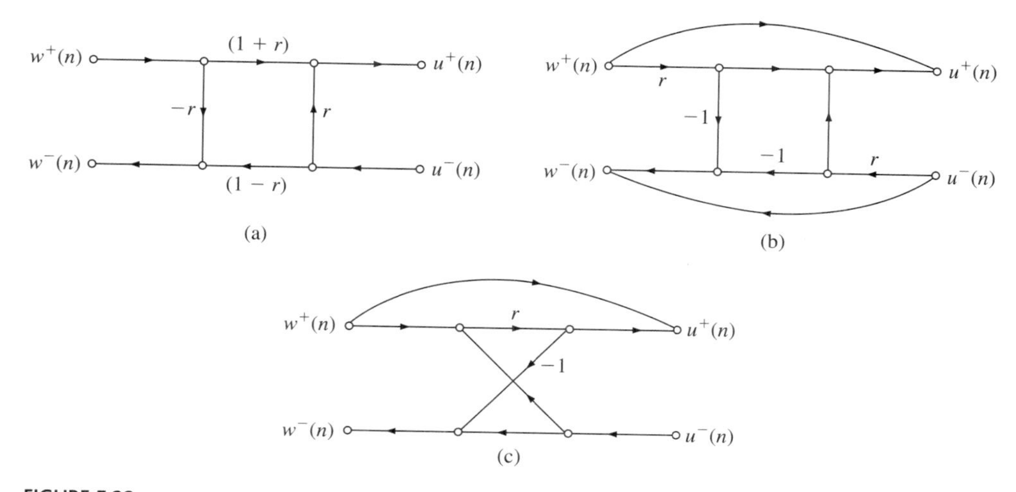

FIGURE 5.38
(a) 4-multiplier representation of lossless tube junction; (b) 2-multiplier configuration; (c) 1-multiplier configuration.

Since the terms $rw^+[n]$ and $ru^-[n]$ occur in both equations, two out of the four multiplications in Eqs. (5.65a) and (5.65b) can be eliminated as shown in Figure 5.38b. Note that this configuration requires two multiplications and four additions. Still another implementation follows from grouping terms involving r, as in

$$u^+[n] = w^+[n] + r\left(w^+[n] + u^-[n]\right), \tag{5.67a}$$

$$w^-[n] = u^-[n] - r\left(w^+[n] + u^-[n]\right). \tag{5.67b}$$

Now, since the term $r[w^+[n] + u^-[n]]$ occurs in both equations, this configuration requires only one multiplication and three additions as shown in Figure 5.38c. This form of the lossless tube model was first obtained by Itakura and Saito [163]. When using the lossless tube model for speech synthesis, the choice of computational structure depends on the speed with which multiplications and additions can be done, and the ease of controlling the computation.

5.2.4 Transfer Function of the Lossless Tube Model

To complete our discussion of lossless tube discrete-time models for speech production, it is instructive to derive a general expression for the transfer function in terms of the reflection coefficients. Equations of the type that we shall derive have been obtained before by Atal and Hanauer [12], Markel and Gray [232], and Wakita [410] in the context of linear predictive analyses of speech. We shall return to a consideration of lossless tube models and their relation to linear predictive analysis in Chapter 9. Our main concern, at this point, is the general form of the transfer function and the variety of other models suggested by the lossless tube model.

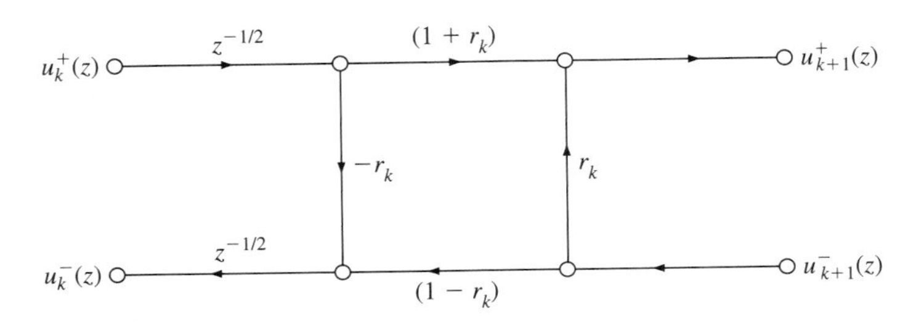

FIGURE 5.39
Flow graph representing the relationship among z-transforms at a junction.

To begin, note that we seek the transfer function

$$V(z) = \frac{U_L(z)}{U_G(z)}. \tag{5.68}$$

To find $V(z)$, it is most convenient to express $U_G(z)$ in terms of $U_L(z)$ and then solve for the ratio above. To do this, we consider Figure 5.39, which depicts a junction in the lossless tube model. The z-transform equations for this junction are

$$U_{k+1}^+(z) = (1+r_k)z^{-1/2}U_k^+(z) + r_k U_{k+1}^-(z), \tag{5.69a}$$

$$U_k^-(z) = -r_k z^{-1}U_k^+(z) + (1-r_k)z^{-1/2}U_{k+1}^-(z). \tag{5.69b}$$

Solving for $U_k^+(z)$ and $U_k^-(z)$ in terms of $U_{k+1}^+(z)$ and $U_{k+1}^-(z)$ gives

$$U_k^+(z) = \frac{z^{1/2}}{1+r_k}U_{k+1}^+(z) - \frac{r_k z^{1/2}}{1+r_k}U_{k+1}^-(z), \tag{5.70a}$$

$$U_k^-(z) = \frac{-r_k z^{1/2}}{1+r_k}U_{k+1}^+(z) + \frac{z^{-1/2}}{1+r_k}U_{k+1}^-(z). \tag{5.70b}$$

Equations (5.70a) and (5.70b) permit us to work backwards from the output of the lossless tube model to obtain $U_G(z)$ in terms of $U_L(z)$.

To make the result more compact, it is helpful to represent the boundary condition at the lips in the same manner as all the junctions in the system. Toward this end, we define $U_{N+1}(z)$ to be the z-transform of the input to a fictitious $(N+1)^{st}$ tube. An equivalent point of view is that the $(N+1)^{st}$ tube is terminated in its characteristic impedance. In any case, $U_{N+1}^+(z) = U_L(z)$ and $U_{N+1}^-(z) = 0$. Then from Eqs. (5.44) and (5.50), we see that if $A_{N+1} = \rho c/Z_L$, we can define $r_N = r_L$.

Now, Eqs. (5.70a) and (5.70b) can be expressed in matrix format:

$$\mathbf{U}_k = \mathbf{Q}_k \mathbf{U}_{k+1}, \tag{5.71}$$

where

$$\mathbf{U}_k = \begin{bmatrix} U_k^+(z) \\ U_k^-(z) \end{bmatrix}, \tag{5.72}$$

and

$$\mathbf{Q}_k = \begin{bmatrix} \dfrac{z^{1/2}}{1+r_k} & \dfrac{-r_k z^{1/2}}{1+r_k} \\ \dfrac{-r_k z^{-1/2}}{1+r_k} & \dfrac{z^{-1/2}}{1+r_k} \end{bmatrix}. \tag{5.73}$$

By repeatedly applying Eq. (5.71), it follows that the variables at the input to the first tube can be expressed in terms of the variables at the output by the matrix product:

$$\mathbf{U}_1 = \mathbf{Q}_1\mathbf{Q}_2 \cdots \mathbf{Q}_N \mathbf{U}_{N+1} \tag{5.74a}$$

$$= \prod_{k=1}^{N} \mathbf{Q}_k \cdot \mathbf{U}_{N+1}. \tag{5.74b}$$

From Figure 5.29 it can be seen that the boundary condition at the glottis can be expressed as

$$U_G(z) = \frac{2}{(1+r_G)} U_1^+(z) - \frac{2r_G}{1+r_G} U_1^-(z), \tag{5.75}$$

which can also be expressed as

$$U_G(z) = \left[\frac{2}{1+r_G}, \quad -\frac{2r_G}{1+r_G} \right] . \mathbf{U}_1. \tag{5.76}$$

Thus, since

$$\mathbf{U}_{N+1} = \begin{bmatrix} U_L(z) \\ 0 \end{bmatrix} = \begin{bmatrix} 1 \\ 0 \end{bmatrix} U_L(z), \tag{5.77}$$

we can write finally,

$$\frac{U_G(z)}{U_L(z)} = \left[\frac{2}{1+r_G}, \quad -\frac{2r_G}{1+r_G} \right] \prod_{k=1}^{N} \mathbf{Q}_k \begin{bmatrix} 1 \\ 0 \end{bmatrix}, \tag{5.78}$$

which is equal to $1/V(z)$.

To examine the properties of $V(z)$, it is helpful to first express $\mathbf{Q}_k$ as follows:

$$\mathbf{Q}_k = z^{1/2}\begin{bmatrix} \dfrac{1}{1+r_k} & \dfrac{-r_k}{1+r_k} \\ \dfrac{-r_k z^{-1}}{1+r_k} & \dfrac{z^{-1}}{1+r_k} \end{bmatrix} = z^{1/2}\hat{\mathbf{Q}}_k. \quad (5.79)$$

Thus, Eq. (5.78) can be expressed as

$$\frac{1}{V(z)} = z^{N/2}\left[\frac{2}{1+r_G}, \quad -\frac{2r_G}{1+r_G}\right]\prod_{k=1}^{N}\hat{\mathbf{Q}}_k\begin{bmatrix}1\\0\end{bmatrix}. \quad (5.80)$$

First, we note that since the elements of the matrices $\hat{\mathbf{Q}}_k$ are either constant or proportional to z^{-1}, the complete matrix product will reduce to a polynomial in the variable z^{-1} of order N. For example, it can be shown (see Problem 5.10) that for $N = 2$,

$$\frac{1}{V(z)} = \frac{2(1 + r_1r_2z^{-1} + r_1r_Gz^{-1} + r_2r_Gz^{-2})z}{(1+r_G)(1+r_1)(1+r_2)}, \quad (5.81)$$

or

$$V(z) = \frac{0.5(1+r_G)(1+r_1)(1+r_2)z^{-1}}{1 + (r_1r_2 + r_1r_G)z^{-1} + r_2r_Gz^{-2}}. \quad (5.82)$$

In general, it can be seen from Eqs. (5.79) and (5.80) that, for a lossless tube model, the transfer function can always be expressed as

$$V(z) = \frac{0.5(1+r_G)\prod_{k=1}^{N}(1+r_k)z^{-N/2}}{D(z)}, \quad (5.83a)$$

where $D(z)$ is a polynomial in z^{-1} given by the matrix

$$D(z) = [1, \; -r_G]\begin{bmatrix}1 & -r_1\\ -r_1z^{-1} & z^{-1}\end{bmatrix}\cdots\begin{bmatrix}1 & -r_N\\ -r_Nz^{-1} & z^{-1}\end{bmatrix}\begin{bmatrix}1\\0\end{bmatrix}. \quad (5.83b)$$

It can be seen from Eq. (5.83b) that $D(z)$ will have the form

$$D(z) = 1 - \sum_{k=1}^{N}\alpha_k z^{-k}; \quad (5.84)$$

i.e., the transfer function of a lossless tube model has a delay corresponding to the number of sections of the model and it has no zeros—only poles. These poles, of course, define the resonances or formants of the lossless tube model.

In the special case $r_G = 1$ ($Z_G = \infty$), the polynomial $D(z)$ can be found using a recursion formula that can be derived from Eq. (5.83b). If we begin by evaluating the matrix product from the left, we will always be multiplying a 1×2 row matrix by a 2×2 matrix until finally we multiply by the 2×1 column vector on the right in Eq. 5.83b. The desired recursion formula becomes evident after evaluating the first few matrix products. We define

$$\mathbf{P}_1 = [1, \ -1]\begin{bmatrix} 1 & -r_1 \\ -r_1 z^{-1} & z^{-1} \end{bmatrix} = \left[(1 + r_1 z^{-1}), \ -(r_1 + z^{-1})\right]. \tag{5.85}$$

If we define

$$D_1(z) = 1 + r_1 z^{-1}, \tag{5.86}$$

then it is easily shown that

$$\mathbf{P}_1 = [D_1(z), \ -z^{-1}D_1(z^{-1})]. \tag{5.87}$$

Similarly, the row matrix $\mathbf{P}_2$ is defined as

$$\mathbf{P}_2 = \mathbf{P}_1 \begin{bmatrix} 1 & -r_2 \\ -r_2 z^{-1} & z^{-1} \end{bmatrix}. \tag{5.88}$$

If the indicated multiplication is carried out, it is easily shown that

$$\mathbf{P}_2 = [D_2(z), \ -z^{-2}D_2(z^{-1})], \tag{5.89}$$

where

$$D_2(z) = D_1(z) + r_2 z^{-2} D_1(z^{-1}). \tag{5.90}$$

By induction it can be shown that

$$\mathbf{P}_k = \mathbf{P}_{k-1} \begin{bmatrix} 1 & -r_k \\ -r_k z^{-1} & z^{-1} \end{bmatrix} \tag{5.91a}$$

$$= [D_k(z), \ -z^{-k}D_k(z^{-1})], \tag{5.91b}$$

where

$$D_k(z) = D_{k-1}(z) + r_k z^{-k} D_{k-1}(z^{-1}). \tag{5.92}$$

Finally, the desired polynomial $D(z)$ is

$$D(z) = \mathbf{P}_N \begin{bmatrix} 1 \\ 0 \end{bmatrix} = D_N(z). \tag{5.93}$$

Thus, it is not necessary to carry out all the matrix multiplies. We can simply evaluate the recursion

$$D_0(z) = 1, \tag{5.94a}$$

$$D_k(z) = D_{k-1}(z) + r_k z^{-k} D_{k-1}(z^{-1}), \quad k = 1, 2, \ldots, N, \tag{5.94b}$$

$$D(z) = D_N(z). \tag{5.94c}$$

The effectiveness of the lossless tube model can be demonstrated by computing the transfer function for the area function data used to compute Figures 5.13–5.16. To do this we must decide upon the termination at the lips and the number of sections to use. In our derivations, we have represented the radiation load as a tube of area A_{N+1} that has no reflected wave. The value of A_{N+1} is chosen to give the desired reflection coefficient at the output. This is the only source of loss in the system (if $r_G = 1$), and thus it is to be expected that the choice of A_{N+1} will control the bandwidths of the resonances of $V(z)$. For example, $A_{N+1} = \infty$ gives $r_N = r_L = 1$, the reflection coefficient for an acoustic short circuit. This, of course, is the completely lossless case. Usually A_{N+1} would be chosen to give a reflection coefficient at the lips which produces reasonable bandwidths for the resonances. An example is presented below.

The choice of number of sections depends upon the sampling rate chosen to represent the speech signal. Recall that the frequency response of the lossless tube model with equal-length sections is periodic; and thus the model can only approximate the vocal tract behavior in a band of frequencies $|F| < 1/(2T)$, where T is the sampling period. We have seen that this requires $T = 2\tau$, where τ is the one-way propagation time in a single section. If there are N sections, for a total length l, then $\tau = l/(cN)$. Since the order of the denominator polynomial is N, there can be at most $N/2$ complex conjugate poles to provide resonances in the band $|F| < 1/(2T)$. Using the above value for τ with $l = 17.5$ cm and $c = 35{,}000$ cm/sec, we see that

$$\frac{1}{2T} = \frac{1}{4\tau} = \frac{Nc}{4l} = \frac{N}{2}(1000) \text{ Hz}. \tag{5.95}$$

This implies that there will be about 1 resonance (formant) per 1000 Hz of frequency for a vocal tract of total length 17.5 cm. For example, if $1/T = 10{,}000$ Hz, then the baseband is 5000 Hz. This implies that N should be 10. A glance at Figures 5.11 through 5.16 confirms that vocal tract resonances seem to occur with a density of about 1 formant per 1000 Hz. Shorter overall vocal tract lengths will have fewer resonances per kilohertz and vice versa.

Figure 5.40 shows an example for $N = 10$ and $1/T = 10$ kHz. Figure 5.40a shows the area function data of Figure 5.13 resampled to give a 10-tube approximation for the vowel /AA/. Figure 5.40b shows the resulting set of 10 reflection coefficients for $A_{11} = 30$ cm^2. This gives a reflection coefficient at the lips of $r_N = 0.714$. Note that the largest reflection coefficients occur where the relative change in area is greatest. Figure 5.40c shows the log magnitude frequency response curves for $r_N = 1$ and $r_N = 0.714$ (dashed curve). A comparison of the dashed curve of Figure 5.40c to Figure 5.13 confirms that

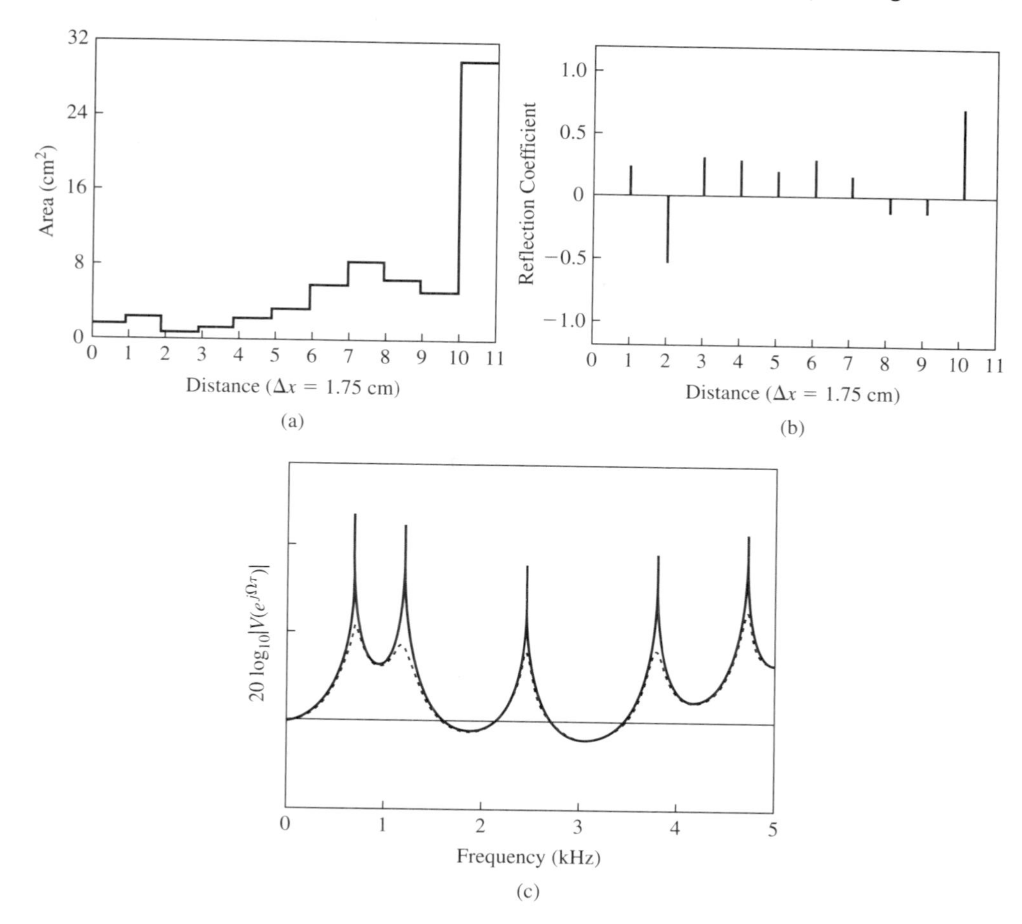

FIGURE 5.40
(a) Area function for 10-section lossless tube terminated with reflectionless section of area 30 cm^2; (b) reflection coefficients for 10-section tube; (c) log magnitude frequency response of 10-section tube; dotted curve corresponds to conditions of (b); solid curve corresponds to short-circuit termination. [Note: Area data of (a) estimated from data given by Fant [101] for the Russian vowel /AA/.]

with appropriate loss at the lip boundary, the frequency response of the lossless tube model is very much like that of the more detailed model discussed in Section 5.1.5.

5.3 DIGITAL MODELS FOR SAMPLED SPEECH SIGNALS

We have seen in Sections 5.1 and 5.2 that it is possible to derive detailed mathematical representations of the acoustics of speech production. Our purpose in surveying this theory is to call attention to the basic features of the speech signal and to show how these features are related to the physics of speech production. We have seen that sound is generated in three ways, and that each mode results in a distinctive type of output. We have also seen that the vocal tract imposes its resonances upon the excitation so as to produce the different sounds of speech. This is the essence of what we have learned so far.

An important idea should now be emerging from our lengthy discussion of physical models. We see that a valid approach to representation of speech signals could be in terms of a "source/system" model such as proposed in Figure 3.11 and as depicted again in Figure 5.23; that is, a linear system whose output has the desired speech-like properties when excited by an appropriate input and controlled by variable parameters that are related to the process of speech production. The model is thus equivalent to the physical model at its input and output terminals, but its internal structure does not mimic the physics of speech production. In particular, we are interested in discrete-time source/system models for representing sampled speech signals.[6]

To produce a speech-like signal, the mode of excitation and the resonance properties of the linear system must change with time. The nature of this time variation can be seen in Section 3.2. In particular, waveform plots such as Figure 3.8 show that the properties of the speech signal change relatively slowly with time. For many speech sounds it is reasonable to assume that the general properties of the excitation and vocal tract remain fixed for periods of 10–40 msec. Thus, a source/system model involves a slowly time-varying linear system excited by an excitation signal that changes back and forth from quasi-periodic pulses for voiced speech to random noise for unvoiced speech.

The lossless tube discrete-time model of the previous section serves as an example. The essential features of that model are depicted in Figure 5.41a. Recall that the vocal tract system was characterized by a set of areas or, equivalently, reflection coefficients. Systems of the form of Figure 5.37c can thus be used to compute the speech output, given an appropriate input. We showed that the relationship between the input

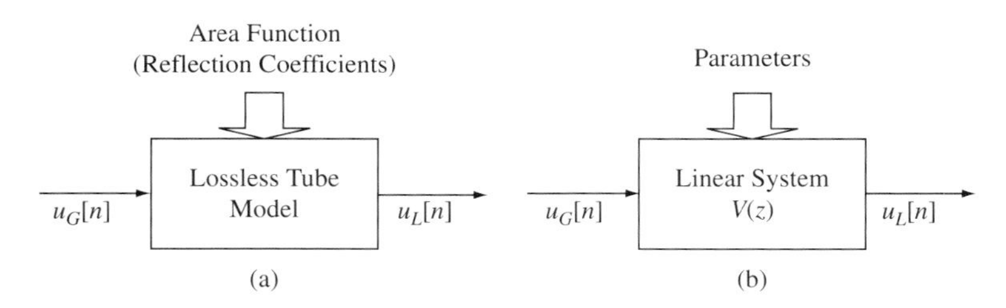

FIGURE 5.41
(a) Block diagram representation of the lossless tube model; (b) source/system model.

[6]Flanagan [104, 105] referred to such models as "terminal analogs" using the term *analog* in the sense of *behaving analogously*. In digital signal processing, the term "analog" has become almost synonymous with "continuous-time." While "discrete-time terminal analog" might be desirable, it seems too awkward and fraught with possibilities for confusion, and therefore, we use the term source/system model.

and output could be represented by a transfer function, $V(z)$, of the form

$$V(z) = \frac{G}{1 - \sum_{k=1}^{N} \alpha_k z^{-k}}, \tag{5.96}$$

where G and $\{\alpha_k\}$ depend upon the reflection coefficients as discussed in Section 5.2.4. (Note that the fixed delay in Eq. (5.83a) has been dropped for practical convenience.) Insofar as the output is concerned, all systems having this transfer function will produce the same output in response to a given input. (This is not strictly true for time-varying systems, but differences can be minimized by careful implementation.) Thus, discrete-time source/system models take the general form of Figure 5.41b. This leads to a consideration of alternative implementations of the vocal tract filter.

In addition to the vocal tract response, a complete source/system model includes a representation of the changing excitation function, and the effect of sound radiation at the lips. In the remainder of this section we examine each of the model components separately, and then combine them into a complete model.

5.3.1 Vocal Tract Modeling

The resonances (formants) of speech correspond to the poles of the transfer function $V(z)$. An all-pole model is a very good representation of vocal tract effects for a majority of speech sounds; however, the acoustic theory tells us that nasals and fricatives require both resonances and anti-resonances (poles and zeros). In these cases, we may include zeros in the transfer function or we may argue that the effect of a zero of the transfer function can be achieved (approximated) by including more poles [12]. (See Problem 5.11.) In most cases this approach is to be preferred.

Since the coefficients of the denominator of $V(z)$ in Eq. 5.96 are real, the roots of the denominator polynomial will either be real or occur in complex conjugate pairs. A typical complex resonance of the physical vocal tract is characterized by the complex frequencies

$$s_k, s_k^* = -\sigma_k \pm j2\pi F_k. \tag{5.97}$$

The corresponding complex conjugate poles in the discrete-time representation would be

$$z_k, z_k^* = e^{-\sigma_k T} e^{\pm j2\pi F_k T} \tag{5.98a}$$

$$= e^{-\sigma_k T} \cos(2\pi F_k T) \pm j e^{-\sigma_k T} \sin(2\pi F_k T). \tag{5.98b}$$

The continuous-time (analog) bandwidth of the vocal tract resonance is approximately $2\sigma_k$ and the center frequency is $2\pi F_k$ [41]. In the z-plane, the radius from the origin to the pole determines the bandwidth; i.e.,

$$|z_k| = e^{-\sigma_k T}, \tag{5.99}$$

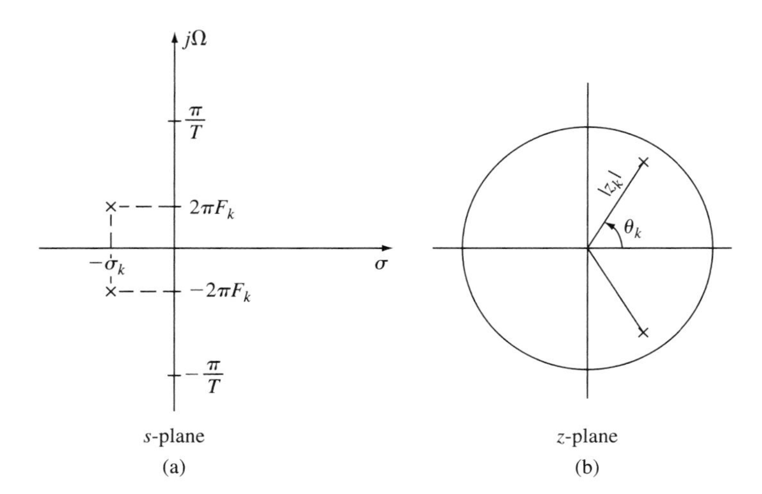

FIGURE 5.42
(a) s-plane and (b) z-plane representations of a vocal tract resonance.

so that the equivalent analog bandwidth is

$$2\sigma_k = -\frac{2}{T}\log_e |z_k|, \tag{5.100a}$$

where T is the sampling period. The z-plane angle, $\theta_k = \angle z_k$, is related to the center frequency as

$$\theta_k = 2\pi F_k T. \tag{5.100b}$$

Thus, if the denominator of $V(z)$ is factored, the corresponding continuous-time formant frequencies and bandwidths can be found using Eqs. (5.100b) and (5.99). As shown in Figure 5.42, the complex natural frequencies of the human vocal tract system are all in the left half of the s-plane since it is a stable system. Thus, $\sigma_k > 0$, and therefore $|z_k| < 1$; i.e., all of the corresponding poles of the discrete-time model must be inside the unit circle as required for stability. Figure 5.42 depicts a typical complex resonant frequency in both the s-plane and the z-plane.

In Section 5.2 we showed how a lossless tube model leads to a transfer function of the form of Eq. (5.96). It can be shown [12, 232] that if the areas of the tube model satisfy $0 < A_k < \infty$, all the poles of the corresponding system function $V(z)$ will be inside the unit circle. Conversely, it can be shown that given a transfer function, $V(z)$, as in Eq. (5.96), a lossless tube model can be found [12, 232]. Thus, one way to implement a given transfer function is to use a ladder structure, as in Figure 5.37c, possibly incorporating one of the junction forms of Figure 5.38. Another approach is to use one of the standard digital filter implementation structures given in Chapter 2. For example, we could use a direct form implementation of $V(z)$ as depicted in Figure 5.43a. Alternatively, we can represent $V(z)$ as a cascade of second-order systems (digital

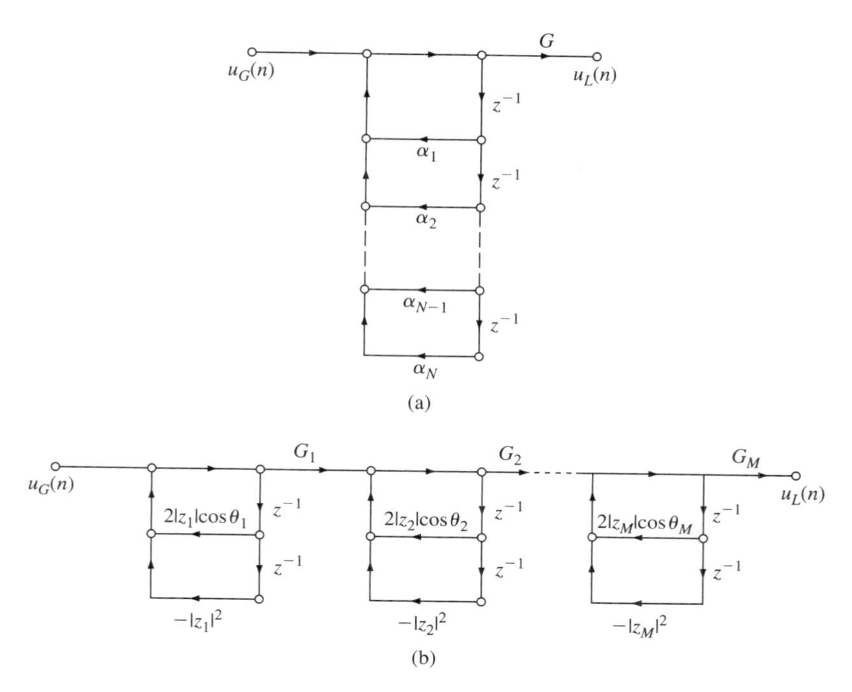

FIGURE 5.43
(a) Direct form implementation of all-pole transfer function; (b) cascade implementation of all-pole transfer function ($G_k = 1 - 2|z_k|\cos\theta_k + |z_k|^2$).

resonators); i.e.,

$$V(z) = \prod_{k=1}^{M} V_k(z), \tag{5.101}$$

where M is the largest integer in $((N+1)/2)$, and

$$V_k(z) = \frac{1 - 2|z_k|\cos(2\pi F_k T) + |z_k|^2}{1 - 2|z_k|\cos(2\pi F_k T)z^{-1} + |z_k|^2 z^{-2}}. \tag{5.102}$$

In this case, the numerator of $V_k(z)$ is chosen so that $V_k(1) = 1$; i.e., the gain of each subsystem is unity at $\omega = 0$. Additional gain could be applied to make the direct form and cascade form completely equivalent. A cascade model is depicted in Figure 5.43b. Problem 5.12 shows a novel way of eliminating multiplications in cascade models. Yet another approach to implementing the system $V(z)$ is to make a partial-fraction expansion of $V(z)$ and thus obtain a parallel form model. This approach is explored in Problem 5.13.

It is interesting to note that cascade and parallel models were first considered as continuous-time (analog) models. In this context there is a serious limitation, since continuous-time second-order systems (resonators) have frequency responses that die away with frequency. This led Fant [101] to derive "higher pole correction" factors that were cascaded with the analog formant resonators to achieve proper high frequency spectral balance. When digital simulations began to be used, Gold and Rabiner [128] observed that digital resonators had, by virtue of their inherent periodicity, the correct high frequency behavior. We have, of course, already seen this in the context of the lossless tube model. Thus no higher pole correction network is required in digital simulations.

5.3.2 Radiation Model

So far we have considered the transfer function $V(z)$ which relates volume velocity at the source to volume velocity at the lips. If we wish to obtain a model for pressure at the lips (as is usually the case), then the effects of radiation must be included. We saw, in Section 5.1.4, that in the analog model, the pressure and volume velocity are related by Eqs. (5.32a) and (5.32b). We desire a similar z-transform relation of the form

$$P_L(z) = R(z)U_L(z). \tag{5.103}$$

It can be seen from the discussion of Section 5.1.4 and from Figure 5.10 that pressure is related to volume velocity by a highpass filtering operation. In fact, at low frequencies, it can be argued that the pressure is approximately the derivative of the volume velocity. Thus, to obtain a discrete-time representation of this relationship, we must use a digitization technique that avoids aliasing. For example, by using the bilinear transform method of digital filter design [270], it can be shown (see Problem 5.14) that a reasonable approximation to the radiation effects is obtained with

$$R(z) = R_0(1 - z^{-1}); \tag{5.104}$$

i.e., a first backward difference. (A more accurate approximation is also considered in Problem 5.14.) The crude "differentiation" effect of the first difference is consistent with the approximate differentiation at low frequencies that is commonly assumed.

This radiation "load" can be cascaded with the vocal tract model as in Figure 5.44, where $V(z)$ can be implemented in any convenient way and the required parameters will, of course, be appropriate for the chosen configuration; e.g., area function (or reflection coefficients) for the lossless tube model, polynomial coefficients for the direct form model, or formant frequencies and bandwidths for the cascade model.

5.3.3 Excitation Model

To complete our source/system model, we must provide the means for generating an appropriate input to the combined vocal tract/radiation system. Recalling that the

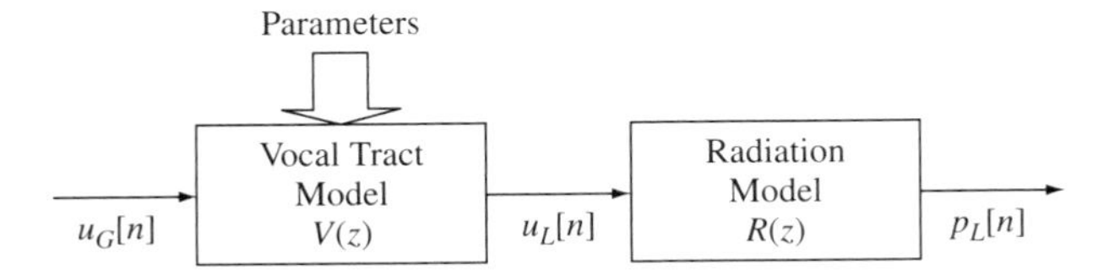

FIGURE 5.44
Source/system model including radiation effects.

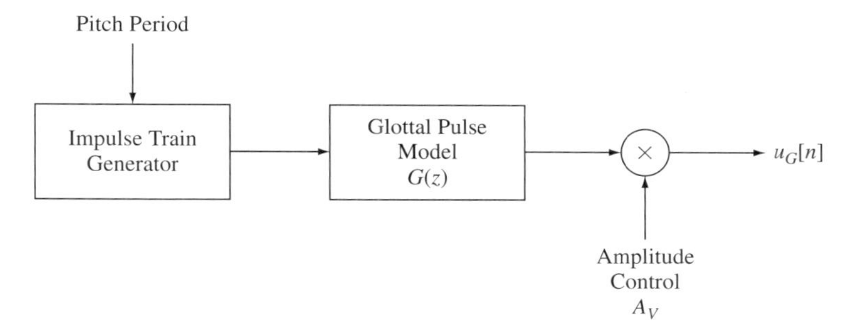

FIGURE 5.45
Generation of the excitation signal for voiced speech.

majority of speech sounds can be classified as either voiced or voiceless, we see that in general terms, what is required is a source that can produce either a quasi-periodic pulse waveform or a random noise waveform.

In the case of voiced speech, the excitation waveform must appear somewhat like the upper waveform in Figure 5.20. A convenient way to represent the generation of the glottal wave is shown in Figure 5.45. The impulse train generator produces a sequence of unit impulses that are spaced by the desired fundamental (pitch) period. This signal, in turn, excites a linear system whose impulse response $g[n]$ has the desired glottal wave shape. A gain control, A_v, controls the intensity of the voiced excitation.

The choice of the form of $g[n]$ is probably not critical as long as its Fourier transform has a significant fall-off at high frequencies. One possibility is a sampled version of the Rosenberg [326] glottal pulse discussed in Section 5.1.7 for which the log magnitude spectrum is shown in Figure 5.21; i.e.,

$$g[n] = g_c(nT) = \begin{cases} \frac{1}{2}[1 - \cos(\pi n/N_1)] & 0 \le n \le N_1 \\ \cos(\pi(n - N_1)/(2N_2)) & N_1 \le n \le N_1 + N_2 \\ 0 & \text{otherwise,} \end{cases} \tag{5.105}$$

where $N_1 = T_1/T$ and $N_2 = T_2/T$. As discussed in Section 5.1.7, this wave shape is very similar in appearance to the glottal pulses in Figure 5.20 that are predicted by detailed models of vocal cord vibration.

Since $g[n]$ in Eq. (5.105) has finite length, its z-transform, $G(z)$, has only zeros. An all-pole model is often more desirable. Good success has also been achieved using a two-pole model for $G(z)$ [232].

For unvoiced sounds, the excitation model is much simpler. All that is required is a source of white noise and a gain parameter to control the intensity of the unvoiced excitation. For discrete-time models, a random number generator provides a source of flat-spectrum noise. The probability distribution of the noise samples does not appear to be critical.

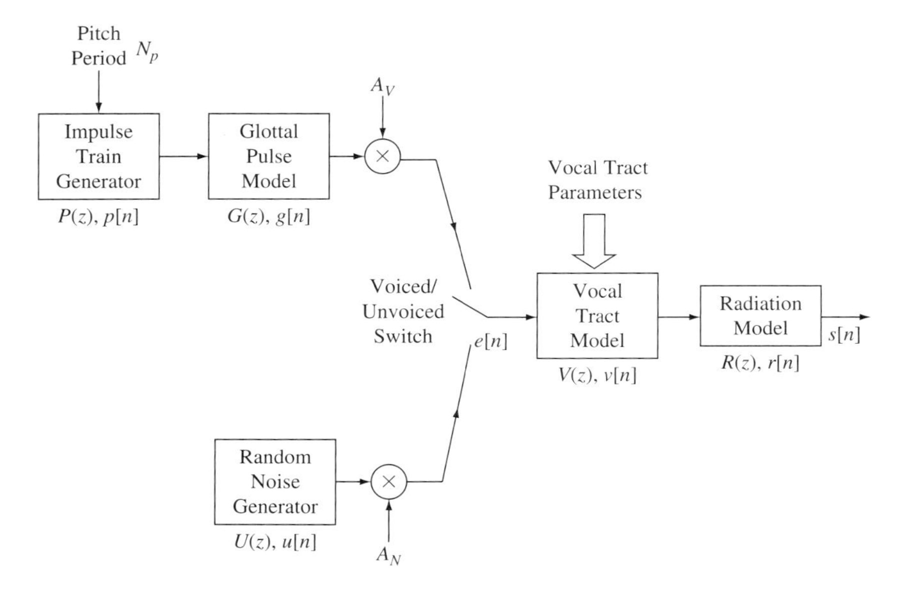

FIGURE 5.46
General discrete-time model for speech production.

5.3.4 The Complete Model

Putting all the ingredients together, we obtain the model of Figure 5.46. By switching between the voiced and unvoiced excitation generators, we can model the changing mode of excitation. The vocal tract can be modeled in a wide variety of ways as we have discussed. In some cases it is convenient to combine the glottal pulse and radiation models into a single system. In fact, we will see in Chapter 9 that in the case of linear predictive analysis, it is convenient to combine the glottal pulse, radiation, and vocal tract components all together and represent them as a single transfer function,

$$H(z) = G(z)V(z)R(z), \tag{5.106}$$

of the all-pole type. In other words, there is much latitude for variation in all the parts of the model of Figure 5.46.

At this point, a natural question concerns the limitations of such a model. Certainly the model is far from the partial differential equations with which we began this chapter. Fortunately none of the deficiencies of this model severely limits its applicability. First, there is the question of time variation of the parameters. In continuant sounds, such as vowels, the parameters change very slowly and the model works very well. With transient sounds, such as stops, the model is not as good but still adequate. It should be emphasized that our use of transfer functions and frequency response functions implicitly assumes that we can represent the speech signal on a "short-time" basis. That is, the parameters of the model are assumed to be constant over time intervals

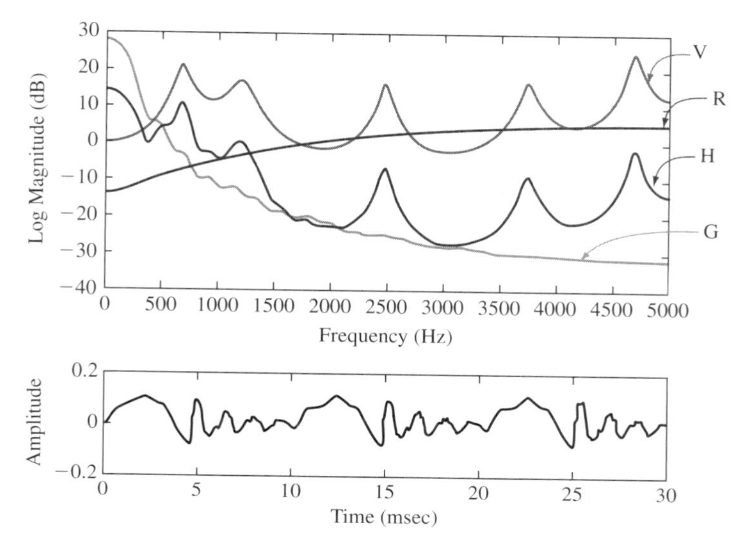

FIGURE 5.47
Frequency response (log magnitude) curves for the various components of the speech model, including the glottal source (G), the vocal tract response (V), the radiation (R), and the overall system response (H). The resulting time waveform of the sound is shown at the bottom of the figure.

typically 10–40 msec long. The transfer function $V(z)$ then really serves to define the structure of a linear system whose parameters vary slowly in time. We shall repeatedly invoke this principle of quasi-stationarity in subsequent chapters. A second limitation is the lack of provision for zeros, as required theoretically for nasals and fricatives. This is definitely a limitation for nasals, but not too severe for fricatives. Zeros can be included in the model, if desired. Third, the simple dichotomy of voiced–unvoiced excitation is inadequate for voiced fricatives. Simply adding the voiced and unvoiced excitations is inadequate since frication is correlated with the peaks of the glottal flow. A more sophisticated model for voiced fricatives has been developed [298] and can be employed when needed. Finally, a relatively minor concern is that the model of Figure 5.46 requires that the glottal pulses be spaced by an integer multiple of the sampling period, T. Winham and Steiglitz [421] have considered ways of eliminating this limitation in situations requiring precise pitch control.

A summary of the components of the speech model and their impact on the resulting speech model log magnitude response (for a voiced speech sound) is shown in Figure 5.47. The upper panel shows four log magnitude frequency response curves corresponding to the glottal source (labeled G in the figure), the vocal tract shape (labeled V), the radiation characteristic (labeled R), and the overall system response (labeled H). The bottom pane in this figure shows the resulting periodic time signal corresponding to the sound produced by the speech model with the selected vocal tract configuration.

5.4 SUMMARY

This chapter has focused upon two main areas: the physics of speech generation in the human vocal tract, and the resulting set of discrete-time models for speech production. Our discussion of the acoustic theory of speech production has been lengthy, but far from complete. Our purpose has been to provide adequate knowledge about the general properties of speech signals so as to motivate and suggest models that are useful for speech processing.

The models discussed in Sections 5.2 and 5.3 will be the basis for our discussion in the remainder of this book. We shall think of these models in two ways. One point of view is called speech analysis; the other is called speech synthesis. In speech analysis we are interested in techniques for estimating the parameters of the model from a natural speech signal that is assumed to be the output of the model. In speech synthesis, we wish to use the model to create a synthetic speech signal by controlling the model with suitable parameters. These two points of view will become intermingled in many cases and will arise in many problem areas. Underlying all our subsequent discussion will be models of the type discussed in this chapter. Having reviewed the subject of digital signal processing in Chapter 2 and the acoustic theory of speech production here, we are now ready to begin to see how digital signal processing techniques can be applied in processing speech signals.

PROBLEMS

5.1. In Section 5.1.7 Rosenberg's [326] continuous-time glottal pulse approximation was defined as

$$g_c(t) = \begin{cases} 0.5[1 - \cos(2\pi t/(2T_1))] & 0 \le t \le T_1 \\ \cos(2\pi(t - T_1)/(4T_2)) & T_1 < t \le T_1 + T_2, \end{cases}$$

where T_1 is the opening time and T_2 is the closing time.

(a) Show that the continuous-time Fourier transform of $g_c(t)$ is

$$G_c(\Omega) = 0.5G_1(\Omega) - 0.25G_1(\Omega - \Omega_1) - 0.25G_1(\Omega + \Omega_1) + 0.5[G_2(\Omega - \Omega_2) + G_2(\Omega + \Omega_2)]e^{-j\Omega T_1},$$

where $\Omega_1 = 2\pi/(2T_1)$ and $\Omega_2 = 2\pi/(4T_2)$ with

$$G_1(\Omega) = \frac{1 - e^{-j\Omega T_1}}{j\Omega},$$

$$G_2(\Omega) = \frac{1 - e^{-j\Omega T_2}}{j\Omega}.$$

(b) Show that if we form a periodic glottal excitation signal

$$u_G(t) = \sum_{k=-\infty}^{\infty} g_c(t - kT_0), \tag{5.107}$$

its continuous-time Fourier transform is

$$U_G(\Omega) = \frac{2\pi}{T_0} \sum_{k=-\infty}^{\infty} G_c(2\pi k/T_0)\delta(\Omega - 2\pi k/T_0); \tag{5.108}$$

i.e., the periodic excitation "samples" the spectrum of a single glottal pulse, creating an impulse spectrum with impulses spaced by the fundamental frequency $\Omega_0 = 2\pi/T_0$.

(c) Use MATLAB to evaluate the equation for $G_c(2\pi F)$ obtained in part (a) over the range $0 \leq F \leq 5000$ for $T_1 = 4$ msec and values of $T_2 = 0.5$, 1, 2, and 4 msec. Plot $20\log_{10}|G_c(2\pi F)|$ for all cases on the same set of axes. What is the effect of decreasing the closing time?

(d) For the case $T_1 = 4$ msec and $T_2 = 2$ msec, sample $g_c(t)$ to obtain $g[n] = g_c(nT)$, where $1/T = 10{,}000$ samples/sec. Use MATLAB's `freqz( )` function to evaluate the discrete-time Fourier transform, $G(e^{j\omega})$, of $g[n]$ at discrete-time frequencies $\omega = 2\pi FT$ corresponding to continuous-time frequencies $0 \leq F \leq 5000$. Plot both $20\log_{10}|G_c(2\pi F)|$ and $20\log_{10}|G(e^{j2\pi FT})|$ on the same axes. What is the effect of frequency-domain aliasing?

5.2. By substitution, show that Eqs. (5.4a) and (5.4b) are solutions to the partial differential equations of Eqs. (5.2a) and (5.2b).

5.3. Note that the reflection coefficients for the junction of two lossless acoustic tubes of areas A_k and A_{k+1} can be written as either

$$r_k = \frac{\dfrac{A_{k+1}}{A_k} - 1}{\dfrac{A_{k+1}}{A_k} + 1},$$

or

$$r_k = \frac{1 - \dfrac{A_k}{A_{k+1}}}{1 + \dfrac{A_k}{A_{k+1}}}.$$

(a) Show that since both A_k and A_{k+1} are positive,

$$-1 \leq r_k \leq 1.$$

(b) Show that if $0 < A_k < \infty$ and $0 < A_{k+1} < \infty$, then $-1 < r_k < 1$.

5.4. In the analysis of the acoustic tube model of the human vocal tract, we have made many assumptions to simplify the solutions. These include the following:

(a) The tube is lossless;

(b) The tube is rigid; i.e., its cross-sectional area does not change with time;

(c) The tube is constructed as a concatenation of a number of uniform sections, each with a different length and cross-sectional area;

(d) The tube's total length is fixed.

State the effects of each of these assumptions on the properties of the resulting system (volume velocity) transfer function of the vocal tract; i.e., had any of the assumptions been eliminated, how would the vocal tract transfer function have changed?

5.5. In determining the effect of the radiation load termination on a lossless tube model, it was assumed that Z_L was real and constant. A more realistic model is given by Eq. (5.32b).

(a) Beginning with the boundary condition

$$P_N(l_N, \Omega) = Z_L \cdot U_N(l_N, \Omega),$$

find a relation between the Fourier transforms of $u_N^-(t+\tau_N)$ and $u_N^+(t-\tau_N)$.

(b) From the frequency-domain relation found in (a) and Eq. (5.32b), show that $u_N^-(t+\tau_N)$ and $u_N^+(t-\tau_N)$ satisfy the ordinary differential equation

$$L_r\left[R_r + \frac{\rho c}{A_N}\right]\frac{du_N^-(t+\tau_N)}{dt} + \frac{\rho c}{A_N}R_r u_N^-(t+\tau_N)$$
$$= L_r\left[R_r - \frac{\rho c}{A_N}\right]\frac{du_N^+(t-\tau_N)}{dt} - \frac{\rho c}{A_N}R_r u_N^+(t-\tau_N).$$

5.6. By substitution of Eq. (5.55) into Eq. (5.54), show that Eqs. (5.53) and (5.54) are equivalent.

5.7. Consider the two-tube model of Figure 5.30. Write the frequency domain equations for this model and show that the transfer function between the input and output volume velocities is given by Eq. (5.56b).

5.8. Consider an ideal lossless tube model for the production of vowels consisting of two sections as shown in Figure P5.8.

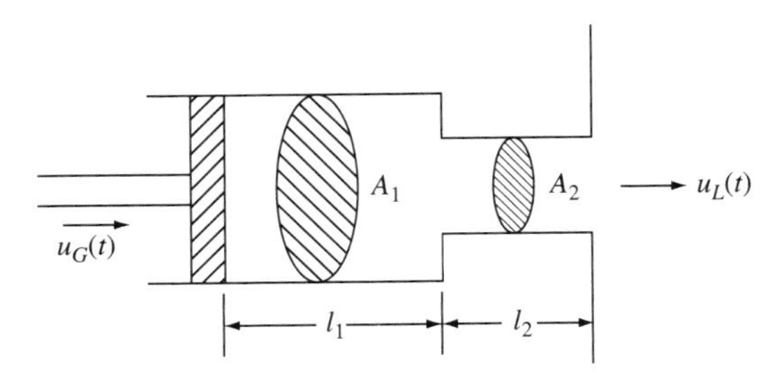

FIGURE P5.8
Two-section ideal lossless tube model.

Assume that the terminations at the glottis and lips are completely lossless. For the above conditions, the system function of the lossless tube model will be obtained from Eq. (5.57) by substituting $r_G = r_L = 1$ and

$$r_1 = \frac{A_2 - A_1}{A_2 + A_1}.$$

(a) Show that the poles of the system are on the $j\Omega$ axis and are located at values of Ω satisfying the equations

$$\cos\left[\Omega(\tau_1 + \tau_2)\right] + r_1 \cos\left[\Omega(\tau_2 - \tau_1)\right] = 0,$$

or equivalently

$$\frac{A_1}{A_2}\tan(\Omega\tau_2) = \cot(\Omega\tau_1),$$

where $\tau_1 = l_1/c$, $\tau_2 = l_2/c$, and c is the velocity of sound.

(b) The values of Ω that satisfy the equations derived in (a) are the formant frequencies of the lossless tube model. By a judicious choice of the parameters l_1, l_2, A_1, and A_2, we can approximate the vocal tract configurations of vowels, and by solving the above equations, obtain the formant frequencies for the model. The following table gives parameters for approximations to several vowel configurations [105]. Solve for the formant frequencies for each case. (Note that the non-linear equations must be solved graphically or iteratively.) Use $c = 35{,}000$ cm/sec.

TABLE P5.1 Parameters of the two-tube model for various vowel approximations.

Vowel	l_1 (cm)	A_1 cm^2	l_2 (cm)	A_2 cm^2
/i/	9	8	6	1
/ae/	4	1	13	8
/a/	9	1	8	7
/Λ /	17	6	0	6

5.9. An acoustic tube of total length $l = 17.5$ cm consists of two uniform sections. Assume that the boundary conditions at the glottis and the lips are $r_L = r_G = 1$; i.e., the equivalent reflection coefficients at the glottis and the lips are equal to 1. Figure P5.9 shows a frequency "line plot" of *some* of the resonant modes of the system transfer function.

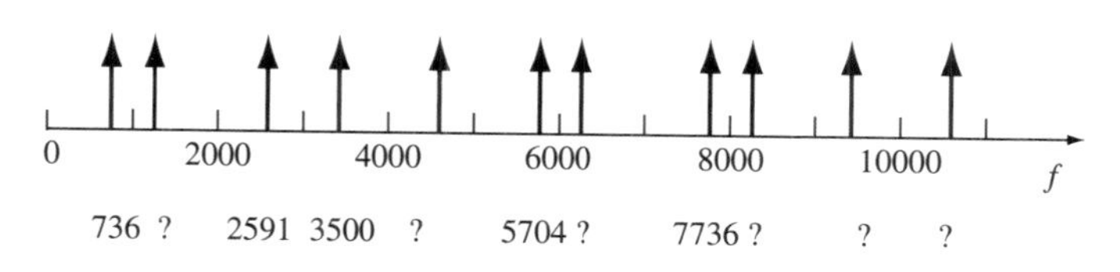

FIGURE P5.9
Locations of some of the resonances of a two-tube model of the vocal tract.

(a) Find the resonant frequencies that are indicated by a /?/ in the figure.

(b) What is the smallest length difference between the two sections?

(c) Estimate the reflection coefficient at the junction between the two uniform sections.

5.10. By substituting the appropriate matrices $\hat{\mathbf{Q}}_1$ and $\hat{\mathbf{Q}}_2$ into Eq. (5.80), show that the transfer function of a two-tube discrete-time vocal tract model is given by Eq. (5.82).

5.11. Show that if $|a| < 1$,

$$1 - az^{-1} = \frac{1}{\sum_{n=0}^{\infty} a^n z^{-n}}$$

and thus, that a zero can be approximated as closely as desired by multiple poles.

5.12. The transfer function of a digital formant resonator is of the form

$$V_k(z) = \frac{1 - 2|z_k|\cos\theta_k + |z_k|^2}{1 - 2|z_k|\cos\theta_k z^{-1} + |z_k|^2 z^{-2}},$$

where $|z_k| = e^{-\sigma_k T}$ and $\theta_k = 2\pi F_k T$.

(a) Plot the locations of the poles of $V_k(z)$ in the z-plane. Also plot the corresponding analog poles in the s-plane.

(b) Write the difference equation relating the output, $y_k[n]$, of $V_k(z)$ to its input, $x_k[n]$.

(c) Draw a digital network implementation of the digital formant network with three multipliers.

(d) By rearranging the terms in the difference equation obtained in part (b), draw a digital network implementation of the digital formant network that only requires two multiplications.

5.13. Consider the system function for a discrete-time vocal tract model

$$V(z) = \frac{G}{\prod_{k=1}^{N}(1 - z_k z^{-1})}.$$

(a) Show that $V(z)$ can be expressed as the partial-fraction expansion

$$V(z) = \sum_{k=1}^{M}\left[\frac{G}{1 - z_k z^{-1}} + \frac{G_k^*}{1 - z_k^* z^{-1}}\right],$$

where M is the largest integer contained in $(N+1)/2$, and it is assumed that all the poles of $V(z)$ are complex. Give an expression for the G_k's in the above expression.

(b) Combine terms in the above partial-fraction expansion to show that

$$V(z) = \sum_{k=1}^{M}\frac{B_k - C_k z^{-1}}{1 - 2|z_k|\cos\theta_k z^{-1} + |z_k|^2 z^{-2}},$$

where $z_k = |z_k|e^{j\theta_k}$. Give expressions for B_k and C_k in terms of G_k and z_k. This expression is the *parallel form* representation of $V(z)$.

(c) Draw the digital network diagram for the parallel form implementation of $V(z)$ for $M = 3$.

(d) For a given all-pole system function $V(z)$, which implementation would require the most multiplications–the parallel form or the cascade form, as suggested in Problem 5.12?

5.14. The relationship between pressure and volume velocity at the lips is given by

$$P(l,s) = Z_L(s)U(l,s),$$

where $P(l,s)$ and $U(l,s)$ are the Laplace transforms of $p(l,t)$ and $u(l,t)$ respectively, and

$$Z_L(s) = \frac{sR_rL_r}{R_r + sL_r},$$

where

$$R_r = \frac{128}{9\pi^2}, \qquad L_r = \frac{8a}{3\pi c};$$

c is the velocity of sound and a is the radius of the lip opening. In a discrete-time model, we desire a corresponding relationship of the form [Eq. (5.103)]

$$P_L(z) = R(z)U_L(z),$$

where $P_L(z)$ and $U_L(z)$ are z-transforms of $p_L[n]$ and $u_L[n]$, the sampled versions of the bandlimited pressure and volume velocity. One approach to obtaining $R(z)$ is to use the bilinear transformation [270], i.e.,

$$R(z) = Z_L(s)\Bigg|_{s = \frac{2}{T}\left[\frac{1-z^{-1}}{1+z^{-1}}\right]}.$$

(a) For $Z_L(s)$ as given above, determine $R(z)$.

(b) Write the corresponding difference equation that relates $p_L[n]$ and $u_L[n]$.

(c) Give the locations of the pole and zero of $R(z)$.

(d) If $c = 35{,}000$ cm/sec, $T = 10^{-4}$ sec^{-1}, and 0.5 cm $< a < 1.3$ cm, what is the range of pole values?

(e) A simple approximation to $R(z)$ obtained above is obtained by neglecting the pole; i.e.,

$$\hat{R}(z) = R_0(1 - z^{-1}).$$

For $a = 1$ cm and $T = 10^{-4}$, find R_0 such that $\hat{R}(-1) = Z_L(\infty) = R(-1)$.

(f) Sketch the frequency responses $Z_L(\Omega)$, $R(e^{j\Omega T})$, and $\hat{R}(e^{j\Omega T})$ as a function of Ω for $a = 1$ cm and $T = 10^{-4}$ for $0 \leq \Omega \leq \pi/T$.

5.15. A simple approximate model for a glottal pulse is given in Figure P5.15a.

(a) Find the z-transform, $G_1(z)$, of this sequence. (Hint: Note that $g_1[n]$ can be expressed as the convolution of the sequence

$$p[n] = \begin{cases} 1 & 0 \leq n \leq N-1 \\ 0 & \text{otherwise} \end{cases}$$

with itself.)

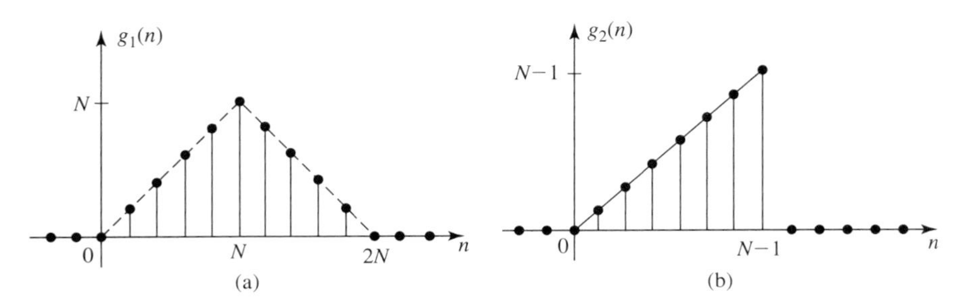

FIGURE P5.15
Glottal pulse approximations.

(b) Plot the poles and zeros of $G_1(z)$ in the z-plane for $N = 10$.

(c) Sketch the magnitude of the Fourier transform of $g_1[n]$ as a function of ω.

Now consider the glottal pulse model $g_2[n]$ as given in Figure P5.15b.

(d) Show that the z-transform, $G_2(z)$, is given by

$$G_2(z) = z^{-1}\sum_{n=0}^{N-2}(n+1)z^{-n}$$

$$= z^{-1}\left\{\frac{1 - Nz^{-(N-1)} + (N-1)z^{-N}}{(1-z^{-1})^2}\right\}.$$

(Hint: Use the fact that the z-transform of $nx[n]$ is $-z\dfrac{dX(z)}{dz}$.)

(e) Show that in general, $G_2(z)$ must have at least one zero outside the unit circle. Find the zeros of $G_2(z)$ for $N = 4$.

5.16. A commonly used approximation to the glottal pulse is

$$g[n] = \begin{cases} na^n & n \geq 0 \\ 0 & n < 0 \end{cases}.$$

(a) Find the z-transform of $g[n]$.

(b) Sketch the Fourier transform, $G(e^{j\omega})$, as a function of ω.

(c) Show how a should be chosen so that

$$20\log_{10}|G(e^{j0})| - 20\log_{10}|G(e^{j\pi})| = 60 \text{ db}.$$

5.17. The shape of the glottal pulse from the vocal cords can be approximated by the impulse response of a second-order filter with system function:

$$G(z) = \frac{az^{-1}}{(1-az^{-1})^2}, \quad 0 < a < 1.$$

(a) Plot the glottal pulse model impulse response, $g[n]$, for $a = 0.95$ and for $a = 0.8$.

(b) Plot the corresponding log magnitude response, $20\log_{10}(|G(e^{j\omega})|$, in dB versus ω (or versus f) for the two values of a used in part (a) of this problem.

(c) The effect of lip radiation can be modeled by a single zero at $z = 1$. Repeat part (b) with the inclusion of this extra zero.

(d) Draw a flow graph representation of the system that models the combined glottal pulse and lip radiation effects.

5.18. Assume we know the system function (transfer function), $V(z)$, of a discrete-time model of the vocal tract; i.e.,

$$V(z) = \frac{G}{D(z)} = \frac{G}{1 - \sum_{k=1}^{N} \alpha_{.k} z^{-k}}.$$

We wish to obtain the areas and reflection coefficients for a lossless tube model, as shown in Figure P5.18, for the case of $N = 3$ sections, with $r_G = 1$ and $r_N = r_L$.

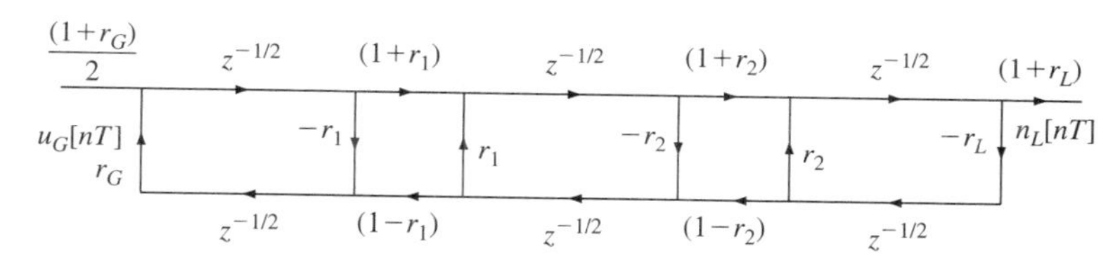

FIGURE P5.18
Lossless tube model for three-section tube model approximation to the vocal tract.

We have seen that the denominator polynomial, $D(z)$, satisfies the recursion:

$$D_0(z) = 1,$$
$$D_k(z) = D_{k-1}(z) + r_k z^{-k} D_{k-1}(z^{-1}), \quad k = 1, 2, \ldots, N,$$
$$D(z) = D_N(z).$$

The goal of this problem is to develop an algorithm for finding the reflection coefficients, and the areas of a lossless tube having a given system function.

(a) Show that r_N is equal to the coefficient of z^{-N} in the denominator of $V(z)$; i.e., $r_N = -\alpha_N$.

(b) Using the recursion in reverse order, show that

$$D_{k-1}(z) = \frac{D_k(z) - r_k D_k(z^{-1})}{1 - r_k^2}, \quad k = N, N-1, \ldots, 2.$$

(c) How would you find r_{k-1} from $D_{k-1}(z)$?

(d) Using the results of parts (b) and (c), state an algorithm for finding all the reflection coefficients, $r_k, k = 1, 2, \ldots, N$, and all the tube areas, $A_k, k = 1, 2, \ldots, N$. Are the A_k's unique?

5.19. (MATLAB Exercise) Use MATLAB to compute and plot the vocal tract log magnitude spectrum and the locations of the formants for a two-tube model of the vocal tract, as shown in Figure 5.30, and as given in Equation (5.56b). Your MATLAB code should accept the input lengths (l_1 and l_2) and areas (A_1 and A_2) of a two-tube model of the vocal tract, along with the reflection coefficients at the glottis (r_G) and at the lips (r_L). Test your code on the following examples:

1. $l_1 = 10, A_1 = 1, l_2 = 7.5, A_2 = 1, r_G = 0.7, r_L = 0.7$
2. $l_1 = 15.5, A_1 = 8, l_2 = 2, A_2 = 1, r_G = 0.7, r_L = 0.7$
3. $l_1 = 9.5, A_1 = 8, l_2 = 8, A_2 = 1, r_G = 0.7, r_L = 0.7$
4. $l_1 = 8.8, A_1 = 8, l_2 = 8.8, A_2 = 1, r_G = 0.7, r_L = 0.7.$

The first example is a sanity check on your code and should provide uniform tube formants of 500, 1500, 2500, 3500, 4500, ... Hz if you used a value of $c = 35{,}000$ cm/sec as the speed of sound. The remaining three examples are close approximations to three of the two-tube configurations in Figure 5.31. What happens to the log magnitude spectral plots if both r_G and r_L are set to 1.0 (rather than 0.7)?

5.20. (MATLAB Exercise) Use MATLAB to compute and plot the vocal tract log magnitude spectrum and the locations of the formants for a three-tube model of the vocal tract. Figure P5.20 shows a complete flow diagram of a three-tube model of the human vocal tract. Assuming a glottal reflection coefficient, r_G, and a lip reflection coefficient, r_L, with the three tubes (beginning at the glottis and ending at the lips) defined by lengths $\{l_1, l_2, l_3\}$, areas $\{A_1, A_2, A_3\}$, and reflection coefficients $\{r_1, r_2\}$, where:

$$r_1 = \frac{A_2 - A_1}{A_2 + A_1},$$

$$r_2 = \frac{A_3 - A_2}{A_3 + A_2}.$$

We can determine the transfer function of the three-tube model by writing down the flow equations at each of the nodes and working our way backwards from the lips to the glottis.

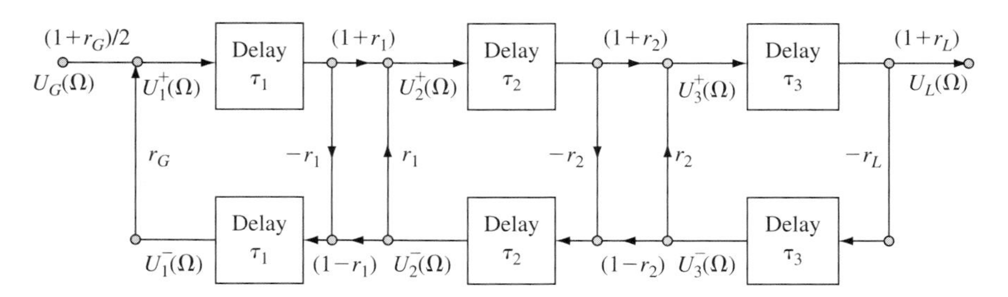

FIGURE P5.20
Complete flow diagram of a three-tube model of the vocal tract.

The resulting three-tube model vocal tract transfer function can be shown to be of the form:

$$V_a(\Omega) = \frac{U_L(\Omega)}{U_G(\Omega)}$$
$$= \frac{0.5(1+r_L)(1+r_1)(1+r_2)(1+r_G)e^{-j\Omega(\tau_1+\tau_2+\tau_3)}}{D(\Omega)},$$

where

$$D(\Omega) = 1 + r_2 r_L e^{-2j\Omega\tau_3} + r_1 r_L e^{-2j\Omega(\tau_2+\tau_3)} + r_1 r_2 e^{-2j\Omega\tau_2}$$
$$+ r_L r_G e^{-2j\Omega(\tau_1+\tau_2+\tau_3)} + r_2 r_G e^{-2j\Omega(\tau_1+\tau_2)}$$
$$+ r_1 r_G e^{-2j\Omega\tau_1} + r_1 r_2 r_L r_G e^{-2j\Omega(\tau_1+\tau_3)}.$$

Your MATLAB code should accept the input lengths (l_1, l_2, and l_3) and areas (A_1, A_2, and A_3) of a three-tube model of the vocal tract, along with the reflection coefficients at the glottis (r_G) and at the lips (r_L). Test your code on the following examples:

1. $l_1 = 5, A_1 = 1, l_2 = 5, A_2 = 1, l_3 = 7.5, A_3 = 1, r_G = 0.7, r_L = 0.7$
2. $l_1 = 7.5, A_1 = 8, l_2 = 8, A_2 = 8, l_3 = 2, A_3 = 1, r_G = 0.7, r_L = 0.7$
3. $l_1 = 15.5, A_1 = 8, l_2 = 1, A_2 = 1, l_3 = 1, A_3 = 1, r_G = 0.7, r_L = 0.7$
4. $l_1 = 5, A_1 = 1, l_2 = 5, A_2 = 8, l_3 = 7.5, A_3 = 4, r_G = 0.7, r_L = 0.7.$

The first example is a sanity check on your code and should provide uniform tube formants of 500, 1500, 2500, 3500, 4500, ... Hz if you used a value of $c = 35{,}000$ cm/sec as the speed of sound. The second and third examples are really two-tube models broken apart into three-tube models by splitting the length of either the first tube (Example 2) or the second tube (Example 3). The resulting formants should agree with those obtained from the two-tube model of Problem 5.19. The remaining example is an arbitrary configuration of three tube lengths and areas.

5.21. (MATLAB Exercise) Use MATLAB to compute and plot the vocal tract log magnitude spectrum and to determine the locations of the formants for a uniform p-tube model of the vocal tract by solving the nodal equations for the forward and backward signals. The concept is simple, namely excite the digitized vocal tract of Figure 5.37, part (c), with an impulse of volume velocity at the glottis, i.e., $u_G[n] = \delta[n]$ and then propagate the impulse (in both time and space) through the set of p uniform tubes until you reach the lips section where the vocal tract output signal $u_L[n]$ is obtained. By iterating the nodal solutions for each time interval n, you obtain the vocal tract impulse response which can be transformed into a log magnitude spectrum and plotted on a logarithmic scale. A simple peak magnitude test can be applied to find the formants and to estimate the formant bandwidths.

The mat files `area_1.mat`, `area_2.mat` and `area_3.mat` provide some of the inputs which are required to test and debug your code. Each of these mat files contains the following parameters:

- `p` = number of vocal tract sections
- `nfft` = size of FFT that should be used to compute log magnitude spectrum
- `c` = speed of sound, nominally 35,000 cm/sec

- `ls` = length of each of the p sections (in cm)
- `fs` = sampling frequency for the simulation
- `area(1:p)` = set of p tube areas from the glottis to the lips
- `source` = reference source for areas(1:p)

By way of example, the mat file area_1.mat contains the following data:

- `p` = 10
- `nfft` = 1024
- `c` = 35000
- `ls` = 1.75
- `fs` = 10000
- `area(1:10)` = [1.5360 2.4000 0.7680 1.3440 2.3000 3.3120 5.9500 8.3500 6.4320 4.9900]
- `source` = 5.40 (source reference figure from book).

The three mat files contain data from Figure 5.40 `(area_1.mat)`, Figure 5.16 `(area_2.mat)` and for a uniform tube `(area_3.mat)`.

In addition to the parameters from the mat files, you will need to choose appropriate values for the glottal resistance (r_G) and for the lip resistance (r_L). It is suggested that you use the values $r_G = r_L = 0.7$ to test your code, and then vary r_G and r_L between 0.7 and 1.0 to see the impact of smaller losses at the glottis and lips. The ultimate test of using $r_G = r_L = 1.0$ leads to log magnitude spectra that are essentially impulses at the formant frequencies (no loss in the system) and these lead to computational issues in solving for formant frequencies and bandwidths.

For each of the three mat files, determine the vocal tract impulse responses and plot the log magnitude spectra. Estimate the formant bandwidths and center frequencies and give an indication of the location of the formants on the plots of the log magnitude spectra.

CHAPTER 6

Time-Domain Methods for Speech Processing

6.1 INTRODUCTION

Our analysis of sound propagation in the human vocal tract in Chapter 5 led to the source/system model for speech production shown in Figure 6.1. We are now ready to begin to see how digital signal processing methods can be applied to speech signals to estimate the properties and parameters of such models. In this chapter we consider signal processing methods that focus on direct manipulation of the time waveform. Chapters 7 and 8 describe methods of speech analysis that are primarily based on Fourier transform methods, while Chapter 9, on linear predictive analysis, unites the ideas of the present chapter with the frequency-domain point of view.

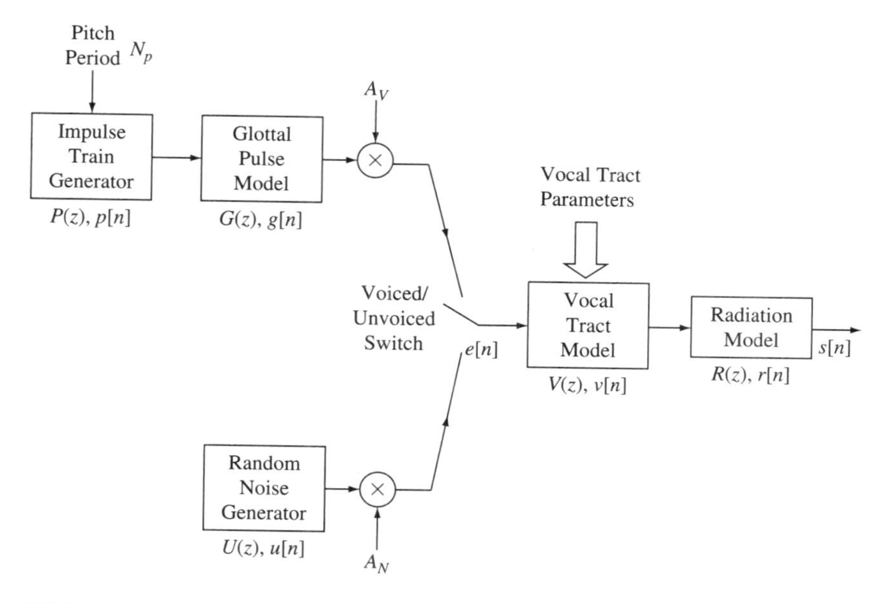

FIGURE 6.1
Model for speech production and synthesis.

In most speech processing applications, the first step usually is to obtain a convenient and useful parametric representation (i.e., other than the samples of the speech waveform) of the information carried by the speech signal. This can be achieved by assuming that the speech signal, $s[n]$, is the output of a parametric synthesis model such as the one shown in Figure 6.1. As discussed in Chapter 5 and represented in Figure 6.1, the information carried by the speech signal includes, but is not limited to, the following:

- the (time-varying) pitch period (in samples), N_p, (or pitch frequency, $F_p = F_s/N_p$, where F_s is the speech sampling frequency), for regions of voiced speech, including possibly the locations of the pitch excitation impulses that define the periods between adjacent pitch pulses
- the glottal pulse model, $g[n]$
- the time-varying amplitude of voiced excitation, A_V
- the time-varying amplitude of unvoiced excitation, A_N
- the time-varying excitation type for the speech signal; i.e., quasi-periodic pitch pulses for voiced sounds or pseudo-random noise for unvoiced sounds
- the time-varying vocal tract model impulse response, $v[n]$, or equivalently, a set of vocal tract parameters that control a vocal tract model
- the radiation model impulse response, $r[n]$ (assumed to be fixed over time).[1]

Usually, the goal of speech analysis is to estimate (as a function of time) parameters of a speech representation such as Figure 6.1, and use them as a basis for an application such as a speech coder, a speech synthesizer, a speech recognizer, etc. In our discussions, it will be convenient to refer to different "representations" of the speech signal as depicted in Figure 6.2. The waveform representation or time-domain representation is simply the sample sequence $s[n]$. We shall refer to any representation derived by digital signal processing operations as an *alternate representation*. If the alternate representation is a parameter or set of parameters of a speech model such as Figure 6.1, then the alternate representation is a *parametric representation*. As shown in Figure 6.2, it is often convenient to think of parametric representations as being

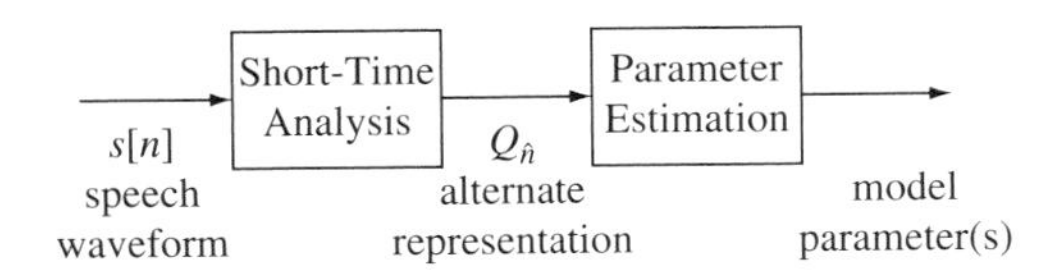

FIGURE 6.2
Digital signal processing for conversion of the speech waveform to an alternate representation that is more suitable for speech analysis; i.e., model parameter estimation.

[1]Often, the effects of the glottal pulse, vocal tract, and radiation are combined (for voiced speech) into one time-varying impulse response.

derived by a two-stage process, with the first stage being the computation of an alternate or intermediate representation from which the desired parameter set is estimated.

We define time-domain processing of speech as performing direct operations on the speech waveform (or a filtered version of the waveform), and frequency-domain processing as performing operations on a Fourier representation of the speech signal. In this chapter we discuss a set of processing techniques that are, by this definition, time-domain methods. That is, the alternate representation methods involve direct operations on the waveform of the speech signal. This is in contrast to the techniques to be described in Chapters 7–9, which we classify as frequency-domain methods since they involve (either explicitly or implicitly) some form of spectrum representation.[2]

Examples of time-domain parametric representations of the speech signal to be discussed in this chapter include short-time (log) energy, short-time zero-crossing (or level-crossing) rate, and short-time autocorrelation. Such representations are attractive because the required digital processing is very simple to implement, and, in spite of this simplicity, the resulting representations provide a useful basis for estimating important parameters of speech models. The required precision of any speech representation is dictated by the intended application as well as the particular information in the speech signal that is to be measured, or, in some cases, made more prominent. For example, the purpose of the digital processing may be to facilitate the determination of whether a particular section of waveform corresponds to speech or background signal (or possibly noise). Or, in a somewhat more complicated situation, we may wish to make a three-way classification as to whether a section of the signal is voiced speech, unvoiced speech, or silence (background signal). In such cases, a representation that discards "irrelevant" information and places the desired features clearly in evidence is to be preferred over a more detailed representation that retains all the inherent information in the speech signal. Other situations, such as digital transmission, may require the most accurate representation of the speech signal that can be obtained with a given set of constraints (e.g., bit rate constraints) so that a waveform can be reconstructed that is perceptually equivalent to the original sampled speech signal. Such applications are generally based on frequency-domain representations such as the short-time Fourier transform or linear predictive representations, which are derived from the short-time autocorrelation function, and will be discussed in later chapters.

We begin this chapter by introducing a general framework for discussing time-dependent (or short-time) processing techniques. The remainder of the chapter focuses on several important examples of short-time processing that are based on direct transformations of the sampled time waveform. Our approach in this chapter, and in Chapters 7–9, is to focus on the fundamental signal processing that leads to the following alternative representations of the speech signal:

- short-time energy
- short-time zero-crossing rate

[2]In all cases, we will assume that the speech signal has been bandlimited and sampled at a rate that is at least twice the Nyquist frequency. Further, we will assume that the resulting samples are finely quantized so that the quantization error is negligible. The effects of quantization are discussed in Chapter 11.

- short-time autocorrelation function
- short-time Fourier transform.

These alternate representations are the basis for almost every speech processing algorithm that we will discuss in subsequent chapters. With the exception of the short-time Fourier transform (Chapter 7), these alternative representations are not sufficient to reconstruct the original time waveform nor are they direct estimates of any of the parameters of speech models such as Figure 6.1. However, most speech processing algorithms are based upon further processing of one or more of these alternative representations. Later, in Chapter 10, we will discuss a number of methods (algorithms) for estimating such features of speech models as voiced/unvoiced classification, pitch detection, and formant frequencies from the various parametric representations discussed in this chapter and in Chapters 7–9. In Chapters 11–14 we will see that these same representations provide the basis for a wide range of algorithms for speech (and audio) coding, synthesis, and recognition.

6.2 SHORT-TIME ANALYSIS OF SPEECH

A plot of waveform samples, along with annotated phoneme regions (at a sampling rate of $F_s = 10{,}000$ samples/sec) representing a speech signal produced by a male speaker is shown in Figure 6.3. This plot serves as a reminder of how the properties of the speech signal change slowly with time, moving from one relatively stationary state (e.g., phoneme) to another. In this example, the excitation mode begins as unvoiced, then switches to voiced, then back to unvoiced, then back to voiced, and finally back to unvoiced. Also observe that there is a significant variation in the peak amplitude of the signal, and there is a steady variation of fundamental (pitch) frequency within voiced regions. In some applications it is useful to be able to define and mark the boundaries between voiced and unvoiced intervals automatically and also to automatically determine the pitch period within the voiced intervals. The fact that these variations are often so evident in a waveform plot such as that of Figure 6.3 suggests that simple time-domain processing techniques should be capable of providing useful estimates of signal features such as signal energy, excitation mode and state (voiced or unvoiced), pitch, and possibly even estimates of vocal tract parameters such as formant frequencies and bandwidths.

The underlying assumption in almost all speech processing systems is that the properties of the speech signal change relatively slowly with time (compared to the detailed sample-to-sample variations of the waveform) with rates of change on the order of 10–30 times/sec. This slow variation corresponds to speech production rates on the order of 5–15 sounds/sec.[3] This assumption leads to a variety of "short-time" processing methods in which short segments of the speech signal are isolated and processed as if they were short segments from a sustained sound with fixed (non-time-varying) properties. This segmentation or "framing" process is illustrated

[3]For the example of Figure 6.3, there are eight distinct phonemes in 0.6 seconds, or a rate of about 13 sounds/sec.

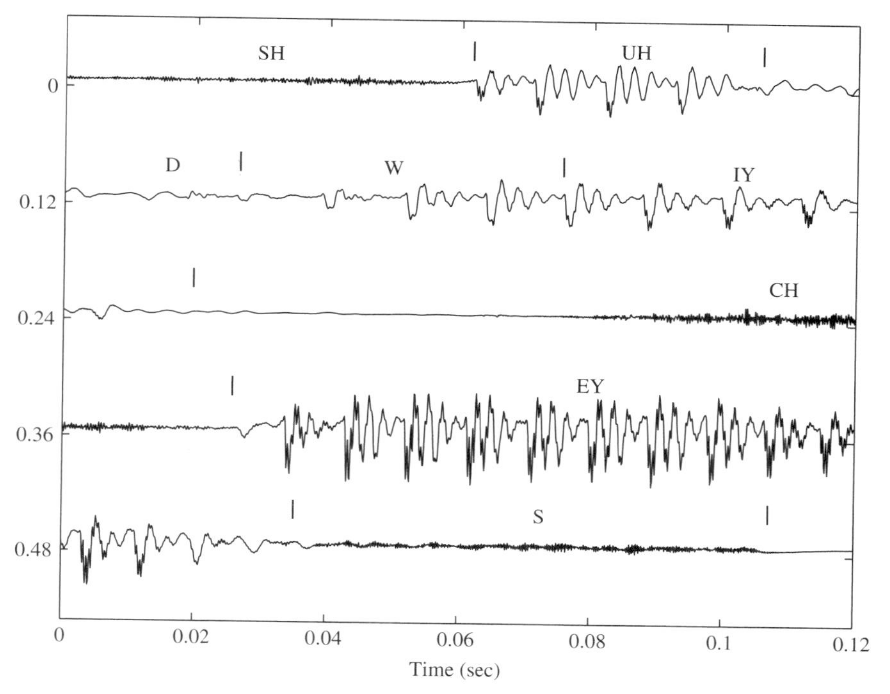

FIGURE 6.3
Waveform of an utterance of /SH UH D - W IY - CH EY S/ ("should we chase"). The sampling rate is 10 kHz, and the samples are connected by straight lines. Each line in the plot corresponds to 1200 speech samples or 0.12 seconds of signal.

in Figure 6.4, which shows 2000 samples of a speech signal ($F_s = 16{,}000$ Hz) together with 641-point Hamming windows[4] positioned at intervals of 240 samples. Each of these shifted windows selects a time interval for analysis. This type of segmentation can be repeated until we reach the end of a sentence or phrase to be processed, or in real-time applications, it could continue indefinitely. The resulting speech segments are generally referred to as *analysis frames*. The result of the processing on each frame can be either a single number, or a set of numbers. Therefore, such processing produces a new time-dependent sequence that can serve as a representation of some relevant property of the speech signal.

A key issue in such short-time processing systems is the choice of segment duration or frame length. The shorter the segment, the less likely that the speech properties of the segment will vary significantly over the segment duration. Thus, the ability to rapidly track abrupt waveform changes (e.g., from unvoiced to voiced which occurs at about sample 840 in Figure 6.4) is best for short segments. However, speech parameters estimated from short segments (on the order of 5–20 msec) have a great deal of "uncertainty" due to the small amount of speech data that is available for processing,

[4]The Hamming window is defined in Eq. (6.4).

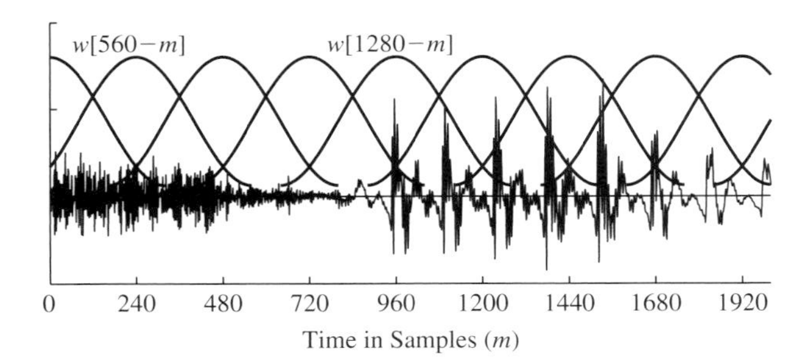

FIGURE 6.4
Example of frames and windows in short-time processing. For this example the sampling rate, F_s, is 16,000 samples/sec, the frame length, L, is 641 samples (equivalent to 40 msec frame duration), and the frame shift, R, is 240 samples (equivalent to 15 msec frame shift), leading to an analysis system with a frame rate of 66.7 frames/sec.

thus leading to highly variable speech parameter estimates. Similarly, speech parameters estimated from medium length segments (20–100 msec) often exhibit a great deal of "uncertainty" because they can encompass multiple sounds (as in the window labeled $w[1280 - m]$ in Figure 6.4), and therefore again provide highly variable estimates of basic speech parameters. Finally, the use of long speech segments (100–500 msec) leads also to inaccuracies because of the large amount of sound change over such a long duration. Furthermore, long segments make it difficult to pinpoint sharp changes (such as voiced/unvoiced transitions). Thus, there is always a degree of "uncertainty" in short-time measurements on speech signals, and we cannot eliminate completely the variability of speech parameter estimates based on finite-duration speech segments. Therefore, a compromise analysis frame duration of between 10 and 40 msec is most often used in speech processing systems.

6.2.1 General Framework for Short-Time Analysis

All the short-time processing techniques that we will discuss in this chapter, as well as the short-time Fourier representation of Chapter 7, can be represented mathematically in the form

$$Q_{\hat{n}} = \sum_{m=-\infty}^{\infty} T(x[m])\tilde{w}[\hat{n} - m], \tag{6.1}$$

where $\tilde{w}[\hat{n} - m]$ is a sliding analysis window and $T(\,\cdot\,)$ represents some operation on the input signal. The quantity $Q_{\hat{n}}$ represents the value (or vector of values) of the alternate short-time representation of a speech signal $x[n]$ at analysis time $\hat{n}$.[5] As depicted in Figure 6.5, the speech signal (possibly after linear filtering to isolate a

[5] In general, we use the notation n and m for discrete indices for sequences, but whenever we want to indicate a specific analysis time, we use the notation $\hat{n}$.

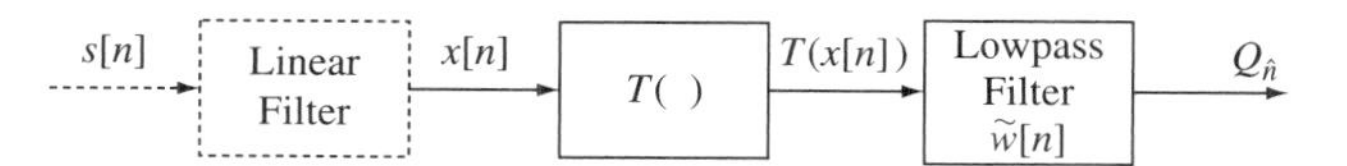

FIGURE 6.5
General representation of the short-time analysis principle.

desired frequency band) is subjected to a transformation, $T(\cdot)$, which may be either linear or non-linear, and which may depend upon some adjustable parameter or set of parameters. The purpose of this transformation is to make some property of the speech signal more prominent. The resulting sequence is then multiplied by a lowpass window sequence, $\tilde{w}[\hat{n}-m]$, positioned at a particular analysis time $\hat{n}$. The multiplication by the lowpass window represents the framing operation; i.e., the focusing of the analysis on the time interval "at or around" the analysis time, $\hat{n}$. The product is then summed over all non-zero values inside the shifted window. This corresponds to averaging the transformed signal over the time interval selected by the window. The combined effect of multiplying by the shifted window and summing the products is equivalent to lowpass filtering by a filter whose impulse response is the window sequence. Often Eq. (6.1) is normalized by dividing by the effective window length, i.e.,

$$L_{\text{eff}} = \sum_{m=-\infty}^{\infty} \tilde{w}[m], \tag{6.2}$$

so that $Q_{\hat{n}}$ is a weighted average. This has the effect of normalizing the frequency response of the effective lowpass filter so that $\tilde{W}(e^{j0}) = 1$.

Usually the window sequence (impulse response) will be of finite duration, but this is not a strict requirement. What is needed is that $\tilde{w}[m]$ be "smooth" and concentrated in time so that it has lowpass filter characteristics. The values $Q_{\hat{n}}$ are therefore a sequence of local weighted averages of values of the sequence $T(x[m])$. Examples of operators of interest for speech analysis are $T(x[m]) = x[m]e^{-j\omega m}$, leading to the short-time Fourier transform, which will be discussed in detail in Chapter 7, and $T(x[m]) = (x[m])^2$, for the short-time energy, which we will discuss in more detail in Section 6.3.

6.2.2 Filtering and Sampling in Short-Time Analysis

Observe that Eq. (6.1) states that $Q_{\hat{n}}$ is the discrete-time convolution of the modified speech signal $T(x[n])$ with the window sequence $\tilde{w}[n]$. Thus, $\tilde{w}[n]$ serves as the impulse response of a linear filter. Although the output of the filter can be computed at the input speech sampling rate by moving the analysis window in steps of 1 sample, typically the window is moved in jumps of $R > 1$ samples as illustrated in Figure 6.4. This corresponds to downsampling the output of the filter by the factor R; i.e., the short-time representation is evaluated at times $\hat{n} = rR$, where r is an integer. The choice of R is not independent of the choice of the window length. Clearly, if the window is of length L samples, then we should choose $R < L$ so that each speech sample is included in at least one analysis segment. This is illustrated in Figure 6.4 where the windows overlap by $L - R = 641 - 240 = 401$ samples. Typically, the analysis windows overlap by more than 50% of the window length.

To be more specific about how the sampling interval (or equivalently, the window shift), R, should be chosen, it will be useful to consider two commonly used window sequences; the rectangular window

$$w_R[n] = \begin{cases} 1 & 0 \le n \le L-1 \\ 0 & \text{otherwise,} \end{cases} \tag{6.3}$$

and the Hamming window

$$w_H[n] = \begin{cases} 0.54 - 0.46\cos(2\pi n/(L-1)) & 0 \le n \le L-1 \\ 0 & \text{otherwise.} \end{cases} \tag{6.4}$$

Time-domain plots of the rectangular and Hamming windows for $L = 21$ are given in Figure 6.6. If we substitute Eq. (6.3) for $w[n]$ in Eq. (6.1), we obtain

$$Q_{\hat{n}} = \sum_{m=\hat{n}-L+1}^{\hat{n}} T(x[m]); \tag{6.5}$$

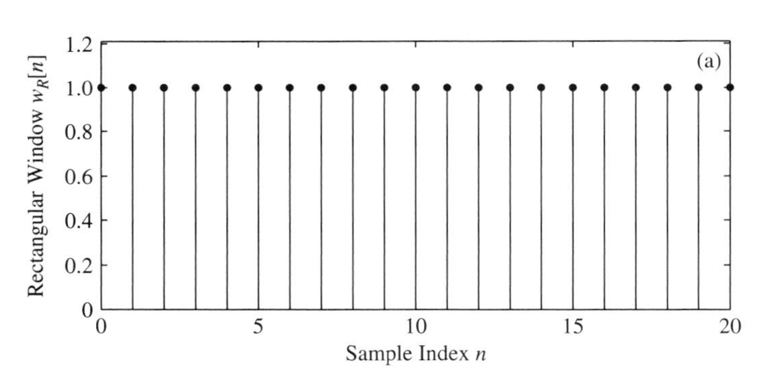

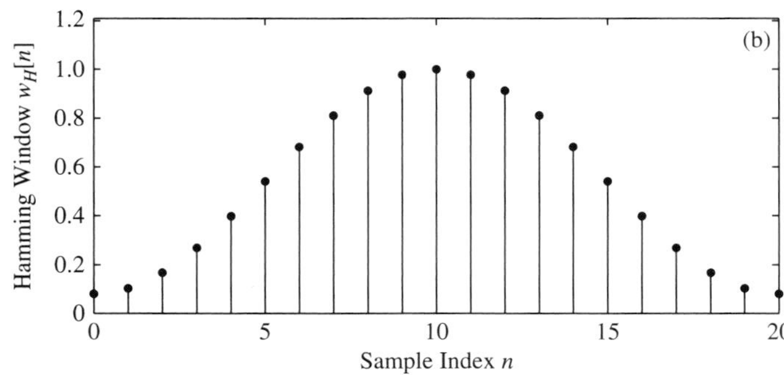

FIGURE 6.6
Plots of the time responses of a 21-point (a) rectangular window; (b) Hamming window.

i.e., the rectangular window corresponds to applying equal weight to all samples in the interval $(\hat{n} - L + 1)$ to $\hat{n}$. If the Hamming window is used, Eq. (6.5) will incorporate the window sequence $w_H[\hat{n} - m]$ from Eq. (6.4) as weighting coefficients, with the same limits on the summation. The frequency response of an L-point rectangular window is easily shown to be (see Problem 6.1)

$$W_R(e^{j\omega}) = \sum_{n=0}^{L-1} e^{-j\omega n} = \frac{\sin(\omega L/2)}{\sin(\omega/2)} e^{-j\omega(L-1)/2}. \tag{6.6}$$

The log magnitude (in dB) of $W_R(e^{j\omega})$ is shown in Figure 6.7a for a 51-point window ($L = 51$). Note that the first zero of Eq. (6.6) occurs at $\omega = 2\pi/L$, corresponding to analog frequency,

$$F = \frac{F_s}{L}, \tag{6.7}$$

where $F_s = 1/T$ is the sampling frequency. Although Figure 6.7a shows that the log magnitude frequency response does not fall off very quickly above this frequency, nevertheless it is often considered to be the cutoff frequency of the lowpass filter

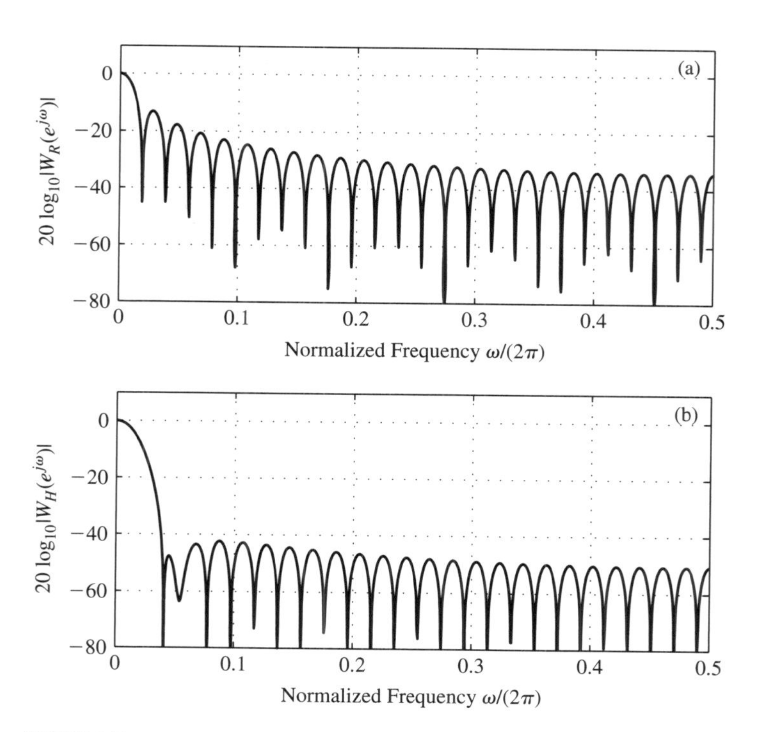

FIGURE 6.7
Fourier transform of (a) 51-point rectangular window; (b) 51-point Hamming window.

corresponding to the rectangular window. Thus for an input sampling frequency of $F_s = 10$ kHz ($T = 0.0001$), the cutoff frequency of an $L = 51$ point rectangular window is approximately $F_s/L = 10{,}000/51 = 196$ Hz. The log magnitude frequency response of a 51-point Hamming window is shown in Figure 6.7b. It can be shown that the nominal cutoff frequency of the Hamming window is approximately $4\pi/L$ in normalized radian units [270] corresponding to a bandwidth of $2F_s/L$ in analog frequency units. Therefore, the bandwidth of a 51-point Hamming window, again with an input sampling frequency of $F_s = 10{,}000$ Hz, is 392 Hz. Figure 6.7b confirms that the bandwidth of the Hamming window is about twice the bandwidth of a rectangular window of the same length. It is also clear that the Hamming window gives much greater attenuation (> 40 dB) outside the passband than the comparable rectangular window (> 14 dB). It is straightforward to show that the attenuation of both these windows is essentially *independent* of the window duration. Thus, increasing the window length, L, simply decreases the bandwidth.[6]

From this discussion it follows that for typical finite-duration analysis windows, the short-time representation $Q_{\hat{n}}$ has a restricted lowpass bandwidth that is inversely proportional to the window length L. Therefore it should be possible to sample $Q_{\hat{n}}$ at a lower rate than the input sampling rate. In particular if we use the nominal cutoff of the Hamming window, it follows from the sampling theorem that the sampling rate of $Q_{\hat{n}}$ should be greater than or equal to $4F_s/L$, or $R \leq L/4$. With its nominal cutoff of F_s/L, the rectangular window would seem to require only half the sampling rate; however, its rather poor frequency selectivity would generally result in significant aliasing at this sampling rate.

We have seen that the sampling rate of a short-time representation is linked to the window length. Recalling our earlier discussion, if L is too small, i.e., on the order of a pitch period or less, $Q_{\hat{n}}$ will fluctuate very rapidly, depending on exact details of the waveform. If L is too large, i.e., on the order of tens of pitch periods, $Q_{\hat{n}}$ will change very slowly and thus will not adequately reflect the changing properties of the speech signal. Unfortunately, this implies that no single value of L is entirely satisfactory because the duration of a pitch period varies from about 20 samples (or 500 Hz pitch frequency at a 10 kHz sampling rate) for a high pitch female or a child, up to 125 samples (80 Hz pitch frequency at a 10 kHz sampling rate) for a very low pitch male. With these shortcomings in mind, a suitable practical choice for L is on the order of 100–400 for a 10 kHz sampling rate (i.e., 10–40 msec duration), and the short-time representation is typically computed with window overlap of 50 to 75% ($R = L/2$ to $R = L/4$).

6.3 SHORT-TIME ENERGY AND SHORT-TIME MAGNITUDE

The energy of a discrete-time signal is defined as

$$E = \sum_{m=-\infty}^{\infty} (x[m])^2. \tag{6.8}$$

[6]A detailed discussion of the properties of windows is not required for the short-time representations of this chapter. Further discussion is given in Chapter 7.

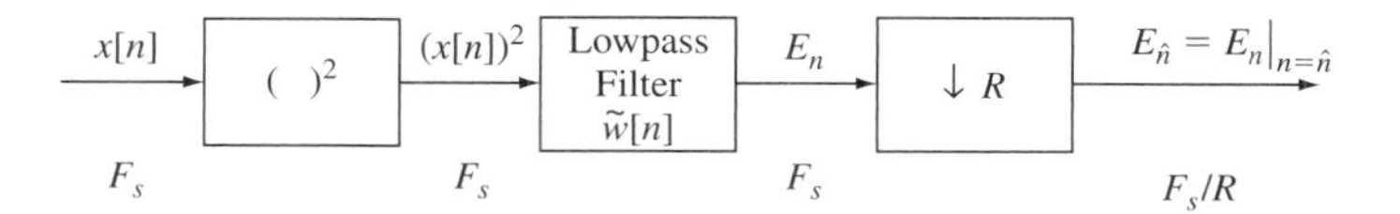

FIGURE 6.8
Block diagram representation of computation of the short-time energy.

The energy of a sequence is a single number that has little meaning or utility as a representation for speech because it gives no information about the time-dependent properties of the speech signal. What is needed is something that is sensitive to the time-varying changes of signal amplitude with time. For this, the short-time energy is much more useful. Our definition of the short-time energy is

$$E_{\hat{n}} = \sum_{m=-\infty}^{\infty} (x[m]w[\hat{n}-m])^2 = \sum_{m=-\infty}^{\infty} (x[m])^2\tilde{w}[\hat{n}-m], \tag{6.9}$$

where $w[\hat{n}-m]$ is a window that is applied directly to the speech samples before squaring, and $\tilde{w}[\hat{n}-m]$ is a corresponding window that can be applied equivalently after squaring. Equation (6.9) can be seen to be in the form of Eq. (6.1) with $T(x[m]) = (x[m])^2$ and $\tilde{w}[m] = w^2[m]$.[7] The computation of the short-time energy representation is depicted in Figure 6.8. It can be seen that the sampling rate of the signal changes from F_s at the input and output of the squaring box to F_s/R at the output of the lowpass filter, $\tilde{w}[n]$.

For an L-point rectangular window, the effective window is $\tilde{w}[m] = w[m]$, and it therefore follows that $E_{\hat{n}}$ is

$$E_{\hat{n}} = \sum_{m=\hat{n}-L+1}^{\hat{n}} (x[m])^2. \tag{6.10}$$

Figure 6.9 depicts the computation of the short-time energy sequence with the rectangular window. Note that as $\hat{n}$ varies, the window literally slides along the sequence of squared values [in general, $T(x[m])$] selecting the interval to be involved in the computation.

Figures 6.10 and 6.11 show the effects of varying the window length L (for the rectangular and Hamming windows, respectively) on the short-time energy representation for an utterance of the text "What she said", spoken by a male speaker. As L increases, the energy contour becomes smoother for both windows. This is because the bandwidth of the effective lowpass filter is inversely proportional to L. These figures illustrate that for a given window length, the Hamming window produces a somewhat smoother curve than does the rectangular window. This is because, even though the

[7]Observe that in the case of the short-time energy, we can choose $w[m] = (w_H[m])^{1/2}$ so that the effective impulse response for short-time analysis is $\tilde{w}[m] = w_H[m]$. Alternatively, if we use $w[m] = w_H[m]$, then the effective analysis impulse response is $\tilde{w}[m] = w_H^2[m]$.

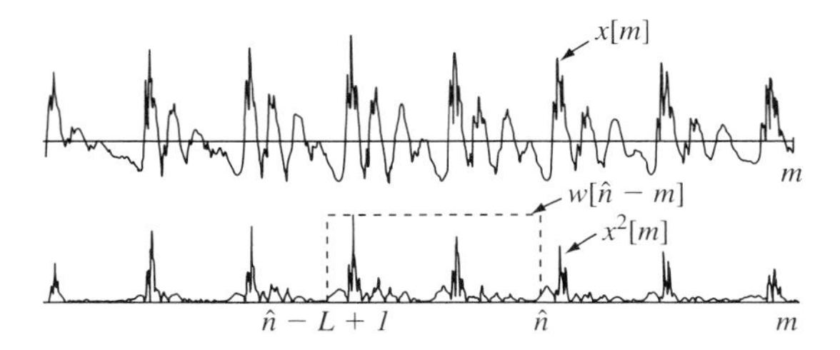

FIGURE 6.9
Illustration of the computation of short-time energy.

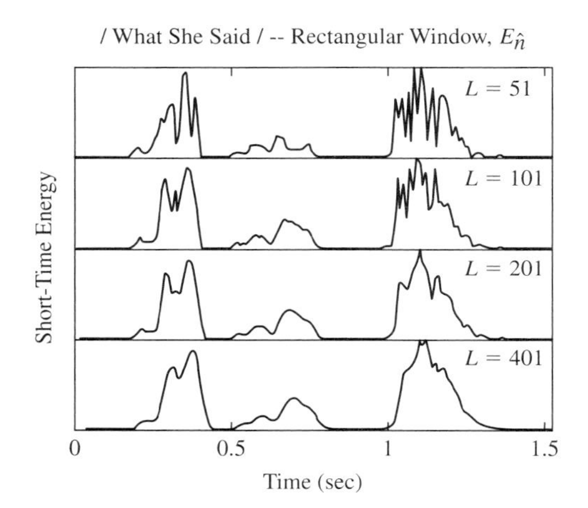

FIGURE 6.10
Short-time energy functions for rectangular windows of length $L = 51$, 101, 201, and 401.

"nominal" cutoff is lower for the rectangular window, the high-frequency attenuation is much greater for the Hamming window filter.

The major significance of $E_{\hat{n}}$ is that it provides a basis for distinguishing voiced speech segments from unvoiced speech segments. As can be seen in Figures 6.10 and 6.11, the values of $E_{\hat{n}}$ for the unvoiced segments are significantly smaller than for voiced segments. This difference in amplitude is enhanced by the squaring operation. The energy function can also be used to locate approximately the time at which voiced speech becomes unvoiced, and vice versa, and, for very high quality speech (high signal-to-noise ratio), the energy can be used to distinguish speech from silence (or background signal). Several speech detection and voiced–unvoiced classification methods are discussed in Section 6.4 and again, in detail, in Chapter 10.

6.3.1 Automatic Gain Control Based on Short-Time Energy

An application of the short-time energy representation is a simple automatic gain control (AGC) mechanism for speech waveform coding (see Chapter 11 for a more

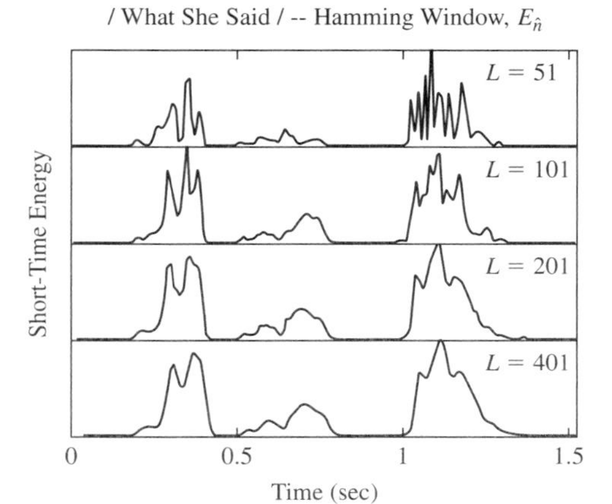

FIGURE 6.11
Short-time energy functions for Hamming windows of length $L = 51$, 101, 201, and 401.

complete discussion). The purpose of an AGC is to keep the signal amplitude as large as possible without saturating or overflowing the allowable dynamic range of a digital representation of the speech samples.

In this type of application, it is of interest to compute the short-time energy at every sample of the input, since we wish to apply the AGC to individual samples. While finite-duration windows can be used for this purpose, it can be more efficient to use an infinite-duration window (impulse response) so that the computation can be recursive. As a simple example, consider the exponential window sequence

$$\tilde{w}[n] = (1-\alpha)\alpha^{n-1}\, u[n-1] = \begin{cases} (1-\alpha)\alpha^{n-1} & n \geq 1 \\ 0 & n < 1, \end{cases} \tag{6.11}$$

so that Eq. (6.9) becomes

$$E_{\hat{n}} = (1-\alpha) \sum_{m=-\infty}^{\hat{n}-1} (x[m])^2. \tag{6.12}$$

Observe that $\tilde{w}[n]$ is defined so that in convolution, $\tilde{w}[\hat{n}-m]$ applies a decreasing weighting to all samples $(x[m])^2$ that are *prior* to $\hat{n}$ but not including the one at analysis time $\hat{n}$. The reason for the one-sample delay that this implies will be discussed below. To obtain the recursive implementation for the short-time energy, it is useful to note that the z-transform of Eq. (6.11) is

$$\tilde{W}(z) = \frac{(1-\alpha)z^{-1}}{1-\alpha z^{-1}}. \tag{6.13}$$

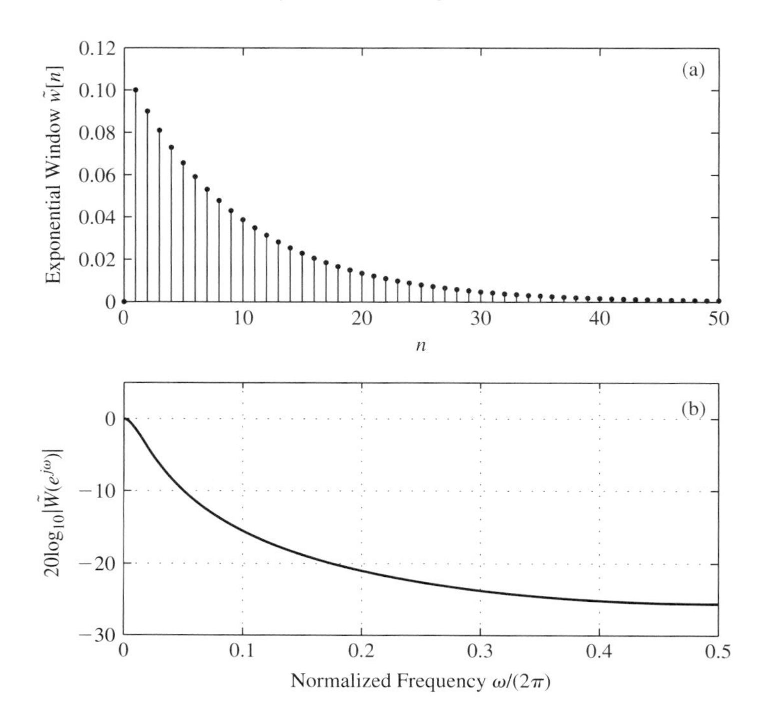

FIGURE 6.12
Exponential window for short-time energy computation using a value of $\alpha = 0.9$: (a) $\tilde{w}[n]$ (impulse response of analysis filter) and (b) $20\log_{10}|\tilde{W}(e^{j\omega})|$ (log magnitude frequency response of the analysis filter).

The discrete-time Fourier transform (DTFT) (frequency response of the analysis filter) is

$$\tilde{W}(e^{j\omega}) = \frac{(1-\alpha)e^{-j\omega}}{1-\alpha e^{-j\omega}}. \tag{6.14}$$

Figure 6.12a shows 51 samples of the exponential window (for a value of $\alpha = 0.9$) and the corresponding DTFT (log magnitude response) is shown in Figure 6.12b. These are the impulse response and frequency response, respectively, of the recursive short-time energy analysis filter. Note that by including the scale factor $(1-\alpha)$ in the numerator, we ensure that $\tilde{W}(e^{j0}) = 1$; i.e., the low frequency gain is around unity (0 dB) irrespective of the value of α. By increasing or decreasing the parameter α, we can make the effective window longer or shorter respectively. The effect on the corresponding frequency response is, of course, opposite. Increasing α makes the filter more lowpass, and vice versa.

To anticipate the need for a more convenient notation, we will denote the short-time energy as $E_n = \sigma^2[n]$, where we have used the index $[n]$ instead of the subscript $\hat{n}$ to emphasize that the time scale of the short-time energy and the input $x[n]$ are the

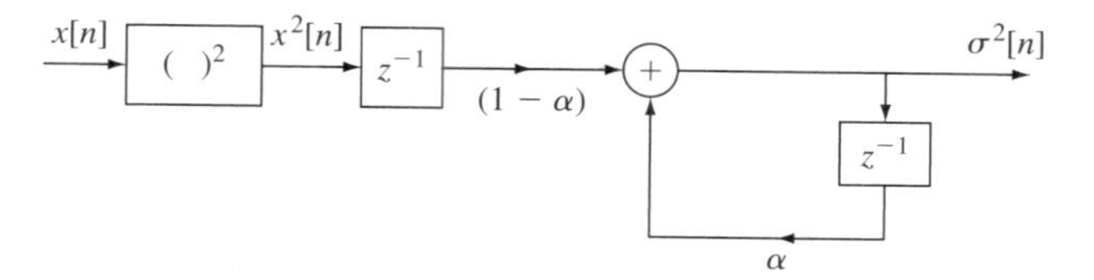

FIGURE 6.13
Block diagram of recursive computation of the short-time energy for an exponential window.

same; i.e., we compute the short-time energy at every input sample time. Furthermore, we use the notation σ^2 to denote the fact that the short-time energy is an estimate of the variance of $x[m]$.[8] Now since $\sigma^2[n]$ is the output of the filter with impulse response $\tilde{w}[n]$ in Eq. (6.11), it follows that it satisfies the recursive difference equation

$$\sigma^2[n] = \alpha\sigma^2[n-1] + (1-\alpha)x^2[n-1]. \tag{6.15}$$

This computation of the short-time energy is depicted in Figure 6.13.

Now we can define an AGC of the form

$$G[n] = \frac{G_0}{\sigma[n]}, \tag{6.16}$$

where G_0 is a constant gain level to which we attempt to equalize the level of all frames, where we have used the square-root of the short-time energy $\sigma^2[n]$ to match the variations to those of the samples $x[n]$. By including the delay of one sample in the definition of $\tilde{w}[m]$, $G[n]$ at time n depends only on the previous samples of $x[n]$. The capability of the AGC control of Eq. (6.16) to equalize the variance (or more precisely the standard deviation) of a speech waveform is illustrated in Figure 6.14, which shows a section of a speech waveform at the top (showing the computed standard deviation, $\sigma[n]$ superimposed on the waveform plot), along with the standard deviation equalized waveform, $x[n] \cdot G[n]$ at the bottom. The value of α was 0.9. With this value of α, the window is short enough to track the local amplitude variations due to the voiced nature of the speech signal, but not the detailed variations within each period, which are due primarily to the formant frequencies. Larger values of α would introduce more smoothing so that the AGC would act over a longer (e.g., syllabic) time scale.

It is also possible to use finite-length windows to compute the short-time energy for AGC. In such cases, the short-time energy is usually computed at a lower sampling rate (every R samples) and the gain computed using Eq. (6.16) is held constant over the corresponding interval of R samples. This approach would introduce a delay of at least

[8]It is easily shown that if $x[m]$ is a zero-mean random signal, then the expected value of Eq. (6.12) is the variance of $x[m]$.

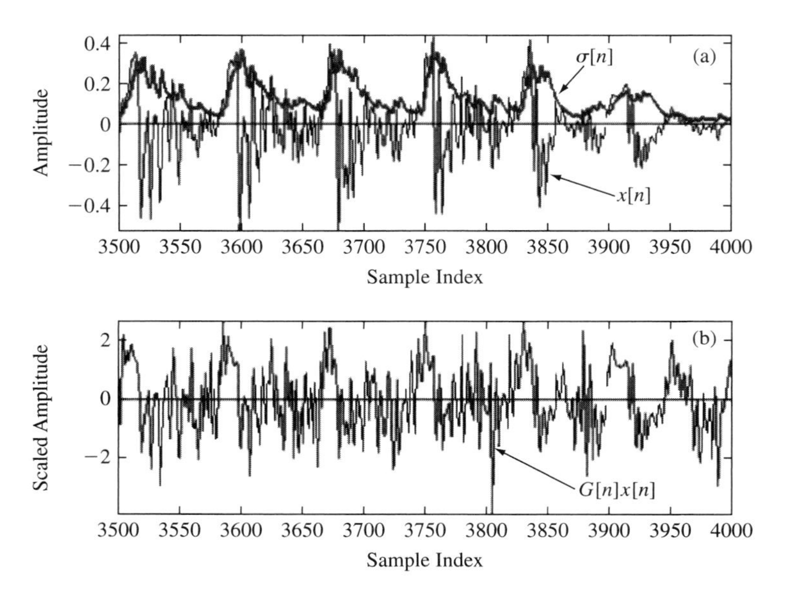

FIGURE 6.14
AGC of speech waveform: (a) waveform and computed standard deviation, $\sigma[n]$; (b) equalized waveform, $x[n] \cdot G[n]$, for a value of $\alpha = 0.9$.

R samples, but it would be useful in applications where block processing is required for other processing tasks.

6.3.2 Short-Time Magnitude

One difficulty with the short-time energy function, as defined by Eq. (6.9), is that it is very sensitive to large signal levels (since they enter the computation as a square), thereby emphasizing large sample-to-sample variations in $x[n]$. In the previous subsection, we addressed this by taking the square-root of the short-time energy. Another way of alleviating this problem is to define a short-time magnitude function,

$$M_{\hat{n}} = \sum_{m=-\infty}^{\infty} |x[m]w[\hat{n}-m]| = \sum_{m=-\infty}^{\infty} |x[m]|\tilde{w}[\hat{n}-m], \tag{6.17}$$

where the weighted sum of absolute values of the signal is computed instead of the sum of squares. (Since the window sequence is normally positive so that $|w[n]| = w[n]$, it can be applied either before or after taking the magnitude of the speech samples.) Figure 6.15 shows how Eq. (6.17) can be implemented as a linear filtering operation on $|x[n]|$. Note that a simplification in arithmetic is achieved by eliminating the squaring operation in the short-time energy computation.

Figures 6.16 and 6.17 show short-time magnitude plots comparable to the short-time energy plots of Figures 6.10 and 6.11. The differences are particularly noticeable in the unvoiced regions. For the short-time magnitude computation of Eq. (6.17), the

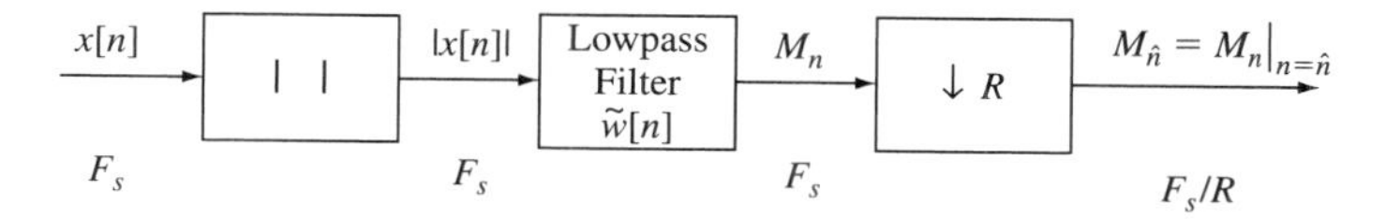

FIGURE 6.15
Block diagram representation of computation of the short-time magnitude function.

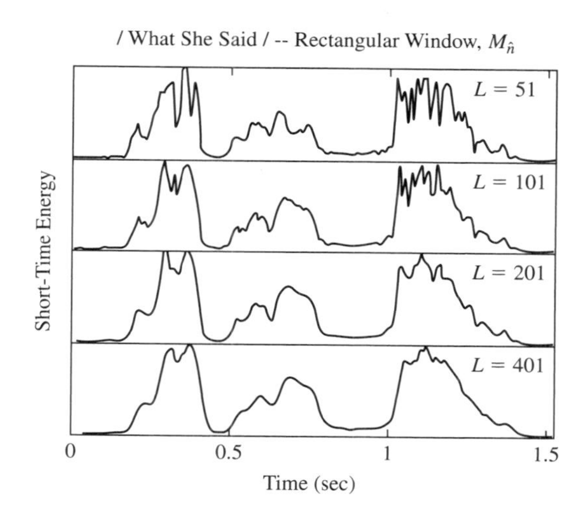

FIGURE 6.16
Short-time magnitude functions for rectangular windows of length $L = 51$, 101, 201, and 401.

dynamic range (ratio of maximum to minimum) is approximately the square-root of the dynamic range for the standard energy computation. Thus the differences in level between voiced and unvoiced regions are not as pronounced as for the short-time energy.

Since the bandwidths of both the short-time energy and the short-time magnitude function are just that of the lowpass filter, it is evident that these functions need not be sampled as frequently as the speech signal. For example, for a window of duration 20 msec, a sampling rate of about 100 samples/sec is adequate. Clearly, this means that much information has been discarded in obtaining these short-time representations. However, it is also clear that information regarding speech amplitude is retained in a very convenient form.

To conclude our comments on the properties of short-time energy and short-time magnitude, it is instructive to point out that the window need not be restricted to rectangular or Hamming form, or indeed to any function commonly used as a window in spectrum analysis or digital filter design. All that is required is that the effective lowpass filter provide adequate smoothing. Thus, we can design a lowpass filter by any of the standard filter design methods [270, 305]. Furthermore, the filter can be either a

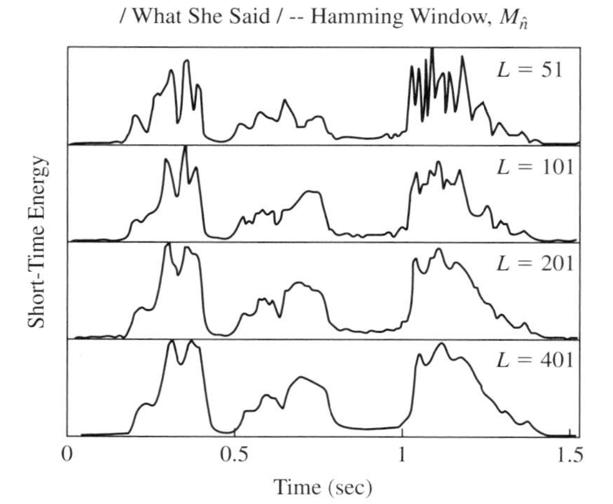

FIGURE 6.17
Short-time magnitude functions for Hamming windows of length $L = 51, 101, 201$, and 401.

finite duration impulse response (FIR) or an infinite duration impulse response (IIR) filter as in the case of the exponential windows.

There is an advantage in having the impulse response (window) be always positive since this guarantees that the short-time energy or short-time magnitude will always be positive. FIR filters (such as the rectangular or Hamming impulse responses) have the advantage that the output can easily be computed at a lower sampling rate than the input simply by moving the window more than one sample between computations. For example, if the speech signal is sampled at $F_s = 10{,}000$ samples/sec, and a window of duration 20 msec (200 samples) is used, the short-time energy can be computed at a sampling rate of about 100 samples/sec, or once every 100 samples at the input sampling rate.

As seen earlier with the AGC example, it is not necessary to use a finite-length window. If we use the exponential window of Eq. (6.11), the short-time magnitude would be

$$M_n = \alpha M_{n-1} + (1 - \alpha)|x[n-1]|, \tag{6.18}$$

where we have again used the appropriate filter normalization factor $(1 - \alpha)$ and included the delay of one sample.

To use Eqs. (6.15) and (6.18), the quantities E_n and M_n must be computed at each sample of the input speech signal, even though a much lower sampling rate would suffice. Sometimes this is required anyway, as in the case of some waveform coding schemes discussed in Chapter 11, and this recursive method is then very attractive. However, when a lower sampling rate will suffice, the non-recursive method may require less arithmetic. (See Problem 6.4.) Another factor of interest is the delay inherent in the lowpass filtering operation. The windows of Eqs. (6.3) and (6.4) have been defined so that they correspond to a delay of $(L-1)/2$ samples. Since they have linear phase, the origin of the energy function can be redefined to take this delay into account. For recursive implementation, the phase is non-linear and therefore the delay cannot be exactly compensated.

6.4 SHORT-TIME ZERO-CROSSING RATE

In the context of discrete-time signals, a zero-crossing is said to occur if successive waveform samples have different algebraic signs. This is illustrated in Figure 6.18. The rate (number of crossings per some unit of time) at which zero-crossings occur is a simple (and often highly reliable) measure of the frequency content of a signal. This is particularly true of narrowband signals. For example, a sinusoidal signal of frequency F_0, sampled at a rate F_s, has F_s/F_0 samples per cycle of the sine wave. Each cycle has two zero-crossings, so that the average rate of zero-crossings per sample is

$$\begin{aligned} Z^{(1)} &= (2) \text{ crossings/cycle } \cdot (F_0/F_s) \text{ cycles/sample} \\ &= 2\frac{F_0}{F_s} \text{ crossings/sample}, \end{aligned} \tag{6.19}$$

and the number of crossings in an interval of M samples is

$$Z^{(M)} = M \cdot (2F_0/F_s) \text{ crossings/}(M \text{ samples}), \tag{6.20}$$

where we use the notation $Z^{(M)}$ to denote the number of crossings per M samples of the waveform. Thus $Z^{(1)}$ is the number of crossings per sample, and $Z^{(100)}$ is the number of zero-crossings per $M = 100$ samples. (The symbol Z, without a superscript, denotes zero-crossings per sample.) The short-time zero-crossing rate gives a reasonable way to estimate the frequency of a sine wave, as we will now show. An alternative form of Eq. (6.20) is

$$F_e = 0.5 F_s Z^{(1)}, \tag{6.21}$$

where F_e is the equivalent sinusoidal frequency corresponding to a given zero-crossing rate, $Z^{(1)}$, per sample. If the signal is a single sinusoid of frequency F_0, then $F_e = F_0$.

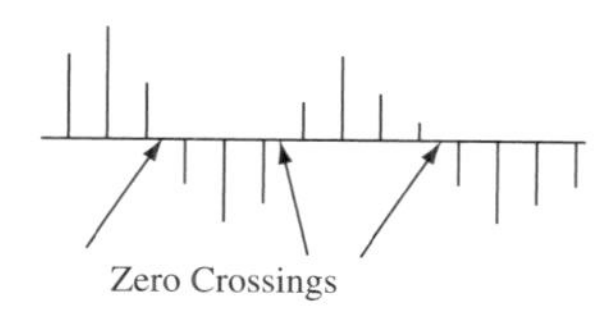

FIGURE 6.18
Plots of waveform showing locations of zero-crossings of the signal.

Thus, Eq. (6.21) can be used to estimate the frequency of a sinusoid, and if the signal is not a sinusoid, Eq. (6.21) can be thought of as an *equivalent sinusoidal frequency* for the signal.

Consider the following examples of zero-crossing rates of sinusoidal waveforms; (We assume a sampling rate of $F_s = 10{,}000$ samples/sec for each of the examples.)

- for a 100 Hz sinusoid ($F_0 = 100$ Hz), with $F_s/F_0 = 10{,}000/100 = 100$ samples per cycle, we get $Z^{(1)} = 2/100 = 0.02$ crossings/sample, or $Z^{(100)} = (2/100) * 100 = 2$ crossings/10 msec interval (or 100 samples);
- for a 1000 Hz sinusoid ($F_0 = 1000$ Hz), with $F_s/F_0 = 10{,}000/1000 = 10$ samples per cycle, we get $Z^{(1)} = 2/10 = 0.2$ crossings/sample, or $Z^{(100)} = (2/10) * 100 = 20$ crossings/10 msec interval (or 100 samples);
- for a 5000 Hz sinusoid ($F_0 = 5000$ Hz), with $F_s/F_0 = 10{,}000/5000 = 2$ samples per cycle, we get $Z^{(1)} = 2/2 = 1$ crossings/sample, or $Z^{(100)} = (2/2) * 100 = 100$ crossings/10 msec interval (or 100 samples).

From the above examples we see that the range of zero-crossing rates ($Z^{(1)}$) for pure sinusoids (or for any other signal) goes from 2 crossings/sample (for a sinusoid with a frequency of half the sampling rate of the system), down to about 0.02 crossings/sample (for a 100 Hz sinusoid). Similarly the zero-crossing rates on a 10 msec interval ($Z^{(100)}$ for $F_s = 10{,}000$) are in the range from 100 (again for a sinusoid with frequency of half the sampling rate of the system) down to 2 for a 100 Hz sinusoid.

There are a number of practical considerations in implementing a representation based on the short-time zero-crossing rate. Although the basic algorithm for detecting a zero-crossing requires only a comparison of signs of pairs of successive samples, special care must be taken in the sampling process. Clearly, the zero-crossing rate is strongly affected by DC offset in the analog-to-digital converter, 60 Hz hum in the signal, and any noise that may be present in the digitizing system. As an extreme example, if the DC offset is greater than the peak value of the signal, no zero-crossings will be detected. (Actually, this could easily occur for unvoiced segments where the peak amplitude is low.) Therefore, care must be taken in the analog processing prior to sampling to minimize these effects. For example, it is often preferable to use a bandpass filter, rather than a lowpass filter, as the anti-aliasing filter to eliminate DC and 60 Hz components in the signal. Additional considerations in the short-time zero-crossing measurement are the sampling period, T, and the averaging interval, L. The sampling period determines the time (and frequency) resolution of the zero-crossing representation; i.e., fine resolution requires a high sampling rate. However, to preserve the zero-crossing information, only 1-bit quantization (i.e., preserving the sign of the signal) is required.

A key question about the use and measurement of zero-crossing rates is the effect of DC offsets on the measurements. Figures 6.19 and 6.20 show the effects of severe offset (much more than might be anticipated in real systems) on the waveforms and the resulting locations and counts of zero-crossings. Figure 6.19 shows the waveforms for a 100 Hz sinusoid with no DC offset (the top panel) and with an offset of 0.75 times the peak amplitude (the bottom panel). It can be seen that, for this case, the locations

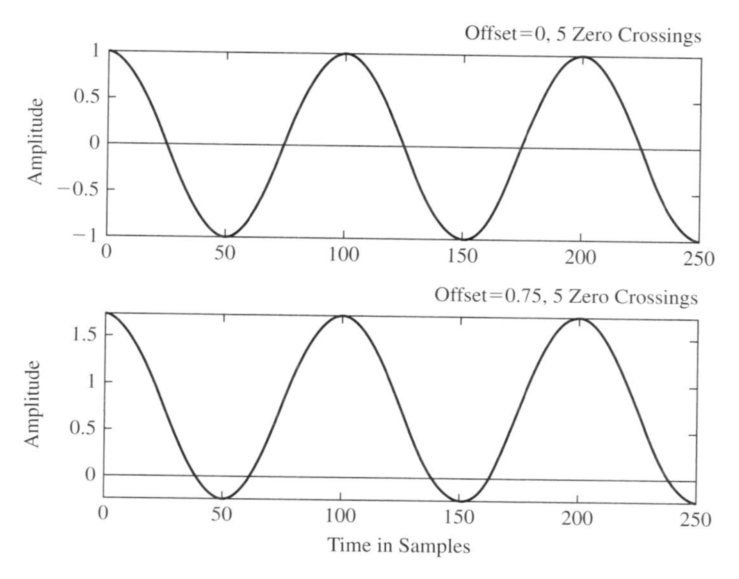

FIGURE 6.19
Plots of waveform for sinusoid with no DC offset (top panel) and DC offset of 0.75 times the peak amplitude (bottom panel).

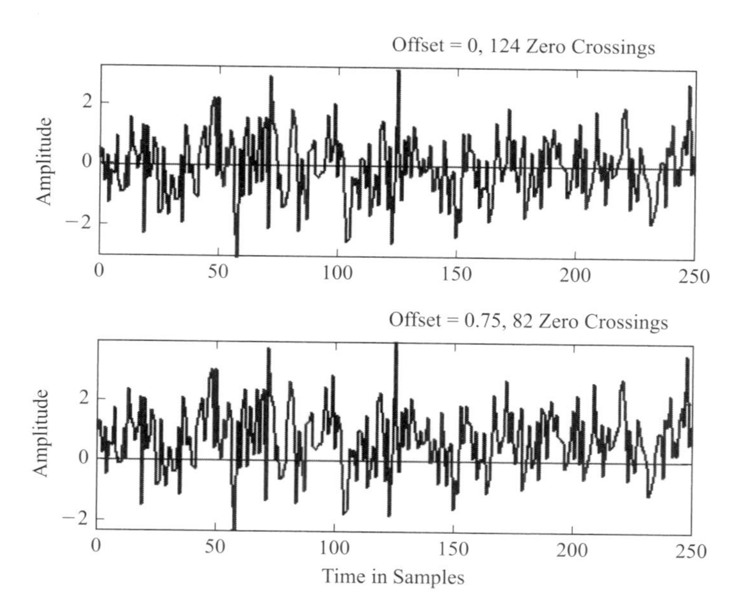

FIGURE 6.20
Plots of waveform for zero-mean, unity variance, Gaussian white noise signal with no offset (top panel), and 0.75 DC offset (bottom panel).

of the zero-crossings change markedly; however, the zero-crossing counts are identical for the $2\frac{1}{2}$ cycles of this sinusoid. Figure 6.20 shows the waveforms for a Gaussian white noise sequence (zero-mean, unit variance, flat spectrum) with no DC offset (the top panel), and with an offset of 0.75 (the bottom panel). Again the locations of the zero-crossings have changed with the DC offset. However in this case, the count of zero-crossings over the 251-sample interval has changed from 124 (for no DC offset) to a value of 82 for the 0.75 DC offset. This shows that for zero-crossing counts to be a useful measure of frequency content, the waveform must be highpass filtered to insure that no such DC component exists in the waveform, prior to calculation of zero-crossing rates.

Because of such practical limitations, a variety of similar representations to the short-time zero-crossing estimate have been proposed. All of the variants introduce some feature intended to make the estimate less sensitive to noise, but each has its own set of limitations. Notable among these is the up-crossing representation studied by Baker [21]. This representation is based upon the time intervals between zero-crossings that occur with positive slope. Baker has applied this representation to phonetic classification of speech sounds [21].

Speech signals are broadband signals and the interpretation of the short-time zero-crossing rate is therefore much less precise. However, rough estimates of spectral properties of the speech signal can be obtained using a representation based on the short-time average zero-crossing rate defined as simply the average number of zero-crossings in a block of L samples. As is evident from Figure 6.20, if we select a block of L samples, all that is required is to check samples in pairs to count the number of times the samples change sign within the block and then compute the average by dividing by L. This would give the average number of zero-crossings/sample. As in the case of the short-time energy, the window can be moved by R samples and the process repeated, thus giving the short-time zero-crossing representation of the speech signal.

The short-time zero-crossing measurement can have the form of the general short-time representation defined in Eq. (6.1) if we define the short-time average zero-crossing rate (per sample) as

$$Z_{\hat{n}} = \frac{1}{2L_{\text{eff}}} \sum_{m=-\infty}^{\infty} |\text{sgn}(x[m]) - \text{sgn}(x[m-1])| \, \tilde{w}[\hat{n}-m], \tag{6.22}$$

where L_{eff} is the effective window length defined in Eq. (6.2), and where the signum (sgn) operator, defined as

$$\text{sgn}(x[n]) = \begin{cases} 1 & x[n] \geq 0 \\ -1 & x[n] < 0, \end{cases} \tag{6.23}$$

transforms $x[n]$ into a signal that retains only the sign of the samples. The terms $|\text{sgn}(x[m]) - \text{sgn}(x[m-1])|$ are equal to 2 when the pair of samples have opposite sign, and 0 when they have the same sign. Thus, each zero-crossing would be represented by a sample of amplitude 2 at the time of the zero-crossing. This factor of 2 is taken into account by the factor of 2 in the averaging factor $1/(2L_{\text{eff}})$ in Eq. (6.22). Typically, the

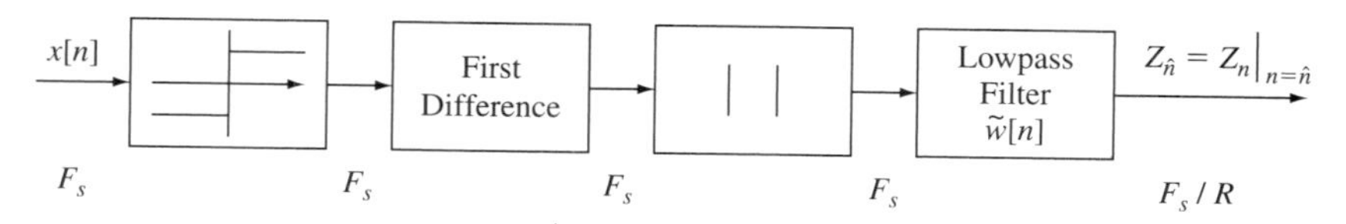

FIGURE 6.21
Block diagram representation of short-time zero-crossings.

window used to compute the average zero-crossing rate is the rectangular window,[9]

$$\tilde{w}[n] = \begin{cases} 1 & 0 \leq n \leq L-1 \\ 0 & \text{otherwise,} \end{cases} \tag{6.24}$$

for which $L_{\text{eff}} = L$. Thus, Eq. (6.22) becomes

$$Z_{\hat{n}} = \frac{1}{2L} \sum_{m=\hat{n}-L+1}^{\hat{n}} |\text{sgn}(x[m]) - \text{sgn}(x[m-1])|. \tag{6.25}$$

The operations involved in the computation of Eq. (6.22) are represented in block diagram form in Figure 6.21. This representation shows that the short-time zero-crossing rate has the same general properties as the short-time energy and the short-time magnitude; i.e., it is a lowpass signal whose bandwidth depends on the shape and length of the window. Equation (6.22) and Figure 6.21 make the computation of $Z_{\hat{n}}$ appear more complex than it really is. As we have said, implementing Eq. (6.25) simply requires counting the zero-crossings and dividing by L. However, the more general mathematical representation of Eq. (6.22) is useful if we wish to use a weighting filter such as the Hamming window, or if we wish to have a recursive implementation similar to Eqs. (6.15) and (6.18). (See Problem 6.5.)

For most applications we are not interested in the zero-crossing rate per sample, but instead we need the rate of zero-crossings per fixed interval of M samples (corresponding to a selected time interval of τ seconds, usually the window length). For a sampling rate of $F_s = 1/T$ and a segment duration of τ seconds, corresponding to an interval of $M = F_s \cdot \tau$ or $M = \tau/T$ samples, all we need to do to modify the rate from $Z^{(1)}$ to $Z^{(M)}$ is multiply the definition of $Z^{(1)}$ by M, that is

$$Z^{(M)} = Z^{(1)} \cdot M; \quad \text{where } M = \tau \cdot F_s = \tau/T.$$

To illustrate how the short-time zero-crossing rate parameter can be properly normalized to a fixed time interval, independent of the sampling rate, consider the example of Table 6.1, whichshows the computational parameters using a 40 msec

[9]If we compute the zero-crossing rate using Eq. (6.22) and using the Hamming window as defined by Eq. (6.4), we can incorporate the factor $1/(2L_{\text{eff}})$ into the definition of the window $\tilde{w}[m]$.

TABLE 6.1 Zero-crossing rates per 10 msec for a 1000 Hz sinusoid at various sampling rates.

F_s	L	$Z_{\hat{n}} = Z^{(1)}$	M	$Z^{(M)}$
8000	320	1/4	80	20
10,000	400	1/5	100	20
16,000	640	1/8	160	20

window length for a 1000 Hz sine wave, with sampling rates of 8000, 10,000, and 16,000 Hz. It can be seen that, for a fixed time interval ($\tau = 10$ msec in the case of Table 6.1), the resulting short-time zero-crossing rate remains the same, independent of sampling rate, by suitably choosing the values of L and M for the different sampling rates. An equivalent way of normalizing the zero-crossing rate is to compute the equivalent sinusoidal frequency using Eq. (6.21).

Now let us see how the short-time zero-crossing rate applies to speech signals. The model for speech production suggests that the energy of voiced speech is concentrated below about 3 kHz because of the spectrum fall-off introduced by the glottal wave, whereas for unvoiced speech, most of the energy is found at higher frequencies. Since high frequency spectral content implies high zero-crossing rates, and low frequency spectral content implies low zero-crossing rates, there is a strong correlation between zero-crossing rate and energy distribution with frequency. Furthermore, the total energy of voiced speech is considerably larger than that of unvoiced speech. Therefore, voiced speech should be characterized by relatively high energy and a relatively low zero-crossing rate, while unvoiced speech will have a relatively high zero-crossing rate and relatively low energy. This is illustrated in Figure 6.22 which shows 2001 samples of a speech signal ($F_s = 16{,}000$ samples/sec) along with the short-time zero-crossing rate (shown by the dashed curve) and short-time energy (shown by the dash-dot curve) for Hamming window analysis windows of lengths $L = 201$ and $L = 401$ samples in Figures 6.22a and b respectively.[10] For this example, the short-time energy and zero-crossing rate were computed at the same sampling rate as the input (i.e., $R = 1$). This figure confirms that for both window lengths, the short-time zero-crossing rate is relatively high for the unvoiced speech interval and the short-time energy is relatively low, with the opposite holding for the voiced speech interval. A close examination of the region of transition from unvoiced to voiced speech, which occurs around sample 900, reveals that the delay of approximately $(L-1)/2$ samples that results from the use of the symmetric Hamming window as a causal analysis filter causes the zero-crossing and energy representations to be shifted to the right by 100 and 200 samples respectively in Figures 6.22a and 6.22b. Also, by normalizing the windows by their effective length, L_{eff}, defined in Eq. (6.2), we see that $Z_{\hat{n}}$ has about

[10]The samples were extracted from an utterance of "... greasy wash ..." having the phonetic representation /G R IY S IY W AA SH/. Samples 0–900 are the end of the fricative /S/, samples 900–1800 are the vowel /IY/, and samples 1800–2000 are the beginning of the glide /W/.

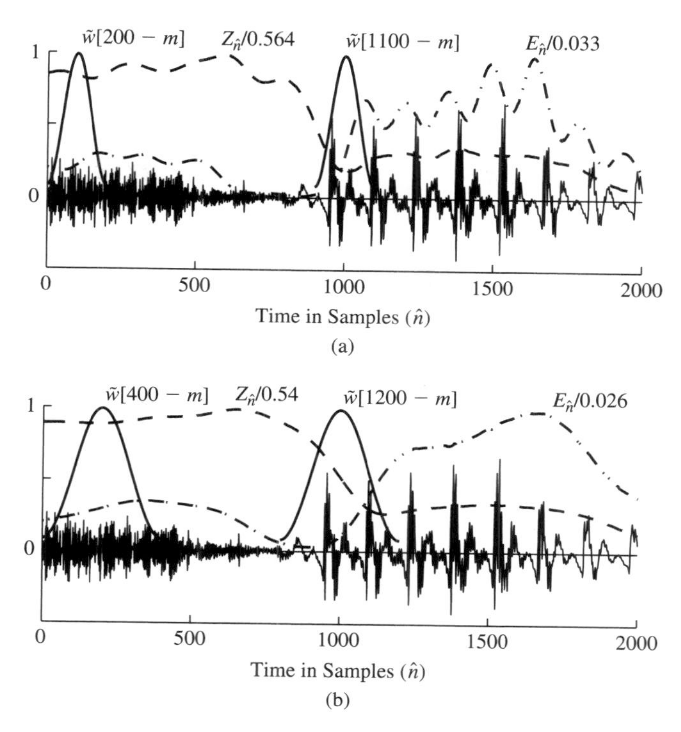

FIGURE 6.22
Illustration of short-time zero-crossings (dashed curve) and short-time energy (dash-dot curve) for 2001 samples of a speech signal ($F_s = 16$ kHz) using (a) a Hamming window with duration $L = 201$ (12.5 msec) and (b) a Hamming window with duration $L = 401$ (25 msec).

the same general size (scale) and temporal contour for both window sizes. This is also true for $E_{\hat{n}}$.

So far, our assertions have not been very precise. We have not said what we mean by *high* and *low* values of short-time zero-crossing rate, and it really is not possible to be precise. However, we can give a meaning to high and low zero-crossing rates in a statistical sense. Figure 6.23 shows histograms of short-time zero-crossing rates obtained using a rectangular window of duration $L = 100$ samples (10 msec intervals for $F_s = 10{,}000$) for both voiced and unvoiced speech segments, along with Gaussian density fits to the histograms. We see that the Gaussian curve provides a reasonably good fit to the zero-crossing rate distribution for both unvoiced and voiced regions. The mean short-time average zero-crossing rate is 49 per 10 msec interval (0.49 per sample) for unvoiced segments and 14 per 10 msec interval (0.14 per sample) for voiced segments. As a reference, the vertical lines in Figure 6.23 show the zero-crossing rates corresponding to a sinusoid at each of the frequencies 1 kHz, 2 kHz, 3 kHz, and 4 kHz. Using Eq. (6.21), the equivalent sinusoidal frequencies of the means of the distributions are $F_v = 0.12 \cdot 5000 = 600$ Hz for the voiced distribution and

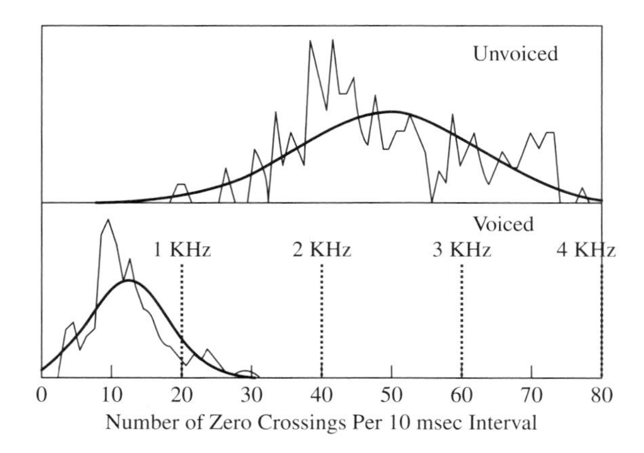

FIGURE 6.23
Distribution of zero-crossing for unvoiced and voiced speech.

$F_u = 0.49 \cdot 5000 = 2450$ Hz for the unvoiced distribution. Since the two distributions overlap, an unequivocal voiced/unvoiced decision is not possible based on short-time zero-crossing rate alone. Nevertheless, a simple comparison to a threshold placed somewhere between 10 and 49 (e.g., 25) can give a reasonably accurate classification.[11]

Note that the plots of short-time zero-crossing rate in Figure 6.22 are consistent with the distributions in Figure 6.23; i.e., $Z_{\hat{n}} \approx 0.54$ zero-crossings per sample for the unvoiced region and $Z_{\hat{n}} \approx 0.16$ zero-crossings per sample for the voiced region. These zero-crossing rates correspond through Eq. (6.21) (with $F_s = 16{,}000$ Hz) to equivalent sinusoidal frequencies of 4320 Hz and 1280 Hz respectively.

Some additional examples of short-time zero-crossing rate measurements are shown in Figure 6.24. In these examples, the duration of the rectangular analysis window is 15 msec, i.e., $L = 150$ samples at a 10 kHz sampling rate. The output was not divided by $L = 150$ so the plots show the number of zero-crossings per 150 samples. Hence the values in Figure 6.24 should be divided by 150 for comparison to those in Figure 6.22, which shows the number of zero-crossings per sample. Furthermore, the values in Figure 6.24 must be multiplied by a factor of 100/150 for fair comparison to the distributions in Figure 6.23. Note that just as in the case of short-time energy and short-time magnitude, the short-time zero-crossing rate can be sampled at a rate much below the sampling rate of the input speech signal. Thus, in Figure 6.24, the output is computed 100 times/sec (window moved in steps of $R = 100$ samples). Note that the upper and lower plots show the regions of the occurrences of the fricatives /S/ and /SH/ very clearly; i.e., the regions of high zero-crossing rate. The middle plot, for which the utterance is mostly voiced speech except for the two instances of the phoneme /H/, which is often partially voiced, is less informative regarding the voiced/unvoiced distinction.

[11]As discussed in Chapter 10, a more formal approach to classification would involve using the estimated probability distributions to set the threshold to minimize the probability of false classification.

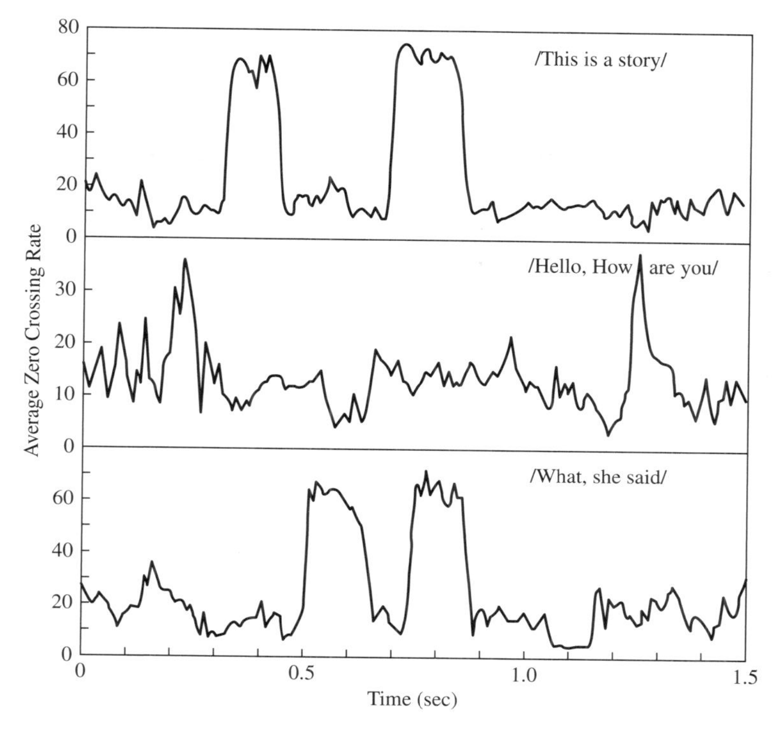

FIGURE 6.24
Short-time zero-crossing rate for three different utterances.

Another application of the zero-crossing representation is as a simple intermediate step in obtaining a frequency-domain representation of speech. The approach involves bandpass filtering of the speech signal in several contiguous frequency bands. Short-time energy and zero-crossing representations are then obtained for the filter outputs. These representations together give a representation that crudely reflects the spectral properties of the signal. Such an approach was initially proposed by Reddy, and studied by Vicens [403] and Erman [99] as the basis for a large-scale speech recognition system.

6.5 THE SHORT-TIME AUTOCORRELATION FUNCTION

The deterministic (or aperiodic) autocorrelation function of a discrete-time signal is defined as [270]

$$\phi[k] = \sum_{m=-\infty}^{\infty} x[m]x[m+k]. \tag{6.26}$$

If the signal is stationary random or periodic, the appropriate definition is [270]

$$\phi[k] = \lim_{N\to\infty} \frac{1}{(2N+1)} \sum_{m=-N}^{N} x[m]x[m+k]. \tag{6.27}$$

In either case, the autocorrelation function representation of the signal is a convenient way of displaying certain properties of the signal. For example, if the signal is periodic (e.g., like voiced speech) with period N_p samples, then it is easily shown that

$$\phi[k] = \phi[k + N_p]; \tag{6.28}$$

i.e., the autocorrelation function of a periodic signal is also periodic with the same period. Other important properties of the autocorrelation function are the following [270]:

1. It is an even function of the lag index; i.e., $\phi[k] = \phi[-k]$.
2. It attains its maximum value at $k = 0$; i.e., $|\phi[k]| \leq \phi[0]$ for all k.
3. The quantity $\phi[0]$ is equal to the total energy [Eq. (6.8)] for deterministic signals or the average power for random or periodic signals.

Thus, the autocorrelation function contains the energy as a special case, but even more important is the convenient way it highlights periodicity in a signal. If we consider Eq. (6.28) together with properties (1) and (2), we see that for periodic signals, the autocorrelation function attains a maximum at samples $0, \pm N_p, \pm 2N_p, \ldots$. That is, *regardless of the time origin of the periodic signal*, the period can be estimated by finding the location of the first maximum of the autocorrelation function (in an appropriate range). This property makes the autocorrelation function an attractive basis for estimating periodicities in all sorts of signals, including speech (if we properly formulate a short-time autocorrelation function). Furthermore, we will see in Chapter 9 that the autocorrelation function contains a great deal of information about the detailed structure of the signal. Thus, it is useful to adapt the definition of the autocorrelation function to obtain a short-time autocorrelation function representation of speech.

Using the same approach that was used to define the other short-time representations that we have discussed in this chapter, we define the short-time autocorrelation function at analysis time, $\hat{n}$, as the deterministic autocorrelation function of the finite-length windowed segment of the speech $(x[m]w[\hat{n}-m])$; i.e.,

$$R_{\hat{n}}[k] = \sum_{m=-\infty}^{\infty} (x[m]w[\hat{n}-m])(x[m+k]w[\hat{n}-k-m]). \tag{6.29}$$

This equation can be interpreted as follows: (1) the window is moved to analysis time $\hat{n}$ to select a segment of speech $x[m]w[\hat{n}-m]$ for values of m in the shifted region of support of $w[\hat{n}-m]$; (2) the selected window-weighted speech segment is multiplied by the window-weighted speech segment shifted left by k samples (for k positive); and (3) the product sequence is summed over all non-zero samples, thereby computing the deterministic autocorrelation of Eq. (6.26) of the windowed segment of speech. Note that the time origin for the summation index in Eq. (6.29) is the time origin of the sequence $x[m]$. The quantity $\hat{n}$ determines the shift of the window, and is therefore the analysis time. The index k, which is the amount of relative shift between the sequences $(x[m]w[\hat{n}-m])$ and $(x[m+k]w[\hat{n}-k-m])$, is called the *autocorrelation lag index*. In general, Eq. (6.26) would be evaluated over a range of lags $0 \le k \le K$, where K is the maximum lag of interest.

Since $R_{\hat{n}}[k]$ is the deterministic autocorrelation function of the windowed sequence $x[m]w[\hat{n}-m]$, $R_{\hat{n}}[k]$ inherits the properties of deterministic autocorrelation functions that were discussed earlier in this section. With the notation established by Eq. (6.29), those properties are restated as follows:

1. The short-time autocorrelation function is an even function of the lag index k; i.e.,

$$R_{\hat{n}}[-k] = R_{\hat{n}}[k]. \tag{6.30a}$$

2. The maximum value of the short-time autocorrelation function occurs at zero lag; i.e.,

$$|R_{\hat{n}}[k]| \le R_{\hat{n}}[0]. \tag{6.30b}$$

3. The short-time energy is equal to the short-time autocorrelation at lag zero; i.e., $E_{\hat{n}} = R_{\hat{n}}[0]$. Specifically, note that

$$R_{\hat{n}}[0] = \sum_{m=-\infty}^{\infty} x^2[m]w^2[\hat{n}-m] = \sum_{m=-\infty}^{\infty} x^2[m]\tilde{w}[\hat{n}-m] = E_{\hat{n}}, \tag{6.30c}$$

where the effective analysis window for the short-time energy is $\tilde{w}[\hat{n}-m] = w^2[\hat{n}-m]$.

As in the case for the short-time energy and the short-time zero-crossing rate, Eq. (6.29) can be manipulated into the general form of Eq. (6.1). Using the evenness property in Eq. (6.30a), we can express $R_{\hat{n}}[k]$ as

$$\begin{aligned} R_{\hat{n}}[k] &= R_{\hat{n}}[-k] \\ &= \sum_{m=-\infty}^{\infty} (x[m]x[m-k])(w[\hat{n}-m]w[\hat{n}+k-m]). \end{aligned} \tag{6.31}$$

If we define the effective window for lag k as

$$\tilde{w}_k[\hat{n}] = w[\hat{n}]w[\hat{n}+k], \tag{6.32}$$

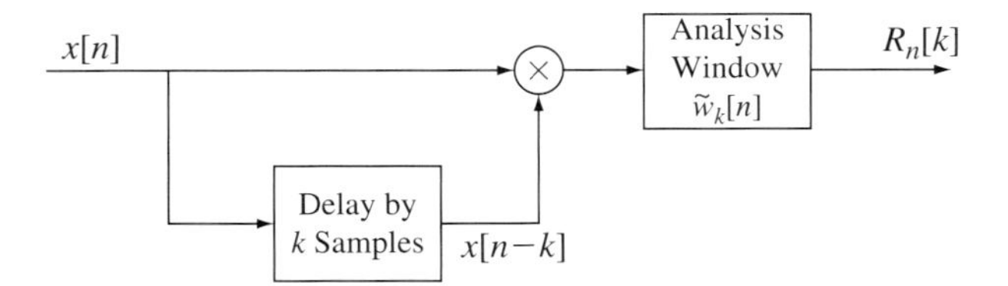

FIGURE 6.25
Block diagram representation of the short-time autocorrelation for lag index k.

then Eq. (6.31) can be written as

$$R_{\hat{n}}[k] = \sum_{m=-\infty}^{\infty} (x[m]x[m-k])\tilde{w}_k[\hat{n}-m], \tag{6.33}$$

which is exactly in the form of Eq. (6.1), with the shift k playing the role of a variable parameter in the non-linear transformation $T(x[m]) = x[m]x[m-k]$. Thus the value at time $\hat{n}$ of the kth autocorrelation lag value can be represented as shown in Figure 6.25 as the output of a linear time invariant (LTI) system with impulse response $\tilde{w}_k[n] = w[n]w[n+k]$, when the input sequence is $x[n]x[n-k]$. This representation reminds us that, when viewed as a function of the analysis time n with k fixed, $R_n[k]$ is a lowpass signal that can be sampled at a lower rate than the sampling rate of the input signal, and, as in the case of the short-time energy, for certain infinite-duration exponential windows, $w[n]$, Eq. (6.33) also can be used to obtain recursive implementations that can be iterated at the sampling rate of the input [24]. (See Problem 6.6.) Since each autocorrelation lag index requires a different impulse response, $\tilde{w}_k[n]$, this approach to computation is generally used when only a few lags are required as in linear predictive analysis (discussed in Chapter 9).

The computation of the short-time autocorrelation function is usually carried out on a frame-by-frame basis with finite-length windows using Eq. (6.29) after replacing $-m$ by $\hat{n}-m$ in Eq. (6.29) and rewriting it in the form

$$R_{\hat{n}}[k] = \sum_{m=-\infty}^{\infty} (x[\hat{n}+m]w'[m])(x[\hat{n}+m+k]w'[k+m]), \tag{6.34}$$

where $w'[m] = w[-m]$.[12] This transformation of summation index is equivalent to redefining the time origin of the summation to be the time origin of the shifted analysis window; i.e., the input sequence is advanced by $\hat{n}$, whereupon $x[\hat{n}+m]$ is multiplied by a window $w'[m]$ in order to select a segment of speech for autocorrelation analysis. Figure 6.26 shows an illustration of the sequences that are involved in the computation of $R_{\hat{n}}[k]$ using a rectangular window for a particular value of k. Figure 6.26a shows $x[m]$

[12]Note that we use m to denote the dummy index of summation in both Eqs. (6.29) and (6.34).

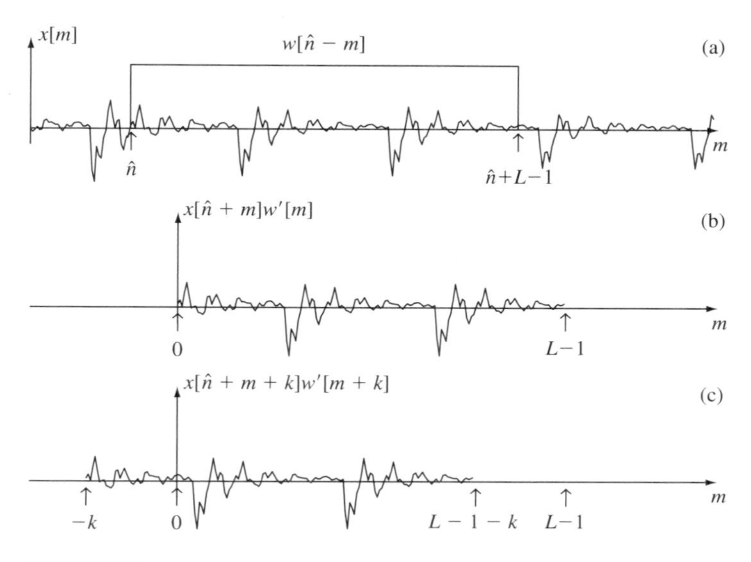

FIGURE 6.26
Illustration of the sequences involved in the computation of the short-time autocorrelation for lag index k: (a) the sequences $x[m]$ and the shifted window sequence $w[\hat{n} - m]$; (b) the windowed sequence after redefining the time origin to the origin of the window $w'[m] = w[-m]$; and (c) the windowed sequence shifted by $k > 0$.

and $w[\hat{n} - m]$ on the same axis as functions of m with $\hat{n}$ fixed. Figure 6.26b shows the windowed segment with time origin redefined to the beginning of the window. This figure shows that the product $(x[\hat{n}+m]w'[m])(x[\hat{n}+m+k]w'[k+m])$ is non-zero only over the interval $0 \leq m \leq L-1-k$. In general, if the window $w'[m]$ is of finite duration (as in Eqs. (6.3) and (6.4)), then the resulting sequence, $x[\hat{n}+m]w'[\hat{n}]$, will be of finite duration, and Eq. (6.34) becomes

$$R_{\hat{n}}[k] = \sum_{m=0}^{L-1-k} (x[\hat{n}+m]w'[m])(x[\hat{n}+m+k]w'[k+m]). \tag{6.35}$$

Note that when the rectangular or Hamming windows of Eqs. (6.3) and (6.4) are used for $w'[m]$ in Eq. (6.35), they correspond to non-causal analysis filters in Eq. (6.33). This is also evident in Figure 6.26a, where we see that $w[\hat{n} - m] = w'[m - \hat{n}]$ is non-zero in the interval *after* $\hat{n}$ if $w'[m]$ is non-zero in the interval $0 \leq m \leq L-1$. For finite-length windows, this poses no problem since suitable delays can be introduced into the processing, even in real-time applications.

The computation of the kth autocorrelation lag using Eq. (6.35) would appear to require L multiplications for computing $x[\hat{n}+m]w'[m]$ (except when the rectangular window is used), and $(L-k)$ multiplications and additions for computing the sum of lagged products. The computation of many lags, as required in estimating periodicity, thus requires a great deal of arithmetic. This can be reduced by taking advantage of

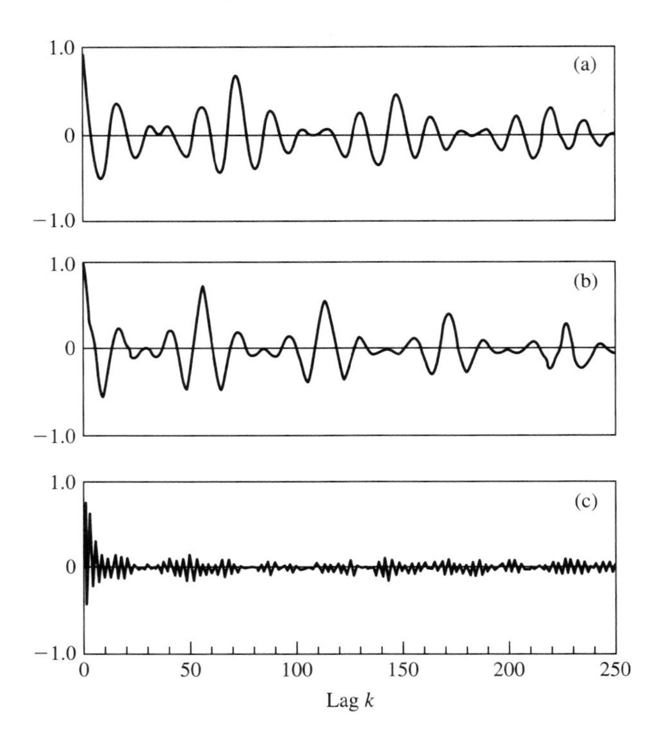

FIGURE 6.27
Autocorrelation function for (a) and (b) voiced speech; and (c) unvoiced speech, using a rectangular window with $L = 401$.

some special properties of Eq. (6.35) [37, 189] or by using the fast Fourier transform (FFT) [270].

Figure 6.27 shows three examples of short-time autocorrelation functions computed using a rectangular window with $L = 401$ for a speech signal sampled at 10 kHz. The short-time autocorrelation was evaluated for lags $0 \leq k \leq 250$.[13] The first two cases are for voiced speech segments and the third is for an unvoiced segment. For the first segment, peaks occur approximately at multiples of 72, indicating a pitch period of 7.2 msec or a fundamental frequency of approximately 140 Hz. Note that even a very short segment of speech differs from a segment of a truly periodic signal. The "period" of the signal changes across a 401 sample interval and also the wave shape varies somewhat from period to period. This is part of the reason that the peaks get smaller for large lags. For the second voiced section (taken at a totally distinct place in the utterance), similar periodicity effects are seen, only in this case the local peaks in the autocorrelation are at multiples of 58 samples, indicating an average pitch period

[13]In this and subsequent plots, the autocorrelation function is normalized by $E_{\hat{n}} = R_{\hat{n}}[0]$. Furthermore, $R_{\hat{n}}[k]/R_{\hat{n}}[0]$ is plotted only for positive values of k. Since $R_{\hat{n}}[-k] = R_{\hat{n}}[k]$, it is not necessary to show values for negative lags.

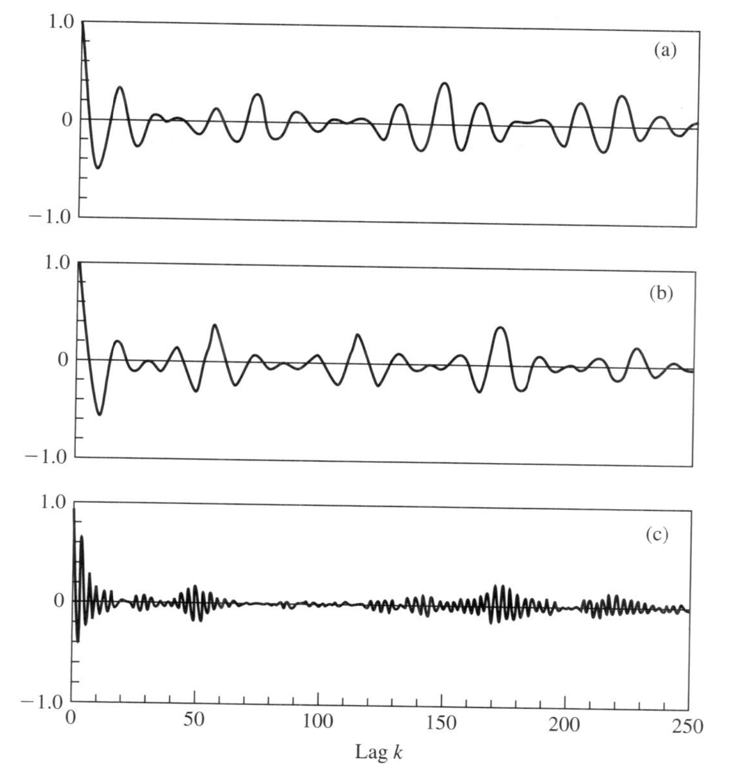

FIGURE 6.28
Autocorrelation functions for (a) and (b) voiced speech; and (c) unvoiced speech, using a Hamming window with $L = 401$.

of about 5.8 msec. Finally for the unvoiced section of speech, there are no strong autocorrelation periodicity peaks, thus indicating a *lack* of periodicity in the waveform. The autocorrelation function for unvoiced speech is seen to be a high frequency noise-like waveform, somewhat like the speech itself.

Figure 6.28 shows the same examples using a Hamming window. By comparing these results to those in Figure 6.27, it can be seen that the rectangular window gives a much stronger indication of periodicity than the Hamming window. This is not surprising in view of the tapering of the speech segment introduced by the Hamming window, which would destroy periodicity even for a perfectly periodic signal.

The examples of Figures 6.27 and 6.28 were for a value of $L = 401$. An important issue is how L should be chosen to give a good indication of periodicity. Again we face conflicting requirements. Because the pitch period of voiced speech changes with time, L should be as small as possible. On the other hand, it should be clear that to get any indication of periodicity in the autocorrelation function, the window must have a duration of at least two periods of the waveform. In fact, Figure 6.26c shows that because of the finite length of the windowed speech segment involved

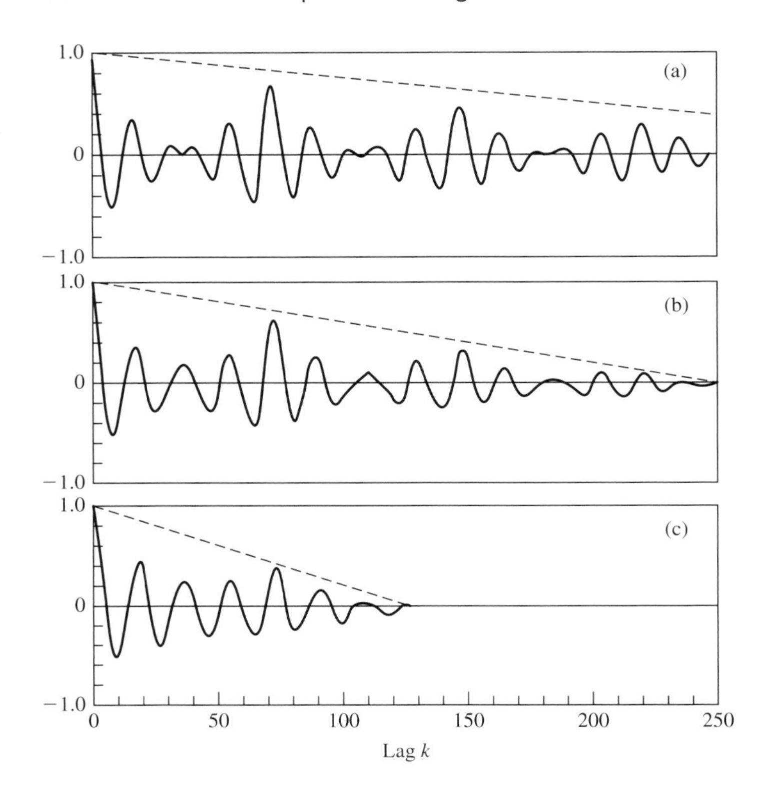

FIGURE 6.29
Autocorrelation function for voiced speech with (a) $L = 401$; (b) $L = 251$; and (c) $L = 125$. Rectangular window used in all cases.

in the computation of $R_{\hat{n}}[k]$, fewer and fewer signal samples are involved in the computation as k increases. (Note that the upper limit of summation in Eq. (6.35) is $L - k - 1$.) This leads to a reduction in amplitude of the correlation peaks as k increases. This can easily be verified for the case of a periodic impulse train (see Problem 6.8), and it is easily demonstrated for speech by example. Figure 6.29 illustrates the effect for rectangular windows of different lengths. The dashed lines are plots of the equation

$$R[k] = 1 - k/L, \qquad |k| < L, \tag{6.36}$$

which is the autocorrelation function of the L-point rectangular window. Clearly, the dashed line fall-off in peak level due to the rectangular window is a good bound on the amplitude of the correlation peaks due to periodicity. In Problem 6.8, it is shown that for a periodic impulse train, the peaks will lie exactly on such a straight line. For the present example, the peaks are further away from the line for $L = 401$ than for the other two cases. This is because the pitch period and wave shape change more across an interval of 401 samples than across the shorter intervals.

Figure 6.29c corresponds to a window length of $L = 125$ samples. Since the pitch period for this example is about 72 samples, not even two complete pitch periods are included in the window. This is clearly a situation to be avoided, but avoiding it is difficult because of the wide range of pitch periods that can be encountered. One approach is to simply make the window long enough to accommodate the longest pitch period, but this leads to undesirable averaging of many periods when the pitch period is short. Another approach is to allow the window length to adapt to match the expected pitch period. Still another approach that allows the use of shorter windows is described in the next section.

6.6 THE MODIFIED SHORT-TIME AUTOCORRELATION FUNCTION

One possible solution to the undesirable tapering of the short-time autocorrelation function for long lags is the use of a "modified" version of the short-time autocorrelation function. This modified short-time autocorrelation function is defined as

$$\hat{R}_{\hat{n}}[k] = \sum_{m=-\infty}^{\infty} x[m]w_1[\hat{n}-m]x[m+k]w_2[\hat{n}-m-k]; \tag{6.37}$$

i.e., we allow the possibility of using different windows for the selected segment and the shifted segment. This expression can be written as

$$\hat{R}_{\hat{n}}[k] = \sum_{m=-\infty}^{\infty} x[\hat{n}+m]\hat{w}_1[m]x[\hat{n}+m+k]\hat{w}_2[m+k], \tag{6.38}$$

where

$$\hat{w}_1[m] = w_1[-m], \tag{6.39a}$$

and

$$\hat{w}_2[m] = w_2[-m]. \tag{6.39b}$$

To accomplish our goal of eliminating the fall-off due to the variable upper limit in Eq. (6.35), we can choose the window $\hat{w}_2[m]$ to include samples outside the non-zero interval of window $\hat{w}_1$. That is, we define two rectangular windows

$$\hat{w}_1[m] = \begin{cases} 1 & 0 \le m \le L-1 \\ 0 & \text{otherwise} \end{cases} \tag{6.40a}$$

and

$$\hat{w}_2[m] = \begin{cases} 1 & 0 \le m \le L-1+K \\ 0 & \text{otherwise,} \end{cases} \tag{6.40b}$$

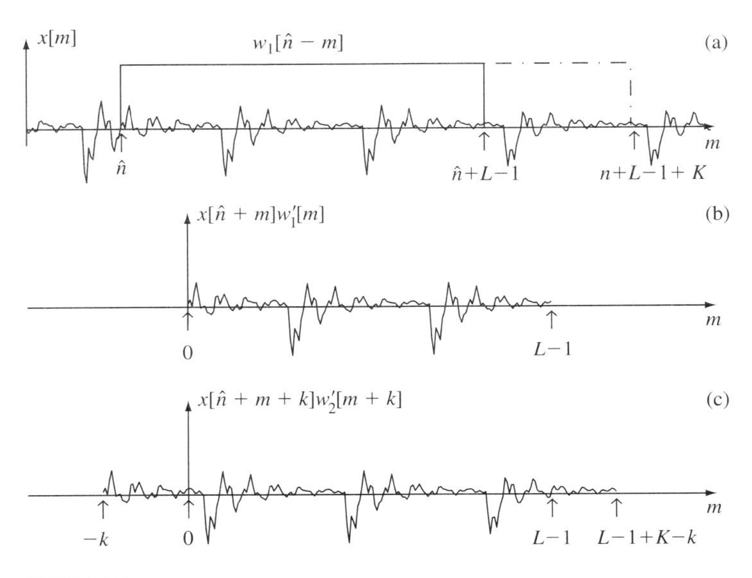

FIGURE 6.30
Illustration of the samples involved in the computation of the modified short-time autocorrelation function.

where K is the greatest lag of interest. Thus, Eq. (6.38) can be written as

$$\hat{R}_{\hat{n}}[k] = \sum_{m=0}^{L-1} x[\hat{n}+m]x[\hat{n}+m+k], \quad 0 \le k \le K; \tag{6.41}$$

i.e., the sum is always over L samples, and samples from outside the interval $\hat{n}$ to $\hat{n}+L-1$ are involved in the computation. The differences in the data involved in the computations of Eqs. (6.35) and (6.41) are depicted in Figure 6.30. Figure 6.30a shows a speech waveform and Figure 6.30b shows a segment of L samples selected by a rectangular window. For a rectangular window, this segment would be used for both terms in Eq. (6.35) and it would be the term $x[\hat{n}+m]\hat{w}_1[m]$ in Eq. (6.41). Figure 6.30c shows the other term in Eq. (6.41). Note that K additional samples are included. These extra samples are shifted into the interval $0 \le m \le L-1$ as k is varied over $0 \le k \le K$.

Equation (6.41) will be referred to as the *modified* short-time autocorrelation function. Strictly speaking, however, it is the *cross-correlation* function for the two different finite-length segments of speech, $x[\hat{n}+m]\hat{w}_1[m]$ and $x[\hat{n}+m]\hat{w}_2[m]$. Thus $\hat{R}_{\hat{n}}[k]$ has the properties of a cross-correlation function, not an autocorrelation function. For example, $\hat{R}_{\hat{n}}[-k] \neq \hat{R}_{\hat{n}}[k]$. Nevertheless, $\hat{R}_{\hat{n}}[k]$ will display peaks at multiples of the period of a periodic signal and it will not display a fall-off in amplitude at large values of k if the signal is perfectly periodic. Figure 6.31 shows the modified autocorrelation functions corresponding to the examples of Figure 6.27. Because for $L = 401$ the effects of waveform variation dominate the tapering effect in Figure 6.27, the two figures look much alike. Figure 6.32 shows the effect of

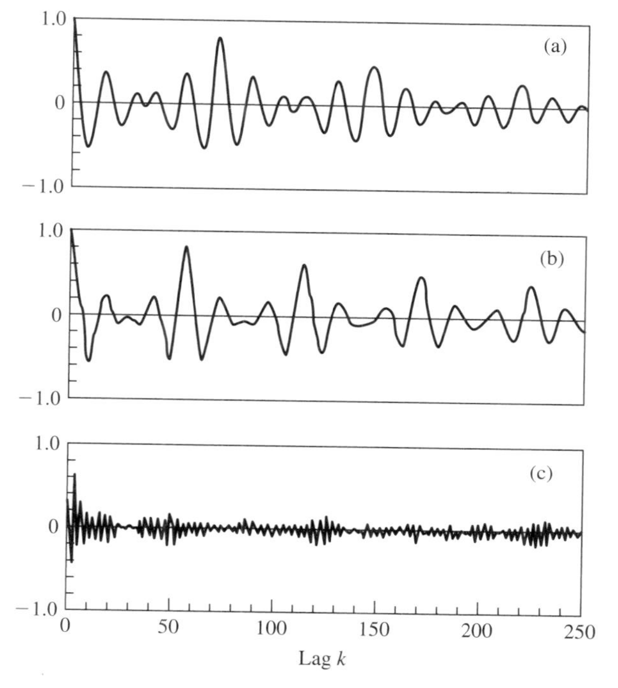

FIGURE 6.31
Modified autocorrelation function for speech segments of Figure 6.27 with $L = 401$.

window length. A comparison of Figure 6.32 with Figure 6.29 shows that the difference is more apparent for smaller values of L. The peaks in Figure 6.32 are less than the $k = 0$ peak only because of deviations from periodicity over the interval $\hat{n}$ to $\hat{n} + N - 1 + K$, which is involved in the evaluation of Eq. (6.31). Problem 6.8 shows that for a perfectly periodic impulse train, all the peaks will have the same amplitude.

6.7 THE SHORT-TIME AVERAGE MAGNITUDE DIFFERENCE FUNCTION

As we have pointed out, the computation of the autocorrelation function involves considerable arithmetic, even using the simplifications presented in Refs. [37, 189]. A technique that eliminates the need for multiplications is based upon the idea that for a truly periodic input of period N_p, the sequence

$$d[n] = x[n] - x[n - k] \tag{6.42}$$

would be zero for $k = 0, \pm N_p, \pm 2N_p, \ldots$. For short segments of voiced speech, it is reasonable to expect that $d[n]$ will be small at multiples of the period, but not identically zero. The short-time average magnitude of $d[n]$ as a function of k should be

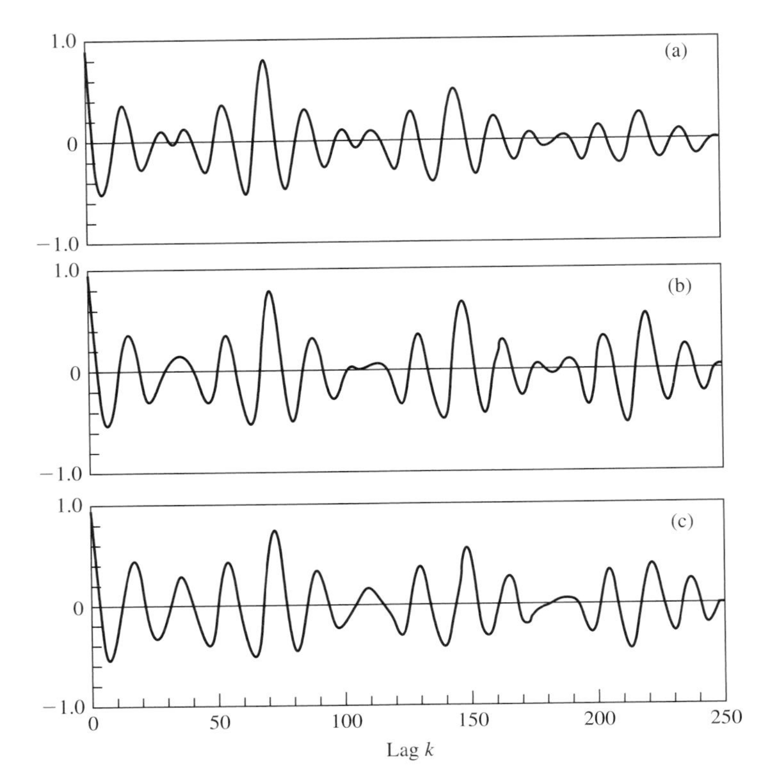

FIGURE 6.32
Modified autocorrelation function for voiced speech for (a) $L = 401$; (b) $L = 251$; and (c) $L = 125$. Examples correspond to Figure 6.29.

small whenever k is close to the period. The short-time average magnitude difference function (AMDF) [331] is thus defined as

$$\gamma_{\hat{n}}[k] = \sum_{m=-\infty}^{\infty} |x[\hat{n}+m]w_1[m] - x[\hat{n}+m-k]w_2[m-k]|. \tag{6.43}$$

Clearly, if $x[\hat{n}]$ is close to being periodic in the interval spanned by the window, then $\gamma_{\hat{n}}[k]$ should dip sharply for $k = N_p, 2N_p, \ldots$. Note that it is most reasonable to choose the windows to be rectangular. If both have the same length, we obtain a function similar to the autocorrelation function of Eq. (6.35). If $w_2[n]$ is longer than $w_1[n]$, then we have a situation similar to the modified autocorrelation of Eq. (6.41). It can be shown [331] that

$$\gamma_{\hat{n}}[k] \approx \sqrt{2}\beta[k](\hat{R}_{\hat{n}}[0] - \hat{R}_{\hat{n}}[k])^{1/2}. \tag{6.44}$$

It is reported that $\beta[k]$ in Eq. (6.44) varies between 0.6 and 1.0 with different segments of speech, but does not change rapidly with k for a particular speech segment [331].

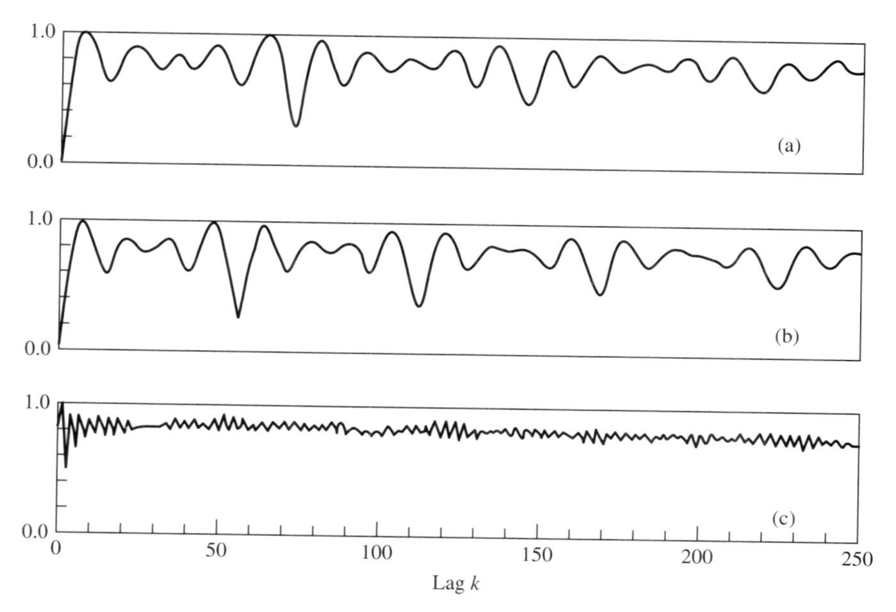

FIGURE 6.33
AMDF function (normalized to 1.0) for the same speech segments as in Figures 6.27 and 6.31.

Figure 6.33 shows the AMDF function for the speech segments of Figures 6.27 and 6.31 for the same length window. The AMDF function does indeed have the shape suggested by Eq. (6.44) and thus, $\gamma_{\hat{n}}[k]$ dips sharply at the pitch period of voiced speech and shows no comparable dips for unvoiced speech.

The AMDF function is implemented with subtraction, addition, and absolute value operations, in contrast to addition and multiplication operations for the autocorrelation function. With floating point arithmetic, where multiplications and additions take approximately the same time, about the same time is required for either method with the same window length. However, for special purpose hardware, or with fixed point arithmetic, the AMDF could have the advantage. In this case multiplications usually are more time consuming, and furthermore, either scaling or a double precision accumulator is required to hold the sum of lagged products. For this reason the AMDF function has been used in numerous real-time speech processing systems.

6.8 SUMMARY

In this chapter we have discussed several representations of speech that are based primarily on processing carried out directly in the time domain, i.e., directly on the waveform samples without any prior analysis. We have discussed these techniques in some detail since they are widely used in speech processing and we feel that a basic understanding of the properties of these techniques is essential to their effective utilization.

We began with a general discussion of the short-time analysis principle, wherein the temporal variations of the properties of the speech waveform are tracked by analyzing short segments of the waveform over which the properties remain relatively constant. Short-time analysis principles will be routinely applied in subsequent chapters. Indeed, in many instances, the short-time analysis principle is so obviously in force that we hardly need to call attention to the fact that the analysis proceeds frame-by-frame.

As illustrations of the short-time analysis techniques we studied short-time formulations of energy, average magnitude, zero-crossings, and autocorrelation function. These are all simple illustrations, but they are all widely used "representations" of the speech signal that have the virtue of highlighting general time-varying features of the speech signal. Future chapters build on these representations and most speech processing systems incorporate them as fundamental analysis techniques.

PROBLEMS

6.1. The rectangular window is defined as

$$w_R[n] = \begin{cases} 1 & 0 \leq n \leq L-1 \\ 0 & \text{otherwise.} \end{cases}$$

The Hamming window is defined as

$$w_H[n] = \begin{cases} 0.54 - 0.46\cos\left[2\pi n/(L-1)\right] & 0 \leq n \leq L-1 \\ 0 & \text{otherwise.} \end{cases}$$

(a) Show that the Fourier transform of the rectangular window is

$$W_R(e^{j\omega}) = \frac{\sin(\omega L/2)}{\sin(\omega/2)} e^{-j\omega(L-1)/2}.$$

(b) Sketch $W_R(e^{j\omega})$ as a function of ω. (Disregard the linear phase factor $e^{-j\omega(L-1)/2}$ in making the sketch.)

(c) Express $w_H[n]$ in terms of $w_R[n]$ and thereby obtain an expression for $W_H(e^{j\omega})$ in terms of $W_R(e^{j\omega})$.

(d) Sketch the individual terms in $W_H(e^{j\omega})$. (Disregard the linear phase factor $e^{-j\omega(L-1)/2}$ that is common to each term.) Your sketch should illustrate how the Hamming window trades frequency resolution for increased suppression of higher frequencies.

6.2. The short-time energy of a sequence $x[n]$ is defined as

$$E_{\hat{n}} = \sum_{m=-\infty}^{\infty} (x[m]w[\hat{n}-m])^2.$$

For the particular choice

$$w[m] = \begin{cases} a^m & m \geq 0 \\ 0 & m < 0, \end{cases}$$

it is possible to find a recurrence formula for $E_{\hat{n}}$.

(a) Find a difference equation that expresses $E_{\hat{n}}$ in terms of $E_{\hat{n}-1}$ and the input $x[n]$.

(b) Draw a digital network diagram of this equation.

(c) What general property must

$$\tilde{w}[m] = w^2[m]$$

have in order that it be possible to find a recursive implementation?

6.3. Using the commutative property of convolution, the short-time energy can be expressed as

$$E_{\hat{n}} = \sum_{m=-L}^{L} \tilde{w}[m]x^2[\hat{n} - m],$$

where $\tilde{w}[n] = \tilde{w}[-n]$ is a symmetric non-causal window. Suppose we wish to compute $E_{\hat{n}}$ at each sample of the input.

(a) Let $\tilde{w}[m]$ be

$$\tilde{w}[m] = \begin{cases} a^{|m|} & |m| \leq L \\ 0 & \text{otherwise.} \end{cases}$$

Find a recurrence relation (i.e., a difference equation) for $E_{\hat{n}}$. (As defined, $\tilde{w}[m]$ is non-causal. Therefore an appropriate delay must be inserted.)

(b) What is the savings in number of multiplications obtained by using the recurrence relation rather than directly computing E_n?

(c) Draw a digital network diagram of the recurrence formula for $E_{\hat{n}}$.

6.4. Suppose that the short-time magnitude is to be estimated every L samples at the input sampling rate. One possibility is to use a finite-length window as in

$$M_{\hat{n}} = \sum_{m=\hat{n}-L+1}^{\hat{n}} |x[m]| w[\hat{n} - m].$$

In this case, $M_{\hat{n}}$ is only computed once for each L samples of the input. Another approach is to use an exponential window for which a recurrence formula can be obtained, e.g.,

$$M_{\hat{n}} = aM_{\hat{n}-1} + |x[\hat{n}]|.$$

In this case $M_{\hat{n}}$ must be computed at each input sample, even though we may only want it every L samples.

(a) How many multiplications and additions are required to compute $M_{\hat{n}}$ once for each L samples with the finite-length window?

(b) Repeat (a) for the recursive definition of $M_{\hat{n}}$.

(c) Under what conditions will the finite-length window be more efficient?

6.5. The short-time zero-crossing rate was defined in Eqs. (6.22)–(6.24) as

$$Z_{\hat{n}} = \frac{1}{2L} \sum_{m=\hat{n}-L+1}^{\hat{n}} |\text{sgn}(x[m]) - \text{sgn}(x[m-1])|.$$

Show that $Z_{\hat{n}}$ can be expressed as

$$Z_{\hat{n}} = Z_{\hat{n}-1} + \frac{1}{2L} \{|\text{sgn}(x[\hat{n}]) - \text{sgn}(x[\hat{n}-1])| \\ - |\text{sgn}(x[\hat{n}-L]) - \text{sgn}(x[\hat{n}-L-1])|\}.$$

6.6. As given by Eq. (6.29), the short-time autocorrelation function is defined as

$$R_{\hat{n}}[k] = \sum_{m=-\infty}^{\infty} x[m]w[\hat{n}-m]x[m+k]w[\hat{n}-k-m].$$

This problem is concerned with the development of some results on recursive computation of the short-time autocorrelation function that were first presented by Barnwell [24].

(a) Show that

$$R_{\hat{n}}[k] = R_{\hat{n}}[-k];$$

i.e., show that $R_{\hat{n}}[k]$ is an even function of k.

(b) Show that $R_{\hat{n}}[k]$ can be expressed as

$$R_{\hat{n}}[k] = \sum_{m=-\infty}^{\infty} x[m]x[m-k]\tilde{w}_k[\hat{n}-m],$$

where

$$\tilde{w}_k[n] = w[n]w[n+k].$$

(c) Suppose that

$$w[n] = \begin{cases} a^n & n \geq 0 \\ 0 & n < 0. \end{cases}$$

Determine the impulse response, $\tilde{w}_k[n]$, for computing the kth lag.

(d) Determine the z-transform of $\tilde{w}_k[n]$ in (c), and from it, obtain a recursive implementation for $R_{\hat{n}}[k]$. Draw a digital network implementation for computing $R_{\hat{n}}[k]$ as a function of $\hat{n}$ for the window of (c).

(e) Repeat parts (c) and (d) for

$$w[n] = \begin{cases} na^n & n \geq 0 \\ 0 & n < 0. \end{cases}$$

6.7. Which of the following functions cannot be a valid autocorrelation function? For all invalid autocorrelations, explain the reason.

(a) $R_x(\tau) = 2e^{-\tau^2}, \ -\infty < \tau < \infty$

(b) $R_x(\tau) = |\tau| e^{-|\tau|}, \ -\infty < \tau < \infty$

(c) $R_x(\tau) = \left(\frac{\sin(\pi\tau)}{\pi\tau}\right)^2, \ -\infty < \tau < \infty$

(d) $R_x(\tau) = 2\frac{\tau^2 + 4}{\tau^2 + 6}, \ -\infty < \tau < \infty$

(e) $R_x(\tau) = [0.2\cos(3\pi\tau)]^3, \ -\infty < \tau < \infty$

6.8. Consider the periodic impulse train

$$x[m] = \sum_{r=-\infty}^{\infty} \delta[m - rN_p].$$

(a) Using Eq. (6.35) with $w'[m]$, a rectangular window whose length, L, satisfies

$$QN_p < L - 1 < (Q+1)N_p,$$

where Q is an integer, find and sketch $R_{\hat{n}}[k]$ for $0 \leq k \leq L - 1$.

(b) How would the result of (a) change if the window is a Hamming window of the same length?

(c) Find and sketch the modified short-time autocorrelation function, $\hat{R}_{\hat{n}}[k]$, given by Eq. (6.41) for the same value of L.

6.9. The long-time autocorrelation function of a stationary random signal or a periodic signal is defined as

$$\phi[k] = \lim_{L\to\infty} \frac{1}{2L+1} \sum_{m=-L}^{L} x[m]x[m+k].$$

The short-time autocorrelation function is defined as

$$R_{\hat{n}}[k] = \sum_{m=0}^{L-|k|-1} x[\hat{n}+m]w'[m]x[\hat{n}+m+k]w'[m+k]$$

and the modified short-time autocorrelation function is defined as

$$\hat{R}_{\hat{n}}[k] = \sum_{m=0}^{L-1} x[\hat{n}+m]x[\hat{n}+m+k].$$

Show whether the following statements are true or false.

(a) If $x[n] = x[n + N_p], -\infty < n < \infty$, then

(i) $\phi[k] = \phi[k + N_p]$, $-\infty < k < \infty$

(ii) $R_{\hat{n}}[k] = R_{\hat{n}}[k + N_p]$, $-(L-1) \leq k \leq L-1$

(iii) $\hat{R}_{\hat{n}}[k] = \hat{R}_{\hat{n}}[k + N_p]$, $-(L-1) \leq k \leq L-1$

(b) If $x[n] = x[n + N_p]$, $-\infty < n < \infty$, then

(i) $\phi[-k] = \phi[k]$, $-\infty < k < \infty$

(ii) $R_{\hat{n}}[-k] = R_{\hat{n}}[k]$, $-(L-1) \leq k \leq L-1$

(iii) $\hat{R}_{\hat{n}}[-k] = \hat{R}_{\hat{n}}[k]$, $-(L-1) \leq k \leq L-1$

(c) If $x[n] = x[n + N_p]$, $-\infty < n < \infty$, then

(i) $\phi[k] \leq \phi[0]$, $-\infty < k < \infty$

(ii) $R_{\hat{n}}[k] \leq R_{\hat{n}}[0]$, $-(L-1) \leq k \leq L-1$

(iii) $\hat{R}_{\hat{n}}[k] \leq \hat{R}_{\hat{n}}[0]$, $-(L-1) \leq k \leq L-1$

(d) If $x[n] = x[n + N_p]$, $-\infty < n < \infty$, then

(i) $\phi[0]$ is equal to the power in the signal

(ii) $R_{\hat{n}}[0]$ is the short-time energy

(iii) $\hat{R}_{\hat{n}}[0]$ is the short-time energy

6.10. **(a)** An analog signal, $x_1(t) = A\cos(200\pi t)$, is sampled at a rate of $F_s = 10{,}000$ Hz, giving the discrete-time signal, $x_1[n]$. Sketch the DTFT, $|X_1(e^{j\omega})|$, versus either ω or f.

(b) A second analog signal, $x_2(t) = B\cos(202\pi t)$, is also sampled at a rate of $F_s = 10{,}000$ Hz, giving the discrete-time signal, $x_2[n]$. Sketch the DTFT, $|X_2(e^{j\omega})|$, versus either ω or f.

(c) Is either of the discrete-time signals (or both), $x_1[n]$ and $x_2[n]$, a periodic signal, and if so, what is the period in samples?

6.11. Consider the signal

$$x[n] = \cos\left(\frac{2\pi}{N_p}n\right), \qquad -\infty < n < \infty.$$

(a) Determine the long-time autocorrelation function, $\phi[k]$, for $x[n]$ [as given in Eq. (6.27)]. Hint: First show that the average over infinite limits is equal to the average over one period of the periodic signal.

(b) Sketch $\phi[k]$ as a function of k.

(c) Determine and sketch the long-time autocorrelation function of the signal

$$y[n] = \begin{cases} 1 & \text{if} \quad x[n] \geq 0, \\ 0 & \text{if} \quad x[n] < 0. \end{cases}$$

6.12. An analog periodic signal of the form:

$$x(t) = 1 + \cos(2\pi(93.75)t)$$

is sampled at a rate of $F_s = 1000$ samples/sec, giving a discrete-time signal of the form:

$$x[n] = 1 + \cos(2\pi(93.75)n/1000) = 1 + \cos(0.1875\pi n).$$

(a) A rectangular window of length $L = 64$ samples is used to compute the modified autocorrelation of $x[n]$ at time $\hat{n} = 0$. Write an expression for the modified autocorrelation, $\hat{R}_0[k]$, of $x[n]$ and sketch its behavior over the region $0 \leq k \leq L$.

(b) Find the non-zero integer lag location of the maximum of the modified autocorrelation function in the interval $0 \leq k \leq L$.

6.13. **(a)** A discrete-time signal, $x_1[n]$, with sampling rate $F_s = 10{,}000$ Hz is defined as:

$$x_1[n] = \begin{cases} 1 & -12 \leq n \leq 12 \\ 0 & \text{otherwise.} \end{cases}$$

Determine the DTFT, $X_1(e^{j\omega})$, of $x_1[n]$. Plot the log magnitude spectrum, $20\log_{10}|X_1(e^{j\omega})|$ versus $f = \omega/2\pi$, with the f scale going from 0 to 0.5.

(b) An analog signal, $x_1(t) = A\cos(2\pi 200t)$, is compressed, giving the analog signal $x_2(t) = \text{comp}(x_1(t))$, where the compression function comp is defined as

$$\text{comp}(x) = \begin{cases} 1 & x \geq 0 \\ 0 & x < 0 \end{cases}$$

and the resulting analog signal, $x_2(t)$, is sampled at a rate of $F_s = 10{,}000$ Hz, giving the discrete-time signal $x_2[n]$. Plot the resulting log magnitude spectrum, $20\log_{10}|X_2(e^{j\omega})|$, versus $f/2\pi$, again from 0 to 0.5.

(c) What special spectral property is shown by the spectral lines corresponding to $X_2(e^{j\omega})$.

6.14. The short-time AMDF of the signal $x[n]$ is defined as [see Eq. (6.43)]

$$\gamma_{\hat{n}}[k] = \frac{1}{L}\sum_{m=0}^{L-1} |x[\hat{n}+m] - x[\hat{n}+m-k]|.$$

(a) Using the inequality [331]

$$\frac{1}{L}\sum_{m=0}^{L-1} |x[m]| \leq \left[\frac{1}{L}\sum_{m=0}^{L-1} |x[m]|^2\right]^{1/2},$$

show that

$$\gamma_{\hat{n}}[k] \leq \left[2(\hat{R}_{\hat{n}}[0] - \hat{R}_{\hat{n}}[k])\right]^{1/2}.$$

This result leads to Eq. (6.44).

(b) Sketch $\gamma_{\hat{n}}[k]$ and the quantity $[2(\hat{R}_{\hat{n}}[0] - \hat{R}_{\hat{n}}[k])]^{1/2}$ for $0 \leq k \leq 200$ for the signal

$$x[n] = \cos(\omega_0 n),$$

with $L = 200$, $\omega_0 = 200\pi/10{,}000$.

6.15. Consider the signal

$$x[n] = A\cos\left(\frac{2\pi}{N_p}n\right)$$

as input to a three-level center clipper that produces an output

$$y[n] = \begin{cases} 1 & x[n] > C_L \\ 0 & |x[n]| \leq C_L \\ -1 & x[n] < -C_L. \end{cases}$$

(a) Sketch $y[n]$ as a function of n for $C_L = 0.5A$, $C_L = 0.75A$, and $C_L = A$.

(b) Sketch the autocorrelation function for $y[n]$ for the values of C_L in (a).

(c) Discuss the effect of the setting of C_L as it approaches A. Suppose that A varies with time such that

$$0 < A[n] \leq A_{\max}.$$

Discuss problems that this can cause if C_L is close to $A_{\max}$.

6.16. (MATLAB Exercise) Write a MATLAB program to plot (and compare) the time and frequency responses of five different L-point windows, namely:

1. Rectangular window:

$$w[n] = \begin{cases} 1 & 0 \leq n \leq L-1 \\ 0 & \text{otherwise.} \end{cases}$$

2. Triangular window:

$$w[n] = \begin{cases} 2n/(L-1) & 0 \leq n \leq (L-1)/2 \\ 2 - 2n/(L-1) & (L+1)/2 \leq n \leq L-1 \\ 0 & \text{otherwise.} \end{cases}$$

3. Hann window:

$$w[n] = \begin{cases} 0.5 - 0.5\cos\left(\frac{2\pi n}{L-1}\right) & 0 \leq n \leq L-1 \\ 0 & \text{otherwise.} \end{cases}$$

4. Hamming window:

$$w[n] = \begin{cases} 0.54 - 0.46 \cos\left(\dfrac{2\pi n}{L-1}\right) & 0 \leq n \leq L-1 \\ 0 & \text{otherwise.} \end{cases}$$

5. Blackman window:

$$w[n] = \begin{cases} 0.42 - 0.5 \cos\left(\dfrac{2\pi n}{L-1}\right) + 0.08 \cos\left(\dfrac{4\pi n}{L-1}\right) & 0 \leq n \leq L-1 \\ 0 & \text{otherwise.} \end{cases}$$

Accept as input the value of the window duration, L, and check that it is an odd integer. Design the five windows and plot their time responses on a common plot. On a separate plot, show the log magnitude responses of all five windows. Compare the effective bandwidths of the five windows along with the peak sidelobe ripple (in dB). (Hint: You may want to consider replotting the log magnitude response over a narrow band between 0 and $5 * F_s/L$ to compare the effective bandwidths of the five windows.)

6.17. (MATLAB Exercise) Write a MATLAB program to analyze a speech file and simultaneously, on one page, plot the following measurements:

1. the entire speech waveform
2. the short-time energy, $E_{\hat{n}}$
3. the short-time magnitude, $M_{\hat{n}}$
4. the short-time zero-crossing, $Z_{\hat{n}}$

Use the speech waveforms in the files `s5.wav` and `should.wav` to test your program. Choose appropriate window sizes (L), window shifts (R), and window type (Hamming, rectangular) for the analysis. Explain your choice of these parameters. (Don't forget to normalize the frequency scale of your analysis depending on the sampling rate of the speech signal in each file.)

6.18. (MATLAB Exercise) Write a MATLAB program to show the effects of window duration on the short-time analysis of energy, magnitude, and zero-crossings. Using the speech file `test_16k.wav`, compute the short-time energy, magnitude, and zero-crossings using frame lengths of $L = 51, 101, 201, 401$ samples using either a Hamming window or a rectangular window. Plot the resulting short-time estimates on a common plot. What effects do you see as the window length shortens or lengthens?

6.19. (MATLAB Exercise) Write a MATLAB program to create an AGC mechanism for a speech waveform. Using either an IIR filter of the form $H(z) = (1-\alpha)z^{-1}/(1-\alpha z^{-1})$ or an FIR filter of the form $h[n] = 1/M, 0 \leq n \leq M-1$, plot a voiced transition of the speech waveform showing the AGC tracking function σ_x superimposed on the speech waveform, and in a separate plot, show the gain equalized speech amplitude, thereby illustrating the effectiveness of AGC control. Consider what would be appropriate values of α or M for both syllabic rate control as well as instantaneous rate control.

6.20. (MATLAB Exercise) Write a MATLAB program to compute the two variants on the short-time autocorrelation of a section of speech namely:

1. short-time autocorrelation
2. modified short-time autocorrelation.

For this exercise you will have to specify the frame length, L, and the maximum number of autocorrelation points, K. Compare the short-time autocorrelation estimates from the short-time autocorrelation and the modified short-time autocorrelation for several voiced regions. Which short-time autocorrelation estimate would be better for pitch period detection algorithms?

Try to estimate a suitable threshold for selecting the peak autocorrelation (in a voiced region) appropriate for pitch detection and, by processing the speech file on a frame-by-frame basis, determine the pitch contour of the speech file based on detecting the peak of the autocorrelation function and deciding if it is above the threshold (voiced speech) or below the threshold (unvoiced speech or background). Which works better for pitch detection, the short-time autocorrelation function or the modified short-time autocorrelation function?

6.21. (MATLAB Exercise) Write a MATLAB program to calculate the AMDF of a speech file and implement a pitch detection algorithm based on using the AMDF on a frame-by-frame basis. Plot the pitch contour of the utterance.

Frequency-Domain Representations

CHAPTER 7

7.1 INTRODUCTION

In many areas of science and engineering, the representation of signals or other functions by sums of sinusoids or complex exponentials leads to convenient solutions to problems and often to greater insight into physical phenomena than is available by other means. Such representations—Fourier representations as they are commonly called—are useful in signal processing for two basic reasons. The first is that for linear systems, it is very convenient to determine the response to a superposition of sinusoids or complex exponentials. The second reason is that the Fourier representations often serve to place in evidence certain properties of the signal that may be obscure or at least less evident in the original signal.

Speech communication research and technology are areas where the concept of a Fourier representation has traditionally played a major role. To see why this is so, it is helpful to recall that the discrete-time model for the production of samples of a steady state speech sound, such as a vowel or fricative, as shown in Figure 7.1, consists of a linear system with system function, $V(z)$, excited by a source which is either periodically varying ($A_V p[n] * g[n]$ for voiced speech) or randomly varying ($A_N u[n]$ for unvoiced speech). A transfer function, $R(z)$, represents radiation of sound at the lips. In general, the spectrum of the output of such a model would be the product of the frequency responses of the vocal tract system, the spectrum of the excitation source, and the spectrum of the model of sound radiation. For voiced speech such as a sustained vowel, we can write the discrete-time Fourier transform (DTFT) expression as:

$$X(e^{j\omega}) = A_V P(e^{j\omega})G(e^{j\omega})V(e^{j\omega})R(e^{j\omega}), \tag{7.1}$$

and for sustained unvoiced speech, assuming white noise excitation with unit power, the power spectrum of the output is

$$\Phi_{xx}(e^{j\omega}) = A_N^2 |V(e^{j\omega})|^2 |R(e^{j\omega})|^2. \tag{7.2}$$

Thus, it is to be expected that the Fourier spectrum of the output would reflect the properties of the excitation, the vocal tract and radiation frequency responses. However, although vowels and fricatives can be sustained for several seconds with little

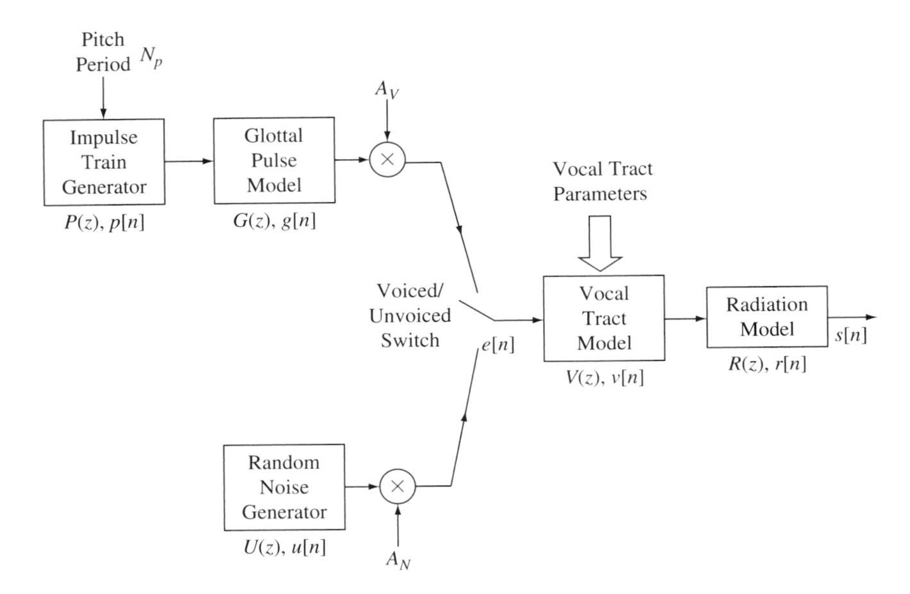

FIGURE 7.1
General discrete-time model of speech production showing explicit sources for voiced and unvoiced speech sounds.

variation, natural speech is continually changing in time. Thus the standard Fourier representations that are appropriate for periodic, transient, or stationary random signals are not directly applicable to the representation of speech signals. We have already seen ample evidence that the short-time analysis principle is a valid approach to speech processing. We have seen, for example, that temporal properties such as energy, zero crossings, and correlation are slowly varying so that they can be assumed to be fixed over time intervals on the order of 10 to 40 msec. We will demonstrate that spectral properties of speech likewise can be assumed to change relatively slowly with time.

In order to study spectral properties of speech signals, we will define a time-varying Fourier transform, which we generally refer to as the *short-time Fourier transform* (STFT). Use of the STFT is therefore termed *short-time Fourier analysis* (STFA). We will also show that the STFT is invertible in the sense that, with certain constraints, we can recover the original sampled signal by a process that we term *short-time Fourier synthesis* (STFS). Indeed, STFA/STFS provides a representation of the speech waveform that can serve as the basis for many types of speech processing including coding and various types of signal enhancement. This is depicted in Figure 7.2, which shows that the processing can be controlled by "side information" extracted by other means from the speech signal.

In the remainder of this chapter, we will show how STFA/STFS is based on conventional discrete-time Fourier analysis. We will also show how it can be interpreted as a bank of linear bandpass filters. This will lead to both theoretical and practical insights into time-dependent Fourier analysis. We will also consider computational techniques based upon fast computation algorithms for the discrete Fourier transform

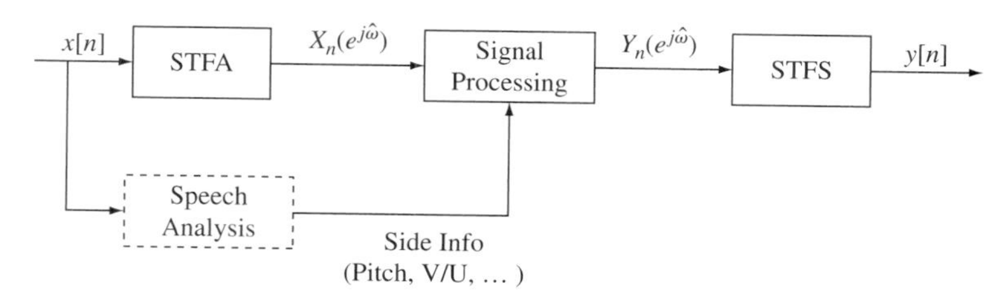

FIGURE 7.2
Model of frequency-domain processing of speech via STFA and STFS methods.

(FFT algorithms). The general concept of short-time Fourier representation is, in fact, the basis for almost everything that follows in this book.

7.2 DISCRETE-TIME FOURIER ANALYSIS

In Chapter 2 we discussed the discrete-time Fourier transform (DTFT) of a discrete-time signal, $x[n]$, and showed that the signal and its DTFT were related by the pair of equations

$$X(e^{j\omega}) = \text{DTFT}\{x[n]\} = \sum_{n=-\infty}^{\infty} x[n]e^{-j\omega n}, \tag{7.3a}$$

$$x[n] = \text{IDTFT}\{X(e^{j\omega})\} = \frac{1}{2\pi}\int_{-\pi}^{\pi} X(e^{j\omega})e^{j\omega n}d\omega, \tag{7.3b}$$

where ω is the normalized frequency variable of $X(e^{j\omega})$ in units of radians.[1] When we wish to clearly specify the discrete-time Fourier transform defined by Eqs. (7.3a) and (7.3b), we shall use the acronym DTFT or the terminology "normal (discrete-time) Fourier transform," to distinguish it from the STFT to be defined below.

We also reviewed the related concept of the discrete Fourier transform (DFT), which is inherently a representation of periodic sequences, but is applicable to finite-length sequences if care is taken to ensure that one period is precisely equal to the desired finite-length sequence. The DFT and its inverse are given by the equations

$$X[k] = \text{DFT}\{x[n]\} = \sum_{n=0}^{N-1} x[n]e^{-j(2\pi k/N)n}, \qquad k = 0, 1, \ldots, N-1, \tag{7.4a}$$

$$x[n] = \text{IDFT}\{X[k]\} = \frac{1}{N}\sum_{k=0}^{N-1} X[k]e^{j(2\pi k/N)n}, \qquad n = 0, 1, \ldots, N-1. \tag{7.4b}$$

[1] Recall that ω is related to analog radian frequency Ω by the equation $\omega = \Omega T = \Omega/F_s$, where $F_s = 1/T$ is the sampling frequency.

The DFT and DTFT can both be used as mathematical representations of a finite-length sequence; specifically, the DFT and the DTFT of a finite-length sequence are related by

$$X[k] = X(e^{j\omega})\Big|_{\omega=(2\pi k/N)} \qquad k = 0, 1, \ldots, N-1,$$

that is, the DFT is a sampled (in frequency) version of the DTFT.

Example 7.1 DFT and DTFT of a Periodic Impulse Train

A particularly useful result concerns the periodic impulse train sequence

$$p[n] = \sum_{r=-\infty}^{\infty} \delta[n - rN], \tag{7.5}$$

which we often invoke in our model for periodic voiced sounds. We will also see that $p[n]$ plays an important role in understanding the process of STFS. From Eq. (7.4a), it follows that the DFT of $p[n]$ is $P[k] = 1$ for $k = 0, 1, \ldots, N-1$. Hence $p[n]$ can also be represented as

$$p[n] = \frac{1}{N} \sum_{k=0}^{N-1} e^{j(2\pi k/N)n}. \tag{7.6}$$

If Eq. (7.6) is evaluated only in $0 \leq n \leq N-1$, we get $\delta[n]$. However, if evaluated outside that interval, Eq. (7.6) repeats periodically with period N as described explicitly by Eq. (7.5). This can be seen by substituting $n+N$ for n in Eq. (7.6). The DTFT can also be used as a frequency-domain representation of the periodic interpretation of $p[n]$ by noting that the DTFT of a complex exponential signal, $e^{j\omega_0 n}$, is simply $2\pi\delta(\omega - \omega_0)$ for $0 \leq \omega < 2\pi$, where $\delta(\ )$ denotes the continuous-variable impulse (or Dirac delta function). Applying this result to Eq. (7.6) gives the DTFT defined over one period in ω as

$$P(e^{j\omega}) = \sum_{k=0}^{N-1} \left(\frac{2\pi}{N}\right) \delta(\omega - (2\pi k/N)), \qquad 0 \leq \omega < 2\pi. \tag{7.7}$$

Substitution of Eq. (7.7) into the inverse DTFT expression in Eq. (7.3b) and evaluating the integral with limits 0 to 2π yields Eq. (7.6), which in turn confirms that $P(e^{j\omega})$ in Eq. (7.7) is the DTFT of the periodic impulse train in Eq. (7.5).

While we might assert that speech signals may go on continuously for hours at a time, they are inherently of finite duration. Even the most loquacious talker must occasionally stop to breathe in. Thus, it is certainly possible to apply conventional discrete-time Fourier analysis to word-length or even sentence-length sequences of

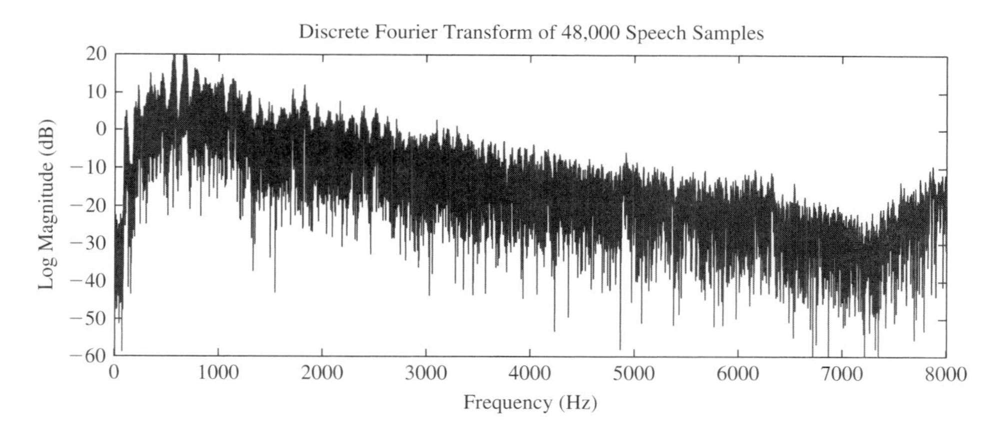

FIGURE 7.3
Log magnitude plot for DFT of 48,000 point section of speech. The speech utterance was "She had your dark suit in greasy wash water all year" and the sampling rate was $F_s = 16{,}000$ samples/sec. (Frequency axis labeled in terms of analog frequency $F_k = kF_s/N$.)

speech samples. Using Eq. (7.4a) we can compute the DFT of any sequence of samples whose length is N or less. The DFT frequencies $\omega_k = 2\pi k/N$ then correspond to analog radian frequencies $\Omega_k = 2\pi kF_s/N$ or corresponding analog cyclic frequencies $F_k = kF_s/N$. Figure 7.3 shows the result of such a computation with 48,000 samples of speech (3 seconds duration at a sampling rate of 16,000 samples/sec). A value $N = 65{,}536$ was used with zero padding in Eq. (7.4a) for computational efficiency.

What can we learn from this application of the DFT to a long sequence of speech samples? Figure 7.3 shows that the DFT oscillates wildly from DFT sample to DFT sample, but our eye is able to pick out some gross features.[2] Observe that the maximum spectrum level occurs below 1000 Hz, with a general fall-off as frequency increases. This can be attributed to the fact that the voiced sounds have the highest amplitude and the filtering effect of the glottal pulse shape in our model for voiced speech production, which tends to suppress high frequencies. We see a set of prominences regularly spaced at intervals of approximately 110 Hz over the range from 0 to 3 kHz. These probably represent the average pitch frequency for this talker. We also might be suspicious that the speech was not adequately lowpass filtered prior to sampling, because of the rising trend around $F_s/2 = 8000$ Hz. Beyond these qualitative observations, not much more can be learned from the long-term Fourier spectrum of an entire-sentence-length signal. A different utterance or different speaker would produce a plot with

[2]This random appearing behavior is also observed when we compute the DFT of a long sequence of random samples. In that context, $|X[k]|^2$ is an estimate of the power density spectrum, and it is called the *periodogram*. Statistical spectrum analysis techniques such as windowing the corresponding autocorrelation function or averaging of short periodograms are used to obtain smooth spectral estimates [270]. These techniques can also be used to compute smooth long-term average spectra for speech.

gross features very similar to Figure 7.3. We need a tool that is much more sensitive to the time-varying properties of the speech signal. This tool is the STFT.

7.3 SHORT-TIME FOURIER ANALYSIS

We define the time-dependent, or short-time, Fourier transform (STFT) as

$$X_{\hat{n}}(e^{j\hat{\omega}}) = \sum_{m=-\infty}^{\infty} w[\hat{n}-m]x[m]e^{-j\hat{\omega}m}, \tag{7.8}$$

where $w[\hat{n}-m]$ is a real window sequence whose purpose is to determine the portion of the input signal that receives emphasis at a particular time index, $\hat{n}$. The time-dependent Fourier transform is a complex function of two variables: the time index, $\hat{n}$, which is discrete, and the frequency variable, $\hat{\omega}$, which is continuous and periodic, with period 2π.[3] A plot showing the domain of the two variables, $\hat{n}$ and $\hat{\omega}$, is given in Figure 7.4 for the range $0 \leq \hat{n} \leq 8$ ($\hat{n}$ is defined for all discrete values but only a few are shown in this figure) and for $0 \leq \hat{\omega} < 2\pi$ (since $\hat{\omega}$ is periodic over intervals of 2π). Alternatively, we could use the range $-\pi < \hat{\omega} \leq \pi$.

An alternative form of Eq. (7.8) is obtained by a change of summation index, which yields the expression

$$\begin{aligned} X_{\hat{n}}(e^{j\hat{\omega}}) &= \sum_{m=-\infty}^{\infty} w[m]x[\hat{n}-m]e^{-j\hat{\omega}(\hat{n}-m)} \\ &= e^{-j\hat{\omega}\hat{n}} \sum_{m=-\infty}^{\infty} x[\hat{n}-m]w[m]e^{j\hat{\omega}m}. \end{aligned} \tag{7.9}$$

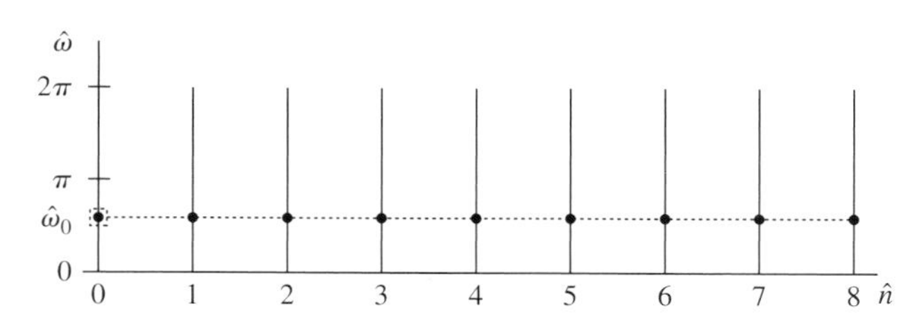

FIGURE 7.4
Domain of variables $\hat{n}$ and $\hat{\omega}$ in the definition of the STFT.

[3]We use $\hat{\omega}$ as the frequency variable of the STFT, to distinguish the STFT analysis frequency from the frequency variable ω, which we reserve for the normal DTFT as in Eq. (7.3a). Similarly, we use $\hat{n}$ for the time index of the STFT to distinguish from n, which is used throughout as a generic time index.

If we define

$$\tilde{X}_{\hat{n}}(e^{j\hat{\omega}}) = \sum_{m=-\infty}^{\infty} x[\hat{n}-m]w[m]e^{j\hat{\omega}m} = \sum_{m=-\infty}^{\infty} x[\hat{n}+m]w[-m]e^{-j\hat{\omega}m}, \tag{7.10}$$

then $X_{\hat{n}}(e^{j\hat{\omega}})$ can be expressed as

$$X_{\hat{n}}(e^{j\hat{\omega}}) = e^{-j\hat{\omega}\hat{n}}\tilde{X}_{\hat{n}}(e^{j\hat{\omega}}). \tag{7.11}$$

The STFT equations can be interpreted in two distinct ways. First, assuming that $\hat{n}$ is fixed, we observe from Eq. (7.8) that $X_{\hat{n}}(e^{j\hat{\omega}})$ is simply the DTFT of the sequence $w[\hat{n}-m]x[m]$, $-\infty < m < \infty$. Therefore, for fixed $\hat{n}$, $X_{\hat{n}}(e^{j\hat{\omega}})$ has the same properties as a normal DTFT. The vertical lines in Figure 7.4 show the regions of support $0 \le \hat{\omega} < 2\pi$ for different values of $\hat{n}$. The second interpretation follows by considering $X_{\hat{n}}(e^{j\hat{\omega}})$ as a function of the time index $\hat{n}$ with $\hat{\omega}$ fixed at, for example, $\hat{\omega}_0$ as in Figure 7.4. This corresponds to points at the intersection of the horizontal dashed line in Figure 7.4 with the vertical lines. (Because $\hat{n}$ is discrete, $X_{\hat{n}}(e^{j\hat{\omega}})$ is not defined at points between.) In this case we observe that Eq. (7.8), Eq. (7.9), and Eq. (7.10) are all in the form of a discrete-time convolution. This interpretation leads us naturally to consider the time-dependent Fourier representation in terms of linear filtering. As we will see, both interpretations lead to useful insights and we will find it worthwhile to examine the time-dependent Fourier transform in detail from both viewpoints.

7.3.1 DTFT Interpretation

Consider $X_{\hat{n}}(e^{j\hat{\omega}})$ as the DTFT of the sequence $w[\hat{n}-m]x[m]$, $-\infty < m < \infty$, for fixed $\hat{n}$. The time-dependent Fourier transform is a function of the time index, $\hat{n}$, which takes on all integer values so as to "slide" the window, $w[\hat{n}-m]$, along the sequence, $x[m]$. This is depicted in Figure 7.5, which shows $x[m]$ and $w[\hat{n}-m]$ as functions of m for several values of $\hat{n}$.

The conditions for the existence of the STFT representation are identical to those for the DTFT; i.e., the sequence must be absolutely summable. For the STFT, the

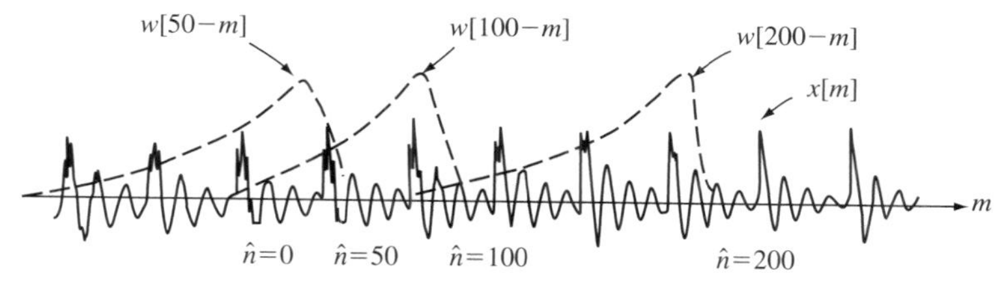

FIGURE 7.5
Plots of $x[m]$ and $w[\hat{n}-m]$ for several values of $\hat{n}$. (Note that the signal and the window are plotted as continuous functions for convenience, even though they are only defined for integer values of m and $\hat{n}-m$.)

sequence $x[m]w[\hat{n}-m]$ must be absolutely summable for all values of $\hat{n}$ [270]. If, as is generally the case, $w[\hat{n}-m]$ is of finite duration and $|x[m]| < \infty$ for all m, then this condition is satisfied for all $\hat{n}$.

As in the case for normal Fourier transforms of discrete-time signals, the STFT is periodic in $\hat{\omega}$ with period 2π. This is easily seen by substituting $\hat{\omega}+2\pi$ into Eq. (7.8). As discussed before, the STFT can be expressed in terms of analog frequencies through the relation $\hat{\omega}=\Omega T$, where T is the sampling period used to obtain the sequence $x[m]$, and Ω denotes analog radian frequency. Also, by making the substitutions $\hat{\omega}=2\pi f$ or $\hat{\omega}=2\pi FT$, we can express the STFT as a function of either normalized cyclic frequency, f, or conventional analog cyclic frequency F (in Hertz) respectively. We will have occasion to use a variety of different frequency variables in equations and figures in this chapter and in the remainder of the book. To avoid confusion, it is important to become familiar with the relationships among the different frequency variables.

The fact that for a given value of $\hat{n}$, $X_{\hat{n}}(e^{j\hat{\omega}})$ has the same properties as a normal DTFT leads to a simple proof that the input sequence $x[m]$ can be recovered *exactly* from the time-varying Fourier transform. Since $X_{\hat{n}}(e^{j\hat{\omega}})$ for fixed $\hat{n}$ is the DTFT of $w[\hat{n}-m]x[m]$, we can use Eq. (7.3b) to write

$$w[\hat{n}-m]x[m] = \frac{1}{2\pi}\int_{-\pi}^{\pi} X_{\hat{n}}(e^{j\hat{\omega}})e^{j\hat{\omega}m}\,d\hat{\omega}. \tag{7.12}$$

That is, we recover the sequence $w[\hat{n}-m]x[m]$ as the inverse DTFT of $X_{\hat{n}}(e^{j\hat{\omega}})$. Note that the integration in Eq. (7.12) could be over any interval of length 2π (e.g., 0 to 2π) since the entire integrand is periodic with period 2π. Now if $w[0]\neq 0$, Eq. (7.12) can be evaluated for $m=\hat{n}$, thereby obtaining

$$x[\hat{n}] = \frac{1}{2\pi w[0]}\int_{-\pi}^{\pi} X_{\hat{n}}(e^{j\hat{\omega}})e^{j\hat{\omega}m}\,d\hat{\omega}. \tag{7.13}$$

Thus, with the rather mild requirement $w[0]\neq 0$, the sequence $x[\hat{n}]$ can be exactly recovered from $X_{\hat{n}}(e^{j\hat{\omega}})$, if $X_{\hat{n}}(e^{j\hat{\omega}})$ is known at each $\hat{n}$ for all values of $\hat{\omega}$ over one complete period. This is an important theoretical result, which, we will see, takes on practical significance with the imposition of a simple additional constraint on the window. Another interesting observation can be made from Eq. (7.12). We see that, in principle, we could recover not just a single value $x[\hat{n}]$ but instead an entire range of $x[m]$ could be recovered by realizing that, so long as the window is positive and non-zero, the sequence $x[m]$ can be recovered as

$$x[m] = \frac{1}{2\pi w[\hat{n}-m]}\int_{-\pi}^{\pi} X_{\hat{n}}(e^{j\hat{\omega}})e^{j\hat{\omega}m}\,d\hat{\omega}, \tag{7.14}$$

for m inside the interval where $w[\hat{n}-m]\neq 0$. To completely reconstruct $x[m]$ for all m, it is only necessary to ensure that values of $\hat{n}$ are chosen so that each value of m is included in the region of support of at least one shifted window $w[\hat{n}-m]$. This demonstrates that it is possible to invert the short-time transform, but it is not a computationally feasible way to do it. Practical algorithms for inverse STFS are discussed later in this chapter.

The STFT, $X_{\hat{n}}(e^{j\hat{\omega}})$, is, in general, a complex-valued function of $\hat{n}$ and $\hat{\omega}$. Therefore, it can be expressed in terms of its real and imaginary parts,[4]

$$X_{\hat{n}}(e^{j\hat{\omega}}) = a_{\hat{n}}(\hat{\omega}) - jb_{\hat{n}}(\hat{\omega}). \tag{7.15}$$

For the case when $x[m]$ and $w[\hat{n}-m]$ are both real, $a_n(\hat{\omega})$ and $b_n(\hat{\omega})$ can be shown to satisfy certain symmetry and periodicity relations (see Problem 7.1) that are the properties of any DTFT. Another representation for $X_{\hat{n}}(e^{j\hat{\omega}})$ is in terms of magnitude and phase as

$$X_{\hat{n}}(e^{j\hat{\omega}}) = |X_{\hat{n}}(e^{j\hat{\omega}})|e^{j\theta_{\hat{n}}(\hat{\omega})}. \tag{7.16}$$

The quantities $|X_{\hat{n}}(e^{j\hat{\omega}})|$ and $\theta_{\hat{n}}(\hat{\omega})$ can readily be related to $a_{\hat{n}}(\hat{\omega})$ and $b_{\hat{n}}(\hat{\omega})$ and vice versa (see Problem 7.2). Additional properties of $a_{\hat{n}}(\hat{\omega})$, $b_{\hat{n}}(\hat{\omega})$, and $X_{\hat{n}}(e^{j\hat{\omega}})$ are emphasized in other problems given at the end of this chapter.

7.3.2 DFT Implementation

While the DTFT interpretation of the STFT yields useful insights, we must rely on the DFT and its fast computation algorithms (FFT) to implement the computation of the STFT as a sequence of Fourier transforms evaluated at a finite discrete set of frequencies $\omega_k = 2\pi k/N$, with $k = 0, 1, \ldots, N-1$. By substituting $(2\pi k/N)$ for ω in Eq. (7.10), we obtain

$$\begin{aligned} \tilde{X}_{\hat{n}}(e^{j(2\pi k/N)}) &= \sum_{m=0}^{L-1} x[\hat{n}+m]w[-m]e^{-j(2\pi k/N)m} \\ &= \tilde{X}_{\hat{n}}[k], \qquad k = 0, 1, \ldots, N-1, \end{aligned} \tag{7.17}$$

where the window is chosen to be an L-point non-causal window such that $w[-m] \neq 0$ only in $0 \leq m \leq L-1$ and $L \leq N$. From Eq. (7.11), it follows that the STFT at time $\hat{n}$ and frequencies $\omega_k = 2\pi k/N$ is

$$X_{\hat{n}}(e^{j(2\pi k/N)}) = e^{-j(2\pi k/N)\hat{n}}\tilde{X}_{\hat{n}}[k], \qquad k = 0, 1, \ldots, N-1. \tag{7.18}$$

Equation (7.17) should be recognized as the DFT of the windowed sequence $\tilde{x}_{\hat{n}}[m] = x[\hat{n}+m]w[-m]$ for $0 \leq m \leq L-1$, and hence $\tilde{X}_{\hat{n}}[k]$ can be computed efficiently by an FFT algorithm if N is a power of two or some other highly composite

[4]Note that $a_{\hat{n}}(\hat{\omega})$ is the real part and $b_{\hat{n}}(\hat{\omega})$ is minus the imaginary part of $X_{\hat{n}}(e^{j\hat{\omega}})$. The negative sign is used for convenience in later discussion.

number [270]. Thus, to compute $\tilde{X}[k]$, we iterate the following steps:

1. Select a set of L samples starting at $\hat{n}$. (For a causal window, we would take sample $\hat{n}$ and the $L-1$ samples preceding $\hat{n}$.)
2. Multiply by the window $w[-m]$ to form $\tilde{x}_{\hat{n}}[m] = x[\hat{n}+m]w[-m]$, for $m = 0, 1, \ldots, L-1$.
3. Compute $\tilde{X}_{\hat{n}}[k]$, the N-point DFT of $\tilde{x}_{\hat{n}}[m]$ using a fast (FFT) algorithm.
4. If magnitude and phase of $X_{\hat{n}}(e^{j(2\pi k/N)})$ are required, use Eq. (7.18). Otherwise, note that $|X_{\hat{n}}(e^{j(2\pi k/N)})| = |\tilde{X}_{\hat{n}}(e^{j(2\pi k/N)})| = |\tilde{X}_{\hat{n}}[k]|$.

Generally, $\tilde{X}_{\hat{n}}[k]$ is not evaluated at each $\hat{n}$, but rather the STFT is sampled in both time and frequency by computing $\tilde{X}_{rR}[k]$, where R is an integer and $R \geq 1$.

7.3.3 Effect of Window on Resolution

So far we have not considered the role of the window, $w[\hat{n}-m]$, beyond its obvious function of selecting the portion of the sequence $x[m]$ to be analyzed. The shape of the window sequence has an important effect on the nature of the time-dependent Fourier transform, and the present viewpoint provides a convenient way to interpret the role of the window sequence, $w[\hat{n}-m]$. If $X_{\hat{n}}(e^{j\hat{\omega}})$ is thought of as the normal DTFT of the sequence $w[\hat{n}-m]x[m]$, and if we assume for the moment that the DTFTs

$$X(e^{j\omega}) = \sum_{m=-\infty}^{\infty} x[m]e^{-j\omega m} \tag{7.19}$$

and

$$W(e^{j\omega}) = \sum_{m=-\infty}^{\infty} w[m]e^{-j\omega m} \tag{7.20}$$

exist, then the DTFT of $w[\hat{n}-m]x[m]$ (for fixed $\hat{n}$) is the convolution of the transforms of $w[\hat{n}-m]$ and $x[m]$ [270]. Since, for fixed $\hat{n}$, the DTFT of $w[\hat{n}-m]$ is $W(e^{-j\omega})e^{-j\omega\hat{n}}$,

$$X_{\hat{n}}(e^{j\hat{\omega}}) = \frac{1}{2\pi}\int_{-\pi}^{\pi} W(e^{-j\omega})e^{-j\omega\hat{n}}X(e^{j(\hat{\omega}-\omega)})d\omega. \tag{7.21}$$

Equation (7.21) states that the Fourier transform of the sequence $x[m]$, $-\infty < m < \infty$ is convolved with the Fourier transform of the shifted window sequence. We have argued that the normal Fourier transform of an entire speech signal is not of great interest even if it exists. However, Eq. (7.21) can be useful if we first recall that the purpose of the window is to emphasize a finite segment of the speech waveform in the vicinity of sample $\hat{n}$, and to deemphasize the remainder of the waveform. Indeed, typical window sequences are such that $w[\hat{n}-m] = 0$ for m outside a finite interval around $\hat{n}$. Insofar as the final result is concerned, then, it is entirely reasonable to assume that the properties of $x[m]$ inside the window persist outside the window. For example, if the speech signal within the window corresponds to a vowel or other voiced sound embedded in a continuously changing speech waveform, we can just as well consider that the windowed sequence $x[m]w[\hat{n} - m]$ arose from windowing a periodic sustained voiced sound. Likewise if the speech within the window is unvoiced, we can assume that the same unvoiced properties exist outside the window. An equally appropriate point of view is that the signal is simply zero outside the window. This would be appropriate for the analysis of transient sounds such as plosives.

Thus, Eq. (7.21) is meaningful if we assume that $X_{\hat{n}}(e^{j\hat{\omega}})$ stands for the STFT (at time $\hat{n}$) of a signal that either continues in the same way outside the window or is zero outside the window.

With this point of view, the properties of the Fourier transform of the window, $W(e^{j\omega})$, become important. It is clear from Eq. (7.21) that for faithful reproduction of the properties of $X(e^{j\omega})$ in $X_{\hat{n}}(e^{j\hat{\omega}})$, the function $W(e^{j\omega})$ should appear as a continuous-variable impulse with respect to $X(e^{j\omega})$; i.e., $W(e^{j\omega})$ should be highly concentrated around $\omega = 0$. We have already discussed the properties of the rectangular and Hamming windows in Chapter 6, where it was shown that the width of the main lobe of $W(e^{j\omega})$ is inversely proportional to the length of the window, whereas the levels of the side-lobes depend mainly on the window shape and are essentially independent of the window length [270]. For an $(L = 2M + 1)$-point rectangular window, the "main lobe" extending from the zero at $\omega = -2\pi/L$ to the zero at $\omega = +2\pi/L$ has a width of approximately $\Delta\omega = 2\pi/M$ (in normalized radian frequency units) or F_s/M (in analog cyclic frequency units), and the peak sidelobes (of the stopband) are only 14 dB below the level of the peak of the main lobe (the passband). Figure 7.6 shows an $(L = 2M + 1)$-point Hamming window defined as

$$w_H[n] = \begin{cases} 0.54 + 0.46\cos(\pi n/M) & -M \le n \le M \\ 0 & \text{otherwise.} \end{cases} \tag{7.22}$$

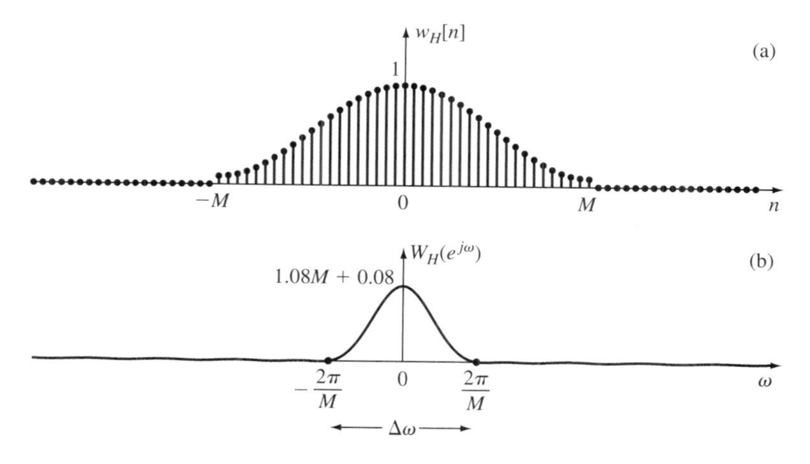

FIGURE 7.6
(a) $(L = 2M + 1)$-point Hamming window and (b) its corresponding DTFT.

in (a) and its DTFT, $W_H(e^{j\omega})$ in (b). As shown in Figure 7.6b, the frequency width of the main lobe is approximately $\Delta\omega = 4\pi/M \approx 8\pi/L$ (in normalized radian frequency units) or $2F_s/M \approx 4F_s/L$ (in analog cyclic frequency units), and the peak side lobes (not visible on the linear scale) are 40 dB or more below the peak of the main lobe (the passband).[5]

The length and shape of the window affect the frequency resolution of the STFT. We can illustrate this by assuming that for voiced speech, the segment within the window is represented by

$$x[n] = h[n] * \sum_{k=-\infty}^{\infty} \delta[n - kN_p] = \sum_{k=-\infty}^{\infty} h[n - kN_p],$$

where $h[n] = A_V g[n] * v[n] * r[n]$ represents the combined effects of the voiced gain, glottal pulse, vocal tract, and radiation in Figure 7.1, and N_p is the pitch period in samples. Using Eq. (7.7), the corresponding DTFT of $x[n]$ is found to be

$$X(e^{j\omega}) = \sum_{k=-\infty}^{\infty} \omega_0 H(e^{jk\omega_0})\delta(\omega - k\omega_0),$$

where $\omega_0 = 2\pi/N_p$ is the fundamental frequency in normalized radian units. That is, the DTFT of a periodic sequence with period N_p samples is comprised of a train of impulse functions spaced by the fundamental frequency, ω_0, where each impulse has a size proportional to the value of $H(e^{j\omega})$ at the harmonic frequency, $k\omega_0$. Substituting this for $X(e^{j\omega})$ in Eq. (7.21) and carrying out the convolution with the impulse functions gives

$$X_{\hat{n}}(e^{j\hat{\omega}}) = \frac{1}{N_p} \sum_{k=-\infty}^{\infty} H(e^{jk\omega_0}) \left[e^{-j(\hat{\omega}-k\omega_0)\hat{n}} W(e^{-j(\hat{\omega}-k\omega_0)}) \right], \quad (7.23)$$

which shows that the STFT of a periodic model of voiced speech consists of a sum of copies of the DTFT of $w[\hat{n} - m]$, which are scaled by the complex number $H(e^{jk\omega_0})/N_p$ and shifted by multiples of the fundamental frequency ω_0. An example of short-time spectral analysis of voiced speech is shown in Figure 7.7. Hamming windows with durations of $L = 251$ samples (narrowband frequency resolution) and $L = 81$ samples (wideband frequency resolution) are shown in Figure 7.7a along with a 251-sample segment of the speech waveform. The corresponding STFTs (magnitudes) are shown in Figure 7.7b. The speech signal, sampled at a rate of $F_s = 8000$ Hz, is from a relatively high-pitched male speaker. Figure 7.7a shows that the pitch period at this

[5] We show the case $L = 2M + 1$, an odd integer, for simplicity. Since the window can be made an even function ($w_H[-n] = w_H[n]$), its DTFT $W(e^{j\omega})$ is real and even. In general, L can be either even or odd, but an even-length window cannot be oriented to be an even function of n.

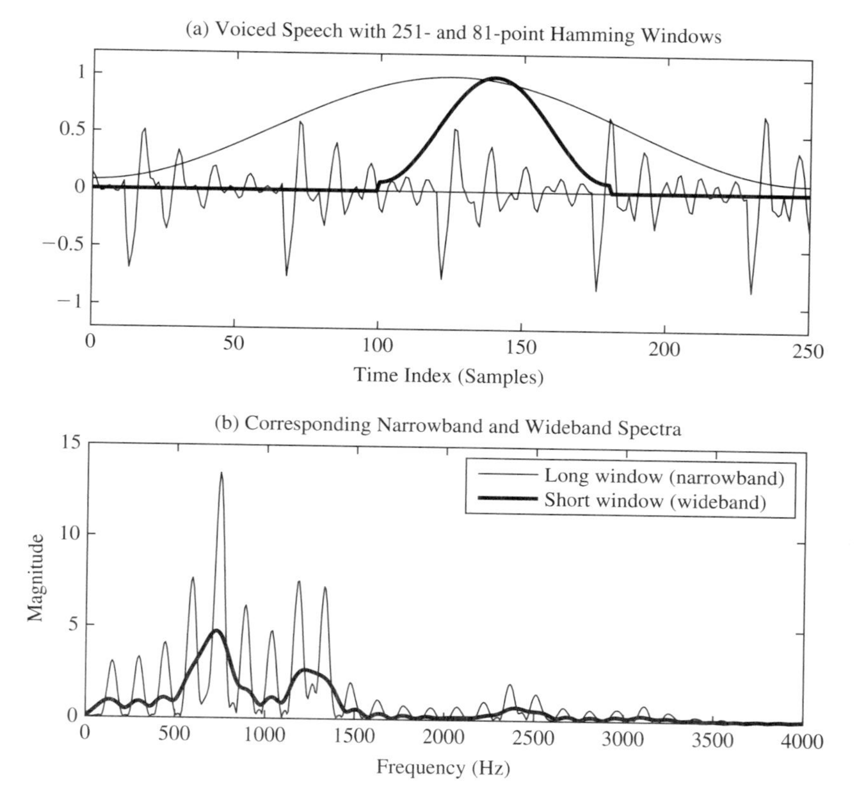

FIGURE 7.7
Windowed speech waveforms for narrowband (long window) and wideband (short window) spectral analysis using $L = 251$ and $L = 81$ sample Hamming windows.

time is approximately 54 samples, implying that the fundamental analog frequency is $F_0 = 8000/54 \approx 148$ Hz. The 251 samples, equivalent to 31.25 msec, span slightly more than four pitch periods, while 81 samples, equivalent to 10 msec, cover only slightly more than one pitch period. This is reflected in the frequency resolution evident in Figure 7.7b. The main lobe width of the longer window is approximately $\Delta F = 2 * 8000/125 = 128$ Hz. Since $\Delta F < F_0$ $(\Delta\omega < \omega_0)$ for the long window, the shifted copies of the DTFT of the window appearing in Eq. (7.23) do not overlap appreciably and we can distinguish copies of $W(e^{-j2\pi FT})e^{-j2\pi F\hat{n}T}$ centered on each multiple of the fundamental frequency. When the window length is several pitch periods long, the STFT analysis is said to be *narrowband*. Note that the narrowband spectrum has about 10 equally spaced peaks in the frequency interval 0 to 1500 Hz, confirming that the fundamental frequency is about 150 Hz.

When the window length is on the order of one pitch period, the analysis is said to be *wideband*; i.e., the frequency resolution is such that individual harmonics of the fundamental frequency are not resolved. The 81-point window yields a wideband analysis as seen in Figure 7.7b. Such analysis resolves individual pitch periods in time, but because the shifted copies of $W(e^{-j2\pi FT})e^{-j2\pi F\hat{n}T}$ have width $\Delta F = 2(8000)/40 = 400$ Hz, they overlap and blur the pitch harmonics together, creating the heavy smooth curve.[6] Since the window is about the same length as a pitch period, it is entirely reasonable that the spectrum does not evince any evidence of periodicity.

Illustrations of the effects of window shape and length are provided by the plots of Figures 7.8–7.10. All of these figures were constructed using speech sampled at a rate of 16,000 samples/sec. In each figure, the baseband 0 to 8000 Hz is shown for the

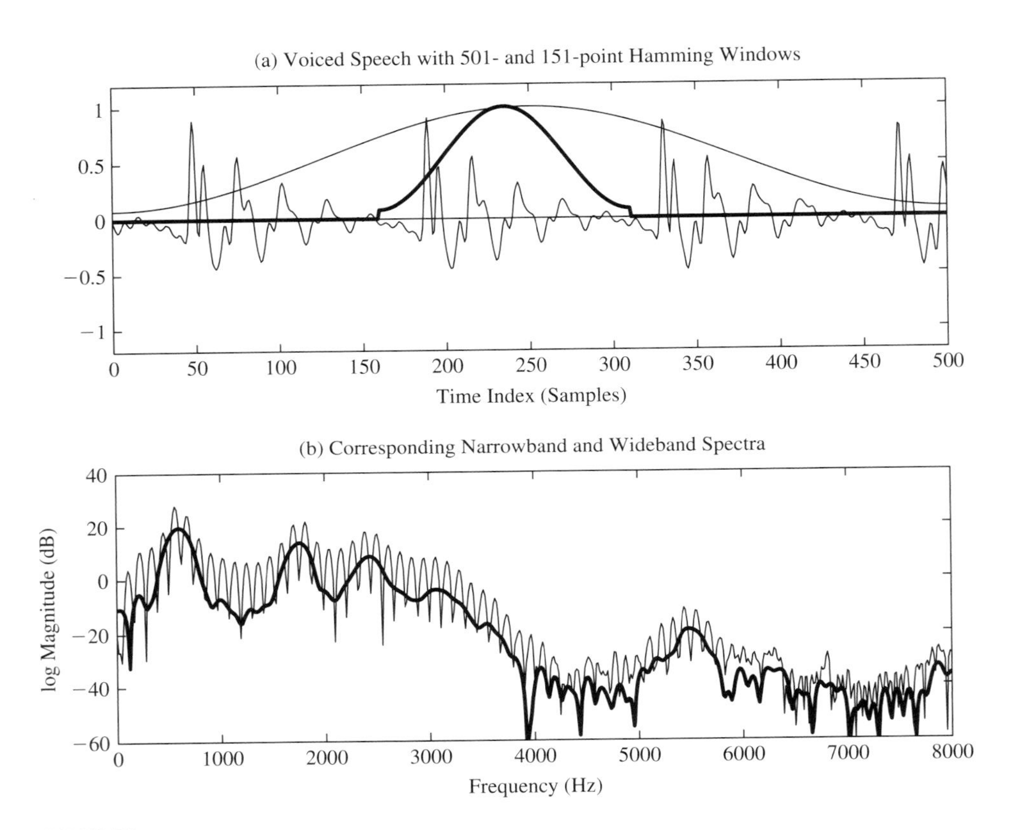

FIGURE 7.8
Spectrum analysis for voiced sound /AE/ as in /HH AE D/. (a) Time waveform (sampling rate 16 kHz) and Hamming windows of length $L = 501$ and $L = 151$ samples. (b) Corresponding spectra.

[6]Note that Figure 7.7b shows $|X_{\hat{n}}(e^{j2\pi FT})|$ after the shifted and scaled copies of $W(e^{-j2\pi FT})e^{-j2\pi F\hat{n}T}$ have been combined by complex addition. This complex addition leads to cancellation and reinforcement between the harmonics.

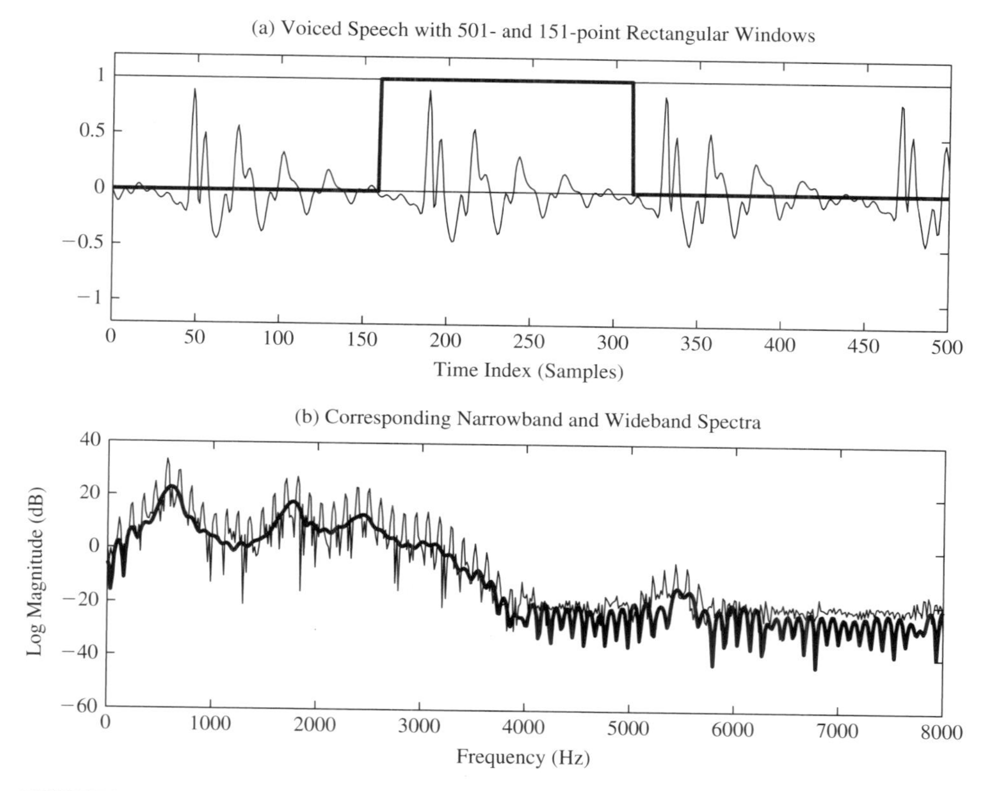

FIGURE 7.9
Spectrum analysis for voiced sound /AE/ as in /HH AE D/. (a) Time waveform (sampling rate 16 kHz) and rectangular windows of length $L = 501$ and $L = 151$ samples. (b) Corresponding spectra.

STFT log magnitude.[7] Figure 7.8 shows results for Hamming windows of 501 samples (31.25 msec duration) and 151 samples (9.375 msec duration) for a section of voiced speech where the pitch period is approximately 140 samples (equivalent to 8.75 msec at the 16 kHz rate). The corresponding fundamental frequency is approximately 114 Hz. These features of the speech signal are clearly reflected in the STFTs. In the case of the narrowband spectrum (window length 501 samples), the width of the main lobe of the DTFT of the window is $\Delta F = 2(16{,}000)/250 = 128$ Hz, which is slightly greater than the harmonic spacing, but the harmonic structure is still clearly in evidence. Each of the "ripples" is due to a shifted version of the DTFT of the window, and the ripple peaks are separated by approximately 114 Hz. The wideband spectrum shows no evidence of

[7]In Figure 7.7 the STFT magnitude was shown because we wanted to illustrate how the shape of the Fourier transform of the window manifests itself in the STFT of voiced speech. However, we generally prefer to plot the log magnitude of the STFT (dB) because it shows fine detail in regions where the spectrum is both small and large.

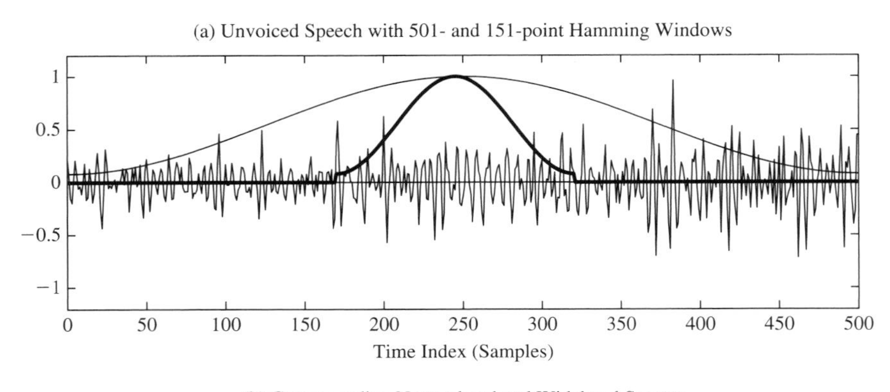

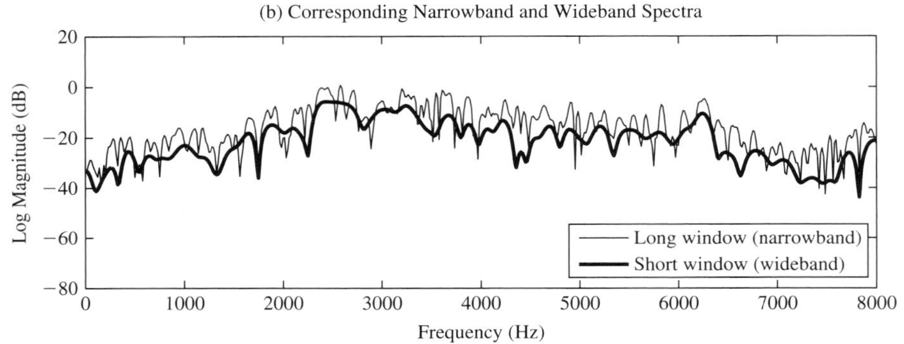

FIGURE 7.10
Spectrum analysis for unvoiced sound /SH/ as in /SH IY/. (a) Time waveform (sampling rate 16 kHz) and Hamming windows of length $L = 501$ and $L = 151$ samples. (The speech signal was multiplied by 10 compared to the voiced segment in Figure 7.8 for plotting.) (b) Corresponding spectra.

the pitch harmonic structure, as expected. However, it does show several prominent peaks corresponding to the formant frequencies for this segment of voiced speech. The first four major peaks of the wideband spectrum are at approximately 600, 1750, 2430, and 3070 Hz. These are quite close to the nominal values for the vowel /AE/ given in Table 3.4. Also note that the spectrum shows a tendency to fall-off at higher frequencies due to the lowpass nature of the glottal pulse spectrum.

Figure 7.9 shows the results for the same segment of the vowel /AE/ but this time with rectangular windows of lengths 501 and 151 samples. A comparison of the spectra of Figures 7.8 (Hamming window) and 7.9 (rectangular window) shows considerable overall similarity in terms of the pitch harmonics, formant structure, and gross spectral shape. Differences in the spectra can also be seen, the most notable being the increased sharpness of the pitch harmonics of Figure 7.9, due to the narrower main lobe of the rectangular window relative to that of the same length Hamming window. (The main lobe width for the rectangular window is half that of the same length Hamming window.) Another difference in the spectra is that the relatively large side lobes of the rectangular window produce a "ragged" or noisy spectrum between the pitch harmonics. This effect occurs because the side lobes of window transforms positioned at adjacent harmonics interact in the space between the harmonics—sometimes reinforcing, sometimes canceling—thereby producing a rather random appearing variation between harmonics. This undesirable "leakage" between adjacent harmonics tends to offset the benefits of the narrower main lobe of the rectangular window. As a result, such windows are rarely used in speech spectrum analysis.

Figure 7.10 shows the effects of window length for a section of unvoiced speech (corresponding to the fricative /SH/) for Hamming window lengths 501 and 151; i.e., the same window conditions as in Figure 7.8. Comparing Figures 7.8 and 7.10 shows that the periodic structure in the narrowband (long window) spectrum is only present for voiced speech. The ragged appearance of the spectrum (for both window lengths) is typical of periodogram analysis of random signals such as unvoiced speech [270]. The shorter (wideband) window produces a smoother estimate of the spectrum while retaining the general overall shape, with a broad peak around 2500 Hz.

The examples of Figures 7.8–7.10 illustrate the basic relationship between the time duration and shape of the window and the properties of the STFT. That is, frequency resolution varies inversely with the length of the window and is also dependent on window shape. When we recall that the purpose of the window is to limit the time interval to be analyzed so that the properties of the waveform do not change appreciably, we see that a compromise is required. In Figure 7.8, for example, it can be seen that the wave shapes of each of the "periods" of the voiced speech signal are slightly different from one another, indicating that the formant frequencies are varying slightly across the 31.25 msec time interval. Similar small variations occur in the pitch period. To track the temporal variation of the formant frequencies, a shorter analysis interval is required. Windows of 10 msec duration, positioned at the beginning and end of the 50 msec interval, would yield distinctly different STFTs. Thus, good temporal resolution requires a short window while good frequency resolution calls for a long window. We will see examples of the use of both types of windows later in this book in our discussion of applications.

We have seen that an interpretation of the short-time Fourier transform as the conventional DTFT of a windowed segment of the speech signal leads to useful insights into both the properties of the time-dependent Fourier representation and the role of the window. Further insight will result from the linear filtering interpretation to be discussed in Section 7.3.5.

7.3.4 Relation to the Short-Time Autocorrelation

An important property of $X_{\hat{n}}(e^{j\hat{\omega}})$ is its relation to the short-time autocorrelation function (STACF) as defined in Chapter 6. Given that $X_{\hat{n}}(e^{j\hat{\omega}})$ is the DTFT of $w[\hat{n}-m]x[m]$ for each value of $\hat{n}$, it is easily seen that

$$S_{\hat{n}}(e^{j\hat{\omega}}) = |X_{\hat{n}}(e^{j\hat{\omega}})|^2 = X_{\hat{n}}(e^{j\hat{\omega}}) \cdot X^{\star}_{\hat{n}}(e^{j\hat{\omega}}), \tag{7.24}$$

is the Fourier transform of

$$R_{\hat{n}}[l] = \sum_{m=-\infty}^{\infty} w[\hat{n}-m]x[m]w[\hat{n}-l-m]x[m+l], \tag{7.25}$$

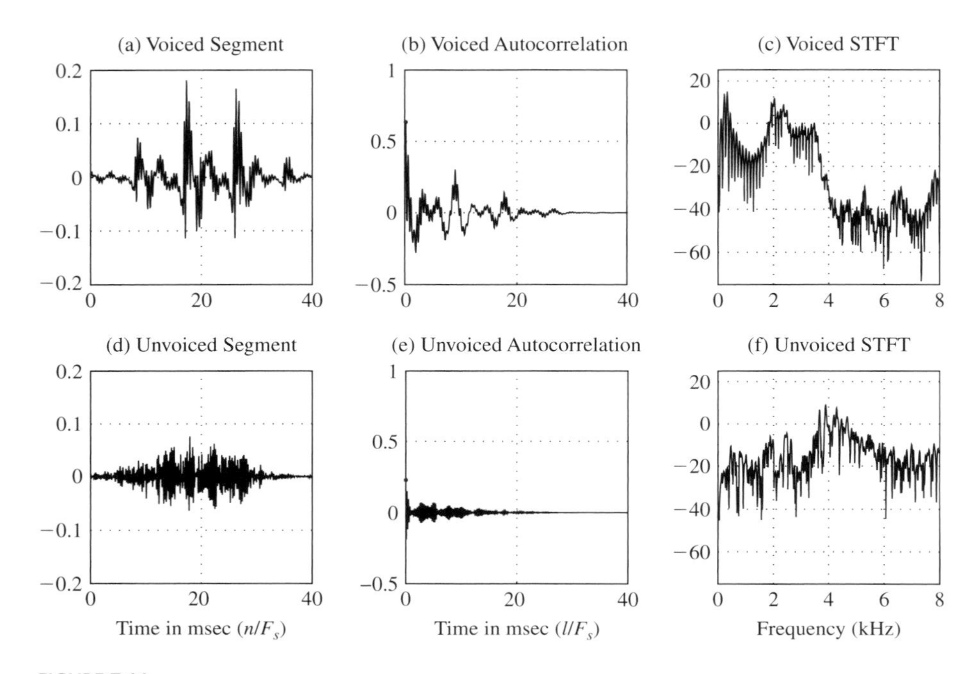

FIGURE 7.11
Short-time autocorrelation and spectrum: (a) windowed speech, (b) STACF, and (c) STFT for a segment of voiced speech; (d) windowed speech, (e) STACF, and (f) STFT for a segment of unvoiced speech.

which is the STACF that we discussed in Chapter 6. Equations (7.24) and (7.25) thus serve to relate the short-time spectrum representation to the STACF. Figure 7.11 shows examples of the windowed speech, the STACF, $R_{\hat{n}}[l]$, and the corresponding STFT, $X_{\hat{n}}(e^{j\hat{\omega}})$, for voiced speech (a,b,c) and unvoiced speech (d,e,f). The window weighting function was a Hamming window of length $L = 641$ samples (equivalent to 40 msec at a 16 kHz sampling rate) and the number of frequency values was $N = 2048$. This choice of N and L avoids time aliasing in the computation of $R_{\hat{n}}[l]$ as the inverse DFT of $X_{\hat{n}}(e^{j2\pi k/N})$. Observe that the pitch ripples in the STFT for voiced speech correspond to the peaks in the STACF at $l/F_s \approx 9$ and 18 msec. Also note that the STACF of unvoiced speech shows little or no correlation at long lags.

7.3.5 Linear Filtering Interpretation

Now we turn our attention to the linear filtering interpretation of the STFT. We begin by repeating Eq. (7.8) as

$$X_n(e^{j\hat{\omega}}) = \sum_{m=-\infty}^{\infty} w[n-m]\left(x[m]e^{-j\hat{\omega}m}\right). \tag{7.26}$$

Note that in Eq. 7.26, we have dropped the ˆ over n. We do this simply to relax our emphasis on the notion that $X_{\hat{n}}(e^{j\hat{\omega}})$ represents the STFT at a particular time $\hat{n}$. This was useful when we wanted to emphasize the DTFT interpretation where it is common

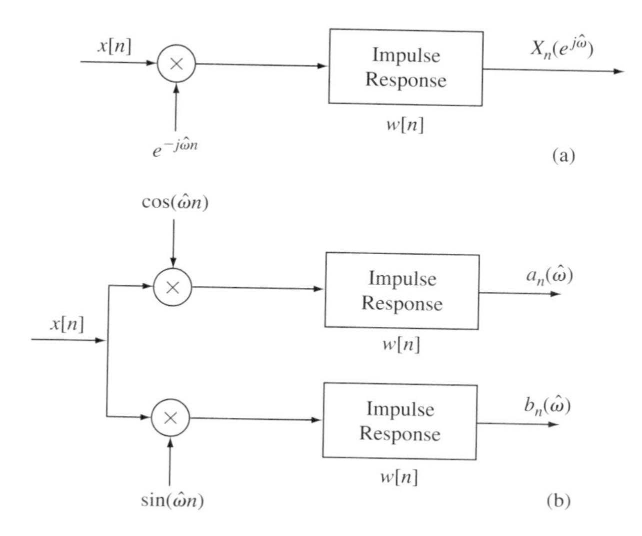

FIGURE 7.12
Linear filtering interpretation of short-time spectrum analysis: (a) complex operations; (b) real operations only.

to move the window in jumps of more than one sample. However, now we want to assume that $\hat{\omega}$ is fixed and think of $X_n(e^{j\hat{\omega}})$ as a one-dimensional time sequence of complex numbers. The notation of Eq. (7.26) is simply intended to convey that the STFT for fixed $\hat{\omega}$ can be viewed as a function of time index n on the same scale as the signal $x[n]$ and to facilitate our discussion of STFA as a linear filtering operation.

It can be seen from Eq. (7.26) that for each fixed value of $\hat{\omega}$, $X_n(e^{j\hat{\omega}})$ is a sequence of complex numbers, and those values are obtained by the convolution of the sequence $w[n]$ with the sequence $(x[n]e^{-j\hat{\omega}n})$. Thus, for a particular value of $\hat{\omega}$, $X_n(e^{j\hat{\omega}})$ can be thought of as the output of the system depicted in Figure 7.12a, where $w[n]$ plays the role of the impulse response of a linear shift-invariant system. Note that even though $x[n]$ is real, the input and output of the linear system are complex, as illustrated in Figure 7.12a. If we express $X_n(e^{j\hat{\omega}})$ as

$$X_n(e^{j\hat{\omega}}) = a_n(\hat{\omega}) - jb_n(\hat{\omega}), \tag{7.27}$$

then the operations required to obtain $a_n(\hat{\omega})$ and $b_n(\hat{\omega})$ are shown in Figure 7.12b, where all the sequences are real.

To see how the system of Figure 7.12a operates to form the STFT at frequency $\hat{\omega}$, it is helpful to again assume that the normal DTFT of $x[n]$ exists, and is denoted $X(e^{j\omega})$. (Recall that we are now considering $\hat{\omega}$ to be a particular value of radian frequency for the STFT.) Then, as a result of the modulation process, the DTFT of the input to the linear filter is $X(e^{j(\omega+\hat{\omega})})$. Thus the DTFT of $x[n]$ at frequency $\hat{\omega}$ is shifted to zero frequency. The DTFT of the output of the filter is $X(e^{j(\omega+\hat{\omega})})W(e^{j\omega})$, so if the filter is a lowpass filter with a very narrow passband, the output of the filter will depend essentially upon $X(e^{j\hat{\omega}})$ alone; i.e., $X_n(e^{j\hat{\omega}}) \approx X(e^{j\hat{\omega}})$. Thus, as in the previous interpretation, $W(e^{j\omega})$ should be non-zero over a very narrow band around zero frequency

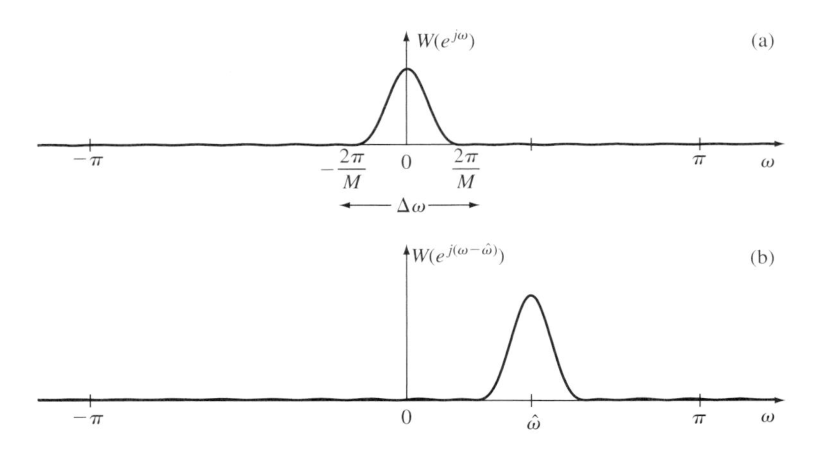

FIGURE 7.13
Filters for STFA based on the Hamming window: (a) lowpass filter frequency response; (b) bandpass filter frequency response for analysis frequency $\hat{\omega}$.

and as small as possible outside this band. This is exactly the same requirement that we arrived at from the point of view of the previous section. Figure 7.13a shows that $W(e^{j\omega})$ for the Hamming window has the desired lowpass property. It is interesting to note that if we make the substitution of variables $\omega = -\omega$ in Eq. (7.21), we obtain

$$X_n(e^{j\hat{\omega}}) = \frac{1}{2\pi}\int_{-\pi}^{\pi} W(e^{j\omega})X(e^{j(\omega+\hat{\omega})})e^{j\omega n}d\omega, \tag{7.28}$$

which is exactly the inverse DTFT of $X(e^{j(\omega+\hat{\omega})})W(e^{j\omega})$. Now, instead of interpreting Eq. (7.28) as a convolution in the frequency domain (smoothing the Fourier transform), we interpret Eq. (7.28) in terms of linear filtering of the frequency-shifted input signal $(x[n]e^{-j\hat{\omega}n})$.

Yet another interpretation of $X_n(e^{j\hat{\omega}})$ in terms of linear filtering is evident from Eq. (7.9), written again here as

$$X_n(e^{j\hat{\omega}}) = e^{-j\hat{\omega}n}\left(\sum_{m=-\infty}^{\infty} x[n-m](w[m]e^{j\hat{\omega}m})\right). \tag{7.29}$$

As shown in Figure 7.14a, $X_n(e^{j\hat{\omega}})$ can also be thought of as the result of modulating $e^{-j\hat{\omega}n}$ with the output of a complex bandpass filter whose impulse response is $w[n]e^{j\hat{\omega}n}$ and whose frequency response is therefore $W(e^{j(\omega-\hat{\omega})})$. If the DTFT $W(e^{j\omega})$ is a lowpass function, as in Figure 7.13a, then the filter in Figure 7.14a will be a bandpass filter whose passband is centered at frequency $\hat{\omega}$, as is shown in Figure 7.13b. Figure 7.14b shows the system of Figure 7.14a in terms of only real quantities. A comparison of Figures 7.12b and 7.14b shows that if both $a_n(\hat{\omega})$ and $b_n(\hat{\omega})$ are required, the implementation of Figure 7.12b is simpler. If, however, only $|X_n(e^{j\hat{\omega}})|$ is required,

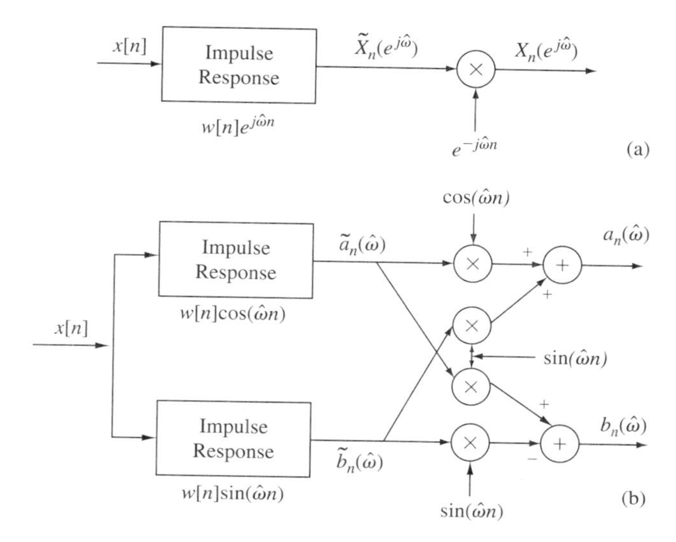

FIGURE 7.14
Another interpretation of short-time spectral analysis in terms of linear filtering: (a) complex operations; (b) real operations only.

implementation with bandpass filters is simpler. To see this, note that from Eqs. (7.11) and (7.15),

$$|X_n(e^{j\hat{\omega}})| = [a_n^2(\hat{\omega}) + b_n^2(\hat{\omega})]^{1/2} \tag{7.30a}$$

$$= |\tilde{X}_n(e^{j\hat{\omega}})| = [\tilde{a}_n^2(\hat{\omega}) + \tilde{b}_n^2(\hat{\omega})]^{1/2}. \tag{7.30b}$$

Figure 7.15a depicts Eq. (7.30a) and Figure 7.15b depicts Eq. (7.30b). The system of Figure 7.15b would, in general, be simpler to implement.

With the point of view that $X_n(e^{j\hat{\omega}})$ at a particular value of $\hat{\omega}$ is the output of a system as depicted in Figure 7.12 or 7.14, we can call on our knowledge of linear systems to help in understanding the properties of the short-time Fourier representation. For example, it is helpful to recall that the impulse response of a discrete-time linear shift-invariant system can be either of finite (FIR) or infinite (IIR) duration. Similarly we may define two classes of windows for STFA.[8] Also, recall that a linear shift-invariant system can be either causal or non-causal, depending upon whether or not its impulse response is zero for $n < 0$. In like manner, we can classify windows as either causal or non-causal. A causal window is one for which

$$w[n] = 0, \quad n < 0, \tag{7.31a}$$

[8]An IIR impulse response can be thought of as an analysis window if it is suitably concentrated in time.

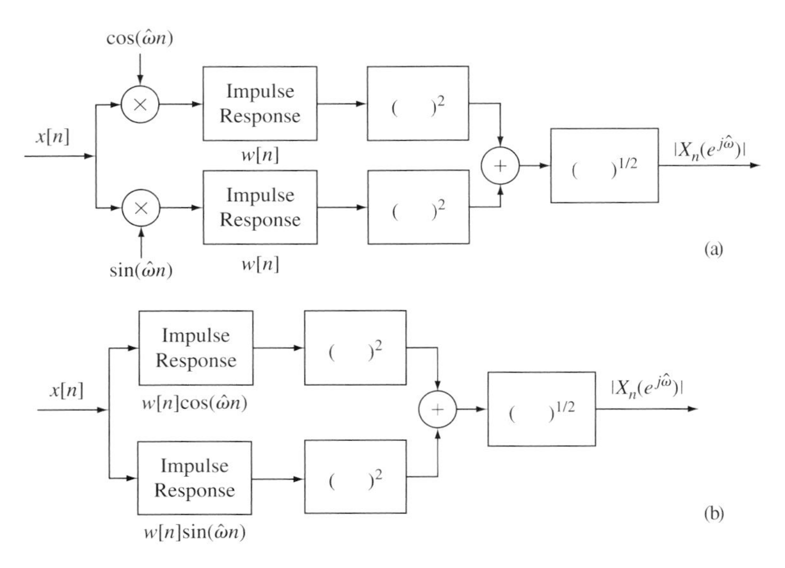

FIGURE 7.15
Two implementations for obtaining the magnitude of the short-time spectrum: (a) using lowpass filters; (b) using bandpass filters.

or equivalently

$$w[n-m]=0, \quad n<m. \tag{7.31b}$$

The Hamming window and the rectangular window are examples of finite duration windows. By appropriate choice of time origin, they can also be defined as either causal or non-causal windows.[9] Such windows are, as we will see, appropriate for use in implementations based upon Figures 7.12 and 7.14 as well as implementations based upon the DFT. Infinite duration windows are also useful, especially when $X_n(e^{j\hat{\omega}})$ is computed using linear filtering as in Figures 7.12 and 7.14. In such cases, we can obtain a recurrence formula that gives $X_n(e^{j\hat{\omega}})$ in terms of values at previous times (see Problem 7.7).

7.3.6 Sampling Rates of $X_n(e^{j\hat{\omega}})$ in Time and Frequency

The STFT is a complex two-dimensional representation of the one-dimensional real speech signal $x[n]$. That is, $X_n(e^{j\hat{\omega}})$ is a function of both the discrete index n which represents time, and continuous normalized radian STFT analysis frequency $\hat{\omega}$. As such, $X_n(e^{j\hat{\omega}})$ is like a (complex-valued) two-dimensional image with one discrete and one continuous dimension. Figure 7.16a shows the region of support for $X_n(e^{j\hat{\omega}})$ in two dimensions. A basic consideration in the digital implementation of systems for

[9]Non-causal computations require buffering in a real-time system.

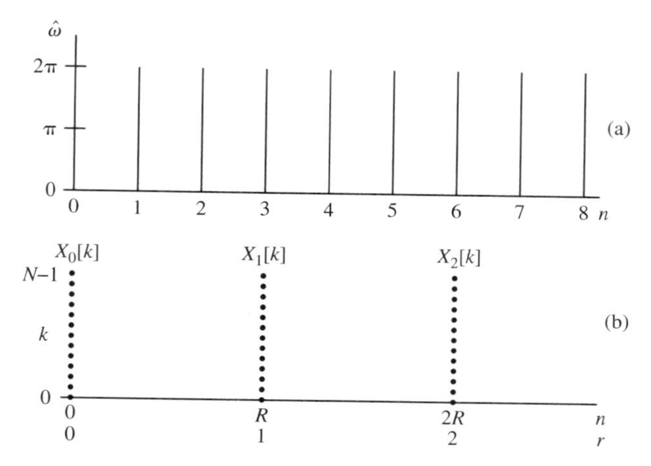

FIGURE 7.16
Domain of STFT variables $\hat{\omega}$ and n for (a) the case with no sampling and (b) the case with sampling based on the bandwidth and timewidth of the lowpass window, $w[n]$.

STFA is the rate at which $X_n(e^{j\hat{\omega}})$ should be sampled in both the time and frequency dimensions to provide an unaliased representation of $X_n(e^{j\hat{\omega}})$ from which $x[n]$ can be exactly recovered. Figure 7.16b shows the discrete region of support when $X_n(e^{j\hat{\omega}})$ is sampled in time with interval R samples and in frequency at frequencies $\omega_k = (2\pi k/N)$ as in

$$X_{rR}(e^{j(2\pi k/N)}) = X_r[k] = \sum_{m=-\infty}^{\infty} x[m]w[rR-m]e^{-j(2\pi k/N)m}, \tag{7.32}$$

where $k = 0, 1, \ldots, N-1$ and r ranges over the integers.[10] The infinite limits in Eq. (7.32) are only meant to imply that the sum is over all m for which $w[rR-m] \neq 0$. For example, for an L-point causal Hamming window, the limits would be finite, ranging from $m = rR - L + 1$ to $m = rR$. It is essential to know how R and N should be chosen so that the speech signal can be reconstructed from its sampled STFT. This question is by no means trivial. Answering it requires careful consideration of the factors entering into the computation of $X_n(e^{j\hat{\omega}})$ in order to arrive at the correct sampling rates in both time and frequency. As will be shown, a complicating factor in the discussion of the choice of the proper sampling rates for $X_n(e^{j\hat{\omega}})$ is that sampling rates lower than the theoretically minimum rate required to avoid aliasing in both dimensions can be used in either the time or the frequency dimensions, and $x[n]$ can still be exactly recovered from the aliased (undersampled) short-time transform. Such undersampled representations are indeed quite useful for applications in which one is only

[10]Note that $\omega_k = (2\pi k/N)$ for $k = 0, 1, \ldots, N-1$ covers the entire range $0 \leq \omega < 2\pi$.

interested in such analysis functions as spectral estimation, pitch and formant analysis, and digital speech spectrograms or for speech and audio coding applications in which minimization of overall bit rate of the system is of prime importance. For applications in which one is interested in obtaining a STFT of the signal, performing some modification on the signal (e.g., linear or non-linear filtering) and then resynthesizing the modified signal, it generally is essential to minimize aliasing in either or both domains.

Sampling Rate of $X_n(e^{j\hat{\omega}})$ in Time

The linear filtering interpretation of the previous section provides the necessary insight for determining the required sampling rate in the time dimension of $X_n(e^{j\hat{\omega}})$. It was shown that for a fixed value of $\hat{\omega}$, $X_n(e^{j\hat{\omega}})$ is the output of a linear filter with impulse response $w[n]$. We have already shown that for most commonly used windows, the DTFT $W(e^{j\omega})$ has the properties of a (non-ideal) lowpass filter frequency response. Let us denote the effective bandwidth of the analysis window as B Hz.[11] Thus the sequence $X_n(e^{j\hat{\omega}})$ (as a function of n with $\hat{\omega}$ fixed) has bandwidth determined by the DTFT of the window, and therefore according to the sampling theorem, $X_n(e^{j\hat{\omega}})$ must be sampled at a rate of at least $2B$ samples/sec to avoid aliasing. As an example, consider an $(L = 2M + 1)$–point Hamming window, as defined in Eq. (7.22). Then the approximate filter bandwidth of $W(e^{j\omega})$ in terms of analog frequencies is[12]

$$B = \frac{2F_s}{L} \quad \text{(Hz)}, \tag{7.33}$$

where F_s is the sampling rate of the signal, $x[n]$, and thus the required sampling rate of $X_n(e^{j\hat{\omega}})$ in the time dimension is $F_s/R = 2B \geq 4F_s/L$ samples/sec or $R \leq L/4$, where R is an integer. In other words, for the Hamming window, in order to avoid aliasing for the sequence of time samples of $X_n(e^{j\hat{\omega}})$ at frequency $\hat{\omega}$, the spacing between analysis window positions must be less than or equal to 25% of the window length. Equivalently, the windows must overlap by at least 75%. Thus for $L = 100$ and $F_s = 10{,}000$ Hz, we get $B = 200$ Hz, so that $X_n(e^{j\hat{\omega}})$ must be evaluated 400 times/sec—i.e., every $R = 100/4 = 25$ samples at the sampling rate of the input.

Sampling Rate of $X_n(e^{j\hat{\omega}})$ in STFT Frequency $\hat{\omega}$

Since $X_{rR}(e^{j\hat{\omega}})$ is periodic in $\hat{\omega}$ with period 2π, it is only necessary to sample over an interval of length 2π.[13] To determine an appropriate finite set of frequencies

[11]Here there is a possibility of confusion of frequency variables. Recall that when $X_n(e^{j\hat{\omega}})$ is viewed as a function of time, the STFT analysis frequency $\hat{\omega}$ is fixed. That is why we have reserved ω to denote the frequency variable associated with the *time variation* of $X_n(e^{j\hat{\omega}})$.

[12]Normally, the term "bandwidth" of a filter means the width of the positive region of support of the frequency response; i.e., the "passband."

[13]Usually we consider the interval $0 \leq \hat{\omega} < 2\pi$ because the frequencies $\omega_k = (2\pi k/N)$, $k = 0, 1, \ldots, N-1$ are conveniently computed using an FFT algorithm.

$\hat{\omega}_k = 2\pi k/N, k = 0, 1, \ldots, N-1$ at which $X_{rR}(e^{j\hat{\omega}})$ must be specified to exactly recover the signal $x[n]$, we use the DTFT interpretation of $X_{rR}(e^{j\hat{\omega}})$.If the window is time-limited, then if $X_{rR}(e^{j\hat{\omega}})$is viewed as a DTFT, its inverse transform is time-limited. In this case the sampling theorem requires that we sample $X_{rR}(e^{j\hat{\omega}})$in the frequency dimension at a rate of at least twice its "time-width." Since the inverse DTFT of $X_{rR}(e^{j\hat{\omega}})$ is the signal $x[m]w[rR-m]$ and this signal is of duration L samples (again due to the finite duration window), then according to the sampling theorem, $X_{rR}(e^{j\hat{\omega}})$ must be sampled (in frequency $\hat{\omega}$) at the set of frequencies

$$\hat{\omega}_k = \frac{2\pi k}{N}, \quad k = 0, 1, \ldots, N-1, \tag{7.34}$$

with $N \geq L$ in order to exactly recover $x[n]$ from $X_{rR}(e^{j\omega_k})$. (See Problem 7.14.) With this set of samples, Eq. (7.32) will be recognized as the N-point DFT of the sequence $x[m]w[rR-m]$, which is assumed to be of finite-length L samples. Thus, for an L-point causal window, the inverse DFT yields

$$\frac{1}{N}\sum_{k=0}^{N-1} X_r[k]e^{j(2\pi k/N)m} = x[m]w[rR-m], \qquad rR-L+1 \leq m \leq rR, \tag{7.35}$$

if $N \geq L$. Otherwise time aliasing will occur; i.e., the inverse DFT evaluated in the interval $rR-L+1 \leq m \leq rR$ will be composed of a sum of shifted (by N) copies of $x[m]w[rR-m]$.

Thus, for the example of a Hamming window of duration $L = 100$ samples, we require $X_{rR}(e^{j\hat{\omega}})$to be evaluated for at least 100 uniformly spaced frequencies.

Total Sampling Rate of $X_n(e^{j\hat{\omega}})$

Based on the above discussion, we can determine the total number of samples of $X_n(e^{j\hat{\omega}})$that must be computed per second to give an unaliased representation of the original signal $x[n]$. The minimum sampling rate of $X_n(e^{j\hat{\omega}})$ in the time dimension is $2B$, where B is the frequency bandwidth of the window, and the minimum number of samples in the frequency dimension is L, the time width of the window. Thus the total sampling rate (SR) of $X_n(e^{j\hat{\omega}})$ is

$$SR = 2B \cdot L, \quad \text{samples/sec.} \tag{7.36}$$

For most practical windows, B can be represented as a multiple of (F_s/L), where F_s is the sampling frequency of $x[n]$; i.e.,

$$B = C_b \frac{F_s}{L} \quad \text{(Hz)}, \tag{7.37}$$

where C_b is the proportionality constant. Thus Eq. (7.36) can be written as

$$SR = 2C_b F_s, \quad \text{samples/sec.} \tag{7.38}$$

The ratio of SR to F_s is therefore

$$\frac{SR}{F_s} = 2C_b. \tag{7.39}$$

The quantity $2C_b$ indicates the "oversampling" ratio of the short-time analysis as compared to the sampling rate used to obtain the sequence $x[n]$.

As an example, if $w[n]$ is a Hamming window, then $2C_b = 4$, whereas if $w[n]$ is a rectangular window [and if the bandwidth B is defined to be the frequency of the first zero of $W(e^{j\omega})$], then $2C_b = 2$. It should be emphasized that the DTFT of the rectangular window is not a very good lowpass filter. Its sidelobes are only 14 dB down from the value at $\omega = 0$ while the sidelobes of the Hamming window are much lower. Thus, it is reasonable to assert that the short-time spectrum representation of $x[n]$ requires on the order of four times as many samples as required to represent the waveform if we wish to avoid aliasing. However, in return for this expansion of sampling rate, we obtain a very flexible representation of the signal from which extensive modifications, in both the time and frequency domains, can be made. (See Section 7.11.)

Summary of Sampling of $X_n(e^{j\hat{\omega}})$

In this section we have discussed the required sampling rates of $X_n(e^{j\hat{\omega}})$ in time and frequency to obtain an unaliased representation. Although the sampling rates that were derived are the minimum rates required to avoid aliasing in both dimensions, there exist special cases in which $X_n(e^{j\hat{\omega}})$ can be undersampled in either dimension, and for which $x[n]$ can be exactly recovered with no aliasing error. Such cases are of practical importance for the implementation of systems in which minimum storage (bit rate) of the representation is of importance; e.g., an analysis/synthesis system, a spectral display, etc. We will discuss how to design and implement such systems later in this chapter. However, before addressing the question of reconstruction of the signal from its STFT, it is useful to turn to the question of how the STFT can be visualized on the graphical displays of computers.

7.4 SPECTROGRAPHIC DISPLAYS

As discussed in Chapter 3, the notion of a time-dependent Fourier representation of speech was prevalent long before the advent of digital speech processing techniques. The sound spectrograph produced a gray scale image, called a spectrogram, on electrically sensitive teledeltos paper. The spectrogram was based on the time-varying average energy in the output of a variable frequency bandpass filter. Through a clever mechanical/electrical design, the darkness of a particular point on the paper represented the short-time energy of the sound at a corresponding particular combination of time and frequency [196]. Therefore a spectrogram was an image of the STFT computed by analog means. Figure 7.17 (the top panel) shows a wideband (short window) spectrogram of the utterance "Every salt breeze comes from the sea." This example illustrates a number of characteristic features of wideband time-dependent spectra. First, observe that at a particular time, the spectrum varies with frequency

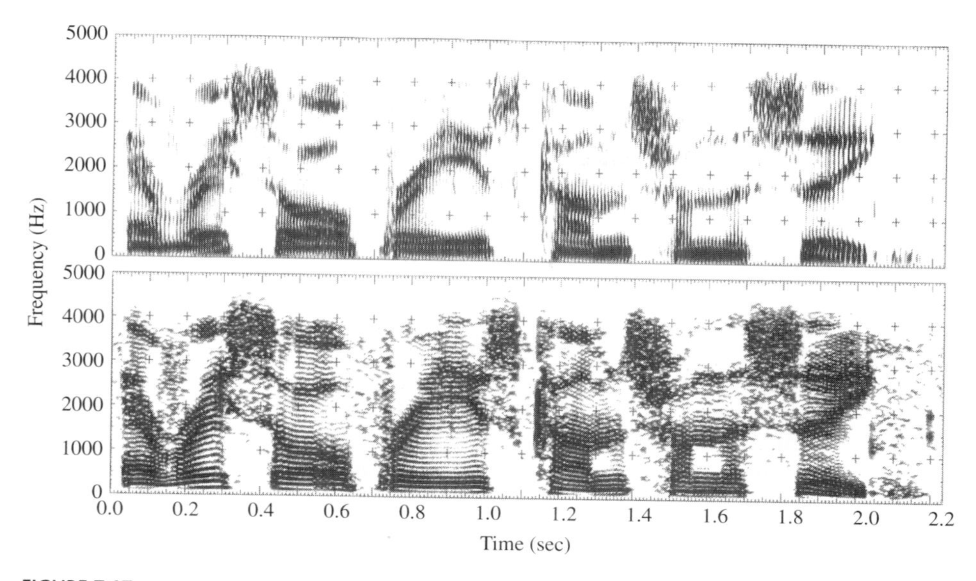

FIGURE 7.17
Wideband and narrowband spectrograms of the sentence "Every salt breeze comes from the sea" made on an analog sound spectrograph machine.

(vertical direction) in a manner suggested by smooth curves plotted with heavy lines in Figures 7.8 and 7.10; i.e., the spectrum consists of a few broad peaks corresponding to the formant frequencies. The horizontally oriented broad bars in the spectrogram show how the formant frequencies vary with time. Another interesting feature of the wideband spectrogram is the vertical striations that appear in regions of voiced speech. These occur because the impulse response of the analyzing bandpass filter (i.e., the spectrum analysis window) is of about the same duration as the pitch period. This is illustrated by the short window in Figure 7.8, where it is positioned slightly to the right of the peak. A further shift to the right would center the short window on the low amplitude part of one of the pitch periods. Thus, the energy in the filter output is maximum when the peak of the impulse response is aligned with the maximum of each individual pitch period. At other times, the output energy is significantly less. This accounts for the finely spaced vertically oriented dark and light regions alternating in time in Figure 7.17a. For unvoiced speech as in Figure 7.10, which is not periodic, the regular vertical striations do not appear and the spectral pattern is much more ragged (random).

Figure 7.17 (the bottom panel) shows a narrowband spectrogram of the same utterance. In this case, the bandwidth of the filter is narrow enough that the individual harmonics are resolved in voiced regions. Thus, while formant frequencies are still in evidence, a cross-section at a particular time is reminiscent of the spectrum for the long windows as in Figures 7.8 and 7.10. No longer is the pattern striated in the voiced regions, since the narrowband impulse response spans several pitch periods; but, rather, the frequency dimension now clearly places in evidence the fundamental frequency and its harmonics. Unvoiced regions are distinguished by a lack of periodicity in the frequency dimension.

The wideband and the narrowband spectrograms display a great deal of information about the properties of a speech utterance. Indeed, when apparatus for displaying such time-dependent Fourier representations first became available, it was hoped that such displays could provide a new "language" for communicating with the deaf. Although this hope was not realized, subsequent research led to the book *Visible Speech* [288], which is still a rich source of information on the short-time spectral and temporal properties of speech. In the years since this early work, many speech researchers have made measurements by hand, on spectrograms, to determine speech parameters such as formant frequencies and fundamental frequency.

Another outgrowth of the invention of the sound spectrograph was the notion that a speaker's identity could be revealed by a detailed analysis of a spectrogram or "voiceprint" of a speech utterance. Although there remains significant debate as to the reliability of voice identification techniques based upon spectrograms [40], these techniques have gained some acceptance in courts of law [50, 140, 147, 293, 324].

The analog hardware sound spectrograph machine was, for a long time, the basic analysis tool in speech research. However, such machines have largely disappeared because spectrograms can be created and displayed much more effectively with modern computer technology. The previous sections of this chapter have shown ways to design and implement time-dependent Fourier representations of much greater sophistication than was ever possible using analog hardware. For example, using the techniques of Section 7.3, we can obtain $|X_{rR}(e^{j\omega_k})|$ with $\omega_k = 2\pi k/N$, which is a two-dimensional representation of the speech signal that is discrete in time (r) and frequency (k). Digital spectrograms are simply displayed by the same technology that is used to display digital representations of photographic images. Generally, all the information is not needed in a display. Often only $20\log_{10}|X_{rR}(e^{j\omega_k})|$ would be displayed, and, since $|X_{rR}(e^{j\omega_k})|$ is even and periodic in k with period N, it is only necessary to display values in the range $0 \leq k \leq N/2$, corresponding to the analog frequency range of $0 \leq F \leq F_s/2$.

In view of the demonstrated usefulness and wide acceptance of the spectrogram as a basic speech analysis tool, it rapidly became clear to the speech research community that a digitally generated spectrogram was undoubtedly more convenient and useful than the original analog spectrogram. In the early days of computing, a TV monitor or CRT display was generally available to output sampled images, and it was obvious that $|X_{rR}(e^{j\omega_k})|$ could be just such a sampled image. A number of researchers investigated such outputs and found that it was possible to roughly duplicate the appearance of analog spectrograms. Indeed, since the teledeltos paper has a gray scale range of only about 12 dB [196], a rather crude quantization of the values of $|X_{rR}(e^{j\hat{\omega}})|$ can be used in the display if the objective is only to duplicate spectrogram appearance. However, most digital image display systems have a much greater dynamic range so that more of the spectral information can be portrayed than with the analog system.

Another advantage of the digital spectrogram is that the spectrum can be conveniently shaped in sophisticated ways to enhance the usefulness of the display. An example is the use of high frequency emphasis to counteract the natural fall-off of the speech spectrum. (This is also used in analog spectrographs.) A simple way to introduce high frequency emphasis is to compute the spectrum of the first difference of the input signal. (See Problem 7.19.) Another more flexible way is to directly shape the spectrum prior to display, and the frequency and time dimensions can be expanded or contracted, as desired [268]. In the 1970s, a hard copy output of images was still rather expensive. This led to techniques involving overstriking on printing devices [365].

Today, a laptop computer and photo printer can produce spectrographic displays whose quality far exceeds that of the best sound spectrograms produced by the original spectrograph machines. Stand-alone waveform processing packages such as Wave Surfer [413] and Praat Speech Analysis [294] provide a range of tools for plotting waveforms, spectrograms, pitch contours, formant contours, etc. MATLAB includes software for creating and plotting spectrograms, and there exists a range of MATLAB toolboxes such as Colea [211] and Voicebox [407] that provide many useful signal processing and graphical capabilities–including the ability to create high quality sound spectrograms. In this section we show examples of spectrograms created using standard MATLAB software tools.

In MATLAB, a spectrogram is basically an image plot of the two-dimensional function:

$$B[k,r] = 20\log_{10}|X_r[k]|. \tag{7.40}$$

Such arrays of data can be stored as a matrix with r, the column index, corresponding to the time scale range $t_r = 0, RT, 2RT, \dots, (N_R - 1)T$ along the x-axis of the image, and k, the row index, corresponding to the frequency scale range $F_k = kF_s/N$, $k = 0, 1, \dots, N/2$ along the y-axis. In Eq. (7.40), the basic short-time spectrum computation is of the form:

$$X_r[k] = X_{rR}(e^{j\frac{2\pi}{N}k}) = \sum_{m=rR}^{rR+L-1} x[m]w[rR-m]e^{-j\frac{2\pi}{N}km}, \tag{7.41}$$

where L is the speech frame duration (in samples), N is the size of the FFT used to compute the spectral slice at time index rR [Eq. (7.41)], $w[m]$ is the L-point (non-causal) window used in the short-time spectral computation, N_R is the number of frames, and R is the shift between frames (in samples). Typically, $B[k,r]$ is clipped at some desired level of dynamic range to adjust the amount of fine detail that is displayed (recall that the original sound spectrogram preserved only a small dynamic range because of limitations on the plotting paper and the plotting technology).

To illustrate the quality of the output, Figure 7.18 shows wideband (at the top) and narrowband (at the bottom) spectrograms of the utterance "This is a test." These were computed using the function call

```
spectrogram(x,L,L-R,N,Fs,'yaxis')
```

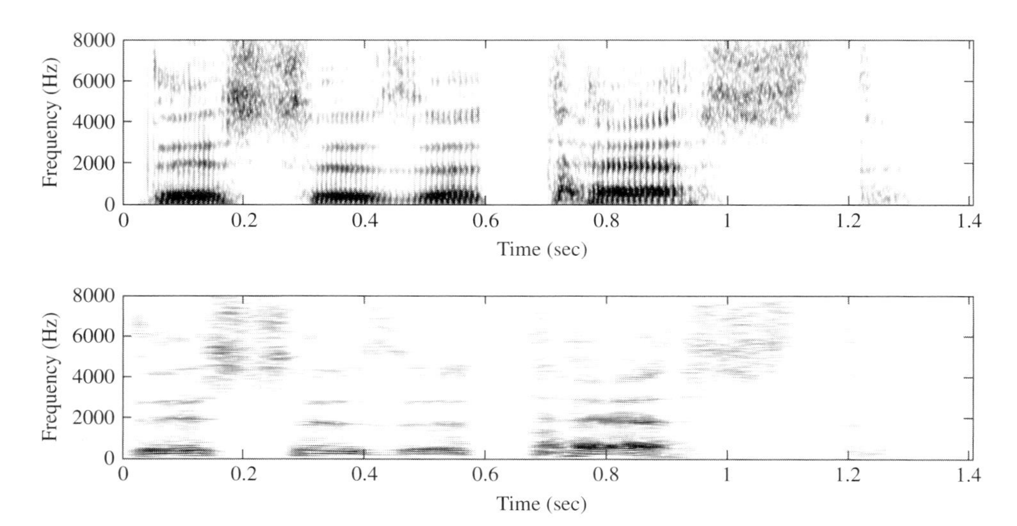

FIGURE 7.18
Wideband (top) and narrowband (bottom) spectrograms of the utterance "This is a test."

in MATLAB.[14] The relevant parameters of the speech and the spectrogram computation are as follows:

- Speech parameters
  - Sampling rate (F_s): 16 kHz
  - Speech duration: 1.406 seconds
  - Speaker: male
  - Utterance: "This is a test"
- Wideband spectrogram parameters
  - Analysis window ($w[n]$): Hamming window (default)
  - Analysis window duration (L): 6 msec (96 samples at the 16 kHz rate)
  - Analysis window shift (R): 10 samples
  - Analysis window overlap ($L - R$): 86 samples
  - Number of analysis sections (N_R): 2250
  - Size of FFT (N): 512
  - Dynamic range of spectral log magnitudes: 40 dB
- Narrowband spectrogram parameters
  - Analysis window ($w[n]$): Hamming window (default)
  - Analysis window duration (L): 60 msec (960 samples at the 16 kHz rate)

[14]MATLAB prefers to specify the frame shift in terms of overlap of successive windows. Thus for a window of length 96 samples, a shift of 10 samples means an overlap of 86 samples.

- Analysis window shift (R): 96 samples
- Analysis window overlap ($L - R$): 864 samples
- Number of analysis sections (N_R): 235
- Size of FFT (N): 1024
- Dynamic range of spectral log magnitudes: 40 dB

The next several examples illustrate the versatility of digital spectrograms, showing the effects of changing the analysis parameters in the following ways:

Figure 7.19 shows a converted gray scale version of a comparison of a gray scale spectrogram (top panel) with two color scale spectrograms (center and bottom panels) of the same utterance. Refer to the endpaper section for the color images. The center panel spectrogram uses a color range of yellow (the lowest level signal log magnitude) to blue (the highest level signal log magnitude), whereas the bottom panel spectrogram uses a color range of blue (lowest level) to red (highest level). The impact of color can be readily seen in these plots. The converted gray scale versions of the color spectrograms are interesting in their own right since they illustrate that the gray

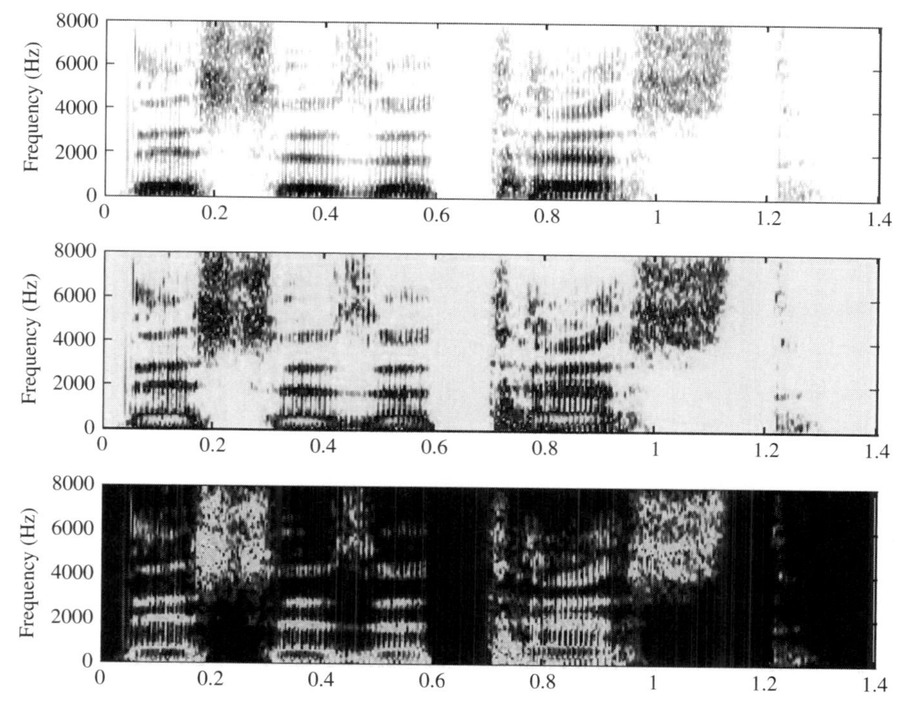

FIGURE 7.19
Comparisons of gray scale wideband spectrogram (top panel) with color wideband spectrograms (middle and bottom panels) of the utterance "This is a test." (Refer to the endpaper section for the corresponding color images.)

scale mapping of spectral amplitude can take various forms which might be useful for highlighting certain features of the spectrogram.

Figure 7.20 shows a comparison of four spectrograms of the utterance "This is a test," where all parameters are the same as used in Figure 7.18 except for L, which is varied from 48 samples (top panel of the figure) to 96 samples (second panel of the figure), to 144 samples (third panel of the figure) and finally to 480 samples (bottom panel of the figure), showing the effect of analysis window size on the resulting spectrogram. Observe that as the analysis window duration gets longer (i.e., from 3 msec in the top panel to 30 msec in the bottom panel), the resolution in frequency becomes sharper, as manifested by the width of the spectral bands around each of the formant frequencies. The first three spectrograms can be classified as wideband spectrograms. The temporal resolution of the individual pitch periods is clear in all three of the upper

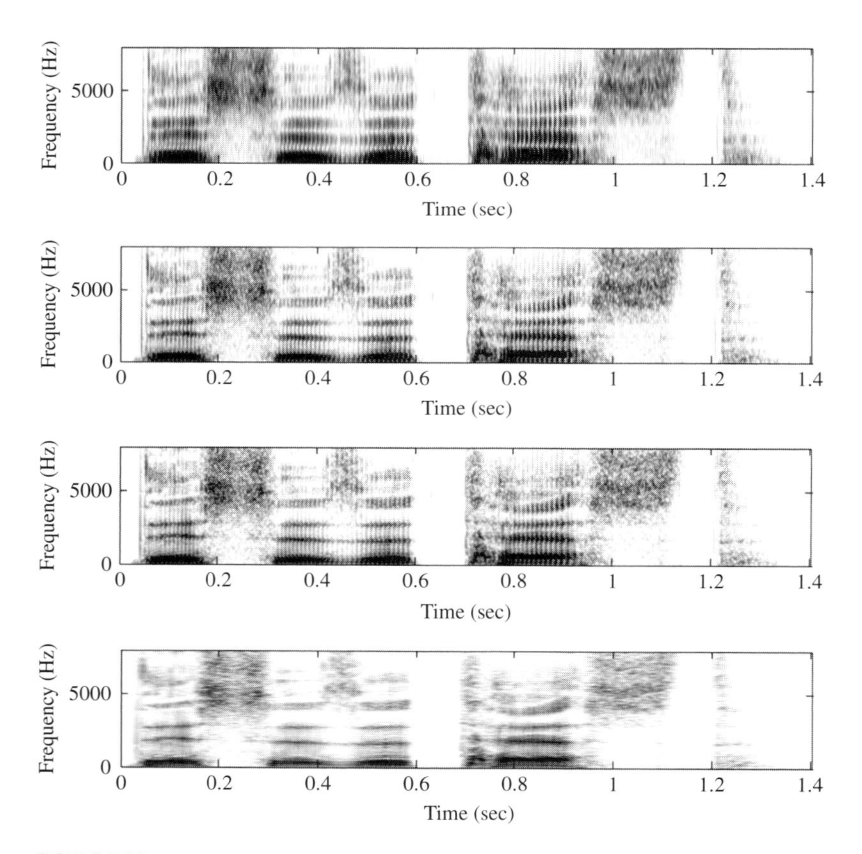

FIGURE 7.20
Comparisons of four spectrograms of the utterance "This is a test." The top panel spectrogram uses a 3 msec (48 samples) analysis window; the second panel spectrogram uses a 6 msec (96 samples) analysis window; the third panel spectrogram uses a 9 msec (144 samples) analysis window; and the bottom panel spectrogram uses a 30 msec (480 samples) analysis window. (All other spectrogram parameters are the same as used in Figure 7.18.)

spectrograms. The fourth spectrogram is a narrowband spectrogram in which each of the harmonics is visible in the voiced regions. Note that the temporal resolution of the individual pitch periods disappears in the fourth spectrogram.

As a final set of examples, Figures 7.21 and 7.22 show comparisons of (color) wideband and narrowband spectrograms (with spectral slices shown at three time slots during the course of the utterance, corresponding roughly to the sounds /IY/, /AE/, and /S/, for a male talker (Figure 7.21) and a female talker (Figure 7.22)). The color versions of these two figures can be found in the endpaper section at the end of this book. The speech for both spectrograms is an utterance of the text "She had your dark suit in," sampled at a 16 kHz rate. Note that both spectrograms are labeled with the approximate locations of the phonemes that occur in the utterance. From either the narrowband spectrograms, or from the narrowband spectral slices, the pitch can be computed as being about 125 Hz for the male talker, and about 300 Hz for the female talker. The impact of the high pitch frequency is seen in the lack of definition of the gross spectral shape for the narrowband slices for the female talker.

7.5 OVERLAP ADDITION METHOD OF SYNTHESIS

The examples of the previous section make it abundantly clear that the STFT has great flexibility in highlighting basic properties of the speech signal. Now we turn to the question of whether the signal can be recovered from the STFT. When using the STFT as a basis for manipulating and processing speech signals, we often need to reconstruct the speech signal from the STFT. This inversion process is called short-time Fourier synthesis or STFS. Recall that the STFT is naturally discrete in the time dimension but continuous in the frequency dimension. In Section 7.3.1, we discussed the invertibility of the STFT by noting that the windowed waveform segment could be recovered if we had $X_n(e^{j\hat{\omega}})$ at all frequencies over a range of 2π radians. However, in order to accomplish this with finite computation, the STFT must be sampled in both time and frequency. For example, if the window is causal with length L samples, the sampled STFT is

$$X_{rR}(e^{j\omega_k}) = \sum_{m=rR-L+1}^{rR} w[rR-m]x[m]e^{-j\omega_k m}, \tag{7.42}$$

where R is the sampling period in time of the STFT, and for uniform sampling in frequency, $\omega_k = 2\pi k/N$, with $k = 0, 1, \ldots, N-1$. One method for reconstructing $x[n]$ from its short-time spectrum is based on the DFT interpretation of the short-time spectrum. Since, as we have already seen, $X_{rR}(e^{j\omega_k})$ can be considered to be the DFT of the sequence $x[m]w[rR-m]$, we can use the inverse DFT to obtain

$$y_r[m] = x[m]w[rR-m] = \frac{1}{N}\sum_{k=0}^{N-1} X_{rR}(e^{j\omega_k})e^{j\omega_k m}. \tag{7.43}$$

If $N \geq L$ so that no time aliasing occurs due to the sampling in the $\hat{\omega}$ variable, $x[m]$ inside the shifted window $w[rR-m]$ can be reconstructed by computing the inverse

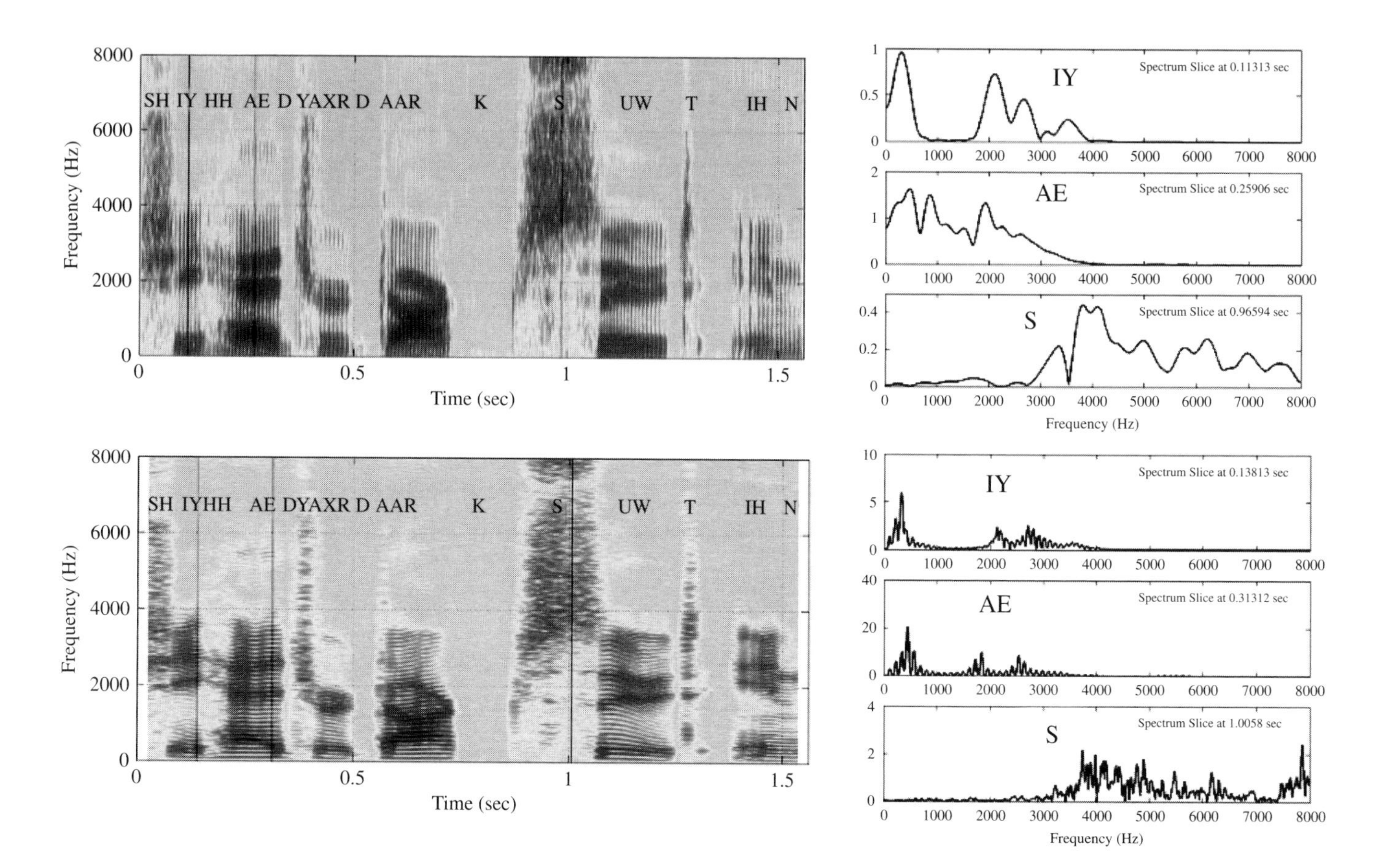

FIGURE 7.21
Spectrograms of a male speaker's utterance of the text "She had your dark suit in." The left column shows wideband and narrowband spectrograms using respectively $L = 80$ sample (5 msec) and $L = 800$ (50 msec) analysis windows, with an 1024-point FFT, and $R = 5$ and $R = 10$ sample shifts between adjacent short-time spectra. The right panel shows wideband and narrowband slices at the three time slots indicated by heavy lines in the sound spectrograms, corresponding to the sounds /IY/, /AE/, and /S/. A full color version of this figure is provided in the endpaper section.

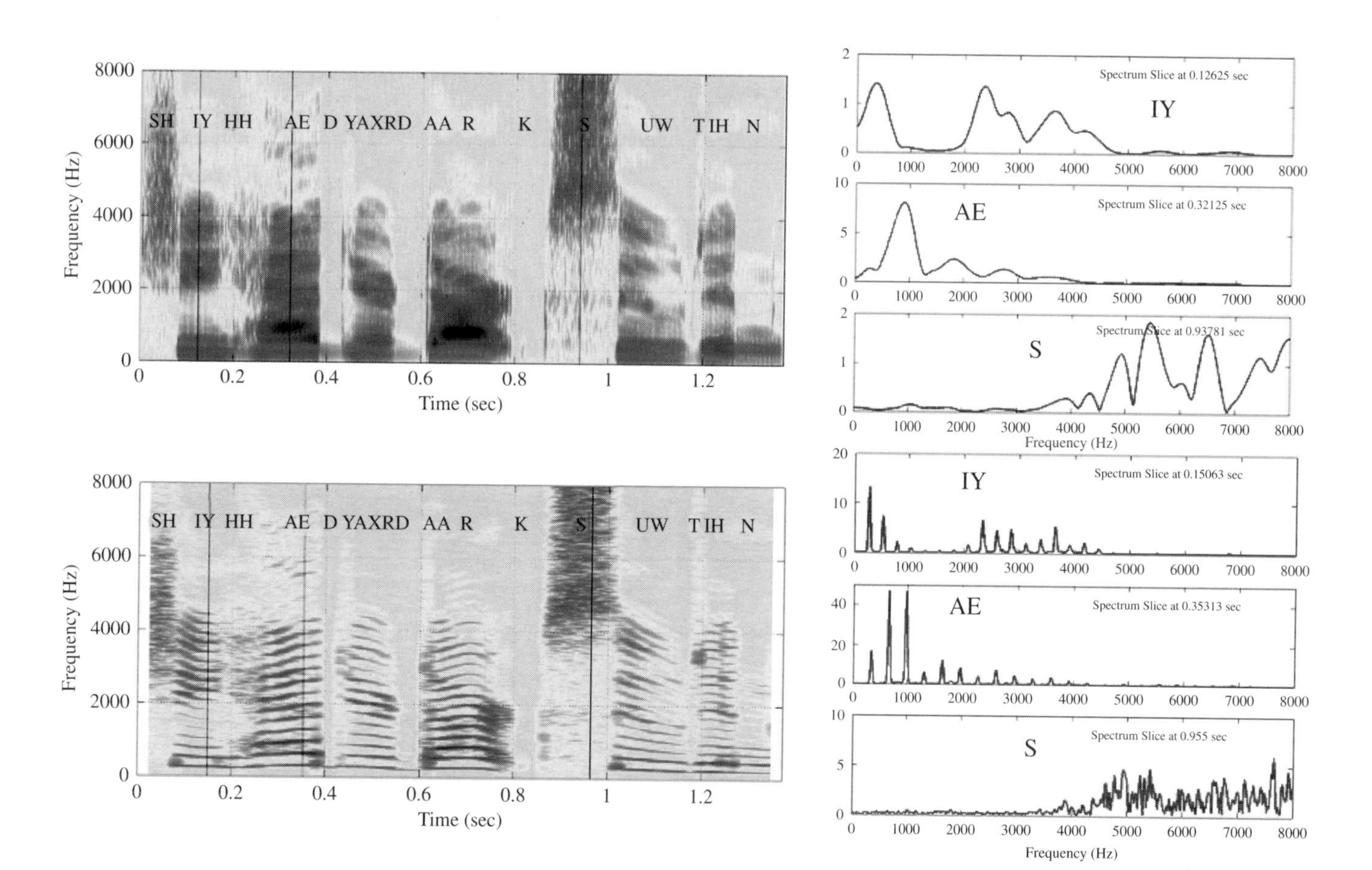

FIGURE 7.22
Spectrograms of a female speaker's utterance of the text "She had your dark suit in." The left column shows wideband and narrowband spectrograms using respectively $L = 80$ sample (5 msec) analysis windows, with an 1024-point FFT, and $R = 5$ and $R = 10$ sample shifts between adjacent short-time spectra. The right panel shows wideband and narrowband slices at the three time slots indicated by heavy lines in the sound spectrograms, corresponding to the sounds /IY/, /AE/, and /S/. A full color version of this figure is provided in the endpaper section.

DFT of $X_n(e^{j\omega_k})$ and dividing by the window (assuming it is strictly non-zero for all values of m). In this manner L signal values of $x[m]$ can be reconstructed for each analysis window (where L is the window duration). Then the window can be moved by L samples and the process repeated. Based on the discussion of Section 7.3.6, it can be seen that this procedure uses an "undersampled in time" representation of $X_n(e^{j\omega_k})$. Using this procedure, we have the constraint

$$R \leq L \leq N. \tag{7.44}$$

The choice $R = L = N$ has the virtue that there would be no expansion of the net sampling rate. However, such a process is not practical since small changes in $X_{rR}(e^{j\omega_k})$ will be amplified by dividing the inverse DFT by the window. Thus although such a procedure is a theoretically valid demonstration that exact reconstruction is possible with $R = L = N$, it has not been found useful for applications requiring reconstruction of the original signal (or a processed version of it). In this section, we present a more robust synthesis procedure similar to the overlap addition (OLA) method for a periodic convolution using DFTs [270].

7.5.1 Conditions for Exact Reconstruction

Assume that the short-time transform is sampled with period R samples in the time dimension and at N frequencies as in Eq. (7.42); i.e., we have $Y_r(e^{j\omega_k}) = X_{rR}(e^{j\omega_k})$ where r is an integer and $\omega_k = 2\pi k/N$ for $0 \leq k \leq N-1$. The OLA method is based upon the synthesis equation

$$y[n] = \sum_{r=-\infty}^{\infty} \left(\frac{1}{N} \sum_{k=0}^{N-1} Y_r(e^{j\omega_k}) e^{j\omega_k m} \right) \Bigg|_{m=n}. \tag{7.45}$$

That is, for a causal window of length L as in Eq. (7.42), to reconstruct the signal, the inverse transform of $Y_r(e^{j\omega_k})$ is computed for each value of r, giving the sequences

$$y_r[m] = x[m]w[rR-m], \quad rR-L+1 \leq m \leq rR. \tag{7.46}$$

Then the signal at time n is obtained by summing the values at time n of all the sequences $y_r[m]$ that overlap at time n; i.e.,

$$y[n] = \sum_{r=-\infty}^{\infty} y_r[n] = x[n] \left(\sum_{r=-\infty}^{\infty} w[rR-n] \right) = x[n]\tilde{w}[n], \tag{7.47a}$$

where $\tilde{w}[n]$ is defined in this case as

$$\tilde{w}[n] = \sum_{r=-\infty}^{\infty} w[rR-n]. \tag{7.47b}$$

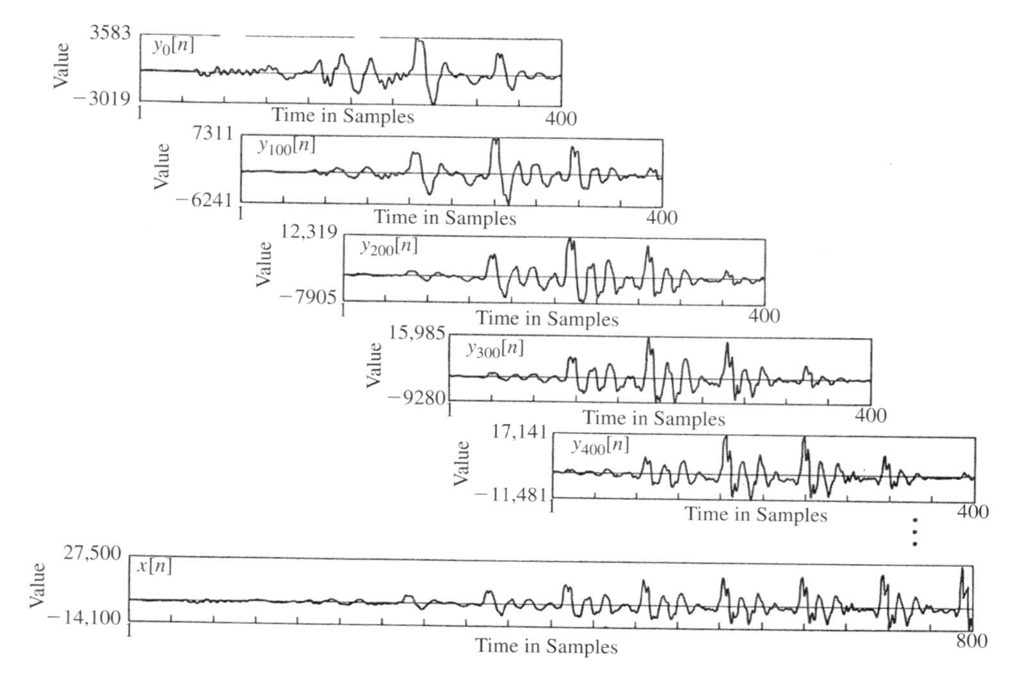

FIGURE 7.23
Illustration of OLA method for STFS of speech.

That is, the reconstructed signal is equal to $x[n]$ multiplied by a periodically time-varying weighting sequence $\tilde{w}[n]$. The condition for exact reconstruction of $x[n]$ is therefore,

$$\tilde{w}[n] = \sum_{r=-\infty}^{\infty} w[rR - n] = C, \tag{7.48}$$

where we will refer to the constant C as the *reconstruction gain*.

The summation process defined by Eq. (7.45) is illustrated in Figure 7.23, which shows five overlapping sections of the speech signal, where each section uses a Hamming window of duration $L = 400$ samples, and adjacent frames are separated by $R = 100$ samples. The summation implied by Eq. (7.47a) is illustrated in this figure by showing at the bottom of the figure the resulting speech waveform that is recovered by overlapping and adding the sections. Observe that since $L = 400$ and $R = 100$, exactly four sections overlap at each synthesis time n.

Note that the sequence $\tilde{w}[n]$ in Eq. (7.47b) is a periodic sequence (with period R) comprised of time-aliased (time-reflected) window sequences. As a simple example, consider a rectangular window $w_{\text{rect}}[n]$ of length L samples. If $R = L$, the windowed segments would simply fit together block-by-block with no overlap. In this case, the condition of Eq. (7.48) is satisfied with $C = 1$ because the shifted windows fit together with no overlap and no gaps. (A simple sketch will confirm this.) If L for the rectangular window is even, and $R = L/2$, a simple analysis or sketch will again verify that the condition of Eq. (7.48) is satisfied with $C = 2$. In fact, for the rectangular window, if $L = 2^\nu$, where ν is an appropriately large integer, the signal $x[n]$ can be perfectly reconstructed from $Y_r[k]$ by the OLA method of Eq. (7.47a) when $L \le N$ and $R = L, L/2, L/4, \ldots, 1$. The corresponding reconstruction gains would be $C = 1, 2, \ldots, L$. While this demonstrates that the OLA method can perfectly reconstruct the original signal for some rectangular windows, and some window spacings R, the rectangular window is rarely used in short-time Fourier analysis/synthesis because of its poor frequency selectivity. Tapered windows such as the Bartlett, Hann, Hamming, and Kaiser windows are more commonly used. Fortunately, these windows, with their superior spectral isolation properties, can also produce perfect or near-perfect reconstruction from the STFT.

Two windows with which perfect reconstruction can be achieved under some conditions are the Bartlett and Hann windows. They are defined in Eqs (7.49) and (7.50) respectively:[15]

Bartlett (triangular)

$$w_{\text{Bart}}[n] = \begin{cases} 1 - |n|/M, & -M \le n \le M \\ 0, & \text{otherwise,} \end{cases} \tag{7.49}$$

Hann

$$w_{\text{Hann}}[n] = \begin{cases} 0.5 + 0.5\cos(2\pi n/M), & -M \le n \le M \\ 0, & \text{otherwise.} \end{cases} \tag{7.50}$$

As these windows are defined, the window length is $L = 2M + 1$, with the two end samples equal to zero.[16] If M is even and $R = M$ for the Bartlett window, it is easily shown by a simple time-domain argument that the condition of Eq. (7.48) is satisfied with $C = 1$ for all n. Figure 7.24a shows overlapping Bartlett windows of length $2M + 1$ (first and last samples zero) when $R = M$. It is clear that these shifted windows sum to the reconstruction gain constant $C = 1$. Figure 7.24b shows the same choice of $L = 2M + 1$ and $R = M$ for the Hann window. Although it is perhaps less obvious from this plot, it is nevertheless true that these shifted windows sum to the constant $C = 1$ for all n.

While the results shown in Figure 7.24, where $R = M$, are intuitively plausible, it is less obvious that the Bartlett and Hann windows for $M = 2^\nu$ can provide perfect reconstruction for values of $R = M, M/2, \ldots, 1$ with corresponding reconstruction

[15]For convenience, they are defined as odd-length symmetric windows. Causal versions are obtained by a delay of M samples. Even-length versions of the causal windows can also be defined. The symmetric Hamming window is defined by Eq. (7.22) and the Hamming window and its DTFT are shown in Figure 7.6.

[16]With these definitions, the actual number of non-zero samples is $2M - 1$ for both the Bartlett and Hann windows, but the inclusion of the zero samples leads to convenient mathematical simplifications.

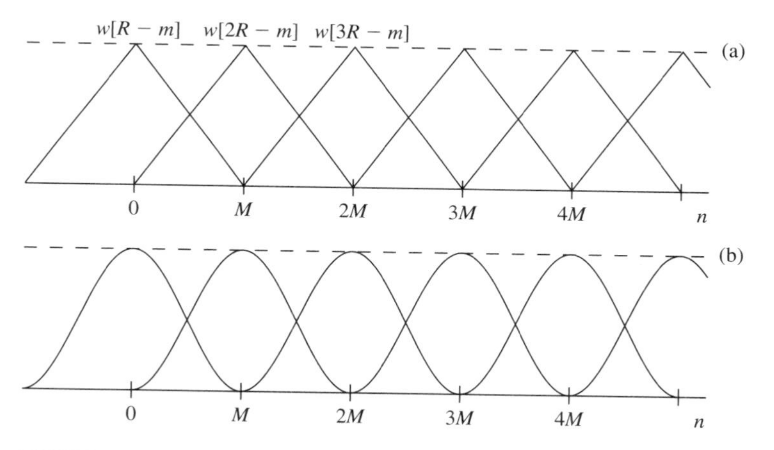

FIGURE 7.24
(a) Shifted $2M + 1$-point Bartlett windows with $R = M$. (b) Shifted $2M + 1$-point Hann windows with $R = M$. The dashed line is the periodic sequence $\tilde{w}[n] = C = 1$ defined by Eq. (7.48).

gains of $C = M/R$. To see this, it is helpful to recall that the envelope sequence $\tilde{w}[n]$ is inherently periodic with period R, so it can be represented by an inverse DFT as

$$\tilde{w}[n] = \sum_{r=-\infty}^{\infty} w[rR - n] = \frac{1}{R}\sum_{k=0}^{R-1} W^*(e^{j(2\pi k/R)})e^{j(2\pi k/R)n}, \tag{7.51}$$

where $W^*(e^{j(2\pi k/R)})$ is the DTFT of the time-reflected window $w[-n]$ sampled at frequencies $(2\pi k/R)$, $k = 0, 1, \ldots, R-1$. It follows from Eq. (7.51) that a condition for perfect reconstruction is

$$|W^*(e^{j(2\pi k/R)})| = |W(e^{j(2\pi k/R)})| = 0, \qquad k = 1, 2, \ldots, R-1, \tag{7.52a}$$

and if Eq. (7.52a) holds, then it follows from Eq. (7.51) that the reconstruction gain is

$$C = \frac{W(e^{j0})}{R}. \tag{7.52b}$$

We can use the result of Eqs. (7.52a) and (7.52b) to verify our intuitive observation about the Bartlett window, which for the definition in Eq. (7.49), has DTFT

$$W_{\text{Bart}}(e^{j\omega}) = \left(\frac{1}{M}\right)\left(\frac{\sin(\omega M/2)}{\sin(\omega/2)}\right)^2. \tag{7.53}$$

From Eq. (7.53), it follows that the Bartlett window Fourier transform has equally spaced zeros at frequencies $2\pi k/M$, for $k = 1, 2, \ldots, M-1$. Therefore, if we choose R so that $2\pi k/R = 2\pi k/M$ or $R = M$, the condition of Eq. (7.52a) is satisfied. Substituting $\omega = 0$ into Eq. (7.53) gives $W_{\text{Bart}}(e^{j0}) = M$, so it follows that perfect reconstruction results with $C = M/R = 1$ if $R = M$. Choosing $R = M$ aligns the frequencies $2\pi k/R$ with all the zeros of $|W_{\text{Bart}}(e^{j\omega})|$ which are at $2\pi k/M$, $k = 1,2,\ldots,M-1$. If M is divisible by 2, we can use $R = M/2$, and the frequencies $2\pi k/R$ will then again align with a subset of the zeros of $|W_{\text{Bart}}(e^{j\omega})|$ and the reconstruction gain will be $C = M/R = 2$. If M is a power of 2, R can be smaller with a concomitant increase in C.

The DTFT $W_{\text{Hann}}(e^{j\omega})$ has zeros equally spaced at integer multiples of π/M for $k = 2,3,\ldots,2M-2$, so a subset of these are at frequencies $2\pi k/M$ for $k = 1,2,\ldots, M-1$. This can be shown by noting that

$$w_{\text{Hann}}[n] = (0.5 + 0.5\cos(\pi n/M))w_r[n], \tag{7.54}$$

where $w_r[n]$ is a rectangular window that ensures that $w_{\text{Hann}}[n] = 0$ for $|n| > M$. Using Eq. (7.54), it is possible to express $W_{\text{Hann}}(e^{j\omega})$ in terms of $W_r(e^{j\omega})$, which, in turn, can easily be expressed in closed form. With this result, it can be verified that $W_{\text{Hann}}(e^{j(\pi k/M)}) = 0$ for $k = 2,3,\ldots,2M-2$. The derivation of the desired relationship for $W_{\text{Hann}}(e^{j\omega})$ is not straightforward, so it is left as a homework exercise (see Problem 7.16). Therefore, exact reconstruction is also possible with the Hann window defined as in Eq. (7.50) or (7.54). The equally spaced zeros of $W_{\text{Bart}}(e^{j\omega})$ and $W_{\text{Hann}}(e^{j\omega})$ are evident in the plots in Figure 7.25.

Figure 7.25 also shows the DTFT of an odd-length symmetric Hamming window, which is closely related to the Hann window, but optimized to minimize the side lobe levels. Because of its superior side lobe structure, the Hamming window may be preferred; however, it is not theoretically possible to achieve perfect reconstruction with the symmetric Hamming window. As a result of the adjustment of the coefficients from

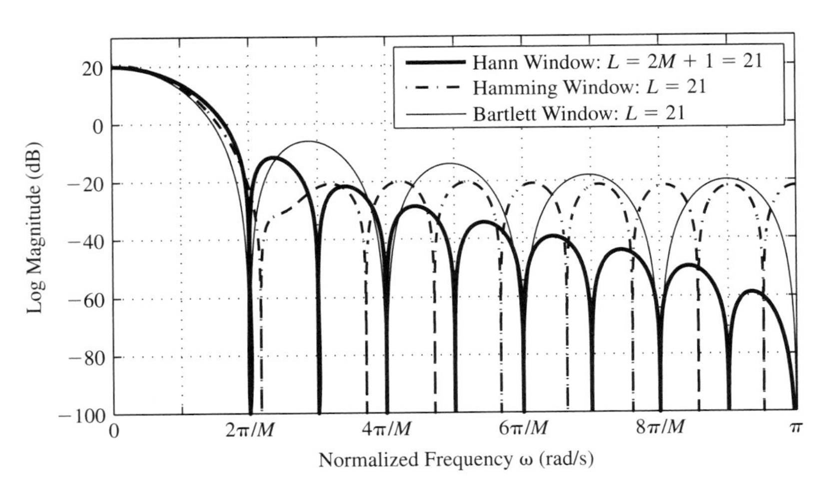

FIGURE 7.25
Discrete-time Fourier transforms of Bartlett, Hann, and Hamming windows for $L = 2M + 1 = 21$ ($M = 10$).

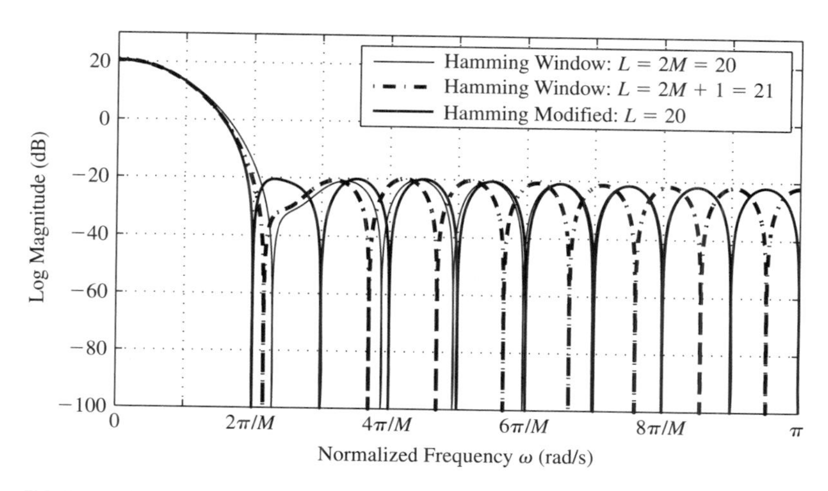

FIGURE 7.26
DTFTs of even-length, odd-length, and modified odd-to-even-length Hamming.

0.5 and 0.5 to 0.54 and 0.46, which minimizes the maximum side lobe level, the zeros of $W_{\text{Hamm}}(e^{j\omega})$ are significantly displaced from the equally spaced zeros of the Hann window, so it is no longer possible to choose R such that the frequencies $2\pi k/R$ fall precisely on zeros of $W_{\text{Hamm}}(e^{j\omega})$. However, as illustrated by Figure 7.25, the maximum side lobe level for frequencies above $2\pi/M$ is approximately -40 dB relative to the magnitude at $\omega = 0$. Thus, the condition of Eq. (7.52a) can be satisfied approximately at each of the frequencies $2\pi k/R$ if we choose $R = M/2$ (or even $R = M$). Figure 7.26 shows a comparison of the DTFTs of an even-length ($L = 20$) and an odd-length ($L = 21$) Hamming window (thin continuous line and dash-dot line respectively). Both windows are defined as causal windows by the equation

$$w_{\text{Hamm}}[n] = \begin{cases} 0.54 - 0.46\cos\left(2\pi n/(L-1)\right) & 0 \le n \le L-1 \\ 0 & \text{otherwise.} \end{cases} \tag{7.55}$$

Note that while the zeros of the even-length Hamming window tend to approach a spacing of $2\pi/M$, neither the even-length nor the odd-length Hamming window as it is normally defined can satisfy the condition of zeros spaced by $2\pi/R$ as required for exact reconstruction. This means that $\tilde{w}[n]$ in Eq. (7.48) will have a slight periodic variation with n. With the Hamming window, an even-length L is preferred so that we can choose the value of the time sampling period R as $L/2$ or $L/4$ corresponding to exactly 50% or 75% overlap respectively. This is, of course, not possible with the odd-length Hamming window. However, a curious result occurs if the odd-length Hamming window is truncated from $L = 2M + 1$ samples to $L = 2M$ samples by simply zeroing the last sample of an odd-length Hamming window from Eq. (7.55). The resulting window is no longer symmetric, but, more importantly, the zeros of the corresponding DTFT shift to the desired multiples of $2\pi/M$. This is illustrated in Figure 7.26, where the heavy continuous line shows the result of truncating the 21-point Hamming window to create a 20-point modified Hamming window. Observe that the side lobe level for the modified Hamming window remains at about 40 dB below the value at $\omega = 0$, while the zeros are exactly at multiples of $2\pi/M$. Thus, the modified odd-to-even length Hamming window can achieve exact reconstruction in OLA synthesis using $R = L/2$ or $R = L/4$, etc.

Equation 7.51 shows that if Eq. (7.52a) is not satisfied exactly, $\tilde{w}[n]$ will tend to oscillate around $C = W(e^{j0})/R$ with period R, thereby imparting a slight amplitude modulation to the reconstructed signal. It is readily shown in general (see Problem 7.15) that if $w[n]$ has a bandlimited Fourier transform and if $X_n(e^{j\omega_k})$ is properly sampled in time, i.e., R is small enough to avoid aliasing in the time dimension, then

$$\sum_{r=-\infty}^{\infty} w[rR - n] \approx \frac{W(e^{j0})}{R}, \tag{7.56}$$

for all n. Examples would include the unmodified Hamming window or the Kaiser window, both of which can achieve very nearly exact reconstruction when R is properly chosen. Thus Eq. (7.47a) becomes

$$y[n] \approx x[n]\frac{W(e^{j0})}{R}, \tag{7.57}$$

showing that the synthesis rule of Eq. (7.45) can lead to nearly exact reconstruction of $x[n]$ (to within a constant multiplier) by adding overlapping sections of the waveform. For the case of the unmodified Hamming window, the nominal reconstruction gain constant is $C = 1.08(M + 1)/R$, and the nominal bandlimit of the window is $2\pi/M$, so sampling the STFT with $2\pi/R = 4\pi/M$ (i.e., $R = M/2$) essentially avoids aliasing due to sampling in the time dimension and the values of $W_{\text{Hamm}}(e^{j2\pi k/R})$ are close to zero. For this reason, it is common to use $R = (L - 1)/4$ when L is odd or $R = L/4$ when L is even; i.e., it is typical to overlap the windows by 75%.

Figures 7.27 and 7.28 illustrate in more detail how the OLA method is implemented for $w[n]$, an L-point Hamming window with $R = L/4$. Figure 7.27 gives a flow chart of the method, assuming the signal $x[n]$ is 0 for $n < 0$. Since a time overlap of 4 to 1 is chosen for the Hamming window, to obtain the correct initial conditions, the first analysis section is positioned to begin at $n = L/4$ as shown in Figure 7.28. The window (assumed to be causal and non-zero for $0 \leq n \leq L - 1$) is used to give the signal $y_r[m] = w[rR - m]x[m]$, which is non-zero for $rR - L + 1 \leq m \leq rR$. This L-point sequence is padded with sufficient zeros to account for the effects of any modifications of the short-time spectrum (as discussed in Section 7.11) and to increase N to a convenient size for fast computation. Then an N-point FFT of the resulting sequence is used to give $Y_r(e^{j\omega_k})$.

To reconstruct the signal at time n, we use Eq. (7.45). Figure 7.28 illustrates the operations implied by Eq. (7.45) for a value of n such that $0 \leq n \leq R - 1$. Note that $y[n]$ for each n in $0 \leq n \leq R - 1$ consists of the sum of four numbers; i.e.,

$$y[n] = x[n]w[R - n] + x[n]w[2R - n] + x[n]w[3R - n] + x[n]w[4R - n]. \tag{7.58}$$

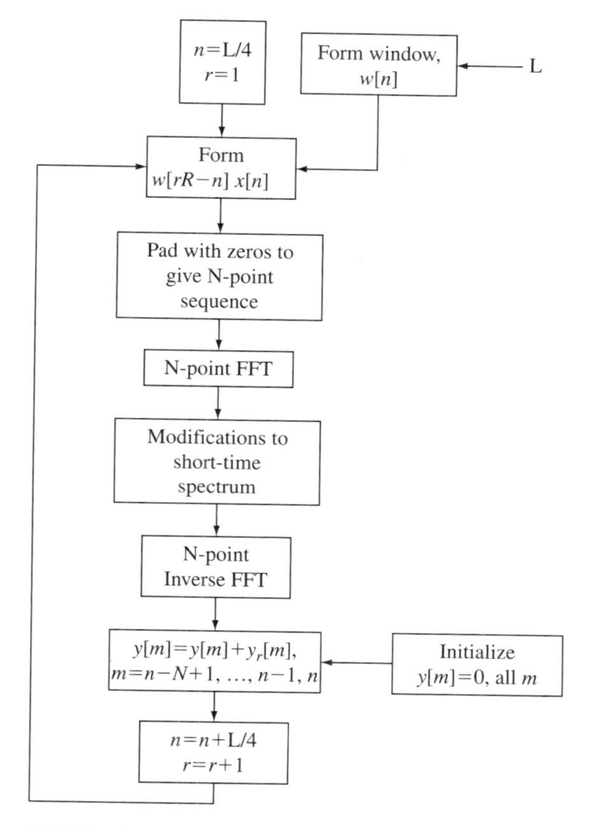

FIGURE 7.27
Flow diagram for OLA method for short-time processing.

For values of n in the next block of samples where $R \leq n \leq 2R-1$, the term $x[n]w[R-n]$ would be replaced by a term $x[n]w[5R-n]$, etc.

7.5.2 Use of a Synthesis Window

As anticipated in Figure 7.27, the OLA synthesis method is often used in applications where the STFT is modified before synthesis. In such applications, the signal sections $y_r[m]$ contain errors and artifacts at the edges of the tapering analysis window. For example, if $N > L$, the values of the signal outside the window may not be zero, and if $N = L$, these values will time alias or "wrap around" into the window region. To mitigate such effects, it is possible to use a synthesis window $w_s[n]$ before overlapping and adding the reconstructed sections. To be specific, Eqs. (7.47a) and (7.47b) are replaced by

$$y[n] = \sum_{r=-\infty}^{\infty} w_s[rR-n]y_r[n] = x[n]\left(\sum_{r=-\infty}^{\infty} w_s[rR-n]w[rR-n]\right) = x[n]\tilde{w}[n], \quad (7.59a)$$

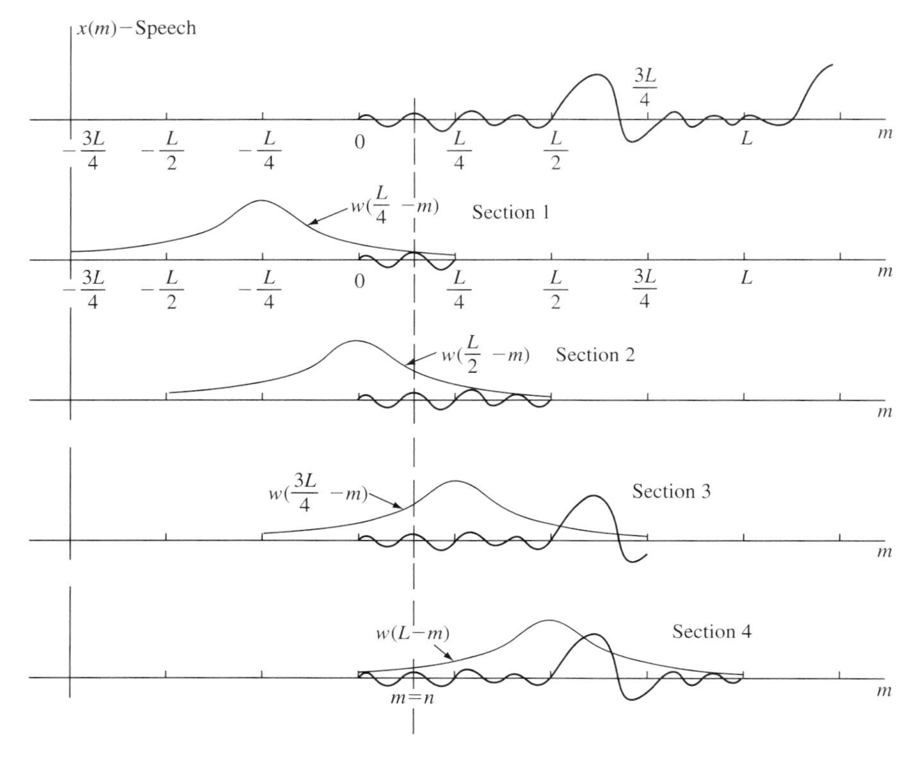

FIGURE 7.28
Reconstruction procedure for $w[n]$ using an L-point Hamming window.

where $\tilde{w}[n]$ is now defined as

$$\tilde{w}[n] = \sum_{r=-\infty}^{\infty} w_s[rR - n]w[rR - n] = \sum_{r=-\infty}^{\infty} w_{\text{eff}}[rR - n]. \tag{7.59b}$$

Now it is the product $w_{\text{eff}}[n] = w_s[n]w[n]$ that must satisfy the exact reconstruction equation. This complicates the choice of window significantly. The window $w[n]$ and the sampling interval R should be chosen to provide a desired degree of spectral selectivity and to avoid aliasing in the time dimension of the STFT. However, $w_s[n]$ cannot be chosen arbitrarily to simply satisfy Eq. 7.59b because we also want it to suppress edge effects due to processing of the STFT. For example, in audio coding, windows are sometimes constructed to have flat regions between tapering edges composed of cosine functions [42].

A detailed discussion of the construction of analysis/synthesis window pairs would take us too far afield. However, it is instructive to note what happens in the case when both the analysis and synthesis windows are Hann windows. Since we have seen that perfect reconstruction is possible with $w_{\text{Hann}}[n]$ given by Eq. (7.50) and the correct

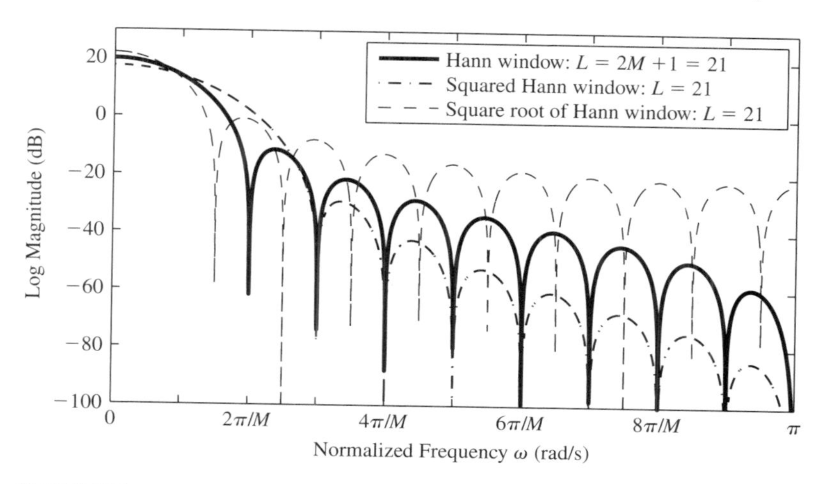

FIGURE 7.29
Illustration of effect of Hann synthesis window.

choice of R, we might be tempted to choose both the analysis and the synthesis window to be $w[n] = w_s[n] = \sqrt{w_{\text{Hann}}[n]}$ so that $w_{\text{eff}}[n] = w[n]w_s[n] = w_{\text{Hann}}[n]$. Figure 7.29 shows the log magnitude of the DTFT of $\sqrt{w_{\text{Hann}}[n]}$ as the thin dashed line. Observe that the main lobe width is narrower but the side lobes of this square-root window are larger than those of the Hann window from which it is derived. Thus, even though the combined analysis/synthesis window would be the Hann window whose log magnitude DTFT is shown by the heavy solid line, the properties of the square-root window are significantly inferior. On the other hand, we could choose $w[n] = w_s[n] = w_{\text{Hann}}[n]$ so that $w_{\text{eff}}[n] = w[n]w_s[n] = (w_{\text{Hann}}[n])^2$. The log magnitude DTFT of the squared Hann window is shown as the dash-dot mid-weight line in Figure 7.29. Observe that the log magnitude response rolls off more slowly than the Hann window and its first zero is at $3\pi/M$. However, the zeros of the DTFT of the squared Hann window are at $\pi k/M$ with $k = 3, 4, \ldots, 2M - 3$, so it is possible, for example if M is even, to choose $2\pi/R = 4\pi/M$ ($R = M/2$) so that exact reconstruction is possible with back-to-back Hann windows of length $L = 2M + 1$. For this condition, it can be shown that if $R = M/2$, the reconstruction gain constant is $C = 3/2$. (See Problem 7.17.)

As in our previous discussion, if we use back-to-back Hamming windows instead of Hann windows, it is not possible to achieve perfect reconstruction, but back-to-back Hamming windows will produce an overall analysis/synthesis with very little variation in the reconstruction function $\tilde{w}[n]$ if we use windows that overlap by 75% ($R = M/2$).

7.6 FILTER BANK SUMMATION METHOD OF SYNTHESIS

As we mentioned in the previous section, it is necessary to sample the STFT in both the time and frequency dimensions in order to obtain efficient and effective computational realizations of short-time Fourier analysis and synthesis. In this section, we shall focus on the implications of sampling in the frequency dimension. Our goal is to show that the requirement $N \geq L$ for exact reconstruction can be relaxed if we choose the window and N properly. In this section, we shall assume that the sampling rate in the time dimension is identical to that of the input signal; i.e., in the notation of the previous section, $R = 1$. In Section 7.7, we will consider the full implications of sampling in the time dimension with $R \neq 1$.

The frequency-sampled STFT is

$$X_n(e^{j\omega_k}) = \sum_{m=-\infty}^{\infty} w[n-m]\left(x[m]e^{-j\omega_k m}\right), \tag{7.60a}$$

where for uniform sampling in frequency, $\omega_k = 2\pi k/N$, with $k = 0, 1, \ldots, N-1$. These are the standard DFT frequencies, so, not surprisingly, the DFT plays a significant role in our discussions even though we shall focus on the linear filtering interpretation.[17] In particular, the STFS equation that we shall employ first is

$$y[n] = \frac{1}{N}\sum_{k=0}^{N-1} X_n(e^{j\omega_k})e^{j\omega_k n}, \tag{7.60b}$$

which, as we pointed out in the previous section, is simply the inverse DFT of $X_n(e^{j\omega_k})$ at the particular time n. In this section, we will study the process of STFS from the point of view of the linear filtering interpretation of the STFT. The method of synthesis that emerges from this interpretation is called the *filter bank summation method* (FBS) of short-time synthesis. Before we explore the details of this method, it is useful to observe that if $N \geq L$, where L is the window length, the inverse DFT produces $w[n-m]x[m]$ for $n-m$ inside the region of support of the window. Therefore, it follows from setting $m = n$ that

$$w[0]x[n] = \frac{1}{N}\sum_{k=0}^{N-1} X_n(e^{j\omega_k})e^{j\omega_k n}. \tag{7.61}$$

That is, Eq. (7.60b) exactly reconstructs $x[n]$ to within a constant multiplier, and if $w[0] > 0$, we can divide by $w[0]$ to obtain $x[n]$. Thus, if we know $X_n(e^{j\omega_k})$ for all n, Eq. (7.60b) is the desired synthesis equation and $w[0]$ is the scale factor on the synthesized output.

[17] It is possible to employ non-uniform sampling where the frequencies ω_k are not equally spaced. However, this requires that the STFT be defined as in

$$X_n(e^{j\omega_k}) = \sum_{m=-\infty}^{\infty} x[m]w_k[n-m]e^{-j\omega_k m},$$

which assumes a different analysis window $w_k[n]$ for each analysis frequency. With this definition we sacrifice the DFT interpretation, and the linear filtering interpretation yields more understanding.

In Section 7.3.5, we showed that when $\hat{\omega}$ is fixed at a frequency ω_k, $X_n(e^{j\omega_k})$ is a lowpass representation of the signal in a band centered at ω_k. When expressed in the form of Eq. (7.60a), $X_n(e^{j\omega_k})$ has the interpretation of lowpass filtering following frequency down-shifting by ω_k. With a change of summation variable, we have the alternative form

$$X_n(e^{j\omega_k}) = e^{-j\omega_k n} \sum_{m=-\infty}^{\infty} x[n-m] \left(w[m] e^{j\omega_k m} \right). \tag{7.62}$$

With the definition

$$h_k[n] = w[n] e^{j\omega_k n}, \tag{7.63}$$

Eq. (7.62) becomes

$$X_n(e^{j\omega_k}) = e^{-j\omega_k n} \left(\sum_{m=-\infty}^{\infty} x[n-m] h_k[m] \right). \tag{7.64}$$

Since the window $w[n]$ has the properties of a lowpass filter, Eq. (7.64) can be interpreted as in Figure 7.14 as a bandpass filter with impulse response $h_k[n]$ followed by frequency down-shifting by modulation with a complex exponential $e^{-j\omega_k n}$. Figure 7.13 shows an example of how the corresponding lowpass and bandpass filter frequency responses are related for the case of a Hamming window for analysis frequency $\hat{\omega}$.

Now, the signal

$$y_k[n] = X_n(e^{j\omega_k}) e^{j\omega_k n} \tag{7.65}$$

is simply one of the terms of the sum in Eq. 7.60b. From Eqs. (7.64) and (7.65), it follows that

$$y_k[n] = \sum_{m=-\infty}^{\infty} x[n-m] h_k[m]. \tag{7.66}$$

That is, $y_k[n]$ is the output of a bandpass filter with impulse response $h_k[n]$ as defined by Eq. (7.63). The operations of Eqs. (7.64) and (7.65) are depicted in Figure 7.30a. Since Eqs. (7.60a) and (7.64) are equivalent, either form for $X_n(e^{j\omega_k})$ can be used in Eq. (7.65), and in both cases, the overall system relating $x[n]$ to $y_k[n]$ is a bandpass filter with impulse response $h_k[n]$. This is depicted in Figure 7.30, where Figure 7.30a depicts Eqs. (7.64) and (7.65) and Figure 7.30b depicts Eqs. (7.60a) and (7.65). Figure 7.30c shows the equivalent bandpass filter for both cases.

The result summarized in Figure 7.30 provides the key to understanding how Eq. (7.60b) provides a practical method for reconstructing the input signal from its time-dependent Fourier transform. We simply implement N bandpass channels of the

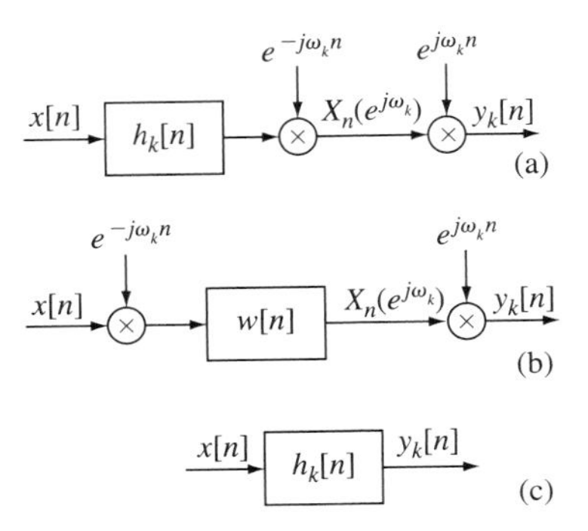

FIGURE 7.30
Methods for implementing the synthesis of a single channel in terms of linear filtering.

form of Eq. (7.65) and sum the outputs; i.e.,

$$y[n] = \frac{1}{N}\sum_{k=0}^{N-1} y_k[n],$$

which is identical to Eq. (7.60b). This motivates the name *FBS* for this approach to STFS.

We have seen that at one value of frequency $\hat{\omega} = \omega_k$, the combination of analysis followed by synthesis can be represented as a bandpass filter centered on the analysis frequency ω_k. Now consider the set of N frequencies $\{\omega_k = 2\pi k/N\}$, $k = 0, 1, \ldots, N-1$, and suppose that the N time sequences $X_n(e^{j\omega_k})$ are available for each frequency. This can be achieved by a "bank" of analysis/synthesis channels of the form of either Figure 7.30a or Figure 7.30b, whose outputs are summed to effect the implementation of Eq. (7.60b). The N bandpass filters of the form of Eq. (7.63) have frequency responses

$$H_k(e^{j\omega}) = W(e^{j(\omega-\omega_k)}), \tag{7.67}$$

as illustrated in Figure 7.13b for a Hamming window. Then if we consider the entire collection of bandpass filters, each having the same input and their outputs added together as in Figure 7.31, the composite frequency response relating $y[n]$ to $x[n]$ is

$$\tilde{H}(e^{j\omega}) = \frac{1}{N}\sum_{k=0}^{N-1} H_k(e^{j\omega}) = \frac{1}{N}\sum_{k=0}^{N-1} W(e^{j(\omega-\omega_k)}). \tag{7.68}$$

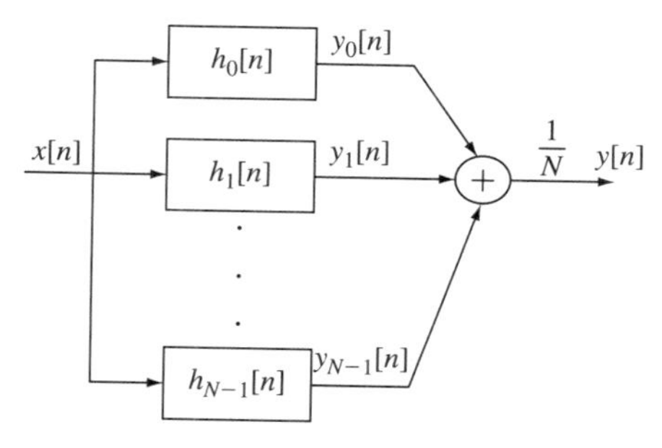

FIGURE 7.31
Equivalent linear system relating $y_k[n]$ and $y[n]$ to $x[n]$.

If $W(e^{j\omega_k})$ is properly sampled in the frequency dimension (i.e., if $N \geq L$ where L is the time duration of the window), then it can be shown that

$$\frac{1}{N}\sum_{k=0}^{N-1} W(e^{j(\omega-\omega_k)}) = w[0], \qquad \text{for all } \omega. \tag{7.69}$$

To derive Eq. (7.69), recall that $W(e^{j\omega_k})$ is the DFT of the window. Therefore, the inverse DFT of $W(e^{j\omega_k})$ is

$$\frac{1}{N}\sum_{k=0}^{N-1} W(e^{j\omega_k})e^{j\omega_k n} = \sum_{r=-\infty}^{\infty} w[n+rN]; \tag{7.70}$$

i.e., $W(e^{j\omega_k})$ corresponds to a time-aliased representation of $w[n]$. (See Problem 7.14.) If $w[n]$ is of duration L samples and $L \leq N$, then

$$w[n] = 0, \quad n < 0, \; n \geq L, \tag{7.71}$$

and no time aliasing occurs due to sampling $W(e^{j\omega})$ in frequency. Therefore, if Eq. (7.70) is evaluated for $n = 0$, we get

$$\frac{1}{N}\sum_{k=0}^{N-1} W(e^{j\omega_k}) = w[0]. \tag{7.72}$$

Equation (7.69) is readily obtained by noting that $W(e^{j(\omega-\omega_k)})$ is a uniformly sampled version of $W(e^{j\omega})$ evaluated at $\omega - \omega_k$ rather than ω_k. According to sampling theory, *any* set of N uniformly spaced samples is adequate. Thus, Eq. (7.69) follows from Eq. (7.72) and the sampling theorem.

It is interesting to note that the FBS method and the OLA method are *duals* of one another; i.e., one depends on a sampling relation in frequency, and the other depends on a sampling relation in time. The FBS requires that the sampling in frequency be such that the window transform obeys the relation

$$\frac{1}{N}\sum_{k=0}^{N-1} W(e^{j(\omega-\omega_k)}) = w[0], \quad \text{for all } \omega, \tag{7.73a}$$

whereas the OLA method requires that the sampling in time be such that the window obeys the relation

$$\sum_{r=-\infty}^{\infty} w[rR-n] = W(e^{j0})/R, \quad \text{for all } n. \tag{7.73b}$$

The duality between Eqs. (7.73a) and (7.73b) is evident.

From Eqs. (7.68) and (7.69), we see that the impulse response of the composite system is

$$\tilde{h}[n] = \frac{1}{N}\sum_{k=0}^{N-1} h_k[n] = w[0]\delta[n], \tag{7.74}$$

and therefore, the composite output $y[n]$ will be $y[n] = w[0]x[n]$ as we showed in Eq. (7.61) under the same assumption $N \geq L$.

We have used the concept of a bank of filters to confirm a very important result that we have already observed from the DFT viewpoint. Under the condition that $w[n]$ has finite duration, L, the sequence $x[n]$ can be reconstructed exactly from the time-dependent Fourier transform sampled in the time dimension at the sampling rate of the input signal and sampled in the frequency dimension at $N \geq L$ equally spaced frequencies over $0 \leq \hat{\omega} < 2\pi$. However, these conditions are far from necessary. There are many ways to achieve exact reconstruction, even when $N < L$.

For example, if the window is the infinite sequence

$$w[n] = \frac{\sin(\pi n/N)}{(\pi n/N)}, \qquad -\infty < n < \infty, \tag{7.75}$$

corresponding to the impulse response of an ideal lowpass filter with cutoff frequency π/N, then the composite frequency response would be constant ($\tilde{H}(e^{j\omega}) = 1$), independent of ω since the resulting bandpass filters $H_k(e^{j\omega}) = W(e^{j(\omega-\omega_k)})$, $k = 0, 1, \ldots, N-1$ would exactly cover the band $0 \leq \omega < 2\pi$. This situation is depicted in Figure 7.32, which shows the composite response for $N = 6$ equally spaced ideal filters each of total width $2\pi/N$. It is clear that exact reconstruction is achieved for this window even though it has infinite length.

To explore this question further and to understand why it is not necessary for the window to have finite length with $L \leq N$, it is helpful to redefine the synthesis

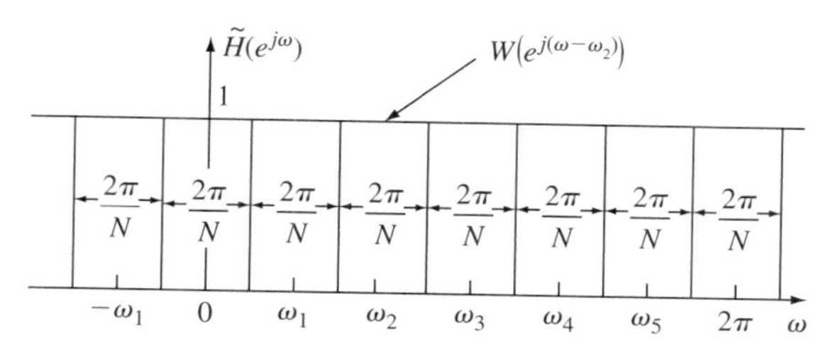

FIGURE 7.32
Composite frequency response for $N = 6$ equally spaced ideal filters.

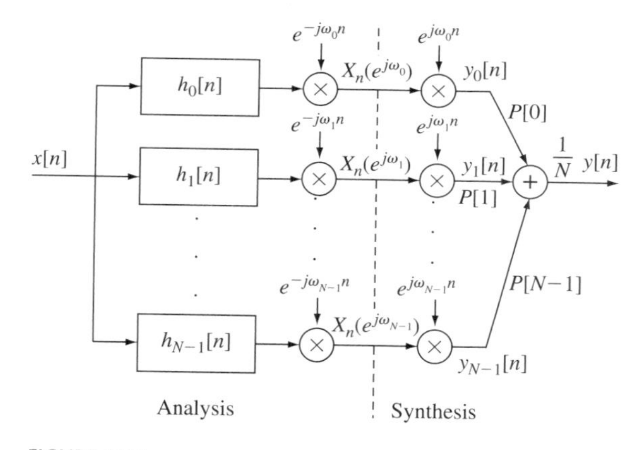

FIGURE 7.33
Analysis and synthesis operations for short-time spectrum analysis.

equation as

$$y[n] = \frac{1}{N} \sum_{k=0}^{N-1} P[k] X_n(e^{j\omega_k n}) e^{j\frac{2\pi}{N}kn}, \tag{7.76}$$

where we have introduced a set of complex gain coefficients, $P[k]$, for each of the N filter bank channels. These complex gain coefficients are shown in Figure 7.33, where the operations of analysis and synthesis implied by Eqs. (7.60b) and (7.62) are depicted.[18] The filters $h_k[n]$ are the equivalent bandpass filters for analysis followed by synthesis. The coefficients $P[k]$ can be used to adjust the magnitude and phase of the

[18]Here we have used the analysis form of Figure 7.30a, but we could just as well have used the form in Figure 7.30b.

individual channels. This adds significant flexibility in the design and implementation of STFA/STFS systems. Now the overall composite impulse response of the filter bank becomes

$$\tilde{h}[n] = \frac{1}{N}\sum_{k=0}^{N-1} P[k]w[n]e^{j\omega_k n} = w[n]\left(\frac{1}{N}\sum_{k=0}^{N-1} P[k]e^{j\omega_k n}\right). \tag{7.77}$$

We define

$$p[n] = \frac{1}{N}\sum_{k=0}^{N-1} P[k]e^{j\omega_k n}, \tag{7.78}$$

to be the inverse DFT of the set of coefficients $P[k]$, $k = 0, 1, \ldots, N-1$. Therefore $\tilde{h}[n]$ can be written as

$$\tilde{h}[n] = w[n]p[n]. \tag{7.79}$$

Since the sequence $p[n]$ is periodic, with period N if evaluated outside the base interval $0 \leq n \leq N-1$, the nature of $w[n]$ outside the base interval can significantly affect the overall impulse response of the filter bank. Specifically, if $P[k] = 1$ for all k we showed in Example 7.1 that $p[n]$ is a periodic train of impulses; i.e.,

$$p[n] = \sum_{r=-\infty}^{\infty} \delta[n - rN]. \tag{7.80}$$

Therefore, $\tilde{h}[n]$ is

$$\tilde{h}[n] = w[n]p[n] = \sum_{r=-\infty}^{\infty} w[rN]\delta[n - rN]; \tag{7.81}$$

i.e., the composite impulse response is simply the window sequence sampled at intervals of N samples. This is depicted in Figure 7.34. Figure 7.34a shows the sequence $p[n]$. Figure 7.34b shows $w[n]$ as given in Eq. (7.75); i.e., the impulse response of an ideal lowpass filter with cutoff frequency π/N. From Eq. (7.75), it follows that for this window, $w[n] = 0$ at integer multiples of N except at $n = 0$. Comparing Figures 7.34a and 7.34b reveals that the product $\tilde{h}[n] = p[n]w[n]$ will be zero everywhere except at $n = 0$ where the product is unity. Thus the composite impulse response is

$$\tilde{h}[n] = \delta[n], \tag{7.82}$$

consistent with our previous frequency-domain argument.

While we cannot implement an ideal lowpass filter, the details of the way $p[n]$ and $w[n]$ interact to produce the composite response suggest a multitude of ways of

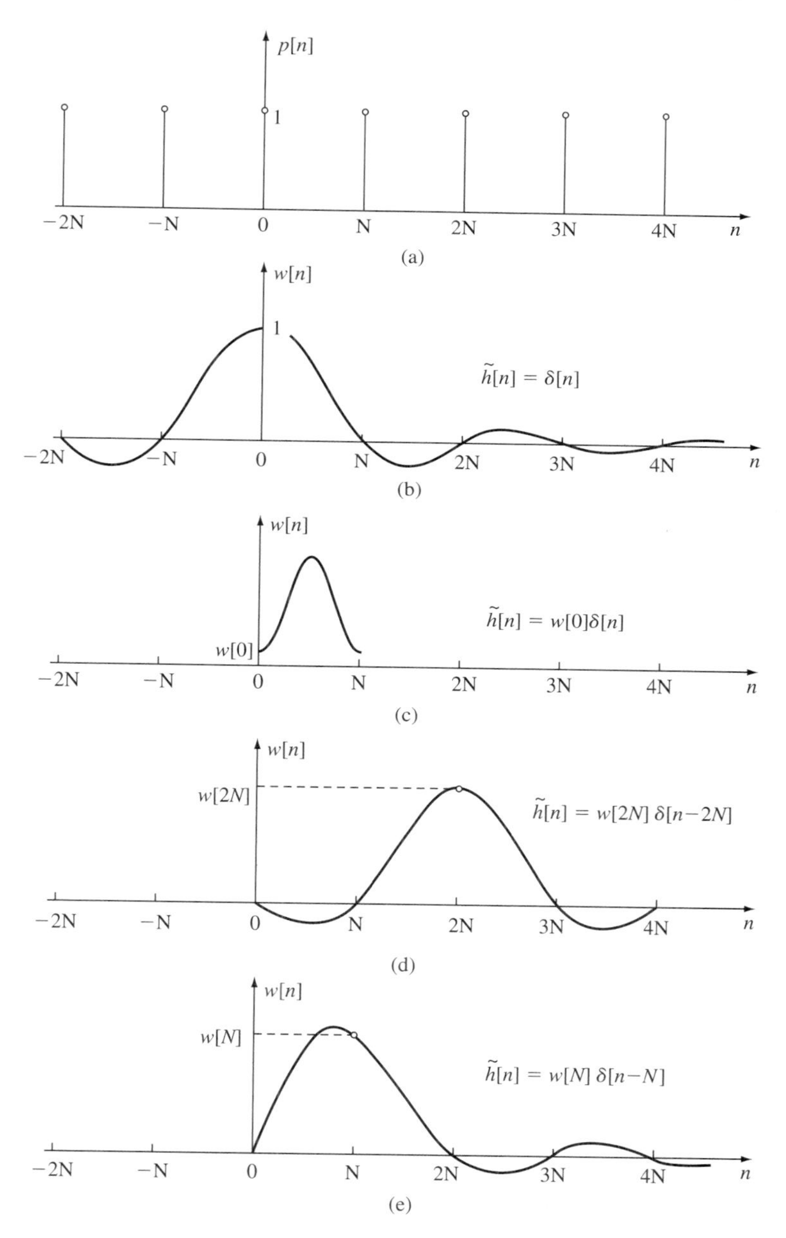

FIGURE 7.34
Typical sequences for $p[n]$ and $w[n]$ for composite filter bank.

choosing $w[n]$ and N so that the signal can be reconstructed exactly from the sampled short-time transform. First of all, note that if $w[n]$ has finite length $L \leq N$ and is causal, then the composite impulse response will be as in Eq. (7.74), thus verifying the previous discussion. Figure 7.34c shows an example of this case. Alternatively, a causal window with length greater than N can be used if $w[n]$ has the properties

$$w[n] = 0 \qquad n = rN \begin{cases} r \neq r_0 \\ r = 0, \pm 1, \pm 2, \ldots \end{cases} \tag{7.83a}$$

so that

$$\tilde{h}[n] = p[n]w[n] = w[r_0 N]\delta[n - r_0 N]. \tag{7.83b}$$

Figure 7.34d shows an example of a finite duration window where $r_0 = 2$. In fact, it is clear that $w[n]$ need be neither time-limited or frequency-limited in order that it be possible to exactly reconstruct a delayed replica of $x[n]$ from $X_n(e^{j\omega_k})$. All that is required is that Eq. (7.83a) hold for $w[n]$ and N. There is no restriction that the window be of finite length, if it has equally spaced zeros. Indeed Figure 7.34e suggests an infinite duration window with the appropriate properties.

The implication of Eq. (7.83b) is that the composite frequency response of an analysis/synthesis system such as Figure 7.33 has a flat magnitude response and linear phase corresponding to a delay of $r_0 N$ samples. That is,

$$\tilde{H}(e^{j\omega}) = \frac{1}{N}\sum_{k=0}^{N-1} P[k]W(e^{j(\omega-\omega_k)}) = w[r_0 N]e^{-j\omega r_0 N}, \tag{7.84}$$

which in turn implies that the output of the analysis/synthesis system is

$$y[n] = w[r_0 N]x[n - r_0 N]. \tag{7.85}$$

Thus, apart from the scale factor, $w[r_0 N]$, and delay of $r_0 N$ samples, the output of the system for time-dependent Fourier analysis and synthesis is an *exact replica* of the input sequence.

We have shown that exact reconstruction of the input is possible using FBS methods with a number of frequency channels less than that required by the sampling theorem and with a causal window that permits the realization of the analysis with causal bandpass or lowpass filters. Thus, an important practical issue is how well we can design digital filters to approximate one of the behaviors shown in Figure 7.34. This issue is discussed extensively in Refs. [345] and [347].

7.7 TIME-DECIMATED FILTER BANKS

We have argued that it should be possible to sample the STFT in the time dimension as well as the frequency dimension since typical windows act like a lowpass filter, and

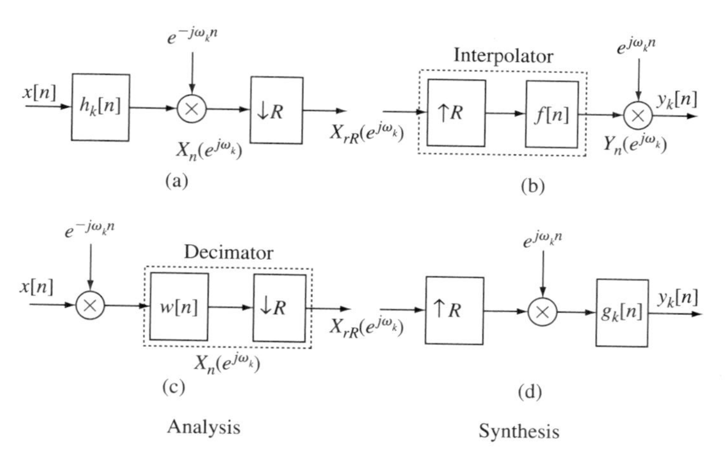

FIGURE 7.35
Implementation of decimation and interpolation in filter bank channels: (a) analysis with bandpass filter followed by frequency down-shift followed by downsampling; (b) synthesis with upsampler followed by lowpass interpolation filter followed by frequency up-shift; (c) analysis with frequency down-shift followed by lowpass filter followed by downsampling; (d) synthesis with upsampling followed by frequency up-shift followed by bandpass filter.

our discussion of the OLA method of synthesis demonstrates that the window can be moved in jumps of $R < L$ samples if $L \leq N$. From the filter bank perspective, the sampling rate for $X_n(e^{j\omega_k})$, thought of as a function of n with ω_k fixed, need only be twice the bandwidth of the window Fourier transform. Assume, therefore, that the kth channel is computed once every R samples of the input. Assuming temporarily that $X_n(e^{j\omega_k})$ is computed at the sampling rate of the input[19] we can modify Figures 7.30a and 7.30b to reflect the fact that $X_n(e^{j\omega_k})$ need only be sampled at a rate of F_s/R by including a downsampler at the output of the analysis as depicted in Figure 7.35a and 7.35c. The downsampler, denoted by $\downarrow R$, simply throws away (or does not compute) $R-1$ samples out of every R at the sampling rate of $x[n]$. Figure 7.35c shows that in this context, the downsampler in combination with the lowpass filter of the window constitutes a *decimator* [270].

From our previous discussion, we have seen that the systems in Figures 7.35a and 7.35c both produce the output $X_{rR}(e^{j\omega_k})$ at the effective sampling rate of F_s/R. To synthesize the output using the FBS method, it is necessary to interpolate $X_{rR}(e^{j\omega_k})$ back to the original sampling rate. Figure 7.35b shows the classical discrete-time interpolator consisting of an upsampler, which inserts $R-1$ zero samples between each sample at the low rate, and then fills in the values by linear filtering with a lowpass filter whose impulse response is denoted $f[n]$. The combination of an upsampler by R followed by a lowpass filter is an *interpolator* [270]. The output of the interpolator is formed by

[19]Usually only the downsampled STFT would be computed in practice.

convolution of $f[n]$ with the upsampled STFT; i.e.,

$$Y_n(e^{j\omega_k}) = f[n] * \sum_{r=-\infty}^{\infty} X_{rR}(e^{j\omega_k})\delta[n - rR]$$
$$= \sum_{r=-\infty}^{\infty} X_{rR}(e^{j\omega_k}) f[n - rR]. \quad (7.86)$$

Figure 7.35d shows an alternative equivalent form, which relies on the fact that the interpolated signal is ultimately frequency up-shifted by ω_k. Thus, we can first upsample by R, then frequency up-shift the result, and finally filter the frequency up-shifted result with the bandpass filter $g_k[n] = f[n]e^{j\omega_k n}$, whose frequency response is $F(e^{j(\omega-\omega_k)})$. The details of decimation and interpolation were reviewed in Chapter 2.

7.7.1 General FBS System with Decimation

Figure 7.36 shows a basic FBS implementation of a STFA/STFS system with N bandpass channels, each using a common lowpass filter (with impulse response $w[n]$) modulated to the set of bandpass frequencies, $\omega_k = 2\pi k/N, k = 0, 1, \ldots, N-1$. The outputs of the analysis channels are downsampled by a factor R, leading to a total sample rate of NF_s/R for the downsampled STFT. Figure 7.36 also shows the possibility of modifications such as quantization in a coding application. The right half of Figure 7.36 shows the synthesis operations, which include interpolation followed by frequency up-shifting (via the set of modulators) to place the channel signals back in their original frequency location.

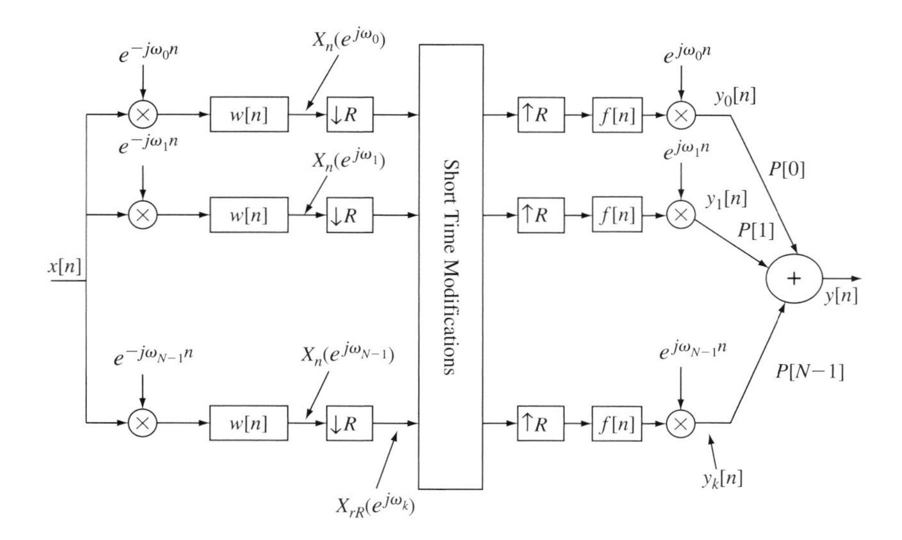

FIGURE 7.36
Full implementation of STFT analysis/synthesis system with channel decimation by a factor of R, channel modification, and channel interpolation by the same factor R.

We have seen in Section 7.6 that if no modifications occur, then it is possible to reconstruct the original signal exactly from the STFT using a back-to-back (no modifications) system like the one shown in Figure 7.33. The question that naturally arises is whether it is possible to choose R, $w[n]$, $f[n]$, N, and the complex gain factors $P[k]$ so that Figure 7.36 is also capable of exact reconstruction of the input. In order to answer this question, we must derive the mathematical description of the signal processing of Figure 7.36 in either the time domain or the frequency domain, assuming no short-time modifications occurred at the interface between the analyzer and the synthesizer.[20]

In the time domain, we have as the output of the analyzer the STFT channel signals

$$X_{rR}[k] = X_{rR}(e^{j\omega_k}) = \sum_{m\in\mathcal{W}_{rR}} w[rR-m]x[m]e^{-j\omega_k m}, \tag{7.87}$$

where $\mathcal{W}_{rR}$ denotes the set of samples within the region of support of $w[rR-m]$. We can then express the composite output of the synthesizer as

$$\begin{aligned} y[n] &= \frac{1}{N}\sum_{k=0}^{N-1} P[k]Y_n(e^{j\omega_k})e^{j\omega_k n} \\ &= \frac{1}{N}\sum_{k=0}^{N-1} P[k]\left[\sum_{r=-\infty}^{\infty} X_{rR}[k]f[n-rR]\right]e^{j\omega_k n}, \end{aligned} \tag{7.88}$$

which, after substituting Eq. (7.87) into Eq. (7.88) and some manipulation, takes the form

$$y[n] = \sum_{m\in\mathcal{W}_{rR}} x[m]\left[\sum_{r=-\infty}^{\infty} w[rR-m]f[n-rR]\right]\frac{1}{N}\sum_{k=0}^{N-1} P[k]e^{\omega_k(n-m)}. \tag{7.89}$$

In the case when $P[k]=1$, $k=0,1,\ldots,N-1$ we can write

$$\frac{1}{N}\sum_{k=0}^{N-1} e^{j\omega_k(n-m)} = \sum_{q=-\infty}^{\infty}\delta[n-m-qN], \tag{7.90}$$

from which it follows that

$$y[n] = \sum_{q=-\infty}^{\infty} x[n-qN]\left[\sum_{r=-\infty}^{\infty} w[rR-n+qN]f[n-rR]\right]. \tag{7.91}$$

[20]We return to the issue of the impact of modifications to the STFT in Section 7.11.

Note that this expression is not in the form of a convolution. In general the input/output relation for an analysis/synthesis filter bank would be time-varying due to aliasing distortions resulting from non-ideal filters. The condition for perfect reconstruction (i.e., $y[n] = x[n]$ for all n) is

$$\sum_{r=-\infty}^{\infty} w[rR - n + qN] f[n - rR] = \begin{cases} 1 & q = 0 \\ 0 & q \neq 0. \end{cases} \tag{7.92}$$

Equation (7.92) is difficult to interpret and generally does not lend much insight into the design of decimated filter banks. In contrast, a frequency-domain analysis leads to a representation that can be split into a set of conditions for perfect reconstruction and a set of conditions to eliminate any aliasing due to the non-ideal filters in the filter bank.

If we begin with the assumption that the input signal has conventional DTFT $X(e^{j\omega})$, we can obtain a relationship between the Fourier transforms of the input and output for the complete filter bank of Figure 7.36. Defining the bandpass filters $h_k[n] = w[n]e^{j\omega_k n}$ and $g_k[n] = f[n]e^{j\omega_k n}$ with corresponding frequency responses $H_k(e^{j\omega}) = W(e^{j(\omega-\omega_k)})$ and $G_k(e^{j\omega}) = F(e^{j(\omega-\omega_k)})$, this relationship takes the form

$$\begin{aligned} Y(e^{j\omega}) &= \frac{1}{N}\sum_{k=0}^{N-1} P[k]G_k(e^{j\omega}) \\ &\quad \times \left[\frac{1}{R}\sum_{l=0}^{R-1} H_k(e^{j(\omega-2\pi l/R)})X(e^{j(\omega-2\pi l/R)}) \right] \end{aligned} \tag{7.93a}$$

$$\begin{aligned} &= X(e^{j\omega})\left(\frac{1}{RN}\sum_{k=0}^{N-1} P[k]G_k(e^{j\omega})H_k(e^{j\omega}) \right) \\ &\quad + \sum_{l=1}^{R-1} X(e^{j(\omega-2\pi l/R)}) \\ &\quad \times \left(\frac{1}{RN}\sum_{k=0}^{N-1} P[k]G_k(e^{j\omega})H_k(e^{j(\omega-2\pi l/R)}) \right) \end{aligned} \tag{7.93b}$$

$$= \tilde{H}(e^{j\omega}) \cdot X(e^{j\omega}) + \text{ aliasing terms.} \tag{7.93c}$$

As suggested by Eq. (7.93c), $Y(e^{j\omega})$ can be expressed as the sum of a term in the form of a product of a linear time invariant (LTI) system with frequency response $\tilde{H}(e^{j\omega})$ with the input signal spectrum and a term involving $R-1$ frequency-shifted copies of the DTFT of the input. Based on this interpretation of Eqs. (7.93a)–(7.93c), we see that the conditions for perfect reconstruction, i.e., $Y(e^{j\omega}) = X(e^{j\omega})$, are the following:

$$\tilde{H}(e^{j\omega}) = \frac{1}{RN}\sum_{k=0}^{N-1} P[k]G_k(e^{j\omega})H_k(e^{j\omega}) = 1; \tag{7.94a}$$

i.e., the equivalent linear filter $\tilde{H}(e^{j\omega})$ must exhibit flat gain and zero phase for all frequencies,[21] and

$$\frac{1}{RN}\sum_{k=0}^{N-1} P[k]G_k(e^{j\omega})H_k(e^{j(\omega-2\pi l/R)}) = 0, \quad l = 1, 2, \ldots, R; \tag{7.94b}$$

i.e., "complete alias cancellation" for each of the frequency-shifted terms.

Figure 7.37 shows a block diagram representation of Eqs. (7.93a)–(7.93c). Note that the lower block accounts for all the aliasing distortion due to the frequency-shifted copies of $X(e^{j\omega})$ that result from the downsampling. The upper part shows a representation of the LTI equivalent analysis/synthesis filter bank with the appropriate bandpass versions of the analysis and interpolation filters back-to-back as suggested by Eq. 7.94a. Note that in each channel, the combination $H_k(e^{j\omega})G_k(e^{j\omega})$ acts in the

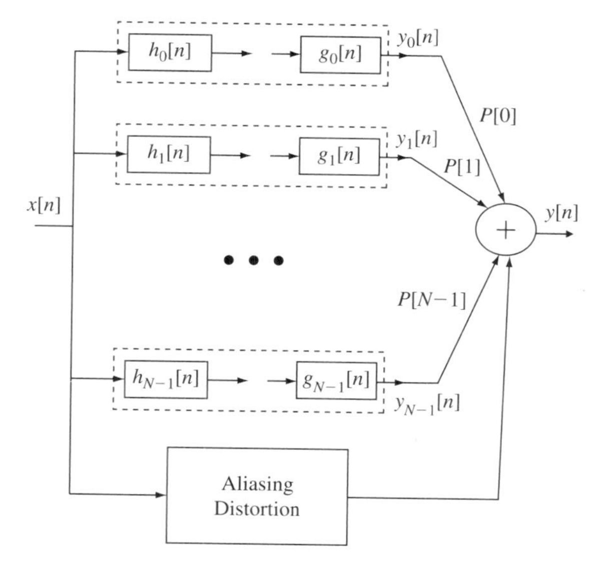

FIGURE 7.37
Back-to-back (no modifications) analysis/synthesis filter bank equivalent overall to Figure 7.36.

[21] In a practical application with causal filters, we would also have to allow for an appropriate time delay between input and output signals.

same way as $H_k(e^{j\omega}) = W(e^{j(\omega-\omega_k)})$ acted alone in Figure 7.33. The impulse responses of the bandpass filters are $h_k[n] = w[n]e^{j\omega_k n}$ and $g_k[n] = f[n]e^{j\omega_k n}$, and their combined effect is described by the impulse response $(w[n] * f[n])e^{j\omega_k n}$. Thus, when decimation/interpolation is introduced into the analysis/synthesis system, it behaves as if the effective analysis window is $w_e[n] = w[n] * f[n]$. Therefore, assuming that the filters and coefficients $P[k]$ are chosen to satisfy Eq. 7.94b so that the aliasing distortion path is eliminated in Figure 7.37, the overall analysis/synthesis system is described by the overall frequency response,

$$\tilde{H}(e^{j\omega}) = \frac{1}{RN} \sum_{k=0}^{N-1} P[k] W_e(e^{j(\omega-\omega_k)}), \tag{7.95a}$$

where $W_e(e^{j\omega}) = W(e^{j\omega})F(e^{j\omega})$, and by the corresponding overall impulse response,

$$\tilde{h}[n] = w_e[n]p[n], \tag{7.95b}$$

where $w_e[n] = w[n] * f[n]$ and

$$p[n] = \frac{1}{RN} \sum_{k=0}^{N-1} P[k] e^{j\omega_k n}. \tag{7.95c}$$

Because of the linearity of Eqs. (7.95a)–(7.95c), it is often convenient to include the gain factor $1/(RN)$ in the definition of the analysis and synthesis windows or in the definition of the coefficients $P[k]$.

To design a decimated analysis/synthesis filter bank system, we need to determine the following:

1. The number of channels N and the decimation/interpolation ratio R. These will usually be fixed by the desired frequency resolution.
2. The window $w[n]$ and the interpolation filter impulse response $f[n]$. These are lowpass filters. Typically they should have good frequency-selective properties such that the bandpass channel responses do not overlap into more than one band on either side of the channel.
3. The complex gain factors $P[k]$, $k = 0, 1, \ldots, N-1$. These constants are important for achieving flat overall response and alias cancellation.

Equations (7.94a) and (7.94b) provide a basis for designing the filters $w[n]$ and $f[n]$ in the decimated analysis/synthesis filter bank system. Added to these two constraints is a third constraint that the filters be "good" lowpass filters as defined by the intended application. Unfortunately, these constraints are not independent. References [345, 347] include discussions of some techniques for achieving flat response with combined filters $w[n] * f[n]$ that meet prescribed specifications. Satisfying Eq. (7.94b) simultaneously is somewhat more problematic since it involves shifted combinations of the bandpass filter responses. A typical approach is to assume that the overlap of frequency responses occurs only between adjacent channels. With that assumption, the constants $P[k]$ can be chosen so that the alias terms cancel in the overlap regions. We will see how this can occur in Section 7.8 where we show how Eqs. (7.94a) and (7.94b) are used to design two-channel filter banks.

7.7.2 Maximally Decimated Filter Banks

So far our discussion has assumed that R, the decimation ratio, and N, the number of frequencies, are different, and our earlier discussion of the OLA method suggests that the condition $R < L \leq N$ should hold. However, our discussion in Section 7.6 showed that we can have $N < L$. In this section, we will show that it is possible to obtain exact reconstruction with $R = N$ and $N < L$. The condition $R = N$ is termed *maximal decimation* because $R = N$ is the largest decimation factor that can be used with an N-channel analysis/synthesis filter bank and still achieve exact reconstruction. Note that with this condition, the number of samples/sec is conserved since the downsampling reduces the sampling rate of each of the N channels to F_s/N.[22]

To derive the maximally decimated filter bank, we set $R = N$ in Figure 7.35a for the analysis part, resulting in Figure 7.38a, and we set $R = N$ in Figure 7.35d to implement the synthesis part, resulting in Figure 7.38b. This leads to the equivalent forms in Figure 7.38c for analysis and Figure 7.38d for synthesis. These equivalences, which are easy to derive, result in both cases from the fact that $e^{\pm j(2\pi/N)rN} = 1$ for any integer r. We see that in addition to expanding the nominal bandwidth of the bandpass channel signal from $\pm\pi/N$ to $\pm\pi$, the downsampling by N accomplishes the frequency down-shifting normally accomplished by modulation with $e^{-j(2\pi/N)kn}$. Similarly, the

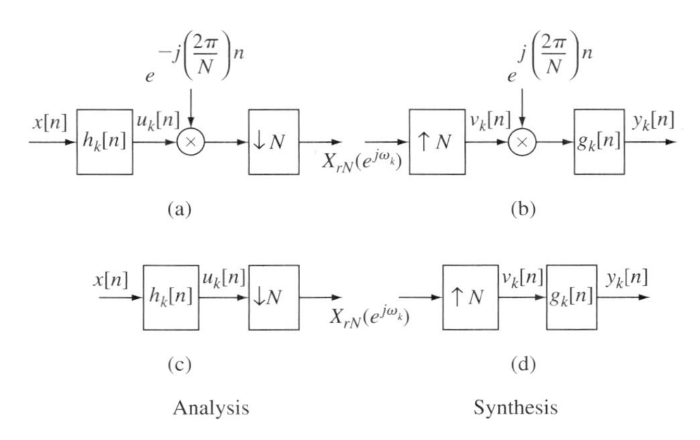

FIGURE 7.38
Maximally decimated analysis and synthesis: (a) analysis with down-sampling by N; (b) synthesis with interpolation by N; (c) equivalent analysis system to (a); (d) equivalent synthesis system to (b).

[22] Although the channel signals are complex for the filter banks we have been discussing, the complex conjugate symmetry between channels ω_k and $2\pi - \omega_k$ renders half the samples redundant.

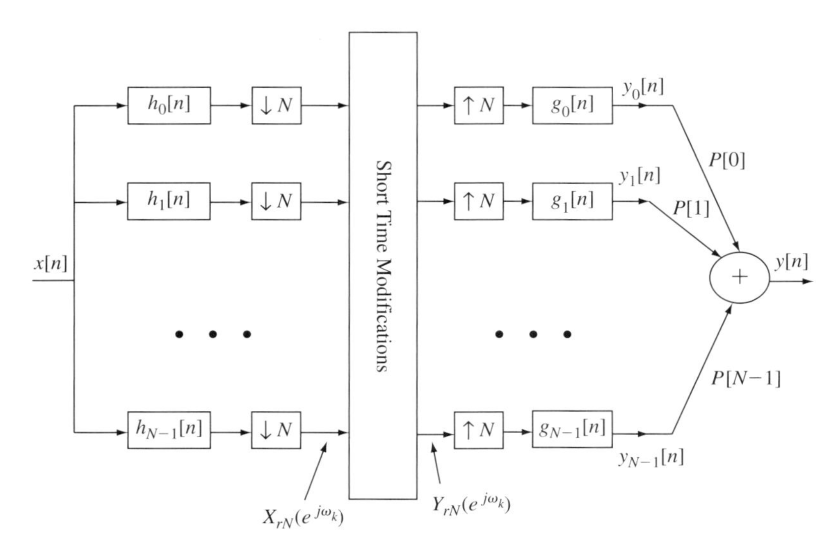

FIGURE 7.39
Maximally decimated analysis/synthesis filter bank showing modulators eliminated.

upsampling operation creates a frequency-compressed copy of the spectrum at each multiple of frequency $\omega_k = (2\pi/N)k$, and thereby effects the frequency shift that was done by modulation by $e^{j(2\pi/N)kn}$ in Figure 7.38b. Using the equivalent analysis and synthesis systems in Figure 7.38 in their appropriate places in Figure 7.36, we obtain the block diagram of Figure 7.39.

While eliminating the modulators in Figure 7.36 yields significant simplification in the maximally decimated analysis/synthesis filter bank, we have yet to demonstrate that this can lead to a useful analysis/synthesis system. Equations (7.94a) and (7.94b) must still be satisfied in order to achieve exact reconstruction at the output of the filter bank. In the next section, we will see examples of how this can be achieved.

7.8 TWO-CHANNEL FILTER BANKS

A two-channel ($N = 2$) maximally decimated analysis/synthesis filter bank system is shown in Figure 7.40. Applications of such systems in subband speech and audio coding will be discussed in Chapter 12. The parameters for this system are $R = N = 2$ and

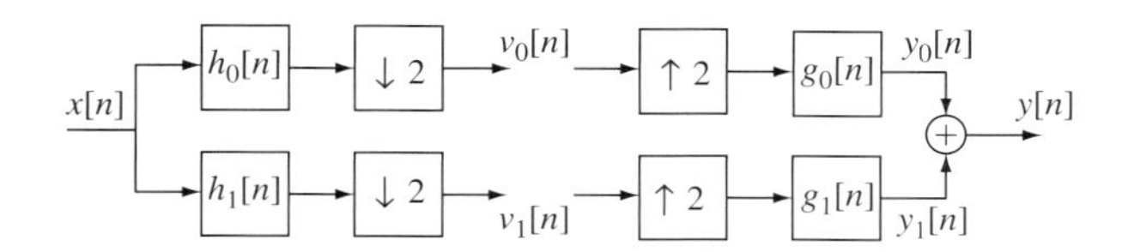

FIGURE 7.40
Two-band maximally decimated analysis/synthesis filter bank.

$\omega_k = \pi k, k = 0, 1$. Applying the frequency-domain analysis of Eq. (7.93b) to the system in Figure 7.40 leads to the following result:

$$Y(e^{j\omega}) = \frac{1}{2}\left[G_0(e^{j\omega})H_0(e^{j\omega}) + G_1(e^{j\omega})H_1(e^{j\omega})\right]X(e^{j\omega}) \tag{7.96a}$$

$$+\frac{1}{2}\left[G_0(e^{j\omega})H_0(e^{j(\omega-\pi)}) + G_1(e^{j\omega})H_1(e^{j(\omega-\pi)})\right]X(e^{j(\omega-\pi)}). \tag{7.96b}$$

Note that we have not included the customary multiplier $1/N$ on the output and we have not explicitly shown the gain factors $P[0]$ and $P[1]$ in Figure 7.40. It is convenient in this special case to assume that $P[0]$ is incorporated into the definition of $G_0(e^{j\omega})$ and $P[1]$ is included in $G_1(e^{j\omega})$. We have taken this into account in making the substitution into Eq. (7.93b).

Perfect or nearly perfect reconstruction can be achieved with this system using overlapping non-ideal filters for which aliasing will occur in the downsampling operations of the analysis filter bank. To see this, note that the second term in the expression for $Y(e^{j\omega})$ (line labeled Eq. (7.96b)), which represents potential aliasing distortion from the downsampling operation, can be eliminated by choosing the filters such that

$$G_0(e^{j\omega})H_0(e^{j(\omega-\pi)}) + G_1(e^{j\omega})H_1(e^{j(\omega-\pi)}) = 0. \tag{7.97a}$$

We called this the *alias cancellation condition* in our general discussion in Section 7.7. The other requirement for perfect reconstruction is

$$\frac{1}{2}\left[G_0(e^{j\omega})H_0(e^{j\omega}) + G_1(e^{j\omega})H_1(e^{j\omega})\right] = 1, \tag{7.97b}$$

which we called the *flat gain condition*. To complete the design of such a two-channel filter bank system, we need to choose the two sets of filters, and we need to specify the gain factors $P[0]$ and $P[1]$ that were assumed to be incorporated in $G_0(e^{j\omega})$ and $G_1(e^{j\omega})$ in Figure 7.40. Either FIR or IIR filters can (in principle) be used in the analysis/synthesis system of Figure 7.40 to provide nearly perfect reconstruction. In the quadrature mirror filter solution to be discussed next, all the filters are based on the analysis lowpass filter, so the design of such filters is based upon finding $H_0(e^{j\omega})$ that is an acceptable lowpass filter approximation while satisfying Eq. (7.97b) to within an acceptable approximation error.

7.8.1 Quadrature Mirror Filter Banks

One approach to the design of a two-band filter bank was proposed by Croisier et al. [74] and applied to subband coding by Galand and Esteban [100, 122]. Their work led to the solution discussed in this section.

The alias cancellation condition in Eq. (7.97a) is satisfied if the filters are chosen as follows:[23]

$$h_1[n] = e^{j\pi n}h_0[n] \Longleftrightarrow H_1(e^{j\omega}) = H_0(e^{j(\omega-\pi)}), \tag{7.98a}$$

$$g_0[n] = 2h_0[n] \Longleftrightarrow G_0(e^{j\omega}) = 2H_0(e^{j\omega}), \tag{7.98b}$$

$$g_1[n] = -2h_1[n] \Longleftrightarrow G_1(e^{j\omega}) = -2H_0(e^{j(\omega-\pi)}). \tag{7.98c}$$

Note that all the filters are based on a single lowpass filter with impulse response $h_0[n]$. This filter should be designed with a nominal cutoff frequency of $\pi/2$ radians and with a narrow transition to a stopband with adequate attenuation to isolate the two bands. The filter $h_1[n] = e^{j\pi n}h_0[n]$ is the complementary highpass filter that covers the remaining frequencies up to π radians. The filters $h_0[n]$ and $h_1[n]$ are termed *quadrature mirror filters (QMF)* since Eq. (7.98a) imposes mirror symmetry about $\omega = \pi/2$. Finally, note that $G_0(e^{j\omega})$ contains a factor of 2. This is equivalent to using a gain constant $P[0] = 2$ on the lowpass channel. Similarly the factor -2 in the definition of $G_1(e^{j\omega})$ is equivalent to using the gain factor $P[1] = -2$ on the highpass channel. The factors of 2 compensate for the factor $1/R = 1/2$ in Eq. (7.93b) that is due to downsampling by 2.

Substituting Eqs. (7.98a)–(7.98c) into Eq. (7.97a) gives

$$2H_0(e^{j\omega})H_0(e^{j(\omega-\pi)}) - 2H_0(e^{j(\omega-\pi)})H_0(e^{j(\omega-2\pi)}) = 0,$$

since $H_0(e^{j(\omega-2\pi)}) = H_0(e^{j\omega})$. This confirms that the alias term cancels for any choice of $H_0(e^{j\omega})$.

Now substituting Eqs. (7.98a)–(7.98c) into Eq. (7.96a) leads to the relation

$$Y(e^{j\omega}) = \left[(H_0(e^{j\omega}))^2 - (H_0(e^{j(\omega-\pi)}))^2\right]X(e^{j\omega}) = \tilde{H}(e^{j\omega})X(e^{j\omega}), \tag{7.99}$$

from which it follows that perfect reconstruction requires

$$\tilde{H}(e^{j\omega}) = [H_0(e^{j\omega})]^2 - [H_0(e^{j(\omega-\pi)})]^2 = e^{-j\omega M}, \tag{7.100}$$

which anticipates the need for delay through the analysis/synthesis system due to the delay of the causal filters that would comprise the analysis and synthesis filter banks.

Suppose that the basic lowpass filter is an FIR system of length L such that $h_0[n] = h_0[(L-1)-n]$ for $0 \le n \le (L-1)$. Such a filter has a generalized linear phase with a delay of $(L-1)/2$ samples, and its frequency response is of the form

$$H_0(e^{j\omega}) = A_0(e^{j\omega})e^{-j\omega(L-1)/2}, \tag{7.101}$$

[23] Recall that the notation $\Longleftrightarrow$ denotes DTFT correspondence.

where $A_0(e^{j\omega})$ is a real function of ω [270]. If Eq. (7.101) is substituted into Eq. (7.100), we obtain

$$\tilde{H}(e^{j\omega}) = \left[(A_0(e^{j\omega}))^2 - e^{j\pi(L-1)}(A_0(e^{j(\omega-\pi)}))^2\right] e^{-j\omega(L-1)} \tag{7.102}$$

as the reconstruction frequency response of the QMF system. Since $A_0(e^{j\omega})$ is a real even function of ω, it follows that $|\tilde{H}(e^{j\omega})| = 0$ at $\omega = \pi/2$ if $(L-1)$ is an even integer, thus making it impossible to satisfy Eq. (7.100). However, if $(L-1)$ is odd (impulse response length L is even), then

$$\tilde{H}(e^{j\omega}) = \left[(A_0(e^{j\omega}))^2 + (A_0(e^{j(\omega-\pi)}))^2\right] e^{-j\omega(L-1)}, \tag{7.103}$$

and the overall analysis/synthesis system will behave as a linear phase FIR system of order $2(L-1)$ with delay of $(L-1)$ samples. Although linear phase is guaranteed for this choice of filters, there remains the problem of ensuring that

$$\left[(A_0(e^{j\omega}))^2 + (A_0(e^{j(\omega-\pi)}))^2\right] = 1, \tag{7.104}$$

but since the left side of Eq. (7.104) is non-negative, a solution is possible. It can be shown [399] that the only computationally realizable filters satisfying Eq. (7.97b) exactly (even with delay) in general (and Eq. (7.104) specifically) are systems with impulse responses of the form $h_0[n] = c_0\delta[n - 2n_0] + c_1\delta[n - 2n_1 - 1]$, where n_0 and n_1 are arbitrarily chosen integers. Such systems cannot provide the sharp frequency selective properties needed in speech and audio coding applications, but to illustrate that such systems can achieve exact reconstruction, consider the simple two-point moving average lowpass filter

$$h_0[n] = \frac{1}{2}(\delta[n] + \delta[n-1]), \tag{7.105a}$$

which has frequency response

$$H_0(e^{j\omega}) = \cos(\omega/2)e^{-j\omega/2}. \tag{7.105b}$$

For this filter, $Y(e^{j\omega}) = e^{-j\omega}X(e^{j\omega})$, as can be verified by substituting Eq. (7.105b) into Eq. (7.99). Recall that the alias cancellation condition is implicitly satisfied by the relationships among the filters as specified by Eqs. (7.98a)–(7.98c).

Thus, the QMF framework with linear phase FIR filters can exactly cancel aliasing and provide overall linear phase (pure time delay), but for higher-order filters that provide the desired separation of the spectrum into two parts, flat magnitude response can only be approximated. Johnston [175] proposed an algorithm for design of the basic lowpass filter that minimizes the deviation of Eq. (7.104) from unity while providing a desired level of stopband attenuation. In speech subband coding, this is necessary to ensure that the quantization error is held within the band in which it was created. In addition to specifying an iterative design algorithm, Johnston [175] gave

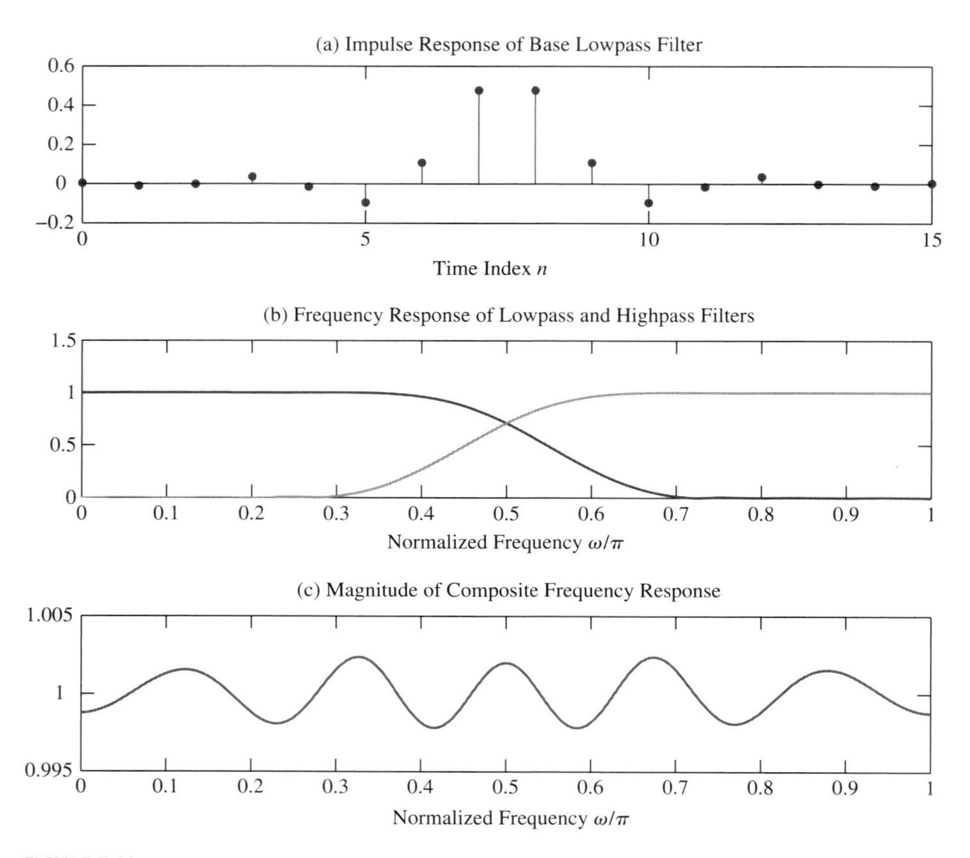

FIGURE 7.41
(a) 16-point impulse response of $h_0[n]$ for QMF system; (b) $|H_0(e^{j\omega})|$ and $|H_1(e^{j\omega})|$ as a function of ω/π; and (c) overall composite frequency response $H(e^{j\omega})$.

tables of impulse response coefficients for filters of lengths $((L-1)+1)$ ranging from 8 to 64. Higher-order filters, of course, give greater frequency selectivity, but recall that alias cancellation does not require high-order filters. It is guaranteed by the structure inherent in the QMF system.

An example of the filters and overall frequency response of a QMF filter bank system is shown in Figure 7.41 for Johnston's 16-point impulse response [175]. Figure 7.41a shows the impulse response ($L = 16$) of the lowpass filter $h_0[n]$, and Figure 7.41b shows the magnitudes of the frequency responses of the lowpass and highpass analysis filters. Note the mirror symmetry between the two filters about the frequency $\omega = \pi/2$. Figure 7.41c shows the magnitude of the frequency response of the overall composite analysis/synthesis system. The error in approximation to unit gain is considerably less than 0.5%. Due to the constraints imposed, the phase of the overall system is linear, corresponding to a delay of $L-1 = 15$ samples.

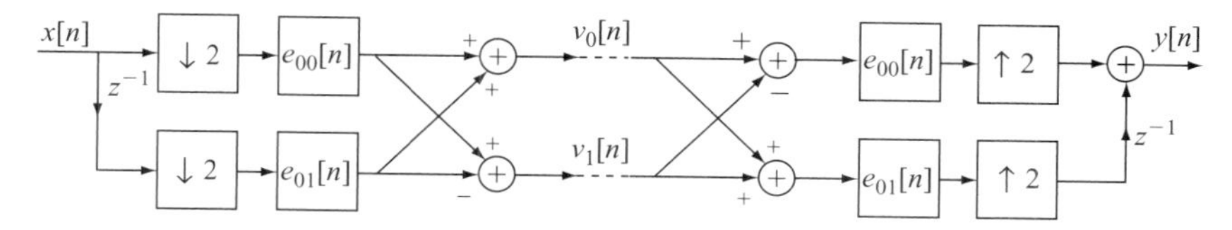

FIGURE 7.42
Polyphase representation of the two-channel analysis and synthesis filter bank of Figure 7.40 .

7.8.2 Polyphase Structure for QMF Filter Banks

Polyphase techniques [270] are multirate signal processing techniques that combine decimation, interpolation, and filtering to obtain efficient implementations of digital signal processing systems. When polyphase techniques are applied to the system in Figure 7.40, the result is as shown in Figure 7.42. The polyphase filters in Figure 7.42 are derived from the QMF filters by the following relations:

$$e_{00}[n] = h_0[2n] \qquad 0 \leq n \leq L/2 - 1, \tag{7.106a}$$

$$e_{01}[n] = h_0[2n+1] \qquad 0 \leq n \leq L/2 - 1. \tag{7.106b}$$

Figures 7.40 and 7.42 are equivalent in the sense that an input $x[n]$ to both systems produces the same output $y[n]$.

The significance of the polyphase form is that it requires only a quarter of the computation required to implement the system in Figure 7.40. If $h_0[n]$ has length L samples, then Figure 7.40, where the filters all operate at the sampling rate of the input, requires $4LF_s$ multiplications/sec to generate the output samples. However, since the impulse responses in Figure 7.42 have length $L/2$, and since the filters operate on the downsampled input, only $4(L/2)F_s/2 = LF_s$ multiplications/sec are required to compute the same output sequence.

Polyphase techniques can be applied to advantage in filter banks with more than two channels. A detailed discussion of polyphase methods can be found in References [71, 399, 402].

7.8.3 Conjugate Quadrature Filters

Quadrature mirror filters can completely eliminate aliasing distortion and can provide exactly linear phase, but for filters longer than $L = 2$, they can only approximate perfectly flat overall magnitude response. Smith and Barnwell [367, 368] and Mintzer [244] showed that exact reconstruction is possible with computationally realizable FIR filters of higher order. The key is to define the highpass analysis filter as

$$h_1[n] = (-1)^n h_0[(L-1) - n] \Leftrightarrow H_1(e^{j\omega}) = -H_0(e^{-j(\omega-\pi)})e^{-j\omega(L-1)}, \tag{7.107a}$$

where L is the length of the impulse response $h_0[n]$ and $\Leftrightarrow$ denotes correspondence of a sequence and its discrete-time Fourier transform. As in the case of QMF filters, $(L-1)$ must be an odd integer. In order to satisfy the alias cancellation condition of

Eq. (7.97a), the corresponding synthesis filters must be

$$g_0[n] = 2h_0[(L-1)-n] \Leftrightarrow G_0(e^{j\omega}) = 2H_1(e^{j(\omega-\pi)}), \tag{7.107b}$$

$$g_1[n] = 2(-1)^{n+1}h_0[n] \Leftrightarrow G_1(e^{j\omega}) = -2H_0(e^{j(\omega-\pi)}), \tag{7.107c}$$

where again, we have incorporated the gain factors $P[0] = -P[1] = 2$ into the definitions of the synthesis filters. Smith and Barnwell termed these filters *conjugate quadrature filters*, but they are now more commonly known as *Smith–Barnwell filters* [368]. If Eqs. (7.107a), (7.107b), and (7.107c) are substituted into the expression for the overall frequency response in Eq. (7.97b), the result is

$$\tilde{H}(e^{j\omega}) = [C_0(e^{j\omega}) + C_0(e^{j(\omega-\pi)})]e^{-j\omega(L-1)}, \tag{7.108a}$$

where

$$c_0[n] = h_0[n] * h_0[-n] \Leftrightarrow C_0(e^{j\omega}) = H_0(e^{j\omega})H_0(e^{-j\omega}). \tag{7.108b}$$

The sequence $c_0[n]$ (called a *product filter* [367, 368])is clearly the deterministic autocorrelation of the lowpass analysis filter $h_0[n]$. Therefore the corresponding DTFT, $C_0(e^{j\omega})$, is real and non-negative. The key to demonstrating that exact reconstruction can be achieved with a causal L-point impulse response is to note that it is possible to use the Parks-McClellan filter design algorithm [270] to obtain a filter with impulse response $c_0[n]$ of length $(2L-1)$ samples and real frequency response $C_0(e^{j\omega}) \geq 0$ such that $C_0(e^{j\omega}) + C_0(e^{j(\omega-\pi)}) = 1$. Then all that remains is to factor $C_0(z) = H_0(z)H_0(z^{-1})$ to obtain $2(L-1)$ zeros. These zeros are in groups of four complex conjugate reciprocals or double zeros on the unit circle or reciprocal real zeros [270]. $(L-1)$ of these zeros are selected to form $H_0(z)$, taking care to select complex conjugate pairs (or single real zeros) to ensure that $h_0[n]$ is real. The basic lowpass filter $h_0[n]$ is obtained by multiplying out the factors of $H_0(z)$. The solution for $h_0[n]$ is not unique. One approach to forming the $H_0(z)$ polynomial is to use only zeros inside or on the unit circle, thus obtaining a minimum-phase solution for $H_0(z)$. Another recommended approach is to select passband zeros alternately from inside and outside the unit circle and one complex pair from each double zero pair on the unit circle. This approach gives approximately linear phase [367, 368].

A length $L = 16$ example (due to Smith and Barnwell [368]) is shown in Figure 7.43. Note that in contrast to the QMF example in Figure 7.41, in this case the overall composite frequency response has absolutely flat amplitude.

We have omitted the details of design of the CQF filters, but they are readily available in Refs. [244, 367, 368]. The important point is that the CQF filters give a constructive proof that perfect reconstruction in a two-channel filter bank is possible with causal FIR filters. This will be important in the next section, where filter banks with more than two channels are considered.

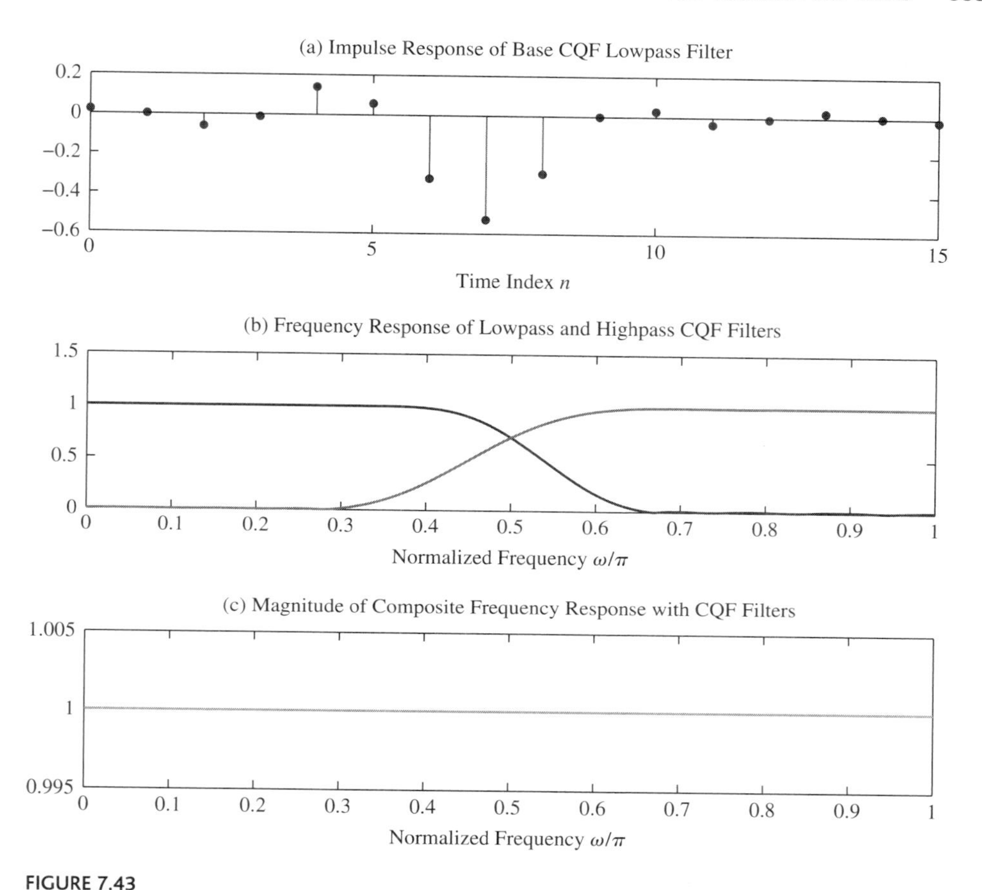

FIGURE 7.43
(a) 16-point impulse response of $h_0[n]$ for a CQF system; (b) $H_0(e^{j\omega})$ and $H_1(e^{j\omega})$ as a function of ω/π; and (c) overall composite frequency response $H(e^{j\omega})$.

7.8.4 Tree-Structured Filter Banks

The QMF and CQF two-channel analysis/synthesis systems divide the speech spectrum into two overlapping frequency bands. The individual channels overlap slightly at the band edges, but this overlap cancels exactly in the synthesized output. The individual channel filters can be made as sharp as desired, and the stopband attenuation can be made as high as desired by increasing the order of the lowpass and corresponding highpass FIR filters. Furthermore, the basic two-channel decomposition can be iterated to create a more refined analysis. For example, the outputs of the lowpass and highpass channels can be subjected to further splitting by additional two-channel filter banks. This leads to what is known as a *tree-structured filter bank*. Although based on the principle of successive splitting of bands into two equal parts, the method has remarkable flexibility for implementing non-uniform spectrum decompositions.

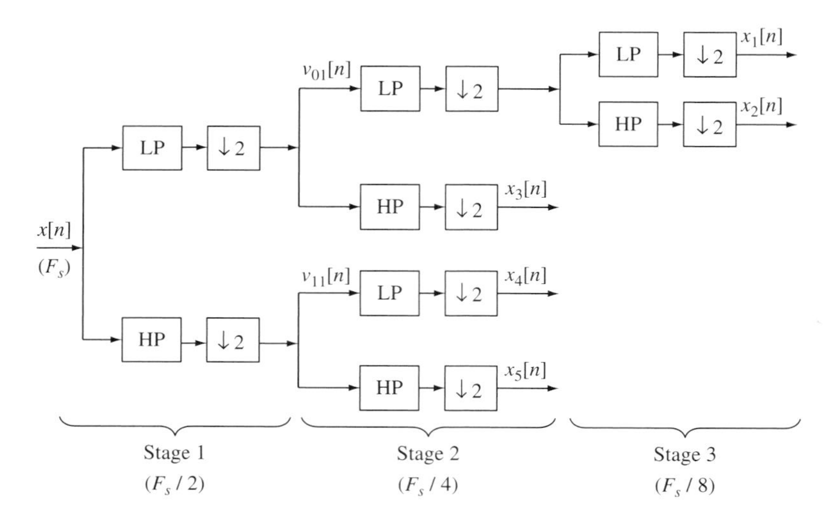

FIGURE 7.44
Tree-structured filter bank.

A simple example is shown in Figure 7.44 which shows three stages of decomposition.[24] Consider a signal $x[n]$ that results from sampling with a sampling rate F_s. The first stage splits the band from 0 to $F_s/2$ into two equal bands each of nominal width $F_s/4$, which after decimation by 2 are effectively sampled at a rate of $F_s/2$. In stage two, each band is further divided into a lowpass and highpass band and after decimation by 2, we obtain four channels each sampled at an effective rate of $F_s/4$. Finally, the lowpass part of the first stage lowpass output is further split into two signals $x_1[n]$ and $x_2[n]$, each of which is effectively sampled at a rate of $F_s/8$. Thus, $x_1[n]$ is the lowpass part of the lowpass part of the lowpass part of the spectrum of the input $x[n]$. Similarly, the signal $x_3[n]$ is the highpass part of the lowpass part of the spectrum of $x[n]$. Thus, by choosing which parts of the divided spectrum to further divide, we can create a non-uniform division of the spectrum. In the case of the example of Figure 7.44, the spectrum is split into five bands. The lowest two bands have nominal width $F_s/16$ and the other three bands have nominal width $F_s/8$. The sum of the bandwidths is equal to the total baseband of $F_s/2$.

It is helpful to visualize the tree-structured analysis pictured in Figure 7.44 in terms of a spectrographic type of diagram as in Figure 7.45, where the frequency axis shows the division into separate bands and the time axis is divided according to the sampling rate of the channel signal for each band. Thus, the two low frequency bands of width $F_s/16$ are sampled at a rate of $F_s/8$ samples/sec so the time division within those two bands is $8/F_s = 8T$, where T is the sampling period associated with $x[n]$.

[24]The two-channel analysis at each level of the tree can also be implemented using the polyphase form of Figure 7.42 for increased efficiency.

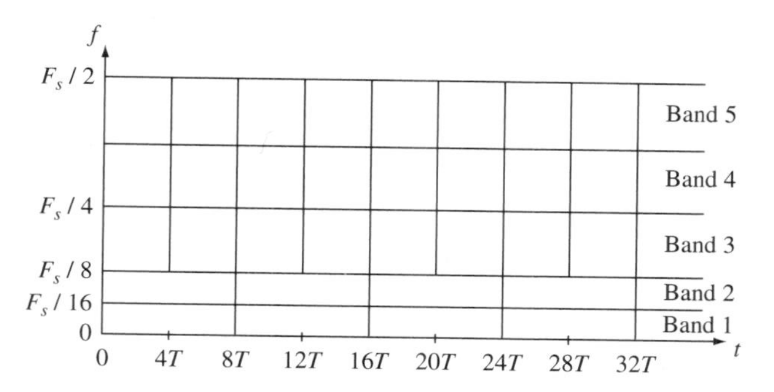

FIGURE 7.45
Sampling pattern for tree-structured filter bank of Figure 7.44.

Similarly, the upper three bands are each of width $F_s/8$, so they are sampled at a rate of $F_s/4$ and the samples for those channels are spaced at time intervals of $4T$. Thus, the time-frequency plane is "tiled" by a regular pattern of patches each of whose length in time is inversely proportional to its width in frequency.

Such non-uniform decompositions of the frequency spectrum are essentially the same as discrete wavelet transforms [49]. Indeed, the vector of outputs in Figure 7.44,

$$\{x_1[n], x_2[n], x_3[n], x_4[n], x_5[n]\},$$

comprises a particular case of the discrete dyadic wavelet transform of $x[n]$. That this is an exactly invertible transformation follows from the fact that perfect reconstruction can be achieved using CQF filters. The synthesis system corresponding to Figure 7.44 would have a mirrored structure as shown in Figure 7.46. For example, signals $x_1[n]$

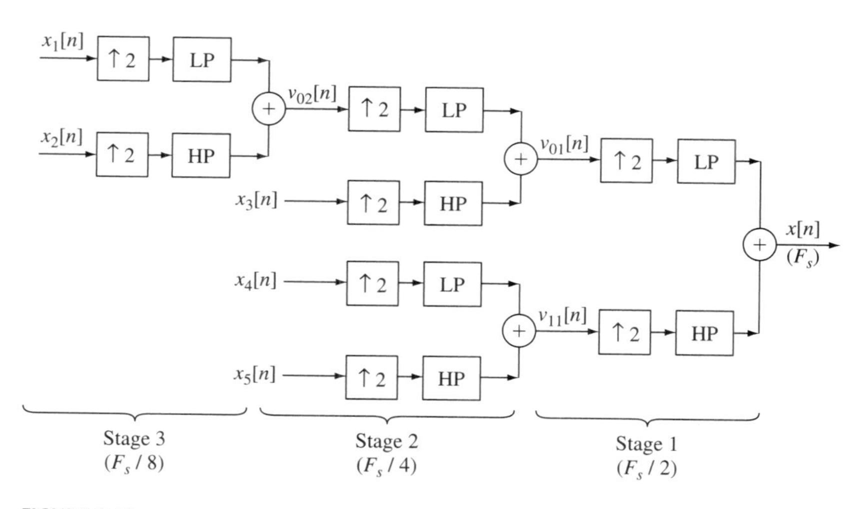

FIGURE 7.46
Tree-structured synthesis corresponding to tree-structured filter bank of Figure 7.44.

and $x_2[n]$ would be upsampled, then filtered by appropriate lowpass and highpass filters, and finally added together to reconstruct the signal $v_{02}[n]$ (exactly if CQFs are used). Then $v_{02}[n]$ and $x_3[n]$ can be used to recover $v_{01}[n]$. Similarly, $x_4[n]$ and $x_5[n]$ can be used to reconstruct $v_{11}[n]$, which, together with the reconstructed $v_{01}[n]$, can be used to reconstruct $x[n]$. It should be noted that the signals $x_1[n]$ and $x_2[n]$ have one more stage of delay than the other three signals. It would be necessary to add this delay to the other three signals before beginning the reconstruction. If CQF filters are used, the reconstruction will be exact since each of the individual stages provides exact reconstruction.

In summary, it is possible to design perfect reconstruction, uniformly or non-uniformly spaced, filter banks using the CQF filters as well as filter banks that provide almost perfect reconstruction using QMF filters. The individual filters have passbands, stopbands, and transition regions that depend on the order $(L-1)$ of the FIR filters. Long filters can provide sharp cutoff and high stopband attenuation at the cost of increased delay through the entire system.

7.9 IMPLEMENTATION OF THE FBS METHOD USING THE FFT

In the preceding section we have shown that it is possible, using causal filters, to design a filter bank whose composite output is identical to the input except for a delay and scale factor. In particular, we saw that FIR filters are particularly well suited to this purpose. Since time-dependent Fourier analysis and synthesis is equivalent to such a filter bank, it is also true that finite duration analysis windows can be used effectively in the design of analysis/synthesis systems. One of the major disadvantages of FIR systems is the extensive computation required to implement them. Fortunately in the particular context of time-dependent Fourier analysis, there are several methods for reducing the computation over that required for straightforward implementation.

7.9.1 FFT Analysis Techniques

Consider a time-dependent Fourier analysis/synthesis system with equally spaced analysis frequencies $\omega_k = 2\pi k/N$, $0 \leq k \leq N-1$. In Section 7.3.6 it was shown that the sequences $X_n(e^{j\omega_k})$ need not be computed at the same rate as the input sampling rate because each sequence $X_n(e^{j\omega_k})$ is effectively the output of a lowpass filter with normalized cutoff frequency π/N. Thus the output can be computed only once for each N consecutive samples of the input. FIR systems are especially suited to this application because it is possible to compute only the desired output samples without computing the intervening $N-1$ samples. With IIR systems, the inherent recursive nature of the implementation requires that all values of the output be computed.

An additional improvement in computational efficiency can be obtained by the use of FFT techniques [346]. To see this, we express the time-dependent Fourier transform as

$$X_n(e^{j\frac{2\pi}{N}k}) = \sum_{m=-\infty}^{\infty} x[m]w[n-m]e^{-j\frac{2\pi}{N}km}, \quad 0 \leq k \leq N-1. \tag{7.109}$$

If $w[m]$ is of finite duration, Eq. (7.109) can be manipulated into the form of a DFT, and therefore an FFT algorithm can be used to compute $X_n(e^{j2\pi k/N})$ for $0 \leq k \leq N-1$. By a substitution of variable of summation, Eq. (7.109) becomes

$$X_n(e^{j\frac{2\pi}{N}k}) = e^{-j\frac{2\pi}{N}kn} \sum_{m=-\infty}^{\infty} x_n[m]e^{-j\frac{2\pi}{N}km}, \tag{7.110}$$

where $k = 0, 1, \ldots, N-1$ and

$$x_n[m] = x[n+m]w[-m] \tag{7.111}$$

is the windowed segment of the signal at time n. The sequence $x_n[m]$ will be non-zero only within the region of support of $w[-m]$. That is, the sequence $x_n[m]$ is obtained by redefining the origin of the sequence $x[m]w[n-m]$ to be at sample n, thus focusing our attention on the sequence in the neighborhood of the time at which $X_n(e^{j2\pi k/N})$ is to be computed. Next, by a substitution $m = q + Nr$, with $0 \leq q \leq N-1$ and $-\infty < r < \infty$, we can express Eq. (7.110) as the double sum

$$X_n(e^{j\frac{2\pi}{N}k}) = e^{-j\frac{2\pi}{N}kn} \sum_{r=-\infty}^{\infty} \left[\sum_{q=0}^{N-1} x_n[q+Nr] \right] e^{-j\frac{2\pi}{N}k(Nr+q)}. \tag{7.112}$$

Since $e^{-j2\pi kr} = 1$, we can interchange the order of summations and obtain

$$X_n(e^{j\frac{2\pi}{N}k}) = e^{-j\frac{2\pi}{N}kn} \sum_{q=0}^{N-1} \left[\sum_{r=-\infty}^{\infty} x_n[q+Nr] \right] e^{-j\frac{2\pi}{N}kq}. \tag{7.113}$$

Now if $u_n[q]$ is defined as the finite-length sequence,

$$u_n[q] = \sum_{r=-\infty}^{\infty} x_n[q+Nr], \quad 0 \leq q \leq N-1, \tag{7.114}$$

it follows that

$$X_n(e^{j\frac{2\pi}{N}k}) = e^{-j\frac{2\pi}{N}kn} \sum_{q=0}^{N-1} u_n[q]e^{-j\frac{2\pi}{N}kq}, \quad k = 0, 1, \ldots, N-1. \tag{7.115}$$

When $L < N$, only the term for $r = 0$ would be non-zero in Eq. (7.114). However, when $L > N$, as is often the case, this corresponds to time aliasing the windowed sequence $x_n[m]$, and if $N < L < \infty$, a finite number of terms would be non-zero. We observe from Eq. (7.115) that $X_n(e^{j2\pi k/N})$ is simply $e^{-j\frac{2\pi}{N}kn}$ times the N-point DFT of the sequence $u_n[q]$. Alternatively, $X_n(e^{j2\pi k/N})$ is the N-point DFT of the sequence $u_n[q]$

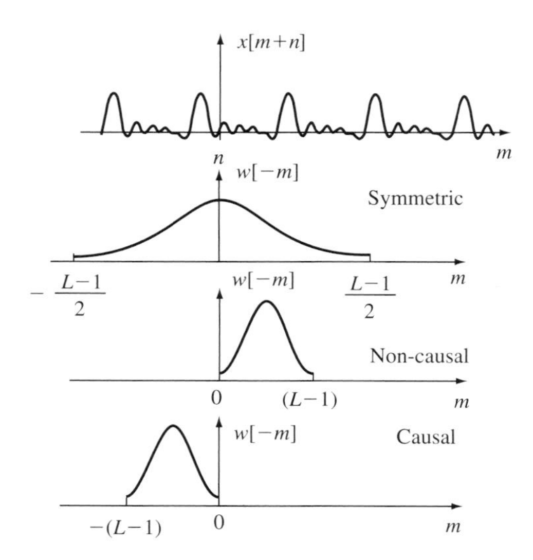

FIGURE 7.47
Plots of the sequences $x[m+n]$ and $w[-m]$.

after a circular shift of n modulo N [270]. That is,

$$X_n(e^{j\frac{2\pi}{N}k}) = \sum_{m=0}^{N-1} u_n[((m-n))_N]e^{-j\frac{2\pi}{N}km}, \tag{7.116}$$

where the notation $((\ ))_N$ means that the integer inside the double set of parentheses is to be interpreted modulo N. Thus, we have succeeded in manipulating $X_n(e^{j2\pi k/N})$ into the form of an N-point DFT of a finite-length sequence that is derived from the windowed input sequence. In summary, the procedure for computing $X_n(e^{j2\pi k/N})$ for $0 \leq k \leq N-1$ is:

1. For a particular time n, form the sequence $x_n[m]$ as in Eq. (7.111) by multiplying $x[m+n]$ by the reversed window sequence $w[-m]$. Figure 7.47 shows $x[m+n]$ and three special cases of $w[-m]$.
2. Break the resulting sequence $x_n[m]$ up into segments of length N samples and add these segments together according to Eq. (7.114) to produce the finite-length sequence $u_n[q]$, $0 \leq q \leq N-1$.
3. Circularly shift $u_n[q]$ by n modulo N to produce $u_n[((m-n))_N]$, $0 \leq m \leq N-1$.
4. Compute the N-point DFT of $u_n[((m-n))_N]$ to produce $X_n(e^{j2\pi k/N})$, $0 \leq k \leq N-1$.

This procedure must be repeated at each value of n at which $X_n(e^{j2\pi k/N})$ is desired; however, as we have already seen in Section 7.7, n can be incremented by more than one sample, so that $X_n(e^{j2\pi k/N})$ would normally be computed for $n = 0, \pm R, \pm 2R, \ldots$;

i.e., at intervals of R samples of the input signal. Recall that this is justified since $X_n(e^{j2\pi k/N})$ is the output of a lowpass filter whose nominal cutoff frequency is π/N radians. Thus, as long as $R \leq N$, the "samples" of $X_n(e^{j2\pi k/N})$ will suffice to reconstruct the input signal.

Note that this method gives $X_n(e^{j2\pi k/N})$ for all values of k. In general, because of the conjugate symmetry of $X_n(e^{j2\pi k/N})$, at most only about half of the channels need be computed. Sometimes the very low frequency and high frequency channels are not implemented. Thus, the question arises as to whether the FFT method is more efficient than direct implementation. To compare, let us assume that we only require $X_n(e^{j2\pi k/N})$ for $1 \leq k \leq M$. Further assume that the window is of length L. Then to obtain the complete set of values of $X_n(e^{j2\pi k/N})$ would require $4LM$ real multiplications and about $2LM$ real additions using the method of Figure 7.33. Assuming a rather straightforward complex FFT algorithm where N is a power of two,[25] it can be shown that the FFT method would require $L + 2N\log_2 N$ real multiplications and $L + 2N\log_2 N$ real additions to obtain all N values of $X_n(e^{j2\pi k/N})$. If we take the number of real multiplications as the basis for comparison, then it follows that the FFT method requires less computation unless

$$M \leq \frac{N\log_2 N}{2L}. \tag{7.117}$$

For example, if $N = 128 = 2^7$, we see that the FFT method is more efficient than the direct method unless $M \leq 3.5$; i.e., for fewer than four channels. Thus, in any application where fine frequency resolution is required, it is almost certain that the FFT method would be most efficient. (Note that if $L > N$, the comparison is even more favorable to the FFT method.)

7.9.2 FFT Synthesis Techniques

The previous discussion of analysis techniques showed that by using an FFT algorithm, we can compute all N equally spaced values of $X_n(e^{j\frac{2\pi}{N}k})$ with an amount of computation that is less than that required to compute M channels using a direct implementation. By rearranging the computations required for synthesis, a similar savings can occur in reconstructing $x[n]$ from values of $X_n(e^{j\frac{2\pi}{N}k})$ that are available at every R samples of $x[n]$, where $R \leq N$ [290].

From Eq. (7.76) with $\omega_k = 2\pi k/N$, the output of the synthesis system is

$$y[n] = \frac{1}{N}\sum_{k=0}^{N-1} Y_n[k]e^{j\frac{2\pi}{N}kn}, \tag{7.118}$$

[25] This does not take advantage of the fact that $u_n((m-n))_N$ is real. This could be used to reduce computation by another factor of 2.

where

$$Y_n[k] = P[k]X_n(e^{j\frac{2\pi}{N}k}), \quad 0 \le k \le N-1. \tag{7.119}$$

Recall that the complex weighting coefficients, $P[k]$, permit adjustments of the magnitude and phase of the channels. If $X_n(e^{j\frac{2\pi}{N}k})$ is available only at integer multiples of R, then the intermediate values must be filled in by interpolation as discussed in Section 7.7. To do this, it is helpful to define the sequence

$$V_n[k] = \begin{cases} P[k]X_n(e^{j\frac{2\pi}{N}k}) & n = 0, \pm R, \pm 2R, \ldots \\ 0 & \text{otherwise.} \end{cases} \tag{7.120}$$

There is a sequence of the above form for each value of k. Now, for each value of k, the intermediate values are filled in by processing the sequence $V_n[k]$ with a lowpass interpolation filter with a nominal cutoff frequency of π/N radians. If we denote the unit sample response of this filter by $f[n]$, and assume that it is symmetrical, with total length $2RQ-1$, then for each value of k, for $0 \le k \le N-1$,

$$Y_n[k] = \sum_{m=n-RQ+1}^{n+RQ-1} f[n-m]V_m[k], \quad -\infty < n < \infty. \tag{7.121}$$

Equation (7.121) together with Eq. (7.118) describes the operations required to compute the output of the synthesis stage when the time-dependent Fourier transform is available at intervals of R samples. This process is illustrated in Figure 7.48, which depicts $V_m[k]$ as a function of m and k. Also, we use n to denote the time index at which we wish to synthesize the output $y[n]$. Remember that m is the time index and k is the frequency index. The solid dots denote points at which $V_m[k]$ is non-zero; i.e., the points at which $X_m(e^{j\frac{2\pi}{N}k})$ is known. The open circles denote points at which $V_m[k]$ is zero, and at which we desire to interpolate the values of $Y_n[k]$. The impulse response of the interpolating filter (for $Q=2$) is shown positioned at a particular time n. Each channel signal is interpolated by convolution with the impulse response of the interpolation filter. As an example, the samples involved in the computation of $Y_3[1]$ are enclosed in a box. In general, the box indicating which samples are involved in computing $Y_n[k]$ will slide along the kth row of Figure 7.48, with the center of the box being positioned at n. Notice that each interpolated value is dependent upon $2Q$ of the known values of $X_n(e^{j\frac{2\pi}{N}k})$. If we assume that M channels are available for synthesis, then it is easily shown that $2(Q+1)M$ real multiplications and $2QM$ real additions are required to compute each value of the output sequence, $y[n]$.

The interpolated values of $Y_n[k]$ could be used in Eq. (7.118) to compute each sample of the output. However, the FFT can also be used to increase the efficiency of the synthesis operation. To see how the output can be computed more efficiently, let

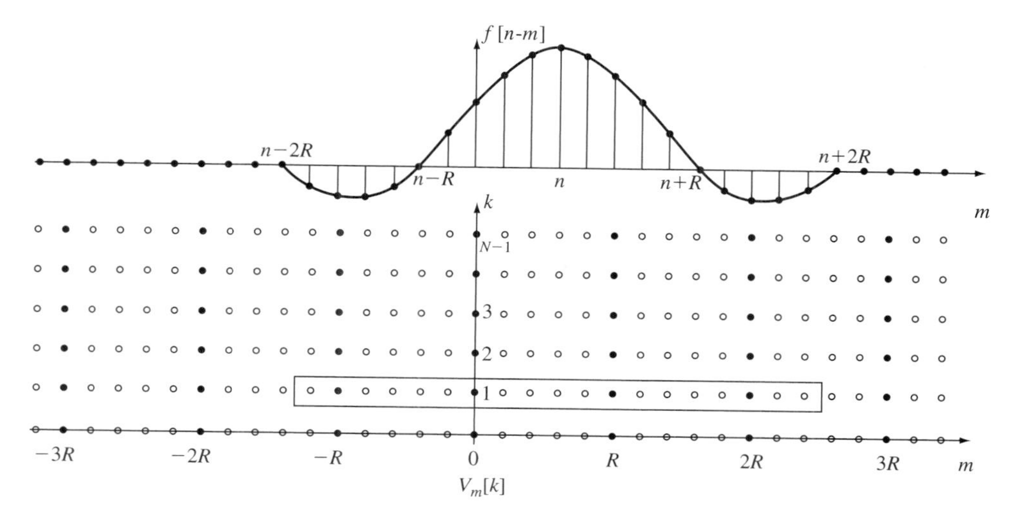

FIGURE 7.48
Samples involved in computing $Y_n[k]$.

us substitute Eq. (7.121) into Eq. (7.118), obtaining

$$y[n] = \frac{1}{N}\sum_{k=0}^{N-1} \sum_{m=n-RQ+1}^{n+RQ-1} f[n-m]V_m[k]e^{j\frac{2\pi}{N}kn}. \tag{7.122}$$

Interchanging the order of summation gives

$$y[n] = \sum_{m=n-RQ+1}^{n+RQ-1} f[n-m]v_m[n], \tag{7.123}$$

where

$$v_m[r] = \frac{1}{N}\sum_{k=0}^{N-1} V_m[k]e^{j\frac{2\pi}{N}kr} \qquad 0 \le r \le N-1. \tag{7.124}$$

Using Eq. (7.120), it can be seen that

$$v_m[r] = \begin{cases} \dfrac{1}{N}\displaystyle\sum_{k=0}^{N-1} P[k]X_m(e^{j\frac{2\pi}{N}k})e^{j\frac{2\pi}{N}kr}, & m = 0, \pm R, \pm 2R, \ldots \\ 0 & \text{otherwise.} \end{cases} \tag{7.125}$$

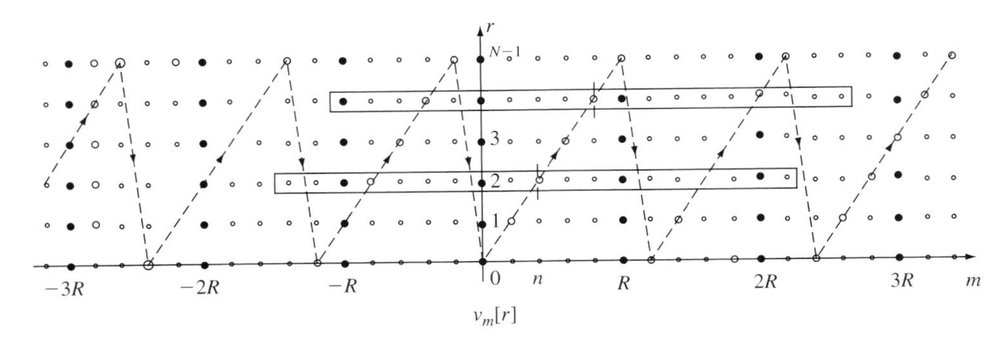

FIGURE 7.49
The interpolation process for $v_m[r]$. (After Portnoff [290]. © [1976] IEEE.)

Thus we see that rather than interpolate the time-dependent Fourier transform, and then evaluate Eq. (7.118), we can instead compute $v_m[r]$ at each time at which $X_n(e^{j\frac{2\pi}{N}k})$ is known (i.e., $m = 0, \pm R, \ldots$), and then interpolate $v_m[r]$ as in Eq. (7.123).

It can be seen that $v_m[r]$ is in the form of an inverse DFT, and thus $v_m[r]$ is periodic in r with period N. Thus, in Eq. (7.123), the index n in the term $v_m[n]$ must be interpreted modulo N. The interpolation process implied by this is depicted in Figure 7.49. The heavy dots in the two-dimensional net of points represent the points at which $v_m[r]$ is non-zero. The remaining points in Figure 7.49 can be interpolated in the same manner described for interpolation of $Y_n[k]$. From Eq. (7.123) and the periodic nature of $v_m[r]$, it can be seen that $y[n]$ is equal to the values of the interpolated sequence along the "saw tooth" pattern in Figure 7.49.

In implementing the synthesis in this manner, the N-point sequences $v_m[r]$, $0 \le r \le N-1$, can be computed for each value of m at which $X_m(e^{j\frac{2\pi}{N}k})$ is known by using an FFT algorithm to perform the inverse DFT computation of Eq. (7.125). Note that a channel can be omitted simply by setting its value equal to zero before computing the inverse DFT. Likewise, if it is desirable to implement a linear phase shift by choosing $P[k] = e^{-j\frac{2\pi}{N}kn_0}$, it is easily shown that the effect is to simply circularly shift the sequences $v_m[r]$. Thus, multiplications by the factor $e^{-j\frac{2\pi}{N}kn_0}$ can be avoided by performing the inverse DFT operation directly upon $X_m(e^{j\frac{2\pi}{N}k})$ and then circularly shifting the result by the desired n_0 samples. After the sequences $v_m[r]$ are obtained, the output can be computed by interpolating $v_m[r]$ as in Eq. (7.123). For each value of $y[n]$, $2Q$ values of $v_m[r]$ are involved. The samples involved for two different values of n are shown enclosed in boxes in Figure 7.49. It should be clear that for R consecutive values of $y[n]$, the values of $v_m[r]$ are obtained from the same $2Q$ columns. Thus it is convenient to compute the output in blocks of R samples.

The amount of computation required to implement a time-dependent Fourier synthesis in the above manner can again be estimated by assuming that N is a power of 2 and that a straightforward complex FFT algorithm is used to compute the inverse transforms called for in Eq. (7.125). For this assumption, the synthesis requires $(2QR + 2N\log_2 N)$ real multiplications and $(2QR - 1 + 2N\log_2 N)$ real additions to compute a group of R consecutive values of the output, $y[n]$. The direct method of synthesis requires $(2Q+1)MR$ real multiplications and $2QMR$ real additions to compute R consecutive samples of the output. If we consider the situation when the direct method requires fewer multiplications than the FFT method, we find that

$$M < \frac{Q + \frac{N}{R}\log_2 N}{Q+1}. \tag{7.126}$$

For typical values of $N = 128$, $Q = 2$ (interpolation over four samples as in Figure 7.49), and $R = N$ (the lowest possible sampling rate for $X_n(e^{j\frac{2\pi}{N}k})$), then we see that the direct method is most efficient only when $M < 3$. Thus for most applications, the FFT offers significant improvements in the computational efficiency of the synthesis operation.

7.10 OLA REVISITED

We have just shown how an FFT algorithm can be used to implement the analysis and synthesis filter banks of the FBS method of STFA/STFS. The technique that we have described is applicable for any choice of N, L, and R as long as $R \leq N$, including the case of maximal decimation, $R = N$. No matter how the parameters are chosen, the window $w[n]$ and the interpolation filter $f[n]$ must be chosen so that their DTFTs satisfy Eqs. (7.94a) and (7.94b) in order to achieve nearly perfect reconstruction.

Although our discussion has been framed in terms of more efficient implementation of filter bank analysis and synthesis, it turns out that the methods proposed also describe the OLA method of analysis and synthesis. To see this, recall that in the OLA method, $N \geq L$; i.e., N is chosen so as to *avoid* time aliasing. This means that in the analysis stage, $u_n[q] = x_n[q]$, $0 \leq q \leq N-1$; i.e., no overlapping of segments of $x_n[q]$ as in Eq. (7.114) is required. Also, recall that in the OLA method, $w[n]$ and R are chosen so that the condition

$$\sum_{r=-\infty}^{\infty} w[rR-n] = \frac{W(e^{j0})}{R}, \qquad \text{for all } n, \tag{7.127}$$

is satisfied. We showed that for a Hamming window, this condition holds if $R = N/4$. Therefore, $X_n(e^{j\frac{2\pi}{N}k})$ for $n = 0, \pm R, \pm 2R, \ldots$, is the DFT of $x_{rR}[((m - rN/4))_N]$; that is before computing the DFT, the windowed sequence $x_{rR}[m]$ is circularly shifted by $rN/4$.

This is illustrated in Figure 7.50, where we assume that the window is non-causal such that $w[-m]$ is non-zero in the interval $0 \leq m \leq L-1$ and $N = L$. (Note that, for convenience in constructing this figure, $N = L = 8$). The window $w[-m]$ is represented at the top of the figure as being comprised of four sections each of length $N/4$ and labeled a, b, c, and d. Below the window is shown the input signal in blocks of $N/4$ samples indexed starting with $r = 0$. The vertical lines of solid dots represent the circularly shifted segments $x_{rR}[((m - rN/4))_N]$. Note that the solid dots are

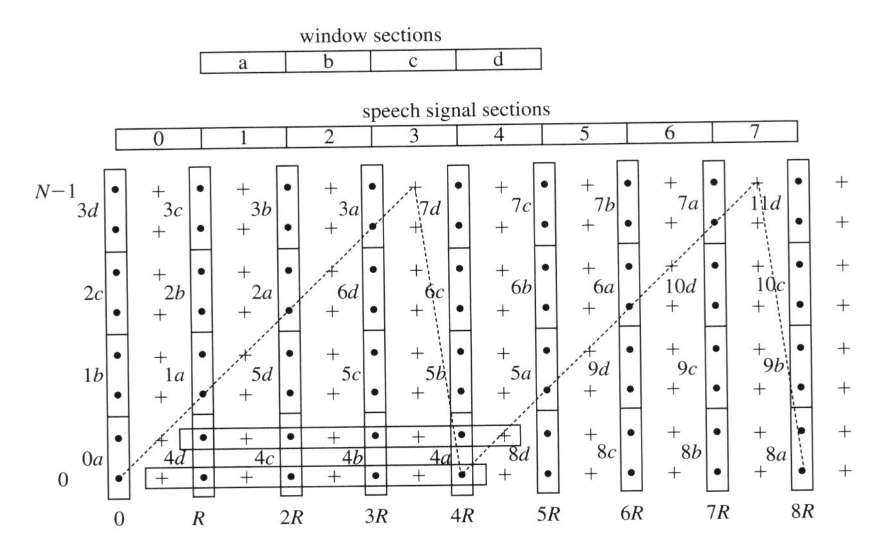

FIGURE 7.50
Representation of indexing in OLA method.

grouped in groups of $N/4$ samples that are labeled with the signal segment number and the window segment number that was applied to it. Thus 5a means signal segment 5 multiplied by window segment a. As the window moves in jumps of $N/4$ samples with respect to the signal, each segment of the input is multiplied by each segment of the window. The circular shifts by multiples of $N/4$ cause these four windowed segments to lie sequentially in time in reverse order at the same vertical level in the diagram.

The values of $X_{rR}(e^{j(2\pi k/N)})$ would be the corresponding DFTs of each of these sequences, and they can be imagined as replacing the circularly shifted sequences at each analysis time $rN/4$. For purposes of synthesis from $X_{rR}(e^{j(2\pi k/N)})$, the circularly shifted windowed segments would be reconstructed by an inverse FFT algorithm, thereby restoring the picture as shown. To interpolate the signal samples, we simply use a rectangular window of length N samples for $f[n]$. This is represented by the boxes positioned at samples $4R$ and $4R+1$, which show that four past windowed and circularly shifted samples are involved in the reconstruction of each sample $y[n]$. This will be seen to be entirely equivalent to our previous interpretation as overlapping, time-shifted windows. Note that with this interpretation, the samples involved in reconstruction are again selected along a sawtooth path through the matrix of windowed signal values.

The conclusion to be drawn from this discussion is that the OLA method and the FBS methods are not fundamentally different ways of doing STFA/STFS. In fact, the OLA method can be thought of as being equivalent to the FBS method if the synthesis window $f[n]$ is a rectangular window, $w[n]$ is a window satisfying Eq. (7.127), $N = L$, and $R = N/4$.

7.11 MODIFICATIONS OF THE STFT

So far, we have discussed many details concerning the STFT representation. In particular, we have looked at the problem of STFA and STFS from two points of view, which we termed the OLA method and the FBS method. Both methods have been shown capable of reconstructing the original signal exactly (to within a scale factor) in the case when the short-time spectrum is properly sampled in both time and frequency. In many applications of the STFT, however, one is interested in making either fixed or time-varying modifications to the short-time spectrum before the speech signal is reconstructed [3, 6]. In this section we show the effects of fixed and time-varying modifications of the short-time spectrum on the resulting synthesis.

7.11.1 Multiplicative Modifications

We have already observed that fixed multiplicative modifications such as

$$Y_n(e^{j\omega_k}) = P[k]X_n(e^{j\omega_k}), \qquad k = 0, 1, \ldots, N-1, \tag{7.128}$$

are useful for obtaining a flat overall frequency response for the analysis/synthesis system or for omitting certain channels from the synthesized output (by setting $P[k]$ to zero for the omitted channels). In this section we consider the effects of more general modifications.

FBS Method

We assume that the inverse DFT of $P[k]$ exists, and we call this sequence $p[n]$ where

$$p[n] = \frac{1}{N}\sum_{k=0}^{N-1} P[k]e^{j\omega_k n}, \tag{7.129}$$

where N is the number of analysis frequencies. The reconstructed signal $y[n]$ from the FBS method is obtained by substituting Eq. (7.128) into Eq. (7.76), giving

$$\begin{aligned}
y[n] &= \frac{1}{N}\sum_{k=0}^{N-1} X_n(e^{j\omega_k})P[k]e^{j\omega_k n} \\
&= \frac{1}{N}\sum_{k=0}^{N-1}\left[\sum_{m=-\infty}^{\infty} w[n-m]x[m]e^{-j\omega_k m}\right] P[k]e^{j\omega_k n} \\
&= \sum_{m=-\infty}^{\infty} w[n-m]x[m]\frac{1}{N}\sum_{k=0}^{N-1} P[k]e^{j\omega_k (n-m)} \\
&= \sum_{m=-\infty}^{\infty} w[n-m]x[m]p[n-m] \\
&= x[n] * \left(w[n]p[n]\right).
\end{aligned} \tag{7.130}$$

Thus, the effect of the fixed spectral modification $P[k]$ is to convolve the signal $x[n]$ with the product of the window $w[n]$ and the periodic sequence $p[n]$. The motivation for making modifications of the form of Eq. (7.128) to the STFT is to effect a linear filtering operation on the signal $x[n]$ with

$$h_p[n] = w[n]p[n], \tag{7.131}$$

the impulse response of the resulting linear filter. Recall that $p[n]$ is a periodic sequence, so that if $w[n]$ is longer than N, there will be a kind of repetitive structure in $h_p[n]$. Thus, for the FBS method, fixed spectral modifications are strongly affected by the window, and only in the case when $p[n]$ is highly concentrated or when a rectangular window is used is it even approximately true that

$$h_p[n] \approx p[n], \quad 0 \le n \le N-1, \tag{7.132}$$

as might be desired.

For time-varying modifications, we model $Y_n(e^{j\omega_k})$ as

$$Y_n(e^{j\omega_k}) = X_n(e^{j\omega_k})P_n[k], \tag{7.133}$$

and we define the time-varying impulse response due to the modification, $p_n[m]$, as

$$p_n[m] = \frac{1}{N}\sum_{k=0}^{N-1} P_n[k]e^{j\omega_k m}. \tag{7.134}$$

Proceeding as before, we solve for $y[n]$, due to the modification, as

$$\begin{aligned}
y[n] &= \frac{1}{N}\sum_{k=0}^{N-1} X_n(e^{j\omega_k})P_n[k]e^{j\omega_k n} \\
&= \frac{1}{N}\sum_{k=0}^{N-1} e^{-j\omega_k n} \sum_{m=-\infty}^{\infty} x[n-m]w[m]e^{j\omega_k m}P_n[k]e^{j\omega_k n} \\
&= \sum_{m=-\infty}^{\infty} x[n-m]w[m]\frac{1}{N}\sum_{k=0}^{N-1} P_n[k]e^{j\omega_k m} \\
&= \sum_{m=-\infty}^{\infty} x[n-m]w[m]p_n[m] \\
&= \sum_{m=-\infty}^{\infty} x[n-m]\left(p_n[m]w[m]\right).
\end{aligned} \tag{7.135}$$

Eq. (7.135) again shows that for the FBS method, the time response of the spectral modification, $p_n[m]$, is weighted by the window before being convolved with $x[n]$.

In this case, since the modification is time-varying, Eq. (7.135) is not a convolution, but a time-varying superposition sum.

OLA Method

Now consider the effect of modifications when the OLA method of synthesis is used. Using the representation of Eq. (7.128) for the modification, we can solve for the reconstructed signal due to a fixed modification by using Eq. (7.45), giving

$$
\begin{aligned}
y[n] &= \sum_{r=-\infty}^{\infty} \frac{1}{N} \sum_{k=0}^{N-1} P[k] X_{rR}(e^{j\omega_k}) e^{j\omega_k n} \\
&= \frac{1}{N} \sum_{r=-\infty}^{\infty} \sum_{k=0}^{N-1} \sum_{l=-\infty}^{\infty} x[l] w[rR - l] e^{-j\omega_k l} P[k] e^{j\omega_k n} \\
&= \sum_{l=-\infty}^{\infty} x[l] \left[\frac{1}{N} \sum_{k=0}^{N-1} P[k] e^{j\omega_k (n-l)} \right] \left[\sum_{r=-\infty}^{\infty} w[rR - l] \right] \\
&= \left(\frac{W(e^{j0})}{R} \right) \sum_{l=-\infty}^{\infty} x[l] p[n - l], \qquad (7.136)
\end{aligned}
$$

or

$$
y[n] = \left(\frac{W(e^{j0})}{R} \right) (x[n] * p[n]) . \qquad (7.137)
$$

Eq. (7.137) shows that $y[n]$ is the convolution of the original signal with the time response of the spectral modification. That is, no window modifications on $p[n]$ are obtained with this method.[26] (The reader should realize that appropriate modifications must be made to the analysis, as shown in Figure 7.27—i.e., padding the signal with a sufficient number of zeros—to prevent aliasing when implementing the analysis and synthesis operations with FFTs.)

For the case of a time-varying modification, we obtain

$$
y[n] = \sum_{r=-\infty}^{\infty} \left[\frac{1}{N} \sum_{k=0}^{N-1} P_r[k] X_{rR}(e^{j\omega_k}) e^{j\omega_k n} \right], \qquad (7.138)
$$

[26]Because of the way Eq. (7.136) is implemented, i.e., by computing the output in blocks of N samples, $p[n]$ in Eq. (7.136) is not periodic but at most only N samples long.

which can be manipulated into the form

$$y[n] = \sum_{l=-\infty}^{\infty} x[l] \sum_{r=-\infty}^{\infty} w[rR - l] \left[\frac{1}{N} \sum_{k=0}^{N-1} P_r[k] e^{j\omega_k (n-l)} \right]. \tag{7.139}$$

Using Eq. (7.134), we get

$$y[n] = \sum_{l=-\infty}^{\infty} x[l] \sum_{r=-\infty}^{\infty} w[rR - l] p_r[n - l]. \tag{7.140}$$

If we let $q = n - l$, or $l = n - q$, then Eq. (7.140) becomes

$$y[n] = \sum_{q=-\infty}^{\infty} x[n - q] \sum_{r=-\infty}^{\infty} p_r[q] w[rR - n + q]. \tag{7.141}$$

If we define $\hat{p}$ by

$$\hat{p}[n - q, q] = \hat{p}[m, q] = \sum_{r=-\infty}^{\infty} p_r[q] w[rR - m], \tag{7.142}$$

then Eq. (7.140) becomes

$$y[n] = \sum_{q=-\infty}^{\infty} x[n - q] \hat{p}[n - q, q]. \tag{7.143}$$

The interpretation of Eq. (7.142) is that for the qth value, $\hat{p}[m, q]$ is the convolution of $p_r[q]$ and $w[r]$. Thus, each coefficient of the time response due to the time-varying modification is smoothed (i.e., lowpass filtered) by the window as illustrated in Figure 7.51. This figure shows a time-varying modification with impulse response $p_n[n]$ defined for $n = 0, 1, \ldots, R - 1$ being smoothed by the window $w[n]$ and applied to the signal $x[n]$. Thus, for the OLA method, any modification is bandlimited by the window but the modification acts as a true convolution on the input signal. This is in direct contrast to the FBS method in which the modifications were *time-limited* by the window, and could change instantaneously.

7.11.2 Additive Modifications

We have been discussing the effects of non-random multiplicative modifications to the short-time spectrum. It is also important to understand the effects of additive, signal independent (random), modifications to the short-time spectrum as might be expected to occur when implementing the analysis with finite precision (i.e., round-off noise), or when quantizing the short-time spectrum, as for a vocoder.

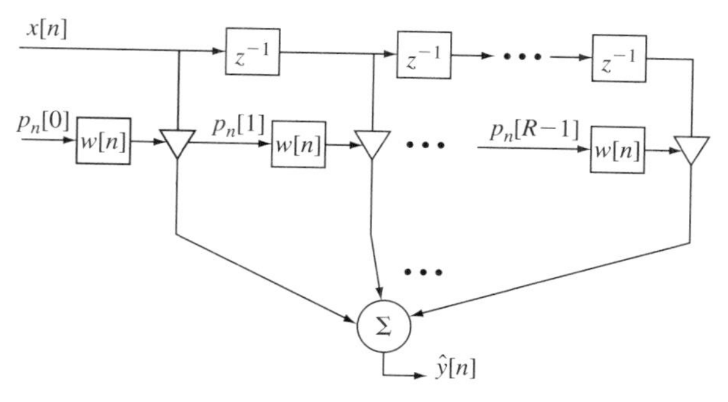

FIGURE 7.51
Implementation of time-varying modification using the OLA method. Each coefficient of the time-varying impulse response $p_n[n], n = 0, 1, \ldots, R-1$ is smoothed by the window, $w[n]$, before being applied to the signal, $x[n]$.

We model such additive modifications to the short-time spectrum as

$$\hat{X}_n(e^{j\omega_k}) = X_n(e^{j\omega_k}) + E_n(e^{j\omega_k}), \tag{7.144}$$

where we define the noise sequence corresponding to $E_n(e^{j\omega_k})$ as

$$e[n] = \frac{1}{N}\sum_{k=0}^{N-1} E_n(e^{j\omega_k})e^{j\omega_k n}. \tag{7.145}$$

(In the case where $e[n]$ is a random noise, then a statistical model for $e[n]$ and $E(e^{j\omega_k})$ is warranted. The results to be presented are not dependent on such a statistical model.)

For the FBS method, the effect of the additive modification of Eq. (7.144) is

$$y[n] = \frac{1}{N}\sum_{k=0}^{N-1} \left[X_n(e^{j\omega_k}) + E_n(e^{j\omega_k})\right] e^{j\omega_k n}, \tag{7.146}$$

which, by linearity, can be put in the form

$$y[n] = x[n] + \frac{1}{N}\sum_{k=0}^{N-1} E_n(e^{j\omega_k})e^{j\omega_k n} \tag{7.147}$$

or

$$y[n] = x[n] + e[n]. \tag{7.148}$$

Thus, an additive spectral modification results in an additive component in the reconstructed signal. The reader should notice that the analysis window has no direct effect on the additive term in the synthesis.

For the OLA method, the effect of the additive modification of Eq. (7.144) is

$$y[n] = \sum_{r=-\infty}^{\infty} \frac{1}{N} \sum_{k=0}^{N-1} \left[X_{rR}(e^{j\omega_k}) + E_{rR}(e^{j\omega_k}) \right] e^{j\omega_k n}, \tag{7.149}$$

which can be put in the form

$$\begin{aligned} y[n] &= x[n] + \sum_{r=-\infty}^{\infty} \left[\frac{1}{N} \sum_{k=0}^{N-1} E_{rR}(e^{j\omega_k}) e^{j\omega_k n} \right] \\ &= x[n] + \sum_{r=-\infty}^{\infty} e_r[n]. \end{aligned} \tag{7.150}$$

Thus, for additive modifications, the resulting synthesis contains a larger additive (noise) signal for the OLA method than for the FBS method due to the overlap between analysis frames. For a Hamming window with a 4-to-1 overlap, the additive term in the synthesis will be on the order of four times larger for the OLA method than for the FBS method. As such, the OLA methods tends to be more sensitive to noise than the FBS method, and thus would be less useful for vocoding applications, etc.

7.11.3 Time-Scale Modifications: The Phase Vocoder

An interesting and novel approach to an analysis/synthesis system based on the short-time spectrum is the phase vocoder. The phase vocoder was originated and intensively investigated by Flanagan and Golden [109] as a means for speech compression, hence the designation "vocoder." The results in this section are based on the work of these investigators. Our interest here is in the basic analysis/synthesis framework of the phase vocoder and in its application to time scale modification of speech.

To understand how the system operates, consider the response of a single channel. For this purpose, it is convenient to represent one back-to-back filter bank channel entirely in terms of real operations as in Figure 7.52. Recalling that we normally choose $\omega_{N-k} = 2\pi - \omega_k$ and $P[k] = |P[k]|e^{j\phi_k} = P^*[N-k]$, we see that the imaginary parts cancel out, leaving only the real parts, which can easily be shown to be equal to

$$\text{Re}[P[k]y_k[n]] = |P[k]||X_n(e^{j\omega_k})| \cos\left[\omega_k n + \theta_n(\omega_k) + \phi_k\right], \tag{7.151}$$

where $\theta_n(\omega_k) = \angle X_n(e^{j\omega_k})$. Signals such as these are summed to produce the composite output. Such signals can be interpreted as discrete cosine waves that are both amplitude and phase modulated by the time-dependent Fourier transform channel signal. The quantity, $|P[k]|$, is generally either one or zero, depending on whether the channel is included or not. The constant phase parameters, ϕ_k, are included to maximize the flatness of the composite response.

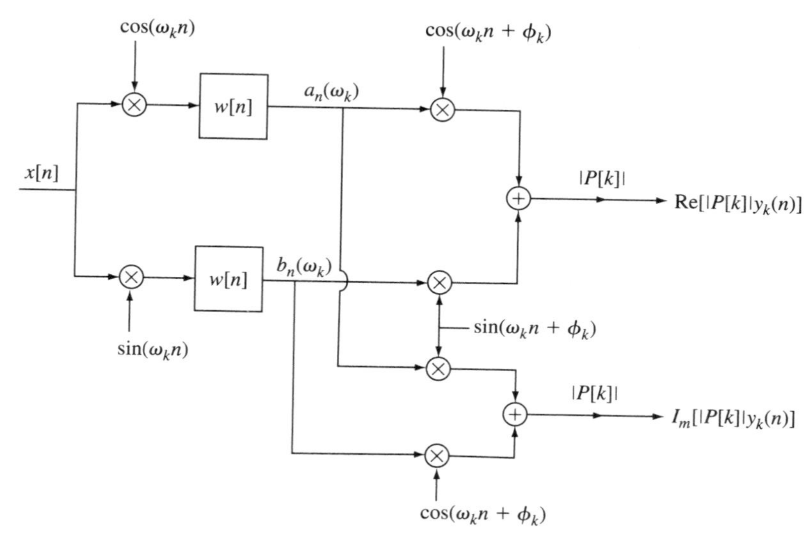

FIGURE 7.52
Implementation of a single channel of the phase vocoder.

A useful interpretation of Eq. (7.151) is possible if we introduce the concept of instantaneous frequency. In order to do this, it is convenient to consider an analog time-dependent Fourier transform

$$X_c(t, \Omega_k) = |X_c(t, \Omega_k)|e^{j\theta_c(t,\, \Omega_k)} \tag{7.152a}$$

$$= a_c(t, \Omega_k) - jb_c(t, \Omega_k), \tag{7.152b}$$

where

$$|X_c(t, \Omega_k)| = \left[a_c^2(t, \Omega_k) + b_c^2(t, \Omega_k)\right]^{1/2} \tag{7.153a}$$

$$\theta_c(t, \Omega_k) = -\tan^{-1}\left[\frac{b_c(t, \Omega_k)}{a_c(t, \Omega_k)}\right]. \tag{7.153b}$$

This time-dependent, continuous-time Fourier transform is defined as

$$X_c(t, \Omega_k) = \int_{-\infty}^{\infty} x_c(\tau)w_c(t-\tau)e^{-j\Omega_k\tau}\,d\tau, \tag{7.154}$$

where $x_c(\tau)$ is the continuous-time waveform of the speech signal, and $w_c(\tau)$ is a continuous-time analysis window or, equivalently, the impulse response of an analog lowpass filter. The quantity

$$\dot{\theta}_c(t, \Omega_k) = \frac{d\theta_c(t, \Omega_k)}{dt} \tag{7.155}$$

is called the phase derivative, and it is the instantaneous frequency deviation of the kth channel from its center frequency, Ω_k. The phase derivative can be expressed in terms of $a_c(t, \Omega_k)$ and $b_c(t, \Omega_k)$ as

$$\dot{\theta}_c(t, \Omega_k) = \frac{b_c(t, \Omega_k)\dot{a}_c(t, \Omega_k) - a_c(t, \Omega_k)\dot{b}_c(t, \Omega_k)}{a_c^2(t, \Omega_k) + b_c^2(t, \Omega_k)}, \tag{7.156}$$

where the raised dot signifies differentiation with respect to t. When dealing with discrete-time signal processing, we assume that $x_c(t)$ and $X_c(t, \Omega_k)$ are bandlimited, and that $X_n(e^{j\omega_k})$ is a sampled version of an analog time-dependent Fourier transform; i.e.,

$$X_n(e^{j\omega_k}) = X_c(nT, \omega_k/T). \tag{7.157}$$

Likewise, the "phase derivative" of $X_n(e^{j\omega_k})$ is defined as a sampled version of $\dot{\theta}_c(t, \Omega_k)$; i.e.,

$$\dot{\theta}_n(\omega_k) = \frac{b_n(\omega_k)\dot{a}_n(\omega_k) - a_n(\omega_k)\dot{b}_n(\omega_k)}{a_n^2(\omega_k) + b_n^2(\omega_k)}, \tag{7.158}$$

where, in this case, $\dot{a}_n(\omega_k)$ and $\dot{b}_n(\omega_k)$ are assumed to be sequences derived by sampling corresponding bandlimited analog derivative signals. These derivative signals can also be obtained by digital filtering of the sequences $a_n(\omega_k)$ and $b_n(\omega_k)$ (see Problem 7.23).

To see why phase derivative signals are of interest, consider the situation where the channel center frequencies are closely spaced. Particularly, consider the case when the pitch is constant and only a single harmonic of the fundamental is in the passband of the kth channel. In this case, we would find that $|X_n(e^{j\omega_k})|$ would reflect the amplitude response of the vocal tract at a frequency of approximately ω_k. The phase derivative would be a constant, which would be equal to the deviation of the harmonic component from the center frequency. Now if the vocal tract response and pitch vary slowly, as occurs during normal speech, it is reasonable to argue that the magnitude and phase derivative will both be slowly varying. Indeed, it is reasonable to argue that the effects of aliasing in sampling the magnitude and phase derivative signals should be less severe, perceptually, than the aliasing effects when the real and imaginary parts of the time-dependent Fourier transform are sampled [109].

It should be noted that for synthesis, $\theta_c(t, \Omega_k)$ is obtained from $\dot{\theta}_c(t, \Omega_k)$ by integration; i.e.,

$$\theta_c(t, \Omega_k) = \int_{t_0}^{t} \dot{\theta}_c(\tau, \Omega_k)d\tau + \theta_c(t_0, \Omega_k). \tag{7.159}$$

This equation suggests that $\theta_n(\omega_k)$, which is a sampled version of $\theta_c(t, \Omega_k)$, should be even smoother and more lowpass than $\dot{\theta}_n(\omega_k)$. Thus, it might be supposed that $\theta_n(\omega_k)$ could be sampled at an even lower rate than $\dot{\theta}_n(\omega_k)$. However, this neglects the fact the $\theta_n(\omega_k)$ is unbounded, and thus unsuitable for quantization in speech coding applications. (This can easily be seen by considering the case of constant pitch.)

A bounded phase can be obtained by computing the principal value; i.e., restricting values of $\theta_n(\omega_k)$ to be in a range 0 to 2π or $-\pi$ to π. Unfortunately, the principal value phase will be discontinuous (i.e., the principal value of $\theta_c(t, \Omega_k)$ will be a discontinuous function of t) and thus it will not be a lowpass signal. The fact that the principal value of the phase is discontinuous does not mean that phase cannot be quantized, since all that is required is that it be possible to reconstruct the corresponding real and imaginary parts of $X_n(e^{j\omega_k})$ at an appropriate sampling rate. Thus, the sampling rate of $\theta_n(\omega_k)$ must be as high as the rate required for $a_n(\omega_k)$ and $b_n(\omega_k)$.

The phase derivative, while appearing to have the advantage of smoothness, is not without disadvantages in an analysis/synthesis system. This can be seen from Eq. (7.159), which shows that in reconstructing $\theta_n(\omega_k)$ from $\dot{\theta}_n(\omega_k)$, we must have an "initial condition." Normally, such initial conditions will not be known and arbitrarily assuming zero initial phase results effectively in an error in the fixed phase angle ϕ_k. This can cause the composite response of the complete analysis/synthesis system to deviate appreciably from the ideal flat magnitude and linear phase, resulting in synthetic speech that may sound quite reverberant.

A phase vocoder analyzer based upon the magnitude and phase derivative is depicted in Figure 7.53. Figure 7.53 shows a single channel of the analysis section for a frequency $0 < \omega_k < \pi$. All other channels have exactly the same form, although they may differ in the details of the decimation and quantization. The operations required to transform $a_n(\omega_k)$ and $b_n(\omega_k)$ into $|X_n(e^{j\omega_k})|$ and $\dot{\theta}_n(\omega_k)$ are depicted in Figure 7.54. One approach to synthesis from magnitude and phase derivative signals is shown in Figure 7.55. This approach involves conversion to real and imaginary parts followed by synthesis as in Figure 7.52. The operations required to convert from magnitude and phase derivative signals to real and imaginary parts are depicted in Figure 7.56. It can be seen that the phase derivative signals must be integrated to produce a phase signal. The cosine and sine of the phase angle are then multiplied by the magnitude function to produce the real and imaginary parts. An alternative approach to synthesis that avoids the conversion process is depicted in Figure 7.57. In this case, the magnitude and phase derivative signals are interpolated with the resulting magnitude and phase sequences

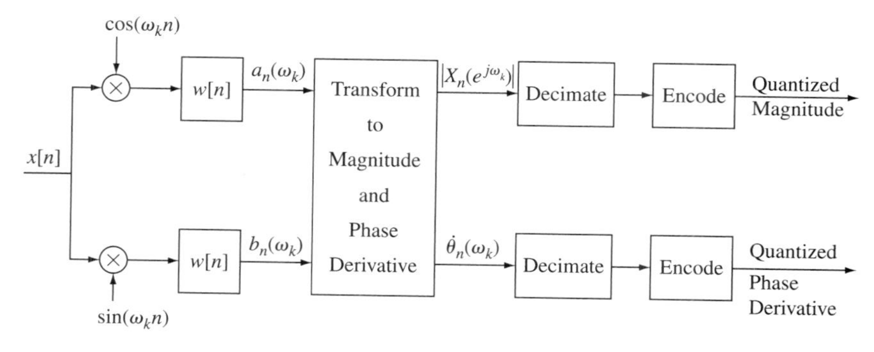

FIGURE 7.53
Complete single channel of a phase vocoder analyzer.

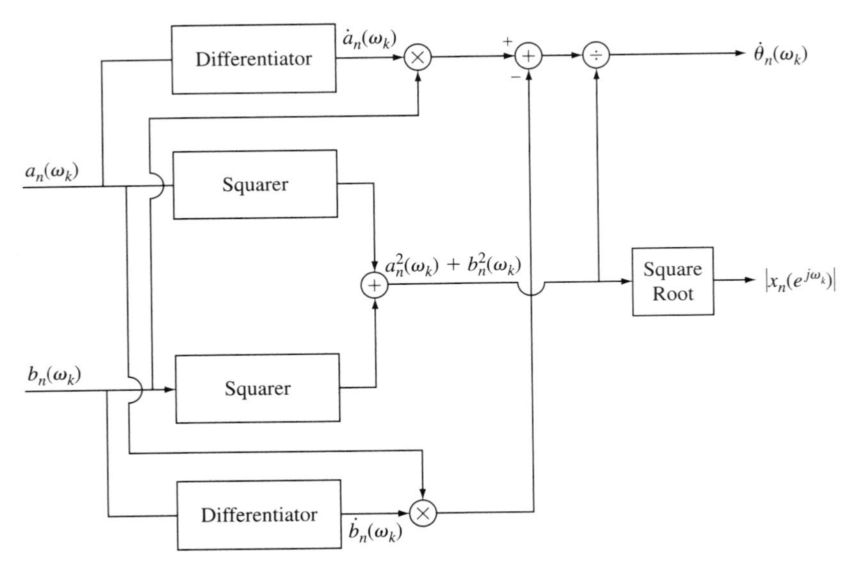

FIGURE 7.54
Conversion from a and b to $\dot{\theta}$ and $|X(e^{j\omega})|$.

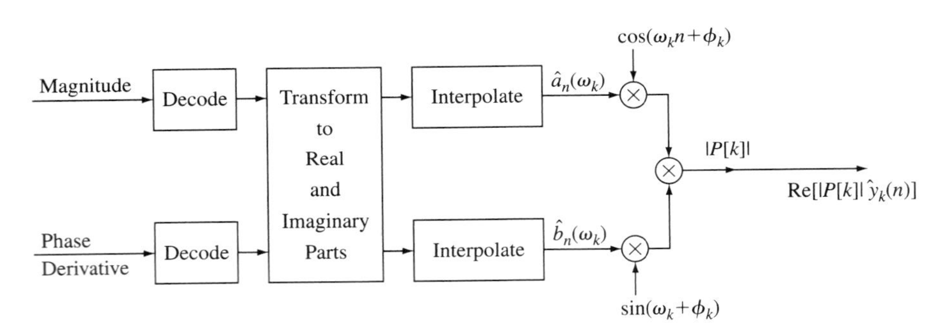

FIGURE 7.55
Implementation of the synthesizer for a single channel of the phase vocoder.

being used to amplitude and phase modulate the sinusoid. Thus, the conversion from magnitude and phase derivative to real and imaginary parts is replaced by the need for a phase modulator. Assuming that the implementation of such a digital phase modulator is not extremely complicated, it is clear that the synthesis scheme of Figure 7.57 is significantly simpler than the scheme depicted in Figures 7.55 and 7.56.

A detailed study of techniques for sampling and quantizing the magnitude and phase derivative signals in a phase vocoder has been carried out by Carlson [52]. In that study, a 28-channel phase vocoder was implemented with a channel spacing of 100 Hz. Linear quantizers were used for the phase derivative parameters and logarithmic quantizers were used for the magnitude parameters. Bits were distributed

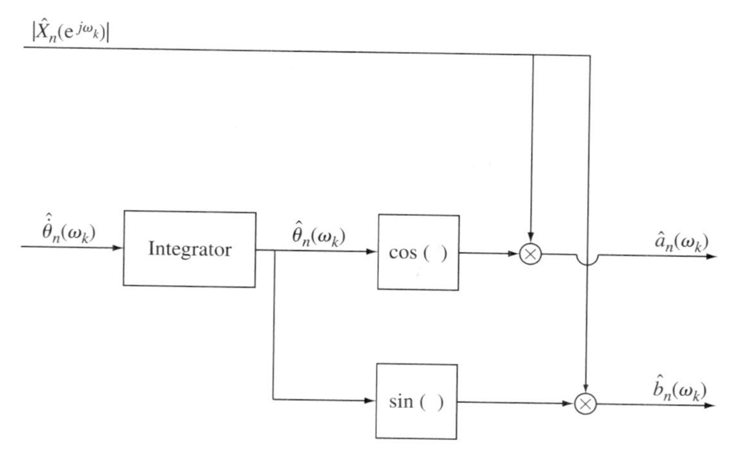

FIGURE 7.56
Conversion from $|X(e^{j\omega})|$ and $\dot{\theta}$ to a and b.

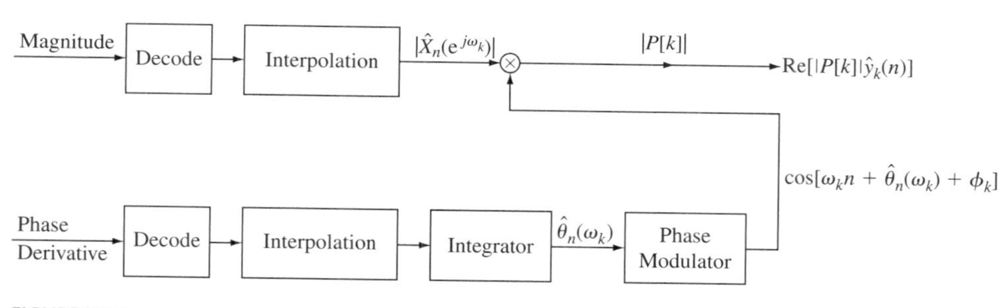

FIGURE 7.57
Alternate form for synthesis.

non-uniformly among the channels, with more bits being allocated to represent the lower channels and fewer bits for the upper channels. Also, more bits were allocated to the phase derivative than to the magnitude signals. By sampling the magnitude and phase derivative signals only 60 times/sec, and using 2 bits for the lowest magnitude channels and 1 bit for the highest channels and 3 bits for the lowest frequency phase derivative channels and 2 bits for the highest frequency channels, a bit rate of 7.2 kbps was achieved. Informal tests showed that speech represented in this way was judged to be comparable in quality to logarithmic pulse code modulation (PCM) representations at two to three times the bit rate.

An additional feature of the phase vocoder and of vocoders in general is increased flexibility for manipulating the parameters of the speech signal. In contrast to waveform representations, where the pattern of variations of the speech signal is represented by a sequence of numbers, vocoders represent the speech signal in terms of parameters more closely related to the fundamental parameters of speech production. For example, as we have already argued, in the case of a phase vocoder with closely

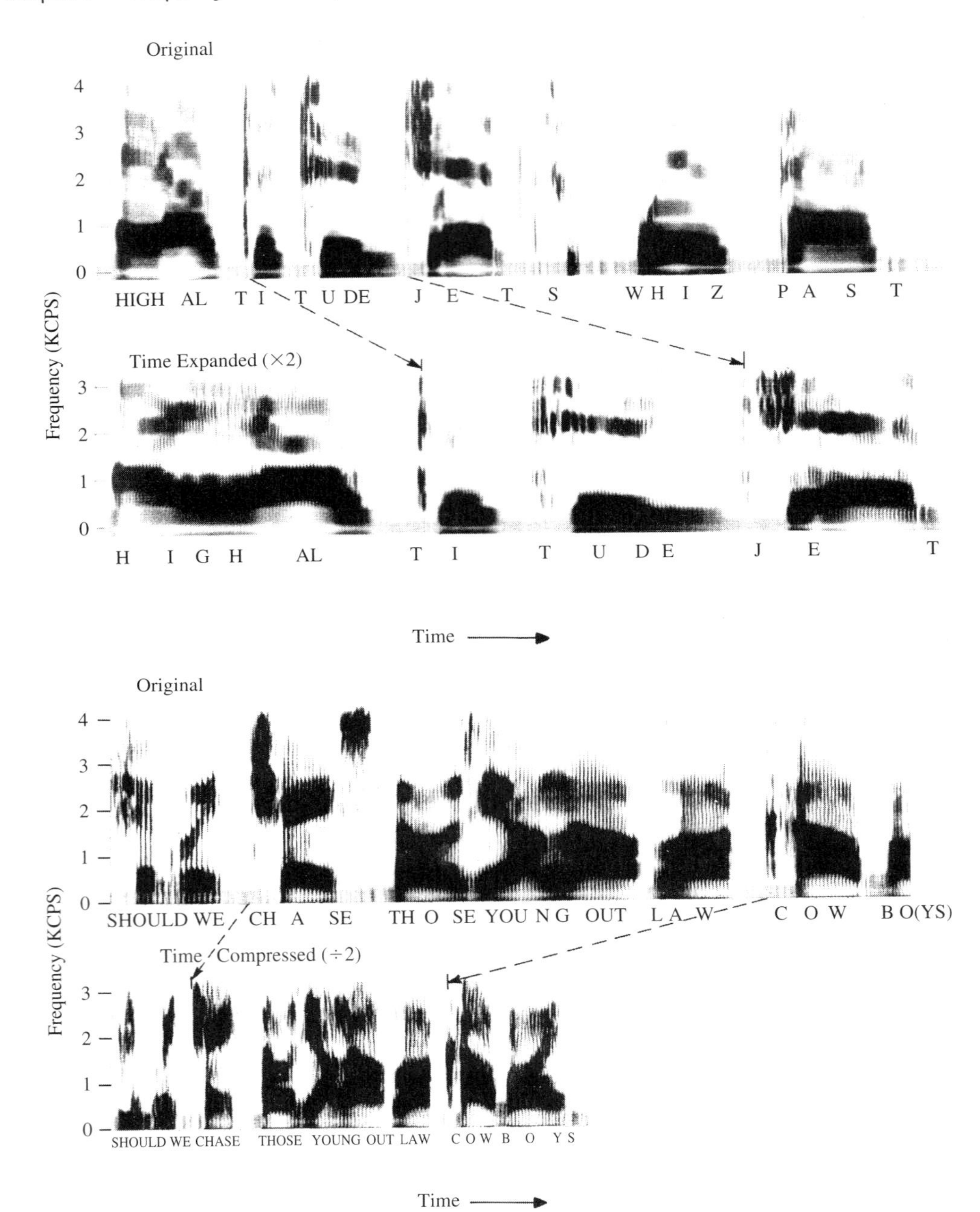

FIGURE 7.58
An example of (a) time expansion and (b) time compression using the phase vocoder. (After Flanagan and Golden [109]. Reprinted with permission of Alcatel-Lucent USA Inc.)

spaced channels, it is reasonable to suppose that the magnitude of the complex channel signals represents mainly information about the vocal tract transfer function, while the phase derivative signals give information about the excitation. A simple example of the manner in which basic speech parameters can be modified using a phase vocoder is suggested by Figure 7.57. Suppose that the phase derivative signal is arbitrarily set to 0, so that the output signal is formed by the product of the magnitude of the STFT, and a cosine of fixed frequency, ω_k. If we assume equally spaced channels, then the composite output will appear as a periodic signal with fundamental frequency equal to the spacing of the channels. Since the magnitude function varies as a function of time, the output will not be periodic, but will be slowly varying. This type of synthesis gives a distinctly monotone output, as would be expected. Alternatively, if the phase derivative signals were allowed to vary randomly, we would expect the resulting speech to sound like whispered speech.

Another more useful application of the flexibility inherent in a phase vocoder system involves the alteration of the time and frequency dimensions of the speech signal as described by Flanagan and Golden [109]. Referring again to Figure 7.57, we recall that the instantaneous frequency of the cosine is $[\omega_k + \dot{\theta}_n(\omega_k)]$. Thus, a frequency-divided signal can be obtained by simply dividing ω_k and $\dot{\theta}_n(\omega_k)$ by a constant q. If each channel is synthesized in this fashion, the result is a frequency-compressed signal, where the frequency scale is compressed by the factor q. The frequency scale of the resulting signal can be restored by recording the signal at one speed and playing it back at q times the speed. Alternatively, one can use a digital-to-analog converter operating at q times the clock frequency of the sampling frequency of the input. In either case, the compression of the time scale counteracts the compression of the frequency scale introduced in the synthesis. The result is a signal with the normal frequency dimension but with a compressed time scale. Similar operations can be applied to expand the time scale. In this case, the center frequency ω_k and the phase angle are multiplied by a factor q and the resulting expanded frequency scale is restored by playing back the output signal at a slower rate. The result in this case is a time-expanded signal with the normal frequency dimensions. Figure 7.58 (due to Flanagan and Golden [109]) shows spectrograms of an example of both time-expanded and time-compressed speech produced using the process just described for a factor $q = 2$.

7.12 SUMMARY

The STFT is both a fundamental concept with far reaching impact on thinking about speech processing and a mathematical and computational tool for representing the speech signal. In this chapter, we have defined the short-time Fourier transform, and illustrated how the concept is well matched to the linear source/system model that evolved out of our discussion in Chapter 5. We showed that the STFT can be thought of either in terms of the DTFT of windowed segments of the speech waveform or in terms of a collection of bandpass filter outputs. Both of these interpretations yield important insights into the properties of the STFT. We considered sampling issues imposed by the need to obtain a computable version of the STFT, and found that with the proper choice of analysis window, frequency sampling increment, and window overlap, it is possible to reconstruct the original speech signal exactly from the STFT. This makes the STFT an extremely flexible representation for the speech signal. We have provided a great deal of detail in some places to demonstrate how the speech signal can be reconstructed and how banks of filters can be designed to implement short-time Fourier analysis and synthesis. In the remainder of this book we will see it appearing either explicitly or implicitly as a representation upon which is based a wide variety of algorithms for extracting speech model parameters, digital coding, spectrogram displays, and manipulation of the speech signal properties.

PROBLEMS

7.1. If the STFT is expressed as

$$X_{\hat{n}}(e^{j\hat{\omega}}) = a_{\hat{n}}(\hat{\omega}) - jb_{\hat{n}}(\hat{\omega}) = |X_{\hat{n}}(e^{j\hat{\omega}})|e^{j\theta_{\hat{n}}(\hat{\omega})},$$

prove that if $x[n]$ is real, then

(a) $a_{\hat{n}}(\hat{\omega}) = a_{\hat{n}}(2\pi - \hat{\omega}) = a_{\hat{n}}(-\hat{\omega})$.

(b) $b_{\hat{n}}(\hat{\omega}) = -b_{\hat{n}}(2\pi - \hat{\omega}) = -b_{\hat{n}}(-\hat{\omega})$.

(c) $|X_{\hat{n}}(e^{j\hat{\omega}})| = |X_{\hat{n}}(e^{j(2\pi-\hat{\omega})}| = |X_{\hat{n}}(e^{-j\hat{\omega}})|$.

(d) $\theta_{\hat{n}}(\hat{\omega}) = -\theta_{\hat{n}}(2\pi - \hat{\omega}) = -\theta_{\hat{n}}(-\hat{\omega})$.

7.2. The STFT is a complex function of $\hat{n}$ and $\hat{\omega}$; i.e.,

$$X_{\hat{n}}(e^{j\hat{\omega}}) = a_{\hat{n}}(\hat{\omega}) - jb_{\hat{n}}(\hat{\omega}) = |X_{\hat{n}}(e^{j\hat{\omega}})|e^{j\theta_{\hat{n}}(\hat{\omega})}.$$

(a) Obtain expressions for $|X_{\hat{n}}(e^{j\hat{\omega}})|$ and $\theta_{\hat{n}}(\hat{\omega})$ in terms of $a_{\hat{n}}(\hat{\omega})$ and $b_{\hat{n}}(\hat{\omega})$.

(b) Obtain expressions for $a_{\hat{n}}(\hat{\omega})$ and $b_{\hat{n}}(\hat{\omega})$ in terms of $|X_{\hat{n}}(e^{j\hat{\omega}})|$ and $\theta_{\hat{n}}(\hat{\omega})$.

7.3. If we define the STFT of the signal $x[n]$ as

$$X_{\hat{n}}(e^{j\hat{\omega}}) = \sum_{m=-\infty}^{\infty} x[m]w[\hat{n} - m]e^{-j\hat{\omega}m},$$

show that the following properties hold:

(a) Linearity: if $v[n] = x[n] + y[n]$, then $V_{\hat{n}}(e^{j\hat{\omega}}) = X_{\hat{n}}(e^{j\hat{\omega}}) + Y_{\hat{n}}(e^{j\hat{\omega}})$.

(b) Shifting property: if $v[n] = x[n - n_0]$, then $V_{\hat{n}}(e^{j\hat{\omega}}) = X_{\hat{n}-n_0}(e^{j\hat{\omega}})e^{-j\hat{\omega}n_0}$.

(c) Scaling property: if $v[n] = \alpha x[n]$, then $V_{\hat{n}}(e^{j\hat{\omega}}) = \alpha X_{\hat{n}}(e^{j\hat{\omega}})$.

(d) Modulation property: if $v[n] = x[n]e^{j\omega_0 n}$, then $V_{\hat{n}}(e^{j\hat{\omega}}) = X_{\hat{n}}(e^{j(\hat{\omega}-\omega_0)})$.

(e) Conjugate symmetry: if $x[n]$ is real, then $X_{\hat{n}}(e^{j\hat{\omega}}) = X^*_{\hat{n}}(e^{-j\hat{\omega}})$.

7.4. If the sequences $x[n]$ and $w[n]$ have DTFTs $X(e^{j\omega})$ and $W(e^{j\omega})$, prove that the STFT

$$X_{\hat{n}}(e^{j\hat{\omega}}) = \sum_{m=-\infty}^{\infty} x[m]w[\hat{n} - m]e^{-j\hat{\omega}m},$$

can be expressed in the form

$$X_{\hat{n}}(e^{j\hat{\omega}}) = \frac{1}{2\pi}\int_{-\pi}^{\pi} W(e^{j\theta})e^{j\theta\hat{n}}X(e^{j(\hat{\omega}+\theta)})d\theta;$$

i.e., $X_{\hat{n}}(e^{j\hat{\omega}})$ is a smoothed spectral estimate of $X(e^{j\omega})$ at frequency $\hat{\omega}$.

7.5. If we define a short-time power density of a signal in terms of its STFT as

$$S_{\hat{n}}(e^{j\hat{\omega}}) = |X_{\hat{n}}(e^{j\hat{\omega}})|^2,$$

and we define the short-time autocorrelation of the signal as

$$R_{\hat{n}}[k] = \sum_{m=-\infty}^{\infty} w[\hat{n}-m]x[m]w[\hat{n}-k-m]x[m+k],$$

then show that if

$$X_{\hat{n}}(e^{j\hat{\omega}}) = \sum_{m=-\infty}^{\infty} x[m]w[\hat{n}-m]e^{-j\hat{\omega}m},$$

$R_{\hat{n}}[k]$ and $S_{\hat{n}}(e^{j\hat{\omega}})$ are related as a DTFT pair—i.e., show that $S_{\hat{n}}(e^{j\hat{\omega}})$ is the Fourier transform of $R_{\hat{n}}[k]$ and vice versa.

7.6. A speech signal is sampled at a rate of 10,000 samples/sec (i.e., $F_s = 10{,}000$). A Hamming window of length L samples is used to compute the STFT of the speech signal. The STFT is sampled in time with period R, and in frequency at $N = 1024$ frequencies.

(a) It can be shown that the main lobe of the Hamming window has a symmetric full width of approximately $8\pi/L$. How should L be chosen if we want the full width of the main lobe to correspond to approximately 200 Hz analog frequency?

(b) How should R be chosen if we wish to compute the STFT every 10 msec?

(c) What is the spacing (in Hz) between sample points in the frequency dimension?

7.7. Suppose that the window sequence, $w[n]$, used in STFA is causal and has a rational z-transform of the form

$$W(z) = \frac{\displaystyle\sum_{r=0}^{N_z} b_r z^{-r}}{1 - \displaystyle\sum_{k=1}^{N_p} a_k z^{-k}}.$$

(a) What properties should $W(z)$ [or equivalently, $W(e^{j\omega})$] have in order that it be suitable for this application?

(b) Obtain a recurrence formula for $X_n(e^{j\hat{\omega}})$ in terms of the signal $x[n]$ and previous values of $X_n(e^{j\hat{\omega}})$.

(c) Consider the case

$$W(z) = \frac{1}{1 - az^{-1}}.$$

How should a be chosen to obtain a frequency resolution of approximately 100 Hz at a sampling rate of 10 kHz?

(d) The value of a required in part (c) suggests that problems may arise in implementing very narrowband time-dependent Fourier analysis recursively. Briefly discuss the nature of these problems.

7.8. The STFT is defined as:

$$X_n(e^{j\hat{\omega}}) = \sum_{m=-\infty}^{\infty} w[n-m]x[m]e^{-j\hat{\omega}m}.$$

Assume that the window used in the computation of the STFT is of the form:

$$w[n] = n\beta^n u[n].$$

(a) What samples of the input sequence, $x[m]$, are involved in the computation of $X_{50}(e^{j\hat{\omega}})$, $X_{100}(e^{j\hat{\omega}})$?

(b) Assuming that $\hat{\omega}$ is fixed, obtain a recurrence formula (difference equation) expressing $X_n(e^{j\hat{\omega}})$ in terms of the input sequence and past values of $X_n(e^{j\hat{\omega}})$.

(c) Plot $|W(e^{j\omega})|$ versus ω for the case where $\beta = 0.9$.

7.9. Prove that

$$\begin{aligned} \frac{1}{N}\sum_{k=0}^{N-1} e^{j(2\pi k/N)n} &= \sum_{r=-\infty}^{\infty} \delta[n-rN] \\ &= \begin{cases} N & n = rN,\ r = 0, \pm 1, \ldots \\ 0 & \text{otherwise.} \end{cases} \end{aligned}$$

In proving this result, make use of the identity

$$\sum_{k=0}^{N-1} \alpha^k = \frac{1-\alpha^N}{1-\alpha}.$$

7.10. A discrete-time signal, $x[n]$, is defined as:

$$\begin{aligned} x[n] = {} & 7\delta[n] + 6\delta[n-1] + 6\delta[n-2] + 7\delta[n-3] \\ & + 3\delta[n-4] + \delta[n-5] \end{aligned}$$

$X(e^{j\omega})$, the DTFT of $x[n]$, is sampled at N points around the unit circle; i.e., at the set of frequencies:

$$\omega_k = \frac{2\pi}{N}k, \quad k = 0, 1, \ldots, N-1,$$

resulting in the sequence

$$\tilde{X}[k] = X(e^{j\omega})|_{\omega=2\pi k/N}.$$

Finally we inverse transform $\tilde{X}[k]$ and obtain the sequence $\tilde{x}[n]$. Determine and plot $\tilde{x}[n]$ for $N = 40$, $N = 10$, $N = 5$, and $N = 3$.

7.11. A discrete-time signal, $x[n]$, has the form

$$x[n] = r^n u[n], \quad |r| < 1.$$

(a) Solve for $X(e^{j\omega})$, the DTFT of $x[n]$.

(b) $X(e^{j\omega})$ is sampled at a set of N uniformly spaced points between $\omega = 0$ and $\omega = 2\pi$, giving the DFT

$$\tilde{X}[k] = X(e^{j\omega})|_{\omega=2\pi k/N}, \quad k = 0, 1, \ldots, N-1.$$

Solve explicitly for $\tilde{x}[n]$, the inverse Fourier transform of $\tilde{X}[k]$.

(c) Plot $\tilde{x}[n]$ for $n = 0, 1, \ldots, 2N-1$ for the case where $N = 5$ and $r = 0.9$.

7.12. A segment of a speech waveform from a sustained vowel is sampled at a sampling rate of $F_s = 8000$ samples/sec. The segment is multiplied by a Hamming window and the log magnitude Fourier transform is computed and is shown in Figure P7.12.

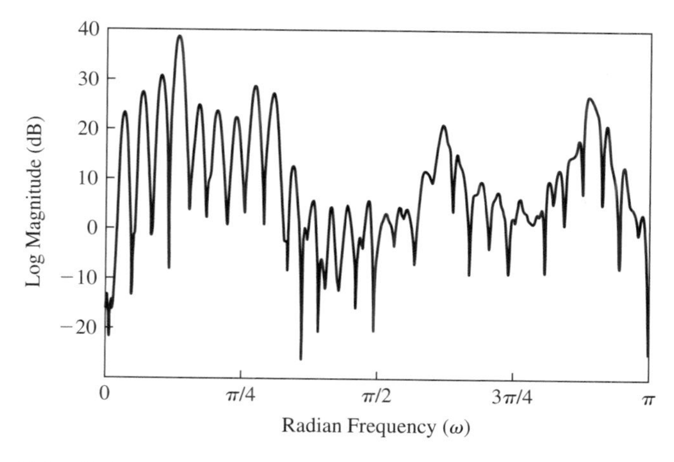

FIGURE P7.12
Log magnitude spectrum of a segment of a speech waveform from a sustained vowel.

(a) Mark the approximate locations of all the formant frequencies and give an estimate of their values in Hz.

(b) Mark on the plot and estimate the value of the fundamental frequency, F_0, in Hz. How did you estimate this value reliably?

(c) Estimate the minimum length of the Hamming window used in this analysis. Explain your reasoning.

7.13. Figure P7.13, part (a), shows a model for speech production that is appropriate for sampled speech waveforms. In answering the questions in this problem, you should assume that the speech signals being analyzed are the output of this model. The different parts of this model will be referred to by their labels (a)–(h) as shown in Figure P7.13, part (a). Figure P7.13, part (b), shows five short-time Fourier transforms (log magnitude spectra) computed from different speech signals at different times and with different window lengths. The five STFTs are identified by their labels, A–E, as shown on the left side of the spectral plots. Assume the speech was sampled at a sampling rate of $F_s = 8000$ samples/sec. Keep in mind that the speech signals being analyzed are assumed to be the output of the model of Figure P7.13, part (a).

(a) Which spectrum(s) in Figure P7.13, part (b), correspond to voiced speech frames? Explain your choice(s).

(b) Which spectrum(s) in Figure P7.13, part (b), correspond to unvoiced speech frames? Explain your choice(s).

(c) Which spectrum(s) in Figure P7.13, part (b), was(were) computed with the longest window(s)? Explain your choice(s).

(d) Which spectrum(s) in Figure P7.13, part (b), was(were) computed with the shortest window(s)? Explain your choice(s).

(e) Which spectrum in Figure P7.13, part (b), most likely corresponds to a female voice? Explain your choice(s).

(f) Estimate the fundamental frequency (in Hz) of the signal in spectrum A in Figure P7.13, part (b). Explain how you reliably estimated the fundamental frequency.

(g) Estimate the first three formant frequencies (in Hz) of the signal in spectrum B in Figure P7.13, part (b).

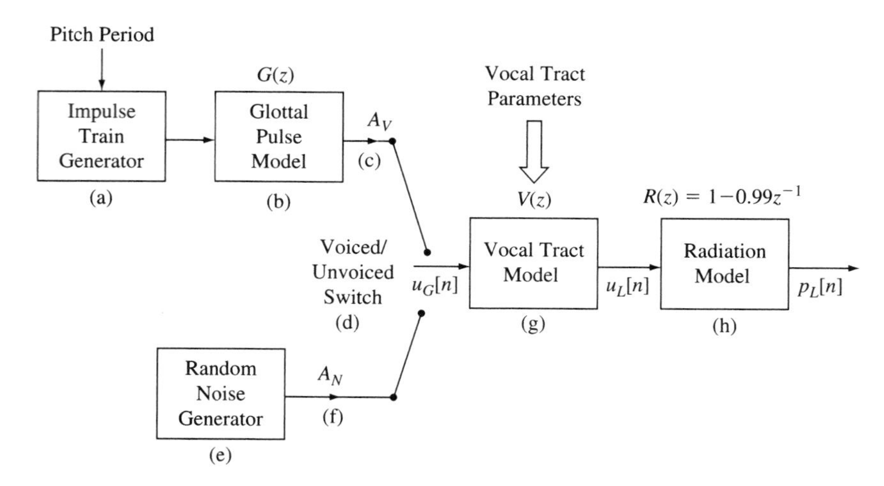

FIGURE P7.13
Part (a)—Discrete-time system model for speech production.

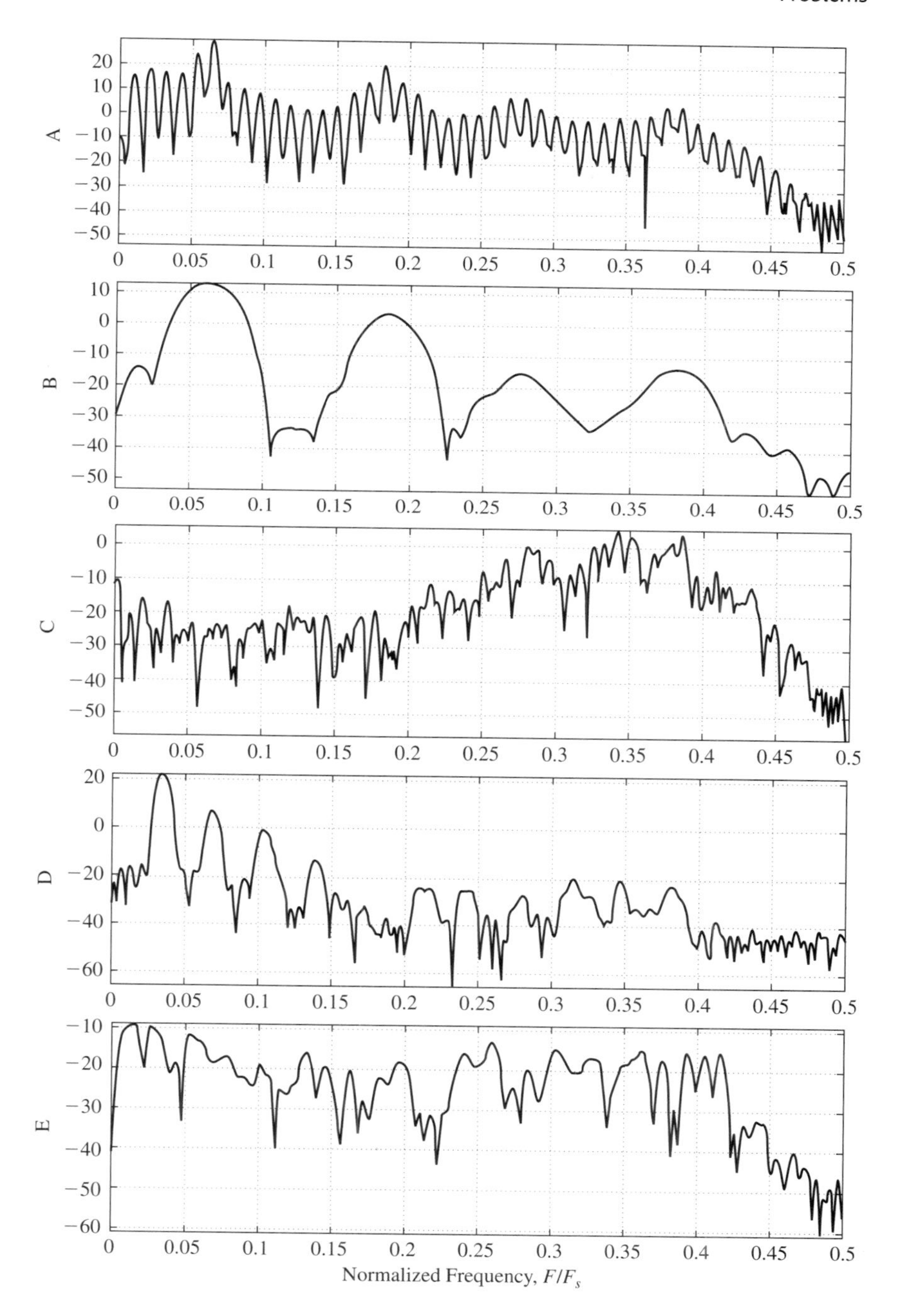

FIGURE P7.13
Part (b)—Log magnitude spectra of different speech frames. (Note the label, A–E, on the left side of the spectral plot which identifies the spectrum under consideration.)

(h) What part of the model in Figure P7.13, part (a), determines the formant frequencies?

(i) What part of the model in Figure P7.13, part (a), determines the spacing of the prominent local peaks in spectrum A in Figure P7.13, part (b)?

(j) What part(s) of the model in Figure P7.13, part (a), if changed, could shift the log magnitude spectra in Figure P7.13, part (b) *down* by 20 dB? What change(s) will accomplish this spectral shift?

(k) How would the curves in Figure P7.13, part (b), be affected if part (h) of the model were removed? How could this be effectively accomplished if we only have $p_L[n]$ available; i.e., how could we get $u_L[n]$ from $p_L[n]$?

7.14. In implementing STFT representations, we employ sampling in both the time and frequency dimensions. In this problem, we investigate the effects of both types of sampling. Consider a sequence $x[n]$ with DTFT

$$X(e^{j\omega}) = \sum_{m=-\infty}^{\infty} x[m]e^{-j\omega m}.$$

(a) If the periodic function $X(e^{j\omega})$ is sampled at frequencies $\omega_k = 2\pi k/N$, $k = 0, 1, \ldots, N-1$, we obtain

$$\tilde{X}[k] = \sum_{m=-\infty}^{\infty} x[m]e^{-j\frac{2\pi}{N}km}.$$

These samples can be thought of as the DFT of the sequence $\tilde{x}[n]$ given by

$$\tilde{x}[n] = \frac{1}{N}\sum_{k=0}^{N-1} \tilde{X}[k]e^{j\frac{2\pi}{N}kn}.$$

Show that

$$\tilde{x}[n] = \sum_{r=-\infty}^{\infty} x[n + rN].$$

(b) What are the conditions on $x[n]$ so that no aliasing distortion occurs in the time domain when $X(e^{j\omega})$ is sampled?

(c) Now consider "sampling" the sequence $x[n]$; i.e., let us form the new sequence

$$y[n] = x[nM]$$

consisting of every M^{th} sample of $x[n]$. Show that the Fourier transform of $y[n]$ is

$$Y(e^{j\omega}) = \frac{1}{M}\sum_{k=0}^{M-1} X(e^{j(\omega-2\pi k)/M}).$$

In proving this result, you may wish to begin by considering the sequence

$$v[n] = x[n]p[n],$$

where

$$p[n] = \sum_{r=-\infty}^{\infty} \delta[n + rM].$$

Then note that $y[n] = v[nM] = x[nM]$.

(d) What are the conditions on $X(e^{j\omega})$ so that no aliasing distortion in the frequency domain occurs when $x[n]$ is sampled?

7.15. Consider a window $w[n]$ with Fourier transform $W(e^{j\Omega T})$ which is bandlimited to the range $0 \leq \Omega \leq \Omega_c$. We wish to show that

$$\sum_{r=-\infty}^{\infty} w[rR - n] = \frac{W(e^{j0})}{R},$$

independent of n if R is a sufficiently small (non-zero) integer.

(a) Let $\hat{w}[r] = w[rR - n]$. Obtain an expression for $\hat{W}(e^{j\Omega T'})$ in terms of R and $W(e^{j\Omega T})$, where T is the sampling rate of $w[n]$, and $T' = RT$ is the sampling rate of $\hat{w}[r]$. (Hint: Recall the problem of decimating a signal by an R to 1 factor or see Problem 7.14c.)

(b) Assuming $W(e^{j\Omega T}) = 0$ for $|\Omega| > \Omega_c$, derive an expression for the maximum value of R (as a function of Ω_c) such that

$$\hat{W}(e^{j0}) = \frac{W(e^{j0})}{R}.$$

(c) Recalling that $\sum_{r=-\infty}^{\infty} \hat{w}[r]e^{-j\Omega T' r} = \hat{W}(e^{j\Omega T'})$, show that if the conditions of part (b) are met, then the relation given at the beginning of this problem is valid.

7.16. The symmetric $(2M + 1)$-point Hann window, often used in OLA analysis/synthesis, was defined in Eq. (7.54) as

$$w_{\text{Hann}}[n] = [0.5 + 0.5\cos(\pi n/M)]w_r[n], \quad \text{(P7.1)}$$

where $w_r[n]$ is a rectangular window that represents the requirement that $w_{\text{Hann}}[n] = 0$ for $|n| > M$. Because the quantity $[0.5 + 0.5\cos(\pi n/M)] = 0$ for $n = \pm M$, it turns out that there are *four* different rectangular windows that will ensure the desired time limiting of $w_{\text{Hann}}[n]$. The one most convenient for our immediate purpose is

$$w_{r1}[n] = \begin{cases} 1 & -M \leq n \leq M - 1 \\ 0 & \text{otherwise.} \end{cases}$$

(a) Show that the DTFT of $w_{r1}[n]$ is

$$W_{r1}(e^{j\omega}) = \left(\frac{1 - e^{-j\omega 2M}}{1 - e^{-j\omega}}\right) e^{j\omega M}.$$

(b) Show that if $w_r[n] = w_{r1}[n]$ in Eq. (P7.1), then

$$W_{\text{Hann}}(e^{j\omega}) = 0.5W_{r1}(e^{j\omega}) + 0.25W_{r1}(e^{j(\omega-\pi/M)}) + 0.25W_{r1}(e^{j(\omega+\pi/M)}),$$

and use the result of part (a) to obtain an equation for $W_{\text{Hann}}(e^{j\omega})$ that depends only on ω and M.

(c) Use the results of (a) and (b) to show that $W_{\text{Hann}}(e^{j2\pi k/M}) = 0$ for $k = 1, 2, \ldots, M-1$. Therefore, perfect reconstruction results if $R = M$ or $R = M/2$ (if $M/2$ is an integer).

(d) Use the results from (a) and (b) to show that $W_{\text{Hann}}(e^{j0}) = M$, and therefore, the reconstruction gain using the Hann window is $C = M/R$.

(e) There are three more rectangular windows that give the same $w_{\text{Hann}}[n]$ when used in Eq. (P7.1). (This simply means that there are three more ways to express $W_{\text{Hann}}(e^{j\omega})$ in terms of ω and M.) Denote these windows as $w_{r2}[n]$, $w_{r3}[n]$, and $w_{r4}[n]$. Determine these windows and determine an equation for each of their DTFTs.

(f) Using the equations derived in parts (a), (b), and (e), write a MATLAB program to evaluate and plot $W_{\text{Hann}}(e^{j\omega})$ for each of the four different rectangular windows for the case $M = 10$. Plot all four plots on the same axes and thereby verify that they are identical and that the zeros of $W_{\text{Hann}}(e^{j\omega})$ are at frequencies $\pi k/M$ for $k = 2, 3, \ldots, 2M-2$, and also therefore, at $2\pi k/M$ for $k = 1, 2, \ldots, M-1$.

7.17. When a Hann window as defined by Eq. (7.54) is used in OLA analysis/synthesis for both the analysis window and an additional synthesis window, the effective window is $w_{\text{eff}}[n] = (w_{\text{Hann}}[n])^2$.

(a) Using Eq. (7.54), show that

$$w_{\text{eff}}[n] = [0.375 + \cos(\pi n/M) + 0.125\cos(2\pi n/M)]w_r[n].$$

(b) Use the result of (a) to obtain an expression for $W_{\text{eff}}(e^{j\omega})$ in terms of $W_r(e^{j\omega})$.

(c) Using $w_r[n] = w_{r1}[n]$ and $W_r(e^{j\omega}) = W_{r1}(e^{j\omega})$ from Problem 7.16, show that $W_{\text{eff}}(e^{j4\pi k/M}) = 0$ for $k = 1, 2, \ldots, (M-2)/2$, and $W_{\text{eff}}(e^{j0}) = 3M/4$. This proves that perfect reconstruction is possible with back-to-back Hann windows for OLA analysis/synthesis with $R = M/2$ and corresponding reconstruction gain of $C = (3M/4)/(M/2) = 3/2$.

7.18. **(a)** Show that the impulse response of the system of Figure P7.18 is

$$h_k[n] = h[n]\cos(\omega_k n).$$

(b) Find an expression for the frequency response of the system of Figure P7.18.

7.19. Emphasis of the high frequency region of the spectrum is often accomplished using a first difference. In this problem we examine the effect of such operations on the STFT.

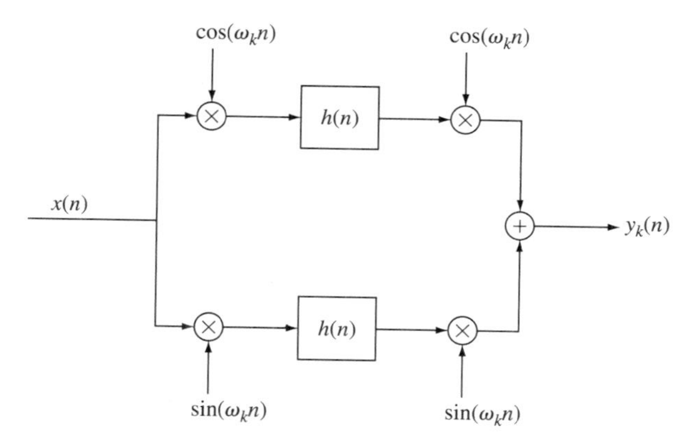

FIGURE P7.18
System implementation.

(a) Let $y[n] = x[n] - x[n-1]$. Show that

$$Y_n(e^{j\hat{\omega}}) = X_n(e^{j\hat{\omega}}) - e^{-j\hat{\omega}} X_{n-1}(e^{j\hat{\omega}}).$$

(b) Under what conditions can we make the approximation

$$Y_n(e^{j\hat{\omega}}) \approx (1 - e^{-j\hat{\omega}}) X_n(e^{j\hat{\omega}}).$$

In general, $x[n]$ may be linearly filtered as in

$$y[n] = \sum_{k=0}^{N-1} h[k] x[n-k].$$

(c) Show that $Y_n(e^{j\hat{\omega}})$ is related to $X_n(e^{j\hat{\omega}})$ by an expression of the form

$$Y_n(e^{j\hat{\omega}}) = X_n(e^{j\hat{\omega}}) * h_{\hat{\omega}}[n].$$

Determine $h_{\hat{\omega}}[n]$ in terms of $h[n]$.

(d) When is it reasonable to expect that

$$Y_n(e^{j\hat{\omega}}) = H(e^{j\hat{\omega}}) X_n(e^{j\hat{\omega}})?$$

7.20. A filter bank with N filters has the following specifications:

1. the center frequencies of the bands are ω_k;
2. the bands are symmetric around $\omega = \pi$, i.e., $\omega_k = 2\pi - \omega_{N-k}$, $P_k = P^*_{N-k}$, $w_k[n] = w_{N-k}[n]$;
3. a channel exists for $\omega_k = 0$.

For both N even and N odd:

(a) sketch the locations of the N filter bands;

(b) derive an expression for the composite impulse response of the filter bank in terms of $w_k[n]$, ω_k, P_k, and N.

7.21. In Section 7.6, we considered the FBS method of short-time analysis and synthesis. There we showed that the overall composite impulse response of the system of Figure 7.33 is

$$\tilde{h}[n] = w[n]p[n] = \sum_{r=-\infty}^{\infty} w[rN]\delta[n - rN].$$

Suppose that the window $w[n]$ is such that the composite impulse response is,

$$\tilde{h}[n] = \alpha_1\delta[n] + \alpha_2\delta[n - N] + \alpha_3\delta[n - 2N].$$

This would result, for example, when the window corresponded to the impulse response of an IIR system. The goal would be to have only a single impulse rather than three. From the point of view of speech processing, the additional impulses represent echoes spaced N samples apart.

(a) Determine the system function $H(e^{j\omega})$ for this example, and show that

$$|H(e^{j\omega})|^2 = (\alpha_2 + (\alpha_1 + \alpha_3)\cos(\omega N))^2 + (\alpha_1 - \alpha_3)^2 \sin^2(\omega N).$$

(b) Show that the phase response of the system can be written as

$$\theta(\omega) = -\omega N + \tan^{-1}\left[\frac{(\alpha_1 - \alpha_3)\sin(\omega N)}{\alpha_2 + (\alpha_1 + \alpha_3)\cos(\omega N)}\right].$$

(c) To determine locations of amplitude maxima and minima, $|H(e^{j\omega})|^2$ is differentiated with respect to ω and the result is set to zero. Show that for $|\alpha_1 + \alpha_3| << |\alpha_2|$, the locations of the maxima and minima are

$$\omega = \pm\frac{k\pi}{N}, \quad k = 0, 1, 2, \ldots.$$

(d) Using the results of part (c), show that the peak-to-peak amplitude ripple (in dB) can be expressed as

$$R_A = 20\log_{10}\left[\frac{|\alpha_2 + \alpha_1 + \alpha_3|}{|\alpha_2 - \alpha_1 - \alpha_3|}\right].$$

(e) Solve for R_A for the cases

(i) $\alpha_1 = 0.1$, $\alpha_2 = 1.0$, $\alpha_3 = 0.2$

(ii) $\alpha_1 = 0.15$, $\alpha_2 = 1.0$, $\alpha_3 = 0.15$.

(iii) $\alpha_1 = 0.1$, $\alpha_2 - 1.0$, $\alpha_3 = 0.1$

(f) By differentiating $\theta(\omega)$ with respect to ω, it can be shown that maxima and minima of θ occur at values of ω satisfying

$$\cos(\omega N) = -\left[\frac{\alpha_1 + \alpha_3}{\alpha_2}\right].$$

Show that the peak-to-peak phase ripple is given by

$$R_p = 2\tan^{-1}\left[\frac{\alpha_1 - \alpha_3}{(\alpha_2^2 - (\alpha_1 + \alpha_3)^2)^{1/2}}\right].$$

(g) Solve for R_p for the cases of part (e). Discuss the effects of varying α_1 and α_3 on R_A and R_p.

7.22. Consider a periodic sequence

$$\tilde{x}[n] = \sum_{r=-\infty}^{\infty} h_v[n + rN_p]$$

representing a voiced speech sound.

(a) Show that the Fourier series of $\tilde{x}[n]$ can be represented as the Fourier series

$$\tilde{x}[n] = \frac{1}{N_p}\sum_{k=0}^{N_p-1} \tilde{X}[k]e^{j\frac{2\pi}{N_p}kn},$$

where the Fourier coefficients $\tilde{X}[k]$ are samples of the Fourier transform of the voiced speech impulse response; i.e.,

$$\tilde{X}[k] = H_v(e^{j\frac{2\pi}{N_p}k}).$$

(See problem 7.14.)

(b) Show that the STFT of $\tilde{x}[n]$ can be expressed as

$$\tilde{X}_n(e^{j\hat{\omega}}) = \frac{1}{N_p}\sum_{k=0}^{N_p-1} H_v(e^{j\frac{2\pi}{N_p}k})W_n(e^{j(\hat{\omega}-2\pi k/N_p)}),$$

where $W_n(e^{j\hat{\omega}})$ is the Fourier transform of $w[n-m]$.

(c) How many *different* values can $\tilde{X}_n(e^{j\hat{\omega}})$ take on for a given frequency, $\hat{\omega}$?

(d) For the rectangular window

$$w[n] = \begin{cases} 1 & 0 \le n \le N_p - 1 \\ 0 & \text{otherwise,} \end{cases}$$

find the function $W_n(e^{j\hat{\omega}})$.

(e) For the rectangular window of length N_p, for what values of n will it be true that

$$\tilde{X}_n(e^{j\frac{2\pi}{N_p}k}) = H_v(e^{j\frac{2\pi}{N_p}k}).$$

7.23. Consider the analysis and synthesis of the signal $x[n] = \cos(\omega_0 n)$. The analysis network is shown in Figure P7.23a for the kth channel.

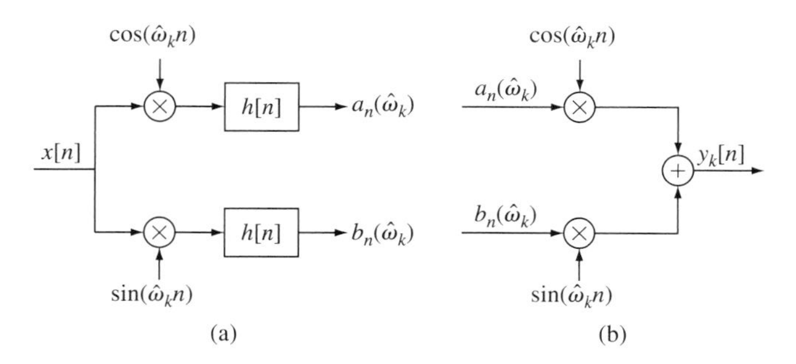

FIGURE P7.23
Signal analysis and synthesis networks.

(a) Determine $a_n(\hat{\omega}_k)$ and $b_n(\hat{\omega}_k)$ for the given input signal.

(b) Assuming $w[n]$ is a narrowband lowpass filter, simplify your expressions for $a_n(\hat{\omega}_k)$ and $b_n(\hat{\omega}_k)$ assuming that $(\omega_0 - \hat{\omega}_k)$ falls within the band of the filter, and that $W(e^{j\omega}) \approx 1$ for such frequencies.

(c) The signals $a_n(\hat{\omega}_k)$ and $b_n(\hat{\omega}_k)$ are combined to give magnitude $M_n(\hat{\omega}_k)$, and phase derivative, $\dot{\phi}_n(\hat{\omega}_k)$. Determine $M_n(\hat{\omega}_k)$ and $\dot{\phi}_n(\hat{\omega}_k)$ for this example.

(d) Show that when using the synthesis network of Figure P7.23b, the output signal is essentially identical to the input signal.

(e) The phase derivative $\dot{\phi}_n(\hat{\omega}_k)$ is computed using the relation

$$\dot{\phi}_n(\hat{\omega}_k) = \frac{b_n(\hat{\omega}_k)\dot{a}_n(\hat{\omega}_k) - a_n(\hat{\omega}_k)\dot{b}_n(\hat{\omega}_k)}{[a_n(\hat{\omega}_k)]^2 + [b_n(\hat{\omega}_k)]^2}.$$

Solve for $\dot{\phi}_n(\hat{\omega}_k)$ for this example, and compare your results with those of part (c).

(f) Now assume that the derivatives of part (e) are computed using a simple first difference; i.e.,

$$\dot{a}_n(\hat{\omega}_k) \approx \frac{1}{T}\left[a_n(\hat{\omega}_k) - a_{n-1}(\hat{\omega}_k)\right],$$

where T is the sampling period in the time dimension. Now solve for $\dot{\phi}_n(\hat{\omega}_k)$ and compare your results with part (c). Under what conditions are they approximately the same?

7.24. (MATLAB Exercise) Write a MATLAB program to perform short-time spectral analysis on a single frame of the speech waveform. Your program should accept the following inputs:

1. the speech filename
2. the starting sample within the speech array for frame analysis
3. the frame length in msec

You will have to convert from the input frame length in msec to the frame length in samples, based on the sampling rate of the digital speech signal being analyzed. Using a Hamming window of the appropriate length, window the selected frame of speech and plot (on a single page) the following signals:

1. the speech waveform for the entire duration of the speech file
2. the windowed speech frame beginning at the desired starting sample
3. the magnitude of the STFT of the selected frame of speech
4. the log magnitude (in dB) of the short-time Fourier transform of the selected frame of speech

Use the following input to test your code:

- speech files: `s5.wav` and `vowel_iy_100hz.wav`
- starting samples: 7000 (for file `s5.wav`) and 1000 (for file `vowel_iy_100hz.wav`)
- frame length: 40 msec for both examples

7.25. (MATLAB Exercise) Write a MATLAB program to perform (and compare the results of) multiple short-time analyses of a section of speech using multiple window lengths. Your program should accept the following inputs:

1. the number of short-time spectral analyses to be compared
2. the speech filename
3. the starting sample within the speech array for analysis of the short-time spectrum
4. the frame lengths in msec for each of the multiple analyses

For each of the spectral analyses you will have to convert from the input frame length in msec to the frame length in samples, based on the sampling rate of the sampled speech signal being analyzed. You should do the short-time analyses using both a Hamming window and a rectangular window so as to be able to contrast the results for these two windows. Your program output should be a plot with the multiple spectral analyses when using a Hamming window, and a separate plot with the multiple spectral analyses using a rectangular window. Each plot should include four sub-plots with the following:

1. the speech waveform for the file being analyzed
2. the waveforms of the speech segments being analyzed, showing the window superimposed over the waveform
3. the magnitude of the short-time transforms, using a different color for each of the analyses
4. the log magnitude (in dB) of the short-time transforms, again using a different color for each of the analyses

Use the following input to test your code:

- speech file: `s5.wav`
- number of analyses: 4
- frame lengths of analyses: 5, 10, 20, 40 msec
- starting sample for analysis: 7000

7.26. (MATLAB Exercise) Write a MATLAB program to create digital spectrograms (both narrowband and wideband spectrograms) from a specified speech file. For this exercise you can either use the MATLAB routine `spectrogram` or write your own code to create the required sequence of short-time spectra for plotting as a spectrogram.

Your program should accept the following parameters for creating a wide range of spectrogram plots:

- the speech filename
- an option to lower the sampling rate from the original rate of the speech to a lower rate, e.g., from 16 kHz to 8 kHz
- the window length (in msec) for the wideband spectrogram
- the window length (in msec) for the narrowband spectrogram
- the FFT length for the wideband spectrogram computation
- the FFT length for the narrowband spectrogram computation
- an option to plot either the log magnitude or the linear magnitude of the short-time spectra
- the dynamic range of the spectrogram image
- an option to display the spectrogram in color or gray scale

Your program should plot both the narrowband and wideband spectrograms on a single page.

Test your program with the following inputs:

- speech filename: `s5.wav`
- resample option: 0 (don't change sampling rate)
- window length for wideband spectrogram: 5 msec
- window length for narrowband spectrogram: 50 msec
- FFT size for wideband spectrogram: 1024
- FFT size for narrowband spectrogram: 1024
- option to use log or linear magnitude: use log magnitude
- dynamic range of spectrogram image: 60 dB
- color option: plot in gray scale and in color (two separate plots)

Once you get your program working well, experiment with utterances at different sampling rates, and vary the analysis parameters to see their effects on the resulting spectrograms.

7.27. (MATLAB Exercise) Write a MATLAB program to plot multiple spectrograms on a single page (for comparison purposes, e.g., to compare synthetic and natural versions of a speech utterance).

Your program should accept, for each spectrogram to be plotted, the set of parameters required to specify the spectrogram properties; i.e., option to resample to specified lower sampling rate, window length (in msec), FFT size, dynamic range, and gray scale/color option. Plot all spectrograms on a single page using the sub-plot option.

Test your program using the following options:

- number of spectrograms to be compared: 2
- speech filename1: `we were away a year ago_suzanne.wav`
- speech filename2: `we were away a year ago_lrr.wav`
- window length1: 5 msec (wideband spectrogram)
- window length2: 5 msec (wideband spectrogram)
- dynamic range: 60 dB
- color option: gray scale plots

7.28. (MATLAB Exercise) Write a MATLAB program to plot the sum of overlapped and added windows in order to verify the window sampling property, namely:

$$\sum_{r=-\infty}^{\infty} w[rR - n] = \frac{W(e^{j0})}{R}.$$

Compute an L-point Hamming window and create an overlap-added array by shifting, overlapping, and adding that window. If L is odd, use $R = (L - 1)/4$, and if L is even, use $R = L/4$. Sum about 50–100 overlap-added sections and plot the resulting window sum. You will see that after a period of $3L/4$ samples (due to initial build-up of the sum), the sum of the OLA windows becomes almost constant, and stays that way until the end, at which there is a decay due to an insufficient number of overlap-added windows at the end of the sequence.

(a) Use $L = 400$. Experiment with values of R (the window shift), including: L=100 (75% overlap), 50 (87.5% overlap), and 25 (93.3% overlap).

(b) Repeat using $L = 401$ and the same values of R.

(c) Modify the 401-point Hamming window by removing the last sample. Then repeat the experiment.

7.29. (MATLAB Exercise) Write a MATLAB program to speed up or slow down a file by an arbitrary factor, α, in the range $0.4 \leq \alpha \leq 2.5$ using the weighted synchronous overlap-add method (WSOLA). The WSOLA method works as follows. A standard OLA system overlaps windows by 75% and adds the results, giving a perfect synthesis. When we speed up ($\alpha > 1$) or slow down ($\alpha < 1$) a waveform, we cannot perfectly add up overlapped windows since the phasing (or time synchrony) of the overlapped and added windows no longer is correct due to the speed-up or slow-down factor, α. Hence we utilize a method of optimization of the phasing of overlapped sections that works in the following manner. We begin with the first frame of speech and we define the "ideal" overlapped section (the one that is a quarter of a window away from the current frame) and the "real" frame; i.e., the one we should be using for speed up or slow down (the one that is separated by α times the normal window shift) but suffers from phase errors. Using the simple expedient of extending the "real" frame by a small but fixed duration at both the beginning and ending of the frame, we then can correlate the "ideal" frame with the "real frame" (over the extended region) and find the location of maximum correlation, from which we extract the frame to be overlap-added to the existing speech. Once we have this best matching frame for OLA, we can now define the next "ideal" frame as the one that is shifted by a quarter of the window from the optimized previous frame, and we can redefine the next "real" frame as the one that is shifted by α times a quarter of the window duration, and iterate the process until we run out of speech. The resulting speeded up or slowed down signal sounds remarkably natural and free of artifacts for a wide range of values of α.

Your program should read in the following parameters:

- the speech filename
- α, the speech speed-up or slow-down factor
- N, the window length in msec
- L, the window shift in msec

- wtype, the window type (rectangular, Hamming, triangular)
- Δ_{max}, the maximum frame offset (in msec) for matches

You should test your program using the following input parameters:

- filename: `s5.wav`
- α: 1 (as a sanity check), 2.2, 0.75
- N: 40 msec
- L: 10 msec
- wtype: 1 (Hamming window)
- Δ_{max}: 5 msec

Listen to the resulting speeded up or slowed down speech file. Experiment with a range of speech (and music) files and a range of speed-up and slow-down factors.

7.30. (MATLAB Exercise) Write a MATLAB program to compute and plot the filter bank modification function:

$$p[n] = \sum_{k=1}^{M} e^{j\frac{2\pi}{N}kn} + \sum_{k=N-M}^{N-1} e^{j\frac{2\pi}{N}kn}$$

$$= \frac{\sin\left(\frac{\pi}{N}(2M+1)\,n\right)}{\sin\left(\frac{\pi}{N}\,n\right)} - 1$$

for the case $N = 15$ (channels in the filter bank) and $M = 2$ (retained channels). Plot the response, $p[n]$, for the range $-15 \leq n \leq 30$ using both the stem and plot functions so that you can see both the actual response (the stem plot) and the envelope of the modification response (the normal line plot).

7.31. (MATLAB Exercise) Write a MATLAB program to compute and plot the impulse and magnitude responses of an $N = 100$ channel filter bank with $M = 30$ active channels using a sixth-order Bessel (maximally flat) lowpass filter prototype. The goal is to show improved filter bank composite response by delaying the modification function, $p[n]$, with respect to the lowpass filter prototype, $w[n]$, thereby providing a solution with reduced log magnitude ripple and reduced ripple in the phase of the composite filter bank response.

Begin by designing the maximally flat (Bessel) lowpass prototype filter using the MATLAB sequence:

```
fs=10000;  % sampling frequency
fc=52;  % lowpass cutoff frequency
norder=6;  % numerator order
dorder=6;  % denominator order
fcn=fc*2/fs;  % normalized lowpass cutoff frequency
[b,a]=maxflat(norder,dorder,fcn);
% b is the numerator polynomial
% a is the denominator polynomial
```

Plot the impulse response ($w[n]$) and frequency (log magnitude and phase) responses of the lowpass prototype filter.

The next step is to compute the modification response, $p[n]$, using the formula

$$p[n] = \frac{\sin\left[\frac{\pi}{N}(2M+1)n\right]}{\sin\left[\frac{\pi}{N}n\right]} - 1$$

and plot this response for the range $-100 \leq n \leq 200$; i.e., over three periods of the periodic response.

Next form the product of $w[n]$ and $p[n]$ (assuming no added delay) and plot the time responses ($w[n] * p[n]$ as well as $w[n]$ as an envelope on the plot) and the log magnitude response. How much ripple is there in the log magnitude response of the composite filter bank?

Finally form the product of $w[n]$ and $p[n - n_0]$ and choose the value of n_0 that gives the minimum ripple in the log magnitude response of the composite filter bank. Again plot the time responses ($w[n] * p[n]$ as well as $w[n]$ as an envelope on the plot) and the log magnitude response. What is the optimum value of n_0 for this filter bank. How much ripple is there in the log magnitude response of the composite filter bank?

7.32. (MATLAB Exercise) Write a MATLAB program to design and implement a bank of 15 uniformly spaced filters that covers the range from 200 to 3200 Hz at a sampling rate of 9600 Hz. Use the windowing method to design each of the filters in the filter bank with 60 dB of attenuation outside the passband and use a Kaiser window to modify the ideal filter response and give a realizable FIR filter.

First design the Kaiser window using the MATLAB code:

```
fs=9600;  % sampling frequency in Hz
nl=175;  % length of Kaiser window in samples
alpha=5.65326;
% design value for bandwidth and ripple
w1=kaiser(nl,alpha);
% w1 is the resulting Kaiser window
```

Next design the set of ideal filters and multiply each filter response by the window and sum over the 15 channels in the filter bank. Assume that the center frequencies of the 15 filters are 300, 500, 700, ..., 3100 Hz, and that the bandwidth of each (ideal) filter is 200 Hz; i.e., the first ideal filter has a passband from 200 to 400 Hz, the second filter has a passband from 400 to 600 Hz, etc. (We see that the ideal filters have no overlap and add up to an ideal flat response between 200 and 3200 Hz.)

Plot the magnitude responses of the 15 individual filters in the filter bank, along with a blow-up of the magnitude response showing the details of the in-band ripple.

7.33. (MATLAB Exercise) Write a MATLAB program to design and implement a bank of four non-uniformly spaced filters that covers the range from 200 to 3200 Hz at a sampling rate of 9600 Hz. Use the windowing method to design each of the filters in the filter bank with 60 dB of attenuation outside the passband and use a Kaiser window to modify the ideal filter response and give a realizable FIR filter.

First design the Kaiser window using the MATLAB code:

```
fs=9600;  % sampling frequency in Hz
nl=301;  % length of Kaiser window in samples
```

```
alpha=5.65326;
% design value for bandwidth and ripple
w1=kaiser(nl,alpha);
% w1 is the resulting Kaiser window
```

Next design the set of ideal filters and multiply each filter response by the window and sum over the four channels in the filter bank. Assume that the center frequencies of the four filters are 300, 600, 1200, and 2400 Hz and that the bandwidths of the (ideal) filters are 200, 400, 800, and 1600 Hz; i.e., the first ideal filter has a passband from 200 to 400 Hz, the second filter has a passband from 400 to 800 Hz, etc. (We see that the ideal filters have no overlap and add up to an ideal flat response between 200 and 3200 Hz.)

Plot the magnitude responses of the four individual filters in the filter bank, along with a blow-up of the magnitude response showing the details of the in-band ripple.

CHAPTER 8

The Cepstrum and Homomorphic Speech Processing

8.1 INTRODUCTION

In 1963, Bogert, Healy, and Tukey published a chapter with one of the most unusual titles to be found in the literature of science and engineering, namely "The Quefrency Alanysis of Time Series for Echoes" [39]. In this paper, the authors observed that the logarithm of the Fourier spectrum of a signal plus an echo (delayed and scaled replica) consists of the logarithm of the signal spectrum plus a periodic component due to the echo. They suggested that further Fourier analyses of the log spectrum could highlight the periodic component in the log spectrum and thus could lead to a new indicator of the occurrence of an echo. Specifically they made the following observation:

> In general, we find ourselves operating on the frequency side in ways customary on the time side and vice versa.

As an aid in formalizing this new point of view, where the time- and frequency-domains are interchanged, Bogert et al. created a number of new terms by transposing letters in familiar engineering terms. For example, to emphasize that the log spectrum was to be viewed as a waveform to be subjected to Fourier analysis, they defined the term "cepstrum" of a signal as the power spectrum of the logarithm of the power spectrum of a signal. Similarly, the term "quefrency" was introduced for the independent variable of the cepstrum. In Reference [39], Bogert et al. proposed a number of other terms with the goal of emphasizing the parallels that arise when the time- and frequency-domains are interchanged. The meanings of a subset of these terms are given in Table 8.1. A few of these terms have survived the test of time and will be used in the discussions of this chapter.

Bogert et al. based their definition of the cepstrum on a rather loose interpretation of the spectrum of an analog signal. In fact, their simulations used discrete-time spectrum estimates based on the discrete Fourier transform as approximations to analog spectra. Since effective application of the cepstrum concept required digital processing, it was necessary, early on, to develop a solid definition of the cepstrum in terms of discrete-time signal theory [271, 342]. For discrete-time signals, a definition that captures the essential features of the original definition is that the *cepstrum* of a signal is the inverse discrete-time Fourier transform (IDTFT) of the logarithm of the magnitude of the discrete-time Fourier transform (DTFT) of the signal. That is, the

TABLE 8.1 Summary of terms introduced by Bogert et al. [39]. (New terms shown italicized.)

Term	Meaning
spectrum	Fourier transform of autocorrelation function
cepstrum	Inverse Fourier transform of logarithm of spectrum
analysis	Determining the spectrum of a signal
alanysis	Determining the cepstrum of a signal
filtering	Linear operation on time signal
liftering	Linear operation on log spectrum
frequency	Independent variable of spectrum
quefrency	Independent variable of cepstrum
harmonic	Integer multiple of fundamental frequency
rahmonic	Integer multiple of fundamental quefrency

cepstrum, $c[n]$, of a signal, $x[n]$, is defined as

$$c[n] = \frac{1}{2\pi} \int_{-\pi}^{\pi} \log |X(e^{j\omega})| e^{j\omega n} d\omega, \tag{8.1a}$$

where the DTFT of the signal is defined as

$$X(e^{j\omega}) = \sum_{n=-\infty}^{\infty} x[n] e^{-j\omega n}. \tag{8.1b}$$

Equation (8.1a) is the definition of the cepstrum that we will use throughout this chapter. Note that $c[n]$, being an IDTFT, is nominally a function of a discrete index n. If the input sequence is obtained by sampling an analog signal, i.e., $x[n] = x_a(nT)$, then it would be natural to associate time with the index n in the cepstrum. Therefore, continuing with the idea of transposing letters in analogous names, Bogert et al. introduced the term "quefrency" for the name of the independent variable of the cepstrum [39]. This new term is often useful in describing the fundamental properties of the cepstrum. For example, we will show that low quefrencies correspond to slowly varying (in frequency) components in the log magnitude spectrum, while high quefrencies correspond to rapidly varying components of the log magnitude spectrum. Isolated peaks at multiples of a quefrency, N_p, in the cepstrum correspond to a periodic component in the log magnitude with period $2\pi/N_p$ in normalized radian frequency, ω, or F_s/N_p in analog frequency. It was the existence of such peaks, when echoes are present in a signal, that led Bogert et al. to consider the cepstrum as a basis for detecting echoes.

One of the first applications of the cepstrum to speech processing was proposed by Noll [252] who recognized that since voiced speech has the same type of repetitive time-domain structure as a signal with an echo, the cepstrum could be used for detecting (estimating) pitch period and voicing. Later in this chapter, we show that the cepstrum has many other applications in speech processing. We begin our discussion of the properties of the cepstrum by generalizing the cepstrum concept. We define the "complex cepstrum" as

$$\hat{x}[n] = \frac{1}{2\pi} \int_{-\pi}^{\pi} \log\{X(e^{j\omega})\} e^{j\omega n} d\omega, \tag{8.2}$$

where $\log\{X(e^{j\omega})\}$ is the complex logarithm as opposed to the logarithm of the magnitude as in Eq. (8.1a), and we use the caret notation, $\hat{x}[n]$, to denote the complex cepstrum of the signal, $x[n]$. We will see that this generalization connects the cepstrum concept to a class of systems for separating the components of a signal obtained as a result of a multiple convolution, and that this leads to wider applicability to the problems of speech processing since convolution is at the heart of our discrete-time model for speech production.

8.2 HOMOMORPHIC SYSTEMS FOR CONVOLUTION

Contemporaneously with the introduction of the cepstrum concept, Oppenheim [266] developed a new theory of systems that was based on the mathematical theory of linear vector spaces. The essence of this theory was that certain operations of signal combination (convolution and multiplication in particular) satisfy the same postulates as does addition in the theory of linear vector spaces. From this observation, Oppenheim showed that classes of non-linear systems could be defined on the basis of a generalized principle of superposition. He termed such systems *homomorphic* systems. Of particular importance for our present discussion is the class of homomorphic systems for which the input and output are combined by convolution.

The principle of superposition, as it is normally expressed for conventional linear systems, is given by

$$\begin{aligned} y[n] = \mathcal{L}\{x[n]\} &= \mathcal{L}\{x_1[n] + x_2[n]\} \\ &= \mathcal{L}\{x_1[n]\} + \mathcal{L}\{x_2[n]\} \\ &= y_1[n] + y_2[n] \end{aligned} \tag{8.3a}$$

and

$$\mathcal{L}\{ax[n]\} = a\mathcal{L}\{x[n]\} = ay[n], \tag{8.3b}$$

where $\mathcal{L}\{\cdot\}$ is the linear operator that represents the system. The principle of superposition of Eq. (8.3a) states that if an input signal is composed of an additive (linear) combination of elementary signals, then the output is an additive combination of the corresponding outputs. This is depicted in Figure 8.1, where the + symbol at the input and output emphasizes that an additive combination at the input produces a corresponding additive combination at the output. Similarly, Eq. (8.3b) states that a scaled input results in a correspondingly scaled output.

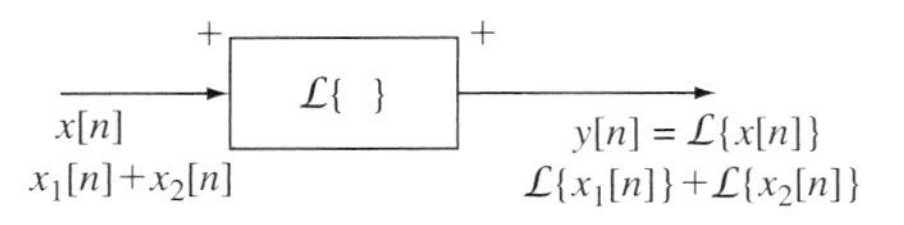

FIGURE 8.1
Representation of a system obeying the superposition principle for addition. Such systems will be termed "conventional linear systems."

As shown in Chapter 2, a direct result of the principle of superposition is that if the system is linear and also time-invariant,[1] the output can be expressed as the convolution sum

$$y[n] = \sum_{k=-\infty}^{\infty} x[k]h[n-k] = x[n] * h[n], \tag{8.4}$$

where $h[n]$ is the impulse response of the system; i.e., $h[n]$ is the output when the input is $\delta[n]$. The $*$ symbol denotes the *operation* of discrete-time convolution; i.e., the transformation of the input sequence $x[n]$, $-\infty < n < \infty$ into the output sequence $y[n]$, $-\infty < n < \infty$.

By analogy with the principle of superposition for conventional linear systems, as described by Eqs. (8.3a) and (8.3b), we can define a class of systems that obey a generalized principle of superposition where addition is replaced by convolution.[2] That is, in analogy to Eq. (8.3a), the output $y[n]$ due to the input $x[n] = x_1[n] * x_2[n]$ is

$$\begin{aligned} y[n] = \mathcal{H}\{x[n]\} &= \mathcal{H}\{x_1[n] * x_2[n]\} \\ &= \mathcal{H}\{x_1[n]\} * \mathcal{H}\{x_2[n]\} \\ &= y_1[n] * y_2[n], \end{aligned} \tag{8.5}$$

where $\mathcal{H}\{\cdot\}$ is the operator that represents the system. An equation similar to Eq. (8.3b), which expresses scalar multiplication in the generalized sense, can also be given [342]; however, the notion of generalized scalar multiplication is not needed for the applications that we will discuss. Systems having the property expressed by Eq. (8.5) are termed "homomorphic systems for convolution." This terminology stems from the fact that such transformations can be shown to be homomorphic transformations in the sense of linear vector spaces [266]. Such systems are depicted as shown

[1] Recall that if an input $x[n]$ to a system produces an output $y[n]$, then the system is time-invariant if an input $x[n-n_0]$ produces the output $y[n-n_0]$.

[2] It can easily be shown that convolution has the same algebraic properties (commutativity and associativity) as addition [271, 272].

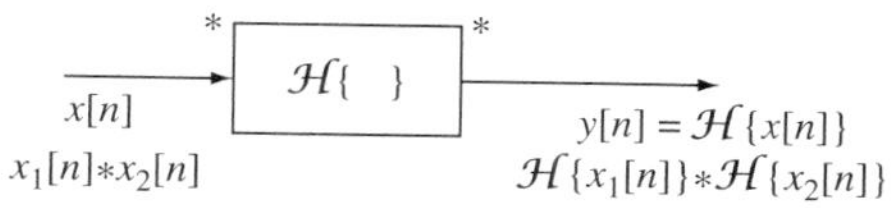

FIGURE 8.2
Representation of a homomorphic system for convolution.

in Figure 8.2, where the operation of convolution is noted explicitly at the input and output of the system through the use of the $*$ symbol.

A *homomorphic filter* is simply a homomorphic system having the property that one component (the desired component) passes through the system essentially unaltered, while the other component (the undesired component) is removed. In Eq. (8.5), for example, if $x_2[n]$ were the undesirable component, we would require that the output corresponding to $x_2[n]$ be a unit sample, $y_2[n] = \delta[n]$, while the output corresponding to $x_1[n]$ would closely approximate $x_1[n]$ so that the output of the homomorphic filter would be $y[n] = x_1[n] * \delta[n] = x_1[n]$. This is entirely analogous to when a conventional linear system is used to separate (filter) a desired signal from an additive combination of the desired signal and noise. In this case the desired result is that the output due to the noise is zero. Thus, the sequence $\delta[n]$ plays the same role for convolution as is played by the zero signal for additive combinations. Homomorphic filters are of interest to us because our goal in speech processing is to separate the convolved excitation and vocal tract components of the speech model.

An important aspect of the theory of homomorphic systems is that any homomorphic system can be represented as a cascade of three homomorphic systems, as depicted in Figure 8.3 for the case of homomorphic systems for convolution [266]. The first system takes inputs combined by convolution and transforms them into an additive combination of corresponding outputs. The second system is a conventional linear system obeying the principle of superposition as given in Eq. (8.3a). The third system is the inverse of the first system; i.e., it transforms signals combined by addition back into signals combined by convolution. The importance of the existence of such a canonic form for homomorphic systems is that the design of such systems reduces to the problem of the design of the central linear system in Figure 8.3. The system $\mathcal{D}_*\{\cdot\}$ is called the *characteristic system for convolution* and it is fixed in the canonic form of Figure 8.3. Likewise, its inverse, called the *inverse characteristic system for convolution*, and denoted $\mathcal{D}_*^{-1}\{\cdot\}$, is also a fixed system. The characteristic system

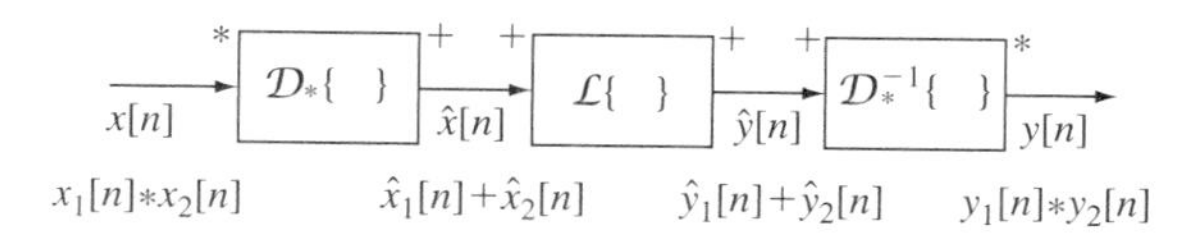

FIGURE 8.3
Canonic form of system for homomorphic deconvolution.

for convolution also obeys a generalized principle of superposition where the input operation is convolution and the output operation is ordinary addition. The properties of the characteristic system are defined as

$$\begin{aligned}\hat{x}[n] = \mathcal{D}_*\{x[n]\} &= \mathcal{D}_*\{x_1[n] * x_2[n]\} \\ &= \mathcal{D}_*\{x_1[n]\} + \mathcal{D}_*\{x_2[n]\} \\ &= \hat{x}_1[n] + \hat{x}_2[n]. \end{aligned} \tag{8.6}$$

Likewise, the inverse characteristic system, $\mathcal{D}_*^{-1}$, is defined as the system having the property

$$\begin{aligned}y[n] = \mathcal{D}_*^{-1}\{\hat{y}[n]\} &= \mathcal{D}_*^{-1}\{\hat{y}_1[n] + \hat{y}_2[n]\} \\ &= D_*^{-1}\{\hat{y}_1[n]\} * D_*^{-1}\{\hat{y}_2[n]\} \\ &= y_1[n] * y_2[n]. \end{aligned} \tag{8.7}$$

8.2.1 Representation by DTFTs

An alternative representation of the canonic form for convolution can be inferred by recalling that the DTFT of $x[n] = x_1[n] * x_2[n]$ is

$$X(e^{j\omega}) = X_1(e^{j\omega}) \cdot X_2(e^{j\omega}); \tag{8.8}$$

i.e., the operation of discrete-time Fourier transformation changes convolution into multiplication, an operation that is also commutative and associative. Thus, in the frequency domain, the characteristic system for multiplication should map multiplication into addition. In the familiar case of real positive numbers, an operation that does this is the logarithm. While the generalization of the logarithm to deal with complex functions such as the DTFT is not straightforward, it can nevertheless be done satisfactorily so that with appropriate attention to mathematical details, it is possible to write

$$\begin{aligned}\hat{X}(e^{j\omega}) &= \log\{X(e^{j\omega})\} = \log\{X_1(e^{j\omega}) \cdot X_2(e^{j\omega})\} \\ &= \log\{X_1(e^{j\omega})\} + \log\{X_2(e^{j\omega})\} \\ &= \hat{X}_1(e^{j\omega}) + \hat{X}_2(e^{j\omega}). \end{aligned} \tag{8.9}$$

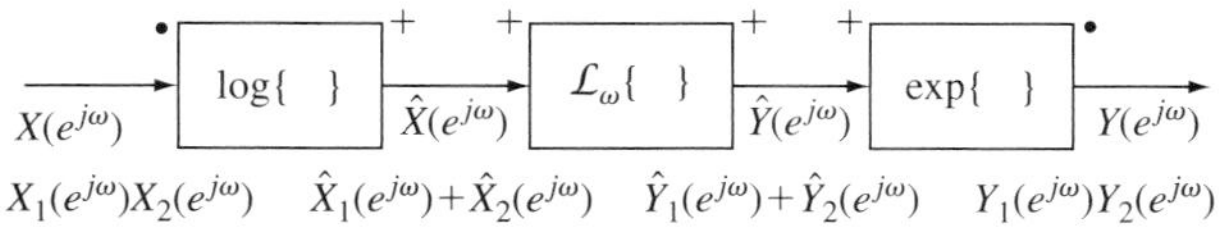

log{ } $\mathcal{L}_\omega$\{ \} exp{ }

$X(e^{j\omega})$ $\hat{X}(e^{j\omega})$ $\hat{Y}(e^{j\omega})$ $Y(e^{j\omega})$

$X_1(e^{j\omega})X_2(e^{j\omega})$ $\hat{X}_1(e^{j\omega})+\hat{X}_2(e^{j\omega})$ $\hat{Y}_1(e^{j\omega})+\hat{Y}_2(e^{j\omega})$ $Y_1(e^{j\omega})Y_2(e^{j\omega})$

FIGURE 8.4
Representation of the canonic form for homomorphic deconvolution in terms of DTFTs.

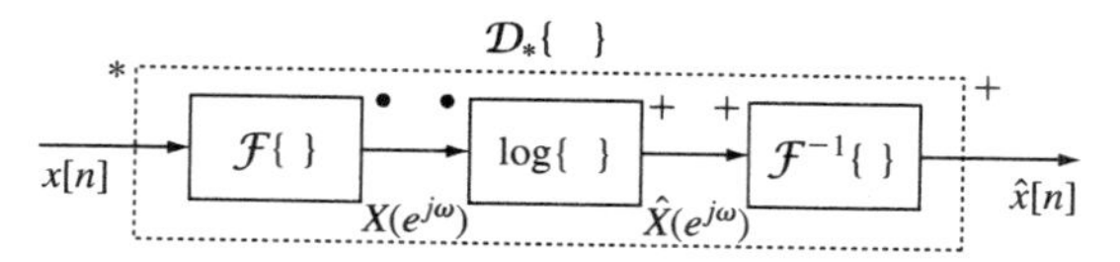

FIGURE 8.5
Representation of the characteristic system for homomorphic deconvolution in terms of DTFT operators (denoted $\mathcal{F}\{\cdot\}$ and $\mathcal{F}^{-1}\{\cdot\}$).

With this in mind, the canonic form for convolution in Figure 8.3 can be represented in the frequency domain as in Figure 8.4, where $\mathcal{L}_\omega\{\cdot\}$ represents a conventional linear operator on the logarithm of DTFTs.

To represent signals as sequences, rather than in the frequency domain as in Figure 8.4, the characteristic system can be represented as depicted in Figure 8.5 where the log function is surrounded by the DTFT operator and its inverse, and where the three operations are defined by the equations

$$X(e^{j\omega}) = \sum_{n=-\infty}^{\infty} x[n]e^{j\omega n}, \tag{8.10a}$$

$$\hat{X}(e^{j\omega}) = \log\{X(e^{j\omega})\}, \tag{8.10b}$$

$$\hat{x}[n] = \frac{1}{2\pi}\int_{-\pi}^{\pi} \hat{X}(e^{j\omega})e^{j\omega n}\,d\omega. \tag{8.10c}$$

In this representation, the characteristic system for convolution is represented as a cascade of three homomorphic systems: the $\mathcal{F}$ operator (DTFT) maps convolution into multiplication, the complex logarithm maps multiplication into addition, and the $\mathcal{F}^{-1}$ operator (IDTFT) maps addition into addition. Similarly, Figure 8.6 depicts the inverse characteristic system for convolution in terms of DTFT operators $\mathcal{F}$ and $\mathcal{F}^{-1}$ and the complex exponential.

The mathematical difficulties associated with the complex logarithm revolve around problems of uniqueness, which are discussed in some detail in Ref. [342]. For our purposes here, it is sufficient to state that an appropriate definition of the complex

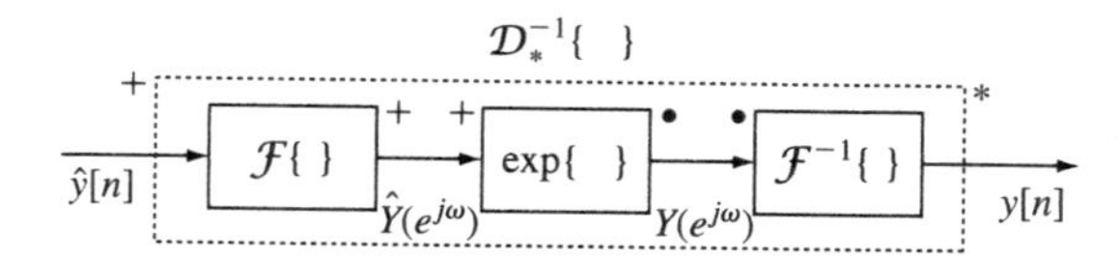

FIGURE 8.6
Representation of the inverse characteristic system for homomorphic deconvolution in terms of DTFT operators.

logarithm is

$$\hat{X}(e^{j\omega}) = \log\{X(e^{j\omega})\} = \log|X(e^{j\omega})| + j\arg\{X(e^{j\omega})\}. \tag{8.11}$$

In this equation, the real part of $\hat{X}(e^{j\omega})$ (i.e., $\log|X(e^{j\omega})|$) causes no particular difficulty. However, problems of uniqueness arise in defining the imaginary part of $\hat{X}(e^{j\omega})$ (i.e., $\arg\{X(e^{j\omega})\}$), which is simply the phase angle of the DTFT $X(e^{j\omega})$. In [342] it is shown that one approach to dealing with the problems of uniqueness of the phase angle is to require that the phase angle be a continuous odd function of ω when the sequence $x[n]$ is real. This condition ensures that

$$\log\{X_1(e^{j\omega})X_2(e^{j\omega})\} = \log\{X_1(e^{j\omega})\} + \log\{X_2(e^{j\omega})\}, \tag{8.12}$$

which is the crucial property to ensure mapping of convolution into addition.

Figure 8.7 illustrates the problem that arises when one tries to properly define the phase angle of the DTFT. It is generally expedient to compute the angle of a complex number as the principal value; i.e., the angle such that $-\pi < \text{ARG}\{X(e^{j\omega})\} \leq \pi$. However, when the principal value is applied to compute the phase angle as a function of ω, a continuous function does not generally result. The principal value phase has discontinuities of size 2π because an angle in the complex plane is always ambiguous to within an integer multiple of 2π. This poses no problem for the complex exponential

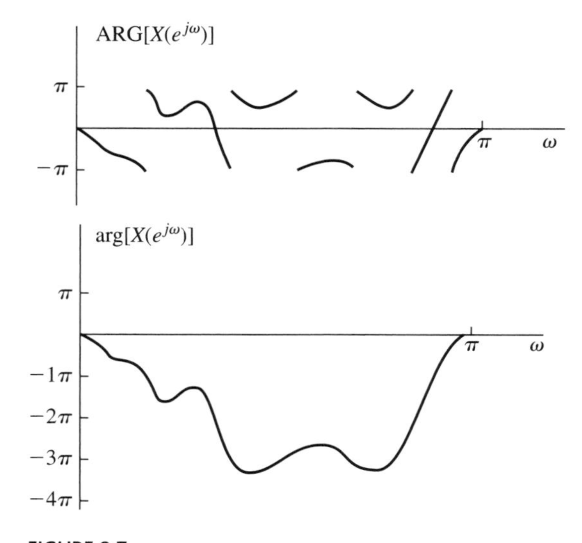

FIGURE 8.7
Phase angle functions ARG and arg.

(inverse of the complex log) since it is defined as

$$\begin{aligned} X(e^{j\omega}) &= \exp\{\log|X(e^{j\omega})| + j\arg\{X(e^{j\omega})\}\} \\ &= e^{\log|X(e^{j\omega})|}e^{j\arg\{X(e^{j\omega})\}} \\ &= |X(e^{j\omega})|e^{j\arg\{X(e^{j\omega})\}}. \end{aligned} \tag{8.13}$$

From Eq. (8.13), it follows that any integer multiple of 2π can be added to the phase $\arg\{X(e^{j\omega})\}$ without a change in the value of the complex exponential.

When computing the phase angle of $X(e^{j\omega})$, the principal value computation only uses the real and imaginary parts of the DTFT at a single frequency ω. For example, in going from ω to $\omega \pm \epsilon$, the complex value of $X(e^{j\omega})$ can jump from the third quadrant of the z-plane to the second quadrant so that an angle that could be defined as slightly more negative than $-\pi$ radians (in order to maintain continuity) would instead be assigned a positive value that is slightly less than π radians (in order to maintain the principal value). Thus, the principal value jumps by 2π whenever the phase angle exceeds π or falls below $-\pi$. This is illustrated in Figure 8.7 where the phase becomes increasingly negative as ω increases so that the principal value phase displays discontinuities at the frequencies at which the phase enters and leaves the band between $-3\pi < \arg\{X(e^{j\omega})\} \leq -\pi$, etc. The phase is said to "wrap around modulo 2π" when the principal value phase is computed. This is seen by comparing the top graph in Figure 8.7 to the bottom graph. The principal value phase clearly does not satisfy the additivity condition; i.e.,

$$\text{ARG}\{X(e^{j\omega})\} \neq \text{ARG}\{X_1(e^{j\omega})\} + \text{ARG}\{X_2(e^{j\omega})\}, \tag{8.14a}$$

since the discontinuity frequencies of the sum will not, in general, be the same as those of either of the two individual principal value functions. If, however, the phase is computed as a continuous function $\arg\{X(e^{j\omega})\}$ as in the lower plot in Figure 8.7, then it follows that since angles add in a product,

$$\arg\{X(e^{j\omega})\} = \arg\{X_1(e^{j\omega})\} + \arg\{X_2(e^{j\omega})\}. \tag{8.14b}$$

Given that it is possible to compute the complex logarithm so as to satisfy Eq. (8.14b), the inverse transform of the complex logarithm of the Fourier transform of the input is the output of the characteristic system for convolution; i.e.,

$$\hat{x}[n] = \frac{1}{2\pi}\int_{-\pi}^{\pi} \left(\log|X(e^{j\omega})| + j\arg\{X(e^{j\omega})\}\right) e^{j\omega n} d\omega. \tag{8.15}$$

If the input signal $x[n]$ is real, then it follows that $\log|X(e^{j\omega})|$ is an even function of ω and $\arg\{X(e^{j\omega})\}$ an odd function of ω. This means that the real and imaginary parts of the complex logarithm in Eq. (8.15) have the appropriate symmetry for $\hat{x}[n]$ to be a real sequence. Furthermore, this means that $\hat{x}[n]$ can be represented as

$$\hat{x}[n] = c[n] + d[n], \tag{8.16a}$$

where $c[n]$ is the inverse DTFT of $\log|X(e^{j\omega})|$ and the even part of $\hat{x}[n]$,

$$c[n] = \frac{\hat{x}[n] + \hat{x}[-n]}{2}, \tag{8.16b}$$

while $d[n]$ is the inverse DTFT of $\arg\{X(e^{j\omega})\}$ and the odd part of $\hat{x}[n]$

$$d[n] = \frac{\hat{x}[n] - \hat{x}[-n]}{2}. \tag{8.16c}$$

As we have already noted, the output of the characteristic system, $\hat{x}[n]$, is called the "complex cepstrum." This is not because $\hat{x}[n]$ is complex, but because the complex logarithm is its basis. As we can see from Eq. (8.16b), the cepstrum $c[n]$ is the *even part* of the complex cepstrum, $\hat{x}[n]$. The odd part of $\hat{x}[n]$, $d[n]$, will be used later in a discussion of a distance measure based on the group delay function $-d\arg\{X(e^{j\omega})\}/d\omega$.

8.2.2 *z*-Transform Representation

We can obtain another mathematical representation of the characteristic system by noting that if the input is a convolution $x[n] = x_1[n] * x_2[n]$, then the z-transform of the input is the product of the corresponding z-transforms; i.e.,

$$X(z) = X_1(z) \cdot X_2(z). \tag{8.17}$$

Again, with an appropriate definition of the complex logarithm, we can write

$$\begin{aligned}\hat{X}(z) &= \log\{X(z)\} = \log\{X_1(z) \cdot X_2(z)\} \\ &= \log\{X_1(z)\} + \log\{X_2(z)\},\end{aligned} \tag{8.18}$$

and the homomorphic systems for convolution can be represented in terms of z-transforms as in Figure 8.8, where the operator $\mathcal{L}_z\{\cdot\}$ is a linear operator on z-transforms. This representation is based upon a definition of the logarithm of a complex product so that it is equal to the sum of the logarithms of the individual terms as in Eq. (8.18).

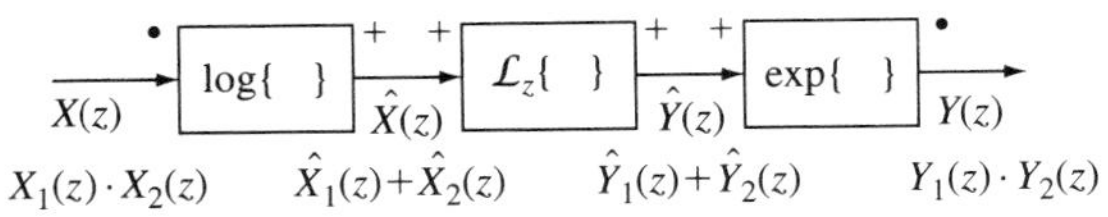

FIGURE 8.8
Representation in terms of z-transforms of the canonic form for homomorphic systems for convolution.

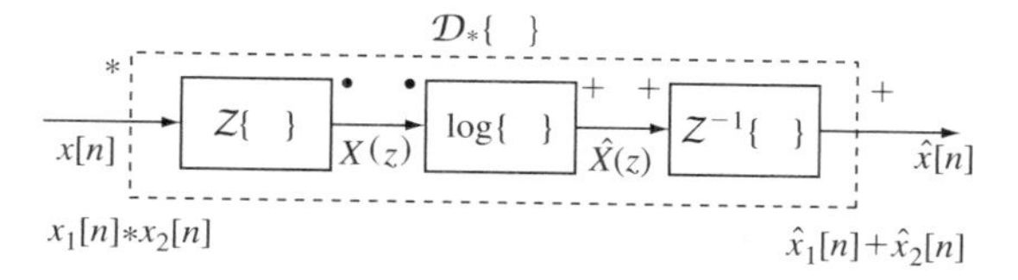

FIGURE 8.9
Representation of the characteristic system for homomorphic deconvolution in terms of z-transform operators.

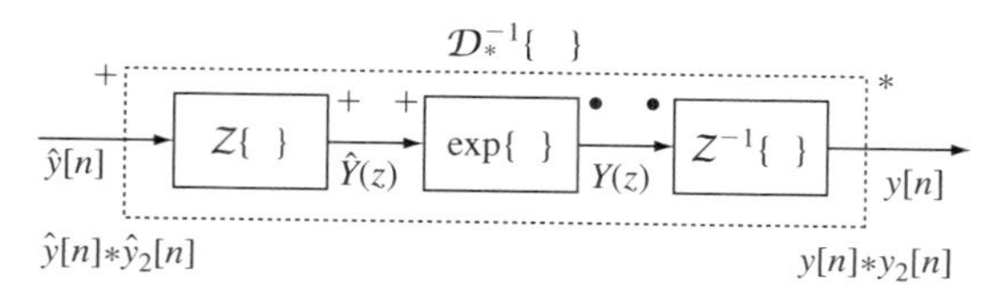

FIGURE 8.10
Representation of the inverse of the characteristic system for homomorphic deconvolution.

To represent signals as sequences, rather than in the z-transform domain as in Figure 8.8, the characteristic system can be represented as depicted in Figure 8.9, where the log function is surrounded by the z-transform operator and its inverse. Similarly, the inverse of the characteristic system can be represented as in Figure 8.10. Figures 8.5 and 8.6 are special cases of Figures 8.9 and 8.10 respectively for the special case $z = e^{j\omega}$; i.e., when the z-transform is evaluated on the unit circle of the z-plane.

The z-transform representation is especially useful for theoretical analysis where the z-transform polynomials are rational functions that can be more easily manipulated than DTFT expressions. This is illustrated in Section 8.2.3. Furthermore, as discussed in Section 8.4.2, if a powerful root solver is available, the z-transform representation is the basis of a method for computing the complex cepstrum.

8.2.3 Properties of the Complex Cepstrum

The general properties of the complex cepstrum can be illustrated by considering the case of signals that have rational z-transforms. The most general form that is necessary to consider for our purpose is

$$X(z) = \frac{A\prod_{k=1}^{M_i}(1 - a_k z^{-1})\prod_{k=1}^{M_o}(1 - b_k^{-1} z^{-1})}{\prod_{k=1}^{N_i}(1 - c_k z^{-1})}, \tag{8.19}$$

which can written as

$$X(z) = \frac{z^{-M_o} A \prod_{k=1}^{M_o} (-b_k^{-1}) \prod_{k=1}^{M_i} (1 - a_k z^{-1}) \prod_{k=1}^{M_o} (1 - b_k z)}{\prod_{k=1}^{N_i} (1 - c_k z^{-1})}, \tag{8.20}$$

where the magnitudes of the quantities, a_k, b_k, and c_k are all less than 1. Thus, the terms $(1 - a_k z^{-1})$ and $(1 - c_k z^{-1})$ correspond to zeros and poles inside the unit circle, and the terms $(1 - b_k^{-1} z^{-1})$ or $(1 - b_k z)$ correspond to zeros outside the unit circle.[3] The representation of Eq. (8.20) suggests factorization into a product of a minimum-phase and a maximum-phase signal [270] as in

$$X(z) = X_{\min}(z) \cdot z^{-M_o} X_{\max}(z), \tag{8.21a}$$

where the minimum-phase component has z-transform

$$X_{\min}(z) = \frac{A \prod_{k=1}^{M_i} (1 - a_k z^{-1})}{\prod_{k=1}^{N_i} (1 - c_k z^{-1})}, \tag{8.21b}$$

with all poles and zeros inside the unit circle, and the z-transform of the maximum-phase component is

$$X_{\max}(z) = \prod_{k=1}^{M_o} (-b_k^{-1}) \prod_{k=1}^{M_o} (1 - b_k z), \tag{8.21c}$$

with all the zeros lying outside the unit circle. The product representation in Eq. (8.21a) suggests the convolution $x[n] = x_{\min}[n] * x_{\max}[n - M_o]$, where the minimum-phase component is causal (i.e., $x_{\min}[n] = 0$ for $n < 0$) and the maximum-phase component is anti-causal (i.e., $x_{\max}[n] = 0$ for $n > 0$). The factor z^{-M_o} represents simply the shift in the time origin by M_o samples that is required to make $x_{\max}[n]$ and the overall sequence, $x[n]$, causal. This factor is generally omitted in computing the complex cepstrum because it is often irrelevant or because it can be dealt with by simply noting the value of M_o. This is equivalent to assuming that $x[n] = x_{\min}[n] * x_{\max}[n]$ and therefore the complex cepstrum is $\hat{x}[n] = \hat{x}_{\min}[n] + \hat{x}_{\max}[n]$;

[3] To simplify our discussion, we do not allow zeros or poles to be precisely on the unit circle of the z-plane. This restriction is not a serious practical limitation, and it does not limit the conclusions we draw about the general properties of the cepstrum and complex cepstrum. Further, we restrict our attention to stable, causal infinite-duration sequences whose poles are all inside the unit circle of the z-plane.

i.e., it is the sum of a part due to the minimum-phase component of the input and a part due to the maximum-phase component.

Under the assumption of Eq. (8.18), the complex logarithm of $X(z)$ is

$$\hat{X}(z) = \log|A| + \sum_{k=1}^{M_0} \log|b_k^{-1}| + \log[z^{-M_o}] + \sum_{k=1}^{M_i} \log(1 - a_k z^{-1})$$
$$+ \sum_{k=1}^{M_o} \log(1 - b_k z) - \sum_{k=1}^{N_i} \log(1 - c_k z^{-1}). \quad (8.22)$$

Observe that we have used only the magnitude of the term $A\prod_{k=1}^{M_o}(-b_k^{-1})$ in Eq. (8.22) when taking the complex logarithm of $X(z)$ in Eq. (8.20). This is because A times the product of the zeros outside the unit circle (which occur in complex conjugate pairs) will always be real if the signal $x[n]$ is real. The algebraic sign of the product can be determined if necessary, but it is usually not used in computing the complex cepstrum. When Eq. (8.22) is evaluated on the unit circle, it can be seen that the term $\log[e^{-j\omega M_o}]$ will contribute only to the imaginary part of the complex logarithm. Since this term only carries information about the time origin, it is generally removed in the process of computing the complex cepstrum [342]. Thus, we will also neglect this term in our discussion of the properties of the complex cepstrum. Using the fact that each of the logarithmic terms can be written as a power series expansion, based on the well known power series,

$$\log(1 - Z) = -\sum_{n=1}^{\infty} \frac{Z^n}{n}, \qquad |Z| < 1, \quad (8.23)$$

it is relatively straightforward to show that the complex cepstrum has the form

$$\hat{x}[n] = \begin{cases} \log|A| + \sum_{k=1}^{M_0} \log|b_k^{-1}| & n = 0 \\ \sum_{k=1}^{N_i} \frac{c_k^n}{n} - \sum_{k=1}^{M_i} \frac{a_k^n}{n} & n > 0 \\ \sum_{k=1}^{M_0} \frac{b_k^{-n}}{n} & n < 0. \end{cases} \quad (8.24)$$

Equation (8.24) displays the following important properties of the complex cepstrum:

1. In general, the complex cepstrum is non-zero and of infinite extent for both positive and negative n, even though $x[n]$ may be causal, or even of finite duration (when $X(z)$ has only zeros).
2. The complex cepstrum is a decaying sequence that is bounded by

$$|\hat{x}[n]| < \beta \frac{\alpha^{|n|}}{|n|}, \quad \text{for} \quad |n| \rightarrow \infty, \tag{8.25}$$

   where α is the maximum absolute value of the quantities a_k, b_k, and c_k, and β is a constant multiplier.[4]
3. The zero-quefrency value of the complex cepstrum (and the cepstrum) depends on the gain constant and the zeros outside the unit circle. Setting $\hat{x}[0] = 0$ (and therefore $c[0] = 0$) is equivalent to normalizing the log magnitude spectrum to a gain constant of

$$A \prod_{k=1}^{M_o} (-b_k^{-1}) = 1. \tag{8.26}$$

4. If $X(z)$ has no zeros outside the unit circle (i.e., all $b_k = 0$),[5] then

$$\hat{x}[n] = 0, \quad \text{for} \quad n < 0. \tag{8.27}$$

   Such signals are called *minimum-phase* signals [270].
5. If $X(z)$ has no poles or zeros inside the unit circle (i.e., all $a_k = 0$ and all $c_k = 0$), then

$$\hat{x}[n] = 0, \quad \text{for} \quad n > 0. \tag{8.28}$$

   Such signals are called *maximum-phase* signals [270].

8.2.4 Some Examples of Complex Cepstrum Analysis

In this section, we give a few examples of complex cepstra of simple signals. We base our work on the power series expansion of Eq. (8.23), which is the basis for the general formula in Eq. (8.24).

[4]In practice, we generally deal with finite-length signals, which are represented by polynomials in z^{-1}; i.e., the numerator in Eq. (8.20). In many cases, the sequence may be hundreds or thousands of samples long. A somewhat remarkable result is that for finite-length sequences of speech samples, almost all of the zeros of the z-transform polynomial tend to cluster around the unit circle, and as the sequence length increases, the roots move closer to the unit circle [152]. This implies that for long, finite-length sequences, the decay of the complex cepstrum is due primarily to the factor $1/|n|$.

[5]We tacitly assume no poles outside the unit circle.

Example 8.1 Decaying Exponential Sequence

Determine the complex cepstrum of the minimum-phase sequence

$$x_1[n] = a^n u[n], \qquad |a| < 1.$$

Solution

We first determine the z-transform of $x_1[n]$ as follows:

$$X_1(z) = \sum_{n=0}^{\infty} a^n z^{-n} = \frac{1}{1 - az^{-1}}, \quad |z| > |a|. \tag{8.29a}$$

Next we solve for $\hat{X}_1(z)$ as:

$$\begin{aligned}\hat{X}_1(z) &= \log[X_1(z)] \\ &= -\log(1 - az^{-1}) \\ &= \sum_{n=1}^{\infty} \left(\frac{a^n}{n}\right) z^{-n}.\end{aligned} \tag{8.29b}$$

The complex cepstrum value $\hat{x}_1[n]$ is simply the coefficient of the term z^{-n} in Eq. (8.29b); i.e.,

$$\hat{x}_1[n] = \frac{a^n}{n} u[n-1]. \tag{8.29c}$$

Example 8.2 Single Zero outside the Unit Circle

Determine the complex cepstrum of the maximum-phase sequence

$$x_2[n] = \delta[n] + b\delta[n+1], \qquad |b| < 1.$$

Solution

In this case the z-transform of $x_2[n]$ is easily shown to be

$$X_2(z) = 1 + bz = bz(1 + b^{-1}z^{-1}). \tag{8.30a}$$

That is, $X_2(z)$ is a single zero outside the unit circle. Next we determine $\hat{X}_2(z)$, obtaining

$$\begin{aligned}\hat{X}_2(z) &= \log[X_2(z)] \\ &= \log(1 + bz) \\ &= \sum_{n=1}^{\infty} \frac{(-1)^{n+1}}{n} b^n z^n.\end{aligned} \tag{8.30b}$$

Again picking the coefficient of z^{-n} for $\hat{x}_2[n]$, we obtain

$$\hat{x}_2[n] = \frac{(-1)^{n+1}b^n}{n}u[-n-1]. \tag{8.30c}$$

Example 8.3 Simple Echo

Determine the complex cepstrum of the sequence

$$x_3[n] = \delta[n] + \alpha\delta[n-N_p], \quad |\alpha| < 1.$$

Discrete convolution of any sequence $x_1[n]$ with this sequence produces a scaled-by-α echo of the first sequence; i.e.,

$$x_1[n] * (\delta[n] + \alpha\delta[n-N_p]) = x_1[n] + \alpha x_1[n-N_p].$$

Solution

The z-transform of $x_3[n]$ is

$$X_3(z) = 1 + \alpha z^{-N_p}, \tag{8.31a}$$

and, assuming that $|\alpha| < 1$, the z-transform of $\hat{x}_3[n]$ is

$$\begin{aligned}\hat{X}_3(z) &= \log\left[X_3(z)\right] \\ &= \log\left(1 + \alpha z^{-N_p}\right) \\ &= \sum_{n=1}^{\infty}\frac{(-1)^{n+1}}{n}\alpha^n z^{-nN_p}.\end{aligned} \tag{8.31b}$$

In this case, $\hat{X}_3(z)$ has only integer powers of z^{-N_p}, so

$$\hat{x}_3[n] = \sum_{k=1}^{\infty}(-1)^{k+1}\frac{\alpha^k}{k}\delta[n-kN_p]. \tag{8.31c}$$

Thus, Eq. (8.31c) shows that the complex cepstrum of $x_3[n] = \delta[n] + \alpha\delta[n-N_p]$ is an impulse train with impulses spaced every N_p samples. It is a useful exercise to work through this example with the assumption $|\alpha| > 1$.

An important generalization of the sequence in Example 8.3 is the sequence

$$p[n] = \sum_{r=0}^{M}\alpha_r\delta[n-rN_p]; \tag{8.32}$$

i.e., a train of impulses with spacing N_p samples. The z-transform of Eq. (8.86) is

$$P(z) = \sum_{r=0}^{M} \alpha_r z^{-rN_p}. \tag{8.33}$$

From Eq. (8.33), it is evident that $P(z)$ is really a polynomial in the variable z^{-N_p} rather than z^{-1}. Thus, $P(z)$ can be expressed as a product of factors of the form $(1 - az^{-N_p})$ and $(1 - bz^{N_p})$, and therefore, the complex cepstrum, $\hat{p}[n]$, will be non-zero only at integer multiples of N_p.

The fact that the complex cepstrum of a train of uniformly spaced impulses is also a uniformly spaced impulse train with the same spacing is a very important result for speech analysis as we will see in Sections 8.3 and 8.5.

A final example shows how the results of the preceding examples can be combined to find the complex cepstrum of a convolution.

Example 8.4 Complex Cepstrum of $x_4[n] = x_1[n] * x_2[n] * x_3[n]$

Determine the complex cepstrum of the sequence $x_4[n] = x_1[n] * x_2[n] * x_3[n]$; i.e., a signal comprised of the convolution of the sequences of Examples 8.1, 8.2, and 8.3.

Solution

We can write the sequence $x_4[n]$ as

$$\begin{aligned} x_4[n] &= x_1[n] * x_2[n] * x_3[n] \\ &= (a^n u[n]) * (\delta[n] + b\delta[n+1]) * (\delta[n] + \alpha\delta[n-N_p]) \\ &= (a^n u[n] + ba^{n+1}u[n+1]) \\ &\quad + \alpha(a^n u[n-N_p] + ba^{n-N_p+1}u[n-N_p+1]). \end{aligned} \tag{8.34a}$$

The complex cepstrum of $x_4[n]$ is the sum of the complex cepstra of the three sequences, giving

$$\begin{aligned} \hat{x}_4[n] &= \hat{x}_1[n] + \hat{x}_2[n] + \hat{x}_3[n] \\ &= \frac{a^n}{n}u[n-1] + \frac{(-1)^{n+1}b^n}{n}u[-n-1] \\ &\quad + \sum_{k=1}^{\infty} \frac{(-1)^{k+1}\alpha^k}{k}\delta[n-kN_p]. \end{aligned} \tag{8.34b}$$

Figure 8.11 shows plots of the input signal, $x_4[n]$, and the resulting complex cepstrum, $\hat{x}_4[n]$, for the case $a = 0.9$, $b = 0.8$, $\alpha = 0.7$, and $N_p = 15$. Figure 8.11a shows the input signal waveform. The decaying signal that corresponds to $x_1[n] * x_2[n]$ and the repetition of that signal scaled by α and delayed by $N_p = 15$

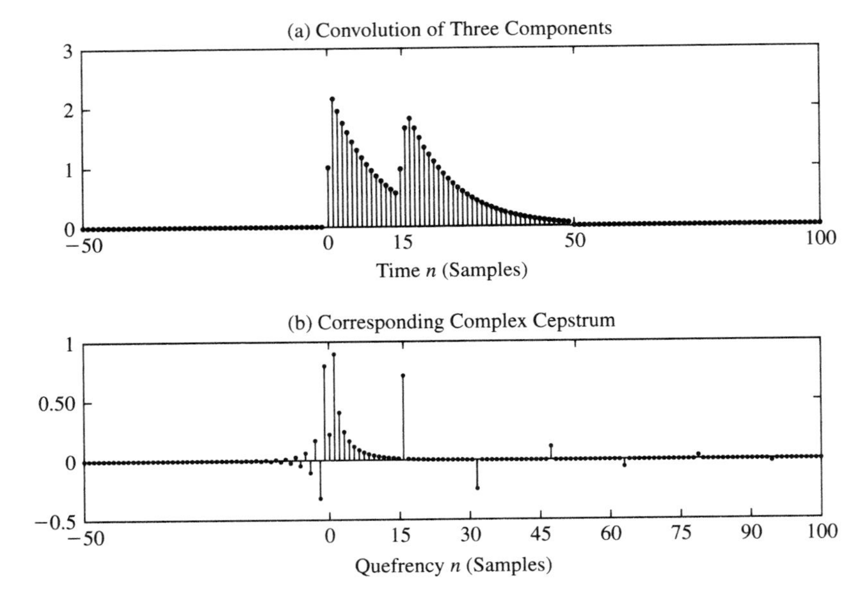

FIGURE 8.11
Plots of the waveform and the resulting complex cepstrum for Example 8.4: (a) the 3-way convolution of the sequences of Examples 8.1, 8.2, and 8.3 and (b) corresponding complex cepstrum.

samples are both clearly visible. Figure 8.11b shows the components of the complex cepstrum, with $\hat{x}[n]$ for $n < 0$ contributed by $x_2[n]$, and $\hat{x}[n]$ for $n \geq 0$ is comprised of the complex cepstra of $x_1[n]$ and $x_3[n]$. In particular, the impulses at multiples of $N_p = 15$ are due to the echoing caused by convolution with $x_3[n]$. Note that the contributions due to the pole at $z = a$ and the zero at $z = -1/b$ die out rapidly as n increases.

8.2.5 Minimum- and Maximum-Phase Signals

A general result for minimum-phase sequences of the form Eq. (8.27) is that they can be completely represented by only the real parts of their DTFTs [270]. Thus, since the real part of the DTFT of the complex cepstrum is $\log|X(e^{j\omega})|$, we should be able to represent the complex cepstrum of minimum-phase signals by the logarithm of the magnitude of the DTFT alone. Remembering that the real part of the Fourier transform is the DTFT of the even part of the sequence, it follows that since $\log|X(e^{j\omega})|$ is the DTFT of the cepstrum, then the cepstrum is the even part of the complex cepstrum; i.e.,

$$c[n] = \frac{\hat{x}[n] + \hat{x}[-n]}{2}. \tag{8.35}$$

It follows from Eqs. (8.27) and (8.35) that since $\hat{x}[n] = 0$ for $n < 0$,

$$\hat{x}_{\text{mnp}}[n] = \begin{cases} 0 & n < 0 \\ c[n] & n = 0 \\ 2c[n] & n > 0, \end{cases} \tag{8.36}$$

where we use the subscript notation mnp for minimum-phase signals, and mxp for maximum-phase signals. Thus, for minimum-phase sequences, the complex cepstrum can be obtained by computing the cepstrum and then using Eq. (8.36).

Similar results can be obtained for maximum-phase signals. In this case, it can be seen from Eqs. (8.28) and (8.35) that, for maximum-phase signals,

$$\hat{x}_{\text{mxp}}[n] = \begin{cases} 0 & n > 0 \\ c[n] & n = 0 \\ 2c[n] & n < 0, \end{cases} \tag{8.37}$$

so again, the complex cepstrum of a maximum-phase signal can be computed from only the $\log|X(e^{j\omega})|$.

8.3 HOMOMORPHIC ANALYSIS OF THE SPEECH MODEL

We have held that a fundamental tenet of digital speech processing is that speech can be represented as the output of a linear, time-varying system whose properties vary slowly with time. This is embodied in the model of Figure 8.12, which emerged from our discussion of the physics of speech production. This leads to the basic principle of speech analysis that assumes that short segments of the speech signal can be modeled as the output of a linear time-invariant system excited either by a quasi-periodic impulse train or a random noise signal. As we have seen repeatedly in previous chapters, the fundamental problem of speech analysis is to reliably and robustly estimate the parameters of the model of Figure 8.12 (i.e., the pitch period control, the shape of the glottal pulse, the gains for the voiced or unvoiced excitation signals, the state of the voiced/unvoiced switch, the vocal tract parameters, and the radiation model, as illustrated in Figure 8.12), and to measure the variations of these model control parameters with time.

Since the excitation and impulse response of a linear time-invariant system are combined by convolution, the problem of speech analysis can also be viewed as a problem in separating the components of a convolution, and therefore, homomorphic systems and the cepstrum are useful tools for speech analysis. In the model of Figure 8.12, the pressure signal at the lips, $s[n]$, for a voiced section of speech is represented as the convolution

$$s[n] = p[n] * h_V[n], \tag{8.38a}$$

where $p[n]$ is the quasi-periodic voiced excitation signal, and $h_V[n]$ represents the combined effect of the vocal tract impulse response $v[n]$, the glottal pulse $g[n]$, the radiation

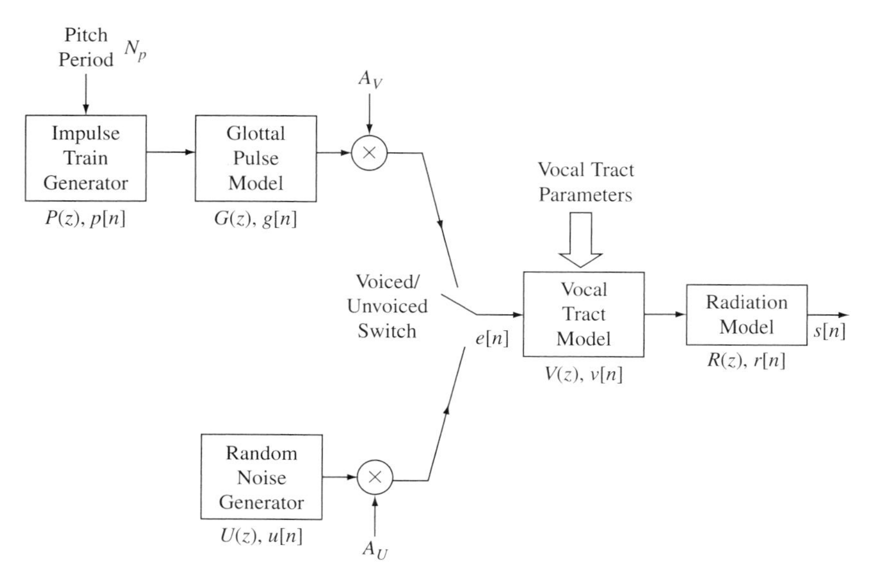

FIGURE 8.12
General discrete-time model of speech production.

load response at the lips, $r[n]$, and the voiced gain, A_V. The effective impulse response, $h_V[n]$, is itself the convolution of $g[n]$, $v[n]$, and $r[n]$, including scaling by the voiced section gain control, A_V; i.e.,

$$h_V[n] = A_V \cdot g[n] * v[n] * r[n]. \tag{8.38b}$$

Recall that when it is not necessary to make a fine distinction, it is common to refer to $h_V[n]$ as simply "the vocal tract impulse response for voiced speech" even though it is only partially determined by the vocal tract configuration.

Similarly, for unvoiced sections of speech, we can represent the pressure signal at the lips as

$$s[n] = u[n] * h_U[n], \tag{8.38c}$$

where $u[n]$ is the (unit-variance) random unvoiced excitation signal and $h_U[n]$ is the convolution of the vocal tract response, $v[n]$, with the radiation load response, $r[n]$, including scaling by the unvoiced section gain control A_U; i.e.,

$$h_U[n] = A_U \cdot v[n] * r[n]. \tag{8.38d}$$

Again, it is generally convenient to refer to $h_U[n]$ as "the vocal tract impulse response for unvoiced speech" even though it is only partially determined by the vocal tract configuration.

8.3.1 Homomorphic Analysis of the Model for Voiced Speech

To illustrate how homomorphic systems and cepstrum analysis can be applied to speech, we apply the results of Section 8.2 to a discrete-time model for a sustained /AE/ vowel with a fundamental frequency $f_0 = 125$ Hz, or a pitch period of 8 msec.

As a model for the glottal pulse, we use the model proposed by Rosenberg [326],

$$g[n] = \begin{cases} 0.5[1 - \cos(\pi(n+1)/N_1)] & 0 \le n \le N_1 - 1 \\ \cos(0.5\pi(n+1-N_1)/N_2) & N_1 \le n \le N_1 + N_2 - 2 \\ 0 & \text{otherwise.} \end{cases} \tag{8.39}$$

This glottal pulse model is illustrated in Figure 8.13a where $g[n]$ in Eq. (8.39) is plotted for $N_1 = 25$ and $N_2 = 10$. This finite-length sequence has a length of 34 samples, so its z-transform, $G(z)$, is a polynomial of the form

$$G(z) = z^{-33} \prod_{k=1}^{33} (-b_k^{-1}) \prod_{k=1}^{33} (1 - b_k z) \tag{8.40}$$

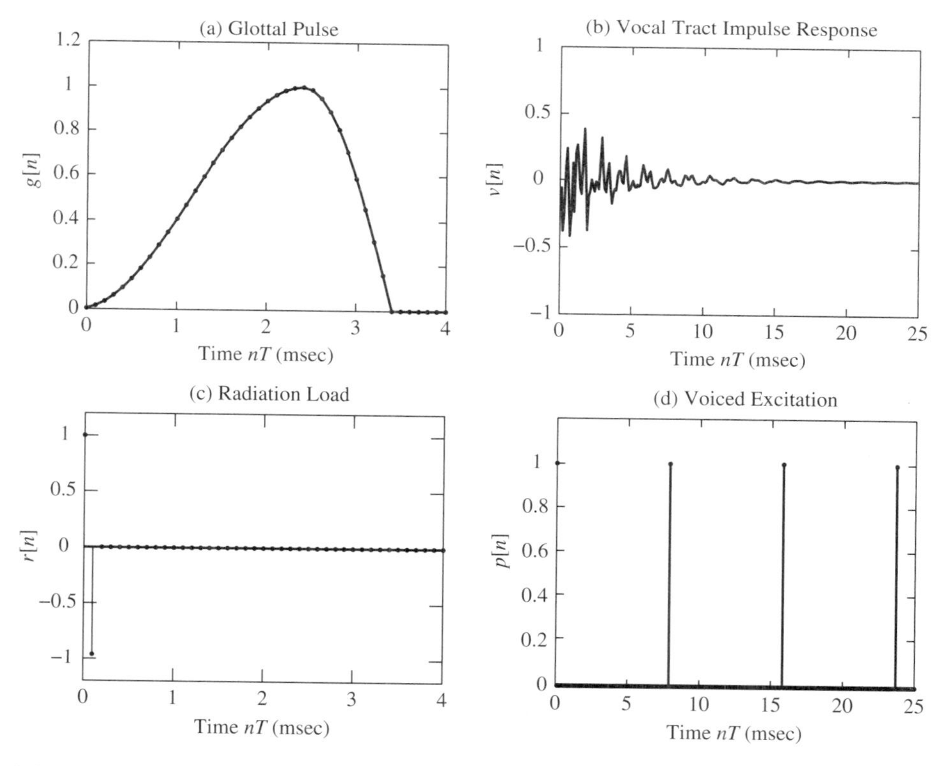

FIGURE 8.13
Time-domain representation of speech model: (a) glottal pulse $g[n]$ as given by Eq. (8.39); (b) vocal tract impulse response, $v[n]$; (c) radiation load impulse response $r[n]$; and (d) periodic excitation $p[n]$.

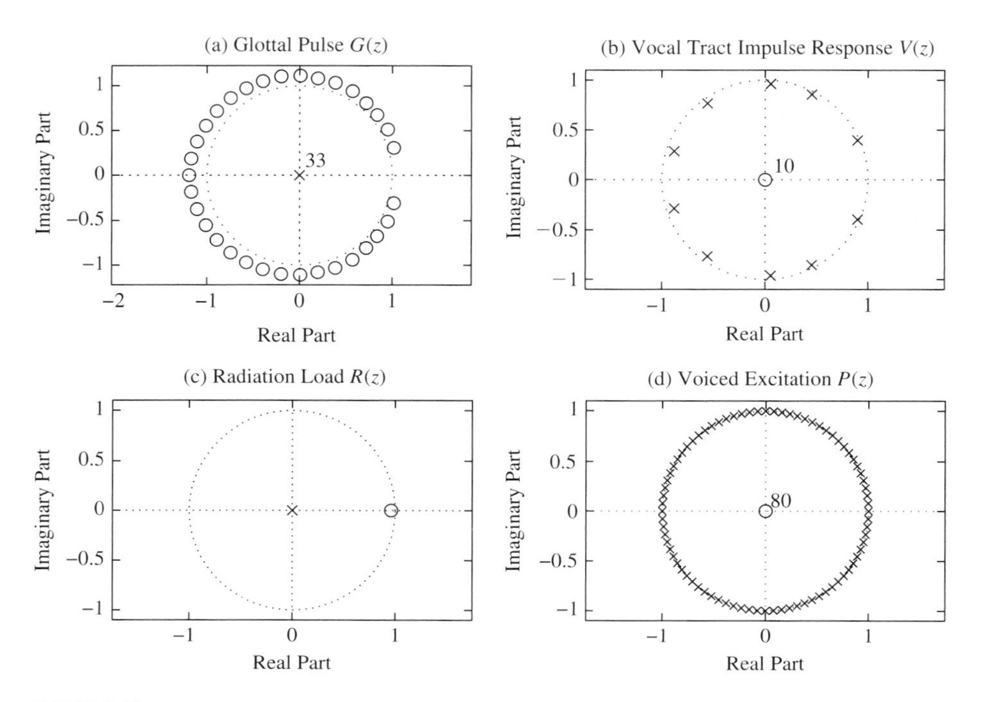

FIGURE 8.14
Pole-zero plots for speech model: (a) glottal pulse $G(z)$; (b) vocal tract system function, $V(z)$; (c) radiation load system function $R(z)$; and (d) z-transform of periodic excitation $P(z)$.

with 33 zeros, which are plotted in Figure 8.14a. Note that all 33 roots of $G(z)$ are outside the unit circle for this example; i.e., this glottal pulse is a *maximum-phase* sequence.

The vocal tract system is specified in terms of its formant frequencies and bandwidths in $V(z)$ given as a product of second-order sections of the form

$$V(z) = \frac{1}{\prod_{k=1}^{5} (1 - 2e^{-2\pi\sigma_k T} \cos(2\pi F_k T) z^{-1} + e^{-4\pi\sigma_k T} z^{-2})}. \tag{8.41}$$

The model for an /AE/ vowel is given in Table 8.2 in terms of the set of analog center frequencies, F_k, and bandwidths, $2\sigma_k$, of the first five formants. All frequencies and bandwidths are specified in Hz. The sampling rate associated with Eq. (8.41) is assumed to be $F_s = 1/T = 10{,}000$ Hz thus providing adequate bandwidth for five formant frequencies. The first 251 samples of $v[n]$ and the 10 poles of $V(z)$ are shown in Figures 8.13b and 8.14b respectively. (Note that the individual signal samples are shown connected by straight lines in this case.)

TABLE 8.2 Formant frequencies (in Hz) and bandwidths (in Hz) for the /AE/ vowel model.

k	F_k (Hz)	$2\sigma_k$ (Hz)
1	660	60
2	1720	100
3	2410	120
4	3500	175
5	4500	250

The model for the radiation load is the simple first difference system

$$R(z) = 1 - \gamma z^{-1}. \tag{8.42}$$

The corresponding impulse response $r[n] = \delta[n] - \gamma\delta[n-1]$ is shown in Figure 8.13c and the single zero is shown in Figure 8.14c for the specific value of $\gamma = 0.96$.

The final component of the model is the periodic excitation $p[n]$. To allow a simple analysis with z-transforms, we define $p[n]$ as the one-sided quasi-periodic impulse train

$$p[n] = \sum_{k=0}^{\infty} \beta^k \delta[n - kN_p], \tag{8.43}$$

which has z-transform

$$P(z) = \sum_{k=0}^{\infty} \beta^k z^{-kN_p} = \frac{1}{1 - \beta z^{-N_p}}. \tag{8.44}$$

Note that $P(z)$ is a rational function of z^{-N_p} due to the even spacing of the assumed excitation sequence. The denominator in Eq. (8.44) has N_p roots at the z-plane locations $z_k = \beta^{1/N_p} e^{j2\pi k/N_p}$, $k = 0, 1, \ldots, N_p - 1$. Figure 8.13d shows the first few impulse samples of $p[n]$ with spacing $N_p = 80$ samples and $\beta = 0.999$, corresponding to a fundamental frequency of $10{,}000/80 = 125$ Hz. Figure 8.14d shows the N_p poles on a circle of radius β^{1/N_p} for the case $\beta = 0.999$. The angular spacing between the poles is $2\pi/N_p$ radians, corresponding to an analog frequency of $10{,}000/N_p = 125$ Hz for the sampling rate of $F_s = 10{,}000$ Hz. This spacing is, of course, equal to the fundamental frequency.

The log magnitudes of the DTFTs corresponding to the sequences in Figure 8.13 and pole-zero plots in Figure 8.14 are shown in Figure 8.15 in corresponding locations. Note that the DTFTs are plotted as $\log_e|\cdot|$ rather than in dB (i.e., $20\log_{10}|\cdot|$) as is common elsewhere throughout this text. To convert the plots in Figure 8.15a–c to dB, simply multiply by $20\log_{10} e = 8.6859$. We see that the spectral contribution due to the glottal pulse is a lowpass component that has a dynamic range of about 6 between $F = 0$ and $F = 5000$ Hz. This is equivalent to about 50 dB spectral fall-off. Figure 8.15b

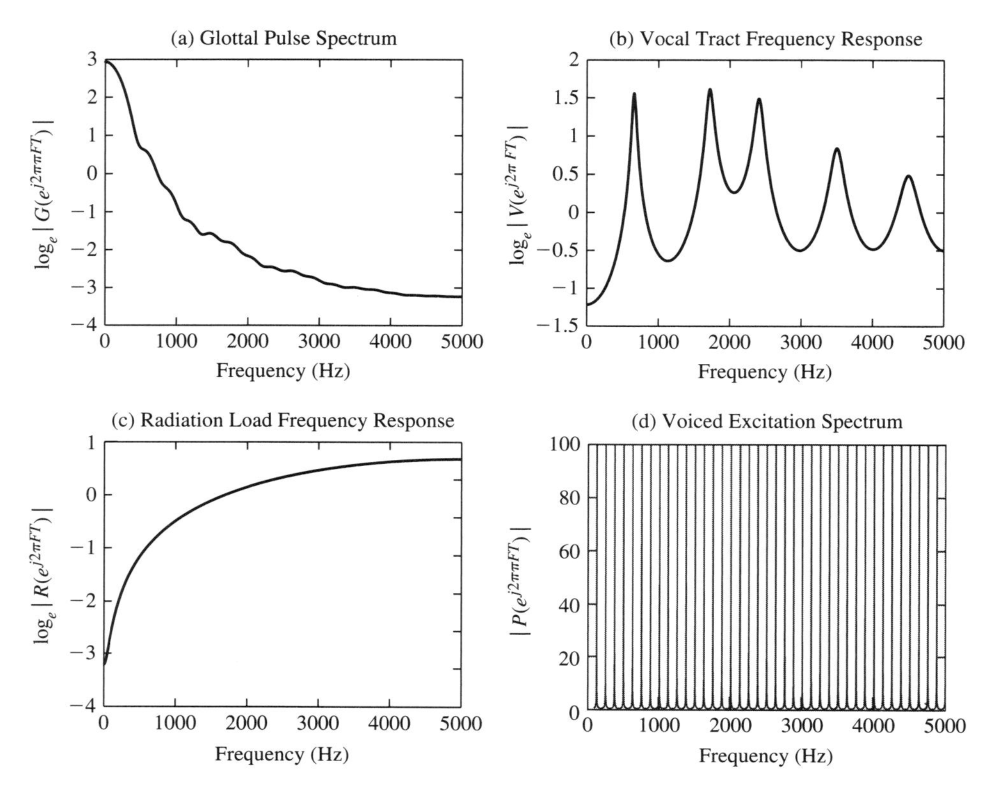

FIGURE 8.15
Log magnitude (base e) of DTFTs: (a) glottal pulse DTFT $\log |G(e^{j\omega})|$; (b) vocal tract frequency response, $\log |V(e^{j\omega})|$; (c) radiation load frequency response $\log |R(e^{j\omega})|$; and (d) magnitude of DTFT of periodic excitation $|P(e^{j\omega})|$.

shows the spectral contribution of the vocal tract system. The peaks of the spectrum are approximately at the locations given in Table 8.2 with bandwidths that increase with increasing frequency. As depicted in Figure 8.15c, the effect of radiation is to give a high frequency boost that partially compensates for the fall-off due to the glottal pulse. Finally, Figure 8.15d shows $|P(e^{j2\pi FT})|$ (not the log) as a function of F. Note the periodic structure due to the periodicity of $p[n]$. The fundamental frequency for $N_p = 80$ is $F_0 = 10{,}000/80 = 125$ Hz.[6]

Now if the components of the speech model are combined by convolution, as defined in the upper branch of Figure 8.12, the result is the synthetic speech signal $s[n]$ which is plotted in Figure 8.16a. The frequency-domain representation is plotted in Figure 8.16b. The smooth heavy line is the sum of parts (a), (b), and (c) of Figure 8.15

[6]In order to be able to evaluate Eq. (8.44) for the plot in Figure 8.15d it was necessary to use $\beta = 0.999$; i.e., the excitation was not perfectly periodic.

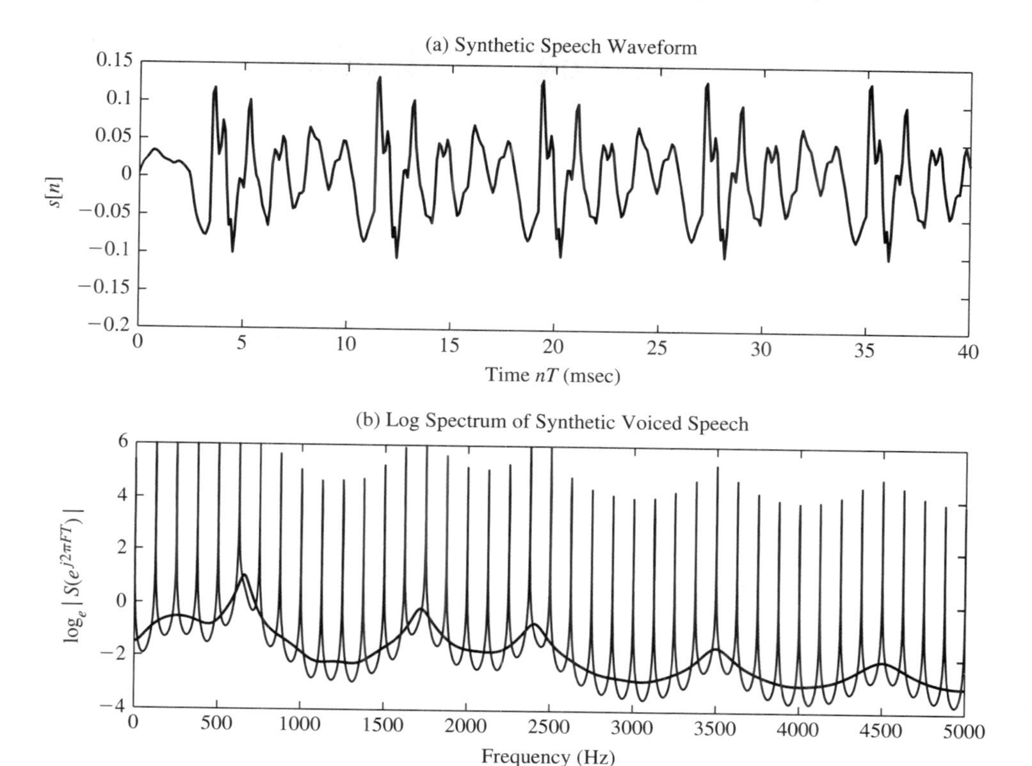

FIGURE 8.16
(a) Output of speech model system for the systems represented in Figures 8.13, 8.14, and 8.15.
(b) Corresponding DTFT.

corresponding to $\log |H_V(e^{j2\pi FT})|$; i.e., the log magnitude of the frequency response corresponding to the impulse response $h_V[n] = Ag[n] * v[n] * r[n]$. The more rapidly varying curve plotted with the thin line includes the log magnitude of the excitation spectrum as well; i.e., this curve represents the logarithm of the magnitude of the DTFT of the output $s[n]$ of the speech model.

Now consider the complex cepstrum of the output of the model. Since the output is $s[n] = h_V[n] * p[n] = A_V g[n] * v[n] * r[n] * p[n]$, it follows that

$$\hat{s}[n] = \hat{h}_V[n] + \hat{p}[n] = \log |A_V|\delta[n] + \hat{g}[n] + \hat{v}[n] + \hat{r}[n] + \hat{p}[n]. \tag{8.45}$$

Using Eq. (8.24), we can use the z-transform representations in Eqs. (8.40), (8.41), (8.42), and (8.44) to obtain the individual complex cepstrum components shown in Figure 8.17. Note that because the glottal pulse is maximum phase, its complex cepstrum satisfies $\hat{g}[n] = 0$ for $n > 0$. The vocal tract and radiation systems are assumed to be minimum-phase so $\hat{v}[n]$ and $\hat{r}[n]$ are zero for $n < 0$. Also note that using the power

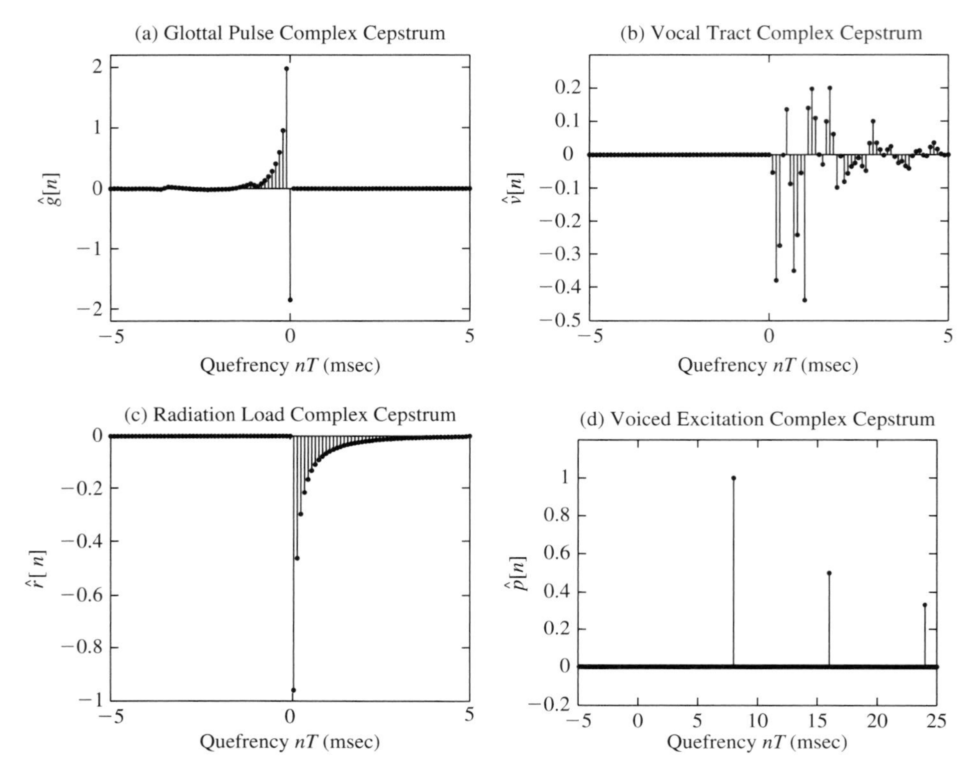

FIGURE 8.17
Complex cepstra of the speech model: (a) glottal pulse $\hat{g}[n]$; (b) vocal tract impulse response $\hat{v}[n]$; (c) radiation load impulse response $\hat{r}[n]$; and (d) periodic excitation $\hat{p}[n]$. (Note differences in amplitude scale among the four plots.)

series expansion in Eq. (8.23) with Eq. (8.44), we obtain

$$\hat{P}(z) = -\log(1 - \beta z^{-N_p}) = \sum_{k=1}^{\infty} \frac{\beta^k}{k} z^{-kN_p}, \tag{8.46}$$

from which it follows that

$$\hat{p}[n] = \sum_{k=1}^{\infty} \frac{\beta^k}{k} \delta[n - kN_p]. \tag{8.47}$$

As seen in Figure 8.17d, the spacing between impulses in the complex cepstrum due to the input $p[n]$ is $N_p = 80$ samples, corresponding to a pitch period of $1/F_0 = 80/10{,}000 = 8$ msec. Note that in Figure 8.17, we have shown the discrete quefrency index in terms of msec; i.e., the horizontal axis shows nT.

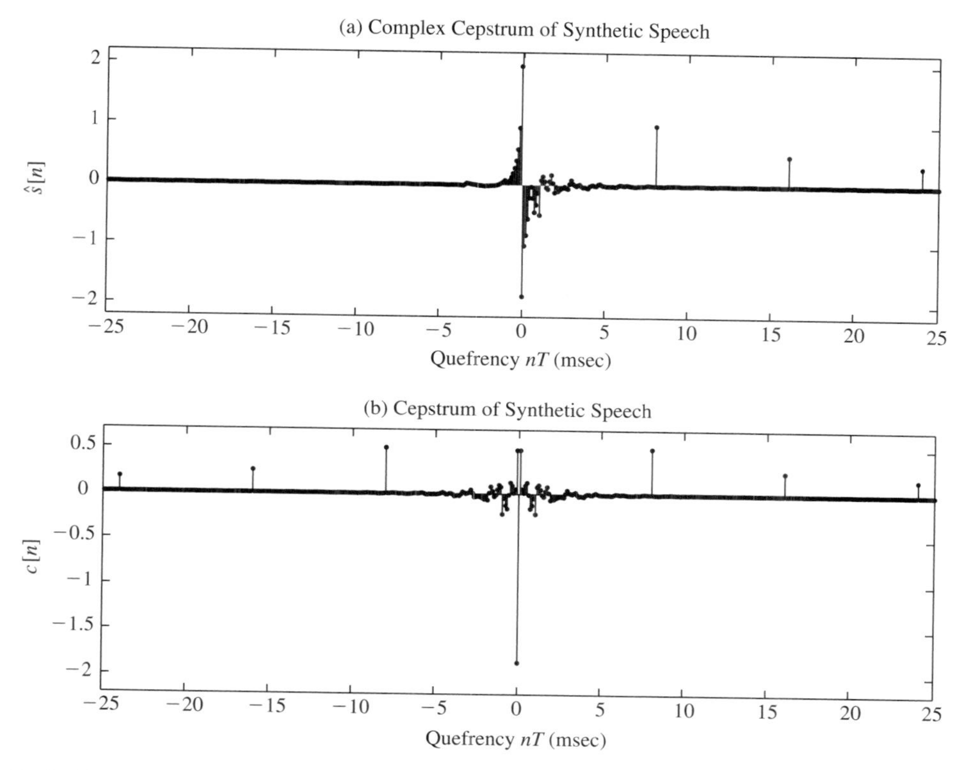

FIGURE 8.18
(a) Complex cepstrum of synthetic speech output. (b) Corresponding cepstrum of synthetic speech output.

According to Eq. (8.45), the complex cepstrum of the synthetic speech output is the sum of all of the complex cepstra in Figure 8.17. Thus, $\hat{s}[n] = \hat{h}_V[n] + \hat{p}[n]$ is depicted in Figure 8.18a. The cepstrum, being the even part of $\hat{s}[n]$, is depicted in Figure 8.18b. Note that in both cases, the impulses due to the periodic excitation tend to stand out from the contributions due to the system impulse response. The location of the first impulse peak is at quefrency N_p, which is the period of the excitation. This impulse would not appear in the cepstrum of an unvoiced speech segment. This is the basis for the use of the cepstrum or complex cepstrum for pitch detection; i.e., the presence of a strong peak signals voiced speech, and its quefrency is an estimate of the pitch period.

Finally, it is worthwhile to connect the z-transform analysis employed in this example to the DTFT representation of the complex cepstrum. This is depicted in Figures 8.19a and 8.19b, which show the log magnitude and continuous phase of the DTFT $S(e^{j2\pi FT})$. These are, of course, the real and imaginary parts of $\hat{S}(e^{j2\pi FT})$, the DTFT of the complex cepstrum, $\hat{s}[n]$. The heavy lines show the contributions to the log magnitude and continuous phase due to the overall system response; i.e., $\hat{H}_V(e^{j2\pi FT}) = \log|H_V(e^{j2\pi FT})| + j\arg\{H_V(e^{j2\pi FT})\}$. The thin lines show the total log

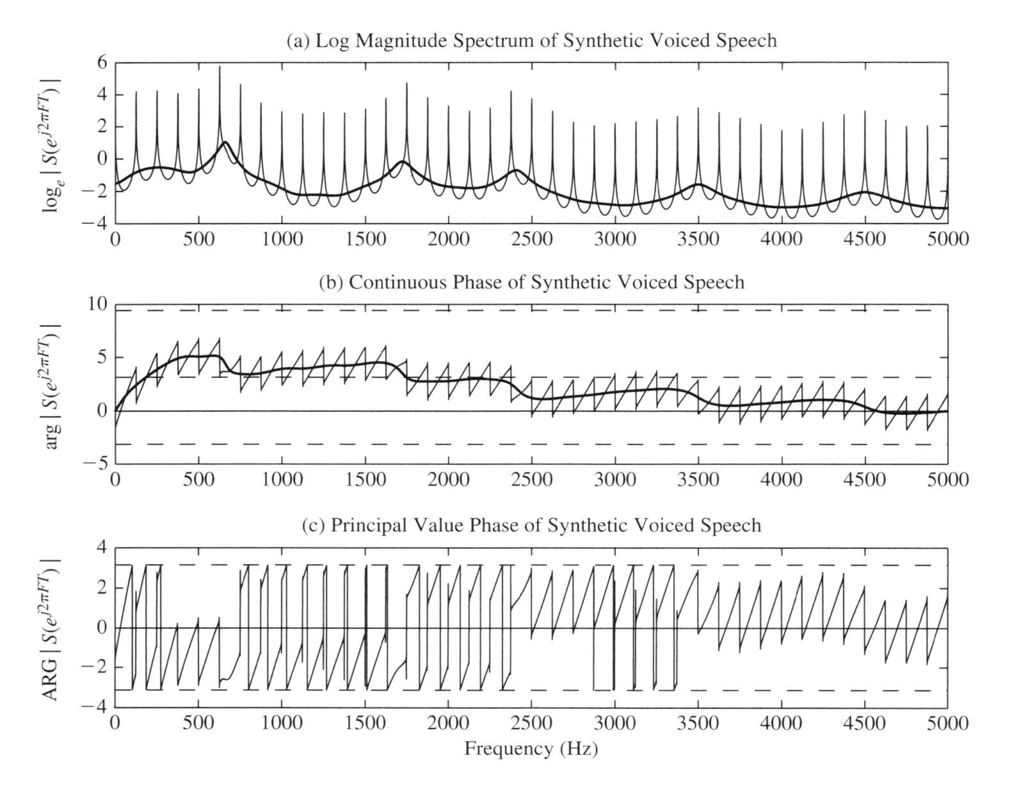

FIGURE 8.19
Frequency-domain representation of the complex cepstrum: (a) log magnitude $\log |S(e^{j2\pi FT})|$ (real part of $\hat{S}(e^{j2\pi FT})$); (b) continuous phase $\arg\{S(e^{j2\pi FT})\}$ (imaginary part of $\hat{S}(e^{j2\pi FT})$); and (c) principal value phase $\text{ARG}\{S(e^{j2\pi FT})\}$. The heavy lines in (a) and (b) represent $\hat{H}_V(e^{j2\pi FT}) = \log |H_V(e^{j2\pi FT})| + j\arg\{H_V(e^{j2\pi FT})\}$.

magnitude and continuous phase of the output of the system. Observe that the excitation introduces a periodic (in F) variation in both the log magnitude and continuous phase that is superimposed upon the more slowly varying components due to the system response. It is this periodic component that manifests itself in the cepstrum as the impulses at quefrencies that are multiples of N_p, and it is this behavior that motivated the original definition of the cepstrum by Bogert et al. [39].

8.3.2 Homomorphic Analysis of the Model for Unvoiced Speech

In Section 8.3.1, we considered an extended example of homomorphic analysis of the discrete-time model of voiced speech production. This analysis is exact for the assumed model, since we were able to determine the z-transforms of each of the convolutional components of the synthetic speech output. A completely similar analysis is not possible for the model for unvoiced speech production since no z-transform representation exists directly for the random noise input signal itself. However, if we employ the autocorrelation and power spectrum representation for the model for unvoiced speech production, we can obtain results similar to those for voiced speech.

Recall that for unvoiced speech, we have no glottal pulse excitation so the model output is $s[n] = h_U[n] * u[n] = v[n] * r[n] * (A_U u[n])$, where $u[n]$ is a unit-variance white noise sequence. The autocorrelation representation of unvoiced speech is therefore

$$\phi_{ss}[n] = \phi_{vv}[n] * \phi_{rr}[n] * (A_U^2 \delta[n]) = A_U^2 \phi_{vv}[n] * \phi_{rr}[n], \tag{8.48}$$

where $\phi_{vv}[n]$ and $\phi_{rr}[n]$ are the deterministic autocorrelation functions of the vocal tract and radiation systems respectively. These are combined by convolution. The z-transform of $\phi_{ss}[n]$ exists and is given by

$$\Phi_{ss}(z) = A_U^2 \Phi_{vv}(z) \Phi_{rr}(z), \tag{8.49}$$

where

$$\Phi_{vv}(z) = V(z)V(z^{-1}) \tag{8.50a}$$

$$\Phi_{rr}(z) = R(z)R(z^{-1}) \tag{8.50b}$$

are deterministic z-transforms representing the shaping of the power spectrum by the vocal tract and radiation respectively. The power spectrum of the synthetic unvoiced speech output would therefore be

$$\Phi_{ss}(e^{j\omega}) = A_U^2 |V(e^{j\omega})|^2 |R(e^{j\omega})|^2. \tag{8.51}$$

From Eq. (8.49) or (8.51), it becomes clear that if we apply the characteristic system for convolution to the convolution of autocorrelation functions in Eq. (8.48), we do not need the complex cepstrum since $\Phi_{ss}(e^{j\omega})$ is real and non-negative, so the cepstrum and complex cepstrum are identical. The cepstrum of an autocorrelation function would be the IDTFT of the logarithm of the Fourier transform of the autocorrelation function; i.e., the cepstrum of the autocorrelation function is the IDTFT of the logarithm of the power spectrum. Thus, the real logarithm of the power spectrum is

$$\hat{\Phi}_{ss}(e^{j\omega}) = 2(\log A_U) + 2\log|V(e^{j\omega})| + 2\log|R(e^{j\omega})|, \tag{8.52}$$

so that the cepstrum (complex cepstrum) of the autocorrelation function is

$$\hat{\phi}_{ss}[n] = 2(\log A_U)\delta[n] + 2(\hat{v}[n] + \hat{v}[-n])/2 + 2(\hat{r}[n] + \hat{r}[-n])/2. \tag{8.53}$$

Therefore, the cepstrum of the autocorrelation function is an even function of quefrency, n, and, because it is based on the power spectrum (which involves squared-magnitudes), it is twice the size of the cepstra of the deterministic components due to the vocal tract and radiation systems. By our assumption, both $v[n]$ and $r[n]$ represent minimum-phase systems, so

$$\hat{h}_U[n] = \begin{cases} \hat{\phi}_{ss}[n]/2 & n \geq 0 \\ 0 & \text{otherwise.} \end{cases} \tag{8.54}$$

However, note that since phase is not included in the power spectrum representation, we cannot distinguish between minimum-phase and maximum-phase components if they were to exist in a more detailed model.

As a simple example, if the vocal tract and radiation systems of the previous section were excited by a white noise input instead of the periodic glottal pulse input, the components of the cepstrum due to $g[n]$ and $p[n]$ would be omitted. The synthetic sound produced would be like a hoarse whisper instead of the voiced vowel /AE/.[7] Thus, the theoretical log power spectrum would be given by Eq. (8.52) and the cepstrum of the autocorrelation function would be given by Eq. (8.53). Using the same values for the parameters of the vocal tract and radiation systems, we obtain the plots in Figure 8.20. In this case, there is no periodic component in the log spectrum and therefore no isolated peak in the cepstrum. As discussed above, the absence of

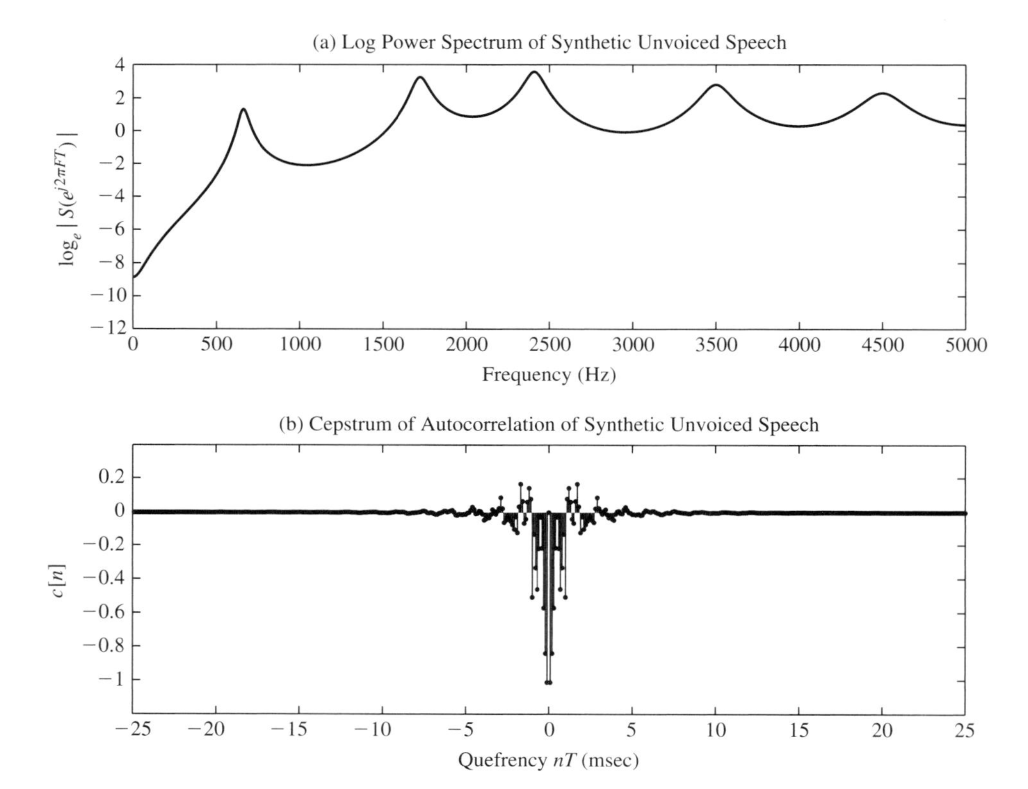

FIGURE 8.20
Homomorphic analysis of unvoiced speech: (a) log magnitude $\log\{\Phi_{ss}(e^{j2\pi FT})\}$; (b) cepstrum of autocorrelation function $\hat{\phi}_{ss}[n]$.

[7] A more realistic model of an unvoiced fricative would differ in the locations and bandwidths of poles and might also include zeros.

such a peak in the expected quefrency range of the pitch period serves as an indicator of unvoiced excitation. Algorithms for pitch detection based on the cepstrum are discussed in Chapter 10.

8.4 COMPUTING THE SHORT-TIME CEPSTRUM AND COMPLEX CEPSTRUM OF SPEECH

In the previous section, we computed exact expressions for the complex cepstrum of the output of a discrete-time model for speech production. This was possible because the synthetic speech signal was created with known systems and known excitations for which z-transform representations could be determined. This model is implicitly assumed in most speech analysis techniques; however, a major difference in practice is that we base our analysis on short segments of a given natural speech signal. We simply assume that the short segment of a natural speech signal could have been a short segment of the model output. Since the speech signal changes continuously with time, we employ a sequence of analyses to track those changes. That is the approach that we shall now develop for homomorphic speech analysis, leading to short-time versions of the cepstrum and complex cepstrum.

8.4.1 Computation Based on the Discrete Fourier Transform

Recall that in Chapter 7, we defined an alternative form of the short-time Fourier transform as

$$\tilde{X}_{\hat{n}}(e^{j\hat{\omega}}) = \sum_{n=0}^{L-1} w[n]x[\hat{n}+n]e^{-j\hat{\omega}n}, \tag{8.55}$$

where $\hat{n}$ denotes the analysis time and $\hat{\omega}$ denotes a short-time analysis frequency.[8] That is, the short-time Fourier transform at analysis time $\hat{n}$ is the DTFT of the finite-length sequence

$$x_{\hat{n}}[n] = \begin{cases} w[n]s[\hat{n}+n] & 0 \le n \le L-1 \\ 0 & \text{otherwise}, \end{cases} \tag{8.56}$$

where $s[n]$ denotes the speech signal and we assume that $w[n]=0$ outside the interval $0 \le n \le L-1$. In this formulation, the time origin of the windowed segment is reset from $\hat{n}$ to the origin of $w[n]$. In our basic definition in Chapter 7 (see Eq. (7.8)), the time origin of the window is shifted to the analysis time $\hat{n}$. This definition facilitates interpretation in terms of linear filtering and filter banks, but for cepstrum analysis, it is preferable to consider the window origin fixed at $n=0$, with the signal samples to be analyzed being shifted into the window as in Eq. (8.56). This allows us to focus

[8] This definition is identical to the alternate definition in Eq. (7.10) except that we have redefined the window by replacing $w[-n]$ by $w[n]$.

on the interpretation of the short-time Fourier transform as simply the DTFT of the finite-length sequence $x_{\hat{n}}[n]$. Since each windowed segment will be processed independently by the techniques of homomorphic filtering, we can simplify our notation by dropping the subscript $\hat{n}$ except where it is necessary to specify the analysis time. Furthermore, there will be no need to distinguish between the DTFT variable ω and the specific short-time analysis frequency variable $\hat{\omega}$ since we will focus on the DTFT interpretation.

Therefore, the representations of the characteristic system for convolution and its inverse, depicted in Figures 8.5 and 8.6 respectively, are the basis for a short-time homomorphic system for convolution if we simply note that the input is the finite-length windowed sequence $x[n] = w[n]s[\hat{n} + n]$. In other words, the short-time characteristic system for convolution is defined by the equations

$$X(e^{j\omega}) = \sum_{n=0}^{L-1} x[n]e^{-j\omega n}, \tag{8.57a}$$

$$\hat{X}(e^{j\omega}) = \log\{X(e^{j\omega})\} = \log|X(e^{j\omega})| + j\arg\{X(e^{j\omega})\}, \tag{8.57b}$$

$$\hat{x}[n] = \frac{1}{2\pi}\int_{-\pi}^{\pi} \hat{X}(e^{j\omega})e^{j\omega n}\,d\omega. \tag{8.57c}$$

Equation (8.57a) is the DTFT of the windowed input sequence defined by Eq. (8.56), Eq. (8.57b) is the complex logarithm of the DTFT of the input, and Eq. (8.57c) is the inverse DTFT of the complex logarithm of the Fourier transform of the input.

As we have already observed, there are questions of uniqueness of this set of equations. In order to clearly define the complex cepstrum with Eqs. (8.57a)–(8.57c), we must provide a unique definition of the complex logarithm of the Fourier transform. To do this, it is helpful to impose the constraint that the complex cepstrum of a real input sequence be also a real sequence. Recall that for a real sequence, the real part of the Fourier transform is an even function and the imaginary part is odd. Therefore, if the complex cepstrum is to be a real sequence, we must define the log magnitude function to be an even function of ω and the phase must be defined to be an odd function of ω. As we have already asserted, a further sufficient condition for the complex logarithm to be unique is that the phase be computed so that it is a continuous periodic function of ω with a period of 2π [271, 342]. Algorithms for the computation of an appropriate phase function typically start with the principal value phase sampled at the discrete Fourier transform (DFT) frequencies as a basis for searching for discontinuities of size 2π. Due to the sampling, care must be taken to locate the frequencies at which the discontinuities occur. A simple approach that generally works well if the phase is densely sampled is to search for jumps (either positive or negative) of size greater than some prescribed tolerance.[9] Once the frequencies where the principal value "wraps around" are found, the appropriate multiples of 2π radians can be

[9]The `unwrap( )` function in MATLAB uses a default tolerance of π, which is reasonable, since jumps of close to π radians can occur because of zeros that are very close to the unit circle.

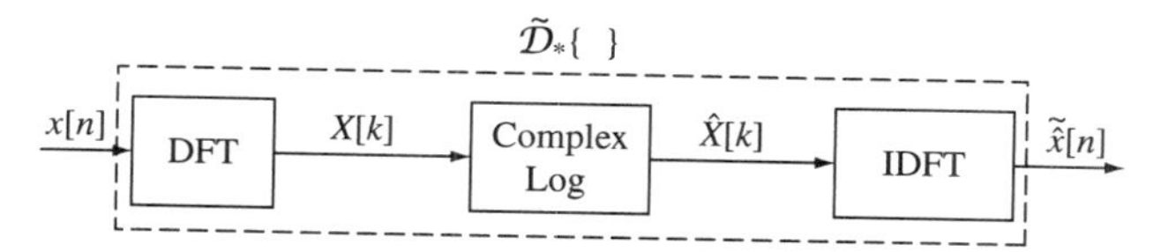

FIGURE 8.21
Computation of the complex cepstrum using the DFT (implementation of the approximate characteristic system for convolution $\tilde{\mathcal{D}}_*\{\cdot\}$).

added or subtracted to produce the "unwrapped phase" [342, 392]. Another method of computing the phase is discussed in Section 8.4.2.

Although Eqs. (8.57a)–(8.57c) can be useful for theoretical analysis, they are not in a form that is useful for computation, since Eq. (8.57c) requires the evaluation of an integral. However, we can approximate Eq. (8.57c) by using the DFT. The DFT of a finite-length sequence is identical to a sampled version of the DTFT of that same sequence [270]; i.e., $X[k] = X(e^{j2\pi k/N})$. Furthermore, the DFT can be efficiently computed by a fast Fourier transform (FFT) algorithm [270]. Thus, one approach that is suggested for computing the complex cepstrum is to replace all of the DTFT operations in Figure 8.5 by corresponding DFT operations. The resulting implementation of the characteristic system is depicted in Figure 8.21 and defined by the equations

$$X[k] = X(e^{j2\pi k/N}) = \sum_{n=0}^{L-1} x[n]e^{-j\frac{2\pi}{N}kn}, \quad 0 \le k \le N-1, \tag{8.58a}$$

$$\hat{X}[k] = \hat{X}(e^{j2\pi k/N}) = \log\{X[k]\}, \quad 0 \le k \le N-1, \tag{8.58b}$$

$$\tilde{\hat{x}}[n] = \frac{1}{N}\sum_{k=0}^{N-1} \hat{X}[k]e^{j\frac{2\pi}{N}kn}, \quad 0 \le n \le N-1, \tag{8.58c}$$

where the window length satisfies $L \le N$. Equation (8.58c) is the inverse discrete Fourier transform (IDFT) of the complex logarithm of the DFT of a finite-length input sequence. The symbol $\tilde{}$ explicitly signifies that the operation denoted $\tilde{\mathcal{D}}_*\{\cdot\}$ in Figure 8.21 and defined by Eqs. (8.58a)–(8.58c) produces an output sequence that is not precisely equal to the complex cepstrum as defined in Eqs. (8.57a)–(8.57c). This is because the complex logarithm used in the DFT calculations is a sampled version of $\hat{X}(e^{j\omega})$ and therefore, the resulting inverse transform is a quefrency-aliased version of the true complex cepstrum (see Refs. [270, 271, 342]). That is, the complex cepstrum computed by Eqs. (8.58a)–(8.58c) is related to the true complex cepstrum by the quefrency-aliasing relation [270]:

$$\tilde{\hat{x}}[n] = \sum_{r=-\infty}^{\infty} \hat{x}[n + rN], \quad 0 \le n < N-1. \tag{8.59}$$

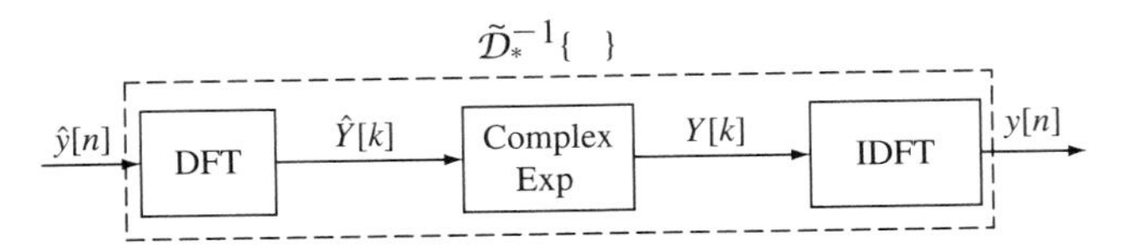

FIGURE 8.22
Implementation of the approximate inverse characteristic system for convolution $\tilde{\mathcal{D}}_*^{-1}\{\cdot\}$ using the DFT.

The inverse characteristic system for convolution is needed for homomorphic filtering of speech. Following our approach above, we obtain this system from Figure 8.6 by simply replacing the DTFT operators by their corresponding DFT computations.

We have observed that the complex cepstrum involves the use of the complex logarithm and that the cepstrum, as it has traditionally been defined, involves only the logarithm of the magnitude of the Fourier transform; that is, the short-time cepstrum, $c[n]$, is given by

$$c[n] = \frac{1}{2\pi}\int_{-\pi}^{\pi} \log|X(e^{j\omega})|e^{j\omega n}d\omega, \qquad -\infty < n < \infty, \tag{8.60}$$

where $X(e^{j\omega})$ is the DTFT of the windowed signal $x[n]$. An approximation to the cepstrum can be obtained by computing the IDFT of the logarithm of the magnitude of the DFT of the finite-length input sequence; i.e.,

$$\tilde{c}[n] = \frac{1}{N}\sum_{k=0}^{N-1} \log|X[k]|e^{j\frac{2\pi}{N}kn}, \qquad 0 \leq n \leq N-1. \tag{8.61}$$

As before, the cepstrum computed using the DFT is related to the true cepstrum computed by Eq. (8.60) by the quefrency-aliasing formula

$$\tilde{c}[n] = \sum_{r=-\infty}^{\infty} c[n+rN], \quad 0 \leq n \leq N-1. \tag{8.62}$$

Furthermore, just as $c[n]$ is the even part of $\hat{x}[n]$, $\tilde{c}[n]$ is the N-periodic even part of $\tilde{\hat{x}}[n]$; i.e.,

$$\tilde{c}[n] = \frac{\tilde{\hat{x}}[n] + \tilde{\hat{x}}[N-n]}{2}. \tag{8.63}$$

Figure 8.23 shows how the computations leading to Eq. (8.62) are implemented using the DFT and the IDFT. In this figure, the operation of computing the cepstrum using the DFT as in Eq. (8.61) is denoted $\tilde{C}\{\cdot\}$.

The effect of quefrency aliasing can be significant for high quefrency components such as those corresponding to voiced excitation. A simple example will illustrate this

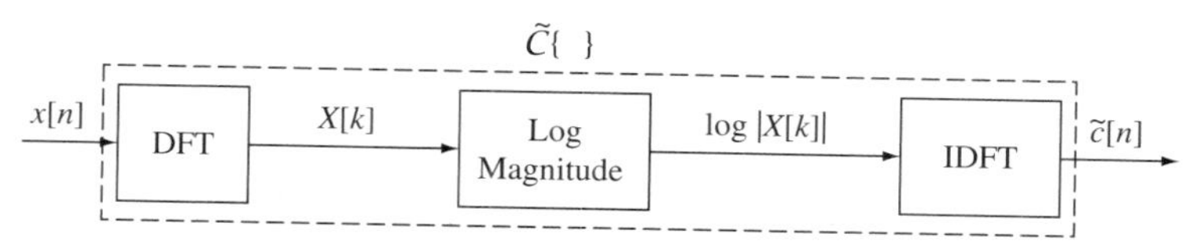

FIGURE 8.23
Computation of the cepstrum using the DFT.

point. Consider a finite-length input $x[n] = \delta[n] + \alpha\delta[n - N_p]$. Its DTFT is

$$X(e^{j\omega}) = 1 + \alpha e^{-j\omega N_p}. \tag{8.64}$$

Using the power series expansion of Eq. (8.23), the complex logarithm can be expressed as

$$\hat{X}(e^{j\omega}) = \log\{1 + \alpha e^{-j\omega N_p}\} = \sum_{m=1}^{\infty} \left(\frac{(-1)^{m+1}\alpha^m}{m}\right) e^{-j\omega m N_p}, \tag{8.65}$$

from which it follows that the complex cepstrum is

$$\hat{x}[n] = \sum_{m=1}^{\infty} \left(\frac{(-1)^{m+1}\alpha^m}{m}\right) \delta[n - mN_p]. \tag{8.66}$$

If, instead of this exact analysis, we compute the complex cepstrum as the inverse N-point DFT of the complex logarithm of the N-point DFT of $x[n]$ (i.e., $\hat{X}[k] = \log\{X[k]\} = \log\{X(e^{j2\pi k/N})\}$), then the resulting N-point sequence $\tilde{\hat{x}}[n]$ will be given by Eq. (8.59) with Eq. (8.66) substituted for $\hat{x}[n]$.

Note the sequence $\hat{x}[n]$ is non-zero for $n = mN_p$ for $1 \leq m < \infty$, so aliasing will produce complex cepstrum values that are out of sequence. In fact, we can show that the non-zero values of $\tilde{\hat{x}}[n]$ are at positions $((mN_p))_N$ for all positive integers m.[10] This is illustrated in Figure 8.24a for the case $N = 256$, $N_p = 75$, and $\alpha = 0.8$. Observe that since $3N_p < N < 4N_p$ for the specific values $N = 256$ and $N_p = 75$, $\hat{x}[N_p]$, $\hat{x}[2N_p]$, and $\hat{x}[3N_p]$ are in their correct positions, but values of $\hat{x}[n]$ for $n \geq 4N_p$ "wrap around" into the base interval $0 \leq n \leq N - 1$. Since $\hat{x}[n] \to 0$ for $n \to \infty$, increasing N will tend to mitigate the effect by allowing more of the impulses to be at their correct position, and, at the same time, ensuring that the aliased samples will have smaller amplitudes due to the $1/|n|$ fall-off. As discussed in Refs. [270, 271, 342, 392], a large value for N (that is, a high rate of sampling of the Fourier transform) is also required for accurate computation of the complex logarithm. However, the use of FFT algorithms makes it feasible to

[10]Following the notation in [270], $((mN_p))_N$ means mN_p taken modulo N.

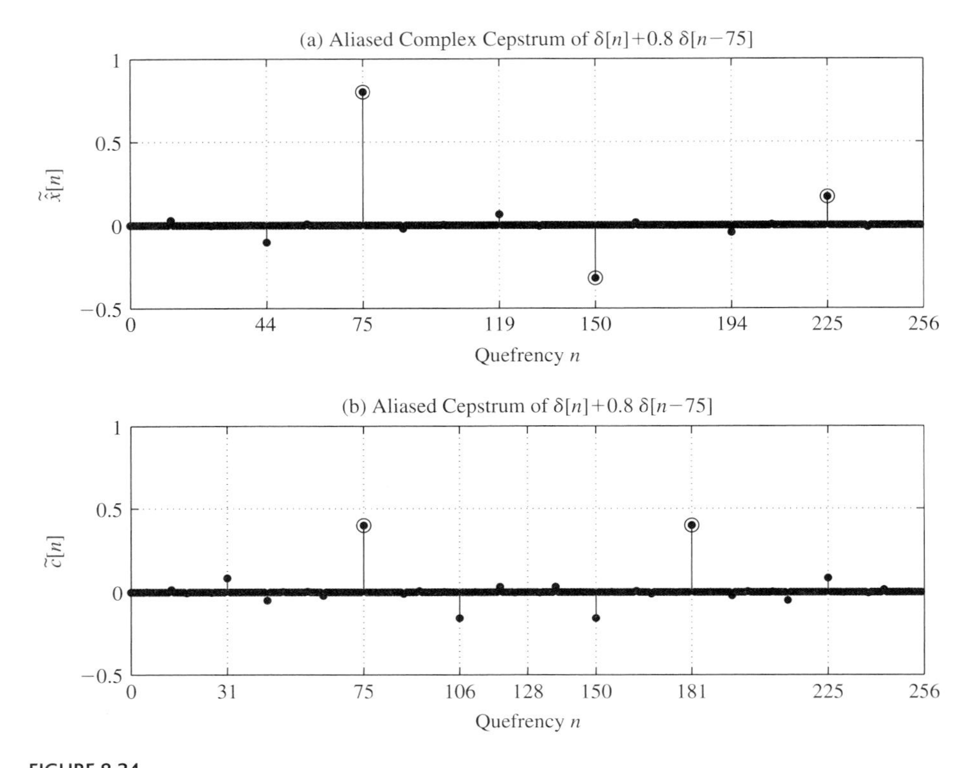

FIGURE 8.24
(a) Quefrency-aliased complex cepstrum; and (b) quefrency-aliased cepstrum. Circled dots are cepstrum values in correct locations.

use reasonably large values of N such as $N = 1024$ or $N = 2048$, so quefrency aliasing need not be a significant problem.

Figure 8.24b shows the aliasing effects for the cepstrum, which in this example is the periodic even part of Figure 8.24a. Since the cepstrum is non-zero for both positive and negative n, the implicit periodicity of the DFT representation causes the negative quefrency samples to be located at positions $((N-n))_N$. Therefore, in this example, only the two circled samples at $n = N_p = 75$ and $n = N - N_p = 256 - 75 = 181$ are in what could be considered their "correct" locations (assuming samples at $128 < n < 256$ to be "negative quefrency" samples). Again, by increasing N, we can mitigate the aliasing effects since the cepstrum approaches zero for large n.

8.4.2 Computation Based on the z-Transform

An approach to computing the complex cepstrum of finite-length sequences that does not require phase unwrapping is suggested by the extended example of Section 8.3. In that example, the z-transforms of all the convolved components of the model had closed-form expressions as rational functions. By factoring the numerator and denominator polynomials, it was possible to compute the complex cepstrum exactly.

Short-time analysis of natural speech signals is based upon finite-length (windowed) segments of the speech waveform, and if a sequence, $x[n]$, has finite length, then its z-transform is a polynomial in z^{-1} of the form

$$X(z) = \sum_{n=0}^{M} x[n]z^{-n}. \tag{8.67a}$$

Such an Mth-order polynomial in z^{-1} can be represented in terms of its roots as

$$X(z) = x[0] \prod_{m=1}^{M_i} (1 - a_m z^{-1}) \prod_{m=1}^{M_o} (1 - b_m^{-1} z^{-1}), \tag{8.67b}$$

where the quantities a_m are the (complex) zeros that lie inside the unit circle (minimum-phase part) and the quantities b_m^{-1} are the zeros that are outside the unit circle (maximum-phase part); i.e., $|a_m| < 1$ and $|b_m| < 1$. We assume that no zeros lie precisely on the unit circle.[11] If we factor a term $-b_m^{-1} z^{-1}$ out of each factor of the product at the far right in Eq. (8.67b), then Eq. (8.67b) can be expressed as

$$X(z) = Az^{-M_o} \prod_{m=1}^{M_i} (1 - a_m z^{-1}) \prod_{m=1}^{M_o} (1 - b_m z), \tag{8.67c}$$

where

$$A = x[0](-1)^{M_o} \prod_{m=1}^{M_o} b_m^{-1}. \tag{8.67d}$$

This representation of a windowed frame of speech can be obtained by using a polynomial rooting algorithm to find the zeros a_m and b_m^{-1} that lie inside and outside the unit circle, respectively, for the polynomial whose coefficients are the sequence $x[n]$.

Computing the Complex Cepstrum

Given the numeric representation of the z-transform polynomial, as in Eqs. (8.67c) and (8.67d), numerical values of the complex cepstrum sequence can be computed

[11] Perhaps not surprisingly, it is rare that a computed root of a polynomial is precisely on the unit circle; however, as previously mentioned, most of the zeros lie close to the unit circle for high-order polynomials.

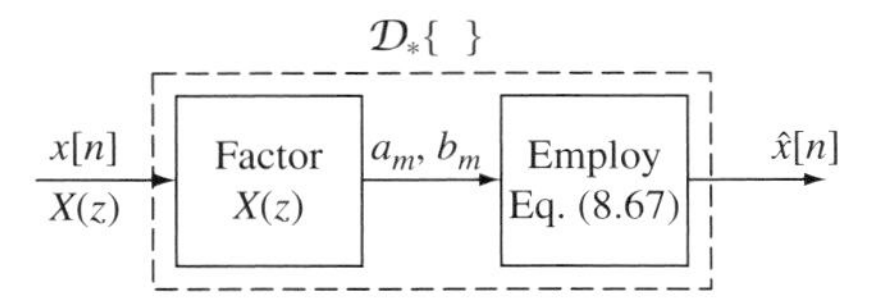

FIGURE 8.25
Computation of the complex cepstrum by polynomial rooting of the z-transform of a finite-length windowed speech segment.

from Eq. (8.24) as

$$\hat{x}[n] = \begin{cases} \log|A| & n = 0 \\ -\displaystyle\sum_{m=1}^{M_i} \frac{a_m^n}{n} & n > 0 \\ \displaystyle\sum_{m=1}^{M_o} \frac{b_m^{-n}}{n} & n < 0. \end{cases} \tag{8.68}$$

If $A < 0$, this fact can be recorded separately, along with the value of M_o, the number of roots that are outside the unit circle. With this information and $\hat{x}[n]$, we have all that is needed to completely characterize the original signal $x[n]$ by its complex cepstrum. This method of computing the complex cepstrum of a windowed speech segment is depicted in Figure 8.25.

This method of computation is particularly useful when $M = M_o + M_i + 1$ is small, but it is not limited to small M. Steiglitz and Dickinson [379] first proposed this method and reported successful rooting of polynomials with degree as high as $M = 256$, which was a practical limit imposed by computational resources readily available at that time. More recently, Sitton et al. [362] have shown that polynomials of order up to 1,000,000 can be rooted accurately using the FFT to implement a systematic grid search. Thus, the complex cepstrum of very long sequences can be computed accurately. The advantages of this method are that there is no quefrency aliasing and no uncertainty about whether a phase discontinuity was either undetected or falsely detected in the phase unwrapping computation of the complex cepstrum.

Phase Unwrapping by Summing Phases of Individual Zeros

If the numerical values of the zeros in Eq. (8.67c) are known, we have seen that the complex cepstrum can be computed using Eq. (8.68) *without* time aliasing and *without* computing the unwrapped phase function. However, if it is DFT samples of the unwrapped phase that we desire, we can compute them from the zeros without first computing the complex cepstrum. To see how this can be done, we write the DTFT in terms of the numerical values of the zeros as

$$X(e^{j\omega}) = Ae^{-j\omega M_0} \prod_{m=1}^{M_i} (1 - a_m e^{-j\omega}) \prod_{m=1}^{M_o} (1 - b_m e^{j\omega}). \tag{8.69}$$

Assuming $A > 0$, the phase of $X(e^{j\omega})$ is

$$\arg [X(e^{j\omega})] = -j\omega M_0 + \sum_{m=1}^{M_i} \text{ARG}\{(1 - a_m e^{-j\omega})\} + \sum_{m=1}^{M_o} \text{ARG}\{(1 - b_m e^{j\omega})\}. \tag{8.70}$$

Our notation in Eq. (8.70) implies that the sum of the terms on the right is the continuous phase function $\arg\{X(e^{j\omega})\}$ that we require in the computation of the complex cepstrum even though the right-hand side terms are principal value phases. This is true because of the following relations for the principal value phases of the individual polynomial factors:

$$-\pi/2 < \text{ARG}\{(1 - ae^{-j\omega})\}] < \pi/2 \text{ if } |a| < 1, \tag{8.71a}$$

$$-\pi/2 < \text{ARG}\{(1 - be^{j\omega})\} < \pi/2 \text{ if } |b| < 1, \tag{8.71b}$$

which hold for $-\pi < \omega \leq \pi$. Furthermore, it can be shown that

$$\text{ARG}\{(1 - ae^{-j\omega})\} = 0 \text{ for } \omega = 0 \text{ and } \pi, \tag{8.72a}$$

$$\text{ARG}\{(1 - be^{j\omega})\} = 0 \text{ for } \omega = 0 \text{ and } \pi. \tag{8.72b}$$

Since $|a_m| < 1$ for $m = 1, \ldots, M_i$ and $|b_m| < 1$ for $m = 1, \ldots, M_o$ in Eq. (8.70), it follows that the principal values of all the terms in that equation can be computed without discontinuities, and therefore the sum will have no discontinuities. Also note that the term $-j\omega M_o$ can be included or not as desired. Generally it would be omitted since its IDTFT can dominate the complex cepstrum values.

To compute the sampled values of the continuous phase at frequencies $2\pi k/N$, we simply need to compute N-point DFTs of the individual terms and then compute and add the principal-value phases of those terms. Specifically, the continuous phase of the DFT $X[k]$ of the finite-length sequence $x[n]$ is

$$\arg\{X[k]\} = \sum_{m=1}^{M_i} \text{ARG}\{(1 - a_m e^{-j2\pi k/N})\} + \sum_{m=1}^{M_o} \text{ARG}\{(1 - b_m e^{-j2\pi (N-k)/N})\}, \quad 0 \leq k \leq N-1. \tag{8.73}$$

The contributions due to the zeros that are inside the unit circle (a_m's) can be computed by using an N-point FFT algorithm to compute the DFTs of the sequences

$\delta[n] - a_m\delta[n-1]$. Similarly the contributions due to the zeros that are outside the unit circle are obtained by computing the DFTs of the sequences $\delta[n] - b_m\delta[n-N+1]$. The computation can be reduced significantly by combining complex conjugate pairs of zeros into second-order factors before computing the DFTs.

If this method is used to compute the DFT-sampled unwrapped phase function, then the time-aliased cepstrum can be computed from

$$\tilde{\hat{x}}[n] = \frac{1}{N}\sum_{k=0}^{N-1}\left(\log|X[k]| + j\arg\{X[k]\}\right)e^{j2\pi kn/N}, \quad 0 \leq n \leq N-1. \tag{8.74}$$

Phase Unwrapping by Time-Aliasing the Complex Cepstrum

Another approach to computing the unwrapped phase function is to first compute the complex cepstrum $\hat{x}[n]$ using Eq. (8.68) over a long time interval such as, for example, $-RN \leq n \leq (R-1)N$, and then compute an approximation to the time-aliased complex cepstrum as in

$$\hat{x}_a[n] = \sum_{r=-R}^{R-1} \hat{x}[n+rN], \qquad 0 \leq n \leq N-1. \tag{8.75}$$

The N-point DFT of $\hat{x}_a[n]$ can be computed with values of N and R chosen jointly so that $\hat{x}_a[n]$ is an accurate approximation to $\tilde{\hat{x}}[n]$. If this is done, $\mathcal{Im}\{\hat{X}_a[k]\}$ will be an accurate approximation for the unwrapped phase function $\arg\{X[k]\}$.

8.4.3 Recursive Computation for Minimum- and Maximum-Phase Signals

In Section 8.2.5 we noted that in the special cases of minimum-phase and maximum-phase signals, the complex cepstrum can be computed using only the magnitude of the DTFT. In this section we show that a recursive time-domain relation is also possible for these cases.

A general non-linear difference equation relation exists between a sequence $x[n]$ and its complex cepstrum $\hat{x}[n]$ if the complex cepstrum exists. This relation can be derived by observing that the derivative of $\hat{X}(z)$ is

$$\frac{d\hat{X}(z)}{dz} = \frac{d}{dz}\left(\log\left[X(z)\right]\right) = \frac{1}{X(z)}\frac{dX(z)}{dz}. \tag{8.76}$$

Using the derivative theorem for z-transforms [270], which states that the z-transform of the sequence $nx[n]$ is $-zdX(z)/dz$, and the fact that multiplication of z-transforms is equivalent to convolution of corresponding sequences, it follows that Eq. (8.76) implies

$$(nx[n]) = x[n] * (n\hat{x}[n]) = \sum_{k=-\infty}^{\infty} k\hat{x}[k]x[n-k]. \tag{8.77}$$

This difference equation is an implicit relation between $x[n]$ and $\hat{x}[n]$ that holds for any sequence $x[n]$ whose complex cepstrum exists. It cannot be used to compute $\hat{x}[n]$ in general because all values of $\hat{x}[n]$ are needed to evaluate the right-hand side of the equation, but in the case of minimum- and maximum-phase sequences, Eq. (8.77) can be specialized to a form that permits recursive computation.

Specifically, consider a minimum-phase signal $x_{\text{mnp}}[n]$ and its corresponding complex cepstrum $\hat{x}_{\text{mnp}}[n]$, which, by definition, have the properties $x_{\text{mnp}}[n] = 0$ and $\hat{x}_{\text{mnp}}[n] = 0$ for $n < 0$. If we impose these conditions, Eq. (8.77) becomes

$$n x_{\text{mnp}}[n] = \sum_{k=0}^{n} k \hat{x}_{\text{mnp}}[k] x_{\text{mnp}}[n-k]. \tag{8.78}$$

If we separate the $k = n$ term from the sum on the right, we obtain

$$n x_{\text{mnp}}[n] = n \hat{x}_{\text{mnp}}[n] x_{\text{mnp}}[0] + \sum_{k=0}^{n-1} k \hat{x}_{\text{mnp}}[k] x_{\text{mnp}}[n-k]. \tag{8.79}$$

Finally, dividing both sides by n and solving for $\hat{x}_{\text{mnp}}[n]$ in Eq. (8.79), we have the result that we are seeking; i.e.,

$$\hat{x}_{\text{mnp}}[n] = \frac{x_{\text{mnp}}[n]}{x_{\text{mnp}}[0]} - \sum_{k=0}^{n-1} \left(\frac{k}{n}\right) \hat{x}_{\text{mnp}}[k] \frac{x_{\text{mnp}}[n-k]}{x_{\text{mnp}}[0]}, \quad n > 0, \tag{8.80}$$

where Eq. (8.80) holds only for $n > 0$ since we cannot divide by 0. Of course, by definition, both $x_{\text{mnp}}[n]$ and $\hat{x}_{\text{mnp}}[n]$ are zero for $n < 0$.

Observe that Eq. (8.80) can be used to compute $\hat{x}_{\text{mnp}}[n]$ for $n > 0$ if we know $\hat{x}_{\text{mnp}}[0]$. To determine $\hat{x}_{\text{mnp}}[0]$, consider the z-transform

$$\hat{X}_{\text{mnp}}(z) = \sum_{n=0}^{\infty} \hat{x}_{\text{mnp}}[n] z^{-n} = \log\left\{\sum_{n=0}^{\infty} x_{\text{mnp}}[n] z^{-n}\right\}, \tag{8.81}$$

from which it follows by the initial value theorem of z-transforms that

$$\lim_{n \to \infty} \hat{X}_{\text{mnp}}(z) = \hat{x}_{\text{mnp}}[0] = \log\{x_{\text{mnp}}[0]\}. \tag{8.82}$$

Therefore, combining Eqs. (8.82) and (8.80), we can finally write the recursion:

$$\hat{x}_{\text{mnp}}[n] = \begin{cases} 0 & n < 0 \\ \log\{x_{\text{mnp}}[0]\} & n = 0 \\ \dfrac{x_{\text{mnp}}[n]}{x_{\text{mnp}}[0]} - \displaystyle\sum_{k=0}^{n-1} \left(\frac{k}{n}\right) \hat{x}_{\text{mnp}}[k] \frac{x_{\text{mnp}}[n-k]}{x_{\text{mnp}}[0]} & n > 0, \end{cases} \tag{8.83}$$

which is a recursive relationship that can be used to implement the characteristic system for convolution $\mathcal{D}_*\{\cdot\}$ if it is known that the input is a minimum-phase signal. The inverse characteristic system can be implemented recursively by simply rearranging Eq. (8.83) to obtain

$$x_{\text{mnp}}[n] = \begin{cases} 0 & n < 0 \\ \exp\{\hat{x}_{\text{mnp}}[0]\} & n = 0 \\ \hat{x}_{\text{mnp}}[n]x_{\text{mnp}}[0] + \sum_{k=0}^{n-1}\left(\frac{k}{n}\right)\hat{x}_{\text{mnp}}[k]x_{\text{mnp}}[n-k] & n > 0. \end{cases} \tag{8.84}$$

A maximum-phase sequence is defined by the properties $x_{\text{mxp}}[n] = 0$ and $\hat{x}_{\text{mxp}}[n] = 0$ for $n > 0$. Starting with Eq. (8.77), we can apply these constraints to obtain the following recursion relation for the complex cepstrum of a maximum-phase sequence:

$$\hat{x}_{\text{mxp}}[n] = \begin{cases} \dfrac{x_{\text{mxp}}[n]}{x_{\text{mxp}}[0]} - \sum_{k=n+1}^{0}\left(\frac{k}{n}\right)\hat{x}_{\text{mxp}}[k]\dfrac{x_{\text{mxp}}[n-k]}{x_{\text{mxp}}[0]} & n < 0 \\ \log\{x_{\text{mxp}}[0]\} & n = 0 \\ 0 & n > 0 \end{cases} \tag{8.85}$$

and by reorganizing Eq. (8.85), we obtain the following recursion for the inverse characteristic system for maximum-phase signals:

$$x_{\text{mxp}}[n] = \begin{cases} \hat{x}_{\text{mxp}}[n]x_{\text{mxp}}[0] + \sum_{k=n+1}^{0}\left(\frac{k}{n}\right)\hat{x}_{\text{mxp}}[k]x_{\text{mxp}}[n-k] & n < 0 \\ \exp\{\hat{x}_{\text{mxp}}[0]\} & n = 0 \\ 0 & n > 0. \end{cases} \tag{8.86}$$

These recursive relations are useful when computing the complex cepstrum of the impulse response of speech models obtained by linear predictive analysis as discussed in Chapter 9.

8.5 HOMOMORPHIC FILTERING OF NATURAL SPEECH

We are now in a position to apply the concepts of the cepstrum and homomorphic filtering to a natural speech signal. Recall that the model for speech production, as shown in Figure 8.12, consists of a slowly time-varying linear system excited by either a quasi-periodic impulse train or by random noise. Thus, it is appropriate to think of a short segment of voiced speech as having been taken from the steady-state output of a linear time-invariant system excited by a periodic impulse train. Similarly, a short segment of unvoiced speech can be thought of as resulting from the excitation of a linear time-invariant system by random noise. The analysis of Section 8.3, which was based on exact z-transform representations of the components of the model, demonstrated that for this convolutional model, there is an interesting separation in the cepstrum between the excitation and the vocal tract impulse response components. The purpose of this section is to demonstrate that similar behavior results if short-time homomorphic analysis methods are employed with natural speech inputs.

8.5.1 A Model for Short-Time Cepstral Analysis of Speech

Following the approach presented in [269], we begin by assuming that over the length (L) of the window, the speech signal $s[n]$ satisfies the convolution equation

$$s[n] = e[n] * h[n], \qquad 0 \le n \le L-1, \tag{8.87}$$

where $h[n]$ is the impulse response of the system from the point of excitation (at the glottis for voiced speech and at a constriction for unvoiced speech) to the radiation at the lips. In this analysis, the impulse response $h[n] = h_U[n]$ models the combined effects of the excitation gain, the vocal tract system, and radiation of sound at the lips for unvoiced speech, while $h[n] = h_V[n]$ contains an additional convolutional component due to the glottal pulse for voiced speech.[12] Furthermore, we assume that the impulse response $h[n]$ is short compared to the length of the window so that the windowed segment can be represented as

$$\begin{aligned} x[n] &= w[n]s[n] = w[n](e[n] * h[n]) \\ &\approx e_w[n] * h[n], \qquad 0 \le n \le L-1, \end{aligned} \tag{8.88}$$

where $e_w[n] = w[n]e[n]$; i.e., any tapering due to the analysis window is incorporated into the excitation as a slowly varying amplitude modulation.

In the case of unvoiced speech, the excitation $e[n]$ would be white noise and $h[n] = h_U[n]$. In the case of voiced speech, $h[n] = h_V[n]$ and $e[n]$ would be a unit impulse train of the form

$$e[n] = p[n] = \sum_{k=0}^{N_w-1} \delta[n - kN_p], \tag{8.89}$$

where N_w is the number of impulses in the window and N_p is the discrete-time pitch period (measured in samples).

For voiced speech, the windowed excitation is

$$e_w[n] = w[n]p[n] = \sum_{k=0}^{N_w-1} w_{N_p}[k]\delta[n - kN_p], \tag{8.90}$$

[12]Note that it is often convenient to incorporate the excitation gain (A_V or A_U in Figure 8.12) into $h[n]$ so that we can assume that $e[n]$ consists of unit impulses for voiced excitation and unit variance white noise for unvoiced excitation.

where $w_{N_p}[k]$ is the "time-sampled" window sequence defined as

$$w_{N_p}[k] = \begin{cases} w[kN_p] & k = 0, 1, \ldots, N_w - 1 \\ 0 & \text{otherwise.} \end{cases} \tag{8.91}$$

From Eq. (8.90), the DTFT of $e_w[n]$ is

$$E_w(e^{j\omega}) = \sum_{k=0}^{N_w-1} w_{N_p}[k]e^{-j\omega kN_p} = W_{N_p}(e^{j\omega N_p}), \tag{8.92}$$

and from Eq. (8.92), it follows that $E_w(e^{j\omega})$ is periodic in ω with period $2\pi/N_p$. Therefore,

$$\hat{X}(e^{j\omega}) = \log\{H_V(e^{j\omega})\} + \log\{E_w(e^{j\omega})\} \tag{8.93}$$

has two components: (1) $\log\{H_V(e^{j\omega})\}$, due to the vocal tract frequency response, which is slowly varying in ω, and (2) $\log\{E_w(e^{j\omega})\} = \log\{W_{N_p}(e^{j\omega N_p})\}$, which is due to the excitation and is periodic with period $2\pi/N_p$.[13] The complex cepstrum of the windowed speech segment $x[n]$ is therefore

$$\hat{x}[n] = \hat{h}_V[n] + \hat{e}_w[n]. \tag{8.94}$$

For voiced speech, the cepstral component due to the excitation has the form[14]

$$\hat{e}_w[n] = \begin{cases} \hat{w}_{N_p}[n/N_p] & n = 0, \pm N_p, \pm 2N_p, \ldots \\ 0 & \text{otherwise.} \end{cases} \tag{8.95}$$

That is, in keeping with the $2\pi/N_p$ periodicity of $\log\{E_w(e^{j\omega})\} = \log\{W_{N_p}(e^{j\omega N_p})\}$, the corresponding complex cepstrum (or cepstrum) has impulses (isolated samples) at quefrencies that are multiples of N_p.

For unvoiced speech, no such periodicity occurs in the logarithm of the DTFT of the windowed unvoiced signal, and therefore no cepstral peaks occur. In fact, the magnitude-squared of the DTFT of a finite segment of a random signal is called the periodogram, and it is well known that the periodogram displays random fluctuations with frequency [270]. As we will see, the low quefrencies are primarily due to $\log|H_U(e^{j\omega})|$, but the high quefrencies represent the random fluctuations in $\log|X(e^{j\omega})|$.

The above analysis suggests that the primary effect of the windowing that is required for short-time analysis is to be found in the cepstrum contributions due to

[13]For signals sampled with sampling rate F_s, this period corresponds to F_s/N_p Hz in cyclic analog frequency.

[14]Note that $\hat{w}_{N_p}[n]$ corresponds to $\log\{W_{N_p}(e^{j\omega})\}$, so through the upsampling theorem [270], $\log\{W_{N_p}(e^{j\omega N_p})\}$ corresponds to Eq. (8.95).

the excitation, with the contributions due to $h_V[n]$ or $h_U[n]$ being very similar to those described in Section 8.3 for exact analysis of the idealized model. Our examples in this section will support this assertion.

8.5.2 Example of Short-Time Analysis Using Polynomial Roots

In Section 8.4.2 we showed that the complex cepstrum of a finite-length sequence can be computed by factoring the polynomial whose coefficients are the signal samples. In short-time cepstral analysis, we can use this approach with the windowed speech segments. As an example, Figure 8.26 shows a voiced segment (multiplied by a 401 sample Hamming window) at the beginning of an utterance by a low-pitch male speaker of the diphthong /EY/ as in the word "shade", i.e., /SH/ /EY/ /D/. For the sampling rate of $F_s = 8000$ Hz, the 401 samples span a time interval of 50 msec. Note that the pitch period in Figure 8.26 is on the order of $N_p = 90$ samples, corresponding to a fundamental frequency of approximately $8000/90 \approx 89$ Hz.

To make an understandable plot with so many samples, the windowed waveform samples must be plotted as a continuous function, but for purposes of analysis, the 401 individual waveform samples serve as the coefficients of a 400th degree polynomial as in Eq. (8.67a), which can be represented in terms of its zeros as in Eqs. (8.67b)–(8.67d). The zeros of the polynomial whose coefficients are plotted in Figure 8.26 are displayed in Figure 8.27. Only the zeros in the upper half of the z-plane are shown, and one zero at $z = 2.2729$ occurs off the plotting range and is not shown.[15] The plot shows that almost all the zeros lie very close to the unit circle. This is a general property of high-order polynomials whose coefficients are random variables [152], and speech samples appear to share similar properties when viewed as coefficients of polynomials. In this example, the number of zeros inside and outside the unit circle are $M_i = 220$ and $M_o = 180$ respectively. No zeros were exactly on the unit circle to within the double precision floating-point accuracy of MATLAB.

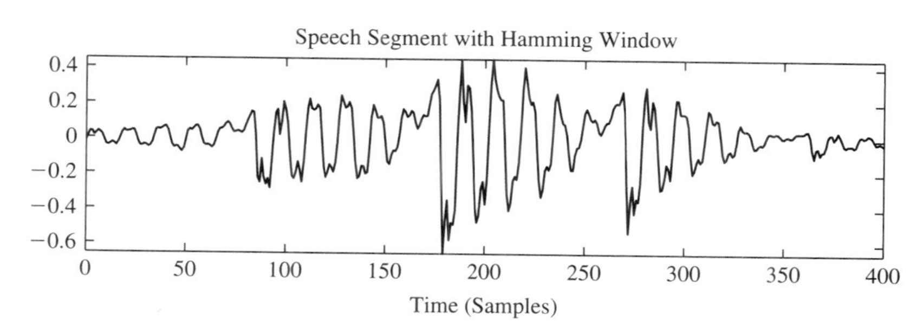

FIGURE 8.26
Windowed time waveform $x[n]$ for homomorphic filtering example.

[15] Since the polynomial coefficients corresponding to speech samples are real numbers, the zeros are either real or in complex conjugate pairs.

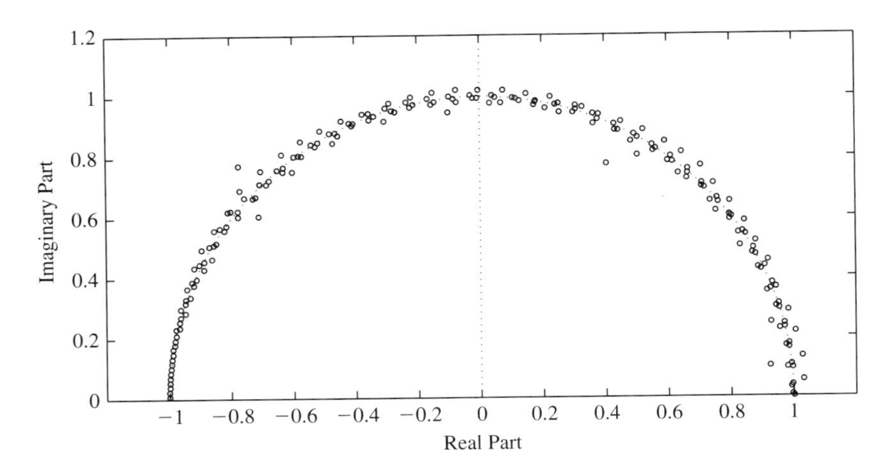

FIGURE 8.27
Zeros of polynomial $X(z)$ (complex conjugate zeros and one zero at $z = 2.2729$ not shown).

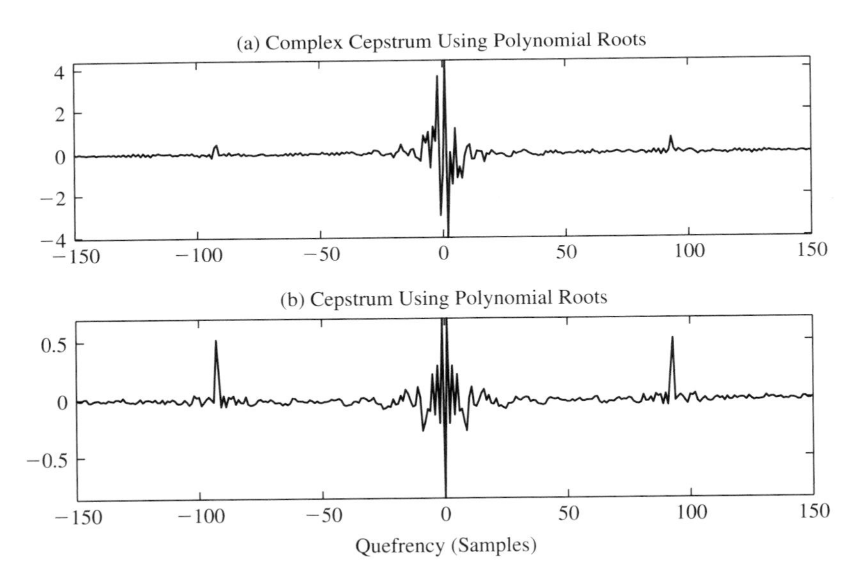

FIGURE 8.28
Cepstrum analysis based on polynomial roots: (a) complex cepstrum $\hat{x}[n]$; (b) cepstrum $c[n]$.

The complex cepstrum computed using Eq. (8.68) from the zeros plotted in Figure 8.27 is shown in Figure 8.28a. The cepstrum for this segment of speech, plotted in Figure 8.28b, is obtained by taking the even part of the sequence in Figure 8.28a. Since the z-transform of the windowed speech segment has zeros both inside and outside the unit circle, the complex cepstrum is non-zero for both positive and negative quefrencies, n. At low quefrencies, this is consistent with the fact that the combined contributions of the vocal tract, glottal pulse, and radiation in $h_V[n]$ will, in general, be non-minimum-phase. Note from Eq. (8.24) that both the complex cepstrum and cepstrum should decay rapidly for large n, and this is evident in Figure 8.28. Also, note that Eq. (8.95) predicts that the contribution to the complex cepstrum due to the periodic excitation will occur at integer multiples of the spacing between impulses; i.e., we should see impulses in the complex cepstrum at multiples of the fundamental period N_p, and this is confirmed by the plots in Figure 8.28. Furthermore, the sizes and locations (positive and negative quefrencies) depend on the window shape and its positioning with respect to the speech waveform.

8.5.3 Voiced Speech Analysis Using the DFT

The z-transform method of computing the short-time complex cepstrum yields the exact values of $\hat{x}[n]$ (within computational error). This requires the rooting of high-degree polynomials and evaluation of Eq. (8.68) for each n. Although this can be computationally demanding for large polynomials, it is made more feasible with root finders based on FFT evaluations of the polynomial [362]. However, it is generally more efficient to use the DFT implementation discussed in Section 8.4.1.

An example of the use of the DFT (FFT) in computation of the short-time complex cepstrum is shown in Figures 8.29–8.34 for the voiced segment of Figure 8.26. Figure 8.29 shows the DFT of the windowed speech segment, i.e., one frame of the short-time Fourier transform.[16] The thin line in Figure 8.29a shows the log magnitude of the DFT. The frequency-periodic component in this function is, of course, due to the time-periodic nature of the input, with the ripples being spaced approximately at multiples of the fundamental frequency. Figure 8.29c shows the discontinuous nature of the principal value of the phase, while the thin line in Figure 8.29b shows the unwrapped phase curve with the discontinuities removed by the addition of appropriate integer multiples of 2π at each DFT frequency. Note that the unwrapped phase curve also displays periodic "ripples" with period equal to the fundamental frequency. Figures 8.29a and 8.29b together comprise the complex logarithm of the short-time Fourier transform; i.e., they are the real and imaginary parts of the Fourier transform of the complex cepstrum, which is shown in Figure 8.30a. Figure 8.30 is virtually identical to Figure 8.28 because the DFT length ($N = 4096$) was large enough to make the time aliasing effects negligible and ensure accurate phase unwrapping. Notice again the peaks at both positive and negative times equal to the pitch period, and notice the rapidly decaying low-time components representing the combined effects of the vocal tract, glottal pulse, and radiation. The cepstrum, which is simply the inverse transform of only the log magnitude with zero imaginary part, is shown in Figure 8.30b. Note that the cepstrum also displays the same general properties as the complex cepstrum, as it should, since the cepstrum is the even part of the complex cepstrum.

[16] For the examples in this section, the DFT length was $N = 4096$. With this value, time aliasing is not a problem in the complex cepstrum, and plots of the DFT are a good approximation to a plot of the DTFT. Thus, for the most part it will not be necessary to distinguish between the "true" complex cepstrum $\hat{x}[n]$ that would be obtained with the DTFT and $\tilde{\hat{x}}[n]$ computed with the DFT.

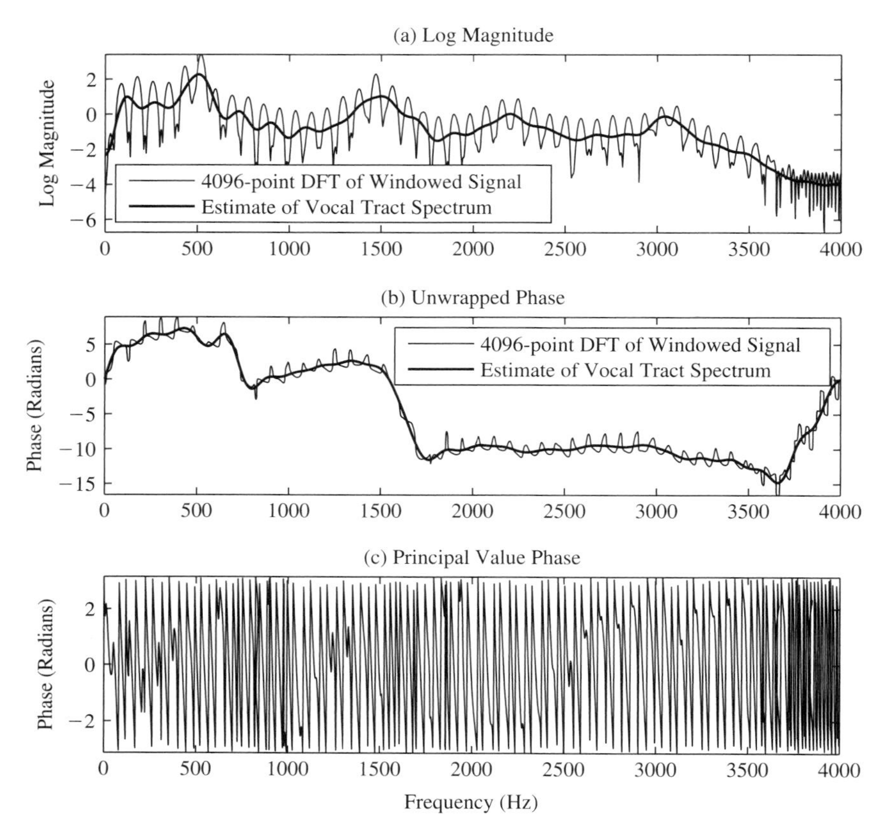

FIGURE 8.29
Homomorphic analysis of voiced speech: (a) Log magnitude of short-time Fourier transform, $\log|X(e^{j\omega})|$. (The heavy line is an estimate of $\log|H_V(e^{j\omega})|$ obtained by lowpass liftering.) (b) Unwrapped phase $\arg\{X(e^{j\omega})\}$. (The heavy line is an estimate of $\arg\{H_V(e^{j\omega})\}$ obtained by lowpass liftering.) (c) Principal value of phase of short-time Fourier transform $\text{ARG}\{X(e^{j\omega})\}$. (All functions in this figure are plotted as a function of analog frequency F; i.e., $\omega = 2\pi FT$ for $F_s = 1/T = 8000$ Hz.)

The plots in Figures 8.29a and 8.29b and the cepstrum plots in Figure 8.30 suggest how homomorphic filtering can be used to separate the excitation and vocal tract components. First, note that the impulses in the complex cepstrum due to the periodic excitation tend to be separated from the low quefrency components. This suggests that the appropriate system for short-time homomorphic filtering of speech is as depicted in Figure 8.31, which shows a segment of speech selected by the window, $w[n]$, with the complex cepstrum computed as discussed in Section 8.4.[17] The desired component

[17]For theoretical analysis, the operators $\mathcal{D}_*\{\cdot\}$ and $\mathcal{D}_*^{-1}\{\cdot\}$ would be represented in terms of the DTFT, but in practice, we would use the operators $\tilde{\mathcal{D}}_*\{\cdot\}$ and $\tilde{\mathcal{D}}_*^{-1}\{\cdot\}$ implemented using the DFT with N large enough to avoid aliasing in the cepstrum.

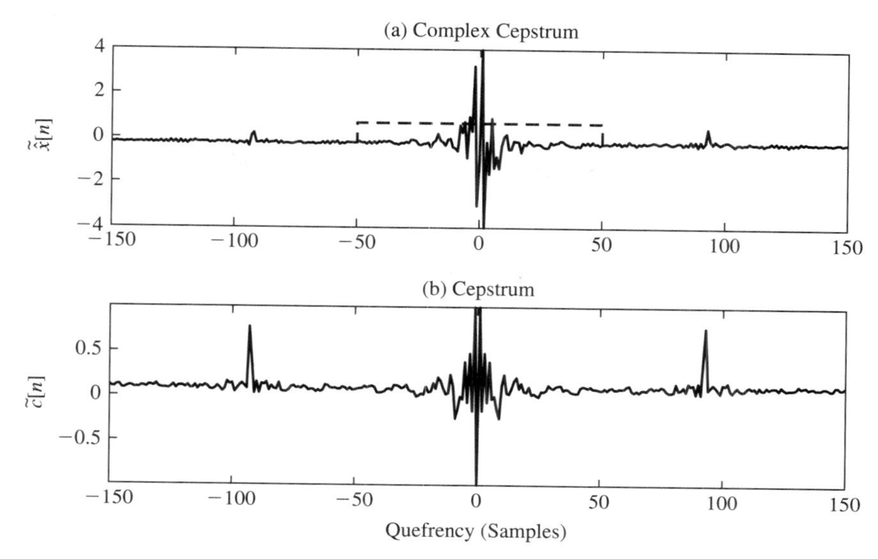

FIGURE 8.30
Homomorphic analysis of voiced speech: (a) complex cepstrum $\tilde{\hat{x}}[n]$; (b) cepstrum $\tilde{c}[n]$. (For comparison to Figure 8.27, the samples of $\tilde{\hat{x}}[n]$ and $\tilde{c}[n]$ were reordered by placing the "negative quefrency" samples from the interval $N/2 < n \leq N-1$ before the samples in the range $0 \leq n < N/2$.)

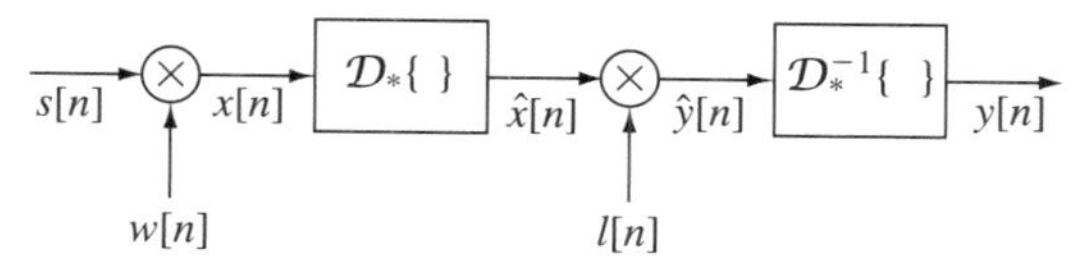

FIGURE 8.31
Implementation of a system for short-time homomorphic filtering of speech.

of the input is selected by what might be termed a "cepstrum window," denoted $l[n]$. This type of filtering is appropriately called "frequency-invariant linear filtering" since multiplying the complex cepstrum by $l[n]$ corresponds to convolving its DTFT, $L(e^{j\omega})$, with the complex logarithm, $\hat{X}(e^{j\omega})$, as in

$$\hat{Y}(e^{j\omega}) = \frac{1}{2\pi}\int_{-\pi}^{\pi} \hat{X}(e^{j\theta})L(e^{j(\omega-\theta)})d\theta. \tag{8.96}$$

This operation, which is simply linear filtering of the complex logarithm of the DTFT, was also called "liftering" by Bogert et al. [39], and therefore $l[n]$ is often called a "lifter." The resulting windowed complex cepstrum is processed by the inverse characteristic system to recover the desired component.

Liftering is illustrated by the thick lines in Figures 8.29a and 8.29b, which show the log magnitude and phase obtained in the process of implementing the inverse characteristic system (i.e., $\hat{Y}(e^{j\omega})$) when $l[n]$ is of the form

$$l_{\text{lp}}[n] = \begin{cases} 1 & |n| < n_{\text{co}} \\ 0.5 & |n| = n_{\text{co}} \\ 0 & |n| > n_{\text{co}}, \end{cases} \tag{8.97}$$

where, in general, n_{co} is chosen to be less than the pitch period, N_p, and in this example, $n_{\text{co}} = 50$ as shown by the dashed line in Figure 8.30a.[18]

When using the DFT implementation, the lifter in Eq. (8.97) must conform to the sample ordering of the DFT; i.e., the negative quefrencies fall in the interval $N/2 < n \leq N-1$ for an N-point DFT. Thus, for DFT implementations, the lowpass lifter has the form,

$$\tilde{l}_{\text{lp}}[n] = \begin{cases} 1 & 0 \leq n < n_{\text{co}} \\ 0.5 & n = n_{\text{co}} \\ 0 & n_{\text{co}} < n < N - n_{\text{co}} \\ 0.5 & n = N - n_{\text{co}} \\ 1 & N - n_{\text{co}} < n \leq N-1. \end{cases} \tag{8.98}$$

For simplicity, we shall henceforth define lifters in DTFT form as in Eq. (8.97), recognizing that the DFT form is always obtained by the process that yielded Eq. (8.98).

The thick lines that are superimposed on the plots of $\log|X(e^{j\omega})|$ and $\arg\{X(e^{j\omega})\}$ in Figures 8.29a and 8.29b show the real and imaginary parts of $\hat{Y}(e^{j\omega})$ corresponding to the liftered complex cepstrum $\hat{y}[n] = l_{\text{lp}}[n]\hat{x}[n]$. By comparing these plots to the corresponding plots with thin lines in Figures 8.29b and 8.29c respectively, it can be seen that $\hat{Y}(e^{j\omega})$ is a smoothed version of $\hat{X}(e^{j\omega})$. The result of the lowpass liftering is to remove the effect of the excitation in the short-time Fourier transform. That is, retaining only the low quefrency components of the complex cepstrum is a way of estimating $\hat{H}_V(e^{j\omega}) = \log|H_V(e^{j\omega})| + j\arg\{H_V(e^{j\omega})\}$, the complex logarithm of the frequency response of the vocal tract system. We see that the smoothed log magnitude function in Figure 8.29a clearly displays formant resonances at about 500, 1500, 2250, and 3100 Hz. Also note that if the lifter $l_{\text{lp}}[n]$ is applied to the cepstrum $c[n]$, the corresponding Fourier representation would be only the smoothed log magnitude shown in Figure 8.29a.

Now if the smoothed complex logarithm in Figures 8.29a and 8.29b is exponentiated to obtain $Y(e^{j\omega}) = \exp\{\hat{Y}(e^{j\omega})\}$, the corresponding time function $y[n]$ would be the waveform shown in Figure 8.32a. This waveform is an estimate of the impulse

[18] A one-sample transition is included in Eq. (8.97). Expanding or omitting this transition completely usually has little effect.

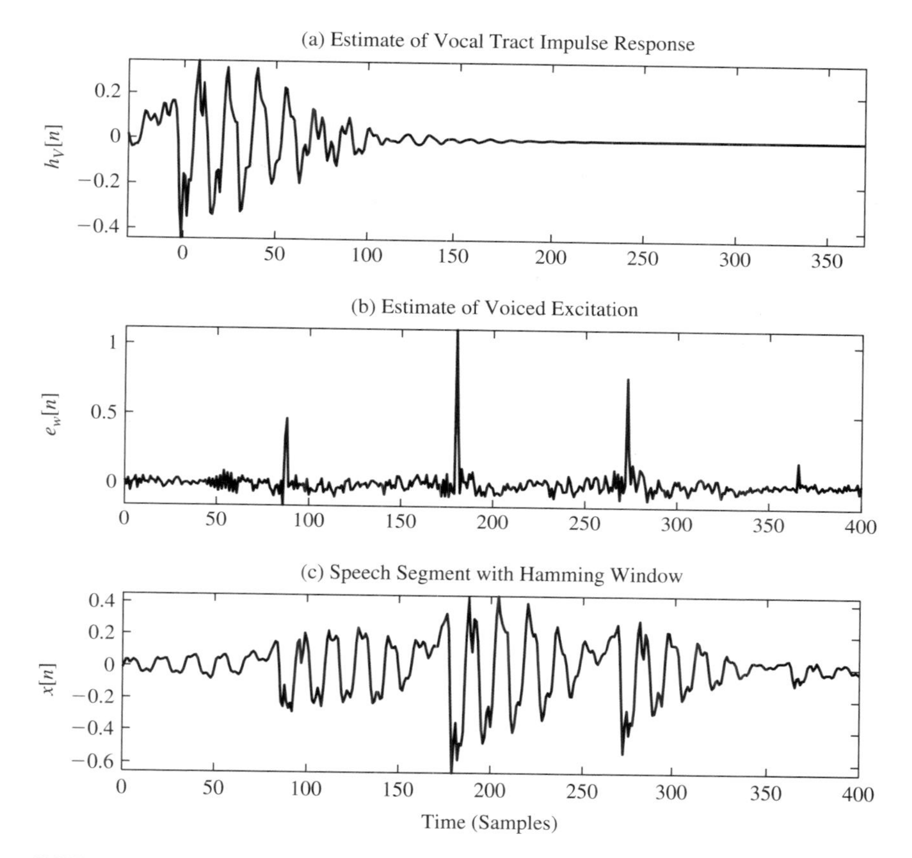

FIGURE 8.32
Homomorphic filtering of voiced speech: (a) estimate of vocal tract impulse response $h_V[n]$; (b) estimate of excitation component $e_w[n]$; and (c) original windowed speech signal.

response $h_V[n]$, including the effects of excitation gain, glottal pulse, vocal tract resonance structure, and radiation.

On the other hand, $l[n]$ can be chosen so as to retain only the high quefrency excitation components using the highpass lifter

$$l_{\text{hp}}[n] = \begin{cases} 0 & |n| < n_{\text{co}} \\ 0.5 & |n| = n_{\text{co}} \\ 1 & |n| > n_{\text{co}}, \end{cases} \tag{8.99}$$

where again, $n_{\text{co}} < N_p$. Figures 8.33a and 8.33b are obtained for $\log |Y(e^{j\omega})|$ and $\arg\{Y(e^{j\omega})\}$, respectively, if $\hat{y}[n] = l_{\text{hp}}[n]\hat{x}[n]$. These two functions would be the log magnitude and phase of an estimate of $E_w(e^{j\omega})$, the DTFT of the window-weighted excitation sequence $e_w[n]$. Note that the output $y[n]$ corresponding to $\hat{y}[n] = l_{\text{hp}}[n]\hat{x}[n]$,

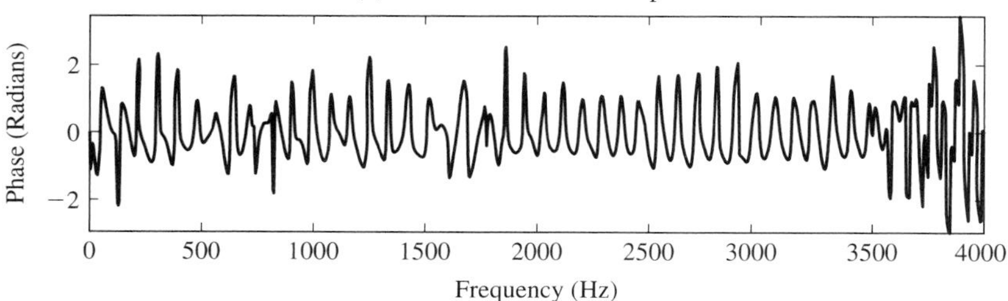

FIGURE 8.33
Homomorphic filtering of voiced speech; (a) estimate of log magnitude and (b) estimate of phase of $E_w(e^{j\omega})$.

which is shown in Figure 8.32b, approximates an impulse train with spacing equal to the pitch period and amplitudes retaining the shape of the Hamming window used to weight the input signal. Thus, with the highpass lifter, $y[n]$ serves as an estimate of $e_w[n]$.

If the same value of n_{co} is used for both the lowpass and highpass lifters, then $l_{\text{lp}}[n] + l_{\text{hp}}[n] = 1$ for all n. Thus, the choice of the lowpass and highpass lifters defines $e_w[n]$ and $h_V[n]$ so that $h_V[n] * e_w[n] = x[n]$; i.e., convolution of the waveforms in Figures 8.32a and 8.32b will result in the original windowed speech signal shown in Figure 8.32c. In terms of the corresponding DTFT, adding the curves in Figures 8.33a and 8.33b to the smooth curves plotted with thick lines in Figures 8.29a and 8.29b respectively results in the rapidly varying curves in Figures 8.29a and 8.29b.

8.5.4 Minimum-Phase Analysis

Since the cepstrum is the inverse DTFT of the logarithm of the magnitude of the DTFT of the windowed speech segment, it is also the even part of the complex cepstrum. If the input signal is known to have the minimum-phase property, we also know that the complex cepstrum is zero for $n < 0$ and therefore, it can be obtained from the cepstrum by the operation in Eq. (8.36), which can be seen to be equivalent to multiplying the cepstrum by a lifter; i.e., $\hat{x}_{\text{mnp}}[n] = l_{\text{mnp}}[n]c[n]$, where

$$l_{\text{mnp}}[n] = \begin{cases} 0 & n < 0 \\ 1 & n = 0 \\ 2 & 0 < n. \end{cases} \tag{8.100a}$$

If, on the other hand, we do not know whether or not the input signal has the minimum-phase property, we can nevertheless *assume* that it does. Then the sequence $\hat{y}[n] = l_{\text{mnp}}[n]c[n]$ would be the complex cepstrum of a signal $y[n]$ whose DTFT would have the same log magnitude as the DTFT of the original signal $x[n]$. If the original signal were not minimum-phase, $\arg\{X(e^{j\omega})\}$ and $\arg\{Y(e^{j\omega})\}$ would differ, but $\log|Y(e^{j\omega})| = \log|X(e^{j\omega})|$.

If we assume that the low quefrencies in the cepstrum are due to the vocal tract system, and we further assume that the vocal tract system is minimum-phase, then we can accomplish the estimation of a minimum-phase vocal tract impulse response by combining Eqs. (8.100a) and (8.97) to obtain

$$l_{\text{mnp}}[n] = \begin{cases} 0 & n < 0 \\ 1 & n = 0 \\ 2 & 0 < n < n_{\text{co}} \\ 1 & n = n_{\text{co}} \\ 0 & n_{\text{co}} < n, \end{cases} \tag{8.100b}$$

which imposes a cutoff quefrency to remove the excitation components in the cepstrum and simultaneously imposes the minimum-phase condition.[19] Note that we have again included a one sample transition, which can be expanded if desired.

For the voiced example of this section, the result of liftering the cepstrum in Figure 8.30b with the lowpass lifter in Eq. (8.100b) with cutoff quefrency $n_{\text{co}} = 50$ is the impulse response shown in Figure 8.34b. From the above discussion, it follows that the log magnitude of the DTFT of the vocal tract impulse response estimate in Figure 8.34b is identical to the log-magnitude of the DTFT of the vocal tract impulse response estimate in Figure 8.32a, since both were obtained with a cutoff quefrency of $n_{\text{co}} = 50$; i.e., the smoothed log magnitude in Figure 8.29a is the log magnitude of the DTFT of the waveforms in Figures 8.34b and 8.32a. In fact, the other two impulse response estimates in Figures 8.34a and 8.34c also have the same log magnitude of their DTFTs. The impulse response in Figure 8.34a corresponds to applying the lifter in Eq. (8.98) to the cepstrum (i.e., without incorporating the minimum-phase condition). This is equivalent to assuming that the phase is zero. The resulting impulse response is an even time sequence and is therefore non-causal. It could be made causal by truncating it symmetrically and introducing sufficient delay. The impulse response in Figure 8.34c is a maximum-phase impulse response that is obtained at the output after multiplying the cepstrum by $l_{\text{mxp}}[n] = l_{\text{mnp}}[-n]$; i.e., by imposing the maximum-phase condition that the complex cepstrum is zero for $n > 0$. The waveform of Figure 8.34c is seen to be a time-reversed version of the minimum-phase impulse response in Figure 8.34b. Again it can be made causal by truncating it and including sufficient delay. The effects of phase on synthetic speech reconstructed from impulse responses derived by homomorphic filtering were studied by Oppenheim [267] using vocal tract impulse

[19]Observe that Eq. (8.100b) reduces to Eq. (8.100a) when $n_{\text{co}} \to \infty$.

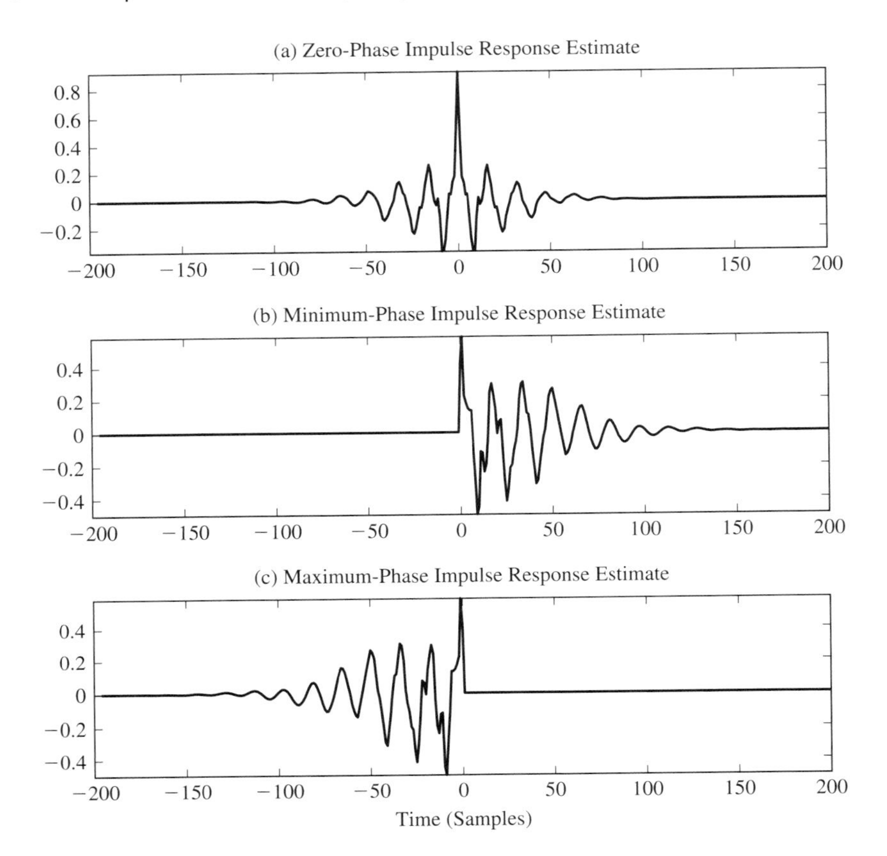

FIGURE 8.34
Homomorphic filtering of voiced speech: (a) zero-phase estimate of $h_V[n]$; (b) minimum-phase estimate of $h_V[n]$; (c) maximum-phase estimate of $h_V[n]$.

responses derived by homomorphic filtering. The use of the cepstrum in speech coding is discussed in more detail in Chapter 11.

8.5.5 Unvoiced Speech Analysis Using the DFT

To complete the illustration of homomorphic analysis of natural speech, consider the example of unvoiced speech given in Figure 8.35. Figure 8.35a shows a waveform segment of the fricative /SH/ multiplied by a 401-point Hamming window. The rapidly varying curve plotted with the thin line in Figure 8.35b is the corresponding log magnitude function $\log|X(e^{j\omega})|$. Figure 8.35c shows the corresponding cepstrum $c[n]$. For consistency, and since we generally do not know in advance whether a particular speech segment is voiced or unvoiced, $c[n]$ for unvoiced speech is computed as the inverse DTFT of $\log|X(e^{j\omega})|$ just as for voiced speech. Note the erratic variation of the log magnitude function (log periodogram). It is clear from Figure 8.35c that, in contrast to the case of voiced speech, the cepstrum of an unvoiced speech segment does

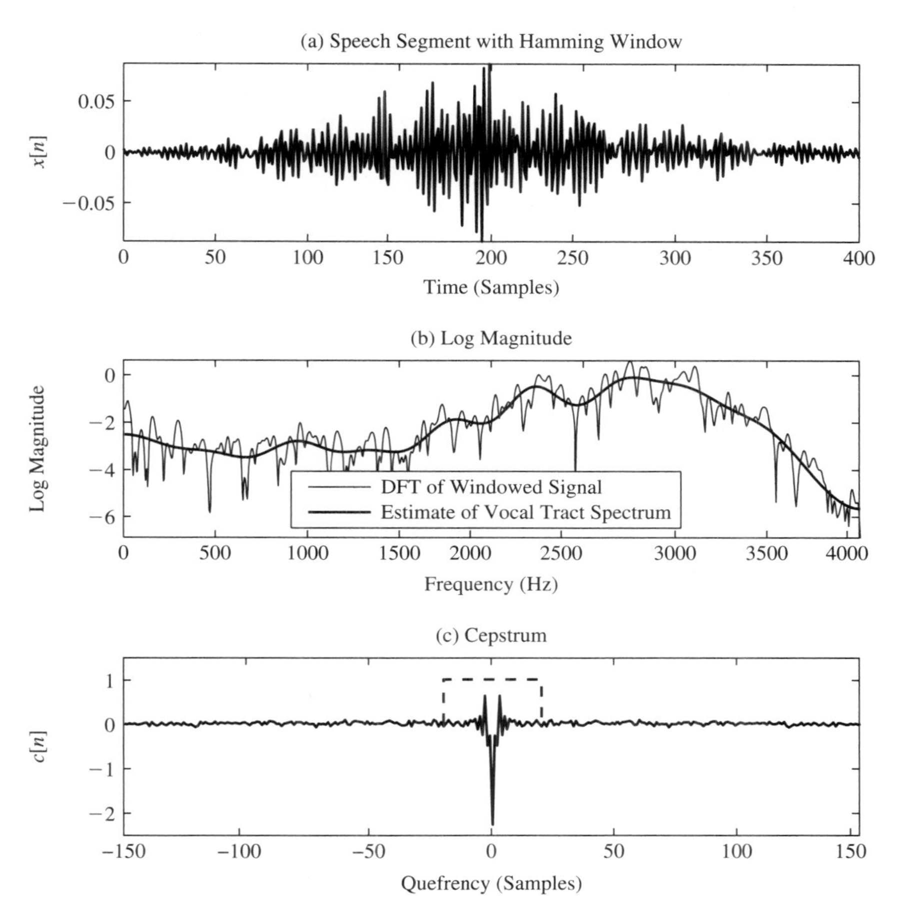

FIGURE 8.35
Homomorphic filtering of unvoiced speech: (a) windowed unvoiced speech segment $x[n]$; (b) log magnitude of short-time Fourier transform $\log|X(e^{j\omega})|$ (heavy line shows estimate of vocal tract spectrum $\log|H_U(e^{j\omega})|$); (c) corresponding cepstrum.

not display any sharp peaks in the high quefrency region. Instead, the high quefrencies represent the rapid random fluctuations in Figure 8.35b. However, the low-time portion of the cepstrum can still be assumed to represent $\log|H_U(e^{j\omega})|$. This is illustrated in Figure 8.35b by the smooth curve plotted with the thick line, which represents the smoothed log magnitude function obtained by applying the lowpass cepstrum window of Eq. (8.97) to the cepstrum of Figure 8.35c with $n_{co} = 20$ as shown.

As in the case of voiced speech, we can compute zero-phase, minimum-phase, or maximum-phase impulse responses as the output of the inverse characteristic system $\mathcal{D}_*^{-1}\{\cdot\}$ with the liftered cepstrum as input. Figure 8.36a shows the zero-phase impulse response corresponding to the smoothed log magnitude in Figure 8.35b obtained using the lifter in Eq. (8.98) with $n_{co} = 20$ (shown superimposed on the cepstrum in

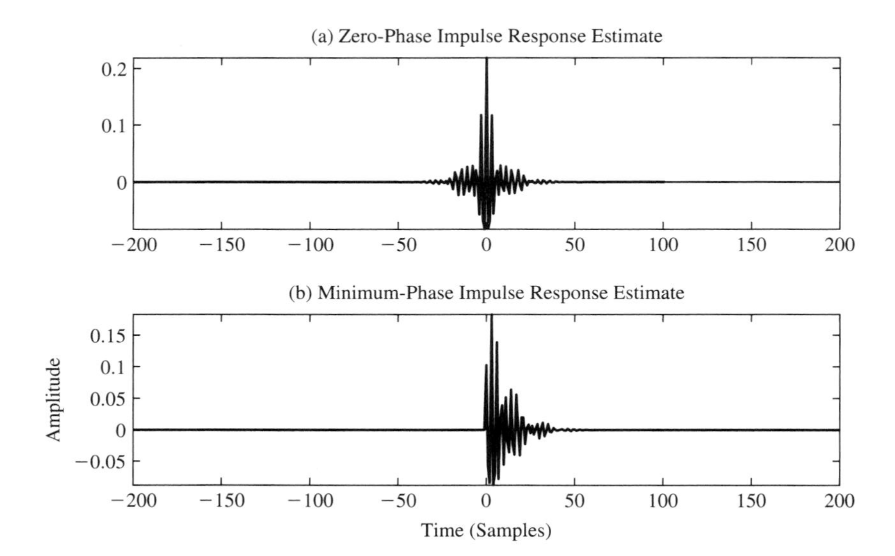

FIGURE 8.36
Homomorphic analysis of unvoiced speech: (a) zero-phase estimate of vocal tract impulse response $h_U[n]$; (b) zero-phase estimate of vocal tract impulse response $h_U[n]$.

Figure 8.36c). Figure 8.36b shows the corresponding minimum-phase impulse response obtained using the lifter of Eq. (8.100b) with $n_{co} = 20$. Not shown in Figure 8.36 is the maximum-phase impulse response, which would simply be a time-reversed version of the impulse response in Figure 8.36b. Observe that the impulse response for the unvoiced example varies quite rapidly in time compared to the impulse responses derived for voiced speech (plotted in Figures 8.32a and 8.34). This is because the log spectrum for the fricative has a broad peak at about 2700 Hz, while the voiced spectrum is concentrated at low frequencies.

8.5.6 Summary Illustration of Short-Time Cepstrum Analysis

The previous discussion and examples show that it is possible to obtain approximations to some of the basic components of the speech waveform by homomorphic filtering. While the examples show that it is possible to separate a windowed speech segment into plausible excitation and vocal tract impulse response components, complete deconvolution of the speech waveform is often not necessary. Rather, we may only require estimates of basic parameters such as pitch period and formant frequencies. For this purpose, the cepstrum is entirely sufficient. Thus, in many speech analysis applications, it is not necessary to compute the unwrapped phase. Notice, for example, by comparing Figure 8.30b to Figure 8.35c that the cepstrum provides a basis for distinguishing between voiced and unvoiced speech, and furthermore, the pitch period of voiced speech is placed clearly in evidence in the cepstrum. Also note that the formant frequencies show up clearly in the smoothed log magnitude estimate of the vocal tract frequency response, which can be obtained by applying the lifter of Eq. (8.98) or (8.100b) to the cepstrum.

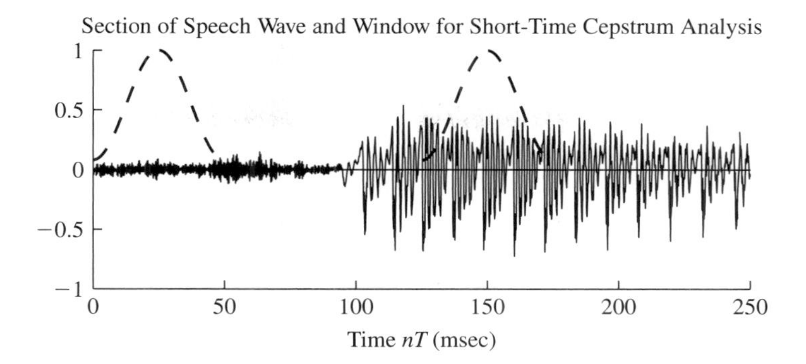

FIGURE 8.37
2001 samples of speech waveform at the transition from the fricative /SH/ to the diphthong /EY/ for an utterance of the word /SH EY D/. The short-time homomorphic analysis of this segment is plotted in Figure 8.38.

As an illustration of the effectiveness of the short-time cepstrum in highlighting the time-varying properties of speech, consider the 2001-point (250 msec) segment of speech shown in Figure 8.37. This figure shows two Hamming windows of length 401 samples (50 msec at an 8 kHz sampling rate). Short-time cepstrum analysis and homomorphic smoothing was performed on this waveform every 100 samples (12.5 msec) to obtain a cepstrum and smoothed log magnitude function for each frame. Figure 8.38 shows 15 consecutive frames of this processing, with the first frame positioned at the beginning of the waveform in Figure 8.37. The window on the left in Figure 8.37 corresponds to the unvoiced example in Section 8.5.5 and to frame 1 in Figure 8.38. The other window, positioned at 1000 samples (12.5 msec) after the beginning of the plot in Figure 8.37, corresponds to the voiced example considered in detail in Section 8.5.3 and to frame 10 in Figure 8.38.

The set of plots in Figure 8.38 confirm all the assertions that we have made about the properties of the cepstrum and about how homomorphic filtering can achieve a degree of separation between the excitation and vocal tract system impulse response. For the given window length (401 samples) and effective spacing between frames (100 samples), the first five frames include only the fricative /SH/. This is evident in the cepstrum, the short-time Fourier transform, and the liftered spectrum in each of the first five frames. Frame 6 begins at 75 msec and ends at 125 msec; thus, it straddles the transition from unvoiced to voiced speech. This is manifested in the cepstrum as a lack of a strong peak because the rapid variations in the short-time spectrum are more random than periodic. However, the formant structure of the voiced part of the window, being of larger amplitude, tends to give the smoothed log spectrum the formant resonance pattern that is characteristic of the later voiced frames. Frames 7–15 contain only voiced speech of the diphthong /EY/, and the corresponding cepstra show a clear peak at the quefrency corresponding to the pitch period. The strong peak in the cepstrum corresponds to the clearly defined periodic structure in the short-time Fourier transform. Subsequent frames contain only voiced speech, and the cepstrum peak moves slightly to the right with increasing frame number, implying a slight drop

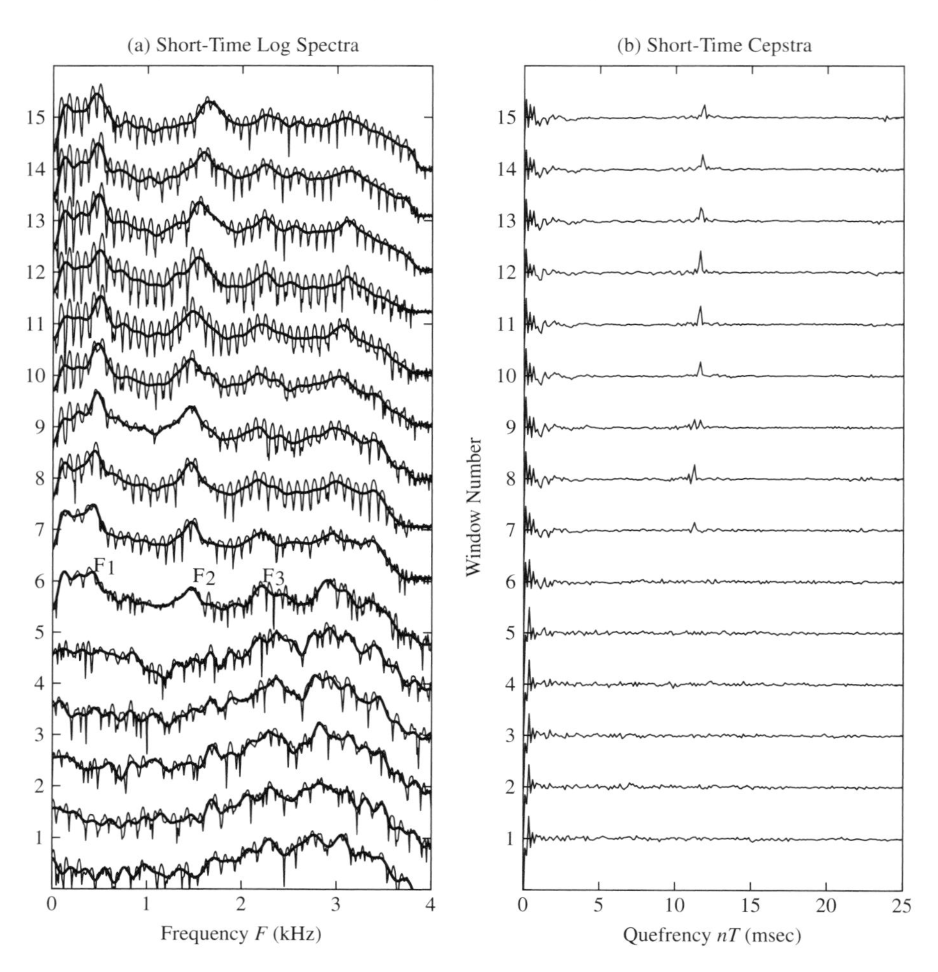

FIGURE 8.38
Short-time homomorphic analysis of the waveform in Figure 8.37 at 15 window positions separated by 100 samples (12.5 msec).

in pitch frequency. The smoothed log spectra in frames 7–15 show how the formant frequencies change with time.

In Chapter 10 we discuss how the short-time cepstrum can be used as the basis for algorithms for pitch and voiced/unvoiced detection and also for estimating the time variation of formant frequencies. In Chapter 11, we will also discuss the role of the cepstrum in speech coding.

8.6 CEPSTRUM ANALYSIS OF ALL-POLE MODELS

In Section 8.4 we showed several ways of computing the short-time cepstrum and complex cepstrum, and in Section 8.5, we illustrated how these methods can be applied to natural speech signals. Another, somewhat indirect, method of cepstrum analysis is based on all-pole modeling of speech using the methods of linear prediction that are discussed in detail in Chapter 9. Although it may seem somewhat awkward to introduce linear predictive models before discussing how they are derived, this should not cause much difficulty, since our focus is simply on the all-pole model that results. Linear predictive analysis is simply a means to compute the model directly from short segments of the speech signal.

For our purposes at this point, it is sufficient to assert that short-time analysis techniques can be applied to estimate a vocal tract system model of the form

$$H(z) = \frac{G}{A(z)} = \frac{G}{1 - \sum_{k=1}^{p} \alpha_k z^{-k}} = \frac{G}{\prod_{k=1}^{p} (1 - z_k z^{-1})}. \tag{8.101}$$

The optimal values of the parameters G and α_k for $k = 1, 2, \dots, p$ can be found by solving a set of linear equations whose coefficients are determined by the autocorrelation function of the windowed speech segment $x[n]$. The parameters z_k are simply the roots of the denominator polynomial $A(z)$ (poles of $H(z)$), which can be found by polynomial rooting. The corresponding impulse response satisfies the difference equation

$$h[n] = \sum_{k=1}^{p} \alpha_k h[n-k] + G\delta[n]. \tag{8.102}$$

An important feature of the all-pole model produced by linear predictive techniques is that the model is minimum phase; i.e., all the poles satisfy $|z_k| < 1$. This means that the complex cepstrum of the impulse response of the all-pole model has the property $\hat{h}[n] = 0$ for $n < 0$. The complex cepstrum (and therefore its even part, the cepstrum) can be determined in several ways. One approach recognizes that Eq. (8.101) is a special case of Eq. (8.20) with $M_i = M_o = 0$ and $N_i = p$. Using Eq. (8.24), it follows that the complex cepstrum of the impulse response of the minimum-phase all-pole model is

$$\hat{h}[n] = \begin{cases} 0 & n < 0 \\ \log G & n = 0 \\ \sum_{k=1}^{p} \dfrac{z_k^n}{n} & n > 0. \end{cases} \tag{8.103}$$

This method of computing the cepstrum of an all-pole model is depicted in the block diagram of Figure 8.39a.

A second approach takes advantage of the recursion formula Eq. (8.83) of Section 8.4.3. Since $h[n]$ is minimum phase, and since it follows from Eq. (8.102) that

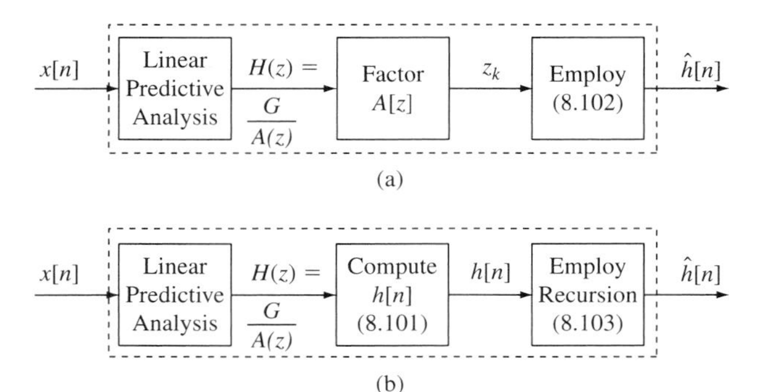

FIGURE 8.39
Computation of the complex cepstrum of the impulse response of an all-pole minimum-phase model of the vocal tract system: (a) polynomial rooting of the denominator of the all-pole system function; (b) recursive computation using Eq. (8.104). (Numbers in parenthesis refer to text equations.)

$h[0] = G$, Eq. (8.83) becomes

$$\hat{h}[n] = \begin{cases} 0 & n < 0 \\ \log G & n = 0 \\ \dfrac{h[n]}{G} - \displaystyle\sum_{k=0}^{n-1} \left(\frac{k}{n}\right) \hat{h}[k] \frac{h[n-k]}{G} & n > 0. \end{cases} \tag{8.104}$$

As depicted in Figure 8.39b, linear predictive analysis can provide the parameters of the difference equation in Eq. (8.102), which can in turn be used to compute any number of samples of the impulse response of the all-pole model. Then Eq. (8.104) can be used to compute the complex cepstrum for as many samples as the available impulse response samples. If it is desired to go from the complex cepstrum of the impulse response of the minimum-phase model back to the impulse response itself, we need only rearrange the terms in Eq. (8.104) to obtain

$$h[n] = \begin{cases} 0 & n < 0 \\ e^{\hat{h}[0]} & n = 0 \\ h[0]\hat{h}[n] + \displaystyle\sum_{k=0}^{n-1} \left(\frac{k}{n}\right) \hat{h}[k] h[n-k] & n > 0. \end{cases} \tag{8.105}$$

This method of computation of the complex cepstrum of the all-pole vocal tract model relies on linear predictive analysis to remove the effects of the excitation. By restricting p to be much less than the pitch period, N_p, linear predictive modeling in a sense accomplishes what the lowpass lifter accomplishes in homomorphic filtering. However, it should be noted that the impulse response obtained by linear predictive methods will differ somewhat from the minimum-phase impulse response obtained by cepstrum liftering as in Section 8.5.4.

8.7 CEPSTRUM DISTANCE MEASURES

Perhaps the most pervasive application of the cepstrum in speech processing is its use in pattern recognition problems such as vector quantization (VQ) and automatic speech recognition (ASR). In such applications, a speech signal is represented on a frame-by-frame basis by a sequence of short-time cepstra. In later discussions in this section, it will be useful to use somewhat more complicated notation. Specifically, we denote the cepstrum of the mth frame of a signal, $x_m[n]$, as $c_m^{(x)}[n]$, where n denotes the quefrency index of the cepstrum. In cases where it is not necessary to distinguish between signals or frames, these additional designations will be omitted as we have done up to this point in this chapter.

Cepstrum-like representations can be obtained in many ways as we have seen. No matter how it is computed, we can assume that the cepstrum vector corresponds to a gain-normalized ($c[0] = 0$) minimum-phase vocal tract impulse response, $h[n]$, having the complex cepstrum

$$\hat{h}[n] = \begin{cases} 2c[n] & 1 \leq n \leq n_{\text{co}} \\ 0 & n < 0. \end{cases} \tag{8.106}$$

(In this discussion, we use $h[n]$ to denote the vocal tract impulse response of the speech model for both voiced and unvoiced speech.) In applications such as VQ or ASR, a test pattern $c[n]$ (vector of cepstrum values defined for $n = 1, 2, \ldots, n_{\text{co}}$), which implicitly corresponds to an impulse response, $h[n]$, is compared against a comparable reference pattern, $\bar{c}[n]$, which implicitly corresponds to a reference impulse response, $\bar{h}[n]$. Such comparisons require a suitable measure of distance (or distortion) between the pair of cepstral vectors. For example, the Euclidean distance applied to the cepstrum would be

$$D = \sum_{n=1}^{n_{\text{co}}} \left| c[n] - \bar{c}[n] \right|^2. \tag{8.107a}$$

Equivalently (based on Parseval's theorem) in the frequency domain, the Euclidean distance (distortion) measure is

$$D = \frac{1}{2\pi} \int_{-\pi}^{\pi} \left| \log |H(e^{j\omega})| - \log |\bar{H}(e^{j\omega})| \right|^2 d\omega, \tag{8.107b}$$

where $\log |H(e^{j\omega})|$ is the log magnitude of the DTFT of $h[n]$ corresponding to the complex cepstrum in Eq. (8.106) or the real part of the DTFT of $\hat{h}[n]$ in Eq. (8.106), and $\log |\bar{H}(e^{j\omega})|$ is the corresponding reference log spectrum. Thus, cepstrum-based comparisons are directly related to comparisons of smoothed short-time log spectra.

The cepstrum offers an effective and flexible representation of speech for pattern recognition problems as we will discuss below.

8.7.1 Compensation for Linear Filtering

Suppose that we have only a linearly filtered version of the speech signal, $y[n] = h_d[n] * x[n]$, instead of $x[n]$. If the analysis window is long compared to the length of the distorting impulse response $h_d[n]$, the short-time cepstrum of one frame of the filtered speech signal, $y[n]$, will be approximately

$$c_m^{(y)}[n] = c_m^{(x)}[n] + c^{(h_d)}[n], \tag{8.108}$$

where $c^{(h_d)}[n]$ will appear more or less the same in each frame. Therefore, if we can estimate $c^{(h_d)}[n]$,[20] which we assume is non-time-varying, we can obtain $c_m^{(x)}[n]$ at each frame from $c_m^{(y)}[n]$ by subtraction; i.e., $c_m^{(x)}[n] = c_m^{(y)}[n] - c^{(h_d)}[n]$. This process is called *cepstrum-mean subtraction*. This property of the cepstrum is extremely attractive in situations where the set of reference patterns, $\bar{c}[n]$, has been obtained under different recording or transmission conditions from those used to acquire the test vectors. In these circumstances, the test vectors can be compensated for the effects of the linear filtering prior to computing the distance measures used for comparison of patterns.

Another approach to removing the effects of linear distortions is to observe that the cepstrum component due to the distortion is the same in each frame. Therefore it can be removed by a simple first difference operation of the form

$$\Delta c_m^{(y)}[n] = c_m^{(y)}[n] - c_{m-1}^{(y)}[n]. \tag{8.109}$$

It is clear that if $c_m^{(y)}[n] = c_m^{(x)}[n] + c^{(h_d)}[n]$, with $c^{(h_d)}[n]$ being independent of m, then $\Delta c_m^{(y)}[n] = \Delta c_m^{(x)}[n]$; i.e., the linear distortion effects are removed. While this idea is simple, more complex processing, to be discussed in Section 8.7.5, is required in most applications.

8.7.2 Weighted Cepstral Distance Measures

In using linear prediction analysis to obtain cepstrum feature vectors for pattern recognition problems, it is observed that there is significant statistical variability due to a variety of factors including short-time analysis window position, bias toward harmonic peaks, and noise [181, 391]. A solution to this problem is to use a weighted cepstral distance measure of the form

$$D = \sum_{n=1}^{n_{co}} (l[n])^2 \left|c[n] - \bar{c}[n]\right|^2, \tag{8.110a}$$

[20]Stockham [381] showed how $c^{(h_d)}[n]$ for such linear distortions can be estimated from the signal, $y[n]$, by time averaging the logarithm of the short-time Fourier transform.

where $(l[n])^2$ is the squared weighting sequence. Equation (8.110b) can be written as the Euclidean distance of liftered cepstra

$$D = \sum_{n=1}^{n_{co}} \left| l[n]c[n] - l[n]\bar{c}[n] \right|^2. \tag{8.110b}$$

Tohkura [391] found that, when averaged over many frames of speech and speakers, cepstrum values, $c[n]$, have zero means and their variances are on the order of $1/n^2$. This suggests that a cepstrum weighting of the form $l[n] = n$ for $n = 1, 2, \ldots, n_{co}$ could be used to equalize the contributions for each term to the cepstrum difference.

Juang et al. [181] observed that the variability due to the vagaries of linear prediction analysis could be lessened by using a bandpass lifter of the form

$$l[n] = \begin{cases} 1 + 0.5n_{co}\sin(\pi n/n_{co}) & n = 1, 2, \ldots, n_{co} \\ 0 & \text{otherwise.} \end{cases} \tag{8.111}$$

Tests of such weighted cepstral distance measures showed consistent performance improvements in automatic speech recognition tasks.

8.7.3 Group Delay Spectrum

The weighted cepstral distance measures of the previous sub-section were given a new interpretation by Itakura and Umezaki [164], who invoked the following basic property of the DTFT:

$$n\hat{h}[n] \Longleftrightarrow j\frac{d\hat{H}(e^{j\omega})}{d\omega}, \tag{8.112}$$

where $\Longleftrightarrow$ denotes the transform relationship between a sequence and its DTFT. An interesting result can be obtained if we represent the complex cepstrum as

$$\hat{h}[n] = c[n] + d[n], \tag{8.113}$$

where $c[n] = \mathcal{E}v\{\hat{h}[n]\}$ is the even part and $d[n] = \mathcal{O}dd\{\hat{h}[n]\}$ is the odd part of the complex cepstrum. Recalling that the DTFT of the complex cepstrum is, by definition, $\hat{H}(e^{j\omega}) = \log|H(e^{j\omega})| + j\arg\{H(e^{j\omega})\}$, it can be shown that the following DTFT relations hold:

$$nc[n] \Longleftrightarrow j\frac{d\log|H(e^{j\omega})|}{d\omega} \tag{8.114a}$$

and

$$nd[n] \Longleftrightarrow -\frac{d\arg\{H(e^{j\omega})\}}{d\omega}. \tag{8.114b}$$

The DTFT expression on the right in Eq. (8.114b) is the group delay function [270] for $H(e^{j\omega})$; i.e.,

$$\text{grd}\{H(e^{j\omega})\} = -\frac{d \arg\{H(e^{j\omega})\}}{d\omega}. \tag{8.115}$$

Now if $h[n]$ is assumed to be obtained by all-pole modeling as discussed in Section 8.6, the complex cepstrum satisfies $\hat{h}[n] = 0$ for $n < 0$. This means that $\hat{h}[n] = 2c[n] = 2d[n]$ for $n > 0$. If we define $l[n] = n$, then the liftered cepstrum distance

$$D = \sum_{m=-\infty}^{\infty} \left| l[m]c[m] - l[m]\bar{c}[m] \right| = \sum_{m=-\infty}^{\infty} \left| l[m]d[m] - l[m]\bar{d}[m] \right| \tag{8.116a}$$

is equivalent to either

$$D = \frac{1}{2\pi} \int_{-\pi}^{\pi} \left| \frac{d \log |H(e^{j\omega})|}{d\omega} - \frac{d \log |\bar{H}(e^{j\omega})|}{d\omega} \right| d\omega, \tag{8.116b}$$

or

$$D = \frac{1}{2\pi} \int_{-\pi}^{\pi} \left| \text{grd}\{H(e^{j\omega})\} - \text{grd}\{\bar{H}(e^{j\omega})\} \right| d\omega, \tag{8.116c}$$

where $H(e^{j\omega})$ is obtained from Eq. (8.101) with $z = e^{j\omega}$. The result of Eq. (8.116b) was also given by Tohkura [391].

Instead of $l[n] = n$ for all n, or the lifter of Eq. (8.111), Itakura proposed the lifter

$$l[n] = n^s e^{-n^2/2\tau^2}. \tag{8.117}$$

This lifter has great flexibility. For example, if $s = 0$, we have simple low quefrency liftering of the cepstrum. If $s = 1$ and τ is large, we essentially have $l[n] = n$ for small n with high quefrency tapering. The effect of liftering with Eq. (8.117) is illustrated in Figure 8.40, which shows in (a) the short-time Fourier transform of a segment of voiced speech along with a linear predictive analysis spectrum with $p = 12$. In Figure 8.40(b) is shown the liftered group delay spectrum for $s = 1$ and τ ranging from 5 to 15 to 30. Observe that as τ increases, the formant frequencies are increasingly emphasized. If larger values of s are used, an even greater enhancement of the resonance structure is observed.

When Itakura and Umezaki [164] tested the group delay spectrum distance measure in an automatic speech recognition system, they found that for clean test utterances, the difference in recognition rate was small for different values of s when $\tau \approx 5$, although performance suffered with increasing s for larger values of τ. This was attributed to the fact that for larger s, the group delay spectrum becomes very sharply peaked and thus more sensitive to small differences in formant locations. However, in test conditions with additive white noise and also with linear filtering

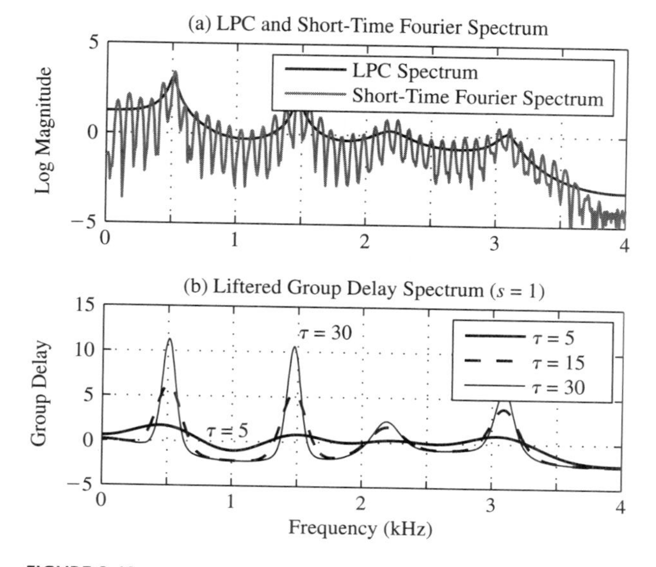

FIGURE 8.40
(a) Short-time Fourier transform and linear predictive coding (LPC) spectrum; (b) liftered group delay spectrum.

distortions, recognition rates improved significantly with $\tau = 5$ and increasing values of the parameter s.

8.7.4 Mel-Frequency Cepstrum Coefficients

As we have seen, weighted cepstrum distance measures have a directly equivalent interpretation in terms of log spectrum distance in the frequency domain. This is significant in light of models for human perception of sound, which are based upon a frequency analysis performed in the inner ear (recall the discussion of Chapter 4). With this in mind, Davis and Mermelstein [82] formulated a new type of cepstrum representation that has come to be widely used and known as the mel-frequency cepstrum coefficients (mfcc).

The basic idea is to compute a frequency analysis based upon a filter bank with approximately critical band spacing of the filters and bandwidths. For a 4 kHz bandwidth, approximately 20 filters are used. In most implementations, a short-time Fourier analysis is done first, resulting in a DFT, $X_m[k]$, for the mth frame. Then the DFT values are grouped together in critical bands and weighted by triangular weighting functions such as those depicted in Figure 8.41. Note that the bandwidths in Figure 8.41 are constant for center frequencies below 1 kHz and then increase exponentially up to half the sampling rate of 4 kHz, resulting in $R = 24$ "filters." The mel-spectrum of the mth frame is defined for $r = 1, 2, \dots, R$ as

$$\text{MF}_m[r] = \frac{1}{A_r} \sum_{k=L_r}^{U_r} \left| V_r[k] X_m[k] \right|^2, \quad r = 1, 2, \dots, R \tag{8.118a}$$

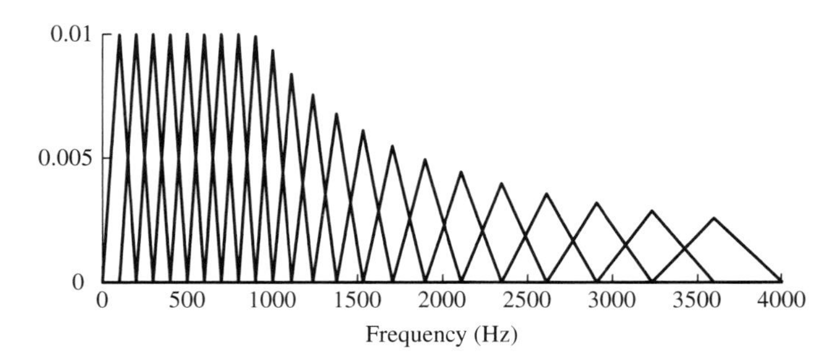

FIGURE 8.41
DFT weighting functions for mel-frequency-cepstrum computations.

where $V_r[k]$ is the weighting function for the rth filter ranging from DFT index L_r to U_r, and

$$A_r = \sum_{k=L_r}^{U_r} |V_r[k]|^2 \tag{8.118b}$$

is a normalizing factor for the rth mel-filter. This normalization is built into the plot of Figure 8.41. It is needed so that a perfectly flat input Fourier spectrum will produce a flat mel-spectrum. For each frame, m, a discrete cosine transform of the logarithm of the magnitude of the filter outputs is computed to form the function $\text{mfcc}_m[n]$ as

$$\text{mfcc}_m[n] = \frac{1}{R} \sum_{r=1}^{R} \log\left(\text{MF}_m[r]\right) \cos\left[\frac{2\pi}{R}\left(r + \frac{1}{2}\right) n\right]. \tag{8.119}$$

Typically, $\text{mfcc}_m[n]$ is evaluated for $n = 1, 2, \ldots, N_{\text{mfcc}}$, where N_{mfcc} is less than the number of mel-filters, e.g., $N_{\text{mfcc}} = 13$ and $R = 24$. Figure 8.42 shows the result of mfcc analysis of a frame of voiced speech in comparison with the short-time spectrum, LPC spectrum, and a homomorphically smoothed spectrum.[21] The large dots are the values of $\log\left(\text{MF}_m[r]\right)$ and the line interpolated between them is a spectrum reconstructed by interpolation at the original DFT frequencies. Note that these spectra are different from one another in detail, but they have, in common, peaks at the formant resonances. At higher frequencies, the reconstructed mel-spectrum, of course, has more smoothing due to the structure of the filter bank.

The mfcc parameters have become firmly established as the basic feature vector for most speech and acoustic pattern recognition problems. For this reason, new and efficient ways of computing $\text{mfcc}[n]$ are of interest. An intriguing proposal is to

[21] The speech signal was pre-emphasized by convolution with $\delta[n] - 0.97\delta[n-1]$ prior to analysis so as to equalize the levels of the formant resonances.

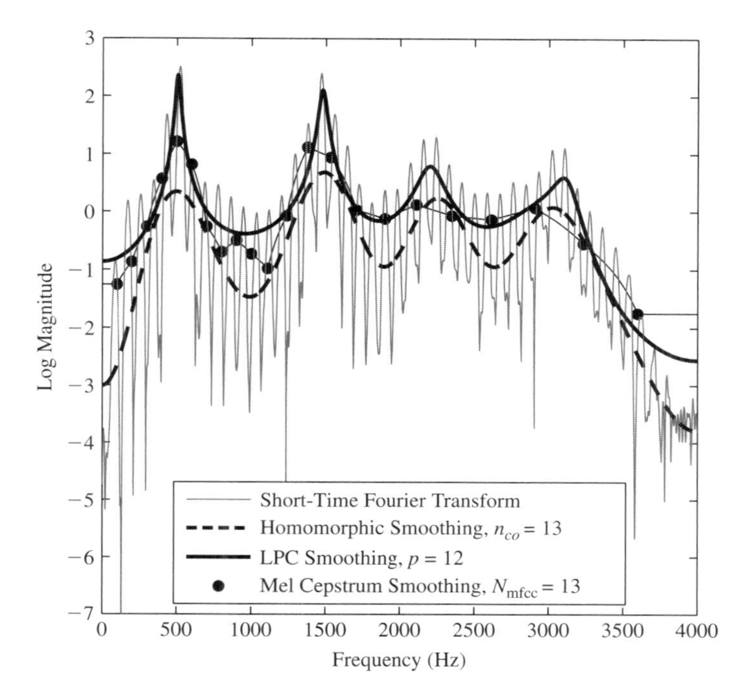

FIGURE 8.42
Comparison of spectral smoothing methods to mel-frequency analysis.

use floating gate electronic technology to implement the filter bank and the DCT computation using microwatts of power [369].

8.7.5 Dynamic Cepstral Features

The set of **mfcc** provide perceptually meaningful and smooth estimates of the speech spectra over time, and have been used effectively in a range of speech processing systems [308]. Since speech is inherently a dynamic signal, changing regularly in time, it is reasonable to seek a representation that includes some aspect of the dynamic nature of the speech signal. As such, Furui [119] proposed the use of estimates of the time derivatives (both first- and second-order derivatives) of the short-term cepstrum. Furui called the resulting parameter sets the delta cepstrum (for the estimate of the first derivative) and the delta-delta cepstrum (for the estimate of the second derivative). Conceptually, the simplest way of computing the delta cepstrum parameters would be as a first difference of cepstral vectors, of the form:

$$\Delta \text{mfcc}_m[n] = \text{mfcc}_m[n] - \text{mfcc}_{m-1}[n]. \tag{8.120}$$

This simple difference is a poor approximation to the first derivative of the mel-cepstral coefficients and is not generally used in practice. Instead the first derivative is often implemented as a least-squares approximation to the local slope (over a region around the current time sample), thereby providing a locally smoothed estimate of the first derivative of the form:

$$\Delta\text{mfcc}_m[n] = \frac{\sum_{k=-M}^{M} k(\text{mfcc}_{m+k}[n])}{\sum_{k=-M}^{M} k^2}. \tag{8.121}$$

It is readily shown that the computation of the delta cepstrum values via Eq. (8.121) is equivalent to fitting the cepstral values of neighboring frames (M frames before frame m and M frames after frame m) by a straight line of slope 1. (It is a simple matter to extend the fitting procedure to a second-order polynomial [308].)

One of the key advantages of using differential parameters such as the delta cepstrum or the delta-delta cepstrum is that the differencing operation removes the effect of simple linear filtering on the parameter values as discussed in Section 8.7.1, thereby making them less sensitive to channel shaping effects that might occur in a speech communication system. This simple cepstral normalization technique has been called cepstral mean subtraction (CMS) or cepstral mean normalization (CMN) and has been widely used in speech recognition systems [151].

8.8 SUMMARY

In this chapter we have presented the basic ideas of homomorphic signal processing as applied to speech. The main idea of homomorphic speech processing is the separation or deconvolution of a segment of speech into a component representing the vocal tract impulse response, and a component representing the excitation source. The way in which such separation is achieved is through liftering or linear filtering of the log spectrum by windowing the cepstrum of the signal. Computational considerations in implementing a homomorphic speech processing system were described. It was shown that several techniques are available for computing the cepstrum and complex cepstrum of a windowed speech signal. These include methods based on polynomial rooting, the DFT, recursive algorithms for minimum-phase signals, and a method based on linear predictive analysis. Detailed examples of homomorphic analysis of synthetic and natural speech were presented. The chapter concluded with a discussion of cepstrum distance measures that are widely used in automatic speech recognition and vector quantization.

PROBLEMS

8.1. The complex cepstrum, $\hat{x}[n]$, of a sequence $x[n]$ is the inverse DTFT of the complex logarithm of the DTFT; i.e.,

$$\hat{X}(e^{j\omega}) = \log|X(e^{j\omega})| + j\arg[X(e^{j\omega})].$$

Show that the (real) cepstrum, $c[n]$, defined as the inverse Fourier transform of the log magnitude, is the even part of $\hat{x}[n]$; i.e., show that

$$c[n] = \frac{\hat{x}[n] + \hat{x}[-n]}{2}.$$

8.2. A linear time-invariant system has the transfer function,

$$H(z) = 8\left[\frac{1 - 4z^{-1}}{1 - \frac{1}{6}z^{-1}}\right].$$

(a) Determine the complex cepstral coefficients, $\hat{h}[n]$, for all n.

(b) Plot $\hat{h}[n]$ versus n for the range $-10 \leq n \leq 10$.

(c) Determine the (real) cepstrum coefficients, $c[n]$, for all n.

8.3. Consider an all-pole model of the vocal tract transfer function of the form

$$V(z) = \frac{1}{\prod_{k=1}^{q}(1 - c_k z^{-1})(1 - c_k^* z^{-1})},$$

where

$$c_k = r_k e^{j\theta_k}.$$

Show that the corresponding cepstrum is

$$\hat{v}[n] = 2\sum_{k=1}^{q} \frac{(r_k)^n}{n} \cos(\theta_k n).$$

8.4. Consider an all-pole model for the combined vocal tract, glottal pulse, and radiation system of the form

$$H(z) = \frac{G}{1 - \sum_{k=1}^{p} \alpha_k z^{-k}}.$$

Assume that all the poles of $H(z)$ are inside the unit circle. Use Eq. (8.85) to obtain a recursion relation between the complex cepstrum, $\hat{h}[n]$, and the coefficients $\{\alpha_k\}$. (Hint: How is the complex cepstrum of $1/H(z)$ related to $\hat{h}[n]$?)

8.5. Consider a *finite-length* minimum-phase sequence $x[n]$ with complex cepstrum $\hat{x}[n]$, and a sequence

$$y[n] = \alpha^n x[n]$$

with complex cepstrum $\hat{y}[n]$.

(a) If $0 < \alpha < 1$, how will $\hat{y}[n]$ be related to $\hat{x}[n]$?

(b) How should α be chosen so that $y[n]$ would no longer be minimum-phase?

(c) How should α be chosen so that $y[n]$ is maximum-phase?

8.6. Show that if $x[n]$ is minimum-phase, then $x[-n]$ is maximum-phase.

8.7. Consider a sequence, $x[n]$, with complex cepstrum $\hat{x}[n]$. The z-transform of $\hat{x}[n]$ is

$$\hat{X}(z) = \log[X(z)] = \sum_{m=-\infty}^{\infty} \hat{x}[m]z^{-m},$$

where $X(z)$ is the z-transform of $x[n]$. The z-transform $\hat{X}(z)$ is sampled at N equally spaced points on the unit circle, to obtain

$$\tilde{\hat{X}}[k] = \hat{X}(e^{j\frac{2\pi}{N}k}), \quad 0 \le k \le N-1.$$

Using the IDFT, we compute

$$\tilde{\hat{x}}[n] = \frac{1}{N}\sum_{k=0}^{N-1} \tilde{\hat{X}}[k]e^{j\frac{2\pi}{N}kn}, \quad 0 \le n \le N-1,$$

which serves as an approximation to the complex cepstrum.

(a) Express $\tilde{\hat{X}}[k]$ in terms of the true complex cepstrum, $\hat{x}[m]$.

(b) Substitute the expression obtained in (a) into the IDFT expression for $\tilde{\hat{x}}[n]$ and show that

$$\hat{x}_p[n] = \sum_{r=-\infty}^{\infty} \hat{x}[n + rN].$$

8.8. Consider the sequence

$$x[n] = \delta[n] + \alpha\delta[n - N_p], \quad |\alpha| < 1.$$

(a) Determine the complex cepstrum of $x[n]$. Sketch your result.

(b) Sketch the cepstrum, $c[n]$, for $x[n]$.

(c) Now suppose that the approximation $\tilde{\hat{x}}[n]$ is computed using Eq. (8.58c). Sketch $\tilde{\hat{x}}[n]$ for $0 \le n \le N-1$, for the case $N_p = N/6$. What if N is not divisible by N_p?

(d) Repeat (c) for the cepstrum approximation $\tilde{c}[n]$ for $0 \le n \le N-1$, as computed using Eq. (8.61).

(e) If the largest impulse in the cepstrum approximation, $\tilde{c}[n]$, is used to detect N_p, how large must N be in order to avoid confusion?

8.9. The cepstrum of a signal $x[n]$, denoted as $c^{(x)}[n]$, has DTFT

$$C^{(x)}(e^{j\omega}) = \log|X(e^{j\omega})|.$$

Liftering is accomplished by

$$c^{(y)}[n] = l[n]c^{(x)}[n].$$

(a) Write an expression relating $C^{(y)}(e^{j\omega})$ to $\log|X(e^{j\omega})|$ and $L(e^{j\omega})$, where $L(e^{j\omega})$ is the DTFT of $l[n]$.

(b) To smooth $\log|X(e^{j\omega})|$, what type of cepstral window, $l[n]$, should be used?

(c) Compare the use of a rectangular cepstral window and a Hamming cepstral window.

(d) What are the considerations in choosing the length of the window?

8.10. In Section 8.5.1 we showed how the window used in short-time Fourier analysis affects cepstrum analysis of voiced speech. For purposes of illustration, a causal window was assumed to be positioned with its origin at the location of one of the pitch impulses. In this problem, we will explore the effect of window positioning. As in Section 8.5.1, assume that the speech signal is

$$s[n] = h[n] * p[n] = h[n] * \sum_{k=-\infty}^{\infty} \delta[n - kN_p].$$

Now assume that $s[n]$ is multiplied by a shifted window $w[n - n_0]$ and assume as in Section 8.5.1 that we can write $s[n]w[n - n_0]$ as

$$x[n] = w[n - n_0]s[n] = h[n] * (w[n - n_0]p[n]) = h[n] * p_w[n],$$

where $p_w[n] = w[n - n_0]p[n]$. That is we absorb the window effect into the excitation sequence.

Furthermore, for this problem, assume that the window is a causal odd-length Hamming window of length $L = 2N_p + 1$ as defined by

$$w[n] = \begin{cases} 0.54 - 0.46\cos\left[2\pi n/(2N_p)\right] & 0 \le n \le 2N_p \\ 0 & \text{otherwise.} \end{cases}$$

(a) Determine the windowed excitation sequences $p_w[n] = w[n - n_0]p[n]$ for the cases $n_0 = 0, N_p/4, N_p/2, 3N_p/4$, and sketch them.

(b) In each of the cases in (a), determine the z-transform $P_w(z)$ and show that all can be placed in the general form

$$P_w(z) = z^{-N_p}(\alpha_0 z^{N_p} + \alpha_1 + \alpha_2 z^{-N_p}).$$

(c) For each of the cases, determine and sketch the complex cepstrum $\hat{p}_w[n]$. (Hint: Use the power series expansion for $\log[P_w(z)]$. You should ignore the term $\log[z^{-N_p}]$ in your analysis.) Which case(s) are minimum-phase? Maximum-phase?

(d) Now consider what happens as n_0 varies from 0 to N_p.

(i) For what values of n_0 will $p_w[n]$ be minimum-phase?
(ii) For what values of n_0 will $p_w[n]$ be maximum-phase?
(iii) For what value(s) of n_0 will the first cepstrum peak be the largest?
(iv) For what value(s) of n_0 will the first cepstrum peak be the smallest?

(e) Generally, the window length would be chosen to be on the order of three or four pitch periods rather than two. In what ways would the results be different for a longer window? In what ways would the results be similar? What happens when the window length is less than two pitch periods?

8.11. (MATLAB Exercise) Write a MATLAB program to compute the cepstrum of the signal:

$$x_1[n] = a^n\, u[n], \quad |a| < 1,$$

in three ways, namely:

1. the analytical solution for the cepstrum of $x_1[n]$
2. the recursion for $\hat{x}_1[n]$ based on $x_1[n]$ being a minimum-phase signal
3. the computational solution for $\hat{x}_1[n]$ using an appropriate-size FFT

Your code should accept an appropriate value for a (typically in the range 0.5 to 0.99), and a size for the FFT to minimize aliasing effects. Plot the three resulting complex cepstra on a common plot, and plot the differences between the complex cepstra for the recursion and the analytical solution (should be essentially zero) and for the differences between the analytical solution and the computational solution.

8.12. (MATLAB Exercise) Write a MATLAB program to compute the complex and real cesptra of the finite duration finite duration signal:

$$x_2[n] = \delta[0] + 0.85\,\delta[n-100], \quad 0 \le n \le 99.$$

Compute the complex and real cepstra using computational methods discussed in this chapter; i.e., use an FFT of appropriate size with phase unwrapping to compute the complex cepstrum and use the log magnitude to compute the real cepstrum. (You might want to compare your results with those obtained using the MATLAB toolbox functions `rcep` and `ccep`.) Plot the signal, the complex cepstrum, and the real cepstrum on a series of three plots, all on a single page.

8.13. (MATLAB Exercise) Write a MATLAB program to compute the complex and real cesptra of the finite duration signal:

$$x_3[n] = \sin\left[\frac{2\pi n}{100}\right] \quad 0 \le n \le 99$$

and zero for all other n.

Compute the complex and real cepstra using computational methods discussed in this chapter; i.e., use an FFT of appropriate size with phase unwrapping to compute the complex cepstrum and use the log magnitude to compute the real cepstrum. (You might want to compare your results with those obtained using the MATLAB toolbox functions `rcep` and `ccep`.) Plot the signal, the complex cepstrum, and the real cepstrum on a series of three plots, all on a single page.

8.14. (MATLAB Exercise) Write a MATLAB program to compute the complex and real cepstra of the following signals:

1. a section of voiced speech
2. a section of unvoiced speech

For each of the two speech signals, plot the signal, the log magnitude spectrum, the real cepstrum, and the appropriately low quefrency liftered log magnitude spectrum. Use the file `test_16k.wav` to test your program. Specify the voiced section as starting at sample 13,000 in the file, and of duration 400 samples. Specify the unvoiced section as starting at sample 3400 in the file, and of duration 400 samples. Use a Hamming window before computing the cepstra of the speech files.

8.15. (MATLAB Exercise) Write a MATLAB program to compute the real cepstrum of a speech file and lifter the cepstrum using both a low quefrency lifter and a high quefrency lifter to illustrate the differences in retrieving a representation of the vocal tract response (the low quefrency lifter) or the source excitation (the high quefrency lifter). Use the speech file `test_16k.wav` with a voiced frame beginning at sample 13,000 (of duration 40 msec) and an unvoiced frame beginning at sample 1000 (again of duration 40 msec). Use a Hamming window to isolate the frame of speech being processed. For both the voiced and unvoiced frames, plot the following quantities:

1. the Hamming window weighted speech section (make sure that your window length of 40 msec reflects the correct number of samples, based on the sampling rate of the speech signal)
2. the log magnitude spectrum of the signal with the cepstrally smoothed log magnitude spectrum (in the case of a low quefrency lifter)
3. the real cepstrum of the speech signal
4. the liftered log magnitude spectrum (for both low quefrency and high quefrency liftering operations)

Make plots for both the voiced and unvoiced sections of speech, and for both the low and high quefrency lifters. Your program should accept the following inputs:

- the speech file being processed
- the starting sample for the speech frame being processed
- the duration of the speech frame (in msec) which must be converted to a frame duration in samples, based on the speech sampling rate
- the size of FFT used in cepstral computation
- the cutoff point of the cepstral lifter (either the high cutoff for a low quefrency lifter, or the low cutoff for a high quefrency lifter)
- the type of lifter (low quefrency or high quefrency)

8.16. (MATLAB Exercise) Write a MATLAB program to demonstrate phase unwrapping for computation of the complex cepstrum of a speech file. You must carefully compensate any linear phase component before attempting to do phase unwrapping. Choose a section of a voiced speech signal and compute its DFT and decompose it into the log magnitude and phase components (this is the wrapped phase). Use the MATLAB toolbox command for phase unwrapping to unwrap the phase. Compute the complex cepstrum, lifter the cepstrum using either a low or high quefrency lifter, and transform the liftered signal back to the frequency domain and ultimately to the time domain. Plot the following quantities on a single page:

1. the original and cepstrally smoothed (in the case of a low quefrency lifter) log magnitude spectra
2. the original (wrapped) phase, the unwrapped phase, and the cepstrally smoothed phase

3. the complex cepstrum calculated with a large size FFT
4. the resulting estimate of the vocal tract response (obtained by low quefrency liftering) or the excitation signal (obtained by high quefrency liftering)

Use the speech file `test_16k.wav` with a starting sample of 13,000 and frame size 40 msec to test your program. Use an FFT size of 4096 samples and a lifter cutoff of 50 samples. Plot the results for both a low quefrency and a high quefrency lifter.

8.17. (MATLAB Exercise) Write a MATLAB program to illustrate the magnitude of aliasing in computing the real cepstrum of a speech signal. Your program should compute the real cepstrum of a frame of speech using an FFT size of 8192 (the standard) and repeat the computation of the exact same frame of speech with FFT sizes of 4096, 2048, 1024, and 512. Compare the resulting cepstra and plot the differences between the real cepstrum computed using an 8192 point FFT with each of the cepstra computed using smaller size FFTs. This exercise illustrates the level of aliasing that is common in speech processing using reasonable-size frames of speech and reasonable-size FFTs. Use the speech file `test_16k.wav` to illustrate the effects of aliasing. Choose a voiced speech frame starting at sample 13,000, of duration 40 msec for this exercise. Plot the 8192 point FFT cepstrum (the standard) in the top panel and the differences between the 8192 point FFT cepstrum and cepstra computed using smaller-size FFTs in the lower panel.

8.18. (MATLAB Exercise) Write a MATLAB program to demonstrate the effects of the cutoff quefrency of the low quefrency lifter on the smoothing of the log magnitude spectrum. Using a voiced section of a speech file, calculate the real cepstrum and apply a low quefrency lifter with cutoff quefrencies of 20, 40, 60, 80, and 100, and show the effects of the low quefrency lifter on the smoothed log magnitude spectrum. Use the file `test_16k.wav` with a voiced frame starting sample of 13,000 and 40 msec frame duration to compute the real cepstrum of the voiced frame. Using a sequence of low quefrency lifters, compute the smoothed log magnitude spectrum and plot the results on top of the curve of the unsmoothed log magnitude spectrum. Repeat for low quefrency cutoffs of 20, 40, 60, 80, and 100.

CHAPTER 9

Linear Predictive Analysis of Speech Signals

9.1 INTRODUCTION

Linear predictive analysis of speech signals is one of the most powerful speech analysis techniques. This method has become the predominant technique for estimating the parameters of the discrete-time model for speech production (i.e., pitch, formants, short-time spectra, vocal tract area functions) and is widely used for representing speech in low bit rate transmission or storage and for automatic speech and speaker recognition. The importance of this method lies both in its ability to provide accurate estimates of the speech parameters and in its relative ease of computation. In this chapter we present the fundamental concepts of linear predictive analysis of speech, and we discuss some of the issues involved in using linear predictive analysis in practical speech applications.

The philosophy of linear prediction is intimately related to the basic speech synthesis model discussed in Chapter 5 where it was shown that a sampled speech signal can be modeled as the output of a linear, time-varying system (difference equation) excited by either quasi-periodic pulses (during voiced speech) or random noise (during unvoiced speech). The difference equation of the speech model suggests that a speech sample can be approximated as a linear combination of p past speech samples. By locally minimizing the sum of the squared differences between the actual speech samples and the linearly predicted samples, a unique set of predictor coefficients can be determined, and by equating the prediction coefficients to the coefficients of the difference equation of the model, we obtain a robust, reliable, and accurate method for estimating the parameters that characterize the linear time-varying system in the speech production model. In the speech processing field, linear prediction was first used in speech coding applications, and the term "linear predictive coding" (or LPC) quickly gained widespread usage. As linear predictive analysis methods became more widely used in speech processing, the term "LPC" persisted and now it is often used as a term for linear predictive analysis techniques in general. Wherever we use the term linear predictive coding or LPC, we intend it to have the general meaning, i.e., not restricted just to "coding."

The techniques and methods of linear prediction have been available in the engineering literature for a long time [38, 417]. One of the earliest applications of the theory of linear prediction was the work of Robinson, who used linear prediction in seismic signal processing [322, 323]. The ideas of linear prediction have been used in the areas of control and information theory under the names of system estimation and system identification. The term "system identification" is particularly descriptive in speech applications since the predictor coefficients are assumed to characterize an all-pole model of the system in the source/system model of speech production.

As applied in speech processing, the term "linear predictive analysis" refers to a variety of essentially equivalent formulations of the problem of modeling the speech signal [12, 161, 218, 232]. The differences among these formulations are often philosophical or in point of view toward the problem of speech modeling. The differences mainly concern the details of the computations used to obtain the predictor coefficients. Thus, as applied to speech, the various (often equivalent) formulations of linear prediction analysis have been:

1. the covariance method [12]
2. the autocorrelation formulation [217, 229, 232]
3. the lattice method [48, 219]
4. the inverse filter formulation [232]
5. the spectral estimation formulation [48]
6. the maximum likelihood formulation [161, 162]
7. the inner product formulation [232].

In this chapter we will examine, in detail, the similarities and differences among the first three basic methods of analysis listed above, since all the other formulations are essentially equivalent to one of these three.

The importance of linear prediction lies in the accuracy with which the basic model applies to speech. Thus a major part of this chapter is devoted to a discussion of how a variety of speech parameters can be reliably estimated using linear prediction methods. Furthermore, some typical examples of speech applications that rely primarily on linear predictive analysis are discussed here, and in Chapters 10–14, to show the wide range of problems to which LPC methods have been successfully applied.

9.2 BASIC PRINCIPLES OF LINEAR PREDICTIVE ANALYSIS

Throughout this book we have repeatedly referred to the basic discrete-time model for speech production that was developed in Chapter 5. The particular form of this model that is most appropriate for the discussion of linear predictive analysis is depicted in Figure 9.1. In this model, the composite spectrum effects are represented by a time-varying digital filter whose steady-state system function is represented by the all-pole rational function

$$H(z) = \frac{S(z)}{GU(z)} = \frac{S(z)}{E(z)} = \frac{1}{1 - \sum_{k=1}^{p} a_k z^{-k}}. \tag{9.1}$$

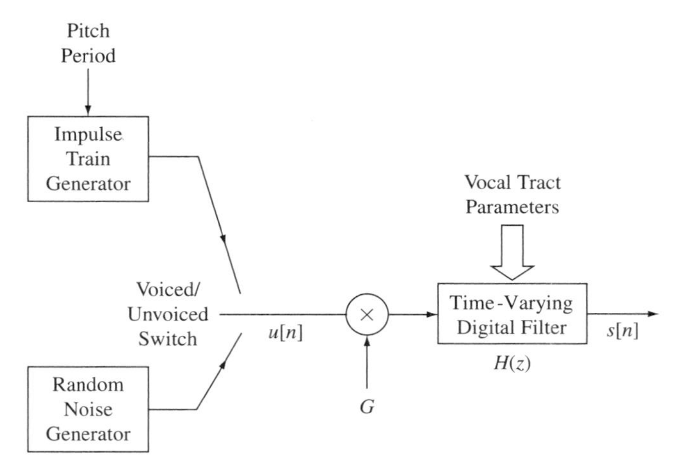

FIGURE 9.1
Block diagram of simplified model for speech production.

We will refer to $H(z)$ as the *vocal tract* system function even though it represents not only the effects of the vocal tract resonances but also the effects of radiation at the lips and, in the case of voiced speech, the spectral effects of the glottal pulse shape. This system is excited by a quasi-periodic impulse train for voiced speech [since the glottal pulse shape is included in $H(z)$] or a random noise sequence for unvoiced speech. Thus, the parameters of this model are:

- Excitation parameters:
  - voiced/unvoiced classification
  - pitch period for voiced speech
  - gain parameter G
- Vocal tract system parameters:
  - coefficients $\{a_k, k = 1, 2, \dots, p\}$ of the all-pole digital filter

These parameters, of course, all vary slowly with time.

The pitch period and voiced/unvoiced classification can be estimated using one of the many methods to be discussed in Chapter 10. As discussed in Chapter 5, this simplified all-pole model is a natural representation for non-nasal voiced sounds. For nasals and fricative sounds, however, the detailed acoustic theory calls for both poles and zeros in the vocal tract transfer function. We will subsequently see that if the digital filter order, p, is high enough, the all-pole model provides a good enough representation for almost all the sounds of speech, including nasal sounds and fricative sounds. The major advantage of this model is that the gain parameter, G, and the filter coefficients, $\{a_k\}$, can be estimated in a very straightforward and computationally efficient way by the method of linear predictive analysis. Furthermore, one of the methods of pitch detection discussed in Chapter 10 is based on using the linear predictor as an inverse filter to extract the error signal, $e[n]$, representative of the excitation signal, $u[n]$.

For the system of Figure 9.1, the model speech samples, $s[n]$, are related to the excitation, $u[n]$, by the simple difference equation

$$s[n] = \sum_{k=1}^{p} a_k s[n-k] + Gu[n]. \tag{9.2}$$

A pth-order linear predictor with prediction coefficients, $\{\alpha_k, k = 1, 2, \ldots, p\}$, is defined as a system whose input is $s[n]$ and whose output is $\tilde{s}[n]$, defined as

$$\tilde{s}[n] = \sum_{k=1}^{p} \alpha_k s[n-k]. \tag{9.3}$$

The system function of this p^{th}-order linear predictor is the z-transform polynomial

$$P(z) = \sum_{k=1}^{p} \alpha_k z^{-k} = \frac{\tilde{S}(z)}{S(z)}. \tag{9.4}$$

$P(z)$ is often referred to as the *predictor polynomial*. The prediction error, $e[n]$, is defined as the difference between $s[n]$ and $\tilde{s}[n]$; i.e.,

$$e[n] = s[n] - \tilde{s}[n] = s[n] - \sum_{k=1}^{p} \alpha_k s[n-k]. \tag{9.5}$$

From Eq. (9.5), it follows that the prediction error sequence is the output of a system whose input is $s[n]$ and whose system function is

$$A(z) = \frac{E(z)}{S(z)} = 1 - P(z) = 1 - \sum_{k=1}^{p} \alpha_k z^{-k}. \tag{9.6}$$

The z-transform, $A(z)$, is a polynomial in z^{-1}. It is often called the *prediction error polynomial*, or equivalently, the *LPC polynomial*.

If the speech signal obeys the model of Eq. (9.2) exactly, so that the model output $s[n]$ is equal to the actual sampled speech signal, then a comparison of Eqs. (9.2) and (9.5) shows that if $\{\alpha_k\} = \{a_k\}$, for all k, then $e[n] = Gu[n]$. Thus, the *prediction error filter*, $A(z)$, is an *inverse filter* for the vocal tract system, $H(z)$, of Eq. (9.1); i.e.,

$$A(z) = \frac{G \cdot U(z)}{S(z)} = \frac{1}{H(z)}. \tag{9.7}$$

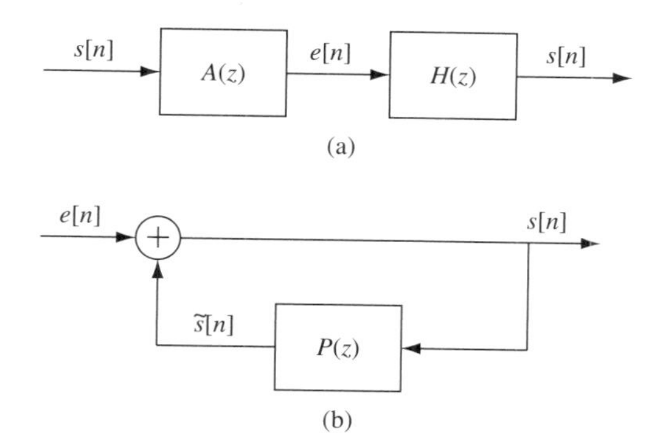

FIGURE 9.2
Signal processing operations for inverse filtering and reconstruction of the speech signal: (a) inverse filtering of $s[n]$, to give the error signal, $e[n]$, and direct filtering of the error signal, $e[n]$, to reconstruct the original speech signal, $s[n]$; (b) implementation of the system $H(z)$ showing the predicted signal $\tilde{s}[n]$.

Figure 9.2 shows the relationships between the signals $s[n]$, $e[n]$, and $\tilde{s}[n]$. The error signal, $e[n]$, is obtained as the output of the inverse filter, $A(z)$, with input $s[n]$, as shown in Figure 9.2a. By assumption, $e[n]$ is the vocal tract excitation, so it should be a quasi-periodic impulse train for voiced speech and random noise for unvoiced speech. The original speech signal, $s[n]$, is obtained by processing the error signal by the all-pole filter, $H(z)$, as also shown in Figure 9.2a. Figure 9.2b shows the feedback processing loop for reconstructing $s[n]$ from $e[n]$; i.e., feedback of the predicted signal, $\tilde{s}[n]$, and addition of the error signal, $e[n]$. Thus we see that $H(z)$ is a pth-order all-pole rational function of the form

$$H(z) = \frac{1}{A(z)} = \frac{1}{1 - P(z)} = \frac{1}{1 - \displaystyle\sum_{k=1}^{p} \alpha_k z^{-k}}, \tag{9.8}$$

and $A(z)$ is the pth-order polynomial in Eq. (9.6).[1] The system whose transfer function is $H(z)$ is often called the *vocal tract model system* or the *LPC model system*.

9.2.1 Basic Formulation of Linear Prediction Analysis Equations

The basic problem of linear predictive analysis is to determine the set of predictor coefficients $\{\alpha_k, k = 1, 2, \dots, p\}$ directly from the speech signal so as to obtain a good estimate of the time-varying spectral properties of the speech signal through the use

[1] As we have stressed before in our discussion of the speech model, the z-transform representations that we have used are not strictly valid for representing the speech model since the system is time-varying. Thus, the z-transform equations and Figure 9.2 are assumed to be valid over only short-time frames, with parameter updates from frame-to-frame.

of Eq. (9.8). Because of the time-varying nature of the speech signal, the predictor coefficients must be estimated by a short-time analysis procedure based on finding the set of predictor coefficients that minimize the mean-squared prediction error over a short segment of the speech waveform. The resulting parameters are then *assumed* to be the parameters of the system function, $H(z)$, in the model for speech production in Figure 9.1.

It may not be immediately obvious that this approach will lead to useful results, but it can be justified in several ways. First, recall that if $\{\alpha_k = a_k,\ k = 1, 2, \dots, p\}$, then $e[n] = Gu[n]$. For voiced speech this means that $e[n]$ would consist of a train of impulses; i.e., $e[n]$ would be small most of the time. For unvoiced speech, the inverse filter flattens the short-time spectrum, thereby creating white noise. Thus, finding α_k's that minimize the prediction error seems consistent with this observation. A second motivation for this approach follows from the fact that if a signal is generated by Eq. (9.2) with non-time-varying coefficients and excited either by a single impulse or by a stationary white noise input, then it can be shown that the predictor coefficients that result from minimizing the mean-squared prediction error (over all time) are identical to the coefficients of Eq. (9.2). A third very pragmatic justification for using the minimum mean-squared prediction error as a basis for estimating the model parameters is that this approach leads to a set of linear equations that can be efficiently solved to obtain the predictor parameters. Perhaps more to the point—the ultimate justification for linear predictive analysis of speech is simply pragmatic. The linear prediction analysis model works exceedingly well.

The short-time total squared prediction error is defined as

$$\mathcal{E}_{\hat{n}} = \sum_m e_{\hat{n}}^2[m] = \sum_m (s_{\hat{n}}[m] - \tilde{s}_{\hat{n}}[m])^2 \tag{9.9a}$$

$$= \sum_m \left(s_{\hat{n}}[m] - \sum_{k=1}^{p} \alpha_k s_{\hat{n}}[m-k] \right)^2, \tag{9.9b}$$

where $s_{\hat{n}}[m]$ is a segment of speech that has been selected in the vicinity of sample $\hat{n}$; i.e.,

$$s_{\hat{n}}[m] = s[m + \hat{n}], \tag{9.10}$$

for m in some finite interval around $\hat{n}$. The range of summation in Eqs. (9.9a) and (9.9b) is temporarily left unspecified, but since we wish to develop a short-time analysis technique, the sum will always be over a finite interval. Also note that to obtain an average (or mean) squared error, we should divide the sum by the length of the speech segment. However, this constant is irrelevant to the set of linear equations that we will obtain and therefore is omitted. We can find the values of α_k that minimize $\mathcal{E}_{\hat{n}}$ in Eq. (9.9b) by setting $\partial \mathcal{E}_{\hat{n}} / \partial \alpha_i = 0$, $i = 1, 2, \dots, p$, thereby obtaining the equations

$$\sum_m s_{\hat{n}}[m-i] s_{\hat{n}}[m] = \sum_{k=1}^{p} \hat{\alpha}_k \sum_m s_{\hat{n}}[m-i] s_{\hat{n}}[m-k], \quad 1 \le i \le p, \tag{9.11}$$

where $\hat{\alpha}_k$ are the values of α_k that minimize $\mathcal{E}_{\hat{n}}$. (Since $\hat{\alpha}_k$ is unique, we will henceforth drop the caret and use the notation α_k to denote the values that minimize $\mathcal{E}_{\hat{n}}$.) If we define

$$\varphi_{\hat{n}}[i,k] = \sum_m s_{\hat{n}}[m-i]s_{\hat{n}}[m-k], \tag{9.12}$$

then Eq. (9.11) can be written more compactly as

$$\sum_{k=1}^{p} \alpha_k \varphi_{\hat{n}}[i,k] = \varphi_{\hat{n}}[i,0], \quad i = 1, 2, \dots, p. \tag{9.13}$$

This set of p equations in p unknowns can be solved efficiently for the unknown predictor coefficients $\{\alpha_k\}$ that minimize the total squared prediction error for the segment $s_{\hat{n}}[m]$.[2] Using Eqs. (9.9b) and (9.11), the minimum mean-squared prediction error can be shown to be

$$\mathcal{E}_{\hat{n}} = \sum_m s_{\hat{n}}^2[m] - \sum_{k=1}^{p} \alpha_k \sum_m s_{\hat{n}}[m]s_{\hat{n}}[m-k], \tag{9.14}$$

and using Eq. (9.12), we can express $\mathcal{E}_{\hat{n}}$ as

$$\mathcal{E}_{\hat{n}} = \varphi_{\hat{n}}[0,0] - \sum_{k=1}^{p} \alpha_k \varphi_{\hat{n}}[0,k], \tag{9.15}$$

where $\{\alpha_k, k = 1, 2, \dots, p\}$ is the set of predictor coefficients satisfying Eq. (9.13). Thus the total minimum error consists of a fixed component, $\varphi_{\hat{n}}[0,0]$, which is equal to the total sum of squares (energy) of the segment $s_{\hat{n}}[m]$, and a component that depends on the predictor coefficients.

To solve for the optimum predictor coefficients, we must first compute the quantities $\varphi_{\hat{n}}[i,k]$ for $1 \leq i \leq p$ and $0 \leq k \leq p$. Once this is done, we only have to solve Eq. (9.13) to obtain the α_k's. Thus, in principle, linear prediction analysis is very straightforward. However, the details of the computation of $\varphi_{\hat{n}}[i,k]$ and the subsequent solution of the equations are somewhat intricate and further discussion is required.

So far we have not explicitly indicated the limits on the sums in Eqs. (9.9a) or (9.9b) and in Eq. (9.11); however, it should be emphasized that the limits on the sum in Eq. (9.11) are identical to the limits assumed for the mean-squared prediction error in Eqs. (9.9a) or (9.9b). As we have stated, if we wish to develop a short-time analysis

[2]While the α_k's are functions of $\hat{n}$ (the time index at which they are estimated), it is cumbersome and generally unnecessary to show this dependence explicitly. It is also advantageous to drop the subscripts $\hat{n}$ on $\mathcal{E}_{\hat{n}}$, $s_{\hat{n}}[m]$, and $\phi_{\hat{n}}[i,k]$ when no confusion will result.

procedure, the limits must be over a finite interval around the analysis time $\hat{n}$. There are two basic approaches to this problem, and we will see below that two methods for linear predictive analysis emerge out of a consideration of the limits of summation and the definition of the waveform segment $s_{\hat{n}}[m]$.

9.2.2 The Autocorrelation Method

One approach to determining the limits on the sums in Eqs. (9.9a) or (9.9b) and Eq. (9.11) is to assume that the waveform segment, $s_{\hat{n}}[m]$, is identically zero outside the interval $0 \leq m \leq L-1$ [217, 222, 232]. This can be conveniently expressed as

$$s_{\hat{n}}[m] = s[m+\hat{n}]w[m], \qquad 0 \leq m \leq L-1, \tag{9.16}$$

where $w[m]$ is a finite length window (e.g., a Hamming window or a rectangular window) that is identically zero outside the interval $0 \leq m \leq L-1$. Figure 9.3a shows the signal $s[n]$ as a function of index n along with a Hamming window $w[n-\hat{n}]$ positioned to begin at $n=\hat{n}$. Redefining the origin of the segment to be at the analysis time $\hat{n}$ by substituting $m = n - \hat{n}$ gives the windowed segment $s_{\hat{n}}[m] = s[m+\hat{n}]w[m]$

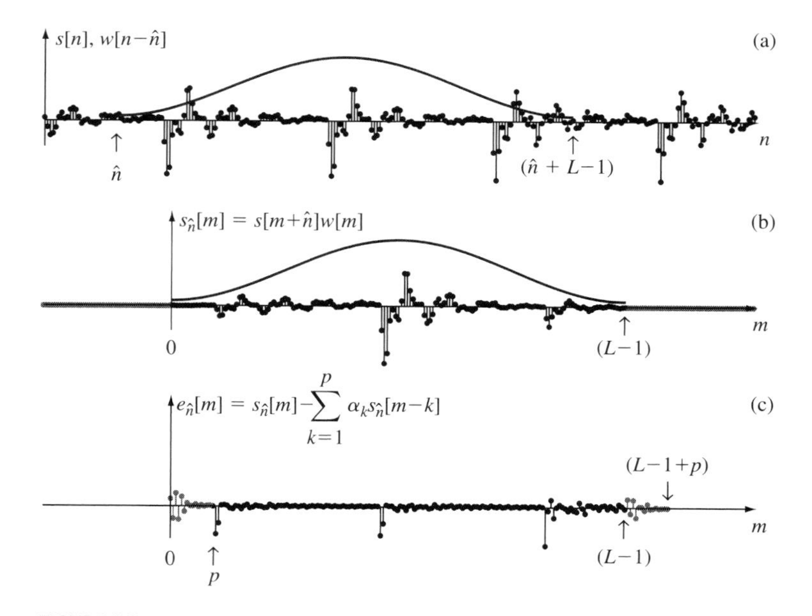

FIGURE 9.3
Illustration of short-time analysis for the autocorrelation method: (a) speech signal $s[n]$ and Hamming window positioned at time $n = \hat{n}$; (b) windowed segment $s_{\hat{n}}[m]$; (c) segment of prediction error signal, $e_{\hat{n}}[m]$, over the range $0 \leq m \leq L-1+p$ obtained using an optimum predictor with $p = 15$.

as shown in Figure 9.3b as a function of m.[3] Figure 9.3c shows the output of an optimum linear predictor of order $p = 15$. The effect of the assumption of Eq. (9.16) on the range of limits of the summation for the expression for $\mathcal{E}_{\hat{n}}$ can be seen by considering Eq. (9.5) and Eqs. (9.9a) or (9.9b). Figure 9.3c shows that if $s_{\hat{n}}[m]$ is non-zero only for $0 \leq m \leq L-1$, then the corresponding prediction error, $e_{\hat{n}}[m]$, for a pth-order predictor, will be non-zero only over the interval $0 \leq m \leq L-1+p$. Thus, for this case, $\mathcal{E}_{\hat{n}}$ is properly expressed as

$$\mathcal{E}_{\hat{n}} = \sum_{m=0}^{L-1+p} e_{\hat{n}}^2[m] = \sum_{m=-\infty}^{\infty} e_{\hat{n}}^2[m], \tag{9.17}$$

where Eq. (9.17) simply relies on the fact that, because of the way that $s_{\hat{n}}[m]$ and $e_{\hat{n}}[m]$ are defined, we can simply indicate that the sum should be over all non-zero values by summing from $-\infty$ to $+\infty$ [217].

Equation (9.5) and Figure 9.3c show that the prediction error is likely to be (relatively) large at the beginning of the interval (specifically $0 \leq m \leq p-1$) because the predictor must predict the (non-zero) signal samples from previous samples that have arbitrarily been set to zero as a result of windowing by the finite duration window. The first $p = 15$ samples are highlighted in Figure 9.3c. Likewise the error is likely to be (relatively) large at the end of the interval (specifically $L \leq m \leq L-1+p$) because the predictor must predict p zero-valued samples outside the window from samples that are non-zero and inside the window. These samples are also highlighted in Figure 9.3c. For this reason, a window, such as a Hamming window, which tapers the segment $s_{\hat{n}}[m]$ to zero at both the beginning and end of the frame is generally used for $w[m]$ in Eq. (9.16) to mitigate these end effects on the error sequence.

This observation explains a basic property of the optimum predictor found using the autocorrelation method. To be more specific, assume that $s_{\hat{n}}[0] \neq 0$. Then it follows that $e_{\hat{n}}[0] = s_{\hat{n}}[0] \neq 0$ as well. Similarly, if $s_{\hat{n}}[L-1] \neq 0$, it follows that $e_{\hat{n}}[L-1+p] = -\alpha_p s_{\hat{n}}[L-1] \neq 0$. Thus, we conclude that $\mathcal{E}_{\hat{n}} > 0$ no matter how we choose p; i.e., the total squared prediction error for the autocorrelation method is always strictly positive. It cannot be zero. This property will be exceedingly important in later discussions.

The limits on the expression for $\varphi_{\hat{n}}[i,k]$ in Eq. (9.12) are identical to those of Eq. (9.17). However, because $s_{\hat{n}}[m]$ is identically zero outside the interval $0 \leq m \leq L-1$, it is simple to show that the expression

$$\varphi_{\hat{n}}[i,k] = \sum_{m=0}^{L-1+p} s_{\hat{n}}[m-i]s_{\hat{n}}[m-k] \quad \begin{cases} 1 \leq i \leq p \\ 0 \leq k \leq p, \end{cases} \tag{9.18a}$$

[3]The time origin of the windowed segment is redefined to be at the beginning of the window. We might also define the analysis time $\hat{n}$ to be the center of the shifted window rather than the beginning of the window. This would lead to a different indexing of the windowed segment.

can be expressed as

$$\varphi_{\hat{n}}[i,k] = \sum_{m=0}^{L-1-(i-k)} s_{\hat{n}}[m]s_{\hat{n}}[m+(i-k)] \quad \begin{cases} 1 \le i \le p \\ 0 \le k \le p. \end{cases} \tag{9.18b}$$

Furthermore, it can be seen that, in this case, $\varphi_{\hat{n}}[i,k]$ is identical to the short-time autocorrelation function of Eq. (6.35) evaluated for $[i-k]$. That is,

$$\varphi_{\hat{n}}[i,k] = R_{\hat{n}}[i-k] \quad \begin{cases} 1 \le i \le p \\ 0 \le k \le p, \end{cases} \tag{9.19}$$

where

$$R_{\hat{n}}[k] = \sum_{m=0}^{L-1-k} s_{\hat{n}}[m]s_{\hat{n}}[m+k]. \tag{9.20}$$

It is this autocorrelation property that motivates the term *autocorrelation method.* The computation of $R_{\hat{n}}[k]$ is covered in detail in Section 6.5 so we will not consider such details again here. Since $R_{\hat{n}}[k]$ is an even function, it follows that the values of $\varphi[i,k]$ required for Eqs. (9.13) are

$$\varphi_{\hat{n}}[i,k] = R_{\hat{n}}[|i-k|] \quad \begin{cases} 1 \le i \le p \\ 0 \le k \le p. \end{cases} \tag{9.21}$$

Therefore Eq. (9.13) can be expressed as

$$\sum_{k=1}^{p} \alpha_k R_{\hat{n}}[|i-k|] = R_{\hat{n}}[i], \quad 1 \le i \le p. \tag{9.22a}$$

Similarly, the expression for the minimum mean-squared prediction error of Eq. (9.15) becomes

$$\mathcal{E}_{\hat{n}} = R_{\hat{n}}[0] - \sum_{k=1}^{p} \alpha_k R_{\hat{n}}[k]. \tag{9.22b}$$

The set of equations given by Eq. (9.22a) can be expressed in matrix form as

$$\begin{bmatrix} R_{\hat{n}}[0] & R_{\hat{n}}[1] & R_{\hat{n}}[2] & \dots & R_{\hat{n}}[p-1] \\ R_{\hat{n}}[1] & R_{\hat{n}}[0] & R_{\hat{n}}[1] & \dots & R_{\hat{n}}[p-2] \\ R_{\hat{n}}[2] & R_{\hat{n}}[1] & R_{\hat{n}}[0] & \dots & R_{\hat{n}}[p-3] \\ \dots & \dots & \dots & \dots & \dots \\ \dots & \dots & \dots & \dots & \dots \\ R_{\hat{n}}[p-1] & R_{\hat{n}}[p-2] & R_{\hat{n}}[p-3] & \dots & R_{\hat{n}}[0] \end{bmatrix} \begin{bmatrix} \alpha_1 \\ \alpha_2 \\ \alpha_3 \\ \dots \\ \dots \\ \alpha_p \end{bmatrix} = \begin{bmatrix} R_{\hat{n}}[1] \\ R_{\hat{n}}[2] \\ R_{\hat{n}}[3] \\ \dots \\ \dots \\ R_{\hat{n}}[p] \end{bmatrix}. \tag{9.23}$$

The $p \times p$ matrix of autocorrelation values is a Toeplitz matrix; i.e., it is symmetric and all the elements along a given diagonal are equal. Furthermore, because it arises out of a least-squares problem, the matrix is also positive-definite. These special properties will be exploited in Section 9.5 to obtain an efficient algorithm for the solution of the set of equations represented by Eq. (9.23).

9.2.3 The Covariance Method

The second basic approach to defining the speech segment, $s_{\hat{n}}[m]$, and the limits on the sum is to fix the interval over which the mean-squared error is computed and then consider the effect on the computation of $\varphi_{\hat{n}}[i, k]$ [12]. That is, we define $\mathcal{E}_{\hat{n}}$ as

$$\mathcal{E}_{\hat{n}} = \sum_{m=0}^{L-1} e_{\hat{n}}^2[m] = \sum_{m=0}^{L-1} \left(s_{\hat{n}}[m] - \sum_{k=1}^{p} \alpha_k s_{\hat{n}}[m-k] \right)^2, \tag{9.24}$$

where we also assume that the prediction error, $e_{\hat{n}}[m]$, is computed over the interval $0 \leq m \leq L-1$ using all samples required by Eq. (9.24). This means that the analysis must be based on $s_{\hat{n}}[m]$ for $-p \leq m \leq L-1$ in order to compute $e_{\hat{n}}[m]$ without edge effects. Therefore, $\varphi_{\hat{n}}[i, k]$ becomes

$$\varphi_{\hat{n}}[i, k] = \phi_{\hat{n}}[i, k] = \sum_{m=0}^{L-1} s_{\hat{n}}[m-i] s_{\hat{n}}[m-k] \quad \begin{cases} 1 \leq i \leq p \\ 0 \leq k \leq p, \end{cases} \tag{9.25}$$

with $s_{\hat{n}}[m] = s[m + \hat{n}]$ for $-p \leq m \leq L-1$. In this case, if we change the index of summation, we can express $\phi_{\hat{n}}[i, k]$ as either

$$\phi_{\hat{n}}[i, k] = \sum_{m=-i}^{L-i-1} s_{\hat{n}}[m] s_{\hat{n}}[m+i-k] \quad \begin{cases} 1 \leq i \leq p \\ 0 \leq k \leq p, \end{cases} \tag{9.26a}$$

or

$$\phi_{\hat{n}}[i, k] = \sum_{m=-k}^{L-k-1} s_{\hat{n}}[m] s_{\hat{n}}[m+k-i] \quad \begin{cases} 1 \leq i \leq p \\ 0 \leq k \leq p. \end{cases} \tag{9.26b}$$

Although the equations look very similar to Eq. (9.18b), note that the limits of summation are not the same. Equations (9.26a) and (9.26b) call for values of $s_{\hat{n}}[m]$ outside the interval $0 \leq m \leq L-1$. Indeed, to evaluate $\phi_{\hat{n}}[i, k]$ for all of the required values of i and k requires that we use values of $s_{\hat{n}}[m]$ in the interval $-p \leq m \leq L-1$. If we are to be consistent with the limits on $\mathcal{E}_{\hat{n}}$ in Eq. (9.24), then we have no choice but to supply the required values. Figure 9.4 shows the signals involved in the computation of $\phi_{\hat{n}}[i, k]$. Figure 9.4a shows the speech waveform $s[n]$ and a rectangular window of length $p + L$ samples positioned to begin at time $\hat{n} - p$ on the time scale n. Figure 9.4b shows the segment of speech selected by the window after redefining the origin of the segment using the substitution $m = n - \hat{n}$. Note that the window extends from $m = -p$

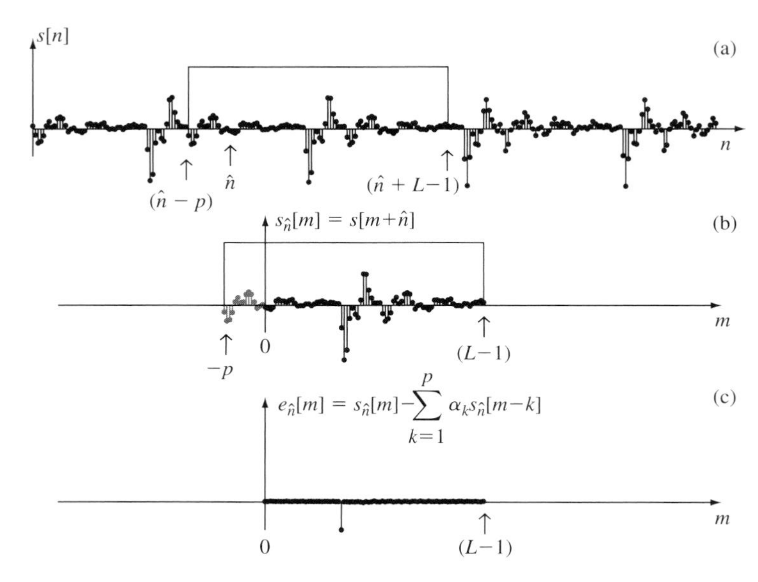

FIGURE 9.4
Illustration of short-time analysis for the covariance method: (a) speech signal $s[n]$ and rectangular window positioned over the range $\hat{n}-p \leq n \leq \hat{n}+L-1$; (b) windowed segment, $s_{\hat{n}}[m]$, defined over the range $-p \leq m \leq L-1$; (c) segment of prediction error signal, $e_{\hat{n}}[m]$, defined over the range $0 \leq m \leq L-1$ using an optimum predictor with $p = 15$.

to $m = L-1$. The p samples in the interval $-p \leq m \leq -1$ are required to properly use the pth-order prediction error filter

$$e_{\hat{n}}[m] = s_{\hat{n}}[m] - \sum_{k=1}^{p} \alpha_k s_{\hat{n}}[m-k] \tag{9.27}$$

to compute $e_{\hat{n}}[m]$ for $0 \leq m \leq L-1$. The prediction error is shown in Figure 9.4c. Note that the prediction error does not display the transient behavior at the beginning and end of the interval as was the case for the prediction error in the autocorrelation method in Figure 9.3c. In this case, it does not make sense to taper the segment of speech to zero at the ends, as in the autocorrelation method, since the necessary signal values are available to properly compute $e_{\hat{n}}[m]$ throughout the interval $0 \leq m \leq L-1$. Thus, the end effects that make it impossible for the total squared prediction error to be zero in the autocorrelation method are avoided by providing the p previous samples for the predictor at the beginning of the segment and only including the samples up to $L-1$. In principle, if the speech signal exactly fits the model, it can be *exactly* predicted (not including excitation impulses as shown in Figure 9.4c) by a linear predictor determined using the covariance method.

Clearly, the correlation function that arises in the covariance method is very similar to what was called the modified autocorrelation function in Chapter 6. As pointed out in Section 6.5, this function is not a true autocorrelation function, but rather, the cross-correlation between two very similar, but not identical, finite length segments of the speech wave. Although the differences between Eqs. (9.26a) and (9.26b) and Eq. (9.18b) appear to be minor computational details, the set of equations

$$\sum_{k=1}^{p} \alpha_k \phi_{\hat{n}}[i,k] = \phi_{\hat{n}}[i,0], \quad i = 1, 2, \ldots, p, \tag{9.28a}$$

has significantly different properties that strongly affect the method of solution and the properties of the resulting optimum predictor. In matrix form, these equations become

$$\begin{bmatrix} \phi_{\hat{n}}[1,1] & \phi_{\hat{n}}[1,2] & \phi_{\hat{n}}[1,3] & \ldots & \phi_{\hat{n}}[1,p] \\ \phi_{\hat{n}}[2,1] & \phi_{\hat{n}}[2,2] & \phi_{\hat{n}}[2,3] & \ldots & \phi_{\hat{n}}[2,p] \\ \phi_{\hat{n}}[3,1] & \phi_{\hat{n}}[3,2] & \phi_{\hat{n}}[3,3] & \ldots & \phi_{\hat{n}}[3,p] \\ \ldots & \ldots & \ldots & \ldots & \ldots \\ \ldots & \ldots & \ldots & \ldots & \ldots \\ \phi_{\hat{n}}[p,1] & \phi_{\hat{n}}[p,2] & \phi_{\hat{n}}[p,3] & \ldots & \phi_{\hat{n}}[p,p] \end{bmatrix} \begin{bmatrix} \alpha_1 \\ \alpha_2 \\ \alpha_3 \\ \ldots \\ \ldots \\ \alpha_p \end{bmatrix} = \begin{bmatrix} \phi_{\hat{n}}[1,0] \\ \phi_{\hat{n}}[2,0] \\ \phi_{\hat{n}}[3,0] \\ \ldots \\ \ldots \\ \phi_{\hat{n}}[p,0] \end{bmatrix}. \tag{9.28b}$$

In this case, since $\phi_{\hat{n}}[i,k] = \phi_{\hat{n}}[k,i]$ [see Eqs. (9.26a) and (9.26b)], the $p \times p$ matrix of correlation-like values is symmetric and positive-definite, but *not* Toeplitz. Indeed, the diagonal elements are related by the equation

$$\begin{aligned} \phi_{\hat{n}}[i+1,k+1] = \phi_{\hat{n}}[i,k] &+ s_{\hat{n}}[-i-1]s_{\hat{n}}[-k-1] \\ &- s_{\hat{n}}[L-1-i]s_{\hat{n}}[L-1-k]. \end{aligned} \tag{9.29}$$

From Eq. (9.15) of our previous discussion, the total squared prediction error for the optimum coefficients that satisfy Eq. (9.28a) is

$$\mathcal{E}_{\hat{n}} = \phi_{\hat{n}}[0,0] - \sum_{k=1}^{p} \alpha_k \phi_{\hat{n}}[0,k]. \tag{9.30}$$

The method of analysis based upon this method of computation of $\phi_{\hat{n}}[i,k]$ has come to be known as the *covariance method* because the matrix of values $\{\phi_{\hat{n}}[i,k]\}$ has the properties of a covariance matrix [222].[4]

9.2.4 Summary

It has been shown that by using different definitions of the segments of the signal to be analyzed, two distinct sets of analysis equations are obtained. For the autocorrelation method, the signal is windowed by an L-point window, and the quantities $\varphi_{\hat{n}}[i,k]$ are

[4]This terminology, which is firmly entrenched, is somewhat confusing since the term "covariance" usually refers to the correlation of a signal after first subtracting its mean.

obtained using a short-time autocorrelation function. The resulting matrix of correlations is a positive-definite Toeplitz matrix, which leads to one type of solution for the predictor coefficients. For the covariance method, the signal is assumed to be known for the set of values $-p \leq m \leq L-1$. Outside this interval, no assumptions need be made about the signal, since these are the only values needed in the computation. The resulting matrix of correlations, in this case, is symmetric and positive-definite, but not Toeplitz. The result is that the two methods of computing the correlations lead to different methods of solution of the analysis equations and to two sets of predictor coefficients with somewhat different properties.

In later sections we will compare and contrast computational details and results for both these techniques as well as for a lattice LPC analysis method, to be discussed in Section 9.5.3. First, however, we will show how the gain, G, in Figure 9.1, can be determined from the prediction error expression.

9.3 COMPUTATION OF THE GAIN FOR THE MODEL

It is reasonable to expect that the speech production model gain, G, could be determined by matching the energy in the signal with the energy of the output of the linear predictive model. This indeed is true when appropriate assumptions are made about the excitation signal to the LPC system [216, 217].

To relate the gain constant, G, to the excitation signal and the error in prediction, we can compare Eq. (9.2), which describes the speech model, and Eq. (9.5), which defines the prediction error.[5] Assuming that the speech signal being analyzed is the output of the model, the excitation signal, $Gu[n]$, can be expressed as

$$Gu[n] = s[n] - \sum_{k=1}^{p} a_k s[n-k], \tag{9.31a}$$

whereas the prediction error signal $e[n]$ is defined as

$$e[n] = s[n] - \sum_{k=1}^{p} \alpha_k s[n-k]. \tag{9.31b}$$

Based on the assumption that $a_k = \alpha_k$, i.e., the predictor coefficients computed from the speech signal are identical to the coefficients of the model, and the output of the model is precisely $s[n]$, it follows that

$$e[n] = Gu[n]. \tag{9.32}$$

That is, the input signal of the model is proportional to the error signal, with the constant of proportionality being the gain constant, G. A detailed discussion of the properties of the prediction error signal is given in Section 9.6.

[5]Note that the gain is also a function of analysis time $\hat{n}$, although we do not show it explicitly.

Since Eq. (9.32) is only approximate (i.e., it is valid to the extent that the ideal and the actual linear prediction parameters are identical), it is generally not possible to solve for G in a reliable way directly from the error signal itself. Instead the more reasonable assumption is made that the total energy of the error signal is equal to the total energy of the excitation input; i.e., assuming the autocorrelation method of analysis,

$$G^2 \sum_{m=0}^{L-1+p} u^2[m] = \sum_{m=0}^{L-1+p} e^2[m] = \mathcal{E}_{\hat{n}}. \tag{9.33}$$

At this point we must make some assumptions about $u[n]$ in order to relate G to the α_k's and the correlation coefficients. There are two cases of interest for the excitation. For voiced speech, it is reasonable to assume $u[n] = \delta[n]$, i.e., the excitation is a unit sample at $n = 0$.[6] For this to be valid requires the assumption that the effects of the radiation load and the glottal pulse shape in the actual excitation for voiced speech can be lumped together with the vocal tract transfer function, and therefore all three of these effects are modeled by the time-varying linear predictor. This requires that the predictor order, p, be large enough to account for the vocal tract and glottal pulse and radiation effects (but smaller than the pitch period for voiced speech). We will discuss the choice of predictor order in Section 9.6. For unvoiced speech, it is convenient to assume that $u[n]$ is white noise having zero mean and unit variance.

Based on these assumptions, we can now determine the gain constant G using Eq. (9.33). For voiced speech, we have input $G\delta[n]$. We denote the resulting output for this particular input as $\tilde{h}[n]$, since it is actually the impulse response corresponding to the system function

$$\tilde{H}(z) = \frac{G}{A(z)} = \frac{G}{1 - \sum_{k=1}^{p} \alpha_k z^{-k}}, \tag{9.34}$$

with gain G incorporated into the system function; i.e., the appropriate input to $\tilde{H}(z)$ is $u[n] = e[n]/G$. From Eq. (9.34), we get the difference equation

$$\tilde{h}[n] = \sum_{k=1}^{p} \alpha_k \tilde{h}[n-k] + G\delta[n]. \tag{9.35}$$

[6] Note that for this assumption to be valid requires that the analysis interval be about the same length as a pitch period. In fact, a similar argument can be constructed assuming that the excitation is a periodic impulse train.

It is straightforward to show (see Problem 9.1) that the deterministic autocorrelation function of $\tilde{h}[n]$, defined as

$$\tilde{R}[m] = \sum_{n=0}^{\infty} \tilde{h}[n]\tilde{h}[m+n] = \tilde{R}[-m], \qquad 0 \leq m < \infty, \tag{9.36}$$

satisfies the relations

$$\tilde{R}[m] = \sum_{k=1}^{p} \alpha_k \tilde{R}[|m-k|], \qquad 1 \leq m < \infty, \tag{9.37a}$$

and

$$\tilde{R}[0] = \sum_{k=1}^{p} \alpha_k \tilde{R}[k] + G^2. \tag{9.37b}$$

Note that Eqs. (9.37a) and (9.22a) have exactly the same form for $m = 1, 2, \ldots, p$, and the coefficients α_k are assumed to be the same in both equations. Therefore, it follows that [216]

$$\tilde{R}[m] = cR_{\hat{n}}[m], \quad 0 \leq m \leq p, \tag{9.38}$$

where c is a constant to be determined. Since the total energies in the signal, $R[0]$, and the impulse response, $\tilde{R}[0]$, must be equal, and since Eq. (9.38) must hold for $m = 0$, it follows that $c = 1$. Therefore, we can use Eqs. (9.22b) and (9.37b) to obtain

$$G^2 = \tilde{R}_{\hat{n}}[0] - \sum_{k=1}^{p} \alpha_k \tilde{R}_{\hat{n}}[k] = R_{\hat{n}}[0] - \sum_{k=1}^{p} \alpha_k R_{\hat{n}}[k] = \mathcal{E}_{\hat{n}}. \tag{9.39}$$

It is interesting to note that Eq. (9.38) together with the requirement that the energy of the impulse response be equal to the energy of the signal requires that the first $p+1$ coefficients of the autocorrelation function of the impulse response of the model be identical to the first $p+1$ coefficients of the autocorrelation function of the speech signal. This condition is called the *autocorrelation matching property* of the autocorrelation method of linear predictive analysis [217, 218].

For the case of unvoiced speech, the correlations are conveniently defined as probability averages (although we use short-time averages in practice). It is assumed that the input is white noise with zero mean and unity variance; i.e.,

$$E\{u[n]u[n-m]\} = \delta[m], \tag{9.40}$$

where $E\{\ \}$ denotes probability average. If we excite the system of Eq. (9.34) with the random input, $Gu[n]$, and denote the output as $\tilde{g}[n]$, then

$$\tilde{g}[n] = \sum_{k=1}^{p} \alpha_k \tilde{g}[n-k] + Gu[n]. \tag{9.41}$$

Since the autocorrelation function for the output is the convolution of the aperiodic autocorrelation function of the impulse response of the linear system with the autocorrelation function of the white noise input, it follows that

$$E\{\tilde{g}[n]\tilde{g}[n-m]\} = \tilde{R}[m] * \delta[m] = \tilde{R}[m], \tag{9.42}$$

where $\tilde{R}[m]$ is given by Eq. (9.36). Therefore, the autocorrelation of the output $\tilde{g}[n]$ can be expressed as

$$\begin{aligned}\tilde{R}[m] &= E\{\tilde{g}[n]\tilde{g}[n-m]\} \\ &= \sum_{k=1}^{p} \alpha_k E\{\tilde{g}[n-k]\tilde{g}[n-m]\} + E\{Gu[n]\tilde{g}[n-m]\}.\end{aligned} \tag{9.43}$$

Since $E\{u[n]\tilde{g}[n-m]\} = 0$ for $m > 0$, i.e., $u[n]$ is uncorrelated with any output signal prior to n, it follows from Eq. (9.43) that

$$\tilde{R}[m] = \sum_{k=1}^{p} \alpha_k \tilde{R}[m-k], \qquad m \neq 0, \tag{9.44}$$

where in this case, $\tilde{R}[m]$ denotes the probability average autocorrelation function. For $m = 0$, Eq. (9.43) becomes

$$\tilde{R}[0] = \sum_{k=1}^{p} \alpha_k \tilde{R}[k] + GE\{u[n]\tilde{g}[n]\}. \tag{9.45}$$

Since it can be shown that $E\{u[n]\tilde{g}[n]\} = G$, Eq. (9.46) reduces to

$$\tilde{R}[0] = \sum_{k=1}^{p} \alpha_k \tilde{R}[k] + G^2. \tag{9.46}$$

As in the aperiodic case before, if we assert that the average power of the output of the model must equal the average power of the speech signal, i.e., $\tilde{R}[0] = R_{\hat{n}}[0]$, then we can argue that

$$\tilde{R}[m] = R_{\hat{n}}[m], \qquad 0 \leq m \leq p, \tag{9.47}$$

and

$$G^2 = R_{\hat{n}}[0] - \sum_{k=1}^{p} \alpha_k R_{\hat{n}}[k] = \mathcal{E}_{\hat{n}}, \tag{9.48}$$

which is identical to the result in Eq. (9.39) that was obtained for the case for the impulse excitation approximation for voiced speech.

When linear predictive analysis is applied to speech, we compute correlation functions of finite-length segments of the speech signal rather than the correlation functions that we assumed in this analysis. Nevertheless, the gain constant G is normally determined from Eqs. (9.39) and (9.48) as the square root of the total squared prediction error.

9.4 FREQUENCY DOMAIN INTERPRETATIONS OF LINEAR PREDICTIVE ANALYSIS

Up to this point we have discussed linear predictive methods mainly in terms of difference equations and correlation functions; i.e., in terms of time domain representations. However, we pointed out at the beginning of this chapter that the coefficients of the linear predictor are *assumed* to be the coefficients of the denominator of the system function that models the combined effects of vocal tract response, glottal wave shape, and radiation. By choosing the order p appropriately, the linear predictive analysis separates these effects from the effects of the excitation. Thus, given the set of predictor coefficients, we can find the frequency response of the model for speech production simply by evaluating the system function $\tilde{H}(z) = G/A(z)$ for $z = e^{j\omega}$ as in

$$\tilde{H}(e^{j\omega}) = \frac{G}{1 - \sum_{k=1}^{p} \alpha_k e^{-j\omega k}} = \frac{G}{A(e^{j\omega})}, \tag{9.49}$$

or $z = e^{j2\pi FT} = e^{j2\pi F/F_s}$ if we wish to use analog frequency F (with sampling period T or sampling frequency $F_s = 1/T$). By including the excitation gain G in the numerator, we obtain a spectrum representation that has the same amplitude scale as the short-time Fourier transform of the windowed speech sample, but with the fine structure due to the excitation removed. If we plot $\tilde{H}(e^{j\omega})$ as a function of frequency,[7] we would expect to see peaks at the formant frequencies just as we have in spectral representations discussed in previous chapters. Thus linear predictive analysis can be viewed as a method of short-time spectrum estimation that incorporates removal of the excitation fine structure. Such techniques are widely applied outside the speech processing field for the purpose of spectrum analysis [382]. In this section we present a frequency

[7] Problem 9.2 is concerned with a method of evaluating $\tilde{H}(e^{j\omega})$ using the FFT.

domain interpretation of the mean-squared prediction error and compare linear predictive techniques to other methods of estimating frequency domain representations of speech.

9.4.1 Linear Predictive Short-Time Spectrum Analysis

In our formulation of linear predictive analysis by the autocorrelation method, we have defined the speech segment to be analyzed as

$$s_{\hat{n}}[m] = s[\hat{n}+m]w[m], \tag{9.50}$$

where $w[m]$ is a window that is non-zero over $0 \leq m \leq L-1$. That is, in performing the short-time linear predictive analysis, we find it convenient to redefine the time origin for analysis to the beginning of the analysis window rather than at $\hat{n}$ as in Eq. (7.8). The discrete-time Fourier transform of this windowed segment is

$$S_{\hat{n}}(e^{j\omega}) = \sum_{m=-\infty}^{\infty} s[\hat{n}+m]w[m]e^{-j\omega m}, \tag{9.51}$$

where the infinite limits simply imply that the sum is over the region of support of the window. Equation (9.51) can be recognized as the alternate form of the short-time Fourier transform that was given in Eq. (7.10). This alternate form differs from Eq. (7.8), the basic definition of the short-time Fourier transform, by a multiplicative factor $e^{j\omega\hat{n}}$, which corresponds to the shift of the time origin for Fourier analysis to the window location rather than at $\hat{n}$ as in Eq. (7.8). This factor does not affect the spectral magnitude, so $|S_{\hat{n}}(e^{j\omega})|$ from Eq. (9.51) is identical to the magnitude of the short-time Fourier transform as defined in Eq. (7.8) of Chapter 7. As shown in Section 7.3.4, $|S_{\hat{n}}(e^{j\omega})|^2$ corresponds to the short-time autocorrelation function, which, as we have seen, is the basis for linear predictive analysis. Therefore, the short-time Fourier transform and $\tilde{H}(e^{j\omega})$ are linked through the short-time autocorrelation function. This is illustrated in Figure 9.5 which shows a windowed voiced speech segment in (a), the corresponding autocorrelation function in (b), and both $20\log_{10}|S_{\hat{n}}(e^{j\omega})|$ and $20\log_{10}|\tilde{H}(e^{j\omega})|$ (in dB) in (c).[8] The equally spaced peaks in the short-time Fourier transform in Figure 9.5c are at multiples of the fundamental pitch frequency (approximately 110 Hz), and they correspond to the peak in the autocorrelation function at about 9 msec (144 samples at a 16 kHz sampling rate), which in turn is the pitch period of the signal in Figure 9.5a. The values of the autocorrelation that were used in determining the linear predictor of order $p = 22$ (corresponding to 1.375 msec) are accented with the heavy line to the left of the dashed vertical line in Figure 9.5b.

Since the values of the autocorrelation function around the pitch peak are not included in the p values used in the linear predictive analysis, the effects of the periodic

[8] By substituting $\omega = 2\pi F/F_s$, we can plot $20\log_{10}|S_{\hat{n}}(e^{j\omega})|$ and $20\log_{10}|\tilde{H}(e^{j\omega})|$ as a function of the analog frequency variable F as in Figures 9.5 and 9.6.

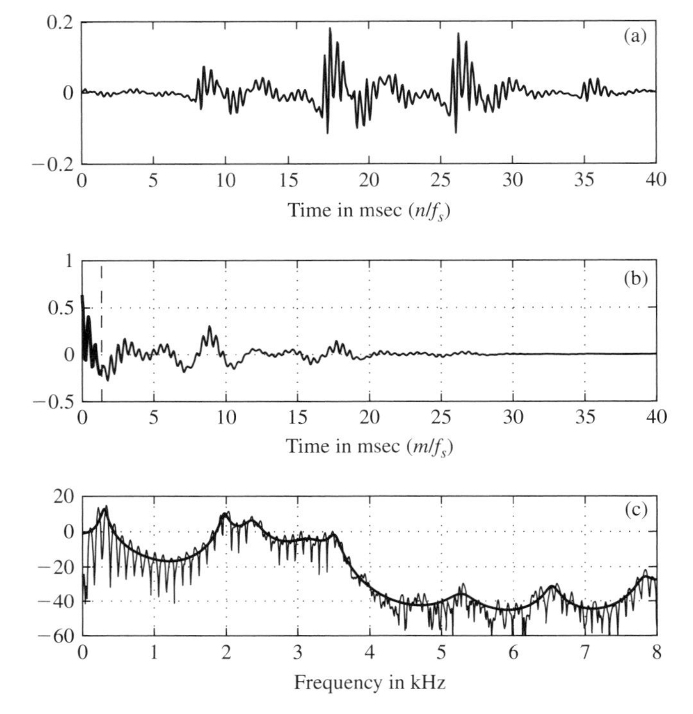

FIGURE 9.5
(a) Voiced speech segment obtained using a Hamming window.
(b) Corresponding short-time autocorrelation function (values ($p = 22$) used in linear predictive analysis are to the left of the dashed vertical line). (c) Corresponding short-time log magnitude Fourier transform and short-time log magnitude linear predictive spectrum plotted on a common scale (with $F_s = 16$ kHz).

excitation are automatically eliminated. This is essentially the same principle that was employed in homomorphic liftering of the log magnitude of the short-time Fourier transform. In linear predictive analysis, the smoothing-by-time-limitation operates on the magnitude-squared indirectly through the linear predictive analysis by basing it on a relatively small number of correlation values. Observe that the linear predictive spectrum has the general shape of the short-time Fourier transform, but it is much smoother. This is because, by using $p = 22$, we have limited the number of resonance peaks for $\tilde{H}(e^{j\omega})$ to a maximum of 11 in the base frequency range $0 \leq \omega \leq \pi$. Most of the prominent peaks of $\tilde{H}(e^{j\omega})$ correspond to the formant resonances of the speech signal, whereas $|S_{\hat{n}}(e^{j\omega})|$ has a general overall shape determined by the vocal tract, glottal pulse shape, and radiation, but the prominent local maxima in $20 \log_{10} |S_{\hat{n}}(e^{j\omega})|$ are at multiples of the pitch frequency.

Figure 9.6 shows an example similar to that of Figure 9.5 for unvoiced speech. As before, Figure 9.6a is the Hamming windowed waveform, Figure 9.6b is the corresponding autocorrelation function, and Figure 9.6c shows the log magnitude of

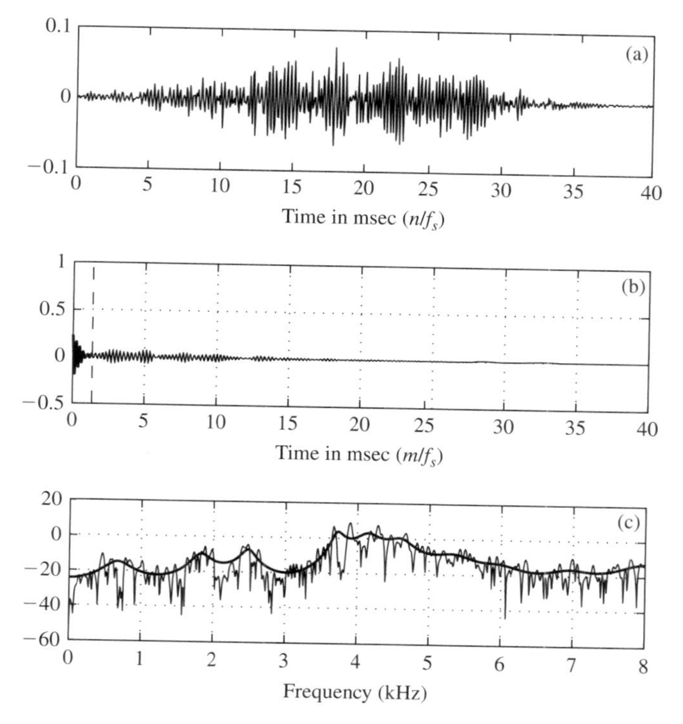

FIGURE 9.6
(a) Unvoiced speech segment obtained using a Hamming window.
(b) Corresponding short-time autocorrelation function (values ($p = 22$) used in linear predictive analysis are to the left of the dashed vertical line). (c) Corresponding short-time log magnitude Fourier transform and short-time log magnitude LPC spectrum plotted on a common scale (with $F_s = 16$ kHz).

the short-time Fourier transform and the log magnitude of the short-time linear predictive spectrum obtained from the $p = 22$ values to the left of the dashed vertical line in Figure 9.6b. The autocorrelation function shows no strong peaks in the expected pitch period range. In this case, the fine structure in the short-time log magnitude spectrum is due to random variability of the periodogram estimate [270]. As before, the linear predictive log magnitude spectrum of order $p = 22$ smoothly follows the general shape of the short-time log magnitude Fourier transform, but the random fluctuations of $20\log_{10}|S_{\hat{n}}(e^{j\omega})|$ are removed by basing the linear predictive spectrum on only $p = 22$ correlation values.

9.4.2 Frequency Domain Interpretation of Mean-Square Prediction Error

The interpretation of the linear predictive spectrum of Section 9.4.1 provides a basis for examining the properties of the prediction error, which, by the assumptions that we have made, is considered to be the excitation to the vocal tract system that is derived from the linear predictor. Recall that for the autocorrelation method, the mean-squared prediction error can be expressed in the time domain as

$$\mathcal{E}_{\hat{n}} = \sum_{m=0}^{L+p-1} e_{\hat{n}}^2[m], \tag{9.52a}$$

or, in the frequency domain (using Parseval's Theorem), as

$$\mathcal{E}_{\hat{n}} = \frac{1}{2\pi}\int_{-\pi}^{\pi} |E_{\hat{n}}(e^{j\omega})|^2 d\omega = \frac{1}{2\pi}\int_{-\pi}^{\pi} |S_{\hat{n}}(e^{j\omega})|^2 |A(e^{j\omega})|^2 d\omega = G^2, \tag{9.52b}$$

where $S_{\hat{n}}(e^{j\omega})$ is the DTFT of the windowed segment of speech $s_{\hat{n}}[m]$, and $A(e^{j\omega})$ is the corresponding prediction error frequency response[9]

$$A(e^{j\omega}) = 1 - \sum_{k=1}^{p} \alpha_k e^{-j\omega k}. \tag{9.53}$$

Using the linear predictive spectrum in the form with gain incorporated

$$\tilde{H}(e^{j\omega}) = \frac{G}{A(e^{j\omega})}, \tag{9.54}$$

Eq. (9.52b) can be expressed as

$$\mathcal{E}_{\hat{n}} = \frac{G^2}{2\pi}\int_{-\pi}^{\pi} \frac{|S_{\hat{n}}(e^{j\omega})|^2}{|\tilde{H}(e^{j\omega})|^2} d\omega = G^2. \tag{9.55}$$

Since the integrand in Eq. (9.55) is positive, it follows that minimizing the total squared prediction error, $\mathcal{E}_{\hat{n}}$, is equivalent to finding gain and predictor coefficients such that the integral of the ratio of the energy spectrum of the speech segment to the magnitude squared of the frequency response of the model linear system is unity. This implies that $|S_{\hat{n}}(e^{j\omega})|^2$ can be interpreted as a frequency-domain weighting function. Implicitly, the linear predictive optimization weights frequencies where $|S_{\hat{n}}(e^{j\omega})|^2$ is large more heavily in determining the gain and predictor coefficients than when $|S_{\hat{n}}(e^{j\omega})|^2$ is small.

To illustrate the nature of the spectral modeling of linear predictive spectra, Figure 9.7 shows a comparison between $20\log_{10}|\tilde{H}(e^{j\omega})|$ and the short-time Fourier spectrum $20\log_{10}|S_{\hat{n}}(e^{j\omega})|$ for a high-pitched voice. The sampling rate of the signal was 8 kHz, and the Hamming window for the autocorrelation linear predictive analysis and the short-time Fourier analysis had length $L = 301$ samples (equivalent to 37.5 msec). The linear predictive spectrum, shown as the heavy solid line, was based on a 12th-order

[9] To show the dependence on analysis time, we should write $A_{\hat{n}}(e^{j\omega})$ and $\tilde{H}_{\hat{n}}(z)$, but we shall continue to suppress the subscript $\hat{n}$ in interest of notational simplicity.

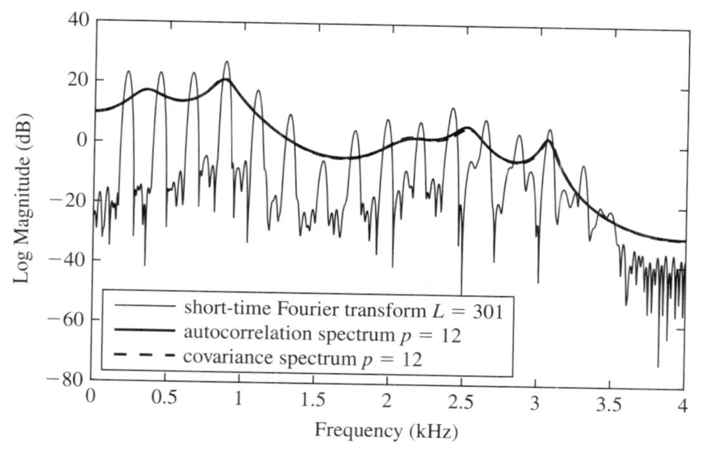

FIGURE 9.7
12th-order linear predictive log magnitude spectra computed by both the autocorrelation and covariance methods compared to the short-time log magnitude Fourier transform of an $L = 301$ sample Hamming windowed speech segment.

predictor ($p = 12$) obtained by the autocorrelation method. The harmonic structure of the signal spectrum is clearly seen in the short-time Fourier transform in this figure since the fundamental frequency is approximately 220 Hz, and the width of the main lobe of the Fourier transform of the 301-sample Hamming window is on the order of 100 Hz. This figure illustrates that the linear predictive spectrum matches the signal spectrum much more closely in the regions of large signal energy (i.e., near the spectrum peaks) than in the regions of low signal energy (i.e., near the spectral valleys). This is to be expected in view of Eq. (9.55) since frequencies where $|S_{\hat{n}}(e^{j\omega})| > |\tilde{H}(e^{j\omega})|$ contribute more to the total error than frequencies where $|S_{\hat{n}}(e^{j\omega})| < |\tilde{H}(e^{j\omega})|$. Thus the linear predictive spectral error criterion favors a good fit near the spectral peaks, whereas the fit near the spectral valleys is nowhere near as good. This can be somewhat problematic for high-pitched voices where the harmonics are widely spaced, since the linear predictive spectrum peaks can be biased toward the widely spaced harmonics of the fundamental frequency.

So far in the discussion of this section, we have assumed that the predictor parameters were computed using the autocorrelation method. This was necessary because only in this case is the Fourier transform of the short-time autocorrelation function equal to the magnitude squared of the short-time Fourier transform of the signal. However, the prediction coefficients and total squared prediction error can also be computed using the covariance method, and these could be used as the basis for the short-time linear predictive spectrum defined by $\tilde{H}(e^{j\omega})$ in Eq. (9.54). If this is done, however, the relationship to the short-time Fourier spectrum is less straightforward, since the covariance method avoids windowing by supplying the previous p signal samples needed to compute the prediction error over the range $0 \leq m \leq L - 1$. Thus, there is no direct relationship between the correlation function used in the covariance method and the short-time Fourier transform as we have defined it. If we use the covariance method to compute a linear predictive spectrum using Eq. (9.54), the denominator polynomial and its roots will be very similar to the polynomial obtained by the autocorrelation method in the case where the analysis intervals for the two methods both span several pitch periods. In this case, however, the gain constant can be significantly different if we simply set $G = \sqrt{\mathcal{E}_{\hat{n}}}$ with $\mathcal{E}_{\hat{n}}$ determined by Eq. (9.22b) for the autocorrelation method and by Eq. (9.30) for the covariance method. This is because the tapering of the window in the autocorrelation method is not present in the covariance method. It is also true that different segment lengths are generally used in the two methods.

To a first approximation, the covariance linear prediction spectrum is comparable to the short-time Fourier transform with a rectangular window where the samples are not tapered to zero. Thus if we wish to compare the linear predictive spectra for the two methods of estimating the parameters, we should multiply the covariance spectrum by the factor

$$U = \frac{1}{L_c} \sum_{n=0}^{L_a-1} (w[n])^2, \tag{9.56}$$

which compensates for the different gains due to possibly different lengths, L_a, for the autocorrelation method and L_c for the covariance method and for the window $w[n]$ in the autocorrelation method. Such normalization is common in spectrum analysis and is used to remove window bias [270, 382]. The dashed curve in Figure 9.7 is the linear predictive spectrum with parameters obtained by the covariance method. For this large window size, the differences between the two linear predictive spectra are insignificant.

One of the significant advantages of the covariance method is that because of the way the correlation function is computed, it is possible to use very short segments of the speech signal. In fact, the covariance method is often used in a pitch synchronous analysis where only the "closed glottis" section of a period is used. This type of analysis is not feasible with the autocorrelation method due to the need to taper the edges of the waveform segments. Figure 9.8 illustrates this point. Included in this figure are the log magnitude responses of the short-time Fourier transform (using $L = 51$), along with the resulting log magnitude responses of both the autocorrelation and the covariance method spectra based on a value of $p = 12$.

The short-time Fourier transform log magnitude response is shown by the thin solid line and is typical of a wideband spectrum corresponding to a short duration window. The dashed curve in Figure 9.8 is the covariance log magnitude spectrum where the window length was 51 samples and the analysis order was $p = 12$. Note that this spectral plot maintains the detailed shape of the linear predictive spectra in Figure 9.7 with a particularly sharp resonance at about 3.2 kHz, while the autocorrelation log magnitude spectrum, shown by the heavy solid line, is much more like the wideband short-time Fourier transform. Thus, the covariance method has the desirable property that it can extract accurate models from very short segments of the speech signal. However, it must be noted that the poles of the covariance-derived vocal tract model

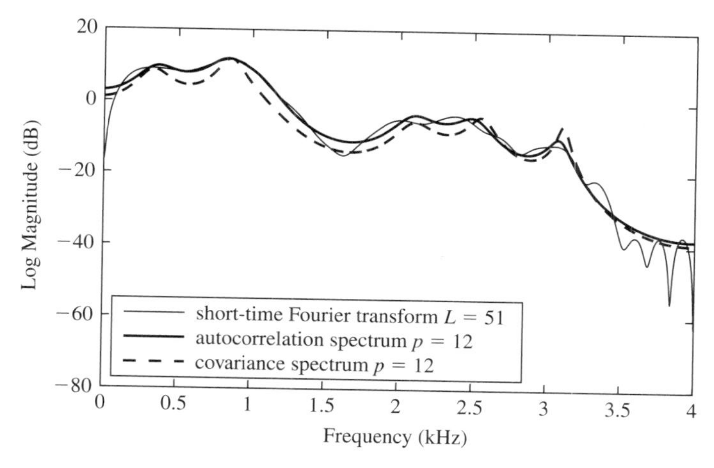

FIGURE 9.8
12-pole linear predictive log magnitude spectra computed by both the autocorrelation and covariance methods compared to the short-time Fourier transform of an $L = 51$ sample Hamming windowed speech segment.

are not *guaranteed* to be within the unit circle; i.e., the resulting system could be unstable. Thus, the autocorrelation method, even with somewhat longer windows, is often preferred for speech processing applications where the vocal tract model is used for synthesis; however, the covariance method is often preferred in applications such as inverse filtering to recover the glottal pulse where it is not necessary to implement the model using a recursive difference equation.

9.4.3 Effect of Model Order p

In Section 9.3 it was shown that the autocorrelation function, $R_{\hat{n}}[m]$, of the segment of speech, $s_{\hat{n}}[m]$, and the autocorrelation function, $\tilde{R}[m]$, of the impulse response, $\tilde{h}[m]$, corresponding to the system function, $\tilde{H}(z)$, are equal for the first $(p+1)$ values. Thus, as $p \to \infty$, the respective autocorrelation functions are equal for all values, and therefore

$$\lim_{p\to\infty} |\tilde{H}(e^{j\omega})|^2 = |S_{\hat{n}}(e^{j\omega})|^2. \tag{9.57}$$

This implies that if p is large enough, the frequency response of the all-pole model, $\tilde{H}(e^{j\omega})$, can approximate the short-time Fourier transform of the signal with arbitrarily small error. This is illustrated in Figure 9.9, which shows the short-time log magnitude spectrum of a segment of voiced speech for a window length of $L = 101$. The dashed curve is the (autocorrelation method) linear predictive log magnitude spectrum with $p = 12$, which is a typical choice for p for a sampling rate of 8 kHz. Note that as in Figure 9.7, the spectral peaks align with the general shape of the short-time spectrum. The heavy line shows the linear predictive log magnitude spectrum for a predictor

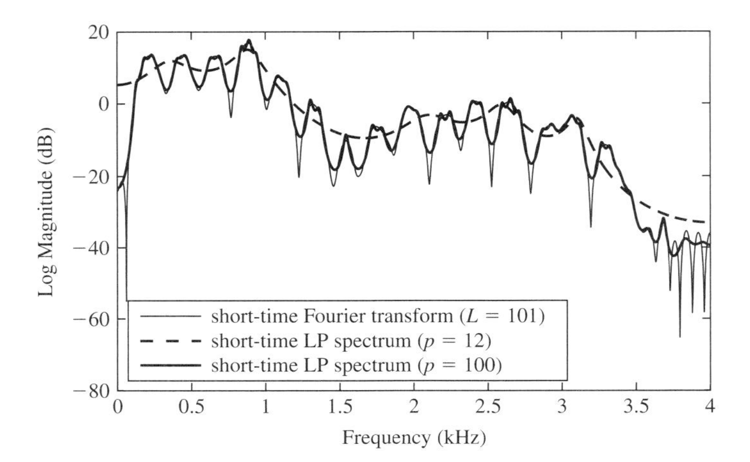

FIGURE 9.9
Short-time Fourier log magnitude spectrum compared to linear predictive log magnitude spectrum for $p = 12$ and p approximately equal to the window length.

order of $p = 100$. Note that the spectrum agrees very closely with the short-time spectrum at the (pitch) peaks of the short-time Fourier spectrum, but even with a predictor order that is equal to the length of the speech segment, the agreement is not perfect at frequencies where $|S_{\hat{n}}(e^{j\omega})|^2$ is small. This is in agreement with our previous argument about the weighting effect of $|S_{\hat{n}}(e^{j\omega})|^2$ in Eq. (9.55). If the order of prediction is further increased, the linear predictive spectrum approaches $|S_{\hat{n}}(e^{j\omega})|^2$ as predicted by Eq. (9.57), but very large values of p are required for a good match at frequencies where $|S_{\hat{n}}(e^{j\omega})|^2$ is close to zero.

Figures 9.10 and 9.11 contribute additional insight into how the linear predictor models the short-time spectrum. Figure 9.10 shows the 12 zeros of the z-transform polynomial $A(z)$ [shown with the symbol $\times$ since they are the poles of $\tilde{H}(z)$]. The 100 zeros of the z-transform polynomial

$$S_{\hat{n}}(z) = \sum_{m=0}^{L-1} s_{\hat{n}}[m] z^{-m}$$

are shown with the symbol $\circ$. Note that $S_{\hat{n}}(z)$ has zeros both inside and outside the unit circle, and some of the zeros of $S_{\hat{n}}(z)$ are very close to the unit circle. It is these zeros that cause the sharp dips in $|S_{\hat{n}}(e^{j\omega})|$ in Figure 9.9. Note that the zeros of $A(z)$, which are the poles of $\tilde{H}(z)$, are all inside the unit circle, as they are constrained to be by the properties of the autocorrelation method. This figure points out that the all-pole system function creates peaks by placing poles close to the unit circle at an angle corresponding to the center frequency of the peak. On the other hand, the all-zero polynomial representation of the segment of speech creates peaks by an *absence* of zeros surrounded by clusters of zeros arrayed close to the unit circle.

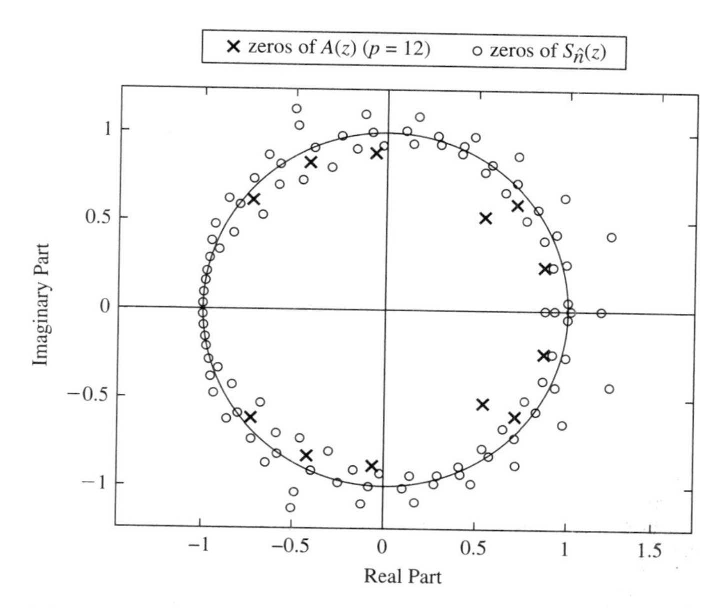

FIGURE 9.10
Zero locations for $S_{\hat{n}}(z)$ and $A(z)$ for $p = 12$.

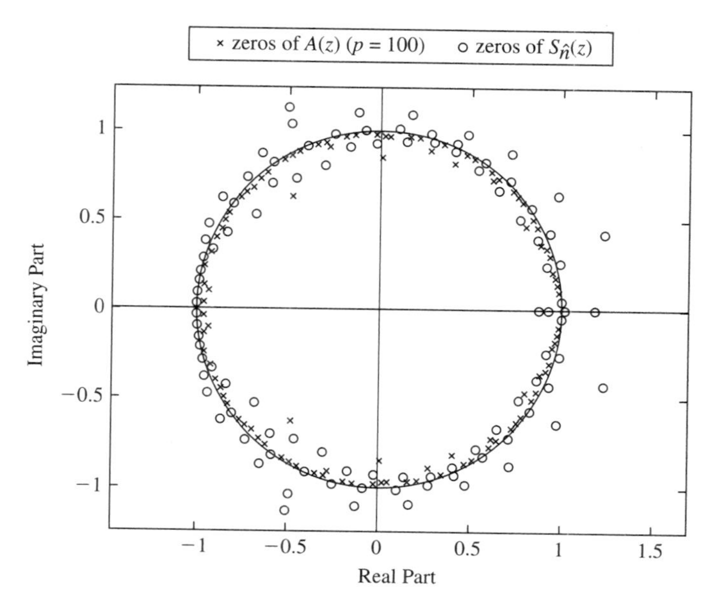

FIGURE 9.11
Zero locations for $S_{\hat{n}}(z)$ and $A(z)$ for $p = 100$.

Additional insight is provided by Figure 9.11, which compares the zeros of $S_{\hat{n}}(z)$ to those of $\tilde{H}(z)$ for $p = 100$. Note in particular how the zeros of $A(z)$ tend to avoid the zeros of $S_{\hat{n}}(z)$. It is interesting to note that even though Eq. (9.57) says that $|\tilde{H}(e^{j\omega})|^2 \to |S_{\hat{n}}(e^{j\omega})|^2$, as $p \to \infty$, it is not necessarily (or generally) true that $\tilde{H}(e^{j\omega}) = S_{\hat{n}}(e^{j\omega})$; i.e., the frequency response of the model need not be the Fourier transform of the signal. This is evident from Figure 9.11, which shows that $S_{\hat{n}}(z)$ need not be minimum-phase, whereas $\tilde{H}(z)$ is required to be minimum-phase since it is the transfer function of an all-pole filter with poles inside the unit circle.

The above discussion suggests that the order p of the linear predictive analysis can effectively control the degree of smoothness of the resulting spectrum. This is illustrated in Figure 9.12, which shows the input speech segment, the log magnitude Fourier transform of that segment, and linear predictive log magnitude spectra for various orders. As p increases, more of the details of the spectrum are preserved. Since our objective is to obtain a representation of only the spectral effects of the glottal pulse, vocal tract, and radiation, we should choose p as discussed before so that the formant resonances and the general spectrum shape are preserved, while keeping p small enough to discard the spectral features primarily due to the excitation.

9.4.4 The Linear Prediction Spectrogram

In Chapter 7 we defined the speech spectrogram to be a gray scale or pseudo-color image plot of the magnitude (or log magnitude in dB) of the short-time Fourier transform. Using Eq. (9.51) as the basis, we obtain

$$|S_r[k]| = \left| \sum_{m=0}^{L-1} s[rR+m]w[m]e^{-j(2\pi/N)km} \right|, \tag{9.58}$$

where the short-time Fourier transform is computed at the set of times, $t_r = rRT$, and the set of frequencies, $F_k = kF_s/N$, where R is the time shift (in samples) between adjacent short-time Fourier transforms, T is the sampling period, $F_s = 1/T$ is the sampling frequency, and N is the size of the discrete Fourier transform used to compute each STFT estimate. As an example, Figure 9.13a shows a wideband gray scale spectrogram (plot of $20 * \log_{10} |S_r[k]|$ as a function of t_r and F_k) of the speech utterance "Oak is strong but also gives shade." The window was a Hamming window with $L = 81$. In order to create this spectrogram plot with no blocking artifacts, the sampling parameters were $R = 3$ and $N = 1000$. The image shows values of $20\log_{10} |S_r[k]|$ within a 40 dB dynamic range.

Similarly, we can define the linear predictive (LP) spectrogram to be an image plot of

$$|\tilde{H}_r[k]| = \left| \frac{G_r}{A_r(e^{j(2\pi/N)k})} \right|, \tag{9.59}$$

where G_r and $A_r(e^{j(2\pi/N)k})$ are the gain and prediction error filter polynomial at analysis time rR. A plot of the LP spectrogram ($20\log_{10} |\tilde{H}_r[k]|$) as a function of t_r and F_k for

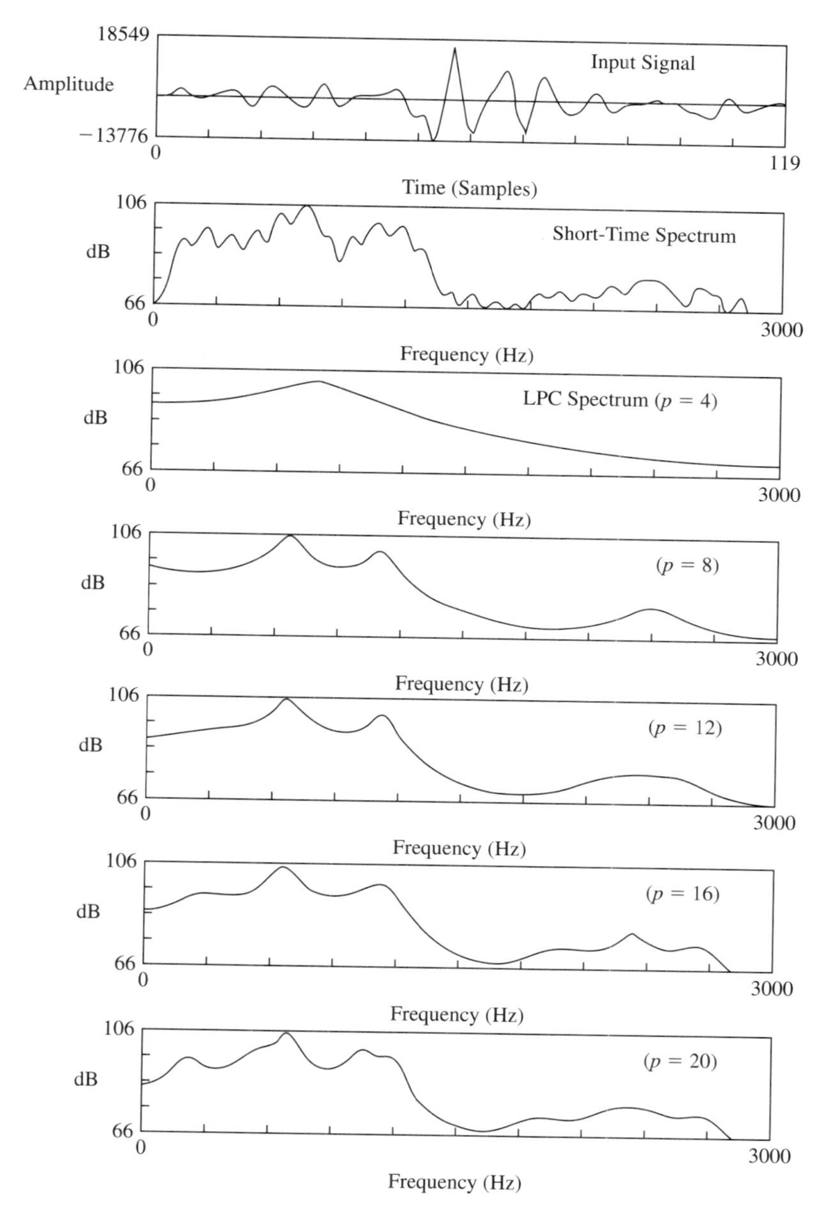

FIGURE 9.12
Linear predictive spectra for /*AA*/ vowel sampled at 6 kHz for several values of predictor order p.

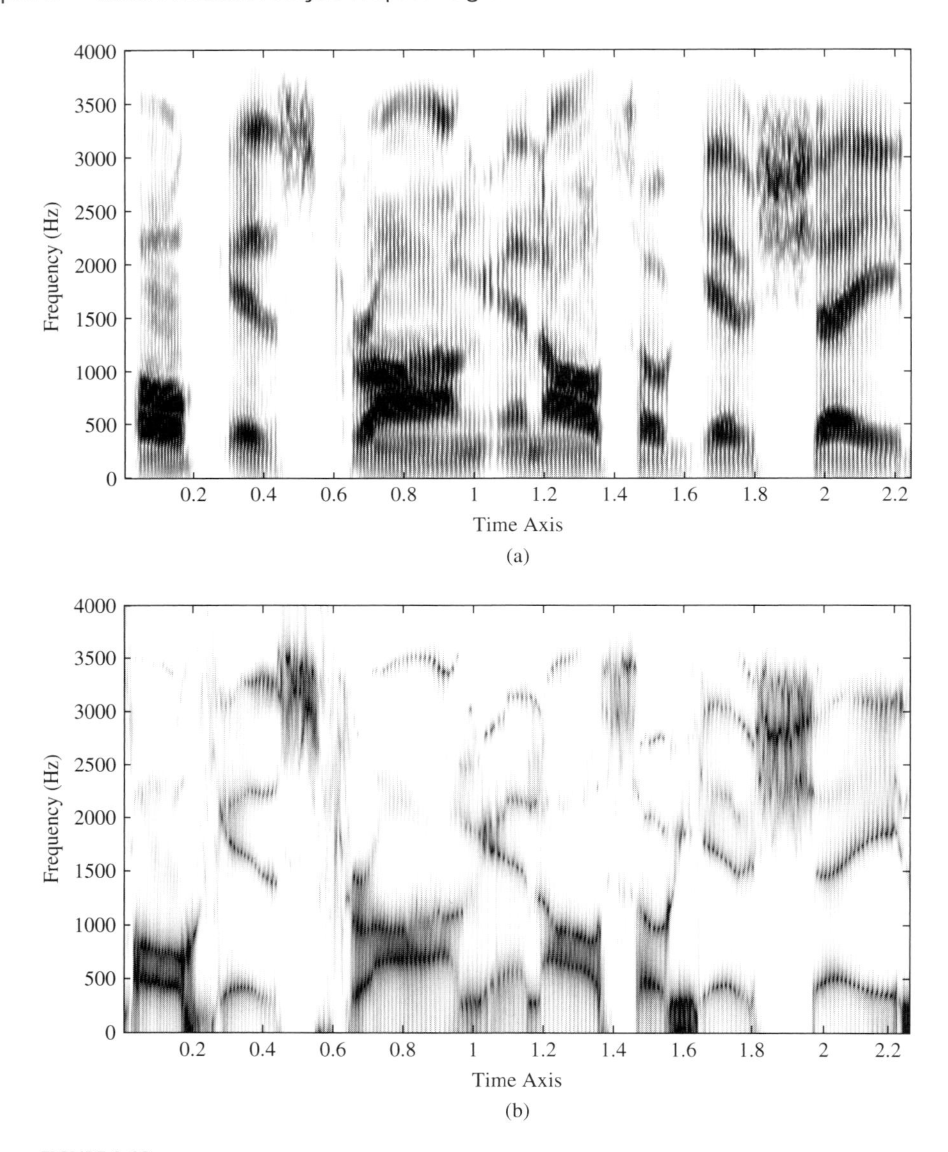

FIGURE 9.13
Spectrograms for an utterance of "Oak is strong and also gives shade.": (a) wideband Fourier spectrogram, Hamming window with $L = 81$; (b) corresponding linear predictive spectrogram.

$R = 3$ and $N = 1000$) for the same utterance as in Figure 9.13a is shown in Figure 9.13b for the same Hamming window of length $L = 81$. The resulting image plot is very similar to the wideband Fourier spectrogram since a low order analysis (e.g., $p = 12$) yields a smooth spectrum at each analysis time that is comparable to wideband short-time Fourier analysis of the signal. However, notice that the peaks of the LP spectrogram (dark regions) are much narrower than those of the Fourier spectrogram.

The wideband LP spectrogram uses a relatively small value of p, and, in principle, it could be based on either long or short segments of the speech waveform. However, in order to provide the same time resolution as a wideband Fourier spectrogram, the analysis window should be short. As we have seen, the covariance method may be preferred in this case. If long time windows are used, much larger values of p (as in Figure 9.9) can be used to capture the pitch structure in order to obtain an LP spectrogram that is comparable to the narrowband Fourier spectrogram.

9.4.5 Comparison to Other Spectrum Analysis Methods

We have already discussed methods of obtaining the short-time spectrum of speech in Chapters 7 and 8. It is instructive to compare these methods with the spectrum obtained by linear predictive analysis.

As an example, Figure 9.14 shows four log spectra of a section of the synthetic vowel /IY/. The first two spectra were obtained using the short-time spectrum method discussed in Chapter 7. For the first spectrum, a section of 400 samples (40 msec) was windowed, and then transformed (using a 2000 point FFT) to give the relatively narrow band spectral analysis shown in Figure 9.14a. In this spectrum the individual harmonics of the excitation are clearly in evidence due to the relatively long duration

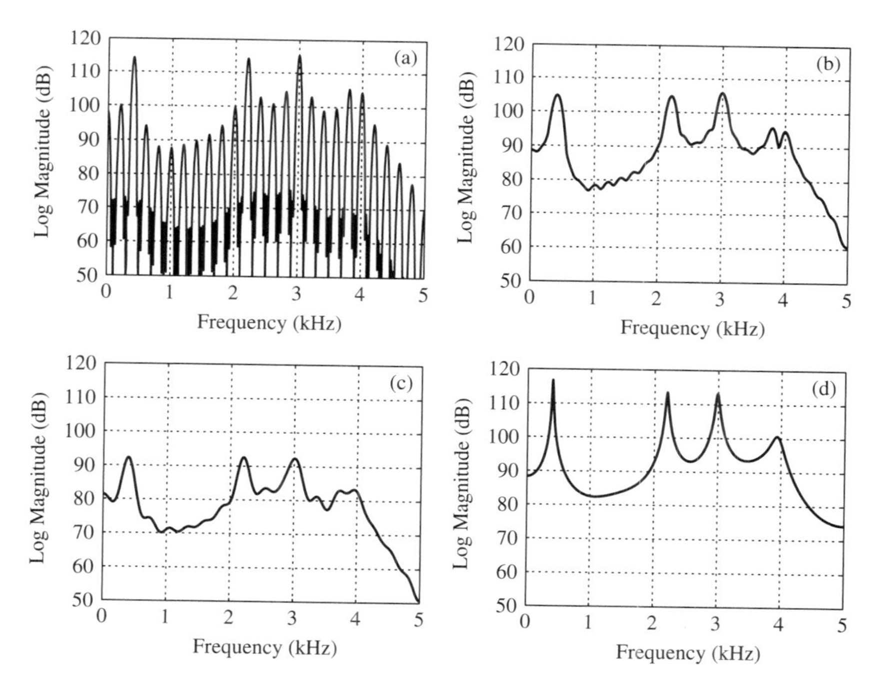

FIGURE 9.14
Spectra of synthetic vowel /IY/: (a) narrowband spectrum using a 40 msec window; (b) wideband spectrum using a 10 msec window; (c) cepstrally smoothed spectrum; (d) LPC spectrum from a 40 msec section using a $p = 12^{th}$-order LPC analysis.

of the window. For the second spectrum shown in Figure 9.14b, the analysis duration was decreased to 100 samples (10 msec), leading to a wideband spectral analysis. In this case, the excitation harmonics are not resolved; instead the overall spectral envelope can be seen. The formant frequencies are clearly in evidence in this spectrum. The third spectrum, shown in Figure 9.14c, was obtained by linear smoothing (liftering) of the log spectrum of a 400 sample segment as described in detail in Chapter 8. For this example the individual formants are again well resolved and are easily measured from the smoothed spectrum using a simple peak picker. However, the bandwidths of the formants are not easily obtained from the homomorphically smoothed spectrum due to the broadening of the resonances by the smoothing processes that have been used in obtaining the final spectrum. Finally the spectrum of Figure 9.14d is the result of a linear predictive analysis using $p = 12$ and a section of $L = 400$ samples (40 msec). A comparison of the linear prediction spectrum to the other spectra shows that the parametric representation appears to represent the formant structure very well with no extraneous peaks or ripples. This is due to the fact that the linear predictive model is very good for vowel sounds if the correct order, p, is used. Since the correct order can be determined by knowing the speech bandwidth, the linear prediction method leads to very good estimates of the spectral properties due to the glottal pulse, vocal tract, and radiation.

Figure 9.15 shows a direct comparison of the log spectra of a voiced section from natural speech obtained by both homomorphic smoothing and linear prediction.

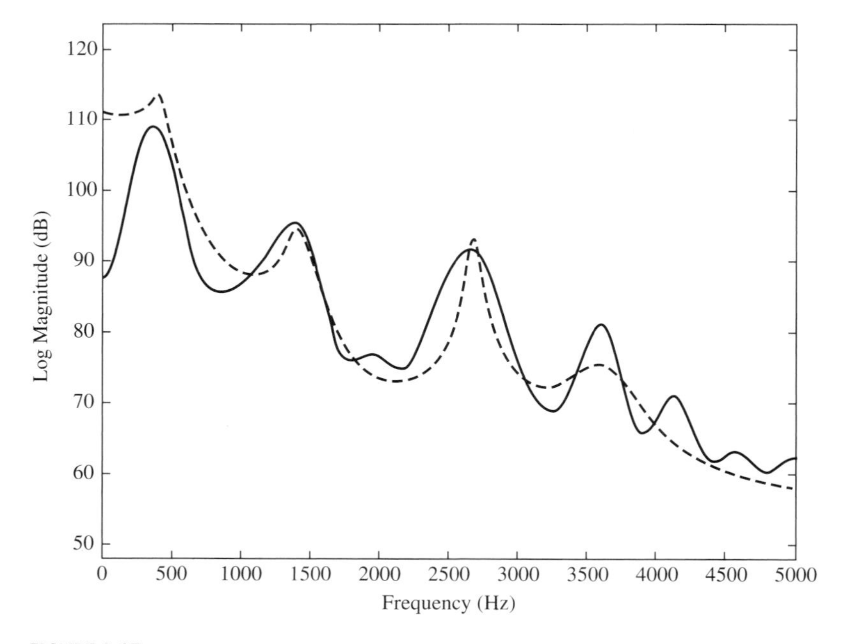

FIGURE 9.15
Comparison of speech spectra obtained by cepstrum smoothing (solid line) and linear prediction analysis (dashed line).

Although the formant frequencies are clearly in evidence in both plots, it can be seen that the LPC spectrum has fewer extraneous local maxima than the homomorphic spectrum. This is because the LPC analysis assumed a value of $p = 12$ so that at most six resonance peaks could occur. For the homomorphic spectrum, no such restriction existed. As noted above, the spectrum peaks from the LPC analysis are much narrower than the spectrum peaks from the homomorphic analysis due to the smoothing of the short-time log spectrum.

9.4.6 Selective Linear Prediction

It is possible to compute a linear predictive spectrum over a selected band of frequencies, rather than uniformly over the entire frequency range. This has been called *selective linear prediction* by Makhoul [216]. This method can be useful where only a portion of the spectrum is of interest. For example, a sampling rate of 20 kHz is required in many speech recognition applications to adequately represent the spectrum of fricatives, but spectral detail in the low frequencies is not required. For voiced sounds, the critical detail is generally in the region from 0 to about 4 or 5 kHz. For unvoiced sounds, the region from 4 to 5 kHz up to 8 or 10 kHz is generally of most importance. Using selective linear prediction, the signal spectrum from 0 to 5 kHz can be modeled by a predictor of order p_1; whereas the region from 5 to 10 kHz can be modeled by a different predictor of order p_2.

The method proposed by Makhoul begins with the computation of the short-time Fourier transform $S_{\hat{n}}(e^{j\omega})$. [In practice, the DFT would be computed at the discrete frequencies $(2\pi k/N)$.] To model only the frequency region from $\omega = \omega_A$ to $\omega = \omega_B$, all that is required is a simple linear mapping of the frequency scale such that $\omega = \omega_A$ is mapped to $\omega' = 0$ and $\omega = \omega_B$ is mapped to $\omega' = \pi$ (i.e., half the sampling frequency in analog terms). The predictor parameters are computed by solving the autocorrelation method equations where the autocorrelation coefficients are obtained from

$$R'[m] = \frac{1}{2\pi}\int_{-\pi}^{\pi} |S_n(e^{j\omega'})|^2 e^{j\omega' m}\, d\omega'. \tag{9.60}$$

Figure 9.16 (due to Makhoul [216]) illustrates the method of selective linear prediction. Figure 9.16a shows the LP spectrum with $p = 28$. Figure 9.16b shows the region from 0 to 5 kHz represented by a 14-pole model ($p_1 = 14$), whereas the region from 5 to 10 kHz is modeled independently by a 5-pole predictor ($p_2 = 5$). It can be seen that at 5 kHz, the model spectra show a discontinuity since there is no constraint that they agree at any frequency. However, the total number of prediction parameters is reduced from 28 to 19 with almost no sacrifice in the quality of the spectral representation.

9.5 SOLUTION OF THE LPC EQUATIONS

Effective implementation of a linear predictive analysis system requires efficient solution of p linear equations for the p unknown predictor coefficients. Because of the special properties of the coefficients in the linear equations, it is possible to solve the equations much more efficiently than is possible using matrix inversion methods

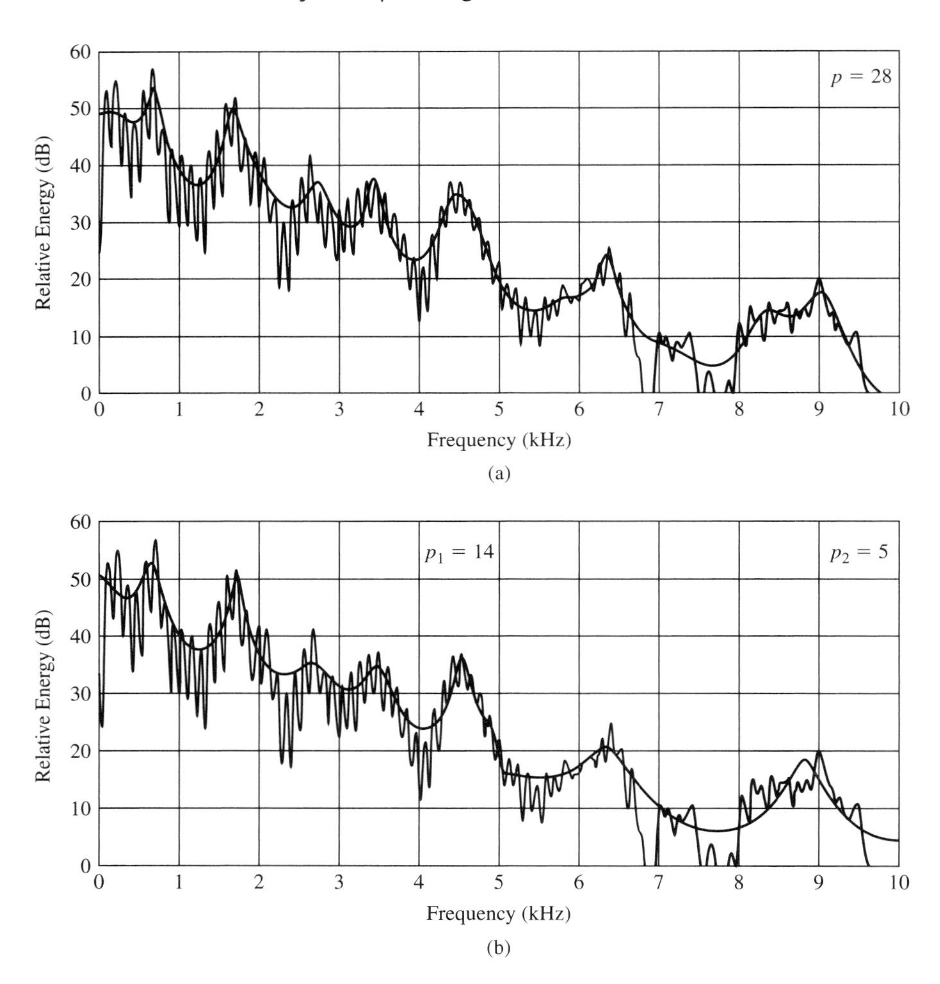

FIGURE 9.16
Application of selective linear prediction: (a) 28-pole LP spectrum. (b) Selective linear prediction with a 14-pole fit to the 0–5 kHz region and a 5-pole fit to the 5–10 kHz region. (After Makhoul [216]. © [1975] IEEE.)

intended for matrices of arbitrary structure. In this section we discuss the details of three methods for obtaining the predictor coefficients, and then we compare and contrast computational requirements of these solutions.

9.5.1 The Cholesky Decomposition

For the covariance method [12], the set of equations that must be solved is of the form:

$$\sum_{k=1}^{p} \alpha_k \phi_{\hat{n}}[i, k] = \phi_{\hat{n}}[i, 0], \qquad i = 1, 2, \dots, p. \tag{9.61}$$

These equations can be represented in matrix notation as

$$\boldsymbol{\Phi}\boldsymbol{\alpha} = \boldsymbol{\psi}, \tag{9.62}$$

where $\boldsymbol{\Phi}$ is a positive-definite symmetric matrix with $(i, j)^{th}$ element $\phi_{\hat{n}}[i, j]$, and $\boldsymbol{\alpha}$ and $\boldsymbol{\psi}$ are column vectors with elements α_i and $\phi_{\hat{n}}[i, 0]$ respectively. As an illustration of the structure of the matrices, consider the case for $p = 4$, which is of the form

$$\begin{bmatrix} \phi_{11} & \phi_{21} & \phi_{31} & \phi_{41} \\ \phi_{21} & \phi_{22} & \phi_{32} & \phi_{42} \\ \phi_{31} & \phi_{32} & \phi_{33} & \phi_{43} \\ \phi_{41} & \phi_{42} & \phi_{43} & \phi_{44} \end{bmatrix} \begin{bmatrix} \alpha_1 \\ \alpha_2 \\ \alpha_3 \\ \alpha_4 \end{bmatrix} = \begin{bmatrix} \psi_1 \\ \psi_2 \\ \psi_3 \\ \psi_4 \end{bmatrix},$$

where, for simplicity of notation, the elements of $\boldsymbol{\Phi}$ are denoted $\phi_{\hat{n}}[i, j] = \phi_{ij}$ and the elements of $\boldsymbol{\psi}$ are denoted $\phi_{\hat{n}}[i, 0] = \psi_i$.

The system of equations given by Eq. (9.61) can be solved efficiently since the matrix $\boldsymbol{\Phi}$ is a symmetric, positive-definite matrix. The resulting method of solution is called the Cholesky decomposition (or sometimes it is called the square root method) [12, 295]. For this method the matrix $\boldsymbol{\Phi}$ is expressed in the form

$$\boldsymbol{\Phi} = \boldsymbol{V}\boldsymbol{D}\boldsymbol{V}^T, \tag{9.63}$$

where $\boldsymbol{V}$ is a lower triangular matrix (whose main diagonal elements are all 1's), and $\boldsymbol{D}$ is a diagonal matrix. The superscript $\boldsymbol{T}$ denotes matrix transpose.[10] For the illustrative case of $p = 4$, Eq. (9.63) takes the form

$$\begin{bmatrix} \phi_{11} & \phi_{21} & \phi_{31} & \phi_{41} \\ \phi_{21} & \phi_{22} & \phi_{32} & \phi_{42} \\ \phi_{31} & \phi_{32} & \phi_{33} & \phi_{43} \\ \phi_{41} & \phi_{42} & \phi_{43} & \phi_{44} \end{bmatrix} = \begin{bmatrix} 1 & 0 & 0 & 0 \\ V_{21} & 1 & 0 & 0 \\ V_{31} & V_{32} & 1 & 0 \\ V_{41} & V_{42} & V_{43} & 1 \end{bmatrix} \begin{bmatrix} d_1 & 0 & 0 & 0 \\ 0 & d_2 & 0 & 0 \\ 0 & 0 & d_3 & 0 \\ 0 & 0 & 0 & d_4 \end{bmatrix} \begin{bmatrix} 1 & V_{21} & V_{31} & V_{41} \\ 0 & 1 & V_{32} & V_{42} \\ 0 & 0 & 1 & V_{43} \\ 0 & 0 & 0 & 1 \end{bmatrix}.$$

By carrying out the matrix multiplications in Eq. (9.63) with the individual matrices expressed in terms of their elements, it can be shown that the diagonal elements ϕ_{ij} of $\boldsymbol{\Phi}$ are related to the elements of $\boldsymbol{D}$ and $\boldsymbol{V}$ by

$$\phi_{ii} = \sum_{k=1}^{i} V_{ik} d_k V_{ik}, \qquad 1 \le i \le p, \tag{9.64}$$

(where $V_{ii} = 1$) and the off-diagonal elements are

$$\phi_{ij} = \sum_{k=1}^{j} V_{ik} d_k V_{jk}, \qquad 2 \le i \le p \text{ and } 1 \le j \le i-1. \tag{9.65}$$

[10] An alternate representation is $\boldsymbol{V}\boldsymbol{V}^T$, where the square roots of the diagonal elements of $\boldsymbol{D}$ in our formulation are incorporated into the diagonal elements of $\boldsymbol{V}$ [295].

From Eq. (9.64) and the fact that the diagonal elements of $\boldsymbol{V}$ are $V_{ii} = 1$ for $1 \leq i \leq p$, it follows that

$$d_1 = \phi_{11}. \tag{9.66}$$

Furthermore, from Eq. (9.65) with $j = 1$, the elements of the first column of $\boldsymbol{\Phi}$ satisfy

$$\phi_{i1} = V_{i1} d_1 V_{11} = V_{i1} d_1, \qquad 2 \leq i \leq p,$$

so

$$V_{i1} = \phi_{i1}/d_1, \qquad 2 \leq i \leq p. \tag{9.67}$$

Using the fact that $V_{jj} = 1$ in Eq. (9.65) allows us to write the diagonal elements of $\boldsymbol{D}$ as,

$$d_i = \phi_{ii} - \sum_{k=1}^{i-1} V_{ik}^2 d_k, \qquad 2 \leq i \leq p, \tag{9.68}$$

and for the remaining off-diagonal elements of $\boldsymbol{V}$,

$$V_{ij} = (\phi_{ij} - \sum_{k=1}^{j-1} V_{ik} d_k V_{jk})/d_j, \quad 2 \leq j \leq i-1. \tag{9.69}$$

Once the matrices $\boldsymbol{V}$ and $\boldsymbol{D}$ have been determined, it is relatively simple to solve for the column vector $\boldsymbol{\alpha}$ in a two-step procedure. From Eqs. (9.62) and (9.63), we get

$$\boldsymbol{V D V^T \alpha} = \boldsymbol{\psi}, \tag{9.70}$$

which can be written as

$$\boldsymbol{V Y} = \boldsymbol{\psi}. \tag{9.71}$$

The example for $p = 4$ would be

$$\begin{bmatrix} 1 & 0 & 0 & 0 \\ V_{21} & 1 & 0 & 0 \\ V_{31} & V_{32} & 1 & 0 \\ V_{41} & V_{42} & V_{43} & 1 \end{bmatrix} \begin{bmatrix} Y_1 \\ Y_2 \\ Y_3 \\ Y_4 \end{bmatrix} = \begin{bmatrix} \psi_1 \\ \psi_2 \\ \psi_3 \\ \psi_4 \end{bmatrix}.$$

From this example it is easily seen that the elements Y_i can be computed using a simple recursion beginning with the initial condition

$$Y_1 = \psi_1, \tag{9.72}$$

and proceeding with

$$Y_i = \psi_i - \sum_{j=1}^{i-1} V_{ij} Y_j, \qquad 2 \leq i \leq p. \tag{9.73}$$

Since, by definition,

$$\boldsymbol{D} \boldsymbol{V}^T \boldsymbol{\alpha} = \boldsymbol{Y}, \tag{9.74}$$

it follows that

$$\boldsymbol{V}^T \boldsymbol{\alpha} = \boldsymbol{D}^{-1} \boldsymbol{Y}. \tag{9.75}$$

The inverse of the diagonal matrix $\boldsymbol{D}$ is easily obtained as the diagonal matrix with reciprocal values $1/d_i$. Again, for the example with $p = 4$, the matrices are

$$\begin{bmatrix} 1 & V_{21} & V_{31} & V_{41} \\ 0 & 1 & V_{32} & V_{42} \\ 0 & 0 & 1 & V_{43} \\ 0 & 0 & 0 & 1 \end{bmatrix} \begin{bmatrix} \alpha_1 \\ \alpha_2 \\ \alpha_3 \\ \alpha_4 \end{bmatrix} = \begin{bmatrix} 1/d_1 & 0 & 0 & 0 \\ 0 & 1/d_2 & 0 & 0 \\ 0 & 0 & 1/d_3 & 0 \\ 0 & 0 & 0 & 1/d_4 \end{bmatrix} \begin{bmatrix} Y_1 \\ Y_2 \\ Y_3 \\ Y_4 \end{bmatrix} = \begin{bmatrix} Y_1/d_1 \\ Y_2/d_2 \\ Y_3/d_3 \\ Y_4/d_4 \end{bmatrix}.$$

Finally, having solved for $\boldsymbol{Y}$ using Eqs. (9.73) and (9.72), Eq. (9.75) can be solved recursively for $\boldsymbol{\alpha}$ beginning with the initial condition

$$\alpha_p = Y_p/d_p, \tag{9.76}$$

and then using the relation

$$\alpha_i = Y_i/d_i - \sum_{j=i+1}^{p} V_{ij} \alpha_j, \quad p - 1 \geq i \geq 1. \tag{9.77}$$

Note that the index i in Eq. (9.77) proceeds backwards from $i = p - 1$ down to $i = 1$.

Equations (9.66), (9.67), (9.68), (9.69), (9.72), (9.73), (9.77), and (9.76), when evaluated in the proper order, are the basis for the Cholesky decomposition method of solving for the prediction coefficients. A pseudo-code description of the complete algorithm incorporating these equations is given in Figure 9.17.

The use of the Cholesky decomposition procedure leads to a very simple expression for the minimum error of the covariance method in terms of the column vector $\boldsymbol{Y}$ and the matrix $\boldsymbol{D}$. We recall that for the covariance method, the prediction error, $\mathcal{E}_{\hat{n}}$, was of the form

$$\mathcal{E}_{\hat{n}} = \phi_{\hat{n}}[0, 0] - \sum_{k=1}^{p} \alpha_k \phi_{\hat{n}}[0, k], \tag{9.78}$$

Cholesky Matrix Inversion Algorithm

```
%%        Find first column of V
```

$d_1 = \phi_{11}$ (9.66)

for $i = 2, 3, \ldots, p$

  $V_{i1} = \phi_{i1}/d_1$ (9.67)

end

%% Find $\boldsymbol{D}$ and remaining columns of $\boldsymbol{V}$

for $j = 2, 3, \ldots, p-1$

$$d_j = \phi_{jj} - \sum_{k=1}^{j-1} V_{jk}^2 d_k \quad (9.68)$$

  for $i = j+1, \ldots, p$

$$V_{ij} = (\phi_{ij} - \sum_{k=1}^{j-1} V_{ik} d_k V_{jk})/d_j \quad (9.69)$$

  end

end

$$d_p = \phi_{pp} - \sum_{k=1}^{p-1} V_{pk}^2 d_k \quad (9.68)$$

%% Find $\boldsymbol{Y} = \boldsymbol{D}\boldsymbol{V}^T\boldsymbol{\alpha}$

$Y_1 = \psi_1$ (9.72)

for $i = 2, 3, \ldots, p$

$$Y_i = \psi_i - \sum_{j=1}^{i-1} V_{ij} Y_j \quad (9.73)$$

end

%% Find $\boldsymbol{\alpha}$ from $\boldsymbol{Y}$

$\alpha_p = Y_p/d_p$ (9.76)

for $i = p-1, p-2, \ldots, 1$

$$\alpha_i = Y_i/d_i - \sum_{j=i+1}^{p} V_{ij}\alpha_j \quad (9.77)$$

end

FIGURE 9.17
Pseudo-code description of the Cholesky matrix inversion algorithm.

or in matrix notation

$$\mathcal{E}_{\hat{n}} = \phi_{\hat{n}}[0, 0] - \boldsymbol{\alpha}^T \boldsymbol{\psi}. \quad (9.79)$$

From Eq. (9.75) we can substitute for $\boldsymbol{\alpha}^T$ the expression $\boldsymbol{Y}^T\boldsymbol{D}^{-1}\boldsymbol{V}^{-1}$, giving

$$\mathcal{E}_{\hat{n}} = \phi_{\hat{n}}[0, 0] - \boldsymbol{Y}^T\boldsymbol{D}^{-1}\boldsymbol{V}^{-1}\boldsymbol{\psi}. \quad (9.80)$$

Using Eq. (9.71) we get

$$\mathcal{E}_{\hat{n}} = \phi_{\hat{n}}[0,0] - \boldsymbol{Y}^T\boldsymbol{D}^{-1}\boldsymbol{Y}, \tag{9.81}$$

or

$$\mathcal{E}_{\hat{n}} = \phi_{\hat{n}}[0,0] - \sum_{k=1}^{p} Y_k^2/d_k. \tag{9.82}$$

Thus the mean-squared prediction error $\mathcal{E}_{\hat{n}}$ can be determined directly from the column vector $\boldsymbol{Y}$ and the matrix $\boldsymbol{D}$. Furthermore Eq. (9.82) can be used to give the value of $\mathcal{E}_{\hat{n}}$ for any value of p up to the value of p used in solving the matrix equations. Thus one can get an idea as to how the mean-squared prediction error varies with the number of predictor coefficients used in the solution.

9.5.2 The Levinson–Durbin Algorithm

For the autocorrelation method [217, 232], the set of equations satisfied by the predictor coefficients is of the form

$$\sum_{k=1}^{p} \alpha_k R[|i-k|] = R[i], \qquad 1 \le i \le p, \tag{9.83}$$

where we have again dropped the subscript $\hat{n}$ for simplicity of notation. It should be understood that the equations in Eq. (9.83) must be solved at each analysis time $\hat{n}$. These equations can be represented in matrix form as

$$\boldsymbol{R}\boldsymbol{\alpha} = \boldsymbol{r}, \tag{9.84}$$

where $\boldsymbol{R}$ is a positive-definite symmetric Toeplitz matrix with $(i,j)^{th}$ element $R[|i-j|]$, and $\boldsymbol{\alpha}$ and $\boldsymbol{r}$ are column vectors with elements α_i, and $r[i] = R[i]$ respectively. By exploiting the Toeplitz nature of the matrix of coefficients, several efficient recursive procedures have been devised for solving this system of equations. The most popular and well known of these methods are the Levinson and Robinson algorithms [232] and the Levinson–Durbin recursive algorithm [217], which we shall now derive.

From Eq. (9.83), the optimum predictor coefficients can be seen to satisfy the set of equations

$$R[i] - \sum_{k=1}^{p} \alpha_k R[|i-k|] = 0, \qquad i = 1, 2, \ldots, p. \tag{9.85a}$$

Furthermore, from Eq. (9.22b), the minimum mean-squared prediction error for a pth-order optimum predictor is given by

$$R[0] - \sum_{k=1}^{p} \alpha_k R[k] = \mathcal{E}^{(p)}. \tag{9.85b}$$

Since Eq. (9.85b) contains the same correlation values as in Eq. (9.85a), it is possible to take them together and write a new set of $p+1$ equations that are satisfied by the p unknown predictor coefficients and the corresponding unknown mean-squared prediction error $\mathcal{E}^{(p)}$. These equations have the form

$$\begin{bmatrix} R[0] & R[1] & R[2] & \cdots & R[p] \\ R[1] & R[0] & R[1] & \cdots & R[p-1] \\ R[2] & R[1] & R[0] & \cdots & R[p-2] \\ \vdots & \vdots & \vdots & \cdots & \vdots \\ R[p] & R[p-1] & R[p-2] & \cdots & R[0] \end{bmatrix} \begin{bmatrix} 1 \\ -\alpha_1^{(p)} \\ -\alpha_2^{(p)} \\ \vdots \\ -\alpha_p^{(p)} \end{bmatrix} = \begin{bmatrix} \mathcal{E}^{(p)} \\ 0 \\ 0 \\ \vdots \\ 0 \end{bmatrix}, \tag{9.86}$$

where we observe that the $(p+1)\times(p+1)$ matrix that we have constructed is also a Toeplitz matrix. It is this set of equations that can be solved recursively by the Levinson–Durbin algorithm. This is done by successively incorporating a new correlation value at each iteration and solving for the next higher-order predictor in terms of the new correlation value and the previously found predictor.

For any order i, the set of equations in Eq. (9.86) can be represented in matrix notation as

$$\boldsymbol{R}^{(i)}\boldsymbol{\alpha}^{(i)} = \boldsymbol{e}^{(i)}. \tag{9.87}$$

We wish to show how the ith solution can be derived from the $(i-1)^{st}$ solution. In other words, given $\boldsymbol{\alpha}^{(i-1)}$, the solution to $\boldsymbol{R}^{(i-1)}\boldsymbol{\alpha}^{(i-1)} = \boldsymbol{e}^{(i-1)}$, we wish to derive the solution to $\boldsymbol{R}^{(i)}\boldsymbol{\alpha}^{(i)} = \boldsymbol{e}^{(i)}$.

First write the equations $\boldsymbol{R}^{(i-1)}\boldsymbol{\alpha}^{(i-1)} = \boldsymbol{e}^{(i-1)}$ in expanded form as

$$\begin{bmatrix} R[0] & R[1] & R[2] & \cdots & R[i-1] \\ R[1] & R[0] & R[1] & \cdots & R[i-2] \\ R[2] & R[1] & R[0] & \cdots & R[i-3] \\ \vdots & \vdots & \vdots & \cdots & \vdots \\ R[i-1] & R[i-2] & R[i-3] & \cdots & R[0] \end{bmatrix} \begin{bmatrix} 1 \\ -\alpha_1^{(i-1)} \\ -\alpha_2^{(i-1)} \\ \vdots \\ -\alpha_{i-1}^{(i-1)} \end{bmatrix} = \begin{bmatrix} \mathcal{E}^{(i-1)} \\ 0 \\ 0 \\ \vdots \\ 0 \end{bmatrix}. \tag{9.88}$$

Next we append a 0 to the vector $\boldsymbol{\alpha}^{(i-1)}$ and multiply by the matrix $\boldsymbol{R}^{(i)}$ to obtain the new set of $(i+1)$ equations:

$$\begin{bmatrix} R[0] & R[1] & R[2] & \cdots & R[i] \\ R[1] & R[0] & R[1] & \cdots & R[i-1] \\ R[2] & R[1] & R[0] & \cdots & R[i-2] \\ \vdots & \vdots & \vdots & \cdots & \vdots \\ R[i-1] & R[i-2] & R[i-3] & \cdots & R[1] \\ R[i] & R[i-1] & R[i-2] & \cdots & R[0] \end{bmatrix} \begin{bmatrix} 1 \\ -\alpha_1^{(i-1)} \\ -\alpha_2^{(i-1)} \\ \vdots \\ -\alpha_{i-1}^{(i-1)} \\ 0 \end{bmatrix} = \begin{bmatrix} \mathcal{E}^{(i-1)} \\ 0 \\ 0 \\ \vdots \\ 0 \\ \gamma^{(i-1)} \end{bmatrix}, \tag{9.89}$$

where, in order to satisfy Eq. (9.89), it must be true that

$$\gamma^{(i-1)} = R[i] - \sum_{j=1}^{i-1} \alpha_j^{(i-1)} R[i-j]. \tag{9.90}$$

It is in Eq. (9.90) that the new autocorrelation value $R[i]$ is introduced. However, Eq. (9.89) is not yet in the desired form $\boldsymbol{R}^{(i)}\boldsymbol{\alpha}^{(i)} = \boldsymbol{e}^{(i)}$. The key step in the derivation is to recognize that due to the special symmetry of the Toeplitz matrix $\boldsymbol{R}^{(i)}$, the equations can be written in reverse order (first equation last and last equation first, etc.) and the matrix for the resulting set of equations is still $\boldsymbol{R}^{(i)}$; i.e.,

$$\begin{bmatrix} R[0] & R[1] & R[2] & \cdots & R[i] \\ R[1] & R[0] & R[1] & \cdots & R[i-1] \\ R[2] & R[1] & R[0] & \cdots & R[i-2] \\ \vdots & \vdots & \vdots & \cdots & \vdots \\ R[i-1] & R[i-2] & R[i-3] & \cdots & R[1] \\ R[i] & R[i-1] & R[i-2] & \cdots & R[0] \end{bmatrix} \begin{bmatrix} 0 \\ -\alpha_{i-1}^{(i-1)} \\ -\alpha_{i-2}^{(i-1)} \\ \vdots \\ -\alpha_1^{(i-1)} \\ 1 \end{bmatrix} = \begin{bmatrix} \gamma^{(i-1)} \\ 0 \\ 0 \\ \vdots \\ 0 \\ \mathcal{E}^{(i-1)} \end{bmatrix}. \tag{9.91}$$

Now Eq. (9.89) is combined with Eq. (9.91) according to

$$\boldsymbol{R}^{(i)} \cdot \left[\begin{bmatrix} 1 \\ -\alpha_1^{(i-1)} \\ -\alpha_2^{(i-1)} \\ \vdots \\ -\alpha_{i-1}^{(i-1)} \\ 0 \end{bmatrix} - k_i \begin{bmatrix} 0 \\ -\alpha_{i-1}^{(i-1)} \\ -\alpha_{i-2}^{(i-1)} \\ \vdots \\ -\alpha_1^{(i-1)} \\ 1 \end{bmatrix} \right] = \left[\begin{bmatrix} \mathcal{E}^{(i-1)} \\ 0 \\ 0 \\ \vdots \\ 0 \\ \gamma^{(i-1)} \end{bmatrix} - k_i \begin{bmatrix} \gamma^{(i-1)} \\ 0 \\ 0 \\ \vdots \\ 0 \\ \mathcal{E}^{(i-1)} \end{bmatrix} \right]. \tag{9.92}$$

Equation (9.92) is now approaching the desired form $\mathbf{R}^{(i)}\boldsymbol{\alpha}^{(i)} = \mathbf{e}^{(i)}$. All that remains is to choose $\gamma^{(i-1)}$ so that the vector on the right has only a single non-zero entry. This requires that the new parameter k_i be chosen as

$$k_i = \frac{\gamma^{(i-1)}}{\mathcal{E}^{(i-1)}} = \frac{R[i] - \sum_{j=1}^{i-1} \alpha_j^{(i-1)} R[i-j]}{\mathcal{E}^{(i-1)}}, \tag{9.93}$$

which ensures cancellation of the last element of the right-hand side vector, and causes the first element to be

$$\mathcal{E}^{(i)} = \mathcal{E}^{(i-1)} - k_i \gamma^{(i-1)} = \mathcal{E}^{(i-1)}(1 - k_i^2). \tag{9.94}$$

For reasons that will be discussed in Section 9.5.3, the parameters, k_i, that arise in the Levinson–Durbin recursive algorithm are called PARCOR (partial correlation) coefficients. They will assume an exceedingly important role in linear predictive analysis.

With this choice of $\gamma^{(i-1)}$, it follows that the vector of ith-order prediction coefficients is

$$\begin{bmatrix} 1 \\ -\alpha_1^{(i)} \\ -\alpha_2^{(i)} \\ \vdots \\ -\alpha_{i-1}^{(i)} \\ -\alpha_i^{(i)} \end{bmatrix} = \begin{bmatrix} 1 \\ -\alpha_1^{(i-1)} \\ -\alpha_2^{(i-1)} \\ \vdots \\ -\alpha_{i-1}^{(i-1)} \\ 0 \end{bmatrix} - k_i \begin{bmatrix} 0 \\ -\alpha_{i-1}^{(i-1)} \\ -\alpha_{i-2}^{(i-1)} \\ \vdots \\ -\alpha_1^{(i-1)} \\ 1 \end{bmatrix}. \tag{9.95}$$

From Eq. (9.95) we can write the set of equations for updating the coefficients as

$$\alpha_j^{(i)} = \alpha_j^{(i-1)} - k_i \alpha_{i-j}^{(i-1)}, \qquad j = 1, 2, \dots, i-1, \tag{9.96a}$$

and

$$\alpha_i^{(i)} = k_i. \tag{9.96b}$$

For a particular order, p, the optimum predictor coefficients are, therefore,

$$\alpha_j = \alpha_j^{(p)}, \qquad j = 1, 2, \dots, p. \tag{9.97}$$

Equations (9.93), (9.96b), (9.96a), (9.94), and (9.97) are the key equations of the Levinson–Durbin algorithm. They are the basis for the recursive algorithm represented in pseudo-code in Figure 9.18, which shows how they are used order-recursively to compute the optimum prediction coefficients as well as the corresponding mean-squared prediction errors and coefficients, k_i, for all linear predictors up to order p.

Note that in the process of solving for the predictor coefficients for a predictor of order p, the solutions for the predictor coefficients of all orders $i < p$ are also obtained; i.e., $\alpha_j^{(i)}$ is the jth predictor coefficient for a predictor of order i. Furthermore, the quantity $\mathcal{E}^{(i)}$ in Eq. (9.94) is the prediction error for a predictor of order i in terms of $\mathcal{E}^{(i-1)}$. The case $i = 0$, which corresponds to no prediction at all, has a total squared prediction error

$$\mathcal{E}^{(0)} = R[0]. \tag{9.98}$$

At each stage of the computation, the total prediction error energy for a predictor of order i is computed as part of the solution for the optimum predictor coefficients.

Levinson–Durbin Algorithm

$\mathcal{E}^{(0)} = R[0]$ (9.98)

for $i = 1, 2, \ldots, p$

$$k_i = \left(R[i] - \sum_{j=1}^{i-1} \alpha_j^{(i-1)} R[i-j] \right) / \mathcal{E}^{(i-1)} \quad (9.93)$$

$\alpha_i^{(i)} = k_i$ (9.96b)

if $i > 1$ then for $j = 1, 2, \ldots, i-1$

$\alpha_j^{(i)} = \alpha_j^{(i-1)} - k_i \alpha_{i-j}^{(i-1)}$ (9.96a)

end

$\mathcal{E}^{(i)} = (1 - k_i^2)\mathcal{E}^{(i-1)}$ (9.94)

end

$\alpha_j = \alpha_j^{(p)} \quad j = 1, 2, \ldots, p$ (9.97)

FIGURE 9.18
Pseudo-code description of the Levinson–Durbin algorithm.

Using Eq. (9.85b), $\mathcal{E}^{(i)}$ for a predictor of order i can be expressed as

$$\mathcal{E}^{(i)} = R[0] - \sum_{k=1}^{i} \alpha_k^{(i)} R[k], \quad (9.99a)$$

or, using Eqs. (9.94) and (9.98), starting with $i = 1$ and working recursively,

$$\mathcal{E}^{(i)} = R[0] \prod_{m=1}^{i} (1 - k_m^2). \quad (9.99b)$$

Also, if the autocorrelation coefficients $R[i]$ are replaced by a set of normalized autocorrelation coefficients, i.e., $r[k] = R[k]/R[0]$, then the solution to the matrix equation remains unchanged. However, the error $\mathcal{E}^{(i)}$ is now interpreted as a normalized error. If we call this normalized error $\mathcal{V}^{(i)}$, then from Eq. (9.99a),

$$\mathcal{V}^{(i)} = \frac{\mathcal{E}^{(i)}}{R[0]} = 1 - \sum_{k=1}^{i} \alpha_k^{(i)} r[k], \quad (9.100)$$

from which it follows that

$$0 < \mathcal{V}^{(i)} \leq 1, \qquad i \geq 0. \quad (9.101)$$

Also, from Eq. (9.99b), $\mathcal{V}^{(p)}$ can be written in the form

$$\mathcal{V}^{(p)} = \prod_{i=1}^{p}(1 - k_i^2), \qquad (9.102)$$

from which it follows that the PARCOR coefficients k_i are in the range

$$-1 < k_i < 1, \qquad (9.103)$$

since we have shown that for the autocorrelation method, $\mathcal{V}^{(p)} = \mathcal{E}^{(p)}/R[0] > 0$ for any value of p. We can use Eq. (9.102) to determine the reduction in normalized error, as a function of predictor order, from the PARCOR coefficients, or equivalently, to determine the set of PARCOR coefficients from the sequence of normalized errors.

The k_i parameters are called by different names in different implementations of the linear prediction method, including reflection coefficients and PARCOR (partial correlation) coefficients. We will see why they are called PARCOR coefficients in the next section when we discuss lattice implementations of the linear predictive analysis.

The inequality of Eq. (9.103) on the parameters, k_i, is important since it can be shown [145, 232] that it is a necessary and sufficient condition for all of the roots of the polynomial $A(z)$ to be inside the unit circle, thereby guaranteeing the stability of the system $H(z)$. A proof of this result will be given later in this chapter. Furthermore, it is possible to show that no such guarantee of stability is available in the covariance method.

9.5.3 Lattice Formulations and Solutions

As we have seen, both the covariance and the autocorrelation methods consist of two steps:

1. computation of a matrix of correlation values
2. solution of a set of linear equations.

These methods have been widely used with great success in speech processing applications. However, another class of linear predictive analysis methods, called *lattice methods* [219], has evolved in which the above two steps are, in a sense, combined into a recursive algorithm for determining the linear predictor parameters directly from the sampled speech signal. To see how these methods are related, it is helpful to begin with the Durbin algorithm solution. First, recall that at the ith-stage of this procedure, the set of coefficients $\{\alpha_j^{(i)}, j = 1, 2, \ldots, i\}$ are the coefficients of the ith-order optimum linear predictor. Using these coefficients, we can define

$$A^{(i)}(z) = 1 - \sum_{k=1}^{i} \alpha_k^{(i)} z^{-k} \qquad (9.104)$$

to be the system function of the ith-order inverse filter (or prediction error filter). If the input to this filter is the segment of the signal, $s_{\hat{n}}[m] = s[\hat{n}+m]w[m]$, then the output would be the prediction error, $e_{\hat{n}}^{(i)}[m] = e^{(i)}[\hat{n}+m]$, where, again dropping the subscript $\hat{n}$ for simplicity of notation,

$$e^{(i)}[m] = s[m] - \sum_{k=1}^{i} \alpha_k^{(i)} s[m-k], \tag{9.105}$$

where henceforth $s[m]$ represents the windowed segment $s_{\hat{n}}[m] = s[\hat{n}+m]w[m]$ since we have suppressed the indication of analysis time. The sequence $e^{(i)}[m]$ is henceforth referred to as the *forward prediction error*, since it is the error in predicting $s[m]$ from i previous samples. In terms of z-transforms, Eq. (9.105) becomes

$$E^{(i)}(z) = A^{(i)}(z)S(z) = \left(1 - \sum_{k=1}^{i} \alpha_k^{(i)} z^{-k}\right) S(z). \tag{9.106}$$

By substituting Eqs. (9.96a) and (9.96b) into Eq. (9.104), we obtain a recurrence formula for $A^{(i)}(z)$ in terms of $A^{(i-1)}(z)$; i.e.,

$$A^{(i)}(z) = A^{(i-1)}(z) - k_i z^{-i} A^{(i-1)}(z^{-1}). \tag{9.107}$$

(See Problem 9.10.) Substituting Eq. (9.107) into Eq. (9.106) results in

$$E^{(i)}(z) = A^{(i-1)}(z)S(z) - k_i z^{-i} A^{(i-1)}(z^{-1})S(z). \tag{9.108}$$

The first term on the right in Eq. (9.108) is obviously the z-transform of the prediction error for an $(i-1)^{st}$-order predictor. The second term can be given a similar interpretation (i.e., as a type of prediction error) if we define the z-transform of a "backward" prediction error $b^{(i)}[m]$ as

$$B^{(i)}(z) = z^{-i} A^{(i)}(z^{-1})S(z). \tag{9.109}$$

It is easily shown that the inverse z-transform of $B^{(i)}(z)$ is

$$b^{(i)}[m] = s[m-i] - \sum_{k=1}^{i} \alpha_k^{(i)} s[m+k-i], \tag{9.110}$$

which we can interpret as predicting sample $s[m-i]$ from the i succeeding samples $\{s[m-i+k], k = 1, 2, \dots, i\}$. It is in this sense that $b^{(i)}[m]$ is the *backward prediction* error sequence. Figure 9.19 shows that the samples involved in computing the forward and backward prediction errors are the same samples. Now, returning to Eq. (9.108), we see that we can express $E^{(i)}(z)$ in terms of $E^{(i-1)}(z)$ and $B^{(i-1)}(z)$ as

$$E^{(i)}(z) = E^{(i-1)}(z) - k_i z^{-1} B^{(i-1)}(z). \tag{9.111}$$

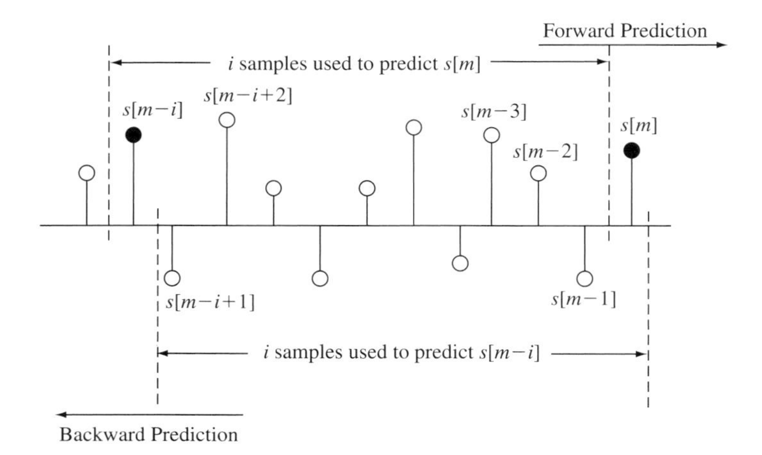

FIGURE 9.19
Illustration of forward and backward prediction using an ith-order predictor.

Therefore, the prediction error sequence $e^{(i)}[m]$ can be expressed as

$$e^{(i)}[m] = e^{(i-1)}[m] - k_i b^{(i-1)}[m-1]. \tag{9.112}$$

Likewise, by substituting Eq. (9.107) into Eq. (9.109), we obtain

$$B^{(i)}(z) = z^{-i}A^{(i-1)}(z^{-1})S(z) - k_i A^{(i-1)}(z)S(z), \tag{9.113}$$

or

$$B^{(i)}(z) = z^{-1}B^{(i-1)}(z) - k_i E^{(i-1)}(z). \tag{9.114}$$

Thus the ith-stage backward prediction error is

$$b^{(i)}[m] = b^{(i-1)}[m-1] - k_i e^{(i-1)}[m]. \tag{9.115}$$

Now Eqs. (9.112) and (9.115) define the forward and backward prediction error sequences for an ith-order predictor in terms of the corresponding prediction errors of an $(i-1)^{th}$-order predictor. Using a zeroth-order predictor is equivalent to using no predictor at all, so we can define the zeroth-order forward and backward prediction error as

$$e^{(0)}[m] = b^{(0)}[m] = s[m], \qquad 0 \leq m \leq L-1. \tag{9.116}$$

If we know k_1 (from the Levinson–Durbin computation), we can compute $e^{(1)}[m]$ and $b^{(1)}[m]$ from $s[m]$ for $0 \leq m \leq L$ using Eqs. (9.112) and (9.115). Then, given $e^{(1)}[m]$

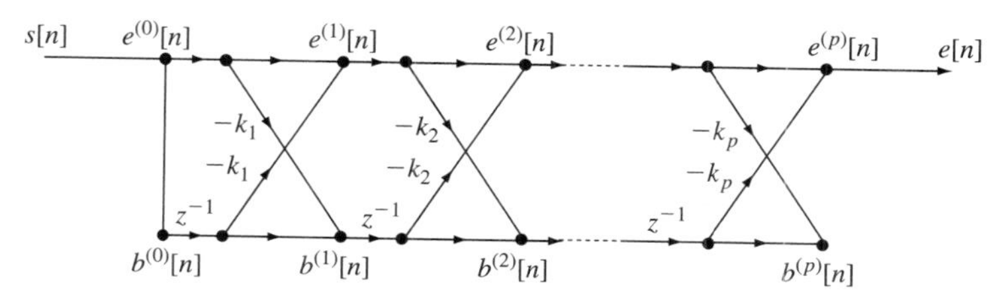

FIGURE 9.20
Signal flow graph of lattice implementation of the prediction error filter $A(z)$.

and $b^{(1)}[m]$, Eqs. (9.112) and (9.115) can be used to compute $e^{(2)}[m]$ and $b^{(2)}[m]$ for $0 \leq m \leq L+1$.[11] This process is depicted in Figure 9.20. Such a structure is called a lattice network. It is clear that if we extend the lattice to p sections, the output of the last upper branch, $e^{(p)}[n]$, will be the forward prediction error $e[n]$ as shown in Figure 9.20. Therefore, Figure 9.20 is a digital network lattice filter implementation of the prediction error filter with transfer function $A(z)$. Instead of the predictor coefficients, however, the coefficients in the lattice filter implementation are the as yet unspecified coefficients k_i.

In summary, the difference equations represented by Figure 9.20 are

$$e^{(0)}[n] = b^{(0)}[n] = s[n], \qquad 0 \leq n \leq L-1, \tag{9.117a}$$

$$e^{(i)}[n] = e^{(i-1)}[n] - k_i b^{(i-1)}[n-1], \qquad 1 \leq i \leq p, \quad 0 \leq n \leq L-1+i, \tag{9.117b}$$

$$b^{(i)}[n] = b^{(i-1)}[n-1] - k_i e^{(i-1)}[n], \qquad 1 \leq i \leq p, \quad 0 \leq n \leq L-1+i, \tag{9.117c}$$

$$e[n] = e^{(p)}[n], \qquad 0 \leq n \leq L-1+p. \tag{9.117d}$$

In computing the output at time n, Eq. (9.117b) is computed first and then Eq. (9.117c) is computed, thereby updating the value of the delayed backward prediction error. This is done for each i up to p. The output is then given by Eq. (9.117d). The cycle depicted in Figure 9.20 is repeated for $n+1, n+2, \ldots$, etc.

All-Pole Lattice Structure

A lattice structure for implementing the all-pole system function $H(z)$ of our discrete-time model for speech production can be developed from the FIR prediction error

[11]For subsequent iterations, the length of the error sequences increases by 1 each time the linear prediction analysis order goes up by 1.

lattice by recognizing that $H(z) = 1/A(z)$; i.e., $H(z)$ is the inverse filter for the FIR system function $A(z)$, and vice versa. To derive the all-pole lattice structure, assume that we are given $e[n]$ and we wish to compute the input $s[n]$. This can be done by working from right to left in Figure 9.20 to invert the computations. More specifically, if we solve Eq. (9.117b) for $e^{(i-1)}[n]$ in terms of $e^{(i)}[n]$ and $b^{(i-1)}[n]$ and leave Eq. (9.117c) as is, we obtain the pair of equations

$$e^{(i-1)}[n] = e^{(i)}[n] + k_i b^{(i-1)}[n-1], \qquad i = 1, 2, \dots, p, \tag{9.118a}$$

$$b^{(i)}[n] = b^{(i-1)}[n-1] - k_i e^{(i-1)}[n], \quad i = 1, 2, \dots, p, \tag{9.118b}$$

which have the flow graph representation shown in Figure 9.21. Note that in this case, the signal flow is from i to $i-1$ along the top of the diagram and from $i-1$ to i along the bottom. Successive connections of p stages of Figure 9.21 with the appropriate k_i in each section takes the input $e^{(p)}[n]$ to the output $e^{(0)}[n]$ as shown in the flow graph of Figure 9.22. Finally, the condition $s[n] = e^{(0)}[n] = b^{(0)}[n]$ at the output of the last stage in Figure 9.22 causes a feedback connection that provides the sequences $b^{(i)}[n]$ that propagate in the reverse direction. Such feedback is, of course, necessary for an IIR system.

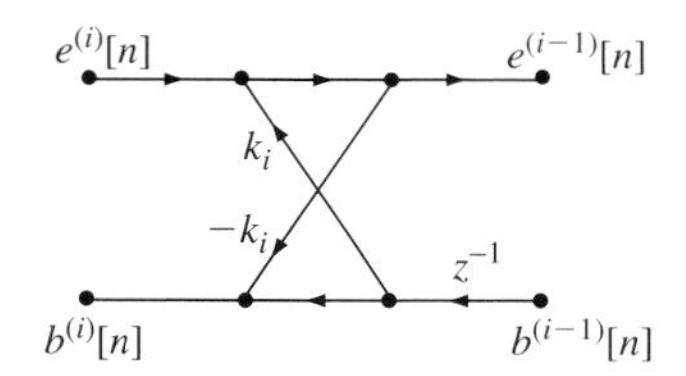

FIGURE 9.21
One stage of computation for inverting the lattice system in Figure 9.20.

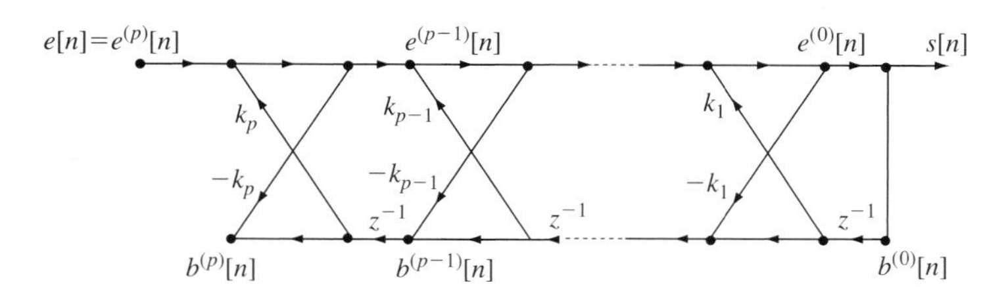

FIGURE 9.22
Signal flow graph representation of an all-pole lattice filter for $H(z) = 1/A(z)$.

The set of difference equations represented by Figure 9.22 is

$$e^{(p)}[n] = e[n], \qquad 0 \le n \le L-1+p, \tag{9.119a}$$

$$e^{(i-1)}[n] = e^{(i)}[n] + k_i b^{(i-1)}[n-1], \qquad i = p, p-1, \dots, 1, \quad 0 \le n \le L-1+i-1, \tag{9.119b}$$

$$b^{(i)}[n] = b^{(i-1)}[n-1] - k_i e^{(i-1)}[n], \qquad i = p, p-1, \dots, 1, \quad 0 \le n \le L-1+i, \tag{9.119c}$$

$$s[n] = e^{(0)}[n] = b^{(0)}[n], \qquad 0 \le n \le L-1. \tag{9.119d}$$

Because of the feedback inherent in Figure 9.22 and the corresponding equations, initial conditions must be specified for all of the node variables associated with delays. Typically, we would specify $b^{(i)}[-1] = 0$ for initial rest conditions. Then, if Eq. (9.119b) is evaluated first, $e^{(i-1)}[n]$ will be available at times $n \ge 0$ for the evaluation of Eq. (9.119c), with the values of $b^{(i-1)}[n-1]$ having been provided by the previous iteration. That is, at a particular time n, Eq. (9.119a) is computed first. Then Eq. (9.119b) for $i = p$ would be used to compute $e^{(p-1)}[n]$ from $e^{(p)}[n]$ and $b^{(i-1)}[n-1]$ from the previous iteration at time $n-1$. Then $b^{(p)}[n]$ can be computed from Eq. (9.119c) using $e^{(p-1)}[n]$ just computed and $b^{(p-1)}[n-1]$ from the previous iteration. This same sequence is then computed for $i = p-1, p-2, \dots$, etc., ending with $s[n] = e^{(0)}[n]$.

Direct Computation of the k_i Parameters

Itakura [161, 162] has shown that, because of the nature of the lattice structure of Figure 9.20, the entire set of coefficients $\{k_i, i = 1, 2, \dots, p\}$ can be computed without computing the predictor coefficients using the Levinson–Durbin algorithm. This is not surprising since the lattice filter structure is an embodiment of the Levinson–Durbin algorithm. There are several possible ways of computing the k_i coefficients, e.g., by minimization of the forward prediction error, by minimization of the backward prediction error, or by some combination of the two minimizations. We begin with the determination of the k_i parameters by minimizing the forward prediction error.

In using the lattice filter structure to compute the k-parameters, we assume as in the autocorrelation method that the signal $s[n]$ is non-zero only in the interval $0 \le n \le L-1$. With this assumption, and the assumption that $k_1, k_2, \dots, k_{i-1}$ have been chosen to minimize the total energies of the forward prediction errors $e^{(1)}[n], e^{(2)}[n], \dots, e^{(i-1)}[n]$, the total energy of the ith-order forward prediction error can be computed as:

$$\mathcal{E}^{(i)} = \sum_{m=0}^{L-1+i} \left[e^{(i)}[m]\right]^2 \tag{9.120}$$

$$= \sum_{m=0}^{L-1+i} \left[e^{(i-1)}[m] - k_i b^{(i-1)}[m-1]\right]^2. \tag{9.121}$$

We can determine the value of k_i that minimizes the total energy of the forward prediction error $e^{(i)}[n]$ by differentiating Eq. (9.121) with respect to the parameter k_i and setting the result equal to zero as follows:

$$\frac{\partial \mathcal{E}^{(i)}}{\partial k_i} = 0 = -2 \sum_{m=0}^{L-1+i} \left[e^{(i-1)}[m] - k_i b^{(i-1)}[m-1] \right] b^{(i-1)}[m-1],$$

$$k_i = \frac{\displaystyle\sum_{m=0}^{L-1+i} \left[e^{(i-1)}[m] \cdot b^{(i-1)}[m-1] \right]}{\displaystyle\sum_{m=0}^{L-1+i} \left[b^{(i-1)}[m-1] \right]^2}. \tag{9.122}$$

In a similar manner we can choose to compute the k_i parameters of the lattice network by minimizing the total energy of the backward prediction error via the following computations:

$$\tilde{\mathcal{E}}^{(i)} = \sum_{m=0}^{L-1+i} \left[b^{(i)}[m] \right]^2 = \sum_{m=0}^{L-1+i} \left[-\tilde{k}_i e^{(i-1)}[m] + b^{(i-1)}[m-1] \right]^2,$$

$$\frac{\partial \tilde{\mathcal{E}}^{(i)}}{\partial \tilde{k}_i} = 0 = -2 \sum_{m=0}^{L-1+i} \left[-\tilde{k}_i e^{(i-1)}[m] + b^{(i-1)}[m-1] \right] e^{(i-1)}[m-1],$$

$$\tilde{k}_i = \frac{\displaystyle\sum_{m=0}^{L-1+i} \left[e^{(i-1)}[m] \cdot b^{(i-1)}[m-1] \right]}{\displaystyle\sum_{m=0}^{L-1+i} \left[e^{(i-1)}[m-1] \right]^2}, \tag{9.123}$$

where $\tilde{k}_i$ is the coefficient that minimizes the total energy of the backward prediction error. It is relatively straightforward to show that the total forward and backward error energies are the same; i.e.,

$$\sum_{m=0}^{L-1+i} \left[e^{(i-1)}[m] \right]^2 = \sum_{m=0}^{L-1+i} \left[b^{(i-1)}[m-1] \right]^2. \tag{9.124}$$

Since the numerators of Eqs. (9.122) and (9.123) are identical, it must be true that $k_i = \tilde{k}_i$. Therefore, we can compute the k_i parameters using either Eq. (9.122) or Eq. (9.123). The expression for k_i can be given a more symmetrical representation as a normalized cross-correlation by combining Eqs. (9.122) and (9.123), giving k_i as the geometric mean:

$$k_i = \frac{\sum_{m=0}^{L-1+i} e^{(i-1)}[m] \cdot b^{(i-1)}[m-1]}{\left\{ \sum_{m=0}^{L-1+i} \left[e^{(i-1)}[m]\right]^2 \sum_{m=0}^{L-1+i} \left[b^{(i-1)}[m-1]\right]^2 \right\}^{\frac{1}{2}}}. \quad (9.125)$$

This is the form of the PARCOR coefficient equation originally given by Makhoul [219] and Itakura [161, 162]. Indeed, it is this expression that motivates the designation PARCOR for partial correlation, since it shows that the lattice structure systematically removes part of the correlation from the input stage-by-stage.

The Burg Method

Another way of computing the k_i parameters is due to Burg [48], who proposed minimizing the *sum* of the lattice filter forward and backward prediction errors over a fixed interval as in the covariance method. This leads to the following equation:

$$\begin{aligned}\hat{\mathcal{E}}^{(i)} &= \sum_{m=0}^{L-1} \left[\left(e^{(i)}[m]\right)^2 + \left(b^{(i)}[m]\right)^2 \right] \\ &= \sum_{m=0}^{L-1} \left(e^{(i-1)}[m] - \hat{k}_i b^{(i-1)}[m-1]\right)^2 \\ &\quad + \sum_{m=0}^{L-1} \left(-\hat{k}_i e^{(i-1)}[m] + b^{(i-1)}[m-1]\right)^2, \end{aligned} \quad (9.126)$$

which is minimized with respect to $\hat{k}_i$ by differentiating and setting the result to 0, giving the following equation:

$$\begin{aligned}\frac{\partial \hat{\mathcal{E}}^{(i)}}{\partial \hat{k}_i} = 0 = &-2 \sum_{m=0}^{L-1} \left(e^{(i-1)}[m] - \hat{k}_i b^{(i-1)}[m-1]\right) b^{(i-1)}[m-1] \\ &-2 \sum_{m=0}^{L-1} \left(-\hat{k}_i e^{(i-1)}[m] + b^{(i-1)}[m-1]\right) e^{(i-1)}[m]. \end{aligned} \quad (9.127)$$

Solving Eq. (9.127) for $\hat{k}_i$) gives

$$\hat{k}_i = \frac{2\sum_{m=0}^{L-1}\left(e^{(i-1)}[m]\cdot b^{(i-1)}[m-1]\right)}{\sum_{m=0}^{L-1}\left(e^{(i-1)}[m]\right)^2 + \sum_{m=0}^{L-1}\left(b^{(i-1)}[m-1]\right)^2}. \tag{9.128}$$

By considering the relation

$$\sum_{m=0}^{L-1}\left(e^{(i-1)}[m] - b^{(i-1)}[m-1]\right)^2 \geq 0,$$

it is straightforward to show that the $\hat{k}_i$ parameters obtained from Eq. (9.128) satisfy the inequality

$$-1 < \hat{k}_i < 1. \tag{9.129}$$

The various expressions for the k_i parameters are all in the form of a normalized cross-correlation function; i.e., the k_i parameters are, in a sense, a measure of the degree of correlation between the forward and backward prediction error sequences. For this reason the parameters k_i are called the *partial correlation* or *PARCOR* coefficients [161, 162]. It is relatively straightforward to verify that Eq. (9.125) is identical to Eq. (9.93) by substituting Eqs. (9.105) and (9.110) in Eq. (9.125).

If Eq. (9.125) replaces Eq. (9.93) in the Durbin algorithm, the predictor coefficients can be computed recursively as before. Thus the PARCOR analysis leads to an alternative to the inversion of a matrix and gives results similar to the autocorrelation method; i.e., the set of PARCOR coefficients is equivalent to a set of predictor coefficients that minimize just the mean-squared *forward* prediction error. More importantly, this approach opens up a whole new class of procedures based upon the lattice configuration of Figure 9.20 [219].

To summarize what we have learned about the lattice method of LPC analysis, the steps involved in determining the predictor coefficients and the k parameters (using the all-zero lattice method of Figure 9.20) are shown in the pseudo-code of Figure 9.23 for the PARCOR definition of the k-parameters. Included in the pseudo-code is the computation of the predictor coefficients and the mean-squared errors, $\mathcal{E}^{(i)}$.

Figure 9.24 shows the lattice algorithm using Burg's definition of the k-parameters. Note that the Burg algorithm is similar to the covariance method in that we must supply p samples prior to the interval $0 \leq n \leq L-1$ in order to compute all p of the prediction error sequences required in the algorithm.

There are clearly several differences in implementation between the lattice method and the covariance and autocorrelation methods discussed earlier. One major difference is that, in the lattice method, the predictor coefficients are obtained directly from the speech samples without an intermediate calculation of an autocorrelation function. At the same time, the method is guaranteed to yield a stable filter without

PARCOR Lattice Algorithm

$$\mathcal{E}^{(0)} = R[0] = \sum_{n=0}^{L-1} (s[n])^2 \quad (1)$$

$$e^{(0)}[n] = b^{(0)}[n] = s[n], \quad 0 \leq n \leq L-1 \quad (2)$$

for $i = 1, 2, \ldots, p$

compute k_i using Eq. (9.125) (3)

compute $e^{(i)}[n], \quad 0 \leq n \leq L-1+i$ using Eq. (9.117b) (4a)
compute $b^{(i)}[n], \quad 0 \leq n \leq L-1+i$ using Eq. (9.117c) (4b)

$$\alpha_i^{(i)} = k_i \quad (5)$$

compute predictor coefficients
if $i > 1$ then for $j = 1, 2, \ldots, i-1$

$$\alpha_j^{(i)} = \alpha_j^{(i-1)} - k_i \alpha_{i-j}^{(i-1)} \quad (6)$$

end
compute mean-squared energy

$$\mathcal{E}^{(i)} = (1 - k_i^2)\mathcal{E}^{(i-1)} \quad (7)$$

end

$$\alpha_j = \alpha_j^{(p)} \quad j = 1, 2, \ldots, p \quad (8)$$

$$e[n] = e^{(p)}[n], \quad 0 \leq n \leq L-1+p \quad (9)$$

FIGURE 9.23
Pseudo-code description of PARCOR lattice algorithm for computing both the k-parameters and the corresponding predictor coefficients.

requiring the use of a window. For these reasons the lattice formulation has become an important and viable approach to the implementation of linear predictive analysis.

9.5.4 Comparison of Computational Requirements

We have already discussed the differences in the theoretical formulations of the covariance, autocorrelation, and lattice formulations of the linear predictive analysis equations. In this section we discuss some of the issues involved in practical implementations of the analysis equations. Included among these issues are computational considerations, numerical and physical stability of the solutions, and the question of how to choose the number of poles and section length used in the analysis. We begin first with the computational considerations involved in obtaining the predictor coefficients from the speech waveform.

The two major issues in the computation of the predictor coefficients are the amount of storage, and the computation (in terms of number of multiplications). Table 9.1 (due to Portnoff et al. [292] and Makhoul [219]) shows the required computation for the covariance, autocorrelation, and lattice methods. In terms of storage,

Burg Lattice Algorithm

$\mathcal{E}^{(0)} = R[0] = \sum_{n=0}^{L-1} (s[n])^2$	(1)
$e^{(0)}[n] = b^{(0)}[n] = s[n], \quad -p \leq n \leq L-1$	(2)
for $i = 1, 2, \ldots, p$	
compute $\hat{k}_i$ using Eq. (9.128)	(3)
compute $e^{(i)}[n], \quad -p+i \leq n \leq L-1$ using Eq. (9.117b)	(4a)
compute $b^{(i)}[n], \quad -p+i \leq n \leq L-1$ using Eq. (9.117c)	(4b)
$\alpha_i^{(i)} = \hat{k}_i$	(5)
compute predictor coefficients	
if $i > 1$ then for $j = 1, 2, \ldots, i-1$	
$\alpha_j^{(i)} = \alpha_j^{(i-1)} - \hat{k}_i \alpha_{i-j}^{(i-1)}$	(6)
end	
compute mean-squared energy	
$\mathcal{E}^{(i)} = (1 - \hat{k}_i^2)\mathcal{E}^{(i-1)}$	(7)
end	
$\alpha_j = \alpha_j^{(p)} \quad j = 1, 2, \ldots, p$	(8)
$e[n] = e^{(p)}[n], \quad 0 \leq n \leq L-1$	(9)

FIGURE 9.24
Pseudo-code description of Burg lattice algorithm for computing both the $\hat{k}$-parameters and the corresponding predictor coefficients.

for the covariance method, the requirements are essentially L_1 locations for the signal segment (data), and on the order of $p^2/2$ locations for the correlation matrix, where L_1 is the number of samples in the analysis window. For the autocorrelation method, the requirements are L_2 locations for both the data and the window, and a number of locations proportional to p for the autocorrelation matrix. For the lattice method the requirements are $3L_3$ locations for the data and the forward and backward prediction errors. For emphasis we have assumed that the L_1 for the covariance method, the L_2 for the autocorrelation method, and the L_3 for the lattice method need not be the same. Thus, in terms of storage (assuming L_1, L_2, and L_3 are comparable), the covariance and autocorrelation methods require less storage than the lattice method.

The computational requirements for the three methods, in terms of multiplications, are shown at the bottom of Table 9.1. For the covariance method, the computation of the correlation matrix requires about $L_1 p$ multiplications, whereas the solution to the matrix equation (using the Cholesky decomposition procedure) requires a number of multiplications proportional to p^3. (Portnoff et al. give an exact figure of $(p^3 + 9p^2 + 2p)/6$ multiplications, p divides, and p square roots [292].) For the

TABLE 9.1 Computational considerations in the LPC solutions.

	Covariance Method (Cholesky Decomposition)	Autocorrelation Method (Durbin Method)	Lattice Method (Burg Method)
Storage			
Data	L_1	L_2	$3L_3$
Matrix	$\sim p^2/2$	$\sim p$	—
Window	0	L_2	—
Computation			
(Multiplications)			
Windowing	0	L_2	—
Correlation	$\sim L_1 p$	$\sim L_2 p$	—
Matrix Solution	$\sim p^3$	$\sim p^2$	$5L_3 p$

autocorrelation method, the computation of the autocorrelation matrix requires about $L_2 p$ multiplications, whereas the solution to the matrix equations requires about p^2 multiplications. Thus if L_1 and L_2 are approximately equal, and with $L_1 \gg p$, $L_2 \gg p$, then the autocorrelation method will require somewhat less computation than the covariance method. However, since in most speech problems, the number of multiplications required to compute the correlation function far exceeds the number of multiplications to solve the matrix equations, the computation times for both these formulations are quite comparable. For the lattice method, a total of $5L_3 p$ multiplications are needed to compute the set of partial correlation coefficients.[12] Thus the lattice method is the least computationally efficient method for solving the LPC equations. However, the other advantages of the lattice method must be kept in mind when considering the use of this method.

9.6 THE PREDICTION ERROR SIGNAL

The prediction error sequence, defined as

$$e[n] = s[n] - \sum_{k=1}^{p} \alpha_k s[n-k], \tag{9.130}$$

is the basis for linear predictive analysis. When the predictor parameters are computed by the Levinson–Durbin algorithm, the prediction error sequence is implicit, but not explicitly computed. In the lattice methods, the prediction error sequences of all orders up to p are actually computed. In any case, it is always possible to compute $e[n]$ using Eq. (9.130). To the extent that the actual speech signal is generated by a system that

[12] Makhoul has discussed a modified lattice method for obtaining the partial correlation coefficients with the same efficiency as the normal covariance method [219].

is well modeled by a time-varying linear predictor of order p, $e[n]$ is equally a good approximation to the excitation source, $e[n] = Gu[n]$. Based on this reasoning, it is expected that the prediction error will be large (for voiced speech) at the beginning of each pitch period. Thus, the pitch period can be determined by detecting the positions of the samples of $e[n]$ that are large, and defining the period as the time difference between pairs of samples of $e[n]$ that exceed a threshold. Alternatively the pitch period can be estimated by performing an autocorrelation analysis on $e[n]$ and detecting the largest peak in the appropriate pitch period range. Another way of understanding why the error signal is valuable for pitch detection is the observation that the spectrum of the error signal is approximately flat; thus the influence of the formants has been eliminated in the error signal.

Examples of the results of linear predictive analyses are given in Figures 9.25–9.28. In each of these figures, the top panel shows the section of

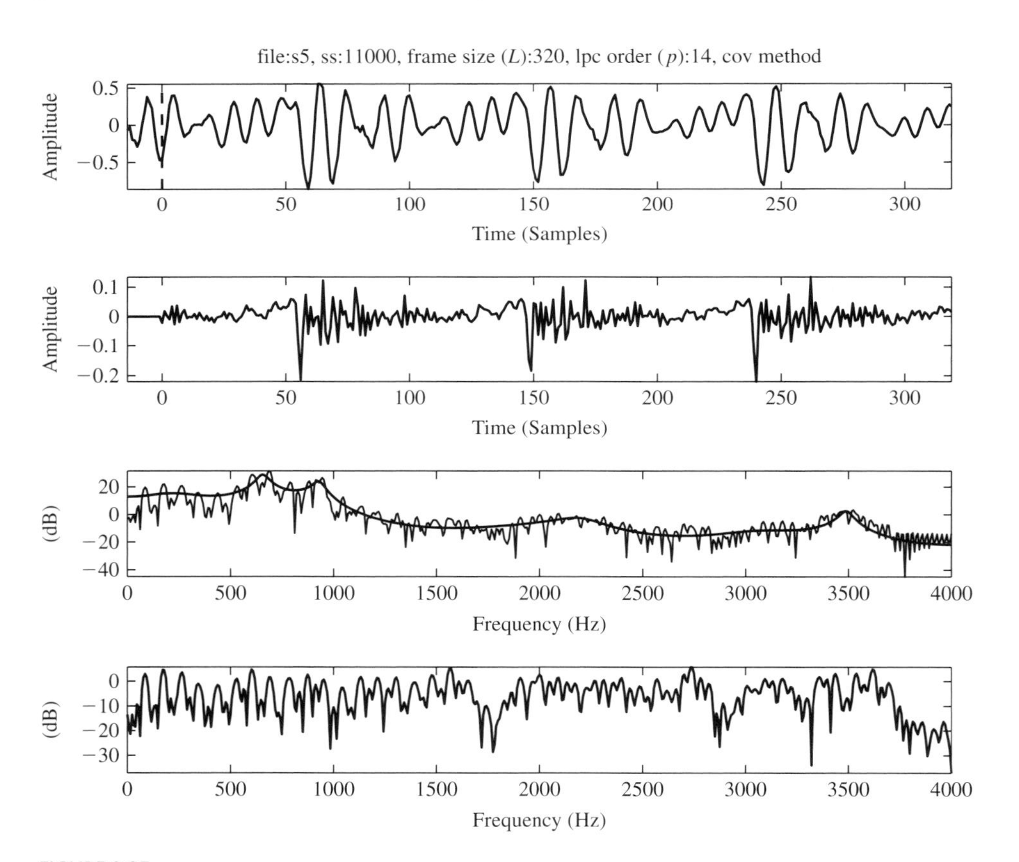

FIGURE 9.25
Typical signals and spectra for the use of the linear predictive covariance method on the speech of a male speaker. The upper panel shows the speech frame being analyzed; the second panel shows the prediction error resulting from a $p = 14$ linear prediction analysis; the third panel shows the log magnitude spectrum of the signal along with the linear prediction model spectrum; the bottom panel shows the log magnitude spectrum of the error signal.

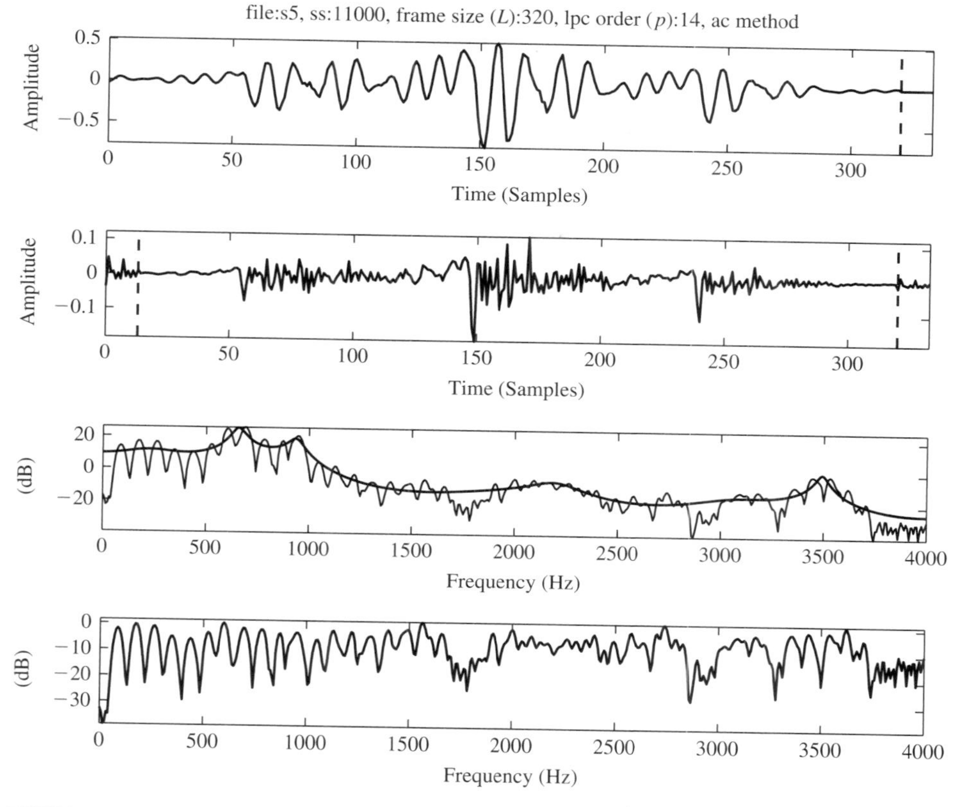

FIGURE 9.26
Typical signals and spectra for the use of the linear predictive autocorrelation methods for a male speaker. The upper panel shows the speech frame being analyzed; the second panel shows the prediction error resulting from a $p = 14$ linear prediction analysis; the third panel shows the log magnitude spectrum of the signal along with the linear prediction model spectrum; the bottom panel shows the log magnitude spectrum of the error signal.

speech being analyzed, the second panel shows the resulting prediction error signal, the third panel shows the log magnitude of the DFT of the original speech signal along with the log magnitude of $\tilde{H}(e^{j\omega T})$ superimposed, and the bottom panel shows the log magnitude spectrum of the error signal. Figures 9.25 and 9.26 are for 40 msec of a voiced sound spoken by a male speaker using the covariance and autocorrelation methods. The error signal is seen to be sharply peaked at the beginning of each pitch period, and the error spectrum is fairly flat, showing a comb effect due to the spectral harmonics. Note the rather large prediction error at the beginning and end of the segment in Figure 9.26 for the autocorrelation method. This is, of course, due to the fact that we are attempting to predict the first p samples of the signal from the zero-valued samples outside the window, and the final p zero-valued signal values from the non-zero signal samples at the end of the windowed speech frame. The

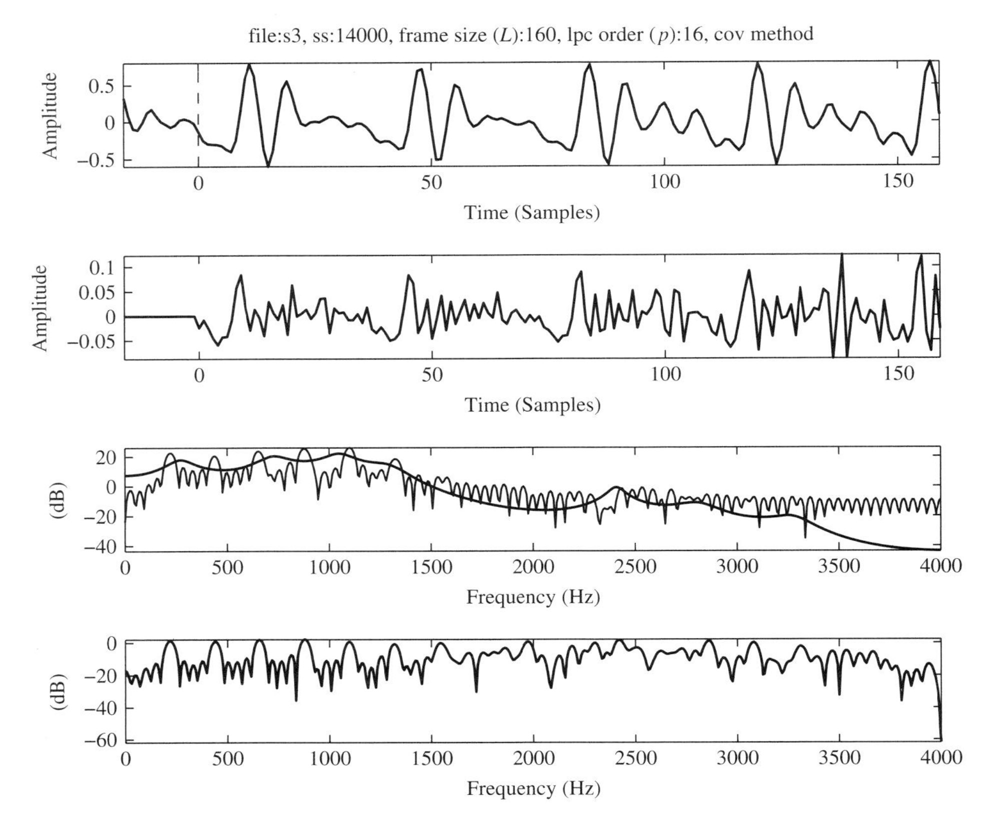

FIGURE 9.27
Typical signals and spectra for the use of the linear predictive covariance method for a female speaker. The upper panel shows the speech frame being analyzed; the second panel shows the prediction error resulting from a $p = 16$ linear prediction analysis; the third panel shows the log magnitude spectrum of the signal along with the linear prediction model spectrum; the bottom panel shows the log magnitude spectrum of the error signal.

tapering effect of the Hamming window is thus not completely effective in reducing this error.

Figures 9.27 and 9.28 show similar results for 20 msec of a voiced sound for a female speaker. For this speaker, approximately 4.5 complete pitch periods are contained within the analysis interval. Thus, in Figure 9.27, the error signal displays a large number of sharp peaks during the analysis interval for the covariance method of analysis. However, the effect of the Hamming window in the autocorrelation methods of Figure 9.28 is to taper the pitch pulses near the ends of the analysis interval; hence the peaks in the error signal due to the pitch pulses are likewise tapered. Note again the rather large errors at the beginning and end of the windowed segment.

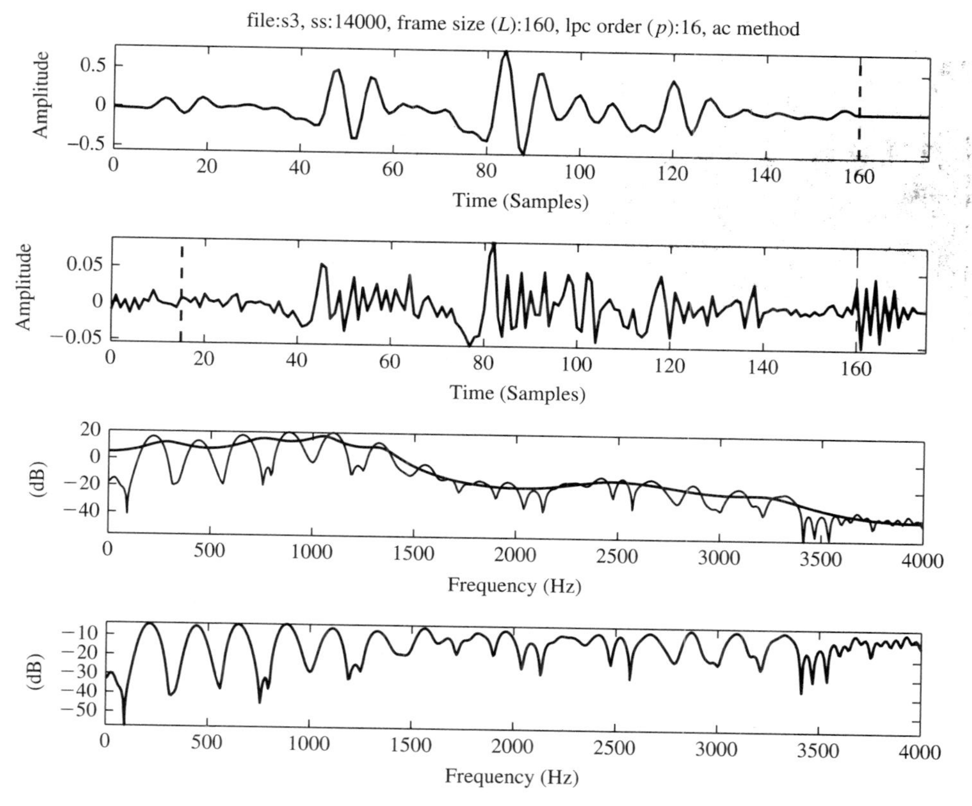

FIGURE 9.28
Typical signals and spectra for the use of the linear predictive autocorrelation method for a female speaker. The upper panel shows the speech frame being analyzed; the second panel shows the prediction error resulting from a $p = 16$ linear prediction analysis; the third panel shows the log magnitude spectrum of the signal along with the linear prediction model spectrum; the bottom panel shows the log magnitude spectrum of the error signal.

The behavior of the error signal, as shown in the preceding figures, suggests that the prediction error would, by itself, be a natural candidate for a signal from which pitch could be detected by simple processing. However, the situation is not quite so clear for other examples of voiced speech. Makhoul and Wolf [222] have shown that for sounds that are not rich in harmonic structure, e.g., liquids like /R/, /L/, or nasals such as /M/, /N/, the peaks in the error signal are not always very sharp or distinct. Additionally, at the junctions between voiced and unvoiced sounds, the pitch markers in the error signal often totally disappear.

In summary, although the error signal, $e[n]$, appears to be a good basis for a pitch detector, it cannot be relied on exclusively for this purpose. In Chapter 10 we discuss one simple pitch detection scheme, called the SIFT algorithm, based upon the prediction error signal.

9.6.1 Alternative Expressions for the Normalized Mean-Squared Error

The normalized mean-squared prediction error for the autocorrelation method is defined as

$$\mathcal{V}_{\hat{n}} = \frac{\sum_{m=0}^{L+p-1} e_{\hat{n}}^2[m]}{\sum_{m=0}^{L-1} s_{\hat{n}}^2[m]}, \tag{9.131a}$$

where $e_{\hat{n}}[m]$ is the output of the prediction error filter corresponding to the speech segment $s_{\hat{n}}[m]$ located at time index $\hat{n}$. For the covariance method, the corresponding definition is

$$\mathcal{V}_{\hat{n}} = \frac{\sum_{m=0}^{L-1} e_{\hat{n}}^2[m]}{\sum_{m=0}^{L-1} s_{\hat{n}}^2[m]}. \tag{9.131b}$$

By defining $\alpha_0 = -1$, the prediction error sequence can be expressed as

$$e_{\hat{n}}[m] = s_{\hat{n}}[m] - \sum_{k=1}^{p} \alpha_k s_{\hat{n}}[m-k] = -\sum_{k=0}^{p} \alpha_k s_{\hat{n}}[m-k]. \tag{9.132}$$

Substituting Eq. (9.132) into Eq. (9.131a) and using Eq. (9.12), it follows that

$$\mathcal{V}_{\hat{n}} = \sum_{i=0}^{p}\sum_{j=0}^{p} \alpha_i \frac{\phi_{\hat{n}}[i,j]}{\phi_{\hat{n}}[0,0]} \alpha_j, \tag{9.133a}$$

and substituting Eq. (9.13) into (9.133a) gives

$$\mathcal{V}_{\hat{n}} = -\sum_{i=0}^{p} \alpha_i \frac{\phi_{\hat{n}}[i,0]}{\phi_{\hat{n}}[0,0]} = 1 - \sum_{i=1}^{p} \alpha_i \frac{\phi_{\hat{n}}[i,0]}{\phi_{\hat{n}}[0,0]}. \tag{9.133b}$$

Still another expression for $\mathcal{V}_{\hat{n}}$ was obtained in the Durbin algorithm; i.e.,

$$\mathcal{V}_{\hat{n}} = \prod_{i=1}^{p} (1 - k_i^2). \tag{9.134}$$

The above expressions are not all equivalent and are subject to interpretation in terms of the details of a given linear predictive method. For example, Eq. (9.134), being based

TABLE 9.2 Expressions for the normalized error.

	Covariance Method	Autocorrelation Method	Lattice Method
$\mathcal{V} = \dfrac{\sum_{m=0}^{L+p-1} e^2[m]}{\sum_{m=0}^{L-1} s^2[m]}$	Valid	Valid	Valid
$\mathcal{V} = \sum_{i=0}^{p}\sum_{j=0}^{p} \alpha_i \dfrac{\phi[i,j]}{\phi[0,0]} \alpha_j$	Valid	Valid*	Not Valid
$\mathcal{V} = -\sum_{i=0}^{p} \alpha_i \dfrac{\phi[i,0]}{\phi[0,0]}$	Valid	Valid*	Not Valid
$\mathcal{V} = \prod_{i=0}^{p} (1 - k_i^2)$	Not Valid	Valid	Valid

*In these cases $\phi[i,j] = R[|i-j|]$.

upon the Durbin algorithm, is valid only for the autocorrelation and lattice methods. Also, since the lattice method does not explicitly require the computation of the correlation functions, Eqs. (9.133a) and (9.133b) do not apply directly to the lattice method. Table 9.2 summarizes the above expressions for normalized mean-squared error and indicates the scope of validity of each expression. (Note that the subscript $\hat{n}$ has been suppressed in the table for simplicity.)

9.6.2 Experimental Evaluation of Values for the LPC Parameters

To provide guidelines to aid in the choice of the linear predictive parameters p and L for practical implementations, Chandra and Lin [53] performed a series of investigations in which they plotted the normalized mean-squared prediction error, for a pth-order predictor versus the relevant parameter for the following conditions:

1. the covariance method and the autocorrelation method
2. synthetic vowel and natural speech
3. pitch synchronous and pitch asynchronous analysis

where the normalized error, $\mathcal{V}$, is defined as in Table 9.2. Figures 9.29–9.34 show the results obtained by Chandra and Lin for the above conditions [53].

Figure 9.29 shows the variation of $\mathcal{V}$ with the order of the linear predictor, p, for a section of a synthetic vowel (/IY/ in /heed/) whose pitch period was 83 samples. The analysis section length L was 60 samples with the window beginning at the beginning of a pitch period—i.e., these results are for a pitch synchronous analysis. For the covariance method, the prediction error decreases monotonically to 0 at $p = 11$, which was the order of the system used to create the synthetic speech. For the autocorrelation

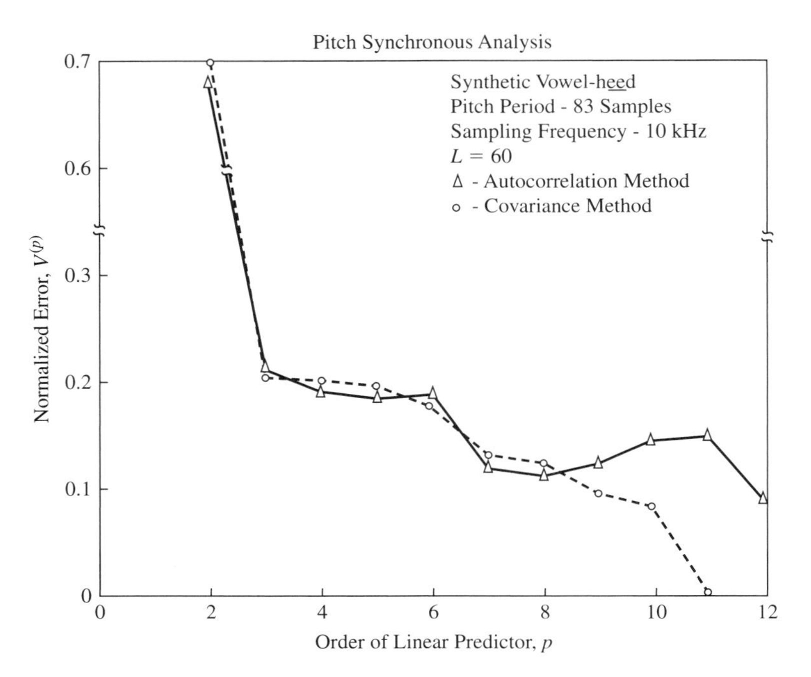

FIGURE 9.29
Variation of prediction error with predictor order, p, for voiced section of a synthetic vowel—pitch synchronous analysis. (After Chandra and Lin [53]. © [1974] IEEE.)

method, the normalized prediction error remains at a value of about 0.1 for values of p greater than about 7. This behavior is due to the fact that for the autocorrelation method with short windows ($L = 60$), the prediction error at the beginning and end of the segment is an appreciable part of the total mean-squared error, and this error can never be exactly zero. This is, of course, not the case with the covariance method, where speech samples from outside the averaging interval are available for prediction.

Figure 9.30 shows the variation of $\mathcal{V}$ with the order of the linear predictor for a pitch asynchronous analysis for a section of speech at the same analysis time $\hat{n}$ as used in Figure 9.29. This time, however, the section length was $L = 120$ samples. For this case the covariance and autocorrelation methods yielded nearly identical values of $\mathcal{V}$ for different values of p. Further, the values of $\mathcal{V}$ decreased monotonically to a value of about 0.1 near $p = 11$. Thus, in the case of an asynchronous linear predictive analysis, at least for the example of a synthetic vowel, both analysis methods yield similar results.

Figure 9.31 shows the variation of $\mathcal{V}$ with L (section length) for a linear predictor of order 12 for the synthetic speech section. As anticipated, for values of L below the pitch period (83 samples), the covariance method gives significantly smaller values of $\mathcal{V}$ than the autocorrelation method. For values of $\mathcal{V}$ at or near multiples of the pitch period, the values of $\mathcal{V}$ show fairly large jumps due to the large prediction error when a pitch pulse is used to excite the system. However, for most values of L on the order of 2 or more pitch periods, both analysis methods yield comparable values of $\mathcal{V}$.

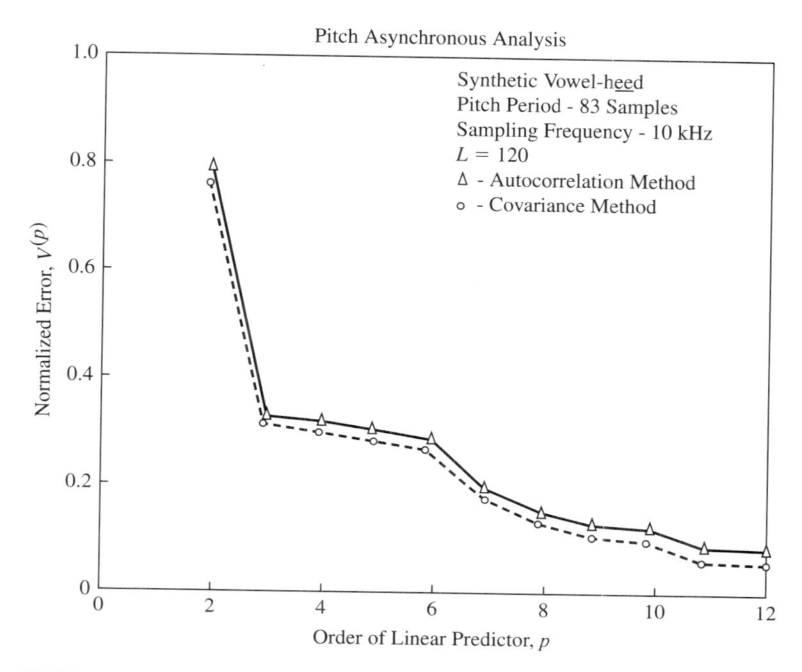

FIGURE 9.30
Variation of prediction error with predictor order, p, for voiced section of a synthetic vowel—pitch asynchronous analysis. (After Chandra and Lin [53]. © [1974] IEEE.)

Figures 9.32–9.34 show a similar set of figures for the case of a section of natural voiced speech. Figure 9.32 shows that the normalized error for the covariance method is significantly lower than the normalized error for the autocorrelation method for a pitch synchronous analysis, whereas Figure 9.33 shows that for a pitch asynchronous analysis, the values of $\mathcal{V}$ are comparable. Finally Figure 9.34 shows how the values of $\mathcal{V}$ vary as L varies for an analysis with $p = 12$. It can be seen that in the region of pitch pulse occurrences, the value of $\mathcal{V}$ for the autocorrelation analysis jumps significantly, whereas the value of $\mathcal{V}$ for the covariance analysis changes only a small amount at these points. Also for large values of L, it is seen that the curves of $\mathcal{V}$ for the two methods approach each other.

9.6.3 Variations of the Normalized Error with Frame Position

We have already shown some properties of the linear predictive normalized error in Section 9.6.2—namely its variation with section length L, and with the number of poles in the analysis, p. There remains one other major source of variability of $\mathcal{V}$—namely its variation with respect to the position of the analysis frame. To demonstrate this variability, Figure 9.35 shows plots of the results of a sample-by-sample (i.e., the window is moved one sample at a time) linear predictive analysis of 40 msec of the vowel sound /IY/, spoken by a male speaker. The sampling rate was $F_s = 10$ kHz.

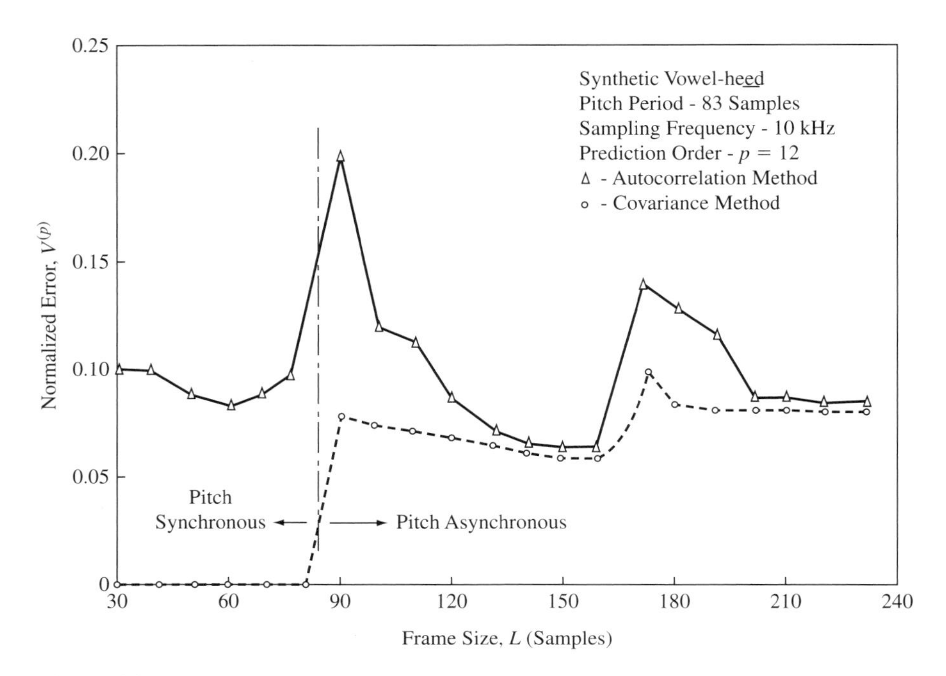

FIGURE 9.31
Variation of prediction error with frame size, for a voiced section of synthetic speech. (After Chandra and Lin [53]. © [1974] IEEE.)

Figure 9.35a shows the signal energy $R_{\hat{n}}[0]$ as a function of $\hat{n}$ (computed at a 10 kHz rate); Figure 9.35b shows the normalized mean-squared error ($\mathcal{V}_{\hat{n}}$) (again computed at a 10 kHz rate) for a 14th-order ($p = 14$) analysis with a 20 msec ($L = 200$) frame size for the covariance method; Figure 9.35c shows the normalized mean-squared error for the autocorrelation method using a Hamming window; and Figure 9.35d shows the normalized mean-squared error for the autocorrelation method using a rectangular window. The average pitch period for this speaker was 84 samples (8.4 msec); thus about 2.5 pitch periods were contained within the 20 msec frame. For the covariance method, the normalized error shows a substantial variation with the position of the analysis frame (i.e., the error is not a smooth function of time). This effect is mainly due to the large peaks in the error signal, $e[n]$, at the beginning of each pitch period, as discussed previously. Thus, in this example, when the analysis frame is positioned to encompass three sets of error peaks, the normalized error is much larger than when only two sets of error peaks are included in the analysis interval. This accounts for the normalized error showing a fairly large discrete jump in level as each new error peak is included in the analysis frame. Each discrete jump of the normalized error is followed by a gradual tapering off and flattening of the normalized error. The exact detailed behavior of the normalized error between discrete jumps depends on details of the signal and the analysis method.

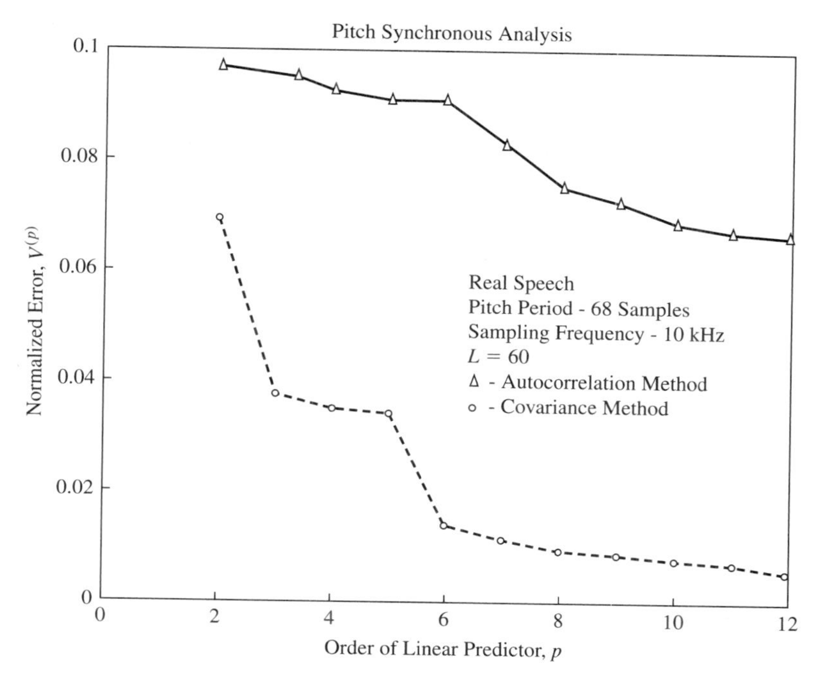

FIGURE 9.32
Variation of prediction error with predictor order, p, for a voiced section of a natural vowel—pitch synchronous analysis. (After Chandra and Lin [53]. © [1974] IEEE.)

Figures 9.35c and 9.35d show a somewhat different behavior of the linear predictive normalized error for the autocorrelation analysis method using a Hamming window and a rectangular window respectively. As seen in this figure, the normalized mean-squared error shows a substantial amount of high frequency variation, as well as a small amount of low frequency and pitch synchronous variation. The high frequency variation is due primarily to the error signal for the first p samples in which the signal is not linearly predictable. The magnitude of this variation is considerably smaller for the analysis using the Hamming window than for the analysis with the rectangular window due to the tapering of the Hamming window at the ends of the analysis window. Another component of the high frequency variation of the normalized error is related to the position of the analysis frame with respect to pitch pulses as discussed previously for the covariance method. However, this component of the error is much less a factor for the autocorrelation analysis than for the covariance method–especially in the case when a Hamming window is used, since new pitch pulses that enter the analysis frame are tapered by the window.

Variations of the type shown in Figure 9.35 have been found for most vowel sounds [302]. The variability with the analysis frame position can be reduced using all-pass filtering and spectral pre-emphasis of the signal prior to linear predictive analysis [302].

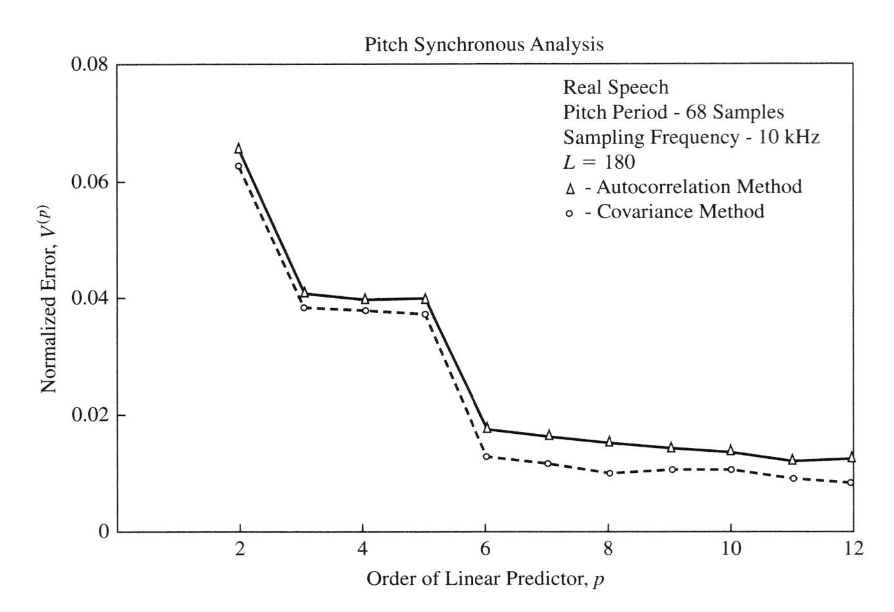

FIGURE 9.33
Variation of prediction error with predictor order for a voiced section of a natural vowel—pitch asynchronous analysis. (After Chandra and Lin [53]. © [1974] IEEE.)

9.7 SOME PROPERTIES OF THE LPC POLYNOMIAL $A(z)$

In this section we examine the following important properties of the prediction error system function polynomial:

- the minimum-phase property of the polynomial, $A(z)$
- the relation between the PARCORs and the root locations of the polynomial $A(z)$
- the relation between the roots of the $A(z)$ polynomial and the formant frequencies of the speech signal spectrum.

9.7.1 The Minimum-Phase Property of the Prediction Error Filter

The system function, $A(z)$, of the optimum prediction error filter cannot have a zero that is outside the unit circle. To prove this assertion, we assume that z_o is a root of $A(z)$ whose magnitude is greater than 1; i.e., $|z_o|^2 > 1$. If this is the case, then we can express the polynomial $A(z)$ as

$$A(z) = (1 - z_o z^{-1}) \cdot A'(z), \tag{9.135}$$

where $A'(z)$ has all the remaining roots of $A(z)$. If $S_{\hat{n}}(e^{j\omega})$ is the discrete-time Fourier transform of the segment of speech on which the linear predictive analysis is based, the minimum mean-squared prediction error at time $\hat{n}$ can be expressed using Parseval's

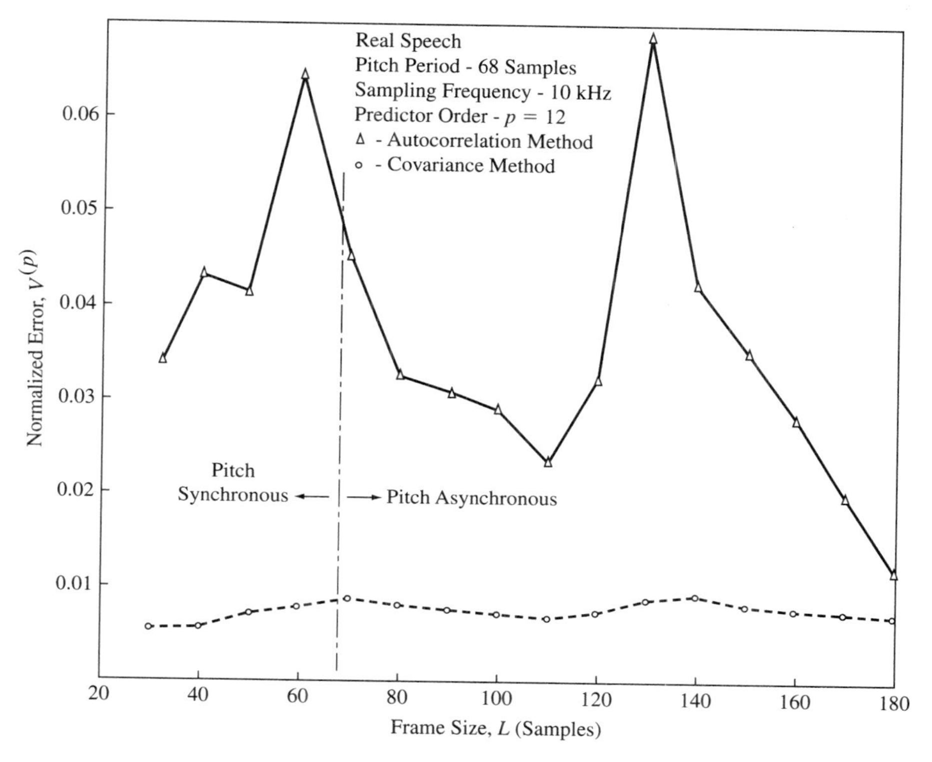

FIGURE 9.34
Variation of prediction error with frame size for a voiced section of natural speech. (After Chandra and Lin [53]. © [1974] IEEE.)

theorem as

$$\begin{aligned}\mathcal{E}_{\hat{n}} &= \sum_{m=-\infty}^{\infty} e_{\hat{n}}[m]^2 \\ &= \frac{1}{2\pi}\int_{-\pi}^{\pi} |A(e^{j\omega})|^2 |S_{\hat{n}}(e^{j\omega})|^2 d\omega \\ &= \frac{1}{2\pi}\int_{-\pi}^{\pi} \left|1 - z_o e^{-j\omega}\right|^2 \left|A'(e^{j\omega})\right|^2 \left|S_{\hat{n}}(e^{j\omega})\right|^2 d\omega. \end{aligned} \tag{9.136}$$

We can rewrite the term $\left|1 - z_o e^{-j\omega}\right|^2$ in the form

$$\left|1 - z_o e^{-j\omega}\right|^2 = |z_o^2| \cdot |1 - (1/z_o^*)e^{-j\omega}|^2. \tag{9.137}$$

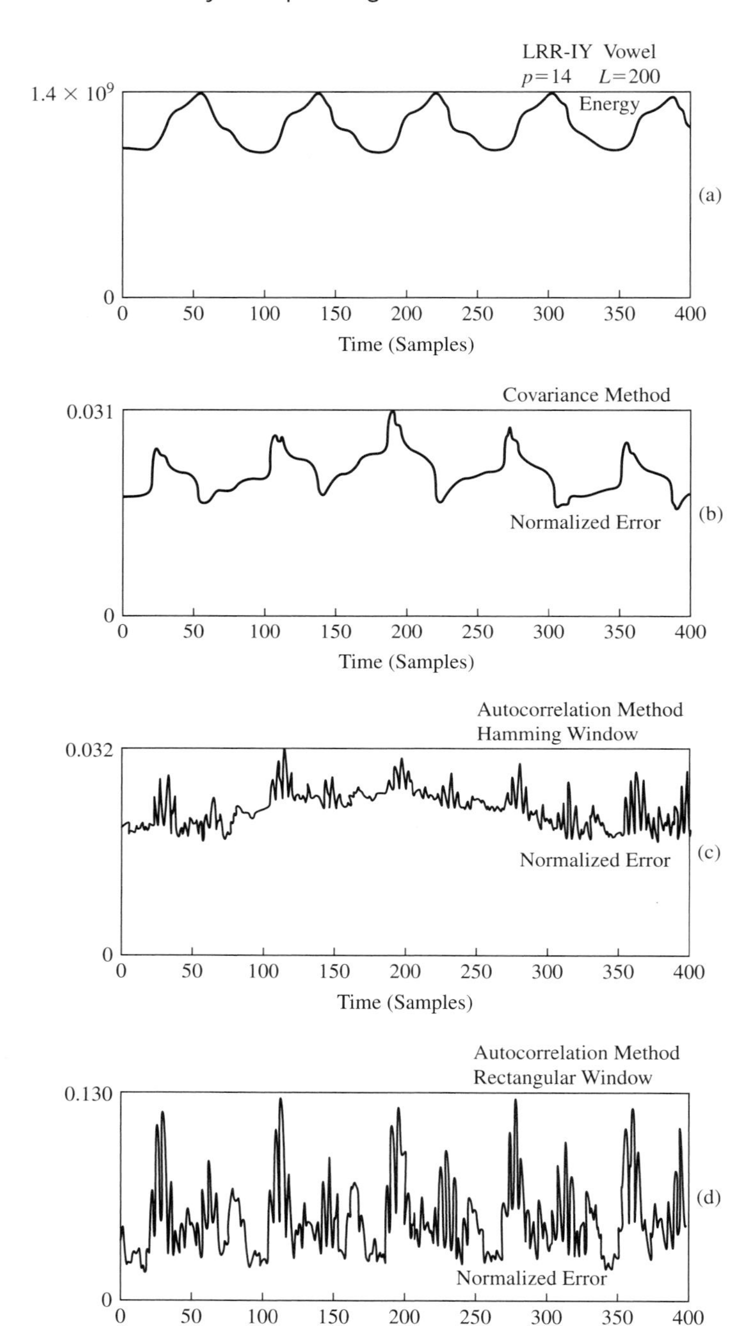

FIGURE 9.35
Prediction error sequences for 200 samples of speech for three linear predictive systems. (After Rabiner et al. [302]. © [1977] IEEE.)

If we substitute this result into Eq. (9.136), we obtain

$$\tilde{\mathcal{E}}_{\hat{n}} = \frac{\mathcal{E}_{\hat{n}}}{|z_o^2|} = \frac{1}{2\pi}\int_{-\pi}^{\pi} \left|1-(1/z_o^*)e^{-j\omega}\right|^2 \left|A'(e^{j\omega})\right|^2 \left|S_{\hat{n}}(e^{j\omega})\right|^2 d\omega, \tag{9.138}$$

thereby showing that $A(z)$ could not be the optimum filter because the FIR system with system function $(1-(1/z_o^*)z^{-1})A'(z)$ has the same order and gives a total squared prediction error, $\tilde{\mathcal{E}}_{\hat{n}}$, which is less than that of $A(z)$ (since $|z_o|^2$ is assumed to be greater than 1).

This proves that the optimum filter polynomial, $A(z)$, cannot have a zero outside the unit circle. If z_o is on the unit circle, Eq. (9.138) still holds since $z_o = 1/z_o^*$, but this does not lead to a contradiction since $|z_o|^2 = 1$. It can be shown, however, that the zeros of $A(z)$ must be strictly *inside* the unit circle. Complete proofs of the minimum-phase property can be found in Refs. [61, 201, 232].

9.7.2 PARCORs and Stability of the LPC Polynomial

If any of the PARCOR coefficients, k_i, have a magnitude greater than or equal to 1, then this implies that there is a root, $z_j^{(i)}$, of the ith-order LPC polynomial, whose magnitude is greater than 1. To show this, we begin by recalling that the Levinson–Durbin recursion states that for the ith-order optimum linear predictor, the total squared prediction error satisfies

$$\mathcal{E}^{(i)} = (1-k_i^2)\cdot\mathcal{E}^{(i-1)} = \prod_{j=1}^{i}(1-k_j^2)\cdot\mathcal{E}^{(0)} > 0. \tag{9.139}$$

Since $\mathcal{E}^{(i)}$ must always be strictly positive, it follows that $|k_i| < 1$, i.e., the PARCOR coefficients, must always be strictly less than 1 in magnitude.

In the derivation of the Durbin recursion, we showed [Eq. (9.107)] that the optimum prediction error polynomial for the ith-stage of processing, $A^{(i)}(z)$, could be expressed in terms of the optimum prediction error polynomial for the $(i-1)^{st}$ stage, as

$$A^{(i)}(z) = A^{(i-1)}(z) - k_i z^{-i} A^{(i-1)}(z^{-1}) = \prod_{j=1}^{i}(1-z_j^{(i)}z^{-1}). \tag{9.140}$$

It follows from Eq. (9.140) that $-k_i$ is the coefficient of the highest order term (z^{-i}) in $A^{(i)}(z)$; i.e., $\alpha_i^{(i)} = k_i$. Therefore we have the result that

$$|k_i| = |\alpha_i^{(i)}| = \prod_{j=1}^{i}|z_j^{(i)}|. \tag{9.141}$$

If $|k_i| \geq 1$, then at least one root of the polynomial $A^{(i)}(z)$ must be outside the unit circle if any other roots are inside. The case $|k_i| = 1$ could occur if all the roots were on the unit circle or if one was outside and another inside. However, we know from Eq. (9.139) that $|k_i| < 1$. Therefore, it follows that the condition $|k_i| < 1$ necessarily holds if $|z_j^{(i)}| < 1$ for $j = 1, 2, \ldots, i$. Since the proof above is valid for all $A^{(i)}(z)$ $i = 1, 2, \ldots, p$, a necessary condition for no roots of $A^{(p)}(z)$ to be outside the unit circle is

$$|k_i| < 1, \quad i = 1, 2, \ldots, p. \tag{9.142}$$

The proof that the condition in Eq. (9.142) is also sufficient to guarantee that the roots of $A^{(p)}(z)$ are strictly inside the unit circle is more involved, but nevertheless true [229].

Thus, Eq. (9.142) is a necessary and sufficient condition for $A(z)$ to be a minimum phase system function and consequently for the all-pole model system $\tilde{H}(z) = G/A(z)$ to be a stable system.

9.7.3 Root Locations for Optimum LP Model

The roots of the prediction error polynomial, $A(z)$, can be interpreted in various ways, including some roots which correspond to the formant frequencies of the vocal tract for the sound being analyzed. A key issue is how to align the roots of the prediction error polynomial with the properties of the LP spectrum so as to assign some roots to the formants and other roots to effects like the radiation model or the glottal excitation pulse shape, or possibly other factors such as transmission effects, etc.

As an example, Figure 9.36 shows a plot of the pole-zero locations of the LP model transfer function,

$$\tilde{H}(z) = \frac{G}{A(z)} = \frac{G}{1 - \sum_{i=1}^{p} \alpha_i z^{-i}} = \frac{G}{\prod_{i=1}^{p}(1 - z_i z^{-1})} = \frac{G z^p}{\prod_{i=1}^{p}(z - z_i)},$$

where the poles of $\tilde{H}(z)$ are indicated by the symbol $\times$ and the pth-order zero is indicated by an "○" with the number 12 (for a system with $p = 12$).

The poles most likely to correspond to formants are the ones closest to the unit circle in Figure 9.36 because these poles produce resonance-like peaks when $\tilde{H}(z)$ is evaluated on the unit circle. This is illustrated by Figure 9.37, which shows the log magnitude frequency response corresponding to the pole-zero plot in Figure 9.36. The prominent peaks in Figure 9.37 each correspond to one of the poles that is close to the unit circle.

The classification into formants and non-formant roots (poles) is illustrated by Table 9.3, which gives values of the root magnitudes and angles (both in degrees, θ, and in hertz, $F = \theta \cdot F_s/360$), along with estimates of the poles that correspond to

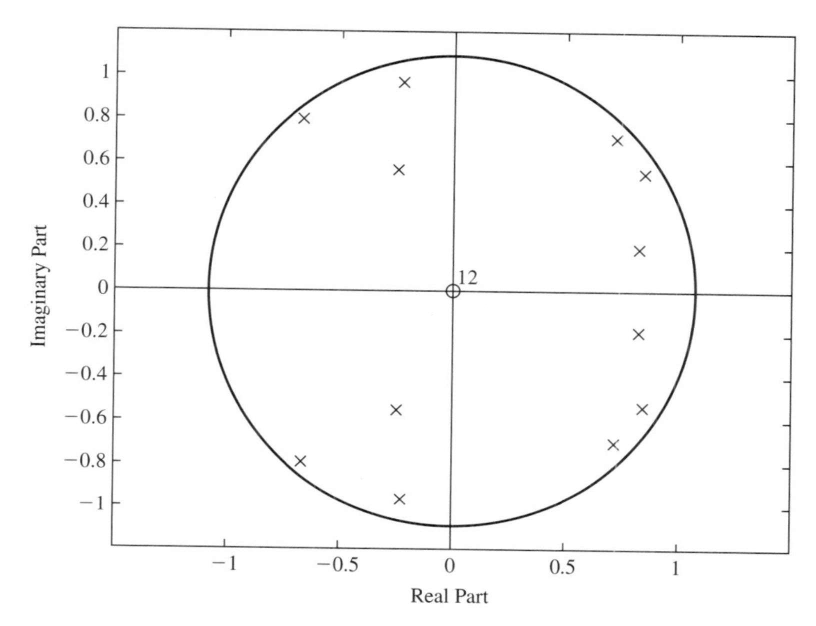

FIGURE 9.36
Plot of the locations of the poles and zeros in the z-plane of the LP vocal tract model system function $\tilde{H}(z)$ for $p = 12$.

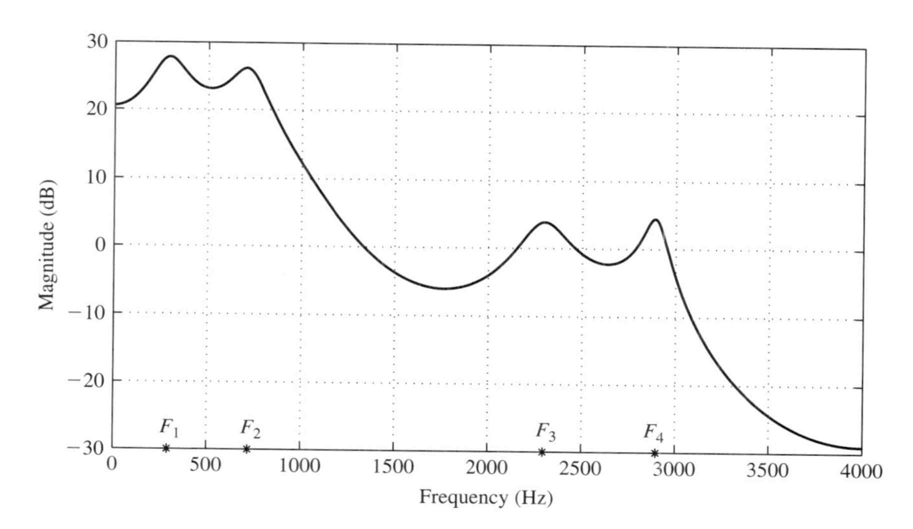

FIGURE 9.37
Plot of the frequency response $\tilde{H}(e^{j\omega})$ corresponding to the poles and zeros shown in Figure 9.36.

TABLE 9.3 Pole locations in the z-plane, identifying poles that are likely to correspond to formants.

Root magnitude	θ root angle(degrees)	F root angle (Hz)	Formant
0.9308	10.36	288	F_1
0.9308	−10.36	−288	F_1
0.9317	25.88	719	F_2
0.9317	−25.88	−719	F_2
0.7837	35.13	976	
0.7837	−35.13	−976	
0.9109	82.58	2294	F_3
0.9109	−82.58	−2294	F_3
0.5579	91.44	2540	
0.5579	−91.44	−2540	
0.9571	104.29	2897	F_4
0.9571	−104.29	−2897	F_4

formants. The poles are labeled in the natural order of increasing frequency (increasing angle in the z-plane). The assignment of root frequencies to formants in Table 9.3 is simply based on the closeness of the (complex) root to the unit circle, with all roots whose magnitude is greater than 0.9 being assigned to a formant frequency, and the remaining roots (whose magnitude is less than 0.9) being assumed to contribute to the representation of the gross spectral shape due to glottal pulse and the radiation components of the combined vocal tract model (or just to inaccuracies in the basic speech model which are not accounted for by the simple all-pole model). The frequencies corresponding to the angle of each pole are indicated by the "$*$" symbol on the frequency axis in Figure 9.37.

The assignment of pole locations from $\tilde{H}(z)$ to individual formants should be based on two criteria, namely the closeness of the complex pole to the unit circle, and the continuity of the pole location over some time interval, since a valid formant frequency at a given time will be manifest over a range of times in the vicinity of the current analysis frame. This concept of the formant frequency being manifested in the spectral section over a period of time is illustrated in Figure 9.38, which shows a wideband spectrogram of the same utterance as in the wideband spectrograms of Figure 9.13, with the estimated formants superimposed on the spectrogram plot. In this case the plot shows all-pole locations (angles of roots of $A(z)$) whose magnitudes are greater than a threshold of 0.9. We can clearly see the continuity of the formants throughout the voiced regions. We also see many instances of isolated pole locations that occur in unvoiced regions (where there are no pitch harmonics) or even in voiced regions. The goal of a successful formant-tracking algorithm is to retain the pole locations that exist over a reasonable range of time and thus correspond to the vocal tract resonances (i.e., the formants) and to eliminate those isolated occurrences that are not sustained over a reasonable time interval or that occur during unvoiced regions.

One way to remove some of the extraneous "formants" is to raise the threshold on the pole magnitude. Figure 9.39 shows a plot similar to Figure 9.38 but, in this

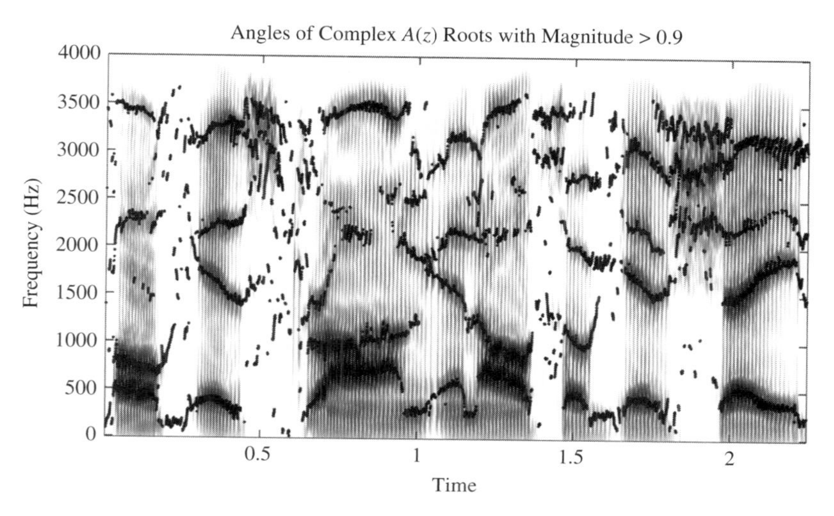

FIGURE 9.38
Wideband spectrogram showing the locations of all complex roots with magnitude greater than a threshold of 0.9. Regions of strong continuity of the root locations are clearly seen, corresponding to formant tracks of the speech. Regions of isolated root locations are also seen.

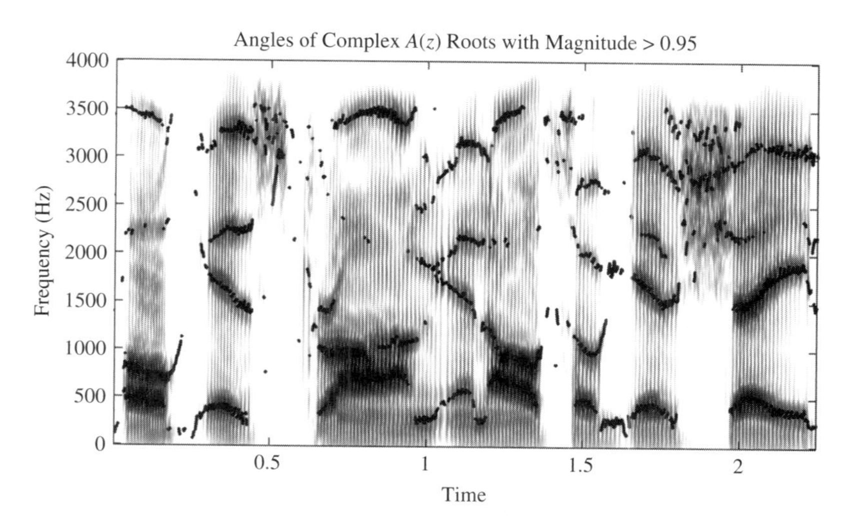

FIGURE 9.39
Wideband spectrogram showing the locations of all complex roots with magnitude greater than a threshold of 0.95.

case, with the threshold for plotting the root magnitudes set to 0.95, instead of 0.9. In this case we see a plot where virtually all of the pole locations align well, with reasonable formant tracks in the spectrogram. However, there are often gaps in the formant tracks where the formants are not defined over regions of time where the signal is clearly voiced and where there should be a well-defined formant track. Hence we see the conflict between using a low threshold (where we get too many false positive indications of formant locations) for assigning roots to formants, and using a higher threshold (where we get too many misses of valid formant locations because the root magnitude was below the threshold for some period of time). What is needed is an algorithm that incorporates more knowledge of the speech model such as temporal continuity of formant motions, expected ranges of formant frequencies, etc. We shall discuss such matters further in Chapter 10.

9.8 RELATION OF LINEAR PREDICTIVE ANALYSIS TO LOSSLESS TUBE MODELS

In Chapter 5 we discussed a model for speech production consisting of a concatenation of N lossless acoustic tubes (each, except the last tube, of length $\Delta x = \ell/N$) as shown in Figure 9.40, where ℓ is the overall length of the vocal tract model. In order to allow such a model to correspond to a discrete-time system, it is necessary to assume that the sampling rate of the input (and output) is related to the desired vocal tract length and number of sections by $F_s = cN/(2\ell)$, where c is the velocity of sound. The reflection coefficients, $\{r_k,\ k = 1, 2, \ldots, N-1\}$, in Figure 9.40b are related to the areas of the lossless tubes by

$$r_k = \frac{A_{k+1} - A_k}{A_{k+1} + A_k}. \tag{9.143}$$

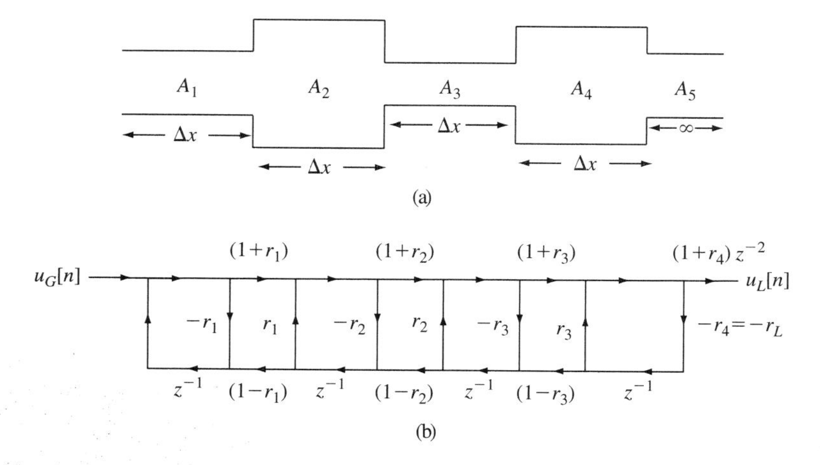

FIGURE 9.40
(a) Lossless tube model terminated in infinitely long tube; (b) corresponding signal flow graph for infinite glottal impedance.

In Section 5.2.4, the transfer function of such a system was derived subject to the condition that the reflection coefficient at the glottis was $r_G = 1$; i.e., the glottal impedance, r_G, was assumed to be infinite, and the reflection coefficient at the Nth section was defined as

$$r_N = r_L = \frac{\rho c/A_N - Z_L}{\rho c/A_N + Z_L}. \tag{9.144}$$

The acoustic impedance Z_L, which models the radiation at the lips, is the characteristic acoustic impedance of an infinite-length tube (to prevent reflections) whose area $A_{N+1} = \rho c/Z_L$ is chosen to introduce an appropriate amount of loss into the system. In Section 5.2.4, the system function of a system such as shown in Figure 9.40 was shown to have the form

$$V(z) = \frac{U_L(z)}{U_G(z)} = \frac{0.5(1 + r_G)\prod_{k=1}^{N}(1 + r_k)z^{-N/2}}{D(z)}, \tag{9.145}$$

where $D(z)$ satisfies the polynomial recursion

$$D_0(z) = 1, \tag{9.146a}$$

$$D_k(z) = D_{k-1}(z) + r_k z^{-k} D_{k-1}(z^{-1}), \quad k = 1, 2, \ldots, N, \tag{9.146b}$$

$$D(z) = D_N(z). \tag{9.146c}$$

All of this is very reminiscent of the discussion of the lattice formulation in Section 9.5.3. Indeed, there it was shown that the polynomial

$$A(z) = 1 - \sum_{k=1}^{p} \alpha_k z^{-k} \tag{9.147}$$

obtained by linear prediction analysis could be obtained by the recursion

$$A^{(0)}(z) = 1, \tag{9.148a}$$

$$A^{(i)}(z) = A^{(i-1)}(z) - k_i z^{-i} A^{(i-1)}(z^{-1}), \quad i = 1, 2, \ldots, p, \tag{9.148b}$$

$$A(z) = A^{(p)}(z), \tag{9.148c}$$

where the parameters $\{k_i\}$ were called the PARCOR coefficients. By comparing Eqs. (9.146a)–(9.146c) and (9.148a)–(9.148c), it is clear that the system function

$$\tilde{H}(z) = \frac{G}{A(z)} \tag{9.149}$$

obtained by linear prediction analysis has the same form as the system function of a lossless tube model consisting of p sections. If we make the associations $N = p$ and

$$r_i = -k_i, \qquad i = 1, 2, \dots, p, \tag{9.150}$$

then it is clear that although the numerators of $V(z)$ and $\tilde{H}(z)$ may differ, the denominators are the same; i.e.,

$$D(z) = A(z). \tag{9.151}$$

Thus, the lossless tube model of Figure 9.40 is equivalent, except for gain, to the all-pole lattice model of Figure 9.22. Using Eqs. (9.143) and (9.150), we get the relation

$$-1 < -k_i = r_i = \left(\frac{A_{i+1} - A_i}{A_{i+1} + A_i}\right) = \left(\frac{\dfrac{A_{i+1}}{A_i} - 1}{\dfrac{A_{i+1}}{A_i} + 1}\right) < 1, \tag{9.152}$$

and it is easy to show that the areas of the equivalent tube model are related to the PARCOR coefficients by

$$A_{i+1} = \left[\frac{1 - k_i}{1 + k_i}\right] A_i > 0 \text{ or equivalently } \frac{A_{i+1}}{A_i} = \left[\frac{1 - k_i}{1 + k_i}\right] > 0. \tag{9.153}$$

Note that each PARCOR coefficient (or reflection coefficient) gives us a ratio between areas of adjacent sections. Thus the areas of the equivalent tube model are not absolutely determined and any convenient normalization will produce a tube model with the same transfer function.

Rather than using area ratios, as expressed in Eq. (9.153), it has been shown that "log area ratios" of the form

$$g_i = \log\left(\frac{A_{i+1}}{A_i}\right) = \log\left(\frac{1 - k_i}{1 + k_i}\right) \tag{9.154}$$

are a more robust representation if the parameters are quantized, since the log area ratio coefficients minimize the spectral sensitivity under uniform quantization. We will examine this spectral sensitivity issue later in this chapter, and also in Chapter 11 when we discuss quantization methods for speech coders.

While the lossless tube model is a convenient way to connect discrete-time signals to the physics of speech production, and good agreement is attained in comparing the output of the discrete-time model with sampled speech, it must be conceded that the "area function" obtained using Eq. (9.153) cannot be said to be the area function of the human vocal tract. The areas are arbitrarily normalized, the human vocal tract varies continuously, and it is not lossless. However, Wakita [410] has shown that if pre-emphasis is used prior to linear predictive analysis to remove the effects due to the glottal pulse and radiation that are relatively constant with time and contribute mainly to the spectral tilt, then the resulting area functions are often very similar to spatially sampled vocal tract configurations that might occur in human speech production.

Based on the above discussion, Wakita [410] demonstrated a simple procedure for estimating the vocal tract tube areas (of a p-section tube approximation to the vocal tract) from the speech waveform, consisting of the following steps:

1. Sample the speech signal at sampling rate $F_s = 1/T = cp/(2\ell)$, where p is the model order and c is the velocity of sound. This gives the input signal, $s[n]$.
2. Remove the effects of the glottal source and the radiation load by using a simple, first order, pre-emphasis system of the form $x[n] = s[n] - s[n-1]$.
3. Compute the PARCOR coefficients on a short-time basis, giving the set $\{k_i, i = 1, 2, \dots, p\}$ for each analysis frame.
4. Assuming $A_1 = 1$ (arbitrary), compute A_2 to A_p using the formula

$$A_{i+1} = \left(\frac{1-k_i}{1+k_i}\right) A_i, \quad i = 1, 2, \dots, p-1. \tag{9.155}$$

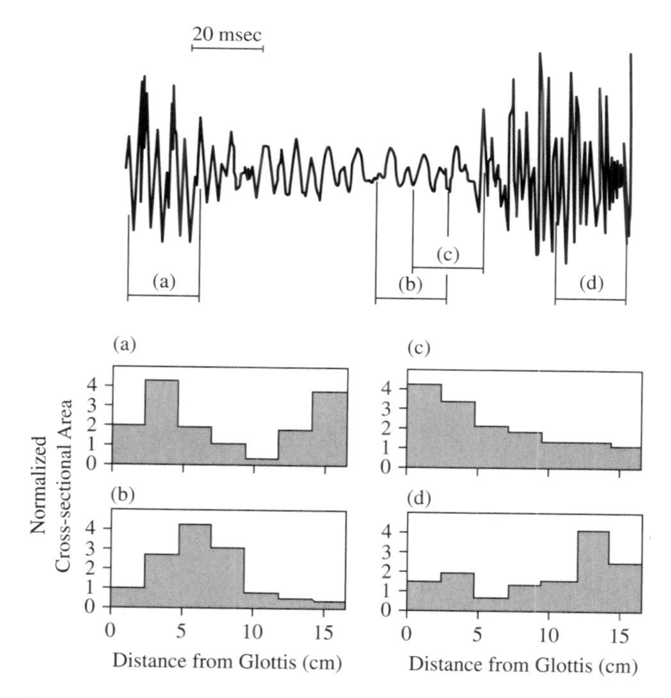

FIGURE 9.41
Simple example of use of Wakita procedure for estimation of tube areas from the speech waveform. Four sections of the nonsense syllable /IY B AA/ are used to give estimates of the vocal tract shape during the /IY/ sound [denoted by (a)], during the /B/ sound [denoted by (b) and (c)], and during the /AA/ sound [denoted by (d)]. (After Wakita [410]. © [1973] IEEE.)

The results of applying the Wakita procedure for estimating tube areas from the speech waveform are shown in Figures 9.41 and 9.42. Figure 9.41 shows four sections of the waveform for the nonsense syllable /IY B AA/ and the resulting estimates of vocal tract areas. The initial speech frame is during the /IY/ sound [denoted as (a) in the figure], where we see that the resulting vocal tract shape (shown at the bottom of the figure) is a typical vocal tract shape for a high front vowel. Speech frames denoted by (b) and (c) occur during the /B/ sound and we see that the resulting estimated vocal tract shapes are essentially closed at the lip end of the vocal tract; i.e., typical for what one might expect for a /B/ sound. Finally the speech frame denoted by section (d) occurs during the /AA/ sound, and the estimated vocal tract shape is again typical of a high mid-vowel vocal tract shape.

Figure 9.42 shows a set of five estimates of vocal tract shape from waveforms of five steady state vowels. Shown in the figure are the log magnitude spectral fits from the LP analysis (on the right side) along with the estimated vocal tract area functions

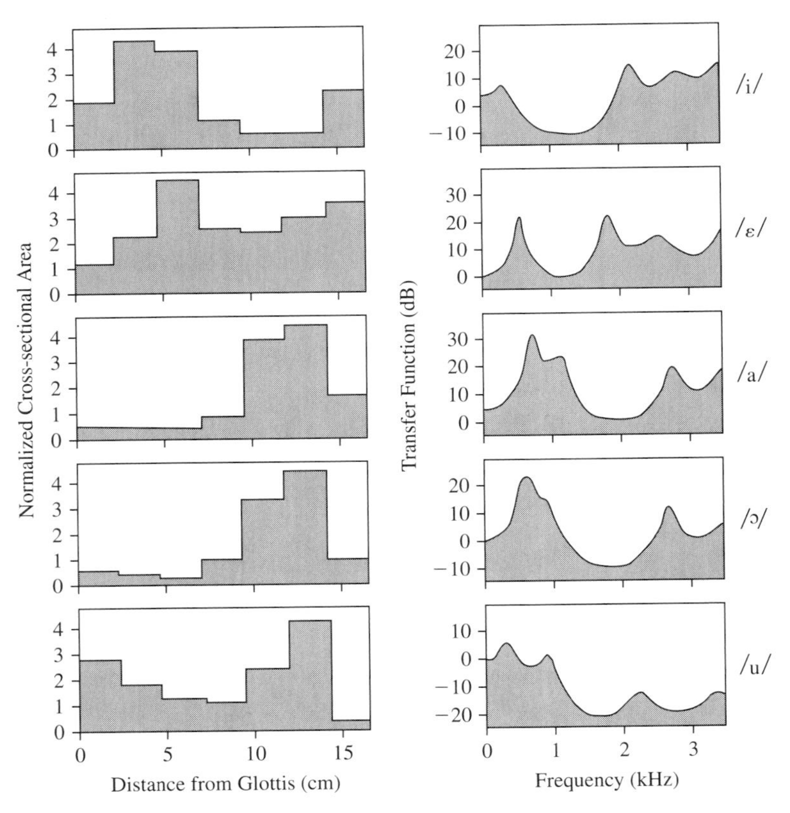

FIGURE 9.42
Use of the Wakita procedure to estimate tube areas for five vowels. Shown in the figure are the estimates of the vocal tract area (on the left side of the figure) along with the resulting LP spectrum (shown on the right side of the figure). (After Wakita [410]. © [1973] IEEE.)

(on the left side) for the five vowels. It can be seen that both the spectral shape and the vocal tract shape agree well with what might be expected for such vowel sounds, showing the efficacy of the procedure, at least for pseudo-steady state sounds.

9.9 ALTERNATIVE REPRESENTATIONS OF THE LP PARAMETERS

Although the set of predictor coefficients, $\{\alpha_k, 1 \leq k \leq p\}$, is often thought of as the basic parameter set of linear predictive analysis, it is straightforward to transform this set of coefficients to a number of other parameter sets, to obtain alternative representations that are often more convenient for applications of linear predictive analysis. In this section we discuss how other useful parameter sets can be obtained directly from LP coefficients [217, 232].

A large number of equivalent LP parameter sets have been proposed for use in various speech systems for coding, synthesis, recognition, verification, enhancement, etc. Table 9.4 lists eight of the most often used parameter sets, and this section discusses how to transform between parameter sets for some of the most common sets of parameters.

We want to be able to transform any LP parameter set to any other LP parameter set in a simple, well-defined manner. For most transformations, such transformations exist and several will be described in this section. Some LP parameter set transformations are not invertible (e.g., we cannot go from an autocorrelation of predictor coefficients back to the predictor coefficients since the inverse transformation is not unique or well defined), and we will note such cases as they occur.

9.9.1 Roots of the Prediction Error Polynomial

Perhaps the simplest alternative to the predictor parameters is the set of roots of the prediction error system function polynomial

$$A(z) = 1 - \sum_{k=1}^{p} \alpha_k z^{-k} = \prod_{k=1}^{p} (1 - z_k z^{-1}), \tag{9.156}$$

TABLE 9.4 LP parameter sets.

Parameter set	Representation
LP coefficients and gain	$\{\alpha_k, 1 \leq k \leq p\}, G$
PARCOR coefficients	$\{k_i, 1 \leq i \leq p\}$
Log area ratio coefficients	$\{g_i, 1 \leq i \leq p\}$
Roots of predictor polynomial	$\{z_k, 1 \leq k \leq p\}$
Impulse response of $\tilde{H}(z)$	$\{\tilde{h}[n], 0 \leq n \leq \infty\}$
LP cepstrum	$\{\hat{\tilde{h}}[n], -\infty \leq n \leq \infty\}$
Autocorrelation of impulse response	$\{\tilde{R}(i), -\infty \leq i \leq \infty\}$
Autocorrelation of predictor polynomial	$\{R_a[i], -p \leq i \leq p\}$
Line spectral pair parameters	$P(z), Q(z)$

which we have already shown have the potential for estimating the formant frequencies. The roots $\{z_k, k = 1, 2, \ldots, p\}$ are a unique representation of $A(z)$ and can be determined using a variety of root-finding algorithms such as the MATLAB function `roots( )`. If conversion of the z-plane roots to the s-plane (analog complex frequency) is desired, this can be achieved by setting

$$z_k = e^{s_k T}, \tag{9.157}$$

where $s_k = \sigma_k + j\Omega_k$ is the s-plane root corresponding to z_k in the z-plane. If $z_k = z_{kr} + jz_{ki}$, then

$$\Omega_k = 2\pi F_k = \frac{1}{T}\tan^{-1}\left[\frac{z_{ki}}{z_{kr}}\right], \tag{9.158}$$

where $T = 1/F_s$ is the sampling period, and

$$\sigma_k = \frac{1}{2T}\log(z_{kr}^2 + z_{ki}^2) = \frac{1}{T}\log|z_k|. \tag{9.159}$$

Equations (9.158) and (9.159) are useful for formant analysis applications, as suggested in Section 9.7.3, and as will be discussed in more detail in Chapter 10.

9.9.2 Impulse Response of $\tilde{H}(z)$, the All-Pole System

The impulse response, $\tilde{h}[n]$, of the all-pole system

$$\tilde{H}(z) = \frac{G}{1 - \sum_{k=1}^{p} \alpha_k z^{-k}} \tag{9.160}$$

can be solved for recursively from the LP coefficients as

$$\tilde{h}[n] = \sum_{k=1}^{p} \alpha_k \tilde{h}[n-k] + G\delta[n], \quad 0 \le n, \tag{9.161}$$

with initial rest conditions $\tilde{h}[n] = 0$ for $n < 0$ assumed. This is useful if we wish to compute a finite set of values of $\tilde{h}[n]$. If we require a closed-form expression for the impulse response, we can determine it from a partial fraction expansion of the form

$$\tilde{H}(z) = \sum_{k=1}^{p} \frac{A_k}{1 - z_k z^{-1}}, \tag{9.162}$$

from which it follows that

$$\tilde{h}[n] = \sum_{k=1}^{p} A_k (z_k)^n u[n]. \tag{9.163}$$

The partial fraction expansion coefficients A_k can be found numerically using the function `residuez( )` in MATLAB.

9.9.3 Autocorrelation of the Impulse Response

As discussed in Section 9.3, it is easily shown (see Problem 9.1) that the autocorrelation function of the impulse response of the all-pole filter defined as

$$\tilde{R}[i] = \sum_{n=0}^{\infty} \tilde{h}[n]\tilde{h}[n-i] = \tilde{R}[-i] \tag{9.164}$$

satisfies the relations

$$\tilde{R}[i] = \sum_{k=1}^{p} \alpha_k \tilde{R}[|i-k|], \quad 1 \leq i \tag{9.165}$$

and

$$\tilde{R}[0] = \sum_{k=1}^{p} \alpha_k \tilde{R}[k] + G^2. \tag{9.166}$$

Equations (9.165) and (9.166) can be used to determine $\tilde{R}[i]$ from the predictor coefficients and vice versa.

9.9.4 Cepstrum

Another alternative to the LP coefficients is the cepstrum of the impulse response of the LP model system. If the model system has transfer function $\tilde{H}(z)$ with impulse response $\tilde{h}[n]$ and complex cepstrum $\hat{\tilde{h}}[n]$, then it can be shown that, using the recursion relationship derived in Chapter 8, for minimum phase signals, $\hat{\tilde{h}}[n]$ can be obtained from the recursion

$$\hat{\tilde{h}}[n] = \alpha_n + \sum_{k=1}^{n-1} \left[\frac{k}{n}\right] \hat{\tilde{h}}[k]\, \alpha_{n-k}, \quad 1 \leq n. \tag{9.167}$$

Similarly we can derive the predictor coefficients from the cepstrum of the impulse response using the relation

$$\alpha_n = \hat{\tilde{h}}[n] - \sum_{k=1}^{n-1} \left[\frac{k}{n}\right] \hat{\tilde{h}}[k] \cdot \alpha_{n-k}, \quad 1 \leq p. \tag{9.168}$$

The cepstrum of the impulse response can also be obtained from the poles of $\tilde{H}(z)$. Using the equations derived in Chapter 8, it can be shown that

$$\hat{\tilde{h}}[n] = \begin{cases} 0 & n < 0 \\ \log(G) & n = 0 \\ \displaystyle\sum_{k=1}^{p} \frac{z_k^n}{n} & n \geq 1. \end{cases} \tag{9.169}$$

9.9.5 Autocorrelation Coefficients of the Predictor Polynomial

Corresponding to the predictor polynomial, or inverse filter,

$$A(z) = 1 - \sum_{k=1}^{p} \alpha_k z^{-k} \tag{9.170}$$

is the impulse response of the inverse filter

$$a[n] = \delta[n] - \sum_{k=1}^{p} \alpha_k \delta[n-k].$$

The autocorrelation function of the inverse filter impulse response is

$$R_a[i] = \sum_{k=0}^{p-i} a[k]a[k+i], \quad 0 \leq i \leq p. \tag{9.171}$$

9.9.6 PARCOR Coefficients

For the autocorrelation method, the predictor coefficients can be obtained from the PARCOR coefficients using the recursion of Figure 9.43, where it is assumed that the PARCOR parameters k_i, $i = 1, 2, \ldots, p$ are available. The algorithm in Figure 9.43 is obtained from the Levinson–Durbin algorithm in Figure 9.18 by replacing the step labeled as Eq. (9.93) by the appropriate value of k_i, which is assumed to be known in advance.

Similarly, if the predictor parameters α_j are known for $j = 1, 2, \ldots, p$, the set of PARCORs can be obtained using a backward recursion of the form given in Figure 9.44.

PARCOR-to-Predictor Algorithm

Given $k_1, k_2, \ldots, k_p$
for $i = 1, 2, \ldots, p$
$\quad \alpha_i^{(i)} = k_i$ Eq. (9.96b)
$\quad$ if $i > 1$ then for $j = 1, 2, \ldots, i-1$
$\quad\quad \alpha_j^{(i)} = \alpha_j^{(i-1)} - k_i \alpha_{i-j}^{(i-1)}$ Eq. (9.96a)
$\quad$ end
end
$\alpha_j = \alpha_j^{(p)} \quad j = 1, 2, \ldots, p$ Eq. (9.97)

FIGURE 9.43
Algorithm for converting from PARCOR (k-parameters) to predictor coefficients.

Predictor-to-PARCOR Algorithm

$a_j^{(p)} = a_j \;\; j = 1, 2, \ldots, p$
$k_p = a_p^{(p)}$ (P.1)
for $i = p, p-1, \ldots, 2$
$\quad$ for $j = 1, 2, \ldots, i-1$

$$a_j^{(i-1)} = \frac{a_j^{(i)} + k_i a_{i-j}^{(i)}}{1 - k_i^2} \qquad \text{(P.2)}$$

$\quad$ end
$\quad k_{i-1} = a_{i-1}^{(i-1)}$ (P.3)
end

FIGURE 9.44
Algorithm for converting from predictor coefficients to PARCOR coefficients.

9.9.7 Log Area Ratio Coefficients

An important set of equivalent parameters that can be derived from the PARCOR parameters is the log area ratio parameters defined as

$$g_i = \log\left[\frac{A_{i+1}}{A_i}\right] = \log\left[\frac{1 - k_i}{1 + k_i}\right], \quad 1 \leq i \leq p. \tag{9.172}$$

These parameters are equal to the log of the ratio of the areas of adjacent sections of a lossless tube equivalent of the vocal tract having the same transfer function as the linear predictive model, as discussed in Section 9.8. The g_i parameters have also been found to be especially useful for quantization [217, 232, 404] because of the relatively flat spectral sensitivity of the g_i's.

The k_i parameters can be directly obtained from the g_i by the inverse transformation

$$k_i = \frac{1 - e^{g_i}}{1 + e^{g_i}}, \quad 1 \leq i \leq p. \tag{9.173}$$

In applications such as coding, the LP parameters must be quantized. The quantized parameters correspond to a different (suboptimal) predictor and therefore to a different model system. Thus, we would prefer a parameter set that can be quantized without dramatically changing properties of the LP solution vector. In particular, it is important to maintain the accuracy of representation of the frequency response $\tilde{H}(e^{j\omega})$. Unfortunately most LP parameter sets are very sensitive to quantization and do not have low spectral sensitivity. The notion of spectral sensitivity is based on the magnitude-squared of the model frequency response, which is assumed to be of the form:

$$|H(e^{j\omega})|^2 = \frac{1}{|A(e^{j\omega})|^2} = P(\omega, g_i), \tag{9.174}$$

where g_i is a parameter that affects P, and the spectral sensitivity is formally defined as:

$$\frac{\partial S}{\partial g_i} = \lim_{\Delta g_i \to 0} \left| \frac{1}{\Delta g_i} \left[\frac{1}{2\pi} \int_{-\pi}^{\pi} \left| \log \frac{P(\omega, g_i)}{P(\omega, g_i + \Delta g_i)} \right| d\omega \right] \right|, \tag{9.175}$$

which measures sensitivity to errors in the g_i parameters.

Figure 9.45 shows the spectral sensitivity when using the PARCOR coefficients directly, i.e., $g_i = k_i$, as the basic representation of the LP parameters. Each curve in the figure is for a particular PARCOR coefficient k_i. As seen in Figure 9.45, the PARCOR coefficient set has very low spectral sensitivity around values of $k_i \approx 0$, but we also see that as $|k_i|$ approaches ± 1, the spectral sensitivity grows dramatically, making this coefficient set useless when the coefficients are heavily quantized (e.g., for use in an LP vocoder, as will be discussed in Chapter 11).

Figure 9.46, on the other hand, shows that the log area ratio parameter set, g_i in Eq. (9.172), has low spectral sensitivity over a very large dynamic range, making them useful as the basis for quantizing LP parameters in vocoders and other systems where the LP parameter sets must be quantized for storage or transmission purposes. A detailed study of the spectral sensitivity of the log area ratio transformation was done by Viswanathan and Makhoul [404].

Another transformation of the k_i parameter set that has been widely used in speech coding is the transformation

$$g_i = \arcsin(k_i), \qquad i = 1, 2, \ldots, p. \tag{9.176}$$

This transformation performs very nearly as well as the log area ratio transformation and has the highly desirable property that $-\pi/2 < g_i < \pi/2$ (since $-1 < k_i < 1$). Note that the log area ratio becomes quite large for values of k_i that are close to 1 in magnitude [11].

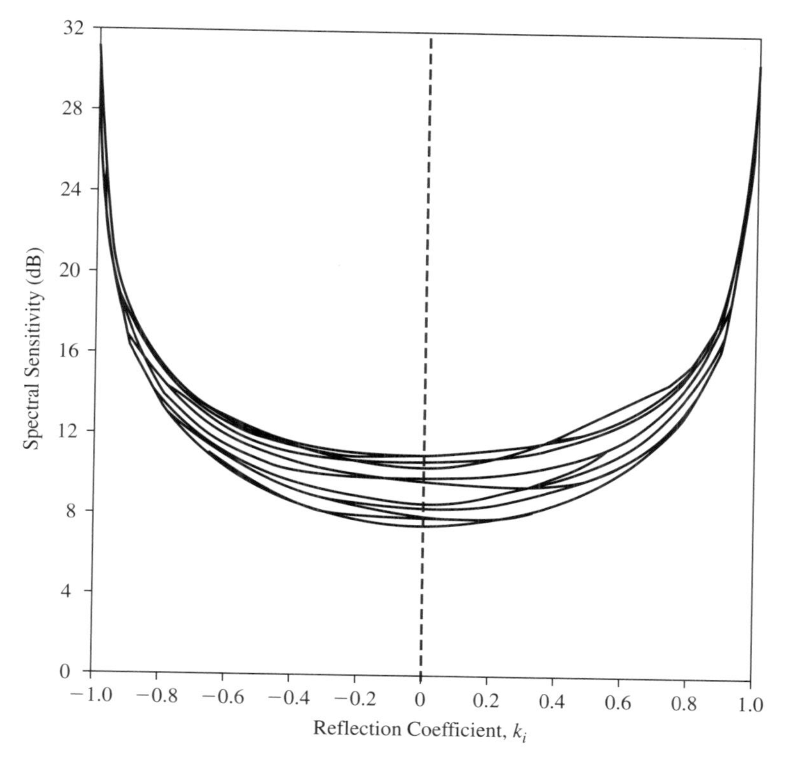

FIGURE 9.45
Plot of spectral sensitivity for k_i parameters. The low sensitivity around $k_i = 0$ is seen, along with the high sensitivity around $|k_i| \approx 1$. (After Viswanathan and Makhoul [404]. © [1975] IEEE.)

9.9.8 Line Spectral Pair Parameters

An alternative set of LP parameters, called the line spectral pair (LSP) parameters, was initially introduced by Itakura [160], and has been extensively studied by Sugamura and Itakura [385], Soong and Juang [373], and Crosmer and Barnwell [75].

The LSP representation is based on an extension of the pth-order all-zero prediction error filter, $A(z)$, which has all of its roots ($z_k, k = 1, 2, \ldots, p$) inside the unit circle ($|z_k| < 1$), and is of the form:

$$A(z) = 1 - \alpha_1 z^{-1} - \alpha_2 z^{-2} - \ldots - \alpha_p z^{-p}. \tag{9.177}$$

We form the reciprocal polynomial, $\tilde{A}(z)$, as

$$\tilde{A}(z) = z^{-(p+1)} A(z^{-1}) = -\alpha_p z^{-1} - \ldots - \alpha_2 z^{-p+1} - \alpha_1 z^{-p} + z^{-(p+1)}, \tag{9.178}$$

where the roots of $\tilde{A}(z)$ ($\tilde{z}_k, k = 1, 2, \ldots, p$) are the inverse of the roots of $A(z)$; i.e., $\tilde{z}_k = 1/z_k, k = 1, 2, \ldots, p$. We can then create an all-pass rational function, $F(z)$, of the

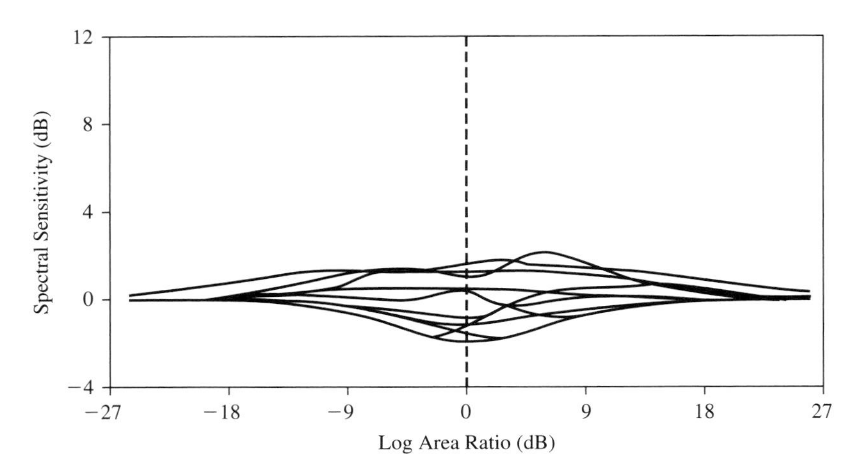

FIGURE 9.46
Plot of spectral sensitivity for log area ratio parameters, g_i. The low sensitivity for virtually the entire range of g_i is seen, making these parameters excellent candidates for use in an LP vocoder. (After Viswanathan and Makhoul [404]. © [1975] IEEE.)

form

$$F(z) = \frac{\tilde{A}(z)}{A(z)} = \frac{z^{-(p+1)}A(z^{-1})}{A(z)}, \tag{9.179}$$

which has the property that $|F(e^{j\omega})| = 1$ for all ω. Finally we create a set of two extended $(p+1)^{st}$-order polynomials:

$$P(z) = A(z) + \tilde{A}(z) = A(z) + z^{-(p+1)}A(z^{-1}), \tag{9.180}$$

$$Q(z) = A(z) - \tilde{A}(z) = A(z) - z^{-(p+1)}A(z^{-1}). \tag{9.181}$$

The polynomials $P(z)$ and $Q(z)$, called the *line spectral pair* (LSP), correspond to lossless vocal tract models with $p+1$ tube approximations to the vocal tract area function. The $(p+1)^{st}$ reflection equivalently occurs at the glottis and it is readily seen from Eqs. (9.180)–(9.181) that the glottal reflection coefficient is $k_G = +1$ (corresponding to an open glottis) for $P(z)$, and $k_G = -1$ (corresponding to a closed glottis) for $Q(z)$.

All of the roots of $P(z)$ and $Q(z)$ occur on the unit circle in the z-plane. We show this for $P(z)$ by realizing that the zeros for $P(z)$ occur when $A(z) = -\tilde{A}(z)$, or equivalently when $F(z) = -1$. Since $|F(e^{j\omega})| = 1$ for all ω, we see that the zeros of $P(z)$ occur when the phase of $F(e^{j\omega})$ satisfies the relation

$$\arg\{F(e^{j\omega_k})\} = \left(k + \frac{1}{2}\right) \cdot 2\pi, \quad k = 0, 1, \ldots, p-1. \tag{9.182}$$

Similarly, for the polynomial $Q(z)$, we see that $Q(z)$ has zeros when $A(z) = \tilde{A}(z)$, or equivalently when $F(z) = +1$. Again, since $|F(e^{j\omega})| = 1$ on the unit circle, the zeros of

$Q(z)$ occur when the phase of $F(e^{j\omega})$ satisfies the relation

$$\arg\{F(e^{j\omega_k})\} = k \cdot 2\pi, \quad k = 0, 1, \ldots, p-1. \tag{9.183}$$

Based on Eqs. (9.182)–(9.183), we see that the zeros of $P(z)$ and $Q(z)$ all fall on the unit circle, and are interlaced with each other [373]. The resulting set of zeros, $\{\omega_k\}$, when ordered in ascending frequency is called the set of *line spectral frequencies* (LSF).This set of LP parameters has been used extensively in low bit rate coding of speech [75]. The property of the LSF parameters that is most important is the guaranteed stability of the LP transfer function, $H(z) = 1/A(z)$, when we quantize the LSF parameters and then convert back to the polynomial $A(z)$ via the relation

$$A(z) = \frac{P(z) + Q(z)}{2}. \tag{9.184}$$

In summary, the properties of the LSP parameters are the following:

- $P(z)$ corresponds to a lossless tube with $p+1$ equal length sections from the lips to the glottis, open at the lips and open ($k_G = k_{p+1} = +1$) at the glottis.
- $Q(z)$ corresponds to a lossless tube with $p+1$ equal length sections from the lips to the glottis, open at the lips and closed ($k_G = k_{p+1} = -1$) at the glottis.

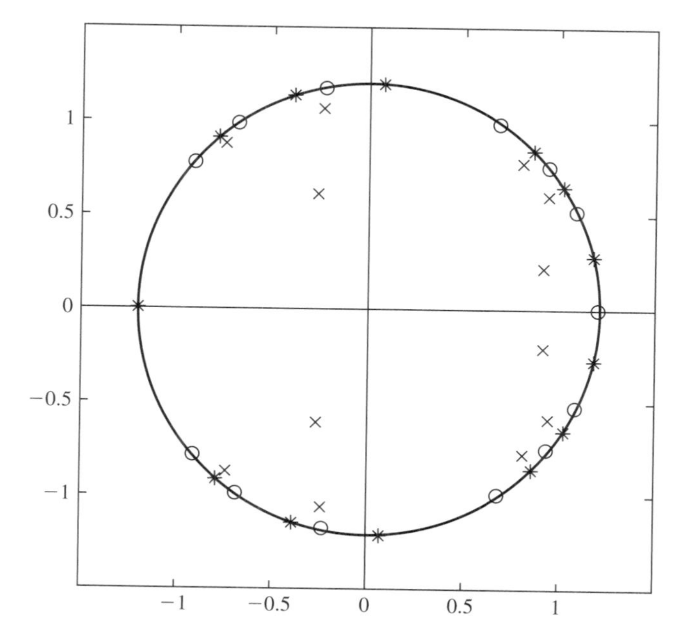

FIGURE 9.47
z-plane plot of the locations of the roots of the polynomials $A(z)$ [marked with $\times$], $P(z)$ [marked with $*$], and $Q(z)$ [marked with O].

- All the roots of $P(z)$ and $Q(z)$ are on the unit circle.
- If p is an even integer, then $P(z)$ has a root at $z = +1$, and $Q(z)$ has a root at $z = -1$.
- A necessary and sufficient condition for $|k_i| < 1, i = 1, 2, \ldots, p$ is that the roots of $P(z)$ and $Q(z)$ alternate on the unit circle in pairs.
- The LSF frequencies get close together when the roots of $A(z)$ are close to the unit circle (i.e., when the roots of $A(z)$ correspond to the formants of the vocal tract response).

To illustrate the properties of the LSP parameters, Figure 9.47 shows a plot of the locations of the roots of a $p = 12^{th}$-order polynomial $A(z)$ (indicated by $\times$), along with the roots of $P(z)$ (indicated by the symbol $*$), and $Q(z)$ (indicated by the symbol $\circ$) in the z-plane. The interlacing property of the roots of $P(z)$ and $Q(z)$ is illustrated by this figure, as is the closeness of the roots of $A(z)$ to the roots of $P(z)$ and $Q(z)$.

9.10 SUMMARY

In this chapter we have studied the technique of linear prediction as it is applied to speech analysis. We have primarily focused on the formulations that provide the most insight into the modeling of the process of speech production. We have discussed the issues involved with implementing these systems and have pointed out the similarities and differences between the basic methods, whenever possible. The technique of linear prediction has come to occupy a dominant position in the set of tools for speech analysis, synthesis, and recognition, as we will see in the remaining chapters of this book.

PROBLEMS

9.1. Consider the difference equation

$$\tilde{h}[n] = \sum_{k=1}^{p} \alpha_k \tilde{h}[n-k] + G\delta[n].$$

The autocorrelation function of $\tilde{h}[n]$ is defined as

$$\tilde{R}[m] = \sum_{n=0}^{\infty} \tilde{h}[n]\tilde{h}[n+m].$$

(a) Show that $\tilde{R}[m] = \tilde{R}[-m]$.

(b) By substituting the difference equation into the expression for $\tilde{R}[-m]$, show that

$$\tilde{R}[m] = \sum_{k=1}^{p} \alpha_k \tilde{R}[|m-k|], \quad m = 1, 2, \ldots, p.$$

9.2. The system function $\tilde{H}(z)$ evaluated at N equally spaced points on the unit circle is

$$\tilde{H}(e^{j\frac{2\pi}{N}k}) = \frac{G}{1 - \sum_{n=1}^{p} \alpha_n e^{-j\frac{2\pi}{N}kn}}, \quad 0 \le k \le N-1.$$

Describe a procedure for using an FFT algorithm to evaluate $\tilde{H}(e^{j\frac{2\pi}{N}k})$.

9.3. An unvoiced speech signal segment can be modeled as a segment of a stationary random process of the form:

$$x[n] = \epsilon[n] - \beta\epsilon[n-1],$$

where $\epsilon[n]$ is a zero mean, unit variance, stationary white noise process and $|\beta| < 1$.

(a) What are the mean and variance of $x[n]$?

(b) What system can be used to *recover* $\epsilon[n]$ from $x[n]$? That is, what is the whitening filter for $x[n]$?

(c) What is the normalized autocorrelation of $x[n]$ at a delay of 1 sample; i.e., what is $r_x[1] = \dfrac{R_x[1]}{R_x[0]}$?

9.4. Equation (9.29) can be used to reduce the amount of computation required to obtain the covariance matrix in the covariance method.

(a) Using the definition of $\phi_n[i,k]$ in the covariance method, prove Eq. (9.29); i.e., show that

$$\begin{aligned}\phi_n[i+1,k+1] = \phi_n[i,k] &+ s_n[-i-1]s_n[-k-1] \\ &- s_n[L-1-i]s_n[L-1-k].\end{aligned}$$

Now suppose that $\phi_n[i,0]$ is computed for $i = 0, 1, 2, \ldots, p$.

(b) Show that the elements on the main diagonal can be computed recursively starting with $\phi_n[0,0]$; i.e., obtain a recurrence formula for $\phi_n[i,i]$.

(c) Show that the elements on the lower diagonals can also be computed recursively beginning with $\phi_n[i,0]$.

(d) How can the elements on the upper diagonal be obtained?

9.5. A speech signal, $s[n]$, is predicted using the predictor:

$$\tilde{s}[n] = \sum_{k=k_0}^{k_1} \alpha_k s[n-k], \quad 1 \le k_0 < k_1 \le p,$$

where p is the usual order of linear predictive analysis for speech. Determine the optimum predictor that minimizes the mean-squared prediction error.

9.6. A speech signal was sampled with a sampling rate of $F_s = 8000$ samples/sec. A 300-sample segment was selected from a vowel sound and multiplied by a Hamming window. From this signal a set of linear prediction error filters

$$A^{(i)}(z) = 1 - \sum_{k=1}^{i} \alpha_k^{(i)} z^{-k},$$

with orders ranging from $i = 1$ to $i = 11$ was computed using the autocorrelation method. This set of predictors is shown in Table 9.6a in a form suggestive of the Levinson-Durbin recursion.

You should be able to answer the following questions by examining the entries in this table.

(a) Determine the z-transform $A^{(4)}(z)$ of the 4th-order prediction error filter. Draw and label the flow graph of the direct form implementation of this system.

(b) Determine the set of k-parameters $\{k_1, k_2, k_3, k_4\}$ for the 4th-order prediction error lattice filter. Draw and label the flow graph of the lattice implementation of this system.

(c) Use the numbers in Table 9.6a to verify numerically that

$$A^{(4)}(z) = A^{(3)}(z) - k_4 z^{-1} A^{(3)}(z^{-1}).$$

(d) The minimum mean-squared prediction error for the 2nd-order predictor is $E^{(2)} = 0.5803$. What is the minimum mean-squared prediction error $E^{(4)}$ for the 4th-order predictor? What is the total energy $R[0]$ of the windowed signal $s[n]$? What is the value of the autocorrelation function $R[1]$?

(e) The minimum mean-squared prediction errors for these predictors form a sequence $\{E^{(0)}, E^{(1)}, E^{(2)}, \ldots, E^{(11)}\}$. This sequence decreases abruptly in going from $i = 0$ to $i = 1$ and then decreases slowly for several orders and then makes a sharp decrease. At what order i would you expect this to occur?

(f) The system function of the 11th-order all-pole model is

$$\tilde{H}(z) = \frac{G}{A^{(11)}(z)} = \frac{G}{1 - \displaystyle\sum_{k=1}^{11} \alpha_k^{(11)} z^{-k}} = \frac{G}{\displaystyle\prod_{i=1}^{11} (1 - z_i z^{-1})}.$$

Table 9.6b gives five of the roots of the 11th-order prediction error filter $A^{(11)}(z)$. State briefly in words where the other six zeros of $A^{(11)}(z)$ are located. Be as precise as possible. If you cannot precisely determine the pole locations, explain where the pole might occur in the z-plane.

(g) Use information given in Table 9.6a and in part (c) of this problem to determine the gain parameter G for the 11th-order all-pole model.

(h) Estimate the first three formant frequencies (in Hz) for this segment of speech. Which of the first three formant resonances has the smallest bandwidth? How is this determined?

(i) Carefully sketch and label a plot of the frequency response of the all-pole model filter for analog frequencies $0 \leq F \leq 4000$ Hz.

TABLE 9.6a Prediction coefficients for a set of linear predictors.

i	$\alpha_1^{(i)}$	$\alpha_2^{(i)}$	$\alpha_3^{(i)}$	$\alpha_4^{(i)}$	$\alpha_5^{(i)}$	$\alpha_6^{(i)}$	$\alpha_7^{(i)}$	$\alpha_8^{(i)}$	$\alpha_9^{(i)}$	$\alpha_{10}^{(i)}$	$\alpha_{11}^{(i)}$
1	0.8328										
2	0.7459	0.1044									
3	0.7273	−0.0289	0.1786								
4	0.8047	−0.0414	0.4940	−0.4337							
5	0.7623	0.0069	0.4899	−0.3550	−0.0978						
6	0.6889	−0.2595	0.8576	−0.3498	0.4743	−0.7505					
7	0.6839	−0.2563	0.8553	−0.3440	0.4726	−0.7459	−0.0067				
8	0.6834	−0.3095	0.8890	−0.3685	0.5336	−0.7642	0.0421	−0.0713			
9	0.7234	−0.3331	1.3173	−0.6676	0.7402	−1.2624	0.2155	−0.4544	0.5605		
10	0.6493	−0.2730	1.2888	−0.5007	0.6423	−1.1741	0.0413	−0.4103	0.4648	0.1323	
11	0.6444	−0.2902	1.3040	−0.5022	0.6859	−1.1980	0.0599	−0.4582	0.4749	0.1081	0.0371

TABLE 9.6b Root locations of 11th-order prediction error filter.

i	$\lvert z_i \rvert$	$\angle z_i$ (rad)
1	0.2567	2.0677
2	0.9681	1.4402
3	0.9850	0.2750
4	0.8647	2.0036
5	0.9590	2.4162

9.7. A sampled speech signal is predicted using the predictor:

$$\tilde{s}[n] = \beta s[n - n_0], \quad 2 \leq n_0 \leq p,$$

where p is the normal predictor order for speech. Find the optimum values of β and $\mathcal{E}_{\hat{n}}$ that minimize the mean-squared prediction error using both the autocorrelation and covariance formulations; i.e., find optimum values of β and $\mathcal{E}_{\hat{n}}$ in terms of the signal correlations.

9.8. An LTI system has the system function:

$$H(z) = \frac{1}{1 - \beta z^{-1}}, \quad |\beta| < 1.$$

The system is excited by an input $x[n] = \delta[n]$ giving the output $y[n] = h[n] = \beta^n u[n]$.

(a) A segment of the output signal is selected at time $\hat{n}$ and windowed with a rectangular window of length N samples to give the segment:

$$s_{\hat{n}}[m] = \begin{cases} \beta^{\hat{n}+m} & 0 \leq m \leq L-1 \\ 0 & \text{otherwise.} \end{cases}$$

(i) Determine a closed-form expression for the autocorrelation function, $R_{\hat{n}}[k]$, for $-\infty < k < \infty$, assuming $\hat{n} \geq 0$.

(ii) Determine the predictor coefficient that minimizes the mean-squared prediction error:

$$\mathcal{E}_{\hat{n}}^{\text{auto}} = \sum_{m=-\infty}^{\infty} (s_{\hat{n}}[m] - \alpha_1 s_{\hat{n}}[m-1])^2.$$

(iii) Show that $\alpha_1 \to \beta$ as $L \to \infty$.

(iv) Compute the optimum prediction error using the formula:

$$\mathcal{E}_{\hat{n}}^{\text{auto}} = R_{\hat{n}}[0] - \alpha_1 R_{\hat{n}}[1]$$

and show that $E_{\hat{n}}^{\text{auto}}$ does <u>not</u> approach zero as $L \to \infty$.

(b) Now assume that $\hat{n} \geq 1$ and that the segment of the output signal is selected at time $\hat{n}$ such that:

$$s_{\hat{n}}[n] = \beta^{m+\hat{n}}, \qquad -1 \leq m \leq L-1.$$

(i) Compute the function, $\phi_{\hat{n}}[i,k]$, for $i=1$, $k=0,1$ as required in the covariance method of linear prediction.

(ii) Compute the predictor coefficient that minimizes the mean-squared prediction error:

$$\mathcal{E}_{\hat{n}}^{\text{cov}} = \sum_{m=0}^{L-1} (s_{\hat{n}}[m] - \alpha_1 s_{\hat{n}}[m-1])^2.$$

(iii) How does α_1 depend on L in this case?

(iv) Now compute the optimum prediction error using the formula:

$$\mathcal{E}_{\hat{n}}^{\text{cov}} = \phi_{\hat{n}}[0,0] - \alpha_1 \phi_{\hat{n}}[0,1].$$

How does $E_{\hat{n}}^{\text{cov}}$ depend on L in this case?

9.9. Linear prediction can be viewed as an optimal method of estimating a linear system, based on a certain set of assumptions. Figure P9.9 shows another way in which an estimate of a linear system can be made.

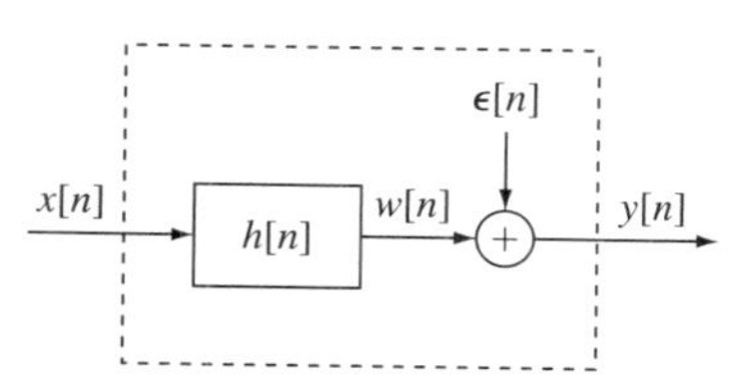

FIGURE P9.9
Output of linear time-invariant system with additive white noise.

Assume that we can observe both $x[n]$ and $y[n]$, and that $\epsilon[n]$ is a white Gaussian noise of zero mean, and variance σ_ϵ^2, and $\epsilon[n]$ is statistically independent of $x[n]$. An estimate of the impulse response of the linear system is desired such that the squared error

$$\mathcal{E} = E[(y[n] - \hat{h}[n] * x[n])^2]$$

is minimized, where $E[\ ]$ denotes expected value and $\hat{h}[n]$, $0 \leq n \leq M-1$ is the estimate of $h[n]$.

(a) Determine a set of linear equations for $\hat{h}[n]$ in terms of the autocorrelation function of $x[n]$ and the cross-correlation function between $y[n]$ and $x[n]$.

(b) How would you implement a solution to the set of equations derived in part (a)? How is this related to the linear predictive methods discussed in this chapter?

(c) Derive an expression for $\mathcal{E}$, the minimum mean-squared error.

9.10. In deriving the lattice formulation, the ith-order prediction error filter was defined as

$$A^{(i)}(z) = 1 - \sum_{k=1}^{i} \alpha_k^{(i)} z^{-k}.$$

The predictor coefficients satisfy

$$\alpha_j^{(i)} = \alpha_j^{(i-1)} - k_i \alpha_{i-j}^{(i-1)}, \quad 1 \le j \le i-1,$$
$$\alpha_i^{(i)} = k_i.$$

Substitute these expressions for $\alpha_j^{(i)}$, $1 \le j \le i$ into the expression for $A^{(i)}(z)$ to obtain

$$A^{(i)}(z) = A^{(i-1)}(z) - k_i z^{-i} A^{(i-1)}(z^{-1}).$$

9.11. The k-parameters in the Burg (lattice) method are obtained as:

$$\hat{k}_i = \frac{2\sum_{m=0}^{L-1} e^{(i-1)}[m] b^{(i-1)}[m-1]}{\sum_{m=0}^{L-1} (e^{(i-1)}[m])^2 + \sum_{m=0}^{L-1} (b^{(i-1)}[m-1])^2}.$$

Show that $|\hat{k}_i| < 1$, $i = 1, 2, \ldots, p$. (Hint: Consider the algebraic difference between the denominator and the numerator, and use this to show that the magnitude of the denominator is always greater than the magnitude of the numerator. Be sure to consider all possibilities for the signs of the numerator and the denominator.)

9.12. Given a section of speech, $s[n]$, which is perfectly periodic with period N_p samples, $s[n]$ can be represented as the discrete Fourier series

$$s[n] = \sum_{k=1}^{M} \left[\beta_k e^{j\frac{2\pi}{N_p}kn} + \beta_k^* e^{-j\frac{2\pi}{N_p}kn} \right],$$

where M is the number of harmonics of the fundamental $(2\pi/N_p)$ that are present. To spectrally flatten the signal (to aid in pitch detection), we desire a signal $y[n]$ of the form

$$y[n] = \sum_{k=1}^{M} \left[e^{j\frac{2\pi}{N_p}kn} + e^{-j\frac{2\pi}{N_p}kn} \right].$$

This problem is concerned with a procedure for spectrally flattening a signal using a combination of linear predictive and homomorphic processing techniques.

(a) Show that the spectrally flattened signal, $y[n]$, can be expressed as

$$y[n] = \frac{\sin\left[\frac{\pi}{N_p}(2M+1)n\right]}{\sin\left[\frac{\pi}{N_p}n\right]} - 1.$$

Sketch this sequence for $N_p = 15$ and $M = 2$.

(b) Now suppose that an linear predictive analysis is done on $s[n]$ using a window that is several pitch periods long and a value p in the linear predictive analysis such that $p = 2M$. From this analysis, the system function

$$H(z) = \frac{1}{1 - \sum_{k=1}^{p} \alpha_k z^{-k}} = \frac{1}{A(z)}$$

is obtained. The denominator can be represented as

$$A(z) = \prod_{k=1}^{p} (1 - z_k z^{-1}).$$

(c) How are the $p = 2M$ zeros of $A(z)$ related to the frequencies present in $s[n]$? The cepstrum $\hat{h}[n]$, of the impulse response, $h[n]$, is defined as the sequence whose z-transform is

$$\hat{H}(z) = \log H(z) = -\log A(z).$$

[Note that $\hat{h}[n]$ can be computed from the α_k's using Eq. (9.167).] Show that $\hat{h}[n]$ is related to the zeros of $A(z)$ by

$$\hat{h}[n] = \sum_{k=1}^{p} \frac{z_k^n}{n}, \quad n > 0.$$

(d) Using the results of (a) and (b), argue that

$$y[n] = n\,\hat{h}[n]$$

is a spectrally flattened signal as desired for pitch detection.

9.13. The "standard" method for obtaining the short-time spectrum of a section of speech is shown in Figure P9.13a. A much more sophisticated, and computationally more expensive, method of obtaining $\log |X(e^{j\omega})|$ is shown in Figure P9.13b.

(a) Discuss the new method of obtaining $\log |X(e^{j\frac{2\pi}{N}k})|$ and explain what the spectral correction system should be.

(b) What are the possible advantages of this new method? Consider the use of windows, the presence of zeros in the spectrum of $x[n]$, etc.

9.14. A causal LTI system has system function:

$$H(z) = \frac{1 - 4z^{-1}}{1 - 0.25z^{-1} - 0.75z^{-2} - 0.875z^{-3}}.$$

(a) Use the Levinson-Durbin recursion to determine whether or not the system is stable.

(b) Is the system minimum phase?

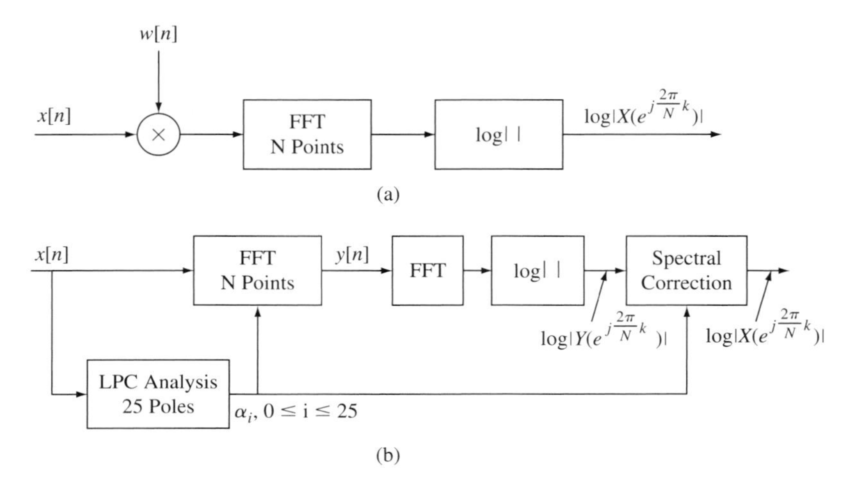

FIGURE P9.13
Methods of obtaining short-time spectrum of section of speech.

9.15. Consider an optimum linear prediction polynomial, of order 3, of the form:

$$A(z) = 1 + 0.5z^{-1} + 0.25z^{-2} + 0.5z^{-3}.$$

(a) Find the PARCOR coefficients associated with the optimum linear prediction polynomial.

(b) Is the linear prediction polynomial stable?

9.16. A speech signal frame of length L samples has energy:

$$R_{\hat{n}}[0] = \mathcal{E}_{\hat{n}}^{(0)} = \sum_{m=0}^{L-1} s_{\hat{n}}^2[m] = 2000.$$

Using the autocorrelation method of analysis on this speech frame, the first three PARCOR coefficients are computed and their values are:

$$k_1 = -0.5,$$
$$k_2 = 0.5,$$
$$k_3 = 0.2.$$

Find the energy of the linear prediction residual, $\mathcal{E}_{\hat{n}}^{(3)} = \sum_{m=0}^{L+2} e_{\hat{n}}^2[m]$ that would be obtained by inverse filtering $s_{\hat{n}}[m]$ by the optimal third-order predictor inverse filter, $A^{(3)}(z)$.

9.17. Linear prediction based on autocorrelation analysis is performed on a segment of speech. The resulting mean-squared errors are found to be:

$$\mathcal{E}^{(0)} = 10 \quad \text{for zeroth-order analysis,}$$

$$\mathcal{E}^{(1)} = 5.1 \quad \text{for first-order analysis,}$$

$$\mathcal{E}^{(2)} = 4.284 \quad \text{for second-order analysis,}$$

$$\mathcal{E}^{(3)} = 4.11264 \quad \text{for third-order analysis.}$$

Find the third order, all-pole, speech model for this frame. (Hint: The solution is not unique, so you must make some assumptions about signs. Any set of assumptions, consistent with the set of mean-squared errors, is acceptable.)

9.18. The *optimum* pitch predictor for voiced speech tries to minimize the average squared error between the speech signal, and the speech signal delayed by a pitch period; i.e.,

$$\mathcal{E}(\beta, N_p) = \sum_m (s[m] - \beta s[m - N_p])^2,$$

where β is the predictor coefficient, and N_p is the candidate pitch period.

(a) Find a formula for the optimum value of β that minimizes $\mathcal{E}(\beta, N_p)$ with N_p held fixed.

(b) Using the optimum value of β, state a simple procedure to find the optimum value of N_p.

(c) Contrast this method for pitch detection with the autocorrelation method (i.e., when would it work well, when might it run into problems).

9.19. Consider the LSF polynomials, $P(z)$ and $Q(z)$ derived from the pth optimum prediction error filter, $A(z)$:

$$P(z) = A(z) + z^{-(p+1)}A(z^{-1}),$$

$$Q(z) = A(z) - z^{-(p+1)}A(z^{-1}).$$

(a) Show that if p is even, then $P(z)$ has a zero at $z = -1$, and $Q(z)$ has a zero at $z = +1$.

(b) Show that the zeros of $P(z)$ and $Q(z)$ must be on the unit circle if $A(z)$ is the system function of an optimum prediction error filter derived by the autocorrelation method.

9.20. If we assume a flat speech spectrum for a degenerative p^{th}-order polynomial, i.e., $A(z) = 1$, show that the resulting LSF are equally spaced over the interval $[0, \pi]$ with the roots of $P(z)$ and $Q(z)$ being interleaved with spacing $\Delta\omega = \pi/(p+1)$.

9.21. Determine the second-order linear prediction inverse filter, $A(z)$, for which the two LSF are 666.67 Hz and 2000 Hz, when $F_s = 8000$ samples/sec. (Note: You may want to use the relationship $\cos(\pi/6) = \sqrt{3}/2$ in your solution.)

9.22. A method for detecting pitch, based on linear predictive analysis, is to use the autocorrelation function of the prediction error signal, $e[n]$. Recall that $e[n]$ can be written as

$$e[n] = x[n] - \sum_{i=1}^{p} \alpha_i x[n-i]$$

and if we define $\alpha_0 = -1$, then

$$e[n] = -\sum_{i=0}^{p} \alpha_i x[n-i],$$

where the windowed signal $x[n] = s[n]w[n]$ is non-zero for $0 \leq n \leq L-1$, and zero everywhere else. (Note that we have omitted the frame subscript $\hat{n}$ to simplify notation.)

(a) Show that the autocorrelation function of $e[n]$, $R_e[m]$, can be written in the form

$$R_e[m] = \sum_{l=-\infty}^{\infty} R_\alpha[l] R_x[m-l],$$

where $R_\alpha[l]$ is the autocorrelation function of the sequence of predictor coefficients, and $R_x[l]$ is the autocorrelation function of $x[n]$.

(b) For a speech sampling rate of 10 kHz, how much computation (i.e., multiplies and adds) is required to evaluate $R_e[m]$ for values of m in the interval 3–15 msec?

9.23. Consider two (windowed) speech sequences $x[n]$ and $\hat{x}[n]$ both defined for $0 \leq n \leq L-1$. (Outside this region the sequences $x[n]$ and $\hat{x}[n]$ are defined to be 0.) We perform an LPC analysis (using the autocorrelation method) of each frame. Thus we obtain autocorrelation sequences $R[k]$ and $\hat{R}[k]$ defined as

$$R[k] = \sum_{n=0}^{L-1-k} x[n]x[n+k], \quad 0 \leq k \leq p,$$

$$\hat{R}[k] = \sum_{n=0}^{L-1-k} \hat{x}[n]\hat{x}[n+k], \quad 0 \leq k \leq p.$$

From the autocorrelation sequences, we solve for the predictor parameters $\boldsymbol{\alpha} = (\alpha_0, \alpha_1, \ldots, \alpha_p)$ and $\hat{\boldsymbol{\alpha}} = (\hat{\alpha_0}, \hat{\alpha_1}, \ldots, \hat{\alpha_p})$ $(\alpha_0 = \hat{\alpha_0} = -1)$.

(a) Show that the prediction (residual) error, defined as

$$\mathcal{E}^{(p)} = \sum_{n=0}^{L-1+p} e^2[n] = \sum_{n=0}^{L-1+p} \left[-\sum_{i=0}^{p} \alpha_i x[n-i] \right]^2$$

can be written in the form

$$E^{(p)} = \boldsymbol{\alpha} \boldsymbol{R}_\alpha \boldsymbol{\alpha}^T,$$

where $\boldsymbol{R}_\alpha$ is a $(p+1)$ by $(p+1)$ matrix. Determine $\boldsymbol{R}_\alpha$.

(b) Consider passing the input sequence $\hat{x}[n]$ through the prediction error system with coefficients $\boldsymbol{\alpha}$, to give the error signal $f[n]$, defined as

$$f[n] = -\sum_{i=0}^{p} \alpha_i \hat{x}[n-i].$$

Show that the mean-squared error, $\mathcal{F}^{(p)}$, defined as

$$\mathcal{F}^{(p)} = \sum_{n=0}^{L-1+p} (f[n])^2$$

can be written in the form

$$\mathcal{F}^{(p)} = \boldsymbol{\alpha} \boldsymbol{R}_{\hat{\alpha}} \boldsymbol{\alpha}^{\boldsymbol{T}},$$

where $\boldsymbol{R}_{\hat{\alpha}}$ is a $(p+1)$ by $(p+1)$ matrix. Determine $\boldsymbol{R}_{\hat{\alpha}}$.

(c) If we form the ratio

$$D = \frac{\mathcal{F}^{(p)}}{\mathcal{E}^{(p)}},$$

what can be said about the range of values of D?

9.24. A proposed measure of similarity between two frames of speech with linear prediction coefficients $\boldsymbol{\alpha}$ and $\hat{\boldsymbol{\alpha}}$, and augmented correlation matrices $\boldsymbol{R}_{\alpha}$ and $\boldsymbol{R}_{\hat{\alpha}}$ (see Problem 9.23) is

$$D(\boldsymbol{\alpha}, \hat{\boldsymbol{\alpha}}) = \frac{\boldsymbol{\alpha} \boldsymbol{R}_{\hat{\alpha}} \boldsymbol{\alpha}^{\boldsymbol{T}}}{\hat{\boldsymbol{\alpha}} \boldsymbol{R}_{\hat{\alpha}} \hat{\boldsymbol{\alpha}}^{\boldsymbol{T}}}.$$

(a) Show that the distance function $D(\boldsymbol{\alpha}, \hat{\boldsymbol{\alpha}})$ can be written in the computationally efficient form

$$D(\boldsymbol{\alpha}, \hat{\boldsymbol{\alpha}}) = \left[\frac{\left(b[0]\hat{R}[0] + 2\sum_{i=1}^{p} b[i]\hat{R}[i]\right)}{\hat{\boldsymbol{\alpha}} \boldsymbol{R}_{\hat{\alpha}} \hat{\boldsymbol{\alpha}}^{\boldsymbol{T}}} \right],$$

where $b[i]$ is the autocorrelation of the $\boldsymbol{\alpha}$ array; i.e.,

$$b[i] = \sum_{j=0}^{p-i} \alpha_j \alpha_{j+i}, \quad 0 \le i \le p.$$

(b) Assume the quantities (i.e., vectors, matrices, scalars) $\boldsymbol{\alpha}$, $\hat{\boldsymbol{\alpha}}$, $\boldsymbol{R}$, $\hat{\boldsymbol{R}}$, $(\hat{\boldsymbol{\alpha}}\, \boldsymbol{R}_{\hat{\alpha}} \hat{\boldsymbol{\alpha}}^{\boldsymbol{T}})$, $\boldsymbol{R}_{\hat{\alpha}}$, $\boldsymbol{b}$ are precomputed–i.e., they are available at the time the distance computation is required. Contrast the computation required to evaluate $D(\boldsymbol{\alpha}, \hat{\boldsymbol{\alpha}})$ using both expressions for D given in this problem.

9.25. (MATLAB Exercise) Write a MATLAB program to compute the optimal set of linear prediction coefficients from a frame of speech using the three LPC analysis methods discussed in this chapter; i.e., the autocorrelation method, the covariance method, and the lattice method. Initially you should use the MATLAB routines `durbin.m`, `cholesky.m`, and `lattice.m` provided on the book website to solve for the optimal LPC solutions. After all LPC analysis methods are working properly, write your own routines for each of the three methods of solution and compare your results with those from the routines provided on the book website. To test your program, use a frame of a steady vowel and a frame of a fricative sound. Hence for each test example, you should input the following:

- speech filename: `ah.wav` for the voiced test frame, `test_16k.wav` for the unvoiced test frame
- starting sample of frame: 3000 for both the voiced and unvoiced test frames
- frame length (in samples): 300 for the voiced frame (where the sampling frequency is 10 kHz), 480 for the unvoiced frame (where the sampling frequency is 16 kHz)
- window type: Hamming for both frames
- LPC analysis order: 12 for both the voiced and unvoiced frames

For each of the analyzed frames, plot (on a single plot) the short-time Fourier transform log magnitude spectrum of the windowed frame of speech and the LPC log magnitude spectra from the three LPC analysis methods. (Don't forget that for the covariance and lattice methods, you also need to preserve $p = 12$ samples before the frame starting sample for computing correlations and error signals.)

9.26. (MATLAB Exercise) Write a MATLAB program to convert either from LPC coefficients (for a pth-order system) to PARCOR coefficients, or from PARCOR coefficients to LPC coefficients. For the routine that converts from LPC coefficients to PARCOR coefficients, use the test calling sequence:

```
p=4;
a=[0.2 0.2 0.2 0.1];
kp=lpccoef_parcor(p,a);
```

Your output should be `kp=[0.4 0.2857 0.2222 0.1]`. For the routine that converts from PARCOR coefficients to LPC coefficients, use the test calling sequence:

```
p=4;
kp=[0.9 -0.5 -0.3 0.1];
a=parcor_lpccoef(p,kp);
```

Your output should be `a=[1.23 -0.0855 -0.42 0.1]`.

9.27. (MATLAB Exercise) Write a MATLAB program to show the effect (on the LPC log magnitude spectrum) of a range of values of p, the LPC system order, on the spectral matching properties of the lattice method of analysis. Your program should select a frame of speech and then perform LPC analysis using the lattice method for a range of values of p; i.e., $p, p+20, p+40, p+60, p+80$. Your program should plot the log magnitude spectrum (in dB) for the original speech frame, along with the LPC log magnitude spectrums for each of the five values of p. Test your program using the speech file `test_16k.wav` with a voiced frame starting sample of 6000, with frame length of 640 samples, with LPC order 16, and using a Hamming window. What happens to the degree of spectral match as p increases?

9.28. (MATLAB Exercise) Write a MATLAB program to show the effect (on the LPC log magnitude spectrum) of a range of values of L, the LPC frame size (in samples), on the spectral matching properties of the lattice method of analysis. Your program should select a frame of speech and then perform LPC analysis using the lattice method for a range of values of L; i.e., 160, 320, 480, 640, and 800. This set of values was chosen for a value of $F_s = 16{,}000$ Hz. What changes would be made if we used a speech file with a sampling rate of $F_S = 10{,}000$ Hz and wanted to keep the same set of frame durations? Your program should plot the log magnitude spectrum (in dB) for the original speech frame, along with the LPC log magnitude spectrums for each of the five values of L. Test your program using the speech file `test_16k.wav` with voiced frame starting sample 6000, with frame length 640 samples, with LPC order 16, and using a Hamming window. What happens to the degree of spectral match as L increases?

9.29. (MATLAB Exercise) Write a MATLAB program to calculate and plot the location of the roots of the prediction error system function polynomial, $A(z)$, and to transform the polynomial $A(z)$ into the LSF polynomials, $P(z)$ and $Q(z)$, defined as:

$$P(z) = A(z) + z^{-p+1}A(z^{-1}),$$

$$Q(z) = A(z) - z^{-p+1}A(z^{-1}),$$

and plot the roots of $P(z)$ and $Q(z)$ on a common plot. Use the section of voiced speech beginning at sample 6000 in speech file `test_16k.wav`, with frame length 640 samples (40 msec at $F_s = 16$ kHz), and with an order of $p = 16$. Compare the root locations of the LPC polynomial, $A(z)$, with the root locations of the LSP polynomials. Do the roots appear where you expect them? To successfully complete this exercise, you will need to transform the LPC polynomial into the LSP polynomials. You should use the MATLAB function `atolsp`, which is provided on the book website, with calling sequence `[P,PF,Q,QF]=atolsp(A,fs)`, where the input and output parameters are defined as:

```
A=prediction error filter coefficients (column vector)
fs=sampling frequency
P=lsp polynomial (column vector)
PF=line spectral frequencies of P (column vector)
Q=lsp polynomial (column vector)
PQ=line spectral frequencies of Q (column vector)
```

You also need to use the MATLAB routine

```
zroots(numerator,denominator)
```

to find and plot the roots of each of the polynomials in the z−plane.

9.30. (MATLAB Exercise) Write a MATLAB program that takes a speech file and computes either a conventional or LPC spectrogram (either gray scale or color scale), and superimposes the roots of the LPC polynomial whose magnitude is greater than some threshold (i.e., it plots the candidates for formant frequencies). Test your program using

the following sets of speech files and parameters for computing the spectrograms and the LPC roots:
Testing Set #1

- speech file: `s5.wav`
- frame size (samples): 400 (50 msec at $F_S = 8$ kHz)
- frame shift (samples): 40 (5 msec at $F_S = 8$ kHz)
- LPC order: 12
- minimum radius for LPC root to be plotted: 0.8
- plot both short-time Fourier transform and LPC spectrograms

Testing Set #2

- speech file: `we were away a year ago_lrr.wav`
- frame size (samples): 800 (50 msec at $F_S = 16$ kHz)
- frame shift (samples): 80 (5 msec at $F_S = 16$ kHz)
- LPC order: 16
- minimum radius for LPC root to be plotted: 0.8
- plot both short-time Fourier transform and LPC spectrograms

9.31. (MATLAB Exercise) Write a MATLAB function that converts from a set of $p = N$ areas of a set of concatenated lossless tubes (with a specified reflection coefficient at the Nth tube output) to a set of reflection coefficients, r_k, $k = 1, 2, \ldots, N$ using Eq. (9.143) for the first $N - 1$ reflection coefficients, and the specified reflection coefficient, r_N, as the Nth reflection coefficient. Also convert from the set of reflection coefficients to the LPC polynomial using the recursion relations of Eqs. (9.146a)–(9.146c). The calling function for this conversion from areas to reflection coefficients and to the LPC polynomial should be:

```
[r,D,G]=AtoV(A,rN)
% A=array of areas (size N)
% rN=reflection coefficient at lips
% r=set of N reflection coefficients
% D=array of LPC polynomial denominator coefficients
% G=numerator of LPC transfer function
```

To test your function, you should write the inverse function that converts from the LPC polynomial coefficients (with an arbitrary area of the first lossless tube model) to the set of reflection coefficients and then to the lossless tube areas, with calling function:

```
[r,A]=VtoA(D,A1)
% D=array of LPC polynomial denominator coefficients
% A1=arbitrary area of first lossless tube
% r=set of N reflection coefficients
% A=array of areas (size N)
```

Test your functions with the set of 10 lossless tube areas (corresponding to a vocal tract shape for an /iy/ vowel):

```
aIY=[2.6, 8, 10.5, 10.5, 8, 5, 0.65, 0.65, 1.3, 3.2]
```

and with the Nth reflection coefficient $r_N = 0.71$.

9.32. (MATLAB Exercise) Write a MATLAB program to synthesize a vowel sound from a given set of areas of an N-tube lossless vocal tract model, along with the reflection coefficient at the end of the Nth tube, r_N. Assume a sampling rate of $F_s = 10$ kHz, and a pitch period of 100 samples. Using the MATLAB function `AtoV` of Problem 9.31, convert from the set of areas to a set of reflection coefficients and then to the LPC polynomial that best matches the given area function. Next convert the LPC transfer function to the impulse response of the LPC polynomial using the MATLAB filter function. To create the synthetic vowel, generate an excitation source of 10 or so pitch periods and then convolve the excitation source with a glottal pulse, then with the LPC impulse response, and finally with a radiation load transfer function in order to generate the vowel output signal. For the glottal source, you should consider using either the Rosenberg pulse (as discussed in Chapter 5) or a simple exponential pulse of the form $g[n] = a^n$ with $a = 0.91$. For the radiation load, you can use a simple transfer function of the form $R(z) = 1 - 0.98z^{-1}$. Plot the log magnitude response (in dB) of the final vowel signal and listen to the sound to verify that you have created a reasonable synthetic vowel.

You can test your MATLAB program with the following area function data for the /IY/ and /AA/ vowel sounds:

/iy/ vowel area function: `aIY=[2.6, 8, 10.5, 10.5, 8, 5, 0.65, 0.65, 1.3, 3.2], rN=0.71`

/aa/ vowel area function: `aAA=[1.6, 2.6, 0.65, 1.6, 2.6, 4, 6.5, 8, 7, 5], rN=0.71`

9.33. (MATLAB Exercise) Write a MATLAB program to analyze a speech file using LPC analysis methods, extract the error signal, and then use the error signal to do an exact reconstruction of the original speech file. Use the speech file `s5.wav` as input with LPC frame size of 320 samples (40 msec at $F_S = 8$ kHz), LPC frame shift of 80 samples (10 msec at $F_S = 8$ kHz), and LPC order $p = 12$. On a single page, plot the following quantities:

- the original speech signal, $s[n]$
- the error signal, $e[n]$
- the resynthesized speech signal, $\hat{s}[n]$, from the excitation signal

Listen to all three files to verify the perfect reconstruction of the speech signal and to hear the characteristics of the error signal for a typical speech file.

9.34. (MATLAB Exercise) Write a MATLAB program to perform LPC analysis on a frame of speech and to display (on a single page) the following quantities:

- the original speech signal for the specified frame
- the LPC error signal for the specified frame
- the signal short-time Fourier transform log magnitude spectrum (dB) along with the LPC spectrum for the specified frame
- the error signal log magnitude spectrum

To test your code, use the file `test_16k.wav` with the following frame parameters:

- starting sample: 6000
- frame size: 640 samples
- LPC order: 12

9.35. (MATLAB Exercise) Write a MATLAB program to perform LPC analysis of speech and plot the average normalized rms error versus LPC order for a range of LPC orders from 1 to 16. Test out your code on several different speech files to see how much variability exists across speech files.

9.36. (MATLAB Exercise) Write a MATLAB program to analyze and synthesize an utterance using the LPC model of Figure P9.36.

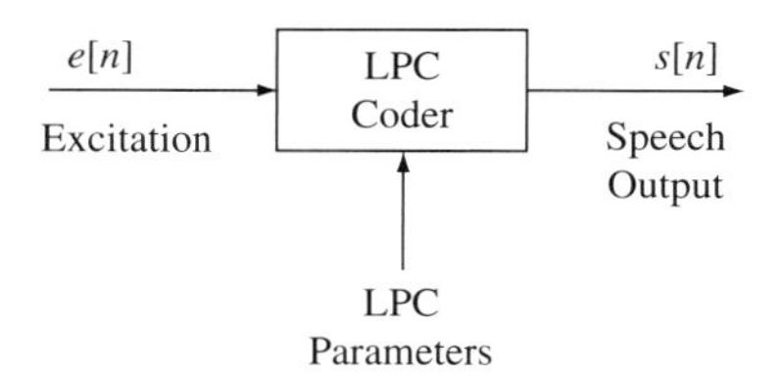

FIGURE P9.36
LPC model of speech synthesis.

In order to synthesize the utterance, you will need to analyze the speech to obtain the LPC parameters at an appropriate update rate, and you will need to create the appropriate excitation signal, again at the same rate as the LPC parameters. To make this problem a lot easier, you will work with the speech file `s5.wav` and you will use the excitation analysis file `pp5.mat`, which contains estimates of the pitch period of the speech (as well as a voiced–unvoiced decision), every 10 msec (i.e., 100 frames/sec). Since the sampling rate of the speech signal is 8 kHz, the frame spacing is 80 samples. A zero value of the pitch period estimate indicates that the speech frame was classified as unvoiced (or silence/background signal), and a non-zero value is the pitch period estimate (in samples at an 8 kHz sampling rate) when the frame is voiced. Using the excitation signal created from the data in `pp5.mat`, and the predictor data that you measure periodically throughout the speech, using conventional LPC analysis methods, the task is to synthesize a synthetic version of sentence `s5.wav` that preserves as much of the intelligibility and naturalness of the original speech as possible. You will save your resulting synthesis in a file called `s5_synthetic.wav`.

One of the challenges of this problem is that you will have to think about how to properly create the excitation signal of Figure P9.36 so that it provides a close approximation to the real speech excitation signal, $e[n]$, from the LPC analysis, and is periodic at the appropriate period for voiced speech and noise-like with the correct gain for unvoiced/silence/background speech. You must think carefully about how to create the excitation signal so that the pitch pulses are separated by the local estimate of pitch period.

In the course of creating this "LPC Vocoder," you should think about the following issues and how they affect the quality of your synthesis:

1. Note that what goes on in a given frame is *not* independent of what happened in previous frames. For example, as the pitch period changes, you will need to know where the last pitch impulse occurred in the previous frame so as to determine the location of the next impulse in the current frame.
2. You can change the vocal tract filter only once per frame, or you can interpolate between frames and change it more often (e.g., at each new pitch period). What parameters can you interpolate and still maintain stability of the resulting synthesis?

3. You don't have to quantize the vocal tract filter and gain parameters, but you should consider doing this if you are satisfied with the quality of your synthesized speech utterance.
4. Listen to your synthetic speech and see if you can isolate the main sources of distortion.
5. Implement a "debug" mode where you can show the "true excitation signal"; i.e., the residual LPC analysis error, $e[n]$, and the synthetic excitation signal that you created, as well as the resulting synthetic speech signal and the original speech signal, for any frame or group of frames of speech. Using this debug mode, see if you can refine your estimates of the key sources of distortion in the LPC Vocoder.
6. Try implementing a pitch detector based on either the speech autocorrelation or the LPC residual signal, and use its output instead of the provided pitch file. What major differences do you perceive in the synthetic speech from the two pitch contours.
7. How could you take advantage of the overlap-add method of short-time analysis and synthesis to simplify your synthesis procedure and make it less sensitive to frame phasing and alignment errors.

You can also consider using the speech files `s1.wav`, `s2.wav`, `s3.wav`, `s4.wav`, `s6.wav` along with pitch period contours `pp1.mat`, `pp2.mat`, `pp3.mat`, `pp4.mat`, `pp6.mat` to provide additional speech files to test your LPC Vocoder code.

CHAPTER 10

Algorithms for Estimating Speech Parameters

10.1 INTRODUCTION

The previous chapters of this book were intended to create a coherent framework of knowledge about the speech signal and about how the speech signal can be represented using discrete-time signal processing concepts. Chapters 3–5 dealt with the acoustic, perceptual, and linguistic foundations of speech processing, showing the links between linguistic descriptions of speech and language and the acoustic realization of sounds, words, and sentences, as well as explaining how the human speech production system is able to create a range of sounds (along with the prosodic aspects of speech such as emphasis, speed, emotion, etc.), and how the human perceptual system is able to recognize the sounds and attach linguistic meaning to human speech. Chapters 6–9 discussed four speech signal representations, namely temporal, spectral, homomorphic, and model-based, and showed the implications of these representations for understanding how the fundamental properties of sound and language are estimated (or measured) from the speech signal. As is the case with almost any representation of a signal, each of the four speech representations was shown to possess certain desirable properties that can enable the creation of algorithms for measuring basic speech properties, and for estimating speech parameters.

The remaining chapters of the book demonstrate how the knowledge framework presented in Chapters 3–9, which culminates in the model for speech production in Figure 10.1, can be applied to solve speech communication problems. The present chapter serves as a transition to the discussion of systems for speech and audio coding, speech synthesis, and speech recognition to be considered in Chapters 11–14. In this chapter, we seek to demonstrate how to use knowledge about the speech signal, along with our understanding of how to best exploit properties of each of the four short-time speech representations, to devise and implement algorithms for measuring and estimating fundamental properties of the speech signal represented in the model of Figure 10.1. The algorithms that we discuss are selected to illustrate how speech processing algorithms are often a combination of fundamental knowledge of the speech signal, digital signal processing theory, statistics, and heuristics. Our discussion is not meant to be exhaustive, but some or even all of the algorithmic features that we discuss can be found in most digital speech processing systems.

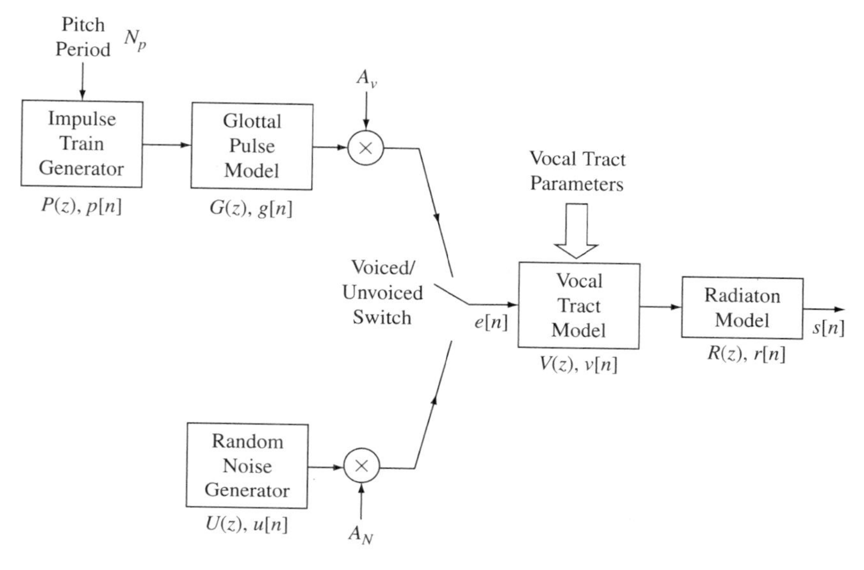

FIGURE 10.1
The basic parametric model for speech production.

We begin with a brief general discussion of a class of non-linear smoothing techniques that can be applied to correct "outlier points" that inevitably result from the short-time analysis methods that are the basis for all speech processing algorithms. Next, we discuss an algorithm for isolating speech from a background signal, i.e., for determining when a speech utterance begins and ends in a specified recording interval. This simple algorithm is based on time-domain measurements. The next algorithm that we describe estimates whether a short-time analysis frame can best be classified as voiced speech, unvoiced speech, or background signal (silence) based on a statistical characterization of the properties of five different parameters estimated from the speech signal. The third set of algorithms we describe are most often referred to as pitch detectors. Their function is to estimate, as a function of time, the pitch period, N_p (or equivalently the pitch frequency), in the model of Figure 10.1 for all voiced regions of the speech signal. Pitch detection has been a very active area of investigation in speech processing, and many pitch detection algorithms, using each of the four representations of speech, have been proposed and studied extensively [144]. In this chapter we briefly describe five different approaches to pitch detection, and discuss some of their properties (e.g., immunity to noise, sensitivity to algorithm assumptions). We end this chapter with a discussion of two methods for estimating the formant frequencies of speech [i.e., the resonance frequencies of the system, $V(z)$, in Figure 10.1] as a function of time—one based on the homomorphic representation of speech, and one based on the linear predictive model of speech.

In summary, the goal of this chapter is to show how the various short-time representations of speech can be used as a basis for algorithms for speech parameter measurement or estimation, and to give a sense as to why some representations are better for some measurements than others. We attempt to justify many of the algorithm steps based on our understanding of speech production and perception, showing the links of practice to theory in such cases.

10.2 MEDIAN SMOOTHING AND SPEECH PROCESSING

The algorithms that we will discuss in this chapter are all based on some sort of short-time representation of the speech signal. As we have pointed out in Chapter 6, short-time analysis generally is based on a window of fixed length, and it is not generally possible to find a window length that is perfect for all frames of speech or for all speakers. Short windows provide responsiveness to changes in the speech signal, but do not provide adequate averaging or spectral resolution. For longer windows, the properties of the speech signal can change significantly across the window duration. In any case, it is highly probable that a window of any length will, for some frames, span both voiced and unvoiced speech. Parameters estimated from such frames often appear as "outliers" relative to values before and after that frame. These "errors" occur because the signal within the window does not perfectly fit the model for either voiced or unvoiced speech, but the estimation of parameters generally assumes that the model is known for the segment under analysis. One approach to dealing with such anomalies is to apply a local smoothing operation that tends to bring the outlying values back into line with neighboring values of that parameter. In this section we show that median filters are a useful tool for this purpose.

In many signal processing applications, a linear filter (or smoother) is used to eliminate the noise-like components of a signal. For some speech processing applications, however, linear smoothers are not completely adequate due to the type of data being smoothed. An example is the pitch period contour (sequence of pitch periods, N_p, in Figure 10.1 as a function of time) estimated using any of the pitch detection algorithms to be described in Section 10.5. An example of a raw pitch period contour is shown in Figure 10.2, which shows an estimate of the pitch period at a rate of 100 frames/sec. The pitch period for voiced speech is normally assumed to be an integer number of samples, and it is common to use this parameter to indicate the

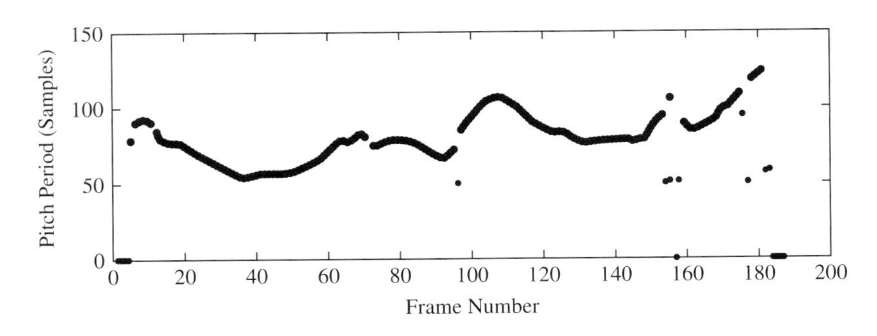

FIGURE 10.2
Example of pitch period contour showing obvious pitch period discontinuities that need to be "smoothed" for use in speech processing systems.

voiced/unvoiced distinction in the model of Figure 10.1 as well. This is done by using the value $N_p = 0$ to signify unvoiced speech. This is illustrated in Figure 10.2, which shows a pitch period contour with a pitch period discontinuity around frame 100, and a region of uncertain pitch period in the vicinity of frames 150–160 and around frame 180. The isolated unvoiced frame is very likely to be an error due to weak voicing. Other than these few discontinuities and irregularities in the pitch period contour, the pitch period estimates follow a smooth curve, which is to be expected because pitch period is a physical quantity that cannot change rapidly. Thus, the pitch period estimates that are out of line are likely to be errors due to weak voicing, which occurs, for example, at the end and/or beginning of regions of voicing or during weakly voiced stop consonants. These errors show up as point (or short region) discontinuities that ideally should be brought back into line with the rest of the pitch period data. An ordinary linear lowpass filter would not only fail to bring the errant pitch period estimates back into line but would severely distort the pitch period contour around the errant points and also at the transition between voiced and unvoiced speech. What is needed is some type of non-linear smoothing algorithm that can preserve signal discontinuities yet still filter out large isolated errors. Although an ideal non-linear smoothing algorithm with these properties does not exist, a non-linear smoother using a combination of running medians and linear smoothing (originally proposed by Tukey [396]) can be shown to have approximately the desired properties [312].

The basic concept of operation of a linear filter is the separation of signals based on their (approximately) non-overlapping frequency content. For non-linear smoothers, it is more appropriate to consider separating signals based on whether they can be considered smooth or rough (noise-like). Thus a signal, $x[n]$, can be considered to be of the form:

$$x[n] = S\{x[n]\} + R\{x[n]\}, \tag{10.1}$$

where $S\{x[n]\}$ is the smooth part of the signal, $x[n]$, and $R\{x[n]\}$ is the rough part of the signal $x[n]$. A non-linear filter that is capable of separating $S\{x[n]\}$ from $R\{x[n]\}$ is the running median of $x[n]$. The output of the causal running median smoother, $M_L\{x[n]\}$, is simply the median of the set of L signal values; i.e., $x[n], x[n-1], \dots, x[n-L+1]$. We generally represent median smoothing in a more illustrative form as

$$y[n] = M_L\{x[n]\} = \text{med}_{m=0}^{L-1}\, x[n-m], \tag{10.2}$$

which shows that an L-point running median filter acts like an L-point finite impulse response (FIR) linear filter in the sense that the output is determined by input samples within a shifting region of support of L samples. However, instead of computing a weighted sum of the samples within the shifted interval $n-L+1 \leq m \leq n$, the median filter simply orders the samples within that interval and outputs the median value. Running medians of length L have the following desirable properties (for smoothing parameter estimates):

1. Median filters do not blur discontinuities in the signal if the signal has no other discontinuities within $L/2$ samples.

2. Median filters will approximately follow low order polynomial trends in the signal. Specifically, an L-point median filter can pull groups of $(L-1)/2$ points back in line with a locally smooth contour.
3. It should be emphasized that median filters, like other non-linear processing algorithms, do not obey the additive part of the principle of superposition; i.e.,

$$M_L\{x_1[n]+x_2[n]\} \neq M_L\{x_1[n]\}+M_L\{x_2[n]\}, \tag{10.3}$$

but they do obey the homogeneity part,

$$M_L\{\alpha x[n]\} = \alpha M_L\{x[n]\}. \tag{10.4}$$

4. The delay of an L-point median filter is $(L-1)/2$ samples. (L is normally an odd integer so that the output value can be associated with the middle of the ordered list of L samples.)
5. Finally, median filters are shift-invariant.

To illustrate how median filters operate on several types of inputs, both with and without significant discontinuities, Figure 10.3 shows a series of plots of four sequences smoothed by a 5-point median filter. The sequence of three sub-plots in each of the four parts of this figure shows the original signal, $x[n]$, the output of the $L=5$ median filter, $M_5\{x[n]\}$, and the difference signal, $d[n]$, between the input signal and the median filter output. (Note the scale difference within the three figures in a group.) The first sequence in Figure 10.3a has no jumps or outlying points of discontinuity. For this sequence we see that the median smoother provides only a small degree of smoothing of signal values (as seen in the plot of the difference signal). The second sequence in Figure 10.3b has a single point discontinuity (as might result from a gross error in estimating a parameter), and, in this case, the median smoother does an excellent job of smoothing out the discontinuity with virtually no significant changes in any of the other signal values. The third sequence in Figure 10.3c has a double discontinuity, where one value is higher and the other is lower than the local trend. In this case both discontinuity points are smoothed back into line with the surrounding points, with no significant changes in all other points of the input sequence. Finally, the fourth sequence in Figure 10.3d is a sequence where the values change discontinuously to a new range. In this case we see that the median smoother is able to follow such a jump without significantly changing the values of any of the sequence points. A linear smoother would smooth the discontinuity, treating it like a step sequence, thereby causing major errors during the region of transition between the two values.

Although running medians generally preserve sharp discontinuities in a signal, they often fail to provide sufficient smoothing of the undesirable noise-like components of a signal. A good compromise is to use a smoothing algorithm based on a combination of running medians and linear smoothing. Since the running medians provide some smoothing, the linear smoother can be a low order system. Usually the linear filter is a symmetrical FIR filter so that delays can be exactly compensated. For

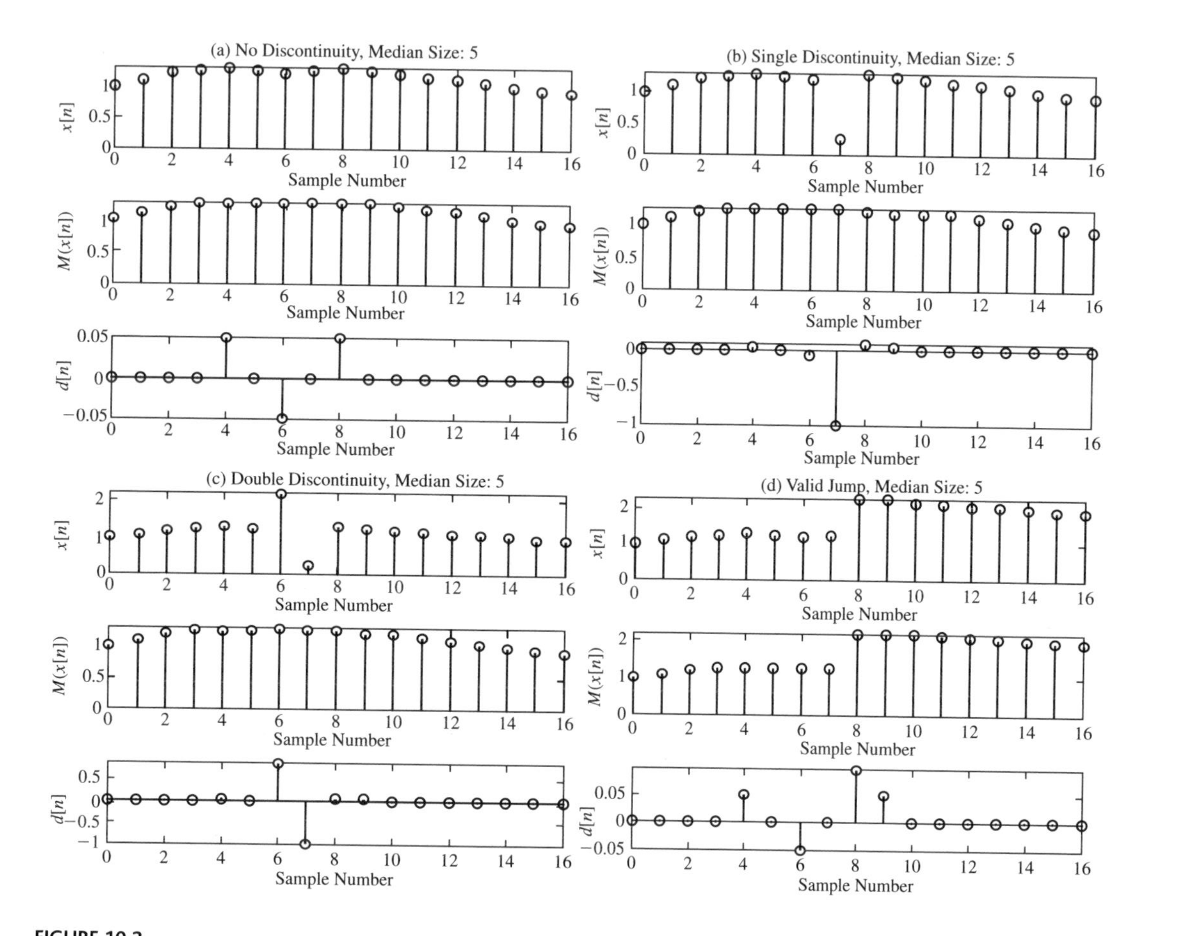

FIGURE 10.3
Illustration of the effects of an $L = 5$ point median smoother on four sequences: (a) a sequence with no strong discontinuity; (b) a sequence with a single point discontinuity; (c) a sequence with a double point discontinuity where one point is high and the other is low; (d) a sequence with a jump in values.

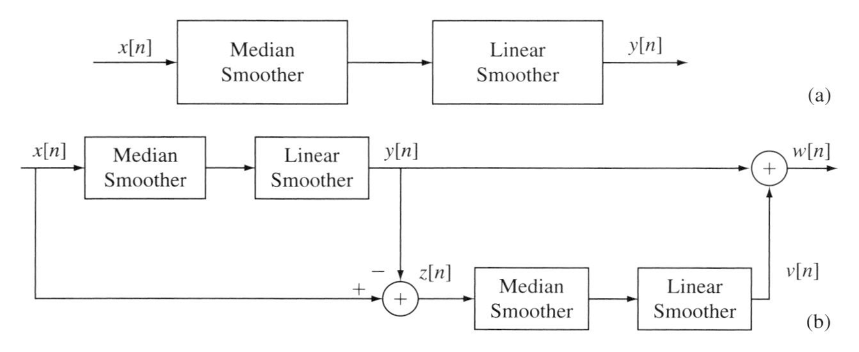

FIGURE 10.4
Block diagram of a non-linear smoothing system. (After Rabiner et al. [312]. © [1975] IEEE.)

example, a filter with impulse response

$$h[n] = \begin{cases} 1/4 & n = 0 \\ 1/2 & n = 1 \\ 1/4 & n = 2 \end{cases} \tag{10.5}$$

having frequency response

$$H(e^{j\omega}) = 0.5(1 + \cos\omega)e^{-j\omega} = \cos^2(\omega/2)e^{-j\omega} \tag{10.6}$$

is generally adequate [396].[1] Figure 10.4a shows a block diagram of a combination smoother based on running medians and linear filtering.

The signal $y[n]$ at the output of the linear smoother is an approximation to the signal $S\{x[n]\}$, i.e., the smooth part of $x[n]$. Since the approximation is not ideal, a second pass of non-linear smoothing is incorporated into the smoothing algorithm as shown in Figure 10.4b. Since

$$y[n] \approx S\{x[n]\}, \tag{10.7}$$

then

$$z[n] = x[n] - y[n] \approx R\{x[n]\}. \tag{10.8}$$

The second pass of non-linear smoothing of $z[n]$ yields a correction signal that is added to $y[n]$ to give $w[n]$, a refined approximation to $S\{x[n]\}$. The signal $w[n]$ satisfies the relation

$$w[n] = S\{x[n]\} + S\{R\{x[n]\}\}. \tag{10.9}$$

[1] Tukey [396] called such a filter a Hann filter because its frequency response magnitude as a function of ω has the form of a Hann window.

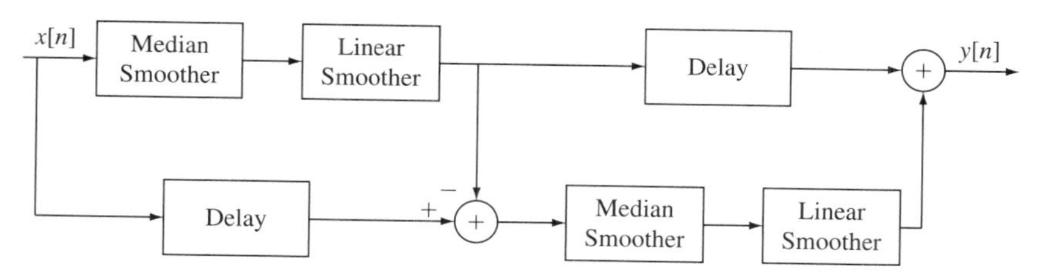

FIGURE 10.5
Non-linear smoothing system with delay compensation. (After Rabiner et al. [312]. © [1975] IEEE.)

If $z[n] = R\{x[n]\}$ exactly, i.e., the non-linear smoother were ideal, then $S\{R\{x[n]\}\}$ would be identically zero and the correction term would be unnecessary.

To implement the non-linear smoother of Figure 10.4 in a realizable system requires accounting for the delays in each path of the smoother. Each median smoother has a delay of $(L-1)/2$ samples, and each linear smoother has a delay corresponding to the impulse response used. For example, a 5-point median filter has a delay of two samples, and the 3-point linear filter in Eq. (10.5) has a delay of one sample. Figure 10.5 shows a block diagram of a realizable version of the smoother of Figure 10.4b.

The final issue concerned with the implementation of the non-linear smoother of Figure 10.5 is the question of how the running median of the signal is defined at the beginning and end of the signal to be smoothed. This is particularly important in applications such as speech recognition where word-length segments of speech are analyzed. Although a variety of approaches are possible, the simple expedient of extrapolating the signal backwards and forwards by assuming the signal remains constant is generally a reasonable solution.

Figure 10.6 shows the results of using several different smoothers on a zero-crossing representation of a speech signal corresponding to an utterance of the 3-digit sequence /seven seven seven/. The input signal (Figure 10.6a) is rough due to the use of a short averaging time. Figure 10.6d shows that the output of the median smoother alone (a 5-point median followed by a 3-point median) has a block-like effect that is characteristic of median filter outputs. This suggests that high frequency components remain in the smoothed output. The output of the linear smoother (a 19-point FIR lowpass filter), shown in Figure 10.6b, is blurred whenever rapid changes occurred in the input signal. The output of the combination smoother [a median of 5 followed by a median of 3 followed by the 3-point filter in Eq. (10.5)], shown in Figure 10.6c, is seen to follow the changes in the input signal well while eliminating most of the noise in the signal.

Figure 10.7a shows an example in which a simple median smoother was used to smooth an estimated pitch period contour having significant errors. As Figure 10.7b shows, the simple median smoother was able to eliminate the gross measurement errors and adequately smooth the pitch period contour, while leaving the voiced/non-voiced transitions virtually unchanged.

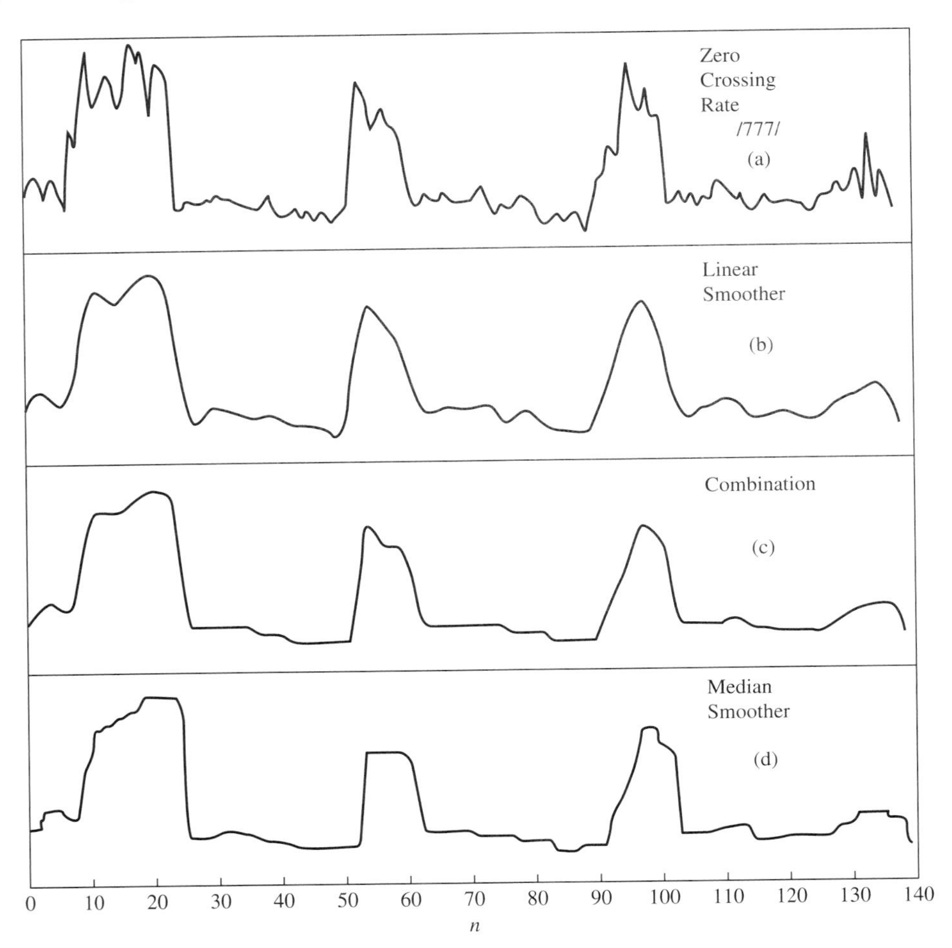

FIGURE 10.6
Example of non-linear smoothing applied to zero-crossing representation. (After Rabiner et al. [312]. © [1975] IEEE.)

10.3 SPEECH-BACKGROUND/SILENCE DISCRIMINATION

The problem of locating the beginning and end of a speech utterance in a background of noise (or other acoustic signals) is important in many areas of speech processing. In particular, in automatic speech recognition, it is essential to locate the time intervals when speech is present so as to avoid trying to classify background sounds as speech inputs. A scheme for locating the beginning and end of a speech signal can also be used to eliminate significant computation by making it possible to process only the parts of an input signal that correspond to speech. Accurately discriminating speech from background noise is difficult, except in the case of extremely high signal-to-noise ratio acoustic environments—e.g., high fidelity recordings made in an anechoic chamber or a soundproof room. For such high signal-to-noise ratio environments, the energy

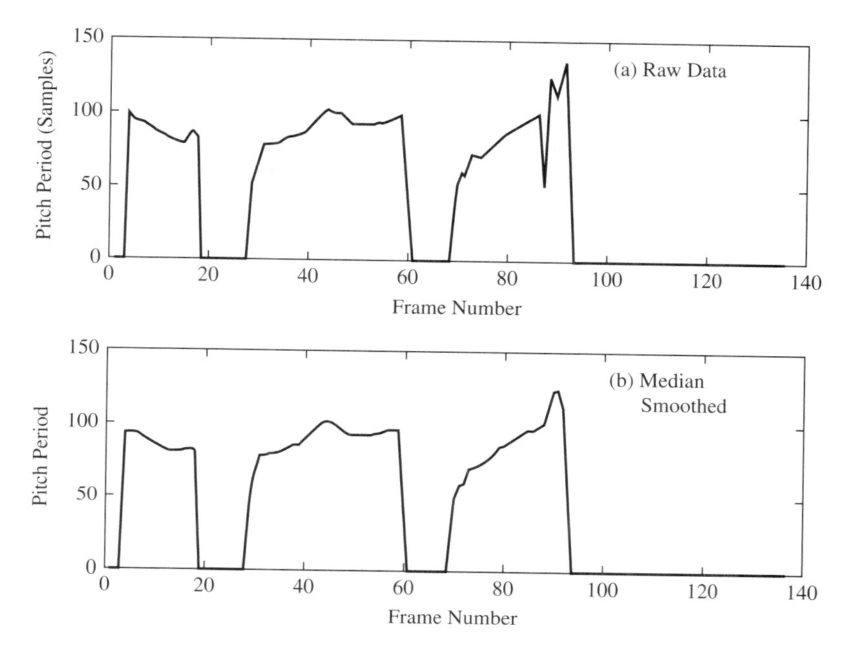

FIGURE 10.7
Example of non-linear smoothing of a pitch contour. (a) The upper plot shows the raw pitch period contour. (b) The lower plot shows the median smoothed pitch period contour.

of the lowest level speech sounds (e.g., weak fricatives) exceeds the background signal energy, and thus a simple short-time energy measurement with an appropriate energy threshold suffices. An example of a speech waveform from this type of recording environment is given in Figure 10.8, which shows a section of a recorded signal in a very low noise environment. Clearly, the beginning of the speech signal is indicated as the point in time where the level of the waveform increases above the background signal level baseline. In this case, the instant where speech begins is easily distinguished and measured. However, such ideal recording conditions are not practical for most applications.

In some sense, the problem is to decide whether a given frame of speech could have been generated by the model of Figure 10.1 or not. The algorithm to be discussed in this section is an example of a heuristically developed method that incorporates simple measurements and basic knowledge of the nature of the speech signal as represented in the model of Figure 10.1. The algorithm is based on two simple time-domain measurements—short-time log energy, which provides a measure of relative signal energy, and short-time zero-crossing rate, which provides a crude measure of concentration of spectral energy in the signal. These representations of speech signals were discussed in Sections 6.3 and 6.4 respectively. Several simple examples illustrate some of the difficulties encountered in locating the beginning and end of a speech utterance in real (noisy) background environments. Figure 10.8 shows an example (the waveform of the signal at the beginning of the word /eight/) for which the low level background

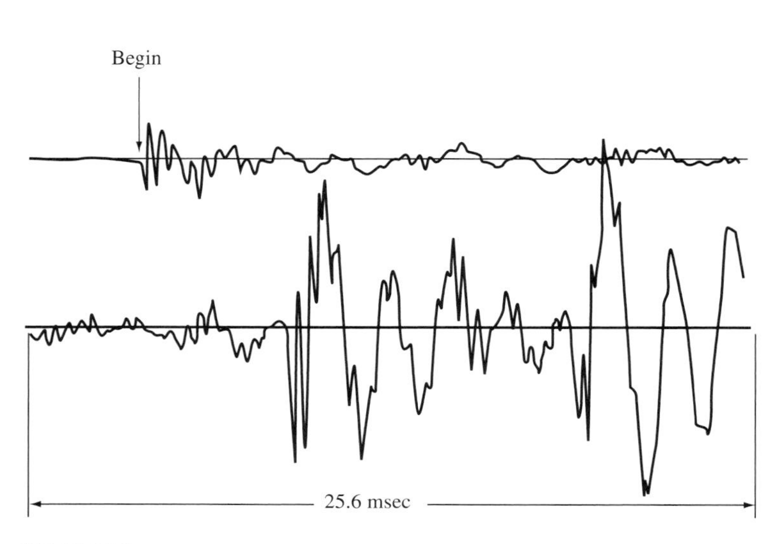

FIGURE 10.8
Waveform for the beginning of the utterance /eight/. (After Rabiner and Sambur [311]. Reprinted with permission of Alcatel-Lucent USA Inc.)

signal is easily distinguished from the level of the speech. In this case a radical change in the waveform amplitude between the background signal and the speech signal is the essential cue to the location of the beginning of the utterance. This change of amplitude would show up clearly in a short-time log energy representation. Figure 10.9 shows the waveform of another example (the beginning of the word /six/) for which, even by casual visual inspection of the waveform, it is easy to locate the beginning of the speech. In this case, the frequency content of the initial /S/ in the word /six/ is radically different from that of the background signal, as seen by the sharp increase in the zero-crossing rate of the waveform. In this case, the speech amplitude at the beginning of the utterance is comparable to the background noise signal amplitude so the short-time log energy alone would not necessarily provide a clear indication of the beginning of speech.

Figure 10.10 shows the waveform of a case in which it is extremely difficult to accurately locate the beginning of the speech signal. This figure shows the waveform for the beginning of the utterance /four/. Since /four/ begins with the weak (low energy) high frequency fricative /F/, it is very difficult to precisely identify the beginning

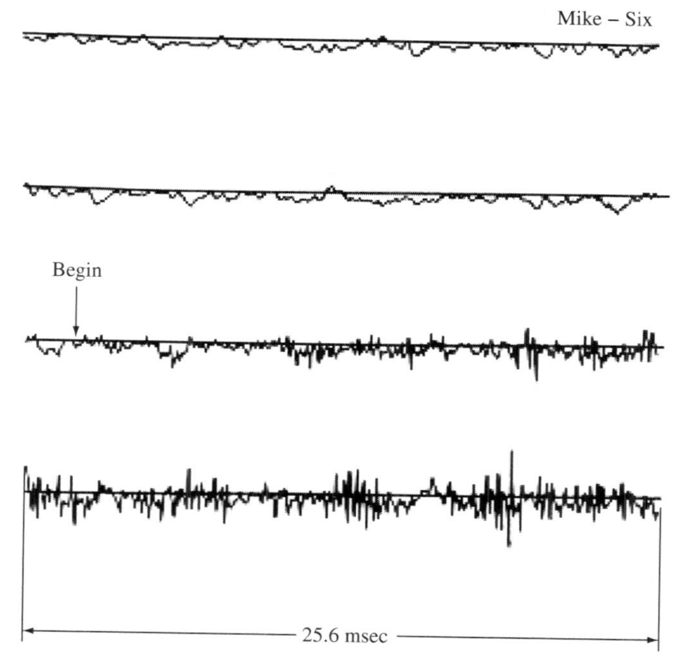

FIGURE 10.9
Waveform for the beginning of the utterance /six/. (After Rabiner and Sambur [311]. Reprinted with permission of Alcatel-Lucent USA Inc.)

point.[2] Although the point marked B in this figure is a good candidate for the beginning of the speech signal, point A is actually the beginning point (based on listening and examining spectrograms).

In general, it is difficult to accurately locate the beginning and end of an utterance for any of the following conditions:

1. weak fricatives (/F/, /TH/, /H/) occurring at the beginning or end of the speech interval;
2. weak plosive bursts (/P/, /T/, /K/) occurring at the beginning or end of the speech interval;
3. nasals occurring at the end of the speech interval (nasals are often devoiced and of reduced levels);
4. voiced fricatives (which become devoiced) occurring at the end of words;
5. trailing off of vowel sounds occurring at the end of an utterance.

[2]This problem is exacerbated by sampling rates below 10 kHz, which require lowpass filter cutoff below the frequency where the spectrum of /F/ is maximum.

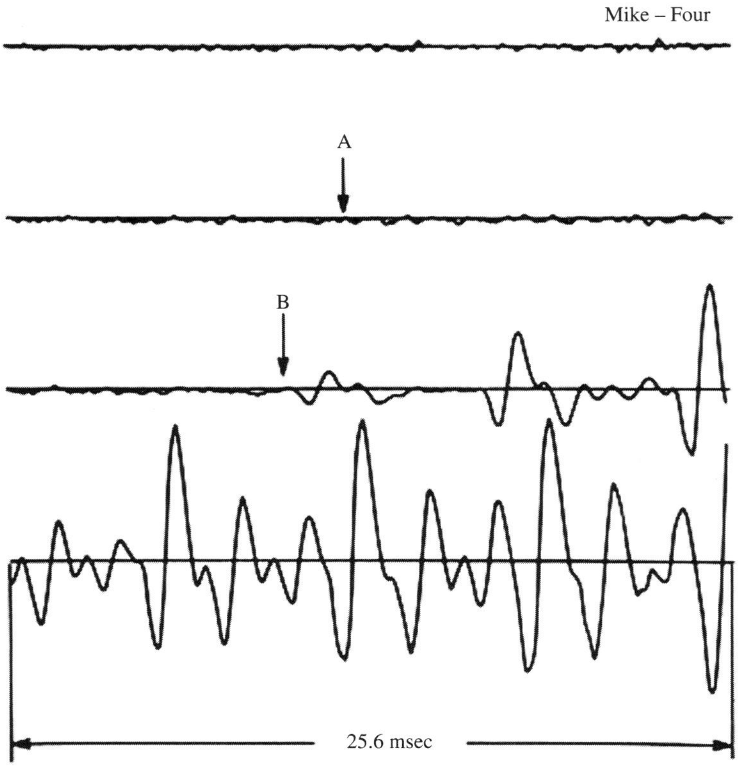

Mike – Four

A

B

25.6 msec

FIGURE 10.10
Waveform for the beginning of the utterance /four/. (After Rabiner and Sambur [311]. Reprinted with permission of Alcatel-Lucent USA Inc.)

In spite of the difficulties posed by the above situations, short-time log energy and short-time zero-crossing rate measurements together can serve as the basis of a reliable algorithm for accurately locating the beginning and end of a speech signal.[3] For the algorithm to be discussed in this section, it is assumed that a speaker utters a single utterance (word, phrase, sentence), without pausing, during a prescribed recording interval, and that the entire recording interval is sampled and stored for subsequent processing. The goal of the speech endpoint detection algorithm is to reliably find the beginning (B) and end (E) of the spoken utterance so that subsequent processing and pattern matching can ignore the surrounding background noise.

A simple algorithm for speech/background detection can be described by reference to Figures 10.11 and 10.12. Figure 10.11 shows the signal processing operations

[3]We could just as easily use short-time magnitude rather than short-time energy in the speech detection algorithm. An example of a speech detection algorithm using short-time magnitude and short-time zero crossings was given by Rabiner and Sambur [311] in the context of an isolated word speech recognition system.

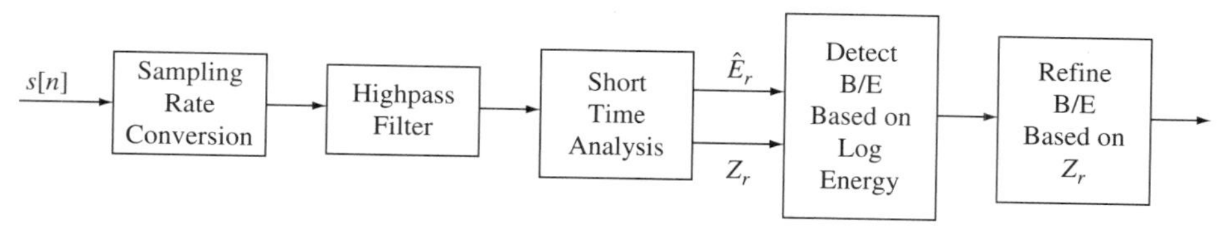

FIGURE 10.11
Block diagram of signal processing operations for speech endpoint detection algorithm.

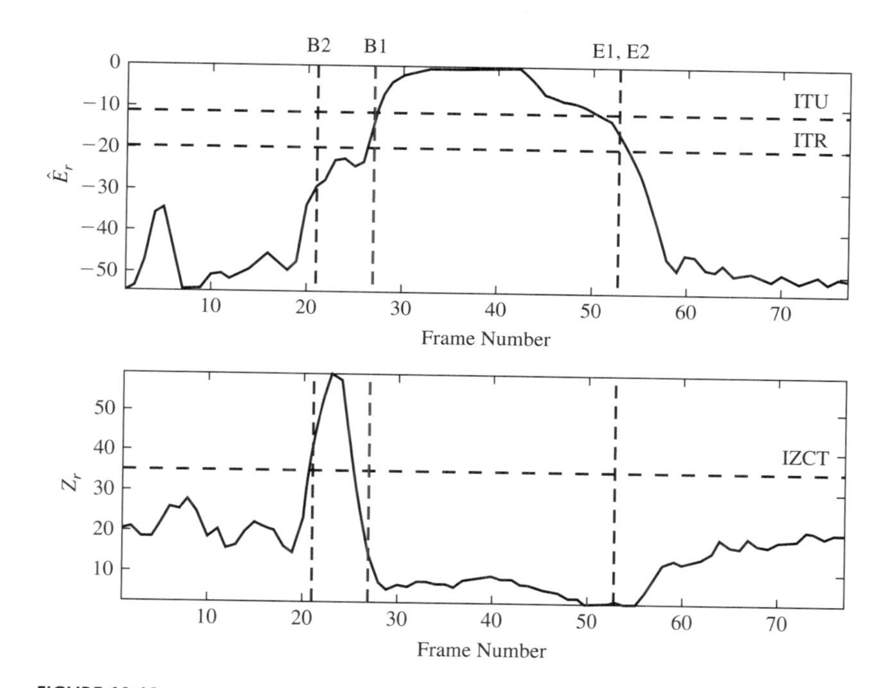

FIGURE 10.12
Typical example of short-time log energy and zero-crossing contours for a word with a fricative at the beginning. The log energy and zero-crossing thresholds are defined as the horizontal lines in the figures; the final detected beginning and ending frames are denoted as B2 and E2 in the log energy plot.

in the speech endpoint detection algorithm. The recorded speech signal, $s[n]$, is first sampling rate converted to a standard sampling rate (10 kHz for this algorithm), and then highpass filtered to eliminate DC offset and hum, using a length 101-point FIR equiripple highpass filter.[4] Next short-time energy and zero-crossing analysis are

[4]The sampling rate conversion step is included in the signal processing because many commonly used databases use non-standard sampling rates, e.g., the TI (Texas Instruments) digits database is sampled at $F_s = 20{,}000$ Hz.

performed using a frame size of 40 msec ($L = 400$ samples at $F_s = 10{,}000$ Hz), with a frame shift of 10 msec ($R = 100$ samples at $F_s = 10{,}000$ Hz), for a basic frame rate of 100 frames/sec. The two short-time parameters, log energy ($\log E_r$) and zero-crossing rate, (Z_r) are computed for each frame, r, of the recorded signal. The short-time parameters are computed as:

$$E_r = \sum_{m=0}^{L-1} (s[(rR+m] \cdot w[m])^2, \quad r = 0, 1, 2, \ldots$$

$$\hat{E}_r = 10 \log_{10} E_r - \max_r (10 \log_{10} E_r),$$

$$Z_r = \frac{R}{2L} \sum_{m=0}^{L-1} |\text{sgn}(s[rR+m] - s[(rR+m-1])|,$$

where $L = 400$, $R = 100$, and $w[m]$ is an L-point Hamming window, and we see that the short-time log energy parameter, $\hat{E}_r$, is normalized to a peak of 0 dB and Z_r is the zero-crossing count per 10 msec interval.

It is assumed that the first 100 msec (10 frames) of the recorded signal contains no speech. This is justified for most applications by the slow reaction time of speakers when asked to begin speaking during a specified time interval, after being cued when that interval begins. The mean and standard deviation of the short-time log energy and zero-crossing rate are computed for this initial 100 msec interval to give a rough statistical characterization of the background signal. These means and standard deviations are denoted as `eavg` and `esig` for the log energy, and `zcavg` and `zcsig` for the zero-crossing rate. Using these measurements, a zero-crossing rate threshold, IZCT, is computed as

$$\text{IZCT} = \max(\text{IF}, \texttt{zcavg} + 3 \cdot \texttt{zcsig}).$$

The quantity IF (which is set to a nominal value of 35) is a global threshold for detecting unvoiced frames (based on long-term statistics of zero crossings for unvoiced sounds). The value of the threshold, IZCT, is raised if the background signal, over the first 100 msec of recording, shows high zero-crossing activity, as estimated from the values of `zcavg` and `zcsig`.

Similarly, for the log energy measurement we define a pair of thresholds, namely ITU, an upper (conservative) threshold, and ITR, a somewhat less conservative threshold for the presence of speech. ITU and ITR are again determined from long-term statistics of the speech and background log energy levels, and again can be modified by the behavior of log energy over the first 100 msec of recording, according to the relations

$$\text{ITU} = \text{constant in the range of } -10 \text{ to } -20 \text{ dB},$$

$$\text{ITR} = \max(\text{ITU} - 10, \texttt{eavg} + 3 \cdot \texttt{esig}).$$

We see from the definitions above that the log energy thresholds are based both on long-term statistics of log energy for speech and non-speech sounds (i.e., the value for ITU and the nominal value for ITR) and the specifics of the log energy of the actual background signal over the first 100 msec of the recording.

Based on the thresholds for $\hat{E}_r$, an initial search is made to find a region of concentration of the log energy contour around the log energy peak. We do this using the following search procedure:

1. Search for a region of concentration of log energy (searching forward from frame 1) to find a frame where the log energy exceeds the lower threshold, ITR; next check the region around the detected frame to make sure that the log energy of adjacent frames extends above the high threshold, ITU, before falling below the low threshold, ITR. If the test fails, reset the initial search frame to the previously detected frame, and rerun the search from that frame forward. Ultimately we find a stable initial frame that defines our initial estimate of the beginning frame of the utterance.[5] We call this initial estimate of beginning frame, B1. (A typical example of the results of each of the steps in the search procedure is illustrated by the appropriate region of Figure 10.12.)
2. Repeat the search of step 1, this time searching backward from the last frame of the utterance. Denote the estimate of ending frame that results from the backward search as E1.
3. Search backward from frame B1 to frame B1−25; count the number of frames where the short-time zero-crossing count exceeds the threshold IZCT; if the number of frames is ≥ 4, the beginning frame is modified to the first frame (lowest index) for which the zero-crossing threshold is exceeded. We call this modified estimate of beginning frame, B2.
4. Search forward from frame E1 to frame E1+25 and count the number of frames where the threshold, IZCT, is exceeded; if the count exceeds the threshold of 4, denote the highest index frame whose count exceeded the threshold as the new ending frame, E2.
5. As a final check, search the local region around the interval [B2, E2] to see if the log energy threshold, ITR, is exceeded; generally this is not the case, but if it occurs, modify the beginning and/or ending frame to match the extended region.

The above search algorithm is heuristic but works extremely well given good estimates of the thresholds, IF for zero crossings and ITU for log energy. The differential correction factor from the first 100 msec of recording is also quite helpful in making small adjustments to the thresholds and is especially useful in adaptive systems when frequent estimates of background signal distributions can be made.

[5]We are guaranteed to find a valid beginning frame in this manner since the peak log energy of 0 dB occurs at some frame within the utterance and this frame satisfies the search constraints.

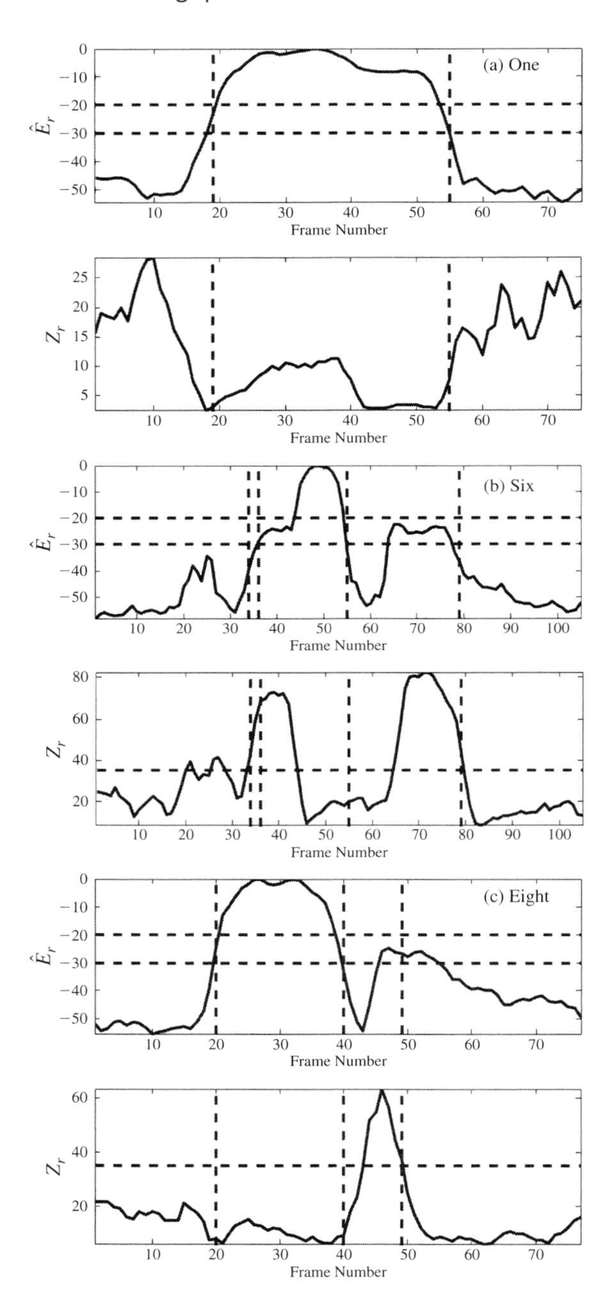

FIGURE 10.13
Typical example of performance of the speech detection method described in this section. Part (a) shows the short-time log energy and zero-crossing contours for the isolated digit /one/, along with the detected beginning and ending frames. Part (b) shows similar results for the digit /six/, and part (c) shows the results for the digit /eight/.

To illustrate the performance of this simple speech detection method, Figure 10.13 shows short-time log energy and zero-crossing plots for three isolated digit utterances. Figure 10.13a shows results for the word /one/. We see that the beginning and ending points were determined solely from the log energy thresholds as the level of zero crossings was too low to modify the results in any manner. For the example in Figure 10.13b, for the digit /six/ (/S/ /IH/ /K/ /S/), we see that both the initial and final endpoints were modified based on the level of zero crossings at the beginning and end due to the strong fricative, /S/, at the beginning and end of the word /six/. It is interesting to note that the log energy of the initial /S/ was sufficiently high that most of the duration of the initial /S/ was already included in the log energy determination of the initial frame; this was not the case for the final /S/ due to the stop gap region for the /K/ sound in /six/. Finally, the example in Figure 10.13c shows a case where the final endpoint was modified because of the high zero-crossing rate in the frication of the release of the /T/ sound from the word /eight/. One of the strengths of this endpoint algorithm is its ability to cross stop regions to look for speech after a stop consonant release.

This application of zero crossings and log energy illustrates the utility of these simple representations when combined with heuristics derived from a practical setting. These representations are particularly attractive since very little arithmetic is required for their implementation.

10.4 A BAYESIAN APPROACH TO VOICED/UNVOICED/SILENCE DETECTION

The previous section described a heuristically derived algorithm for distinguishing between background and speech. Such an algorithm is useful for determining the beginning and end of speaking in applications where word- or sentence-length segments are to be isolated from a longer sequence of recorded signal for further processing. In some applications, such as speech coding, it is of interest to make similar distinctions on a frame-by-frame basis. Again referring to the model of Figure 10.1, the problem is to decide whether the model applies to a given frame of a signal, and if so, to determine the position of the V/U switch. In this section we describe a simple statistical algorithm for classifying individual frames of a signal as being voiced speech, unvoiced speech, or silence (background signal). This discussion will show how statistics can replace heuristics and provide a systematic approach to the design of such algorithms. The algorithm to be discussed is based on a multivariate Gaussian distribution of a set, $\mathbf{X} = [x_1, x_2, x_3, x_4, x_5]$ of five parameters that can be extracted by short-time analysis methods, namely:

1. x_1, the short-time log energy of the signal;
2. x_2, the short-time zero-crossing rate of the signal per 10 msec (100 samples at a 10 kHz sampling rate) interval;
3. x_3, the short-time autocorrelation coefficient at unit sample delay;
4. x_4, the first predictor coefficient of a pth-order linear predictor;
5. x_5, the normalized energy of the prediction error of a pth-order linear predictor.

$\mathbf{X}$ is called the feature vector and is used for making the voiced-unvoiced-silence (VUS) decision on a frame-by-frame basis.

The choice of these five parameters was based on their proven ability to distinguish among voiced sounds, unvoiced sounds, and background sounds [13] for a wide range of speech recording environments. The five features were measured on contiguous blocks of $L = 400$ samples (40 msec of signal, for a 10 kHz sampling rate), with adjacent blocks spaced by $R = 100$ samples (i.e., 10 msec frame shift). A highpass filter with low frequency cutoff of 200 Hz was used to eliminate any low frequency hum, DC offset, or noise components that might have been introduced in the recording process. In the definitions below, we focus on a single 400-sample frame that is indexed as $\tilde{x}[m] = x[m + \hat{n}] \cdot w[m]$, $m = 0, 1, 2, \dots, L-1$, where $w[m]$ is an L-point Hamming window. Note that in the definitions below, we have suppressed the indication of the frame index or analysis time, $\hat{n}$, to simplify the notation. The five parameters are defined for each frame by the following:

1. short-time log energy, $\hat{E}$, defined as:

$$\hat{E} = 10\log_{10}\left(\epsilon + \frac{1}{L}\sum_{m=0}^{L-1}\tilde{x}^2[m]\right),$$

   where $L = 400$ is the number of samples in each frame and $\epsilon = 10^{-5}$ is a small value used to prevent the computation of the logarithm of 0 when the weighted speech samples are all zero; i.e., $\tilde{x}[m] = 0$, $m = 0, 1, \dots, L-1$;
2. short-time zero-crossing count, Z, defined as the number of zero crossings per 10 msec interval;
3. normalized short-time autocorrelation coefficient at unit sample delay, C_1, defined as

$$C_1 = \frac{\displaystyle\sum_{m=1}^{L-1}\tilde{x}[m]\tilde{x}[m-1]}{\sqrt{\left(\displaystyle\sum_{m=0}^{L-1}\tilde{x}^2[m]\right)\left(\displaystyle\sum_{m=1}^{L-1}\tilde{x}^2[m-1]\right)}}$$

4. first predictor coefficient, α_1, of a $p = 12$ pole linear predictive analysis using the covariance method;
5. normalized log prediction error, $\hat{\mathcal{V}}^{(p)}$, defined as

$$\hat{\mathcal{V}}^{(p)} = 10\log_{10}\left(\epsilon + R[0] - \sum_{k=1}^{p}\alpha_k R[k]\right) - 10\log_{10} R[0],$$

where

$$R[k] = \frac{1}{L} \sum_{m=0}^{L-1-k} \tilde{x}[m]\tilde{x}[m+k], \qquad k = 1, 2, \dots, p$$

are the values of the autocorrelation of the speech samples and the α_k's are the predictor coefficients of the pth-order autocorrelation linear predictive analysis.

The training set used for estimating the probability distributions of the above set of features was comprised of about 20 utterances recorded in a quiet background noise environment. Regions of voiced speech, unvoiced speech, and background signal were manually determined and marked by listening and visual inspection of waveform plots and saved in separate training data sets for each of the three decision classes. A total of 760 frames were manually classified as voiced speech, 220 frames were classified as unvoiced speech, and 524 frames were classified as background signal. Using this manually segmented set of training material, one-dimensional distributions (histograms) for each of the five measured parameters were estimated for frames of voiced speech, unvoiced speech, and silence. The distribution means and variances were also calculated in order to compare plots of the feature histograms (i.e., the one-dimensional distributions) to Gaussian fits to these histograms. The resulting set of plots are shown in Figure 10.14 for each of the five analysis features.

By examining the plots of Figure 10.14, we see that the short-time log energy measurement provides excellent separation of voiced frames from either unvoiced or silence frames, as would be expected from our understanding of the significance of short-time log energy from Chapter 6. Similarly we see that the short-time zero-crossing feature parameter separates a large percentage of the frames of the unvoiced feature set from silence and voiced frames. Interestingly the first autocorrelation coefficient of the signal shows a separation between the distributions for voiced frames and silence frames from unvoiced frames, and the first prediction coefficient appears to separate silence fairly well from voiced and unvoiced frames. Finally we see that the normalized log prediction error separates voiced frames from unvoiced and silence frames.

Overall we see that no single measurement, in fact no set of measurements, is perfect in separating voiced frames from unvoiced frames and from silence frames. However, given a choice of features and estimates of the corresponding distributions, the statistical approach allows us to make decisions in a systematic way that allows us to minimize the probability of error.

As an example of this approach, we shall consider a system based on a simple Bayesian framework [91, 390], with the three decision classes denoted ω_i, $i = 1, 2, 3$,

1. class 1, ω_1, representing the silence class (denoted as value 1 on plots),
2. class 2, ω_2, representing the unvoiced class (denoted as value 2 on plots),
3. class 3, ω_3, representing the voiced class (denoted as value 3 on plots).

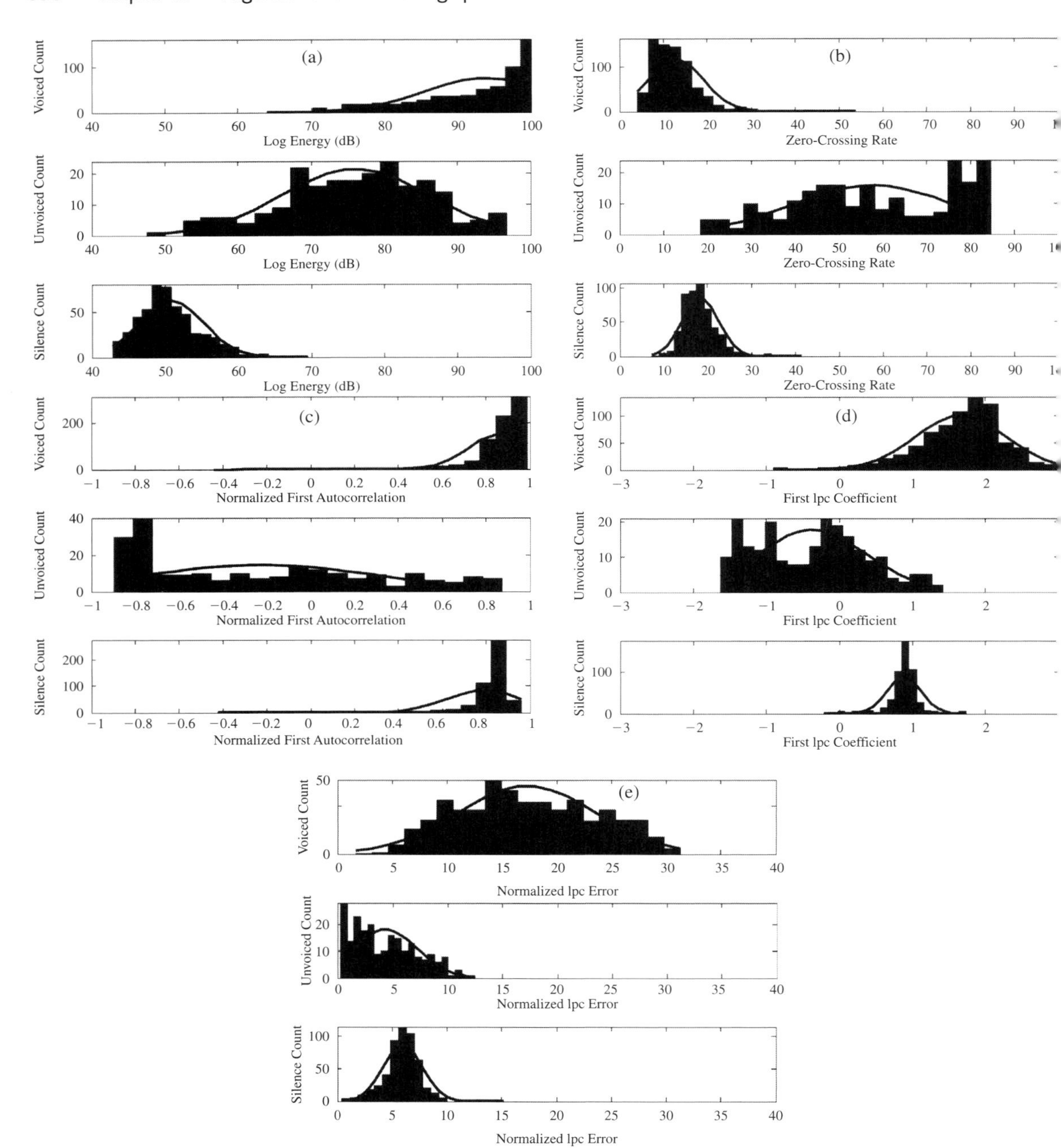

FIGURE 10.14
Measured histograms for the five speech parameters used for VUS estimation, along with simple Gaussian fits to the data. Part (a) shows results for the log energy parameter; part (b) is for the zero-crossing rate; part (c) is for the normalized first autocorrelation; part (d) is for the first predictor coefficient, and part (e) is for the normalized mean-squared prediction error.

The three classes are modeled by Gaussian probability densities based on fits to the 5-dimensional feature vector, $\mathbf{X}$, with means and diagonal covariances, $\mathbf{m}_i$ and $\mathbf{W}_i$, for each class, ω_i, of the form

$$\mathbf{m}_i = E[\mathbf{X}] \text{ for all } \mathbf{X} \text{ in class } \omega_i, \tag{10.10}$$

$$\mathbf{W}_i = E[(\mathbf{X} - \mathbf{m}_i)(\mathbf{X} - \mathbf{m}_i)^T] \text{ for all } \mathbf{X} \text{ in class } \omega_i. \tag{10.11}$$

The statistical approach to VUS detection for a given frame is to choose the class for which the (a posteriori) probability of class i, given the measurement vector $\mathbf{X}$, is the greatest. Using Bayes Theorem [91, 390], we can express the a posteriori probability density function as

$$p(\omega_i|\mathbf{X}) = \frac{p(\mathbf{X}|\omega_i) \cdot P(\omega_i)}{p(\mathbf{X})}, \tag{10.12}$$

where $P(\omega_i)$ is the a priori probability of class i, $p(\mathbf{X}|\omega_i)$ is the conditional probability of $\mathbf{X}$ given ω_i, and

$$p(\mathbf{X}) = \sum_{i=1}^{3} p(\mathbf{X}|\omega_i) \cdot P(\omega_i) \tag{10.13}$$

is the total probability density of $\mathbf{X}$. The Gaussian distribution assumed for the feature vector $\mathbf{X}$ of each of the classes is of the form

$$p(\mathbf{X}|\omega_i) = \frac{1}{(2\pi)^{5/2}|\mathbf{W}_i|^{1/2}} e^{-\frac{1}{2}(\mathbf{X}-\mathbf{m}_i)\mathbf{W}_i^{-1}(\mathbf{X}-\mathbf{m}_i)^T}, \tag{10.14}$$

where $|\mathbf{W}_i|$ is the determinant of $\mathbf{W}_i$.[6]

The Bayes decision rule maximizes the a posteriori probability of class ω_i, as given by Eq. (10.12), which is equivalent to maximization of the monotonic discriminant functions

$$g_i(\mathbf{X}) = \log p(\omega_i|\mathbf{X}), \qquad i = 1, 2, 3.$$

Substituting Eq. (10.12) gives the expression to be maximized as

$$g_i(\mathbf{X}) = \log p(\mathbf{X}|\omega_i) + \log P(\omega_i) - \log p(\mathbf{X}). \tag{10.15}$$

We can disregard the term $\ln p(\mathbf{X})$ since it is independent of the class, ω_i. Substituting the Gaussian distribution in Eq. (10.14) into Eq. (10.15), we get

$$g_i(\mathbf{X}) = -\frac{1}{2}(\mathbf{X} - \mathbf{m}_i)\mathbf{W}_i^{-1}(\mathbf{X} - \mathbf{m}_i)^T + \log P(\omega) + c_i, \tag{10.16}$$

[6]In general, the covariance matrices $\mathbf{W}_i$ will reflect the correlation among the variables of $\mathbf{X}$. However, in the results reported in Ref. [13], the covariance matrices are assumed to be diagonal; i.e., the correlation among the features is neglected.

where

$$c_i = -\frac{5}{2}\ln(2\pi) - \frac{1}{2}\log|\mathbf{W}_i|.$$

The term c_i and the a priori class probability term $\log P(\omega_i)$ (which is difficult to estimate) in Eq. (10.16) are considered to be a "bias" toward class i, and generally can be neglected with little effect on classification accuracy. Neglecting these terms, we can convert the maximization problem to a minimization problem by reversing the sign (and ignoring the factor of 1/2), giving the final decision rule of the form:

Decide class ω_i if and only if

$$d_i(\mathbf{X}) = (\mathbf{X} - \mathbf{m}_i)\mathbf{W}_i^{-1}(\mathbf{X} - \mathbf{m}_i)^T \leq d_j(\mathbf{X}) \;\; \forall \; j \neq i. \tag{10.17}$$

In addition to the simple Bayesian decision rule in Eq. (10.17), a measure based on the relative decision scores was created, thereby providing a confidence measure for how much the class decision could be relied on. The confidence measure was defined as

$$P_i = \frac{d_j d_k}{d_1 d_2 + d_1 d_3 + d_2 d_3}, \qquad i = 1, 2, 3, \text{ and } j, k \neq i, j \neq k. \tag{10.18}$$

The quantities P_i satisfy $0 \leq P_i < 1$. When d_i is very small relative to the other two distances, the other distances will be large and P_i will approach 1. When the distances are all about the same, the confidence score approaches 1/3, which is the equivalent of guessing with no reliable class information coming from the feature vector, $\mathbf{X}$.

Using a relatively small amount of training data, Atal and Rabiner showed that the VUS classification decision, as described above, performed surprisingly well on both the training set (i.e., the exact same data used to train the classifier) and on an independent test set of new utterances and new talkers [13]. The performance accuracy of the VUS estimation algorithm, as obtained by Atal and Rabiner [13], is shown in Table 10.1. The performance on the training and testing sets was comparable, indicating that the classification algorithm was fairly robust in deciding among the three classes.

Of course there will always be errors in classifying individual frames into these three categories, but many of these errors can be corrected using a non-linear smoothing algorithm of the type discussed in Section 10.2 (i.e., a median smoother) on the

TABLE 10.1 Accuracy of VUS estimation algorithms for both training and testing sets. (After Atal and Rabiner [13]. © [1976] IEEE.)

Class	Training set	Count	Testing set	Count
1 (S)	85.5%	76	96.8%	94
2 (U)	98.2%	57	85.4%	82
3 (V)	99%	313	98.9%	375

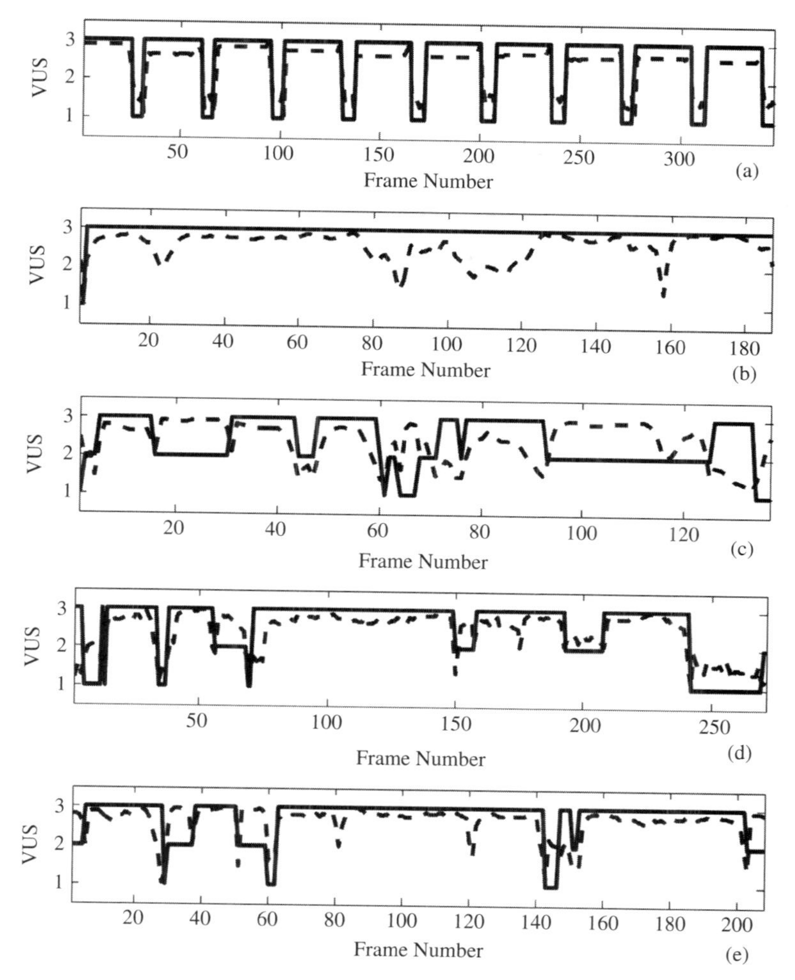

FIGURE 10.15
Plots of VUS classification and confidence scores (scaled up by a factor of 3) for five utterances: (a) synthetic vowel sequence; (b) all-voiced utterance; (c)–(e) speech utterances with a mixture of regions of voiced speech, unvoiced speech, and silence. The solid line indicates decision and the dashed line indicates the corresponding confidence score (multiplied by 3 for plotting).

time sequence of the classifications so that individual frame errors can be spotted and corrected if the surrounding frames are all classified correctly.

Figure 10.15 shows plots of how the VUS classifier worked on five utterances, including the following speech:

1. a synthetic speech sequence of 10 vowels, each separated by a noisy section of background signal, as shown in part (a);
2. the all-voiced utterance, "We were away a year ago", as shown in part (b);

3. the utterance, "This is a test", with alternate regions of voiced, unvoiced, and silence (background), as shown in part (c);
4. the utterance, "Oak is strong and often gives shade", again with alternate regions of voiced, unvoiced, and silence, as shown in part (d);
5. the utterance, "Should we chase those young outlaw cowboys", again with alternate regions of voiced, unvoiced, and silence, as shown in part (e).

Each panel in Figure 10.15 shows the resulting VUS contour (using the code V=3, U=2, S=1) as the solid line in each plot, along with the confidence score contour (where the confidence score was scaled up by a factor of 3 to fall in the range [0–3]). For the synthetic vowel sequence of part (a), the VUS contour is essentially error free, with alternate voiced and silence regions. The scaled confidence scores during voiced regions are almost at the value 3.0; the confidence scores are lower during the background regions but this is due to the specifics of the noise background that was inserted between vowel sounds. For the all-voiced utterance of part (b) of Figure 10.15, again the VUS algorithm perfectly segmented the utterance into a brief region of silence (the first frame) followed by the remaining frames being correctly classified as voiced. The scaled confidence scores for this utterance were again very close to the value of 3.0 for most of the voiced region, sometimes dipping lower at frames within the voiced region (this is particularly noticeable around frame 160, during the /G/ of the word "ago"). For the utterances of parts (c) to (e) of Figure 10.15, the VUS classification algorithm correctly classified most of the frames during regions of V, U, and S, as seen by the steady regions in each of the plots. Single and sometimes double frame errors occurred a few times in these three examples, but no major errors resulted from these small discontinuities. Again we see that the scaled confidence of the classification decision is mostly close to the highest possible value (3 in the plots), dipping below the high confidence region for a few classification decisions; most, if not all, of the errors could easily be repaired using a median smoothing algorithm.

A complete analysis of the effectiveness of each of the five features used in the classifier showed the following:

- For distinguishing between voiced and unvoiced frames, the most effective *single* parameter was the first autocorrelation coefficient, and the second most effective parameter was the log normalized prediction error energy.
- For distinguishing between unvoiced and silence frames, the most effective *single* parameter was the short-time log energy, and the second most effective parameter was the short-time zero-crossing count.
- For distinguishing between voiced and silence frames, the most effective *single* parameter was the short-time log energy, and the second most effective parameter was the log normalized prediction error energy.

The key result is that no single parameter (or even pair of parameters) performed anywhere near as well as the group of five parameters chosen for this task.

The algorithm that we have just described addresses what is perhaps the simplest pattern recognition problem in speech processing. Nevertheless, it illustrates the power of statistical approaches, which are used with great effectiveness in the much more difficult problem of speech recognition. We will discuss more sophisticated statistical techniques in Chapter 14. The basic Bayesian framework can be applied in many ways. In the VUS problem, we can choose different features to compute from the speech signal. Our choice might be influenced by a need to reduce computation or we might seek other parameters where the VUS states are more clearly separated. More extensive training sets would improve performance over a wider range of speakers and signal acquisition and transmission situations. Another modification would be to design a scheme for adapting the algorithm to different speakers and/or recording situations.

10.5 PITCH PERIOD ESTIMATION (PITCH DETECTION)

The problem of pitch period estimation, or pitch detection, can be viewed as an extension of the VUS classification problem in the sense that pitch detection is only necessary for voiced speech, so it is natural to begin by detecting the state of voicing. If we have determined that a frame of speech is voiced, then to determine the excitation model in Figure 10.1, we need an algorithm that can determine the pitch period, N_p, associated with that voiced speech frame. Thus, at each frame, a pitch detector must make a decision between voiced and non-voiced frames (i.e., unvoiced or silence/background frames), and if voiced, estimate the pitch period N_p. Often, the voiced/non-voiced decision is encoded in N_p by setting $N_p = 0$ for non-voiced frames. Alternatively, a more complete analysis system could return a V, U, or S indication along with an estimate of N_p for voiced frames.

Pitch period estimation (or equivalently, fundamental frequency estimation) is one of the most important problems in speech processing. Pitch detectors are essential components of vocoders [105], speaker identification and verification systems [7, 328], and aids-to-the-handicapped [207]. Because of its importance, many solutions to the problem of pitch period estimation have been proposed and widely studied [90, 303]. It is safe to say that all of the proposed schemes have their limitations, and that no presently available pitch detection scheme can be expected to give perfect pitch period estimates across a wide range of speakers, applications, and operating environments.

The goal of a pitch detector (or a pitch period estimation method) is to reliably and accurately estimate the (time-varying) pitch period of the speech waveform during voiced sections of speech. What makes this difficult is that voiced speech is not really a periodic signal as assumed in Figure 10.1, but instead is quasi-periodic, with a period that changes (usually slowly) over time. Further, because of slow changes in vocal tract shape (over time), the speech waveform changes form from period to period, making reliable and accurate identification of the pitch period even more difficult. Finally, during the initiation and termination of voicing, the pitch period is often ill-defined since the glottal excitation is building up, or breaking down, and the waveform has few properties that enable accurate identification or estimation of the pitch period.

10.5.1 Ideal Pitch Period Estimation

The "ideal" input waveform for reliable pitch period detection would be either a perfectly periodic impulse train (as shown in Figure 10.16) or a pure sine wave at the pitch

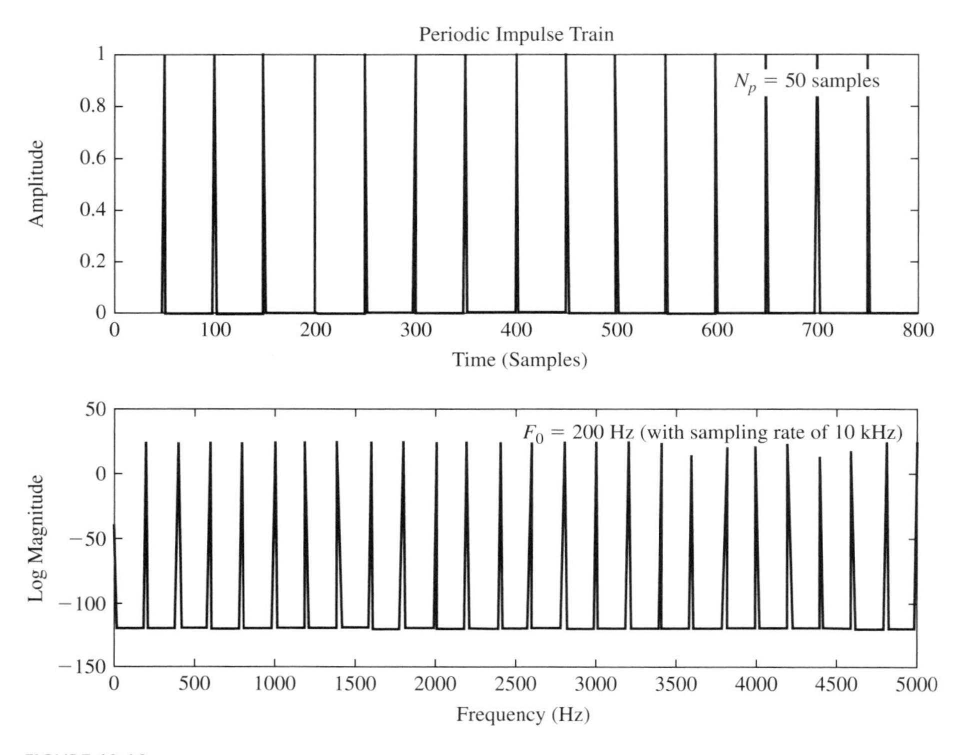

FIGURE 10.16
"Ideal" input signal for speech pitch detector: (top) periodic impulse train with period $N_p = 50$ samples; (bottom) log magnitude spectrum of periodic impulse train with frequency spacing of $F_0 = F_s/N_p =$ 200 Hz (for a 10 kHz sampling rate).

frequency (the inverse of the pitch period) as shown in Figure 10.17. The perfectly periodic impulse train has a flat (log) magnitude spectrum that is also perfectly periodic, as shown at the bottom of Figure 10.16, making pitch detection a trivial process in the time or frequency domains. The sinusoidal signal of Figure 10.17 has a period that is easy to measure in the time domain, and its spectrum consists of a single harmonic at the fundamental frequency, as shown at the bottom of Figure 10.17, thus again making pitch estimation a trivial exercise in the time or frequency domains.

In general, speech does not look like the time-domain plots of either the "ideal" impulse train, or the pure sine wave; i.e., the spectrum is neither flat with many harmonics as in Figure 10.16 nor impulsive as in Figure 10.17. This is illustrated in Figures 10.18 (for a synthesized vowel sound) and 10.19 (for a section of a steady vowel from real speech). In these cases, one approach to pitch period estimation is to try to "flatten" the speech spectrum to resemble the spectral shape of a periodic impulse train, thereby giving an approximation to the ideal periodic impulse train. Another approach to pitch period estimation is to filter the waveform to remove all harmonics except the fundamental, thereby creating a signal that resembles a pure sine wave. Unfortunately it is very difficult to filter out all harmonics except the fundamental without accurately

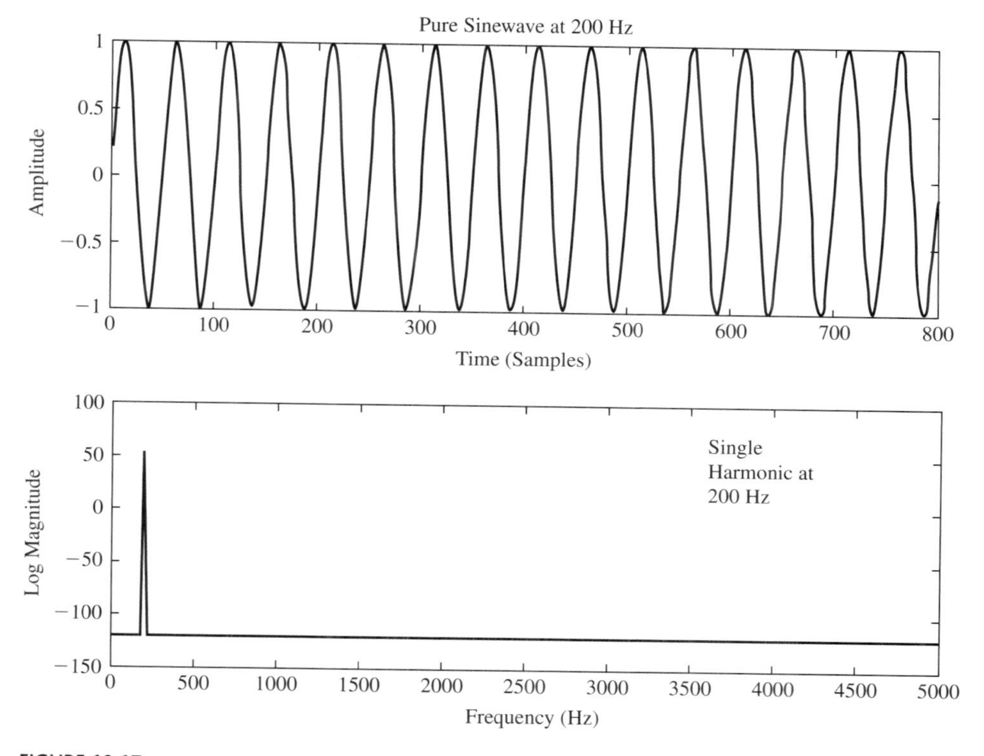

FIGURE 10.17
"Ideal" input signal for speech pitch detector: (top) pure sinusoid at period 50 samples; (bottom) log magnitude spectrum consisting of a single impulse at frequency 200 Hz.

knowing the fundamental (and that is, after all, the goal of pitch detection), so such schemes generally fail badly in practice.

The example waveform in Figure 10.18 is for an ideal synthetic vowel with a constant 100 Hz pitch frequency (or a pitch period of 100 samples at a 10 kHz rate). The "perfect" periodicity of the waveform can be seen in the upper plot; the lower plot shows the perfect (non-flat) harmonic structure of the log magnitude spectrum, clearly showing the vocal tract spectral shape in the lower (or upper) envelope of the log magnitude spectrum. In this case, virtually any of several pitch period estimation methods would do an excellent job of estimating the pitch period or the fundamental frequency from the time-domain or log magnitude frequency-domain plots for this synthetic vowel. Finally, the waveform and log magnitude spectrum of a "real world" vowel (with as constant a pitch as can be readily produced by a non-singer) is shown in Figure 10.19. The top plot (the waveform) shows the period-by-period differences in the signal (hence the term "quasi-periodic"), and the bottom plot shows the lack of regularly spaced, well defined spectral peaks at the "harmonics" of the pitch period. Estimation of pitch period (or fundamental frequency) from the time-domain or frequency-domain plot is difficult, even for such a simple example.

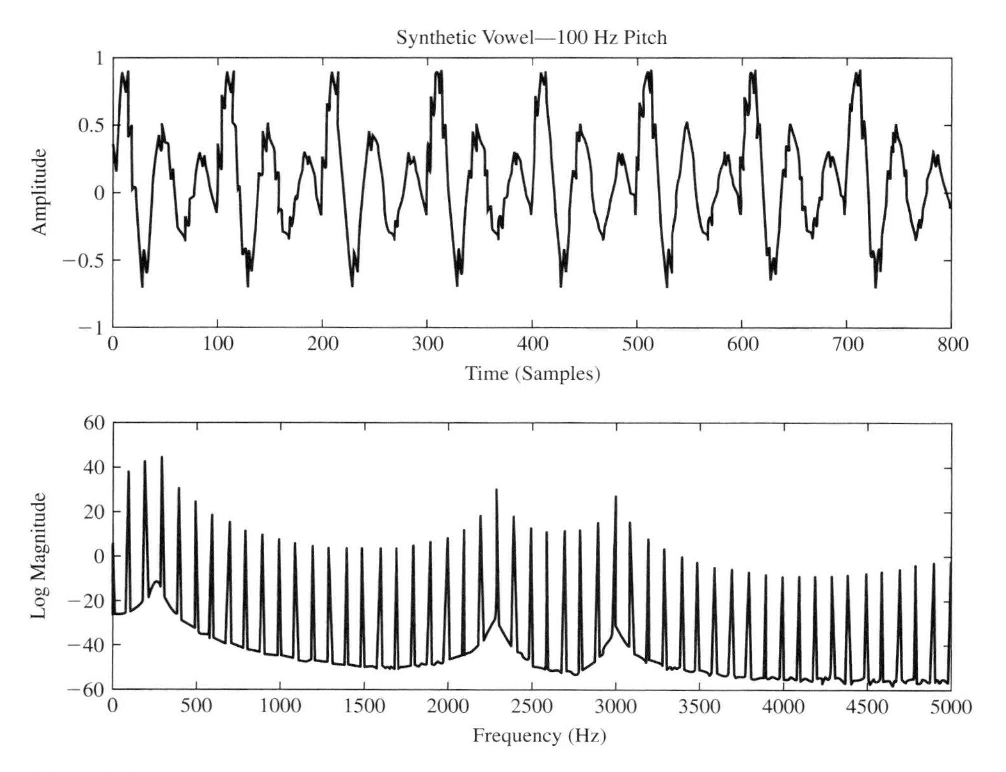

FIGURE 10.18
Waveform and log magnitude spectrum of an ideal synthetic vowel with a pitch period of 100 samples and a pitch frequency of 100 Hz (assuming a 10 kHz sampling rate): (top) waveform of perfectly periodic synthetic vowel; (bottom) log magnitude spectrum of perfectly periodic synthetic vowel.

In the following sub-sections, we present a series of algorithms for pitch detection based on speech measurements from the four representations of speech properties that were discussed in Chapters 6–9, namely time-domain measurements, frequency-domain measurements, cepstral domain measurements, and finally linear prediction-based measurements. As will become clear, most of the approaches rely on transformations of the speech signal aimed at turning it into a signal that is rich in harmonics as in Figure 10.16 or one that has only one or two harmonics such as in Figure 10.17. We begin with a discussion of a simple time-domain parallel processing approach.

10.5.2 Pitch Period Estimation Using a Parallel Processing Approach

In this section we discuss a variation on a particular pitch detection scheme first proposed by Gold [126] and later modified by Gold and Rabiner [129]. Our reasons for discussing this particular pitch detector in this chapter are: (1) it has been used successfully in a wide variety of applications; (2) it is based on purely time-domain processing concepts; (3) it can be implemented to operate very quickly on a general purpose

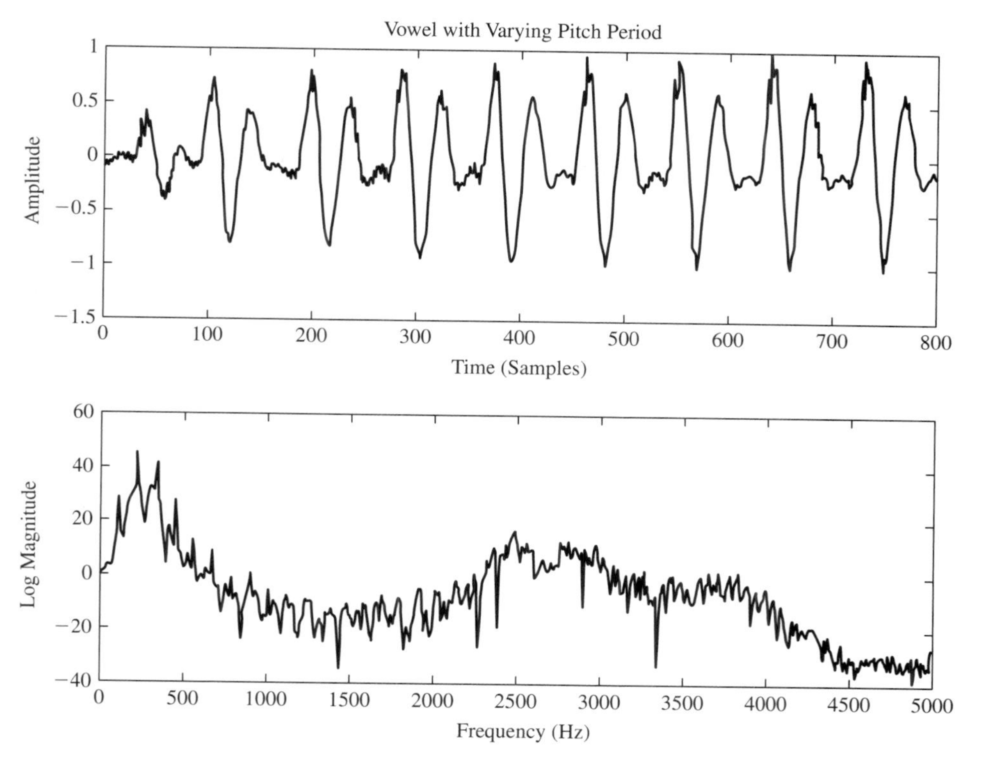

FIGURE 10.19
Waveform and log magnitude spectrum for a "real world" vowel with time-varying pitch period and time-varying vowel tract shape: (top) waveform of vowel showing variations in period and shape of waveform over time; (bottom) log magnitude spectrum showing lack of pronounced harmonic structure.

computer or it can be easily constructed in digital hardware; and (4) it illustrates the use of the basic principle of parallel processing in speech processing.

The basic principles of this pitch period estimation scheme are as follows:

1. The speech signal is processed so as to create a number of impulse trains which approximate the ideal of a periodic impulse train, and which retain the periodicity of the original signal and discard features which are irrelevant to the pitch detection process.
2. This processing permits very simple pitch detectors to be used to estimate the period of each impulse train.
3. The estimates of several of these simple pitch detectors are logically combined to enable estimation of the period of the speech waveform.

The particular scheme proposed by Gold and Rabiner [129] is depicted in Figure 10.20. The speech waveform is sampled at a 10 kHz rate, which allows the period

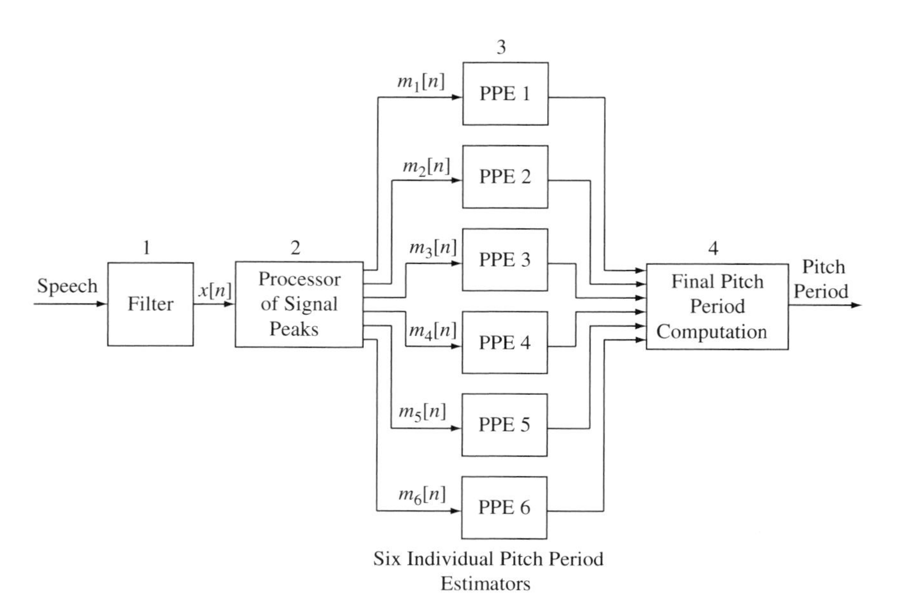

FIGURE 10.20
Block diagram of a parallel processing time-domain pitch detector.

to be determined to within 1 sample or $T = 10^{-4}$ sec. The speech is lowpass filtered with a cutoff of about 900 Hz to produce a relatively smooth (lowpass) waveform. A bandpass filter passing frequencies between 100 and 900 Hz can also be used to remove DC offset and 60 Hz noise in some applications. (This filtering can be done either with an analog filter before sampling or with a digital filter after sampling.)

Following the lowpass or bandpass filtering, the "peaks and valleys" (local maxima or positive peaks and local minima or negative peaks) are located (as shown in Figure 10.21) and, from their locations and amplitudes, several impulse trains (six in Figure 10.20) are derived from the filtered signal. Each impulse train consists of positive impulses occurring at the location of either the positive peaks or the negative peaks (valleys). The six cases defined by Gold and Rabiner [129] are given below:

1. $m_1[n]$: At the location of each positive peak, an impulse equal in amplitude to the peak amplitude occurs. (If the peak amplitude is negative, the impulse amplitude is set to zero.)
2. $m_2[n]$: At the location of each positive peak, an impulse equal to the difference between the positive peak amplitude and the preceding negative peak amplitude occurs. (If the positive peak amplitude is negative, the impulse amplitude is set to zero.)
3. $m_3[n]$: At the location of each positive peak, an impulse equal to the difference between the positive peak amplitude and the preceding positive peak amplitude occurs. (If this difference is negative, the impulse amplitude is set to zero.)

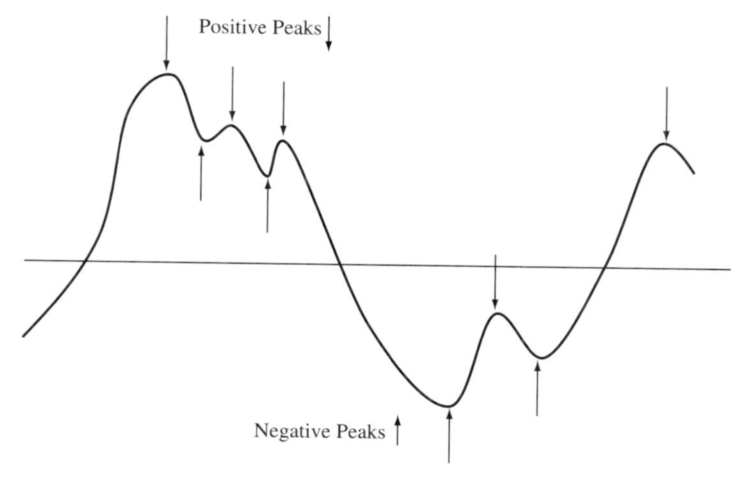

FIGURE 10.21
Locations of positive and negative peaks in the filtered waveform.

4. $m_4[n]$: At the location of each negative peak, an impulse equal to the negative of the amplitude at that negative peak occurs. (If the resulting amplitude is negative, the impulse amplitude is set to zero.)
5. $m_5[n]$: At the location of each negative peak, an impulse equal to the negative of the amplitude at the valley plus the amplitude at the preceding peak occurs. (If the resulting amplitude is negative, the impulse amplitude is set to zero.)
6. $m_6[n]$: At the location of each negative peak, an impulse equal to the negative of the amplitude at the valley plus the amplitude at the preceding local negative peak occurs. (If this difference is negative, the impulse amplitude is set to zero.)

Figures 10.22 and 10.23 show two waveform examples—one a pure sine wave and one a weak fundamental plus a strong second harmonic—together with the resulting impulse trains as defined above. Clearly the impulse trains have the same fundamental period as the original input signals, although $m_5[n]$ of Figure 10.23 is close to being periodic with half the fundamental period. The purpose of generating these impulse trains is to make it simple to estimate the period on a short-time basis. The operation of the simple pitch period estimators is depicted in Figure 10.24. Each impulse train is processed by a time-varying non-linear system (called a peak detecting exponential window rundown circuit in Ref. [126]). When an impulse of sufficient amplitude is detected in the input, the output is reset to the value of that impulse and then held for a blanking interval, $\tau[n]$ seconds—during which no additional pulse can be detected. At the end of the blanking interval, the output begins to decay exponentially. When an impulse exceeds the level of the exponentially decaying output, the process is repeated. The rate of decay and the blanking interval are dependent upon the most recent estimates of pitch period [129]. The length of each pulse is an estimate of the pitch period. The pitch

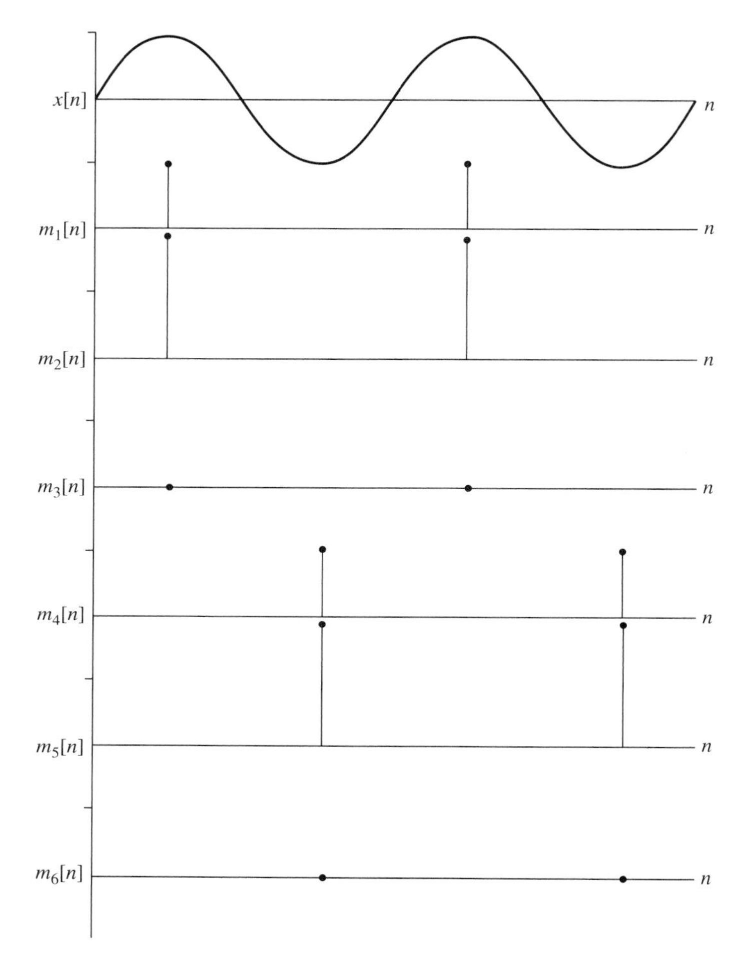

FIGURE 10.22
Input (sinusoid) and corresponding impulse trains generated from the peaks and valleys.

period is estimated periodically (e.g., 100 times/sec) by measuring the length of the pulse spanning the current sampling interval.

This technique is applied to each of the six impulse trains, thereby obtaining six estimates of the pitch period. These six estimates are combined with the two most recent estimates for each of the six pitch detectors. These estimates are then compared, and the value with the most occurrences (within some tolerance) is declared the pitch period at that time. This procedure produces very good estimates of the period of voiced speech. For unvoiced speech there is a distinct lack of consistency among the estimates. When this lack of consistency is detected, the speech is classified as unvoiced. The entire process is repeated periodically to produce an estimate of the pitch period and voiced/unvoiced classification about 100 times/sec.

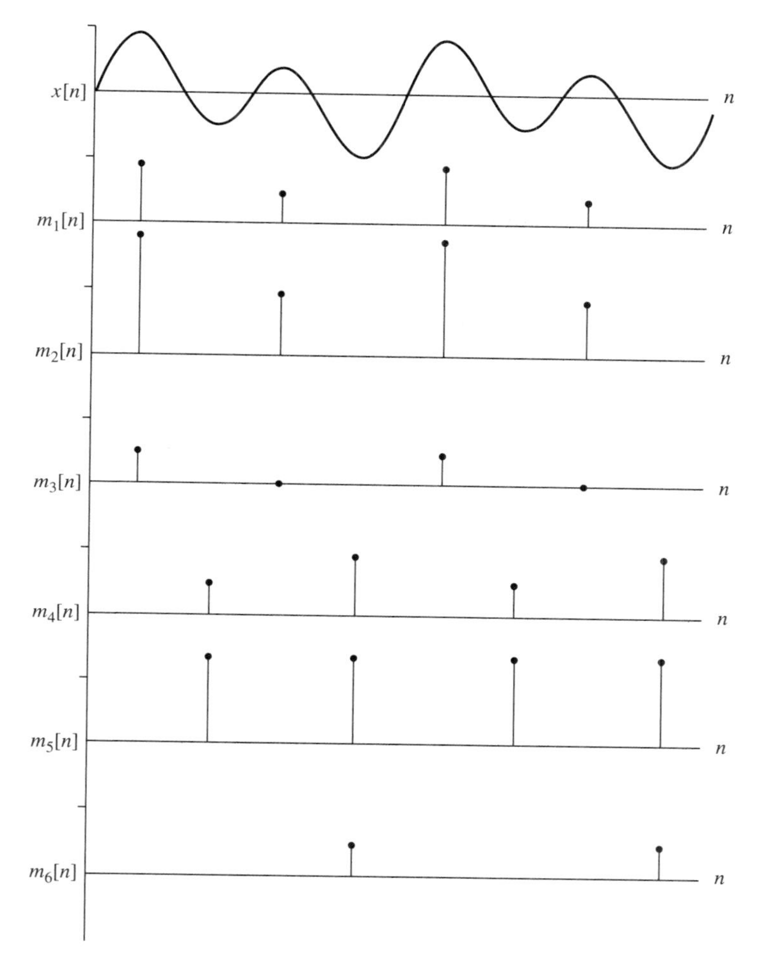

FIGURE 10.23
Input (weak fundamental and second harmonic) and corresponding impulse trains generated from the peaks and valleys.

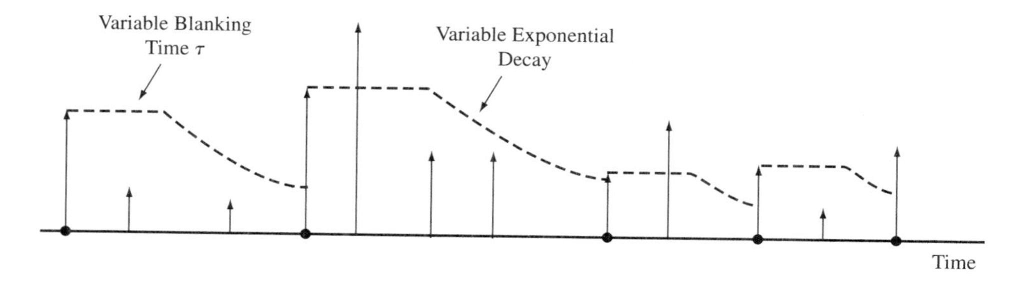

FIGURE 10.24
Basic operation of each individual pitch period estimator of the time-domain pitch detector. (After Gold and Rabiner [129].)

The time-domain parallel processing pitch detector algorithm can be summarized as follows:

1. Filter speech to retain the 100–900 Hz region (adequate for all ranges of pitch; eliminates extraneous signal harmonics).
2. Find all positive and negative peaks in the waveform.
3. At each positive peak:
   - determine peak amplitude pulse (positive pulses only)
   - determine peak-valley amplitude pulse (positive pulses only)
   - determine peak-previous peak amplitude pulse (positive pulses only).
4. At each negative peak:
   - determine peak amplitude pulse (negative pulses only)
   - determine peak-valley amplitude pulse (negative pulses only)
   - determine peak-previous peak amplitude pulse (negative pulses only).
5. Filter pulses with an exponential (peak detecting) window to eliminate false positives and negatives that are far too short to be pitch pulse estimates.
6. Determine pitch period estimate as the time between remaining major pulses in each of the six elementary pitch period detectors.
7. Vote for best pitch period estimate by combining the three most recent estimates for each of the six pitch period detectors.
8. Clean up errors using some type of non-linear (e.g., median) smoother.

The performance of this pitch detection scheme is illustrated by Figure 10.25, which shows the output for a sample of synthetic speech. The advantage of using synthetic speech is that the true pitch periods are known exactly (since they were artificially generated), and thus a measure of the accuracy of the algorithm can be obtained. The disadvantage of synthetic speech is that it is generated according to a simple model, and therefore may not show any of the unusual properties of natural speech. In any case, testing with synthetic speech has shown that most of the time, the method tracks the pitch period to within two samples or less. Furthermore it has been observed that at the initiation of voicing (i.e., the first 10–30 msec of voicing), the speech is often classified as unvoiced. This result is due to the decision algorithm which requires about three pitch periods before a reliable pitch decision can be made—thus a delay of about two pitch periods is inherently built into the method. In a comparative study of pitch detection algorithms carried out under a wide range of conditions with natural speech, this method compared well with other pitch estimation methods that have been proposed [303].

In summary, the details of this particular method are not so important as the basic principles that are introduced. (The details are available in Ref. [129].) First, note that the speech signal was processed to obtain a set of impulse trains that retain only the essential feature of periodicity (or lack of periodicity). Because of this simplification in the structure of the signal, a very simple pitch estimator suffices to produce good

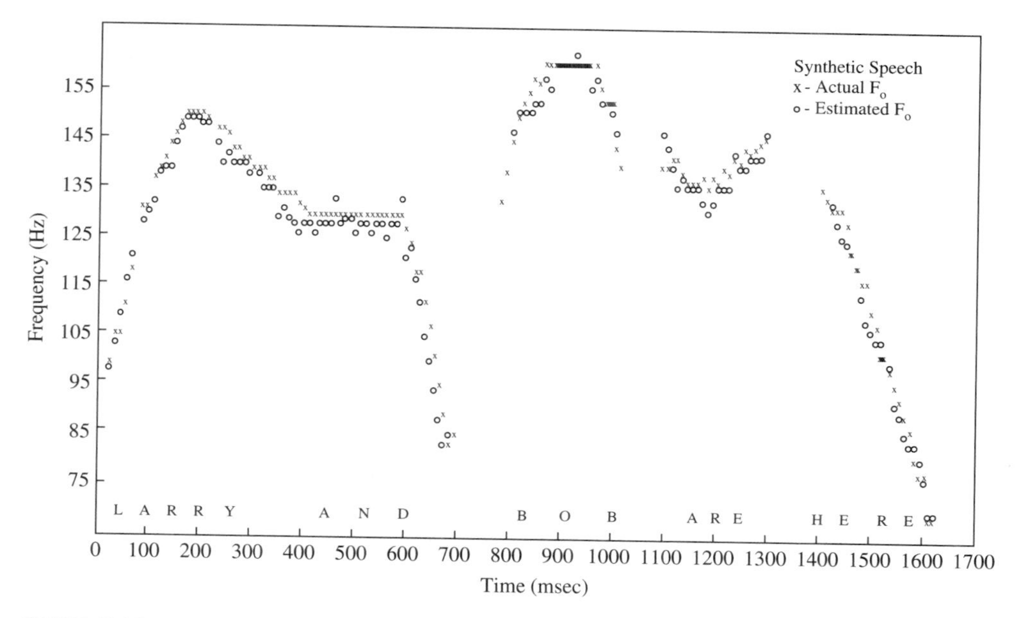

FIGURE 10.25
Comparison between actual and estimated pitch frequency for a synthetic utterance. (After Gold and Rabiner [129]. © [1969] Acoustical Society of America.)

estimates of the pitch period. Finally, several estimates are combined to increase the overall reliability of the estimate. Thus, signal processing simplicity is achieved at the expense of increased logical complexity in estimating the desired feature of the speech signal. Because the logical operations are carried out at a much lower rate (e.g., 100 times/sec) than the signal processing, this results in an overall speed up in processing. A similar approach was used by Barnwell et al. [25] in designing a pitch detector in which the outputs of four simple zero-crossing pitch detectors were combined to produce a reliable estimate of pitch.

10.5.3 Autocorrelation, Periodicity, and Center Clipping

As demonstrated in Section 6.5, the short-time autocorrelation function is defined in Eq. (6.34) as

$$R_{\hat{n}}[k] = \sum_{m=0}^{L-1-|k|} (x[\hat{n}+m]w[m])(x[\hat{n}+m+k]w[k+m]), \tag{10.19}$$

where $w[m]$ denotes a window defined over $0 \leq m \leq L-1$. This short-time representation reflects the periodicity of the input into a convenient form that is not sensitive to the time origin of the signal. Thus, the short-time autocorrelation is frequently used as a basis for determining the pitch period as a function of time. In this section we discuss several issues that arise when the short-time autocorrelation is used in this way as a pitch detector.

One of the major limitations of the autocorrelation representation is that, in a sense, it retains *too much* of the information in the speech signal. [We saw in Chapter 9 that the low-time autocorrelation values ($R[k]$, $0 \leq k \leq p$) are sufficient to accurately estimate the vocal tract transfer function.] As a result, we noted in Figure 6.29, for example, that the autocorrelation function had many peaks. Most of these peaks can be attributed to the damped oscillations of the vocal tract response which are responsible for the shape of each period of the speech wave. In Figures 6.29a and 6.29b, the peak at the pitch period had the greatest amplitude; however, in Figure 6.29c, the peak at about $k = 15$ was actually greater than the peak at $k = 72$. This occurs, in this case, because the window is short compared to the pitch period, but rapidly changing formant frequencies can also create such a confusing situation, as we will see later. Clearly, in cases when the autocorrelation peaks due to the vocal tract response are bigger than those due to the periodicity of the vocal excitation, the simple procedure of picking the largest peak in the autocorrelation function and designating the location of that largest peak as the pitch period indicator will fail.

To avoid this problem, it is again useful to process the speech signal so as to make the periodicity more prominent while suppressing other distracting features of the signal. This was the approach followed in Section 10.5.2 to permit the use of a very simple pitch detector. Techniques that perform this type of operation on a signal are sometimes called "spectrum flatteners" since their objective is to remove the effects of the vocal tract transfer function, thereby bringing each harmonic to the same amplitude level as in the case of a periodic impulse train as depicted in Figure 10.16. Numerous spectrum flattening techniques have been proposed; however, a technique called "center clipping" works well in the present context [370].

In the scheme proposed by Sondhi [370], the center-clipped speech signal is obtained by a non-linear transformation,

$$y[n] = C[x[n]], \tag{10.20}$$

where $C[\ \ ]$ is shown in Figure 10.26. The operation of the center clipper is depicted in Figure 10.27. A segment of speech to be used in computing an autocorrelation function is shown in the upper plot. For this segment, the maximum amplitude, A_{max}, is found and the clipping level, C_L, is set equal to a fixed percentage of A_{max}.[7] (Sondhi [370] used 30%.) Figure 10.26 shows that for samples above C_L, the output of the center clipper is equal to the input minus the clipping level. For samples below the clipping level, the output is zero. The lower plot in Figure 10.27 shows the output for the above input. In contrast to the scheme of Section 10.5.2, where peaks were converted to impulses,

[7] A more sophisticated method sets the clipping threshold to a fixed percentage of the smaller of the maxima obtained from the first and last third of the analysis frame, thereby accounting for buildup and decay of speech during a given frame.

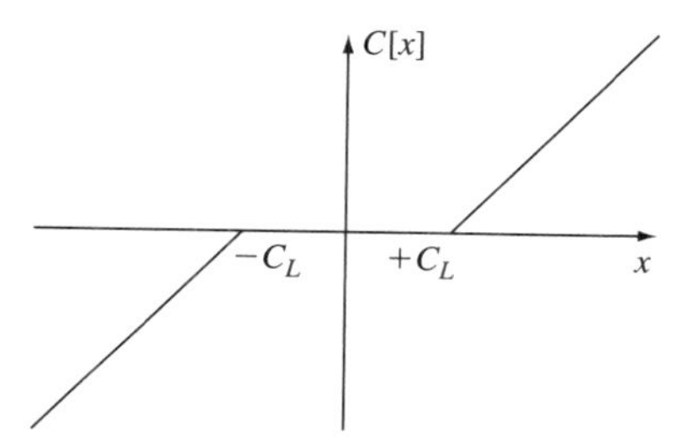

FIGURE 10.26
Center clipping function.

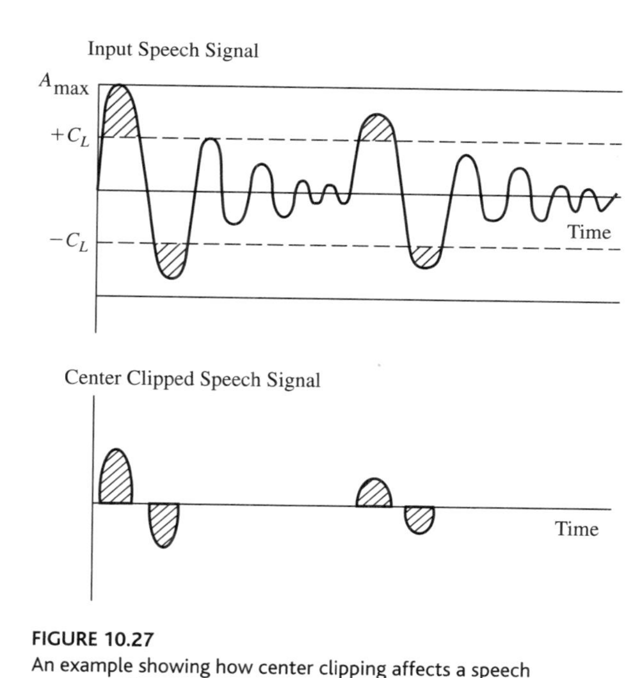

FIGURE 10.27
An example showing how center clipping affects a speech waveform. (After Sondhi [370]. © [1968] IEEE.)

in this case, the peaks are converted to short pulses consisting of the part of each peak that exceeds the clipping level.

Figure 10.28 illustrates the effect of the center clipping operation on the computation of the autocorrelation function. Figure 10.28a shows a 400 sample segment (40 msec at $F_s = 10$ kHz) of voiced speech. Note that there is a strong peak at the pitch period in the autocorrelation function for this segment, which is shown on the right. However, it is also clear that there are many peaks that can be attributed to the damped oscillations of the vocal tract. Figure 10.28b shows the corresponding center-clipped signal where the clipping level was set as shown by the dashed lines in Figure 10.28a.

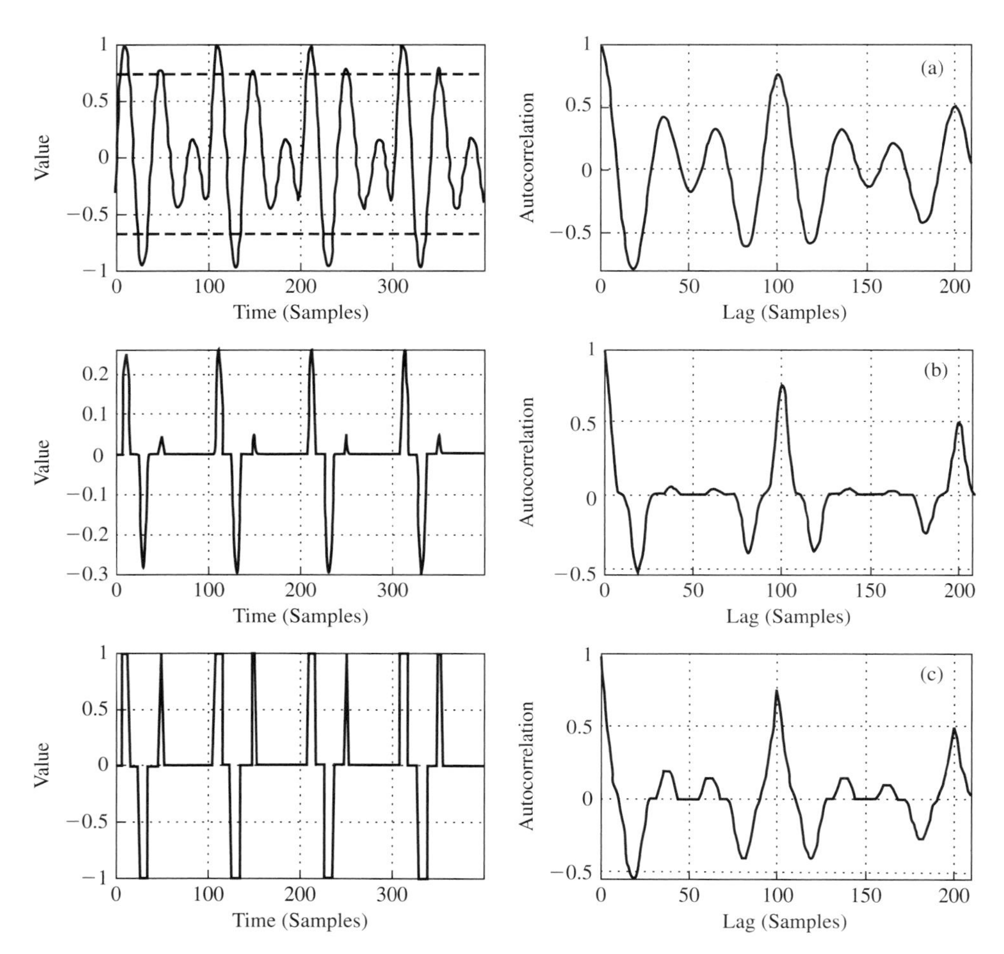

FIGURE 10.28
Example of waveforms and autocorrelation functions: (a) no clipping (dashed lines showing clipping level of 70%); (b) center clipped at 70% threshold; (c) 3-level center clipped. (All correlation functions normalized to 1.0.)

(In this case the clipping level was set at 70% of the maximum signal level in the speech frame.) We see that all that remains in the clipped waveform are several pulses spaced at the original pitch period. The resulting autocorrelation function has considerably fewer extraneous peaks to create confusion.

We will return to part (c) of Figure 10.28 soon. However, first let us examine the effect of the clipping level. Clearly, for high clipping levels, fewer peaks will exceed the clipping level and thus fewer pulses will appear in the output, and therefore, fewer extraneous peaks will appear in the autocorrelation function. This is illustrated by Figure 10.29, which shows the autocorrelation functions for the segment of speech corresponding to Figure 10.28, for decreasing clipping levels. Clearly, as the clipping level is decreased, more peaks pass through the clipper and thus the autocorrelation

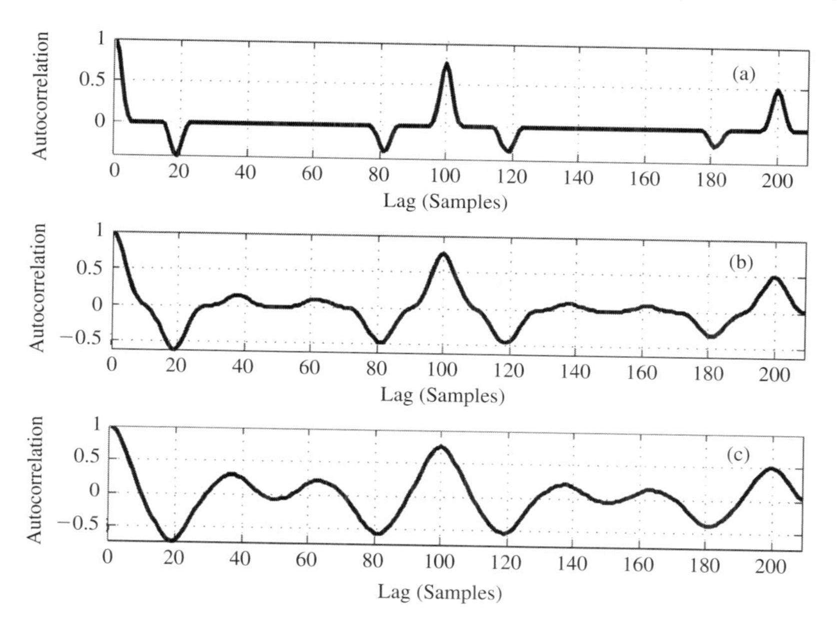

FIGURE 10.29
Autocorrelation functions of center-clipped speech using $L = 401$: (a) C_L set at 90% of maximum; (b) 60%; (c) 30%. (Speech segment same as for Figure 10.28.)

function becomes more complex. (Note that a clipping level of zero corresponds to Figure 10.28a.) The implication of this example is that the clearest indication of periodicity is obtained for the highest possible clipping level. However, the clipping level should not be set too high because the amplitude of the signal can vary appreciably across the duration of the speech segment (e.g., at the beginning or end of voicing). If the clipping level is set at a high percentage of the maximum amplitude across the whole segment, it is possible that much of the waveform will fall below the clipping level and be lost. For this reason Sondhi's original proposal was to set the clipping level at 30% of the maximum amplitude. A procedure which permits a greater percentage (60–80%) to be used is to find the peak amplitude in both the first third and the last third of the segment and set the clipping level at a fixed percentage of the minimum of these two maximum levels.

The problem of extraneous peaks in the autocorrelation function can be alleviated by center clipping prior to computing the autocorrelation function. However, another difficulty with the autocorrelation representation (that remains even with center clipping) is the large amount of computation that is required. A simple modification of the center clipping function leads to a great simplification in computation of the autocorrelation function with essentially no degradation in utility for pitch detection [90]. This modification is shown in Figure 10.30. As indicated there, the output of the clipper is $+1$ if $x[n] > C_L$ and -1 if $x[n] < -C_L$. Otherwise the output is zero. This function is called a 3-level center clipper. Figure 10.28c shows the output of the 3-level center clipper for the input segment of Figure 10.28a. Note that

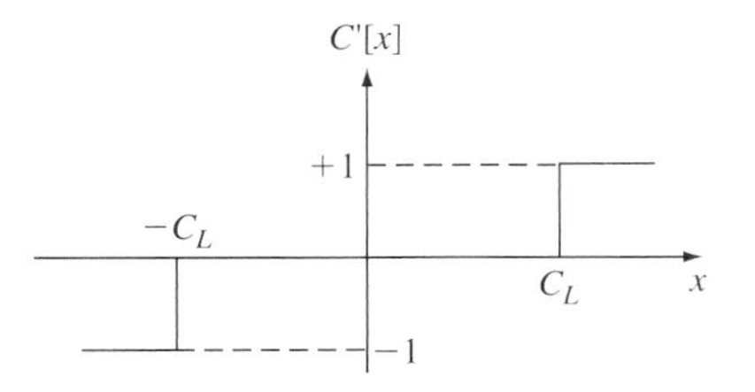

FIGURE 10.30
3-Level center clipping function.

although this operation tends to emphasize the importance of peaks that just exceed the clipping level, the autocorrelation function is very similar to that of the center clipper of Figure 10.28b. That is, most of the extraneous peaks are eliminated, and a clear indication of periodicity is retained.

The computation of the autocorrelation function for a 3-level center-clipped signal is particularly simple. If we denote the output of the 3-level center clipper as $y[n]$, then the product terms $y[\hat{n}+m]y[\hat{n}+m+k]$ in the autocorrelation function

$$R_{\hat{n}}[k] = \sum_{m=0}^{L-1-k} y[\hat{n}+m]y[\hat{n}+m+k] \tag{10.21}$$

can have only three different values:

$$y[\hat{n}+m]y[\hat{n}+m+k] = \begin{cases} 0 & \text{if} \quad y[\hat{n}+m]=0 \quad \text{or} \quad y[\hat{n}+m+k]=0 \\ +1 & \text{if} \quad y[\hat{n}+m]=y[\hat{n}+m+k] \\ -1 & \text{if} \quad y[\hat{n}+m]\neq y[\hat{n}+m+k]. \end{cases}$$

Thus, in hardware terms, all that is required is some simple combinatorial logic and an up-down counter to accumulate the autocorrelation value for each value of k.

While the autocorrelation function effectively represents periodicity in a waveform in a convenient form, it is not a foolproof measure when applied to segments of voiced speech, which are only quasi-periodic. This is illustrated by Figure 10.31. The general format of this figure shows a waveform segment on the left side and a corresponding correlation function on the right. Figure 10.31a shows a 480-sample (30 msec) Hamming windowed speech segment for a male speaker. Figure 10.31b shows the corresponding autocorrelation function $R[k]$ for $k=0,1,\ldots,200$. Note the linear fall-off of the peaks. The heavy vertical line at $k=89$ marks the location of the largest peak in the range $48\leq k\leq 200$, corresponding to a pitch period range of 3 to 12.5 msec or fundamental frequency range of 80–333 Hz. The value 89 is the correct pitch period for the segment in Figure 10.31a. However, note that the peak at lag $k=45$ is higher than the one at $k=89$. If we were to lower the lag range only slightly, that peak location would be chosen as the period. Indeed, the peaks of the autocorrelation function are roughly at multiples of 45. The reason for this is evident from Figure 10.32, which shows the short-time Fourier transform (STFT) corresponding to Figures 10.31a and b.

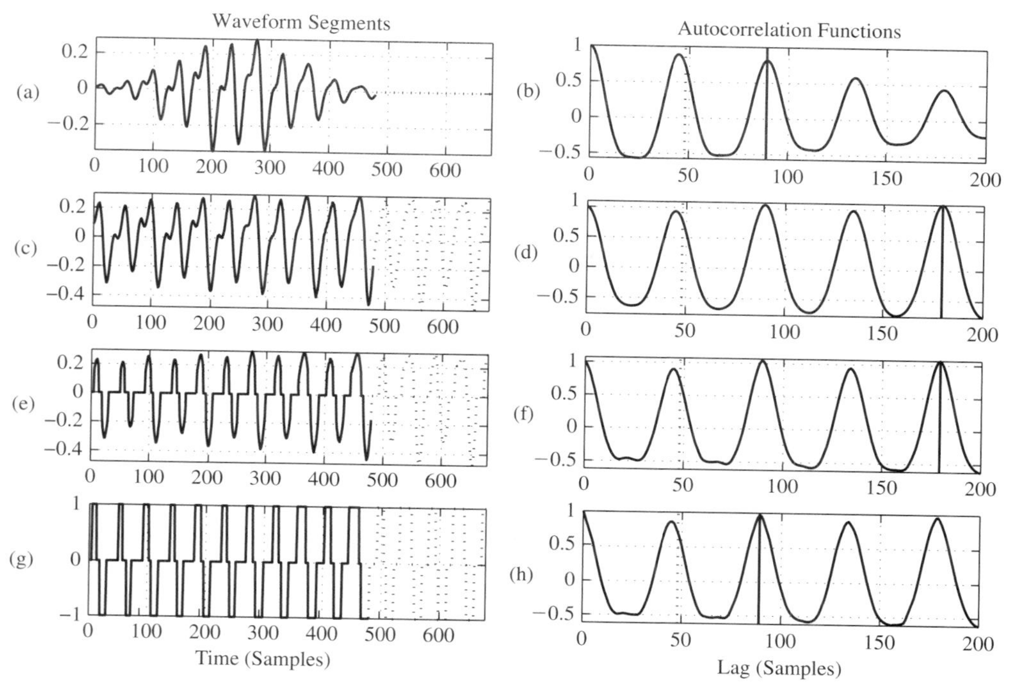

FIGURE 10.31
Illustration of doubling errors in autocorrelation functions: (a,b) autocorrelation function with Hamming window; (c,d) modified autocorrelation function; (e,f) modified autocorrelation function with center clipping; (g,h) modified autocorrelation function with 3-level center clipping.

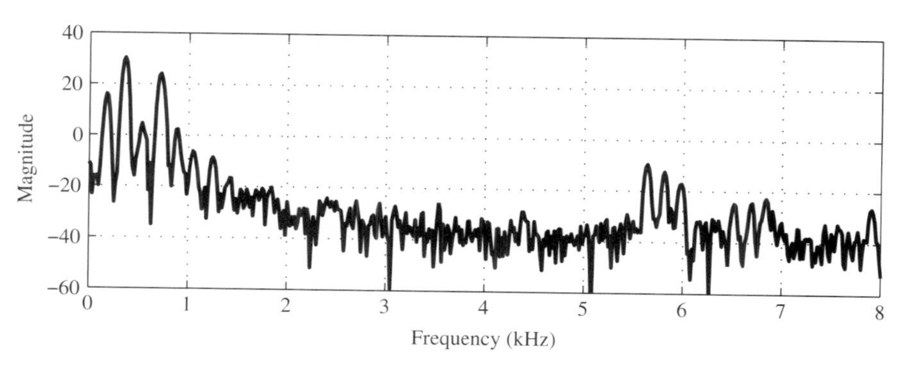

FIGURE 10.32
STFT corresponding to Figure 10.31a.

Observe that this segment of speech has essentially only three strong frequency components at frequencies 180, 360, and 720 Hz. From the point of view of a periodic signal model, the signal has a weak fundamental, with the second and fourth harmonics being much stronger, and the third harmonic essentially missing. This causes the signal to appear to have a fundamental frequency of 360 Hz instead of 180 Hz. In the autocorrelation function, a fundamental frequency of 180 Hz corresponds to a period of $16{,}000/180 \approx 89$ samples, but the two strong components at 360 and 720 suggest a fundamental period of $16{,}000/360 \approx 45$ samples. Thus, the autocorrelation function has peaks at multiples of 45, not just multiples of 89. The missing third harmonic exacerbates this condition.

Figures 10.31c and 10.31d show the waveform and modified autocorrelation function. In this case, note the rectangular window and the extra 200 samples on the right required for computation of the modified autocorrelation function for lags up to 200 samples. Also note that the peaks in the modified autocorrelation function do not fall off with increasing k, and in this case, the maximum peak is at $k = 178$, which is twice the true pitch period. This occurs because the speech signal is not perfectly periodic across the entire window interval. This is not uncommon with the modified autocorrelation function. Figures 10.31e and 10.31f show the center-clipped waveform and modified autocorrelation function. The clipping threshold is set at 60% of the smaller of the peak amplitude during the first third and during the last third of the analysis frame. Again, the maximum peak in the range $56 \leq k \leq 200$ occurs at $k = 178$. Finally, Figures 10.31g and 10.31h show the 3-level center-clipped waveform and modified autocorrelation function when the clipping threshold is set at 60% of the peak amplitude within the analysis interval. In this case, the maximum peak in the pre-set lag interval occurs at $k = 89$.

Another example is shown in Figure 10.33. In this case, the pitch period of the waveform in Figure 10.33a is 161 samples; however, the largest peak in the range $48 \leq k \leq 200$ for the autocorrelation function in Figure 10.33b occurs at $k = 81$. The reason is again clear from Figure 10.34, which shows the STFT corresponding to Figures 10.33a and 10.33b. In this case, the true fundamental is about 100 Hz, as indicated by the 10 peaks in the range 0–1000 Hz in Figure 10.34. This corresponds to the period of 161 samples for a sampling rate of $F_s = 16{,}000$ (16,000/161 = 99.4 Hz). In this example, the peak at $k = 81$ samples is slightly higher than the peak at $k = 161$. While the STFT in Figure 10.34 shows a much richer harmonic structure, the second and fourth harmonics are the strongest.

The above examples illustrate that while the various combinations of center clipping and autocorrelation measures generally give accurate indications of the pitch period, they cannot be relied on to give a foolproof estimate of the pitch period. Therefore, a complete pitch detection algorithm based on autocorrelation will generally involve a variety of logical tests to detect doubling or halving and post-processing using non-linear filtering to eliminate outlying estimates due to gross errors. An example is discussed in the next section.

10.5.4 An Autocorrelation-Based Pitch Estimator

We conclude our discussion of the use of autocorrelation in pitch period estimation by considering the details of a pitch estimation algorithm that was implemented in digital hardware [90] at a time when it was difficult to achieve real-time operation. Thus, the basis for the algorithm was the 3-level center-clipped autocorrelation. For our purposes here, however, emphasis will be on the details of the algorithm, which are not specific

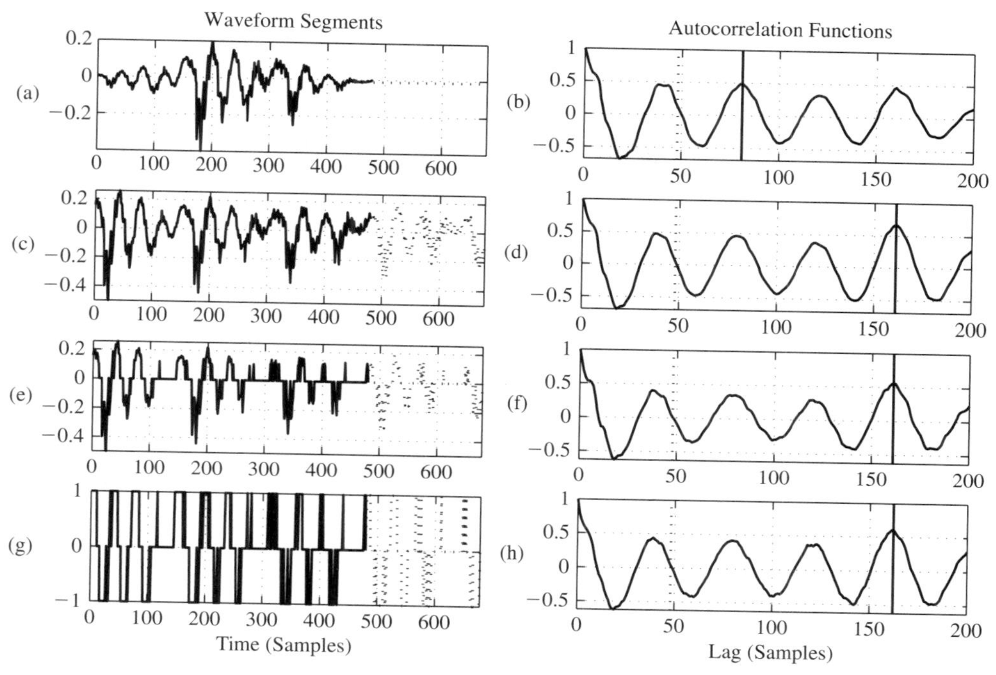

FIGURE 10.33
Illustration of doubling errors in autocorrelation functions: (a,b) autocorrelation function with Hamming window; (c,d) modified autocorrelation function; (e,f) modified autocorrelation function with center clipping; (g,h) modified autocorrelation function with 3-level center clipping.

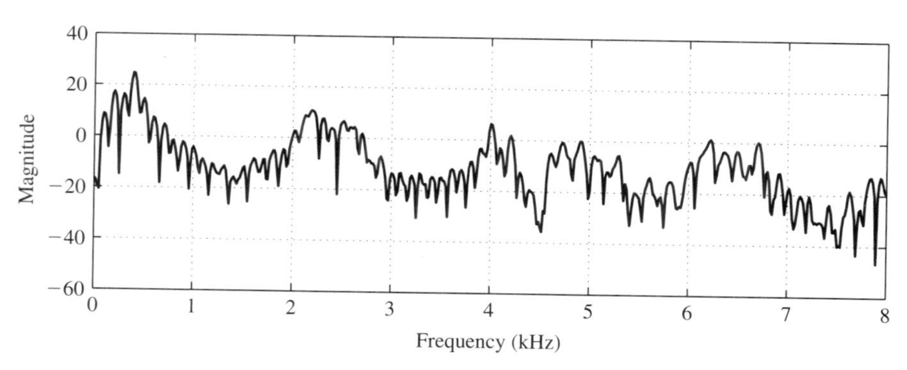

FIGURE 10.34
STFT corresponding to Figure 10.33a.

to any hardware or software implementation. The system is depicted in Figure 10.35 and summarized by the following statements:

1. The speech signal is filtered with a 900 Hz lowpass analog filter and sampled at a rate of 10 kHz.

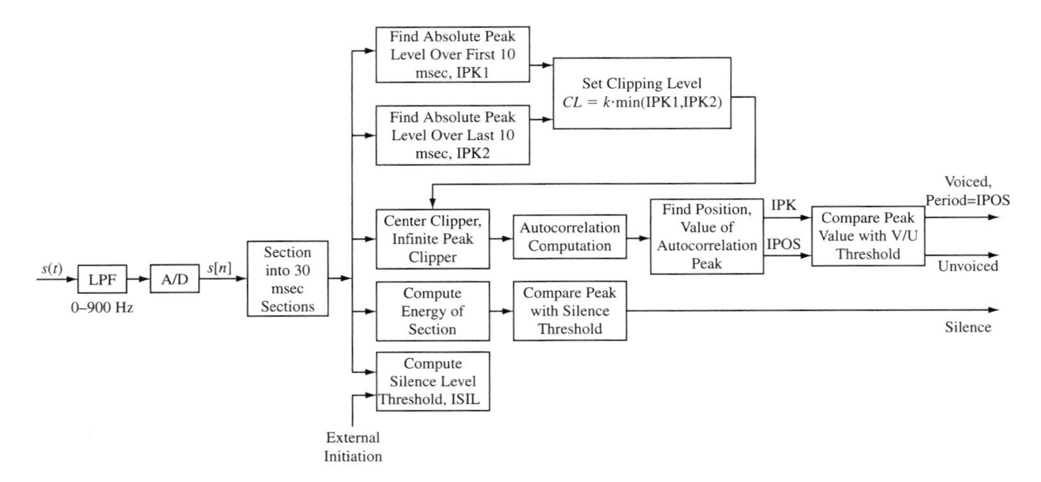

FIGURE 10.35
Block diagram of clipping autocorrelation pitch detector. (After Dubnowski et al. [90]. © [1976] IEEE.)

2. Segments of length 30 msec (300 samples) are selected at 10 msec intervals. Thus, the segments overlap by 20 msec.
3. The short-time average magnitude, Eq. (6.17), is computed with a 100 sample rectangular window. The peak signal level in each frame is compared to a threshold determined by measuring the peak signal level for 50 msec of background noise. If the peak signal level is above the threshold, signifying that the segment is speech, not noise, then the algorithm proceeds as outlined in the next steps; otherwise the segment is classed as silence and no further action is taken.
4. The clipping level is determined as a fixed percentage (e.g., 68%) of the minimum of the maximum absolute values in the first and last 100 samples of the speech segment.
5. Using this clipping level, the speech signal is processed by a 3-level center clipper and the correlation function is computed over a range spanning the expected range of pitch periods.
6. The largest peak of the autocorrelation function is located and the peak value is compared to a fixed threshold (e.g., 30% of $R_{\hat{n}}[0]$). If the peak value falls below the threshold, the segment is classified as unvoiced and if it is above, the pitch period is defined as the location of the largest peak.

This is essentially the algorithm that was implemented in digital hardware [90]; however, there is considerable latitude for variation in the details. For example, steps 4 and 5 could be altered to use the center clipper of Figure 10.26 and standard arithmetic for the autocorrelation computation, or center clipping could be completely eliminated. Still another possibility is to use the AMDF function, as explained in Chapter 6 (and thus search for dips instead of peaks), either with or without some form of center clipping.

Note that this algorithm combines VUS detection with estimation of the pitch period. A more effective, but more computationally expensive, approach would be to use the VUS classifier of Section 10.4 and then apply an autocorrelation-based pitch detector only to the frames classified as voiced.

Figure 10.36 shows the outputs (pitch contours) of three variants of the above algorithm. Figure 10.36a is the pitch contour obtained using the autocorrelation of the speech signal without clipping. Note the scattering of points that are obviously errors due to the fact that a peak at a short lag was greater than the peak at the pitch period. Also note that the pitch period averages between 100 and 150 samples, so that the inherent fall-off of the autocorrelation function causes significant attenuation of the peak at the pitch period. Thus, peaks in the autocorrelation function due to the vocal tract response are likely to be greater than those due to periodicity. Figures 10.36b and 10.36c are respectively for the cases when center clipping and 3-level center clipping are used with the autocorrelation function. Clearly, most of the errors have been eliminated by the inclusion of clipping, and furthermore, there is no significant difference between the two results. A few obvious errors remain in both pitch contours. These errors can be removed by the non-linear smoothing method that was discussed in Section 10.2. In the example shown in Figure 10.36d, all the isolated errors have been removed by the non-linear smoothing.

10.5.5 Pitch Detection in the Spectral Domain

We have seen in Chapter 7 and in the examples in Figures 10.32 and 10.34 that in a narrowband short-time Fourier representation, the excitation for voiced speech is manifested in sharp peaks that occur at integer multiples of the fundamental frequency. This fact has served as the basis of a number of pitch detection schemes based on the STFT. In this section we briefly discuss an example of a pitch detector that illustrates both the basic concepts of using the short-time spectrum for pitch detection and the flexibility afforded by digital processing methods. The example shows that many possibilities exist for exploiting the time-dependent Fourier representation in determining excitation parameters. (Another example is suggested by Problem 10.3.)

One approach to pitch detection in the spectral domain involves the computation of the *harmonic product spectrum* [352], which is defined as

$$P_{\hat{n}}(e^{j\omega}) = \prod_{r=1}^{K} |X_{\hat{n}}(e^{j\omega r})|^2. \tag{10.22}$$

Taking the logarithm gives the *log harmonic product spectrum*:

$$\hat{P}_{\hat{n}}(e^{j\omega}) = 2\sum_{r=1}^{K} \log |X_{\hat{n}}(e^{j\omega r})|. \tag{10.23}$$

The function $\hat{P}_{\hat{n}}(e^{j\omega})$ is seen to be a sum of K frequency-compressed replicas of $\log |X_{\hat{n}}(e^{j\omega})|$. The motivation for using the function of Eq. (10.23) is that for voiced speech, compressing the frequency scale by integer factors should cause harmonics

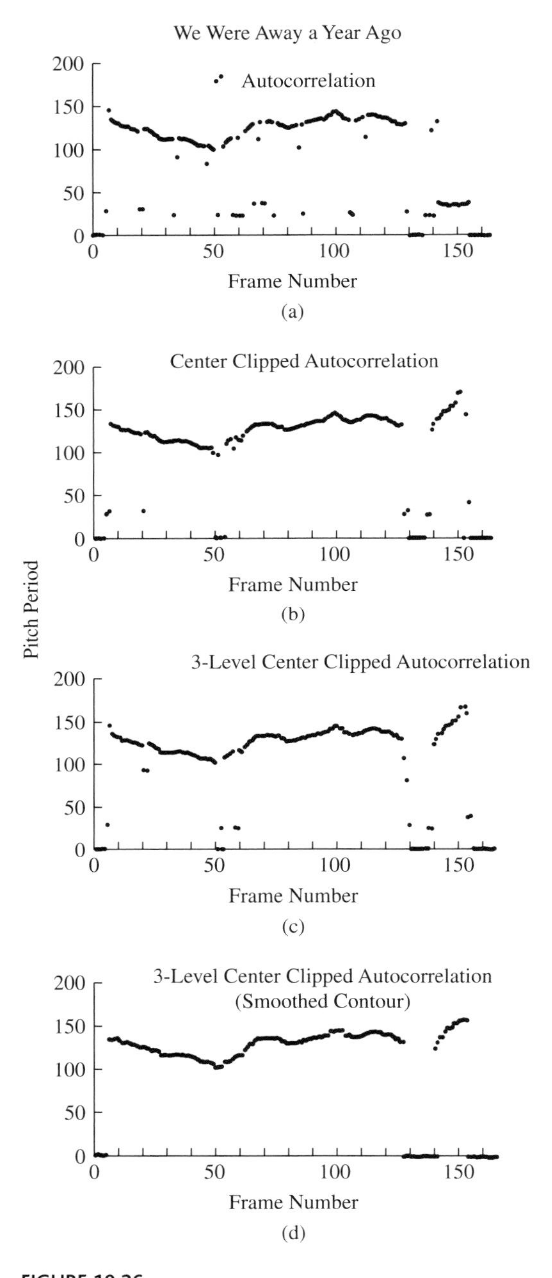

FIGURE 10.36
Autocorrelation pitch detector outputs: (a) no clipping; (b) center clipping (Figure 10.26); (c) 3-level center clipping (Figure 10.30); (d) non-linearly smoothed output from (c). (After Rabiner [300]. © [1977] IEEE.)

of the fundamental frequency to coincide at the fundamental frequency (e.g., for a compression of the frequency scale by a factor of 2, the second harmonic will coincide with the fundamental; for a compression by a factor of 3, the third harmonic will coincide with the fundamental, etc.). At frequencies between the harmonics, some of the frequency-compressed harmonics will coincide, but only at the fundamental will there *always* be reinforcement.

This process of compressing the frequency scale is depicted schematically in Figure 10.37, which shows the log magnitude spectrum with no compression ($r = 1$) in the top panel, with a compression of 2-to-1 ($r = 2$) in the middle panel, and with a compression of 3-to-1 ($r = 3$) in the bottom panel. As is easily seen in this figure, for the continuous function $|X_{\hat{n}}(e^{j2\pi FTr})|$, the peak at F_0 becomes sharper as r increases; thus, the sum of Eq. (10.23) will have a sharp peak at F_0, with possibly some lesser peaks elsewhere. The resulting harmonic product spectra and the log harmonic product spectra have been found to be especially resistant to independent, additive noise, since the contributions of the noise to $X_{\hat{n}}(e^{j\omega})$ have no coherent structure when viewed as a function of frequency. Therefore, in Eq. (10.23), the noise components in $X_{\hat{n}}(e^{j\omega r})$ also tend to add incoherently. For the same reason, unvoiced speech will not exhibit a peak in $\hat{P}_{\hat{n}}(e^{j\omega})$. Another important point is that a peak at the fundamental frequency need not be present in $|X_{\hat{n}}(e^{j\omega})|$ for there to be a peak in $\hat{P}_{\hat{n}}(e^{j\omega})$, since the compressed harmonics of the fundamental frequency will add coherently at the (missing) fundamental frequency, much as shown above. Thus, the harmonic product and log harmonic product spectrum are attractive for operation on highpass filtered speech such as telephone speech, which often is missing the fundamental frequency component for low-pitched voices.

An example of the use of the harmonic product spectrum and the log harmonic product spectrum for pitch detection is shown in the "waterfall" plot of Figure 10.38. The input speech from a female talker was sampled at a 10 kHz sampling rate, and every 10 msec the signal was multiplied by a 40 msec Hamming window (400 samples). Then values of $X_{\hat{n}}(e^{j2\pi k/N})$ were computed using a fast Fourier transform (FFT) algorithm with $N = 4000$. Figures 10.38a and 10.38b show a sequence of log harmonic product spectra and harmonic product spectra, respectively, for the case $K = 5$ in Eqs. (10.23) and (10.22). The pitch frequency peak in the range of 180–240 Hz for this female talker is obvious in the harmonic product spectra plots of part (b) of this figure. The peak appears less distinct in the log harmonic product spectrum. This is simply due to the logarithm operation, which compresses the amplitude scale. It is clear, from this figure, that a rather simple pitch estimation algorithm could be designed with the harmonic product spectrum as input. Such an algorithm has been studied and the superior noise resistance of the resulting algorithm has been cited by Noll [253].

10.5.6 Homomorphic System for Pitch Detection

In Chapter 8, we showed that the cepstrum of voiced speech displays an impulsive peak at a quefrency equal to the pitch period. This is illustrated in Figure 10.39, which shows cepstrum examples from Chapter 8. Figure 10.39a shows the cepstrum for a voiced segment (also see Figure 8.30a) and Figure 10.39b shows the cepstrum of an unvoiced

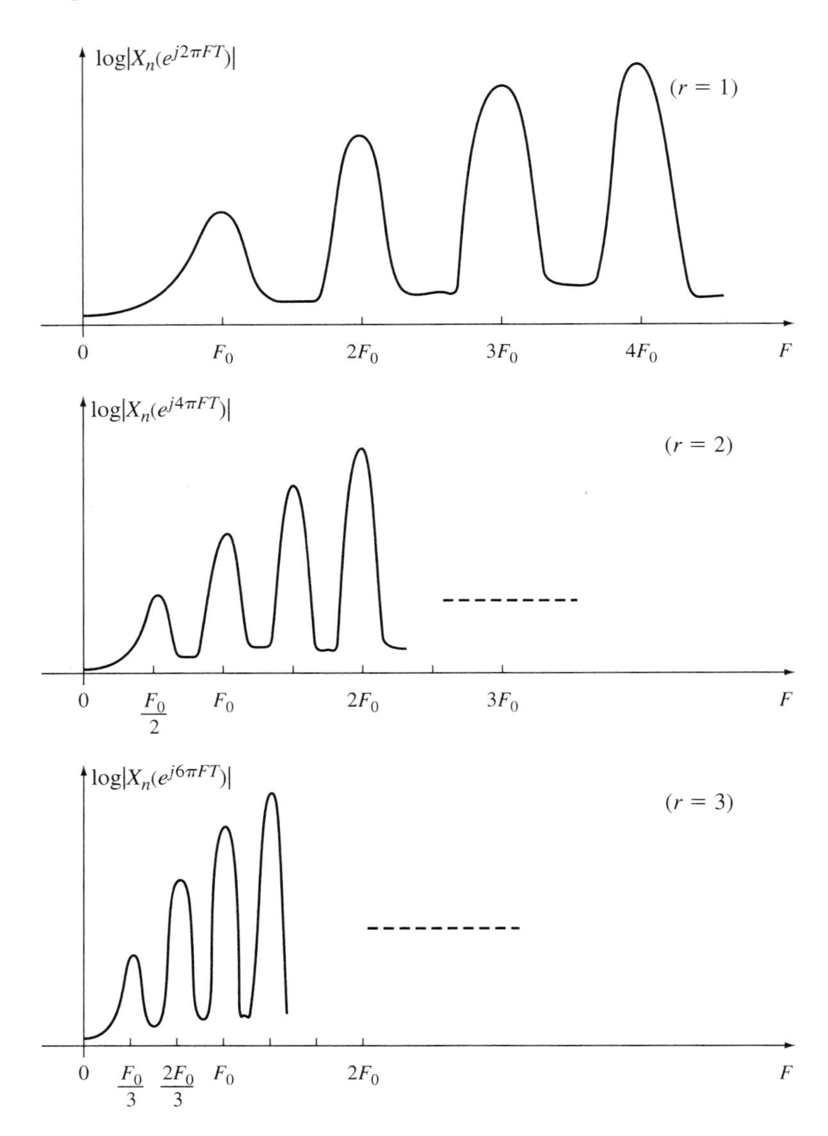

FIGURE 10.37
Representations of terms in the log harmonic product spectrum.

segment (see Figure 8.35c). These plots suggest that the cepstrum could be the basis for pitch estimation algorithms, in the style of algorithms based on the autocorrelation function. We observe that for voiced speech, there is a peak in the cepstrum at the fundamental period of the input speech segment. No such peak appears in the cepstrum for unvoiced speech. These properties of the cepstrum can be used as a basis for determining whether a speech segment is voiced or unvoiced and for estimating the fundamental period of voiced speech.

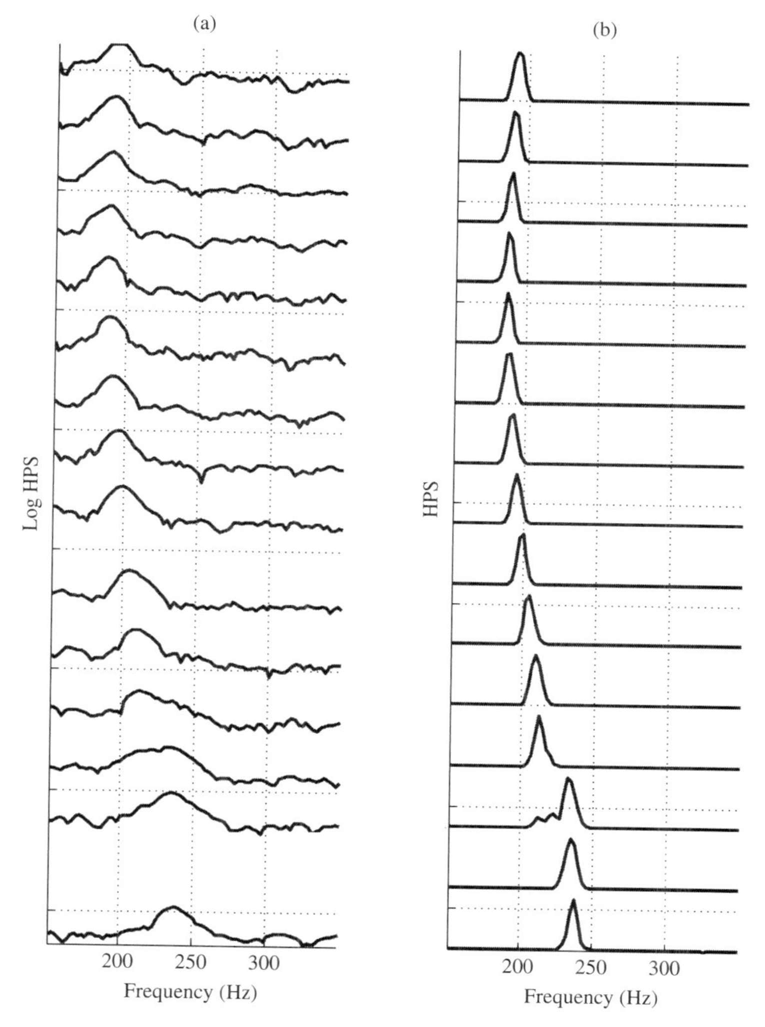

FIGURE 10.38
Log harmonic product spectra and harmonic product spectra for a female talker: part (a) shows the sequence of log harmonic product spectra and part (b) shows the corresponding sequence of harmonic product spectra.

The cepstrum definition combines several features of the center-clipped autocorrelation function. The log operation flattens the spectrum of the signal, and, as shown in Chapter 8, the low-quefrency cepstrum values fall off as $1/|n|$, thus causing less interference between the vocal tract and the high-quefrency excitation components of the cepstrum. For voiced speech, the cepstrum peaks occur at multiples of the pitch period, rN_p, but peaks at these locations also are reduced by a factor $1/|r|$. For these reasons, the cepstrum can be a more reliable indicator of periodicity than the autocorrelation function even with center clipping.

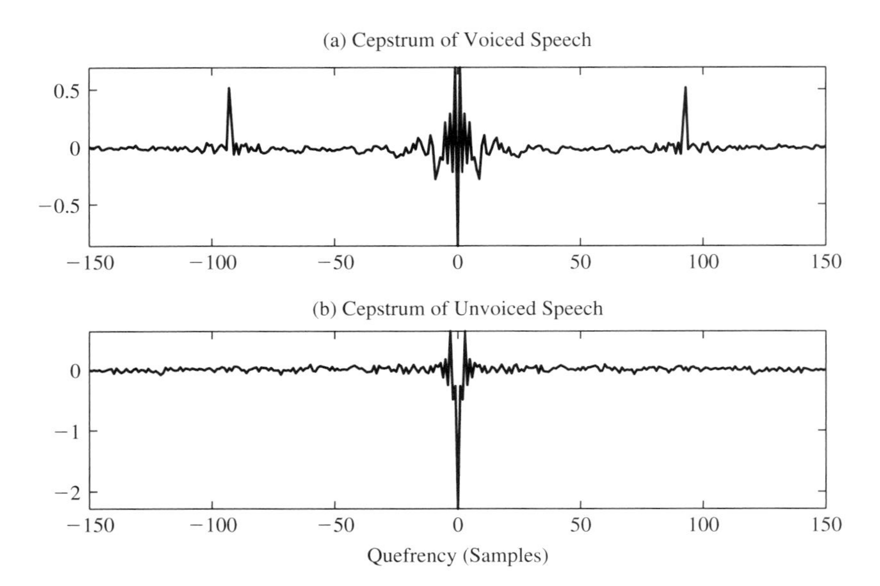

FIGURE 10.39
Cepstra for (a) voiced speech (Figure 8.30a) and (b) unvoiced speech (Figure 8.35c).

The general algorithm for cepstrum pitch estimation is much the same as the autocorrelation-based algorithm discussed in Section 10.5.4. The cepstrum, computed as discussed in Section 8.4, is searched for a peak in the vicinity of the expected pitch period. If the cepstrum peak is above a pre-set threshold, the input speech segment is likely to be voiced, and the position of the peak is a good estimate of the pitch period. If the peak does not exceed the threshold, it is likely that the input speech segment is unvoiced. The time variation of the model of excitation and the pitch period can be estimated by computing a time-dependent cepstrum based upon a time-dependent Fourier transform. Typically, the cepstrum is computed once every 10–20 msec since the excitation parameters do not change rapidly in normal speech.

Figures 10.40 and 10.41 again show waterfall plot examples of short-time log magnitude spectra and short-time cepstra for a male and a female talker. In these examples, the sampling rate of the input was 10 kHz. A 40 msec (400 samples) Hamming window weighted frame of speech was moved in jumps of 10 msec; i.e., log spectra on the left and corresponding cepstra on the right are computed at 10 msec intervals. Clear indications of the pitch period (in quefrency or samples) can be seen from Figure 10.40 throughout the 350 msec region of the plot. The pitch period is seen to rise during the first 10 or so frames, then fall over the next 20 frames, and is fairly steady over the last 5 or so frames. Figure 10.41 shows similar results for a female talker. In this case, the pitch period only changes a small amount (order of a few percentage points) over the entire 350 msec interval in the plots. Again we see that the cepstral peak at the pitch period is the dominant signal level for the entire voiced interval. It is clear that the cepstrum has the potential to form the basis for a highly accurate pitch detection algorithm.

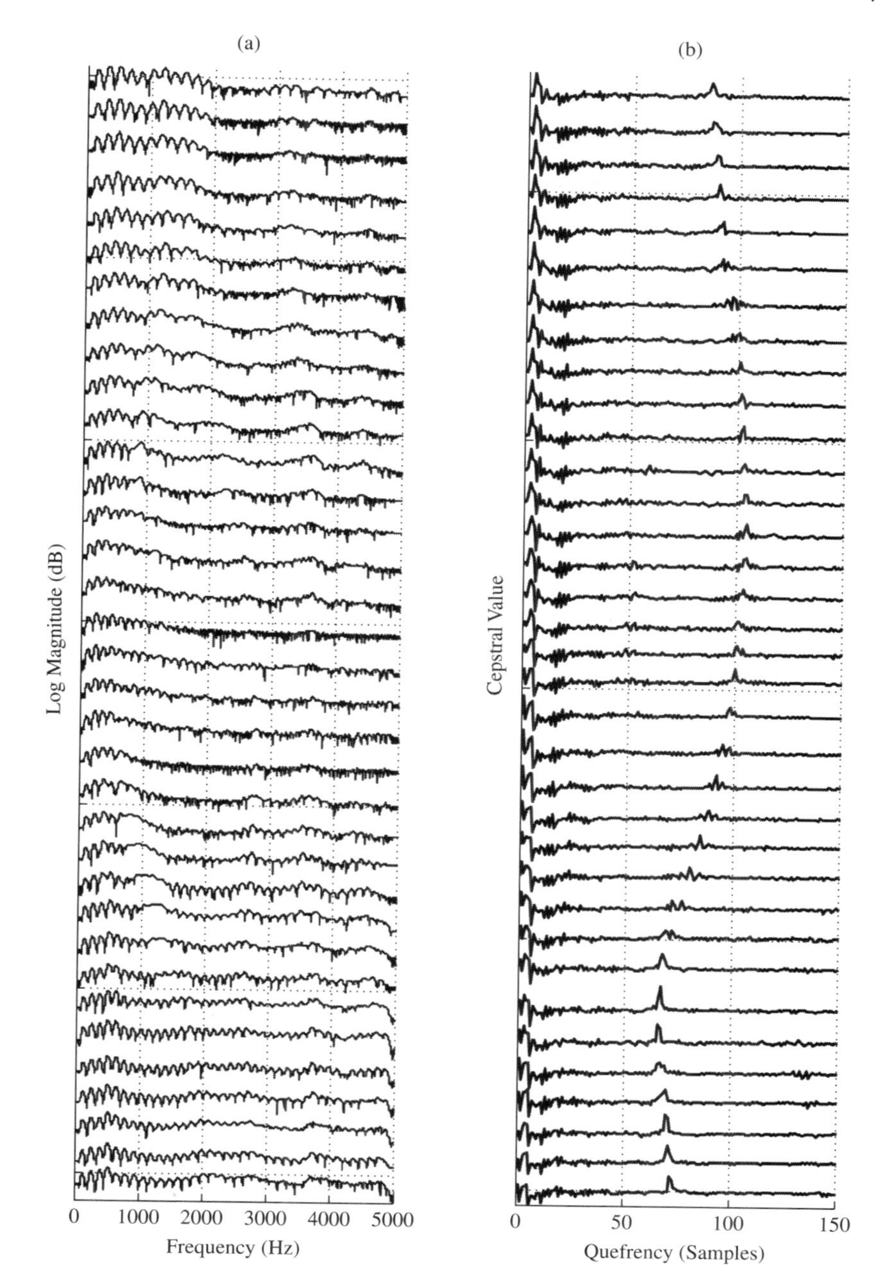

FIGURE 10.40
Series of log spectra and cepstra for a male speaker.

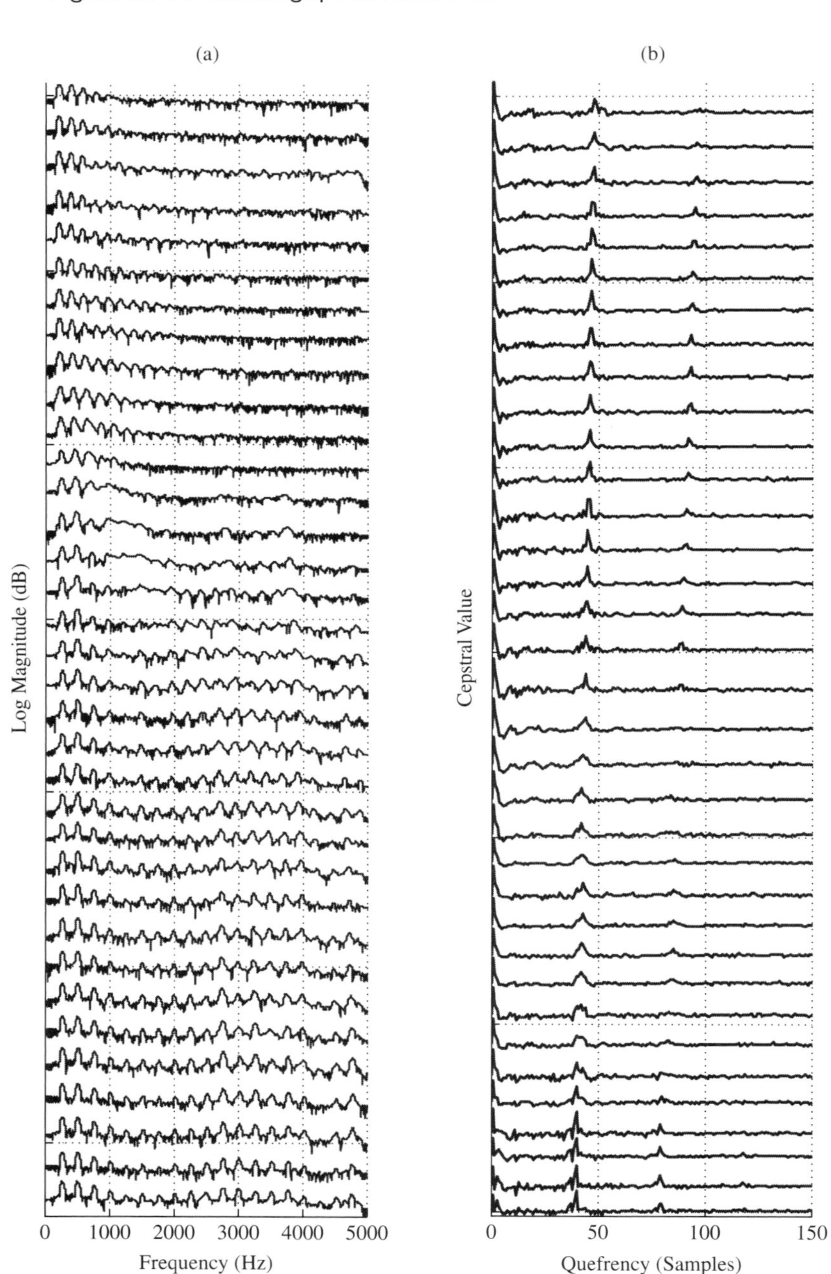

FIGURE 10.41
Series of log spectra and cepstra for a female speaker.

These two examples, although impressive in the prominence with which the pitch information is displayed, may lead us to assume that an overly simplistic algorithm will produce high quality estimates of pitch and voicing. As is often the case in speech analysis algorithms, there are numerous special cases and trade-offs that must be considered in designing a cepstrum pitch detection algorithm; however, a variety of schemes based upon the cepstrum have been used with success. Noll [252] has given a flowchart of one such algorithm. Rather than give the details of any one procedure here, it is more useful to call attention to some of the essential difficulties in using the cepstrum for pitch detection.

First, the presence of a strong peak in the cepstrum in the range 3–20 msec is a very strong indication that the input speech segment is voiced. However, the absence of a peak or the existence of a low level peak is not necessarily a strong indication that the input speech segment is unvoiced. That is, the strength of or even the existence of a cepstrum peak for voiced speech is dependent on a variety of factors, including the length of the window applied to the input signal and the formant structure of the input signal. It is easily shown (see Problem 10.3) that the maximum height of the "pitch peak" is unity. This can be achieved only in the case of absolutely identical pitch periods. This is, of course, highly unlikely in natural speech, even in the case of a rectangular window which encloses exactly an integer number of periods. Rectangular windows are rarely used due to the inferior spectrum estimates that result, and in the case of, for example, a Hamming window, it is clear that both window length and the relative positions of the window and the speech signal will have a considerable effect upon the height of the cepstrum peak. As an extreme example, suppose that the window is less than two pitch periods long. Clearly it is not reasonable to expect any strong indication of periodicity in the spectrum, or the cepstrum, in this case. Thus, the window duration is usually set so that, taking into account the tapering of the data window, at least two clearly defined periods remain in the windowed speech segment. For low pitched male speech, this requires a window on the order of 40 msec in duration. For higher pitched voices, proportionately shorter windows can be used. It is, of course, desirable to maintain the window as short as possible so as to minimize the variation of speech parameters across the analysis interval. The longer the window, the greater the variation from beginning to end and the greater will be the deviation from the model upon which the analysis is based. One approach to maintaining a window that is neither too short or too long is to adapt the window length based upon the previous (or possibly average) pitch estimates [300, 344].

Another way in which the cepstrum can fail to give a clear indication of periodicity is if the spectrum of the input signal is severely bandlimited. An extreme example is the case of a pure sinusoid. In this case there is only one peak in the log spectrum. If there is no periodic oscillation in the log spectrum, there will be no peak in the cepstrum. In speech, some sounds such as voiced stops are extremely bandlimited, with no clearly defined harmonic structure at frequencies above a few hundred hertz. An example of this situation is the signal in the example of Figure 10.31a, whose log spectrum is shown in Figure 10.32. Figure 10.42a shows the autocorrelation function (repeated from Figure 10.31b) for this segment of speech. Note that it shows a confusing array of peaks at approximate multiples of 45 samples, although the correct period is 89. The corresponding cepstrum is shown in Figure 10.42b. In this case, although

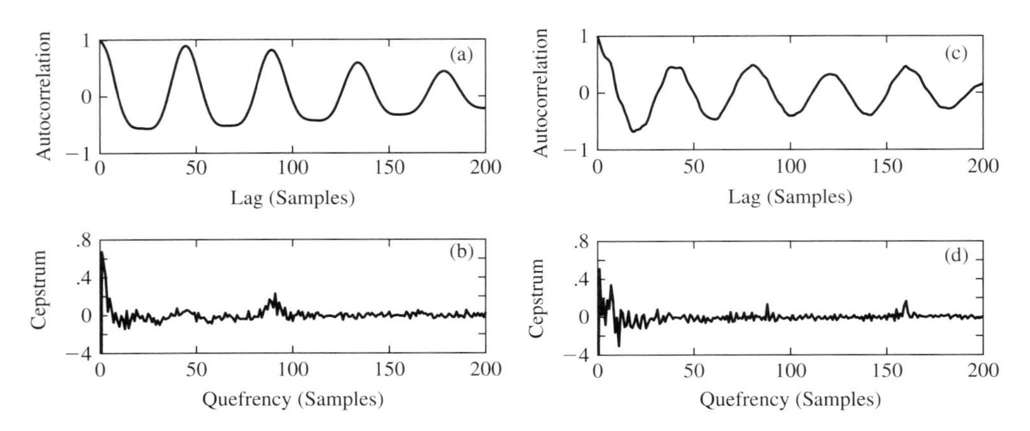

FIGURE 10.42
Comparison of autocorrelation and cepstrum for two frames of speech: (a) autocorrelation and (b) cepstrum for the waveform in Figure 10.31a; (c) autocorrelation and (d) cepstrum for the waveform in Figure 10.33a.

the peak in the cepstrum at quefrency 89 is weak, it still stands out in the interval 48 to 200. (Recall that when center clipping and the modified autocorrelation function are used, the peak occurs at lag 178 instead of the correct value of 89.) In many such cases, the peak is not weak, but fortunately, for all but the shortest pitch period, the pitch peak occurs in a region where the other cepstrum components have died out appreciably. Therefore, a rather low threshold can be used in searching for the pitch peak (e.g., on the order of 0.1). As another example, Figures 10.42c and d show the autocorrelation function (repeated from Figure 10.33b) and cepstrum respectively for the waveform segment in Figure 10.33a. In this case, the autocorrelation peak at lag 80 was the highest although the correct pitch period was 160. Note in Figure 10.42d that the maximum cepstrum peak is at the correct value, although there is a lesser amplitude peak around 80.

With an appropriate window length, the location and amplitude of the cepstrum peak together provide a reliable pitch and voicing estimate for most speech frames. In the cases where the cepstrum fails to clearly display the pitch and voicing, the reliability can be improved by the addition of other information such as zero-crossing rate and energy, and by forcing the pitch and voicing estimates to vary smoothly [344]. The extra logic required to take care of special cases often requires considerable code in software implementations, but this part of a cepstrum pitch detection scheme is a small portion of the total computational effort and is well worthwhile.

10.5.7 Pitch Detection Using Linear Prediction Parameters

In Section 9.6, we showed that the linear prediction residual signal, $e[n]$, is essentially white noise when the input speech signal is unvoiced, and it displays periodicity properties when the input is voiced. As illustrated in Figures 9.25–9.28, with an appropriate choice of the predictor order p, the output for voiced speech often looks very much like the idealized quasi-periodic impulse train of the model for

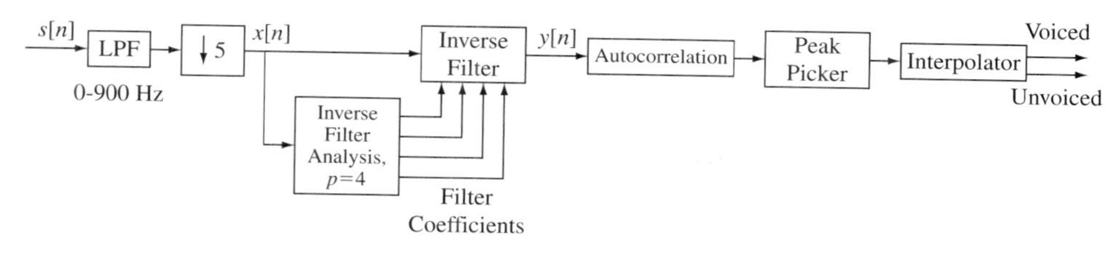

FIGURE 10.43
Block diagram of the SIFT algorithm for pitch detection.

speech production. Thus, linear predictive inverse filtering can be used as the spectral flattening operation preceding autocorrelation analysis. This is the basis for a pitch estimation algorithm proposed by Markel [227], who called it the SIFT (simple inverse filtering tracking) method. A similar method was proposed by Maksym [223].

Figure 10.43 shows a block diagram of the SIFT algorithm. The input signal $s[n]$ is lowpass filtered with a cutoff frequency of about 900 Hz, and then the sampling rate (nominally 10 kHz) is reduced to 2 kHz by decimation by 5 (i.e., four out of every five samples are dropped at the output of the lowpass filter). The decimated output, $x[n]$, is then analyzed using the autocorrelation method with a value of $p = 4$ for the filter order. A fourth-order filter is sufficient to model the signal spectrum in the frequency range 0–1 kHz because there will generally be only 1–2 formants in this range. The signal $x[n]$ is then inverse filtered to give $y[n]$, a signal with an approximately flat spectrum.[8] Thus the purpose of the linear predictive analysis is to spectrally flatten the input signal, similar to the clipping methods discussed earlier in this chapter. The short-time autocorrelation of the inverse filtered signal is computed and the largest peak in the appropriate range is chosen as the pitch period. However, due to the down sampling, the time resolution is only to the nearest $5T$ sec. To obtain additional resolution in the value of the pitch period, the autocorrelation function is interpolated in the region of the maximum value. An unvoiced classification is chosen when the level of the autocorrelation peak (suitably normalized) falls below a prescribed empirically determined threshold.

Figure 10.44 illustrates some typical waveforms, and log magnitude spectra, obtained at several points in the analysis. Figure 10.44a shows one frame of the input waveform. Figure 10.44b shows the corresponding log magnitude spectrum together with the reciprocal of the log magnitude spectrum of the inverse filter. For this example, there appears to be a single formant in the range of 250 Hz. Figure 10.44c shows the log magnitude spectrum of the residual error signal (the output of the inverse filter), whereas Figure 10.44d shows the time waveform of the residual error signal (at the output of the inverse filter). Finally, Figure 10.44e shows the normalized autocorrelation of the residual error signal. A pitch period of 20 samples (10 msec) is clearly in evidence in the autocorrelation function.

[8]The output $y[n]$ is simply the prediction error for the fourth-order predictor.

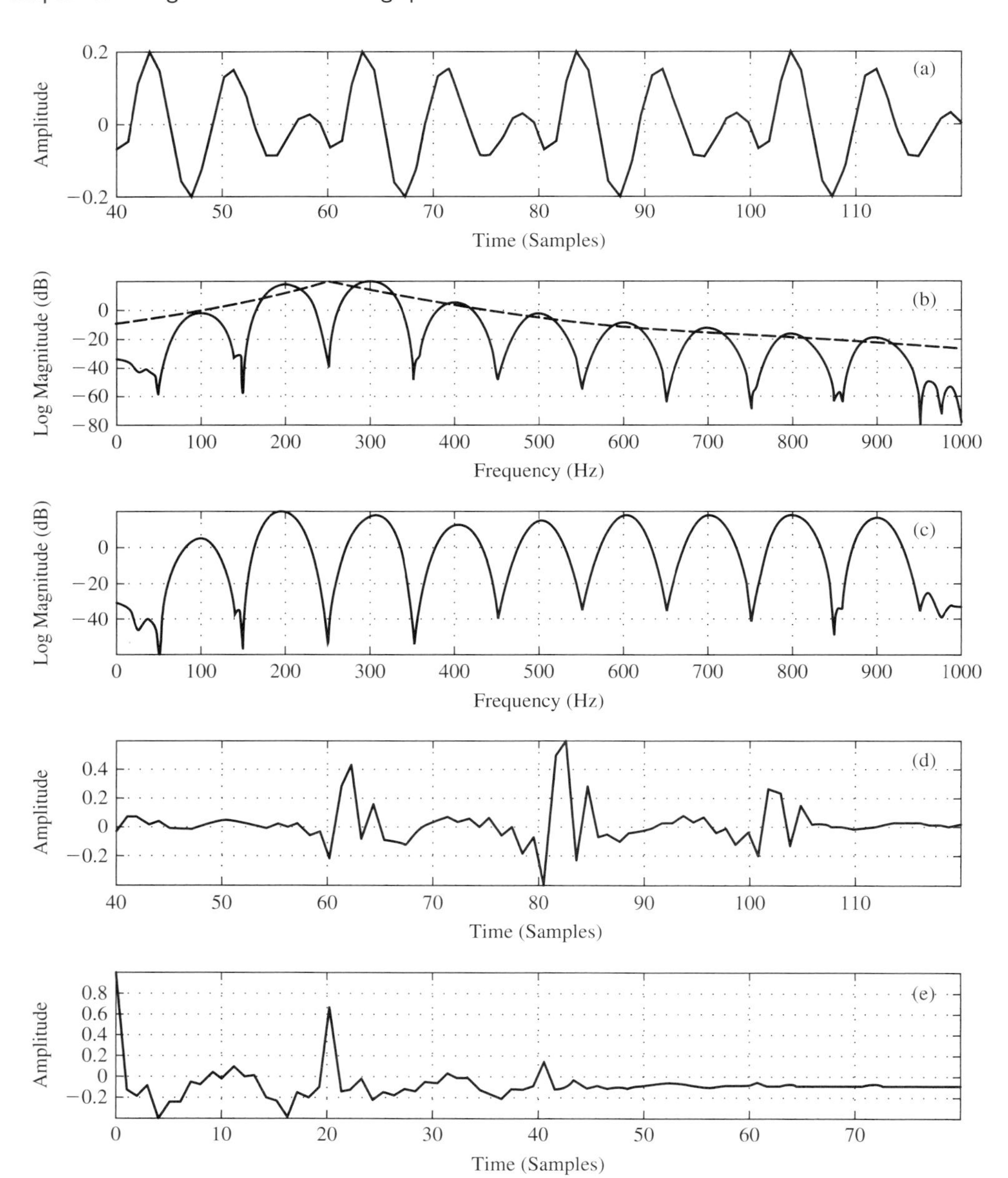

FIGURE 10.44
Typical signals from the SIFT algorithm: (a) a frame of the downsampled signal; (b) the log magnitude spectrum of the original signal (solid curve) and the $p = 4$ LPC spectral fit (dashed curve); (c) the log magnitude spectrum of the residual error signal; (d) the residual error signal; (e) the autocorrelation of the residual error signal showing pitch period peaks at 20 and 40 samples. (Note that the sampling rate is 2000 Hz in all parts of this figure.)

The SIFT algorithm uses linear predictive analysis to provide a spectrally flattened signal to facilitate pitch detection. To the extent that this spectral flattening is successful, the method performs well. However, for high pitched speakers (such as children), the spectral flattening is generally unsuccessful due to the lack of more than one pitch harmonic in the band from 0 to 900 Hz (especially for telephone line inputs). For such speakers and transmission conditions, other pitch detection methods may be more applicable.

10.6 FORMANT ESTIMATION

So far, in this chapter, we have focused on the left hand side of the discrete-time model for speech production in Figure 10.1; i.e., on algorithms for detecting when speech is present, voiced/unvoiced detection, and estimation of the pitch period for voiced speech. The right-hand side of the speech production model represents the vocal tract transmission function as well as sound radiation at the lips. It is of interest, of course, to develop algorithms for estimating the parameters of vocal tract models directly from sampled natural speech signals. We have shown that the technique of homomorphic filtering (Chapter 8) can be used to estimate the impulse response of the vocal tract system, and linear predictive analysis (Chapter 9) is a very effective tool for estimating the parameters of an all-pole model for the vocal tract filter. As we will see in Chapters 11, 13, and 14, these methods yield representations of the vocal tract system that are useful in speech coding, speech recognition, and speech synthesis. However, in some cases, it may be useful to obtain a representation of the vocal tract system in terms of formant frequencies, which, along with vocal tract area functions, are often considered the most fundamental characterization of the vocal tract system. Figure 10.45 shows a slightly reconfigured version of the basic model for speech production. As can be seen, this model aims at parsimony, with separate highly constrained channels for voiced and unvoiced speech, involving only three formant frequencies for voiced

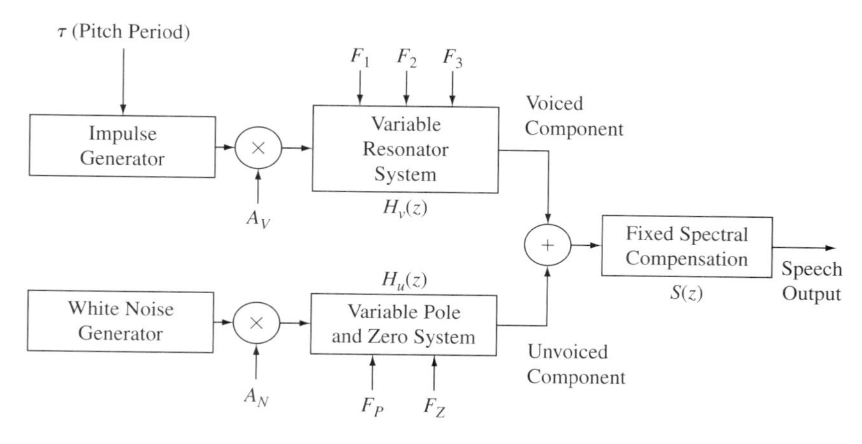

FIGURE 10.45
Discrete-time model for speech production based on formant frequency parameters.

speech and a pole and zero frequency for unvoiced speech. The details of an algorithm for estimating the time-varying parameters in Figure 10.45 are described in the next sub-section.

10.6.1 Homomorphic System for Formant Estimation

From the examples of Section 8.5, we have seen that it is reasonable to assume that the low-time (low quefrency) part of the cepstrum corresponds primarily to the vocal tract, glottal pulse, and radiation systems, while the high-time (high quefrency) part is due primarily to the excitation. As discussed in Section 10.5.6, this property of the cepstrum is exploited in pitch detection and voicing estimation by searching only the high-time portion for peaks. The examples of Section 8.5 also suggest ways of using the cepstrum to estimate vocal tract response parameters. Specifically, recall that the "smoothed" log magnitude functions of Figures 8.33a and 8.33d can be obtained by windowing the cepstrum. These smoothed log spectra display the resonant structure of the particular input speech segment; i.e., the locations of the peaks in the spectrum correspond approximately to the formant frequencies. This suggests that formants can be estimated by locating the peaks in the cepstrally smoothed log spectra.

The model in Figure 10.45 represents voiced speech by a set of parameters consisting of pitch period (N_p), voiced excitation amplitude (A_V), and the lowest three formant frequencies (F_1, F_2, F_3); similarly, unvoiced speech can be represented by the parameter set consisting of unvoiced excitation amplitude (A_N), and a single complex zero (F_Z) and complex pole (F_P). Additional fixed compensation accounts for the high frequency properties of the speech signal and ensures proper spectral balance. The steady state form of the vocal tract transfer function for voiced speech is represented as

$$H_V(z) = \prod_{k=1}^{4} \frac{1 - 2e^{-\alpha_k T} \cos(2\pi F_k T) + e^{-2\alpha_k T}}{1 - 2e^{-\alpha_k T} \cos(2\pi F_k T) z^{-1} + e^{-2\alpha_k T} z^{-2}}. \tag{10.24}$$

This equation describes a cascade of digital resonators having unity gain at zero frequency so that the speech amplitude depends only on the amplitude control, A_V. The first three formant frequencies, F_1, F_2, and F_3, are assumed to vary with time. Since higher formants usually do not vary significantly with time, a fourth resonance is fixed at $F_4 = 4000$ Hz, and a sampling rate of $F_s = 10$ kHz is assumed. The formant bandwidths $\{\alpha_k, k = 1, 2, 3, 4\}$ are also fixed at average values for speech. Cascaded with the vocal tract systems is an additional fixed spectral compensation system having the system function,

$$R(z) = \frac{(1 - e^{-aT})(1 + e^{-bT})}{(1 - e^{-aT} z^{-1})(1 + e^{-bT} z^{-1})}, \tag{10.25}$$

which approximates the glottal pulse and radiation contributions, with a and b chosen to provide a good overall spectral match. Representative values are $a = 400\pi$ and $b = 5000\pi$. More accurate values for a given speaker can be determined from a long-term average spectrum for that speaker.

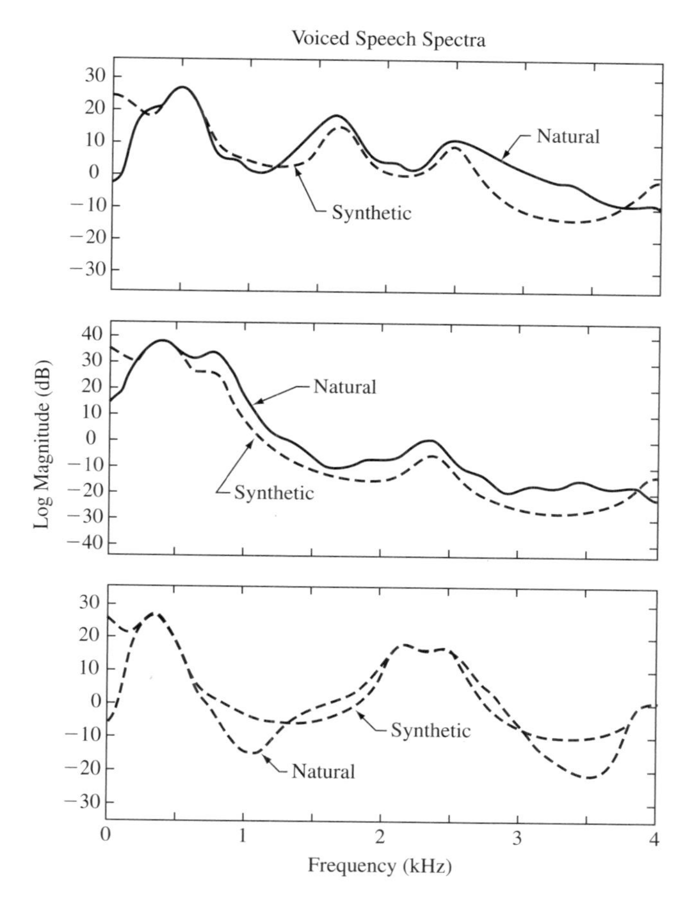

FIGURE 10.46
Comparisons between cepstrally smoothed log spectra and the speech model spectra for voiced speech.

Figure 10.46 shows that close matches can be obtained between a cepstrally smoothed spectrum and the frequency response of the model of Eq. (10.24). In these examples, F_1, F_2, and F_3 in Eq. (10.24) were estimated as the frequency locations of the first three peaks (local maxima) in the cepstrally smoothed spectrum.

For unvoiced speech, the vocal tract contributions are simulated by a simpler system, consisting of a single complex pole (F_P) and a single complex zero (F_Z) and having the steady state transfer function

$$H_U(z) = \frac{(1 - 2e^{-\beta T}\cos(2\pi F_P T) + e^{-2\beta T})(1 - 2e^{-\beta T}\cos(2\pi F_Z T)z^{-1} + e^{-2\beta T}z^{-2})}{(1 - 2e^{-\beta T}\cos(2\pi F_P T)z^{-1} + e^{-2\beta T}z^{-2})(1 - 2e^{-\beta T}\cos(2\pi F_Z T) + e^{-2\beta T})}, \tag{10.26}$$

where F_P is selected as the frequency location of the largest peak of the smoothed log spectrum above 1000 Hz. Choosing F_Z according to the empirically derived formula

$$F_Z = (0.0065F_P + 4.5 - \Delta)(0.014F_P + 28) \tag{10.27a}$$

with

$$\Delta = 20\log_{10}\left|H(e^{j2\pi F_P T})\right| - 20\log_{10}|H(e^{j0})| \tag{10.27b}$$

ensures that the approximate relative amplitude relationship between high and low frequencies is preserved [108]. That this rather simple model can preserve the essential spectral features of the unvoiced speech spectrum is illustrated in Figure 10.47, which shows comparisons between the cepstrally smoothed log spectrum and the model defined by Eqs. (10.26)–(10.27b). In this case, F_P was chosen as the frequency location of the highest peak in the cepstrally smoothed spectrum.

All of the indicated parameters, of course, vary with time. A method for estimating these parameters is based on the computation of a cepstrally smoothed log magnitude function once every 10–20 msec [108, 344]. The peaks of the log spectrum are located and a voicing decision is made based on the level of the cepstrum peak. If the speech segment is voiced, the pitch period is estimated from the location of the cepstral peak over an appropriate range of values and the first three formant frequencies are estimated from the set of the locations of the peaks in the log spectrum using logic based upon the model for speech production [108, 344]. In the case of unvoiced speech, the pole is set at the location of the highest peak in the log spectrum and the zero located so that the relative amplitude between low and high frequencies is preserved [108].

An illustration of estimation of pitch and formant frequencies for voiced speech is given in Figure 10.48. The left-hand half of the figure shows a sequence of cepstra computed at 20 msec intervals. On the right, the log magnitude spectrum is plotted with the corresponding cepstrally smoothed log spectrum superimposed. The lines connect the peaks that were selected by the algorithm described in Ref. [344] as the first three formant frequencies. Figure 10.48 shows that two formant frequencies occasionally come so close together that there are no longer two distinct peaks. These situations can be detected and the resolution can be improved by evaluating the z-transform of the windowed speech segment on a contour that lies inside the unit circle around the frequency region of the merged peaks. This evaluation is facilitated by a spectrum analysis algorithm, called the chirp z-transform (CZT) [315].

Another approach to formant estimation from cepstrally smoothed log spectra was proposed by Olive [262], who used an iterative procedure similar to the analysis-by-synthesis method to find a set of poles for a transfer function to match the smoothed log spectrum with minimum squared error.

Estimates of formant frequency trajectories are exceedingly useful in many applications including linguistic and phonetic studies, speaker verification, and speech synthesis studies. For example, speech can be synthesized from the formant and pitch data estimated as described above by simply controlling the model of Figure 10.45 with the estimated parameters. An example of speech synthesized using this model is

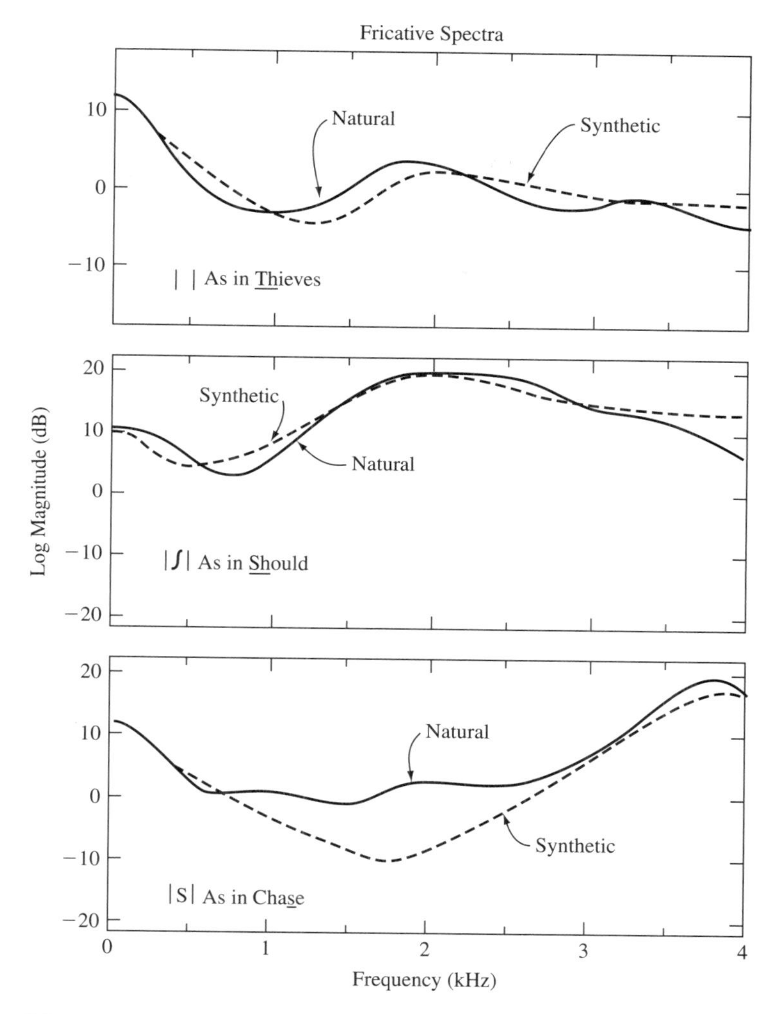

FIGURE 10.47
Comparisons between cepstrally smoothed log spectra and the speech model spectra for unvoiced speech.

shown in Figure 10.49. The upper part of the figure shows the parameters estimated from the natural speech utterance whose spectrogram is given in Figure 10.49b. Figure 10.49c shows a spectrogram of synthetic speech created by controlling the model of Figure 10.45 with the parameters of Figure 10.49a. It is clear that the essential features of the signal are well preserved in the synthetic speech. Indeed, even though the model is very crude, the synthetic speech is very intelligible and retains many of the identifying features of the original speaker. In fact, the pitch and formant frequencies estimated by this procedure formed the basis for extensive experiments in speaker verification [327].

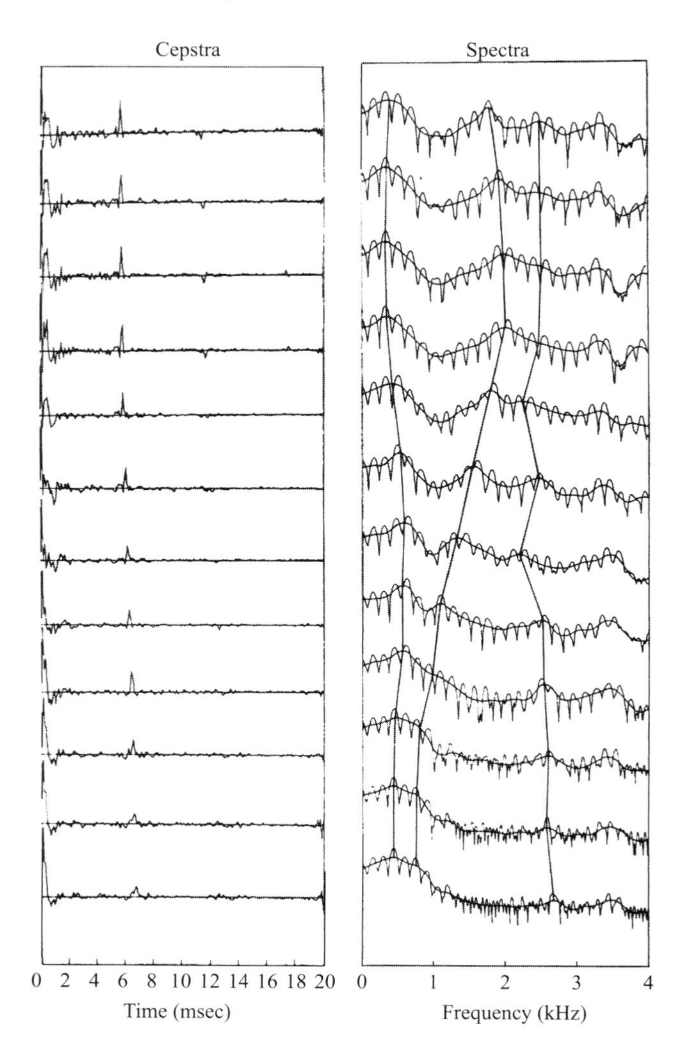

FIGURE 10.48
Automatic formant estimation from cepstrally smoothed log spectra. (After Schafer and Rabiner [344]. © [1970] Acoustical Society of America.)

An important property of the representation that we have been discussing is that the information rate can be very low. As we will discuss in more detail in Chapter 11, an alternate representation to the samples of the speech waveform can be in terms of a model such as Figure 10.45 if an algorithm exists for estimating the model parameters as a function of time. Then by controlling the model with those estimates, samples of a reconstructed waveform can be synthesized. Such a complete system is called an analysis/synthesis system, or equivalently, a *vocoder*. A complete formant analysis/synthesis system (or formant vocoder) based upon the model of Figure 10.45 is shown

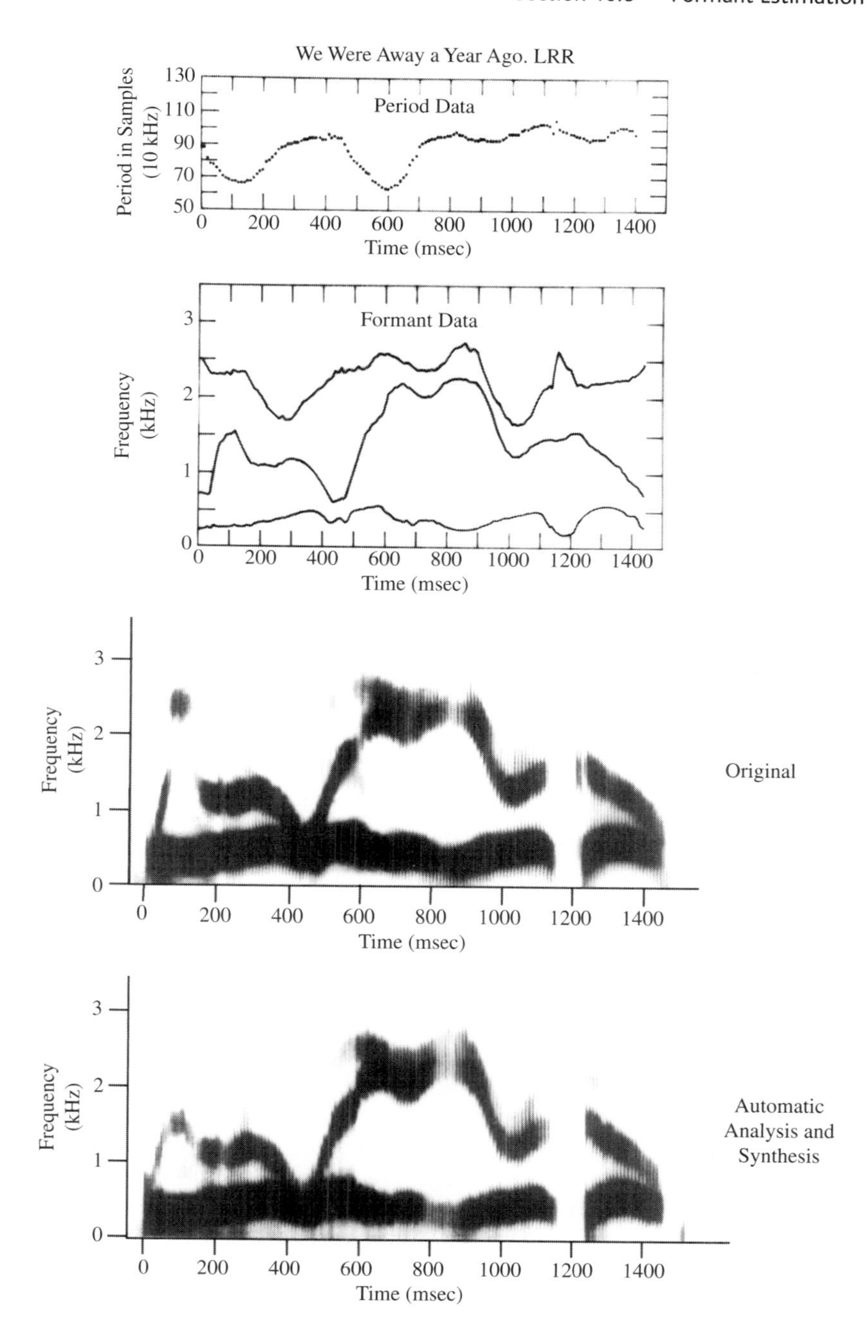

FIGURE 10.49
Automatic analysis and synthesis of "We were away a year ago": (a) pitch period and formant data as plotted by computer; (b) wideband spectrogram of original speech; (c) wideband spectrogram of synthetic speech generated from the data in (a). (After Schafer and Rabiner [344].)

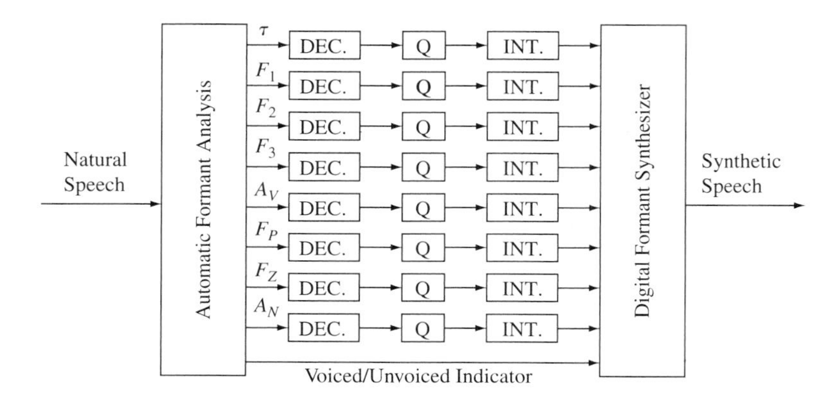

FIGURE 10.50
Block diagram of a formant vocoder.

in Figure 10.50. The model parameters are estimated 100 times/sec and lowpass filtered to remove noise. The sampling rate is reduced to twice the filter cutoff frequency and the parameters are quantized. For synthesis, each parameter is interpolated back to a rate of 100 samples/sec and supplied to a synthesizer as depicted in Figure 10.45.

A perceptual study was performed to determine appropriate sampling rates and quantization parameters for the formant vocoder in Figure 10.50 [329]. The analysis and synthesis sections were first connected directly to produce synthetic reference signals. Then the parameters were lowpass filtered to determine the lowest bandwidth for which no perceptual difference could be detected between synthesis with the filtered and unfiltered parameters. It was found that the bandwidth could be reduced to about 16 Hz with no noticeable change in quality. The filtered parameters could then be sampled at about 33 Hz (3-to-1 decimation). Then an experiment was performed to determine the required information rate. The formant and pitch parameters were quantized with a linear quantizer (adjusted to the range of each parameter), and the amplitude parameters were quantized with a logarithmic quantizer.[9] A summary of the results of the perceptual test is given in Table 10.2. Using a sampling rate of 33 samples/sec and the numbers of bits per parameter as given in Table 10.2, it was found that for the all-voiced utterances employed in the experiment, no degradation in quality over the unquantized synthesis occurred for a total bit rate of about 600 bits/sec. (Note that an additional one-bit voiced/unvoiced parameter transmitted 100 times/sec was required to adequately represent voiced/unvoiced transitions.)

The results of this experiment help to establish a low end target for speech compression systems. While the experiment shows that good quality representation of speech signals can be achieved at rates close to our original estimate in Chapter 3 of a few hundred bits per second, it is difficult to achieve this type of performance robustly across a wide range of speakers and speech sounds. Several "formant vocoders" have

[9]Both linear and logarithmic quantization are discussed in detail in Sections 11.4 and 11.4.2.

TABLE 10.2 Results of perceptual evaluation of a formant vocoder [329]. © [1971] Acoustical Society of America.

Parameter	Required bits/sample
τ	6
F_1	3
F_2	4
F_3	4
$\log[A_V]$	2

been studied [148, 236, 359], but with the widespread availability of adequate digital transmission or storage capacity, other approaches to speech quantization to be discussed in Chapter 11 have become firmly established.

10.6.2 Formant Analysis Using Linear Prediction Parameters

Linear predictive analysis of speech has several advantages, and some disadvantages, when applied to the problem of estimating the formants for voiced sections of speech. Formants can be estimated from the predictor parameters in one of two ways. The most direct way is to factor the predictor polynomial and decide which roots correspond to formants, and which roots correspond to spectral shaping poles [226, 228]. The alternative way of estimating formants is to obtain the spectrum, and choose the formants by a peak picking method similar to the one discussed in Section 10.6.1.

A distinct advantage inherent in the linear predictive method of formant analysis is that the formant center frequency and bandwidth can be determined accurately by factoring the predictor polynomial. (See Section 9.7.3.) Since the predictor order p is chosen a priori, the maximum possible number of complex conjugate poles that can be obtained is $p/2$. Thus the labeling problem inherent in deciding which poles correspond to which formants is less complicated for the linear predictive method since there are generally fewer poles to choose from than for comparable methods of obtaining the spectrum such as cepstral smoothing. Finally extraneous poles are generally easily isolated in the linear predictive analysis since their bandwidths are often very large, compared to what would be expected for bandwidths typical of speech formants. Figure 10.51 shows an example that illustrates that the pole locations do indeed give a good representation of the formant frequencies [12]. The upper panel of Figure 10.51 shows a narrowband spectrogram of the utterance, “This is a test”, with formants estimated from a $p = 16^{th}$-order linear predictive analysis overlaid on a narrowband spectrogram of the spoken utterance; the lower panel shows the roots by themselves. It is clear that the roots preserve the formant structure extremely well in voiced regions of speech and thus form the basis for a high quality formant estimation procedure. Another example showing the relation of the prediction polynomial roots to the formants was given in Figure 9.38 of Section 9.7.3.

The disadvantage inherent with the use of linear predictive analysis in formant tracking is that an all-pole model is used to model the speech spectrum. For sounds

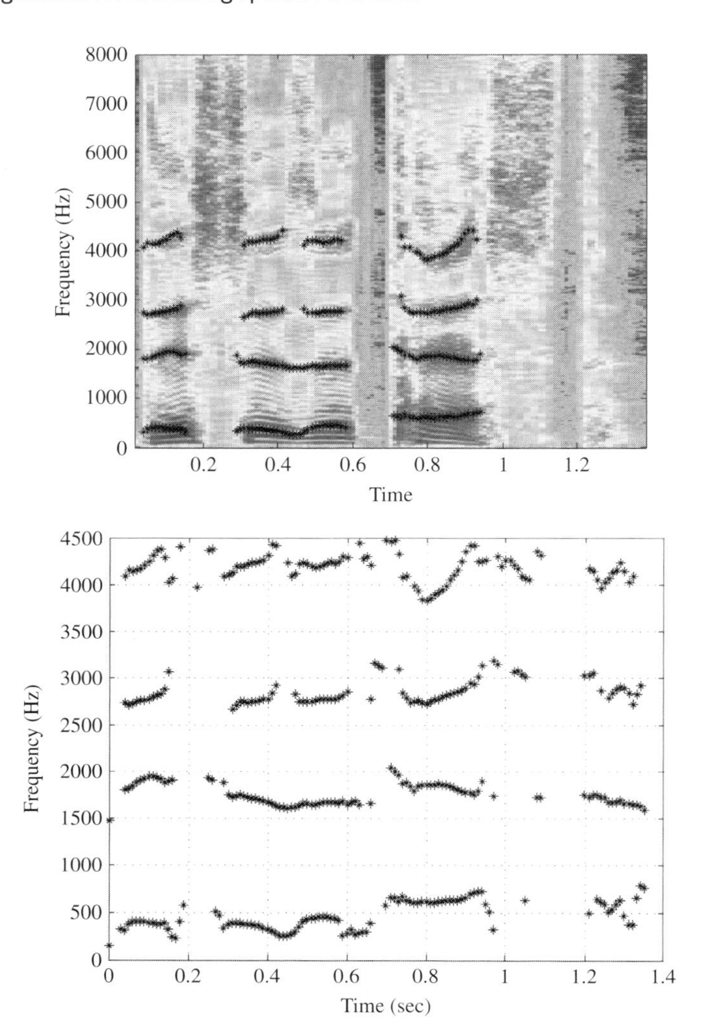

FIGURE 10.51
(a) Spectrograms of original speech utterance, "This is a test", with formants estimated from $p = 16^{th}$-order roots of prediction error filter overlaid; (b) polynomial roots.

such as nasals and nasalized vowels, although the analysis is adequate in terms of its spectral matching capabilities, the physical significance of the roots of the predictor polynomial is unclear when the physical model calls for both poles and zeros. It is not clear how the zeros of the physical model, which are manifest in the short-time spectrum, influence the pole locations in the all-pole linear predictive model. Another difficulty with the analysis is that although the bandwidth of the root is readily determined, it is generally not clear how it is related to the actual formant bandwidth. This is because the measured bandwidth of the root has been shown to be sensitive to the frame duration, frame position, and method of analysis.

With these advantages and disadvantages in mind, several methods have been proposed for estimating formants from linear prediction-derived spectra using peak picking methods, and from the predictor polynomial by factoring methods. Once the candidates for the formants have been chosen, the techniques used to label these candidates, i.e., the assigning of a candidate to a particular formant, are similar to those used for any other analysis method. These include reliance on formant continuity, a need for spectral pre-emphasis to minimize the possibility of close formants merging, and the use of an off-the-unit circle contour for evaluating the linear prediction spectrum, thereby sharpening the spectral peaks. A discussion of the various methods for estimating formant frequencies and bandwidths is given by Markel [226, 228], Atal and Hanauer [12], Makhoul and Wolf [222], and McCandless [237].

10.7 SUMMARY

In this chapter we have discussed some examples of the way that DSP techniques have been employed in digital speech processing. Our purpose was to show how the basic representations of the speech signal, which are based on DSP concepts, have been used in creating algorithms for estimating the parameters of speech models. We began with the simple problem of VUS classification of frames of speech samples. The algorithms discussed combined heuristics guided by knowledge of the speech signal properties with statistical methods for pattern analysis. Next we considered the difficult and pervasive problem of pitch period estimation for voiced speech, where we described algorithms based on time-domain, frequency-domain, cepstrum, and linear predictive representations of the speech signal. The chapter concluded with a discussion of the use of homomorphic filtering and linear predictive analysis in estimating the formant frequencies of speech. Many of the principles illustrated by the specific examples in this chapter will appear again in the remaining chapters of this book, where the broad areas of speech and audio coding, speech synthesis, and speech recognition are considered.

PROBLEMS

10.1. (MATLAB Exercise) The purpose of this MATLAB exercise is to compare linear, median, and combination smoothers on short-time zero-crossing estimates. Using the speech file `test_16k.wav`, compute the short-time zero-crossing rate (per 10 msec of speech) using a frame length of 10 msec and a frame shift of 5 msec. Plot the resulting short-time estimate of zero-crossing rate per 10 msec interval. (It should exhibit a lot of high frequency variation since the estimation interval is short.) Now design a lowpass filter to preserve the low frequency band until about $f = 0.1F_s$, or about 1.6 kHz for this 16 kHz sampled signal. Design a lowpass filter to remove the band from $f = 0.2 * F_s$ to $f = 0.5 * F_s$. (Hint: You should find that you get about 40 dB out-of-band rejection using a filter length of $L = 41$ samples and this should be adequate for this exercise.) Use this filter to smooth the short-time zero-crossings rate contour and plot the resulting smoothed curve as a sub-plot on the page with the original zero-crossings rate contour. Now median-smooth the original zero-crossings rate contour using a combination of a running median of seven samples followed by a running median of five samples. Plot the resulting curve as the third sub-plot. Finally use a combination smoother (of the type discussed in this chapter) and plot the resulting smoothed contour. What differences do you see in the four contours? Which contour best preserves the most salient characteristics of the original short-time zero-crossing rate contour?

10.2. To demonstrate how a parallel processing pitch detector can combine several independent pitch detectors, each with a fairly high error probability, and give a highly reliable result, consider the following idealized situation. Assume there are seven *independent* pitch detectors, each having a probability p of correctly estimating pitch period, and probability $1-p$ of incorrectly estimating pitch period. The decision logic is to combine the seven pitch estimates in such a way that an overall error is made only if four or more of the individual pitch detectors make an error.

(a) Derive an explicit expression for the probability of error of the parallel processing pitch detector in terms of p. [Hint: Consider the result of each pitch detector a Bernoulli trial with probability $(1-p)$ of making an error and probability p of no error.]

(b) Sketch a curve showing the overall error probability as a function of p.

(c) For what value of p is the overall error probability less than 0.05?

10.3. A proposed pitch detector consists of a bank of digital bandpass filters with lower cutoff frequencies given as

$$F_k = 2^{k-1}F_1, \quad k = 1, 2, \dots, M,$$

and upper cutoff frequencies given as

$$F_{k+1} = 2^k F_1, \quad k = 1, 2, \dots, M.$$

This choice of cutoff frequencies gives the filter bank the property that if the input is periodic with fundamental frequency F_0, such that

$$F_k < F_0 < F_{k+1},$$

then the filter outputs of bands 1 to $k-1$ will have little energy, the output of band k will contain the fundamental frequency, and bands $k+1$ to M will contain one or more harmonics. Thus, by following each filter output by a detector which can detect pure tones, a good indication of pitch can be obtained.

(a) Determine F_1 and M such that this method would work for pitch frequencies from 50 to 800 Hz.

(b) Sketch the required frequency response of each of the M bandpass filters.

(c) Can you suggest simple ways to implement the tone detector required at the output of each filter?

(d) What types of problems would you anticipate in implementing this method using non-ideal bandpass filters?

(e) What would happen if the input speech were bandlimited from 300 to 3000 Hz; e.g., telephone line input? Can you suggest improvements in these cases?

10.4. (MATLAB Exercise: Rule-Based Isolated Word Speech Detector) Write a MATLAB program to detect (isolate) a spoken word in a relatively quiet and benign acoustic background using simple rules based on frame-based measurements of log energy and zero-crossing rate. The speech files that to be used for this exercise can be found on the book website in the zip file `ti_isolated_unendpointed_digits.zip`. Included in this zip file are two tokens of each of the waveforms of 11 digits (namely zero-nine plus oh), embedded in a relatively quiet acoustic background in the file format `{1-9,O,Z}{A,B}.waV`. Thus the file `3B.waV` contains the second token of the spoken digit /3/, and the file `ZA.waV` contains the first token of the spoken digit /zero/, etc.

The goal of this exercise is to implement a simple, rule-based, speech endpoint detector which uses frame-based measurements of log energy (on a dB scale) and zero-crossing rate per 10 msec interval, to detect the region of the speech signal and separate it from the background signal.

The steps to follow in programming this algorithm are the following:

1. Read in any one of the speech files.
2. Convert the input signal sampling rate from $F_s = 20{,}000$ samples/sec to a rate of $F_s = 8000$ samples/sec.
3. Design an FIR bandpass filter to eliminate DC, 60 Hz hum, and high frequency noise (at the sampling frequency, $F_s = 8000$ Hz), using the MATLAB filter design function `firpm`. The filter design characteristics should include a stopband corresponding to $0 \leq |F| \leq 100$ Hz, a transition band corresponding to $100 \leq |F| \leq 200$ Hz, and a passband corresponding to $200 \leq |F| \leq 4000$ Hz. Filter the speech file using the resulting FIR filter.
4. Define the frame analysis parameters, namely:
   - `NS=frame duration in msec, L=NS*8` is the frame duration in samples based on a sampling rate of 8000 Hz;
   - `MS=frame offset in msec, R=MS*8` is the frame offset in samples based on a sampling rate of 8000 Hz.
5. Compute the log energy and zero-crossing rates (per 10 msec interval) for all the frames of the entire file, i.e., every `R` samples.
6. Compute the mean and standard deviation of both the log energy and zero-crossing rate over the first 10 frames of the file (assumed to be purely background signal); call these parameters `eavg, esig` for the mean and standard deviation of the log energy parameter, and `zcavg, zcsig` for the mean and standard deviation of the zero-crossing rate parameter.
7. Define the endpoint detector parameters:
   - `IF=35`—absolute threshold for zero-crossing rate
   - `IZCT=max(IF,zcavg+3*zcsig)`—variable zero-crossing rate threshold based on background signal statistics
   - `IMX=max(eng)`—absolute peak of log energy parameter
   - `ITU=IMX-20`—high threshold on log energy parameter
   - `ITL=max(eavg+3*esig, ITU-10)`—low threshold on log energy parameter.
8. Search the signal interval to find the peak of log energy (this is assumed to be at or near the center of the speech interval), and then search the log energy contour to find the main energy concentration, where the log energy falls to the level `ITU` on both sides of the peak. Next search to find the points where the log energy falls below `ITL` and this defines the extended main energy contour of the spoken word. Next the region of the word is expanded based on the zero-crossing rate parameters exceeding the threshold `IZCT` for at least four consecutive frames in the regions adjacent to the

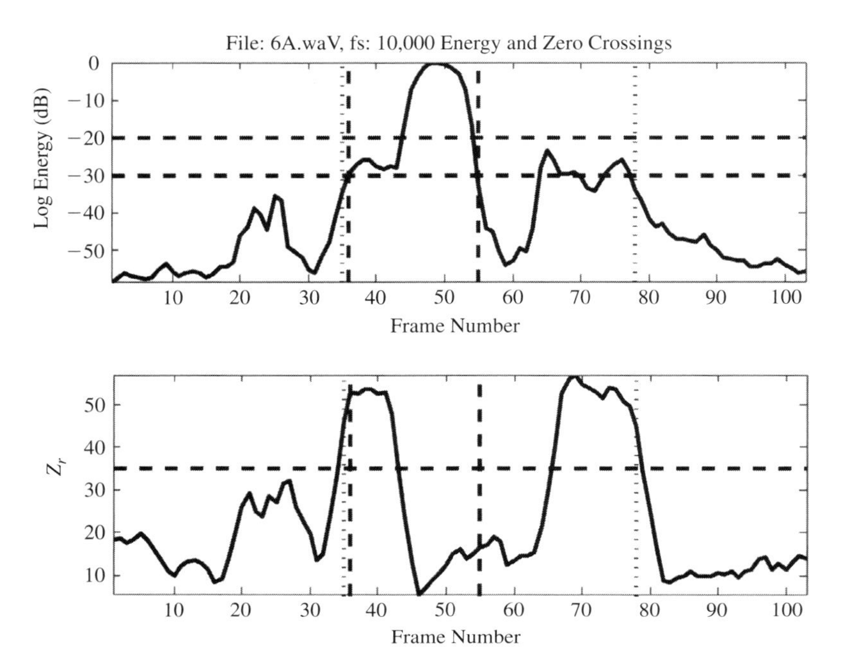

FIGURE P10.4
Endpoint detector regions for spoken word /six/ showing log energy and zero-crossing rate contours.

current estimate of word boundaries. The word boundaries are adjusted to include the entire region where the zero-crossing rate exceeds the threshold, once the minimum duration threshold has been crossed.

To illustrate the behavior of the speech endpoint detector, do the following:

1. Plot the log energy and zero-crossing rate contours on two separate plots on the same page, and show the endpoint detector decisions based solely on log energy speech detection (use dashed lines on the plots) and based on both log energy and zero-crossing rates (use dotted lines on the plots). (You may also want to show dashed lines on the plots to show where the various thresholds turned out to occur.) Figure P10.4 shows a typical plot for the spoken word /six/, where we have used dashed lines to show the log energy boundary points, and dotted lines to show the zero-crossing rate boundary points, along with dashed lines for the log energy and zero-crossing rate thresholds.
2. Listen to the detected speech region to determine whether any part of the speech is missing, or whether any extraneous background signal is included within the speech boundaries.

10.5. (MATLAB Exercise: Bayesian Isolated Word Speech Classifier) Write a MATLAB program to classify frames of signal as either non-speech (Class 1) or speech (Class 2) frames using a Bayesian statistical framework. The feature vector for frame classification consists of short-time measurements of log energy and zero-crossing rate (per 10 msec

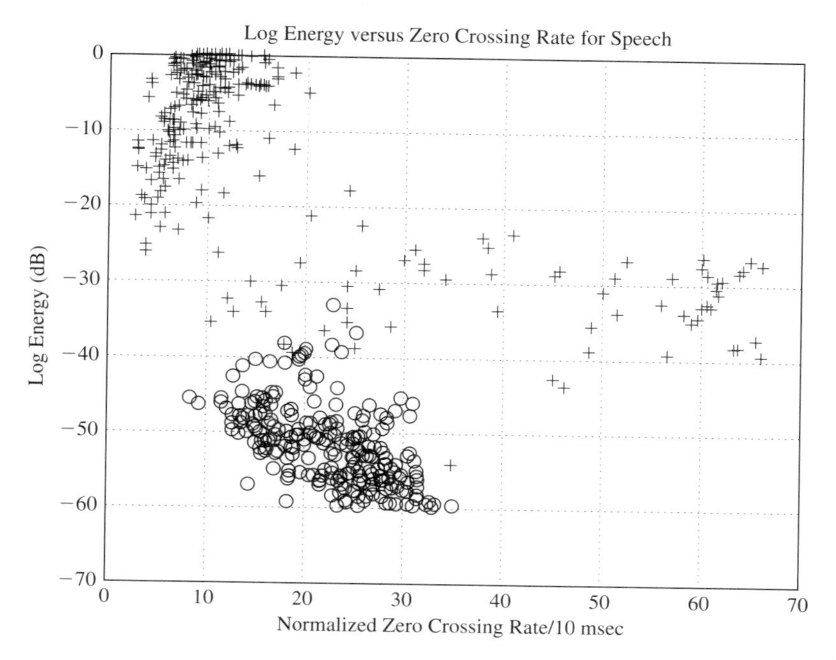

FIGURE P10.5(a)
Plots of log energy versus zero-crossing rate for both non-speech (circles) and speech (crosses) training data.

interval), and the two components of the feature vector are modeled using a simple Gaussian fit, with no correlation between the two feature vector components.

The speech files to be used for this exercise are the same ones used in Problem 10.4 and can be found on the book website in the zip file `ti_isolated_unendpointed_digits.zip`.

Included in this zip file are two tokens of each of the waveforms of 11 digits (zero-nine plus oh), embedded in a relatively quiet acoustic background in the file format `{1-9,O,Z}{A,B}.waV`. Thus the file `3B.waV` contains the second token of the spoken digit /3/, and the file `ZA.waV` contains the first token of the spoken digit /zero/, etc. The two possible outcomes of the Bayesian analysis are `speech` and `nonspeech`. Labeled training data, which can be found on the book website, are contained in the files `nonspeech.mat` and `speech.mat`. These `mat` files contain arrays of the two parameters (log energy and zero-crossing rate/10 msec interval) for the classes "non-speech" (`logen` array and `zcn` array) and "speech" (`loges` array and `zcs` array).

The first thing to do is train the Bayesian classifier; i.e., determine the means and standard deviations for a Gaussian representation of the feature vectors for both Class 1 (non-speech) and Class 2 (speech) frames. A plot of the locations of the feature vectors for both Class 1 and Class 2 tokens, from the training sets, is shown in Figure P10.5(a), where feature vectors for Class 1 tokens are shown as circles and feature vectors for Class 2 tokens are shown as pluses. The distribution for Class 1 tokens is compact, whereas the distribution for Class 2 tokens is spread out.

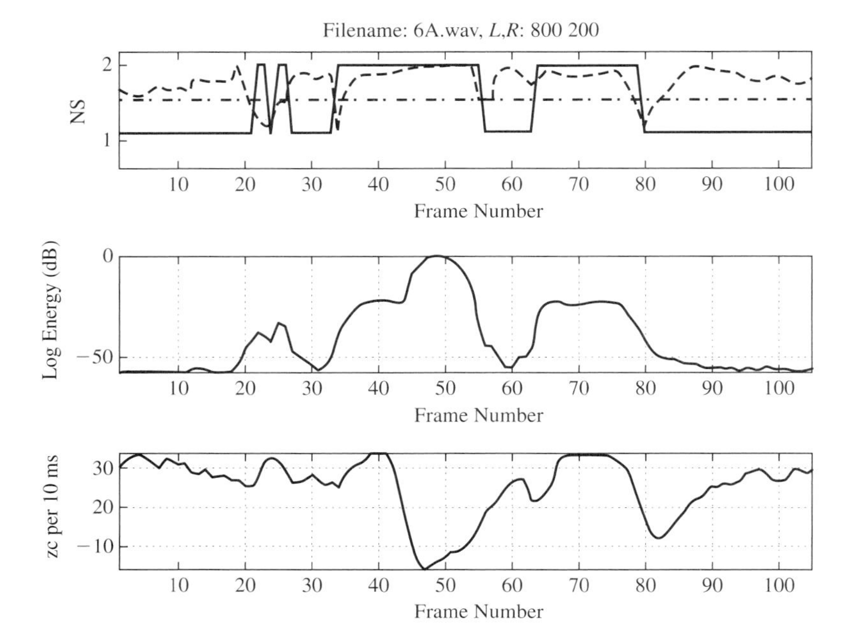

FIGURE P10.5(b)
Bayesian classification regions for spoken word /six/ showing class decision (1=non-speech, 2=speech) and confidence score (in the top panel with the horizontal dashed-dot line showing the 75% confidence level), log energy contour (middle panel), and zero-crossing rate contour (bottom panel).

For testing the Bayesian classifier, use the isolated digit files and score each frame parameter set, $\mathbf{X}$, against the Gaussian distributions for both Class 1 and Class 2 data using the distance measure:

$$d_j = d(\mathbf{X}, N(\boldsymbol{\mu}_j, \boldsymbol{\sigma}_j)) = \sum_{k=1}^{2} (\mathbf{X}(k) - \boldsymbol{\mu}_j(k))^2 / \boldsymbol{\sigma}_j(k)^2, \quad j = 1, 2,$$

where j is the class number ($j = 1$ for non-speech, $j = 2$ for speech), and $\boldsymbol{\mu}_j$ and $\boldsymbol{\sigma}_j$ are the means and standard deviations of the two classes. The chosen class is the one with the smallest distance score. We can attach a measure of confidence to the classification score by forming the confidence measure:

$$c_1 = \frac{\left(\frac{1}{d_1}\right)}{\left(\frac{1}{d_1} + \frac{1}{d_2}\right)} = \frac{d_2}{d_1 + d_2},$$

$$c_2 = \frac{\left(\frac{1}{d_2}\right)}{\left(\frac{1}{d_1} + \frac{1}{d_2}\right)} = \frac{d_1}{d_1 + d_2},$$

where a confidence score of 0.5 represents total lack of confidence in the classification decision, and a confidence score approaching 1 represents total confidence in the classification decision.

The results of doing this Bayesian classification on the isolated digit file `6A.waV` are shown in the top panel of Figure P10.5(b), which shows the frame classification scores, along with the confidence measure and a dashed line for a confidence threshold of 0.75, along with plots of log energy (middle panel) and zero-crossing rate (bottom panel). Here we see that the voiced regions of the digit /6/ were correctly classified as Class 2 (speech) data, and most of the background (along with the stop gap in /6/) was classified as Class 1 (background). However, due to a spurious click prior to the beginning of the word /6/, we see that there is a short region that was (incorrectly) classified as speech.

Test your classification algorithm on several of the isolated digit files to see what types of errors are made, and how they correlate with the computed confidence scores.

10.6. (MATLAB Exercise: Bayesian VUS Classifier) Write a MATLAB program to classify frames of signal as either silence/background (Class 1), unvoiced speech (Class 2), or voiced speech (Class 3) using the Bayesian statistical framework discussed in Section 10.4. The feature vector for frame classification, $\mathbf{X}$, consists of short-time measurements of the five parameters defined in Section 10.4.

A set of training files, `training_files_VUS.zip`, containing hand-selected examples of feature vectors from each of the three classes is available on the book website. These training files are labeled `silence.mat`, `unvoiced.mat`, and `voiced.mat`, with each `.mat` file containing the training set of vectors for that class.

Assume that the training data is Gaussian distributed and that each of the five components of the feature vectors are independent. Hence we need to estimate, from the training data, the means and standard deviations of each of the five feature parameters and for each of the three classes of sound. We denote the mean and standard deviation for the jth class as $\boldsymbol{\mu}_j$ and $\boldsymbol{\sigma}_j$, where $j=1$ denotes Class 1 (silence/background), $j=2$ denotes Class 2 (unvoiced speech), and $j=3$ denotes Class 3 (voiced speech), and each mean and standard deviation is a 5-element vector.

For each test set vector, $\mathbf{X}$, being analyzed, use the simplest Bayesian classification rule, namely compute, for each class, j, the distance measure:

$$d_j = \sum_{k=1}^{5} (\mathbf{X}(k) - \boldsymbol{\mu}_j(k))^2/(\boldsymbol{\sigma}_j(k))^2, \quad j = 1, 2, 3$$

where we recognize that since the components of the feature vector are treated as being independent of each other, we can sum up the weighted distances. Also compute a measure of confidence, C_j, in the classification score, as:

$$C_j = \frac{\left(\dfrac{1}{d_j}\right)}{\left(\dfrac{1}{d_1} + \dfrac{1}{d_2} + \dfrac{1}{d_3}\right)}, \quad j = 1, 2, 3.$$

It can be shown that the higher the confidence score for the class having the minimum distance, the greater the assurance that the algorithm has chosen the correct class.

The testing speech files to be used for this exercise can be found on the book website in the zip file `testing_files_VUS.zip`. Included in this zip file are several files

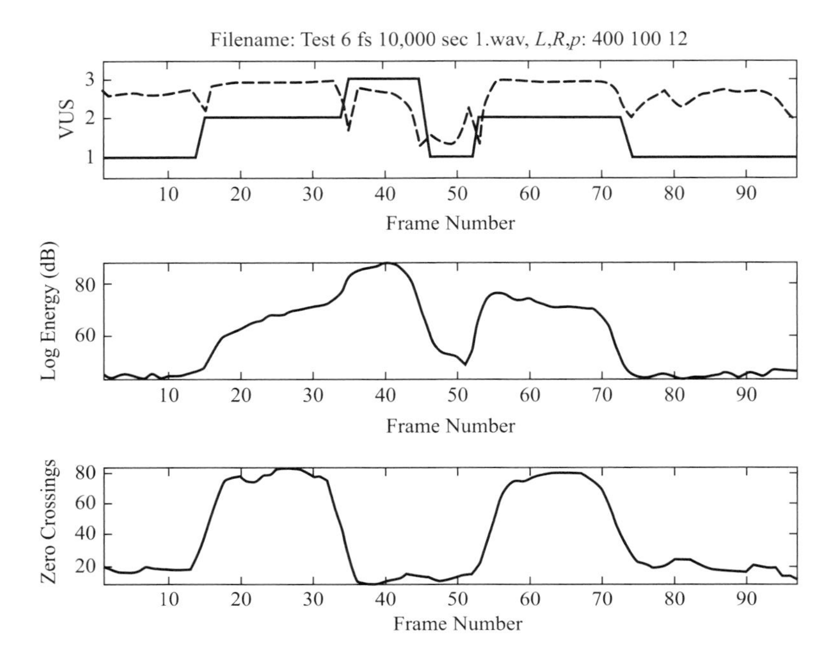

FIGURE P10.6
Bayesian classification regions for spoken word /six/ showing class decision (1=silence, 2=unvoiced, 3=speech) and confidence score (dashed line in the top panel), log energy contour (middle panel), and zero-crossing rate contour (bottom panel).

containing isolated digits, and two full sentence files. Plot, for each test file analyzed, the following:

1. the resulting classification decisions for all the frames in the test file using the code 1 for silence/background classification, 2 for unvoiced speech classification, and 3 for voiced speech classification; also plot (as a dashed line) the confidence scores for all frames (scaled by 3 so that the peak confidence score is 3);
2. the log energy (in dB) contour of the test signal utterance;
3. the zero-crossing rate (per 10 msec interval) contour of the test signal utterance.

Figure P10.6 shows such a plot for the file containing the isolated digit /6/. The upper panel shows the classification scores and the confidence levels (scaled by 3-to-1 for convenience in plotting and plotted as a dashed line); the middle panel shows the log energy contour, and the lower panel shows the zero-crossings contour. Clearly, the algorithm provides very high accuracy in making this 3-level classification decision on the isolated digit /6/.

10.7. (MATLAB Exercise: Autocorrelation-Based Pitch Detector) Program an autocorrelation-based pitch detector using the modified autocorrelation function. Compare the results using both the original speech file and those obtained from a bandpass filtered version of the speech file.

The steps to follow for implementing the pitch detector are the following:

1. specify whether the talker is a male or female (the program uses the talker gender to set ranges for the pitch period contour);
2. read in the speech file (including determining the speech sampling rate, `fs`);
3. convert the sampling rate to a standard value of `fsout=10000` Hz for this exercise;
4. design and implement a bandpass filter to eliminate DC offset, 60 Hz hum, and high frequency (above 1000 Hz) signal, using the design parameters:
   - stopband from 0 to 80 Hz
   - transition band from 80 to 150 Hz
   - passband from 150 to 900 Hz
   - transition band from 900 to 970 Hz
   - stopband from 970 to 5000 Hz
   - filter length of `n=301` samples
5. save both the full band speech and the bandpass filtered speech files for processing and comparison;
6. play both the original and bandpass filtered speech files to be sure that the filtering worked properly;
7. block the signal into frames of length `L=400` samples (corresponding to 40 msec in duration), with frame shift of `R=100` samples (corresponding to 10 msec shift duration);
8. compute the frame-by-frame modified correlation between the frames of signal specified as:

$$s_1[n] = [s[n], s[n+1], \ldots, s[n+L-1]],$$

$$s_2[n] = [s[n], s[n+1], \ldots, s[n+L+\text{pdhigh}-1],$$

   where `n` is the starting sample of the current frame, and `pdhigh` is the longest anticipated pitch period (based on the gender of the speaker) and is specified as:

$$\texttt{pdhigh} = \begin{cases} \texttt{fsout}/75 & \text{for males} \\ \texttt{fsout}/150 & \text{for females} \end{cases}$$

   (Hint: Use the MATLAB function `xcorr` to compute the modified correlation between $s_1[n]$ and $s_2[n]$ as it is considerably faster than any other implementation);
9. search from `pdlow` to `pdhigh` to find the maximum of the modified autocorrelation function (the putative pitch period estimate for the current frame), along with the value of the modified autocorrelation at the maximum (the confidence score), using the range estimate of:

$$\texttt{pdlow} = \begin{cases} \texttt{fsout}/200 & \text{for males} \\ \texttt{fsout}/300 & \text{for females,} \end{cases}$$

   do all operations on both the original file and the bandpass filtered speech file;

10. convert the confidence score (the value of the modified autocorrelation at the maximum) to a log confidence, set a threshold at 0.75 of the maximum value of the log confidence score, and set the pitch period to zero for all frames whose log confidence scores fell below the threshold;

11. plot the resulting pitch period contour along with the log confidence scores for both the original speech file and the bandpass filtered speech file; how do these contours compare?

12. use a 5-point median smoother to smooth the pitch period scores as well as the confidence scores;

13. plot the median smoothed pitch period scores along with the median smoothed confidence scores;

14. save the computed pitch period and confidence scores in the file `out_autoc.mat`.

Which processing works best; i.e., using the full band original speech file or using the bandpass filtered speech file? How much difference do you observe in the resulting pitch period contours?

10.8. (MATLAB Exercise: Log Harmonic Product Spectrum-Based Pitch Detector) Program a pitch detector based on the log harmonic product spectrum method as discussed in Section 10.5.5. The log harmonic product spectrum of a signal, at time n, is defined as:

$$\hat{P}_{\hat{n}}(e^{j\omega}) = 2\sum_{k=1}^{K} \log_{10} |X_{\hat{n}}(e^{j\omega k})|,$$

where K is the number of frequency-compressed replicas of $\log_{10} |X_{\hat{n}}(e^{j\omega})|$ that are summed up to give $\hat{P}_{\hat{n}}(e^{j\omega})$.

The steps to follow are:

1. read in the speech filename and the gender of the talker; load the speech file and ascertain the speech file sampling rate, `fs`; convert the signal to the standard sampling rate of `fsout=10000` Hz;

2. equalize the speech spectrum using a simple FIR filter of the form $H(z) = 1 - z^{-1}$;

3. play both the original file and the spectrum equalized file to verify that the filtering worked properly;

4. define the signal processing parameters for calculation of the STFT and the log harmonic product spectrum, namely:

- `nfft=4000`; size of FFT for STFT
- `L=400`; frame length in samples at a 10,000 Hz sampling rate
- `R=100`; frame shift in samples at a 10,000 Hz sampling rate
- `K=10`; number of replicas of frequency-compressed STFTs added to produce the log harmonic product spectrum
- `coffset=275`; the offset from the peak value of $\hat{P}_n(e^{j\omega})$ (over the entire utterance) for deciding when to classify the frame as non-voiced

5. perform a frame-based STFT analysis over the entire duration of the signal; use `N` sample frames, padded with `nfft-N` zeros; use a Hamming window of duration `N`

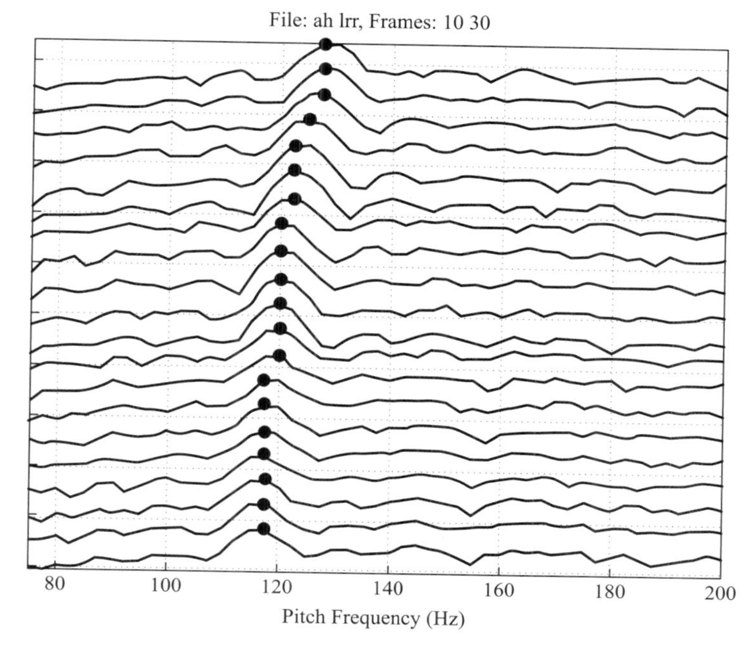

FIGURE P10.8
Plots of $\hat{P}_{\hat{n}}(e^{j\omega})$ for a range of speech frames. The maximum value of $\hat{P}_{\hat{n}}(e^{j\omega})$ is noted by a filled circle.

samples and (for each frame) calculate:

$$\hat{P}_{\hat{n}}(e^{j\omega}), \quad 2\pi f_L \leq \omega \leq 2\pi f_H,$$

$$f_L = \begin{cases} 75 & \text{for males} \\ 150 & \text{for females,} \end{cases}$$

$$f_H = \begin{cases} 200 & \text{for males} \\ 300 & \text{for females,} \end{cases}$$

6. find the peak location (the putative pitch frequency) and the peak amplitude (called the confidence score) and save both values in arrays (with one value per frame); convert the frame-based pitch frequency array to a frame-based pitch period array (how is this done?);
7. using a confidence threshold based on the maximum value of the confidence score minus the confidence offset, set the pitch period to zero for all frames whose confidence scores fall below the confidence threshold;
8. use a 5-point median smoother to smooth the pitch period array and the confidence score array;
9. plot the median-smoothed pitch period contour; on the same plot, show the median-smoothed confidence contour.

It will be helpful to create a debug mode capability whereby a region of frames is specified and the MATLAB program plots the function $\hat{P}_{\hat{n}}(e^{j\omega})$ in a waterfall-type plot, with the location of the maximum (for each frame in the debug set) being noted by a large filled circle (or any desired symbol). Figure P10.8 shows such a waterfall plot for 20 frames of voiced speech from a male speaker.

10.9. (MATLAB Exercise: Cepstrum-Based Pitch Detector). Program a pitch detector based on the (real) cepstrum of a speech signal. The real cepstrum is defined as the inverse FFT of the log magnitude spectrum of the signal, and the pitch period (for voiced speech sections) is found as the location of the peak of the cepstrum over the range of pitch periods appropriate for the gender of the speaker. A variation on standard pitch detectors is proposed in this exercise where the primary cepstral peak as well as a secondary cepstral peak (along with their peak amplitudes) are used as the basis for the pitch period decision.

The steps to follow in implementing the cepstrum-based pitch detector are similar to the ones used in the previous exercise (Problem 10.8), with the following exceptions:

- The signal processing parameters define the size of the FFT used to measure the spectrum and the cepstrum as `nfft=4000` to minimize aliasing.
- A threshold on the ratio of the primary cepstral peak (within the designated region of pitch periods) to the secondary cepstral peak is specified as `pthr1=4` and is used to define a region of "certainty (high confidence)" about pitch period estimates.
- The pitch period range for searching the cepstrum for the pitch peak is specified as the range $\texttt{nlow} \leq \texttt{n} \leq \texttt{nhigh}$, where the low and high of the pitch period range (as well as the cepstral search range) (for both males and females) is specified as:

$$\texttt{nlow} = \begin{cases} 40 & \text{for males} \\ 28 & \text{for females,} \end{cases}$$

$$\texttt{nhigh} = \begin{cases} 167 & \text{for males} \\ 67 & \text{for females.} \end{cases}$$

The process for finding primary and secondary cepstral peaks (within the specified range) is as follows:

1. Locate the maximum of the cepstrum over the specified range (`p1`), and record the quefrency at which the maximum occurs (`pd1`).
2. Zero the cepstrum over a range of ± 4 quefrencies around the maximum location found in the previous step, thereby eliminating the possibility that the secondary cepstral maximum will essentially be the same as the primary maximum.
3. Locate the secondary maximum of the processed cepstrum; record its quefrency (`pd2`), and its value (`p2`).

Figure P10.9(a) illustrates the results of the cepstral peak detection process. This figure shows a sequence of cepstral frames, displayed as a waterfall plot, with the primary cepstral peak indicated by a darkly shaded circle, and the secondary cepstral peak indicated by a lightly shaded circle. The continuity of pitch period, over frames, is clearly seen in parts of this figure.

The next step in the process is to define "reliable" regions of voiced speech. These regions are identified as those frames whose ratio of primary cepstral maximum to

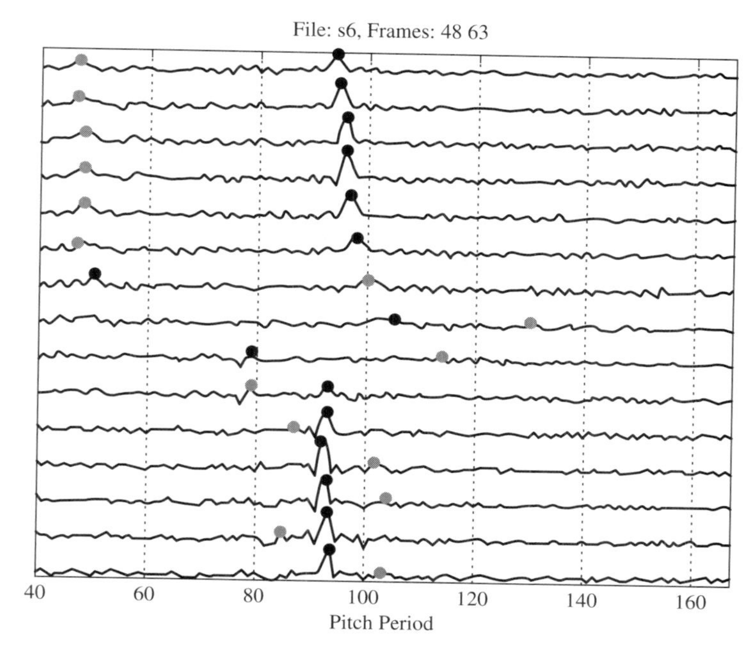

FIGURE P10.9(a)
Plots of sequence of cepstra in a waterfall display showing frame-by-frame cepstra with primary peak indicated by a darkly shaded circle, and secondary peak indicated by a lightly shaded circle.

secondary cepstral maximum values exceeds the pre-set threshold of `pthr1=4.0`. Each of these reliable regions of voiced speech are extended by searching the neighboring regions (i.e., the frames prior to and following the reliable regions) and detecting adjacent frames whose primary or secondary pitch periods are within $\pm10\%$ of the pitch period at the boundary frames. Whenever both putative pitch period estimates (i.e., primary and secondary) exceed the 10% difference threshold, the local search for additional adjacent frames is terminated. The next region of reliable voiced speech frames is searched and extended in a similar manner as above. This process is continued until all reliable regions have been extended to include neighboring pitch estimates that fall within the search criteria.

The final step in the signal processing is median smoothing of the pitch period contour using a 5-point median smoother.

Figure P10.9(b) shows plots of the resulting pitch period contour, along with the confidence scores (cepstral peak levels), for the waveform in the file `s6.wav`. It can be seen that, for the most part, regions of high cepstral value provide reliable pitch period estimates.

10.10. (MATLAB Exercise: LPC SIFT-Based Pitch Detector). Program a pitch detector based on the SIFT method, as discussed in Section 10.5.7, and as implemented in the block diagram of Figure 10.43. The SIFT algorithm for pitch detection filters the speech to a 900 Hz bandwidth, down-samples to a 2 kHz sampling rate, performs a low order ($p = 4$) linear predictive analysis, and inverse filters the speech to provide an estimate of a spectrally flat error signal, with pitch being preserved throughout the processing.

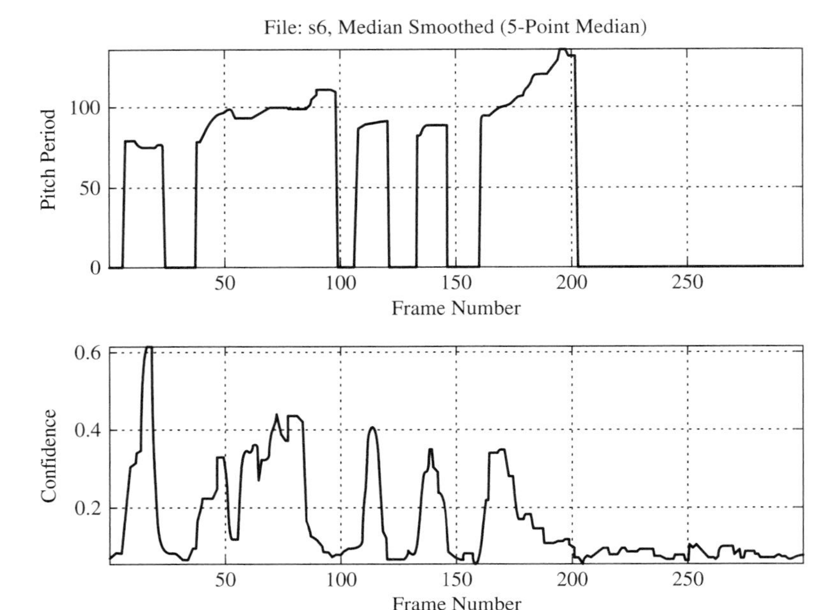

FIGURE P10.9(b)
Plots of pitch period estimates and cepstral magnitudes at the pitch peaks for the utterance in file `s6.wav`.

A standard autocorrelation analysis of the flattened error signal is performed and a peak picker selects the maximum peak (in a suitable range of pitch periods) as well as a secondary peak (again in the range of pitch periods). An interpolator increases the sampling rate of the autocorrelation to 10 kHz, thereby providing much higher precision to the peak locations (potential pitch period estimates), and the primary and secondary peaks are re-estimated using the high rate autocorrelation function. Finally a threshold on the normalized autocorrelation peaks is utilized to determine regions of high confidence in the pitch period estimate, and a pitch period interpolation process is used to extend the "reliable" regions forward and backward, based on pitch period consistency at the boundaries of the reliable regions. A median smoother, with a value of `L=5`, is used to remove isolated fluctuations in pitch period estimates.

The detailed steps of the LPC SIFT pitch detector are as follows:

1. Read in the speech file and the speaker gender.
2. Resample the speech to a sampling rate of `fs=10000` Hz.
3. Design an FIR bandpass filter (slight modification to the original SIFT algorithm) with the following specifications:
   - stopband from 0 to 60 Hz
   - transition band from 60 to 120 Hz
   - passband from 120 to 900 Hz
   - transition band from 900 to 960 Hz
   - stopband from 960 to 5000 Hz
   - filter length of 401 samples.

4. Filter the speech and compensate for the delay of 200 samples.
5. Decimate the speech by a factor of 5-to-1 to give a new sampling rate of 2000 Hz.
6. Perform frame-based analysis of the speech using frames of `L=80` samples (40 msec at a 2 kHz sampling rate) with frame offset of `R=20` samples (10 msec at a 2 kHz sampling rate); use an `N` sample Hamming window to weight each frame and perform `p=4` LPC analysis of each frame.
7. Inverse filter the frame of speech (using the derived LPC inverse filter) to give the LPC error signal.
8. Autocorrelate the error signal providing the autocorrelation function $R_n[k]$ using the MATLAB function `xcorr`.
9. Search the region $\texttt{pdmin} \leq k \leq \texttt{pdmax}$ to find the location and value of the largest (normalized) autocorrelation peak, and repeat the search (suitably nulling out the first peak) to find the location and value of the second largest autocorrelation peak, where `pdmin` and `pdmax` are suitably defined minimum and maximum pitch periods appropriate for the gender of the talker; i.e.,

$$\texttt{pdmin} = \begin{cases} 40 & \text{for males} \\ 28 & \text{for females,} \end{cases}$$

$$\texttt{pdmax} = \begin{cases} 167 & \text{for males} \\ 67 & \text{for females.} \end{cases}$$

10. Interpolate the autocorrelation function by a factor of 5-to-1 (to a sampling rate of 10 kHz) and refine the location of the primary and secondary autocorrelation peaks.
11. Determine regions of high reliability (of pitch period estimate) based on frames where the ratio of primary autocorrelation peak to secondary autocorrelation peak exceeds a pre-set threshold of 3.0.
12. Extend the high reliability regions (both prior to and after the high reliability regions) based on how closely the pitch period estimates just outside the boundary match the pitch period estimates at the boundary; if a frame pitch period match is within $\pm 10\%$ of the boundary value, a new boundary value is chosen and the search continues; the search/extension process for a single region of high reliability is terminated when both the primary and secondary peak locations (the pitch period estimates) are more than 10% away from the local boundary frame pitch period; the extension process is repeated until all of the regions of high reliability have been examined and extended based on the local matches to the boundary frames.
13. Non-linearly smooth the resulting pitch period contour using a median smoother with a duration of 5 frames.

As an example, Figure P10.10 shows a waterfall plot of 20 frames of the normalized autocorrelation function showing the locations (and values) of the primary autocorrelation peaks (dark filled circles) and the secondary autocorrelation peaks (lightly filled circles) from the utterance `s5.wav`. It can be seen that the primary peak is much stronger than the secondary peak, and further that the primary peaks form a smooth pitch period contour over the duration of the frames shown in this plot.

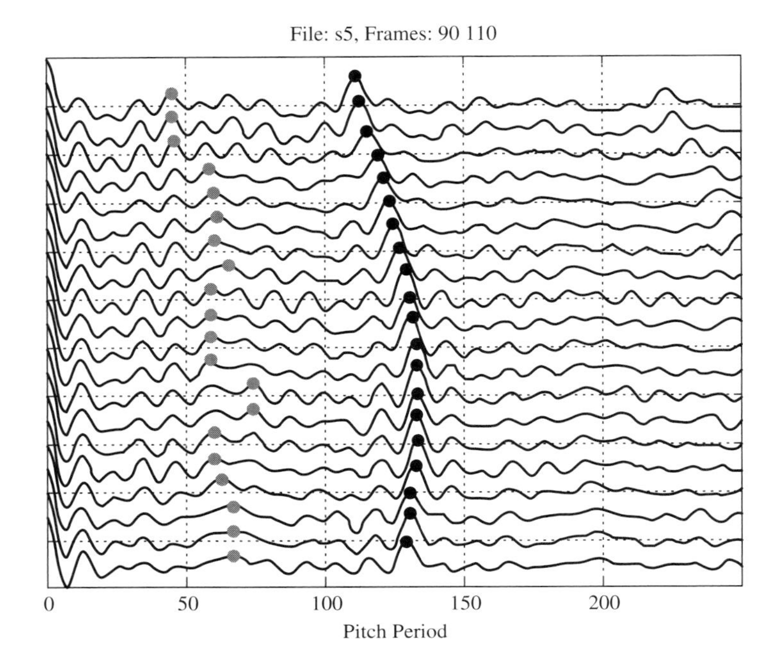

FIGURE P10.10
Plots of normalized autocorrelations of the LPC error signal over a 20 frame region for the utterance `s5.wav`.

10.11. (MATLAB Exercise: Linear Prediction-Based Formant Estimator). Program an algorithm for detecting and estimating formants of (voiced) speech based on finding the roots of the linear prediction polynomial $A(z)$. The detailed steps that must be performed in order to implement a linear prediction-based formant estimator are the following:

1. Read in the speech file to be processed (and determine the sampling rate `fs` of the speech signal).
2. Read in the speech processing parameters for frame-based LPC analysis, namely:
   - `L`: the frame size (either in samples or msec); typically we use 40 msec frames with 640/400/320 samples per frame for sampling rates of 16,000/10,000/8000 samples/sec
   - `R`: the frame shift (either in samples or msec); typically we use 10 msec frame shifts with 160/100/80 sample shifts for sampling rates of 16,000/10,000/8000 samples/sec
   - `p`: the LPC analysis order; typically we use analysis orders of 16/12/10 for sampling rates of 16,000/10,000/8000 samples/sec
   - `rthresh`: threshold on the magnitude of any complex root to be included in the formant estimation algorithm.
3. Perform LPC analysis using the autocorrelation method.
4. Find the roots of the resulting p^{th}-order LPC polynomial. Filter out all real roots and all complex roots whose magnitude falls below the specified threshold, `rthresh`; eliminate any root whose (equivalent) frequency ($f = \angle(\text{degrees}) * F_S/180$ is greater than 4500 Hz (unlikely to find a significant formant above this frequency).

5. Sort the resulting roots and store them in an array that is indexed by the frame number.
6. Determine runs (sequences of frames with the same number of putative formants) whose length is five or more frames; record the length and the starting frame of each run.
7. Process the putative formant array by extending each of the runs backwards (from the beginning of each run to the end of the previous run) and forwards (from the end of each run to the beginning of the next run) and, using a simple Euclidean distance measure, determine the alignment between putative formants in a given frame with those of the previous or following frame, thereby creating a second array of aligned formant tracks with deaths of some tracks (no logical continuation path) and births of new tracks (no logical initial path); hence every frame where the number of putative formants is less than the number of formants for the succeeding frame will have one or more empty slots after formant alignment (and vice versa).
8. Plot the resulting formant tracks and compare these results with the raw putative formant data from the root finding routine.

The formant estimation procedure described above will not work on all inputs (think of the problems that occur for female speakers, children, rapidly spoken speech, sounds with rapid changes within an utterance, etc.). However this exercise serves as a good

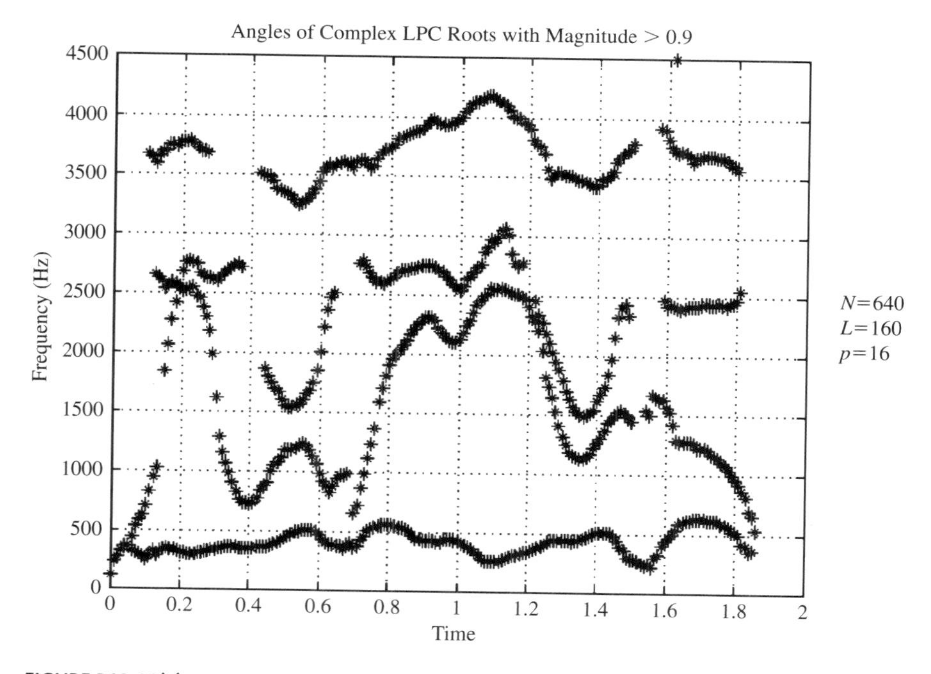

FIGURE P10.11(a)
Plot of complex root angles (scaled to frequency) as a function of the speech analysis frame showing putative formants for the utterance `we were away a year ago_lrr.wav`.

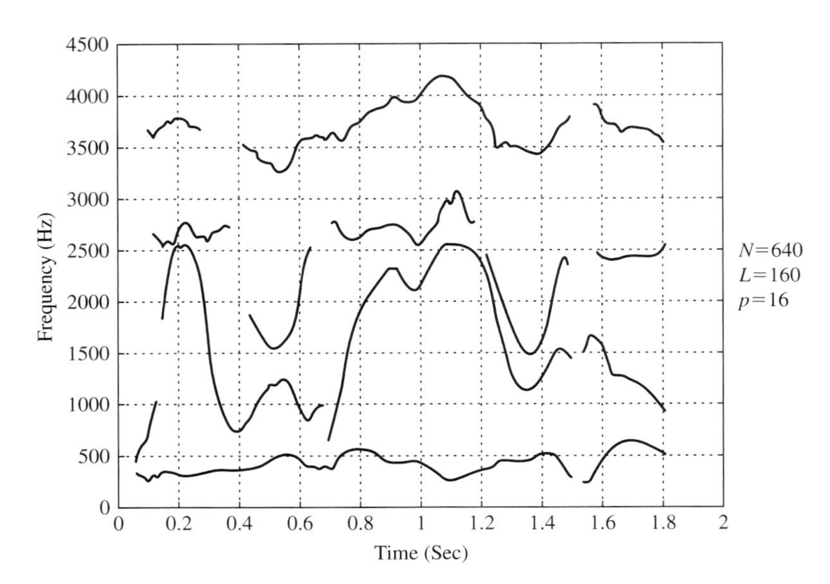

FIGURE P10.11(b)
Plot of estimated formants for the utterance
`we were away a year ago_lrr.wav`.

introduction to the many issues that must be attacked before a reliable formant track can be obtained for an arbitrary sentence.

Test your resulting formant estimation algorithm on the following three sentences:

1. `ah_lrr.wav`, a simple steady vowel spoken by a male talker (hence the formants do not vary much during the course of the sound); sampling rate of 10,000 samples/sec
2. `we were away a year ago_lrr.wav`, a sentence with a fairly smooth formant track that exists throughout most of the utterance (since it is an all-voiced sentence); male speaker with a sampling rate of 16,000 samples/sec
3. `test_16k.wav`, a sentence with very distinct voiced and unvoiced regions; spoken by a male speaker at a sampling rate of 16,000 samples/sec.

As an example, Figures P10.11(a)–P10.11(b) show plots of the complex root angles (suitably converted to frequency) versus frame number for the sentence `we were away a year ago_lrr.wav` along with the estimated formants using the algorithm described in this problem. It can readily be seen that there are stray putative formants (not linked to any reasonable formant track) along with complicated situations where the formants get very close to each other. The resulting formant track of Figure P10.11(b) is a fairly good track for this all-voiced utterance, but, unfortunately, not representative of the problems that must be faced for a general purpose formant tracking method.

CHAPTER 11

Digital Coding of Speech Signals

Watson, if I can get a mechanism which will make a current of electricity vary its intensity as the air varies in density when sound is passing through it, I can telegraph any sound, even the sound of speech.—A.G. Bell [47]

11.1 INTRODUCTION

Bell's simple but profound idea, so important in the history of human communication, seems commonplace today. The basic principle embodied in his great invention, the telephone, is fundamental to a multitude of devices and systems for recording, transmitting, or processing of speech signals wherein the speech signal is represented by measuring and reproducing the amplitude fluctuations of the acoustic waveform of the speech signal. This has always been true for analog systems, but it is also the case for digital systems, where at some point, the speech waveform is often represented by a sequence of numbers (i.e., the set of speech samples) that specifies the pattern of amplitude fluctuations.

The basic digital representation of the speech waveform is depicted in Figure 11.1, which shows the operations of analog-to-digital (A-to-D) and digital-to-analog (D-to-A) conversion. As illustrated, the analog speech waveform, $x_a(t)$, converted from acoustic to electrical form by a microphone and thought of as a continuous function of a continuous-time variable, is sampled periodically in time (with sampling period, T) to produce a sequence of samples, $x[n] = x_a(nT)$. These samples, were they actually available, would take on a continuum of values. However, as shown in Figure 11.1, it is necessary to quantize them to a finite set of values in order to obtain a digital representation. Thus, the output, $\hat{x}[n]$, of an A-to-D converter is discrete in both time and amplitude, and the discrete amplitudes, which might have an intrinsic amplitude scale in terms of sound pressure or volts, are encoded into dimensionless digital codewords $c[n]$.[1] In a similar manner, the D-to-A converter, shown

[1]The codewords in A-to-D conversion are generally two's-complement digital numbers so that they can be used directly in digital computations. Also observe that we use the ^ notation in this chapter to denote quantization, while in earlier chapters, the ^ was used to denote the complex cepstrum of a signal. In general, the meaning of ^ will be clear from the context.

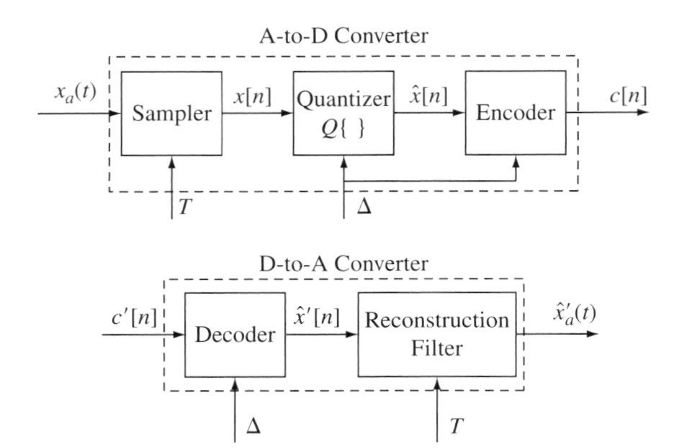

FIGURE 11.1
General block diagram depicting digital waveform representations.

at the bottom of Figure 11.1, converts received codewords, $c'[n]$, into digital samples, $\hat{x}'[n]$, and ultimately to an approximation of the original signal, $\hat{x}'_a(t)$, via a lowpass reconstruction filter. We use the $'$ to denote that the received codewords may differ from the transmitted codewords.

Figure 11.1 is a convenient conceptualization of the process of A-to-D conversion, or sampling and quantization. However, it will soon become clear that if we think of the quantization block more generally, it is also a representation of an entire class of speech coders known as "waveform coders," which are designed to preserve the details of the waveform of the speech signal. While it may not always be possible to separate a given representation into two distinct stages, the two basic features, sampling and quantization, are inherent in all the waveform coding schemes that we will discuss in this chapter. In most waveform coding systems, the sequence $\hat{x}[n]$ has a sampling rate equal to the sampling rate of the input sequence $x[n]$. In A-to-D conversion, the quantized samples are treated as an approximation to the samples $x[n]$. In other systems, the quantized representation must be converted by additional processing to "quantized waveform samples." In waveform coding systems the information rate, I_w, of the digital representation is simply

$$I_w = B \cdot F_s,$$

where B is the number of bits used to represent each sample and $F_s = 1/T$ is the number of samples/sec. The representation of a speech waveform by sampling and quantizing the samples is commonly known as PCM for *pulse code modulation*. The basic principles of PCM were laid out in 1948 in a classic paper by Oliver, Pierce, and Shannon [265].

In a second class of digital speech coding systems, known variously as analysis/synthesis systems, model-based systems, hybrid coders, or *vocoder* systems, detailed waveform properties are not necessarily preserved, but instead the speech coder

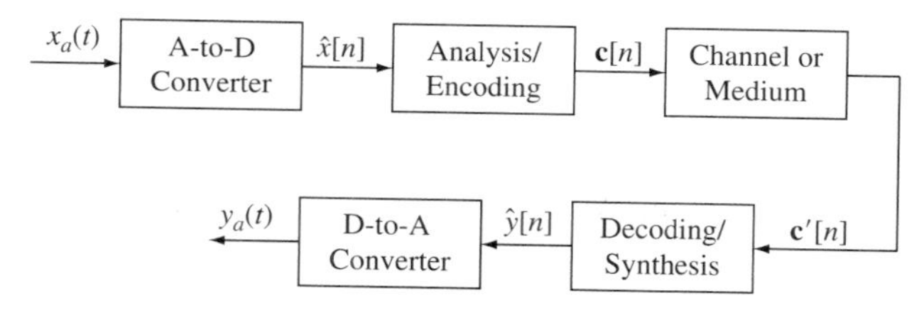

FIGURE 11.2
Block diagram representation of the generic processing in a general analysis/synthesis system speech coder.

attempts to estimate the parameters of a model for speech production that serves as the basis for the digital representation of a given speech signal. The goal in such coders is not to preserve the waveform, but rather to preserve intelligibility and the perceived quality of reproduction from the digital representation. A block diagram representation of the processing for such speech analysis/synthesis systems is given in Figure 11.2. Starting with the analog (continuous-time) signal, $x_a(t)$, the speech signal is first converted to digital format using a high-accuracy (usually at least 16 bits/sample) A-to-D converter, giving the discrete-in-time and discrete-in-amplitude signal, $\hat{x}[n]$. Assuming that the sampled signal is the output of a synthesis model, it is then analyzed using one or more of the speech representations presented in Chapters 6–9 to determine the parameters of the synthesis model such that the model output will be a perceptually acceptable approximation to the original signal. The resulting parameters are encoded for transmission or storage. The result of the analysis and encoding processing is the data parameter vector, $\mathbf{c}[n]$, which would normally be computed at a lower sampling rate than the sampling rate of the original sampled and quantized speech signal, $\hat{x}[n]$. If the "frame rate" of the analysis is F_{fr}, the total information rate for such model-based coders, I_m, is

$$I_m = B_c \cdot F_{fr},$$

where B_c is the total number of bits required to represent the parameter vector $\mathbf{c}[n]$. The output of the channel through which the encoded signal is transmitted is denoted as $\mathbf{c}'[n]$. At the synthesizer, the channel signal is decoded and a synthesis model is used to reconstruct a quantized approximation, $\hat{y}[n]$, to the original digital signal which is then converted back to analog format using a D-to-A converter, giving the processed signal, $y_a(t)$. In such analysis/synthesis systems, a fundamental goal is to represent the speech signal with fewer bits/sec ($I_m < I_w$) and with a perceived quality of reproduction comparable to or equal to the original input signal.

Analysis/synthesis systems, of the type shown in Figure 11.2, can be implemented using a range of the speech representations presented in this book, including:

1. the short-time Fourier transform (STFT) (both filter bank summation and overlap add implementations);
2. the cepstrally smoothed spectrum, with an excitation function obtained from a cepstrum-based pitch detector (or any other method of pitch detection);

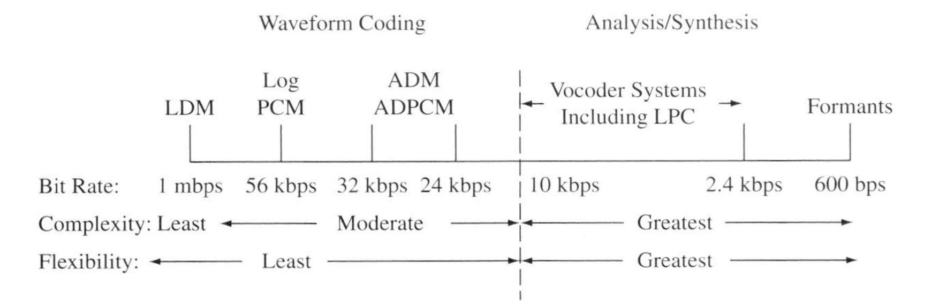

FIGURE 11.3
Representation of the characteristics of waveform codings and analysis/synthesis coders along the dimensions of bit rate, complexity, and flexibility.

3. the linear predictive smoothed spectrum, with an excitation function obtained from coding of the resulting error sequence, including both waveform coding as well as analysis/synthesis coding methods (to be discussed later in this chapter).

A useful way of comparing the two types of speech coders is given in Figure 11.3. This figure compares waveform coding methods and analysis/synthesis coding methods along three dimensions, namely overall system bit rate, computational complexity, and flexibility of the resulting speech coder in terms of being able to manipulate the resulting speech, and thereby slow down or speed up the speech, change the pitch or intensity, modify the frequency spectrum, etc. From Figure 11.3, we see that the class of waveform coders generally has the following properties:

- They require high bit rates, operating in the broad range from about 16 kbps (the lowest coding rate) to upwards of 1 mbps.
- Their complexity varies from low (virtually no computation per speech sample) to moderate.
- They have virtually no flexibility to modify speech signal properties.

In contrast, the class of analysis/synthesis coders has the following properties:

- They can be implemented at bit rates as low as 600 bps and as high as 10 kbps or higher. Hence, this class of coders is a natural candidate for use in low bit rate transmission systems, especially cellular and VoIP systems.
- Their computational complexity is much higher than most waveform coders. However, the amount of computation is small compared to the raw computation power of modern signal processing chips, so this issue is not one of concern for most applications. Such coders often operate on a frame basis, which requires buffering. This can lead to significant delay in real-time systems.
- Their flexibility is high, enabling these coders to modify the time or frequency scales of the processed signal, thereby enabling their use in systems for speeding up speech (e.g., reading machines for the blind), in systems for enhancing speech

(e.g., correction of the frequency spectrum for speech recorded in a helium–oxygen mixture, as might be used by deep sea divers), and as a basis for speech synthesis.

In this chapter we discuss both waveform coders and model-based coders in some detail. We begin by discussing the process of sampling of speech signals. Then we discuss a variety of schemes for quantizing the samples of speech signals. Later sections of the chapter focus on model-based coders. A third class of speech coders is based on the principle of quantizing the short-time Fourier transform of the speech signal. We call these frequency-domain speech coders, and since they share many features with coders designed especially for audio (singing and instrumental music), we defer their discussion to Chapter 12.

11.2 SAMPLING SPEECH SIGNALS

As discussed in our review of digital signal processing in Chapter 2, unquantized samples of an analog signal are a unique signal representation if the analog signal is bandlimited and if the sampling rate is at least twice the Nyquist (highest) frequency of the signal spectrum. Since we are concerned with sampled speech signals, we need to consider the spectral properties of the speech signal. We recall from the discussion of Chapter 3 that according to the steady state models for the production of vowel and fricative sounds, speech signals are not inherently bandlimited, although the short-time spectra of voiced speech sounds do tend to fall off rapidly at high frequencies. Figure 11.4 shows short-time spectra of some typical speech sounds. It is observed that for the voiced sounds such as the vowel spectrum shown in Figure 11.4a, the spectral levels of the high frequencies can be as much as 40 dB below the low frequency peak of the spectrum. On the other hand, for unvoiced sounds as shown in Figure 11.4b, the spectrum is peaked around 8 kHz with a fall-off above due to the anti-aliasing filter. Thus, to accurately represent all speech sounds would require a sampling rate greater than 20 kHz. In many applications, however, this sampling rate is not required. For example, if the sampling operation is a prelude to the estimation of the first three formant frequencies of voiced speech, we need only consider the portion of the spectrum up to about 3.5 kHz. Therefore, if the speech is filtered by a sharp cutoff analog filter prior to sampling, so that the Nyquist frequency is 4 kHz, then a sampling rate of 8 kHz is possible. As another example, consider speech that has been transmitted over a telephone line. Figure 11.5 shows a typical frequency response curve for a telephone line transmission path. It is clear from Figure 11.5 that telephone transmission has a bandlimiting effect on speech signals, and indeed, a Nyquist frequency of 4 kHz (or lower) is commonly assumed for "telephone speech."

An important point that is often overlooked in discussions of sampling is that the speech signal may be corrupted by wideband random noise prior to A-to-D conversion. In such cases, the "speech signal plus noise" combination should be filtered with an analog lowpass filter that cuts off sharply above the desired highest frequency to be preserved for the speech signal, so that images of the high frequency noise are not aliased into the baseband.

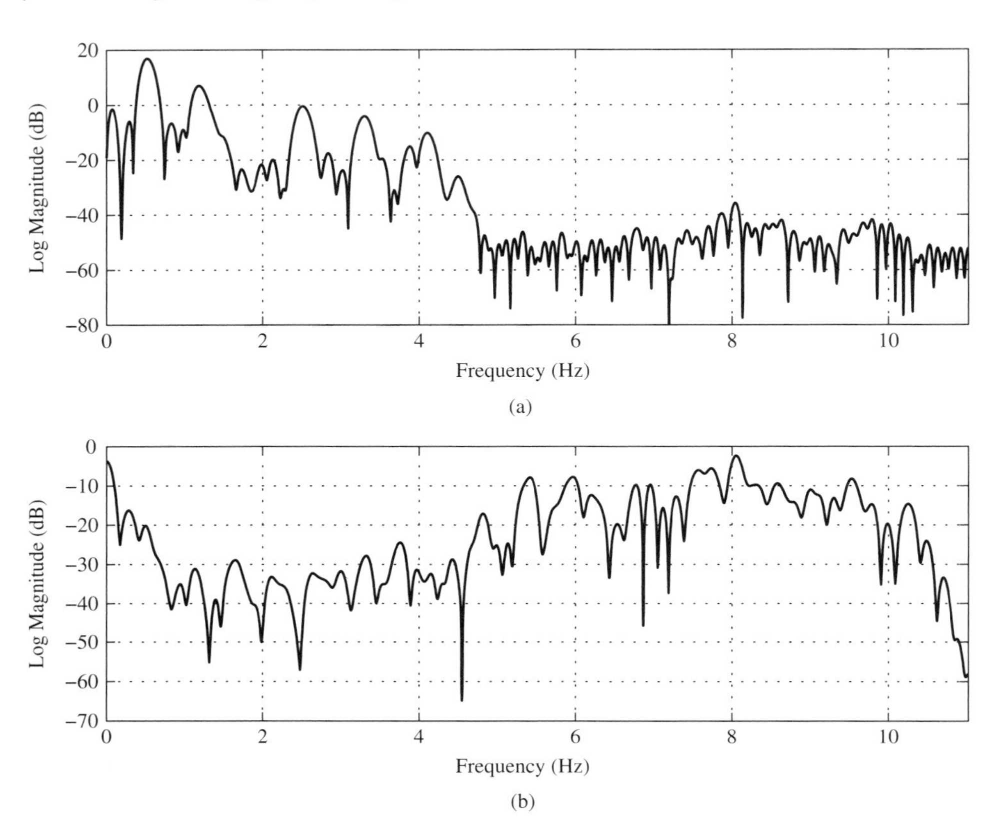

FIGURE 11.4
Wide-band short-time Fourier transforms for a speech signal sampled with $F_s = 22050$ Hz: (a) vowel sound; (b) fricative sound.

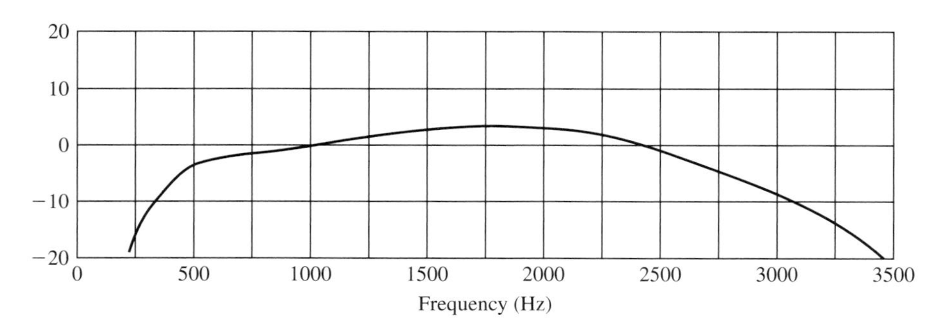

FIGURE 11.5
Frequency response of a typical telephone transmission path. (After BTL, Transmission Systems for Communication, p. 73. Reprinted with permission of Alcatel-Lucent USA Inc.)

11.3 A STATISTICAL MODEL FOR SPEECH

In discussing digital waveform representations, it is often convenient to assume that speech waveforms can be represented by an ergodic random process. Although this is a gross simplification, we will see that a statistical point of view often yields useful results, thereby justifying the use of such a model.

11.3.1 Autocorrelation Function and Power Spectrum

If we assume that an analog speech signal, $x_a(t)$, is a sample function of a continuous-time random process, then the sequence of samples $x[n] = x_a(nT)$ derived by periodic sampling can likewise be thought of as a sample sequence of a discrete-time random process; i.e., each sample $x[n]$ is considered to be a realization of a random variable associated with sample time n, and governed by a set of probability distributions. In communication system analysis, an adequate characterization of the analog signal often consists of a first-order probability density, $p(x)$, for computing the mean and average power, together with the autocorrelation function of the random process, which is defined as

$$\phi_a(\tau) = E[x_a(t)x_a(t+\tau)], \tag{11.1}$$

where $E[\ \cdot\ ]$ denotes the expectation (probability or time average) of the quantity within the brackets. For ergodic random processes, the expectation operator can be defined as a probability average or as a time average over all time, and it is common to estimate averages by averaging over finite time intervals rather than approximations based on estimated probability distributions.[2]

The analog power spectrum, $\Phi_a(\Omega)$, is the Fourier transform of $\phi_a(\tau)$; i.e.,

$$\Phi_a(\Omega) = \int_{-\infty}^{\infty} \phi_a(\tau)e^{-j\Omega\tau}d\tau. \tag{11.2}$$

The discrete-time signal obtained by sampling the random signal $x_a(t)$ has an autocorrelation function defined as

$$\begin{aligned}\phi[m] &= E[x[n]x[n+m]] \\ &= E[x_a(nT)x_a(nT+mT)] = \phi_a(mT).\end{aligned} \tag{11.3}$$

[2]Computation of correlation functions by probability averages requires knowledge of second-order (joint) distributions at times separated by τ, for a range of τ.

Thus, since $\phi[m]$ is a sampled version of $\phi_a(\tau)$, the power spectrum of $\phi[m]$ is given by

$$\Phi(e^{j\Omega T}) = \sum_{m=-\infty}^{\infty} \phi[m]e^{-j\Omega Tm} = \frac{1}{T}\sum_{k=-\infty}^{\infty} \Phi_a\left(\Omega + \frac{2\pi}{T}k\right). \tag{11.4}$$

Equation (11.4) shows that the random process model of speech also suggests that aliasing can be a problem if the speech signal is not bandlimited prior to sampling.

The probability density function for the sample amplitudes, $x[n]$, is the same as for the amplitudes, $x_a(t)$, since $x[n] = x_a(nT)$. Thus averages such as mean and variance are the same for the sampled signal as for the original analog signal.

When applying statistical notions to speech signals, it is necessary to estimate the probability density and correlation function (or the power spectrum) from speech waveforms. The probability density is estimated by determining a histogram of amplitudes for a large number of samples; i.e., over a long time interval. Davenport [79] made extensive measurements of this kind, and Paez and Glisson [275], using similar measurements, have shown that a good approximation to measured speech amplitude densities is a gamma distribution of the form

$$p(x) = \left(\frac{\sqrt{3}}{8\pi\sigma_x|x|}\right)^{1/2} e^{-\frac{\sqrt{3}|x|}{2\sigma_x}}. \tag{11.5}$$

A somewhat simpler approximation is the Laplacian density

$$p(x) = \frac{1}{\sqrt{2}\sigma_x} e^{-\frac{\sqrt{2}|x|}{\sigma_x}}. \tag{11.6}$$

Figure 11.6 shows a measured density for speech, along with gamma and Laplacian densities, all of which have been normalized so that the mean ($\overline{x}$) is zero and the variance (σ_x^2) is unity. The gamma density is clearly a better approximation than the Laplacian density, but both are reasonably close to the actual measured speech density. Figure 11.6 shows that the probability density function for speech is highly concentrated around 0, implying that small speech amplitudes are much more likely than large speech amplitudes. (We will see how we can exploit the shape of the speech density function later in this chapter.)

The autocorrelation function and power spectrum of speech signals can be estimated by standard time-series analysis techniques, which are discussed for example in Ref. [270]. An estimate of the autocorrelation function of an ergodic random process can be obtained by estimating the time average autocorrelation function from a long (but finite) segment of the signal. For example, the definition of the short-time autocorrelation function [Eq. (6.35) of Chapter 6] can be slightly modified to give the estimate

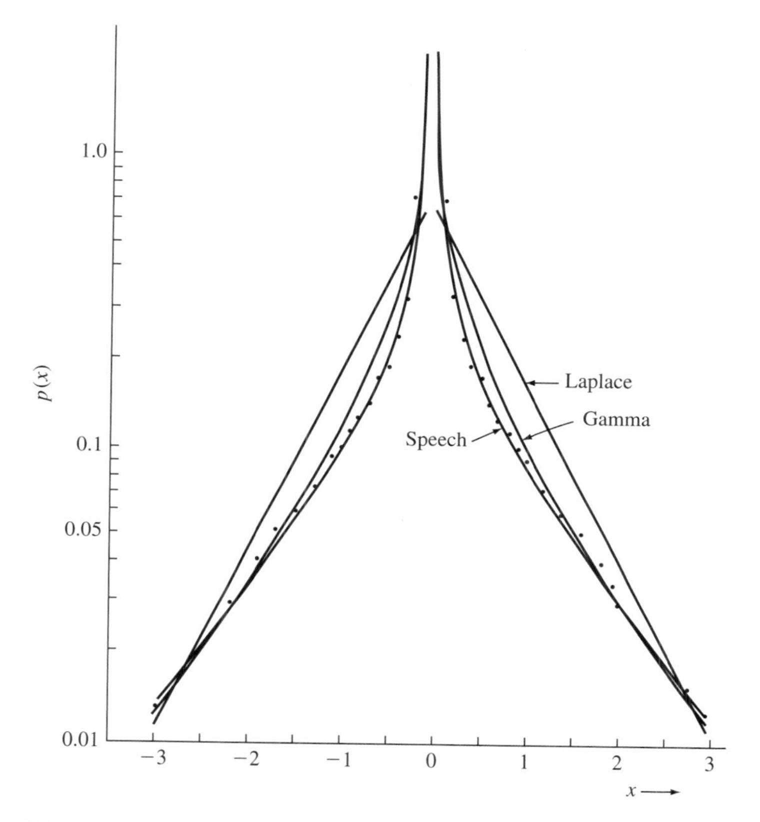

FIGURE 11.6
Speech histogram values (dots) and theoretical gamma and Laplace probability densities. (After Paez and Glisson [275]. © [1972] IEEE.)

of the long-time average autocorrelation function as

$$\hat{\phi}[m] = \frac{1}{L} \sum_{n=0}^{L-1-m} x[n]\, x[n+m], \quad 0 \leq |m| \leq L-1, \tag{11.7}$$

where L is a large integer [305]. An example of such an estimate is shown in Figure 11.7 for an 8 kHz sampling rate [254]. The upper curve shows the normalized autocorrelation $\hat{\rho}[m] = \hat{\phi}[m]/\hat{\phi}[0]$ for lowpass filtered sampled speech signals and the lower curve is for bandpass filtered speech. The shaded region around each curve shows the variation of the estimate due to different speakers. The correlation is high between adjacent samples ($\hat{\rho}[1] > 0.9$) and it decreases significantly for greater sample spacing. Also evident is the fact that lowpass filtered speech is more highly correlated than bandpass filtered speech.

The power spectrum can be estimated in a variety of ways. Before digital computers were employed in the study of speech, the spectrum was estimated by measuring

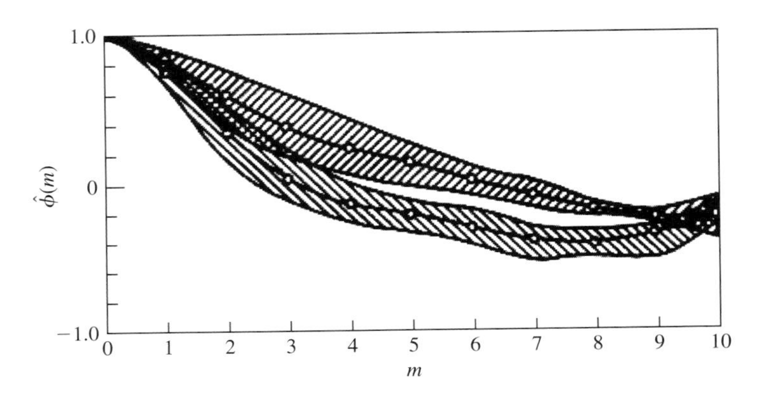

FIGURE 11.7
Autocorrelation functions of speech signals; upper curves for lowpass speech, lower curves for bandpass speech. (After Noll [254].)

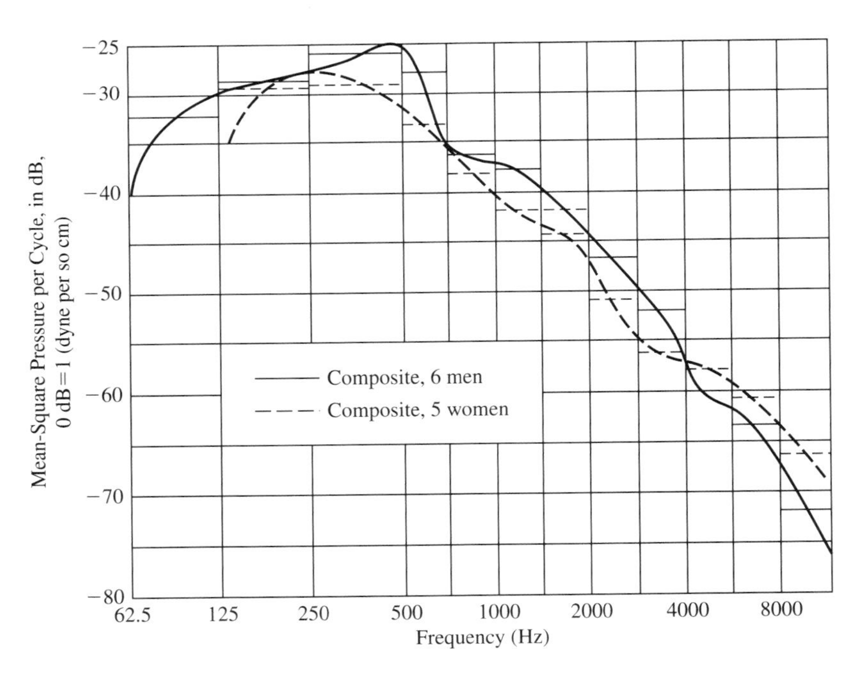

FIGURE 11.8
Long-time power density spectrum for continuous speech. (After Dunn and White [95]. © [1940] Acoustical Society of America.)

the average output of a set of bandpass filters [95]. Figure 11.8 shows such an example where the power was averaged over about a minute of continuous speech. This early result indicates that the average power spectrum is peaked at about 250–500 Hz, and that above this frequency, the spectrum falls off at about 8–10 dB/octave.

With sampled speech signals, one approach to the estimation of the long-term average power spectrum is to first estimate the speech correlation function, $\hat{\phi}[m]$, as in Eq. (11.7) and then after applying a symmetric window $w[m]$, compute its N-point discrete-time Fourier transform [270]; i.e.,

$$\hat{\Phi}(e^{j(2\pi k/N)}) = \sum_{m=-M}^{M} w[m]\,\hat{\phi}[m]\,e^{-j(2\pi k/N)m}, \qquad k = 0, 1, \ldots, N-1. \tag{11.8}$$

An example of this method of spectrum estimation applied to speech is given in Figure 11.9 for $w[m]$ a Hamming window [256]. The upper panel shows the estimate of the long-term correlation of the sampled speech signal for the first 20 lags at a sampling

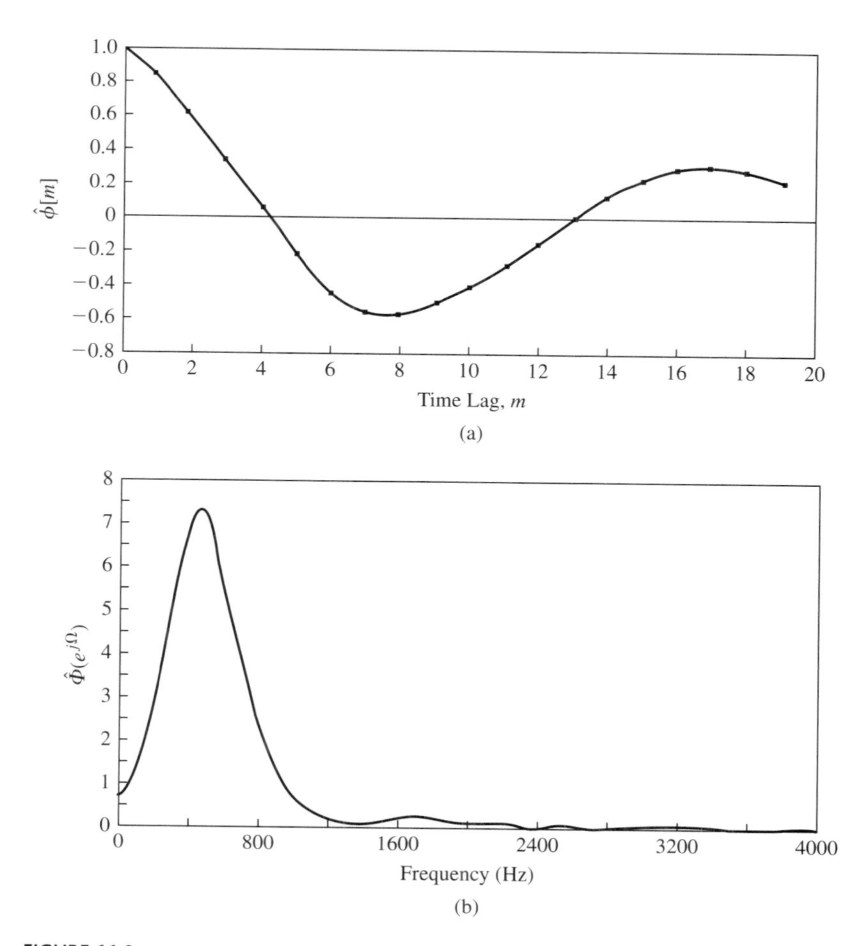

FIGURE 11.9
(a) Autocorrelation function and (b) power density estimates for speech. (After Noll [256]. Reprinted with permission of Alcatel-Lucent USA Inc.)

rate of 8 kHz; the lower panel shows the resulting power spectrum estimate (on a linear magnitude scale) over the frequency range $0 \leq F \leq 4000$ kHz, where the power spectrum estimate was computed using Eq. (11.8) and plotted as a function of $F_k = kF_s/N$. We again see that the power spectrum estimate of Figure 11.9 displays a prominent spectral magnitude peak around 500 Hz, as seen previously in Figure 11.8.

Another common approach to estimating the power density spectrum of a signal is the method of averaged periodograms [270, 305, 416]. The periodogram is defined as

$$P(e^{j\omega}) = \frac{1}{LU}\left|\sum_{n=0}^{L-1} x[n]w[n]e^{-j\omega n}\right|^2, \tag{11.9}$$

where $w[n]$ is an L-point analysis window such as the Hamming window, and

$$U = \frac{1}{L}\sum_{n=0}^{L-1}(w[n])^2 \tag{11.10}$$

is a normalizing constant used to compensate for the tapering of the window. In practice we use the discrete Fourier transform (DFT) to compute the periodogram as:

$$P(e^{j(2\pi k/N)}) = \frac{1}{LU}\left|\sum_{n=0}^{L-1} x[n]w[n]e^{-j(2\pi/N)kn}\right|^2, \tag{11.11}$$

where N is the size of the DFT and $k = 0, 1, \ldots, N-1$.

It is tempting to suppose that if L is very large, we will obtain a result that is representative of the long-term spectrum characteristics of speech. However, an example will demonstrate that this is not the case. The periodogram estimate in Figure 11.10 is based on the computation of Eq. (11.11) with $L = 88{,}200$ (4s at a sampling rate of 22,050 samples/sec) and $N = 131{,}072$. The periodogram of a long segment of speech hints at something that we have seen in Figures 11.8 and 11.9, with the spectrum peak in the range of about 500 Hz and a fall-off in spectral level of about 40 dB between the low frequency spectral peak and the spectrum level at 10 kHz. However, it is also evident that the periodogram varies wildly across frequency, thus rendering it virtually useless for modeling the general long-term spectral shape for speech. Increasing the length L of the segment only leads to more variation [270].

The key to using the periodogram for spectrum estimation is to note that the variability can be reduced by averaging short periodograms computed at intervals over a long segment of the speech signal [416]. This sort of long-term average power spectrum estimate has an interesting interpretation in terms of the short-time Fourier transform as given in sampled form in Eq. (7.41), which is repeated here for an L-point window as

$$X_r[k] = X_{rR}(e^{j(2\pi k/N)}) = \sum_{m=rR}^{rR+L-1} x[m]w[rR-m]e^{-j(2\pi k/N)m}, \tag{11.12}$$

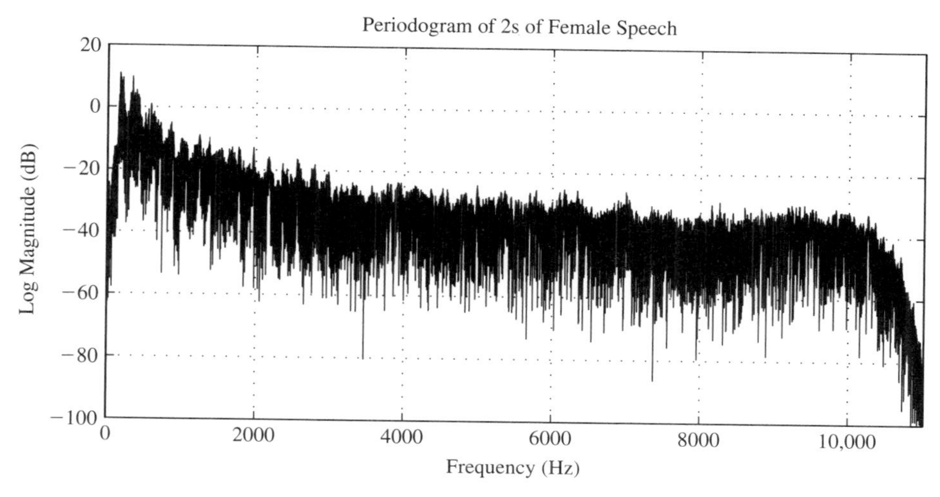

FIGURE 11.10
Example of a periodogram obtained from an 88,400 sample (4s of speech at a sampling rate of 22,050 samples/sec) section of speech from a female speaker.

where $k = 0, 1, \ldots, N-1$. Recall that the spectrogram is simply an image plot of the two-dimensional function $B[k,r] = 10\log_{10}|X_r[k]|^2$. That is, except for lacking the constant $1/(LU)$, the spectrogram is simply a plot of the logarithm of a sequence of periodograms. Figure 11.11 shows wideband spectrograms ($L = 100$ equivalent to 4.54 msec with $F_s = 22{,}050$ Hz) for male and female speakers. Suppose that we have a long sequence of speech of length N_s samples. The average periodogram is obtained as in

$$\tilde{\Phi}(e^{j(2\pi/N)k}) = \frac{1}{KLU}\sum_{r=0}^{K-1} |X_{rR}(e^{j(2\pi k/N)})|^2, \quad k = 0, 1, \ldots, N-1, \tag{11.13}$$

where K is the number of windowed segments contained in the sequence of length N_s. Welch [416] showed that the best compromise between independence of the individual periodogram estimates and overall reduction of variance of the spectrum estimate occurs when $R = L/2$. This value of R was used in forming the image in Figure 11.11 where, although overlap of 90% or more is usually used for most effective viewing as an image, it can be seen that 50% overlap yields a spectrogram that very effectively shows the time-varying spectral properties of speech. The average periodogram is seen from Eq. (11.13) to be simply the average over time with each frequency $(2\pi k/N)$ held fixed; i.e., along horizontal lines in the image such as the line at 4000 Hz. Figure 11.12 shows the resulting long-term spectrum estimates for the male and female speech shown in Figure 11.11. Figure 11.10 is the DFT of the entire 4s utterance for the female speaker. The average spectrum estimate in Figure 11.12 is much smoother, clearly quantifying the trend that is only suggested in Figure 11.10. The curves in Figure 11.12 illuminate several features of the speech signals that were used in the measurements. For example, the region from 10 to 11.025 kHz is the transition band of a digital filter that was used

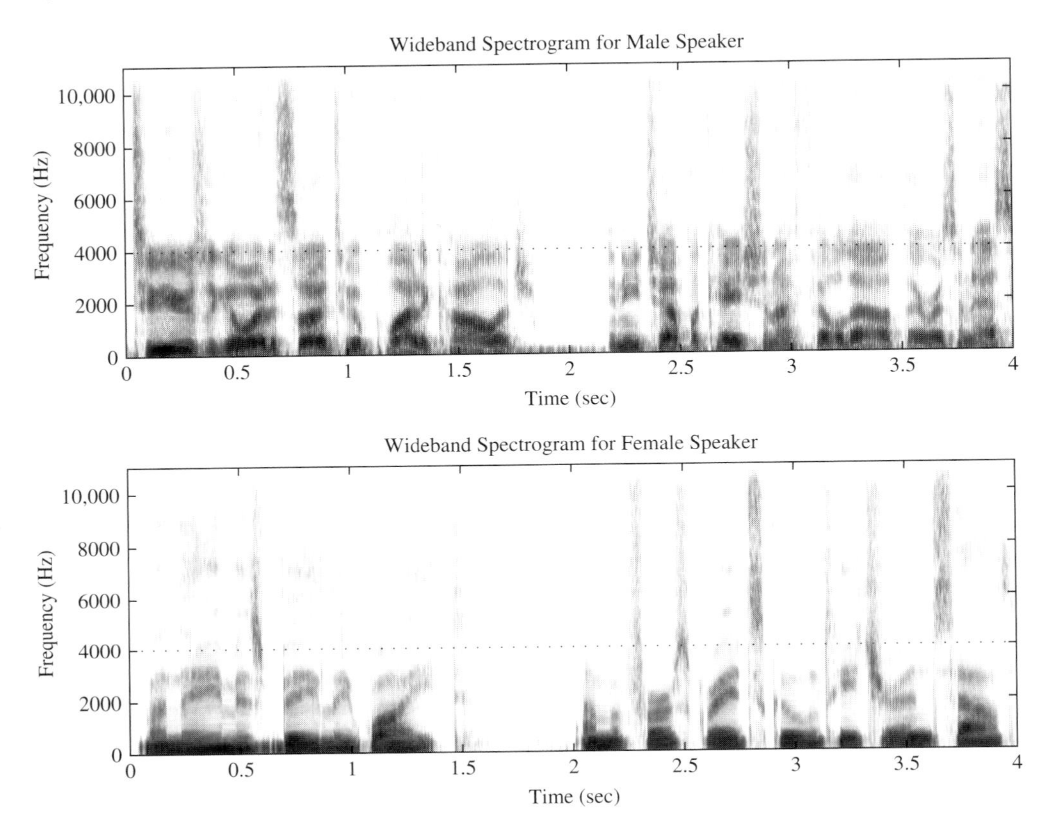

FIGURE 11.11
Spectrograms for male and female speech using a window length of 100 samples (4.54 msec) with 50% overlap.

prior to downsampling the signal from an original sampling rate of 44.1 kHz. While the spectral shapes for the male and female speakers are quite different, we could find similar differences between two speakers of the same sex. The details of the spectrum shapes are generally reflective of speaker characteristics, but it is difficult to attribute particular peaks or valleys to particular characteristics of the speaker. Furthermore, as illustrated by the high-frequency roll-off, the overall spectral shape can be greatly affected by the frequency response of microphones, filters, and electronic systems that precede the sampling of the signal.

11.4 INSTANTANEOUS QUANTIZATION

As we have already pointed out, it is convenient to consider the processes of sampling and quantization separately, with the effects of sampling being understood through the sampling theorem. We now turn to a consideration of quantization of the samples of a speech signal. We begin by assuming that a speech waveform has been lowpass filtered and sampled at a suitable rate, giving a sequence whose sample values are denoted $x[n]$.

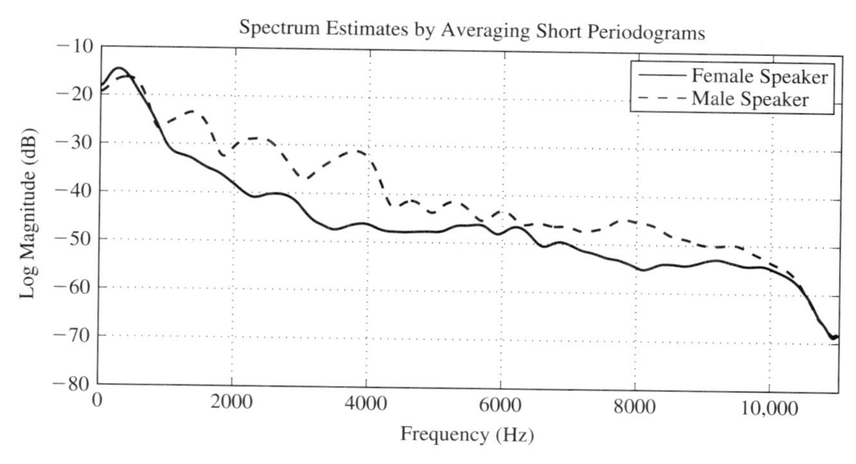

FIGURE 11.12
Example of the power density spectrum obtained from averaging periodograms (plotted as a spectrogram in Figure 11.11) over the entire 4 sec duration. Long-term average log magnitude spectra are given for both a male and a female speaker.

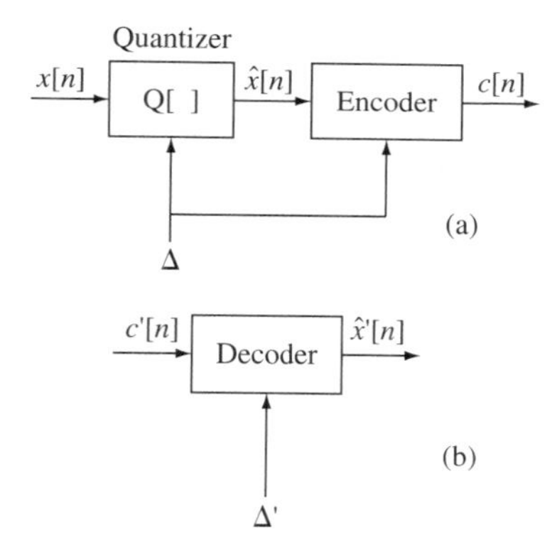

FIGURE 11.13
Process of quantization and coding: (a) encoder; (b) decoder.

Furthermore, we assume that the samples are known with infinite precision; i.e., they are real numbers. In much of our discussion in this chapter, it will be convenient to view this sequence of samples as a discrete-time random process. In order to transmit the sequence of samples over a digital communication channel, or to store them in digital memory or to use them as the input to a digital signal processing algorithm, the sample values must be quantized to a finite set of amplitudes so that they can be represented by a finite set of dimensionless symbols (binary numbers). This process of quantization and coding is depicted in Figure 11.13. Just as it is conceptually useful to separate sampling and quantization into two distinct steps, it is likewise useful to separate the process of representing the samples, $x[n]$, by a finite set of symbols, $c[n]$, into two stages: a quantization stage that produces a sequence of quantized amplitudes $\hat{x}[n] = Q[x[n]]$, and an encoding stage that represents each quantized sample by a codeword, $c[n]$. This is depicted in Figure 11.13a. (The quantity Δ in Figure 11.13a represents the quantization step size for the quantizer.) Likewise, it is convenient to define a decoder that takes a sequence of codewords, $c'[n]$, and transforms it back into a sequence of quantized samples, $\hat{x}'[n]$, as depicted in Figure 11.13b.[3] If the codewords, $c'[n]$, are the same as the codewords, $c[n]$, i.e., no errors have been introduced (by a transmission channel, for example), then the output of the decoder is identical to the quantized samples; i.e., $\hat{x}'[n] = \hat{x}[n]$. In cases where $c'[n] \neq c[n]$, the processes of encoding and quantization have lost information and, as a result, $\hat{x}'[n] \neq \hat{x}[n]$.

In most cases, it is convenient to use binary numbers to represent the quantized samples. With B-bit binary codewords, it is possible to represent 2^B different quantization levels. The information capacity, I, required to transmit or store the digital waveform representation is therefore:

$$I_w = B \cdot F_s = \text{Bit rate in bps}, \tag{11.14}$$

where F_s is the sampling rate (i.e., samples/sec) and B is the number of bits/sample. Thus, for a sampling rate of $F_s = 8000$ samples/sec and $B = 8$ bits/sample as is common in digital coding for telephony, $I = 64{,}000$ bps. On the other hand, the standard values for digital audio waveform coding are $F_s = 44{,}100$ samples/sec and $B = 16$ bits/sample, which gives a total bit rate of $I = 705{,}600$ bps.[4]

It is generally desirable to keep the bit rate as low as possible while maintaining a required level of perceived quality of reproduction of the speech signal. For a given speech bandwidth, F_N, the minimum sampling rate is fixed by the sampling theorem (i.e., $F_s = 2F_N$). Therefore the only way to reduce the bit rate is to reduce the number of bits/sample. For this reason we now turn to a discussion of a variety of techniques for efficiently quantizing a signal at a range of bit rates.

Electronic systems such as amplifiers and A-to-D converters have a finite dynamic range over which they operate linearly. We will denote the maximum allowable amplitude for such a system as $X_{\max}$. In using such systems with speech inputs, it is important to ensure that the signal samples $x[n]$ satisfy the constraint

$$|x[n]| \leq X_p \leq X_{\max}, \tag{11.15}$$

where $2X_p$ is the peak-to-peak range of input signal amplitudes and $X_{\max}$ is the peak-to-peak operating range of the system. For analysis purposes, it may be convenient to assume that both X_p and $X_{\max}$ are infinite, and thereby neglect the effects of clipping.

[3]Sometimes the decoding operation is called "requantization" because the quantized sequence is recovered from the codeword sequence.

[4]The total bit rate for digital audio is usually 1,411,200 bps since a 2-channel (left–right) audio characterization is used for stereo sounds.

We do this, for example, when we assume a particular form for the probability density function of amplitudes of $x[n]$ such as the gamma or Laplacian distribution, both of which have infinite ranges. As a compromise between theoretical models and reality, it is often convenient to assume that the peak-to-peak range of the speech signal values, $2X_p$, is proportional to the standard deviation of the signal, σ_x. For example, if we assume a Laplacian density, it is easy to show (see Problem 11.2) that only 0.35% of the speech samples fall outside the range

$$-4\sigma_x \leq x[n] \leq 4\sigma_x, \tag{11.16}$$

and fewer than 0.002% fall outside the range $\pm 8\sigma_x$.

The amplitudes of the signal samples are quantized by dividing the entire amplitude range into a finite set of amplitude ranges, and assigning the same (quantized) amplitude value to all samples falling in a given range. This quantization operation is shown in Figure 11.14 for an 8-level (3-bit) quantizer, and the sets of input ranges, output levels, and assigned 3-bit codewords for this particular quantizer are shown in Table 11.1. The boundaries of the input ranges are called *decision levels*, and the output levels are called *reconstruction levels*.

For example, we see from Table 11.1 that for all values of $x[n]$ between decision levels x_1 and x_2, the output of the quantizer is set at the reconstruction level $\hat{x}[n] = Q[x[n]] = \hat{x}_2$. The two "saturation regions" of the coder are denoted in Figure 11.14. For these saturation regions, the high and low ranges of these quantization levels must account for signals that exceed the expected maximum value, $X_{\max}$, and hence the maximum is set to ∞ for the fourth positive region, and the minimum is set to $-\infty$ for the fourth negative region. For such cases, the potential quantization error is essentially

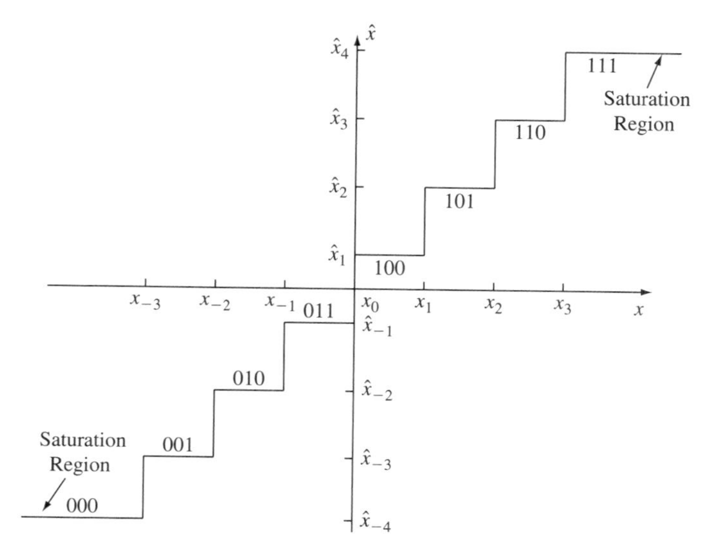

FIGURE 11.14
Input–output characteristic of a 3-bit quantizer.

TABLE 11.1 Quantization ranges, output levels, and selected codewords for a $B = 3$ bit (8-level) quantization method.

Input Range	Output Level	Codeword
$0 = x_0 < x[n] \leq x_1$	$\hat{x}_1$	100
$x_1 < x[n] \leq x_2$	$\hat{x}_2$	101
$x_2 < x[n] \leq x_3$	$\hat{x}_3$	110
$x_3 < x[n] < \infty$	$\hat{x}_4$	111
$x_{-1} < x[n] \leq x_0 = 0$	$\hat{x}_{-1}$	011
$x_{-2} < x[n] \leq x_{-1}$	$\hat{x}_{-2}$	010
$x_{-3} < x[n] \leq x_{-2}$	$\hat{x}_{-3}$	001
$-\infty < x[n] \leq x_{-3}$	$\hat{x}_{-4}$	000

unbounded (at least in theory), and we must pay attention to the potential effects of such saturation regions on overall quantizer performance.

Each of the eight quantizer levels in Figure 11.14 and Table 11.1 is labeled with a 3-bit binary codeword that serves as a symbolic representation of that amplitude level. For example, in Figure 11.14, the coded output for a sample whose amplitude is between x_1 and x_2 would be the binary number 101. The particular labeling scheme in Figure 11.14 is arbitrary. Any of the eight factorial possible labeling schemes is a possibility; however, there are often good reasons for the choice of a particular scheme.

11.4.1 Uniform Quantization Noise Analysis

The quantization ranges and levels may be chosen in a variety of ways depending on the intended application of the digital representation. When the digital representation is to be processed by a digital signal processing system, the quantization levels and ranges are generally distributed uniformly. To define a uniform quantizer using the example of Figure 11.14, we set

$$x_i - x_{i-1} = \Delta \tag{11.17}$$

and

$$\hat{x}_i - \hat{x}_{i-1} = \Delta, \tag{11.18}$$

where Δ is the quantization step size. Two common uniform quantizer characteristics are shown in Figure 11.15 for the case of eight quantization levels. Figure 11.15a shows the case where the origin is in the middle of a rising part of the staircase-like function. This class of quantizers is called the "mid-riser" class. Likewise Figure 11.15b shows an example of the "mid-tread" class of quantizers where the origin is in the middle of a flat part of the staircase-like function. For the case where the number

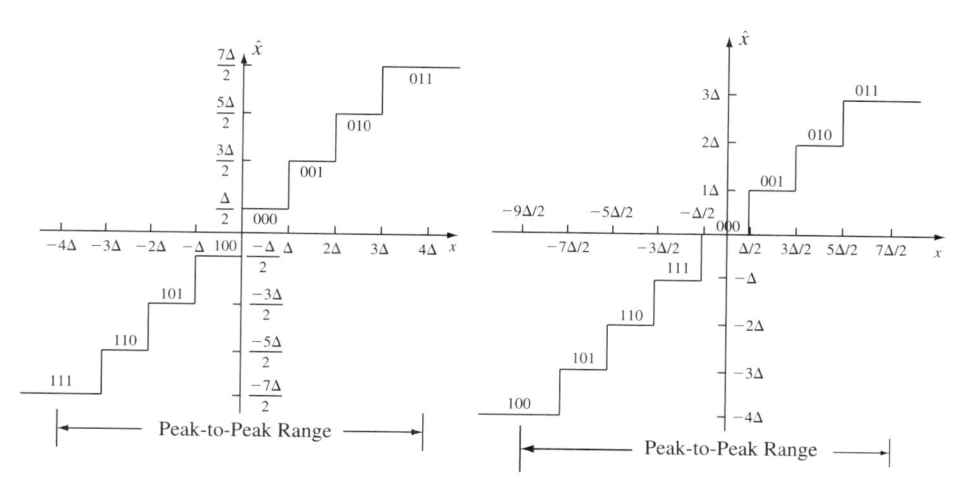

FIGURE 11.15
Two common uniform quantizer characteristics: (a) mid-riser; (b) mid-tread.

of levels is a power of 2, as is convenient for a binary coding scheme, the mid-riser quantizer in Figure 11.15a has the same number of positive and negative levels, and these are symmetrically positioned about the origin. In contrast, the mid-tread quantizer in Figure 11.15b has one more negative level than positive; however in this case, one of the quantization levels is zero while there is no zero level in the mid-riser case. Codeword assignments are shown in Figure 11.15 in the manner of Figure 11.14. In this case, the codewords have been assigned so as to have a direct numerical significance. For example, in Figure 11.15a, if we interpret the binary codewords as a sign-magnitude representation, with the left most bit being the sign bit, then the quantized samples are related to the codewords by the relationship

$$\hat{x}[n] = \frac{\Delta}{2}\text{sign}(c[n]) + \Delta c[n], \tag{11.19}$$

where $\text{sign}(c[n])$ is equal to $+1$ if the first bit of $c[n]$ is 0, and -1 if the first bit of $c[n]$ is 1. Similarly, we can interpret the binary codewords in Figure 11.15b as a 3-bit two's complement representation, in which case the quantized samples are related to the codewords by the relationship

$$\hat{x}[n] = \Delta\, c[n]. \tag{11.20}$$

This latter method of assignment of codewords to quantization levels is most commonly used when the sequence of samples is to be processed by a signal processing algorithm that is implemented with two's complement arithmetic (as on most computers and DSP chips), since the codewords can serve as a direct numerical representation of the sample values.

For uniform quantizers (as shown in Figure 11.15) there are only two parameters, namely the number of levels and the quantization step size, Δ. The number of levels is generally chosen to be of the form 2^B so as to make the most efficient use of B-bit binary codewords. Together, Δ and B must be chosen so as to cover the range of input samples. If we assume that the peak-to-peak range of the quantizer is $2X_{\max}$, then (assuming a symmetrical probability density function for $x[n]$) we should set

$$2X_{\max} = \Delta\, 2^B. \tag{11.21}$$

In discussing the effect of quantization, it is helpful to represent the quantized samples $\hat{x}[n]$ as

$$\hat{x}[n] = x[n] + e[n], \tag{11.22}$$

where $x[n]$ is the unquantized sample and $e[n]$ is the quantization error. With this definition, it is natural to think of $e[n]$ as additive noise. It can be seen from both Figures 11.15a and 11.15b that if Δ and B are chosen as in Eq. (11.21), then

$$-\frac{\Delta}{2} \leq e[n] < \frac{\Delta}{2}. \tag{11.23}$$

For example, if we choose the peak-to-peak range of $x[n]$ to be $8\sigma_x$ and if we assume a Laplacian probability density function (as discussed in Section 11.3), then only 0.35% of the samples will fall outside the range of the quantizer. The clipped samples will incur a quantization error in excess of $\pm\Delta/2$; however, their number is so small that it is common to assume a range on the order of $8\sigma_x$ and neglect the infrequent large errors in theoretical calculations [169].

The results of an experiment on quantization of a simple signal are given in Figure 11.16. Figure 11.16a shows the unquantized (computed with 64-bit floating point arithmetic) sampled signal $x[n] = \sin(0.1n) + 0.3\cos(0.3n)$.[5] Figure 11.16b shows the digital waveform using a 3-bit quantizer; part (c) shows the 3-bit quantization error, $e[n] = \hat{x}[n] - x[n]$, which can be computed exactly since both $x[n]$ and $\hat{x}[n]$ are known. Finally part (d) shows the quantization error for an 8-bit quantizer. The peak-to-peak range of the quantizer in both cases was set at ± 1. The dashed lines in Figure 11.16a are the decision levels for the 3-bit quantizer; i.e., $\pm\Delta/2$, $\pm 3\Delta/2$, $\pm 5\Delta/2$, and $-7\Delta/2$ as shown in Figure 11.15b. The sample values in Figure 11.16b are the reconstruction levels for the 3-bit quantizer; i.e., 0, $\pm\Delta$, $\pm 2\Delta$, $\pm 3\Delta$, and -4Δ. In Figures 11.15c and 11.15d, the dashed lines denote the range $-\Delta/2 \leq e[n] < \Delta/2$ for the 3-bit and 8-bit quantizers respectively; i.e., $\pm\Delta/2 = \pm 1/8$ and $\pm\Delta/2 = \pm 1/256$. Note that in the case of 3-bit quantization, the error signal in Figures 11.15c displays long intervals where the error looks like the negative of the unquantized signal. Also, in regions corresponding to the positive peaks, the signal is clipped and the error exceeds the

[5]The frequencies were chosen to ensure that $x[n]$ was not periodic since this special case would produce a periodic error signal.

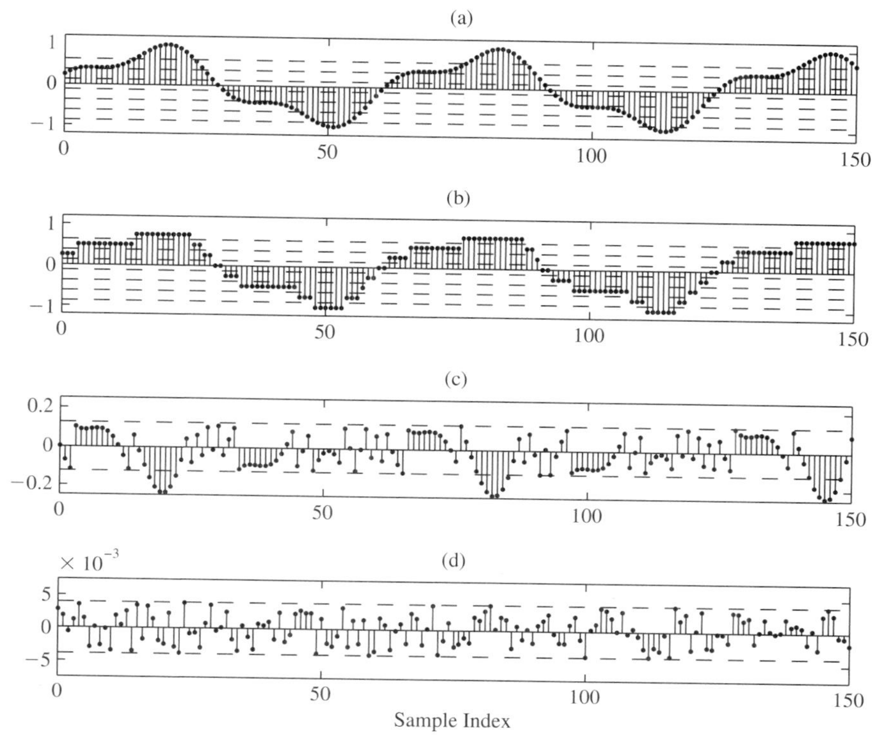

FIGURE 11.16
Waveform plots of the results of quantization of the signal $x[n] = \sin(0.1n) + .3\cos(0.3n)$ using both a 3-bit and an 8-bit quantizer: (a) the unquantized signal $x[n]$; (b) the result of 3-bit quantization, $\hat{x}[n]$; (c) the quantization error $e[n] = \hat{x}[n] - x[n]$ for 3-bit quantization; and (d) the quantization error for 8-bit quantization. [Note that the vertical scales in (c) and (d) are different from the scales of (a) and(b).]

bounds $\pm\Delta/2$. In other words, the error $e[n]$ and the signal $x[n]$ are obviously correlated in these regions. On the other hand, the 8-bit quantization error appears to vary randomly within the bounds $\pm\Delta/2$, and visually there is no obvious correlation with $x[n]$. While it is clear that $x[n]$ and $e[n]$ are certainly not independent, because quantization is a deterministic operation, we shall see that a model for quantization error that assumes that they are uncorrelated yields accurate and useful results and insights.

In a practical speech processing application, we know neither $x[n]$ nor $e[n]$, but only the quantized value $\hat{x}[n]$. However, a quantitative description of the effect of quantization can be obtained using random signal analysis applied to the quantization error. If the number of bits in the quantizer is reasonably high and no clipping occurs, the quantization error sequence, although it is completely determined by the signal amplitudes, nevertheless behaves in the same way as a random signal with the following properties [33, 138, 419, 420]:

1. The noise samples appear to be[6] uncorrelated with the signal samples; i.e.,

$$E[x[n]\, e[n+m]] = 0, \quad \text{for } all\ m. \tag{11.24}$$

2. Under certain assumptions, notably smooth input probability density functions and high rate quantizers, the noise samples appear to be uncorrelated from sample to sample; i.e., $e[n]$ acts like a white noise sequence with autocorrelation:

$$\phi_e[m] = E[e[n]\, e[n+m]] = \sigma_e^2 \delta[n] = \begin{cases} \sigma_e^2, & m = 0 \\ 0 & \text{otherwise.} \end{cases} \tag{11.25}$$

3. The amplitudes of the noise samples are uniformly distributed across the range $-\Delta/2 \leq e[n] < \Delta/2$, resulting in zero noise mean ($\overline{e} = 0$) and an average noise power of $\sigma_e^2 = \Delta^2/12$.

These simplifying assumptions allow a linear analysis that yields accurate results if the signal is not too coarsely quantized. Under these conditions, it can be shown that if the output levels of the quantizer are optimized, then the quantizer error will be uncorrelated with the quantizer output (however, not the quantizer input, as commonly stated). These results can easily be shown to hold in the simple case of a uniformly distributed memoryless input, and Bennett has shown how the result can be extended to inputs with smooth densities if the bit rate is assumed high [33].

We can verify experimentally the assumption that the quantization error samples are approximately uniformly distributed by measuring histograms of the quantization noise. The histogram of a sequence is a plot of the number of samples that lie in a set of "bins" that divide the amplitude range into equal increments. Therefore, if the signal samples are truly from a uniform distribution, each bin should contain approximately the same number of samples; that number would be the total number of samples divided by the number of bins.[7] If the histogram is based on a finite number of samples, the actual measured values will vary since they are random variables too. However, as the number of samples grows large, the variance of the estimate decreases.

Figure 11.17a shows a 51-bin histogram of 173,056 samples of the quantization noise for a speech input from a male speaker. The box drawn with a thin line shows the number of samples to be expected in each bin ($173{,}056/51 \approx 3393$) if the noise samples were in fact uniformly distributed over $-1/2^3 \leq e_3[n] < 1/2^3$. Clearly the assumption of uniform distribution is not valid for 3-bit quantization. In fact, the distribution is similar, but not identical, to the distribution of the speech samples, which have a

[6] By "appear to be" we mean that measured correlations are small. Bennett has shown that the correlations are only small because the error is small and, under suitable conditions, the correlations are equal to the negative of the error variance [33, 138]. The condition "uncorrelated" often implies independence, but in this case, the error is a deterministic function of the input and hence it cannot be independent of the input.

[7] If the distribution is not uniform, then the number of samples in a bin would be proportional to the area of the probability density over the bin interval.

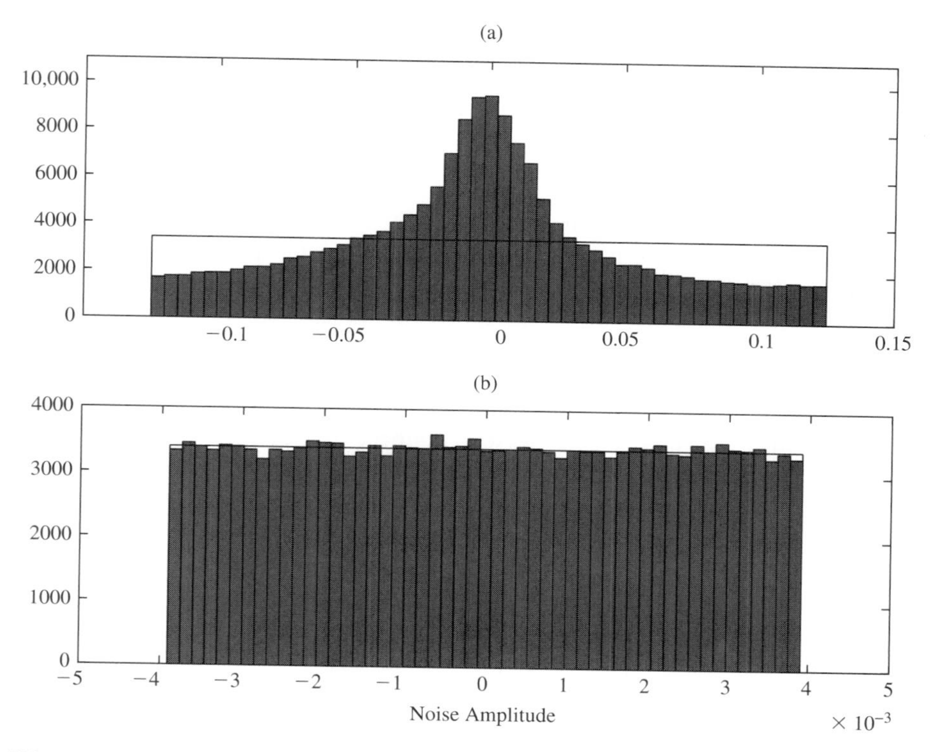

FIGURE 11.17
Fifty-one-bin histograms of 173,056 samples of quantization noise for uniform quantizers: (a) 3-bit; (b) 8-bit.

Laplacian-like distribution. However, the histogram for 8-bit quantization shown in Figure 11.17b is much more consistent with the assumption of uniform distribution over the range $-1/2^8 \leq e_8[n] < 1/2^8$. Measurements for other values of B confirm that a reasonably uniform distribution is obtained for as low as $B = 6$, and the uniform distribution assumption gets increasingly better as B increases.

The other major assumption of our approximate analysis of quantization noise is that its autocorrelation satisfies Eq. (11.25) and therefore its power spectrum is

$$\Phi_e(e^{j\omega}) = \sigma_e^2, \quad \text{for all } \omega, \tag{11.26}$$

where σ_e^2 is the average power of the white noise.[8] If the peak-to-peak quantizer range is assumed to be $2X_{\max}$, then, for a B-bit quantizer, we get

$$\Delta = \frac{2X_{\max}}{2^B}. \tag{11.27}$$

[8]Observe that $\frac{1}{2\pi}\int_{-\pi}^{\pi} \sigma_e^2 d\omega = \sigma_e^2$.

If we assume a uniform amplitude distribution for the noise, [i.e., $p(e) = 1/\Delta$ over the range $-\Delta/2 \leq e < \Delta/2$], we obtain (see Problem 11.1)

$$\sigma_e^2 = \frac{\Delta^2}{12} = \frac{X_{\max}^2}{(3)2^{2B}}. \tag{11.28}$$

To confirm the assumption that the quantization noise sequence behaves like white noise, we can use the techniques of power spectrum estimation discussed in Section 11.3. Figure 11.18 shows the long-term power spectrum of the speech of a male speaker (same as in Figure 11.12). The speech signal was originally quantized with 16-bit precision and sampling rate $F_s = 22{,}050$ Hz. This signal was further quantized using different quantizers and the quantization error computed as the difference between the quantized signal and the original. Also shown are spectrum estimates for the measured quantization noise sequences. It can be seen that the measured spectrum is quite flat for 10-bit quantization noise. Similarly, the spectrum is very flat for 8-bit and 6-bit quantization noise with only a slight coloration at very low frequencies.

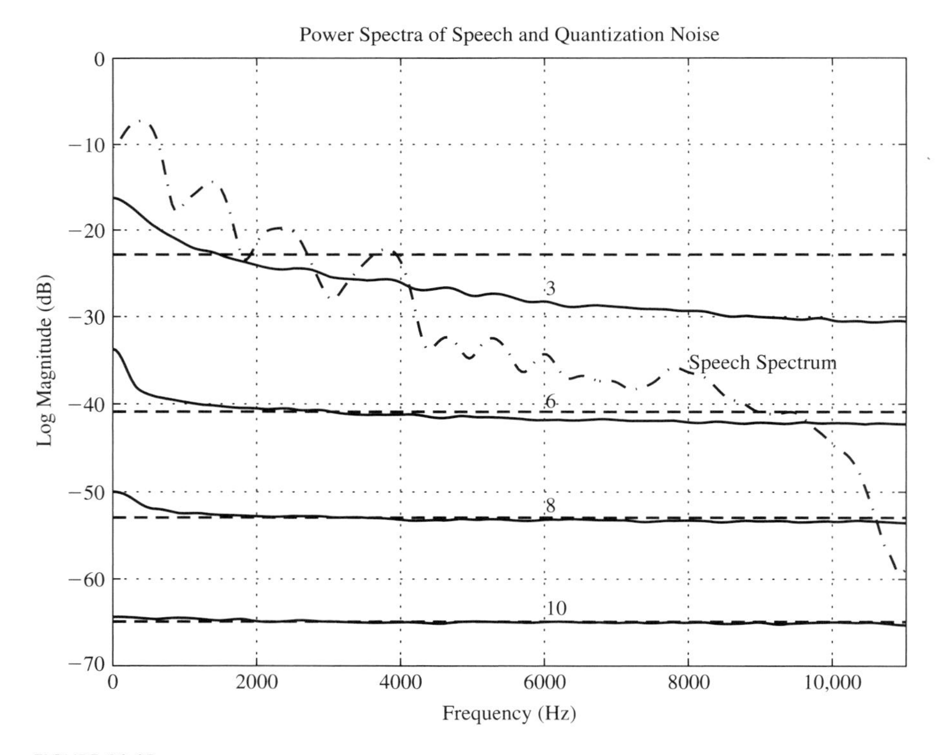

FIGURE 11.18
Spectra of quantization noise for uniform quantizers with 10-, 8-, 6-, and 3-bits/sample (solid lines), along with the power spectrum of the speech signal being quantized (dash-dot line). Dashed lines are theoretical power spectra for white noise, with average power corresponding to that of 10-, 8-, 6-, and 3-bit quantization noise.

On the other hand, the 3-bit quantization noise tends toward the shape of the speech spectrum. Using Eq. (11.28) with $X_{\max} = 1$, the white noise model predicts that the spectrum of the noise (in dB) should be

$$10 \log_{10} \sigma_e^2 = -6B - 4.77, \quad \text{for all } \omega. \tag{11.29}$$

The dashed lines in Figure 11.18 show this function for the values $B = 10, 8, 6, 3$ corresponding to the measured spectra. The model fits very well for all except for the 3-bit case. Note also, as predicted by Eq. (11.29), the measured noise spectra for $B = 10, 8, 6$ differ by approximately 12 dB for all ω.

Armed with confidence in the flat-spectrum model for the quantization noise, it is possible to relate the variance of the noise to the signal variance and the parameters of the quantizer. For this purpose, it is convenient to compute the signal-to-quantization noise ratio, defined as[9]

$$\text{SNR}_Q = \frac{\sigma_x^2}{\sigma_e^2} = \frac{E[x^2[n]]}{E[e^2[n]]} \approx \left(\frac{\sum_n x^2[n]}{\sum_n e^2[n]} \right), \tag{11.30}$$

where σ_x^2 and σ_e^2 are the average powers of the signal and quantization noise respectively. (The quantities σ_x and σ_e are the rms values of the input signal and quantization noise.) Substituting Eq. (11.28) into Eq. (11.30) gives

$$\text{SNR}_Q = \frac{\sigma_x^2}{\sigma_e^2} = \frac{(3)2^{2B}}{\left(\dfrac{X_{\max}}{\sigma_x} \right)^2}, \tag{11.31}$$

or, expressing the signal-to-quantizing noise ratio in dB units, we get

$$\begin{aligned} \text{SNR}_Q(\text{dB}) &= 10 \log_{10} \left[\frac{\sigma_x^2}{\sigma_e^2} \right] \\ &= 6B + 4.77 - 20 \log_{10} \left[\frac{X_{\max}}{\sigma_x} \right]. \end{aligned} \tag{11.32}$$

In fixed quantizers, $X_{\max}$ is a fixed parameter of the quantizer; however, in most speech processing applications, the signal level, represented by σ_x, can vary dramatically with speaker and signal acquisition conditions. Equation (11.32) provides a convenient representation of the fundamental trade-off that arises in quantization of any signal. The quantization noise power is proportional to the step size squared, which

[9]Note we are assuming that $x[n]$ has zero mean. If this is not the case, the mean value of $x[n]$ should be subtracted from the signal prior to signal-to-noise ratio calculations.

is inversely proportional to 2^B. The size of the noise is therefore independent of the input signal. If the signal power goes down because a talker has a weak voice or because an amplifier gain is improperly set, the signal-to-quantization noise ratio decreases. This effect is summarized by the term

$$-20\log_{10}\left[\frac{X_{\max}}{\sigma_x}\right]$$

in Eq. (11.32). The other term in Eq. (11.32) is $6B + 4.77$, which causes SNR_Q to change by 6 dB for a change of 1 in B. This results in the commonly referred to "6 dB per bit" rule.

The speech signal used in the experiment depicted in Figure 11.18 was again used in an experiment to confirm the validity of the result in Eq. (11.32). Fixing the quantizer scale such that $X_{\max} = 1$, the original 16-bit signal, whose maximum value was 1, was multiplied by scaling constants that caused the ratio $X_{\max}/\sigma_x$ to vary over a wide range.[10] The scaled signals were quantized with 14-, 12-, 10-, and 8-bit uniform quantizers and the signal-to-quantization noise ratio was computed for each value of B and each value of $X_{\max}/\sigma_x$. The results are shown in Figure 11.19 as the solid curves. The dashed curves in Figure 11.19 are plots of Eq. (11.32) for the same values of B. There is excellent agreement between the dashed curves in Figure 11.19 and the measurements over a range of $X_{\max}/\sigma_x$ from about 8 to over 100. Note also that the curves are vertically offset from one another by the 12 dB predicted by the 2-bit differences between the quantizers. Also note that the curves verify that with B fixed, if $X_{\max}/\sigma_x$ changes by a factor of 2, SNR_Q(dB) changes by 6 dB. However, for the test signal, when $X_{\max}/\sigma_x$ is below about 8, more and more of the signal falls outside the quantizer range (the region of saturation) and the curves of SNR_Q fall precipitously. The deviation from the model for small values of $X_{\max}/\sigma_x$ is significant. Figure 11.20 shows a plot of the percentage of clipped samples as a function of $X_{\max}/\sigma_x$ for the 8-bit quantizer in Figure 11.19. Note that no clipping occurred for the tested speech waveform for values of $X_{\max}/\sigma_x$ greater than about 8.5. As $X_{\max}/\sigma_x$ decreases (σ_x increases), the percentage of clipped samples increases. This increase corresponds to the rapid decrease in signal-to-noise ratio (SNR) as seen in Figure 11.19. Note that Figure 11.20 shows that only a small percentage of clipped samples is sufficient to decrease the signal-to-quantization noise ratio dramatically.

Since the maximum signal value X_p must always be greater than the rms value σ_x, it is necessary to adjust the size of the signal to avoid excessive clipping. To quantify this adjustment, it is helpful to assume that $X_{\max} = C\sigma_x$, where C will depend on the nature of the distribution of signal samples. For example, if $x[n]$ is from a sampled sine wave, we know that $\sigma_x = X_p/\sqrt{2}$, so it follows that if $C = \sqrt{2}$, then $X_{\max} = X_p$ and no clipping will occur. Therefore, for sinusoidal signals, the curves

[10] σ_x was measured from the scaled signal as $\sigma_x = \left(\frac{1}{N}\sum_{n=0}^{N-1}(x[n])^2\right)^{1/2}$, where $N = 173{,}056$ samples.

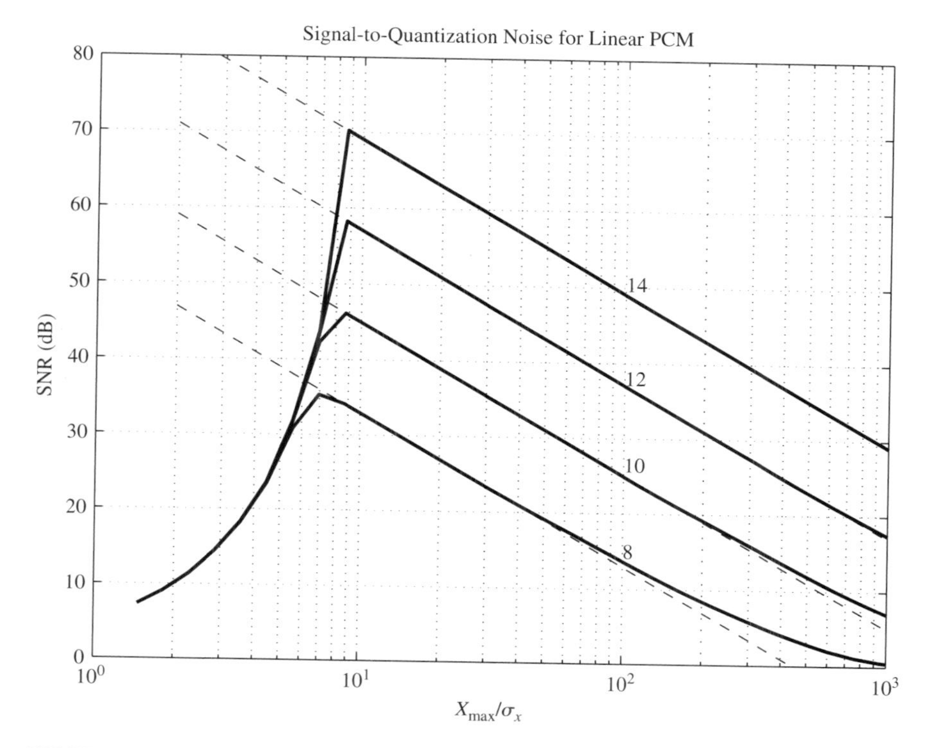

FIGURE 11.19
Solid lines show variation of SNR_Q(dB) as a function of X_{max}/σ_x for a uniform quantizer with $B = 14, 12, 10$, and 8 bits/sample. Dashed curves are plots of Eq. (11.32). Quantizer saturation is in evidence for values of $X_{max}/\sigma_x < 9$.

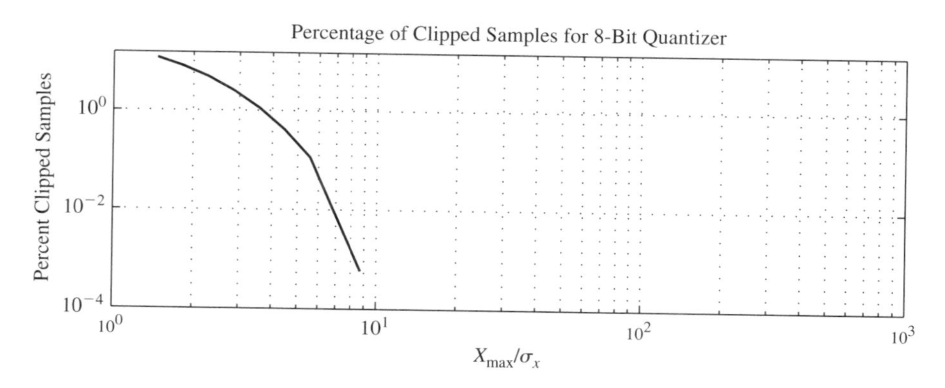

FIGURE 11.20
The percentage of clipped samples as a function of X_{max}/σ_x for 8-bit quantization of a speech signal.

of Figure 11.19 would continue to increase linearly with decreasing $X_{\max}/\sigma_x$ until $X_{\max}/\sigma_x = \sqrt{2}$, in which case $X_p = X_{\max}$; i.e., when the peak signal value equals the maximum amplitude for which no clipping occurs. Sine waves have a much different distribution of amplitudes from the distribution of speech samples, which is much more concentrated at low amplitudes as indicated by the measurement results shown in Figure 11.19, which imply that clipping begins to occur when $X_{\max}/\sigma_x = C = 8$ or even greater values. Since clipping produces large errors, it does not take many clipped samples to reduce the signal-to-quantizing noise ratio significantly. On the other hand, the perceptual effect may be small if samples are infrequently clipped.

In setting the size of the signal relative to $X_{\max}$ for the quantizer, we must balance the increase in SNR obtained by increasing σ_x against the reduction due to clipping. A common choice is $X_{\max} = 4\sigma_x$ [169]. With this choice, Eq. (11.32) becomes

$$\text{SNR}_Q(\text{dB}) = 6B - 7.2, \tag{11.33}$$

which again shows that each bit in the codeword contributes 6 dB to the SNR. For a 16-bit quantizer, we get $\text{SNR}_Q \approx 89$ dB for this assumption.

The expressions for SNR_Q of Eqs. (11.32) and (11.33) are valid subject to the following assumptions:

1. The input signal fluctuates in a complicated manner so that a statistical model for the noise sequence is valid.
2. The quantization step size is small enough to remove any possibility of signal-correlated patterns in the noise waveform.
3. The range of the quantizer is set so as to match the peak-to-peak range of the signal so that very few samples are clipped, but yet the full range is utilized. Note that Figures 11.19 and 11.20 show that $X_{\max}/\sigma_x = 4$ would not guarantee that clipping would not occur for the test speech signal used to create those figures.

For speech signals, the first two assumptions hold up very well when the number of quantizer levels is reasonably large, say greater than 2^6. However, the third assumption is less valid for speech signals in an application setting, since the signal energy can vary as much as 40 dB among speakers and with the transmission environment. Also, for a given speaking environment, the amplitude of the speech signal varies considerably from voiced speech to unvoiced speech, and even within voiced sounds. Since Eq. (11.33) assumes a given range of amplitudes, if the signal fails to achieve that range, it is as if fewer quantization levels are available to represent the signal; i.e., as if fewer bits were used. For example, it is evident from Eq. (11.32) that if the input actually varies only one half the range $\pm X_{\max}$, the SNR is reduced by 6 dB. Likewise, on a short-time basis, the rms value of an unvoiced segment can be 20 to 30 dB below the rms value for voiced speech. Thus, the "short-time" SNR can be much less during unvoiced segments than during voiced segments.

The implication of Figure 11.19 is that in order to maintain a fidelity of representation with uniform quantization that is acceptable perceptually, it is necessary to use many more bits than might be implied by the previous analysis in which we have assumed that the signal is stationary (i.e., its statistics do not vary over time, or across speakers, or for different transmission conditions, etc.). For example, whereas Eq. (11.33) suggests that $B = 7$ would provide about $\text{SNR}_Q = 36$ dB for a value of $X_{\max}/\sigma_x = 4$, which would most likely provide adequate quality in a communications system, it is generally accepted that more than 11 bits are required to provide a high quality representation of speech signals with a uniform quantizer due to the non-stationarity of the speech signal.

For all of the above reasons, it would be very desirable to have a quantizing system for which the SNR was essentially independent of signal level. That is, rather than the error being of *constant variance*, independent of signal amplitude, as for uniform quantization, it would be desirable to have constant *relative* error. This can be achieved using a non-uniform distribution of quantization levels or by adapting the step size of the quantizer. We investigate non-uniform quantizers in Sections 11.4.2 and 11.4.3 and adaptive quantization in Section 11.5.

11.4.2 Instantaneous Companding (Compression/Expansion)

To achieve constant relative error, the quantization levels for the speech signal must be logarithmically spaced. Equivalently, the logarithm of the input can be uniformly quantized. This representation of logarithmic quantization is depicted in Figure 11.21, which shows the input amplitudes being compressed by the logarithm function prior to quantization and encoding, and being expanded by the exponential function after decoding. The combination of compressing followed by expanding is called *companding*. To see that this processing leads to the desired insensitivity to signal amplitude, assume that

$$y[n] = \log |x[n]|. \tag{11.34}$$

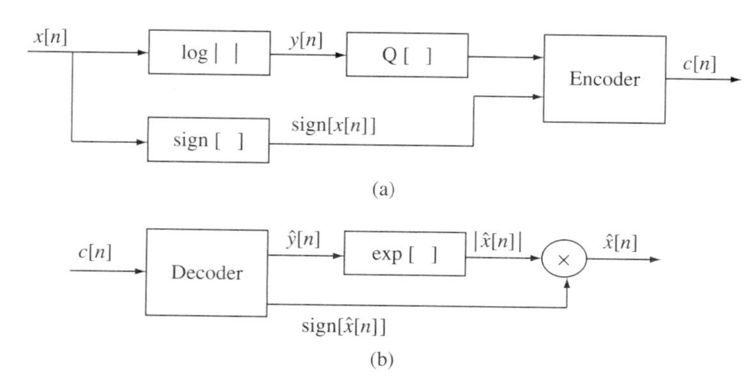

FIGURE 11.21
Block diagram of a logarithmic quantizer: (a) encoder and (b) decoder.

The inverse transformation (expansion) is

$$x[n] = \exp(y[n]) \cdot \text{sign}(x[n]), \tag{11.35}$$

where

$$\text{sign}(x[n]) = \begin{cases} +1 & x[n] \geq 0 \\ -1 & x[n] < 0. \end{cases}$$

The quantized log magnitude can be represented as

$$\begin{aligned} \hat{y}[n] &= Q[\log|x[n]|] \\ &= \log|x[n]| + \epsilon[n], \end{aligned} \tag{11.36}$$

where we assume again that the resulting error sample, $-\Delta/2 \leq \epsilon[n] < \Delta/2$, is white noise that is uncorrelated with the signal sample, $\log|x[n]|$. The inverse of the quantized log magnitude is

$$\begin{aligned} \hat{x}[n] &= \exp(\hat{y}[n])\text{sign}(x[n]) \\ &= |x[n]|\,\text{sign}(x[n])\exp(\epsilon[n]) \\ &= x[n]\exp(\epsilon[n]). \end{aligned} \tag{11.37}$$

If $\epsilon[n]$ is small, we can approximate $\hat{x}[n]$ in Eq. (11.37) by

$$\hat{x}[n] \approx x[n](1+\epsilon[n]) = x[n] + \epsilon[n]\,x[n] = x[n] + f[n], \tag{11.38}$$

where $f[n] = x[n]\epsilon[n]$. Note that $x[n]$ and $f[n]$ are clearly not independent. In fact, this is the desired result; the noise amplitude is scaled by the signal amplitude. However, since $\log|x[n]|$ and $\epsilon[n]$ are assumed to be uncorrelated, $x[n]$ and $\epsilon[n]$ also can be assumed to be uncorrelated, so that

$$\sigma_f^2 = \sigma_x^2 \cdot \sigma_\epsilon^2, \tag{11.39}$$

and therefore,

$$\text{SNR}_Q = \frac{\sigma_x^2}{\sigma_f^2} = \frac{1}{\sigma_\epsilon^2}. \tag{11.40}$$

Thus, the resulting SNR is independent of the signal variance, σ_x^2, and therefore depends only upon the step size of the quantizer of the logarithmically compressed signal. The analysis leading to Eq. (11.40) implicitly assumes quantization with no minimum or maximum limits to cause clipping. If we assume that, to fix the step size, the dynamic range of the quantizer is $\pm X_{\max}$ as before and that no clipping occurs, then σ_ϵ^2 is given by Eq. (11.28) and we can write the signal-to-quantization noise ratio in dB as

$$\text{SNR}_Q(\text{dB}) = 6B + 4.77 - 20\log_{10}[X_{\max}], \tag{11.41}$$

which again shows no dependence on the input signal level.

This type of logarithmic compression and uniform quantization is not really practical since $\log|x[n]| \to -\infty$ when $|x[n]| \to 0$; i.e., the dynamic range (ratio between the largest and the smallest values) of the output of a logarithmic compressor is infinite, and therefore an infinite number of quantization levels would be required. The above analysis, however impractical it may be, suggests that it would be desirable to obtain an approximation to a logarithmic compression characteristic that does not diverge when $|x[n]| \to 0$. A more practical compressor/expander system for pseudo-logarithmic compression is shown in Figure 11.22. This type of system, investigated in detail by Smith [366], is called a μ-law compressor. The μ-law compression function is defined as

$$\begin{aligned} y[n] &= F[x[n]] \\ &= X_{\max}\left(\frac{\log\left[1+\mu\dfrac{|x[n]|}{X_{\max}}\right]}{\log[1+\mu]}\right)\cdot \text{sign}[x[n]]. \end{aligned} \tag{11.42}$$

Figure 11.23 shows a family of μ-law functions for different values of μ. It is clear that using the function of Eq. (11.42) avoids the problem of small input amplitudes since $y[n]=0$ when $|x[n]|=0$. Note that $\mu=0$ corresponds to no compression; i.e.,

$$y[n] = x[n], \qquad \text{when } \mu = 0, \tag{11.43}$$

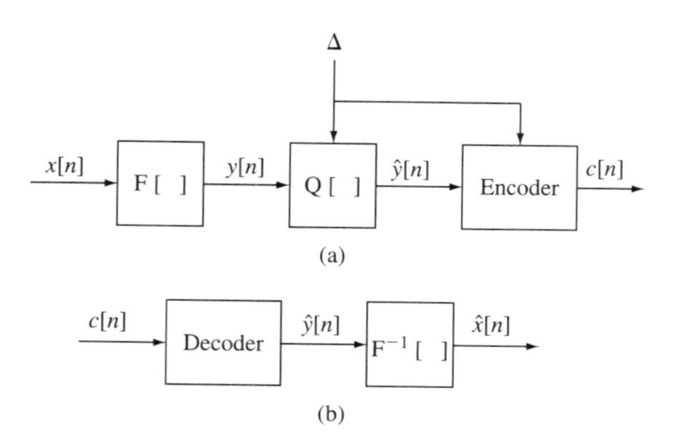

FIGURE 11.22
Block diagram of a compressor/expander system for quantization: (a) the system for compression; (b) the system for expansion.

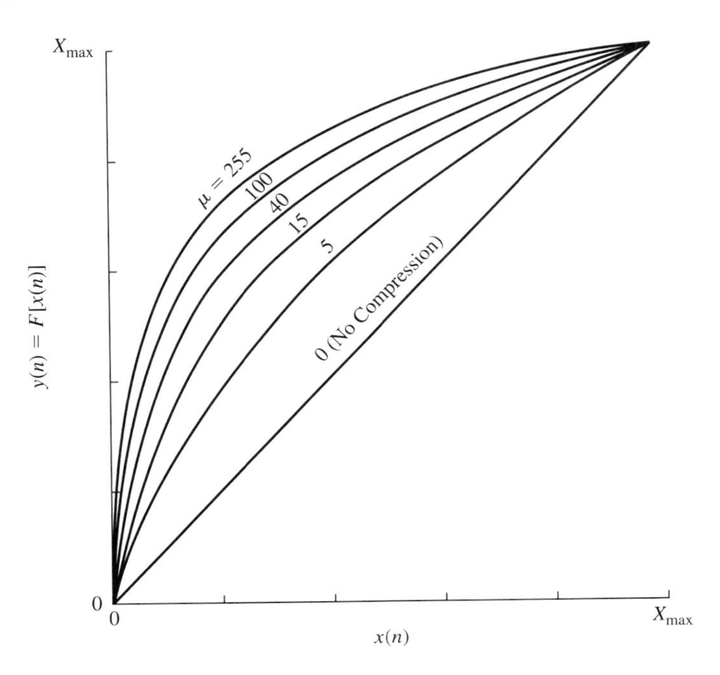

FIGURE 11.23
Input–output relations for μ-law compression characteristics. (After Smith [366]. Reprinted with permission of Alcatel-Lucent USA Inc.)

so the quantization levels remain uniformly spaced for $\mu = 0$. However, for large values of μ, and for large $|x[n]|$, we have

$$|y[n]| \approx X_{max} \frac{\log\left[\mu \dfrac{|x[n]|}{X_{max}}\right]}{\log(\mu)} \qquad \text{for large } \mu. \tag{11.44}$$

That is, μ-law compression for large μ is essentially true logarithmic compression over a large range of amplitudes. Therefore, except for very low amplitudes, the μ-law compression curve gives a very good approximation to the logarithm function required to give constant relative error after compression/quantization/expansion. Figure 11.24 shows the effective set of quantization levels for the case $\mu = 40$ and eight quantization levels. (The quantizer characteristic is anti-symmetric about the origin and a mid-riser quantizer is normally assumed.)

The effect of the μ-law compressor on a speech waveform is illustrated in Figure 11.25, which shows 3000 samples of a speech waveform in (a), along with the corresponding waveform of the μ-law compressed signal ($\mu = 255$) in (b). It is easily seen that the μ-law compressed signal utilizes the full dynamic range ± 1 much more effectively than the original speech signal. Figure 11.25c shows the quantization error $f[n] = x[n]\epsilon[n]$ for the case when the output of the compressor is quantized with an 8-bit quantizer. Note that because the quantization error is introduced before the

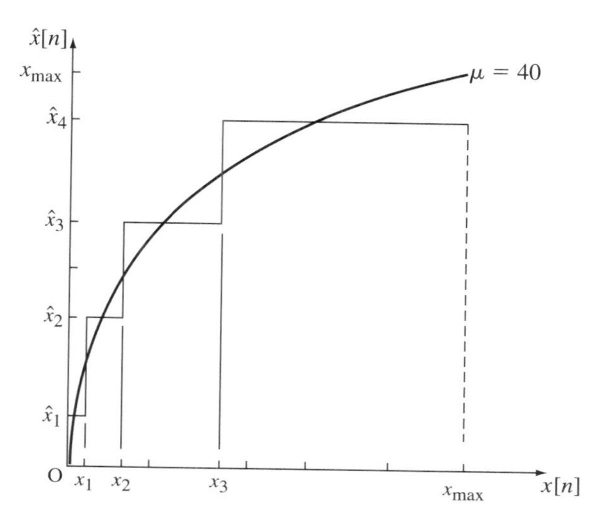

FIGURE 11.24
Distribution of quantization levels for a μ-law 3-bit quantizer with $\mu = 40$.

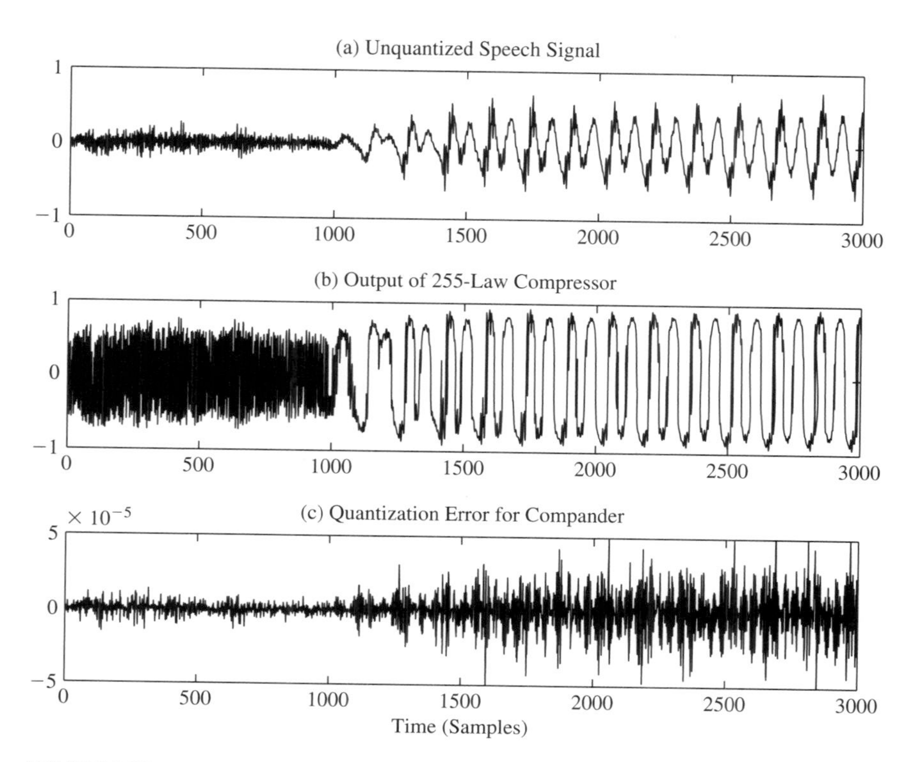

FIGURE 11.25
μ-law companding: (a) original speech signal; (b) 255-law compressed signal; and (c) quantization error for 8-bit quantization of compressed signal followed by expansion.

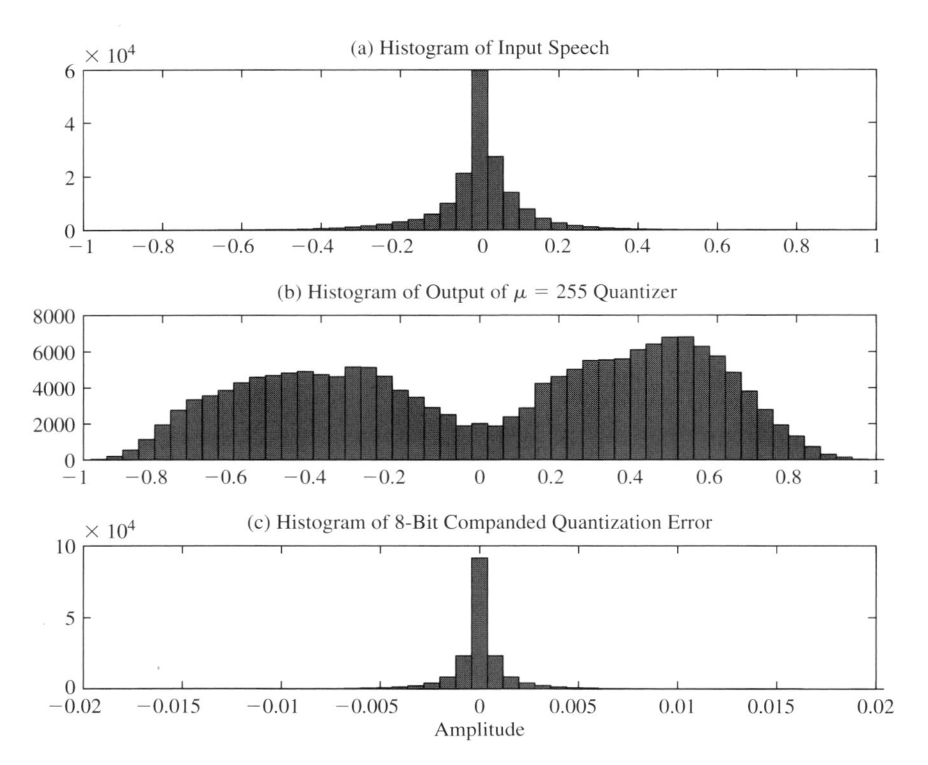

FIGURE 11.26
Amplitude histograms for μ-law companded quantization: (a) input speech signal; (b) μ-law compressed speech signal using $\mu = 255$; and (c) quantization noise at output of μ-law expander.

expansion by exponentiation, the size of the error tracks the size of the signal, thereby providing the desired constant relative error quantization noise.

A plot of the amplitude histogram of speech along with the amplitude histogram of μ-law compressed speech is given in Figure 11.26, for a value of $\mu = 255$. The original speech histogram in (a) is similar to a Laplacian density; however, the histogram of the μ-law compressed speech is much flatter, with relatively fewer small amplitudes. The histogram of the expanded 8-bit quantization error (i.e., of $f[n]$) in (c) shows that its distribution is very similar to that of the input speech signal, which was, of course, the goal of the compression/quantization/expansion operation.

Employing the same kind of assumptions that were used to analyze the uniform quantization case, Smith [366] derived the following formula for the signal-to-quantizing noise ratio for a μ-law quantizer:

$$\begin{aligned}\text{SNR}_Q(\text{dB}) = 6B + 4.77 - 20\log_{10}\left[\log\left(1+\mu\right)\right] \\ -10\log_{10}\left[1 + \left(\frac{X_{\max}}{\mu\sigma_x}\right)^2 + \sqrt{2}\left(\frac{X_{\max}}{\mu\sigma_x}\right)\right].\end{aligned} \tag{11.45}$$

This equation, when compared to Eq. (11.32), indicates a much less severe dependence of $\text{SNR}_Q(\text{dB})$ upon the quantity $(X_{\max}/\sigma_x)$, but it falls short of achieving constant $\text{SNR}_Q(\text{dB})$ as given in Eq. (11.41) for quantization with ideal logarithmic companding. We see that if $X_{\max} = 1$ in Eq. (11.41), ideal logarithmic companding (ignoring clipping) gives $\text{SNR}_Q(\text{dB}) = 6B + 4.77$, which is the same as the first two terms in Eq. (11.45). For example, the ideal logarithmic quantization could be expected to give $\text{SNR}_Q(\text{dB}) = 52.77$ for 8-bit quantization. However, the third term in Eq. (11.45) gives a reduction of $20\log_{10}[\log(256)] = -14.88$ dB. Therefore, with $B = 8$, the first three terms in Eq. (11.45) total 37.89 dB. In other words, 255-law compression sacrifices at least 14.88 dB over pure logarithmic compression. The last term in Eq. (11.45) is very insensitive to changes in $X_{\max}/\sigma_x$ because of the division by μ. For example, when $X_{\max}/\sigma_x = \mu$, SNR_Q in Eq. (11.45) is only reduced by 5.33 dB. Thus, we can expect that 255-law companded 8-bit quantization should maintain $\text{SNR}_Q(\text{dB}) \approx 38$ dB for values of $X_{\max}/\sigma_x \approx 10$ with a drop of about 5 dB as $X_{\max}/\sigma_x$ increases up to 255.

The same speech signal used for the measurements in Figure 11.19 was used in a similar experiment to verify the validity of Eq. (11.45). Figures 11.27 and 11.28 show

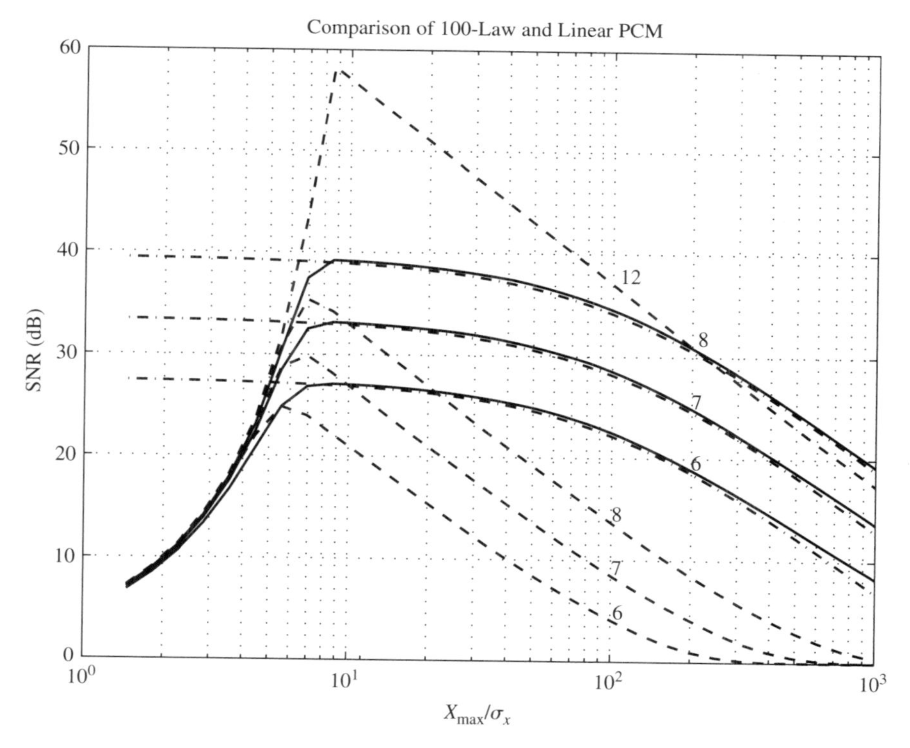

FIGURE 11.27
Measured $\text{SNR}_Q(\text{dB})$ as a function of $X_{\max}/\sigma_x$ for 100-law companding and $B = 6$, 7, and 8-bit quantizers (solid lines) and $B = 6$, 7, 8, and 12-bit uniform quantizers (dashed lines). The dash-dot lines show values computed from Eq. (11.45).

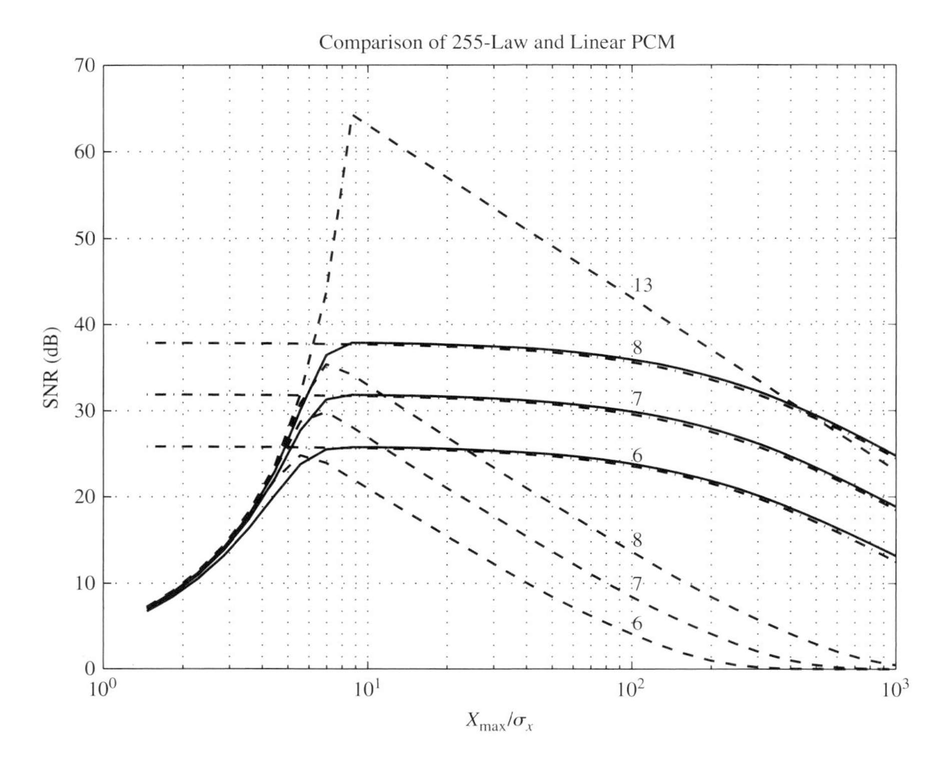

FIGURE 11.28
$\text{SNR}_Q(\text{dB})$ as a function of X_{max}/σ_x for 255-law companding and $B = 6$, 7, and 8-bit quantizers (solid lines) and $B = 6$, 7, 8, and 13-bit uniform quantizers (dashed lines). The dash-dot lines show values computed from Eq. (11.45).

plots of Eq. (11.45) (dash-dot lines) and the results of experimental measurements of $\text{SNR}_Q(\text{dB})$ for $\mu = 100$ and 255 respectively.

Several important points are true for both cases. The fit between the theoretical result in Eq. (11.45) and the measurements is excellent across a wide range, and, as predicted by Eq. (11.45), the curves are separated vertically by 6 dB since B differs by 1 between the curves. Also, the curves for μ-law quantization are quite flat over a wide range. As already pointed out, $\text{SNR}_Q(\text{dB})$ is within about 5 dB up to $X_{max}/\sigma_x = \mu$, and, from Figure 11.27, it is clear that for the case $\mu = 100$, $\text{SNR}_Q(\text{dB})$ remains within 2 dB of the maximum attainable for

$$8 < \frac{X_{max}}{\sigma_x} < 35, \tag{11.46}$$

and from Figure 11.28, we see that for $\mu = 255$, $\text{SNR}_Q(\text{dB})$ is within 2 dB of maximum for

$$8 < \frac{X_{max}}{\sigma_x} < 85. \tag{11.47}$$

However, Eq. (11.45) shows that the term $20\log_{10}(\log(1+\mu))$ causes a reduction in the maximum achievable SNR_Q(dB) that increases with increasing μ. The difference in this quantity for $\mu = 100$ and $\mu = 255$ is about 1.6 dB. Larger values of μ would yield flatter curves that would be further reduced in maximum achievable SNR_Q(dB). However, the reduction is a rather small price to pay for increased dynamic range.

As in the case of uniform quantization, the fit between the theoretical result of Eq. (11.45) and the measured results on speech is excellent for values of $X_{\max}/\sigma_x > 8$. Below this value, clipping occurs and the theoretical model no longer applies. Note that the precipitous drop occurs at about the same value as for the uniform quantization measurements shown in Figure 11.19.

The dashed curves in Figures 11.27 and 11.28 are measured SNR_Q(dB) as a function of $X_{\max}/\sigma_x$ for uniform quantizers operating on the same input signal. For both $\mu = 100$ and $\mu = 255$, the dashed curves fall below the corresponding solid curves in the useful operating range $X_{\max}/\sigma_x > 8$. An additional curve for uniform quantization is shown in each case. Figure 11.27 shows that to achieve the performance of an 8-bit 100-law quantizer over the range $8 < X_{\max}/\sigma_x < 200$ requires at least a 12-bit uniform quantizer. Similarly, Figure 11.28 shows that to achieve the performance of an 8-bit 255-law quantizer over the range $8 < X_{\max}/\sigma_x < 400$ requires at least a 13-bit uniform quantizer. Thus, by using μ-law companding, we achieve the dynamic range of 13-bit uniform quantization with only 8 bits/sample.

Since SNR_Q(dB) is well above 30 dB over a wide range, the quality of reproduction is adequate for many communications applications. As a result, 8-bit μ-law companded waveform coding at a sampling rate of 8 kHz has been standardized as ITU-T G.711. This type of waveform coding has a bit rate of 64 kbps and it is often known synonymously as 64 kbps log-PCM. This bit rate was settled on early in the history of digital speech coding, and therefore many systems that multiplex speech signals tend to aggregate data streams in multiples of 64 kbps. The G.711 standard also specifies an alternative compander definition called "A-law" which is very similar to the μ-law function [172, 401]. The A-law compression function is defined by the following equation:

$$y[n] = \begin{cases} \dfrac{Ax[n]}{1+\log A} & -\dfrac{X_{\max}}{A} \leq x[n] \leq \dfrac{X_{\max}}{A} \\ X_{\max}\dfrac{1+\log(A|x[n]|/X_{\max})}{1+\log A}\text{sign}(x[n]) & \dfrac{X_{\max}}{A} \leq |x[n]| \leq X_{\max}. \end{cases} \tag{11.48}$$

The A-law compressor function for $A = 87.56$ is very similar to the μ-law function for $\mu = 255$. These values are specified in the G.711 standard, which realizes both the μ-law and A-law companders using piecewise linear approximations that facilitate encoding and decoding in hardware and software implementations [401].

11.4.3 Quantization for Optimum SNR

μ-law compression is an example of non-uniform quantization. It is based on the intuitive notion of constant relative error. A more rigorous approach is to design a non-uniform quantizer that minimizes the mean-squared quantization error and thus maximizes the SNR_Q. To do this analytically, it is necessary to know the probability distribution of the signal sample values so that the most probable samples, which for speech are the low amplitude samples, will incur less error than the least probable samples. To apply this idea to the design of a non-uniform quantizer for speech requires the assumption of an analytical form for the probability distribution or some algorithmic approach based on measured distributions.

The fundamentals of optimum quantization were established by Lloyd [210] and Max [235]. For this reason, quantizers with decision and reconstruction levels that maximize signal-to-quantization noise ratio are called *Lloyd–Max quantizers*. Paez and Glisson [275] gave an algorithm for designing optimum quantizers for assumed Laplace and gamma probability densities, which are useful approximations to measured distributions for speech. Lloyd [210] gave an algorithm for designing optimum non-uniform quantizers based on sampled speech signals.[11] Optimum non-uniform quantizers can improve the SNR by as much as 6 dB over μ-law quantizers with the same number of bits. However, little or no improvement in perceived quality of reproduction results.

The variance of the quantization noise is

$$\sigma_e^2 = E[e^2[n]] = E[(\hat{x}[n] - x[n])^2], \tag{11.49}$$

where $\hat{x}[n] = Q[x[n]]$. Generalizing from the example of Figure 11.14, we observe that, in general, we have M quantization levels labeled $\{\hat{x}_{-M/2}, \hat{x}_{-M/2+1}, \dots, \hat{x}_{-1}, \hat{x}_1, \dots, \hat{x}_{M/2}\}$, assuming M is an even number. The quantization (reconstruction) level associated with the interval x_{j-1} to x_j is denoted $\hat{x}_j$. For a symmetric, zero mean amplitude distribution, it is sensible to define the central boundary point $x_0 = 0$, and if the density function is non-zero for large amplitudes, such as the Laplacian or gamma densities, then the extremes of the outer range are set to $\pm\infty$; i.e., $x_{\pm M/2} = \pm\infty$. With this assumption, we get

$$\sigma_e^2 = \int_{-\infty}^{\infty} e^2\, p_e(e)\, de. \tag{11.50}$$

Figure 11.29 shows a plot of a small region of a typical probability distribution for speech. Some typical decision boundaries x_k and reconstruction levels $\hat{x}_k$ for a non-uniform quantizer are also shown in Figure 11.29. This figure shows that since

$$e = \hat{x} - x, \tag{11.51}$$

the contribution to the average error variance for samples quantized to a given reconstruction level, $\hat{x}_k$, is due solely to those samples that fall in the interval $x_{k-1} \leq x < x_k$.

[11] Lloyd's work was initially published in a Bell Laboratories Technical Note with portions of the material having been presented at the Institute of Mathematical Statistics Meeting in Atlantic City, New Jersey, in September 1957. Subsequently this pioneering work was published in the open literature in March 1982 [210].

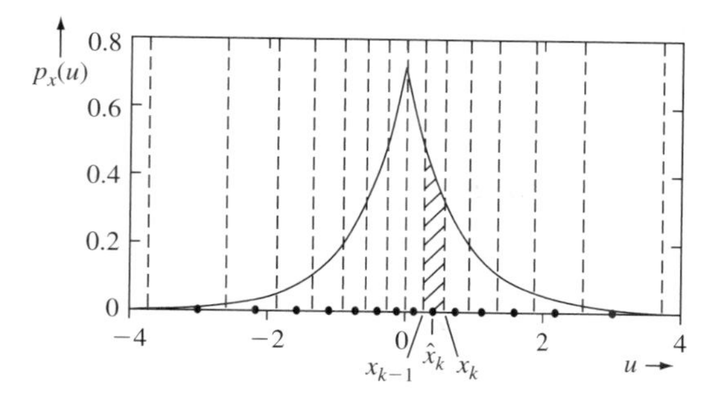

FIGURE 11.29
Illustration of quantization for maximum SNR, where x_i denotes a decision level and $\hat{x}_i$ denotes a reconstruction level. (After Vary and Martin [401].)

Therefore we only need to know $p_x(x)$ over that interval to compute the contribution of those samples to the total average variance of the error. Thus, we can make a transformation of variables in Eq. (11.50), of the form

$$p_e(e) = p_e(\hat{x} - x) = p_{x|\hat{x}}(x|\hat{x}) \triangleq p_x(x), \tag{11.52}$$

which results in

$$\sigma_e^2 = \sum_{i=-\frac{M}{2}+1}^{\frac{M}{2}} \int_{x_{i-1}}^{x_i} (\hat{x}_i - x)^2\, p_x(x)\, dx. \tag{11.53}$$

(Note that this formulation of the noise variance includes the errors due to clipping or "overload.") If $p_x(x) = p_x(-x)$, then the optimum quantizer characteristic will be anti-symmetric so that $\hat{x}_i = -\hat{x}_{-i}$ and $x_i = -x_{-i}$. Therefore the quantization noise variance (noise power) is

$$\sigma_e^2 = 2\sum_{i=1}^{\frac{M}{2}} \int_{x_{i-1}}^{x_i} (\hat{x}_i - x)^2\, p_x(x)\, dx. \tag{11.54}$$

To choose the sets of parameters $\{x_i\}$ and $\{\hat{x}_i\}$ so as to minimize σ_e^2 (maximize SNR), we differentiate σ_e^2 with respect to each parameter and set the derivatives equal to zero, obtaining the equations [235]

$$\int_{x_{i-1}}^{x_i} (\hat{x}_i - x) p_x(x) dx = 0, \qquad i = 1, 2, \ldots, \frac{M}{2}, \tag{11.55a}$$

$$x_i = \frac{1}{2}(\hat{x}_i + \hat{x}_{i+1}), \quad i = 1, 2, \ldots, \frac{M}{2} - 1. \tag{11.55b}$$

Also, we assume

$$x_0 = 0, \tag{11.56a}$$

$$x_{\pm\frac{M}{2}} = \pm\infty. \tag{11.56b}$$

Equation (11.55b) states that the optimum boundary points lie halfway between the $M/2$ reconstruction levels. Equation (11.55a) can be solved for $\hat{x}_i$ as

$$\hat{x}_i = \frac{\int_{x_{i-1}}^{x_i} x p_x(x)dx}{\int_{x_{i-1}}^{x_i} p_x(x)dx}, \tag{11.57}$$

which shows that the optimum location of the reconstruction level $\hat{x}_i$ is at the centroid of the probability density over the interval x_{i-1} to x_i. These two sets of equations must be solved simultaneously for the $M-1$ unknown parameters of the quantizer. Since these equations are generally non-linear, closed form solutions can only be obtained in some special cases. Otherwise an iterative procedure must be used. Such an iterative procedure was given by Max [235]. Paez and Glisson [275] have used this procedure to solve for optimum boundary points for the Laplace and gamma probability density functions.

In general, the solution of Eqs. (11.55a)–(11.55b) will result in a non-uniform distribution of quantization levels. Only in the special case of a uniform amplitude density will the optimum solution be uniform; i.e.,

$$\hat{x}_i - \hat{x}_{i-1} = x_i - x_{i-1} = \Delta. \tag{11.58}$$

We can, however, constrain the quantizer to be uniform [even if $p_x(x)$ is not uniform] and solve for the value of the step size, Δ, that gives minimum quantization error variance and, therefore, maximum SNR_Q. In this case

$$x_i = \Delta \cdot i, \tag{11.59}$$

$$\hat{x}_i = \frac{(2i-1)\Delta}{2}, \tag{11.60}$$

and Δ satisfies the equation

$$\sum_{i=1}^{\frac{M}{2}-1} (2i-1) \int_{(i-1)\Delta}^{i\Delta} \left[\left[\frac{2i-1}{2} \right] \Delta - x \right] p(x)dx$$

$$+ (M-1) \int_{[\frac{M}{2}-1]\Delta}^{\infty} \left[\left[\frac{M-1}{2} \right] \Delta - x \right] p(x)dx = 0. \tag{11.61}$$

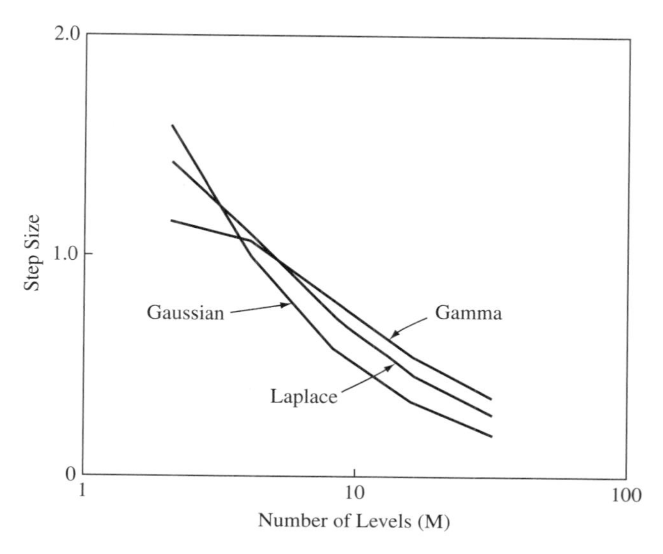

FIGURE 11.30
Optimum step sizes for a uniform quantizer for Laplace, gamma, and Gaussian density functions having unit variance. (Data from [235]. © [1960] IEEE.)

If $p(x)$ is known or assumed (e.g., Laplacian), then the integrations can be performed to yield a single equation that can be solved on a computer using iterative techniques by varying Δ until the optimum value is obtained. Figure 11.30 shows the optimum step size for a uniform quantizer for gamma and Laplacian densities [275] and a Gaussian density [235]. It is clear that, as expected, the step size decreases roughly exponentially with increasing numbers of levels. The details of the curves are attributable to the differences in the shape of the density functions.

Tables 11.2 and 11.3 show optimum quantizer parameters for Laplacian and gamma densities [275]. (Note that these numbers are derived assuming unit variance. If the variance of the input is σ_x^2, then the numbers in the tables should be multiplied by σ_x.) Figure 11.31 shows a 3-bit quantizer for a Laplacian density. It is clear from this figure that the quantization levels get further apart as the probability density decreases. This is consistent with intuition, which suggests that the largest quantization errors should be reserved for the least frequently occurring samples. A comparison of Figures 11.24 and 11.31 shows a similarity between the μ-law quantizer and the optimum non-uniform quantizer. Thus, the optimum non-uniform quantizers might be expected to have improved dynamic range. This is, in fact, true, as discussed in Ref. [275].

Although optimum quantizers yield minimum mean-squared error when matched to the variance and amplitude distribution of the signal, the non-stationary nature of the speech communication process leads to less than satisfactory results. The simplest manifestation of this occurs in transmission systems during periods when no one is talking; i.e., the so-called "idle channel" condition. In this case the input to the

TABLE 11.2 Optimum quantizers for signals with Laplace density ($m_x = 0, \sigma_x^2 = 1$). (After Paez and Glisson [275]. © [1972] IEEE.)

N	**2**		**4**		**8**		**16**		**32**	
i	x_i	$\hat{x}_i$	x_i	$\hat{x}_i$	x_i	$\hat{x}_i$	x_i	$\hat{x}_i$	x_i	$\hat{x}_i$
1	∞	0.707	1.102	0.395	0.504	0.222	0.266	0.126	0.147	0.072
2			∞	1.810	1.181	0.785	0.566	0.407	0.302	0.222
3					2.285	1.576	0.910	0.726	0.467	0.382
4					∞	2.994	1.317	1.095	0.642	0.551
5							1.821	1.540	0.829	0.732
6							2.499	2.103	1.031	0.926
7							3.605	2.895	1.250	1.136
8							∞	4.316	1.490	1.365
9									1.756	1.616
10									2.055	1.896
11									2.398	2.214
12									2.804	2.583
13									3.305	3.025
14									3.978	3.586
15									5.069	4.371
16									∞	5.768
MSE	0.5		0.1765		0.0548		0.0154		0.00414	
SNR dB	3.01		7.53		12.61		18.12		23.83	

TABLE 11.3 Optimum quantizers for signals with gamma density ($m_x = 0, \sigma_x^2 = 1$). (After Paez and Glisson [275]. © [1972] IEEE.)

N	**2**		**4**		**8**		**16**		**32**	
i	x_i	$\hat{x}_i$	x_i	$\hat{x}_i$	x_i	$\hat{x}_i$	x_i	$\hat{x}_i$	x_i	$\hat{x}_i$
1	∞	0.577	1.205	0.302	0.504	0.149	0.229	0.072	0.101	0.033
2			∞	2.108	1.401	0.859	0.588	0.386	0.252	0.169
3					2.872	1.944	1.045	0.791	0.429	0.334
4					∞	3.799	1.623	1.300	0.630	0.523
5							2.372	1.945	0.857	0.737
6							3.407	3.798	1.111	0.976
7							5.050	4.015	1.397	1.245
8							∞	6.085	1.720	1.548
9									2.089	1.892
10									2.517	2.287
11									3.022	2.747
12									3.633	3.296
13									4.404	3.970
14									5.444	4.838
15									7.046	6.050
16									∞	8.043
MSE	0.6680		0.2326		0.0712		0.0196		0.0052	
SNR dB	1.77		6.33		11.47		17.07		22.83	

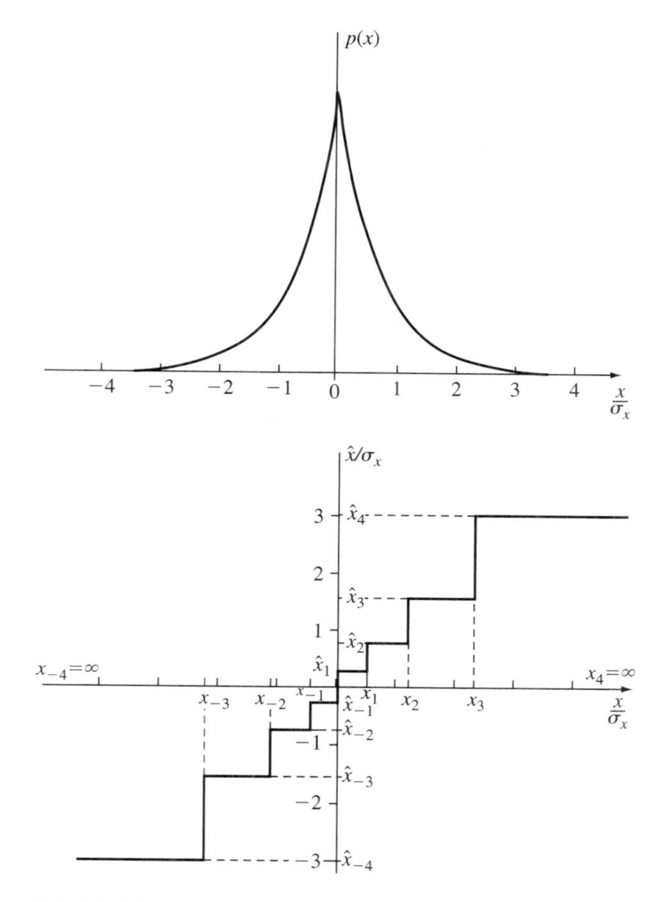

FIGURE 11.31
Density function and quantizer characteristic for Laplace density function and 3-bit quantizer.

quantizer is very small (assuming low noise) so that the output of the quantizer will jump back and forth between the lowest magnitude quantization levels. For a symmetric quantizer, such as Figure 11.15a, if the lowest quantization levels are greater than the amplitude of the background noise, the output noise of the quantizer will be greater than the input noise. For this reason, optimum quantizers of the minimum mean-squared error type are not practical when the number of quantization levels is small. Table 11.4 [275] shows a comparison of the smallest quantizer levels for several uniform and non-uniform optimum quantizers, as compared to a μ-law quantizer with $\mu = 100$. It can be seen that the μ-law quantizer would produce much lower idle channel noise than any of the optimum quantizers. For larger values of μ, the smallest quantization level would be even smaller. (If $\mu = 255$, the minimum quantization level is 0.031.) For this reason, μ-law quantizers are used in practice, even though they provide somewhat lower SNR_Q than optimum designs.

TABLE 11.4 SNRs for 3-bit quantizers. (After Noll [255].)

Non-uniform Quantizers	SNR_Q (dB)	Smallest Level ($\sigma_{x=1}$)
μ-law ($X_{max} = 8\sigma_x$, $\mu = 100$)	9.5	0.062
Gaussian	14.6	0.245
Laplace	12.6	0.222
Gamma	11.5	0.149
Speech	12.1	0.124
Uniform Quantizers	**SNR_Q (dB)**	**Smallest Level ($\sigma_{x=1}$)**
Gaussian	14.3	0.293
Laplace	11.4	0.366
Gamma	11.5	0.398
Speech	8.4	0.398

11.5 ADAPTIVE QUANTIZATION

As we have seen in the previous section, we are confronted with a dilemma in quantizing speech signals. On the one hand, we wish to choose a quantization step size large enough to accommodate the maximum peak-to-peak range of the signal. On the other hand, we would like to make the quantization step small so as to minimize the quantization noise. This is compounded by the non-stationary nature of the speech signal and the speech communication process. The amplitude of the speech signal can vary over a wide range, depending on the speaker, the communication environment, and within a given utterance, from voiced to unvoiced segments. Although one approach to accommodating these amplitude fluctuations is to use a non-uniform quantizer, an alternate approach is to adapt the properties of the quantizer to the level of the input signal. In this section, we discuss some general principles of adaptive quantization, and in later sections, we show examples of adaptive quantization schemes in conjunction with linear prediction. When adaptive quantization is used directly on samples of the input system, it is called adaptive PCM or simply, APCM.

The basic idea of adaptive quantization is to let the step size Δ (or in general the quantizer levels and ranges) vary so as to match the time-varying variance (short-time energy) of the input signal. This is depicted schematically in Figure 11.32a. An alternative point of view, depicted in Figure 11.32b, is to consider a fixed quantizer characteristic preceded by a time-varying gain that tends to keep the signal variance constant. In the first case, the step size should increase and decrease with increases and decreases of the variance of the input. In the case of a non-uniform quantizer, this would imply that the quantization levels and ranges would be scaled linearly to match the variance of the signal. In the second point of view, which applies without modification to both uniform and non-uniform quantizers, the gain changes inversely with changes in the variance of the input, so as to keep

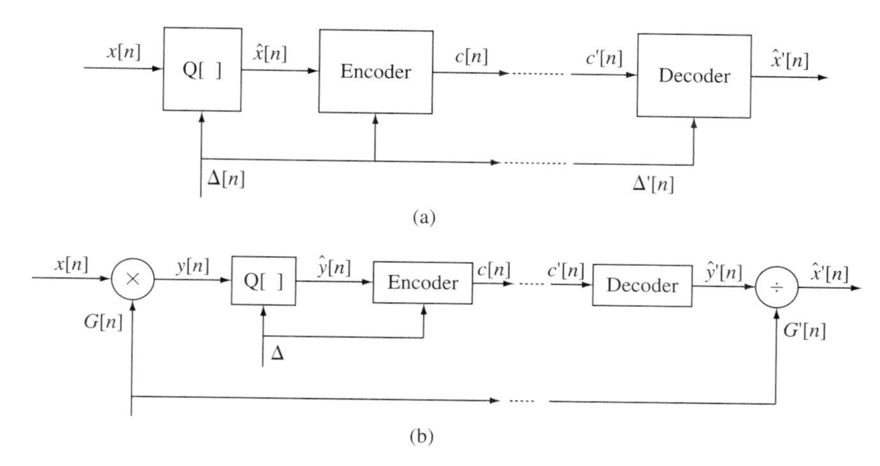

FIGURE 11.32
Block diagram representation of adaptive quantization: (a) variable step size representation; (b) variable gain representation.

the variance of the quantizer input relatively constant. In either case, it is necessary to obtain an estimate of the time-varying amplitude properties of the input signal.

In discussing the time-varying properties of speech signals, it is necessary first to consider the time scale over which changes take place. In the case of amplitude changes, we will refer to changes occurring from sample-to-sample or rapid changes within a few samples as *instantaneous* changes. General trends in the amplitude properties, as for example the peak amplitude in an unvoiced interval or in a voiced interval, remain essentially unchanged for relatively long-time intervals. Such slowly varying trends are referred to as *syllabic* variations, implying that they occur at a rate comparable to the syllable rate in speaking. In discussing adaptive quantizing schemes, it will likewise be convenient to classify them according to whether they are slowly adapting or rapidly adapting; i.e., syllabic or instantaneous.

Two types of adaptive quantizers are commonly considered. In one class of schemes, the amplitude or variance of the input is estimated from the input itself. Such schemes are called *feed-forward adaptive quantizers*. In the other class of adaptive quantizers, the step size is adapted on the basis of the output of the quantizer, $\hat{x}[n]$, or equivalently, on the basis of the output codewords, $c[n]$. These are called *feedback adaptive quantizers*. In general, the adaptation time of either class of quantizers can be either syllabic or instantaneous.

11.5.1 Feed-Forward Adaptation

Figure 11.33 depicts a general representation of the class of feed-forward quantizers. We assume, for convenience in the discussion, that the quantizer is uniform so that it is sufficient to vary a single step size parameter. It is straightforward to generalize this discussion to the case of non-uniform quantizers. The step size $\Delta[n]$, used

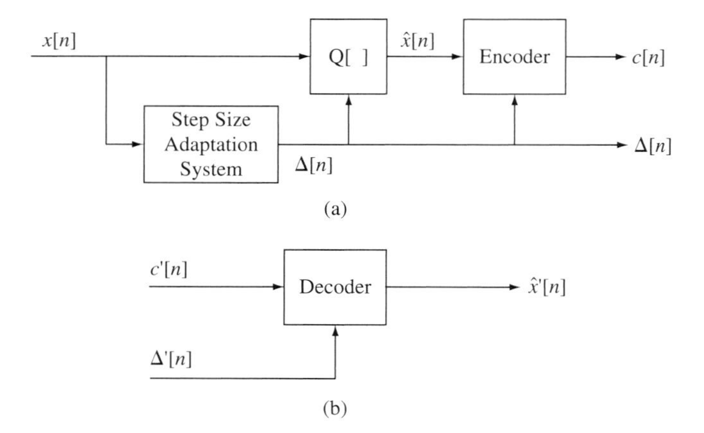

FIGURE 11.33
General representation of feed-forward quantizers: (a) encoder; (b) decoder.

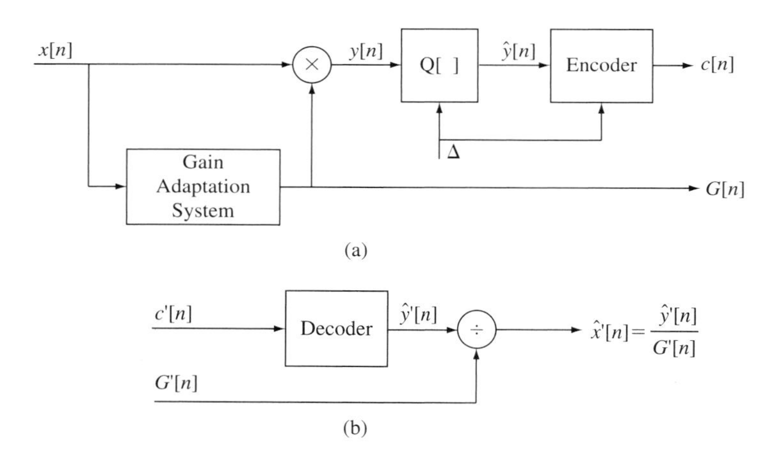

FIGURE 11.34
General feed-forward adaptive quantizer with a time-varying gain: (a) encoder; (b) decoder.

to quantize the sample $x[n]$ in Figure 11.33a, must be available at the receiver in Figure 11.33b. Thus, the codewords $c[n]$ and the step size $\Delta[n]$ together represent the sample $x[n]$. If $c'[n] = c[n]$ and $\Delta'[n] = \Delta[n]$, then $\hat{x}'[n] = \hat{x}[n]$; however, if $c'[n] \neq c[n]$ or $\Delta'[n] \neq \Delta[n]$, e.g., if there are errors in transmission, then $\hat{x}'[n] \neq \hat{x}[n]$. The effect of errors will depend upon the details of the adaptation scheme. Figure 11.34 shows the general feed-forward adaptive quantizer represented in terms of a time-varying gain. In this case, the codewords $c[n]$ and the gain $G[n]$ together represent the quantized samples.

To see how feed-forward schemes work, it is helpful to consider some examples. Most systems of this type attempt to obtain an estimate of the time-varying variance. Then the step size or quantization levels are made proportional to the standard deviation, or the gain applied to the input can be made inversely proportional to the standard deviation.

A common approach is to assume that the variance is proportional to the short-time energy, which is defined as the output of a lowpass filter with input, $x^2[n]$. That is,

$$\sigma^2[n] = \frac{\sum_{m=-\infty}^{\infty} x^2[m]\, h[n-m]}{\sum_{m=0}^{\infty} h[m]}, \tag{11.62}$$

where $h[n]$ is the impulse response of the lowpass filter and the denominator term is a normalization factor so that the term $\sigma^2[n]$ is an estimate of the signal variance in the case of a zero-mean signal, $x[n]$. For a stationary input signal, it can easily be shown that the expected value of $\sigma^2[n]$ is proportional to the variance, σ_x^2 [see Problem 11.10].

A simple example is

$$h[n] = \begin{cases} \alpha^{n-1} & n \geq 1 \\ 0 & \text{otherwise,} \end{cases} \tag{11.63}$$

with normalization factor

$$\sum_{m=0}^{\infty} h[m] = \sum_{m=1}^{\infty} \alpha^{m-1} = \sum_{m=0}^{\infty} \alpha^m = \frac{1}{1-\alpha}. \tag{11.64}$$

Using this in Eq. (11.62) gives

$$\sigma^2[n] = \sum_{m=-\infty}^{n-1} (1-\alpha) \cdot x^2[m]\, \alpha^{n-m-1}. \tag{11.65}$$

It can be shown that $\sigma^2[n]$ in Eq. (11.65) also satisfies the difference equation,

$$\sigma^2[n] = \alpha\, \sigma^2[n-1] + (1-\alpha) \cdot x^2[n-1]. \tag{11.66}$$

(For stability we require $0 < \alpha < 1$.) The step size in Figure 11.33a would therefore be of the form

$$\Delta[n] = \Delta_0\, \sigma[n], \tag{11.67}$$

or the time-varying gain in Figure 11.34a would be of the form[12]

$$G[n] = \frac{G_0}{\sigma[n]}, \tag{11.68}$$

The choice of the parameter α controls the effective interval that contributes to the variance estimate. Figure 11.35 shows an example as used in adaptive quantization in a differential PCM system [26]. Figure 11.35a shows the standard deviation estimate superimposed upon the waveform for the case $\alpha = 0.99$. Figure 11.35b shows the product $y[n] = x[n]G[n]$. For this choice of α, the dip in amplitude of $x[n]$ is clearly not compensated by the time-varying gain. Figure 11.36 shows the same waveforms for the case $\alpha = 0.9$. In this case the system reacts much more quickly to changes in the input amplitude. Thus the variance of $y[n] = G[n]x[n]$ remains relatively constant even through the rather abrupt dip in amplitude of $x[n]$. In the first case, with $\alpha = 0.99$, the time constant (time for weighting sequence to decay to e^{-1}) is about 100 samples (or 12.5 msec at an 8 kHz sampling rate). In the second case, with $\alpha = 0.9$, the time constant is only nine samples, or about 1 msec at an 8 kHz sampling rate. Thus, it would be reasonable to classify the system with $\alpha = 0.99$ as syllabic and the system with $\alpha = 0.9$ as instantaneous.

As is evident from Figures 11.35a and 11.36a, the standard deviation estimate, $\sigma[n]$, and its reciprocal, $G[n]$, are slowly varying functions as compared to the original speech signal. The rate at which the gain (or step size) control signal must be sampled depends upon the bandwidth of the lowpass filter. For the cases shown in Figures 11.35 and 11.36, the frequencies at which the filter gain is down by 3 dB are about 13 and 135 Hz, respectively, for a speech sampling rate of 8 kHz. The gain function (or step size) as used in Figures 11.34 or 11.33 must be sampled and quantized before transmission. It is important to consider the lowest possible sampling rate for the gain, since the information rate of the complete digital representation of the speech signal is the sum of the information rate of the quantizer codeword output and the information rate of the gain function.

To permit quantizing and because of constraints of physical implementations, it is common to limit the range of variation of the gain function or the step size. That is, we define limits on $G[n]$ and $\Delta[n]$ of the form

$$G_{\text{min}} \le G[n] \le G_{\text{max}} \tag{11.69}$$

or

$$\Delta_{\text{min}} \le \Delta[n] \le \Delta_{\text{max}}. \tag{11.70}$$

It is the ratio of these limits that determines the dynamic range of the system. Thus, to obtain a relatively constant SNR, over a range of 40 dB, requires $G_{\text{max}}/G_{\text{min}} = 100$ or $\Delta_{\text{max}}/\Delta_{\text{min}} = 100$.

[12]The constants Δ_0 and G_0 would account for the gain of the filter.

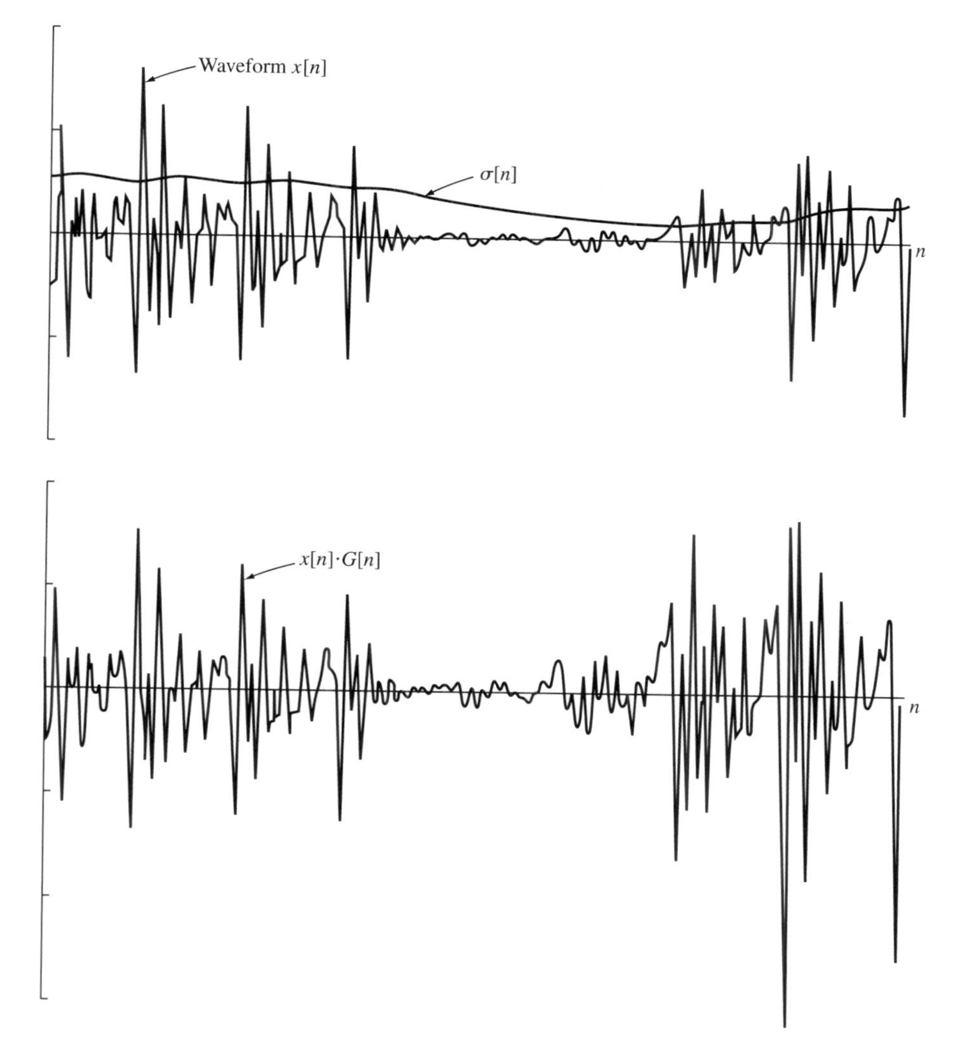

FIGURE 11.35
Example of variance estimate using Eq. (11.66): (a) waveform $x[n]$ and standard deviation estimate $\sigma[n]$ for $\alpha = 0.99$; (b) product of time-varying gain and waveform. (After Barnwell et al. [26].)

An example of the improvement in SNR_Q that can be achieved by adaptive quantization is given in a comparative study by Noll [255].[13] He considered a feed-forward

[13]This technique was also studied by Crosier [73]. He used the term "block companding" to describe the process of evaluating the gain (or step size) every M samples.

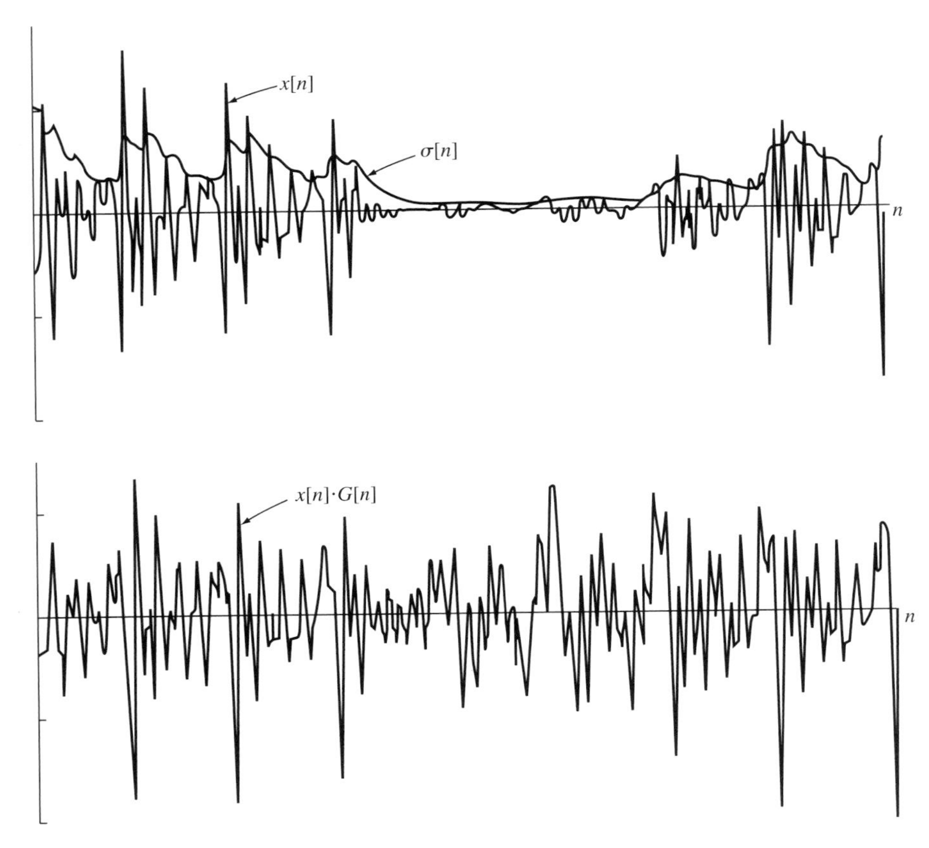

FIGURE 11.36
Variance estimate using Eq. (11.66): (a) $x[n]$ and $\sigma[n]$ for $\alpha = 0.9$; (b) $x[n] \cdot G[n]$.

scheme in which the variance estimate was

$$\sigma^2[n] = \frac{1}{M} \sum_{m=n}^{n+M-1} x^2[m]. \tag{11.71}$$

The gain or step size is evaluated and transmitted every M samples. In this case the system requires a buffer of M samples to permit the quantizer gain or step size to be determined in terms of the samples that are to be quantized rather than in terms of past samples as in the previous example.

Table 11.5 shows a comparison of various 3-bit quantizers with a speech input of known variance.[14] The first column lists the various quantizer types; e.g., a Gaussian quantizer is an optimum quantizer assuming a Gaussian distribution with variance equal to that of the speech signal. The second column gives the SNRs with no

[14]The results in this table are for quantization of actual speech signals.

TABLE 11.5 Adaptive 3-bit quantization with feed-forward adaptation. (After Noll [255].)

Non-uniform Quantizers	Non-adaptive SNR$_Q$ (dB)	Adaptive (M = 128) SNR$_Q$ (dB)	Adaptive (M = 1024) SNR$_Q$ (dB)
μ-law ($\mu = 100$, $X_{\max} = 8\sigma_x$)	9.5	–	–
Gaussian	7.3	15.0	12.1
Laplace	9.9	13.3	12.8
Uniform Quantizers			
Gaussian	6.7	14.7	11.3
Laplace	7.4	13.4	11.5

adaptation. The third and fourth columns give the SNRs for step size adaptation based upon the variance estimate of Eq. (11.71) with $M = 128$ and $M = 1024$, respectively. It can be readily seen that the adaptive quantizer achieves up to 5.6 dB better SNR$_Q$ for the speech signals used in the test. Similar results can be expected with other speech utterances, with slight variations in all the numbers. Thus, it is evident that adaptive quantization achieves a definite advantage over fixed non-uniform quantizers. An additional advantage which is not illustrated by the numbers in Table 11.5 is that by appropriately choosing $\Delta_{\min}$ and $\Delta_{\max}$, it is possible to achieve the improvement in SNR$_Q$, while maintaining low idle channel noise and wide dynamic range. This is true, in general, for most well-designed adaptive quantization systems. The combination of all these factors makes adaptive quantization an attractive alternative to instantaneous companding or minimum mean-squared error quantization.

11.5.2 Feedback Adaptation

The second class of adaptive quantizer systems is depicted in Figures 11.37 and 11.38, where it is noted that the variance of the input is estimated from the quantizer output or equivalently from the codewords. As in the case of feed-forward systems, the step size and gain are proportional and inversely proportional, respectively, to an estimate of the standard deviation of the input as in Eqs. (11.67) and (11.68). Such feedback adaptive schemes have the distinct advantage that the step size or gain need not be explicitly retained or transmitted since these parameters can be derived at the decoder directly from the sequence of codewords. The disadvantage of such systems is increased sensitivity to errors in the codewords, since such errors imply not only an error in the quantizer level, but also in the step size.

One simple approach for calculating the signal variance based on the quantized output is to apply the signal variance estimate of Eq. (11.62) directly to the quantizer output; i.e.,

$$\sigma^2[n] = \frac{\displaystyle\sum_{m=-\infty}^{\infty} \hat{x}^2[n]\, h[n-m]}{\displaystyle\sum_{m=0}^{\infty} h[m]}. \tag{11.72}$$

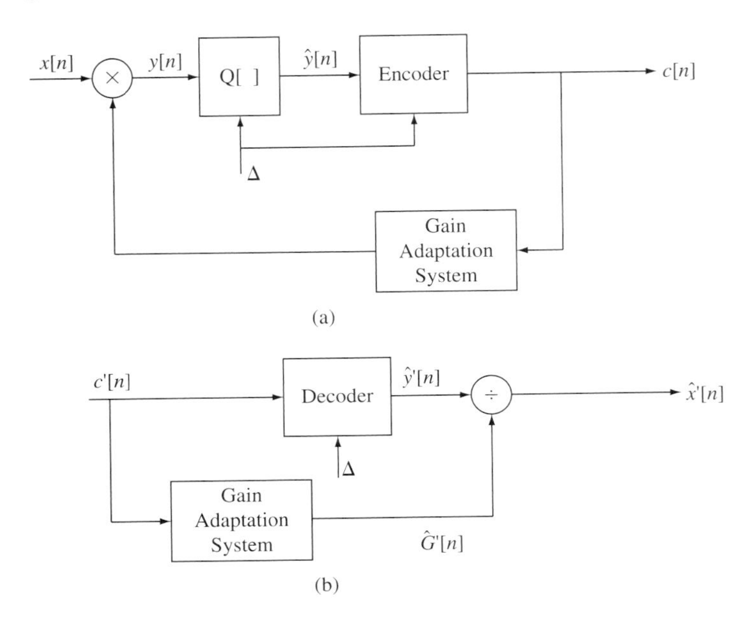

FIGURE 11.37
General feedback adaptation of the time-varying gains: (a) encoder; (b) decoder.

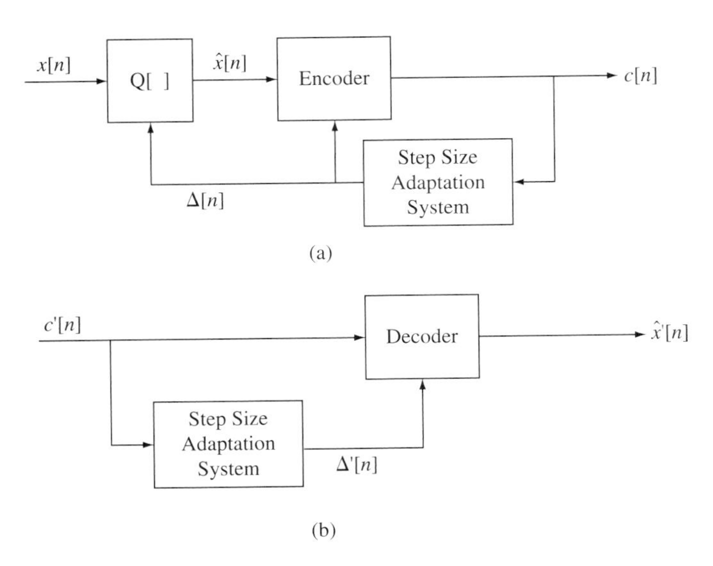

FIGURE 11.38
General feedback adaptation of the step size: (a) encoder; (b) decoder.

In this case, however, it will not be possible to use buffering to implement a non-causal filter. That is, the variance estimate must be based only on "past" values of $\hat{x}[n]$ since the present value of $\hat{x}[n]$ will not be available until after the quantization has occurred, which in turn must be after the variance has been estimated. For example, we could use a filter whose impulse response is

$$h[n] = \begin{cases} \alpha^{n-1} & n \geq 1 \\ 0 & \text{otherwise,} \end{cases} \tag{11.73}$$

as in Eq. (11.65). Alternatively the filter could have an impulse response

$$h[n] = \begin{cases} 1 & 1 \leq n \leq M \\ 0 & \text{otherwise,} \end{cases} \tag{11.74}$$

so that

$$\sigma^2[n] = \frac{1}{M} \sum_{m=n-M}^{n-1} \hat{x}^2[m]. \tag{11.75}$$

The feedback adaptive system based on Eq. (11.75) was studied by Noll [255] who found that with suitable adjustment of the constants Δ_0 or G_0 in Eqs. (11.67) or (11.68), an SNR on the order of 12 dB could be obtained for a 3-bit quantizer with a window length of only two samples. Larger values of M produced only slightly better results.

A somewhat different approach, based on Figure 11.38, was studied extensively by Jayant [168]. In this method, the step size of a uniform quantizer is adapted at each sample time by the formula

$$\Delta[n] = P\,\Delta[n-1], \tag{11.76}$$

where the step size multiplier, P, is a function only of the magnitude of the previous codeword, $|c[n-1]|$. This quantization scheme is depicted in Figure 11.39 for a 3-bit uniform quantizer. With the choice of codewords in Figure 11.39, if we assume the most significant bit is the sign bit and the rest of the word is the magnitude, then

$$\hat{x}[n] = \frac{\Delta[n]\,\text{sign}(c[n])}{2} + \Delta[n]\,c[n], \tag{11.77}$$

where $\Delta[n]$ satisfies Eq. (11.76). Note that since $\Delta[n]$ depends upon the previous step size and the previous codeword, the sequence of codewords is all that is required to represent the signal. As a practical consideration, it is necessary to impose the limits

$$\Delta_{\min} \leq \Delta[n] \leq \Delta_{\max}. \tag{11.78}$$

As mentioned before, the ratio $\Delta_{\max}/\Delta_{\min}$ controls the dynamic range of the quantizer.

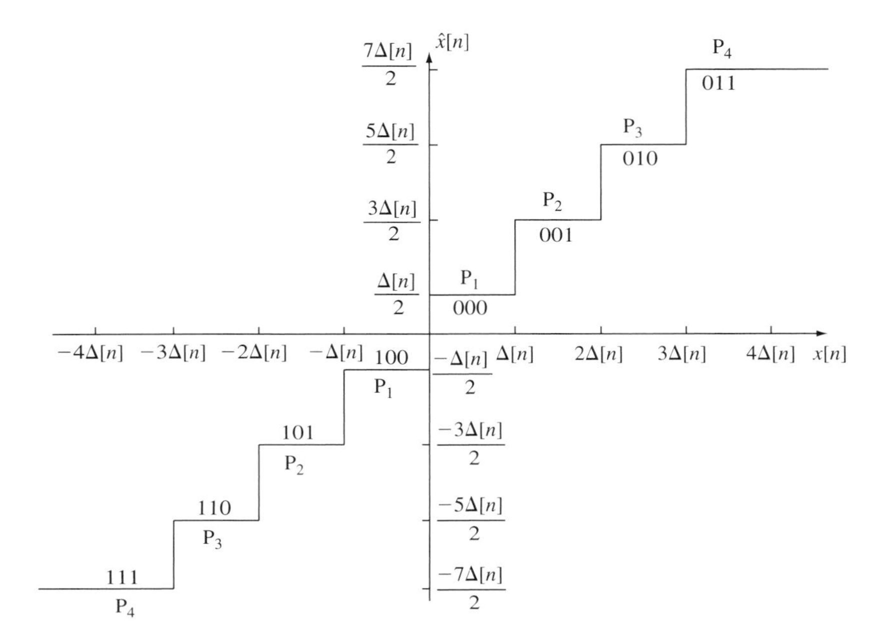

FIGURE 11.39
Input–output characteristic of a 3-bit adaptive quantizer.

The manner in which the multiplier in Eq. (11.76) should vary with $|c[n-1]|$ is intuitively clear. If the previous codeword corresponds to either the largest positive or largest negative quantizer level, then it is reasonable to assume that the quantizer is, or is in danger of being overloaded and, thus, that the quantizer step size is too small. In this case, then, the multiplier in Eq. (11.76) should be greater than 1 so that $\Delta[n]$ increases. Alternatively, if the previous codeword corresponds to either the smallest positive or smallest negative level, then it is reasonable to decrease the step size by using a multiplier less than 1. The design of such a quantizer involves the choice of multipliers to correspond to each of the 2^B codewords for a B-bit quantizer. Jayant [168] approached this problem by finding a set of step size multipliers that minimize the mean-squared quantization error. He was able to obtain theoretical results for Gaussian signals, and, using a search procedure, he obtained empirical results for speech. The general conclusions of Jayant's study are summarized in Figure 11.40, which shows the approximate way in which the step size multipliers should depend upon a quantity denoted Q, defined as

$$Q = \frac{1 + 2|c[n-1]|}{2^B - 1}. \tag{11.79}$$

The shaded region in Figure 11.40 represents the variation in the multipliers to be expected as the input statistics change or as B changes. The specific multiplier values should follow the general trend of Figure 11.40, but the specific values are not overly

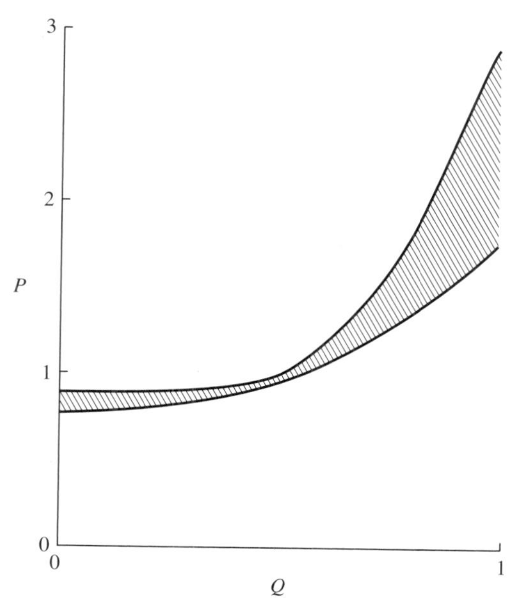

FIGURE 11.40
General shape of optimal multiplier function in speech quantization for $B > 2$. (After Jayant [168]. Reprinted with permission of Alcatel-Lucent USA Inc.)

TABLE 11.6 Step size multipliers for adaptive quantization methods. (After Jayant [168]. Reprinted with permission of Alcatel-Lucent USA Inc.)

	Coder Type	
B	**PCM**	**DPCM**
2	0.6, 2.2	0.8, 1.6
3	0.85, 1, 1, 1.5	0.9, 0.9, 1.25, 1.75
4	0.8, 0.8, 0.8, 0.8 1.2, 1.6, 2.0, 2.4	0.9, 0.9, 0.9, 0.9, 1.2, 1.6, 2.0, 2.4
5	0.85, 0.85, 0.85, 0.85 0.85, 0.85, 0.85, 0.85, 1.2, 1.4, 1.6, 1.8, 2.0, 2.2, 2.4, 2.6	0.9, 0.9, 0.9, 0.9, 0.95, 0.95, 0.95, 0.95, 1.2, 1.5, 1.8, 2.1, 2.4, 2.7, 3.0, 3.3

critical. It is important, however, that the multipliers be such that step size increases occur more vigorously than step size decreases. Table 11.6 shows the sets of multipliers obtained by Jayant [168] for $B = 2, 3, 4$ and 5.

The improvement in SNR that is obtained using this mode of adaptive quantization is shown in Table 11.7. The multipliers of Table 11.6 were used with

TABLE 11.7 Improvements in SNR using optimum step size multipliers for feedback adaptive quantization. (After Jayant [168]. Reprinted with permission of Alcatel-Lucent USA Inc.)

B	Logarithmic PCM with μ-law ($\mu = 100$) Quantization	Adaptive PCM with Uniform Quantization
2	3 dB	9 dB
3	8 dB	15 dB
4	15 dB	19 dB

$\Delta_{\max}/\Delta_{\min} = 100$. Table 11.7 shows a 4–7 dB improvement over μ-law quantization. A 2–4 dB improvement was also noted over non-adaptive optimum quantizers. In another study, Noll [256] noted SNRs for 3-bit μ-law and adaptive quantizers of 9.4 and 14.1 dB respectively. In this experiment the multipliers were {0.8, 0.8, 1.3, 1.9} in contrast to those used by Jayant, which are seen from Table 11.6 to be {0.85, 1, 1, 1.5}. The fact that such different multipliers can produce comparable results lends support to the contention that the values of the multipliers are not particularly critical.

11.5.3 General Comments on Adaptive Quantization

As the discussion of this section suggests, there are almost unlimited possibilities for adaptive quantization schemes. Most reasonable schemes will yield SNRs that exceed the SNR of μ-law quantization, and with a suitable ratio, $\Delta_{\max}/\Delta_{\min}$, the dynamic range of an adaptive quantizer can be fully comparable to that of μ-law quantization. Also, by choosing $\Delta_{\min}$ to be small, the idle channel noise can be made very small. Thus adaptive quantization has many attractive features. However, it is unreasonable to expect that further sophistication of quantizer adaptation alone will yield dramatic savings in bit rate since such techniques simply exploit our knowledge of the amplitude distribution of the speech signal. Thus, we turn our attention in Section 11.7 to exploiting the sample-to-sample correlation through the techniques of differential quantization.

11.6 QUANTIZING OF SPEECH MODEL PARAMETERS

Our discussion, so far in this chapter, has focused on quantization of the samples of the speech waveform. This is the simplest method of obtaining a digital representation of speech, but it is neither the most flexible nor the most efficient in terms of data rate. As we will see in the remainder of the chapter, the key to increased flexibility and efficiency of a digital coding scheme is to take advantage of our knowledge of the discrete-time models for speech production developed in Chapter 5 and all the analysis tools that we have developed in Chapters 6–9 for estimating the parameters of the models. Figure 11.41 depicts a generalized representation of the speech model showing

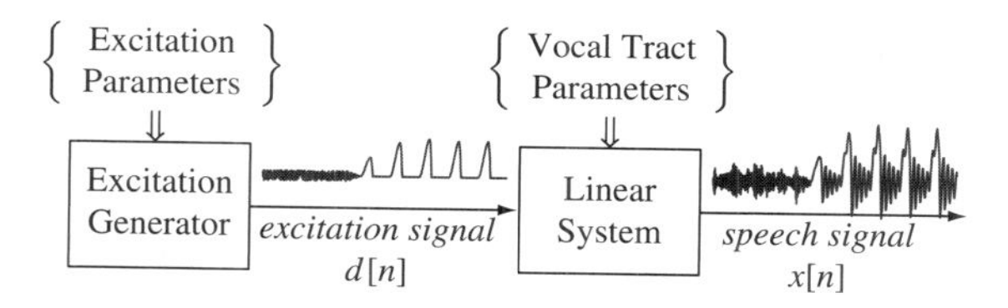

FIGURE 11.41
General source/system model for speech production.

a means for generating an excitation signal which switches between unvoiced (noise-like) and voiced (quasi-periodic) states. The resulting excitation signal is the input to a time-varying linear system that models the resonance properties of the vocal tract and other factors such as glottal pulse shape and radiation effects. The key point of this figure is that the excitation and the vocal tract system are characterized by sets of parameters that can be estimated from a speech signal by techniques such as linear predictive analysis or homomorphic filtering. Assuming that the parameters are estimated as a function of time, so as to track the changing nature of the speech signal, then by using the estimated parameters to control the model, we can *synthesize* an output signal that is an approximation to the original speech signal. The quality and data rate of this approximation depends on many factors including:

- the capability of the model to represent the speech signal;
- the existence of efficient computational algorithms for estimating the parameters of the model;
- the accuracy of the estimates of the parameters using the basic analysis algorithms; and
- the ability to quantize the parameters so as to obtain a low data rate digital representation that will yield a high quality reproduction of the speech signal.

Discrete-time speech models are used in speech coding in two fundamentally different ways. One class of coders can be termed *closed-loop coders* because, as shown in Figure 11.42a, the speech model is used in a feedback loop wherein the synthetic speech output is compared to the input signal, and the resulting difference is used to determine the excitation for the vocal tract model, etc. Examples of this class of coders include differential PCM (both adaptive and fixed) and analysis-by-synthesis coders such as code excited linear prediction (CELP). We shall discuss these systems in later sections of this chapter. In the second class of coders, called *open-loop coders*, and also discussed later, the parameters of the model are estimated directly from the speech signal with no feedback as to the quality of the representation of the resulting synthetic speech. These systems are often called voice coders or *vocoders*. In both types of speech coders, it is necessary to estimate the parameters of the model and quantize those parameters to obtain a digital representation that can be used for efficient digital storage or transmission. In the remainder of this section, we will consider techniques for quantizing the parameters of the speech model assuming that they

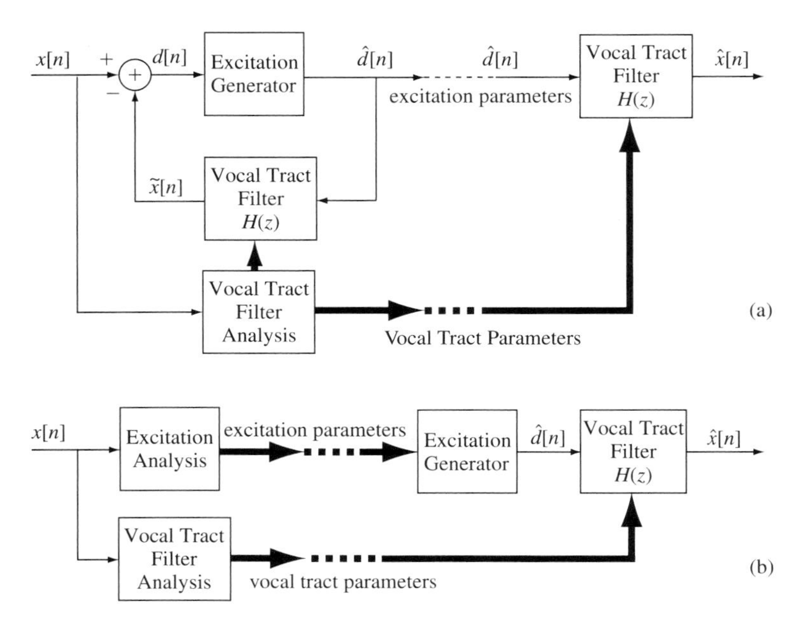

FIGURE 11.42
Use of the speech model in coding: (a) closed-loop coder; (b) open-loop coder.

have been accurately estimated by the techniques that we have studied in previous chapters.

11.6.1 Scalar Quantization of the Speech Model

One approach to quantization of the speech model is to treat each parameter of the model separately as a number to be represented with a fixed number of bits as discussed in Section 11.4. To do this, it is necessary to determine the statistics of that parameter; i.e., its mean, variance, minimum and maximum value, and distribution. Then an appropriate quantizer can be designed for that parameter. Each parameter can have a different bit allocation. As a simple example, consider the pitch period as part of the excitation description. Usually, this parameter is determined in samples at the sampling rate of the input signal. For an 8 kHz sampling rate, for example, the pitch period typically ranges from as low as 20 samples (i.e., 400 Hz pitch frequency) up to 150 samples (i.e., 53.3 Hz pitch frequency) for very low-pitched speakers. Since all periods within such a range must be accurately represented, the pitch period is generally quantized with 128 values (7-bits) uniformly over the expected range of pitch periods. Often the value of zero is used to signal unvoiced excitation. Another parameter of the excitation would be the excitation amplitude. This parameter might be quantized with a non-uniform quantizer such as μ-law. Typically 4–5 bits might be allocated to the amplitude. For an analysis frame rate of 100 frames/sec, the total data rate for the pitch period would be 700 bps, and if 5 bits were used for amplitude, an additional 500 bps would be added to the excitation information.

The vocal tract system in a speech model is usually estimated by linear predictive methods, and therefore the basic representation of this part of the model is a set of p predictor coefficients. As discussed in Chapter 9, the predictor can be represented in many equivalent ways, almost all of which are preferable to the predictor coefficients themselves for quantization. While such representations as log-area coefficients, cepstrum coefficients, and line-spectrum frequencies (LSF) are sometimes used, the PARCOR coefficients (reflection coefficients) are also frequently chosen as the basis for quantization. The PARCOR coefficients can be quantized with a non-uniform quantizer or transformed with an inverse sine or hyperbolic tangent function to flatten their statistical distribution and then quantized with a fixed uniform quantizer. Each coefficient can be allocated a number of bits contingent on its importance in accurately representing the speech spectrum [404]. Figure 11.43 shows an example of the effect of quantization of the linear predictive vocal tract model where $p = 20$. In this example, each PARCOR coefficient was assumed to be in the range $-1 < k_i < 1$ for $i = 1, 2, \ldots, p$. In fact, the higher-order coefficients have a much more restricted range [404] and could easily be represented with fewer bits than the lower-order PARCORs. For Figure 11.43, each PARCOR coefficient was transformed by an inverse sine (arcsin) function into the range $-\pi/2 < \arcsin(k_i) < \pi/2$ and then quantized with

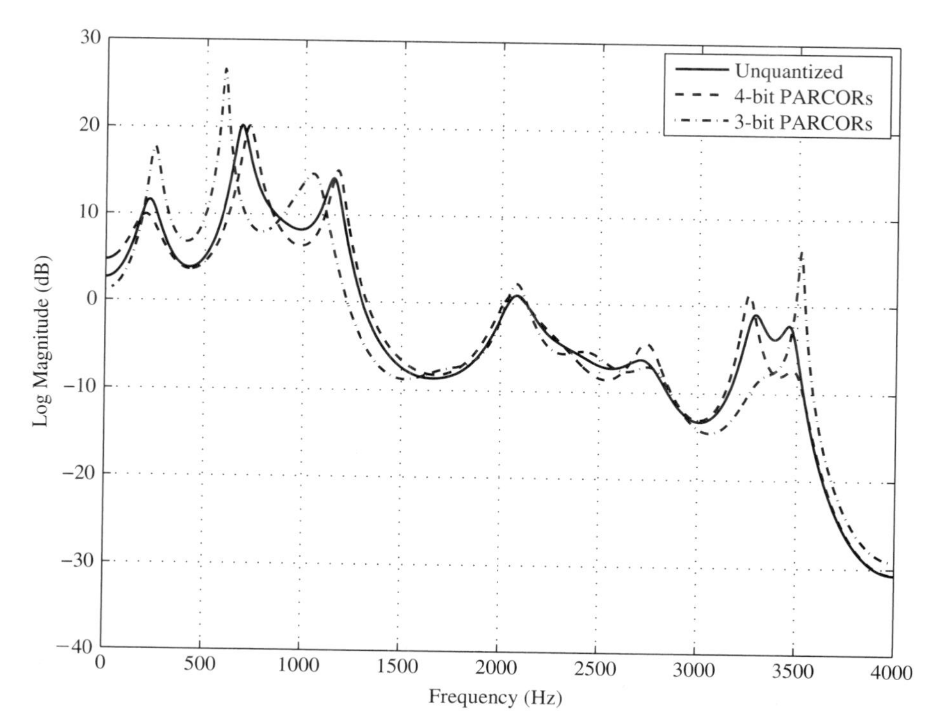

FIGURE 11.43
Illustration of effect of quantization of PARCOR coefficients.

both a 4-bit uniform quantizer and a 3-bit uniform quantizer. Figure 11.43 shows the frequency response, $20 \log_{10} |H(e^{j2\pi F/F_s})|$, of the vocal tract filter with unquantized coefficients (solid line) compared to the frequency response with 4-bit (dashed line) and 3-bit (dash-dot line) quantization of the 20 PARCOR coefficients. For 4-bit quantization, the effect is seen to be only a slight shift of the formant resonances, while a significant shift of frequency and level is observed for 3-bit quantization. More effective quantization would use different numbers of bits for each coefficient. For example, a speech coding system reported by Atal [9], which will be discussed in detail in Section 11.10.3, used a 20th-order predictor and quantized the resulting set of PARCOR coefficients after transformation with an inverse sine function. The number of bits per coefficient ranged from 5, for each of the lowest two PARCORs, down to 1 each for the six highest-order PARCORs, yielding a total of 40 bits/frame [9]. If the parameters are updated 100 times/sec, then a total of 4000 bps for the vocal tract predictor information is required. To save bit rate with modest sacrifice in reproduction quality, it is possible to update the predictor 50 times/sec for a total bit rate contribution of 2000 bps [9]. Another way to save bits is to use a lower order model; however, as we noted in Chapter 5, a sampling rate of 8 kHz requires at least $p = 10$ for accurate representation of the formant structure.

While the above discussion focused on quantization of the PARCOR coefficients, other representations of the vocal tract filter would yield similar results for the total number of bits/frame. If we combine the estimates for excitation and vocal tract components of the model, we arrive at approximately 5000 bps. This would be a significant investment in data rate. Whether it is a worthwhile contribution to the quality of reproduction would depend on how the speech model is incorporated into the specific coder.

11.6.2 Vector Quantization

The technique of vector quantization, or VQ [137], is particularly effective in coding the parameters of the vocal tract model. The basic principle of VQ is depicted in Figure 11.44. VQ can be applied in any context where a set of parameters naturally

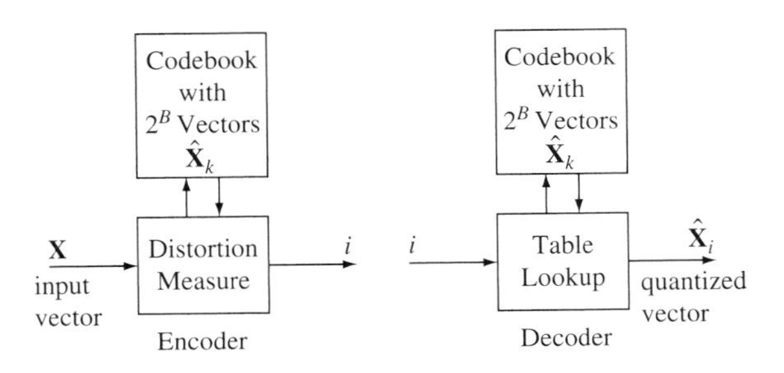

FIGURE 11.44
Vector quantization.

groups together into a vector (in the sense of linear algebra),[15] and is therefore ideally suited to coding sets of parameters such as predictor coefficients, PARCOR coefficients, cepstrum values, log area ratios, or line spectrum frequencies as commonly used to represent the vocal tract system function.

The basic idea of VQ is simple to state, namely that it is generally more efficient to code a "block of scalars" as a vector rather than coding each of the scalars individually using simple quantizers of the type discussed earlier in this chapter. Furthermore, an "optimal VQ" procedure can be designed based on some type of mean-squared distortion metric, leading to very efficient VQ methods. As depicted in Figure 11.44, in VQ, a vector, $\mathbf{X}$, to be quantized is compared exhaustively to a "codebook" populated with representative vectors of that type. The index i of the closest vector, $\hat{\mathbf{X}}_i$, to $\mathbf{X}$, according to a prescribed distortion measure, is returned from the exhaustive search. This index i then represents the quantized vector in the sense that if the codebook is known at the decoder, the corresponding quantized vector $\hat{\mathbf{X}}_i$ can be looked up. If the codebook has 2^B entries, the index i can be represented by a B bit number.

A simple example of how VQ can be applied to a simple waveform coding system is shown in Figure 11.45. If $x[n]$ is a sequence of scalar samples of a speech signal, we can define a vector sequence as:

$$\mathbf{X}[n] = [x[2n], x[2n+1]];$$

i.e., we take the samples in pairs as depicted by the solid dots at $n = 20$ and $n = 50$ in the waveform plot in Figure 11.45a. The lower plots each show 73,728 individual vectors (corresponding to different times in the sequence) plotted as gray dots in a two-dimensional graph, along with the results of clustering these (training set vectors) into 1 [part(b)], 2 [part(c)], 4 [part(d)], and 8 [part(e)] vectors. (Since we have many points that cluster together, only the outlying points stand out in isolation.) We know that speech has considerable sample-to-sample correlation, and this is confirmed by the grouping of the vectors around the 45 degree line. As seen in Figure 11.45e, for example, if the eight vectors denoted by the black * symbols provide acceptable accuracy in representing the complete set of speech vectors, each of the gray dots could be represented by the 3-bit index of the closest vector (*) to it. The result would be the vector quantized speech waveform with an average data rate of 3 bits/2 samples or 1.5 bits/sample. While this example illustrates the basic principles of VQ and is easy to understand, it does not lead to a useful approach to speech coding since far too many vectors would be required for accurate representation of the speech waveform by vector quantized pairs of speech samples. However, the following example is more representative of how VQ is used in coding parameters of the speech model.

Figure 11.46 (due to Quatieri [297]) gives a simple example that illustrates the potential advantages of a vector quantizer over a scalar quantizer for the vocal tract system parameters. For this artificial example, it is assumed that there are only four

[15] Independent quantization of individual speech samples or individual model parameters is called scalar quantization.

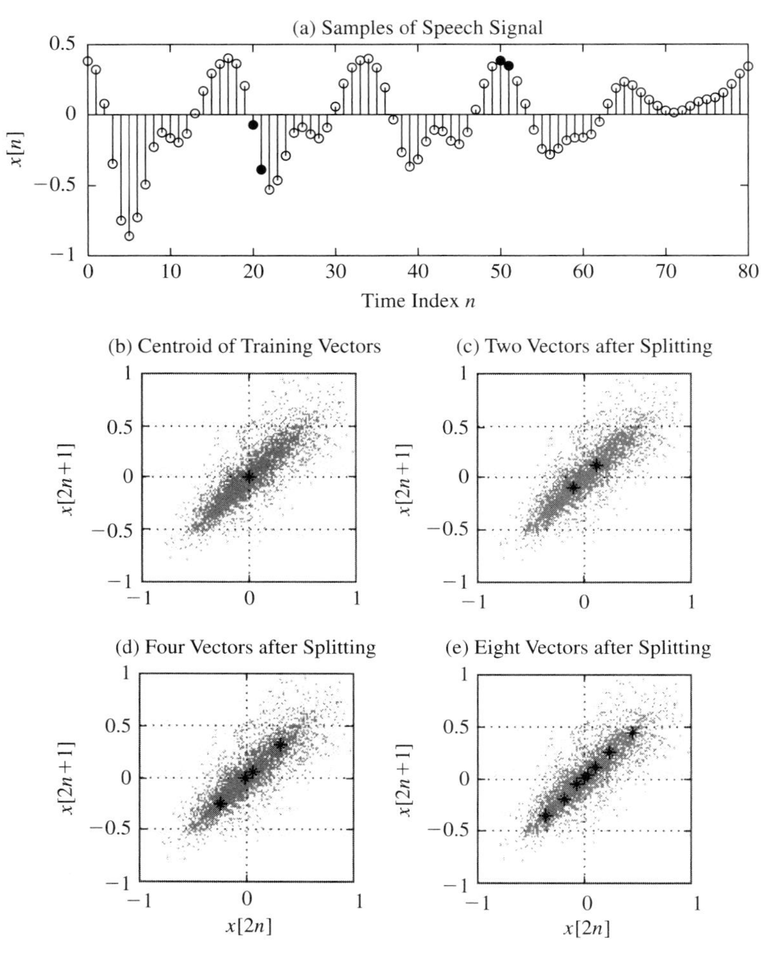

FIGURE 11.45
Illustration of VQ for pairs of speech samples: (a) speech samples showing pairs; (b–e) 73,728 pairs of samples plotted in gray; 1, 2, 4, and 8 pattern vectors (codewords) derived from test samples.

possible shapes that the human vocal tract can take, and that each vocal tract shape is characterized by two formant resonances. The spectra corresponding to the four vocal tract shapes are shown at the left in Figure 11.46, and each spectrum is analyzed into a set of four reflection coefficients, yielding vectors of the form:

$$\mathbf{X} = [k_1, k_2, k_3, k_4].$$

There are only four possible analysis vectors, corresponding to the four vocal tract shapes, so if we code spectral shape using a codebook with the four spectral shapes,

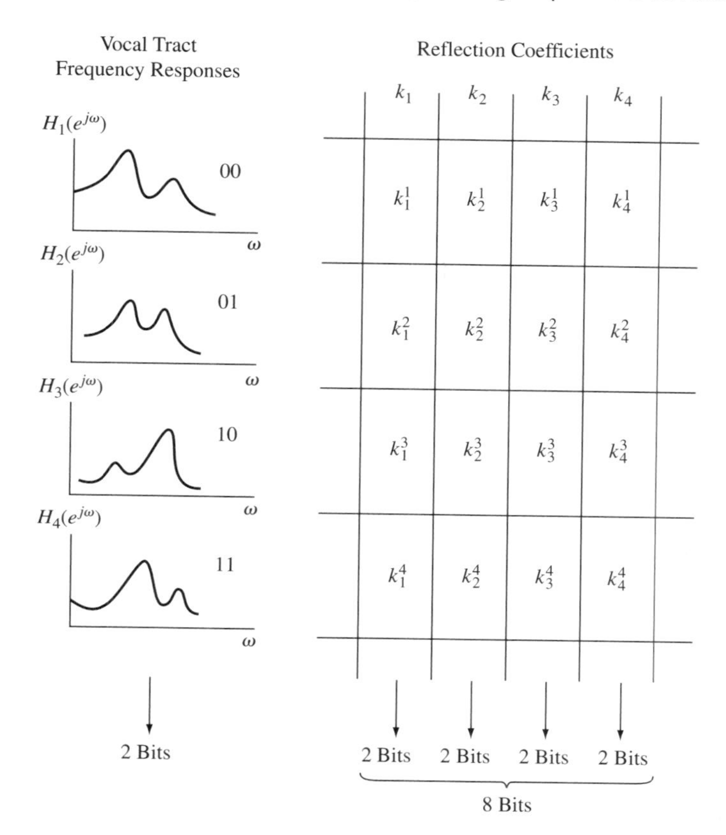

FIGURE 11.46
Example illustrating the power of a VQ codebook over scalar quantization in the case where there are only four possible vectors (2-bit code) but there are 256 combinations of the scalar components of the vector (8-bit code). (After Quatieri [297].)

we need only a 2-bit code to choose arbitrarily any of the four spectral vectors. If, instead, we choose to use a scalar quantizer for each of the four components of $\mathbf{X}$, i.e., k_1, k_2, k_3, k_4, then we would need a 2-bit scalar quantizer for each reflection coefficient (since again there are only four possible values of each reflection coefficient), or a total of 8 bits for the four scalar components of $\mathbf{X}$. Hence, in this highly stylized case, the use of a vector quantizer is four times more efficient than the use of a set of scalar quantizers. Of course, this is entirely due to the artificially high correlation among the four reflection coefficients. If the reflection coefficients were totally independent of each other, there would be virtually no gain by using a vector quantizer. Fortunately, for most speech analysis/synthesis systems, there is a reasonable amount of correlation among the components of the speech analysis vector, $\mathbf{X}$, so the use of a vector quantizer is highly desirable.

11.6.3 Elements of a VQ Implementation

In order to build a VQ codebook and implement a VQ analysis procedure,[16] we need the following components:

1. A large training set of analysis vectors, $\mathcal{X} = \{\mathbf{X}_1, \mathbf{X}_2, \dots, \mathbf{X}_L\}$. The training set provides an implicit estimate of the probability density function of $\mathbf{X}$ and thereby enables the codebook design procedure to choose optimal codeword locations that minimize overall distortion in coding an arbitrary set of analysis vectors (the testing set). If we denote the size of the VQ codebook as $M = 2^B$ vectors (i.e., a B-bit codebook), then we require $L >> M$ so as to be able to find the best set of M codebook vectors in a robust manner. In practice, it has been found that L should be in the range $10 \cdot M \leq L \leq 100 \cdot M$ in order to reliably and robustly train a VQ codebook [182].
2. A measure of distance between a pair of analysis vectors so as to be able to cluster the training set vectors as well as to associate or classify arbitrary vectors (the test set) into unique codebook entries. We denote the distance between two analysis vectors, $\mathbf{X}_i$ and $\mathbf{X}_j$, as:
$$d(\mathbf{X}_i, \mathbf{X}_j) = d_{ij}.$$
3. A centroid computation procedure and a centroid splitting procedure. On the basis of the partitioning that classifies the L training set vectors into M clusters, we choose the M codebook vectors as the centroid of each of the M clusters. We will see that once we have obtained a stable codebook with M vectors, we can begin the search for higher order codebooks by splitting single or multiple centroids based on a centroid splitting procedure, thereby enabling the design of larger codebooks, based on optimal designs of smaller codebooks.
4. A classification procedure for arbitrary analysis vectors that chooses the codebook vector closest (in the sense of the distance measure) to the input vector, providing the codebook index of the resulting codeword. This classification procedure is often referred to as the nearest-neighbor labeling and is essentially a quantizer that accepts, as input, a speech analysis vector and provides, as output, the codebook index of the codebook vector that best matches the input (in a minimum distance sense).

A block diagram of the basic VQ training and classification structure is shown in Figure 11.47. The training procedure, shown at the top of the figure, consists of the K-means clustering algorithm, which maps the training set of vectors into an M-element codebook. The testing procedure, shown at the bottom of the figure, consists of a nearest-neighbor search of the codebook, based on the distance measure appropriate to the analysis vector, $\mathbf{X}$. We now provide more detail about each of the VQ elements.

[16]Most of the material in this section comes directly from Rabiner and Juang [308], with some small modifications.

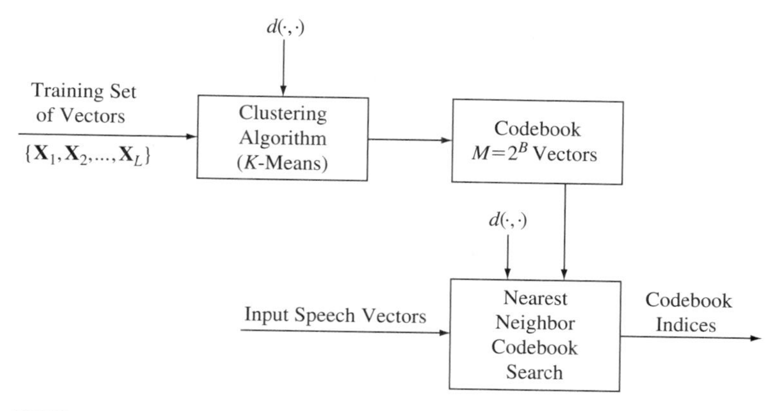

FIGURE 11.47
Block diagram of the basic VQ training and classification structure. (After Rabiner and Juang [308].)

The VQ Training Set

To properly train the VQ codebook, a training set of $L \geq 10M$ vectors should span the anticipated range of:

- talkers, including ranges in age, accent, gender, speaking rate, speaking levels, and any other relevant variables;
- speaking conditions, such as quiet rooms, automobiles, noisy work places;
- transducers and transmission systems, including a range of microphones telephone handsets, cell phones, speaker phones, etc.;
- speech, including carefully recorded material, conversational speech, and telephone query material.

The more narrowly focused the training set, the smaller the quantization error in representing the analysis vectors with a fixed size codebook. However, such narrowly focused training sets are not broadly applicable to a wide range of applications, and thus the training set should be as broad as possible.

The Distance Measure

The distance measure appropriate to the analysis vector $\mathbf{X}$ depends critically on the components of the analysis vector. Thus if $\mathbf{X}_i = [x_i^1, x_i^2, \ldots, x_i^R]$ is a log spectral vector, then a possible distance measure would be an L_p log spectral difference; i.e.,

$$d(\mathbf{X}_i, \mathbf{X}_j) = \left[\sum_{k=1}^{R} |x_i^k - x_j^k|^p \right]^{1/p}$$

where $p = 1$ is a log magnitude spectral difference, $p = 2$ is a root mean-squared log spectral difference, and $p = \infty$ is a peak log spectral difference. Similarly, if

$\mathbf{X}_i = [x_i^1, x_i^2, \ldots, x_i^R]$ is a cepstrum vector, then the distance measure might well be a cepstral distance of the form:

$$d(\mathbf{X}_i, \mathbf{X}_j) = \left[\sum_{k=1}^{R} (x_i^k - x_j^k)^2 \right]^{1/2}.$$

A range of distance measures has been used for various speech analysis vectors and further examples can be found in Ref. [308].

Clustering the Training Vectors

The way in which a set of L training vectors can be optimally clustered into a set of M codebook vectors is based on the generalized Lloyd algorithm (also known as the K-means clustering algorithm) [210], and consists of the following steps:

1. **Initialization**: Arbitrarily choose M vectors (initially out of the training set of L vectors) as the initial set of codewords in the codebook.
2. **Nearest-Neighbor Search**: For each training vector, find the codeword in the current codebook that is closest (in terms of the chosen distance measure), and assign that vector to the corresponding cell (associated with the closest codeword).
3. **Centroid Update**: Update the codeword in each cell to the centroid of all the training vectors assigned to that cell in the current iteration. (We discuss the computation of the centroid in the next section.)
4. **Iteration**: Repeat steps 2 and 3 until the average distance between centroids at consecutive iterations falls below a pre-set threshold.

Figure 11.48 illustrates the result of designing a VQ codebook by showing the partitioning of a two-dimensional space into distinct regions, each of which is represented by a centroid vector (as denoted by the x in each cell). The shape (i.e., the cell boundaries) of each partitioned cell is highly dependent on the distance measure and the statistics of the vectors in the training set.

Centroid Computation

The procedure for the computation of the centroid of a set of assignment vectors is critical to the success of a VQ design procedure. Assume we have a set of V vectors, $\mathcal{X}^c = \{\mathbf{X}_1^c, \mathbf{X}_2^c, \ldots, \mathbf{X}_V^c\}$, all assigned to cluster c. The centroid of the set $\mathcal{X}^c$ is defined as the vector $\bar{\mathbf{Y}}$ that minimizes the average distortion, i.e.,

$$\bar{\mathbf{Y}} \triangleq \min_{\mathbf{Y}} \frac{1}{V} \sum_{i=1}^{V} d(\mathbf{X}_i^c, \mathbf{Y}). \tag{11.80}$$

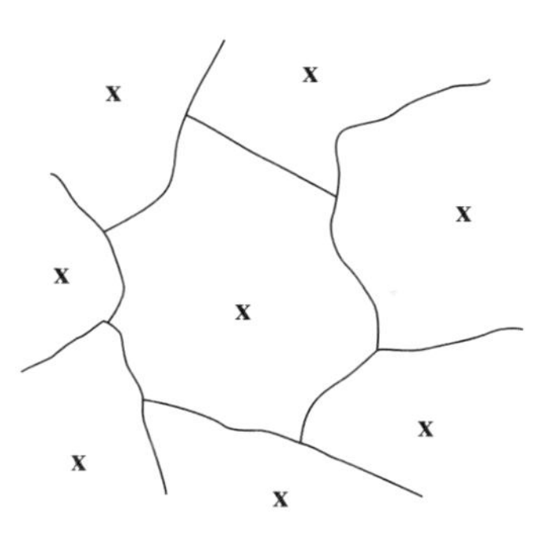

FIGURE 11.48
Partitioning of a two-dimensional vector space into VQ cells with each cell represented by a centroid vector (denoted by **x**). (After Rabiner and Juang [308].)

The solution to Eq. (11.80) is highly dependent on the choice of the distance measure. When both $\mathbf{X}_i^c$ and $\mathbf{Y}$ are measured in a K-dimensional space with the L_2 norm (Euclidean distance), the centroid is the mean of the vector set,

$$\bar{\mathbf{Y}} = \frac{1}{V}\sum_{i=1}^{V} \mathbf{X}_i^c. \tag{11.81}$$

When using an L_1 distance measure, the centroid is the median vector of the set of vectors assigned to the given class.

Vector Classification Procedure

The classification procedure for an arbitrary test set of vectors (i.e., vectors not used to train the codebook) requires a full search through the codebook to find the "best" (minimum distance) match. This is how the codebook is used to quantize the vectors of speech model parameters. Thus if we denote the codebook vectors of an M-vector codebook as $\hat{\mathbf{X}}_i$, for $1 \leq i \leq M$, and we denote the vector to be classified (and quantized) as $\mathbf{X}$, then the index, i^*, of the best codebook entry is

$$i^* = \arg \min_{1\leq i\leq M} d(\mathbf{X}, \hat{\mathbf{X}}_i). \tag{11.82}$$

Binary Split Codebook Design

Although the iterative method for designing a VQ codebook works well, it has been found that the process of designing an M-vector codebook in stages often works even better. This binary splitting process works by first designing a 1-vector codebook, then using a splitting technique on the codewords to initialize the search for a 2-vector codebook, solving for the optimum 2-vector codebook using the K-means clustering algorithm, and continuing the splitting process until the desired M-vector codebook is obtained. The resulting procedure (assuming M is a power of 2) is called the binary split algorithm, and is formally implemented by the following procedure:

1. Design a 1-vector codebook; the single vector in the codebook is the centroid of the entire set of training vectors (so essentially minimal computation is involved in this step).
2. Double the size of the codebook by splitting each current codebook vector, $\mathbf{Y}_m$, according to the rule:

$$\mathbf{Y}_m^+ = \mathbf{Y}_m(1 + \epsilon),$$

$$\mathbf{Y}_m^- = \mathbf{Y}_m(1 - \epsilon),$$

   where m varies from 1 to the current size of the codebook, and ϵ is a splitting parameter (typically ϵ is chosen in the range $0.01 \leq \epsilon \leq 0.05$).
3. Use the K-means clustering algorithm to get the best set of centroids for the split codebook (i.e., the codebook of twice the size).
4. Iterate steps 2 and 3 until a codebook of size M is designed.

In the simple example of vector quantizing pairs of speech samples, Figures 11.45b–e show the first four iterations of the binary split codebook design.

Note that the binary split algorithm could be easily modified to a unity split algorithm whereby only a single VQ cell is split at each iteration (usually the one with the largest average distance from the training set vectors assigned to that cell) and the codebook grows a single cell at a time. Such procedures are often used for small codebook designs or cases when the codebook size is not a power of 2.

Figure 11.49 shows a flow diagram of the binary split VQ codebook generation method. The box labeled "classify vectors" is the nearest-neighbor search procedure, and the box labeled "find centroids" is the centroid update procedure of the K-means algorithm. The box labeled "compute D (distortion)" sums the distances of all training vectors in the nearest-neighbor search so as to determine whether the procedure has converged (i.e., $D' = D$ of the previous iteration, or equivalently, the current set of centroids is essentially identical to the set of centroids at the previous iteration).

Figure 11.50 shows the resulting set of spectral shapes corresponding to codebook vectors in an $M = 64$ codebook. Ideally it would be expected that the individual spectral shapes would correspond to the set of distinct sounds of the language; i.e., the set of phonemes that we defined in Chapter 3 of this book. Since the phonemes occur with unequal probabilities, we might also expect to have multiple spectral shapes for the most common phonemes. By examining the spectral shapes corresponding to the $M = 64$ codebook, we see that, to first order, the shapes seem to reflect a range of spectral shapes for vowels and consonants (especially the most prominent vowels and consonants), along with a few spectra corresponding to background signals (which often occur a significant number of times in any reasonable training set of vectors).

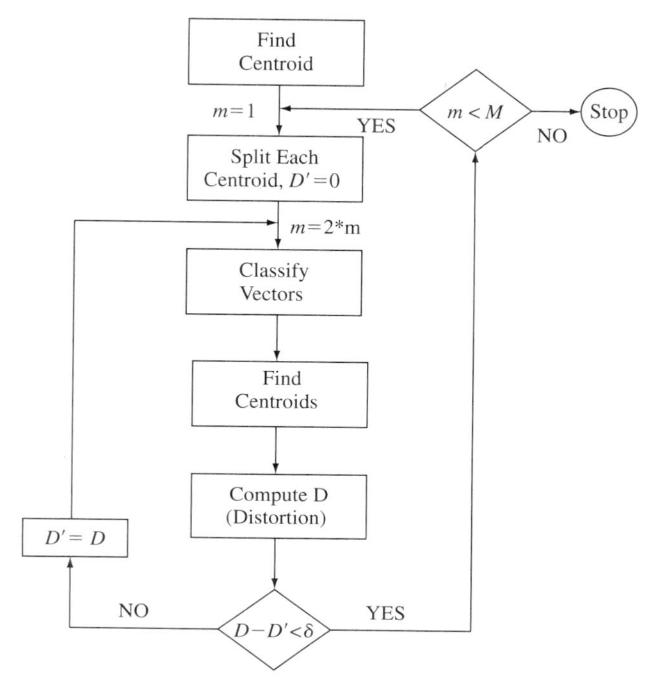

FIGURE 11.49
Flow diagram of binary split codebook generation algorithm. (After Rabiner and Juang [308].)

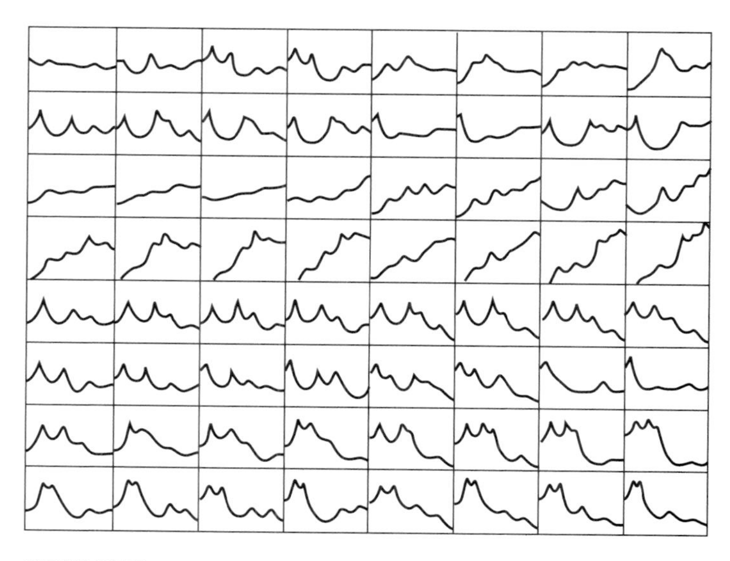

FIGURE 11.50
Spectral shapes corresponding to codebook vectors in an $M = 64$ codebook.

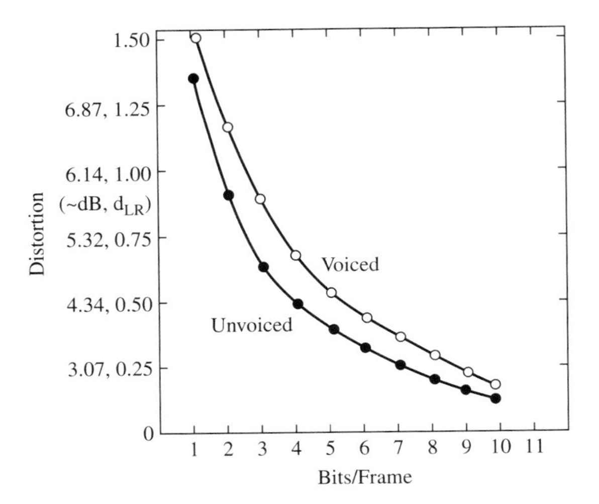

FIGURE 11.51
Codebook distortion versus codebook size (measured in bits/frame) for both voiced and unvoiced speech frames. (After Juang et al. [182].)

To illustrate the effect of codebook size (i.e., the number of codebook vectors) on the average training set distortion, Figure 11.51 shows experimentally measured values of distortion (as measured on a dB scale and on a likelihood ratio scale) versus codebook size (as measured in bits/frame, B) for vectors of both voiced and unvoiced speech. The figure shows that very significant reductions in distortion are achieved in going from a codebook size of 1 bit (2 vectors) to about 7 bits (128 vectors) for both voiced and unvoiced speech. Beyond this point, reductions in distortion are much smaller.

The technique of VQ has found widespread use in coding of parameters of the speech model. Our discussion has emphasized the use of VQ for coding the parameters of the speech model as part of a digital speech coding system. VQ is also widely used as a basis for selecting features in automatic speech recognition systems. We will discuss this application in Chapter 14, and much more detail can be found in Ref. [308]. In the remainder of this chapter, we shall see many examples of how the speech model is built into a speech coding system. In most such applications, VQ can be an effective way of representing the model parameters.

11.7 GENERAL THEORY OF DIFFERENTIAL QUANTIZATION

Figure 11.9a shows that there is considerable correlation between adjacent speech samples, and indeed, even at sampling rates as low as 8 kHz, the correlation is significant even between samples that are several sampling intervals apart. This high correlation implies that, in an average sense, the signal does not change rapidly from sample to sample. This is also evident in the example of Figure 11.45. Therefore, the difference between adjacent samples should have a lower variance than the variance of the signal itself. That this is so can be easily verified (see Problem 11.13). More generally, we showed in Chapter 9 that correlation in speech signals is the basis for linear predictive analysis. Indeed, we motivated our discussion of linear predictive analysis of speech by observing the close relationship between optimum linear predictors and the discrete-time linear system model for speech production that we derived from acoustic principles. From our previous discussions, we know that the output of an optimum linear predictor has lower variance than the input. This is illustrated by Figure 11.52, where the upper plot is a segment of a speech signal and the lower plot is the output of a linear prediction error filter $A(z)$ (with $p = 12$) that was derived from the given segment by the techniques of Chapter 9. Note that the prediction error sequence amplitude of Figure 11.52 is about a factor of 2.5 lower than the amplitude of the signal itself, which means that for a fixed number of bits, this segment of the prediction error could be quantized with a smaller step size than the waveform itself. If the unquantized prediction error (residual) segment is used as input to the corresponding inverse of the prediction error filter, $H(z) = 1/A(z)$, a very close approximation to the waveform of the original segment of speech would result. This is the basis for asserting that with proper choice of the predictor order p, $H(z)$ represents the vocal tract, and the prediction error plays the role of the excitation to the vocal tract system. However, if the prediction error is quantized, the quantization error is also filtered by $H(z)$, generally leading to unsatisfactory reconstruction of the speech signal.

While direct quantization of the output of a linear predictor does not lead to a practical method of speech coding, the basic principle of variance reduction by linear prediction is the basis for the general differential quantization scheme depicted

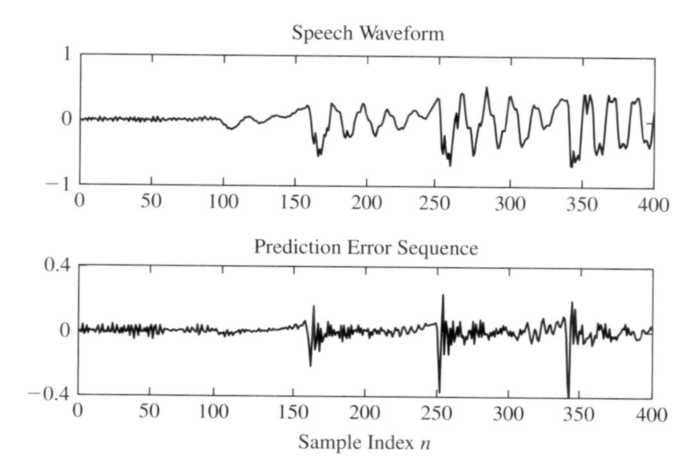

FIGURE 11.52
Illustration of reduction of variance by linear prediction. The upper plot shows a 400-sample section of the speech waveform; the lower plot shows the resulting prediction error sequence.

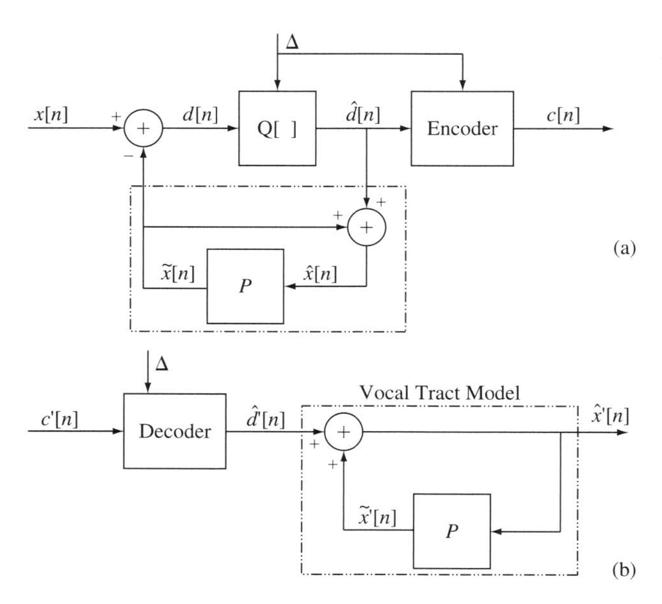

FIGURE 11.53
General differential quantization scheme: (a) encoder; (b) decoder.

in Figure 11.53 [78, 241]. In this system the input to the quantizer is the difference signal

$$d[n] = x[n] - \tilde{x}[n], \tag{11.83}$$

which is the difference between the unquantized input sample, $x[n]$, and an estimate, or prediction, of the input sample, which is denoted as $\tilde{x}[n]$. This predicted value is the output of a predictor system P, whose input, $\hat{x}[n]$, is a quantized version of the input signal, $x[n]$. This is clearly an example of employing the speech production model in a closed-loop coder as defined in Figure 11.42a of Section 11.6. The difference signal is like the prediction error signal, since it is the amount by which the predictor fails to exactly predict the input, but it is not the same as the prediction error signal discussed in Chapter 9 or in Figure 11.52, because the prediction is based on the signal $\hat{x}[n]$, not on $x[n]$. Temporarily leaving aside the question of how the estimate, $\tilde{x}[n]$, is obtained, we note that it is the difference signal that is quantized, rather than the input. The quantizer could be either fixed or adaptive, uniform or non-uniform, but in any case, its parameters should be adjusted to match the variance of $d[n]$. The quantized difference signal can be represented as

$$\hat{d}[n] = d[n] + e[n], \tag{11.84}$$

where $e[n]$ is the resulting quantization error. According to Figure 11.53a, the quantized difference signal is added to the predicted value, $\tilde{x}[n]$, to produce a quantized version of the input; i.e.,

$$\hat{x}[n] = \tilde{x}[n] + \hat{d}[n]. \tag{11.85}$$

Substituting Eqs. (11.83) and (11.84) into Eq. (11.85), we see that

$$\hat{x}[n] = x[n] + e[n]. \tag{11.86}$$

That is, independent of the properties of the system labeled P, the signal $\hat{x}[n]$ differs from the input only by the quantization error of the difference signal. We therefore refer to $\hat{x}[n]$ as the "quantized input" even though the input is not directly quantized by the quantizer. If the prediction is good, the variance of $d[n]$ will be smaller than the variance of $x[n]$, so that a quantizer with a given number of levels can be adjusted to give a smaller quantization error than would be possible when quantizing the input directly.

Note that it is the quantized difference signal, not the quantized input, that is coded for transmission or storage. The system for reconstructing the quantized input $\hat{x}[n]$ from the codewords is implicit in Figure 11.53a. This system, depicted in Figure 11.53b, involves a decoder to reconstruct the quantized difference signal, from which the quantized input is reconstructed using the same predictor as used in Figure 11.53a. Clearly, if $c'[n]$ is identical to $c[n]$, then $\hat{x}'[n] = \hat{x}[n]$, which differs from $x[n]$ only by the quantization error incurred in quantizing $d[n]$.

If we treat $x[n]$ as a random signal, we can determine the signal-to-quantizing noise ratio of the system of Figure 11.53 as,

$$\text{SNR} = \frac{E[x^2[n]]}{E[e^2[n]]} = \frac{\sigma_x^2}{\sigma_e^2}, \tag{11.87}$$

which can be written as

$$\text{SNR} = \frac{\sigma_x^2}{\sigma_d^2} \cdot \frac{\sigma_d^2}{\sigma_e^2} = G_P \cdot \text{SNR}_Q, \tag{11.88}$$

where SNR_Q is the signal-to-quantization noise ratio of the quantizer; i.e.,

$$\text{SNR}_Q = \frac{\sigma_d^2}{\sigma_e^2}, \tag{11.89}$$

and the quantity

$$G_P = \frac{\sigma_x^2}{\sigma_d^2} \tag{11.90}$$

is defined as the *prediction gain* due to the differential configuration.

The quantity SNR_Q is dependent upon the particular quantizer that is used, and, given knowledge of the properties of $d[n]$, SNR_Q can be maximized by using the techniques of the previous sections. The quantity G_P, if greater than unity, represents the gain in SNR that is due to the differential scheme. Clearly, our objective should be to maximize G_P by an appropriate choice of the predictor system, P. For a given input signal, σ_x^2 is a fixed quantity so that G_P can be maximized by minimizing the denominator of Eq. (11.90); i.e., by minimizing the variance of the difference signal.

To proceed, we need to specify the nature of the predictor, P. One approach that is well motivated by our previous discussion of the model for speech production, and extensively studied in Chapter 9, is to use a linear predictor. That is, $\tilde{x}[n]$ is a linear combination of past quantized values as in

$$\tilde{x}[n] = \sum_{k=1}^{p} \alpha_k \hat{x}[n-k]. \tag{11.91}$$

The predicted value is therefore the output of a finite impulse response filter whose system function is

$$P(z) = \sum_{k=1}^{p} \alpha_k z^{-k} = 1 - A(z), \tag{11.92}$$

and whose input is the reconstructed (quantized) signal $\hat{x}[n]$. We also note that the reconstructed signal is the output of a system whose system function is

$$H(z) = \frac{1}{1 - \displaystyle\sum_{k=1}^{p} \alpha_k z^{-k}} = \frac{1}{A(z)}, \tag{11.93}$$

and whose input is the quantized difference signal $\hat{d}[n]$. In Chapter 9, we associated this system function with the transmission properties of the vocal tract. With this point of view, the predictor system represents the knowledge of the speech spectrum that is built into the quantization process by incorporating the predictor. The sequence, $\hat{d}[n]$, corresponds roughly to the excitation sequence in the speech production model. Thus, the differential quantization system of Figure 11.53 has a built-in discrete-time model of speech production. We will see that this is a useful interpretation for speech coding systems ranging from systems where the predictor is fixed and based on long-term correlation measurements to more complex systems where the predictor is adapted to track the time-varying correlation/spectral properties of the speech signal.

Although the results of Chapter 9 are available to be applied here, we shall repeat the derivation of the optimum linear predictor. This is because we want to minimize the variance of the difference signal in Figure 11.53 rather than the prediction error as was done in Chapter 9. We shall see that the feedback of the quantization error leads to some significant differences that we should keep in mind in applying linear prediction in differential quantization systems. Specifically, the variance of the difference signal is[17]

$$\sigma_d^2 = E[d^2[n]] = E[(x[n] - \tilde{x}[n])^2]$$
$$= E\left[\left(x[n] - \sum_{k=1}^{p} \alpha_k \hat{x}[n-k]\right)^2\right]. \tag{11.94}$$

In order to choose a set of predictor coefficients $\{\alpha_k\}$, $1 \le k \le p$, that minimize σ_d^2, we follow the approach of Chapter 9 and differentiate σ_d^2 with respect to each parameter and set the derivatives equal to zero, thereby obtaining the following set of p equations with p unknown coefficients, α_k, $k = 1, 2, \cdots, p$:

$$\frac{\partial \sigma_d^2}{\partial \alpha_i} = E\left[-2\left(x[n] - \sum_{k=1}^{p} \alpha_k \hat{x}[n-k]\right) \cdot \hat{x}[n-i]\right]$$
$$= 0, \qquad 1 \le i \le p. \tag{11.95}$$

Equation (11.95) can be written in the more compact form

$$E[(x[n] - \tilde{x}[n])\,\hat{x}[n-i]] = E[d[n]\,\hat{x}[n-i]] = 0, \quad 1 \le i \le p, \tag{11.96}$$

from which we make the important observation that if the predictor coefficients satisfy Eq. (11.95) so as to minimize σ_d^2, then the difference signal (prediction error) is uncorrelated with (i.e., orthogonal to) past values of the predictor input $\hat{x}[n-i]$, $1 \le i \le p$.

Assuming the additive noise approximation for the quantizer ($\hat{x}[n] = x[n] + e[n]$), Eqs. (11.96) can be expanded into the set of p equations

$$E[x[n-i]\,x[n]] + E[e[n-i]\,x[n]] = \sum_{k=1}^{p} \alpha_k\, E[x[n-i]\,x[n-k]]$$
$$+\sum_{k=1}^{p} \alpha_k E[e[n-i]\,x[n-k]]$$
$$+\sum_{k=1}^{p} \alpha_k E[x[n-i]\,e[n-k]]$$
$$+\sum_{k=1}^{p} \alpha_k E[e[n-i]\,e[n-k]], \tag{11.97}$$

[17]Note that we are using the notation of expected value to denote the averaging. In practical applications $E[\,]$ would be replaced by a finite time average in order to estimate the predictor coefficient from the speech signal.

where $1 \leq i \leq p$. Now, if the quantization is fine enough, we can assume that $e[n]$ is uncorrelated with $x[n]$ and that $e[n]$ is a stationary white noise sequence; i.e.,

$$E[x[n-i]\, e[n-k]] = 0, \quad \text{for all } i \text{ and } k, \tag{11.98}$$

and

$$E[e[n-i]e[n-k]] = \sigma_e^2\, \delta[i-k]. \tag{11.99}$$

Using these assumptions, Eq. (11.97) can be simplified to

$$\phi[i] = \sum_{k=1}^{p} \alpha_k \left(\phi[i-k] + \sigma_e^2\, \delta[i-k] \right), \quad 1 \leq i \leq p, \tag{11.100}$$

where $\phi[i]$ is the autocorrelation function of $x[n]$. Dividing both sides of the above equations by σ_x^2 and defining the normalized autocorrelation as

$$\rho[i] = \frac{\phi[i]}{\sigma_x^2}, \tag{11.101}$$

we can express Eq. (11.100) in matrix form as

$$\boldsymbol{\rho} = \boldsymbol{C}\,\boldsymbol{\alpha}, \tag{11.102a}$$

where

$$\boldsymbol{\rho} = \begin{bmatrix} \rho[1] \\ \rho[2] \\ \cdot \\ \cdot \\ \cdot \\ \rho[p] \end{bmatrix} \tag{11.102b}$$

and

$$\boldsymbol{C} = \begin{bmatrix} (1 + \frac{1}{\text{SNR}}) & \rho[1] & \dots & \rho[p-1] \\ \rho[1] & (1 + \frac{1}{\text{SNR}}) & \dots & \rho[p-2] \\ \cdot & \cdot & \cdot & \cdot \\ \cdot & \cdot & \cdot & \cdot \\ \rho[p-1] & \rho[p-2] & \dots & (1 + \frac{1}{\text{SNR}}) \end{bmatrix}, \tag{11.102c}$$

with $\text{SNR} = \sigma_x^2/\sigma_e^2$, and

$$\boldsymbol{\alpha} = \begin{bmatrix} \alpha_1 \\ \alpha_2 \\ \cdot \\ \cdot \\ \cdot \\ \alpha_p \end{bmatrix}. \tag{11.102d}$$

Thus the vector of optimum predictor coefficients is obtained as the solution of the matrix equation [Eq. (11.102a)]; i.e.,

$$\boldsymbol{\alpha} = \boldsymbol{C}^{-1}\boldsymbol{\rho}. \tag{11.103}$$

The matrix $\boldsymbol{C}^{-1}$ can be computed by a variety of numerical methods, including methods that take advantage of the fact that $\boldsymbol{C}$ is a Toeplitz matrix (see Chapter 9). However, Eq. (11.102a) cannot be solved in the most general case since the diagonal terms of the matrix $\boldsymbol{C}$ depend on the SNR, which is equal to σ_x^2/σ_e^2 [see Eq. (11.102c)]; but SNR depends on the coefficients of the linear predictor, which in turn depend upon SNR through Eq. (11.102a). One possibility is to neglect the term $1/\text{SNR}$ in Eq. (11.102c), in which case, the solution is straightforward since it reduces to minimizing the mean-squared prediction error based on $x[n]$. For example, if $p = 1$, and $1/\text{SNR}$ is neglected, then $\alpha_1 = \rho[1]$. However, such an assumption is unnecessary for the case $p = 1$ since Eq. (11.103) can be directly solved to give

$$\alpha_1 = \frac{\rho[1]}{1 + \dfrac{1}{\text{SNR}}}, \tag{11.104}$$

from which it follows that $\alpha_1 < \rho[1]$.

In spite of the difficulties in solving explicitly for the predictor coefficients, it is possible to obtain an expression for the optimum G_P in terms of the α_i's. To do this, we solve for σ_d^2 by rewriting Eq. (11.94) in the form

$$\begin{aligned} \sigma_d^2 &= E[(x[n] - \tilde{x}[n])\,(x[n] - \tilde{x}[n])] \\ &= E[(x[n] - \tilde{x}[n])\,x[n]] - E[(x[n] - \tilde{x}[n])\,\tilde{x}[n]]. \end{aligned} \tag{11.105}$$

Using Eq. (11.96) it is straightforward to show that for the optimum predictor coefficients, the second term in the above equation is zero; i.e., the predicted value is also uncorrelated with the prediction error (see Problem 11.15). Thus we can write

$$\begin{aligned} \sigma_d^2 &= E[(x[n] - \tilde{x}[n])\,x[n]] \\ &= E[x^2[n]] - E\left[\sum_{k=1}^{p} \alpha_k (x[n-k] + e[n-k])\,x[n]\right]. \end{aligned} \tag{11.106}$$

Using the assumptions of uncorrelated signal and noise, we obtain

$$\sigma_d^2 = \sigma_x^2 - \sum_{k=1}^{p} \alpha_k \phi[k] = \sigma_x^2 \left[1 - \sum_{k=1}^{p} \alpha_k \rho[k]\right]. \tag{11.107}$$

Thus, from Eq. (11.90),

$$(G_P)_{\text{opt}} = \frac{1}{1 - \sum_{k=1}^{p} \alpha_k \rho[k]}, \tag{11.108}$$

where the α_k's are the optimum coefficients satisfying Eq. (11.102a).

For the case $p = 1$, we can examine the effects of using a suboptimum value of α_1 on the quantity $G_P = \sigma_x^2/\sigma_d^2$. From Eq. (11.108), we get

$$(G_P)_{\text{opt}} = \frac{1}{1 - \alpha_1 \rho[1]}. \tag{11.109}$$

If we choose an arbitrary value for α_1, then by repeating the derivation leading to Eq. (11.107), we get

$$\sigma_d^2 = \sigma_x^2[1 - 2\alpha_1\rho[1] + \alpha_1^2] + \alpha_1^2 . \sigma_e^2. \tag{11.110}$$

Solving for σ_x^2/σ_d^2 gives

$$(G_P)_{\text{arb}} = \frac{1}{1 - 2\alpha_1\rho[1] + \alpha_1^2\left[1 + \dfrac{1}{\text{SNR}}\right]}. \tag{11.111}$$

The term α_1^2/SNR represents the increase in variance of $d[n]$ due to the feedback of the error signal $e[n]$. It can be shown (see Problem 11.16) that Eq. (11.111) can be rewritten in the form:

$$(G_P)_{\text{arb}} = \frac{1 - \dfrac{\alpha_1^2}{\text{SNR}_Q}}{1 - 2\alpha_1\rho[1] + \alpha_1^2}, \tag{11.112}$$

for any value of α_1 (including the optimum value). Thus, for example, if $\alpha_1 = \rho[1]$ [which is suboptimum according to Eq. (11.104)] then

$$(G_P)_{\text{subopt}} = \frac{1 - \dfrac{\rho^2[1]}{\text{SNR}_Q}}{1 - \rho^2[1]} = \left[\frac{1}{1 - \rho^2[1]}\right]\left[1 - \frac{\rho^2[1]}{\text{SNR}_Q}\right]. \tag{11.113}$$

Thus the gain in prediction obtained without the quantizer, $1/(1-\rho^2[1])$, is reduced by the second factor in Eq. (11.113) due to feedback of the error signal.

To obtain the optimum gain, Eq. (11.112) can be differentiated with respect to α_1 to give

$$\frac{d(G_P)}{d\alpha_1} = 0, \tag{11.114}$$

which can be solved directly for the optimum value of α_1.[18]

To illustrate the improvement to be gained by predictive quantization, assume that we can neglect the term 1/SNR in Eq. (11.102c). For a first-order predictor, Eq. (11.104) becomes $\alpha_1 = \rho[1]$ and the gain due to prediction is

$$(G_P)_{\text{opt}} = \frac{1}{1-\rho^2[1]}, \tag{11.115}$$

from which it follows that so long as $\rho[1] \neq 0$, there will be some improvement due to prediction. We have already seen (Figure 11.7) typical correlation functions for lowpass and bandpass filtered speech sampled at 8 kHz [254]. The shaded regions in this figure indicate the range of variation of $\rho[n]$ over four speakers, and the central curve is the average for these four speakers. We see from these curves that a reasonable assumption is that for lowpass filtered speech, sampled at the Nyquist rate,

$$\rho[1] > 0.8, \tag{11.116}$$

which implies that

$$(G_P)_{\text{opt}} > 2.77 \;\; (\text{or } 4.43 \text{ dB}). \tag{11.117}$$

Noll [254] used the data shown in Figure 11.7 to compute $(G_P)_{\text{opt}}$ as a function of p for a 55-second segment of speech that was both lowpass and bandpass filtered. The results are depicted in Figure 11.54.[19] The shaded region shows the amount of variation obtained for four speakers, with the central curve representing the average over four speakers. It is clear that, even with the simplest predictor, it is possible to realize about a 6 dB improvement in SNR. This is, of course, equivalent to adding an extra bit to the quantizer. Note also that in no case does the gain reach 12 dB, which would be required to achieve the effect of adding 2 bits to the quantizer word length (quadrupling the number of levels). An alternative point of view is that differential quantization permits a reduction in bit rate, while keeping the SNR the same. The price paid is increased complexity in the quantization system.

[18] We are indebted to Professor Peter Noll for helpful comments on this analysis.

[19] Again the sampling rate was 8 kHz; thus nT is a multiple of 125 μsec.

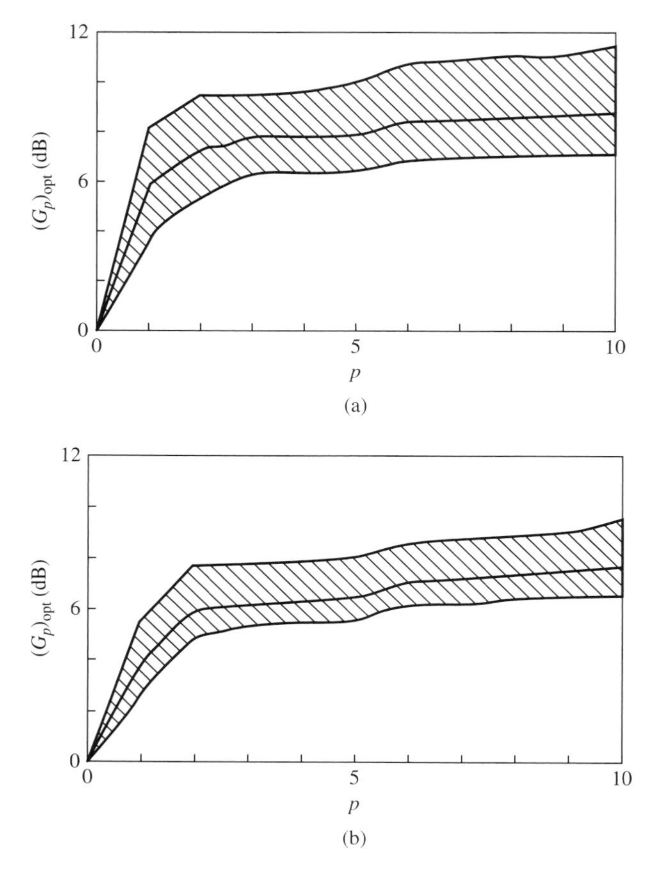

FIGURE 11.54
Optimum SNR gain $(G_P)_{\text{opt}}$ versus number of predictor coefficients: (a) lowpass filtered speech; (b) bandpass filtered speech. (After Noll [256]. Reprinted with permission of Alcatel-Lucent USA Inc.)

Some basic principles of application of the differential quantization scheme emerge from a consideration of Figures 11.7 and 11.54. First, it is clear that differential quantization can yield improvements over direct quantization. Second, the amount of improvement is dependent upon the amount of correlation. Third, a fixed predictor cannot be optimum for all speakers and for all speech material. These facts have led to a variety of schemes based on the basic configuration of Figure 11.53. These schemes combine a variety of fixed and adaptive quantizers and/or predictors to achieve improved quality or lowered bit rate. We will now discuss several examples that represent the range of possibilities.

11.8 DELTA MODULATION

The simplest application of the concept of differential quantization is in *delta modulation* (abbreviated DM) [349–375]. In this class of systems, the sampling rate is chosen to be many times the Nyquist rate for the input signal. As a result, adjacent samples become highly correlated. This is evident from the discussion of Section 11.3, where we showed that the autocorrelation of the sequence of samples is a sampled version of the analog autocorrelation; i.e.,

$$\phi[m] = \phi_a(mT). \tag{11.118}$$

Given the properties of autocorrelation functions, it is reasonable to expect the correlation to increase as $T \to 0$. Indeed, we expect that except for strictly uncorrelated signals,

$$\phi[1] \to \sigma_x^2 \text{ as } T \to 0. \tag{11.119}$$

This high degree of correlation implies that as T tends to zero, we should be better able to predict the input from past samples and, as a result, the variance of the prediction error should be low. Therefore, because of the high gain due to the differential configuration, a rather crude quantizer can provide acceptable performance. Indeed, delta modulation systems employ a simple 1-bit (2-level) quantizer. Thus, the bit-rate of a delta modulation system is simply equal to the sampling rate; i.e.,

$$I_{\text{DM}} = F_s.$$

11.8.1 Linear Delta Modulation

The simplest delta modulation system is depicted in Figure 11.55. In this case the quantizer has only two levels and the step size is fixed. The positive quantization level is represented by $c[n] = 0$ and the negative by $c[n] = 1$. Thus, $\hat{d}[n]$ is

$$\hat{d}[n] = \begin{cases} \Delta & \text{if } c[n] = 0 \\ -\Delta & \text{if } c[n] = 1. \end{cases} \tag{11.120}$$

Figure 11.55 incorporates a simple first-order fixed predictor for which the prediction gain is approximately

$$G_P = \frac{1}{1 - \rho^2[1]}. \tag{11.121}$$

In terms of our interpretation of the predictor system as a model for the vocal tract system in the model for speech production, we see that the simple first-order fixed

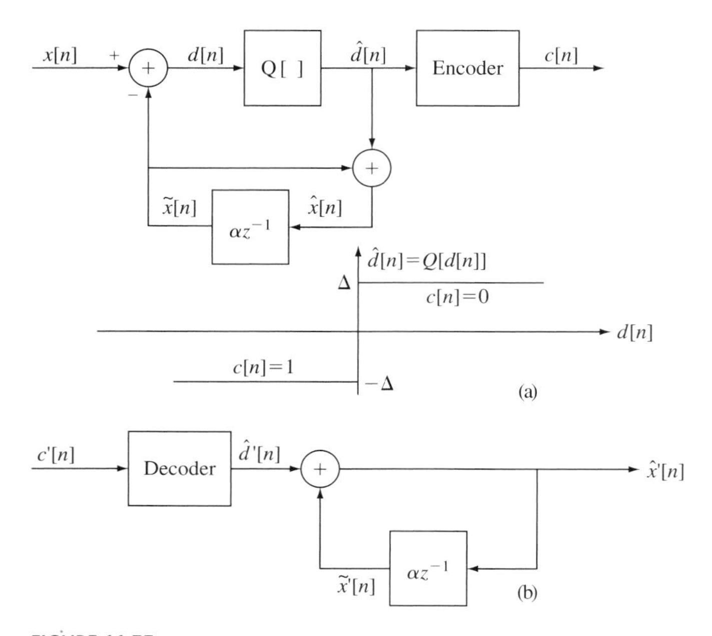

FIGURE 11.55
Block diagram of a linear delta modulation system: (a) encoder; (b) decoder. Not shown is a lowpass filter for removing high frequency noise introduced by the quantization.

system can only model the general long-term spectral shape of the speech spectrum. Nevertheless, as $\rho[1] \to 1$ with increasing F_s, $G_P \to \infty$. This result can be viewed only in qualitative terms, however, since the assumptions under which the expression for G_P was derived tend to break down for such crude quantization.

The effect of quantization error can be observed from Figure 11.56a, which depicts an analog waveform $x_a(t)$ and resulting samples, $x[n]$, $\tilde{x}[n]$, and $\hat{x}[n]$, for a given sampling period, T, and assuming α (the feedback multiplier) is set to 1.0. Figure 11.55a shows that, in general, $\hat{x}[n]$ satisfies the difference equation

$$\hat{x}[n] = \alpha\hat{x}[n-1] + \hat{d}[n]. \tag{11.122}$$

With $\alpha \approx 1$, this equation is the digital equivalent of integration, in the sense that it represents the accumulation of positive and negative increments of (small) magnitude Δ. We also note that the input to the quantizer is

$$d[n] = x[n] - \hat{x}[n-1] = x[n] - x[n-1] - e[n-1]. \tag{11.123}$$

Thus except for the quantization error in $\hat{x}[n-1]$, $d[n]$ is a first backward difference of $x[n]$, which can be viewed as a digital approximation to the derivative of the input and the inverse of the digital integration process. If we consider the maximum slope of the waveform, it is clear that in order for the sequence of samples $\{\hat{x}[n]\}$ to increase as fast

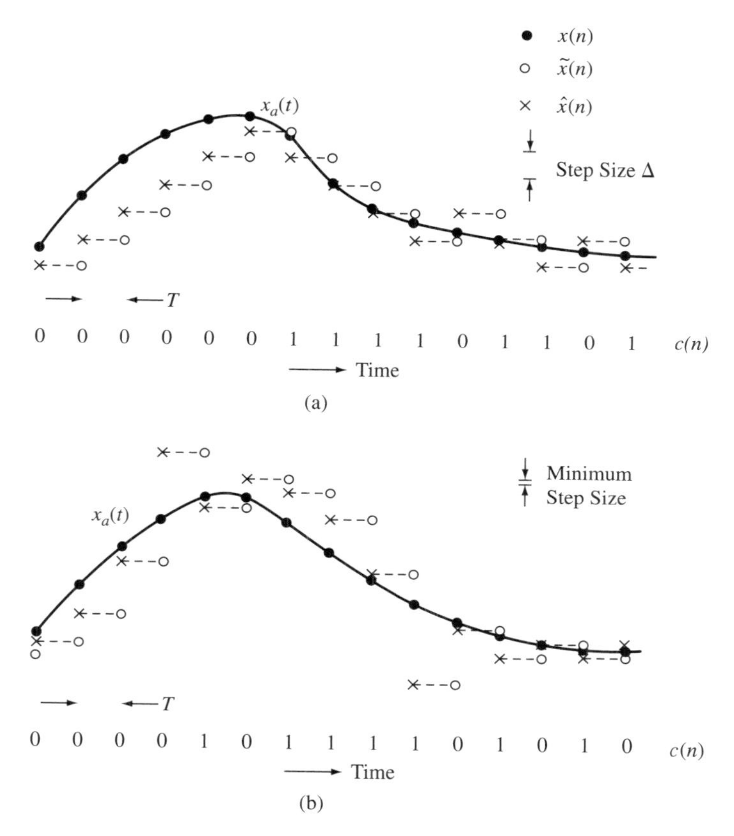

FIGURE 11.56
Illustration of delta modulation: (a) fixed step size; (b) adaptive step size.

as the sequence of samples $\{x[n]\}$ in a region of maximum slope of $x_a(t)$, we require

$$\frac{\Delta}{T} \geq \max\left|\frac{dx_a(t)}{dt}\right|. \tag{11.124}$$

Otherwise, the reconstructed signal will fall behind, as shown on the left side of Figure 11.56a. This condition is called "slope overload" and the resulting quantization error is called *slope overload distortion* (noise). Note that since the maximum slope of $\hat{x}[n]$ is fixed by the step size, increases and decreases in the sequence $\hat{x}[n]$ tend to occur along straight lines. For this reason, fixed (non-adaptive) delta modulation is often called *linear delta modulation* (abbreviated LDM) even though such systems are clearly non-linear in the system theory sense.

The step size, Δ, also determines the peak error when the slope is very small. For example, it is easily verified that when the input is zero (idle channel condition), the output of the quantizer will be an alternating sequence of 0's and 1's, in which case the reconstructed signal $\hat{x}[n]$ will alternate about zero (or some constant level) with a peak-to-peak variation of Δ. This latter type of quantization error, depicted on the right in Figure 11.56a, is called *granular noise*.

Again there is a need to have a large step size to accommodate a wide dynamic range, while a small step size is required for accurate representation of low level signals. In this case, however, we are concerned with the dynamic range and amplitude of the difference signal (or derivative of the analog signal). The choice of step size that minimizes the mean-squared quantization error must therefore be a compromise between slope overload and granular noise.

Figure 11.57, which is from a detailed study of delta modulation by Abate [1], shows SNR as a function of a normalized step size variable defined as $\Delta/(E[(x[n]-x[n-1])^2])^{1/2}$, with oversampling index $F_0 = F_s/(2F_N)$ as a parameter, where F_s is the sampling rate of the delta modulator, and F_N is the Nyquist frequency of the signal. Note that the bit rate is

$$\text{Bit rate } = F_s \cdot (1 \text{ bit}) = F_s = F_0 \cdot (2F_N). \tag{11.125}$$

Thus the oversampling index plays the role of the number of bits/sample for a multi-bit quantizer with sampling at the Nyquist rate. These curves are for flat-spectrum bandlimited Gaussian noise. Somewhat higher SNR values are obtained for speech since there is greater correlation; however, the shape of the curves is much the same. It can be seen from Figure 11.57 that for a given value of F_0, the SNR curve has a rather sharp peak, with values of Δ above the location of the peak corresponding to granular noise and values below the peak location corresponding to slope overload. Abate [1] gives the empirical formula

$$\Delta_{\text{opt}} = \{E(x[n]-x[n-1])^2\}^{1/2} \log(2F_0), \tag{11.126}$$

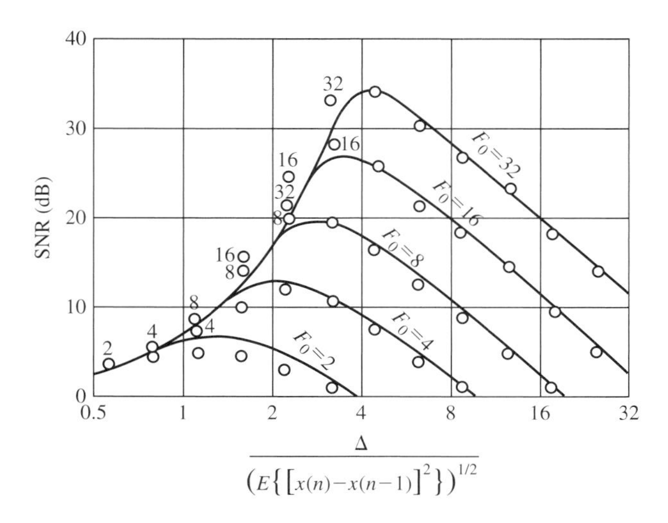

FIGURE 11.57
SNR for delta modulators as a function of the normalized step size. (After Abate [1]. © [1967] IEEE.)

for the optimum step size; i.e., for the location of the peak of the SNR curve for a given value of F_0. Figure 11.57 also shows that the optimum SNR increases at the rate of about 8 dB for each doubling of F_0. Since doubling F_0 is equivalent to doubling F_s, we note that doubling the bit rate increases the SNR by 8 dB. This is in contrast to PCM where if we double the bit rate by doubling the number of bits/sample, we achieve a 6 dB increase for *each added bit*; thus the increase of SNR with bit rate is much more dramatic for PCM than for LDM.

Another important feature of the curves of Figure 11.57 is the sharpness of the peak of the SNR curve, which implies that the SNR is very sensitive to the input level. [Note that $E[(x[n] - x[n-1])^2] = 2\sigma_x^2(1 - \rho[1])$.] Thus, it can be seen from Figure 11.57 that to obtain an SNR of 35 dB for a Nyquist frequency of 3 kHz would require a bit rate of about 200 kbps. Even at this rate, however, this quality can only be maintained over a rather narrow range of input levels if the step size is fixed. To achieve toll quality, i.e., quality comparable to 7- or 8-bit log-PCM, for speech at a sampling rate of 8 kHz, requires much higher bit rates.

The main advantage of LDM is its simplicity. The system can be implemented with simple analog and digital integrated circuits, and since only a 1-bit code is required, no synchronization of bit patterns is required between transmitter and receiver. The limitations on the performance of LDM systems stem mainly from the crude quantization of the difference signal. In view of our previous discussion of adaptive quantization, it is natural to suppose that adaptive quantization schemes would greatly improve the performance of a delta modulator. Of greatest interest are simple adaptive quantization schemes that improve performance but do not greatly increase the complexity of the system.

11.8.2 Adaptive Delta Modulation

LDM systems must use high oversampling ratios to allow a small step size and corresponding small quantization error. Another possibility is to adapt the step size depending on the slope of the signal. Most *adaptive delta modulation* (ADM) systems are of the feedback type in which the step size for the 2-level quantizer is adapted from the output codewords. The general form of such systems is shown in Figure 11.58. Such schemes maintain the advantage that no synchronization of bit-patterns is required since, in the absence of errors, the step size information can be derived from the codeword sequence at both the transmitter and the receiver.

In this section we will illustrate the use of adaptive quantization in delta modulation through the discussion of two specific adaptation algorithms. Many other possibilities can be found in the literature [348, 375].

The first system that we will discuss was studied extensively by N. S. Jayant [167]. Jayant's algorithm for ADM is a modification of the quantization scheme discussed in Section 11.5.2, where the step size obeys the rule

$$\Delta[n] = M\Delta[n-1], \tag{11.127a}$$

$$\Delta_{\min} \leq \Delta[n] \leq \Delta_{\max}. \tag{11.127b}$$

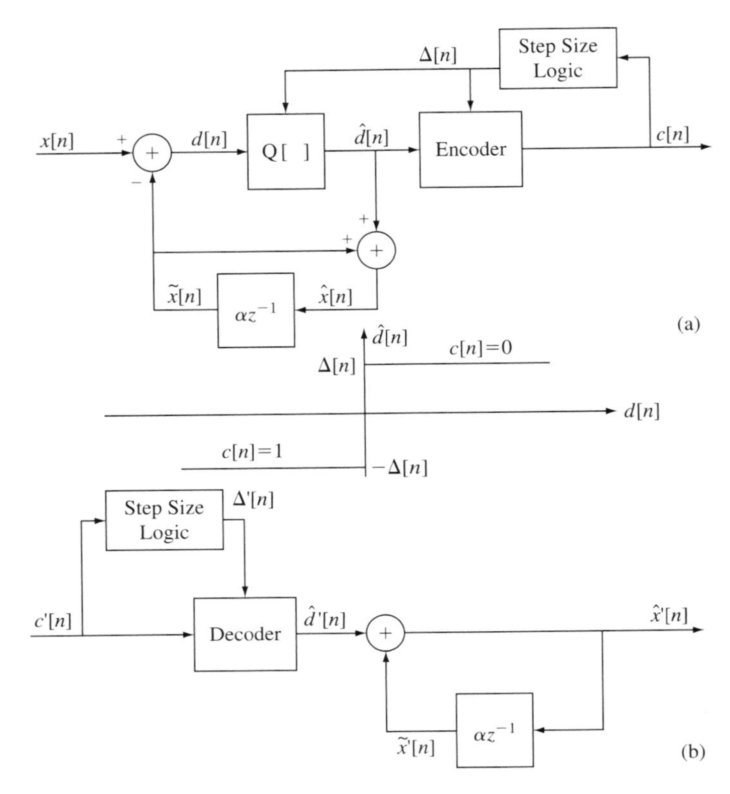

FIGURE 11.58
Delta modulation with adaptive step size: (a) encoder; (b) decoder.

In this case, the multiplier is a function of the present and the previous codewords, $c[n]$ and $c[n-1]$. This is possible since $c[n]$ depends only on the sign of $d[n]$, which is given by

$$d[n] = x[n] - \alpha\hat{x}[n-1]. \tag{11.128}$$

Thus the sign of $d[n]$ can be determined before the determination of the actual quantized value $\hat{d}[n]$, which must await the determination of $\Delta[n]$ from Eq. (11.127a). The algorithm for choosing the step size multiplier in Eq. (11.127a) is

$$\begin{aligned} M &= P > 1 \quad \text{if } c[n] = c[n-1], \\ M &= Q < 1 \quad \text{if } c[n] \neq c[n-1]. \end{aligned} \tag{11.129}$$

This adaptation strategy is motivated by the bit patterns observed in LDM. For example, in Figure 11.56a, periods of slope overload are signaled by runs of consecutive 0's or 1's. Periods of granularity are signaled by alternating sequences of the form ...0 1 0 1 0 1.... Figure 11.56b shows how the waveform in Figure 11.56a would be quantized by an adaptive delta modulator of the type described in Eqs. (11.127a) and (11.129). For convenience, the parameters of the system are set at $P = 2$, $Q = 1/2$, $\alpha = 1$, and the minimum step size is shown in the figure. It can be seen that the region of large positive slope still causes a run of 0's but in this case, the step size increases exponentially so as to follow the increase in the slope of the waveform. The region of granularity to the right in the figure is again signaled by an alternating sequence of 0's and 1's, but in this case, the step size falls rapidly to the minimum (Δ_{min}) and remains there as long as the slope is small. Since the minimum step size can be made much smaller than that required for optimum performance of a linear delta modulator, granular noise can be greatly reduced. Likewise the maximum step size can be made larger than the maximum slope of the input signal so as to reduce slope overload noise.

The parameters of this ADM system are P, Q, Δ_{min}, and Δ_{max}. The step size limits should be chosen to provide the desired dynamic range for the input signal. The ratio $\Delta_{\text{max}}/\Delta_{\text{min}}$ should be large enough to maintain a high SNR over a desired range of input signal levels. The minimum step size should be as small as is practical so as to minimize the idle channel noise. Jayant [167] has shown that for stability, i.e., to maintain the step size at values appropriate for the level of the input signal, P and Q should satisfy the relation

$$PQ \leq 1. \tag{11.130}$$

Figure 11.59 shows the results of a simulation for speech signals with $PQ = 1$ for three different sampling rates. It is evident that the maximum SNR is obtained for $P = 1.5$; however, the peak of all three curves is very broad, with SNR being within a few dB of

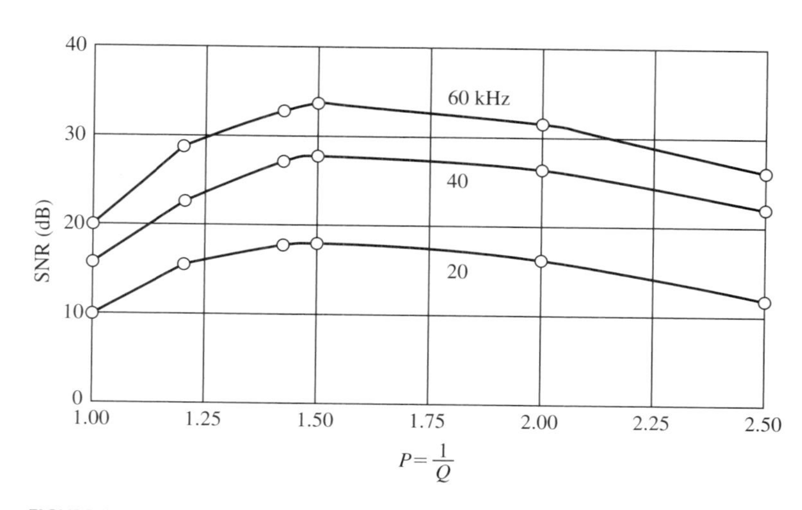

FIGURE 11.59
SNRs of an adaptive delta modulator as functions of P. (After Jayant [167]. Reprinted with permission of Alcatel-Lucent USA Inc.)

the maximum for

$$1.25 < P < 2. \tag{11.131}$$

The results of Figure 11.59 are replotted in Figure 11.60 to compare ADM ($P = 1.5$) to LDM ($P = 1$) and log-PCM. It is noted that with $P = 1/Q$, the condition $P = 1 = 1/Q$ implies no adaptation at all; i.e., LDM. The SNR values for this condition and for $P = 1.5$ are plotted as a function of bit rate in Figure 11.60. Also shown there is maximum SNR for $\mu = 100$ (log-PCM) as a function of bit rate [as computed from Eq. (11.45)], assuming sampling at the Nyquist rate ($F_s = 2F_N = 6.6$ kHz).

Figure 11.60 shows that ADM is superior to LDM by 8 dB at 20 kbps and the SNR advantage increases to 14 dB at 60 kbps. For LDM we observe about a 6 dB increase with a doubling of sampling rate (and bit rate), whereas with ADM, the corresponding increase is 10 dB. Comparing ADM and log-PCM, we note that for bit rates below 40 kbps, ADM outperforms log-PCM. For higher bit rates, log-PCM has a higher SNR. For example, Figure 11.60 shows that the ADM system requires about 56 kbps to achieve the same quality as 7-bit log-PCM, having a bit rate of about $7 \times 6.6 = 46.2$ kbps.

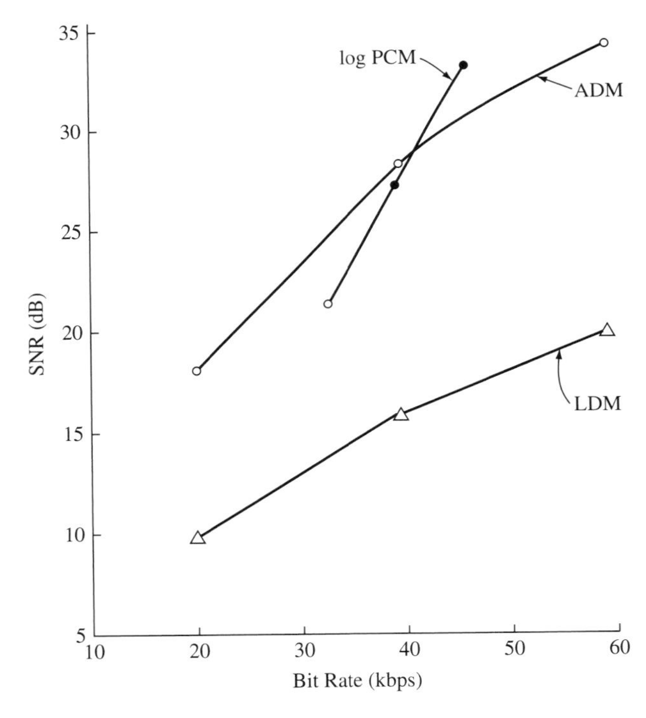

FIGURE 11.60
SNR versus bit rate for three coding schemes using a 6.6 kHz sampling rate.

The improved quality of the ADM system is achieved with only a slight increase in complexity over LDM. Since the step size adaptation is done using the output bit stream, the ADM system retains the basic simplicity of delta modulation systems; i.e., no codeword framing is required. Thus, for many applications, ADM may be preferred to log-PCM even at the expense of slightly higher information rate.

Another example of adaptive quantization in delta modulation is known as *continuously variable slope delta modulation* (CVSD). (A system of this type was first proposed by Greefkes [139].) This system is again based upon Figure 11.58, with the step size logic being defined by the equations

$$\Delta[n] = \beta\Delta[n-1] + D_2 \quad \text{if } c[n] = c[n-1] = c[n-2] \tag{11.132a}$$

$$= \beta\Delta[n-1] + D_1 \quad \text{otherwise,} \tag{11.132b}$$

where $0 < \beta < 1$ and $D_2 \gg D_1 \gg 0$. In this case, the minimum and maximum step sizes are inherent in the recurrence formula for $\Delta[n]$. (See Problem 11.18.)

The basic principle is, as before, to increase the step size in response to patterns in the bit stream that indicate slope overload. In this case, a run of three consecutive 1's or three consecutive 0's causes an increment D_2 to be added to the step size. In the absence of such patterns, the step size decays (because $\beta < 1$) until it reaches $\Delta_{min} = D_1/(1-\beta)$. Thus the step size will increase during slope overload conditions and will decrease otherwise. Again, Δ_{min} and Δ_{max} can be chosen to provide the desired dynamic range and low granular noise during idle channel conditions. This basic adaptation scheme can be adjusted to be either syllabic or instantaneous, since the parameter β controls the speed of adaptation. If β is close to 1, the rate of build-up and decay of $\Delta[n]$ is slow, whereas if β is much less than 1, the adaptation is much faster.

This system has been used in situations requiring low sensitivity to channel errors, with speech quality requirements below those required for commercial communication channels. In this situation, the parameters of the system are adjusted to provide syllabic adaptation. Also the predictor coefficient, α, is set at a value considerably less than 1 so that the effect of channel errors dies out quickly. Of course, the price paid for insensitivity to errors is decreased quality when no errors occur. A major advantage of the ADM system in this situation is that it has sufficient flexibility to provide effective trade-offs between quality and robustness.

11.8.3 Higher-Order Predictors in Delta Modulation

For simplicity, most LDM and ADM systems use a first-order fixed predictor of the form

$$\tilde{x}[n] = \alpha\hat{x}[n-1], \tag{11.133}$$

as shown in Figure 11.58. In this case, the reconstructed signal satisfies the difference equation

$$\hat{x}[n] = \alpha\hat{x}[n-1] + \hat{d}[n], \tag{11.134}$$

which is characterized by the system function

$$H_1(z) = \frac{1}{1 - \alpha z^{-1}}. \tag{11.135}$$

This, we have suggested, is the digital equivalent of an integrator (if $\alpha = 1$). When $\alpha < 1$, it is sometimes called a "leaky integrator."

The results shown in Figure 11.54 suggest[20] that for delta modulation systems, a greater SNR is possible with a second-order predictor; i.e., with

$$\tilde{x}[n] = \alpha_1 \hat{x}[n-1] + \alpha_2 \hat{x}[n-2]. \tag{11.136}$$

In this case

$$\hat{x}[n] = \alpha_1 \hat{x}[n-1] + \alpha_2 \hat{x}[n-2] + \hat{d}[n], \tag{11.137}$$

which is characterized by

$$H_2(z) = \frac{1}{1 - \alpha_1 z^{-1} - \alpha_2 z^{-2}}. \tag{11.138}$$

It has been shown empirically [76] that second-order prediction gives improved performance over first-order prediction when the poles of $H_2(z)$ are both real; i.e.,

$$H_2(z) = \frac{1}{(1 - az^{-1})(1 - bz^{-1})}, \quad 0 < a, b < 1. \tag{11.139}$$

This is often called "double integration." Improvements over first-order prediction can be as high as 4 dB, depending on speaker and speech material [76].

Unfortunately, the use of higher order prediction in ADM systems is not just a simple matter of replacing the first-order predictor with a second-order predictor, since the adaptive quantization algorithm interacts with the prediction algorithm. For example, the idle channel condition will be signaled by different bit patterns depending on the order of the predictor. For a second-order predictor, the bit pattern for the idle channel condition might be ... 010101 ... or ... 00110011 ..., depending upon the choice of α_1 and α_2 and the past state of the system before the input became zero. This clearly calls for an adaptation algorithm based upon more than two consecutive bits if the step size is to fall to its minimum value for idle channel conditions.

The design of ADM systems with high order predictors has not been extensively studied. Whether the added complexity in both the predictor and the quantizer could be justified would depend upon the amount of improvement in quality that could be

[20] To be more specific one would have to know exact values of the speech autocorrelation function for lags much less than 125 μsec (corresponding to the higher sampling rates of delta modulation systems) to calculate high order prediction gains.

obtained. The use of multi-bit quantizers of the type discussed in Section 11.5 simplifies the design somewhat at the expense of the need for framing the bit stream. We now turn to a discussion of differential quantization using multi-bit quantizers.

11.8.4 LDM-to-PCM conversion

In order to obtain a high quality representation of speech, a linear delta modulator employs a very high sampling rate and a simple 1-bit quantizer. Such systems are easy to implement using a combination of simple analog and digital components. For example Figure 11.61 shows a representation of an early integrated circuit implementation of an LDM system. It consists of an analog comparator circuit to create a difference signal, a flip-flop to sense the polarity of the difference signal, and an integrator to reconstruct a (predicted) signal to compare to the input. The gain of the integrator controls the effective step size, and the clocking of the flip-flop accomplishes the sampling in time. This simple combination of analog and digital circuitry is all that is required to implement a delta modulator. The output of the flip-flop is a pulse train that corresponds to the sequence of 1-bit binary codewords that represent the input. Circuits of this type can be implemented as integrated circuits with very high sampling rates $F_s' \gg 2F_N$, thereby achieving digital coding at high SNRs at low cost [22]. The price of this simplicity is the extremely high data rate ($I_{\text{DM}} = F_s'$) required for high quality reproduction of the speech signal.

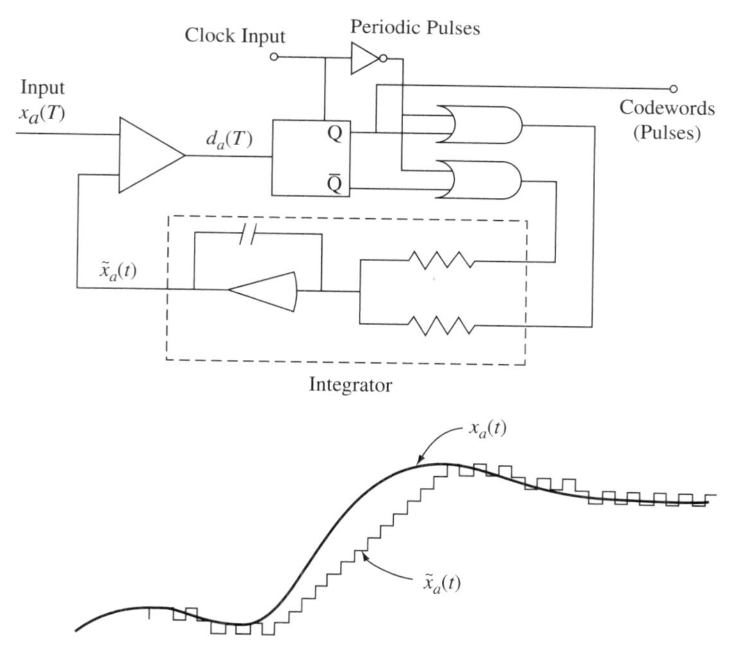

FIGURE 11.61
Circuit implementation of a linear delta modulator. (After Baldwin and Tewksbury [22]. © [1974] IEEE.)

The bit rate can be reduced, however, by using digital signal processing techniques to convert the LDM codewords (1's and 0's) into another, more efficient representation such as PCM or ADPCM at lower sampling rates. One of the most important conversions is from LDM to uniform PCM, since uniform PCM is required whenever further numerical processing of the samples of an analog waveform is desired.

A system for converting the LDM representation to a PCM representation is depicted in Figure 11.62. A PCM representation $\hat{s}[n]$ is first reconstructed from the LDM bit stream at the LDM sampling rate $F_s' = 1/T' = M/T$ by a digital implementation of the LDM decoder. This is followed by lowpass filtering and reduction in sampling rate to the Nyquist rate $F_s = 1/T = 2F_N$. The first step is accomplished as discussed in Section 11.8.1 by decoding the 1's and 0's into increments with values $\pm\Delta$, and then digitally accumulating (instead of integrating as in analog implementations) the resulting positive and negative increments to obtain quantized (PCM) samples of $\hat{s}[n] = x_a(nT') + e[n]$ at the LDM sampling rate $F_s' = 1/T'$. Thus, an LDM-to-PCM converter is essentially an accumulator or up-down counter whose output is filtered and sampled in discrete-time. As Figure 11.62 shows, an LDM system followed by LDM-to-PCM converter effectively constitutes an analog-to-PCM (A-to-D) converter whose implementation is almost completely digital.

If the LDM sampling rate is high enough to avoid slope overload and the granular noise is also low, the quantization error will be small and the reconstruction error $e[n]$ can be modeled as white noise with power $\sigma_e^2 \sim \Delta^2$. As shown in Figure 11.63a, the spectrum of $\hat{s}[n] = x_a(nT') + e[n]$ includes both the speech spectrum, occupying only the band $|F| \leq F_N$, and the quantization noise, which is spread throughout the band $|F| \leq F_s'/2$, where $F_s' = 2MF_N$ is the LDM sampling frequency. That is, oversampling by a factor of M compared to Nyquist sampling allows the total noise power, σ_e^2, to spread over a much wider band of frequencies than will ultimately be preserved after downsampling by a factor of M. Therefore, before reducing the sampling rate to the Nyquist rate for the input, the quantization noise in the band from the Nyquist frequency up to one-half the LDM sampling frequency can be removed by a lowpass digital filter, whose effective cutoff frequency is the desired Nyquist frequency

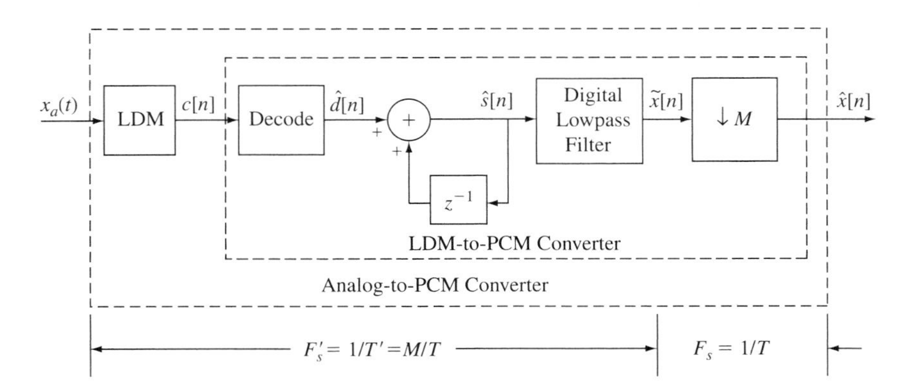

FIGURE 11.62
Analog-to-PCM converter using an LDM system and an LDM-to-PCM converter.

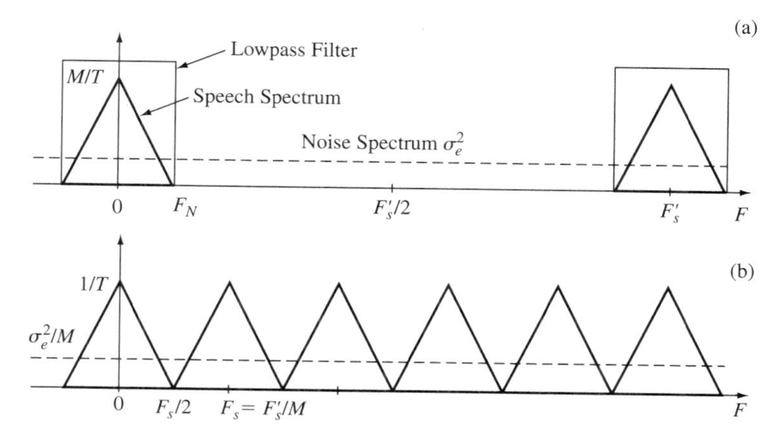

FIGURE 11.63
Representation of spectra of speech and quantization noise in analog-to-LDM-to-PCM conversion: (a) spectra for $\hat{s}[n]$ at LDM sampling rate F_s'; (b) spectra for $\hat{x}[n]$ at PCM sampling rate after filtering and downsampling by M.

of the speech signal, and which operates at the LDM sampling rate [133].[21] (See Section 2.5.3 for a discussion of downsampling.) The spectrum of the output of the downsampler, $\hat{x}[n]$, is shown in Figure 11.63b. Recalling that the normalized radian frequency at the LDM sampling rate is $\omega = 2\pi F/F_s'$, the total noise power in the filter output $\tilde{x}[n] = x_a(nT') + f[n]$ is given by

$$\sigma_f^2 = \frac{1}{2\pi}\int_{-(2\pi F_N/F_s')}^{(2\pi F_N/F_s')} \sigma_e^2 d\omega = \frac{1}{2\pi}\int_{-\pi/M}^{\pi/M} \sigma_e^2 d\omega = \frac{\sigma_e^2}{M}.$$

The output of the filter is computed once per M samples, where M is the ratio of the LDM sampling frequency to the desired PCM sampling frequency. The result after downsampling is the output signal

$$\hat{x}[n] = \tilde{x}[Mn] = x_a(nMT') + f[nM] = x_a(nT) + g[n].$$

Since the average noise power is unchanged in downsampling (i.e., $\sigma_g^2 = \sigma_f^2 = \sigma_e^2/M$) [270], the signal-to-quantization noise ratio in dB for the output of the LDM-to-PCM converter is

$$\text{SNR}_{\text{PCM}}(\text{dB}) = 10\log_{10}\left(\frac{\sigma_x^2}{\sigma_f^2}\right) = 10\log_{10}\left(\frac{M\sigma_x^2}{\sigma_e^2}\right)$$

[21] When LDM is used alone as a digital coding scheme, the lowpass filtering would usually be implemented with an analog filter.

$$= 10\log_{10}\left(\frac{\sigma_x^2}{\sigma_e^2}\right) + 10\log_{10} M$$

$$= \text{SNR}_{\text{LDM}} + 10\log_{10} M. \qquad (11.140)$$

That is, the SNR of the LDM system is increased by the amount $10\log_{10} M$ dB. This means that the filtering and downsampling contributes 3 dB (effectively one-half of 1 bit) to the SNR of the LDM-to-PCM system for each factor of 2 in M. For example, if the oversampling ratio is $M = 256$, 24 dB (5 bits) of SNR_{PCM} is contributed by the filtering and downsampling.

In Section 11.8.1 the results of a digital simulation of LDM were shown in Figure 11.57 (where M was denoted F_0). The simulations included a simulation of the analog lowpass filter that would normally follow an LDM decoder, so Figure 11.57 is also representative of the performance of a discrete-time LDM-to-PCM converter. Figure 11.57 shows that the optimum value of the signal-to-quantization noise ratio of an LDM system (including the simulated analog lowpass filtering) increases a total of about 8 dB for each doubling of M. Equation 11.140 shows that part of this increase is due to the removal of the out-of-band quantization noise and the remaining part must be due to the increased correlation that results from oversampling.

It is clear that large values of M can, in principle, produce high values of SNR_{PCM}. However, as also noted in Section 11.8.1, the optimum SNR is only achieved when the LDM step size is closely matched to the signal variance, and the performance of the LDM system is very sensitive to input signal level. This is why ADM systems are generally preferred in spite of added complexity.

11.8.5 Delta-Sigma A-to-D Conversion

The use of oversampling to reduce quantization noise is not restricted to LDM-to-PCM conversion. In principle, the LDM system and its discrete-time decoder can be replaced by any sampler/quantizer whose decoder can be implemented by digital computation. For example, if we oversample and quantize the samples using a uniform multi-bit quantizer, then we can get the same improvement in SNR of $10\log_{10} M$ dB by lowpass filtering the quantized signal to the Nyquist frequency and downsampling by M. This type of improvement in SNR can be enhanced by the technique of feedback noise shaping, which is employed in systems known as delta-sigma converters [350]. Figure 11.64 shows a block diagram of a discrete-time representation of a first-order delta-sigma converter. As in the case of LDM systems, the comparator at the input and the integrator represented by the accumulator system would both be implemented by analog integrated components. However, for our purposes, the discrete equivalent representation leads to a simpler analysis. Assuming that z-transforms exist for all signals in Figure 11.64,[22] we further assume that $\hat{S}(z) = \tilde{S}(z) + E(z)$, where $E(z)$ stands for the

[22]This is an assumption of convenience that is at odds with our random signal models for speech and quantization noise. However, the algebraic structures revealed by the z-transform analysis are valid for any signal, and lead to useful insights.

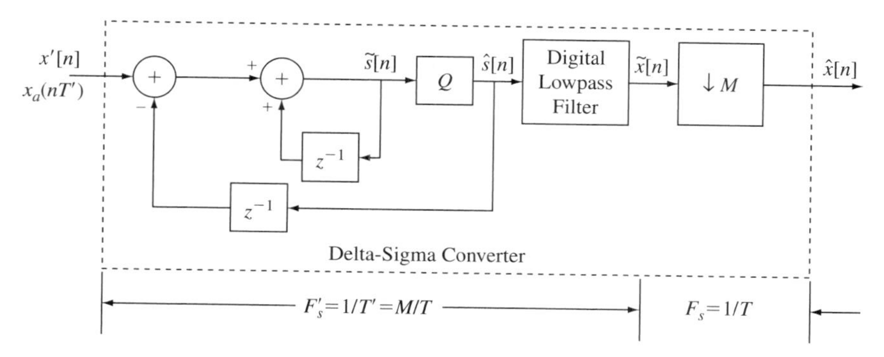

FIGURE 11.64
Discrete-time representation of a delta-sigma A-to-D converter.

quantization noise introduced by the quantizer labeled Q. Now consider the following algebraic manipulation based on the diagram in Figure 11.64:

$$\begin{aligned}\hat{S}(z) &= \tilde{S}(z) + E(z) \\ &= z^{-1}\tilde{S}(z) + X'(z) - z^{-1}\hat{S}(z) + E(z) \\ &= X'(z) + E(z) - z^{-1}(\hat{S}(z) - \tilde{S}(z)) \\ &= X'(z) + (1 - z^{-1})E(z),\end{aligned} \tag{11.141}$$

from which we conclude that

$$\hat{s}[n] = x'[n] + (e[n] - e[n-1]) = x'[n] + f[n]. \tag{11.142}$$

Thus, the signal $\hat{s}[n]$ contains the desired input signal $x'[n] = x_a(nT')$ plus the first difference of the quantization noise introduced by the quantizer. If the quantizer noise has a flat spectrum with noise power σ_e^2, then it is easily shown that the power spectrum of the frequency-shaped quantization noise is

$$\Phi_f(e^{j2\pi F/F_s'}) = \sigma_e^2[2\sin(\pi F/F_s')]^2. \tag{11.143}$$

Figure 11.65 shows $\Phi_f(e^{j2\pi F/F_s'})$ as a function of analog frequency F along with the corresponding power spectrum of the white quantization noise and the lowpass filter for reducing the sampling rate by a factor M. Note that the noise feedback greatly amplifies the high frequencies of the quantization noise, but the low frequencies of the noise in the speech band $|F| \leq F_N$ are also greatly attenuated. Thus, the noise power in the output of the lowpass filter will be much less than the value σ_e^2/M that resulted in the oversampled LDM case. Specifically, with noise shaping, the output of the lowpass filter will be of the form $\tilde{x}[n] = x_a(nT) + f[n]$, and the average power of the noise is

$$\sigma_f^2 = \frac{1}{2\pi}\int_{-(2\pi F_N/F_s')}^{(2\pi F_N/F_s')} \sigma_e^2[2\sin(\omega/2)]^2 d\omega = \frac{1}{2\pi}\int_{-\pi/M}^{\pi/M} \sigma_e^2[2\sin(\omega/2)]^2 d\omega. \tag{11.144}$$

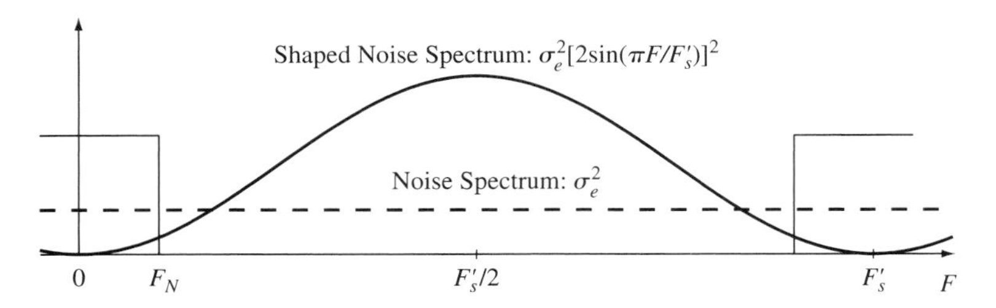

FIGURE 11.65
Illustration of noise shaping in delta-sigma conversion.

Now, if M is large, we can make the small angle approximation $2\sin(\omega/2) \approx \omega$, which will be accurate over the interval of integration $|\omega| \leq \pi/M$. Therefore Eq. (11.144) becomes

$$\sigma_f^2 = \frac{1}{2\pi}\int_{-\pi/M}^{\pi/M} \sigma_e^2 \omega^2 d\omega = \frac{\pi^2}{3M^3}\sigma_e^2. \tag{11.145}$$

After downsampling by M, the output has the form

$$\hat{x}[n] = x_a(nMT') + f[nM] = x_a(nT) + g[n],$$

which means that the signal-to-quantization noise ratio of the delta-sigma converter has the form

$$\text{SNR}_{\Delta\Sigma} = \text{SNR}_Q(\text{dB}) - 5.17 + 30\log_{10} M, \tag{11.146}$$

where $\text{SNR}_Q(\text{dB})$ is the signal-to-quantization noise ratio of the quantizer Q in Figure 11.64. It follows from Eq. (11.146) that with noise shaping, each doubling of the oversampling ratio M increases $\text{SNR}_{\Delta\Sigma}$ by 9 dB, or the equivalent of 1.5 bits. For example if $M = 256$, as is common in delta-sigma systems, the term $30\log_{10} M$ adds 72 dB to $\text{SNR}_{\Delta\Sigma}$. Thus, a 1-bit quantizer can be used, with most of the total SNR being contributed by the combination of noise shaping and lowpass filtering. The basic first-order noise shaping system can be iterated by replacing the quantizer by a first-order delta-sigma system, thereby obtaining a second-order delta-sigma converter [270, 350]. In this case the output noise spectrum before filtering is of the form

$$\Phi_f(e^{j2\pi F/F_s'}) = \sigma_e^2[2\sin(\pi F/F_s')]^4,$$

which, after making the same small angle approximation, gives

$$\sigma_f^2 = \frac{\pi^4}{5M^5}\sigma_e^2, \tag{11.147}$$

for the average noise power of the output of the lowpass filter. With this result, the SNR of the downsampled output of the second-order delta-sigma converter takes the form

$$\text{SNR}_{\Delta\Sigma} = \text{SNR}_Q(\text{dB}) - 12.9 + 50\log_{10} M. \tag{11.148}$$

This indicates that the signal-to-quantization noise increases by 15 dB for each doubling of M. Higher order delta-sigma converters can be formed using the same idea, leading to even greater enhancements of SNR [350].

The concept of delta-sigma A-to-D conversion makes it possible to build very accurate A-to-D converters using a combination of very simple analog circuits coupled to sophisticated but inexpensive digital signal processing. Low-cost A-to-D and D-to-A conversion together with inexpensive digital computation is what makes digital speech processing economically feasible in a wide range of applications.

11.9 DIFFERENTIAL PCM (DPCM)

Any system of the form shown in Figure 11.53 could be called a differential PCM (DPCM) system. Delta modulators, as discussed in the previous section, for example, could also be called 1-bit DPCM systems. Generally, however, the term "differential PCM" is reserved for differential quantization systems in which the quantizer has more than two levels.

As is clear from Figure 11.54, DPCM systems with fixed predictors can provide from 4 to 11 dB improvement in SNR over direct quantization (PCM). The greatest improvement in SNR occurs in going from no prediction to first-order prediction, with somewhat smaller additional gains in SNR resulting from increasing the predictor order up to 4 or 5, after which little additional gain results. As pointed out in Section 11.7, this gain in SNR implies that a DPCM system can achieve a given SNR using one fewer bit than would be required when using the same quantizer directly on the speech waveform. Thus, the results of Sections 11.4 and 11.5 can be applied to obtain a reasonable estimate of the performance that can be obtained for a particular quantizer used in a differential configuration. For example, for a differential PCM system with a uniform fixed quantizer, the SNR would be approximately 6 dB greater than the SNR for a quantizer with the same number of levels acting directly on the input. Differential schemes behave in much the same manner as direct PCM schemes; i.e., the SNR increases by 6 dB for each bit added to the codewords, and the SNR shows the same dependence upon signal level. Similarly, the SNR of a μ-law quantizer improves by about 6 dB when used in a differential configuration and at the same time, its characteristic insensitivity to input signal level is maintained.

Figure 11.54 shows a wide variation of prediction gain with speaker and with bandwidth. Similar wide variations are observed among different speech utterances. All of these effects are characteristic of speech communication. No single set of predictor coefficients can be optimum for a wide variety of speech material or a wide range of speakers.

This variation of performance with speaker and speech material, together with variations in signal level inherent in the speech communication process, makes adaptive prediction and adaptive quantization necessary to achieve the best performance over a wide range of speakers and speaking situations. Such systems are called adaptive differential PCM (ADPCM) systems. We first discuss the use of adaptive quantization with fixed prediction, and then discuss the use of adaptive prediction.

11.9.1 DPCM with Adaptive Quantization

The discussion of adaptive quantization in Section 11.5 can be applied directly to the case of DPCM. As Section 11.5 pointed out, there are two basic approaches to the control of adaptive quantizers, namely feed-forward adaptation and feedback adaptation.

Figure 11.66 shows how a feed-forward-type adaptive quantizer is used in an ADPCM system [256]. Generally, in schemes of this type, the quantizer step size would be proportional to the variance of the input to the quantizer, which in this case is $d[n]$. However, since the difference signal $d[n]$ is directly related to (but smaller than) the input, it is reasonable to control the step size from the input, $x[n]$, as depicted in Figure 11.66. This allows block adaptation of the step size, which would not be possible if $\Delta[n]$ were adapted based on $d[n]$ since $\hat{d}[n]$ is required to compute $d[n]$. The codewords $c[n]$ and a quantized $\Delta[n]$ therefore comprise the representation of $x[n]$. Generally, the sampling rate for $\Delta[n]$ is much lower than F_s, the sampling rate of both $x[n]$ and $c[n]$. The total bit rate would be of the form

$$I = BF_s + B_\Delta F_\Delta,$$

where B is the number of bits for the quantizer and F_Δ and B_Δ are respectively the sampling rate and number of bits per sample for the step size data.

Several algorithms for adjusting the step size were given in Section 11.5.1. The discussion of Section 11.5.1 indicates that such adaptation procedures can provide about 5 dB improvement in SNR over standard μ-law non-adaptive PCM. This

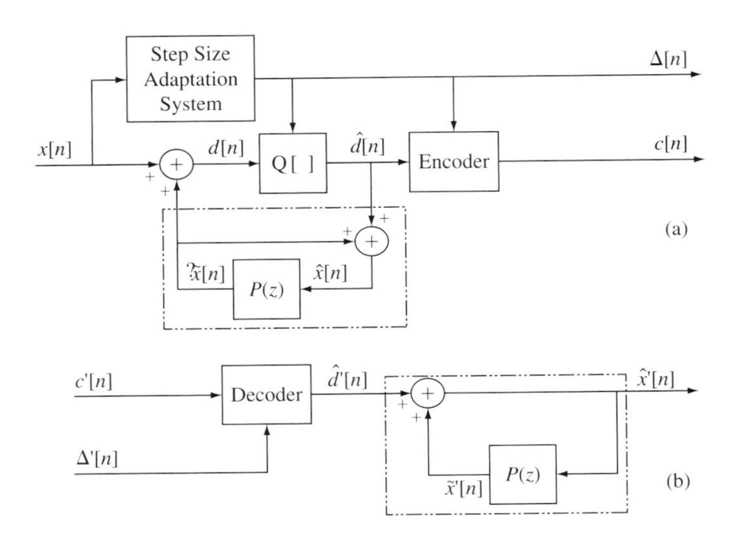

FIGURE 11.66
ADPCM system with feed-forward adaptive quantization: (a) encoder; (b) decoder.

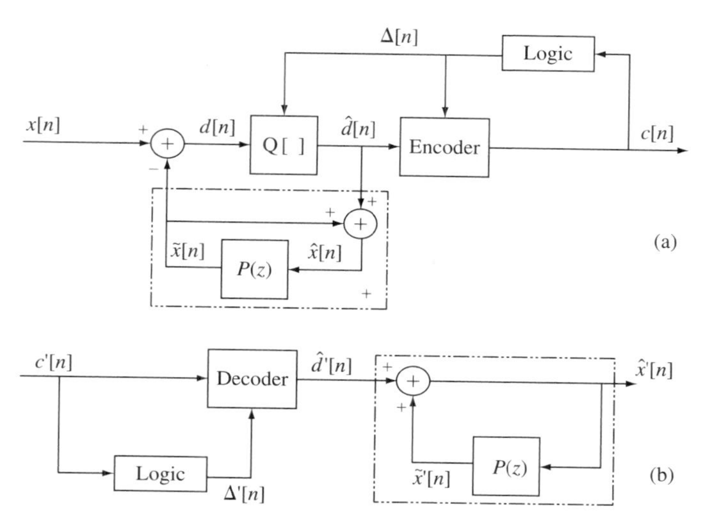

FIGURE 11.67
ADPCM system with feedback adaptive quantization: (a) encoder; (b) decoder.

improvement, coupled with the 6 dB that can be obtained from the differential configuration with fixed prediction, means that ADPCM with feed-forward adaptive quantization should achieve an SNR that is 10–11 dB greater than could be obtained with a fixed quantizer with the same number of levels.

Figure 11.67 shows how a feedback-type adaptive quantizer can be used in an ADPCM system [77]. If, for example, the adaptation strategy described by Eqs. (11.76)–(11.78) is used, we can again expect an improvement of 4–6 dB over a fixed μ-law quantizer with the same number of bits. Thus, both the feed-forward and feedback adaptive quantizers can be expected to achieve about 10–12 dB improvement over a fixed quantizer with the same number of levels.

In either case, the quantizer adaptation provides improved dynamic range as well as improved SNR. The main advantage of the feedback control is that the step size information is derived from the codeword sequence, so that no additional step size information need be transmitted or stored. Therefore, the bit rate of a DPCM system with a B-bit feedback adaptive quantizer is simply $I = BF_S$. This, however, makes the quality of the reconstructed output more sensitive to errors in transmission. With feed-forward control, the codewords and the step size together serve as the representation of the signal. Although this increases the complexity of the representation, there is the possibility of transmitting the step size with error protection, thereby significantly improving the output quality for high error rate transmission [170, 257].

11.9.2 DPCM with Adaptive Prediction

So far we have considered only fixed predictors, which can only model the long-term character of the speech spectrum. We have seen that even with higher-order fixed predictors, differential quantization can provide, under the best circumstances, about 10–12 dB improvement in SNR. Furthermore the amount of improvement is a function

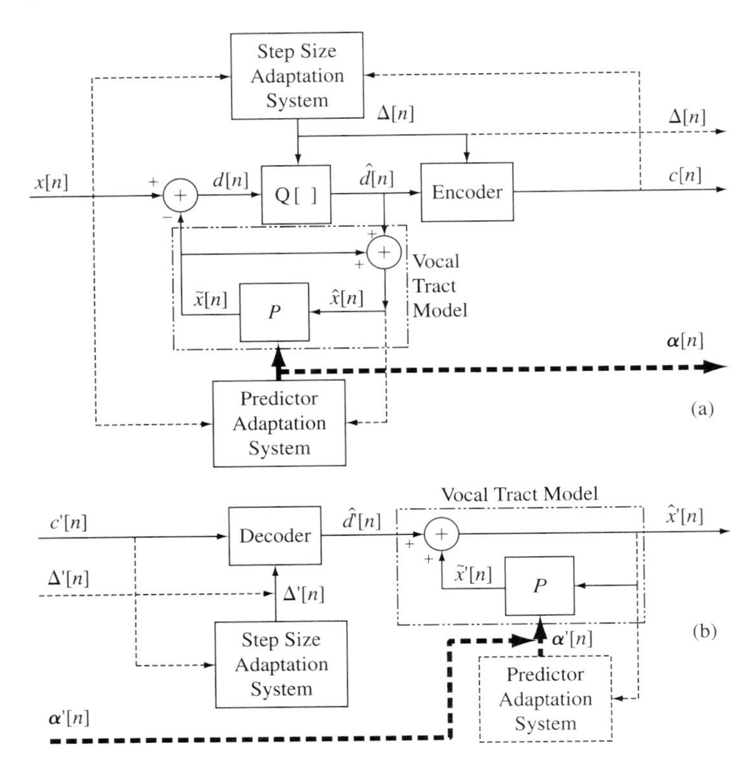

FIGURE 11.68
ADPCM system with both adaptive quantization and adaptive prediction: (a) encoder; (b) decoder.

of speaker and of speech material. In order to effectively cope with the non-stationarity of the speech communication process, it is natural to consider adapting the predictor as well as the quantizer to match the temporal variations of the speech signal [15]. A general adaptive DPCM system with both adaptive quantization and adaptive prediction is depicted in Figure 11.68. The dashed lines representing the adaptation and auxiliary parameters are meant to indicate that both the quantizer adaptation and the predictor adaptation algorithms can be either of the feed-forward or the feedback type. If feed-forward control is used for the quantizer or the predictor, then $\Delta[n]$ or the predictor coefficients, $\boldsymbol{\alpha}[n] = \{\alpha_k[n]\}$ (or both) are also required in addition to the codewords, $c[n]$, to complete the representation of the speech signal. In general, the bit rate of an ADPCM system would be of the form

$$I_{\text{ADPCM}} = BF_s + B_\Delta F_\Delta + B_P F_P, \tag{11.149}$$

where B is the number of bits for the quantizer, B_Δ is the number of bits for encoding the step size with frame rate F_Δ, and B_P is the total number of bits allocated to the predictor coefficients with frame rate F_P (often, $F_P = F_\Delta$). If either the quantizer or the predictor is a feedback adaptive system, the corresponding terms will be omitted from Eq. (11.149).

The predictor coefficients are assumed to be time dependent so that the predicted value is

$$\tilde{x}[n] = \sum_{k=1}^{p} \alpha_k[\hat{n}]\, \hat{x}[n-k], \tag{11.150}$$

where $\hat{n}$ is the frame time. In adapting the set of predictor coefficients, $\boldsymbol{\alpha}[\hat{n}]$, it is common to assume that the properties of the speech signal remain fixed over short time intervals (frames). The predictor coefficients are therefore chosen to minimize the mean-squared prediction error over a short time interval. It is in this case that the notion that the ADPCM system incorporates a model for speech production has its greatest significance. For feed-forward control, the predictor adaptation is based upon measurements on blocks of the input signal, $x[n]$, rather than on the quantized signal $\hat{x}[n]$. That is, the predictor coefficients for a given block to be quantized are estimated from the unquantized signal using the techniques of Chapter 9. These coefficients are then applied in Figure 11.68 to quantize the given block of samples. This is equivalent to neglecting the term 1/SNR due to quantization noise feedback in the analysis of Section 11.7. Using the same type of manipulations that were used to derive Eqs. (11.100) and (11.102a), and neglecting the effect of quantization error, we showed in Chapter 9 that the optimum predictor coefficients satisfy the equations,

$$R_{\hat{n}}[i] = \sum_{k=1}^{p} \alpha_k[\hat{n}]\, R_{\hat{n}}[i-k], \quad i = 1, 2, \ldots, p, \tag{11.151}$$

where $R_{\hat{n}}[i]$ is the short-time autocorrelation function [Eq. (6.29)] at time $\hat{n}$ and is computed as

$$R_{\hat{n}}[i] = \sum_{m=-\infty}^{\infty} x[m]w[\hat{n}-m]x[i+m]w[\hat{n}-m-i], \quad 0 \le i \le p, \tag{11.152}$$

and $w[\hat{n}-m]$ is a window function that is positioned at sample $\hat{n}$ of the input sequence. A rectangular window, or one with much less abrupt tapering of the data (e.g., a Hamming window of length L), can be used. Generally, the speech signal is high-frequency pre-emphasized with a linear filter whose transfer function is of the form $(1-\beta z^{-1})$. Such pre-emphasis helps to regularize the computation of the predictor coefficients. This is particularly important in fixed-point implementations where the numerical dynamic range is limited. A typical value for β is 0.4, which corresponds to raising the spectrum at $\omega = \pi$ by about 7 dB relative to the spectral level at low frequencies. In our discussion, $x[n]$ is assumed to be the output of such a filter. If pre-emphasis is used prior to ADPCM coding, then de-emphasis should be done with the inverse filter $1/(1-\beta z^{-1})$ following the ADPCM decoder.

The adaptive predictor can also be computed from previous samples of the quantized signal $\hat{x}[n]$. If this is done, it is not necessary to transmit the predictor coefficients since, in the absence of transmission errors, they can be computed from past samples of the reconstructed signal at the receiver. Thus, $R_{\hat{n}}[i]$ in Eq. (11.151) would be replaced by

$$R_{\hat{n}}[i] = \sum_{m=-\infty}^{\infty} \hat{x}[m]w[\hat{n}-m]\hat{x}[m+i]w[\hat{n}-m-i], \quad 0 \leq i \leq p. \tag{11.153}$$

In this case, the window must be causal; i.e., $w[n] = 0$ for $n < 0$; i.e., the estimate of predictor coefficients must be based upon *past* quantized values rather than *future* values which cannot be obtained until the predictor coefficients are available. Barnwell [24] has shown how to compute the autocorrelation function recursively using infinite duration impulse response (IIR) systems that weight past samples with exponentially decaying windows. As in the case of adaptive quantizer control, the feedback mode has the advantage that only the quantizer codewords need be transmitted. Feedback control of adaptive predictors, however, has not been widely used due to the inherent sensitivity to errors and the inferior performance that results from basing the control upon a noisy input. An interesting approach to feedback control was considered by Stroh [383], who studied a gradient scheme for adjusting the predictor coefficients.

Since the parameters of speech vary rather slowly, it is reasonable to adjust the predictor parameters, $\boldsymbol{\alpha}[\hat{n}]$, infrequently. For example, a new estimate may be computed every 10–20 msec, with the values being held fixed between estimates; i.e., the frame rate is approximately $F_P = 50$ to 100 Hz. The window duration may be equal to the interval between estimates or it may be somewhat larger, in which case, successive segments of speech overlap. As defined by Eq. (11.152), the computation of the correlation estimates required in Eq. (11.151) would require the accumulation of L samples of $x[n]$ in a buffer before computing $R_{\hat{n}}[i]$. The set of coefficients, $\boldsymbol{\alpha}[\hat{n}]$, satisfying Eq. (11.151) are used in the configuration of Figure 11.68a to quantize the input during the interval of L samples beginning at sample $\hat{n}$. Thus, to reconstruct the input from the quantizer codewords, we also need the predictor coefficients (and possibly the quantizer step size) as depicted in Figure 11.68b.

In order to quantitatively express the benefits of adaptive prediction, Noll [256] examined the dependence of the predictor gain, G_P, upon predictor order for both fixed and adaptive predictors. Figure 11.69[23] shows the quantity

$$10\log_{10}[G_P] = 10\log_{10}\left[\frac{E[x^2[n]]}{E[d^2[n]]}\right] \tag{11.154}$$

as a function of predictor order, p, for both fixed and adaptive prediction. The lower curve, obtained by computing a long-term estimate of the autocorrelation for a given speech utterance and solving for the set of predictor coefficients satisfying

[23]The results in this figure are for a single speaker. In addition, no error feedback was included in the system that was studied.

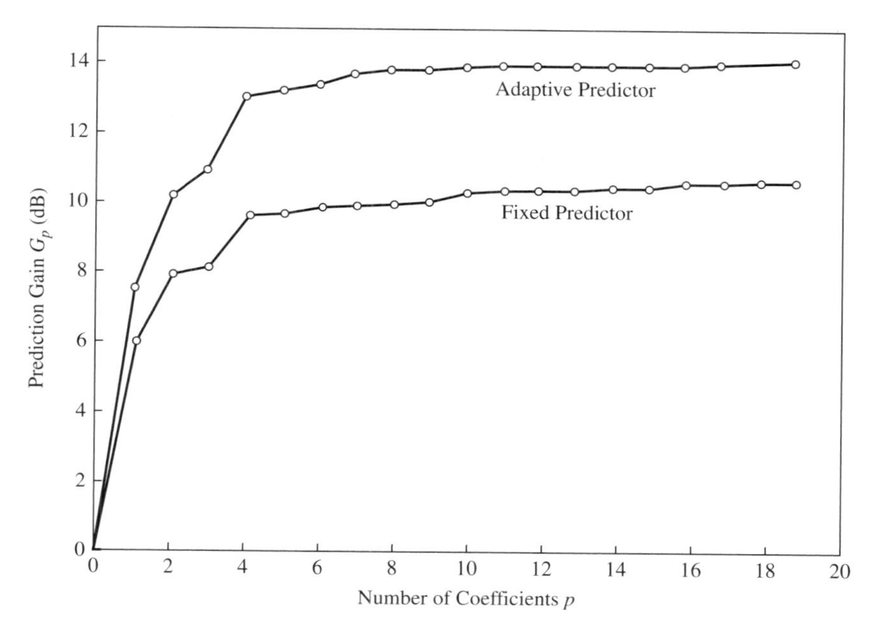

FIGURE 11.69
Prediction gains versus number of predictor coefficients for one female speaker (band from 0 to 3200 Hz). (After Noll [256]. Reprinted with permission of Alcatel-Lucent USA Inc.)

Eq. (11.102a), shows a maximum gain of about 10.5 dB. Note that this curve is within the range shown in Figure 11.54. The upper curve was obtained by finding the value of window length, L, and the predictor coefficients, $\boldsymbol{\alpha}[\hat{n}]$, that maximized G_P frame-by-frame across the entire utterance for a fixed value of p. This maximum value is plotted for each value of p. In this case, the maximum gain is about 14 dB. Based on these measurements, Noll [256] suggested that reasonable upper bounds on the performance of DPCM systems with fixed and adaptive prediction are 10.5 and 14 dB, respectively. Not evident in the curves of Figure 11.69 is the fact that the optimum fixed predictor is likely to be very sensitive to speaker and speech material, whereas the adaptive prediction scheme is inherently less sensitive.

11.9.3 Comparison of ADPCM Systems

DPCM systems are generally considered to be waveform coders since the reconstructed quantized signal can be modeled by $\hat{x}[n] = x[n] + e[n]$. In comparing digital waveform coding systems, it is convenient to use signal-to-quantization noise as a criterion. However, the ultimate criterion for systems that are to be used for voice communication is a perceptual one. The question of how good the coded speech sounds in comparison to the original unquantized speech is often of paramount importance. Unfortunately this perceptual criterion is often the most difficult to quantify and there is no unified set of results that we can refer to. Thus, in this section, we will briefly summarize the results of objective SNR measurements for a variety of speech coding systems, and then summarize a few perceptual results that appear to be particularly illuminating.

Noll [256] performed a very useful comparative study of digital waveform coding schemes. He considered the following systems:

1. $\mu = 100$ log-PCM with $X_{\max} = 8\sigma_x$ (PCM);
2. adaptive PCM (optimum Gaussian quantizer) with feed-forward control (PCM-AQF);
3. differential PCM with first-order fixed prediction and adaptive Gaussian quantizer with feedback control (DPCM1-AQB);
4. adaptive DPCM with first-order adaptive predictor and adaptive Gaussian quantizer with feed-forward control of both the quantizer and the predictor (window length 32) (ADPCM1-AQF);
5. adaptive DPCM with fourth-order adaptive predictor and adaptive Laplacian quantizer, both with feed-forward control (window length 128) (ADPCM4-AQF);
6. adaptive DPCM with 12th-order adaptive predictor and adaptive gamma quantizer, both with feed-forward control (window length 256) (ADPCM12-AQF).

In all these systems, the sampling rate was 8 kHz and the quantizer word length, B, ranged from 2 to 5 bits/sample. Thus the bit rate for the codeword sequence, $c[n]$, ranges from 16 to 40 kbps. Signal-to-quantizing noise ratios for all the systems are plotted in Figure 11.70. Note that the systems are ordered roughly from least complex on the left to most complex on the right, and SNR follows the same pattern. Also, only for the systems denoted PCM and DPCM1-AQB is the sequence of codewords, $c[n]$, the complete representation. All the other coders require additional bits (not shown) for quantizer step size and predictor coefficients. The curves of Figure 11.70 display a number of interesting features. First, note that the lowest curve corresponds to the use of a 2-bit quantizer, and moving upward from one curve to the next corresponds to adding 1 bit to the quantizer word length. Thus, the curves are displaced from one another by roughly 6 dB. Notice also the sharp increase in SNR with the addition of both fixed prediction and adaptive quantization, and note that almost no gain results from adapting a simple first-order predictor. (Compare DPCM1-AQB with ADPCM1-AQF.) However, it also is clear that higher order adaptive prediction does offer significant improvements.

For telephone transmission, it is generally accepted that acceptable speech quality is obtained with a μ-law quantizer with 7–8 bits/sample. From Eq. (11.45), it can be seen that 7-bit $\mu = 100$ PCM would have an SNR of about 33 dB. On the basis of Figure 11.70, it would appear that comparable quality could be obtained using a 5-bit quantizer with adaptive quantization and adaptive prediction. In practice, there is strong evidence to suggest that the perceived quality of ADPCM coded speech is better than a comparison of SNR values would suggest. In a study of an ADPCM system with fixed prediction and feedback control of an adaptive quantizer, Cummiskey et al. [77] found that listeners preferred ADPCM coded speech to log-PCM coded speech with higher SNR. The results of a preference test are given in Table 11.8, where

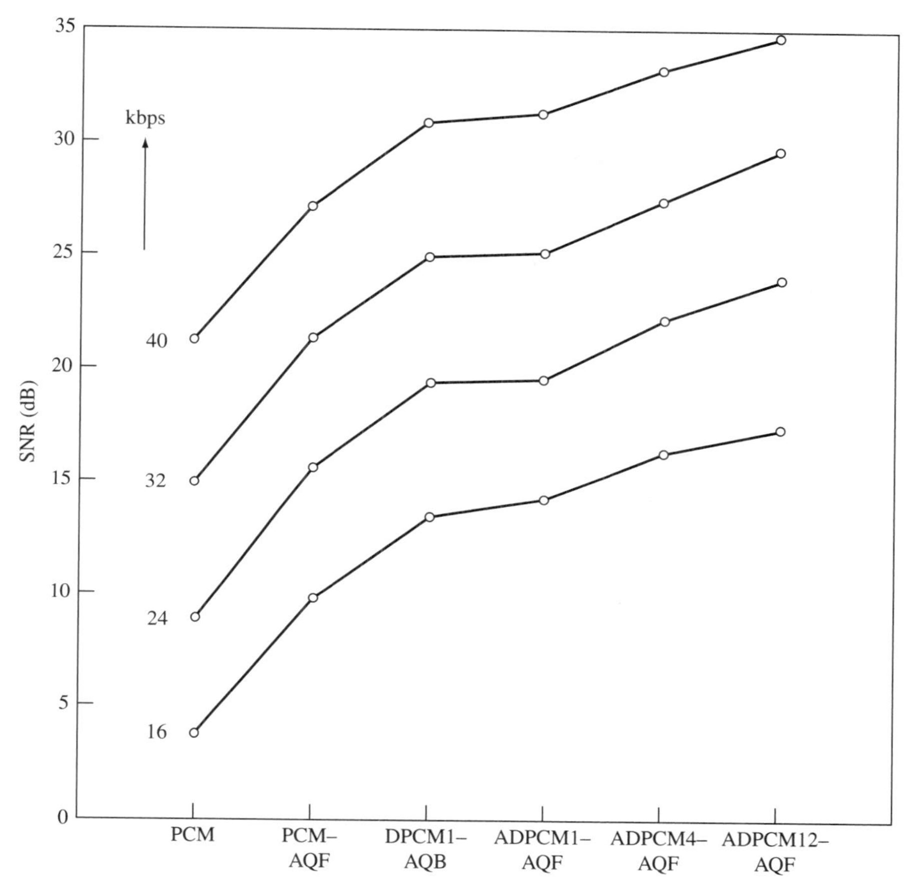

FIGURE 11.70
SNR values for quantization with 2 bits/sample (16 k/sec) up to 5 bits/sample (40 kbps). Code: AQF—adaptive quantizer, feed-forward; AQB—adaptive quantizer, feed backward; ADPCMr—ADPCM system with r^{th}-order predictor. (After Noll [256]. Reprinted with permission of Alcatel-Lucent USA Inc.)

TABLE 11.8 Comparison of objective and subjective performance of ADPCM and log-PCM. (After Cummiskey, Jayant, and Flanagan [77]. Reprinted with permission of Alcatel-Lucent USA Inc.)

Objective Rating (SNR)	Subjective Rating Preference
7-bit log-PCM	7-bit log-PCM (High Preference)
6-bit log-PCM	4-bit ADPCM
4-bit ADPCM	6-bit log-PCM
5-bit log-PCM	3-bit ADPCM
3-bit ADPCM	5-bit log-PCM
4-bit log-PCM	4-bit log-PCM (Low preference)

the PCM system is like system 1 of Noll's study and the ADPCM system is like system 3 of Noll's study. Table 11.8 shows that 4-bit ADPCM is preferred to 6-bit log-PCM. Recalling that the SNR improvement for ADPCM with fixed prediction and adaptive quantization is expected to be on the order of 10–12 dB, or roughly 2 bits, it is not surprising that the systems would be comparable, but in fact, the 4-bit ADPCM was preferred to the 6-bit log-PCM even though the SNR of 4-bit ADPCM was somewhat lower.

In their study of adaptive prediction, Atal and Schroeder [15] found that their ADPCM system with a 1-bit adaptive quantizer and complex adaptive predictor yielded coded speech whose quality was slightly inferior to 6-bit log-PCM. The estimated bit rate for this system was about 10 kbps, in contrast to the 40 kbps required for 6-bit PCM at a sampling rate of 6.67 kHz. Especially in this case, the subjective quality was greater than would be expected from a consideration of the SNR.

A precise explanation of this phenomenon is difficult to obtain; however, it is reasonable to conjecture that it is due to a combination of such factors as better idle channel performance of the adaptive quantizer and greater correlation between the quantization noise and the signal [256].

11.10 ENHANCEMENTS FOR ADPCM CODERS

The basic ADPCM system incorporates the basic linear predictive speech model, and for this reason, ADPCM coding is also called *linear predictive coding*, or simply LPC. It is also called adaptive predictive coding (APC). In Section 11.9, we showed that the incorporation of the linear predictive model for speech production in a feedback path around the quantizer improves quantization accuracy significantly. We have already seen that there are many possible combinations of adaptive quantizers and predictors. Over the past decades since the classic paper of Atal and Schroeder [15], many improvements have been proposed to the basic system. In this section, we discuss several of these improvements.

11.10.1 Pitch Prediction in ADPCM Coders

So far our examples have assumed implicitly that the linear predictor is configured with model order p chosen so as to incorporate a representation of the vocal tract filter (i.e., the formant information) into the quantized signal representation. With this point of view, the difference signal in a predictive coder, being similar to (but not identical to) the prediction error signal, retains the character of the excitation signal of the model for speech production. This means that the quantized difference signal will switch back and forth between a quasi-periodic state and a random noise state. Thus, as was illustrated by Figure 11.52, more redundancy could be removed from the difference signal in voiced intervals. Indeed, if perfect prediction were achieved, the difference signal, $d[n]$, would be almost completely uncorrelated (white noise). It can be seen from Figure 11.69 that little additional prediction gain results from increasing the order of prediction beyond 4 or 5. However, for $F_s = 8$ kHz, even a relatively large order such as $p = 20$ only corresponds to a prediction interval of 2.5 msec, which is

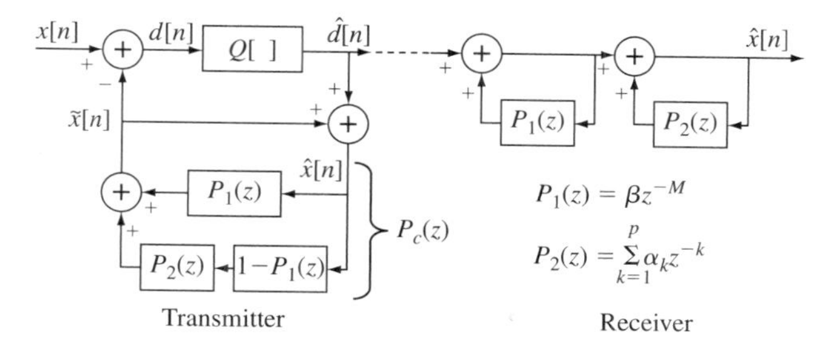

FIGURE 11.71
Two-stage predictor used in a differential quantization method, where the first stage, P_1, is a pitch predictor and the second stage, P_2, is a predictor tuned to the vocal tract response [15, 16].

much smaller than a typical pitch period. One approach to exploitation of the long-term correlation due to pitch was proposed by Atal and Schroeder [15, 16], who used the two-stage predictor, shown in Figure 11.71.[24] The first stage consists of a "pitch predictor" specified by the z-transform system function

$$P_1(z) = \beta z^{-M}, \tag{11.155}$$

where M is chosen to be the pitch period of the current frame (in samples), and β is a gain constant that allows for variations from period to period. For 100 Hz pitch frequency, with a sampling rate of $F_s = 8{,}000$ samples/sec, $M = 80$ samples. For frames of silence (background noise) or unvoiced speech, the values chosen for M and β are essentially irrelevant. (Note that the pitch predictor has no effect if $\beta = 0$.)

The second stage predictor of Figure 11.71 is a linear predictor of the form

$$P_2(z) = \sum_{k=1}^{p} \alpha_k z^{-k}. \tag{11.156}$$

Since p is chosen to capture only the formant resonance structure of speech, we will refer to this predictor as the "vocal tract predictor." The inverse systems for these two predictors appear in cascade in the decoder system on the right in Figure 11.71. The combined system function is of the form:

$$H_c(z) = \left(\frac{1}{1 - P_1(z)}\right)\left(\frac{1}{1 - P_2(z)}\right) = \frac{1}{1 - P_c(z)}. \tag{11.157}$$

The combined (two-stage) prediction error filter has the transfer function

$$1 - P_c(z) = [1 - P_1(z)][1 - P_2(z)] = 1 - [1 - P_1(z)]P_2(z) - P_1(z). \tag{11.158}$$

[24]Note that in this diagram, we have omitted an explicit indication of the parameters that must be transmitted to the decoder, allowing us to focus on the signal relationships.

Therefore, the combined vocal tract/pitch predictor has system function

$$P_c(z) = [1 - P_1(z)]P_2(z) + P_1(z). \tag{11.159}$$

As shown in Figure 11.71, this combined predictor is implemented as a parallel combination of two predictors $[1 - P_1(z)]P_2(z)$ and $P_1(z)$. The corresponding time-domain expression for the predicted signal, $\tilde{x}[n]$, as a function of $\hat{x}[n]$ is

$$\tilde{x}[n] = \beta\hat{x}[n-M] + \sum_{k=1}^{p}\alpha_k(\hat{x}[n-k] - \beta\hat{x}[n-k-M]), \tag{11.160}$$

where the predictor parameters β, M, and $\{\alpha_k\}$ are all adapted at the predictor frame rate $F_P < F_s$. As before, we determine all the prediction parameters by analysis of $x[n]$ rather than $\hat{x}[n]$, which is not known until the predictors are set. Therefore we define the combined prediction error as

$$\begin{aligned} d_c[n] &= x[n] - \tilde{x}[n] \\ &= v[n] - \sum_{k=1}^{p}\alpha_k v[n-k], \end{aligned} \tag{11.161}$$

where

$$v[n] = x[n] - \beta x[n-M] \tag{11.162}$$

is the prediction error of the pitch predictor. The joint computation of the values of β, M, and $\{\alpha_k\}$ that minimize the variance of $d_c[n]$ is not straightforward. For this reason, Atal and Schroeder [15] considered a suboptimum solution in which the variance of $v[n]$ is first minimized, and then the variance of $d_c[n]$ is minimized subject to the previously determined values of β and M.

The mean-squared prediction error for the pitch predictor is

$$E_1 = \left\langle (v[n])^2 \right\rangle = \left\langle (x[n] - \beta x[n-M])^2 \right\rangle, \tag{11.163}$$

where $\langle\ \rangle$ denotes the operation of averaging over a finite frame of speech samples. Recall that in Chapter 9, we discussed two distinct ways of carrying out this averaging, one leading to the autocorrelation method and the other to the covariance method. For the pitch predictor, the covariance-type averaging is preferred because it does not have the windowing effects inherent in the autocorrelation method.

Following the approach of differentiating E_1 with respect to β with M fixed, it can be shown that the coefficient that minimizes the pitch prediction error is

$$(\beta)_{\text{opt}} = \frac{\langle x[n]x[n-M]\rangle}{\langle (x[n-M])^2\rangle}. \tag{11.164}$$

Substituting this value of β into Eq. (11.163) gives the minimum mean-squared pitch prediction error as

$$(E_1)_{\text{opt}} = \left\langle (x[n])^2 \right\rangle \left(1 - \frac{(\langle x[n]x[n-M] \rangle)^2}{\langle (x[n])^2 \rangle \langle (x[n-M])^2 \rangle} \right), \tag{11.165}$$

which is minimized when the normalized covariance

$$\rho[M] = \frac{\langle x[n]x[n-M] \rangle}{(\langle (x[n])^2 \rangle \langle (x[n-M])^2 \rangle)^{1/2}} \tag{11.166}$$

is maximized. Therefore, we can find the optimum pitch predictor by first searching for M that maximizes Eq. (11.166) and then computing $(\beta)_{\text{opt}}$ from Eq. (11.164).

A major issue with the pitch prediction of Eq. (11.162) is the assumption that the pitch period is an integer number of samples. An alternative (somewhat more complicated) pitch predictor is of the form

$$P_1(z) = \beta_{-1} z^{-M+1} + \beta_0 z^{-M} + \beta_1 z^{-M-1} = \sum_{k=-1}^{1} \beta_k z^{-M-k}. \tag{11.167}$$

This representation of the pitch periodicity handles non-integer pitch periods via interpolation around the nearest integer pitch period value. As in the simpler case, M is chosen as the location in the pitch period range of the peak of the covariance function in Eq. (11.166). Then the three prediction coefficients are found by solving a set of linear equations derived by minimizing the variance of the pitch prediction error

$$v[n] = x[n] - \beta_{-1}x[n-M+1] - \beta_0 x[n-M] - \beta_1 x[n-M-1]$$

with M fixed. The optimum solution involves the covariance values $\langle x[n-M-k]\, x[n-M-i] \rangle$ for $i, k = -1, 0, 1$. The solution gives a prediction error filter polynomial $1 - P_1(z)$ of order $M+1$. Since this is the denominator of the inverse system function $H_1(z) = 1/(1 - P_1(z))$ in the receiver side of Figure 11.71, we must be careful to ensure that all the roots of $1 - P_1(z)$ are inside the unit circle. This is not guaranteed by the covariance method, so additional steps must be taken to ensure stability. In the case of the simpler pitch predictor of Eq. (11.155), stability of the corresponding inverse system is easily guaranteed by ensuring that $\beta < 1$.

Given M and β, the sequence $v[n]$ can be determined and its autocorrelation computed for $i = 0, 1, \ldots, p$, from which the prediction coefficients, $\{\alpha_k\}$, can be obtained from Eq. (11.151) with $R_{\hat{n}}[i]$ being the short-time autocorrelation of the sequence $v[\hat{n}]$.[25]

[25] An alternative would be to first compute the vocal tract prediction coefficients from $x[n]$ and then estimate M and β from the prediction error signal from the vocal tract predictor.

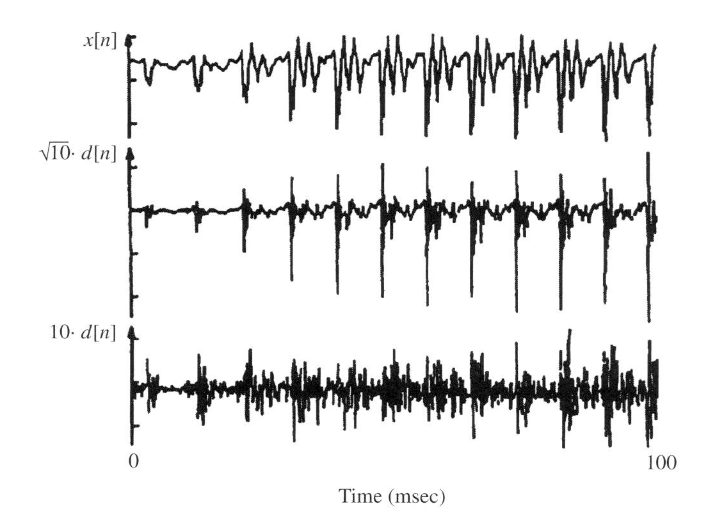

FIGURE 11.72
Examples of waveforms of a speech signal, $x[n]$, the vocal tract prediction error, $\sqrt{10} \cdot d[n]$, and the pitch and vocal tract prediction error, $10 \cdot d[n]$ [15]. Reprinted with permission of Alcatel-Lucent USA Inc.

Figure 11.72 illustrates the potential of the 2-level prediction system of Figure 11.71, showing plots of the input signal, $x[n]$, the error signal resulting from a single stage optimum linear (vocal tract) predictor, $P_2(z)$ (denoted $\sqrt{10} \cdot d[n]$), and the error signal resulting from a two-stage predictor (denoted $10 \cdot d[n]$). The scale factors on $d[n]$ show that, in this example, the prediction gains for the vocal tract predictor and the pitch predictor are approximately the same. Further we see that the single stage vocal tract predictor difference signal shows very sharp pulses due to the pitch excitation, whereas the two-stage prediction error retains vestiges of the pitch excitation signal, but is much less impulsive.

In order to represent speech based on the processing shown in Figure 11.71, it is necessary to transmit or store the quantized difference signal, the quantizer step size (if feed-forward control is used), and the (quantized) predictor coefficients. In the original work of Atal and Schroeder, a 1-bit quantizer was used for the difference signal and the step size was adapted every 5 msec (33 samples at a 6.67 kHz sampling rate), so as to minimize the quantization error. Likewise, the predictor parameters were also estimated every 5 msec. Although no explicit SNR data was given, it was suggested that high quality reproduction of the speech signal could be achieved at bit rates on the order of 10 kbps. Jayant [169] asserts that using reasonable quantization of the parameters, SNR gains of 20 dB are possible over PCM when both long-term and short-term predictors are used.

Unfortunately, no careful study (such as Noll's [256] study discussed in Section 11.9.3) of the limits of performance of adaptive prediction, including pitch parameters, has been done. However, it is apparent that schemes such as this represent one extreme of complexity of digital waveform coding systems. On the other end of the scale would be LDM with its simple quantization process and unstructured stream of 1-bit binary codewords. The choice of quantization scheme depends on a variety of factors, including the desired bit rate, desired quality, coder complexity, and complexity of the digital representation.

11.10.2 Noise Shaping in DPCM Systems

As we have seen, the reconstructed signal at the output of an ADPCM encoder/decoder is $\hat{x}[n] = x[n] + e[n]$, where $e[n]$ is the quantization noise. Since $e[n]$ generally has a flat spectrum, the quantization noise is especially audible in spectral regions of low intensity, e.g., the regions between formants. For very fine quantization, this is not a major problem since the level of the noise spectrum is low. However, for low bit rate systems, the level of the noise is often significantly greater than the level of the signal in parts of the spectrum, resulting in distortion that is clearly audible. This has led to the invention of differential quantization methods that shape the quantization noise to match the speech spectrum, so as to take advantage of the fact that loud sounds can mask weaker sounds at adjacent frequencies [16].

Figure 11.73a shows a block diagram of the basic operations of an ADPCM coder/decoder. To derive the noise shaping configuration in Figure 11.73c, we shall first

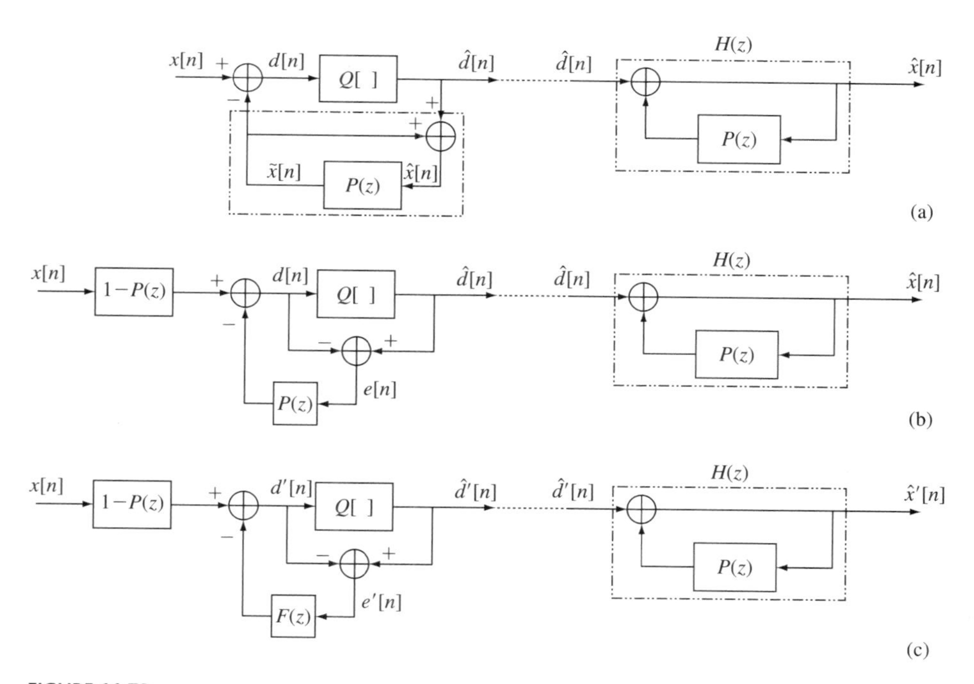

FIGURE 11.73
Block diagram of an alternate representation of a DPCM encoder and decoder showing explicit shaping of the signal and the noise: (a) block diagram of the basic operations of an ADPCM coder/decoder; (b) equivalent organization of the basic operations of part (a); (c) replacement of $P(z)$ by the noise-shaping filter, $F(z)$.

show that the configuration of Figure 11.73b is completely equivalent to Figure 11.73a. To do so, we assume that z-transforms exist for all signals in Figure 11.73a.[26] Assuming that $\hat{X}(z) = X(z) + E(z)$, it follows that $d[n]$ in Figure 11.73a has z-transform

$$\begin{aligned} D(z) &= X(z) - P(z)\hat{X}(z) \\ &= [1 - P(z)]X(z) - P(z)E(z). \end{aligned} \tag{11.168}$$

Now in Figure 11.73b, $E(z) = \hat{D}(z) - D(z)$ is explicitly computed and fed back through the predictor $P(z)$, so it follows that Figure 11.73b is precisely the block diagram representation of Eq. (11.168). Furthermore, since $\hat{D}(z) = D(z) + E(z)$, it also follows that

$$\hat{D}(z) = [1 - P(z)]X(z) + [1 - P(z)]E(z),$$

and therefore the z-transform of the reconstructed signal is

$$\begin{aligned} \hat{X}(z) = H(z)\hat{D}(z) &= \left(\frac{1}{1 - P(z)}\right)\hat{D}(z) \\ &= \left(\frac{1}{1 - P(z)}\right)([1 - P(z)]X(z) + [1 - P(z)]E(z)) \\ &= X(z) + E(z). \end{aligned} \tag{11.169}$$

Thus, we have shown that feeding back the quantization error through the predictor $P(z)$ as in Figure 11.73b ensures that the reconstructed signal $\hat{x}[n]$ differs from $x[n]$ by the quantization error $e[n]$ incurred in quantizing the difference signal $d[n]$.

Equation (11.169) provides the key to shaping the quantization noise spectrum. If we simply replace $P(z)$ in Figure 11.73b by a different system function $F(z)$ in the lower block of Figure 11.73c, we obtain instead of Eq. (11.169), the following equation for the reconstructed signal:

$$\begin{aligned} \hat{X}'(z) = H(z)\hat{D}'(z) &= \left(\frac{1}{1 - P(z)}\right)\hat{D}'(z) \\ &= \left(\frac{1}{1 - P(z)}\right)([1 - P(z)]X(z) + [1 - F(z)]E'(z)) \\ &= X(z) + \left(\frac{1 - F(z)}{1 - P(z)}\right)E'(z). \end{aligned} \tag{11.170}$$

[26] As before, we justify this assumption of convenience that is at odds with our random signal models for speech and quantization noise on the grounds that it reveals algebraic structures that are valid for any signal.

Therefore, we have shown that if $x[n]$ is coded by the system on the transmitter side in Figure 11.73c, then the z-transform of the reconstructed signal at the receiver is

$$\hat{X}'(z) = X(z) + \hat{E}'(z), \tag{11.171}$$

where the quantization noise, $\hat{E}'(z)$, of the reconstructed speech signal is related to the quantization noise $E'(z)$ introduced by the quantizer by

$$\hat{E}'(z) = \left(\frac{1 - F(z)}{1 - P(z)}\right) E'(z) = \Gamma(z)E'(z). \tag{11.172}$$

The system function

$$\Gamma(z) = \frac{1 - F(z)}{1 - P(z)} \tag{11.173}$$

is the effective noise shaping filter of the ADPCM coder/decoder system. By an appropriate choice of $F(z)$, the noise spectrum can be shaped in many ways. Some examples are as follows:

1. If we choose $F(z) = 0$ and we assume that the noise has a flat spectrum, then the noise and the speech spectrum have the same shape. Setting $F(z) = 0$ in Figure 11.73c corresponds to direct quantization of the prediction error sequence, so the quantization noise is directly shaped by $H(z)$ in the receiver. This type of coding has been called open-loop ADPCM and D*PCM [256].
2. If we choose $F(z) = P(z)$, the equivalent system is the standard DPCM system of Figure 11.73a, where $\hat{E}'(z) = E'(z) = E(z)$ and where the noise spectrum is flat and independent of the signal spectrum.
3. If we choose $F(z) = P(\gamma^{-1}z) = \sum_{k=1}^{p} \alpha_k \gamma^k z^{-k}$, we shape the noise spectrum so as to "hide" the noise beneath the spectral peaks of the speech signal.

The third option above is the one most often used. It shapes the noise spectrum into a modified version of the speech spectrum where the noise is suppressed at high frequencies at the cost of slight elevation of the noise spectrum at low frequencies. Each zero of $[1 - P(z)]$ [pole of $H(z)$ and $\Gamma(z)$] is paired with a zero of $[1 - F(z)]$ as in Figure 11.74, where we see that a zero of $\Gamma(z)$ is at the same angle as one of the poles, but with a radius that is divided by γ, where $\gamma = 1.2$ in this figure.

If we assume that the quantization noise has a flat spectrum and the noise power is $\sigma_{e'}^2$, then the power spectrum of the shaped noise would be

$$\Phi_{e'}(e^{j2\pi F/F_s}) = \left| \frac{1 - F(e^{j2\pi F/F_s})}{1 - P(e^{j2\pi F/F_s})} \right|^2 \sigma_{e'}^2. \tag{11.174}$$

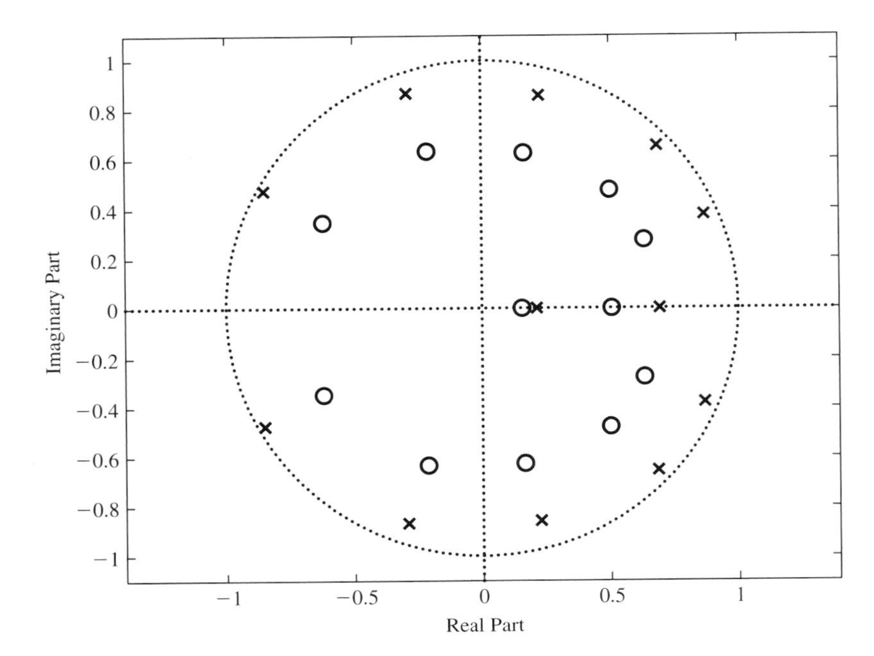

FIGURE 11.74
Locations of poles and zeros of a noise shaper with $\gamma = 1.2$.

Figure 11.75 shows $\Phi_{e'}(e^{j2\pi F/F_s})$ in dB as a function of $F(z)$ for the three cases that we have considered. Also shown as the dotted curve is the linear predictive estimate of the short-time speech spectrum $|G/(1 - P(e^{j2\pi F/F_s}))|^2$ in dB. The dash-dot curve (horizontal line) shows the case $F(z) = P(z)$, which is equivalent to the standard DPCM system. In this case, $\Phi_{e'}(e^{j2\pi F/F_s}) = \sigma_{e'}^2$. Observe that in this case, the noise spectrum level is higher than the speech spectrum in the bands $2700 < F < 3200$ and $3500 < F < 4000$. The dashed curve denotes the noise spectrum when $P(z) = 0$, i.e., this is the open-loop case when there is no noise feedback. In this case, the noise spectrum has the same shape as the speech spectrum, which means that the noise spectrum is lowered at high frequencies but greatly amplified at low frequencies. Unless the quantizer noise power $\sigma_{e'}^2$ is very low, the amplified low frequency noise will be very perceptible. Finally, the solid curve denotes the noise spectrum when $F(z) = P(z/1.37)$. Observe that this choice gives a compromise between the other two cases. The low frequencies of the noise spectrum are raised somewhat and the high frequencies are significantly lowered compared to the flat quantizer noise spectrum. Observe that the frequency response of the noise shaping filter has the interesting property that the average value of $20\log_{10}|\Gamma(e^{j2\pi F/F_s})|$ is zero [16].

It is important to point out that noise shaping with $F(z) \neq P(z)$ will always *decrease* the SNR compared to the standard ADPCM configuration [16, 172]. However, by redistributing the noise as illustrated in Figure 11.75, the decrease in traditional SNR is more than compensated by an increase in the "perceptual SNR."

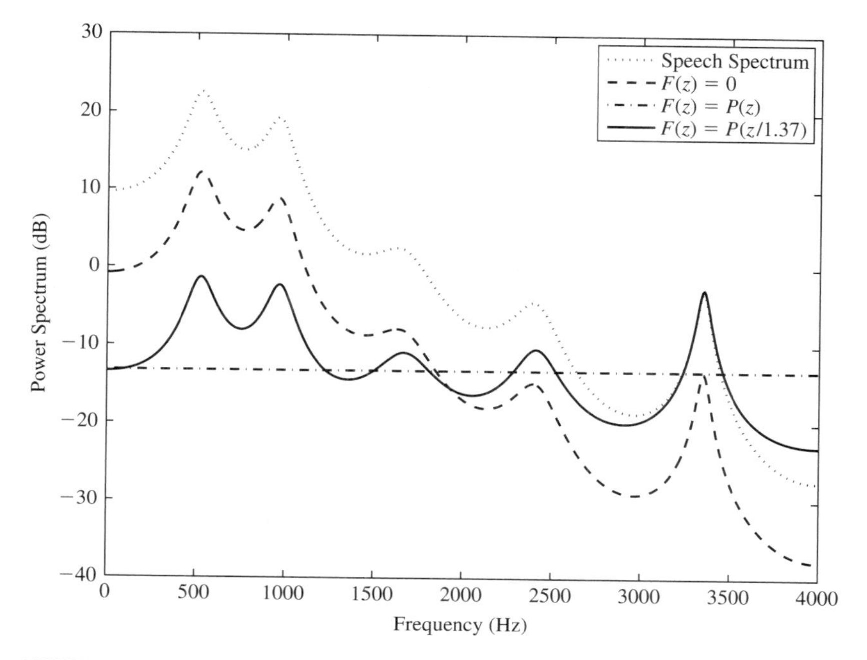

FIGURE 11.75
Spectra of a speech signal and three types of noise shaping.

The use of noise shaping to mask the quantization noise and enable it to "hide" underneath the spectral peaks of the speech signal leads to a noise signal that has more energy than the usual white noise signal from DPCM systems, but which is less audible and annoying due to the effects of tonal masking, and thus this noise masking technique is used in a range of speech coding systems based on predictive coding methods. We return to this topic in the section on analysis/synthesis methods of speech coding, later in this chapter.

11.10.3 A Fully Quantized Adaptive Predictive Coder

We have discussed many aspects of ADPCM coding of speech including adaptive quantization, adaptive (vocal tract) prediction, adaptive pitch prediction, and perceptually motivated adaptive noise shaping. All these topics were brought together in a structure proposed by Atal and Schroeder [17] and further developed by Atal [9]. A block diagram of their proposed system is shown in Figure 11.76. Note that this diagram does not explicitly show that the predictors, noise shaping filter, and quantizer are adapted at a frame rate of 100 Hz. That is, the operations of the diagram are performed on successive 10 msec blocks of samples. The input, $x[n]$, is the pre-emphasized speech signal. The predictor, $P_2(z)$, is the short-term predictor that represents the vocal tract transfer function. The signal, $v[n]$, is the short-term prediction error derived directly from the

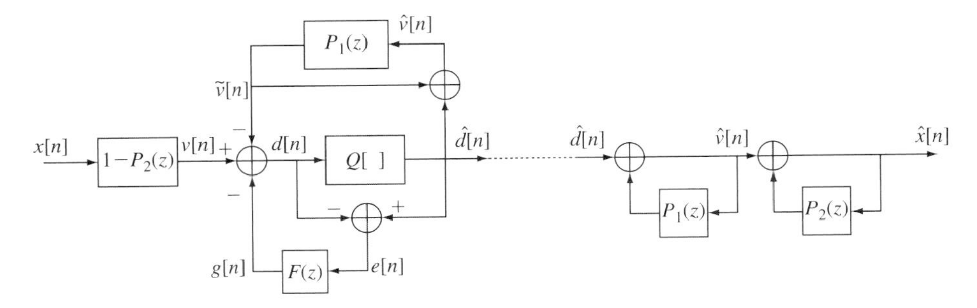

FIGURE 11.76
ADPCM encoder/decoder employing pitch prediction and noise shaping.

input as

$$v[n] = x[n] - \sum_{k=1}^{p} \alpha_k x[n-k].$$

If this signal was the input to the vocal tract filter, $H_2(z) = 1/(1 - P_2(z))$, the output would be the original speech signal. The goal of the encoder is to obtain a quantized representation of this excitation signal from which an approximation, $\hat{x}[n]$, can be reconstructed using $H_2(z)$. The output, $\tilde{v}[n]$, of the pitch predictor $P_1(z)$,

$$\tilde{v}[n] = \sum_{k=-1}^{1} \beta_k \hat{v}[n-M-k],$$

and the noise feedback signal $g[n]$ are both subtracted from $v[n]$ to form the difference signal $d[n] = v[n] - g[n] - \tilde{v}[n]$, which is the input to the quantizer. Both predictors are updated every 10 msec. The quantization noise $e[n] = \hat{d}[n] - d[n]$ is filtered by the system $F(z) = P_2(z/1.37)$, producing the noise feedback signal $g[n]$ as discussed in Section 11.10.2.[27]

The decoder for this system is shown on the right-hand side of Figure 11.76. For convenience, the quantized difference signal is shown connecting directly from the encoder to the decoder. In a practical implementation, the quantized difference signal would be coded efficiently as a sequence of codewords $c[n]$ which, together with the quantization step size information, prediction coefficients, and pitch period estimate, would comprise the representation of $\hat{d}[n]$. The reconstructed $\hat{d}[n]$ is the input to the inverse of the pitch prediction error filter $H_1(z) = 1/(1 - P_1(z))$ whose output is $\hat{v}[n]$. This is the quantized input to the vocal tract system $H_2(z) = 1/(1 - P_2(z))$, whose output is the reconstructed speech signal $\hat{x}[n]$.

[27] Atal and Schroeder [17] recommended peak-limiting the output of the filter $F(z)$ in order to maintain stability.

The total bit rate for an ADPCM coder of this type is expressed by the following equation:

$$I_{\text{ADPCM}} = BF_s + B_\Delta F_\Delta + B_P F_P, \tag{11.175}$$

where B is the number of bits for the output quantized difference signal, B_Δ is the number of bits for encoding the step size with frame rate F_Δ, and B_P is the total number of bits allocated to the predictor coefficients (both long- and short-term) with frame rate F_P. The step size and predictor information, which is often termed the "side information," typically requires 3–4 kbps for high quality coding. We discuss the issue of quantizing the side information at the end of this section. We will focus first on quantization of the difference signal, which, as we have mentioned, plays a role very similar to the excitation signal in the model for speech production, and as it arises in the ADPCM framework, it accounts for the greatest share of the bit rate. Since the difference signal is effectively sampled at the sampling rate of the input signal, even a one bit quantizer will require F_s bps. For typical ADPCM coders $F_s = 8000$ Hz, so this is a significant limit to how low the bit rate can be for an ADPCM coder.

Figure 11.77 illustrates the operation of the encoder of Figure 11.76 with a 2-level ($B = 1$ bit) adaptive quantizer. The scale at the bottom of the figure shows the frame boundaries for $F_\Delta = F_P = 100$ Hz. In each frame, the predictors $P_1(z)$ and $P_2(z)$ are estimated from the input signal by the techniques discussed earlier in this chapter and in Chapter 9. Figure 11.77a shows the partial difference signal $v[n] - \tilde{v}[n]$ which can be computed from the input signal and used as a basis for determining the step size for the block. The quantizer input cannot be used for this purpose since (because of the noise feedback) it is not available until the signal is quantized. Figure 11.77b shows the difference signal $d[n]$, which differs from Figure 11.77a by the quantization noise that is fed back through $F(z)$ and the pitch predictor. Figure 11.77c shows the output of the 2-level quantizer. The effect of the block-to-block variation of the step size is evident. Figure 11.77d shows the signal $\hat{v}[n]$, which is the output of the inverse of the pitch prediction error filter. This signal should be compared to Figure 11.77e, which shows the original unquantized vocal tract prediction error. The goal of the coder is to make $\hat{v}[n]$ as much like $v[n]$ as possible so that the reconstructed pre-emphasized speech signal $\hat{x}[n]$ in Figure 11.77f is as much like (in both the waveform and auditory senses) the original input $x[n]$.

This example reminds us that the problem with 2-level quantization (as we have seen before with delta modulation) is the difficulty of accurately representing a signal with such a wide dynamic range as the one in Figure 11.77b. Any compromise results in peak clipping for high amplitude samples and excessive granularity for the low amplitude samples. The effect in this example is clearly evident in the comparison of $\hat{v}[n]$ in Figure 11.77d with $v[n]$ in Figure 11.77e. The result of this inability to accurately represent the amplitude range of the difference signal is that the reproduction quality of the ADPCM coder with a 2-level quantizer falls short of "transparency" even if the side information is not quantized.

As a solution to this problem, Atal and Schroeder [17] proposed a multi-level quantizer combined with center clipping as depicted in Figure 11.78. This quantizer is motivated by the empirical observation that it is perceptually most important to

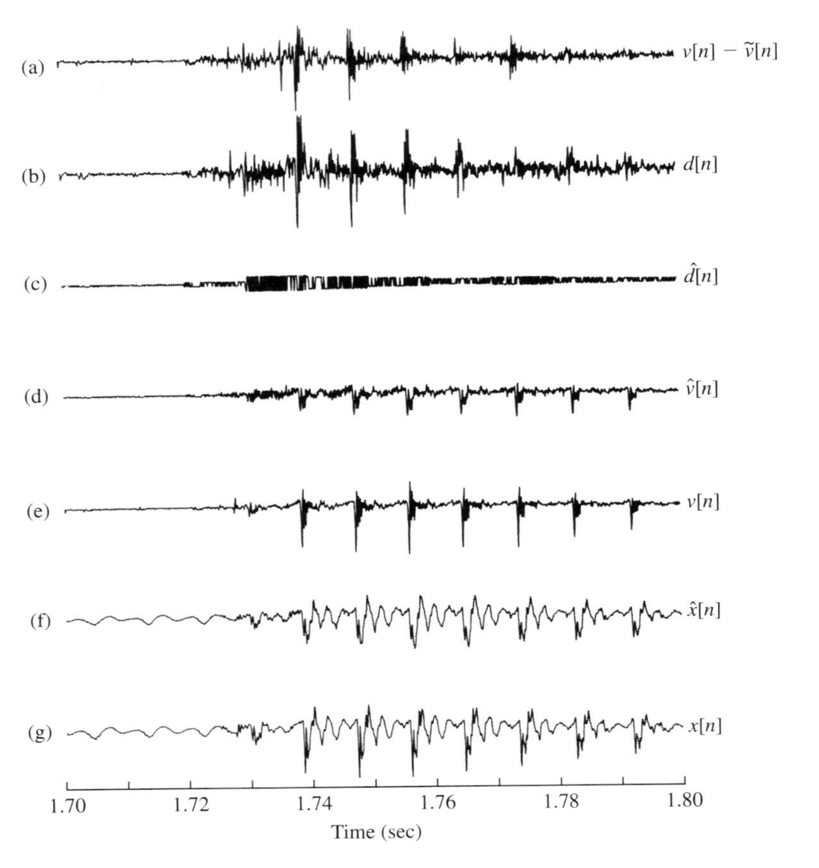

FIGURE 11.77
Waveforms of signals in an ADPCM coder with adaptive 2-level quantizer: (a) partial difference signal $v[n] - \tilde{v}[n]$; (b) difference signal (quantizer input) $d[n]$ with noise feedback; (c) quantizer output $\hat{d}[n]$; (d) reconstructed short-term prediction residual $\hat{v}[n]$; (e) original unquantized short-term prediction residual $v[n]$; (f) reconstructed pre-emphasized speech signal $\hat{x}[n]$; (g) original pre-emphasized input speech signal $x[n]$. The signals $v[n]$ and $\hat{v}[n]$ in (d) and (e) are amplified by 6 dB relative to the speech signal $x[n]$. The signals in (a)–(c) are amplified by 12 dB relative to $x[n]$. (After Atal and Schroeder [17]. © [1980] IEEE.)

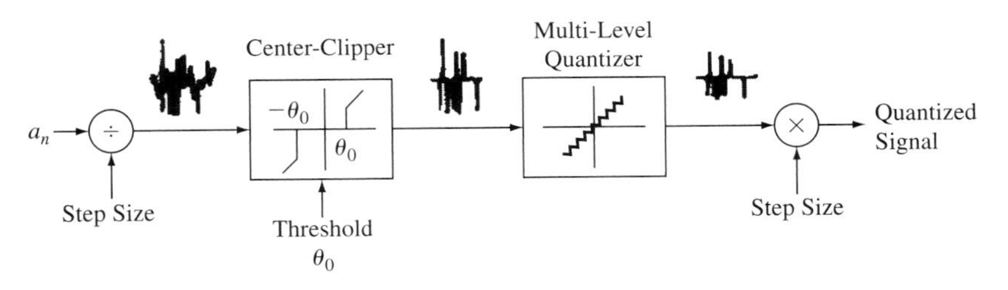

FIGURE 11.78
Three-level quantizer with center clipping.

accurately represent the large amplitude samples of the difference signal. Figure 11.78 represents the quantizer as a normalization by the step size followed by center clipping with threshold θ_0, which is, in turn, followed by a normalized multi-level quantizer. The quantized difference signal is then obtained by reapplying the step size. The center clipping operation has the effect of setting to zero all samples of $d[n]$ falling below the threshold θ_0. By adjusting this threshold to a high value, it is possible to force a large percentage of samples to be zero. This leads to a sparse difference signal that can be coded at average bit rates below F_s bps.

Figure 11.79 shows an identical set of waveforms to those in Figure 11.77 except with a multi-level center-clipped quantizer as described in detail in Ref. [17]. Figure 11.79a again shows the signal $v[n] - \tilde{v}[n]$ derived from the input by the vocal tract and pitch predictors. This waveform is identical to that of Figure 11.77a.

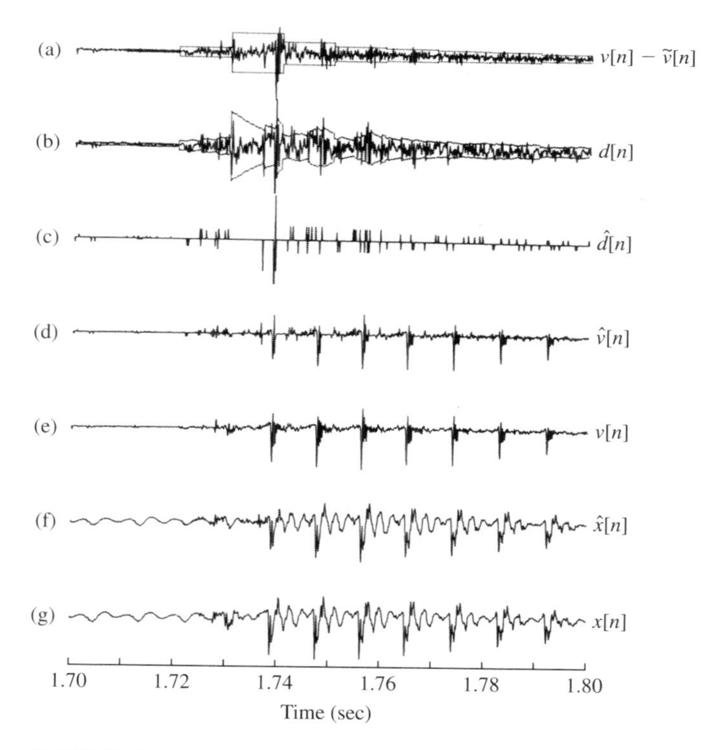

FIGURE 11.79
Waveforms of signals in an ADPCM coder with adaptive multi-level center clipping quantizer: (a) partial difference signal $v[n] - \tilde{v}[n]$; (b) difference signal (quantizer input) $d[n]$ with noise feedback; (c) quantizer output $\hat{d}[n]$; (d) reconstructed short-term prediction residual $\hat{v}[n]$; (e) original unquantized short-term prediction residual $v[n]$; (f) reconstructed pre-emphasized speech signal $\hat{x}[n]$; (g) original pre-emphasized input speech signal $x[n]$. The signals $v[n]$ and $\hat{v}[n]$ in (d) and (e) are amplified by 6 dB relative to the speech signal $x[n]$. The signals in (a)–(c) are amplified by 12 dB relative to $x[n]$. (After Atal and Schroeder [17]. © [1980] IEEE.)

The dotted lines in this case represent initial estimates of the clipping threshold θ_0 to produce a prescribed percentage of zero samples in the quantized difference signal. As before, this cannot be determined from the difference signal, which is not available until the quantizer is determined. It was found that setting a fixed threshold at the beginning of the processing of a frame resulted in significant frame-to-frame variation in the number of non-zero samples. An algorithm given in Ref. [17] avoids large variations by making the center clipping threshold vary from sample to sample. An illustration of how the center clipping threshold varies is shown in Figure 11.79b by the dotted curve superimposed on the difference signal $d[n]$. The multi-level quantizer proposed in Ref. [17] has 15 levels placed symmetrically about zero. The step size is determined by dividing twice the peak amplitude in the frame by 14. Because the center clipping removes as much as 90% of the low-amplitude samples, only the top three positive and negative levels are ever used to quantize the difference signal. Figure 11.79c shows that the quantized difference signal has mostly zero samples, with the non-zero samples concentrated where the unquantized difference signal is large. The reconstructed signal, $\hat{v}[n]$ in Figure 11.79d, shows much better reconstruction of the peaks than the 2-level quantizer in Figure 11.77d.

Estimates of probabilities of usage of the quantization levels showed that the entropy of the signal in Figure 11.79c was on the order of 0.7 bits/sample, which would imply an average data rate of 5600 bps for the excitation signal. However, this is a long-term average value, so in order to take advantage of the fact that the zero level is much more probable than any other to reduce the bit rate of the difference signal, it would be necessary to code blocks of samples together. Such block coding could increase the coding delay by as much as 10 frames (100 msec) in order to assure a constant data rate [17].

To complete the discussion of the coder, we note that Atal [9] used a 16th-order predictor and quantized the resulting set of PARCOR coefficients after transformation with an inverse sine function. The number of bits per coefficient ranged from 5 for each of the lowest two PARCORs down to 1 each for the two highest-order PARCORs, for a total of 46 bits/frame. If the parameters are updated 100 times/sec, then a total of 4600 bps for the short-delay predictor information is required. To save bit rate, it is possible to update the short-delay predictor 50 times/sec for a total bit rate contribution of 2300 bps [9]. The long delay predictor has a delay parameter M, which requires 7 bits to cover the range of pitch periods $20 \leq M \leq 147$ to be expected with 8 kHz sampling rate of the input. The long-delay predictor in the system used delays of $M-1$, M, $M+1$ and three coefficients (gains) each quantized to 4 or 5 bit accuracy. This gave a total of 20 bits/frame and added 2000 bps to the overall bit rate. Finally the rms value of the difference signal requires 6 bits at a frame rate of 100 frames/sec or a total of 600 bps. Thus, the side information requires $4600 + 2000 + 600 = 7200$ bps if a frame rate of 100 is used for all parameters or 4900 bps if the PARCORs are updated every 50 msec. When added to the bit rate for the excitation, the total bit rate with a 2-level quantizer would be either 12,900 or 15,200 bps, and with the multi-level quantizer encoded to take advantage of the low entropy of the difference signal, the total data rate would be either 10,500 or 12,800 bps. The conclusion from this study is that telephone quality reproduction can be achieved with the system of Figure 11.76 at bit rates around 10 kbps.

There is much to be learned from the discussion of the system of Figure 11.76. The basic structure affords great flexibility in creating sophisticated adaptive predictive coders. It can be called a waveform coder since it does tend to preserve the waveform of the speech signal. However, it uses noise shaping to attempt to exploit features of auditory perception, and in using the center clipping multi-level quantizer, it discards information that is important in preserving the detailed wave shape but less important for perception. By building in the adaptive vocal tract and pitch predictors, it could also be termed a model-based coder. In fact it might best be called a *model-based closed-loop waveform coder* that derives its excitation signal by a quantizer inside a feedback loop. The sparse difference signal created by the simple adaptive quantizer suggests that other approaches might lead to even more effective ways of determining the excitation signal for the vocal tract filter. Indeed, this led to the development of the multipulse (MPLPC) and code excited (CELP) linear predictive coders to be discussed in the next section of this chapter.

11.11 ANALYSIS-BY-SYNTHESIS SPEECH CODERS

Our previous discussions in Sections 11.7–11.10 have focused on various embellishments and refinements of the basic DPCM system that we introduced in Section 11.7. We have seen that by incorporating linear prediction, we have, in a very real sense, incorporated the discrete-time model for speech production, and by doing this, we have been able to lower the data rate for high quality telephone bandwidth speech reproduction from 128 kbps (for 16-bit uniformly quantized samples) to about 10 kbps (for the most sophisticated ADPCM systems like the one discussed in Section 11.10.3). To further reduce the data rate requires different approaches. In this section we study a technique called *analysis-by-synthesis* (A-b-S) coding, which is closely related to (and motivated by) the technique for quantization of the difference signal in a predictive coder that was presented in Section 11.10.3. There it was shown that the key to reducing the data rate of a closed-loop adaptive predictive coder was to force the coded difference signal (input to the vocal tract model) to be better represented at low data rates while maintaining very high quality at the output of the decoder synthesizer.

Figure 11.80 shows a block diagram of a somewhat different approach that bears a great resemblance to the adaptive differential PCM coders that we have studied. However, instead of a quantizer for generating the excitation signal for the vocal tract filter, this system substitutes an optimization process (denoted as Error Minimization in Figure 11.80) whereby the excitation signal, $d[n]$, is constructed based on minimization of the mean-squared value of the synthesis error, $d[n] = x[n] - \tilde{x}[n]$, defined as the difference between the input signal $x[n]$ and the synthesized signal $\tilde{x}[n]$. By careful design of the optimization process, the excitation signal, $d[n]$, created by the optimization process more closely approximates the key features of the ideal excitation signal while maintaining a structure that is easy to encode. A key feature of the A-b-S system is the incorporation into the analysis loop of a perceptual weighting filter, $w[n]$, of the type discussed earlier in this chapter, so as to take advantage of the masking effects in speech perception.

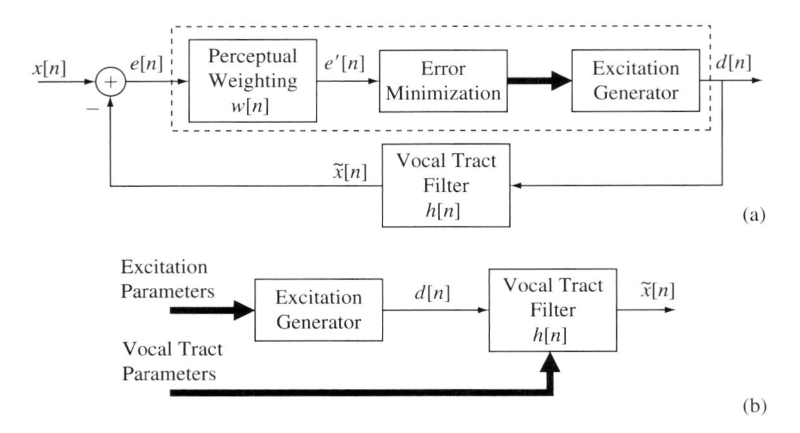

FIGURE 11.80
A-b-S coding: (a) closed-loop coder for speech incorporating a perceptual weighting filter; (b) corresponding decoder (synthesizer).

The operations depicted in Figure 11.80a are applied iteratively to blocks of speech samples. The basic operation of each iteration of the closed-loop analysis-synthesis system of Figure 11.80 is as follows:

1. At the beginning of the iteration process (and only once for each full set of loop iterations), the speech signal, $x[n]$, is used to generate an optimum p^{th}-order linear predictive vocal tract filter, of the form:

$$H(z) = \frac{1}{1 - P(z)} = \frac{1}{1 - \sum_{i=1}^{p} \alpha_i z^{-i}}. \tag{11.176}$$

2. The difference signal, $d[n] = x[n] - \tilde{x}[n]$, based on an initial estimate of the speech signal, $\tilde{x}[n]$ (which itself is based on an initial estimate of the error signal), is perceptually weighted by a speech-adaptive filter of the form:

$$W(z) = \frac{1 - P(z)}{1 - P(\gamma z)}. \tag{11.177}$$

The frequency response (as a function of γ) of the weighting filter is shown in Figure 11.81, where we see that as γ approaches 1, the weighting is essentially flat and independent of the speech spectrum (i.e., no perceptual weighting), and as γ approaches 0, the weighting becomes the inverse frequency response of the vocal tract filter.

3. The error minimization box and the excitation generator create a sequence of error signals that iteratively (once per iteration) improve the match to the weighted error signal by using one of several techniques to be described in this

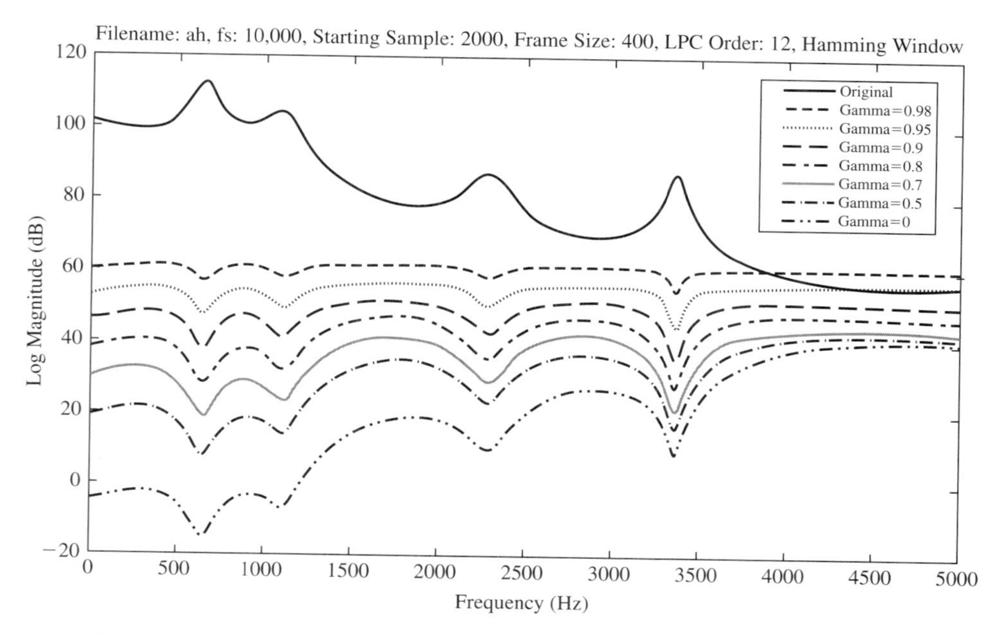

FIGURE 11.81
Plots of the log magnitude of the vocal tract model frequency response for a frame of a speech signal, along with six weighting filter responses, corresponding to values of the weighting parameter, $\gamma = 0.98, 0.95, 0.9, 0.8, 0.7, 0.5, 0$. (After Kondoz [197].)

section, such as by using multiple pulses per frame with variable amplitudes (multipulse LPC or MPLPC), or by selecting an excitation vector from a codebook of Gaussian white noise vectors (CELP) and appropriately scaling that excitation vector.

4. The resulting excitation signal, $d[n]$, which is an improved estimate of the actual prediction error signal at each loop iteration, is used to excite the vocal tract model system. The processing is iterated until the resulting error signal meets some criterion for stopping the closed-loop iterations.

It should be noted that the perceptual weighting filter of Eq. (11.177) is often modified to the form:

$$W(z) = \frac{1 - P(\gamma_1 z)}{1 - P(\gamma_2 z)}, \quad \text{where } 0 \leq \gamma_1 \leq \gamma_2 \leq 1, \tag{11.178}$$

so as to make the perceptual weighting be less sensitive to the detailed frequency response of the vocal tract filter. Further, the vocal tract filter is often preceded by a (long-term) pitch prediction filter of the type discussed earlier in this chapter. (We will return to the use of pitch prediction filters in the closed-loop system later in this section.) Finally, it is simple to show that the perceptual weighting filter can be moved from its position (after the difference signal is computed) to a position outside the closed-loop (so as to process the speech signal only one time per frame). However, the perceptual weighting filter then has to be combined with the vocal tract filter so that the synthesized signal is properly weighted. This presents no problem and, in fact, substantially reduces the overall loop computations.

We now analyze the closed-loop system of Figure 11.80 for two coding methods, namely MPLPC and CELP. First, however, we discuss the general principles of A-b-S systems and show how we construct the excitation signal by optimization using a series of "basis-like" functions.

11.11.1 General Principles of A-b-S Coding of Speech

The key problem that we are attempting to solve using A-b-S methods is to find a representation of the excitation for the vocal tract filter that produces high quality synthetic output while maintaining a structured representation that makes it easy to code the excitation at low data rates. Another approach, to be discussed in Section 11.12, is to construct an excitation in an open-loop manner using a combination of periodic pulses and random noise. The excitation produced by an A-b-S system is more robust and accurate than the simple set of pitch pulses (for voiced speech sounds) or Gaussian noise (for unvoiced sounds or background) that is conventionally used in the LPC vocoder discussed in Section 11.12. The reasons for this are clear; namely that the ideal model of speech production is just not good enough for high quality speech to be generated. The model breaks down for nasal sounds, for fricatives, for nasalized vowels, and for many other sounds. Further, the idealization of pure pitch pulse excitation for voiced sounds produces a buzzy effect, showing that the spectrum of voiced sounds is not only a sequence of harmonics at multiples of the pitch frequency, but instead also contains a noise-like component that is simply not well represented by the ideal speech model. As such, we seek a method of representing the excitation, $d[n]$, in a more complete and robust manner. The beauty of the closed-loop systems that incorporate the speech production model is that there is no need for a hard delineation between the "vocal tract model" and the "excitation model." What counts is the quality of the synthesized output.

Assume we are given a set of Q "basis functions" of the form

$$\mathcal{F}_\gamma = \{f_1[n], f_2[n], \cdots, f_Q[n]\}, \tag{11.179}$$

where each basis function, $f_i[n]$, $i = 1, 2, \ldots, Q$ is defined over the frame interval $0 \leq n \leq L-1$ and is zero outside this interval. Our goal is to build up an optimal excitation function for each frame, in stages, by adding a new weighted basis function at each iteration of the A-b-S process. Thus, at each iteration of the A-b-S loop, we select the basis function from the set $\mathcal{F}_\gamma$ which maximally reduces the perceptually weighted mean-squared error, $\mathcal{E}$ (between the synthetic waveform and the original speech signal), which is of the form:

$$\mathcal{E} = \sum_{n=0}^{L-1} \left[(x[n] - d[n] * h[n]) * w[n]\right]^2, \tag{11.180}$$

where $h[n]$ and $w[n]$ are the vocal tract impulse response and the perceptual weighting filters, respectively, for a particular frame of the speech signal.[28]

We denote the optimal basis function at the kth iteration as $f_{\gamma_k}[n]$, giving the excitation signal component $d_k[n] = \beta_k f_{\gamma_k}[n]$, where β_k is the optimal weighting coefficient for basis function $f_{\gamma_k}[n]$ at iteration k. The A-b-S iteration continues adding scaled basis components until the perceptually weighted error falls below some desired threshold, or until a maximum number of iterations, N, is reached, giving the final excitation signal, $d[n]$, as

$$d[n] = \sum_{k=1}^{N} \beta_k f_{\gamma_k}[n]. \tag{11.181}$$

We solve for the optimal excitation signal of Eq. (11.181) at each iteration stage using an iterative method based on the signal processing shown in Figure 11.82a.[29] We

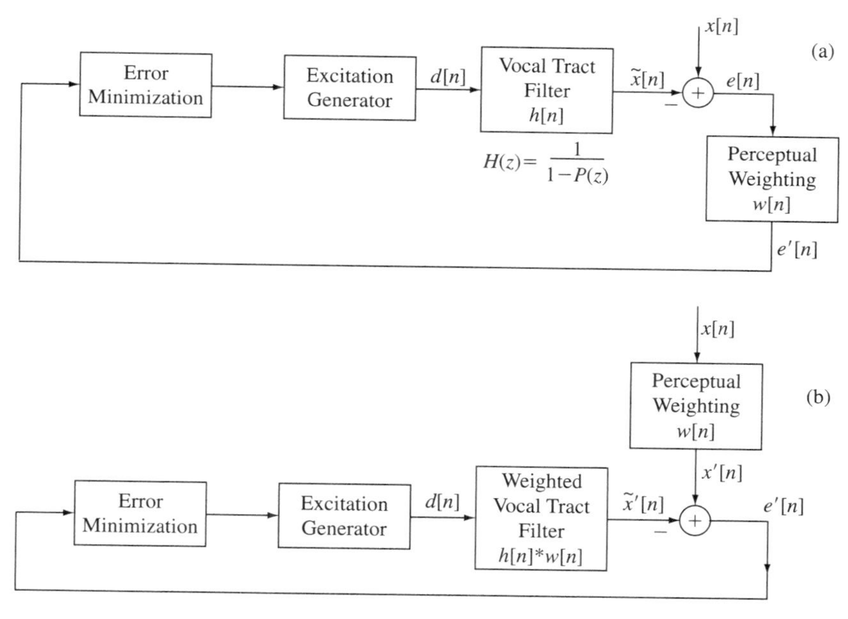

FIGURE 11.82
(a) Closed-loop coder of Figure 11.80 redrawn to match conventional representations; (b) reformulation of the A-b-S structure in terms of perceptually weighted input and impulse response.

[28]Note that we do not show the frame index explicitly.

[29]Figure 11.82a, which is identical to Figure 11.80, is the conventional way to draw the A-b-S system. Figure 11.80 was used to highlight the close relationship of the A-b-S system to other differential coding systems we have studied.

begin by assuming that $d[n]$ is known up to the current frame (i.e., up to $n = 0$) for simplicity, and we initialize our 0th estimate of the excitation, $d_0[n]$ as,

$$d_0[n] = \begin{cases} d[n] & n < 0 \\ 0 & 0 \leq n \leq L-1. \end{cases} \tag{11.182}$$

We form the initial estimate of the speech signal as

$$y_0[n] = \tilde{x}_0[n] = d_0[n] * h[n], \tag{11.183}$$

where $h[n]$ is our estimate of the vocal tract impulse response. Since $d_0[n] = 0$ in the frame $0 \leq n \leq L-1$, $y_0[n]$ consists of the decaying oscillations from the previous frame(s), which, of course, must be computed and stored for use in coding the present frame. We complete the initial (0th) iteration by forming the perceptually weighted difference signal as:

$$e'_0[n] = (x[n] - y_0[n]) * w[n] \tag{11.184a}$$

$$= x'[n] - y'_0[n] = x'[n] - d_0[n] * h'[n], \tag{11.184b}$$

where

$$x'[n] = x[n] * w[n], \tag{11.185a}$$

$$h'[n] = h[n] * w[n], \tag{11.185b}$$

are respectively the perceptually weighted input signal and the perceptually weighted impulse response. Note that throughout this discussion, we will use the notation $'$ to denote "perceptually weighted." The diagram of Figure 11.82a can be redrawn as Figure 11.82b, showing that the operations of Eqs. (11.185a) and (11.185b) can be done once per frame and then used at each stage of the iterative search for $d[n]$, thereby saving significant computation.

We now begin the kth iteration of the A-b-S loop, where $k = 1, 2, \ldots, N$. In each iteration we select one of the $\mathcal{F}_\gamma$ basis set (call this $f_{\gamma_k}[n]$) and determine the associated amplitude β_k to minimize the error in representing $x[n]$. This gives

$$d_k[n] = \beta_k \cdot f_{\gamma_k}[n], \quad k = 1, 2, \ldots, N. \tag{11.186}$$

To determine the value β_k that reduces the error the most, assuming a chosen basis sequence $f_{\gamma_k}[n]$, we form the new perceptually weighted error as

$$e'_k[n] = e'_{k-1}[n] - \beta_k f_{\gamma_k}[n] * h'[n] \tag{11.187a}$$

$$= e'_{k-1}[n] - \beta_k y'_k[n], \tag{11.187b}$$

where we recall that γ_k is temporarily assumed to be known. We now can define the total mean-squared residual error for the kth iteration as:

$$\mathcal{E}_k = \sum_{n=0}^{L-1} (e'_k[n])^2 = \sum_{n=0}^{L-1} (e'_{k-1}[n] - \beta_k y'_k[n])^2, \tag{11.188}$$

i.e., the error at the kth iteration is what remains after subtracting the new contribution to the output from the previous error sequence in the frame. Since we assume that γ_k is known, we can find the optimum value of β_k by differentiating $\mathcal{E}_k$ with respect to β_k, giving the following:

$$\frac{\partial \mathcal{E}_k}{\partial \beta_k} = -2\sum_{n=0}^{L-1} (e'_{k-1}[n] - \beta_k y'_k[n]) \cdot y'_k[n] = 0. \tag{11.189}$$

We can then solve for the optimum β_k as:

$$\beta_k^{\text{opt}} = \frac{\displaystyle\sum_{n=0}^{L-1} e'_{k-1}[n] \cdot y'_k[n]}{\displaystyle\sum_{n=0}^{L-1} (y'_k[n])^2}. \tag{11.190}$$

Substituting Eq. (11.190) into Eq. (11.188) leads (after some manipulation) to the following expression for the minimum mean-squared error:

$$\mathcal{E}_k^{\text{opt}} = \sum_{n=0}^{L-1} (e'_{k-1}[n])^2 - (\beta_k^{\text{opt}})^2 \sum_{n=0}^{L-1} (y'_k[n])^2. \tag{11.191}$$

The only remaining task is to find the optimum function $f_{\gamma_k}[n]$ (for each k), and we do this by searching through all possible f_{γ_k}'s and picking the one that maximizes $\sum_{n=0}^{L-1} (y'_k[n])^2$, where we recall that $y'_k[n] = f_{\gamma_k}[n] * h'[n]$. After we have found the complete set of optimum f_{γ_k}'s and the associated β_k's for all N iterations, our final results are the set of relations:

$$\tilde{x}'[n] = \sum_{k=1}^{N} \beta_k f_{\gamma_k}[n] * h'[n] = \sum_{k=1}^{N} \beta_k \cdot y'_k[n] \tag{11.192a}$$

$$\mathcal{E}_N = \sum_{n=0}^{L-1} (x'[n] - \tilde{x}'[n])^2 = \sum_{n=0}^{L-1} \left(x'[n] - \sum_{k=1}^{N} \beta_k \cdot y'_k[n] \right)^2 \tag{11.192b}$$

$$\frac{\partial \mathcal{E}_N}{\partial \beta_j} = -2\sum_{n=0}^{L-1}\left(x'[n] - \sum_{k=1}^{N}\beta_k y'_k[n]\right) y'_j[n] = 0, \tag{11.192c}$$

where the re-optimized β_k's satisfy (for $k = 1, 2, \dots, N$) the relation:

$$\sum_{n=0}^{L-1} x'[n] y'_j[n] = \sum_{k=1}^{N}\beta_k \left(\sum_{n=0}^{L-1} y'_k[n] \cdot y'_j[n]\right). \tag{11.193}$$

The result of the optimization is, for each frame, the set of coefficients, β_k, and the excitation function indices, γ_k, for $k = 1, 2, \dots, N$. This information, together with the parametric representation of the vocal tract filter (e.g., quantized PARCOR coefficients), comprises the representation of the speech signal. At the receiver, a table of the excitation sequences $f_{\gamma_k}[n]$ is used with the β_ks to recreate the excitation, and the output is reconstructed according to

$$\tilde{x}[n] = \left(\sum_{k=1}^{N}\beta_k f_{\gamma_k}[n]\right) * h[n]. \tag{11.194}$$

We now have all the equations we need to implement the closed-loop A-b-S speech coding system. What remains is to specify the "basis-like" function sets. There are many possibilities for these functions. We shall discuss two of these sets of basis functions in detail, namely the set used for MPLPC and the set used for CELP.

The "basis-like" functions proposed by Atal and Remde [14] for MPLPC coding are of the form:

$$f_\gamma[n] = \delta[n - \gamma] \quad 0 \le \gamma \le (Q - 1 = L - 1); \tag{11.195}$$

i.e., the "basis-like" functions consist of delayed impulses within the frame of the speech signal, (the region $0 \le n \le L - 1$). Thus for the MPLPC solution, the excitation signal is represented as a series of impulses (of varying amplitudes β_k).

A second set of "basis-like" functions, proposed by Schroeder and Atal [353], is based on a codebook of white Gaussian noise vectors, and is of the form:

$$f_\gamma[n] = \text{vector of white Gaussian noise}, \quad 1 \le \gamma \le Q = 2^M, \tag{11.196}$$

where each of the white Gaussian noise vector is defined over the entire speech frame ($0 \le n \le L - 1$), and there are 2^M such vectors in an M-bit codebook, where M typically is on the order of 10 (i.e., 1024 Gaussian noise vectors in the codebook). We will see, later in this section, that such random noise codebooks are most effective when used in conjunction with a pitch predictor that helps to form the regular (pitch) pulses during voiced speech regions of the signal.

A third set of "basis-like" functions, proposed by Rose and Barnwell [325], is the set of shifted versions of parts of the previously formed excitation sequence; i.e.,

$$f_\gamma[n] = d[n - \gamma] \quad \Gamma_1 \le \gamma \le \Gamma_2. \tag{11.197}$$

In the following sections we give more details on multipulse excitation and code-excited excitation for linear predictive coders. The interested reader is referred to Ref. [325] for more details on the "self-excited" LP coder.

11.11.2 Multipulse LPC

A multipulse coder uses the system of Figure 11.82b, with basic excitation components given by Eq. (11.195). The error minimization block, at each iteration, minimizes the expression in Eq. (11.188), where $n = 0$ is the beginning of the current frame. For the simple impulse functions, this gives

$$\mathcal{E} = \sum_{n=0}^{L-1} (x'[n] - \sum_{k=1}^{N} \beta_k h'[n - \gamma_k])^2, \quad (11.198)$$

i.e., the mean-squared error between the speech signal and the predicted speech signal based on an N-impulse representation of the excitation signal, $d[n]$, over the current speech frame ($0 \leq n \leq L - 1$).

Figure 11.83 illustrates the procedure for determining the optimum impulse locations. A pulse is placed at every location in the current speech frame and processed by the vocal tract filter (labeled as linear filter in the figure). This is easy to accomplish because convolution with shifted impulses simply shifts the impulse response. Based on the residual error obtained by using the pulse at a given location, the energy of the residual is calculated and normalized by the energy of the output of the vocal tract filter (when excited by the single pulse) and the resulting peak location is selected as the optimum pulse location. The optimum weighting coefficient, β_k, is obtained using the computation of Eq. (11.190). This process is iterated by finding the best single pulse

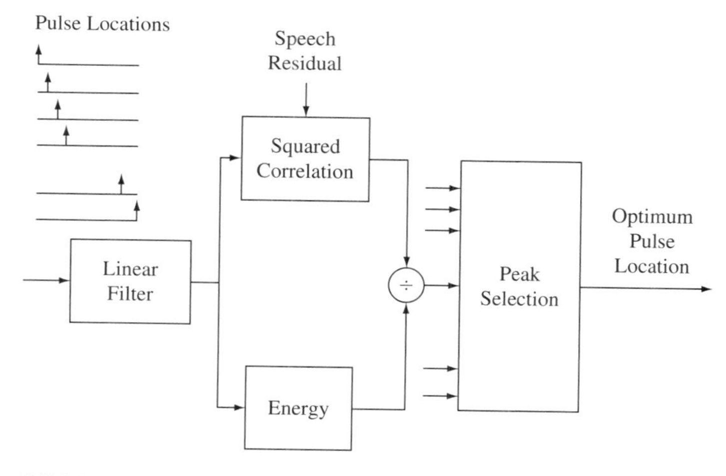

FIGURE 11.83
Block diagram illustrating the procedure for determining the optimum pulse location in multipulse analysis. (After Atal and Remde [14]. © [1982] IEEE.)

solution (both location and amplitude), subtracting out the effect of this pulse from the speech waveform, and iterating the process to find subsequent best pulse locations and amplitudes. This iterative process is continued until the desired minimum error is obtained or the maximum number of impulses in a given frame, N, has been used. It has been found that about eight impulses for each 10 msec interval of speech gives a synthetic speech signal that is perceptually close to the original speech signal. For a 10 kHz sampling rate, the uncoded excitation signal consists of 100 samples of each 10 msec interval. Hence MPLPC has the ability to reduce the number of parameters to be coded by more than an order of magnitude (from 100 waveform samples per 10 msec interval to eight impulses). Depending on the quantization method, we would expect the multipulse process to be about 10 times more efficient than a simple waveform coding method, operating directly on the speech waveform.

The results of the multipulse analysis process are illustrated in Figure 11.84, which shows signal waveforms from the first four iterations of the A-b-S loop. The original speech signal (over the analysis frame) is shown at the top of the figure for iterations $k = 0, 1, 2, 3, 4$. The excitation signal, at each iteration, is shown below the speech waveform, followed by the synthetic speech generated using the current coded excitation, followed finally by the error between the original speech signal and the synthetic speech signal. Initially (at iteration $k = 0$), the excitation signal is zero and the synthetic speech is the extended output due to the excitation up through the previous frame. Since the impulse response of the vocal tract filter generally is longer than a single frame of speech, it extends to the following frame and must be accounted for in the computation of the optimum excitation "basis-like" functions. The initial error signal is the difference between the speech signal in the current frame and the tail of the output due to the previous frame.

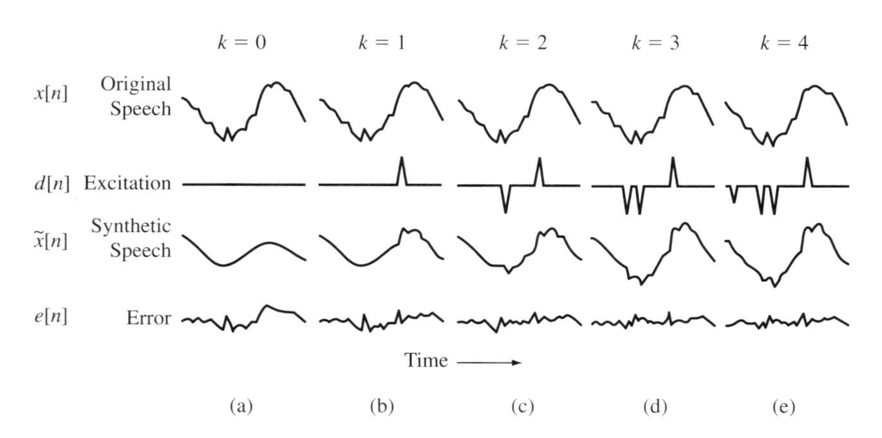

FIGURE 11.84
Illustration of the multipulse analysis process. The upper plots show the original speech waveform over the analysis frame; the next layer shows the excitation signal for the first four iterations of the A-b-S loop; the next layer shows the synthetic speech signal generated with k pulses, and the lowest layer shows the resulting error signal over the first four iterations. (After Atal and Remde [14]. © [1982] IEEE.)

At the first iteration ($k = 1$), the multipulse excitation analysis (as shown in Figure 11.83 and explained in the previous section) finds the location and amplitude of the best single pulse, as shown under the $k = 1$ column of Figure 11.84, and the resulting synthetic speech and the error signal both change in response to this initial excitation signal impulse. The process is iterated for $k = 2, 3, 4$ and optimum pulse locations and amplitudes are determined, and the synthetic speech gets perceptually closer in quality to the original speech signal, with the error function energy becoming smaller and smaller.

At the end of the search, the MPLPC method re-optimizes the amplitudes of all the pulses obtained from the search, retaining the pulse locations as fixed by the search. The re-optimization is straightforward and is simply the direct application of the optimization formula of Eq. (11.193) to the final solution. Although this is a sub-optimal procedure (the method could have re-optimized all pulse amplitudes at each iteration), it has been shown to yield results that are comparable to the optimized procedure, with significantly less computational complexity.

Figure 11.85 gives examples of how well the MPLPC coding method works on two segments of speech. For each of the speech segments, the figure shows:

- the original speech signal;
- the resulting synthetic speech signal;
- the multipulse excitation signal;
- the resulting error signal (the difference between the original speech signal and the synthetic speech signal derived from the multipulse representation of the excitation).

As seen in Figure 11.85, with a sufficient number of pulses (typically 8 per 10 msec interval or 800 pulses/sec), the resulting synthetic speech is essentially indistinguishable from the original speech. Using simple scalar quantizers with 9 bits per pulse (both

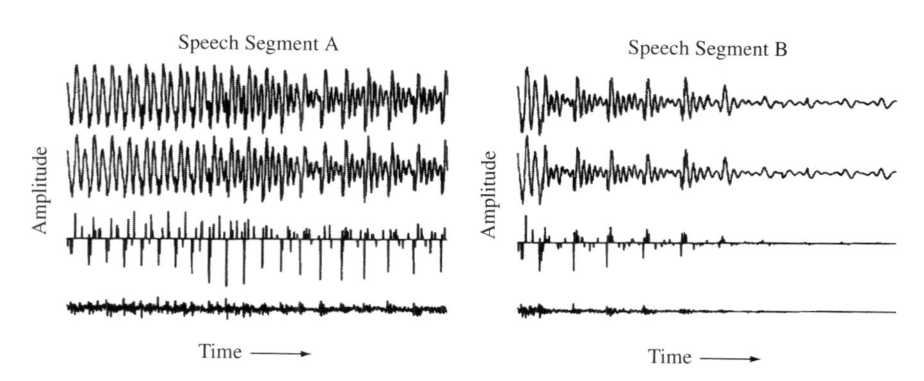

FIGURE 11.85
Examples of MPLPC coding of two segments of speech. For each segment, there are plots of the original speech signal (the upper plot), followed by the synthetic speech signal, the set of pulses of the excitation signal, and finally the error signal. (After Atal and Remde [14]. © [1982] IEEE.)

location and amplitude included in the 9 bits), a total of 7200 bps would be needed to code the excitation signal pulses. We have already shown that we need about 2400 bps to code the vocal tract filter coefficients (using whatever equivalent parameter set that is desired), giving a total bit rate of about 9600 bps for a full MPLPC system.

Various techniques have been proposed for reducing the bit rate of MPLPC systems, including coding the pulse locations differentially (since they are constrained to occur within a single frame and are naturally ordered in location), and to normalize the pulse amplitudes to reduce the dynamic range of the amplitude parameters. Other simplifications include using a long-term pitch predictor in conjunction with the vocal tract filter, thereby decomposing the correlation of speech into a short-time component (used to provide spectral estimates) and a long-time component (used to provide pitch period estimates). Hence, in the A-b-S processing loop, we first remove the long-time correlation by using the pitch prediction filter, followed by removing the short-time correlation using the vocal tract filter. The side result is that by removing the pitch pulses from the multipulse analysis, there are fewer large pulses to code, and we can code the speech at somewhat lower bit rates—typically on the order of 8000 bps (as opposed to 9600 bps for the original MPLPC system).

Figure 11.86 shows the resulting A-b-S system for MPLPC coding, using a long-term pitch predictor of the form:

$$\hat{B}(z) = 1 - bz^{-M}, \tag{11.199}$$

where M is the pitch period (as estimated by an open-loop process pitch detector), and using a short-term vocal tract predictor of the form:

$$\hat{A}(z) = 1 - \sum_{k=1}^{p} \alpha_k z^{-k}, \tag{11.200}$$

where p is the linear predictive analysis order and $\{\alpha_k\}$ is the prediction coefficient set as determined using an open-loop analysis. The closed-loop residual generator is

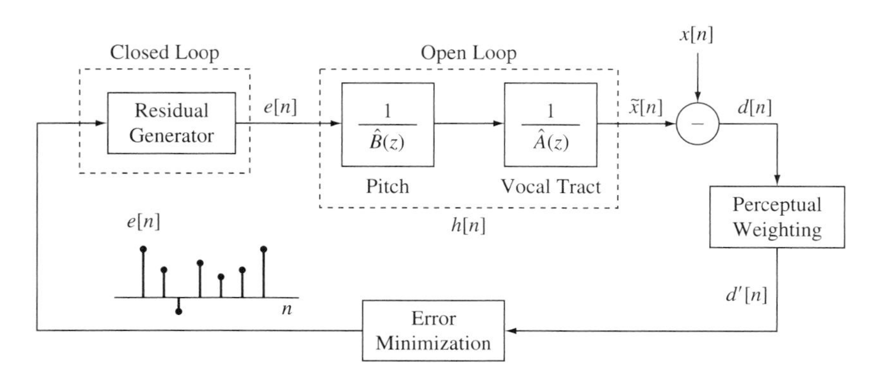

FIGURE 11.86
Block diagram of MPLPC A-b-S processing. (After Atal and Remde [14]. © [1982] IEEE.)

explicitly marked in Figure 11.86. The system of Figure 11.86 has been shown to yield high quality speech at bit rates of between 8000 and 9600 bps [14].

11.11.3 Code Excited Linear Prediction (CELP)

We have just shown that the excitation to a vocal tract model derived by linear predictive analysis, $d[n]$, could be approximated by a series of variable amplitude and variable location pulses, suitably optimized to minimize the perceptually weighted error of the approximation to the speech signal. The basic idea behind CELP coding is to represent the residual after both long-term (pitch period) and short-term (vocal tract) prediction on each frame by an optimally selected codeword (sequence) from a VQ codebook (suitably scaled in amplitude), rather than by multiple pulses [10]. A typical VQ codebook consists of a set of 2^M codewords (for an M-bit codebook) of length 40 samples (corresponding to each 5 msec analysis frame at an 8 kHz sampling rate). The VQ codebook can either be deterministic, i.e., derived from an appropriate training set of residual error vectors, or stochastic, i.e., generated randomly from white Gaussian random numbers with unit variance. Deterministic codebooks have been found to have robustness problems due to channel mismatch conditions, so they have generally been avoided in CELP coders. Stochastic codebooks are motivated by the observation that the cumulative amplitude distribution of the residual from the long-term pitch predictor output is roughly identical to a Gaussian distribution function with the same mean and variance, as illustrated in Figure 11.87. Typical CELP VQ stochastic codebooks are ($M = 10$)-bit codebooks; i.e., there are 1024 Gaussian random codewords, each of length 40 samples.

Block diagrams of a general CELP encoder and decoder are given in Figures 11.88 and 11.89. The iterative process begins as in the MPLPC case by subtracting the tail of the output due to the excitation signal prior to the current block. For each of the excitation VQ codebook vectors, the following operations occur:

- The codebook vector is scaled by a gain estimate, yielding the excitation signal $d[n]$.
- The excitation signal $d[n]$ is used to excite the inverse of the long-term pitch prediction error filter (labeled as the long-term synthesis filter in Figure 11.88) and then the short-term vocal tract model system (labeled as the short-term synthesis filter in Figure 11.88), yielding the estimate of the speech signal, $\tilde{x}[n]$, for the current codebook vector.
- The error signal, $e[n]$, is generated as the difference between the speech signal, $x[n]$, and the estimated speech signal, $\tilde{x}[n]$.
- The difference signal is perceptually weighted and the resulting mean-squared error is calculated.

The above set of processing steps is iterated for each of the 2^M codebook vectors, and the codebook vector yielding the minimum weighted mean-squared error is chosen as the best representation of the excitation for the 5 msec frame being analyzed. The process of searching for the best codeword is illustrated in Figure 11.90, where the linear filter refers to the combination of the long-term pitch predictor and the short-term

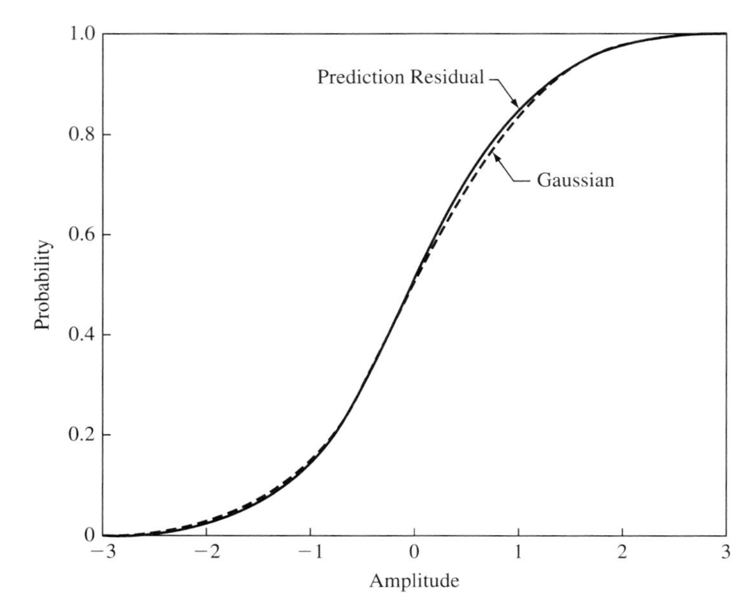

FIGURE 11.87
Cumulative amplitude distributions for prediction residual after pitch prediction and a corresponding Gaussian distribution function with the same mean and variance. (Courtesy of B. S. Atal [10].)

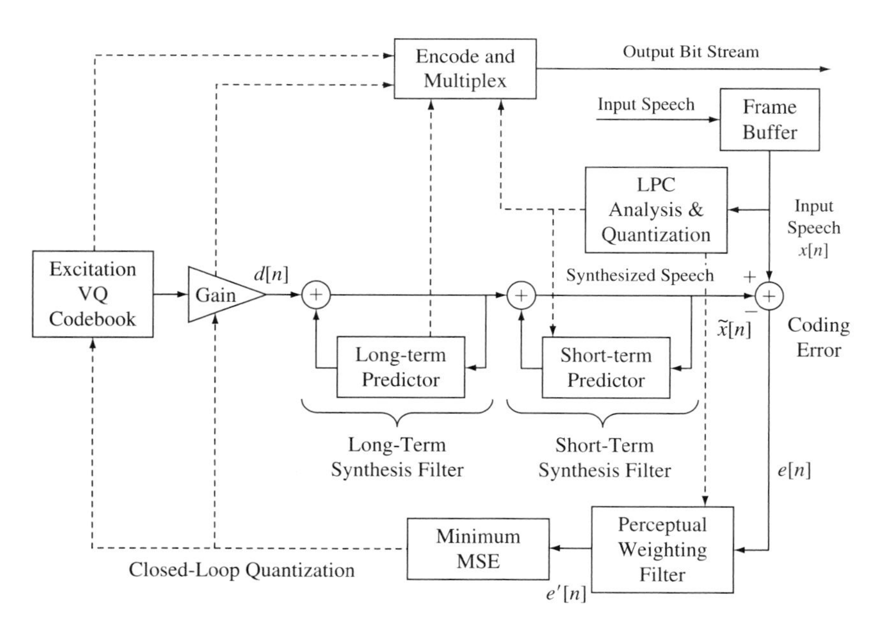

FIGURE 11.88
Block diagram of general CELP encoder.

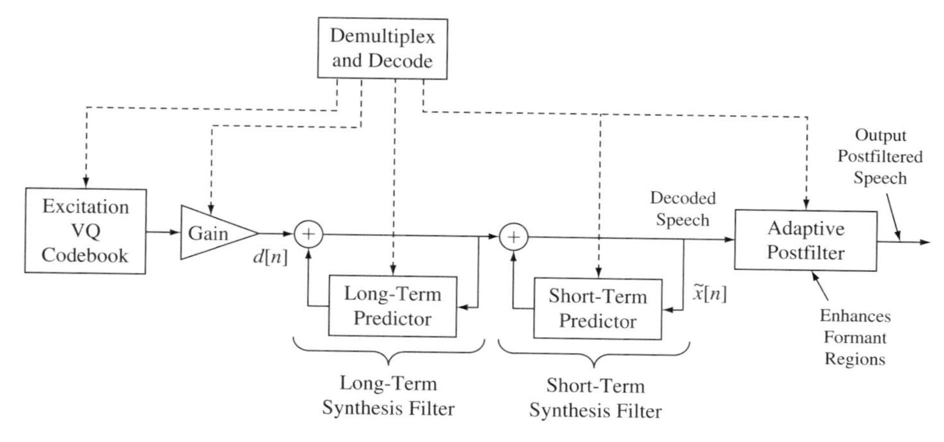

FIGURE 11.89
Block diagram of general CELP decoder.

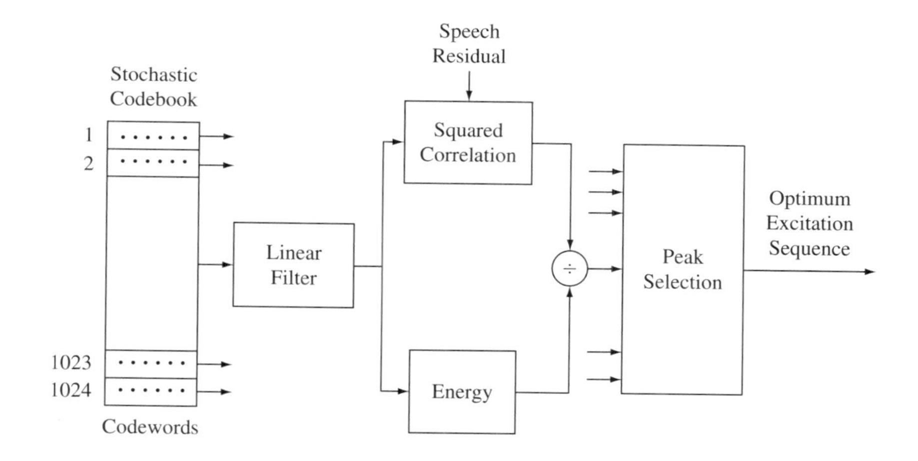

FIGURE 11.90
Search procedure for determining the best stochastic codeword match to the speech residual for the current 5 msec frame.

vocal tract predictor, and the boxes labeled "squared correlation" and "energy" compute the normalized correlation and the energy of the match between the speech residual and each codeword in the CELP VQ codebook. The division computes the optimum given coefficient β. The peak selection chooses the codeword with the highest normalized squared correlation to the speech residual.

The signal processing operations of the CELP decoder of Figure 11.89 consist of the following steps (for each 5 msec frame of speech):

1. Select the appropriate codeword for the current frame from a matching excitation VQ codebook (which exists at both the encoder and the decoder).

2. Scale the codeword sequence by the gain of the frame, thereby generating the excitation signal, $e[n]$.
3. Process $e[n]$ by the long-term synthesis filter (the pitch predictor) and the short-term vocal tract filter, giving the reconstructed speech signal, $\tilde{x}[n]$.
4. Process the reconstructed speech signal by an adaptive postfilter whose function is to enhance the formant regions of the speech signal, and thus to improve the overall quality of the synthetic speech from the CELP system.

The use of an adaptive postfilter was originally proposed by Chen and Gersho [56, 57], who observed that at low bit rates (4000–8000 bps), where the average quantizing noise power was relatively high, it was difficult to adequately suppress the noise below the masking threshold at all frequencies and the resulting encoded signal thus sounded noisier than would have been predicted by auditory masking experiments at higher SNRs. As such, Chen and Gersho [56, 57] proposed an adaptive postfilter of the form:

$$H_p(z) = (1 - \mu z^{-1}) \frac{\left[1 - \sum_{k=1}^{p} \gamma_1^k \alpha_k z^{-k}\right]}{\left[1 - \sum_{k=1}^{p} \gamma_2^k \alpha_k z^{-k}\right]}, \tag{11.201}$$

where the $\{\alpha_k\}$ are the prediction coefficients of the current frame, and μ, γ_1, and γ_2 are the tunable parameters of the adaptive postfilter. Typical ranges of the adaptive postfilter parameters are:

- range of μ of $0.2 \leq \mu \leq 0.4$;
- range of γ_1 of $0.5 \leq \gamma_1 \leq 0.7$;
- range of γ_2 of $0.8 \leq \gamma_2 \leq 0.9$.

It has been found that the postfilter attenuates the spectral components in the valleys without distorting the speech. [The use of the simple highpass filter, via the term $(1 - \mu z^{-1})$, provides a spectral tilt and helps to minimize the distortion due to the adaptive postfilter]. Figure 11.91 shows plots of a typical voiced sound STFT spectrum, frequency response of vocal tract system function, and the frequency response of the postfilter [197]. It can be seen that the spectral tilt almost makes the postfilter frequency response flat (as compared to the falling spectrum of the signal and the vocal tract filter response), with small peaks at the formants and small dips at the valleys. A well-designed adaptive postfilter has been shown to be capable of reducing the background noise of CELP coders and making the speech sound more natural ("smoother") than without the adaptive postfilter, especially at low bit rates. The effectiveness of the adaptive postfilter is well established, and virtually all CELP coders utilize this method to improve synthesis quality.

Before showing some typical results of CELP coding of speech, it is worthwhile making a few comments about VQ codebooks and how they are populated and searched. Initial CELP systems used a Gaussian codebook with random vectors (zero

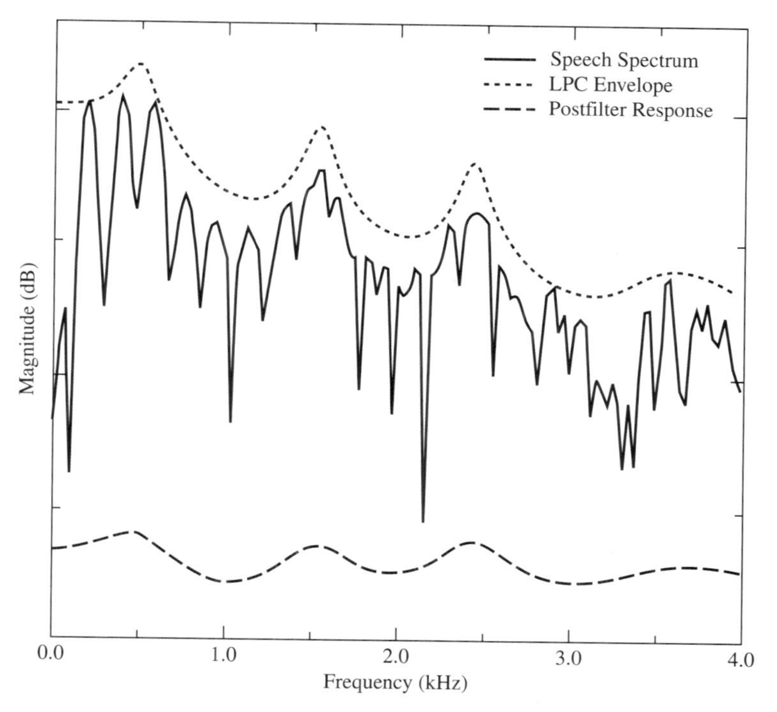

FIGURE 11.91
Typical spectra of a section of voiced speech using an adaptive postfilter, including the STFT spectrum, the frequency response of vocal tract system function, and the frequency response of the adaptive postfilter. (After Kondoz [197].)

mean, unit variance) for each codeword. This codebook was shown to be impractical because of the large memory required to store the random vectors, and the huge amount of computation to process each vector to determine the best fit to each frame of speech. A more practical solution was to populate the codebook from a one-dimensional array of Gaussian random numbers, where most of the samples between adjacent codewords were common (identical). Such overlapping codebooks typically use shifts of one or two samples, and provide large complexity reductions for storage and computation of optimal codebook vectors for a given frame. An example of two codeword vectors from a highly overlapped stochastic codebook is given in Figure 11.92, which shows two codewords that are identical except for a shift of two samples (the initial two codeword samples in the plot at the bottom of Figure 11.92).

To reduce the computation using stochastic codebooks even further, the amplitudes of the codeword samples are often center-clipped to a 3-level codeword (± 1 and 0). Interestingly, subjective listening tests have shown that the resulting speech quality is generally improved when using center-clipped codebooks as the source of the excitation vectors. Finally, it should be clear that CELP codewords can be considered to be a special case of MPLPC where the excitation is vector quantized using a codebook.

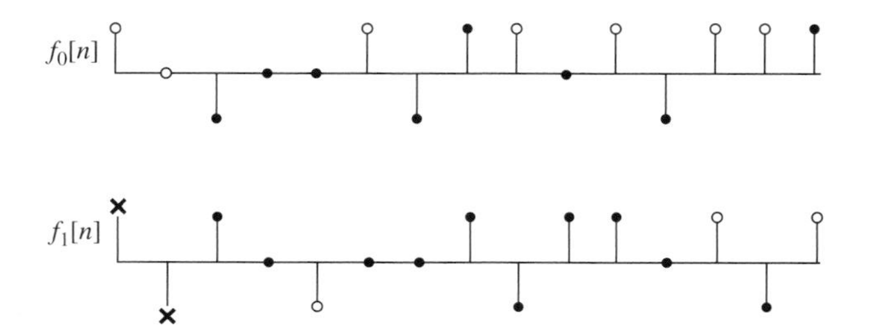

FIGURE 11.92
Example of stochastic codeword vectors which differ from each other by a two-sample shift.

We have not explicitly described the search procedure for choosing which codebook entry optimizes the match to the excitation signal. There have been developed two methods for significantly reducing the computational load of the full search of the codebook vectors. The first method uses a singular value decomposition approach to transform the convolution operation into a reduced computation set of multiplications. The second method uses DFT methods to convert convolutions to products which can be computed in significantly less time. Suffice to say that very efficient methods have been developed for the codebook optimization to be a practical and useful coding technique [182].

Figures 11.93 and 11.94 show typical waveforms and spectra derived from a CELP coding system. The waveform plots in Figure 11.93 show 100 msec sections of the original speech waveform [part (a)], the synthetic speech output from the CELP decoder [part (b)], the short-term prediction residual [part (c), amplified by a factor of 5], the reconstructed short-term prediction residual [part (d), again amplified by a factor of 5], the prediction residual after long-term pitch prediction [part (e), amplified by a factor of 15], and the coded residual from a 10-bit random codebook [part (f), again amplified by a factor of 15]. The first observation is how closely the synthetic speech output matches the original speech, showing the capability of the CELP coding approach. The second observation is that the prediction residual (after pitch prediction) and the coded residual (using the 10-bit codebook) look to be very different, but clearly have common statistical properties, and thus work very well for matching the perceptually relevant properties of the speech spectrum.

The spectral comparison shown in Figure 11.94 shows the very strong spectral matches around the peaks of the formants, with rather significant spectral differences in the valleys. Using the perceptual weighting on the spectral difference would show that the regions of difference are highly masked, and thus contribute little to any loss in fidelity of the resulting synthetic speech.

11.11.4 CELP Coder at 4800 bps

A CELP coder, operating at a bit rate of 4800 bps, has been standardized by the U.S. Government as Federal Standard FS-1016. Block diagrams of the FS-1016 CELP encoder and decoder are shown in Figures 11.95 and 11.96, respectively [51]. The

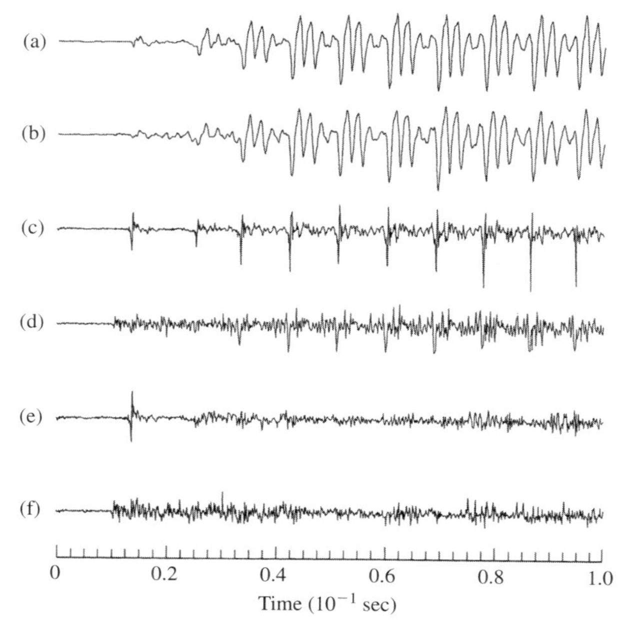

FIGURE 11.93
Typical waveforms from a CELP coder: (a) original speech; (b) synthetic speech output; (c) LPC prediction residual; (d) reconstructed LPC residual; (e) prediction residual after pitch prediction; (f) coded residual from a 10-bit random codebook. (Parts (c) and (d) are amplified five times for display purposes; parts (e) and (f) are amplified 15 times for display purposes.) (After Atal [10].)

encoder uses a stochastic codebook with 512 codewords (9-bit codebook), and an adaptive codebook with 256 codewords, to estimate the long-term correlation (the pitch period). The adaptive codebook is simply 256 previous samples of the excitation signal $d[n]$. Each codeword in the stochastic codebook is sparsely populated with ternary valued samples (-1, 0, $+1$), and with codewords overlapped and shifted by two samples, thereby enabling a fast convolution solution for selection of the optimum codeword for each frame of speech. The linear predictive analyzer uses a frame size of 30 msec and a predictor of order $p = 10$ using the autocorrelation method with a Hamming window. The 30 msec frame is broken into four sub-frames (7.5 msec in duration) and the adaptive and stochastic codewords are updated every sub-frame, whereas the linear predictive analysis is only performed once every full frame.

The following sets of features are produced by the coding system of Figure 11.95:

1. the linear predictive spectral parameters [coded as a set of 10 line spectral pair (LSP) parameters] for each 30 msec frame;
2. the codeword index γ_a and gain G_a of the adaptive codebook vector for each 7.5 msec sub-frame;

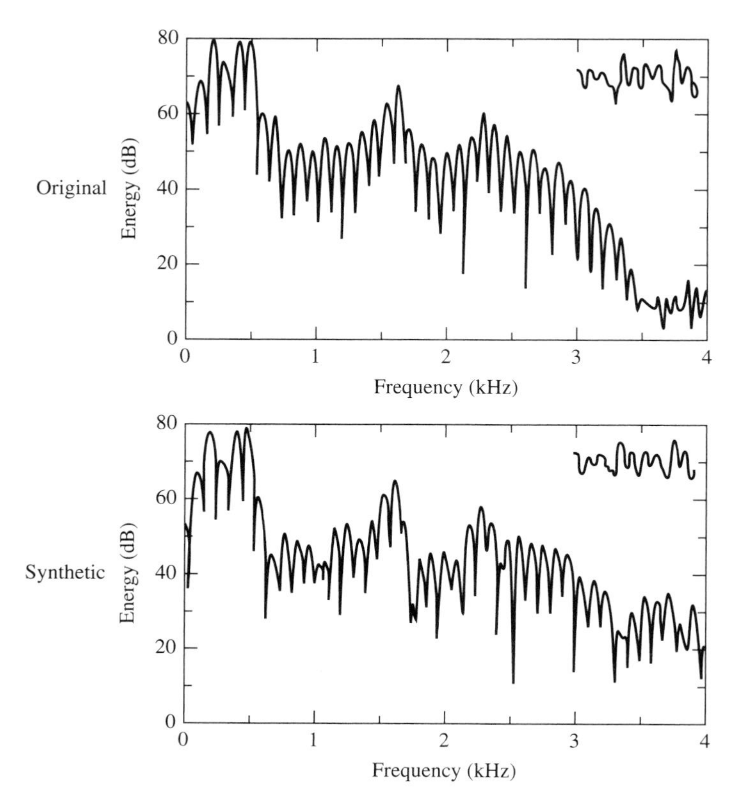

FIGURE 11.94
Original and synthetic speech spectra for a short voiced speech segment from the CELP coder. (After Atal [10].)

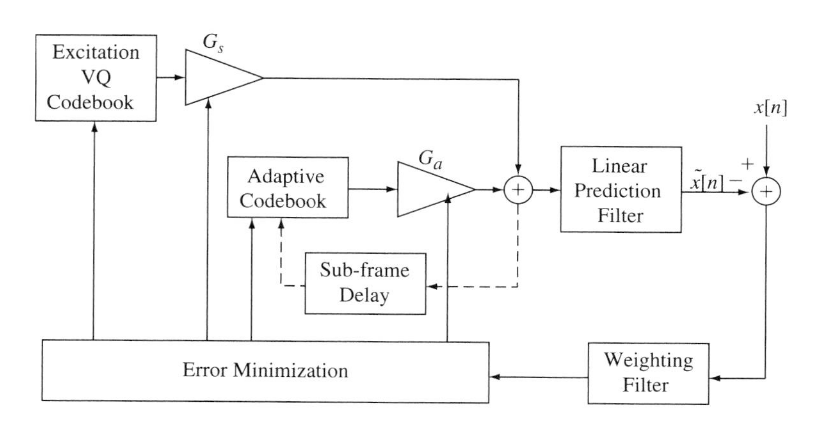

FIGURE 11.95
FS-1016 speech encoder. (After Campbell et al. [51].)

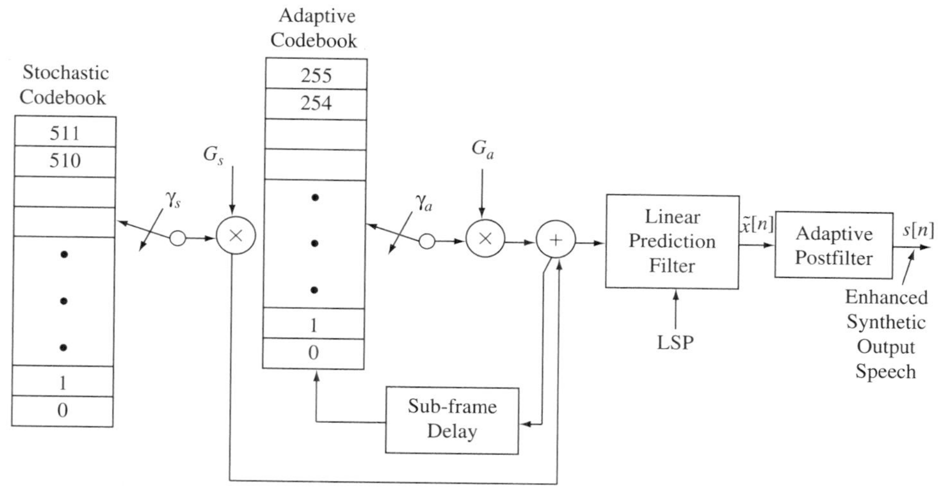

FIGURE 11.96
FS-1016 speech decoder. (After Campbell et al. [51].)

3. the codeword index γ_s and gain G_s of the stochastic codebook vector for each 7.5 msec sub-frame.

To code these parameters to a total bit rate of 4800 bps, the bits are allocated to the three sets of features as shown in Table 11.9. On the basis of bits allocated to every 30 msec frame, we see from the table that 34 bits are allocated to the 10 LSP features, 48 bits to the adaptive codebook codeword (the pitch delay) and the gain associated with the pitch delay (non-uniformly over the four sub-frames), 56 bits are allocated to the stochastic codebook codeword and gain (14 bits per sub-frame), and 6 bits are allocated to bookkeeping functions of the coder (synchronization, future use, Hamming window parity). Thus a total of 144 bits are allocated to each 30 msec frame, giving a grand total of 4800 bps for the coder.

We have discussed the basic form of CELP coding. Because of its remarkable ability to produce high quality synthetic speech approximations to natural speech signals, at bit rates from 16 kbps and lower, a wide variety of CELP variations have appeared in the literature, and for completeness, we list several of these variants:

- **ACELP**: algebraic code excited linear prediction
- **CS-ACELP**: conjugate-structure **ACELP**
- **VSELP**: vector-sum excited LPC
- **EVSELP**: enhanced **VSELP**
- **PSI-CELP**: pitch synchronous innovation-code excited linear prediction
- **RPE-LTP**: regular pulse exciting-long-term prediction
- **MP-MLQ**: multipulse-maximum likelihood quantization.

TABLE 11.9 Bit allocation for FS-1016 4800 bps CELP coder.

Parameter	Sub-frame 1	2	3	4	Frame
LSP1					3
LSP2					4
LSP3					4
LSP4					4
LSP5					4
LSP6					3
LSP7					3
LSP8					3
LSP9					3
LSP10					3
Pitch delay γ_a	8	6	8	6	28
Pitch gain G_a	5	5	5	5	20
Codeword index γ_s	9	9	9	9	36
Codeword gain G_s	5	5	5	5	20
Future expansion					1
Hamming parity					4
Synchronization					1
Total					144

Many of these variations on the CELP theme have found their way into speech coder standards created by the International Telecommunication Union (ITU) and other standards bodies. We give a partial list in Section 11.13.1.

11.11.5 Low Delay CELP (LD-CELP) Coding

The delay associated with any coder is the time taken by the input speech sample to be processed, transmitted, and decoded at the receiver, plus any transmission delay. The major components of coding delay include:

1. buffering delay at the encoder, namely the length of the frame analysis window (order of 20–40 msec for most linear predictive coders);
2. processing delay at the encoder, namely the time needed to compute all coder parameters and encode them for transmission over a channel;
3. buffering delay at the decoder, namely the time to collect all parameters for a frame of speech samples;
4. processing delay at the decoder, namely the time to compute a frame of output using the speech synthesis model.

For most CELP coders, the first three delay components are comparable, and are typically on the order of 20–40 msec for each component, for a total delay of about

60–120 msec. The fourth delay component generally takes less time and typically adds about 10–20 msec. Thus the total delay of a CELP coder (irrespective of transmission delay, interleaving of signals or forward error correction protection methods, as might be used in a practical speech communication system), is about 70–130 msec.

For many applications, the delay associated with traditional CELP coding is too large. The major cause of this large delay is the use of a forward adaptation method for estimating the short-term and long-term predictors (i.e., the vocal tract filter and the pitch prediction filter). Generally, the use of backward adaptation methods for estimating short-term and long-term prediction parameters produces poor quality speech. However, Chen showed how a backward adaptive CELP coder could be made to perform as well as the conventional forward adaptive CELP coder [55] at bit rates of 16 and 8 kbps. A block diagram of the resulting "low delay" CELP coder, operating at a rate of 16 kbps, is shown in Figure 11.97.

The key properties of the low delay, backward adaptive, CELP coder are the following:

- In the conventional CELP coder, the predictor parameters, the gain, and the codebook excitation are all transmitted to the receiver; in the LD-CELP coder, only the excitation sequence is transmitted. The two traditional predictors (the long-term pitch predictor and the short-term vocal tract filter) are combined into one high order (50th-order) predictor whose coefficients are updated by performing a linear predictive analysis on the previously quantized speech signal.
- The excitation gain is updated by using the gain information embedded in the previously quantized excitation.
- The LD-CELP excitation signal, at 16 kbps, is represented at a rate of 2 bits/sample for an 8 kHz sampling rate. Using a codeword length of five samples, each excitation vector is coded using a 10-bit codebook. (In practice, a 3-bit gain codebook and a 7-bit shape codebook are used in a product representation).
- A closed-loop optimization procedure is used to populate the shape codebook using the same weighted error criterion as is used to select the best codeword in the CELP coder.

The resulting LD-CELP system achieves a delay of less than 2 msec and has been shown to produce speech whose quality exceeded that of 32 kbps ADPCM. This structure is the basis for the ITU G.728 LD-CELP standard.

11.11.6 Summary of A-b-S Speech Coding

In this section we have shown that by using the closed-loop analysis methods of an A-b-S approach to represent the excitation signal in LPC, we were able to derive an excitation signal approximation that produced very good quality synthetic speech while being efficient to code. Although a wealth of A-b-S techniques have been proposed and studied in the literature, we concentrated our discussion on two methods, namely MPLPC and CELP, both of which have been extensively studied and shown to be very effective in representing speech efficiently at bit rates from 2400 bps to 16 kbps.

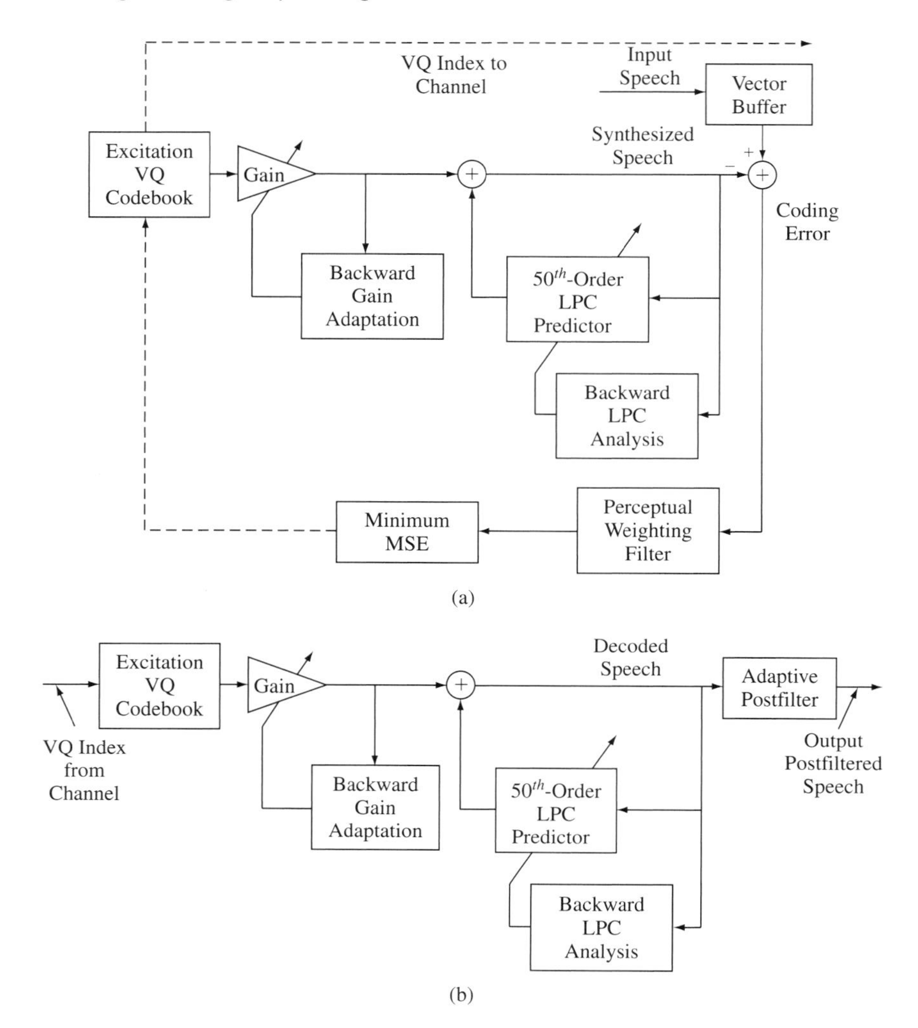

FIGURE 11.97
Block diagram of LD-CELP encoder and decoder operating at a bit rate of 16 kbps. Part (a) shows the encoder operations; part (b) shows the decoder operations. (After Chen [55]. © [1990] IEEE.)

11.12 OPEN-LOOP SPEECH CODERS

So far in this chapter we have studied closed-loop systems that incorporate the discrete-time model of speech. For the most part, we have assumed that the vocal tract part of the model was estimated using short-term linear predictive analysis on a frame-by-frame basis. We have seen that placing the vocal tract model in a feedback loop can lead, either through adaptive quantization or optimization, to an excitation signal for the vocal tract filter that has many of the characteristics that, based on the physical model for speech production, we have posed for the excitation signal in the model.

We have also seen that these closed-loop approaches lead to an excitation signal that may be difficult to code and requires a significant data rate for transmission. In this section we have come at last to perhaps the oldest approach to speech coding, namely the direct and independent estimation of the excitation and the vocal tract filter. These types of open-loop systems have been called vocoders (voice coders) since the original work of Homer Dudley [92]. In this section we discuss several open-loop digital vocoder systems.

11.12.1 The Two-State Excitation Model

Figure 11.98 shows a simplified version of the source/system model for speech that we have inferred from our study of the physics of speech production. As we have discussed in the previous chapters, the excitation model can be very simple. Unvoiced sounds are produced by exciting the system with white noise, and voiced sounds are produced by a periodic impulse train excitation, where the spacing between impulses is the pitch period, $N_p = P_0/F_s$. We shall refer to this system as the *two-state excitation model*. The slowly time-varying linear system models the combined effects of vocal tract transmission, radiation at the lips, and, in the case of voiced speech, the lowpass frequency shaping of the glottal pulse. The V/UV (voiced/unvoiced excitation) switch produces the alternating voiced and unvoiced segments of speech, and the gain parameter, G, controls the level of the filter output. When values for V/UV decision, G, N_p, and the parameters of the linear system are supplied at periodic intervals (frames), then the model becomes a speech synthesizer. When the parameters of the model are estimated directly from a speech signal, the combination of estimator and synthesizer becomes a vocoder or, as we prefer, an *open-loop analysis/synthesis speech coder*. A general block diagram of the class of open-loop coders is given in Figure 11.99.

Pitch, Gain, and V/UV Detection

The fundamental frequency of voiced speech can range from well below 100 Hz for low-pitched male speakers to over 250 Hz for high-pitched voices of women and children. The fundamental frequency varies slowly, with time, more or less at the same rate as the vocal tract motions. It is common to estimate the pitch period, N_p, at a frame

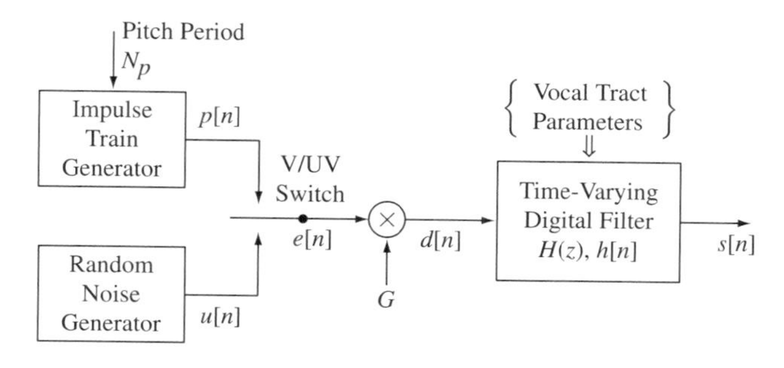

FIGURE 11.98
Two-state excitation model for speech synthesis.

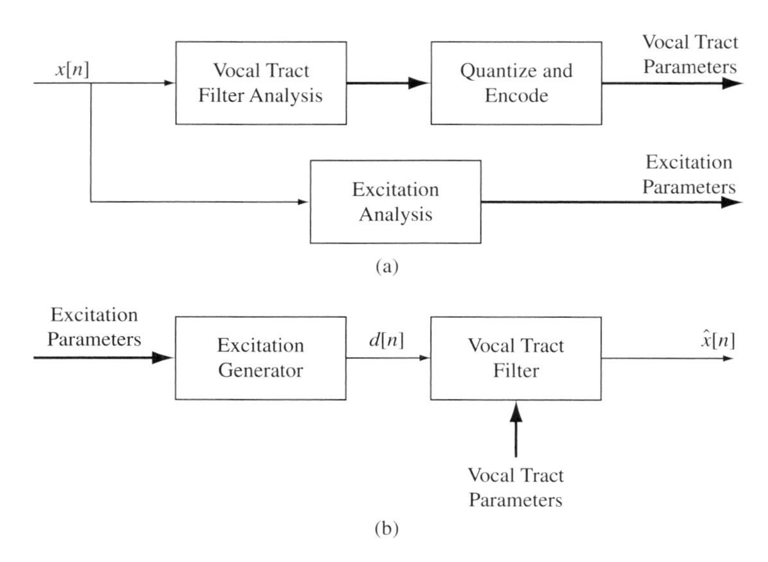

FIGURE 11.99
General block diagram for the class of open-loop coders: (a) analysis/encoding stage; (b) the synthesis/decoding stage.

rate of about 50 to 100 times/sec. To do this, short segments of speech are analyzed to detect periodicity (signaling voiced speech) or aperiodicity (signaling unvoiced speech or background signal).

In previous chapters, most especially in Chapter 10, we have discussed several approaches to pitch detection. One of the simplest, yet most effective, operates directly on the time waveform by locating corresponding peaks and valleys in the waveform, and measuring the times between the peaks and valleys [129]. The short-time autocorrelation function is also useful for this purpose since it shows a peak at the pitch period for voiced speech with no such peak shown for unvoiced speech. In pitch detection applications of the short-time autocorrelation, it is common to pre-process the speech by a spectrum flattening operation such as center clipping [300] or inverse filtering [227]. This pre-processing tends to enhance the peak at the pitch period for voiced speech while suppressing the local correlation due to formant resonances.

Another approach to pitch detection is based on the cepstrum. As discussed in Chapters 8 and 10, a strong peak in the expected pitch period range signals voiced speech, and the location of the peak is the pitch period. Similarly, lack of a peak in the expected range signals unvoiced speech or background signal [252].

The gain parameter, G, is also found by analysis of short segments of speech. It should be chosen so that the short-time energy of the synthetic output matches the short-time energy of the input speech signal. For this purpose, the autocorrelation function value, $\phi_{\hat{n}}[0]$, at lag 0, or the cepstrum value $c_{\hat{n}}[0]$ at quefrency 0 can be used to determine the energy of the segment of the input signal.

For digital coding applications, the pitch period (N_p), V/UV, and gain (G) must be quantized. Typical values are 7 bits for pitch period ($N_p = 0$ signals UV) and 5 bits for G. For a frame rate of 50 frames/sec, this totals 600 bps, which is well below the bit rate used to encode the excitation signal in closed-loop coders such as ADPCM or CELP. Since the vocal tract filter can be coded as in ADPCM or CELP, much lower total bit rates are common in open-loop systems. This comes at a large cost in the loss of quality of the synthetic speech output, however.

Vocal Tract System Estimation

The vocal tract system in the synthesizer of Figure 11.98 can take many forms. The primary methods that have been used have been homomorphic filtering and linear predictive analysis as discussed in Chapters 8 and 9, respectively.

Homomorphic filtering can be used to extract a sequence of impulse responses from the sequence of cepstra that result from short-time cepstrum analysis. Thus, one cepstrum computation can yield both an estimate of pitch and the vocal tract impulse response. In Chapter 8 we discussed an analysis/synthesis system based on homomorphic filtering. In the original homomorphic vocoder, the impulse response was digitally coded by quantizing each cepstrum value individually (scalar quantization) [267]. The impulse response, reconstructed from the quantized cepstrum at the synthesizer, is simply convolved with the excitation created from the quantized pitch, voicing, and gain information; i.e., $s[n] = Ge[n] * h[n]$.[30] In a more recent application of homomorphic analysis in an A-b-S framework [62], the cepstrum values were coded using VQ, and the excitation derived by A-b-S as described in Section 11.11. In still another approach to digital coding, homomorphic filtering was used to remove excitation effects in the short-time spectrum and then three formant frequencies were estimated from the smoothed spectra (see the discussion in Chapter 10). The formant frequencies can be used to control the resonance frequencies of a synthesizer comprised of a cascade of second-order section IIR digital filters [344]. Such a speech coder is called a *formant vocoder*.

Linear predictive analysis can also be used to estimate the vocal tract system for an open-loop coder with two-state excitation [12]. We discuss the details of the LPC vocoder in the next subsection.

11.12.2 The LPC Vocoder

When linear predictive analysis is used in a system like Figure 11.99a to estimate the vocal tract filter in an open-loop vocoder, the system is called a linear predictive coder (LPC) or somewhat redundantly, an LPC vocoder. For each frame of speech, a linear prediction analysis is performed (using any of the methods presented in Chapter 9) and the resulting LPC parameter set (or some equivalent variant like the LSFs or the log area ratios) is quantized and encoded. Simultaneously, the speech excitation parameters are determined using some type of pitch detector. The processing of the decoder consists simply of a generator that generates the excitation signal, $d[n] = Ge[n]$, from

[30] Care must be taken at frame boundaries. For example, the impulse response can be changed at the time a new pitch impulse occurs, and the resulting output can overlap into the next frame.

the excitation parameters, and uses it to excite a model of the vocal tract, derived from the coded set of vocal tract parameters estimated in the encoder of Figure 11.99.

In the simplest LPC vocoder systems, the analysis is performed at a rate on the order of 100 frames/sec, using a linear predictive analysis of order $p = 12$, resulting in a set of 15 parameters/frame [i.e., 12 prediction coefficients (or an equivalent parameter set such as log area ratios or LSF values), pitch period, voicing decision, gain]. These 15 parameters must be quantized and coded for transmission over a channel, as illustrated in Figure 11.99. Ideally we would estimate the probability density function for each parameter to be coded and then use the appropriate scalar quantizer for that parameter. As a baseline, we use a simple coding scheme for the LPC vocoder which allocates bits to the 15 parameters as follows:

- V/UV switch position—1 bit/frame, 100 bps
- pitch period—6 bits/frame, uniformly quantized, 600 bps
- gain—5 bits/frame, non-uniformly quantized, 500 bps
- linear prediction parameters (e.g., bandwidth and center frequency for each of 6 complex poles)—5 bits/parameter, non-uniformly quantized, 12 parameters, 6000 bps.

This leads to an LPC vocoder representation requiring a total of 7200 bps. Using this quantization scheme there is essentially no loss in quality of the decoded signal as compared to the signal synthesized with unquantized parameters. The total bit rate of this simple implementation of an LPC vocoder, namely 7200 bps, is below that of almost all of the waveform coding schemes described earlier in this chapter. However, the quality of the synthetic speech from the LPC vocoder is limited by the requirement that the excitation signal consist of either individual pitch pulses or random noise. This is why A-b-S vocoders are generally preferred even at the cost of higher data rates. There are many ways of reducing the overall bit rate of the LPC vocoder, below the 7200 bps rate of the previous section, without seriously degrading overall speech quality or intelligibility from what can be achieved with unquantized parameters. In this section we outline just a few of these methods.

The areas where we can reduce the LPC vocoder bit rate without seriously degrading the quality of the coded signal include the following:

1. Perform logarithmic encoding of the pitch period and gain signals.
2. The difficulty in achieving accurate V/UV detection and pitch period estimation across a range of speakers.
3. Use the set of PARCOR coefficients, $\{k_i\}$, $i = 1, 2, \ldots, p$, to derive the set of log area ratios, $\{g_i\}, i = 1, 2, \ldots, p$, which can be coded using a uniform quantizer with lower spectral sensitivity (than that of the LPC poles) and fewer bits/coefficient for coding.
4. Utilize a VQ codebook to represent the LPC analysis feature vector as a 10-bit index into a codebook of 1024 vectors. Codebooks based on sets of PARCOR parameters have been utilized for this purpose in existing LPC vocoders.

Based on these coding refinements, LPC vocoders using 4800 bps have been built with virtually the same quality as the LPC vocoder using 7200 bps. Further, by reducing the frame rate from 100 frames/sec down to 50 frames/sec, LPC vocoder rates have been halved down to 2400 bps, with almost the same quality of the resulting synthetic speech as the 4800 bps vocoder. As an extreme case, an LPC vocoder was built that achieved a total bit rate of 800 bps using a 10-bit codebook of PARCOR vectors, and a 44.4 frames/sec analysis rate, allotting just 8 total bits/frame for pitch period, voiced/unvoiced decision, and gain, and using a 2-bit synchronization code between frames. Again the quality of the LPC vocoder at a total bit rate of 800 bps was comparable to that of the LPC vocoder at the 2400 bps total rate [422].

The U.S. Government created an LPC Vocoder standard, known as LPC-10 (or often denoted as LPC-10e for the updated standard), operating at a bit rate of 2400 bps, with the following specifications:

- analysis frame rate of 44.44 frames/sec
- covariance linear predictive analysis using $p = 10$
- AMDF (average magnitude difference function) pitch detector, as described in Chapter 6
- PARCOR coefficients for quantization
  - 5-bit scalar coding of PARCOR features, k_1–k_4
  - 4-bit scalar coding of PARCOR features, k_5–k_8
  - 3-bit scalar coding of PARCOR feature, k_9
  - 2-bit scalar coding of PARCOR feature, k_{10}
- 7-bit scalar coding of pitch period
- 5-bit scalar coding of amplitude
- single frame synchronization bit

Much experimentation has been done to improve the quality of an LPC vocoder, but unfortunately the following inherent problems limit the quality of the output synthesized by the decoder:

1. the inadequacy of the basic source/filter speech production model, especially for nasal sounds, for unvoiced consonants, and for sounds with zeros in the transfer function (e.g., nasalized vowels)
2. the idealization of the excitation source as either a quasi-periodic pitch pulse train or random noise
3. the lack of accounting for parameter correlation using a one-dimensional scalar quantization method (of course the use of a vector quantizer alleviates this problem, somewhat, but there remain issues in designing a uniformly applicable codebook that is independent of speaker, accent, transducer, and speaking environment).

In summary, the open-loop LPC representation can provide intelligible speech of quality that is acceptable for some communication applications. The data rate can be lowered to about 2400 bps with little degradation in quality from what can be achieved with no quantization of the model parameters [422].

11.12.3 Residual-Excited LPC

In Section 11.7 we presented Figure 11.52 as motivation for the use of the source/system model in digital speech coding. This figure shows an example of inverse filtering of the speech signal using a prediction error filter, where the prediction error (or residual) is significantly smaller and less lowpass in nature. The speech signal can be reconstructed from the residual by passing it through the vocal tract system $H(z) = 1/A(z)$. None of the methods discussed so far attempt to directly code the prediction error signal in an open-loop manner. ADPCM and A-b-S systems derive the excitation to the synthesis filter by a feedback process. The two-state model attempts to construct the excitation signal by direct analysis and measurements on the input speech signal. Systems that attempt to code the residual signal directly are called residual-excited linear predictive (RELP) coders.

Direct coding of the residual faces the same problem as in ADPCM or CELP: the sampling rate of the prediction residual is the same as that of the input, and accurate coding could require several bits per sample. Figure 11.100 shows a block diagram of an RELP coder [398]. In this system, which is quite similar to earlier voice-excited vocoders (VEV) [354, 405], the problem of reducing the bit rate of the residual signal is attacked by reducing its bandwidth to about 800 Hz, lowering the sampling rate, and coding the samples with adaptive quantization. ADM was used in Ref. [398], but APCM could be used if the sampling rate is lowered to 1600 Hz by decimation. The 800 Hz band is wide enough to contain several harmonics of the highest pitched voices.

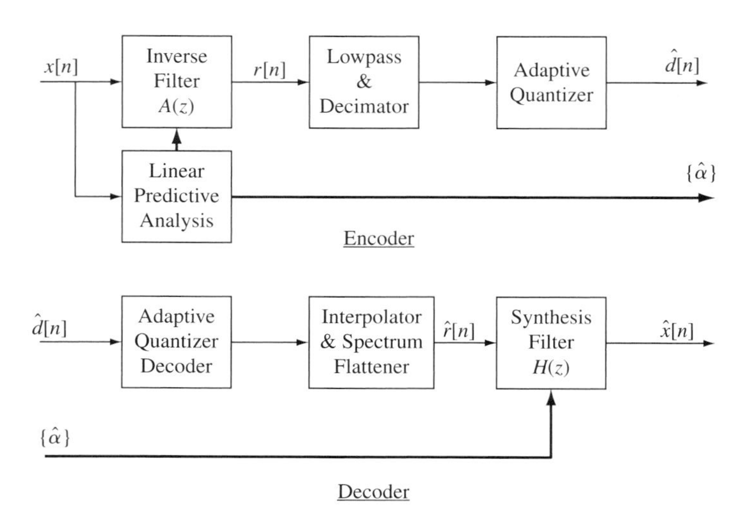

FIGURE 11.100
RELP encoder and decoder.

The reduced bandwidth residual is restored to full bandwidth prior to its use as an excitation signal by a non-linear spectrum flattening operation, which restores higher harmonics of voiced speech. White noise is also added according to an empirically derived recipe. In the implementation of Ref. [398], the sampling rate of the input was 6.8 kHz and the total bit rate was 9600 kbps, with 6800 bps devoted to the residual signal. The quality achieved at this rate was not significantly better than the LPC vocoder with a two-state excitation model. The principal advantage of this system is that no hard V/UV decision must be made, and no pitch detection is required. While this system did not become widely used, its basic principles can be found in subsequent open-loop coders that have produced much better speech quality at bit rates around 2400 bps.

11.12.4 Mixed Excitation Systems

While two-state excitation allows the bit rate to be quite low, the quality of the synthetic speech output leaves much to be desired. The output of such systems is often described as "buzzy," and in many cases, errors in estimating pitch period or voicing decision cause the speech to sound unnatural if not unintelligible. The weaknesses of the two-state model for excitation spurred interest in a mixed excitation model where a hard decision between V and UV is not required. Such a model was first proposed by Makhoul et al. [221] and greatly refined by McCree and Barnwell [240].

Figure 11.101 depicts the essential features of the mixed-excitation linear predictive (MELP) coder proposed by McCree and Barnwell [240]. This configuration was developed as the result of careful experimentation, which focused one-by-one on the sources of distortions manifest in the two-state excitation coder such as buzziness and tonal distortions. The main feature is that impulse train excitation and noise excitation are added instead of switched. Prior to their addition, they each pass through a multi-band spectral shaping filter. The gains in each of five bands are coordinated between the two filters so that the spectrum of $e[n]$ is flat. This mixed excitation helps to model short-time spectral effects such as "devoicing" of certain bands during voiced speech. In some situations, a "jitter" parameter, ΔP, is invoked to better model voicing transitions. Other important features of the MELP system are lumped into the block labeled

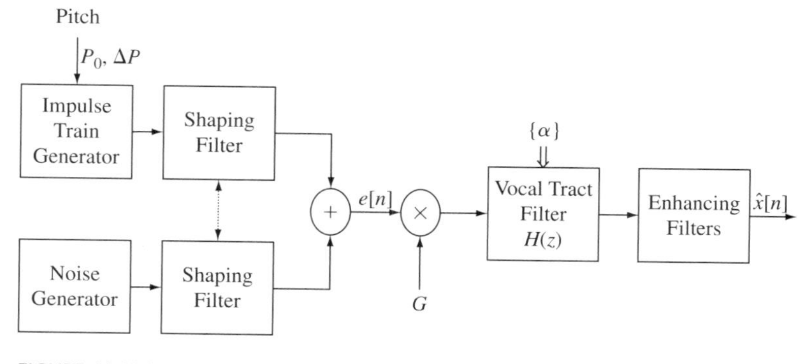

FIGURE 11.101
MELP decoder.

"enhancing filters." This represents adaptive spectrum enhancement filters used to enhance formant regions,[31] and a spectrally flat "pulse dispersion filter" whose purpose is to reduce "peakiness" due to the minimum-phase nature of the linear predictive vocal tract system.

These modifications to the basic two-state excitation LPC vocoder produce marked improvements in the quality of reproduction of the speech signal. Several new parameters of the excitation must be estimated at analysis time and coded for transmission, but these add only slightly to either the analysis computation or the bit rate [240]. The MELP coder is said to produce speech quality at 2400 bps that is comparable to CELP coding at 4800 bps. In fact, its superior performance led to a new Department of Defense standard in 1996 and subsequently to MIL-STD-3005 and NATO STANAG 4591, which operates at 2400, 1200, and 600 bps.

11.13 APPLICATIONS OF SPEECH CODERS

Speech coders have found practical applications in a wide range of systems, including the following:

- network coders—used for transmission and storage of speech within the Public Switched Telephony Network (PSTN)
- international networks—used for transmission of speech via undersea cables and satellites; also used for trans-coding of speech between different digital standards
- wireless—used as the basic speech coder for second and third generation digital wireless networks throughout the world
- privacy and secure telephony—used to encode speech to remain private (free of eavesdropping) for short periods of time (days) or to be encrypted for long-term (years) security applications
- IP networks—used to packetize speech for voice-over-IP (VoIP) digital network integration with other data types including images, video, text, binary files, etc.
- storage—for voice mail, answering machines, announcement machines and any other long-term needs for preserving speech material.

11.13.1 Standardization of Speech Coders

Figure 11.102 illustrates several attributes of coders used in the range of applications outlined above. This figure shows several speech coding algorithms that have become standardized for use in network applications, mobile radio, voice mail, or secure voice. For each of the coding methods, Figure 11.102 shows the designation of the standard, the bit rate of the coder, the class of application, and the quality of the coding output synthesis on the mean opinion score (MOS) quality scale.

[31] Such postfilters are also used routinely in CELP coders as discussed earlier in this chapter.

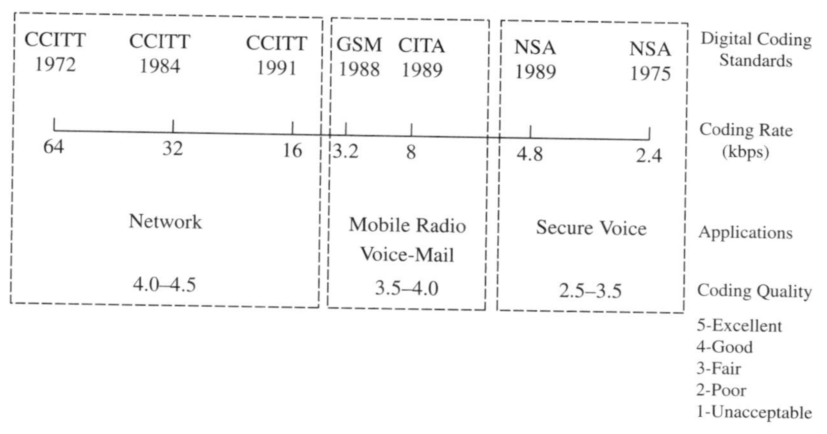

FIGURE 11.102
Illustration of the range of several speech coder attributes.

The highest quality applications of speech coders are those that run in the public telephony network (labeled network in Figure 11.102) including:

- a μ-law PCM instantaneous coder which is designated as the 1972 standard CCITT G.711 (where CCITT was an International standards organization that is now known as the International Telecommunication Union or ITU), running at a bit rate of 64 or 56 kbps (8 kHz sampling rate, 8 or 7 bits/sample), with an MOS score of about 4.3 and an implementation complexity of 0.01 MIPS (million instructions per second);
- an ADPCM adaptive coder which is designated as the 1984 set of CCITT standards G.721 and G.726/G.727 standards, running at a range of bit rates between 16 and 40 kbps, with an MOS score of about 4.1 and a complexity of 2 MIPS;
- an LD-CELP predictive coder which is designated as the 1991 G.728 CCITT standard, running at a bit rate of 16 kbps with an MOS score of 4.1 and a complexity of about 30 MIPS.

In the category of mobile radio and voice mail applications, Figure 11.102 shows two coders, namely:

- a full rate cellular coder used in the GSM (group special mobile) system, based on the regular pulse excitation with long-term prediction coder, designated as the 1988 ETSI (European Telecommunications Standards Institute) standard GSM 6.10, running at a bit rate of 13.2 kbps, with an MOS score of about 3.9 and a complexity of about 6 MIPS;
- a full rate cellular coder used in the North American TDMA (time division multiple access) standard, based on VSELP, designated as the 1989 Cellular Standard IS54, running at a rate of 8 kbps, with an MOS score of about 3.5, and a complexity of about 14 MIPS.

TABLE 11.10 Characteristics of a range of speech coders.

Coding standard	Year	Bit rate (kbps)	MOS	MIPS	Frame size (msec)
Linear PCM	1948	128	4.5	0	0.125
G.711 μ-law PCM	1972	64, 56	4.3	0.01	0.125
G.721 ADPCM	1984	32	4.1	2	0.125
G.722 ADPCM	1984	48/56/64	4.1	5	0.125
G.726/G.727 ADPCM	1990	16/24/32/40	4.1	2	0.125
G.728 LD-CELP	1992	16	4.0	30	0.625
G.729 CS-ACELP	1996	8	4.0	20	10
G.723.1 MPC-MLQ	1995	6.3/5.3	4.0/3.7	11	10
GSM FR RPLPC/LTP	1987	13	3.7	6	22.5
GSM HR VSELP	1994	5.6	3.5	14	22.5
IS-54 VSELP	1989	8	3.6	14	22.5
IS-96 QCELP	1993	1.2/2.4/4.8/9.6	3.5	15	22.5
FS-1015 LPC10(e)	1984	2.4	2.3	7	22.5
FS-1016 CELP	1991	4.8	3.0	16	30/7.5
NSA MELP	1996	2.4	3.2	40	22.5

In the category of secure voice are the two U.S. Government standard coders:

- a low bit rate secure coder used in U.S. Government encryption systems, designated as the 1989 standard FS-1016 (Federal Standard), running at a bit rate of 4.8 kbps, with an MOS score of 3.2, and a complexity of about 16 MIPS;
- a low rate secure coder again used in U.S. Government encryption systems, designated as the 1975 standard LPC-10(e) or known more formally as FS-1015, running at a bit rate of 2.4 kbps, with an MOS score of 2.3, and a complexity of about 7 MIPS.

The coders shown in Figure 11.102 are just a subset of the multiplicity of coders that have been proposed and studied, many of which are utilized in a range of applications. Table 11.10 gives a summary of the characteristics of the range of speech coders that have been standardized over the past 35 years.

These standards often continue to evolve under the same number. Often the evolution is to wider bandwidth coders. An example is the G.722 subband ADPCM coder to which has been added a transform coder (G.722.1) and an adaptive multi-rate coder AMR-WB (G.722.2).

11.13.2 Speech Coder Quality Evaluation

The SNR measure is often an adequate measure of performance for waveform coders, although it is not sensitive to perceptual effects on speech quality. However, it is not appropriate for model-based speech coders since they operate on blocks of speech and do not follow the waveform on a sample-by-sample basis. Thus, differences between the original and the coded waveforms are likely to be large even when the two are perceptually close. For these reasons, several subjective measures of (user-perceived) speech quality and robustness have been developed. These are appropriate to both waveform coding and analysis/synthesis approaches to speech coding. We have shown that there are two types of speech coders, namely:

- Waveform coders that approximate the speech waveform on a sample-by-sample basis, including PCM, DPCM, and ADPCM coders. These waveform coders produce a reconstructed signal which converges toward the original signal with decreasing quantization error (i.e., higher bit rate coding);
- Model-based (or parametric) open/closed-loop coders, including LPC, MPLPC and CELP coders. These frame-based (or block-based) coders produce a reconstructed signal that does not converge to the original signal with decreasing quantization error (i.e., higher bit rates). Instead the model-based coders converge to a model-constrained maximum quality which is related to the model inaccuracy in representing an arbitrary speech signal.

Figure 11.103 illustrates the dichotomy between the reconstructed signal quality of waveform coders and model-based or parametric coders with increasing bit rate. We see that waveform-based coders continue to improve in quality as the bit rate increases, whereas model-based coders level off in quality at a level significantly below that of waveform coders, and do not increase beyond some limit as the bit rate increases.

For waveform coders, we have established the validity of the SNR measurement, which computes the ratio between the (uncoded) signal variance, σ_x^2, and the

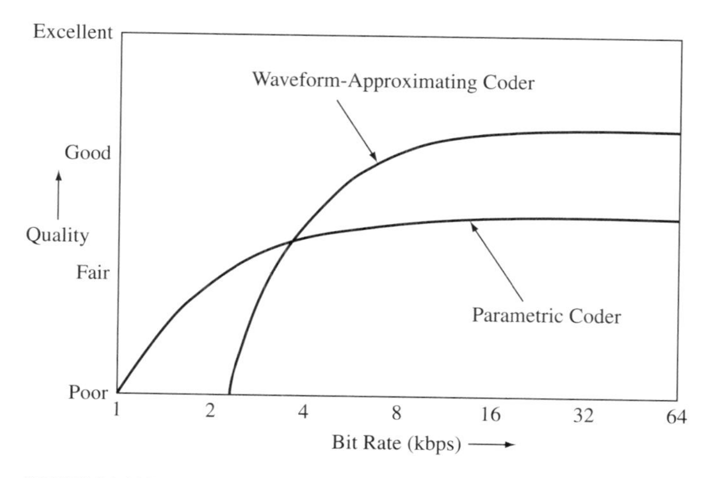

FIGURE 11.103
Typical behavior of the speech quality of waveform coders and parametric coders with increasing bit rates.

quantization error variance, σ_e^2 (measured in dB), over the duration of the signal; i.e.,

$$\text{SNR} = 10\log_{10}\frac{\sigma_x^2}{\sigma_e^2} = 10\log_{10}\left(\frac{\sum_{n=0}^{N-1}(s[n])^2}{\sum_{n=0}^{N-1}(s[n]-\hat{s}[n])^2}\right), \tag{11.202}$$

where $s[n]$ is the unquantized waveform, $\hat{s}[n]$ is the quantized waveform, and N is the number of samples in the waveform. A variant on the "global" SNR of Eq. (11.202) is the so-called "segmental" SNR, defined as:

$$\text{SNR}_{\text{SEG}} = \frac{1}{K}\sum_{k=1}^{K}\text{SNR}_k, \tag{11.203}$$

where SNR_k is the SNR of Eq. (11.202) measured over consecutive (or overlapping) frames of 10–20 msec duration, and K is the total number of frames in the speech signal. The segmental SNR is effective for speech since the waveform variance, σ_x^2, varies so much through voiced, unvoiced, and background regions, and the averaging of Eq. (11.203) provides a better estimate of waveform SNR than that obtained from Eq. (11.202).

For model-based coders, the concept of measuring SNR makes little sense since the resulting waveform does not match the signal on a sample-by-sample basis. Hence the resulting SNR calculations would have little meaning for such coders.

Thus, for model-based coders, we need a completely different measure of speech quality. The measure that has been proposed, and widely accepted, is the MOS, which is a subjective rating test. This test rates the subjective quality of speech on a 5-point scale, where the scores have the following meanings:

- MOS score of 5—excellent, transparent speech quality
- MOS score of 4—good, toll speech quality
- MOS score of 3—fair, communications speech quality
- MOS score of 2—poor, synthetic speech quality
- MOS score of 1—bad, unacceptable speech quality.

MOS testing requires a varied group of listeners, a lot of training, and the need to anchor (especially at the high quality end) the speech material being evaluated so that the listeners are able to determine the range of speech quality of the material being tested and evaluated. When done carefully and properly, the resulting MOS scores are quite repeatable and provide an excellent measure of subjective speech quality.

Figure 11.104 shows MOS scores for telephone bandwidth speech coders, for both waveform coders and model-based coders, operating at bit rates from 64 kbps down to 2.4 kbps. It can be seen that the highest MOS scores are in the range of 4.0–4.2 (slightly above Good on the rating scale), since all the coders being evaluated in this figure are

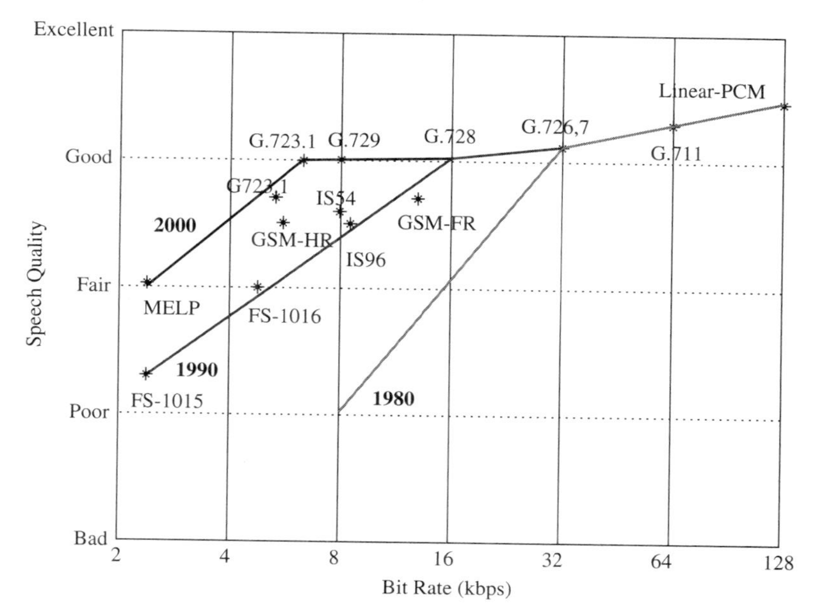

FIGURE 11.104
Speech coder subjective quality (MOS) scores for a range of coders operating at bit rates between 128 and 2.4 kbps. The trend of MOS scores is shown at the end of 1980, 1990, and 2000. (After Cox (private correspondence).)

telephone bandwidth coders, and it is well known that the perceived quality of telephone bandwidth speech (bandwidth of 3.2 kHz) is below that of wideband speech with a bandwidth of 7 kHz. We also see that over a bit rate from 64 kbps down to almost 6 kbps, there exist individual coders which achieve comparable MOS scores (on the order of 4.0), showing the wide range of bit rates over which we can design high quality speech coders. Below 6 kbps, we see a precipitous drop in speech quality (from Good to Fair) as the bit rate falls to 2.4 kbps, where the MELP coder is the highest quality at this time. This figure also shows the rate of progress in creating high speech quality coders over time. We see that in 1980, we could only maintain good speech quality (MOS score of 4 or higher) at 64 kbps with the G.711 (μ-law PCM) coder. By 1990 we could maintain good speech quality down to 16 kbps with the addition of G.726 and G.727 ADPCM coders at 32 kbps, and the G.728 LD-CELP coder at 16 kbps. Finally, by 2000, coders with good quality scores included G.729 at 8 kbps and G.723.1 at 6 kbps. There clearly remain challenges at bit rates below 6 kbps and it is still not clear whether a high quality speech coder can be designed for operation at such low rates.

11.14 SUMMARY

This chapter has presented a detailed discussion of digital speech coding methods. We have seen that a wide variety of approaches, both waveform matching and model-based, is possible. We have made no attempt to cover all the systems that have been proposed, but have chosen instead to emphasize basic principles. Further references on this topic can be found in the review paper by Jayant [169], in the collection of reprints edited by Jayant [171], and in the excellent speech coding presentations by Jayant and Noll [172], Kondoz [197], Quatieri [297], and Atal [10].

PROBLEMS

11.1. The uniform probability density function is defined as

$$p(x) = \begin{cases} \dfrac{1}{\Delta} & |x| < \Delta/2 \\ 0 & \text{otherwise.} \end{cases}$$

Determine the mean and variance of the uniform distribution.

11.2. Consider the Laplacian probability density function

$$p(x) = \frac{1}{\sqrt{2}\sigma_x} e^{-\sqrt{2}|x|/\sigma_x}.$$

Determine the probability that $|x| > 4\sigma_x$.

11.3. Let $x[n]$, the input to a linear shift-invariant system, be a stationary, zero mean, white noise process. Show that the autocorrelation function of the output is

$$\phi[m] = \sigma_x^2 \sum_{k=-\infty}^{\infty} h[k]\, h[k+m],$$

where σ_x^2 is the variance of the input and $h[n]$ is the impulse response of the linear system.

11.4. A speech signal, $s[n]$, is received at two microphones and is modeled as if it were corrupted by two independent Gaussian noise sources, $e_1[n]$ and $e_2[n]$, giving input signals of the form:

$$x_1[n] = s[n] + e_1[n],$$
$$x_2[n] = s[n] + e_2[n].$$

The noise sources are characterized by means and variances of the form:

$$E(e_1[n]) = 0,$$
$$E(e_2[n]) = 0,$$
$$E(e_1[n]^2) = \sigma_{e_1}^2,$$
$$E(e_2[n]^2) = \sigma_{e_2}^2.$$

The two input signals are combined linearly to form the signal, $r[n]$, as:

$$r[n] = ax_1[n] + (1-a)x_2[n],$$

$$= s[n] + ae_1[n] + (1 - a)e_2[n],$$
$$= s[n] + e_3[n].$$

Find the value of a that minimizes the variance of $e_3[n]$. What is the minimum variance of $e_3[n]$ and how does it compare to the variances of $e_1[n]$ and $e_2[n]$?

11.5. A proposed system for "oversampling" speech in order to reduce the effect of quantization noise in sampling an analog signal is shown in Figure P11.5.

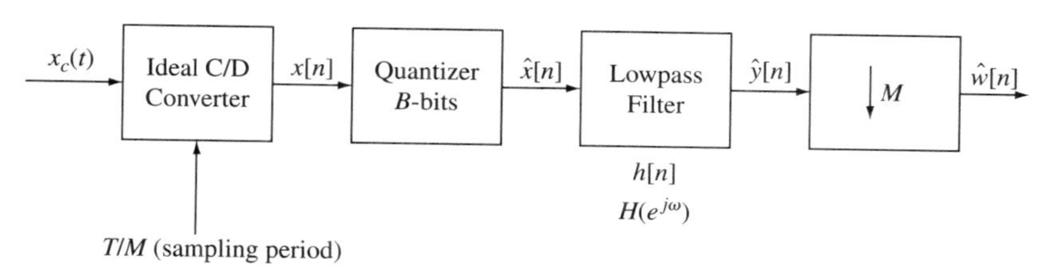

FIGURE P11.5
Block diagram of oversampling speech system.

Assume the following about the signals in Figure P11.5:

1. $x_c(t)$ is bandlimited so that $X_c(j\Omega) = 0$ for $|\Omega| \geq \pi/T$, and the ideal C/D converter oversamples $x_c(t)$ giving $x[n] = x_c(nT/M)$.
2. The quantizer produces output samples rounded to B-bit precision.
3. M is an integer greater than 1.
4. The filter response is:

$$H(e^{j\omega}) = \begin{cases} 1 & |\omega| \leq \pi/M \\ 0 & \pi/M < |\omega| \leq \pi. \end{cases}$$

We represent the signals in the system as:

$$\hat{x}[n] = x[n] + e[n],$$
$$\hat{y}[n] = x[n] * h[n] + e[n] * h[n] = y[n] + f[n],$$
$$\hat{w}[n] = \hat{y}[nM] = w[n] + g[n],$$

where $f[n]$ and $g[n]$ are the quantization noise components of $\hat{y}[n]$ and $\hat{w}[n]$, respectively.

(a) Derive an expression for $w[n]$ in terms of $x_c(t)$.

(b) Derive expressions for the noise power at each point in the system; i.e., σ_e^2, σ_f^2, and σ_g^2.

(c) Determine M so that the noise power, σ_f^2, is "one bit better" than the noise power σ_e^2.

11.6. Consider the design of a high quality digital audio system. The specifications are: 60 dB SNR must be maintained over a range of peak signal levels of 100 to 1. The useful signal bandwidth must be at least 8 kHz.

(a) Draw a block diagram of the basic components needed for A/D and D/A conversion.

(b) How many bits are required in the A/D and D/A converter?

(c) What are the main considerations in choosing the sampling rate? What types of analog filters should be used prior to the A/D converter and following the D/A converter? Estimate the lowest sampling rate that would be possible in a practical system.

(d) How would the specifications and answers change if the objective was only to maintain a telephone quality representation of speech?

11.7. Consider the system shown in Figure P11.7 for sampling speech and converting to a digital format.

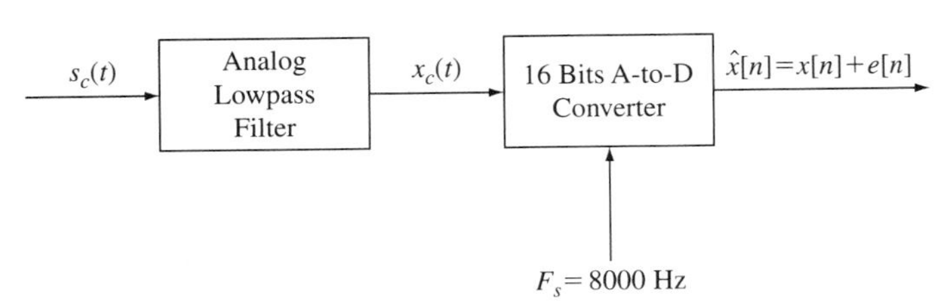

FIGURE P11.7
Block diagram of speech sampling system.

For a given input signal, $s_c(t)$, the signal-to-quantizing noise ratio at the output is:

$$\text{SNR} = 10\log_{10}\left(\frac{\sigma_x^2}{\sigma_e^2}\right) = 89 \text{ dB}.$$

(a) If the sampling rate is changed from $F_s = 8000$ to $F_s = 16{,}000$ Hz, while leaving all other conditions in Figure P11.7 the same, what is the new SNR of the system?

(b) What is the SNR when the input signal is changed so that the input to the analog lowpass filter in Figure P11.7 is $0.1s_c(t)$?

(c) If the A-to-D converter is changed from a 16-bit converter to a 12-bit converter, while leaving all other conditions the same, what is the new SNR?

11.8. A speech signal is bandlimited by an ideal lowpass filter, sampled at the Nyquist rate, quantized by a uniform B-bit quantizer, and converted back to an analog signal by an ideal D/A converter, as shown in Figure P11.8a. Define $y[n] = x[n] + e_1[n]$, where $e_1[n]$ is the quantization error. Assume that the quantization step size is $\Delta = 8\sigma_x/2^B$ and that B is large enough so that we can assume the following:

1. $e_1[n]$ is stationary;
2. $e_1[n]$ is uncorrelated with $x[n]$;
3. $e_1[n]$ is a uniformly distributed white noise sequence.

We have seen that, under these conditions, the signal-to-quantizing noise ratio is:

$$\text{SNR}_1 = \frac{\sigma_x^2}{\sigma_{e_1}^2} = \frac{12}{64} \cdot 2^{2B}.$$

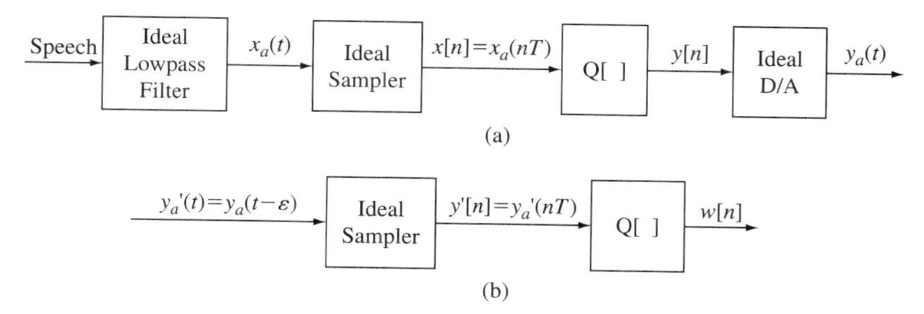

FIGURE P11.8
Block diagram of signal processing system.

Now assume that the analog signal $y_a(t)$ is sampled again at the Nyquist rate and quantized by an identical B-bit quantizer, as shown in Figure P11.8b. (Assume that $0 < \epsilon < T$; i.e., the two sampling systems are not exactly synchronized in time.) Assume that $w[n] = y'[n] + e_2[n]$, where $e_2[n]$ has identical properties to $e_1[n]$.

(a) Show that the overall SNR is

$$\text{SNR}_2 = \frac{\text{SNR}_1}{2}.$$

(b) Generalize the result of (a) to N stages of A/D and D/A conversion.

11.9. Although it is common to treat the quantization error as being independent of the signal $x[n]$, it can easily be shown that this assumption breaks down for a small number of quantization levels.

(a) Show that $e[n] = \hat{x}[n] - x[n]$ is *not* statistically independent of $x[n]$. ($\hat{x}[n]$ is the quantized signal.) Hint: Represent $\hat{x}[n]$ as

$$\hat{x}[n] = \left[\frac{x[n]}{\Delta}\right] \cdot \Delta + \frac{\Delta}{2},$$

where $[\,\cdot\,]$ denotes the "greatest integer in," i.e., the greatest integer less than or equal to the quantity within the brackets. Also represent $x[n]$ as

$$x[n] = \left[\frac{x[n]}{\Delta}\right] \cdot \Delta + x_f[n] = x_i[n] + x_f[n],$$

where $x_i[n]$ is the integer part of $x[n]$, and $x_f[n]$ is the fractional part of $x[n]$. Then $e[n]$ can be determined as a function of $x[n]$. Argue that they cannot be exactly statistically independent.

(b) Under what conditions is the approximation that $x[n]$ and $e[n]$ are statistically independent valid?

(c) Figure P11.9 shows a method which has been suggested for making $e[n]$ and $x[n]$ statistically independent, even for a small number of quantization levels. For this case $z[n]$ is a pseudo-random, uniformly distributed, white noise sequence with

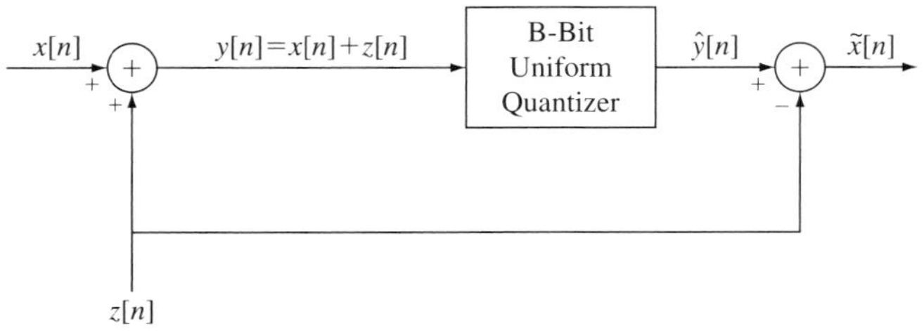

FIGURE P11.9
System for making $e[n]$ and $x[n]$ statistically independent.

probability density function

$$p(z) = \frac{1}{\Delta} \qquad -\frac{\Delta}{2} \leq z \leq \frac{\Delta}{2}.$$

Show that, in this case, the quantization error $e[n] = x[n] - \hat{y}[n]$ *is* statistically independent of $x[n]$ for all values of B. (The noise sequence, $z[n]$, being added to the signal is called *dither noise*). Hint: Look at the range of values for $e[n]$ for ranges of $y[n]$.

(d) Show that the variance of the quantization error at the output of the B-bit quantizer is greater than the variance of the quantization error for the undithered case–i.e., show that

$$\sigma_{e_1}^2 > \sigma_e^2,$$

where

$$e_1[n] = x[n] - \hat{y}[n]$$

and

$$e[n] = x[n] - \hat{x}[n].$$

(e) Show that by simply subtracting the dither noise $z[n]$ from the quantizer output, the variance of the quantization error $e_2[n] = x[n] - (\hat{y}[n] - z[n])$ is the same as the variance of the undithered case; i.e., $\sigma_{e_2}^2 = \sigma_e^2$.

11.10. A common approach to estimating the signal variance is to assume that it is proportional to the short-time energy of the signal, defined as

$$\sigma^2[n] = \frac{\sum_{m=-\infty}^{\infty} x^2[m]w[n-m]}{\sum_{m=0}^{\infty} w[m]},$$

where $w[n]$ is the short-time analysis window.

(a) Show that if $x[n]$ is stationary with zero mean and variance σ_x^2, then $E[\sigma^2[n]]$ is proportional to σ_x^2.

(b) For

$$w[n] = \begin{cases} \alpha^n & n \geq 0 \quad (|\alpha| < 1) \\ 0 & n < 0 \end{cases}$$

and for

$$E[x^2[m]\, x^2[l]] = \begin{cases} B & m = l \\ 0 & m \neq l, \end{cases}$$

determine the variance of $\sigma^2[n]$ as a function of B and α.

(c) Explain the behavior of the variance of $\sigma^2[n]$ of part (b) as α varies from 0 to 1.

11.11. Consider the adaptive quantization system shown in Figure P11.11a. The 2-bit quantizer characteristic and codeword assignment is shown in Figure P11.11b. Suppose the step size is adapted according to the following rule:

$$\Delta[n] = M\, \Delta[n-1],$$

where M is a function of the previous codeword $c[n-1]$ and

$$\Delta_{\min} \leq \Delta[n] \leq \Delta_{\max}.$$

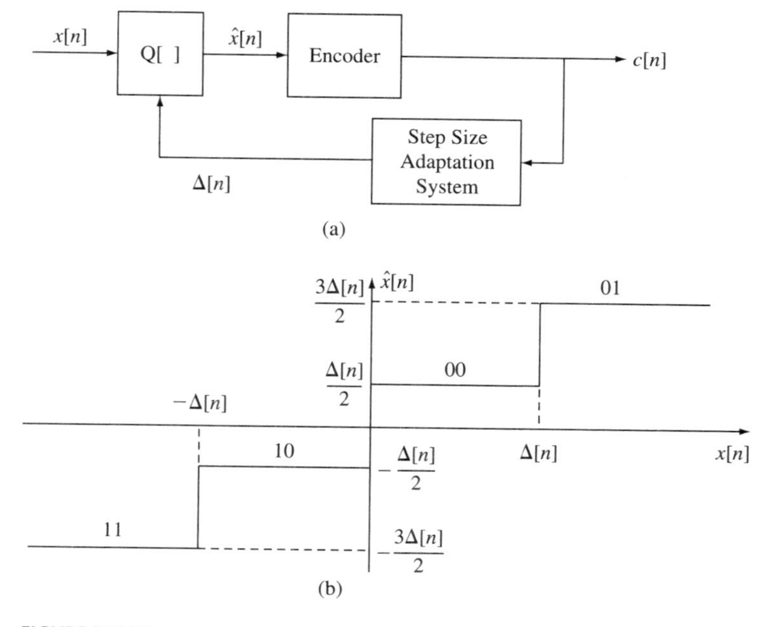

FIGURE P11.11
Block diagram of adaptive quantization system.

Furthermore suppose that

$$M = \begin{cases} P & \text{if } c[n-1] = 01 \text{ or } 11 \\ 1/P & \text{if } c[n-1] = 00 \text{ or } 10. \end{cases}$$

(a) Draw a block diagram of the step size adaptation system.

(b) Suppose that

$$x[n] = \begin{cases} 0 & n < 5 \\ 20 & 5 \le n \le 13 \\ 0 & 13 < n. \end{cases}$$

Assume that $\Delta_{min} = 2$ and $\Delta_{max} = 30$ and $P = 2$. Make a table of values of $x[n]$, $\Delta[n]$, $c[n]$, and $\hat{x}[n]$ for $0 \le n \le 25$. (Assume that at $n = 0$, $\Delta[n] = \Delta_{min} = 2$, and $c[n] = 00$).

(c) Plot the samples $x[n]$ and $\hat{x}[n]$ on the same coordinate scale.

11.12. Consider the 2-bit adaptive quantizing system of Problem 11.11. In this case, however, the step size adaptation algorithm is:

$$\Delta[n] = \begin{cases} \beta\Delta[n-1] + D & \text{if } \sum_{k=1}^{M} LSB[c[n-k]] \ge 2 \\ \beta\Delta[n-1] & \text{otherwise,} \end{cases}$$

where $LSB[c[n-k]]$ means "least significant bit" of the codeword $c[n-k]$ and M is the number of codewords considered.

(a) Draw a block diagram of the step size adaptation system.

(b) In this case the maximum step size is built into the algorithm for adaptation. Find Δ_{max} in terms of β and D. (Hint: Consider the step response of the first equation of this problem.)

(c) Again suppose that

$$x[n] = \begin{cases} 0 & n < 5 \\ 20 & 5 \le n \le 13 \\ 0 & 13 < n. \end{cases}$$

Also suppose that $M = 2$, $\beta = 0.8$, and $D = 6$. Make a table of values of $x[n]$, $\Delta[n]$, $c[n]$, and $\hat{x}[n]$ for $0 \le n \le 25$. (Assume that at $n = 0$, $\Delta[n] = 0$, and $c[n] = 00$.) Plot the samples $x[n]$ and $\hat{x}[n]$ on the same coordinate system.

(d) Find the value of β such that the time constant of the step size adaptation system is 10 msec.

11.13. Consider the first-order linear predictor

$$\tilde{x}[n] = \alpha x[n-1],$$

where $x[n]$ is a stationary, zero mean signal.

(a) Show that the prediction error

$$d[n] = x[n] - \tilde{x}[n]$$

has variance

$$\sigma_d^2 = \sigma_x^2\,(1 + \alpha^2 - 2\alpha\phi[1]/\sigma_x^2).$$

(b) Show that σ_d^2 is minimized for

$$\alpha = \phi[1]/\sigma_x^2 = \rho[1].$$

(c) Show that the minimum prediction error variance is

$$\sigma_d^2 = \sigma_x^2\,(1 - \rho^2[1]).$$

(d) Under what conditions will it be true that $\sigma_d^2 < \sigma_x^2$?

11.14. Given a sequence $x[n]$ with long-term autocorrelation $\phi[m]$, show that the difference signal

$$d[n] = x[n] - x[n - n_0]$$

has lower variance than the original signal $x[n]$ as long as there is some correlation between $x[n]$ and $x[n - n_0]$. (Assume $x[n]$ has zero mean value.)

(a) State the conditions on $\phi[n_0]$ such that

$$\sigma_d^2 \le \sigma_x^2.$$

(b) If $d[n]$ is formed as

$$d[n] = x[n] - \alpha x[n - n_0],$$

where

$$\alpha = \frac{\phi[n_0]}{\phi[0]},$$

state the conditions on $\phi[n_0]$ such that

$$\sigma_d^2 \le \sigma_x^2.$$

11.15. Using Eqs. (11.91) and (11.96), prove the assertion that for the optimum predictor coefficients

$$E[(x[n] - \tilde{x}[n])\,\tilde{x}[n]] = E[d[n]\tilde{x}[n]] = 0;$$

i.e., that the optimum prediction error is uncorrelated with the predicted signal.

11.16. Consider the difference signal

$$d[n] = x[n] - \alpha_1 \hat{x}[n-1],$$

where $\hat{x}[n]$ is the quantized signal in a differential coder.

(a) Show that

$$\sigma_d^2 = \sigma_x^2 \left[1 - 2\alpha_1 \rho[1] + \alpha_1^2\right] + \alpha_1^2 \sigma_e^2.$$

(b) Using the result of part (a), show that

$$G_P = \frac{\sigma_x^2}{\sigma_d^2} = \frac{1 - \dfrac{\alpha_1^2}{\text{SNR}_Q}}{1 - 2\alpha_1 \rho[1] + \alpha_1^2},$$

where

$$\text{SNR}_Q = \frac{\sigma_d^2}{\sigma_e^2}.$$

11.17. Consider the differential quantization system shown at the top of Figure P11.17. Assume the following:

$$\hat{d}[n] = d[n] + e[n],$$

$$P(z) = \sum_{k=1}^{p} \alpha_k z^{-k}.$$

(a) Show that $\hat{s}[n] = s[n] + e[n]$; i.e., the quantization error of the reconstructed speech signal is the same as the quantization error of the difference signal.

(b) Using the result of part (a), obtain an expression for $d[n]$ in terms of $s[n]$ and $e[n]$ only.

Now consider the quantization system shown at the bottom of Figure P11.17. Assume that:

$$F(z) = \sum_{k=1}^{p} \beta_k z^{-k}.$$

(c) Obtain an expression for $d'[n]$ in terms of $s[n]$ and $e'[n]$ only.

(d) How should $F(z)$ be chosen so that $d'[n] = d[n]$ and thus $\hat{d}'[n] = \hat{d}[n]$, $e'[n] = e[n]$, and $\hat{s}'[n] = \hat{s}[n]$.

(e) If $F(z)$ is not chosen as in part (d), show that $\hat{s}'[n]$ satisfies the difference equation:

$$\hat{s}'[n] = s[n] + \sum_{k=1}^{p} \alpha_k \hat{e}'[n-k] + e'[n] - \sum_{k=1}^{p} \beta_k e'[n-k],$$

where $\hat{e}'[n] = \hat{s}'[n] - s[n]$ is the noise in the reconstructed signal.

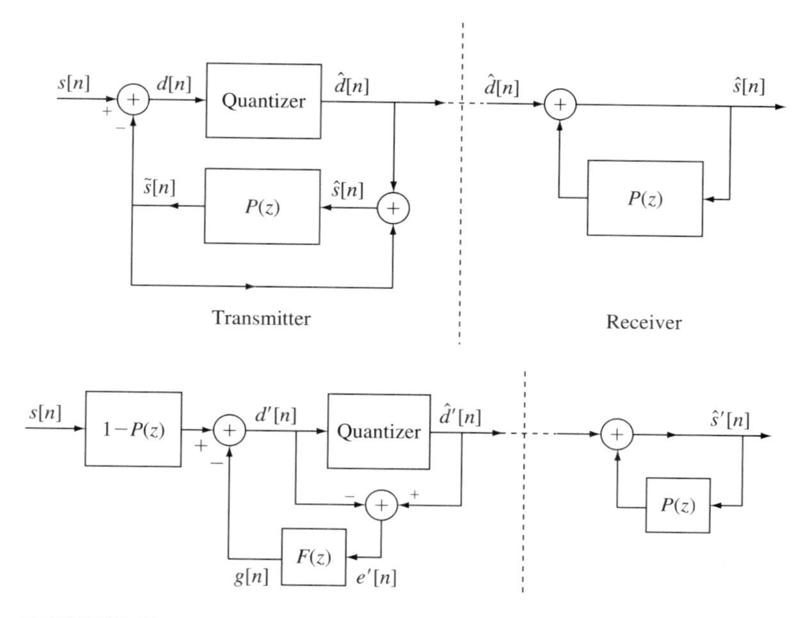

FIGURE P11.17
Block diagrams of two differential quantization systems.

(f) Show that:

$$\frac{\hat{E}'(z)}{E'(z)} = \frac{1 - F(z)}{1 - P(z)},$$

where $\hat{E}'(z)$ and $E'(z)$ are the z-transforms of the output noise and the quantizer noise, respectively.

(g) Recall that the quantizer noise has a flat power spectrum. The result of part (f) suggests that the output noise can be shaped by choosing $F(z)$ appropriately. It can be shown that it is not possible to reduce the overall noise power in the output; however, it is possible to redistribute the noise power in frequency. From what you know about speech and speech perception, how would you choose $F(z)$?

11.18. In the CVSD adaptive delta modulator, the step size adaptation algorithm is

$$\Delta[n] = \begin{cases} \beta\Delta[n-1] + D_2 & \text{if } c[n] = c[n-1] = c[n-2] \\ \beta\Delta[n-1] + D_1 & \text{otherwise,} \end{cases}$$

where $0 < \beta < 1$ and $0 < D_1 \ll D_2$.

(a) The maximum step size is attained if the input to the step size filter is constant at D_2 as would occur in a prolonged period of slope overload. Find $\Delta_{\max}$ in terms of D_2 and β.

(b) The minimum step size would be attained if the pattern $c[n] = c[n-1] = c[n-2]$ does not occur for a prolonged period as in the idle channel condition. Find $\Delta_{\min}$ in terms of D_1 and β.

11.19. Consider the adaptive delta modulator of Figure P11.19a. The 1-bit quantizer is given in Figure P11.19b. The step size is adapted according to the following rule:

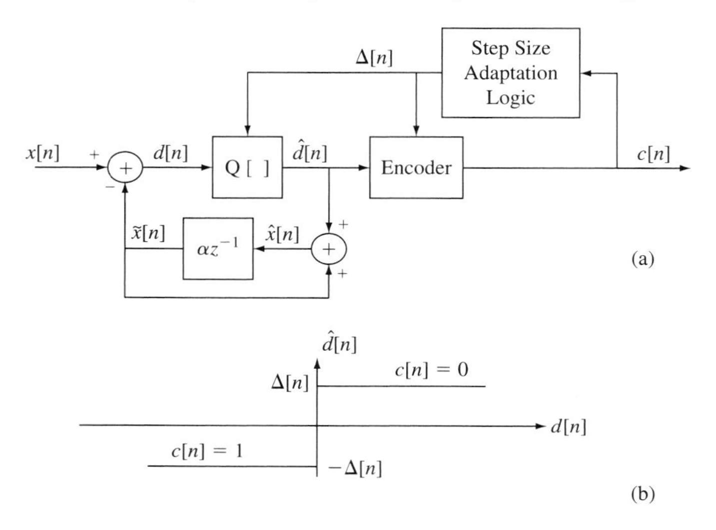

FIGURE P11.19
Block diagram of adaptive delta modulator.

$$\Delta[n] = M\Delta[n-1],$$

where

$$\Delta_{\min} \leq \Delta[n] \leq \Delta_{\max}$$

and the step size multiplier is given by

$$M = \begin{cases} P & \text{if } c[n] = c[n-1] \\ 1/P & \text{if } c[n] \neq c[n-1]. \end{cases}$$

(a) Draw a block diagram of the step size logic.

(b) Suppose that

$$x[n] = \begin{cases} 0 & n < 5 \\ 20 & 5 \leq n \leq 13 \\ 0 & 13 < n. \end{cases}$$

Assume $\Delta_{\min} = 1$, $\Delta_{\max} = 15$, $\alpha = 1$, and $P = 2$. Make a table of values of $x[n]$, $\tilde{x}[n]$, $d[n]$, $\Delta[n]$, $\hat{d}[n]$, and $\hat{x}[n]$ for $0 \leq n \leq 25$. Assume that at $n = 0$, $x[0] = 0$, $\tilde{x}[0] = 1$, $d[0] = -1$, $\Delta(0) = \Delta_{\min} = 1$, and $\hat{d}[0] = -1$. Plot $x[n]$ and $\hat{x}[n]$ for $0 \leq n \leq 25$.

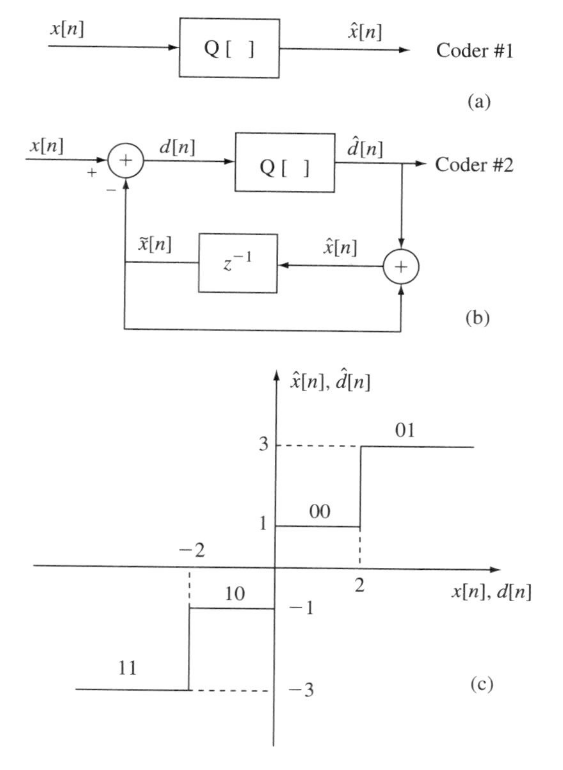

FIGURE P11.20
Block diagrams of two speech coders along with the input–output characteristic of a 2-bit quantizer.

11.20. Consider the two coders shown in Figure P11.20a and b. Each coder uses a 2-bit quantizer with the input–output characteristic shown in Figure P11.20c. Consider the idle channel case, i.e., the case where $x[n]$ is a low level noise. For simplicity we assume $x[n]$ is of the form

$$x[n] = 0.1 \cos(\pi n/4).$$

(a) For $0 \leq n \leq 20$, make a plot of $\hat{x}[n]$ for both coders.

(b) For which coder would the "idle channel noise" be more objectionable in a real communications system? Why?

11.21. Consider the PCM-to-ADPCM code conversion system of Figure P11.21. The PCM coded signal, $y[n]$, can be represented as

$$y[n] = x[n] + e_1[n],$$

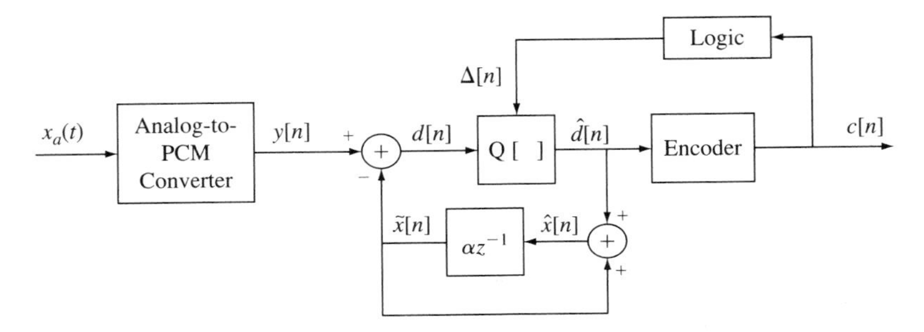

FIGURE P11.21
PCM-to-ADPCM code converter.

where $x[n] = x_a(nT)$, and $e_1[n]$ is the quantization error in the PCM representation. The quantized ADPCM signal, $\hat{y}[n]$, can be represented as

$$\hat{y}[n] = y[n] + e_2[n],$$

where $e_2[n]$ is the ADPCM quantization error.

(a) Assuming that the quantization errors $e_1[n]$ and $e_2[n]$ are uncorrelated, show that the overall SNR is

$$\text{SNR} = \frac{\sigma_x^2}{\sigma_{e_1}^2 + \sigma_{e_2}^2}.$$

(b) Show that SNR can be expressed as

$$\text{SNR} = \frac{\text{SNR}_1}{1 + \dfrac{1 + \text{SNR}_1}{\text{SNR}_2}},$$

where $\text{SNR}_1 = \sigma_x^2/\sigma_{e_1}^2$ and $\text{SNR}_2 = \sigma_y^2/\sigma_{e_2}^2$.

(c) If $\text{SNR}_1 = \text{SNR}_2 \gg 1$, show that we lose only 3 dB of signal-to-quantization noise ratio (in dB) in the conversion from PCM to ADPCM.

11.22. We showed in this chapter that for a logarithmic instantaneous compander, the quantized speech sample can be approximated as:

$$\hat{x}[n] \approx x[n] + f[n],$$

where

$$f[n] = x[n] \cdot \epsilon[n],$$

where $\epsilon[n]$ is the quantization noise resulting from quantization of the log-encoded speech sample.

(a) Assuming that $x[n]$ and $\epsilon[n]$ are independent, zero mean processes with variances σ_x^2 and σ_ϵ^2, show that:

$$\bar{f} = E(f[n]) = 0.$$

(b) Also show that:

$$\overline{\sigma_f^2} = E(f^2) = \sigma_x^2 \cdot \sigma_\epsilon^2.$$

11.23. We have discussed a number of vocoder systems in this book, namely

1. channel vocoder
2. serial formant vocoder
3. parallel formant vocoder
4. homomorphic vocoder
5. phase vocoder
6. LPC vocoder.

Theoretically speaking, how would you order the quality of the output of these vocoders? Explain your ordering completely. The issues that should be discussed include dependence on a model, information lost in analysis, necessity for pitch tracking, etc.

11.24. (MATLAB Exercise) The goal of this MATLAB exercise is to verify some aspects of the statistical model of speech, as discussed in Section 11.3. In this exercise you will use three concatenated speech files with the following content:

1. `out_s1_s6.wav`—the concatenation of six individual speech files (with beginning and ending silence regions removed), namely `s1.wav, s2.wav, s3.wav, s4.wav, s5.wav, s6.wav`
2. `out_male.wav`—the concatenation of four individual male speech files (with beginning and ending silence regions removed), namely `s2.wav, s4.wav, s5.wav, s6.wav`
3. `out_female.wav`—the concatenation of two individual female speech files (with beginning and ending silence regions removed), namely `s1.wav, s3.wav`

(a) Treating the large speech file as a source of statistical speech samples, determine the mean and variance of the speech signal, along with the minimum and maximum values. Create an amplitude histogram (using the MATLAB `hist` routine), and plot the histogram of the speech amplitudes using 25 bins for the histogram. Experiment with other numbers of bins for the histogram.

(b) Using the m-file `pspect.m` from the book website, compute an estimate of the long-term average power spectrum of speech using the signal in the long concatenated file `out_s1_s6.wav`. Experiment with a range of window durations (e.g., 32/64/128/256/512) to determine the effect of window size on the smoothness of the power spectrum. Plot the power spectrum (in dB units, labeling the frequency axis appropriately), for the five window durations given above, on a common plot.

(c) Modify the MATLAB function `pspect( )` to give the autocorrelation function and the corresponding time axis as additional outputs. What condition should you impose on Nfft and Nwin in order that the computed correlation not be time-aliased? Plot the correlation functions corresponding to the power spectrums that you measured in part (b)—i.e., five autocorrelations on a single plot.

(d) Repeat part (b) for the window duration of 32 samples using both the male speech file (out_male.wav) and the female speech file (out_female.wav). Plot both power spectrums (again in dB units) on a single plot. How do the power spectrums for male and female speech compare?

11.25. (MATLAB Exercise) The goal of this MATLAB exercise is to experiment with the process of uniform quantization of a speech signal. This exercise uses the MATLAB function (from the book website):

```
X=fxquant(s,bits,rmode,lmode)
```

where

- s is the input speech signal (to be quantized)
- bits is the total number of bits (including the sign bit) of the quantizer
- rmode is the quantization mode, where rmode is one of 'round' for rounding to the nearest level, 'trunc' for 2's complement truncation, or 'magn' for magnitude truncation
- lmode is the overflow/underflow handler, where lmode is one of 'sat' for a saturation limiter, 'overfl' for a 2's complement overflow, 'triangle' for a triangle limiter, and 'none' for no limiter
- X is the output (quantized) speech signal

(a) Create a linearly increasing input vector going from -1 to $+1$ in increments of 0.001, i.e., xin=-1:0.001:1, and use the MATLAB function fxquant to plot the non-linear quantizer characteristic for the conditions bits=4, rmode='round', lmode='sat'. Repeat this calculation and plot for the case of rmode='trunc'. What is the range of values for $e[n]$ when truncation is used instead of rounding?

(b) Use fxquant( ) to quantize the speech samples of the file s5.wav, between sample 1300 and sample 18,800 (the speech file is exactly zero outside this range), with parameters:

```
rmode='round', lmode='sat'.
```

Experiment with different numbers of bits for the quantizer, namely 10 bits, 8 bits, and 4 bits. For each of these quantizers, compute the quantization error sequences and use the program striplot( ) to plot the first 8000 of these error sequences with 2000 samples per line. What are the important differences between the error plots as a function of the number of bits in the quantizer? Make histograms of the quantization noise samples for each of these bit rates. Do the histograms look like they fit the white noise model (i.e., with a uniform amplitude distribution)?

(c) Compute the power spectrum of the quantization noise sequences for each of the quantizers of part (b) using the file pspect( ). Plot these spectra on the same plot as the power spectrum of the original (unquantized) speech samples. Do the noise spectra support the white noise assumption? What is the approximate difference (in dB) between the noise spectra for 10- and 8-bit quantization?

11.26. The short-time SNR of a signal, $s[n]$, is defined as:

$$\text{SNR}[n] = 10\log_{10}\left[\frac{\sum_{m=0}^{L-1}(s[m+n])^2}{\sum_{m=0}^{L-1}(s[m+n]-\hat{s}[m+n])^2}\right],$$

where $\hat{s}[n]$ is the sequence of quantized samples corresponding to $s[n]$. Figure P11.26 shows the first 4000 samples of $s[n]$ from the speech file `this_8k.wav` with a sampling rate of $F_S = 8000$ samples/sec.

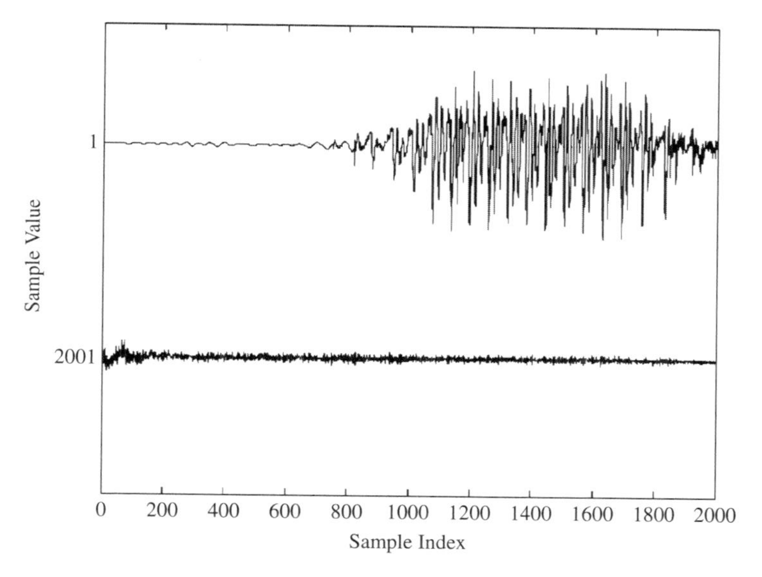

FIGURE P11.26
Plot of first 4000 samples of speech file "test_8k.wav."

Use the function `fxquant( )` described in Problem 11.25 to quantize $s[n]$ using an 8-bit uniform quantizer with the peak value of the speech signal set so that $X_{\max} = 4\sigma_s$, where σ_s is the long-term average rms of the speech signal. The function `fxquant( )` assumes that $X_{\max} = 1$, so you will have to appropriately scale the input samples.

(a) Assuming a value of $L = 100$, use MATLAB to plot SNR$[n]$ as a function of n for the range $0 \le n \le 3999$.

(b) Assuming a value of $L = 600$, use MATLAB to plot SNR$[n]$ as a function of n for the range $0 \le n \le 3999$.

(c) How would the plots change if the quantizer was a μ-law quantizer instead of a uniform quantizer?

11.27. (MATLAB Exercise) The goal of this MATLAB exercise is to experiment with the process of μ-law compression of speech. Using the MATLAB function `y=mulaw(x,mu)`, where:

- `x` is the input signal
- `mu` is the compression parameter
- `y` is the μ-law compressed signal

Do the following:

(a) Create a linearly increasing input vector (−1:0.001:1) and use it, with the above function `mulaw( )` to plot the μ-law characteristic for $\mu = 1, 20, 50, 100, 255, 500$, all on the same plot. (Note that the value $\mu = 255$ is a standard value used in landline telephony.)

(b) Using the segment of speech, from sample 1300 to sample 18,800, from the file `s5.wav`, and a value of $\mu = 255$, plot the output waveform $y[n]$ of the μ-law compressor. Observe how the low amplitude samples are increased in magnitude. Plot a histogram of the output samples and compare it to the "white noise" histogram from uniform quantization.

(c) A block diagram of a full μ-law quantization method is shown in Figure P11.27. To implement this system, you are asked to write an m-file for the inverse of the μ-law compressor. This m-file should have the following calling sequence and parameters:

```
function x=mulawinv(y,mu)
% function for inverse mulaw for Xmax=1
% x=mulawinv(y,mu)
% y=input column vector
% mu=mulaw compression parameter
% x=expanded output vector
```

Use the technique used in `mulaw( )` to set the signs of the samples. Test the inverse system by applying it directly to the output of `mulaw( )` without quantization; e.g., try

```
v=mulawinv(mulaw(2,255),255);
```

In this case, `v` and `x` should be essentially identical.

(d) The MATLAB statement

```
yh=fxquant(mulaw(x,255),6,'round','sat');
```

implements a 6-bit μ-law quantizer. Thus the μ-law compressed samples are coded by a uniform quantizer with 6 bits. When the coded samples are used in a signal

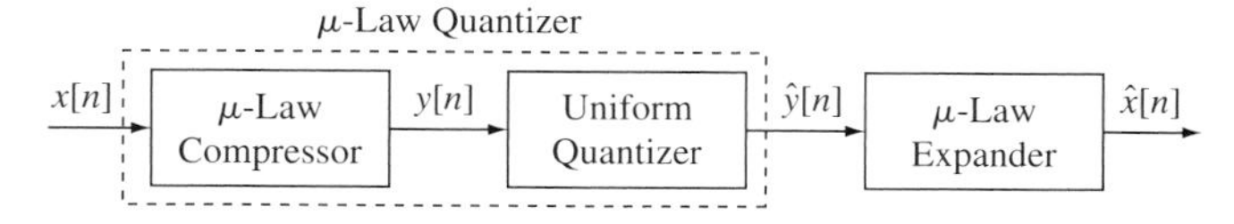

FIGURE P11.27
Representation of μ-law quantization.

processing computation, or when a continuous-time signal needs to be reconstructed, the uniformly coded samples of the μ-law quantizer must be expanded. Hence, the quantization error is likewise expanded, so that to determine the quantization error, it is necessary to compare the output of the inverse system to the original samples. Thus the quantization error is `e=mulaw(yh,255)-x;`. Using uniform quantizers with 10, 8, and 4 bits, compute and plot the first 8000 samples of the resulting quantization error, plot a histogram of the quantization error amplitudes, and plot the power spectrums of the resulting quantization errors, all on common plots.

11.28. (MATLAB Exercise) The goal of this MATLAB exercise is to compare uniform and μ-law quantizers on the basis of their SNRs.

A convenient definition of a waveform (with duration L samples) SNR is:

$$\text{SNR} = 10\log\left(\frac{\displaystyle\sum_{n=0}^{L-1}(x[n])^2}{\displaystyle\sum_{n=0}^{L-1}(\hat{x}[n]-x[n])^2}\right).$$

Note that the division by L, as required for averaging, cancels in the numerator and denominator.

It was shown in this chapter that the SNR for a uniform B-bit quantizer was of the form

$$\text{SNR} = 6B + 4.77 - 20\log_{10}\left(\frac{X_{\text{max}}}{\sigma_x}\right),$$

where X_{max} is the clipping level of the quantizer (1 for the files in these exercises), and σ_x is the rms value of the input signal amplitude. We see that the SNR increases 6 dB per bit added to the quantizer word length. Furthermore we see that if the signal level is decreased by a factor of 2, the SNR decreases by 6 dB.

(a) Write an m-file to compute the SNR, given the unquantized and quantized versions of the signal. The calling sequence and parameters of this m-file should be:

```
function [s_n_r,e]=snr(xh,x);
% function for computing SNR
% [s_n_r,e]=snr(xh,x)
% xh=quantized signal
% x=unquantized signal
% e=quantization error signal (optional)
% s_n_r=snr in dB
```

Use your SNR function to compute the SNRs for uniform quantization with 8 and 9 bits. Do the results differ by the expected amount?

(b) An important consideration in quantizing speech is that signal levels can vary with speakers and with transmission/recording conditions. Write a program to plot the measured SNR for uniform and μ-law quantization as a function of $1/\sigma_x$. Vary the signal level by multiplying the input speech signal, `s5.wav`, by the factors `[2^(0:-1:-12)]` and measure the SNR for all 13 cases. Plot the resulting SNR data on a semi-log plot. Repeat the calculation for 10, 9, 8, 7, and 6 bit uniform quantizers and for $\mu = 100,255,500$ over the same range.

Your plots should show that the μ-law quantizer maintains a constant SNR over an input amplitude range of about 64:1. How many bits are required for a uniform quantizer to maintain at least the same SNR as does a 6-bit, μ-law quantizer over the same range?

11.29. (MATLAB Exercise) The goal of this MATLAB exercise is to demonstrate the process of adaptive quantization using either an IIR filter or an FIR filter. As discussed in this chapter, there are two ways of adapting the quantizer, namely gain adaptation or step size adaptation. Both adaptive methods require the estimation of the signal standard deviation, $\sigma[n]$. Assuming that the signal has a mean of zero (hence you need to subtract any DC component of the signal prior to variance computation), the signal variance, $\sigma^2[n]$, can be computed using an IIR filter of the form:

$$H(z) = \frac{(1-\alpha)z^{-1}}{(1-\alpha z^{-1})}$$

as:

$$\sigma^2[n] = \alpha\sigma^2[n-1] + (1-\alpha)\cdot x^2[n-1],$$

or, using a rectangular window FIR filter of length M samples, as:

$$\sigma^2[n] = \frac{1}{M}\sum_{m=n-M+1}^{n} x^2[m].$$

Using an IIR filter, with values of α of 0.9 and 0.99, and using the speech file `s5.wav`, calculate the standard deviation, $\sigma[n]$, of the speech signal and plot the following:

1. both the speech samples, $x[n]$, and the superimposed standard deviation samples, $\sigma[n]$, for the sample range of $2700 \leq n < 6700$ on the upper plot of a figure with two plots
2. the gain equalized speech samples, $x[n]/\sigma[n]$, for the sample range of $2700 \leq n < 6700$ on the lower plot of the same figure

(a) How do the two plots (i.e., the plot using $\alpha = 0.9$ and the plot using $\alpha = 0.99$) compare, especially in terms of rate of adapting to the changing speech level and ability to equalize the local dynamic range of the speech signal?

(b) Repeat the above exercise using an FIR filter (rectangular window) of duration $M = 10$ and $M = 100$ samples, and replot the same waveforms. How do the results using the FIR window compare to the results using the IIR filter?

11.30. (MATLAB Exercise) The goal of this MATLAB exercise is to implement an ADPCM speech coder using the table of adaptive step size multipliers (Table 11.6) proposed by Jayant [168]. A block diagram of the encoder and the decoder for this MATLAB exercise is given in Figure P11.30.

Use the following parameter values for this coder implementation:

- 4-bit coder with multipliers of [0.9 0.9 0.9 0.9 1.2 1.6 2.0 2.4]

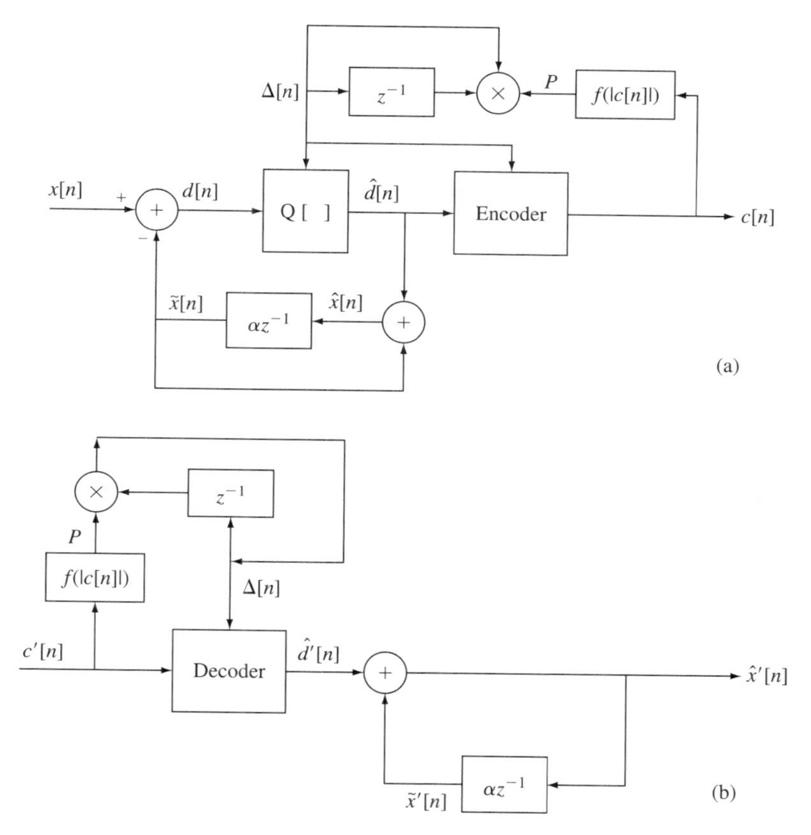

FIGURE P11.30
Block diagram of ADPCM encoder (a) and decoder (b).

- predictor value, $\alpha = 0.6$
- range of step sizes, `deltamin=16`, `deltamax=1600`–i.e., a range of 100-to-1 in step sizes

Write separate routines for the encoder and the decoder and process each of the speech input files `s1.wav`, `s2.wav`, `s3.wav`, `s4.wav`, `s5.wav`, `s6.wav`. Determine the resulting SNRs for each of these files by subtracting the decoded speech from the original speech to give the error signal, and using the conventional SNR calculation. Express the SNR in decibels (logarithmic units). Listen to the decoded speech files and describe the quality of the resulting coded speech at the coded bit rate of 32 kbps (4-bits/sample coding at a sampling rate of 8000 samples/sec speech).

11.31. (MATLAB Exercise) The goal of this MATLAB exercise is to design a set of 11 VQ codebooks, one for each of the 11 digits of English (zero-nine plus oh), using the design algorithm discussed in this chapter. You are asked to design codebooks with from 4 code vectors (i.e., a 2-bit codebook) to 64 code vectors (i.e., a 6-bit codebook). The feature vectors to be coded are sets of 12 cepstral coefficients, measured from a set of isolated digit utterances spoken by 113 different talkers (both male and female talkers from the TI digits database [204]). Each raw feature vector file has about 10,000 or so sets of 12 cepstral coefficients, so there are more than sufficient number of vectors to train codebooks up to size 64.

The files containing the raw cepstral vectors are labeled as `cc_tidig_endpt_[1-9ZO].mat` and are found on the book website, and can be loaded with the MATLAB command `load 'cc_tidig_endpt_1'` (for the file associated with the digit "one") to give the array `c(# vectors,12)`, which contains the raw cepstral coefficient feature vector array for the digit "one." Once you have loaded the cepstral feature vectors, you must run the VQ design procedure. This process must be iterated for all 11 digits (1–9 plus "oh" and "zero").

The VQ design procedure, as outlined in this chapter, consists of the following steps:

Step 1 Determine the centroid of the entire set as the average vector of cepstral coefficients. This is effectively the 1-element codebook.

Step 2 Create a codebook of double the current size. To do this, you need to split the current set of codewords by creating two codewords for every existing codeword. This is done by scaling the existing codeword set by factors of $(1+\epsilon)$ and $(1-\epsilon)$, where ϵ is a codebook design parameter (begin with a value of $\epsilon = 0.001$ and experiment with this splitting factor once your program works properly).

Step 3 Map each cepstral coefficient feature vector in the training set to the new set of codewords by measuring the Euclidean distance of each feature vector to each codeword, and assigning the feature vector to the codeword that gives the minimum distance (distortion).

Step 4 Determine the new centroids of all vectors assigned to each codeword, and iterate the process (Steps 3 and 4) until the change in total distortion (as measured by the normalized sum of Euclidean distances of all feature vectors mapped to codebook centroids) becomes small, at which point the procedure is considered to have converged.

Steps 2–4 are iterated for codebooks of size 2, 4, 8, 16, 32, and 64, and the results of each codebook size should be saved in files of the form `vq_cc_tidig_endpt_vqsize_digits.mat`, where `vqsize` is 4, 8, 16, 32, and 64 for the five codebooks, and `digit` is [1-9ZO].

For each digit, plot the logarithm of the average distortion of each codebook as a function of the size of the codebook (in bits) on a single plot and show how the average distortion falls with increased codebook size from sizes 1 to 64 codewords per codebook.

11.32. (MATLAB Exercise) Write a MATLAB program to show the error weighting filter characteristics, $W(e^{j\omega})$, for a voiced speech frame as a function of the weighting parameter, γ, where the error weighting filter has the form:

$$W(z) = \frac{1 - A(z)}{1 - A(z/\gamma)}$$

$$= \frac{1 - \displaystyle\sum_{k=1}^{p} \alpha_k z^{-k}}{1 - \displaystyle\sum_{k=1}^{p} \alpha_k \gamma^k z^{-k}},$$

which has a frequency response of the form:

$$W(e^{j\omega}) = \frac{1 - \sum_{k=1}^{p} \alpha_k e^{-j\omega k}}{1 - \sum_{k=1}^{p} \alpha_k \gamma^k e^{-j\omega k}}.$$

Plot (on a single sheet) the LPC log magnitude spectrum (in dB) along with the error weighting log magnitude spectrum for a range of values of γ including $\gamma = [0, 0.5, 0.7, 0.8, 0.9, 0.95, 0.98]$. Test your program using the speech file `test_16k.wav` with voiced frame starting sample 6000, with frame length 640 samples, with LPC order 16 and using a Hamming window. What can you say about the effect of γ as it varies between 0 and 0.98.

11.33. (MATLAB Exercise) Write a MATLAB program to plot and compare the error weighting characteristics and the combined LPC with error weighting characteristics of a frame of voiced speech as a function of the weighting parameter, β. The LPC-based error weighting filter was shown to be of the form:

$$W(e^{j\omega}) = \frac{G(1 - \sum_{k=1}^{p} \alpha_k e^{-j\omega k})}{1 - \sum_{k=1}^{p} \alpha_k \beta^k e^{-j\omega k}}$$

and the combined frequency response of the LPC inverse filter and the error weighting was shown to be of the form:

$$W_C(e^{j\omega}) = \frac{G}{1 - \sum_{k=1}^{p} \alpha_k \beta^k e^{-j\omega k}}.$$

Using the range of values $\beta = 0.2{:}0.1{:}1$ for a frame of voiced speech, plot the log magnitude spectra of the original speech frame along with the weighting filter log magnitude responses (for the entire range of β) and the combined log magnitude spectra (again for the entire range of β). Use as the voiced frame the speech samples of `test_16k.wav` starting at sample 6000, using an analysis frame length of 640 samples, an LPC order of 16, and a Hamming window.

CHAPTER

12

Frequency-Domain Coding of Speech and Audio

12.1 INTRODUCTION

Frequency-domain coders are based on the principles of short-time Fourier analysis and synthesis as established in Chapter 7. The basic equation that defines the discrete short-time Fourier transform (STFT) is

STFT $$X_n(e^{j\omega_k}) = \sum_{m=-\infty}^{\infty} w_k[n-m]x[m]e^{-j\omega_k m}, \quad k = 0, 1, \ldots, M-1, \tag{12.1a}$$

where $X_n(e^{j\omega_k}) = |X_n(e^{j\omega_k})|e^{j\theta_n(\omega_k)}$ and $w_k[n]$ is the analysis window for the kth frequency.[1] Recall from our discussion in Chapter 7 that Eq. (12.1a) has the following equivalent interpretations: (1) as frequency down-shifting by ω_k followed by filtering with the lowpass impulse response $w_k[n]$; or (2) as bandpass filtering with impulse response $w_k[n]e^{j\omega_k n}$ followed by down-shifting by ω_k. Using either interpretation, the kth frequency value, $X_n(e^{j\omega_k})$, is a lowpass signal as a function of the time index n that represents the frequency spectrum in the band around ω_k. The short-time Fourier inverse transform (synthesis equation) is defined by

$$\hat{x}[n] = \sum_{k=0}^{M-1} P[k]\hat{X}_n(e^{j\omega_k})e^{j\omega_k n}, \tag{12.1b}$$

where $P[k] = |P[k]|e^{j\phi_k}$ is a complex scaling factor for the kth channel that can be used to ensure nearly perfect reconstruction of the input signal; i.e., $\hat{x}[n] = x[n]$ when $\hat{X}_n(e^{j\omega}) = X_n(e^{j\omega})$.

Figure 12.1 shows a general representation of how short-time Fourier analysis/synthesis is used in speech coding. The basic principle is simply to use the fact that $X_n(e^{j\omega_k})$ for fixed ω_k is a lowpass signal that can be sampled in the time dimension

[1] The reader will recall that we sometimes use the notation $\hat{n}$ to denote a specific time at which a short-time analysis quantity is computed, and we use the notation n or m when the variable is used as a running discrete time index.

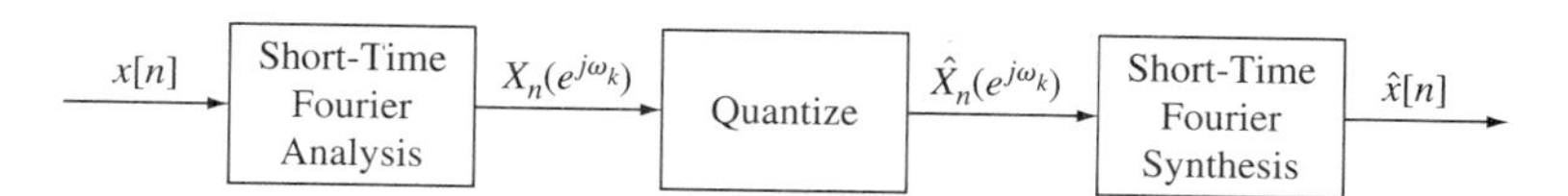

FIGURE 12.1
Digital coding of the STFT.

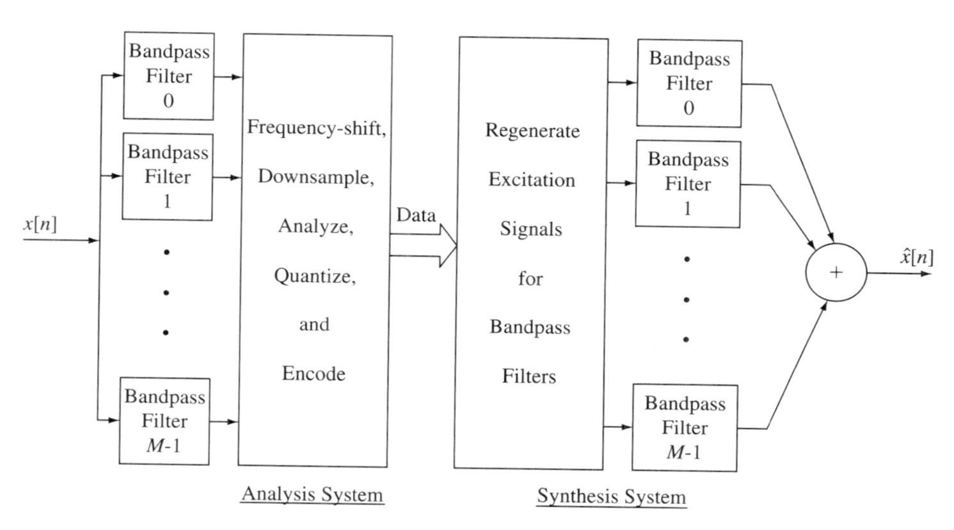

FIGURE 12.2
General frequency-domain coder for speech and audio.

at a lower rate than the input signal. When the time-sampled values of $X_n(e^{j\omega_k})$ are quantized, giving $\hat{X}_n(e^{j\omega_k})$, an approximation, $\hat{x}[n]$, can be recovered from $\hat{X}_n(e^{j\omega_k})$ by interpolation back to the input sampling rate followed by short-time Fourier synthesis as in Eq. (12.1b).

Figure 12.2 depicts the general nature of the large class of frequency-domain coding systems in more detail. On the far left of the diagram in Figure 12.2 is a set (bank) of M bandpass analysis filters. As discussed in detail in Chapter 7, with proper choice of the bandpass filters, the collective set of outputs, which we call the *channel signals*, can be precisely the same as the time-dependent Fourier transform $X_n(e^{j\omega})$ evaluated at a set of frequencies ω_k, $k = 0, 1, \ldots, M - 1$ as in Eq. (12.1a). Other filter sets can implement other spectral transformations. In general, the collective set of bandpass filter outputs is input to a processing stage labeled "frequency-shift, downsample, analyze, quantize, and encode," whose output is labeled "data." This simply suggests that the bandpass filter outputs can be used in many ways to derive a representation (the data) of the speech signal from which a "coded" signal, $\hat{x}[n]$, can be reconstructed by a bank of synthesis filters as depicted on the right half of the diagram in Figure 12.2.

This chapter will show that, depending on how the filter banks are set up and what operations are performed on the channel signals, the diagram of Figure 12.2 can represent many types of frequency-domain coders. Our discussion will begin with the channel vocoder, since it is the antecedent of all speech coders and therefore worthy of special mention even though it is rarely used today. The remainder of our discussion will focus on more recent developments in transform coding and subband coding because they are the most important examples of the use of frequency-domain methods in modern speech and audio coding.

12.2 HISTORICAL PERSPECTIVE

The oldest form of speech coding system is the channel vocoder, which was invented by Homer Dudley in the 1930s [92]. At the time of its invention, all communication signal processing systems were composed of analog filters, analog modulators, and other analog system components such as rectifiers and envelope detectors. Dudley's goal with the channel vocoder was to compress the analog bandwidth of speech so that more than one speech signal could be simultaneously transmitted over the existing 3.2 kHz bandwidth of telephone lines. He coined the word vocoder (*vo*ice *coder*) for his system, and that term has persisted to this day as a generic term for systems that "compress" the speech signal. Dudley demonstrated that it was possible to reduce the bandwidth while maintaining intelligibility of the speech signal, but the quality of the output of the channel vocoder was far from perfect. Nevertheless, Dudley's channel vocoder was a key milestone in speech communication science and engineering. As digital signal processing implementations of speech communication systems became an active topic of research in the 1960s, Dudley's basic principles were adopted in the design of the first digital speech compression systems [105, 130, 131, 351], and they remain at the core of modern speech science and technology.

12.2.1 The Channel Vocoder

The channel vocoder was a frequency-domain coder that attempted to incorporate properties of the speech production model into the analysis and synthesis configuration. From a modern perspective, it was a hybrid between what we are calling, in this chapter, a frequency-domain coder and the model-based coders that we discussed in Chapter 11. For simplicity, a number of approximations were introduced into the implementation of time-dependent Fourier analysis and synthesis. These approximations were forced on Dudley by the analog hardware that was available to him at the time, and later, the early digital channel vocoder researchers had to make do with limited computational power. In the latter case, an equally important limitation was that digital signal processing was a nascent field, with much knowledge about digital filters and discrete-time Fourier analysis yet to be uncovered.

To see, from a modern perspective, how the digital channel vocoder fits into the framework of Figure 12.2, note that since $x[n]$ and $\hat{x}[n]$ are real, it follows from Eq. (12.1b) that the reconstruction formula can be expressed as

$$\begin{aligned}\hat{x}[n] &= \mathcal{R}e\left\{\sum_{k=0}^{M-1} P[k]\hat{X}_n(e^{j\omega_k})e^{j\omega_k n}\right\} \\ &= \sum_{k=0}^{M-1} \hat{x}_k[n], \qquad (12.2a)\end{aligned}$$

where

$$\hat{x}_k[n] = |P[k]| \cdot |\hat{X}_n(e^{j\omega_k})| \cos(\omega_k n + \theta_n(\omega_k) + \phi_k). \tag{12.2b}$$

and

$$\hat{X}_n(e^{j\omega_k}) = |\hat{X}_n(e^{j\omega_k})| e^{j\theta_n(\omega_k)} \tag{12.2c}$$

Equation (12.2b) shows that each channel in short-time Fourier analysis/synthesis can be represented as a cosine of nominal frequency ω_k that is amplitude modulated by $|P[k]| \cdot |\hat{X}_n(e^{j\omega_k})|$ and phase modulated by $\theta_n(\omega_k) + \phi_k$. Furthermore, recall that the basic short-time Fourier analysis can be implemented by a bank of bandpass filters as suggested in Figure 12.2. To obtain $X_n(e^{j\omega_k})$, the output of the kth bandpass filter is down-shifted in frequency to the baseband, and then the resulting lowpass signal is sampled at a low rate determined by the bandwidth of the bandpass filter.

In the original channel vocoder, the combination of bandpass filtering and frequency shifting to baseband was approximated using the system shown in Figure 12.3. An approximation to $|X_n(e^{j\omega_k})|$ was obtained by envelope detection on the output of a bandpass filter with center frequency ω_k. The kth channel of the channel vocoder used a bandpass filter with impulse response such as $w_k[n]\cos(\omega_k n)$, where $w_k[n]$ is a lowpass filter impulse response. The bandpass filter was followed by a full-wave rectifier (the magnitude block) and a lowpass filter to smooth and bandlimit the magnitude output. The full-wave rectifier and lowpass filter served as an approximate envelope detector (AM demodulator), which in turn approximated the frequency down-shifting of the STFT. The processing of Figure 12.3 was the basic analysis component for each frequency band in the channel vocoder. The analyzer consisted of a bank of such channels, with analysis frequencies distributed either uniformly or non-uniformly across the speech band of interest as shown in Figure 12.4 assuming a digital implementation. Dudley's early implementation used analog components.

Of course, the speech signal cannot be represented by the amplitude spectrum alone. Without the short-time phases, $\theta_n(\omega_k) + \phi_k$ in Eq. (12.2b), the synthesized speech would contain only amplitude modulated sinusoids with frequencies equal to the center frequencies of the bandpass filters. If the filters were equally spaced, the synthesized output would sound much like continuously voiced speech with fixed (monotone) pitch, which is a highly undesirable result.

Rather than attempt to estimate the time-varying phase for each channel, a channel vocoder had an additional analysis component for determining the mode

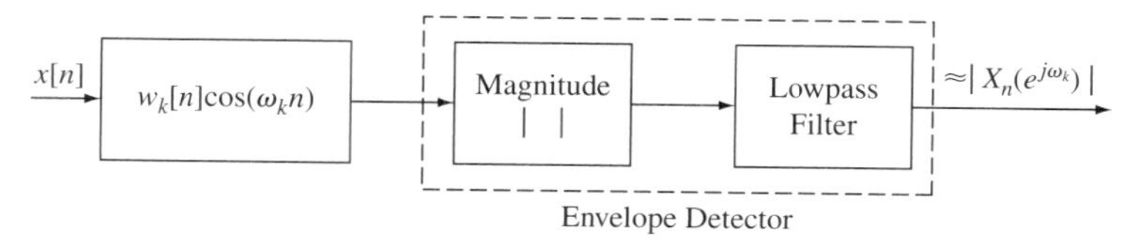

FIGURE 12.3
Method for approximating the short-time spectrum.

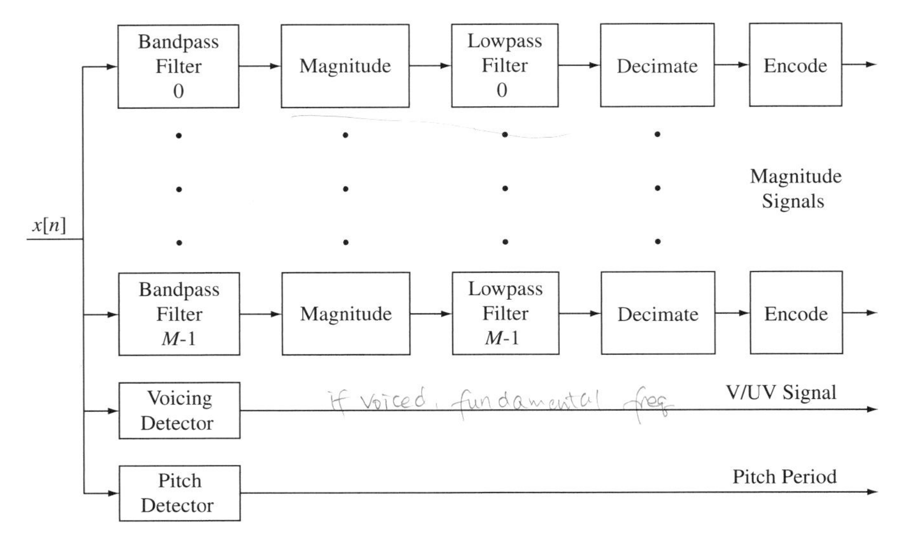

FIGURE 12.4
Block diagram of a digital channel vocoder analyzer.

of excitation, i.e., voiced or unvoiced, and if voiced, the fundamental frequency (or equivalently the pitch period) of the speech signal. These pitch and voicing parameters were estimated approximately every 10 msec so as to track the time-varying excitation properties of the speech signal. As depicted in Figure 12.4, the resulting excitation information, together with the amplitude channel signals, formed the representation for the speech signal. In a digital implementation, these parameters were sampled and quantized for transmission or storage.

The synthesizer used in a channel vocoder is depicted in Figure 12.5. The basic principle of channel vocoder synthesis can be simply stated. The channel signals controlled the amplitude of the contribution of a particular channel to the total output, while the excitation signals controlled the detailed spectral structure within the passband of a given channel. The voiced/unvoiced signal simply served to select an appropriate excitation generator; i.e., random white noise for unvoiced speech or a periodic pulse generator for voiced speech, with the fundamental frequency of the pulse generator being controlled by the pitch period data. Thus, the composite output spectrum was built up out of individual bands in which the amplitude within a given frequency band was roughly constant over short-time intervals. In fact, a particular band retained the frequency selective shaping properties of the bandpass filter used for synthesis of that band. If the excitation was voiced, then the output was composed of contiguous bands in which the fine spectral structure was characteristic of periodicity, while if the excitation was unvoiced, the spectrum was more or less flat across each band.

The resulting speech often sounded highly reverberant because of complete lack of control over the merging together of adjacent bands, which could result in nulls in the composite frequency response. This can be seen from Figure 12.6 (due to Flanagan [105]), which shows a comparison of a spectrogram of a speech input signal with a spectrogram of the corresponding output of a 15-channel vocoder. Note the blocky

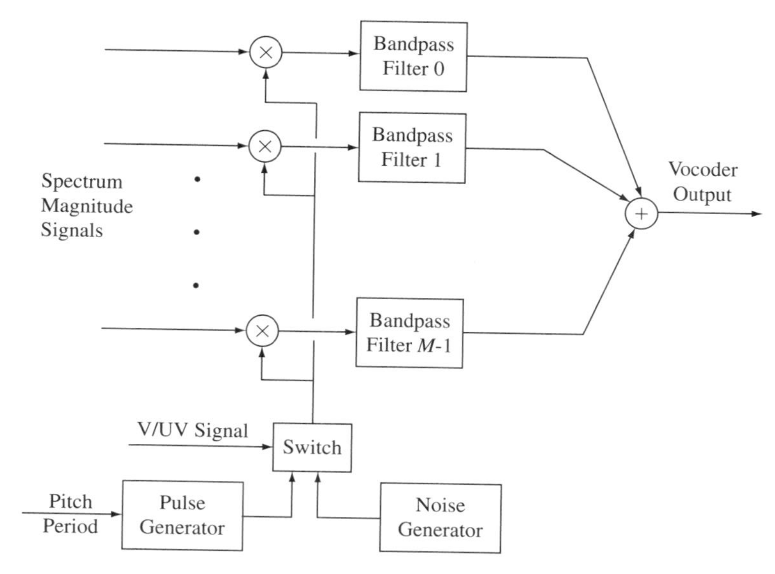

FIGURE 12.5
Block diagram of channel vocoder synthesizer.

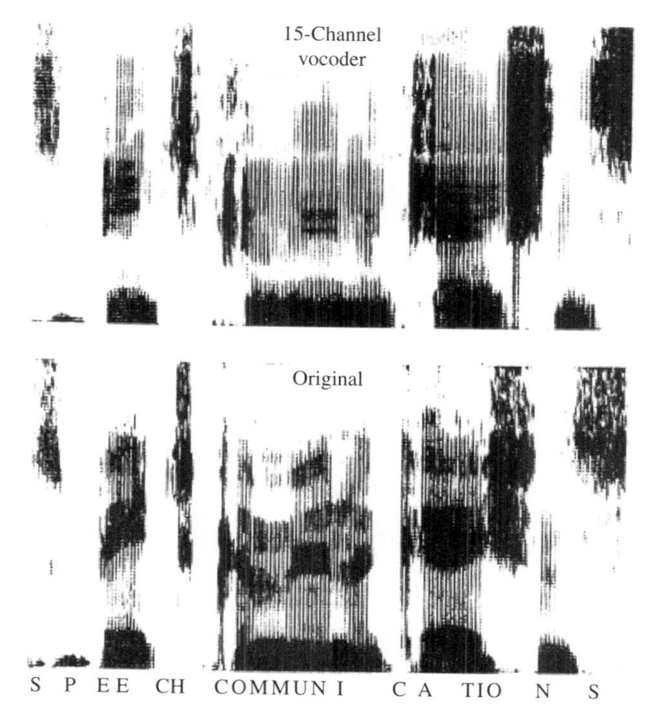

FIGURE 12.6
An example of a 15-channel vocoder. (After Flanagan [105].)

appearance of the spectrogram of the vocoder output. This results from the coarse spacing of channels. The formant information appears highly quantized and formant frequency variations are drastically altered in some cases. Digital channel vocoders typically operate in the range 1200 to 2400 bps, with roughly 600 bps devoted to the pitch and voicing information and the remaining information devoted to the channel signals. A major contribution to the reduction in bit rate that was achieved with a channel vocoder was the direct representation of the pitch and voicing information instead of the short-time phase. This, however, was one of the weaknesses of the channel vocoder system, since accurate detection of pitch and voicing is generally a difficult task. (Recall the discussion in Chapter 10.)

One feature of the channel vocoder was that, because it incorporated the speech production model, it permitted modification of the speech signal. This was because the excitation and vocal tract information were represented separately. It was easy to see, for example, how pitch could be changed independently of the vocal tract information (channel signals). For example, if the pulse generator always produced the same fundamental frequency, i.e., the pitch information is not utilized, then monotone speech would be produced. If no pulse generator excitation was used, but rather the excitation was always random noise, then the output sounded somewhat like whispered speech. Independent variations of the time and frequency scale could also be achieved using a channel vocoder simply by appropriate independent scaling of the center frequencies of the bandpass filters and the pitch period.

While channel vocoders could code speech intelligibly at bit rates on the order of 1200 to 2400 bps the fundamental weaknesses of the system could not be overcome by using more bits to quantize the parameters. In addition to the difficulties with voiced/unvoiced (V/UV) decisions and pitch detection, it is clear that the channel vocoder synthesizer was, in a sense, simply obtaining a crude representation of the spectrum envelope by using the smoothed outputs of a bank of bandpass filters. While this limitation might be overcome by using many bandpass filter channels (with concomitant increase in bit rate), it has turned out that there are much better ways of determining the spectrum envelope and applying it to the excitation and much better ways of determining the excitation. (See Chapter 11 on Model-Based Coding.) This does not mean, however, that frequency-domain representations cannot be used successfully for speech and audio coding. Indeed, as we shall see, the frequency-domain framework is well suited to incorporating knowledge of auditory masking into the coding process. However, taking full advantage of perceptual effects requires much more sophisticated digital signal processing than was used in channel vocoders.

12.2.2 The Phase Vocoder

The channel vocoder established the principle of building a speech coder around a short-time frequency analysis of the speech signal, but, as discussed above, discarding the phase and replacing it with information extracted according to the source/system model of the speech waveform is a fundamental limitation that cannot be overcome by more sophisticated filter banks or even more accurate pitch and voicing detection. As the techniques of digital filter design and digital signal processing progressed during the late 1960s and 1970s, it became clear that discarding the phase was neither desirable nor necessary.

The first significant attempt to utilize the phase of the STFT was reported by Flanagan and Golden [109], who called their system the *phase vocoder*. Their investigation was based on a digital simulation of a system that they defined and analyzed as if it were an analog system. From this point of view, they were led naturally to base their system on the time derivative of the phase rather than the phase itself. There were two reasons for this. First, the principal value phase limited to $[-\pi, \pi)$ or $[0, 2\pi)$ would contain discontinuities, and therefore would require a very wide analog bandwidth for transmission. Second, if the phase were unwrapped to a continuous function of time, it would be unbounded and therefore also unsuitable for transmission. This led them to compute a sampled version of the time derivative of the short-time phase for each channel [109]. By considering an equivalent bandlimited continuous-time signal, it was possible to define a phase derivative signal $\dot{\theta}_n(\omega_k)$ that was approximately the instantaneous frequency variation around the channel frequency ω_k. Using this approach, they were able to demonstrate that the bandwidth of an analog speech signal could be compressed by a factor of 2 without serious degradation [109]. They showed by their simulations that the phase derivative was bounded, and thus would be amenable to quantization in a digitally coded system. Although they did not fully implement and evaluate a digital speech coder based on the phase derivative, they estimated that a bit rate of under 10,000 bps would be possible for acceptable quality of speech reconstituted from quantized magnitude and phase derivative signals [109].

A detailed study of techniques for sampling and quantizing the magnitude and phase derivative signals in a phase vocoder was carried out by Carlson [52]. In that study, a 28-channel phase vocoder was implemented with a channel spacing of 100 Hz. Uniform quantizers were used for the phase derivative parameters and logarithmic quantizers were used for the magnitude parameters. Bits were distributed non-uniformly among the channels, with more bits being allocated to represent the lower channels and fewer bits for the upper channels. Also, more bits were allocated to the phase derivative than to the magnitude signals. By sampling the magnitude and phase derivative signals only 60 times/sec, and using 2 bits for the low frequency magnitude signals and 1 bit for the high frequency magnitude signals, and 3 bits for the low frequency phase derivative channels and 2 bits for the high frequency channels, a bit rate of 7200 bps was achieved. Informal tests found that speech represented in this way was comparable in quality to logarithmic PCM (pulse code modulation) representations at two to three times the bit rate [52].

As described in Chapter 7, computation of the phase derivative leads to a representation that allows manipulation of both the time and frequency dimensions of the speech signal so that, for example, the speech signal can be speeded up or slowed down without affecting the pitch and formants [109]. Although the phase vocoder has not been widely used for coding, this feature of the phase vocoder has made it attractive in areas such as music synthesis [247] and speech manipulation [297].

12.2.3 An Early Effort to Digitally Code the STFT

The phase vocoder was an important advance in understanding that discrete-time short-time Fourier analysis could be the basis for achieving bit-rate reduction without sacrifice of speech quality. Its success stimulated much research on short-time Fourier analysis/synthesis, which led to a more detailed understanding of discrete short-time Fourier analysis and multi-rate systems.

One of the earliest efforts to incorporate some of the new understanding of short-time Fourier analysis/synthesis into coding of speech was reported by Schafer and Rabiner [346]. In this work, the complex STFT was computed at equally spaced frequencies $\omega_k = 2\pi/128$ with an input sampling rate of $F_s = 12{,}194$ Hz. The effect of channel sampling rate on the quality of unquantized analysis/synthesis was explored along with several approaches to quantization of the channel signals. For example, one approach sampled the channel signals at a rate of 500 samples/sec and encoded the real and imaginary parts of $X_n(e^{j\omega_k})$ by 1-bit adaptive delta modulation. This corresponded to a bit rate of 28 kbps, since only 28 out of a possible 65 channels were coded for transmission. The 28 channels covered the band from about 100 to 2700 Hz. In another experiment, the same 28 channel signals were computed at a rate of 100 samples/sec and the log magnitude and phase angle were quantized with 3 bits and 4 bits respectively for channels 1 through 10, and with 2 bits and 3 bits respectively for channels 11 through 28.[2] The resulting bit rate was 16 kbps, with a "slight degradation" observed in the output of the decoder.

While work such as Ref. [346] demonstrated that the short-time Fourier representation had the potential to be the basis for speech coding, the initial results were not compelling. Nevertheless, interest grew in the frequency-domain approach due to the fact that it was known that (1) virtually perfect reconstruction could be achieved in the absence of quantization and (2) the frequency-domain approach appeared to be the ideal venue for incorporating the knowledge of auditory masking that was being established in parallel research efforts. Although advances in signal processing and the incorporation of perceptual models have resulted in the ability to encode speech with high quality at moderate bit rates, production-model-based coders can produce comparable quality at lower bit rates if the input is restricted to be a speech signal with good signal-to-noise ratio (SNR). However, as we will see, frequency-domain coders are dominant in the area of audio coding, where the highest quality reproduction of all types of sound (not only speech) is the goal.

12.3 SUBBAND CODING

Figure 12.7 shows a general block diagram of a subband coder with M channels, each channel consisting of a bandpass filter, a downsampler, a quantizer and encoder, a decoder, and a bandpass interpolator. The bandpass filters have passbands that cover the band of frequencies that are to be preserved in the coded signal. Usually the total band is roughly the entire baseband of the sampled input signal; i.e., frequencies 0 to $F_s/2$, one-half the sampling frequency of the input signal. The individual bandpass filter bandwidths are often all the same, although non-uniform bandwidths have obvious correlates with models for hearing. Since the channel signals are narrowband, they

[2] The "discontinuities" of the principal value phase are not a concern in a digital coding system since the principal value phase is sufficient with the magnitude to reconstruct the real and imaginary parts of the STFT as needed for reconstitution of the speech signal.

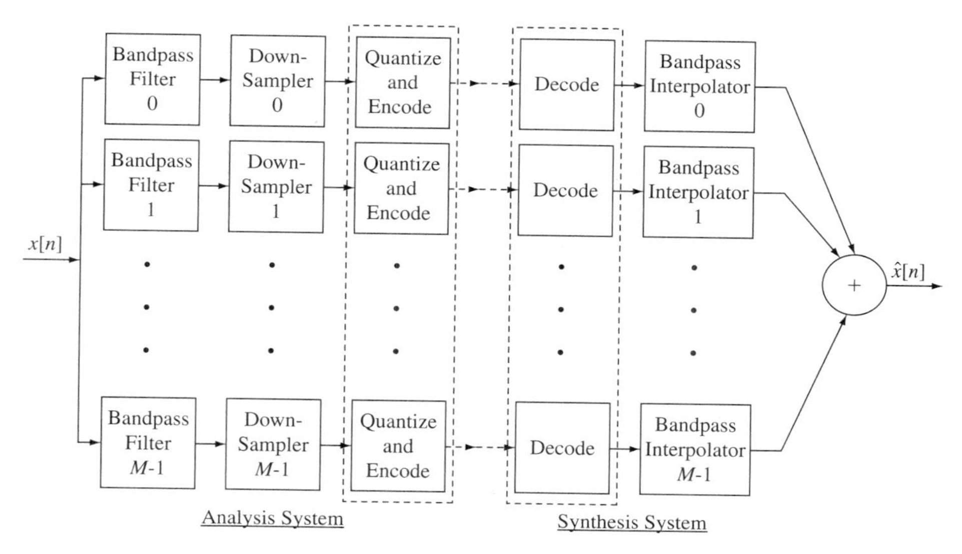

FIGURE 12.7
General block diagram of a subband coder.

can be sampled at a lower rate than the input sampling rate. In the particular case of bandpass filters with equal bandwidths spanning the entire baseband of the input speech signal, the downsampling by simple decimation of the channel signals, as suggested in Figure 12.7, implicitly implements the required frequency down-shift and creates a full band channel signal at the sampling rate F_s/M.[3]

In subband coders, the channel signals are quantized either separately or jointly by quantization techniques of the type that we have discussed primarily in Chapter 11. The quantized signals are also encoded for digital transmission or storage.

To reconstruct the coded speech signal $\hat{x}[n]$, the quantized channel signals are decoded and then fed as inputs to a set of "bandpass interpolators." Each such bandpass interpolator comprises an upsampler (by the corresponding decimation factor for the channel) followed by a bandpass filter that is generally identical in shape to the analysis bandpass filter. This bandpass interpolation operation creates channel signals at the original sampling rate and restores the channel signals to their appropriate passbands. If the filters are carefully designed, and the outputs of the downsamplers in the analysis system are connected directly [no quantization and perfect (lossless) encoding and decoding] to the corresponding inputs of the bandpass interpolators, then it is possible for the output, $\hat{x}[n]$, to be virtually identical to the input, $x[n]$ [71]. Such perfect reconstruction filter bank systems are the basis for the wide-ranging class of speech and audio coders called *subband coders*.

[3]It is easily shown that downsampling of a bandpass filtered signal is equivalent to modulation to DC followed by downsampling of the lowpass signal. (See Problem 12.1.)

When the outputs of the filter bank are quantized, encoded, and decoded, the output $\hat{x}[n]$ is not equal to the input, but by careful design of the quantizer, the output can be perceptually indistinguishable from the input. The goal of such coders is, of course, for the total composite bit rate[4] to be as low as possible while maintaining perceptually high quality.

In contrast to the channel vocoder and the linear predictive (LPC) vocoders discussed in this chapter and in Chapter 11, which incorporate the speech production model into the quantization process, the filter bank structure facilitates incorporation of features of the speech perception model. As discussed in Chapter 4, the basilar membrane, in effect, performs a frequency analysis of the sound impinging on the eardrum. The coupling between points on the basilar membrane results in the masking effects that were mentioned in Chapter 4. This masking was incorporated in some of the coding systems discussed so far, but only in a rudimentary manner. In Section 12.6, we will see that the subband coding framework allows the incorporation of models for masking into the quantization process.

We begin our discussion of subband coding by considering a two-band system based on ideal frequency-selective filters. Although this is the simplest possible subband coder, with little practical value, it demonstrates all the basic principles of subband coder operation, and it is straightforward to generalize to the M-channel case.

12.3.1 Example of an Ideal Two-Band Subband Coder

To illustrate the principles of subband coding, we consider a comparison between direct quantization of the sampled speech signal $x[n]$, as depicted in Figure 12.8, and a two-band subband coder as shown in Figure 12.9. Recall that the non-linear operation of quantization, as illustrated in Figure 12.8a. $\hat{x}[n]$ can be approximated by an additive noise model as depicted on the right in Figure 12.8b, where, for a B-bit uniform quantizer, the step size is $\Delta = 2X_m/2^B$, with $2X_m$ denoting the peak-to-peak range of amplitudes allowed by the quantizer without clipping. As we noted in Chapter 11, X_m, which is a property of the quantizer, is normally set proportional to the rms value of the input signal; i.e., $X_m = C\sigma_x$. For speech, a value of $C = 4$ guarantees that clipping of samples will be a rare event. When B is reasonably large and X_m is set to encompass the dynamic range of $x[n]$, we have seen in Chapter 11 that the quantization noise samples $e[n] = \hat{x}[n] - x[n]$ behave as if the following is true:

1. the noise is uncorrelated with the input,
2. the noise samples themselves are uncorrelated (white noise), and
3. the noise samples are uniformly distributed over the range $-\Delta/2 \leq e[n] < \Delta/2$.

With the above assumptions, the quantization noise has a flat spectrum over $-\pi$ to π, with total average noise power equal to $\sigma_e^2 = \Delta^2/12 = 2^{-2B}X_m^2/3$. For direct

[4]The total bit rate will be the sum of the products of the sampling rates of the downsampled channel signals times the number of bits allocated to each channel.

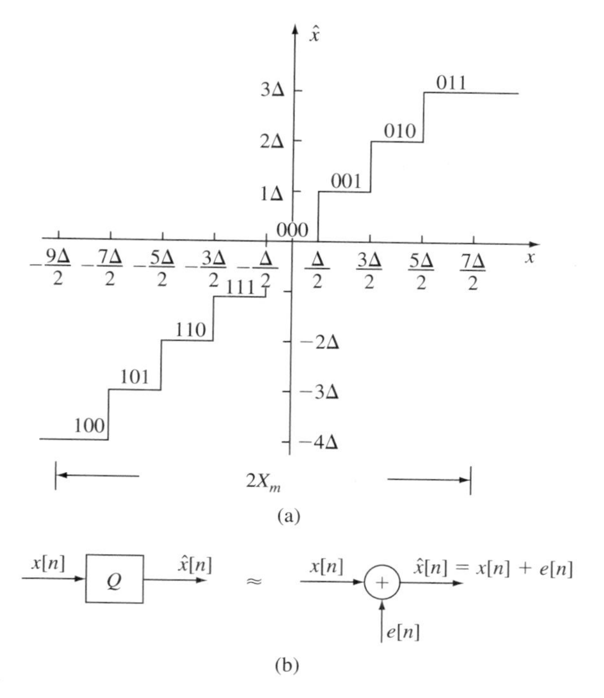

FIGURE 12.8
Additive noise approximation for quantization.

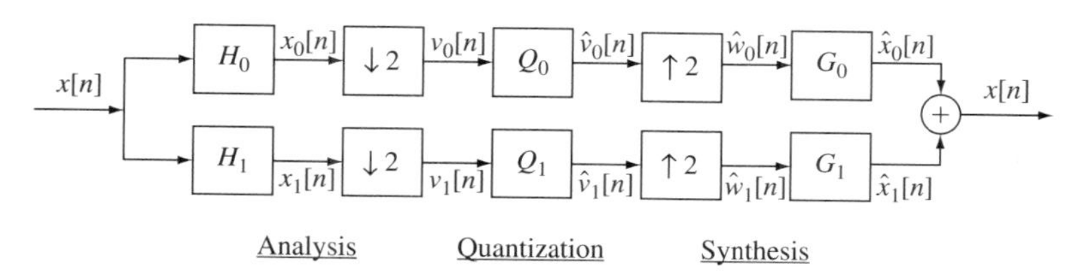

FIGURE 12.9
Block diagram of direct quantization of a two-band subband coder.

quantization, the SNR of the quantizer is therefore of the form

$$\mathrm{SNR}_Q = \frac{\sigma_x^2}{\sigma_e^2} = \frac{3 \cdot 2^{2B}}{C^2}. \tag{12.3}$$

Now consider the two-band subband coder in Figure 12.9. Assume that the filters are ideal filters as depicted in Figure 12.10, so that the spectrum of the input is split into two equal-width non-overlapping bands. Note that Figure 12.10 defines the ideal filters over the band of discrete frequencies $-\pi < \omega < \pi$, but the frequency responses are periodic, with period 2π, because they are discrete-time filters. The input filters

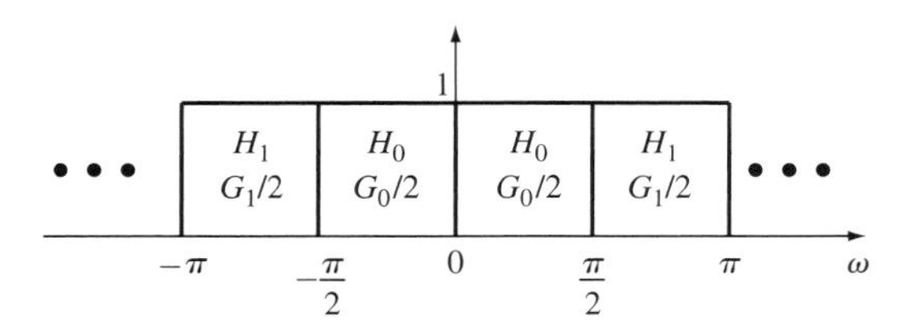

FIGURE 12.10
Ideal filters for two-band subband coder.

$H_0(e^{j\omega})$ and $H_1(e^{j\omega})$ have unit gain, while the output filters have a gain of 2 as required for 1:2 interpolation; i.e., $G_0(e^{j\omega}) = 2H_0(e^{j\omega})$ and $G_1(e^{j\omega}) = 2H_1(e^{j\omega})$.

Without quantization (i.e., $\hat{v}_0[n] = v_0[n]$ and $\hat{v}_1[n] = v_1[n]$), the back-to-back analysis/synthesis system gives perfect reconstruction; i.e., $\hat{x}[n] = x[n]$. To show this, it is useful to assume that $x[n]$ has a discrete-time Fourier transform (DTFT) $X(e^{j\omega})$. We can then trace the progress of the input signal through the system of Figure 12.9. The equations for the system (without quantizers) are

$$X_i(e^{j\omega}) = H_i(e^{j\omega})X(e^{j\omega}), \qquad i = 0, 1, \tag{12.4a}$$

$$V_i(e^{j\omega}) = \frac{1}{2}X_i(e^{j\omega/2}) + \frac{1}{2}X_i(e^{j(\omega-2\pi)/2}), \qquad i = 0, 1, \tag{12.4b}$$

$$\hat{V}_i(e^{j\omega}) = V_i(e^{j\omega}), \qquad i = 0, 1, \quad \text{no quantization} \tag{12.4c}$$

$$\hat{W}_i(e^{j\omega}) = \hat{V}_i(e^{j\omega 2}) = V_i(e^{j\omega 2}), \qquad i = 0, 1, \tag{12.4d}$$

$$\hat{X}_i(e^{j\omega}) = G_i(e^{j\omega})\hat{W}_i(e^{j\omega}), \qquad i = 0, 1, \tag{12.4e}$$

$$\hat{X}(e^{j\omega}) = \hat{X}_0(e^{j\omega}) + \hat{X}_1(e^{j\omega}). \tag{12.4f}$$

To trace the input through the system of Figure 12.9, begin with Eq. (12.4f) and successively substitute the previous equations until $\hat{X}(e^{j\omega})$ is expressed in terms of $X(e^{j\omega})$. The result is the following equation:

$$\begin{aligned}\hat{X}(e^{j\omega}) &= \frac{1}{2}\left[G_0(e^{j\omega})H_0(e^{j\omega}) + G_1(e^{j\omega})H_1(e^{j\omega})\right]X(e^{j\omega}) \\ &\quad + \frac{1}{2}\left[G_0(e^{j\omega})H_0(e^{j(\omega-\pi)}) + G_1(e^{j\omega})H_1(e^{j(\omega-\pi)})\right]X(e^{j(\omega-\pi)}) \\ &= H(e^{j\omega})X(e^{j\omega}) + \tilde{H}(e^{j\omega})X(e^{j(\omega-\pi)}).\end{aligned} \tag{12.5}$$

The coefficient of $X(e^{j\omega})$ in Eq. (12.5) is

$$H(e^{j\omega}) = \frac{1}{2}\left[G_0(e^{j\omega})H_0(e^{j\omega}) + G_1(e^{j\omega})H_1(e^{j\omega})\right], \tag{12.6a}$$

and the coefficient of $X(e^{j(\omega-\pi)})$ is

$$\tilde{H}(e^{j\omega}) = \frac{1}{2}\left[G_0(e^{j\omega})H_0(e^{j(\omega-\pi)}) + G_1(e^{j\omega})H_1(e^{j(\omega-\pi)})\right]. \tag{12.6b}$$

Using these equations, it follows that $\hat{X}(e^{j\omega}) = X(e^{j\omega})$ if

$$\tilde{H}(e^{j\omega}) = \frac{1}{2}\left[G_0(e^{j\omega})H_0(e^{j\omega}) + G_1(e^{j\omega})H_1(e^{j\omega})\right] = 1, \tag{12.7a}$$

and if

$$\left[G_0(e^{j\omega})H_0(e^{j(\omega-\pi)}) + G_1(e^{j\omega})H_1(e^{j(\omega-\pi)})\right] = 0. \tag{12.7b}$$

Equations (12.7a) and (12.7b) together are the conditions for perfect reconstruction of the input signal. A graphical argument easily verifies that these conditions are satisfied exactly for the ideal filters depicted in Figure 12.10.

To determine the SNR of a two-channel subband coder, we assume a random input signal[5] with total signal power σ_x^2. Figure 12.11 shows an illustrative example of two-band coding of such a signal.[6] For purposes of illustration, a "speech-like" power spectrum is shown in Figure 12.11a. The frequency responses of the channel filters are shown by the dotted lines in Figure 12.11a. Figures 12.11b and 12.11c show the spectra of the lowpass and highpass channels respectively. In this example, the signal powers in the two bands are clearly different, which is a key point. Because the filters are ideal, the two channel signals can be assumed to be independent so that the signal power is the sum of the powers in the two channels; i.e., $\sigma_x^2 = \sigma_{x_0}^2 + \sigma_{x_1}^2$. For convenience later, we represent the distribution of power between the two channels by defining $\sigma_{x_0}^2 = \alpha\sigma_x^2$ and $\sigma_{x_1}^2 = (1-\alpha)\sigma_x^2$, where $0 \le \alpha \le 1$. The decimated channel signals $v_i[n] = x_i[2n]$ have autocorrelation functions

$$\phi_{v_i}[k] = E\{v_i[n]v_i[k+n]\} = E\{x_i[2n]x_i[2k+2n]\} = \phi_{x_i}[2k], \tag{12.8}$$

which means that $\sigma_{v_i}^2 = \sigma_{x_i}^2$ for $i = 0, 1$. Furthermore, the power spectra of the decimated channel signals are

$$S_{v_i}(e^{j\omega}) = \frac{1}{2}S_{x_i}(e^{j\omega/2}) + \frac{1}{2}S_{x_i}(e^{j(\omega-2\pi)/2}), \quad i = 0, 1. \tag{12.9}$$

For the example of Figure 12.11a, $S_{v_0}(e^{j\omega})$ and $S_{v_1}(e^{j\omega})$ are shown in Figures 12.11d and 12.11e respectively.

[5]While Eqs. (12.7a) and (12.7b) were derived assuming that $x[n]$ is a deterministic signal, they also apply to random signal models as well.

[6]Note carefully that the normalized frequency range in all parts of the figure is $-\pi \le \omega < \pi$. However, in parts (d) and (e), the underlying sampling rate is $F_s/2$, while in all other parts, the sampling rate is F_s.

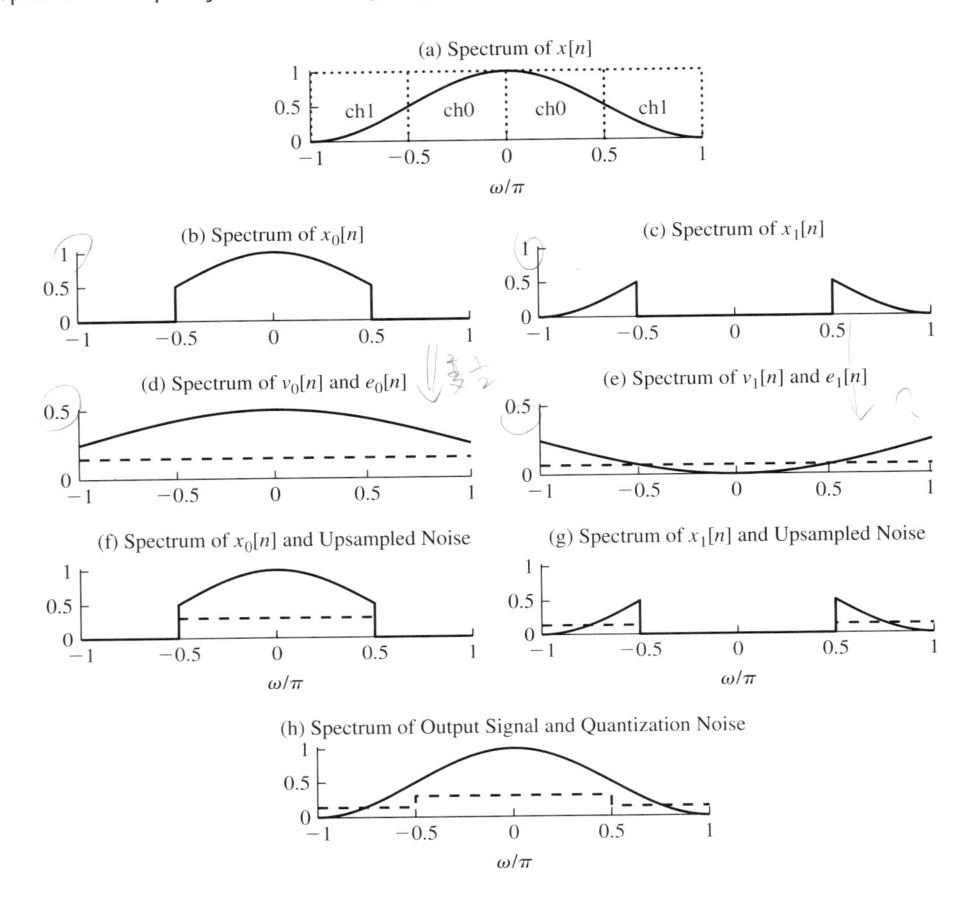

FIGURE 12.11
Illustration of spectra for two-band subband coder.

We can analyze the performance of the subband coder by replacing the quantizers Q_0 and Q_1 by additive noise approximations $e_0[n]$ and $e_1[n]$ [172, 399]. These noise sources propagate linearly through their corresponding channels and are superimposed at the output. This is one of the important features of subband coding: to the extent that the subband filters are good approximations to ideal filters, quantization noise incurred in a given channel affects only that same frequency band in the reconstructed output. Assume that the two quantizers have B_0 and B_1 bits respectively. The sampling rates of the two quantized channel signals $\hat{v}_0[n]$ and $\hat{v}_1[n]$ are both $F_s/2$ because of the decimation by 2. Therefore, the total bit rate of the subband representation is

$$I_{SBC} = B_0\frac{F_s}{2} + B_1\frac{F_s}{2} = \left(\frac{B_0 + B_1}{2}\right) F_s. \tag{12.10}$$

The bit rate of direct quantization of the sampled signal $x[n]$ as depicted in Figure 12.8b is simply $I_Q = BF_s$. One useful objective comparison is to equate the bit rates, i.e., set $B = (B_0 + B_1)/2$, and then determine how performance depends upon the distribution of the bits between the two channels. One convenient measure of performance is, of course, SNR at the output of the decoder.

Now, assume the quantizers are adjusted according to $X_{m_i} = C_i\sigma_{x_i}$ for $i = 0,1$. Then the noise powers at the output of the quantizers are $\sigma_{e_i}^2 = 2^{-2B_i}C_i^2\sigma_{x_i}^2/3$. The flat spectra of the two white noise sources are represented schematically in Figures 12.11d and 12.11e by the dashed lines. The upsampler followed by the filter precisely undoes the effect of the downsampling of the input, creating an additive component $x[n]$ in the output $\hat{x}[n]$. The noise sources $e_0[n]$ and $e_1[n]$ create corresponding noise signals $f_0[n]$ and $f_1[n]$ respectively at the output of the interpolators. The channel signals propagate to the output independently, with the results that $\sigma_{\hat{x}}^2 = \sigma_x^2 + \sigma_f^2$, where $\sigma_x^2 = \sigma_{x_0}^2 + \sigma_{x_1}^2$ and $\sigma_f^2 = \sigma_{e_0}^2 + \sigma_{e_1}^2$. The SNR at the output of the subband coder synthesizer is, therefore,

$$\text{SNR}_{SBC} = \frac{\sigma_x^2}{\sigma_{e_{SBC}}^2} = \frac{\sigma_x^2}{\sigma_{e_0}^2 + \sigma_{e_1}^2}. \tag{12.11}$$

By a simple manipulation, we can write SNR_{SBC} in the form

$$\text{SNR}_{SBC} = \left(\frac{\sigma_e^2}{\sigma_{e_0}^2 + \sigma_{e_1}^2}\right) \cdot \left(\frac{\sigma_x^2}{\sigma_e^2}\right) = G_{SBC} \cdot \text{SNR}_Q, \tag{12.12}$$

where the quantity G_{SBC} has the obvious definition and is called the subband coding gain. Assuming the same scale factor $C = C_i$ for all quantizers, substituting appropriate expressions for the quantization noise powers σ_e^2, $\sigma_{e_0}^2$, and $\sigma_{e_1}^2$, and recalling that we have assumed that $\sigma_{x_0}^2 = \alpha\sigma_x^2$ and $\sigma_{x_1}^2 = (1-\alpha)\sigma_x^2$, we obtain

$$G_{SBC} = \frac{2^{-2B}C^2\sigma_x^2}{2^{-2B_0}C^2\sigma_{x_0}^2 + 2^{-2B_1}C^2\sigma_{x_1}^2} = \frac{2^{-(B_0+B_1)}}{\alpha 2^{-2B_0} + (1-\alpha)2^{-2B_1}}, \tag{12.13}$$

where we have also used the fact that for comparison to direct quantization, $2B = (B_0 + B_1)$. Figure 12.12 shows plots of Eq. (12.13) for $B_0 + B_1 = 8$ and several values of α. Several important points are evident from this figure. First, note that if $\alpha = 0.5$, $\sigma_{x_0}^2 = \sigma_{x_1}^2 = 0.5\sigma_x^2$.[7] As shown by the solid line in Figure 12.12, the maximum subband coding gain in this case is 1 (0 dB) and it occurs when $B_0 = B_1$. Thus, if the spectrum is flat (or the signal power divides equally), subband coding has no SNR advantage over direct quantization of $x[n]$. At the other extreme is the case where $\alpha = 1$ (dashed

[7]One way that this could occur is if the power spectrum of $x[n]$ is perfectly flat; however, all that is assumed is that the total power divides equally between the low and high frequency bands. Further division of the bands could alter the power distribution if the spectrum were not flat.

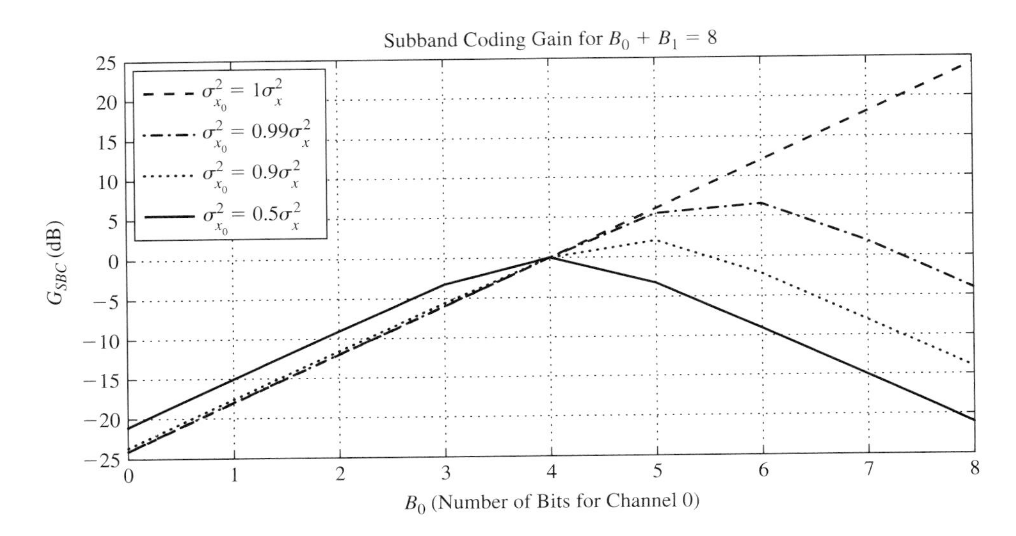

FIGURE 12.12
Coding gain for a two-band subband coder.

line in Figure 12.12). In this case, there is no power in the high frequency band, and by using all 8 bits to quantize the low band with sampling rate $F_s/2$, we gain $4 \times 6 = 24$ dB over quantizing $x[n]$ with sampling rate F_s and $B = 8/2 = 4$ bits/sample. Obviously, this is simply a matter of having over-sampled the input signal. The more interesting cases are when $\alpha = 0.9$ or $\alpha = 0.99$, where most, but not all, power is in the low band. These cases are shown by the dotted and dash-dot curves respectively in Figure 12.12. In these cases, the maximum value of G_{SBC} is greater than 1, and the maximum gain occurs when $B_0 > B_1$. The reason for the unequal distribution of bits is evident from Eq. (12.13). With the numerator fixed [$(B_0 + B_1)$ fixed], and if $\alpha > (1 - \alpha)$, then reducing the size of the denominator to increase G_{SBC} requires $B_0 > B_1$. It can be shown that the subband gain is maximum when B_0 and B_1 are chosen so that the quantization noise power is the same in both bands [172]; i.e., the same noise level results due to both subbands having the same scale factor ($C_0 = C_1 = C$). This means that the SNR would be lower in the high frequency band. This is in contrast to the choice $B_0 = B_1 = B$, where the SNR is the same in both bands.

We have seen that for signals like speech, which have a non-flat spectrum, the signal-to-quantization-noise ratio can be improved by decomposing the signal into disjoint frequency bands that are quantized differently. While our example, which was based on the assumption of ideal filters, suggests significant gains can result from subband coding of speech, to realize such gains in a practical coder, the following two issues must be addressed:

- the design of practical filter banks for analysis and synthesis
- the design of the quantization algorithm that maximizes perceived quality.

We discussed the design of filter banks in Chapter 7. In particular, in Section 7.8, we gave an extensive discussion of the design of two-channel filter banks and showed how such two-channel perfect reconstruction systems can be used in a tree structure to create filter banks with more than two channels with either uniform or non-uniform bandwidths. With Chapter 7 as background, we can proceed to consider the issue of quantization of the channel signals.

12.3.2 Quantizers for Subband Coding

The channel signals in a subband coder are quantized before being sent to the synthesis filter bank. This adds error (quantization noise) to the channel signals that will propagate to the output through the interpolation filter of the synthesis stage. In the example of the subband coder with ideal filters discussed in Section 12.3.1, the quantization noise was confined completely to the channel in which it was generated. This will be true in a practical setting to the extent that the filters have high attenuation in their stopbands and to the extent that the overlap region is small. While aliased signal components cancel completely in a properly designed filter bank, this is not true of the noise components since they are added after the decimation. Thus, reasonably sharp cutoff filters are desirable to minimize leakage of noise into the overlap region of the filters.

In its simplest form, as illustrated in Section 12.3.1, the subband coder simply quantizes the individual channel signals independently. If we use a fixed quantizer, as illustrated in Section 12.3.1, we have seen that if the spectrum of the input signal falls off at high frequencies, as in the case of speech signals, then the SNR can be improved by using different quantizers for each band. The results of Section 12.3.1 can be generalized to filter banks comprised of M equal bandwidth channels, which could be implemented as a tree structure as described in Section 7.8. In such a case, each of the channel signals is sampled at a rate of F_s/M so that the total bit rate will be the sum of the bit rates. If each channel is allocated the same number of bits, we have the same SNR in each channel. On the other hand, as in the two-channel example, the subband coding gain, G_{SBC}, is maximized by distributing bits to match the average power of the channels. Furthermore, it can be shown that if the spectrum falls off monotonically, the subband coding gain increases with increasing number of channels [172].[8] As in the two-channel case, the optimum improvement in SNR occurs when the quantization noise power is about the same in each band [172].

Generally, we want to use subband coding to reduce the bit rate while maintaining a given level of perceptual quality. To gain the most in performance, it is best to use some sort of adaptive quantization scheme. The simplest approach is to distribute the bits among the channels, allocating the fewest number of bits to the channels expected to have lowest power. Then an adaptive PCM quantizer such as Jayant's one-word memory system (see Section 11.5.2) can be used to ensure that the quantization error size tracks the size of the channel signal. Therefore bands with lower signal power will have smaller step sizes and commensurately lower quantization noise. Subbands with

[8] Increasing the number of channels leads to decreased time resolution.

higher power will have larger step sizes and therefore larger quantization noise. Even though the larger signal in the band will mask the noise, it is still better to use more bits for the high power bands.

Still another approach is to allocate the bits dynamically among the channels [68, 317]. This requires an algorithm for deciding where the bits can do the most perceptual good. Yet another approach is to apply vector quantization (see Section 11.6.2) to blocks of samples of the channel signals. Cox et al. [68] used a four-channel filter bank and formed 4 by 4 blocks (4 channels by 4 time samples in each channel) that were vector quantized.

There are many possibilities for quantizing the channel signals in a subband coder. In the next section we describe the performance of a simple five-channel coder designed for speech. Later, we study a subband coder designed for high-quality audio coding, where the quantization algorithm is grounded in perceptual masking theory.

12.3.3 An Example of Subband Speech Coding

The use of subband coding for speech signals originated in about 1976 [72, 100]. At this time, the theory of quadrature mirror filter banks (QMF) was developed and shown to provide nearly perfect reconstruction from subband signals [74]. A study by Daumer [80] comparing several digital speech coding algorithms included a subband coder (supplied by R. E. Crochiere) as represented in Table 12.1.

With a sampling rate of $F_s = 8000$ samples/sec, the filter bank was implemented by the tree structure of Figure 7.44 and the sampling structure depicted in Figure 7.45. The bits were allocated to optimize subjective performance. The quantizers for the channel signals were ADPCM (adaptive differential pulse code modulation) coders as described by Cummiskey et al. [77]. The ADPCM coders had fixed first-order predictors, with prediction coefficient values of $\{-0.71, -0.28, -0.31, 0.26, -0.64\}$ for channels 1–5 respectively. The Jayant adaptive quantizer with one-word memory was used for the quantizer in each ADPCM system.

Note that the 16 and 24 kbps coders did not use (allocate any quantization bits to) channel #5 as a means of lowering the overall bit rate at the sacrifice of signal bandwidth. Also note that more bits are allocated to the lower frequency channels (where most of the signal power is found in speech) rather than to the higher frequency

TABLE 12.1 Subband coder designs for 16, 24, and 32 kbps.

Band	Decimation Factor From 8 kHz	Band Edges (Hz)	Subband Sampling Rates (Hz)	(bits/sample) for I (kbps) = 16	24	32
1	8	0–500	1000	4	5	5
2	8	500–1000	1000	4	5	5
3	4	1000–2000	2000	2	4	4
4	4	2000–3000	2000	2	3	4
5	4	3000–4000	2000	0	0	3

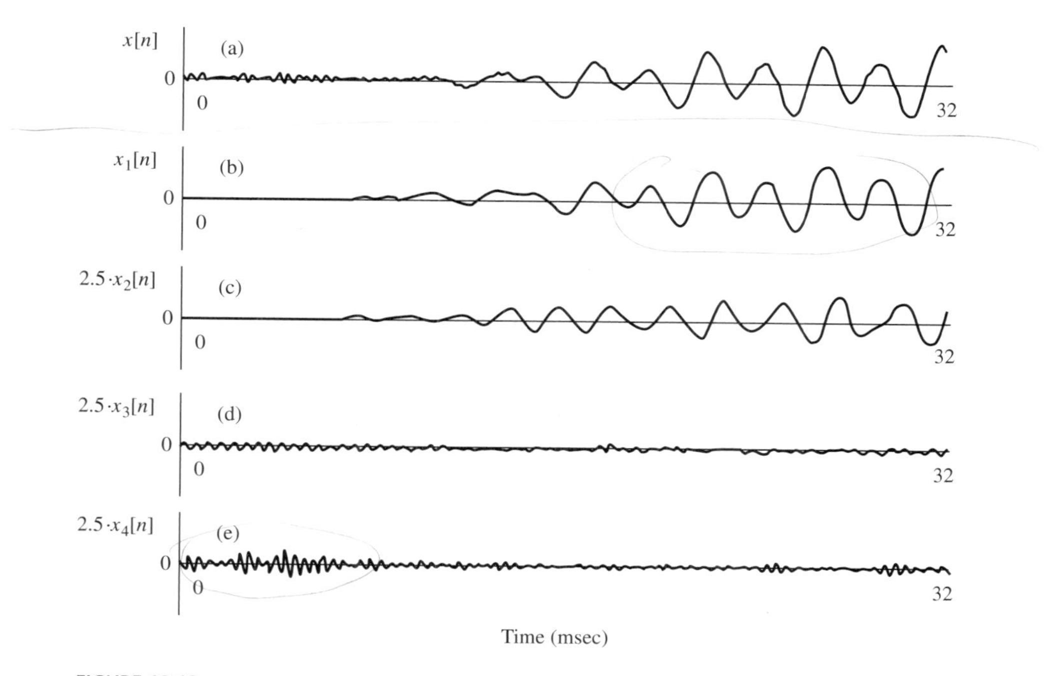

FIGURE 12.13
Waveforms of (a) original and (b–e) channel signals for a five-band subband coder. Jayant [172] shows these waveforms and attributes them to Cox's unpublished work.

channels. Figure 12.13 shows waveforms of the input and the first four channels. The plots show corresponding 32 msec segments of the signal; i.e., 256 samples of the input signal, 32 samples of channels 1 and 2, and 64 samples of channels 3 and 4. Observe that at the beginning of the interval, the speech is unvoiced and most of the energy appears in the upper channels, while at the end of the interval, the speech is voiced and the energy shifts to the lower channels. By using ADPCM with adaptive quantization, it is possible to track the varying signal amplitude and also to take advantage of the correlation that remains in the channel signals.

In Daumer's perceptual tests, the MOS scores were 3.11, 3.93, and 4.25 for 16, 24, and 32 kbps subband coders respectively. As a reference, 64 kbps $\mu = 255$ log-PCM encoding had an MOS score of 4.44. Thus, the 32 kbps subband coder is essentially indistinguishable from the 64 kbps log-PCM reference, and the 24 kbps subband coder was also judged to provide "good" quality. Subband coding is seen to provide significant reductions in bit rate with negligible sacrifice in perceptual quality. Note, however, that the quality drops dramatically in going from 24 to 16 kbps.

12.4 ADAPTIVE TRANSFORM CODING

As mentioned in Section 12.3.2, the subband coding gain increases as the number of channels increases (with corresponding decrease in channel width). Therefore, it is desirable to use a larger number of channels than we have discussed so far. Consider a subband coder with a filter bank that divides the input signal into M contiguous bands each of width π/M radians/sec in normalized frequency [$F_s/(2M)$ in analog frequency]. This can be accomplished with a full tree structure if M is a power of 2, or with a carefully designed M channel filter bank as discussed in Section 12.3. The result would be a set of M channel signals that are sampled in time at a rate of F_s/M. Thus, every MT seconds, we obtain a vector of M channel values. This set of M values represents M samples of the input signal.

If M is large and the filter bank is implemented by digital filtering, the computation cost can be prohibitive. However, another way of looking at this situation suggests a more practical alternative. Because of the downsampling, it is only necessary to compute the channel signals once every M samples of the input. In essence, blocks of speech signal samples of length corresponding to the length of the filter impulse response are transformed into a frequency-domain representation of size M, and the block boundaries are effectively moved in jumps of M samples by the downsampling operation. In the case of a filter bank analysis, the linear transformation is implemented through convolution with the impulse responses of the analysis filters, and the decimation by M produces a set of output samples every MT seconds. This is easy to do with FIR filters. However, there are many invertible linear transformations that could be applied on a block-by-block basis.[9] For example, we could use a linear transform such as the discrete cosine transform (DCT) to transform blocks of L samples selected at intervals of M samples. The blocks would overlap if $L > M$, as is usually the case. Figure 12.14 shows an illustration of how this can be done. The trapezoidal windows overlap by a small amount from block to block, but because of their shape,

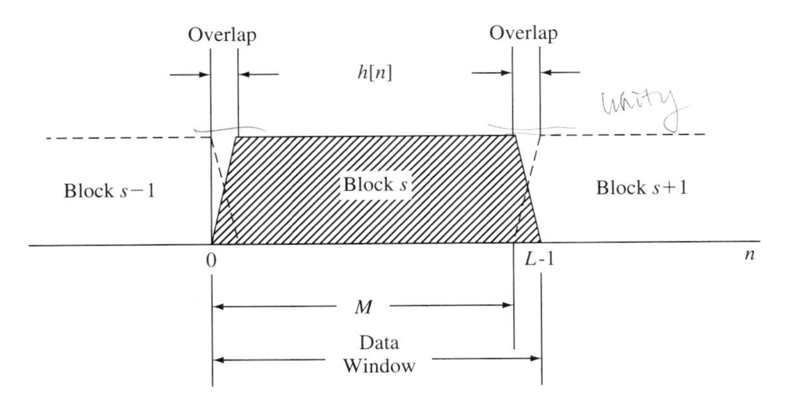

FIGURE 12.14
Illustration of block processing for speech coding [393].

[9]It is generally true that just as a downsampled filter bank analysis can be implemented by an appropriately defined transform; it is also true that transform analysis can be implemented by an appropriate filter bank. Generally the transform analysis is used when there is a reduction in computation complexity, which is the usual case with trigonometric kernel functions.

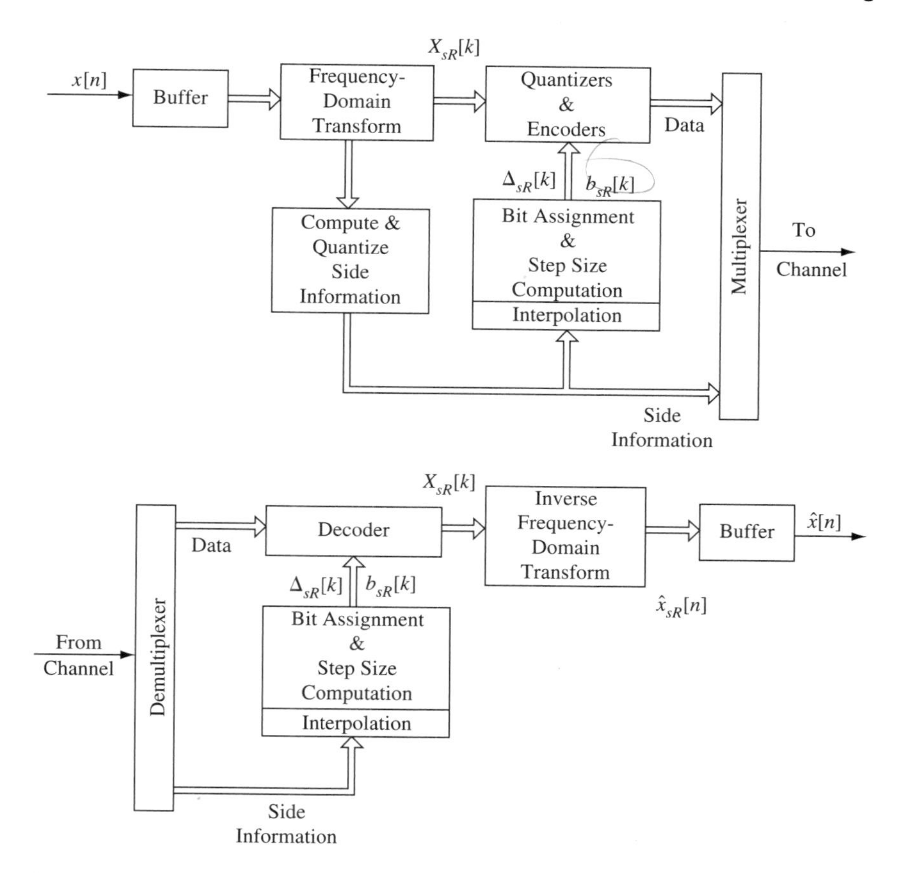

FIGURE 12.15
Adaptive transform coder and decoder [393].

their weighting sums to unity in the overlap region thus satisfying the OLA condition for exact reconstruction.

Thinking about frequency-domain coding in this way leads to the system depicted in Figure 12.15. In this approach, segments of the speech waveform are selected and weighted by a window such as the trapezoidal window of Figure 12.14 to form a block of samples of the form $x_{sR}[n] = w[sR - n]x[n]$, where s is the block index, R is the number of samples between blocks, and $w[sR - n]$ is the analysis window shifted to sample sR. A signal $\hat{x}[n]$ can be reconstructed from the blocks by overlapping and summing the blocks; i.e.,

$$\hat{x}[n] = \sum_{s=-\infty}^{\infty} x_{sR}[n] = \sum_{s=-\infty}^{\infty} w[sR - n]x[n] = x[n] \sum_{s=-\infty}^{\infty} w[sR - n]. \tag{12.14}$$

If the analysis window is properly chosen so that

$$\sum_{s=-\infty}^{\infty} w[sR-n] = 1, \tag{12.15}$$

then it follows that $\hat{x}[n] = x[n]$. The trapezoidal window of Figure 12.14 clearly satisfies the condition of Eq. (12.15).

In transform coding, the blocks $x_{sR}[n]$ are represented by an invertible linear transformation such as the DFT, the DCT, or the Karhunen-Loeve transform. (If the DFT is used, we have simply what we have called the short-time Fourier transform.) The transform is denoted $X_{sR}[k] = \mathcal{T}\{x_{sR}[n]\}$, where k is the index of the transform values. Since the transform is invertible, we can transmit it and then reconstruct the identical signal blocks as $x_{sR}[n] = \mathcal{T}^{-1}\{X_{sR}[k]\}$, which can be overlapped and added as in Eq. (12.14) to reconstruct the original signal.

Now if the transform values are quantized, we reconstruct the signal blocks with some error; i.e., $\hat{x}_{sR}[n] = \mathcal{T}^{-1}\{\hat{X}_{sR}[k]\}$. Using Eq. (12.14) to reconstruct the coded signal will result in a signal that differs from $x[n]$.

The key to transform coding is to quantize $X_{sR}[k]$ effectively so that the reconstructed signal $\hat{x}[n]$ differs imperceptibly from the original signal $x[n]$. The details of algorithms for coding the transforms are given in the fundamental papers of Zelinsky and Noll [425] and Tribolet and Crochiere [393] and in the text by Jayant and Noll [172]. The basic approach is depicted in Figure 12.15, which shows $b_{sR}[k]$, the number of bits of the quantizer for channel k, and $\Delta_{sR}[k]$, the corresponding step size for channel k (at frame s) as controls for the individual adaptive quantizers. These quantizer parameters are computed based on "side information" in the form of spectrum shape information that is coded and transmitted with the quantized transform values. This side information acts like the gain in an adaptive PCM coder for each analysis channel. Since the algorithm for deriving bit assignments and step sizes can be available at the receiver side, the side information is sufficient to allow $b_{sR}[k]$ and $\Delta_{sR}[k]$ to be derived at the receiver. Figure 12.16 shows an example [393] for one frame of speech. The upper plot shows the DCT (rapidly varying curve) and a smoothed estimate of the energy of the spectral bands. This energy distribution is sampled and quantized as the side information. Based on this information, the bits are assigned to each DCT value as shown in the middle plot. Note that many of the DCT values are given 0 bits, meaning that they are not transmitted. This is reflected in the bottom plot, which shows the DCT that is reconstructed from the quantized information. (The dotted curve is the original DCT for comparison.)

Tribolet and Crochiere [393] compared ATC to subband coding (SBC) and two forms of ADPCM. The results are shown in Figure 12.17 where ADPCM-F means fixed first-order predictor and ADPCM-V means eighth-order adaptive predictor. The different coders were rated on a 9-point scale with anchor signals that defined 1 as noisy and highly degraded and 9 being indistinguishable from the original sampled speech signal. The ATC coder (with 256-point transforms) was rated superior to all the others at all bit rates, and the subband coder was superior to the ADPCM systems at 16 and 24 kbps. Observe in particular that the curves for ATC and SBC are almost parallel,

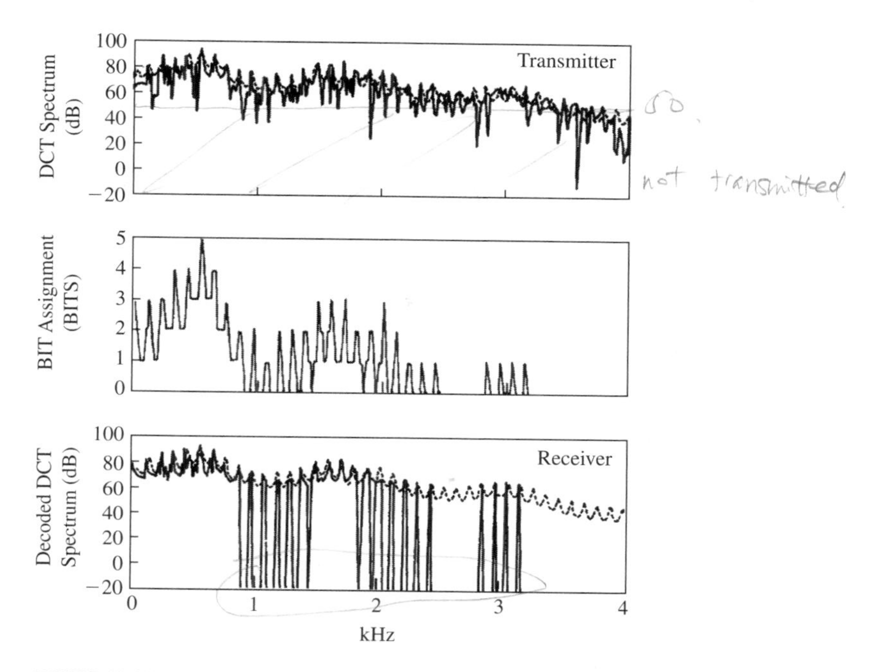

FIGURE 12.16
Illustration of bit allocation in adaptive transform coding (ATC) [393].

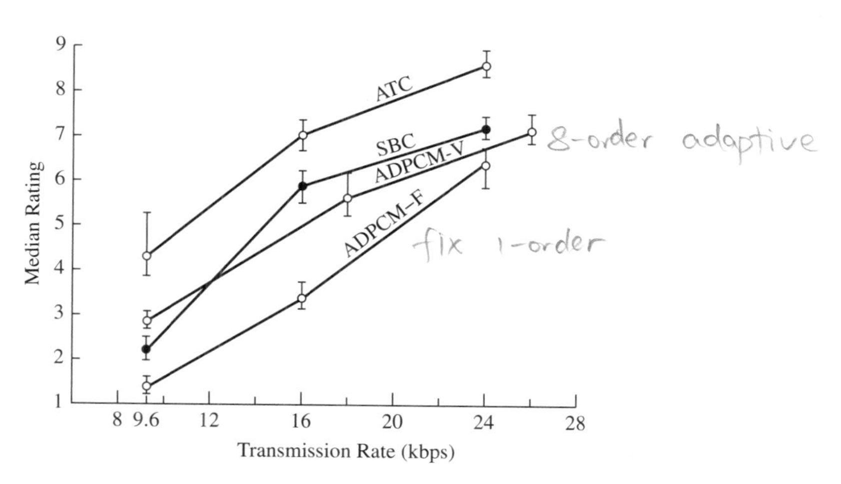

FIGURE 12.17
Median opinion scores for comparing speech coders [393–395].

which shows the perceptual gain of using a larger number of subbands (implied by the high order transform).

12.5 A PERCEPTION MODEL FOR AUDIO CODING

The key to effective coding of the channel signals in both subband coding and transform coding is to incorporate the knowledge of perceptual masking effects into the quantization algorithm. In many speech coders, the perceptual model only serves as a rough guide to the allocation of bits among the channels; e.g., noise feedback is used to shape quantization noise in a fixed spectrum-dependent manner. However, in coders designed for wideband audio signals, known results on auditory masking phenomena are explicitly incorporated into the quantization process. In this section, we illustrate how this can be done by describing a particular model for masking effects as employed in the ISO/IEC 11172-3 (MPEG-1) standard [156].

To illustrate how masking can be applied in coding of speech and audio signals, we consider the system depicted in Figure 12.18. This system is, in essence, the signal processing framework for a transform coder, and is very much like the system introduced by Johnston [176]. First the STFT is computed with relatively high frequency resolution (long time window). Then the STFT is analyzed to determine which (weaker) frequency components will be masked by other (stronger) components in the spectrum. It can be shown that it is possible to drastically modify (quantize) the short-time spectrum without audible distortion by simply setting to zero value (i.e., removing) all masked components. Such an experiment demonstrates that by taking account of the short-time threshold of audibility, it is possible to quantize a short-time spectral representation with minimal audible distortion. This is the basis for a large class of frequency-domain audio coders as used in the popular MP3 and AAC audio compression players.

12.5.1 Short-Time Analysis and Synthesis

The STFT in Figure 12.18 is computed at intervals of R samples and is defined as:

$$X_r[k] = \sum_{n=rR}^{rR+L-1} x[n]w[rR-n]e^{-j(2\pi k/N)n}, \quad 0 \le k \le N-1, -\infty < r < \infty, \qquad (12.16)$$

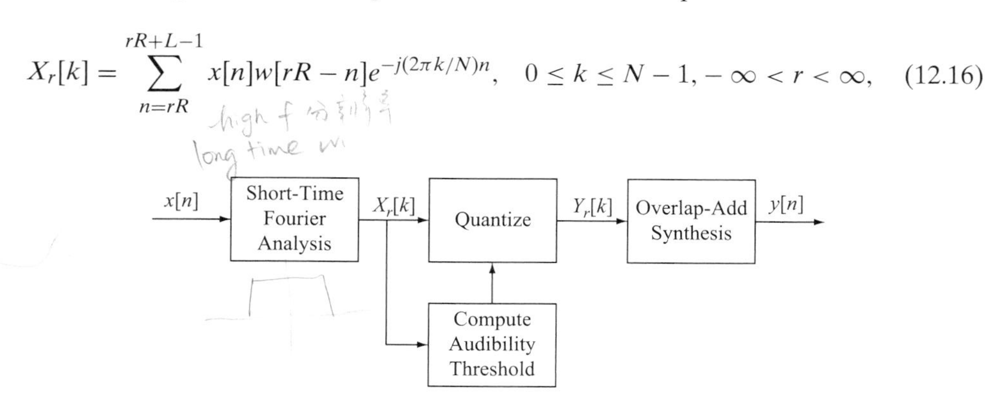

FIGURE 12.18
An experiment to illustrate the effects of masking in speech and audio coding.

where $w[n]$ is an L-point non-causal Hann window.[10] In general, the window length L is less than or equal to the DFT length N.

$$w[-n] = \begin{cases} 0.5(1 - \cos(2\pi n/L)) & 0 \le n \le L-1 \\ 0 & \text{otherwise.} \end{cases} \tag{12.17}$$

For sampling rate F_s, the discrete analysis frequencies $\omega_k = (2\pi k/N)$ in Eq. (12.16) are equivalent to the continuous-time frequencies $\Omega_k = 2\pi k F_s/N$. The MPEG standard [156] specifies $N = 512$ or 1024 for implementing the psychoacoustic model. Equation (12.16) can be evaluated efficiently at the N DFT frequencies by a fast Fourier transform (FFT) algorithm. If $X_r[k]$ is unmodified, the L-point windowed segments $x_r[n] = x[n]w[rR - n]$ can be recovered exactly by the inverse DFT; i.e.,

$$x_r[n] = x[n]w[rR - n] = \frac{1}{N}\sum_{k=0}^{N-1} X_r[k]e^{j(2\pi k/N)n}, \quad rR \le n \le rR + N - 1, \tag{12.18}$$

which can also be evaluated efficiently by FFT methods.

The overlap-add (OLA) method of synthesis shown in Figure 12.18 is used to reconstruct a signal $y[n]$ from the windowed segments in Eq. (12.18):

$$y[n] = \sum_{r=-\infty}^{\infty} x_r[n] = \sum_{r=-\infty}^{\infty} x[n]w[rR - n] = x[n]\sum_{r=-\infty}^{\infty} w[rR - n]. \tag{12.19}$$

If $X_r[k]$ is not modified, it was shown in Section 7.5 that with windows such as the Hann window, $y[n] = x[n]$. Clearly, the OLA synthesis can produce the original input for all n, for the window $w[n]$, if R is chosen such that

$$\tilde{w}[n] = \sum_{r=-\infty}^{\infty} w[rR - n] = 1, \quad \text{for all } n. \tag{12.20}$$

As shown in Section 7.5, a condition that guarantees $\tilde{w}[n] =$ constant for all n is

$$W(e^{j2\pi k/R}) = 0 \quad \text{for } k = 1, 2, \ldots, R-1, \tag{12.21}$$

where $W(e^{j\omega})$ is the DTFT of the window. If Eq. (12.21) holds, then it also follows that $\tilde{w}[n] = W(e^{j0})/R$ for all n. The DTFT of the N-point Hann window as defined in

[10] Observe that even though we view this as an L-point window, it has only $L-1$ non-zero samples, since $w[0] = 0$. In fact $w[n]$ can be thought of as an $(L-1)$-point symmetric window with a shift of 1 or an $(L+1)$-point symmetric window with $w[0] = w[-L] = 0$. The symmetry of the window makes it easy to compensate for delay. The non-causal definition is used so that Eq. (12.16) is easily related to the DFT of a finite-length sequence where the indexing is usually from 0 to $L-1$.

Eq. (12.17) is

$$W(e^{j\omega}) = \left(\frac{\sin(\omega L/2)}{\sin(\omega/2)} + \frac{\sin((\omega - 2\pi/L)L/2)}{\sin((\omega - 2\pi/L)/2)} + \frac{\sin((\omega + 2\pi/L)L/2)}{\sin((\omega + 2\pi/L)/2)} \right) e^{j\omega L/2}, \tag{12.22}$$

from which it follows that

$$W(e^{j\omega}) = \begin{cases} L/2 & \omega = 0 \\ 0 & \omega = \pm 4\pi/L, \pm 6\pi/L, \ldots, \pm\pi. \end{cases} \tag{12.23}$$

Thus, if $R = L/2$, $L/4$, $L/8$, etc., then $\tilde{w}[n] = 0.5L/R$ for all n, and therefore, if the STFT is unmodified, $y[n] = (0.5L/R)x[n]$.

The important point to be illustrated now is that even if the STFT is modified by quantization, the corresponding synthesized signal $y[n]$ can be perceptually indistinguishable from the original signal $x[n]$. To see how this result can be achieved, we must explain how to modify the discrete STFT (using results from psychoacoustics and perception studies) without significantly altering the resulting audio quality.

12.5.2 Review of Critical-Band Theory

We begin by reviewing some results from Chapter 4 (Section 4.3), where we found that many auditory phenomena can be explained by assuming that the auditory system analyzes sounds into frequency bands called *critical bands*. Figure 12.19 shows the relationship between critical bands and frequency [428]. The dots show center frequencies of critical bands selected to cover most of the frequency bands from 0 to 15,500 Hz, which in turn covers most of the baseband for an audio sampling rate

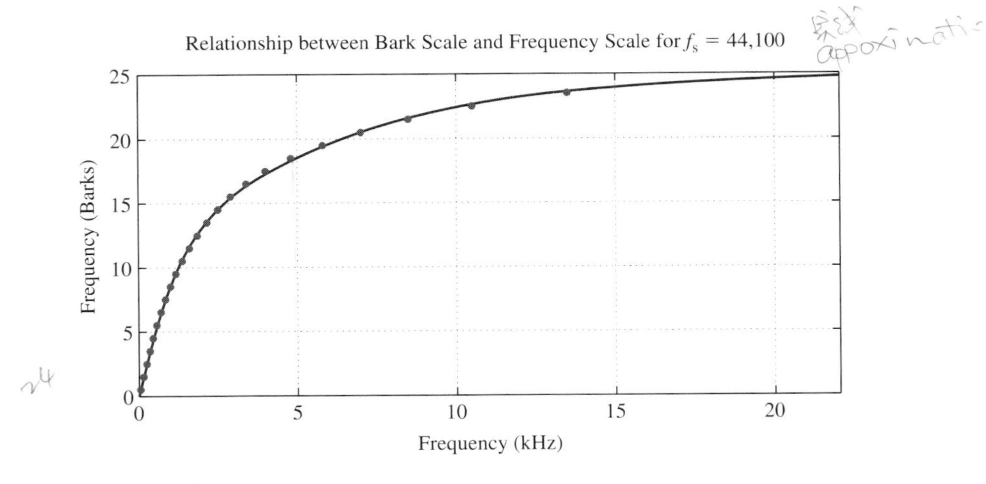

FIGURE 12.19
Relationship between critical bands (in Barks) and frequency (in kHz) [428].

of $F_s = 44{,}100$ Hz. These frequencies represent averages of measurements for many subjects. The solid curve is a graph of the equation

$$Z(f) = 13 \arctan (0.76f) + 3.5 \left(\arctan (f/7.5)\right)^2, \tag{12.24}$$

which is a good approximation to the experimental data [428]. Up to about 500 Hz, the critical bands are about 100 Hz wide. Above 500 Hz they become increasingly wider. By starting with the first critical band which spans the range from 0 to 100 Hz and centered on 50 Hz, the critical bandwidths can be added so as to cover an auditory band of interest. A set of 24 contiguous ideal critical-band-wide filters covers the range from 0 to 15,500 Hz as shown in Figure 12.19. Figure 12.19 and Eq. (12.24) define a critical-band rate scale which is given a unit called "Bark." One Bark is equivalent to one critical bandwidth, with each critical band centered on 1/2 Bark intervals as shown in Figure 12.19. Table 4.2 gives values for the critical band frequencies and their Bark equivalents.

As a first step toward relating the STFT to psychoacoustics, we note that Figure 12.19 and Eq. (12.24) represent a non-linear mapping of the frequency axis into the critical-band scale. This mapping can be applied to the discrete-frequency variable of the STFT in Eq. (12.16) to compute the equivalent Bark values; i.e.,

$$Z[k] = 13 \arctan (0.76kF_s/N) + 3.5 \left(\arctan ((kF_s/N)/7.5)\right)^2, \quad 0 \leq k \leq N-1. \tag{12.25}$$

Figure 12.20 shows an example of one frame of the STFT of a trumpet sound. The upper plot shows $P[k] = C + 20\log_{10}|X_r[k]|$ for one value of r (a single frame) plotted as a function of cyclic frequency $f_k = kF_s/N$ for $N = 512$. As will be discussed in Section 12.5.4, the constant C is chosen to calibrate the short-time Fourier spectrum on an assumed sound pressure level (SPL) scale. The lower plot shows the same $P[k] = C + 20\log_{10}|X_r[k]|$ plotted as a function of $Z[k]$. Note the compression of the high frequency part of the scale. This means that the lower critical bands have many fewer DFT samples than higher critical bands, which are much wider in frequency.

12.5.3 The Threshold in Quiet

As discussed in Section 4.4 of Chapter 4, the auditory threshold in quiet is determined by presenting pure tones of varying amplitude to a subject and asking the subject to indicate when they hear the sound. By an iterative level-adjustment process, it is possible to converge on the minimum SPL at which a tone of a given frequency can be heard. Recall that SPL is defined as $SPL = 20\log_{10}(p/p_0)$, where p is the rms sound pressure and $p_0 = 20 \times 10^{-6}$ Pascal is approximately the rms sound pressure at the threshold of audibility in the most sensitive range of hearing (2–5 kHz). One of the most important properties of human hearing is that the threshold of audibility varies with frequency. Measuring the dependence of the threshold of audibility on frequency requires very careful calibration and low noise conditions. The variation with frequency of the average threshold in quiet, $T_q(f)$, for listeners with acute hearing is

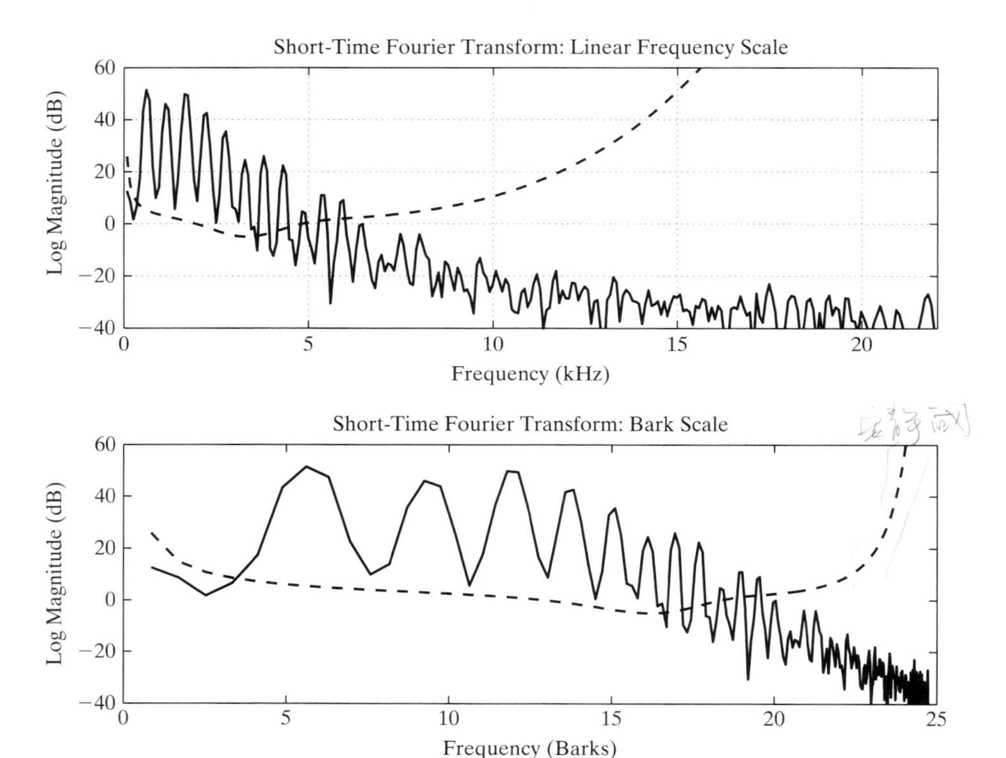

FIGURE 12.20
Example of STFT plotted versus ordinary cyclic frequency (upper plot) and on the Bark scale (lower plot). The dashed lines show the threshold of audibility (loudness level) versus frequency.

well represented by the expression [389]

$$T_q(f) = 3.64(f/1000)^{-0.8} - 6.5e^{-0.6(f/1000-3.3)^2} + 10^{-3}(f/1000)^4 \quad \text{(dB SPL)}. \quad (12.26)$$

The threshold of audibility is plotted (as a dashed curve) as a function of frequency in the upper plot of Figure 12.20 and as a function of critical-band frequency in the lower plot. Note that the threshold curve dips slightly below 0 dB in the most sensitive range (i.e., 2–5 kHz).

12.5.4 SPL Calibration of the STFT

In order to apply the audibility threshold to digital audio coding, it is necessary to calibrate the digital representation of an audio signal in terms of SPL; i.e., we need to convert the STFT into an "SPL spectrum." Without some assumptions, this is impossible to do since the sample values of the STFT have no intrinsic units. The same set of signal samples can produce sounds of widely varying intensities simply by altering the gain of the playback system. For this reason, it is common to create an arbitrary reference point by assuming that a 4 kHz sampled tone with amplitude ± 1 quantization level will create an SPL after playback that is just audible; i.e. about 0 dB SPL [374].[11]

The first step toward calibrating the short-time spectrum in terms of SPL is to recall that the average power of a continuous-time sinusoid $s(t) = A\cos(2\pi f_0 t)$ is $A^2/2$. Therefore, a sampled sinusoid $s[n] = \tilde{A}\cos(2\pi f_0 n/F_s)$ should also have average power $\tilde{A}^2/2$.[12] A further assumption is that the samples $s[n]$ are quantized to $B = 16$ bits. If we think of these B-bit numbers as fractions, then the maximum value of $\tilde{A}$ is 1, and the lowest amplitude is 1/32,768. We seek a conversion factor between the digital number $\tilde{A}$ and the corresponding amplitude A of the analog signal. An estimate of the average power based on a block of N samples is defined as

$$\text{ave. pwr} =< (s[n])^2 >= \frac{1}{N}\sum_{n=0}^{N-1}(s[n])^2 = \frac{1}{N^2}\sum_{k=0}^{N-1}|S[k]|^2, \tag{12.27}$$

where the second expression in Eq. (12.27) follows from Parseval's theorem for the DFT [270]. If $f_0/F_s = k_0/N$, where k_0 is an integer (i.e., k_0/N is exactly a DFT frequency), then the DFT of $s[n]$ is

$$S[k] = \begin{cases} \tilde{A}N/2 & k = k_0, N - k_0 \\ 0 & \text{all other } k \text{ in } 0 \le k \le N-1, \end{cases} \tag{12.28}$$

from which it is easily seen that

$$< (s[n])^2 >= (|S[k_0]|^2 + |S[N-k_0]|^2)/N^2 = 2|S[k_0]|^2/N^2 = \tilde{A}^2/2. \tag{12.29}$$

Note that the total power of a sine wave is the sum of the powers of the positive and negative frequency complex exponential components.

The above discussion suggests that if we start with an audio signal $s[n]$, we can normalize the block of samples by dividing by N, giving $x[n] = s[n]/N$ so that $|X[k_0]|^2 = |S[k_0]|^2/N^2$. Therefore $2|X[k_0]|^2$ will be equal to the total average power at discrete frequency k_0.

Another factor to be accounted for is the effect of the window (an N-point Hann window in this case) used in short-time Fourier analysis. For the sinusoid at DFT frequency k_0/N, it can be shown [42, 156] that if the Hann window varies slowly with respect to the oscillations of the sinusoidal signal, the average power will be reduced

[11] An alternative assumption, which gives comparable results, is to assume that a full-scale 16-bit-quantized sinusoid is reproduced at $SPL = 6 * 16 = 96$ dB [42, 156].

[12] In general, a computed estimate based on the average of N squared sample values will be close to $\tilde{A}^2/2$ if the averaging is over several cycles of the sine wave.

by a constant factor of 3/8 (i.e., independent of N):

$$\frac{1}{N}\sum_{n=0}^{N-1}(x[n]w[n])^2 \approx \frac{1}{N}\sum_{n=0}^{N-1}(x[n])^2\frac{1}{N}\sum_{n=0}^{N-1}(w[n])^2 = (3/8)\frac{1}{N}\sum_{n=0}^{N-1}(x[n])^2. \qquad (12.30)$$

Considering all of the above factors, and assuming that $x[n] = s[n]/N$, where $|s[n]| \leq 1$, we can write the short-time SPL spectrum in dB as

$$P_r[k] = C + 20\log_{10}|X_r[k]|, \qquad (12.31)$$

where the term C is a constant (in dB) that converts from the numeric scale of the STFT to SPL and r is the current frame number. In other words, C can include the effect of the window and assumed playback level. Typically, C is on the order of 90 dB [42, 374]. This is consistent with a full-scale playback level of approximately 90 dB SPL and also with a least significant bit amplitude being barely audible.

The short-time SPL spectrum shown in Figure 12.20 was computed according to Eq. (12.31) with $C = 20\log_{10}(2^{15}) = 90.309$ dB [374]. Observe that with this assumed playback level, none of the frequencies above about 6 kHz will be audible. Also note that increasing C (equivalent to turning up the gain) will shift the spectrum up with respect to the threshold in quiet, thereby making more of the high frequency components audible.

12.5.5 Review of Masking Effects

It is well established that certain signals can make other signals inaudible [428]. This phenomenon is called *masking* and was discussed briefly in Section 4.4 of Chapter 4. Figure 12.21 illustrates masking of a tone by another tone. The dashed curve is the threshold of audibility in quiet, and the solid curves represent the amount that the threshold of audibility is raised for frequencies in the vicinity of a masking tone of frequency 1 kHz and SPL (L_M) as indicated on the curve. For example, the curve labeled

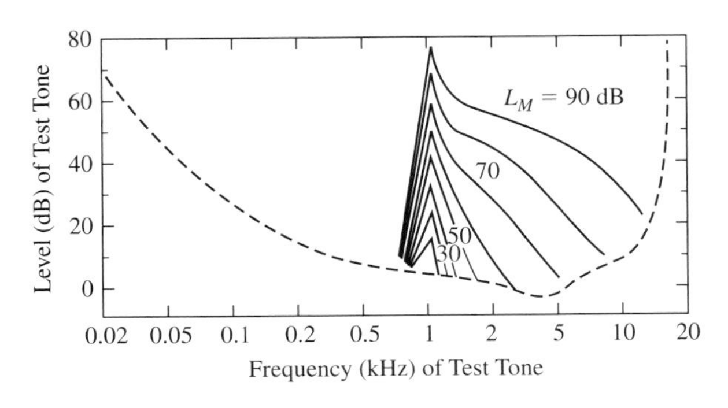

FIGURE 12.21
Level of test tone masked by 1 kHz tones of different level as a function of frequency of the test tone [428].

$L_M = 90$ dB shows that for a 1 kHz masker of 90 dB SPL, a simultaneous 2 kHz tone must have SPL of over 50 dB in order to be heard as a separate tone. It is said that the masking effect "spreads" both above and below the frequency of the masker tone, and the amount of threshold shift (as a function of frequency for the masked tone) is represented by a "spreading function." As shown by the curves of Figure 12.21, the spreading function decreases sharply for frequencies below the masker frequency (less spreading) and less sharply for frequencies above the masker frequency (more spreading). Also, below the masking frequency, the spreading function becomes sharper with increasing SPL of the masker, while the opposite is true for frequencies above the masking frequency.

In another class of masking experiments, it is found that tones can be masked by noise. Figure 12.22 shows an example of this phenomenon. In this example the masker is a critical-band-wide noise centered at 1 kHz.[13] The SPL of the noise masker (L_{CB}) is indicated on the curves, which show the threshold shift for masking tones of frequencies above and below 1 kHz. Note that the curves have the same general shape as those in Figure 12.21, but the peak masking effect is less in the tone-masking-tone case; i.e., noise is a more effective masker than tones.

Masking is exploited in audio coding by estimating the masking effect from the short-time spectrum of the audio signal and distributing quantization bit levels according to the degree of masking at each spectral component. The approach used in the MPEG audio standard is to identify signal components in the short-time spectrum that can have a significant masking effect. Then an approximation to the masking effect of each component is determined from an approximation to the masking curves shown in Figures 12.21 and 12.22. The individual masking effects are then combined to obtain a

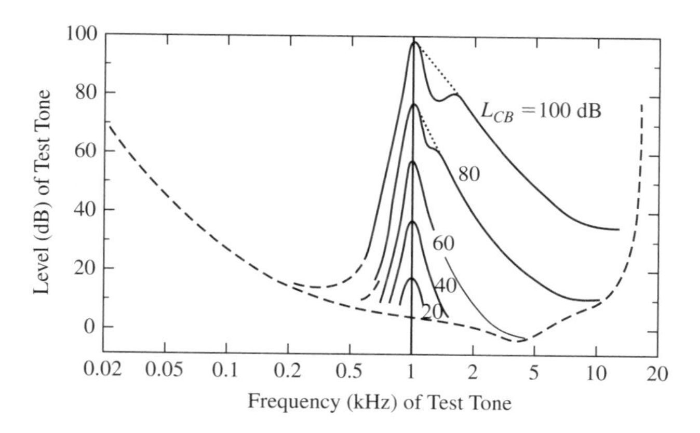

FIGURE 12.22
Level of just-masked test tone by critical-band-wide noise (with center frequency of 1 kHz) as a function of the frequency of the test tone [428].

[13] The critical bandwidth centered on 1 kHz is 920 to 1080 Hz. This critical band is at 9 on the Bark scale.

threshold of audibility that applies during the time interval spanned by the window. In the MPEG standard coder, this threshold is used to guide the quantization of a filter bank representation of the signal.

Psychoacoustic Model 1, as provided by the ISO/IEC 11172-3 (MPEG-1) standard [156], is based on approximations to the spreading functions illustrated in Figures 12.21 and 12.22. These approximations are specified most conveniently on the critical-band scale. However, note that we wish to obtain a masking function defined as a function of normal cyclic frequency so that it can be applied to the DTFT at the set of frequencies $f_k = kF_s/N$. Normally we would refer to these frequencies by the index k. The corresponding Bark scale frequency would be obtained from Eq. (12.24) as $Z[k] = Z(kF_s/N)$. Now suppose that $P_{TM}[k_{tm}] = P_r[k_{tm}]$ represents the power (SPL in dB) at discrete frequency k_{tm} in frame r of an STFT component that has been identified as a tonal masker (TM). Then the MPEG Psychoacoustic Model 1 specifies that the masking contribution at maskee frequency k due to the masker located at k_{tm} is

$$T_{TM}[k, k_{tm}] = P_{TM}[k_{tm}] - 0.275Z[k_{tm}] + S[k, k_{tm}] - 6.025 \quad \text{(dB SPL)}, \qquad (12.32)$$

where the spread of the masking effect from discrete masker frequency k_{tm} to surrounding maskee frequencies k is approximated by the following equation:

$$S[k, k_{tm}] = \begin{cases} 17\Delta Z[k, k_{tm}] - 0.4P_{TM}[k_{tm}] + 11 & -3 \le \Delta Z[k, k_{tm}] < -1 \\ (0.4P_{TM}[k_{tm}] + 6)\Delta Z[k, k_{tm}] & -1 \le \Delta Z[k, k_{tm}] < 0 \\ -17\Delta Z[k, k_{tm}] & 0 \le \Delta Z[k, k_{tm}] < 1 \\ (0.15P_{TM}[k_{tm}] - 17)\Delta Z[k, k_{tm}] & \\ -0.15P_{TM}[k_{tm}] & 1 \le \Delta Z[k, k_{tm}] < 8, \end{cases} \qquad (12.33)$$

where $\Delta Z[k, k_{tm}] = Z[k] - Z[k_{tm}]$ is the Bark scale frequency with origin shifted to the masker Bark frequency. Examples of Eq. (12.33) with Eq. (12.32) are shown in Figure 12.23 for a masker at $Z[k_{tm}] = 10$ (equivalent to 1175 Hz). Although Figures 12.21 and 12.23 are plotted on different frequency scales, it should be clear that Eq. (12.33) is a reasonable approximation to the spreading effect.

In the MPEG standard Psychoacoustic Model 1, the spreading function for noise maskers (NM) is assumed, for convenience, to be identical to Eq. (12.33). However, the threshold for noise masking is defined as

$$T_{NM}[k, k_{nm}] = P_{NM}[k_{nm}] - 0.175Z[k_{nm}] + S[k, k_{nm}] - 2.025 \quad \text{(dB SPL)}. \qquad (12.34)$$

In this case $P_{NM}[k_{nm}]$ represents the total noise power in a critical-band-wide set of frequencies around the frequency k_{nm}.

12.5.6 Identification of Maskers

In order to apply Eqs. (12.32) and (12.34) to determine the overall threshold shift due to masking, it is necessary to identify all the maskers in the STFT. Since most audio signals have complicated spectral properties, this process must be heuristic. It is

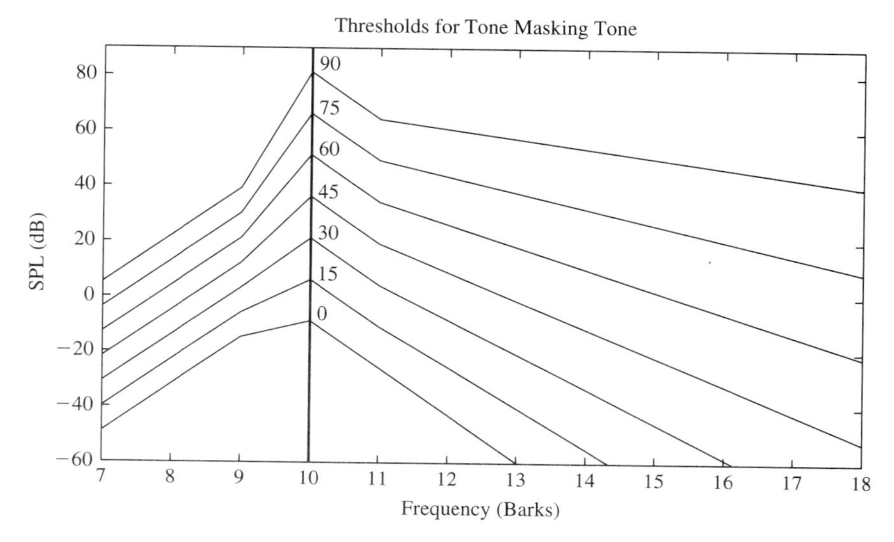

FIGURE 12.23
Thresholds as computed using the MPEG Psychoacoustic Model 1 [156, 374].

based on the assumptions that (1) narrow spectral peaks indicate tonal maskers, and (2) the remaining components not deemed to be tonal maskers in a given critical band together comprise a noise masker in that critical band.

A tonal masker is identified by testing each value of $P_r[k]$ defined by Eq. (12.31) for $0 < k < N/2$ to see if it satisfies the conditions

$$\left\{ \begin{array}{c} P_r[k] > P_r[k \pm 1], \\ P_r[k] > P_r[k \pm \Delta_k] + 7 \text{ dB}, \end{array} \right\} \tag{12.35}$$

where Δ_k defines a set of indices on either side of k such that

$$\begin{array}{lll} \Delta_k = 2 & \text{for } 2 < k < 63 & (170 < f < 5500 \text{ Hz}), \\ \Delta_k = 2, 3 & \text{for } 63 \leq k < 127 & (5500 < f < 11{,}025 \text{ Hz}), \\ \Delta_k = 4, 5, 6 & \text{for } 127 \leq k \leq 256 & (11{,}025 < f < 22{,}050 \text{ Hz}). \end{array} \tag{12.36}$$

That is, $P_r[k]$ is tested to see if it is the maximum of a narrow spectral peak. The frequency ranges in Eq. (12.36) correspond to the given DFT ranges for a sampling rate $F_s = 44{,}100$ Hz. With this set of criteria, peaks can be broader in the higher critical bands. If a particular index k_{tm} is selected as the location of a tonal masker, a new SPL is associated with that masker according to

$$P_{TM}[k_{tm}] = 10 \log_{10} \left(10^{0.1 P_r[k_{tm}-1]} + 10^{0.1 P_r[k_{tm}]} + 10^{0.1 P_r[k_{tm}+1]} \right). \tag{12.37}$$

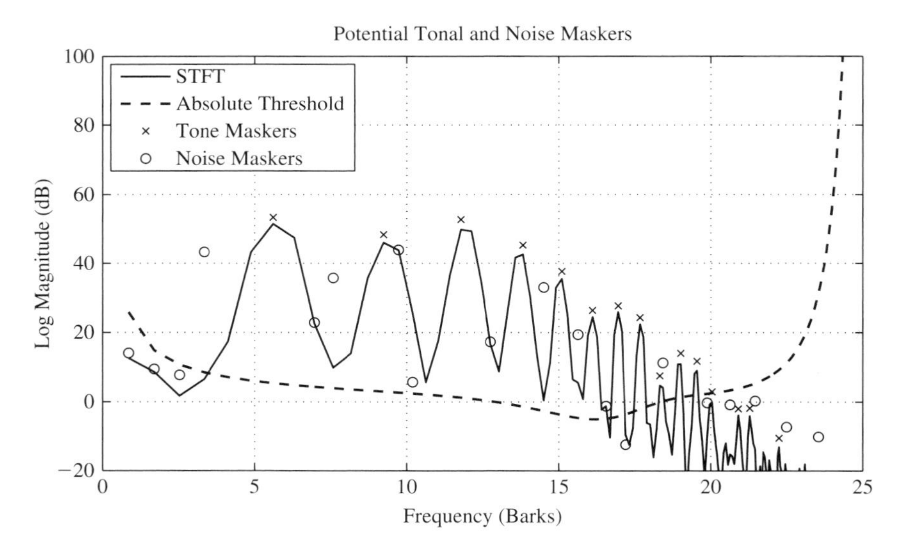

FIGURE 12.24
Tonal and noise masker candidates.

That is, power from three adjacent frequencies is combined to form a single tonal masker centered at k_{tm}. The "$\times$" symbols in Figure 12.24 denote the frequencies and SPL for the tonal maskers selected by this algorithm for the frame of the audio of Figure 12.20.

The remaining spectrum values are assumed to contribute to noise maskers. Specifically, for each critical band, a single noise masker is formed from all the STFT values that are not involved in the definition of a tonal masker; i.e., all the frequencies not included in an interval of Δ_{k_m} around the center frequency of the tonal masker. We will use K_b to denote the set of noise frequencies in critical band b. The powers of the noise masking components in a critical band are added to give the associated SPL of the noise in that critical band; i.e.,

$$P_{NM}[\bar{k}_{nm}] = 10\log_{10}\left(\sum_{k\in K_b} 10^{0.1P[k]}\right), \tag{12.38}$$

where $\bar{k}_{nm}$ denotes the DFT index closest to the geometric mean of the DFT indices within critical band b. The "$\circ$" symbols in Figure 12.24 denote the frequencies and SPL for the noise maskers selected by the above algorithm.

As can be observed in Figure 12.24, the above analysis of the SPL spectrum gives a large number of maskers, some of which are not true maskers and should be eliminated. For example, any tonal or noise masker that is below the threshold in quiet can have no effect. Thus, in Figure 12.24, all the tonal and noise maskers with frequencies above 20 Bark can be eliminated, as can the noise maskers at about 1, 2, 3, and 17 Bark. Secondly, if two candidate maskers are close together (within 0.5 Bark),

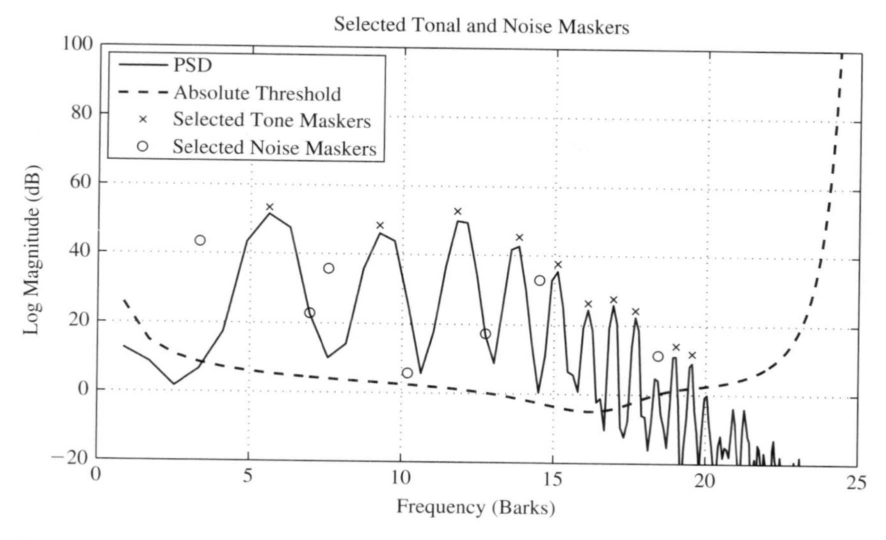

FIGURE 12.25
Tonal and noise masker candidates retained.

only the stronger masker is retained so as not to overestimate the masking effects. Figure 12.25 shows the maskers that are retained from among the original candidate set of Figure 12.24. A comparison of Figures 12.24 and 12.25 shows that three of the noise maskers and one tonal masker above the threshold in quiet have been removed because they were too close to stronger maskers.

Now, given the location (on the Bark scale) and SPL of the maskers, we can use Eqs. (12.32) with (12.33) to compute the contribution of each tonal masker to the threshold of audibility. Similarly, we can use Eqs. (12.34) and (12.33) to compute the contribution of each noise masker to the audibility threshold. Finally, all the contributions are added (in power) to the threshold in quiet to yield the global audibility threshold; i.e.,

$$T_g[k] = 10\log_{10}\left(10^{0.1T_q[k]} + \sum_{k_{tm}} 10^{0.1T_{TM}[k,\, k_{tm}]} + \sum_{\bar{k}_{nm}} 10^{0.1T_{NM}[k,\, \bar{k}_{nm}]}\right). \tag{12.39}$$

Figure 12.26 shows the maskers from Figure 12.25 and the corresponding contributions (in dB) to the overall global threshold of audibility of the three components

$$T_q[k] \qquad \text{(dashed line)},$$

$$10\log_{10}\left(\sum_{k_{tm}} 10^{0.1T_{TM}[k,\, k_{tm}]}\right) \qquad \text{(dotted line)},$$

$$10\log_{10}\left(\sum_{\bar{k}_{nm}} 10^{0.1T_{NM}[k,\, \bar{k}_{nm}]}\right) \qquad \text{(solid line)}.$$

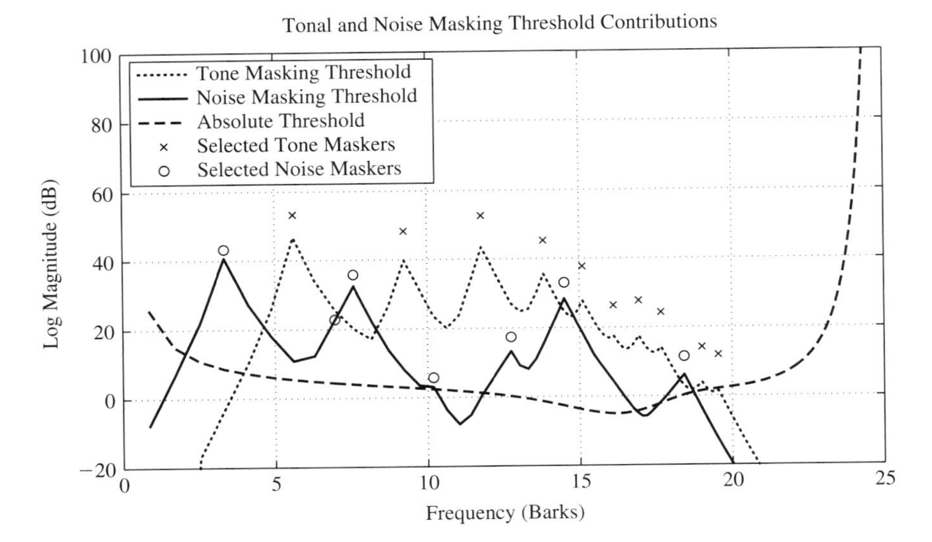

FIGURE 12.26
Tonal and noise masking threshold contributions.

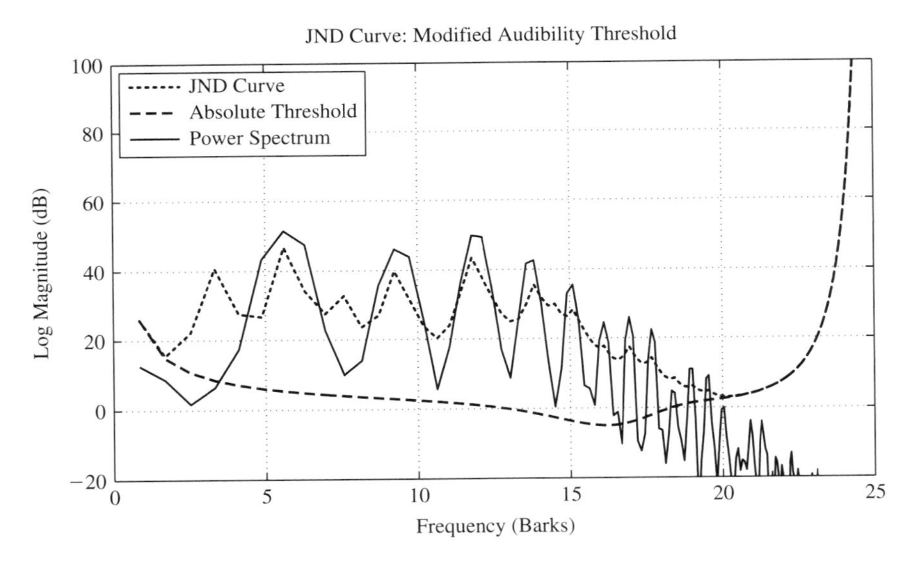

FIGURE 12.27
Just noticeable difference curve.

The power values corresponding to the dB values in Figure 12.26 are added in Eq. (12.39) to give $T_g[k]$, which, for the example SPL spectrum of the trumpet sound, is shown as the solid gray line superimposed on the spectrum shown in Figure 12.27.

12.5.7 Quantization of the STFT

Note that the plot in Figure 12.27 suggests that only the spectral components that exceed the threshold will be audible. Thus, if the playback level is set as assumed in the model, it should be possible to set all frequency components that lie below the estimated global threshold to zero, with no perceptible difference. For this reason, the global audibility threshold is also referred to as the just noticeable difference or JND for the audio signal in the analysis frame. Furthermore, the remaining spectral components can be quantized so long as the quantization error does not exceed the global threshold.

The upper plot in Figure 12.28 shows the original STFT (on the Bark scale) along with the estimated global threshold (dotted line). The lower plot shows the result of setting all 227 components below the threshold to zero and quantizing the remaining spectral components to 6 bits. This results in the interrupted curve composed of the regions that exceed the threshold. The other curve is the difference between the original STFT and the quantized spectrum. Note that the spectral error is simply equal to the original spectrum wherever the spectrum was below the threshold. This is obvious in the region above 20 Bark. The error is the error due to 6-bit quantization for those components above the threshold. The quantization algorithm was adaptive in a very

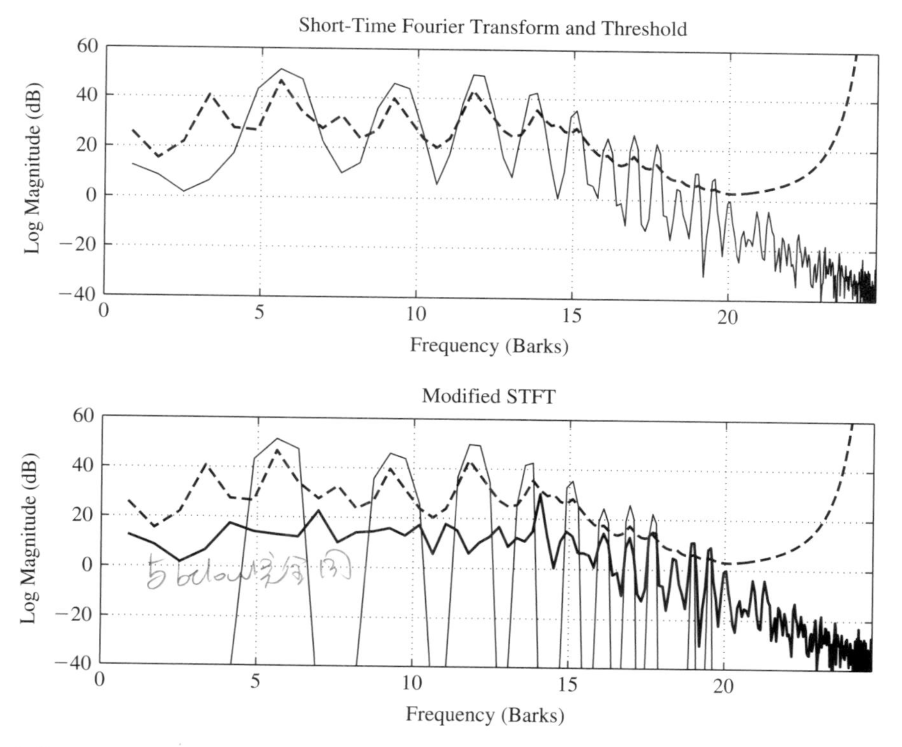

FIGURE 12.28
Illustration of quantization of STFT to 6 bits for components exceeding the threshold and 0 bits for those (209 out of 257) below.

limited sense. A single overall scale factor was computed from the maximum absolute value of all 257 of the real and imaginary parts of the STFT. After normalization by the scale factor, the real and imaginary parts were quantized with a 6-bit quantizer with a dynamic range of ± 1. Observe that the quantization error falls below the global threshold everywhere except for the spectral peaks around 18–19 Bark. Since the quantization error is well below the threshold under most of the major peaks, these regions could in principle be quantized with fewer bits, while more bits are required in regions where the spectrum is close to the threshold. A full-fledged transform coder such as the one proposed by Johnston [176] would use a more sophisticated quantization algorithm.

As a demonstration of the effect of this simple modification, the STFT was computed with a DFT length of $N = 512$ and the Hann window was moved in steps of $R = N/16$, which satisfies the condition for exact reconstruction if the STFT is not modified. For each frame, the threshold of audibility was estimated as described above, and all "inaudible" frequency components were set to zero. The remaining spectral components were quantized to 6 bits. A signal was synthesized from the modified STFT by the OLA method. The results for the trumpet sound are shown in Figure 12.29,

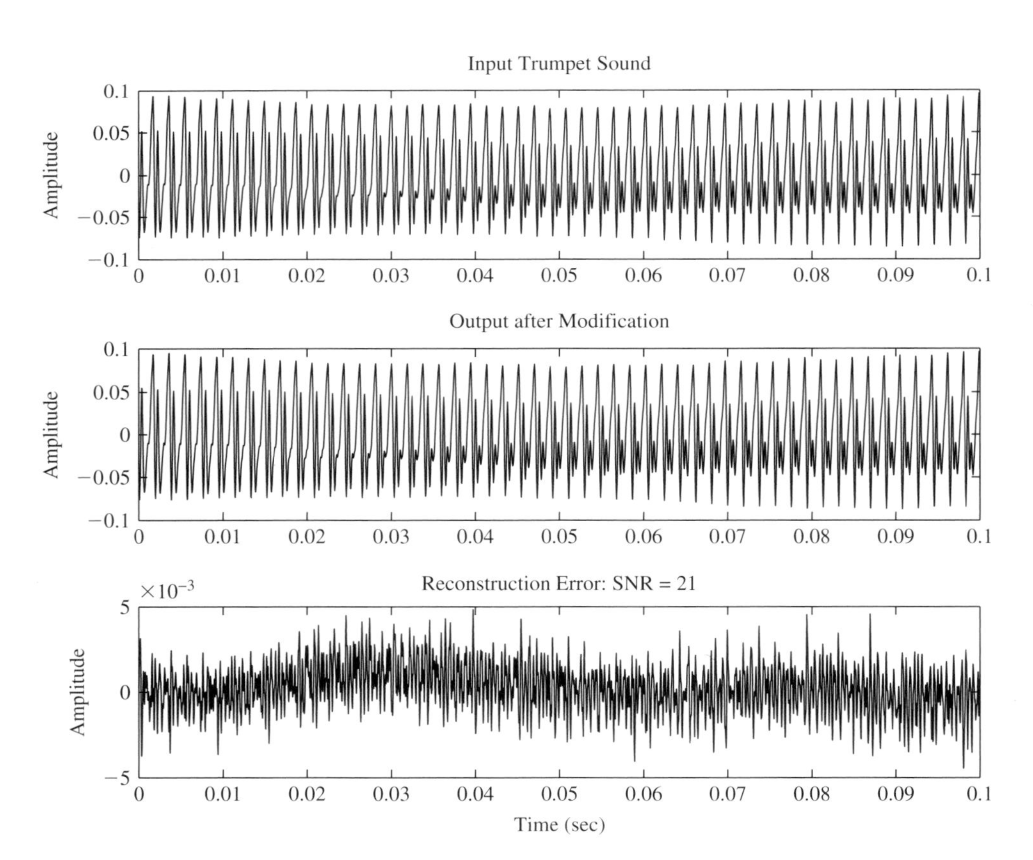

FIGURE 12.29
Illustration of quantization of the STFT of an audio signal.

where the top plot shows the original signal, the middle plot is the synthesized output, and the bottom plot is the difference between the input and output. It is interesting to note that over the entire 9 seconds duration, 208 or more out of 257 DFT values in each frame were set to zero. The SNR, defined as the ratio of the signal power to error power, was 21 dB. Careful listening (by experienced listeners with so-called "golden" ears) revealed no perceptible difference between the original and the modified signals, even though the SNR was only 21 dB.

12.6 MPEG-1 AUDIO CODING STANDARD

The MPEG-1 Psychoacoustic Model 1 of the previous section is primarily aimed at coding of high-quality audio entertainment signals such as instrumental and vocal music, although it applies equally well to speech signals and combined speech and music signals as required in digital movies. In order to encompass such a wide range of sound sources, it is common to use sampling rates such as 32, 44.1, 48, and 96 kHz. Furthermore, the goal in audio coding is to maintain "transparency"; i.e., to ensure that the coded signal is perceptually indistinguishable from the original acoustic source signal. Therefore, the lowest bit rates of interest for audio signals are generally much higher than those that provide acceptable narrowband speech quality for communication applications. Recall that 16-bit sampling at a sampling rate of 44.1 kHz requires a bit rate of around 705 kbps.[14] A typical bit rate for MPEG-1 stereo coding is 128 kbps, corresponding to a compression ratio of about 11. On the other hand, speech coders achieve lower bit rates and higher compression ratios by limiting the bandwidth of the speech signal. For example 64 kbps log-PCM quality can be achieved using an 8 kHz sampling rate at about 8 kbps or lower using analysis/synthesis coding for a compression ratio of about 8-to-1.

Since no particular type of acoustic signal is assumed, audio coders rely primarily on exploiting perceptual models to lower the bit rate. The perception model described in detail in Section 12.5 is the Psychoacoustic Model 1 from the MPEG-1 ISO/IEC 11172-3 international standard [156]. In the standard, this model is given as an example, but the standard really only specifies the precise details of the data stream and the decoder. This leaves great flexibility in the coding side while ensuring that if the data stream meets the specification of the standard, then an MPEG-compliant decoder can reconstruct an audio signal from the data stream. Figure 12.30 depicts the structure of the decoder and a recommended structure for the encoder. The MPEG-1 encoder/decoder is a subband coder in which the quantization of the subband channel signals is controlled by a psychoacoustic model such as the one we have just discussed. The standard specifies how the quantized bandpass filter inputs are to be represented in the data stream, but leaves flexibility in how the subband signals are acquired and quantized.

[14] With two-channel (stereo) coding, requires 705 kbps for each channel or a total overall bit rate of 1.41 mbps for stereo audio coding on high quality CDs.

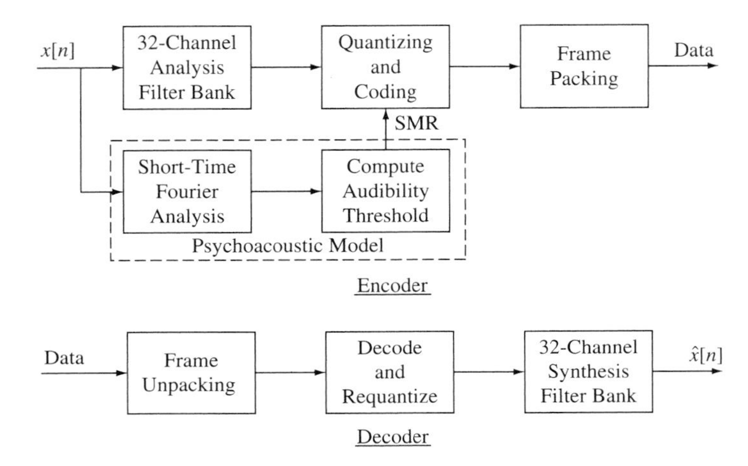

FIGURE 12.30
Block diagram of MPEG-1 encoder and decoder.

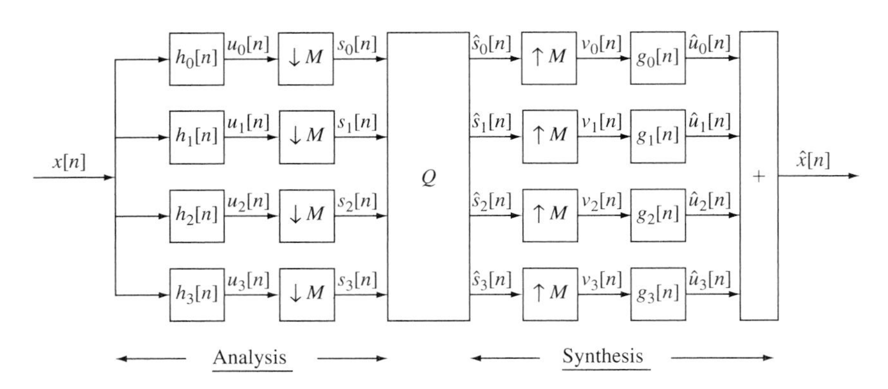

FIGURE 12.31
Filter bank system for analysis and synthesis of audio signals.

In the remainder of this section, we discuss some of the details of the MPEG-1 audio coding standard. The MPEG-1 standard has three "layers" (I, II, and III) which yield increasingly improved quality at lower bit rates, at the cost of increasing complexity in the encoding and decoding processes. MPEG-1 Layer III, which can be expected to provide the highest quality at the lowest bit rate, is, in fact, the popular "MP3" coder that is widely used for representing music in digital form. We will focus the current discussion on Layer I since parts of this layer are common to all the layers, and because it clearly illustrates the basic principles of perceptual audio coding.

12.6.1 The MPEG-1 Filter Bank

The MPEG-1 standard for audio coding [156, 279] is built upon a filter bank for synthesizing an audio signal from a set of $M = 32$ bandpass channel signals. The right half of Figure 12.31 depicts such a filter bank for the special case of $M = 4$ (for simplicity in drawing) channels. A complete system diagram of the MPEG-1 system would require an extension to 32 channels. As depicted in Figure 12.31, the encoder structure includes an analysis filter bank that produces a set of ($M = 32$) subband signals that (with no quantization) can be synthesized by the synthesis filter bank into an almost perfect replica of the original signal. The complete coder/decoder system also includes means (block labeled Q in Figure 12.31) for quantizing the channel signals to obtain data compression.

The MPEG-1 standard does not use a tree-structured decomposition as is common for speech subband coders. Instead, an approximate generalization of the QMF two-channel filter bank is employed, which is often termed a "pseudo-QMF" (PQMF) filter bank. In the standard [156], the analysis and synthesis filter banks are specified by flow charts that show computation algorithms for their implementation. The implementation of the MPEG-1 filter bank involves many sophisticated signal processing concepts, and thus, it serves as an excellent example of filter bank design and implementation. To solidify the connection to the more detailed discussion in Chapter 7, we review the basic signal processing functions that define the MPEG-1 analysis and synthesis filter banks.

The MPEG-1 analysis filter bank is a set of $M = 32$ bandpass filters whose impulse responses are of the form

$$h_k[n] = 2h[n]\cos(\omega_k n - \phi_k) \qquad \begin{cases} n = 0, 1, \ldots, L-1 \\ k = 0, 1, \ldots, M-1, \end{cases} \tag{12.40}$$

where L is the length of the lowpass filter impulse response, $h[n]$, and M is the number of bandpass filters. The center frequencies of the bandpass filters are

$$\omega_k = \frac{2\pi}{4M}(2k+1), \qquad k = 0, 1, \ldots, M-1, \tag{12.41}$$

and the phase shift of the cosine term is $\phi_k = \omega_k M/2$. Equation (12.40) states that the impulse responses of the bandpass analysis filters are obtained by cosine-modulating a prototype lowpass filter impulse response. The multiplicative factor of 2 in Eq. (12.40) compensates for a gain factor of 1/2 introduced by the cosine modulation. As we will show, the non-ideal bandpass frequency responses that result when $h[n]$ is a causal FIR impulse response will necessarily have regions of overlap in the frequency domain, but due to careful design and phasing, this overlap does not present any problems for audio coding.

The equations describing the analysis filter bank system are

$$s_k[n] = \sum_{m=0}^{511} x[n32 - m]2h[m]\cos\left[\frac{2\pi}{128}(2k+1)(m-16)\right], \qquad k = 0, 1, \ldots, 31, \tag{12.42}$$

where we assume an $L = 512$-point impulse response $h[n]$ and $M = 32$ channels in the filter bank.

The outputs of the bandpass filters are decimated by a factor of M (i.e., $s_k[n] = u_k[nM]$). In practice, this is achieved by computing the filter outputs only once for every M samples of the input signal. Thus, for every M samples of the input, the system produces one sample for each of the M bandpass filters so that the total number of samples/sec remains the same.[15] The decimation operation serves the dual purpose of reducing the sampling rate and down-shifting the spectrum of each channel to the baseband for the reduced sampling rate F_s/M.

In audio coding, the bandpass filter outputs are quantized and encoded for transmission or storage. This quantization and coding is represented by the symbol Q in the diagram of Figure 12.31, and the signals $\hat{s}_k[n]$ represent the decoded requantized bandpass channel signals. As we have seen, for subband coders, data compression occurs if, on average, fewer bits/sample are used to represent the 32 bandpass channel signals than are needed for the 32 input signal samples.

For reconstruction of the audio signal, the decoded bandpass filter signals are upsampled, giving

$$v_k[n] = \sum_{r=-\infty}^{\infty} \hat{s}_k[r]\delta[n - rM] = \begin{cases} \hat{s}_k[n/M] & n = 0, M, 2M, \ldots \\ 0 & \text{otherwise}, \end{cases} \tag{12.43}$$

and then interpolated by a second bank of bandpass filters. This restores the channel signal to its original frequency range. The outputs of these filters are summed to produce the overall reconstructed signal. The impulse responses of the synthesis filters are

$$g_k[n] = 2Mh[n]\cos(\omega_k n - \psi_k) \qquad \begin{cases} n = 0, 1, \ldots, L-1 \\ k = 0, 1, \ldots, M-1, \end{cases} \tag{12.44}$$

where the phase shift is $\psi_k = -\phi_k$. The factor $2M$ in Eq. (12.44) compensates for the combined effects of cosine modulation in the filter bank design and decimation by M at the outputs of the analysis filter bank. Using Eq. (12.43), the interpolated outputs are

$$\hat{u}_k[n] = \sum_{r=-\infty}^{\infty} \hat{s}_k[r]g_k[n - rM], \quad k = 0, 1, \ldots, M-1. \tag{12.45}$$

The composite output of the synthesis filter bank, for any time instant n, is computed in blocks of $M = 32$ samples. To see how this is done, let n_0 denote the time index of a particular set of subband samples $\hat{s}_k[n_0]$, $k = 0, 1, \ldots, M-1$. Each of these samples allows the computation of $M = 32$ samples of the corresponding interpolated output signal $\hat{u}_k[n]$ for $n = n_0 M + m$ and $m = 0, 1, \ldots, M-1$. Specifically, since the synthesis filters are causal FIR systems with impulse response length $L = 512$, only

[15]Note that in Figure 12.31, n is used as the time index for all signals. If $F_s = 1/T$ is the sampling rate at the input side, then n denotes sampling time $t_n = nT$. On the other hand, the sampling rate at the output of the decimator is F_s/M, and the index n denotes sample time $t_n = nMT$. We consistently indicate the time index of the stream of signal samples as $[n]$. When a set of values is associated with time index n, we will indicate this by a subscript as in the case $s_k[n]$, $k = 0, 1, \ldots, M-1$, which denotes one output sample at time n for each of the M decimated bandpass filter outputs.

a finite number ($L/M = 16$) of past values of the subband signals will contribute to the computation of the output at any time n. Therefore the equation that describes the computation of $M = 32$ samples of the output of the synthesis filter bank for $n_0M \leq n < (n_0 + 1)M$ is

$$\hat{x}[n_0M + m] = \sum_{k=0}^{31} \left(\sum_{r=0}^{15} \hat{s}_k[n_0 - r] \, (64h[m + r \cdot 32]) \cos\left[\frac{2\pi}{128}(2k + 1)(m + r \cdot 32 + 16) \right] \right), \quad m = 0, 1, \ldots, 31. \tag{12.46}$$

This equation would be applied repeatedly to compute the output in groups of 32 samples. As in the case of Eq. (12.42), this equation describes the system in a mathematical sense, but other forms of the equation are more useful for computational implementations [156]. Filter bank analysis/synthesis systems can, in some cases, be designed and implemented so that in the absence of any quantization of the channel signals ($\hat{s}_k[n] = s_k[n]$), the output $\hat{x}[n]$ will be exactly equal to the input $x[n]$. Such analysis/synthesis systems are called *perfect reconstruction* systems. In the case of the MPEG-1 audio standard, the reconstruction condition is relaxed slightly, leading to systems that give *nearly perfect reconstruction*. In any case, the design of an analysis/synthesis system requires:

1. design of the filters to ensure (nearly) perfect reconstruction,
2. implementation of the filters, decimation, upsampling, and interpolation operations.

The Prototype Lowpass Filter

As discussed more fully in Chapter 7, the prototype lowpass impulse response for a filter bank is given implicitly by the analysis window. For the MPEG-1 standard the analysis window is given in Table C.1 of the standards document [156]. Figure 12.32a shows the impulse response of the prototype lowpass filter. The prototype lowpass impulse response has length $L = 513$, but note, however, that $h[0] = h[512] = 0$. Thus, the impulse response really has only 511 non-zero samples. However, as the plot in Figure 12.32 shows, if the sample $h[0] = 0$ is paired with a sample $h[512] = 0$ (which is true for an FIR system with $L = 513$), then the impulse response is even-symmetric about $n = 256$ such that $h[512 - n] = h[n]$ for $n = 0, 1, 2, \ldots, 512$. Furthermore, $h[256]$ is a unique sample; i.e., $h[512 - 256] = h[256]$. This implies that the prototype lowpass impulse response is a type I FIR linear phase system [270] of effective length 513. Alternatively, $h[n]$ can be thought of as a type I FIR filter of length 511 with an extra sample of delay. An important consequence of either viewpoint is that the frequency response of the prototype lowpass filter has the form

$$H(e^{j\omega}) = A(e^{j\omega})e^{-j\omega L/2}, \tag{12.47}$$

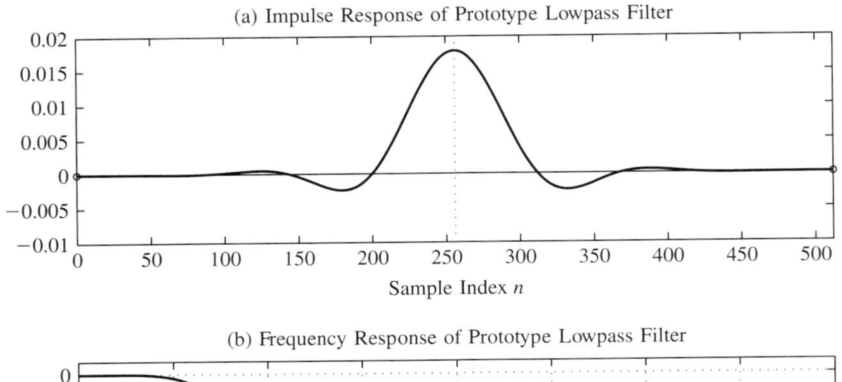

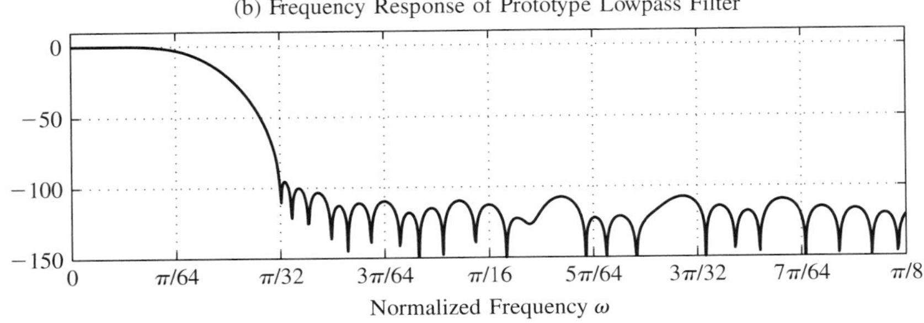

FIGURE 12.32
The MPEG-1 window and corresponding lowpass filter response functions.

where $A(e^{j\omega})$ is a real and even function of ω. Therefore, the analysis system has a time delay of $(L-1)/2 = 256$ samples.

Figure 12.32b shows the log-magnitude of the frequency response of the prototype lowpass filter; i.e., $20\log_{10}|A(e^{j\omega})|$. Observe that the gain in the passband is nominally 1 (0 dB), with the gain being -3 dB at $\omega = \pi/64$. The stopband cutoff frequency is approximately $\omega_s = \pi/32$, so the transition band extends from approximately $\omega_p = \pi/64$ to $\omega_s = \pi/32$. Recall that the multiplier 2 in Eq. (12.40) compensates for the cosine modulation so that the bandpass filters also have unity gain in their passbands. Finally, note that the gain of the filter is well below -90 dB for all frequencies above $\omega_s = \pi/32$. This is crucial for minimizing the interaction of the bandpass channels.

The Bandpass Analysis and Synthesis Filters

Since the impulse responses of the bandpass analysis filters are given by Eq. (12.40), the frequency response of the kth channel is

$$H_k(e^{j\omega}) = e^{-j\phi_k}H(e^{j(\omega-\omega_k)}) + e^{j\phi_k}H(e^{j(\omega+\omega_k)}), \qquad k = 0, 1, \ldots, M-1, \tag{12.48}$$

where the center frequencies of the passbands are

$$\omega_k = \frac{2\pi}{128}(2k+1), \qquad k = 0, 1, \ldots, M-1, \tag{12.49}$$

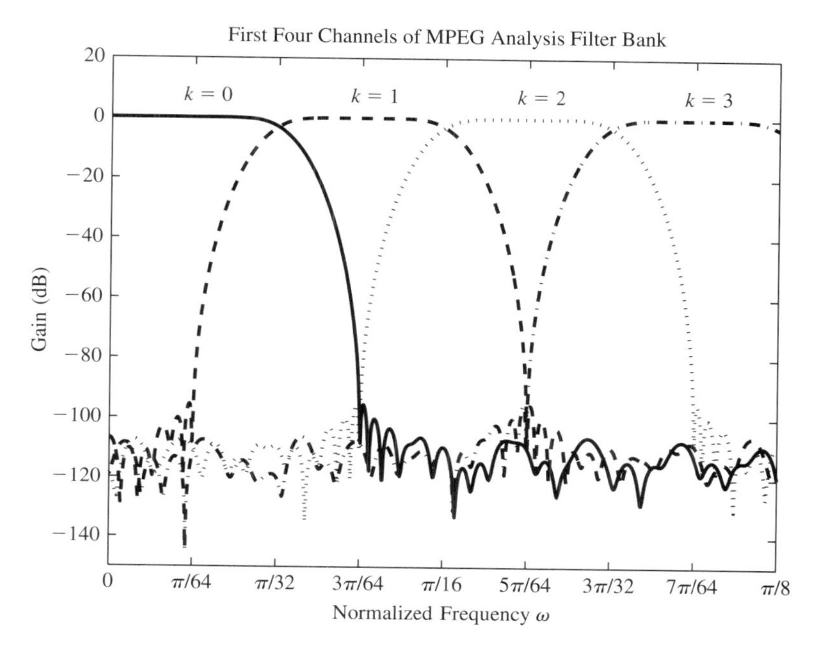

FIGURE 12.33
The first four bandpass filters of the MPEG-1 analysis filter bank.

and the phase angle is $\phi_k = \omega_k M/2$. Thus, the frequency response of the kth channel consists of two frequency-shifted copies of the frequency response of the prototype lowpass filter centered at $\pm\omega_k$. Similarly, from Eq. (12.44), the synthesis filters can be shown to have frequency responses

$$G_k(e^{j\omega}) = Me^{-j\psi_k} H(e^{j(\omega-\omega_k)}) + Me^{j\psi_k} H(e^{j(\omega+\omega_k)}). \tag{12.50}$$

Figure 12.33 shows the log-magnitude frequency responses of the first four channels of the analysis filter bank plotted over the range $0 \le \omega \le \pi/8$. This figure shows that the transition bands of a given filter overlap with the transition bands of adjacent filters, but only the stopbands (which have gain below −90 dB) of non-adjacent filters overlap with the passband and transition bands of a given filter.

Overall Frequency Response of Analysis/Synthesis System

The overall analysis/synthesis system for MPEG-1 audio coding involves two filter banks whose bandpass filters are based upon the same prototype lowpass impulse response, but which have different gains and different phase responses. The phase is controlled by the phase factors ϕ_k for the analysis filters and ψ_k for the synthesis filters, which play an important role in ensuring nearly perfect reconstruction in the absence of quantization.

Taking into account the combined effects of filtering, decimation, upsampling, interpolation filtering, and summing, the frequency-domain representation of the overall system in Figure 12.31 can be shown to be

$$\hat{X}(e^{j\omega}) = \sum_{r=0}^{M-1} \tilde{H}_r(e^{j\omega})X(e^{j(\omega-2\pi r/M)})$$
$$= \tilde{H}_0(e^{j\omega})X(e^{j\omega}) + \sum_{r=1}^{M-1} \tilde{H}_r(e^{j\omega})X(e^{j(\omega-2\pi r/M)}), \quad (12.51)$$

where

$$\tilde{H}_r(e^{j\omega}) = \frac{1}{M}\sum_{k=0}^{M-1} G_k(e^{j\omega})H_k(e^{j(\omega-2\pi r/M)}). \quad (12.52)$$

The two terms on the right in the expanded form of Eq. (12.51) show that there are two important considerations. The first term gives the contribution to the output of the input $X(e^{j\omega})$, and the second sum of the remaining terms gives the contributions of all the aliasing terms introduced by the decimation. In order to avoid aliasing altogether in the output, the bandpass filters must have bandwidths no wider than $\pi/M = \pi/32$. However, as Figure 12.33 shows clearly, the bandwidths of the channels are extended about $\pi/64$ on either side of the desired bandwidth of $\pi/32$. Figure 12.33 also shows, however, that in the overlap regions between channels, the adjacent filter transition regions are symmetrically shaped. With some algebra and invoking the assumption that only adjacent terms interact, it can be shown that, to a high degree of accuracy, the alias terms completely cancel out and the overall system from input to synthesized output is described by the equation

$$\hat{X}(e^{j\omega}) = e^{-j\omega(L-1)}X(e^{j\omega}). \quad (12.53)$$

That is, the overall system has unit gain and a time delay of $L-1 = 512$ samples. This delay is comprised of $L/2 = 256$ samples delay from both the analysis and the synthesis filter bank.

The phase-shift factors ϕ_k and ψ_k are critical for the simultaneous achievement of nearly constant $|\tilde{H}_0(e^{j\omega})|$ and nearly zero values for each of the alias terms $\tilde{H}_r(e^{j\omega})$. In the MPEG-1 standard, the phase shifts of the analysis and synthesis bandpass filters are $\phi_k = \omega_k M/2$ and $\psi_k = -\phi_k = -\omega_k M/2$. This choice of the phases was used by Rothweiler [333, 334] and Nussbaumer and Vetterli [260], but other choices of the phase shifts also lead to nearly perfect reconstruction [234].

12.6.2 Quantization of Channel Signals

In MPEG-1 Layer I, the decimated filter bank outputs are coded in groups of 12 samples. This is equivalent to $12 \times 32 = 384$ samples at the original sampling rate. As depicted in Figure 12.30, a psychoacoustic model such as the model discussed in Section 12.5 is employed to determine the threshold of audibility for each frame of 384 audio samples. As we saw in the previous section, this involves computing the STFT in the form of a 512-point DFT of a Hann windowed segment of the input audio signal. To best represent the properties of the signal in the coding frame, the Hann window should be centered on the group of 384 samples that correspond to the block of 12 decimated bandpass filter output samples. Due to the delay of the analysis filters, which is 256 samples, the Hann window should be offset by $256 + (512 - 384)/2 = 320$ samples at the input sampling rate in order to align samples 64 through 447 of the Hann window of the psychoacoustic model with the block of 12 samples at the outputs of each of the bandpass analysis filters. Maskers are identified in the short-time spectrum and a global threshold of audibility for the coding frame is determined by the process discussed in Section 12.5. This threshold, which is defined across the DFT frequencies $(2\pi k/N)$, $k = 0, 1, \ldots, N-1$,[16] is used to obtain the masking threshold in each of the 32 bands of the analysis filter bank. For a sampling rate of $F_s = 44{,}100$ Hz, the nominal width of the analysis bands is $\Delta f_{\text{sb}} = 22{,}050/32 = 689$ Hz. Thus, the lower channels encompass multiple critical bands. For this reason, the minimum threshold within each band is selected as the threshold in that band. That is, the masking level in the subband is

$$T_{sb}[k_{\text{sb}}] = \min_{k \in K_{k_{\text{sb}}}} \{T_g[k]\}, \qquad k_{\text{sb}} = 0, 1, \ldots, M, \tag{12.54a}$$

where k is the DFT index, k_{sb} denotes the index of the subband, and

$$K_{k_{\text{sb}}} = \{8k_{\text{sb}}, 8k_{\text{sb}} + 1, \ldots, 8k_{\text{sb}} + 7\} \tag{12.54b}$$

represents the set of DFT indices that are contained in the k_{sb}^{th} subband. Figure 12.34 shows the global threshold $T_g[k]$ or JND curve (thin solid line) for the example frame used as illustration in Section 12.5. The global threshold $T_{\text{sb}}[k_{\text{sb}}]$ for the 32-band MPEG-1 filter bank is shown as the heavy dashed line. Note that the threshold is shown only for the first 26 channels. The remaining eight channels have thresholds that are outside the convenient plotting range. In the higher channels, the threshold in quiet determines the threshold since the signal has virtually no power in the upper channels.

From the filter bank global threshold $T_{\text{sb}}[k_{\text{sb}}]$ and the short-time spectrum $P[k]$, it is possible to compute the ratio of the signal power to the masking threshold. The signal-to-mask ratio (SMR) defined as

$$\text{SMR}[k_{\text{sb}}] = \max_{k \in K_{k_{\text{sb}}}} \{P[k]\} - T_{\text{sb}}[k_{\text{sb}}] \tag{12.55}$$

is the final output of the psychoacoustic model.[17] The heavy solid line in Figure 12.34 shows the SMR for the example that we have been following. Note that $\text{SMR}[k_{\text{sb}}] > 0$ implies that the signal components in band k_{sb} are unmasked; i.e., they are above the threshold of audibility. When $\text{SMR}[k_{\text{sb}}] < 0$, the signal components in band k_{sb}

[16]To save computation, the global masking threshold may be computed only at a subset of DFT frequencies.

[17]Note that all quantities in Eq. (12.55) are in dB, so a ratio corresponds to a difference.

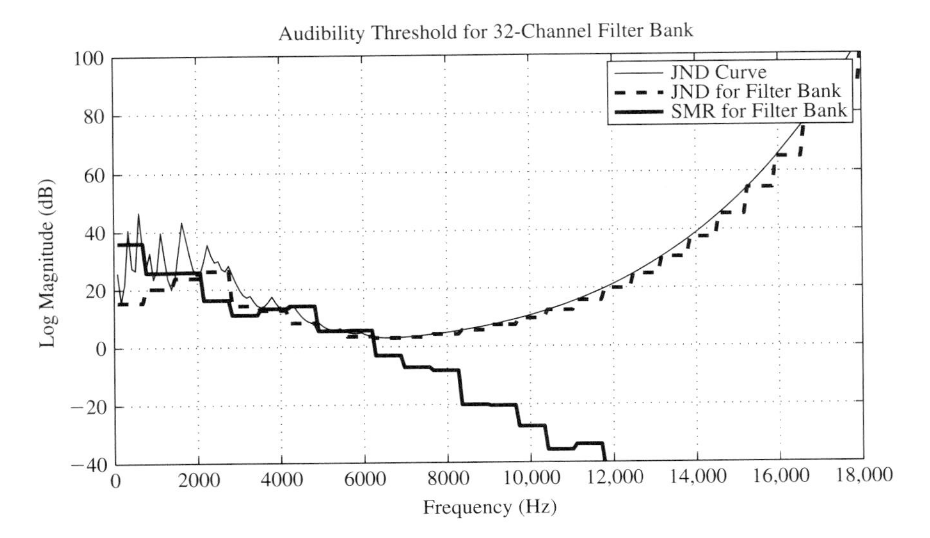

FIGURE 12.34
JND curve and signal-to-mask ratio (SMR) for 32-channel filter bank.

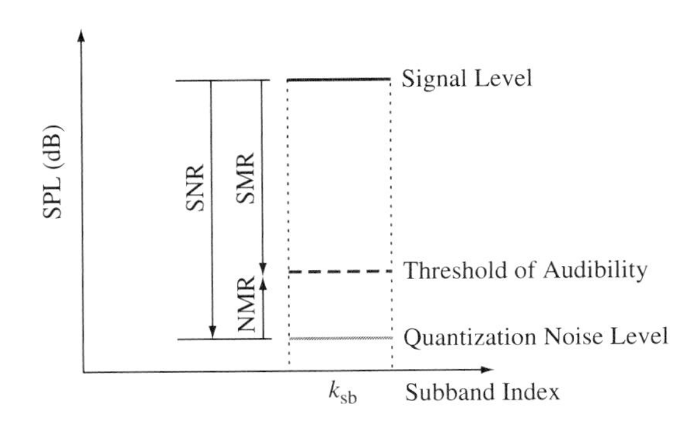

FIGURE 12.35
Illustration of relationship between signal, noise, and masking level in a single subband.

are masked, and they can be set to zero (given 0 bits/sample) with no perceptual consequence.

The SMR is used to ensure that the quantization noise in each channel lies below the masking threshold. Toward this end, the noise-to-mask ratio is defined as

$$\text{NMR}[k_{sb}] = \text{SNR}[k_{sb}] - \text{SMR}[k_{sb}], \tag{12.56}$$

where $\text{SNR}[k_{sb}]$ is the signal-to-quantization-noise ratio in subband k_{sb}. The relationship among all the terms in Eqs. (12.55) and (12.56) is depicted in Figure 12.35, which shows a typical situation where the signal in band k_{sb} is unmasked and quantized so that the quantization noise level is below the threshold of audibility.

When quantizing the audio signal representation to a fixed bit rate, a fixed number of bits is allocated to each frame of 384 samples (12 subband samples). For example, if the sampling rate is $F_s = 44{,}100$ Hz, and we wish to achieve a bit rate of approximately 128 kbps, then the effective bit rate per sample is $128{,}000/44{,}100 = 2.9025 \approx 3$ bits/sample. This means that $3 * 384 = 1152$ bits are available to encode each block of 12 subband samples. A number of bits are set aside for a header (32 bits), which contains among other things, information such as a synchronization word, the sampling rate, bit rate, stereo mode, and layer. Bits may also be set aside for an optional cyclic redundancy code (16 bits) for error checking and optional ancillary data. The remaining bits are used to represent the 12 subband samples as a block in terms of (1) the number of bits/sample for each subband, (2) a scale factor for each subband, and (3) the quantized samples.

The first step in quantization of the 12 samples is to determine a single overall scale factor based on the maximum absolute value of the 12 samples to be quantized. A table lookup procedure prescribed by the MPEG-1 standard [156] is used to select one of 63 possible scale factors. The 12 samples are divided by the selected scale factor to ensure that all normalized samples have a magnitude less than 1. The scale factor is represented with 6 bits as an index to the table specified in the standard. Thus, for single channel mode, the maximum total number of bits required to transmit the scale factors is $6 * 32 = 192$ bits/frame.

The samples are quantized with a normalized symmetric mid-tread quantizer whose levels are represented by a sign-magnitude code with the leading bit indicating the sign bit. The code $100\ldots$ is not used since it would represent -0 in this quantization system. The number of bits/sample (bit allocation) ranges from $B = 0$ to 15, excluding $B = 1$ bit (because the smallest symmetric mid-tread quantizer has three levels). The bit allocation is coded with a 4-bit number whose value is 1 less than the number of bits/sample; i.e., $B_{\text{alloc}} = B - 1$.[18] Thus, the total number of bits for bit allocation information is $4 * 32 = 128$ bits/frame. A value of $B_{\text{alloc}} = 0$ signifies that all 12 subband samples are assumed to be zero, and in this case, a scale factor is not transmitted for that subband. Values $B_{\text{alloc}} = B - 1 = 1, 2, \ldots, 14$ correspond to quantizers with $2^B - 1$ levels for $B = B_{\text{alloc}} + 1 = 2, 3, \ldots, 15$ bits/sample. The MPEG-1 standard provides a table of SNRs corresponding to these symmetric mid-tread quantizers with an assumption of a full-scale sine wave input. This table is used to estimate the SNR for a given choice of B_{alloc}.

After the scale factors have been tentatively determined for each subband, the bits must be distributed among the subbands so as to minimize the noise-to-mask ratio in Eq. (12.56). This is done by an iterative process that starts by setting all bit allocations to zero and assuming no bits are needed to transmit the scale factors. In each iteration, bits are allocated to the subband with the highest noise-to-mask ratio until all the bits have been allocated. The noise-to-mask ratio in Eq. (12.56) is computed

[18]Note also that $B_{\text{alloc}} = 15$ is not allowed.

from the signal-to-mask ratio from the psychoacoustic model and from the SNR value obtained from the table for a proposed bit allocation. Once a subband has been allocated bits, it is necessary to also set aside 4 bits for the scale factor for that subband. The details of the bit allocation algorithm are provided in the MPEG-1 standard [156] and in the text by Bosi and Goldberg [42].

The final step is to quantize the subband samples. All 12 samples are divided by the scale factor to normalize them before they are quantized by the normalized quantizer corresponding to the number of bits allocated to that channel. All the bits for coding an audio frame are packed into a data frame whose structure is specified by the MPEG-1 standard [42, 156]. The data frames are concatenated to produce a stream of bits that represent the audio signal.

In order to decode the MPEG-1 representation, it is necessary to identify the beginning of a frame using the synchronization word. Then, for each frame, the bits are interpreted according to the rules of the standard and the quantized subband samples can be reconstructed from the coded information. These quantized subband signals are the input to some efficient implementation of the synthesis filter bank Eq. (12.46), which reconstructs the audio signal. An example implementation of the synthesis filter bank is given in the MPEG-1 standard [156].

The MPEG-1 standard specifies sampling rates of 32, 44.1, and 48 kHz, with bit rates ranging from 32 to 224 kbps per channel in increments of 32 kbps. Stereo (or 2-channel) coding is supported in the standard. Layer I coding provides transparent quality (mean opinion score 4.7) at bit rates of 192 kbps and above [42, 258]. MPEG-1 Layer I is the simplest of the standardized audio coders, and therefore its compression performance is the least effective. Nevertheless, it was used as the basis for 192 kbps per channel coding in digital compact cassette (DCC) audio recording [42]. It is worth noting that the MPEG-1 Layer I encoder is significantly more complex than the decoder due to the need for computing a perception model at the coder but not at the decoder. This asymmetry is typical of perceptual audio coders and is attractive for applications where decoding (playback) occurs much more frequently than encoding.

In Sections 12.5 and 12.6, we discussed Layer I of the MPEG-1 audio coding standard as an illustration of how perception models can be incorporated into digital audio coding systems. The discussion aimed to show how perceptual models are combined with sophisticated digital signal processing toward the goal of compressing the bit rate of audio signals. Implementing an MPEG-1-compliant audio coder involves many details that either were not discussed or were only mentioned in passing. An example is the fact that the MPEG-1 standard allows for joint coding of stereo signals. For implementation, it would be necessary to follow the standard [156] without deviation. However, the above discussion should make it much easier to interpret the standards document and to understand why each step is taken.

12.6.3 MPEG-1 Layers II and III

The MPEG-1 standard has additional layers (II and III) which add complexity in order to reduce the bit rate [42, 258, 374] and maintain high quality. The Layer I filter banks are used as the basis for all layers. Because of this, a fully-compliant MPEG-1 decoder is capable of decoding all layers of MPEG-1 coding. Instead of quantizing the subband outputs in blocks of 12 samples as in Layer I, Layers II and III collect three subgroups of 12 samples for additional processing. This means that the frame size in Layers II and III is (in effect) $3 * 384 = 1152$ samples at the input sampling rate. In return for the extra delay that this introduces, it becomes possible to share scale factor data between groups in Layer II, leading to about a 50% reduction in the bits required for scale factor information. The Level II psychoacoustic model uses a 1024-point DFT, which provides finer frequency resolution for calculating the SMR. Level II also uses a more complicated arrangement of quantizers, but its basic structure is the same as the Level I coder. The added sophistication yields transparent quality at about 128 kbps per channel (MOS score of 4.6) [42, 258]. MPEG-1 Level II coding has been used in digital audio broadcasting and digital video broadcasting [42].

The MPEG-1 Layer III achieves additional improved performance at bit rates as low as 64 kbps per channel. As depicted in Figure 12.36, this is due to a number of additional features including a refined frequency analysis of the filter bank outputs, more sophisticated bit allocation, and non-uniform quantization with subsequent Huffman coding. Level III also uses a much more sophisticated psychoacoustic model involving two 1024-point DFT calculations. Furthermore, in Level III encoding, groups of 36 filter bank output samples are further transformed by a modified DCT (MDCT) transformation with dynamically changing windows and 50% overlap in the time domain. Furthermore, the transform length can be adapted to handle segments of the signals with different stationarity characteristics. This provides finer frequency sampling $(44{,}100/(18 * 32) = 38.28$ Hz), which is smaller than a critical band over the entire frequency range. This splitting of the frequency analysis into a filter bank followed by a transform analysis makes Level III a hybrid of subband and transform coding.

Non-uniform quantization is used to take advantage of the fact that large transform values can mask quantization error, and Huffman coding provides efficient coding of the quantization indices.

High perceptual quality can be achieved at bit rates as low as 64 kbps (MOS score 3.7), and joint stereo coding at 128 kbps achieved MOS scores above 4 [258]. Applications of MPEG-1 Layer III coding include transmission over ISDN lines and the Internet. As mentioned before, MPEG-1 Layer III is the MP3 format widely used for transmitting audio over the Internet and playback with inexpensive portable players.

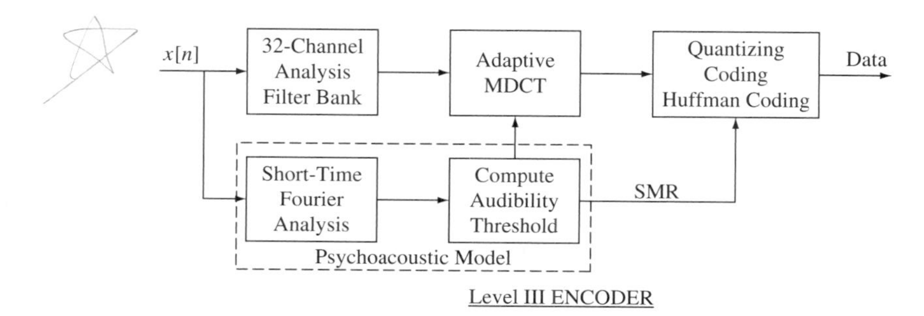

FIGURE 12.36
Operations in MPEG-1 Layer III encoding.

12.7 OTHER AUDIO CODING STANDARDS

Since the early development of perceptual audio coding and its subsequent standardization in MPEG-1, there has been continuous effort to improve quality and lower bit rate. This has led to many variations on the theme of incorporating perception models into algorithms for quantization of a frequency-domain representation. The result has been numerous standardized coders in the style of the MPEG-1 coder/decoder system. In this section, we will give a brief survey of some of these standards. More detailed discussions of the nature of these and other frequency-domain coders are given in the texts by Bosi and Goldberg [42] and Spanias et al. [374].

MPEG-2 and MPEG-2 AAC [157, 158]

The MPEG-2 LSF (low sampling frequency) audio coder standard was developed to extend the MPEG-1 coder to lower sampling rates. The MPEG-2 BC (backwards compatible) standard extended MPEG-1 to multichannel (more than two) in a way that made it backward compatible with MPEG-1. MPEG-2 AAC (advanced audio coder) relaxed the backward compatibility constraint in order to achieve greater compression and higher quality for multichannel audio signals.

MPEG-4 [159]

The MPEG-4 standard includes coders for audio as in MPEG-1 and MPEG-2 along with a framework for representing other "media objects" such as speech, synthetic audio, and text-to-speech. MPEG-4 provides a complete suite of encoder/decoders to tackle different acoustic signals at diverse bit rates, channel configurations, and network situations. It has parametric coders like the Harmonic and Individual Lines plus Noise (HILN) coder, general audio coders like AAC and TWIN-VQ, a hybrid coder HVXC, etc. Even within AAC, MPEG-4 provides additional coding tools such as perceptual noise substitution. Scalability is another feature that provides robustness to different network situations and diverse user bit rates (downloadable rates). Almost all the coders, i.e., AAC, TWIN-VQ, CELP, can be used in scalable configurations. Recently MPEG-4 extended its range to include "scalable to lossless audio."

Dolby AC-3 [19, 81]

The AC-3 coding system was developed by the Dolby Corporation specifically for coding multichannel sound for entertainment. This system shares many characteristics with the other audio coders discussed in this chapter.

12.8 SUMMARY

In this chapter we have provided an introduction to the use of frequency-domain representations as a basis for speech and audio coding. Our goal was to show how models for auditory perception could be incorporated into coders based on a frequency analysis of the signal. We discussed subband coding in general and gave some examples of its application to speech coding. However, the major focus in this chapter has been the illustration of how discrete-time spectrum analysis can be used to estimate masking effects, and how the resulting knowledge can be applied in subband coding of audio signals. In spite of the lengthy and involved discussion, we only scratched the surface of the subject of audio coding. However, our discussion will serve as a good introduction to other texts devoted completely to the subject of audio coding [42, 374].

PROBLEMS

12.1. This problem is concerned with the use of downsampling to frequency shift bandpass signals to a lowpass band.

(a) First consider a complex bandpass signal represented as $x_k[n] = x_0[n]e^{j\omega_k n}$, where $\omega_k = 2\pi k/N$ (with k an integer) and the real signal $x_0[n]$ has a lowpass DTFT such that $X_0(e^{j\omega}) \neq 0$ only in the band $|\omega| < \pi/N$. Determine the relationship between $X_k(e^{j\omega})$ and $X_0(e^{j\omega})$ and sketch a "typical" DTFT, $X_k(e^{j\omega})$, for $-\pi \leq \omega < \pi$.

(b) Next consider the downsampled signal $v_k[n] = x_k[nN]$. Show that, because of the ideal band limitation of $X_k(e^{j\omega})$, $V_k(e^{j\omega}) = \frac{1}{N}X_0(e^{j\omega/N})$, for $-\pi \leq \omega < \pi$.

(c) Next consider a real bandpass signal represented as $x_k[n] = x_0[n]\cos(\omega_k n + \phi_k)$, where $\omega_k = 2\pi k/N$ (with k an integer) and the signal $x_0[n]$ has a lowpass DTFT such that $X_0(e^{j\omega}) \neq 0$ only in the band $|\omega| < \pi/N$. Determine the relationship between $X_k(e^{j\omega})$ and $X_0(e^{j\omega})$ and sketch a "typical" DTFT, $X_k(e^{j\omega})$, for $-\pi \leq \omega < \pi$.

(d) Finally, consider the downsampled signal $v_k[n] = x_k[nN]$, where $x_k[n]$ is defined as in part (c). Show that $V_k(e^{j\omega}) = \frac{\cos(\phi_k)}{N}X_0(e^{j\omega/N})$. Describe what happens in the case $\phi_k = \pi/2$.

(e) How would the above results be affected if $X_k(e^{j\omega})$ is only approximately bandlimited; i.e., if $X_k(e^{j\omega})$ extends beyond the band $\omega_k - \pi/N < |\omega| < \omega_k + \pi/N$?

12.2. The MPEG standard for audio coding [156, 279] is built upon an analysis/synthesis filter bank for decomposing an audio signal into a set of $N = 32$ bandpass channels, as depicted in Figure P12.2a for the special case of four (for simplicity in drawing) channels. In the MPEG standard, the analysis and synthesis filter banks are specified by flow charts that show computation algorithms for their implementation. Some of the details of the system are discussed in Section 12.6.1. However, since these algorithms incorporate numerous detailed properties and features of the filters and the filter bank structure, it is far from

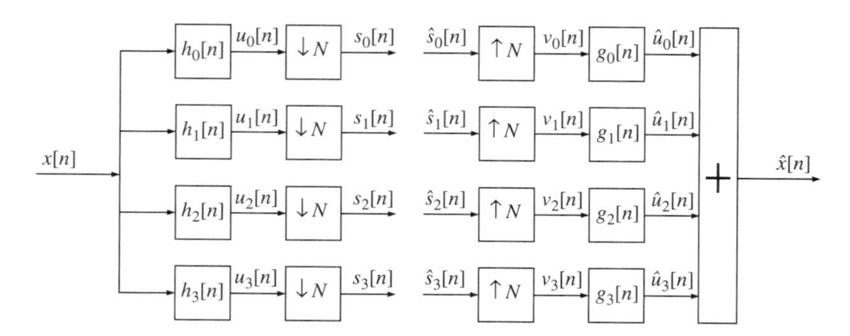

FIGURE P12.2a
Filter bank system for analysis and synthesis of audio signals.

obvious what filter bank is specified and how each step in the algorithm contributes to the implementation. Therefore the purpose of this problem is to explore in depth the nature of the MPEG audio standard filter bank. This lengthy problem is structured to guide the student through the steps of analysis necessary to arrive at the implementation flow charts given in the standard.

The Analysis and Synthesis Filter Banks:

The MPEG analysis filter bank is a set of $N = 32$ bandpass filters, whose impulse responses are of the form

$$h_k[n] = 2h[n]\cos(\omega_k n - \phi_k), \qquad k = 0, 1, \ldots, N-1, \tag{P12.1}$$

where $h[n]$ is the 512-point impulse response of a prototype lowpass filter; the center frequencies of the bandpass filters are

$$\omega_k = \frac{2\pi}{4N}(2k+1) = \frac{\pi}{64}(2k+1), \qquad k = 0, 1, \ldots, N-1, \tag{P12.2}$$

and the phase shifts are $\phi_k = \omega_k N/2$. Equation (P12.1) states that the impulse responses of the bandpass analysis filters are obtained by cosine-modulating a prototype lowpass filter impulse response. The multiplicative factor of 2 in Eq. (P12.1) compensates for a gain factor of 1/2 introduced by the cosine modulation.

(a) Show that the frequency responses of the bandpass channels are

$$H_k(e^{j\omega}) = e^{-j\phi_k}H(e^{j(\omega-\omega_k)}) + e^{j\phi_k}H(e^{j(\omega+\omega_k)}), \qquad k = 0, 1, \ldots, N-1, \tag{P12.3}$$

where the center frequencies of the passbands are given by Eq. (P12.2), and $H(e^{j\omega})$ is the DTFT of $h[n]$.

Also note from Figure P12.2a that the outputs of the bandpass filters are decimated by a factor of N (i.e., $s_k[n] = u_k[nN]$). In practice, this is achieved by computing the filter outputs only once for every N samples of the input signal. Thus, for every N samples of the input, the system produces one sample for each of the N bandpass filters so that the total number of samples/sec remains the same.[19]

(b) Show that the equations describing the analysis filter bank system are

$$s_k[n] = \sum_{m=0}^{511} x[n32 - m]2h[m]\cos[\omega_k(m-16)], \qquad k = 0, 1, \ldots, 31. \tag{P12.4}$$

[19]Note that in Figure P12.2a, n is used as the time index for all signals. If $F_s = 1/T$ is the sampling rate at the input side, then n denotes sampling time $t_n = nT$. On the other hand, the sampling rate at the output of the decimator is F_s/N, and the index n denotes sample time $t_n = nNT$. We consistently indicate the time index of the stream of signal samples as $[n]$ or $[m]$. When a set of values is associated with time index n we will indicate this by a subscript as in the case $s_k[n]$, $k = 0, 1, \ldots, N-1$, which denotes one output sample at time n for each of the N decimated bandpass filter outputs.

In audio coding, the bandpass filter outputs are quantized and encoded for transmission or storage. This coding is represented by the gap in the diagram of Figure P12.2a, and the signals $\hat{s}_k[n]$ represent the quantized bandpass channel signals. Data compression occurs if, on average, fewer bits/sample are used to represent the 32 bandpass channel signals than are needed for the 32 input signal samples. To study the reconstruction ability of the filter bank as in this problem, we assume $\hat{s}_k[n] = s_k[n]$.

For reconstruction of the audio signal, the decoded bandpass filter signals are upsampled by N giving the upsampled signals $v_k[n]$, which are interpolated by a second bank of bandpass filters. The outputs of these filters are summed to produce the overall reconstructed signal. The impulse responses of the synthesis filters are

$$g_k[n] = 2Nh[n]\cos(\omega_k n - \psi_k), \qquad k = 0, 1, \ldots, N-1. \tag{P12.5}$$

In the MPEG standard, the phase shifts of the analysis and synthesis bandpass filters are $\phi_k = \omega_k N/2$ and $\psi_k = -\phi_k = -\omega_k N/2$. This choice of the phases was shown by Rothweiler [333, 334] and Nussbaumer and Vetterli [260] to make it possible to reconstruct the input signal almost perfectly. The factor $2N$ in Eq. (P12.5) compensates for the combined effects of cosine modulation and decimation by N.

(c) Show that the synthesis filters have frequency responses

$$G_k(e^{j\omega}) = Ne^{-j\psi_k}H(e^{j(\omega-\omega_k)}) + Ne^{j\psi_k}H(e^{j(\omega+\omega_k)}). \tag{P12.6}$$

(d) Beginning with $v_k[n]$ expressed as

$$v_k[n] = \sum_{r=-\infty}^{\infty} \hat{s}[r]\delta[n - rN],$$

show that the output of the synthesis filter bank for any output time n is given by the equation

$$\hat{x}[n] = \sum_{k=0}^{31}\left(\sum_{r=-\infty}^{\infty} \hat{s}_k[r](64h[n - r32])\cos[\omega_k(n - r32 + 16)]\right). \tag{P12.7}$$

(e) Since the synthesis filters are causal FIR systems of length 512 samples, only a finite number of past values of the subband signals will contribute to the computation of the output at a given time n. In particular, assume that $r_0 32 \le n < (r_0+1)32$; i.e., assume that the r_0^{th} group of subband samples $\hat{s}_k[r_0]$, $k = 0, 1, \ldots, 31$ has just been received. The sample n can be represented as $n = r_0 32 + m$, where $0 \le m \le 31$ is the index within the next block of output samples to be reconstructed. Show that the equation that describes the computation of the r_0^{th} block of 32 samples of the output of the synthesis filter bank for $0 \le m \le 31$ is

$$\hat{x}[m] = \sum_{k=0}^{31}\left(\sum_{r=0}^{15} \hat{s}_k[r_0 - r](64h[m + r32])\cos[\omega_k(m + r32 + 16)]\right), \tag{P12.8}$$

$$m = 0, 1, \ldots, 31.$$

This equation would be applied repeatedly to compute an output block of 32 samples of $\hat{x}[m]$ from the present (r_0) group and 15 past groups of 32 subband channel signals.

The Prototype Lowpass Filter:

With appropriate choice of the prototype lowpass filter $h[n]$, analysis/synthesis systems based on Eqs. (P12.4) and (P12.8) can be designed and implemented so that, in the absence of any quantization of the channel signals $(\hat{s}_k[n] = s_k[n])$, the analysis/synthesis filter bank of Figure P12.2a is a *nearly perfect reconstruction* system such that $\hat{x}[n] = x[n - 16N]$.

The prototype impulse response for the MPEG audio standard filter bank is given implicitly by the analysis window in Table C.1 of the standard document [156]. This data is contained in the MATLAB file `mpeg_window.mat` available on the book website. The tabulated analysis window as given in the standard was obtained from the prototype impulse response, $h[n]$, by scaling by a factor of 2 and changing the sign of odd numbered blocks (starting from block zero) of 64 samples. The reason for this will become clear later. Figure P12.2b(a) shows a plot of the MPEG analysis window $c[n]$, Figure P12.2b(b) shows the actual impulse response $h[n]$, and P12.2b(c) shows the corresponding lowpass frequency response $|H(e^{j\omega})|$ in dB.

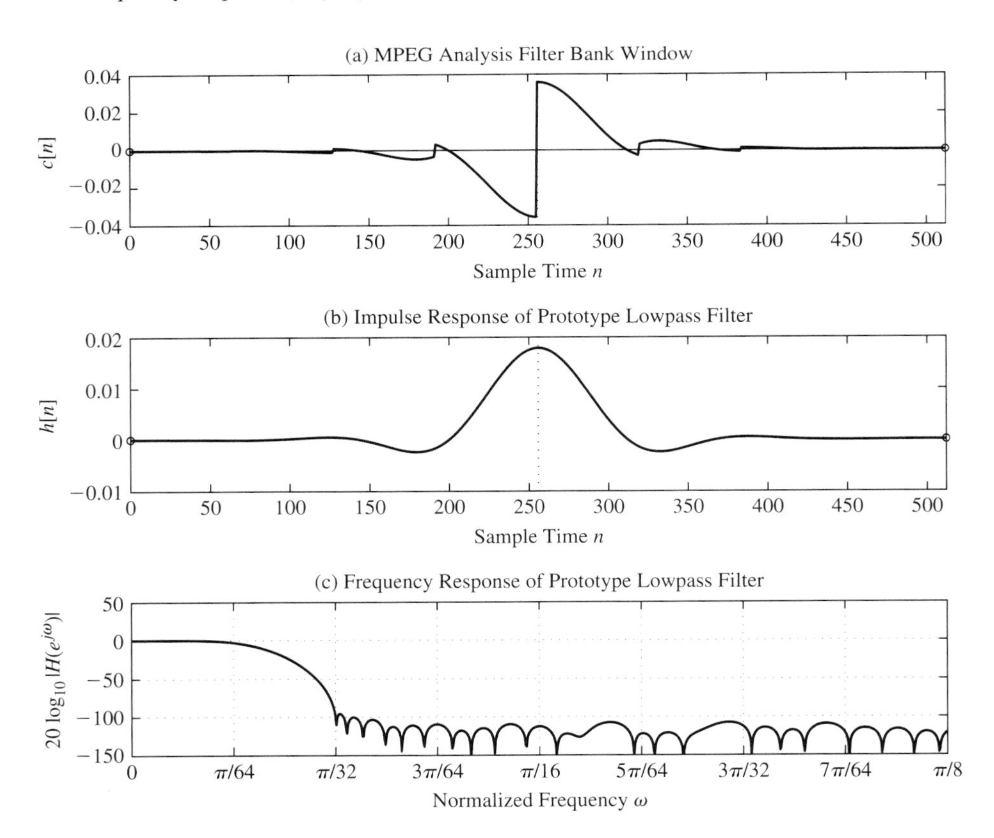

FIGURE P12.2b
The MPEG window and corresponding lowpass filter response functions.

(f) Load the MATLAB file `mpeg_window.mat` and verify the following: $h[0] = 0$ and $h[512 - n] = h[n]$ for $n = 0, 1, \ldots, 512$. Observe that $h[n]$ can be thought of as a type I FIR filter of length 513 samples [270]. What does this imply about the phase response of $H(e^{j\omega})$? What is the time delay of the lowpass prototype filter? The length $L = 513$ can be represented as $L = 2pN + 1$ where $p = 8$. That is, the implied length of the impulse response is 16 times the number of bandpass channels plus one. This plays an important role in the efficient implementation of the filter bank.

(g) Use MATLAB's `freqz( )` function to compute the frequency response of the prototype lowpass filter and construct the plot of Figure P12.2c. (The numerical frequency axis labels are sufficient for the frequency response plot.)

(h) Verify that the gain in the passband is nominally 1 (0 dB), with the gain being -3 dB at $\omega = \pi/64$. Also verify that the gain is less than -90 dB for $\pi/32 \le \omega \le \pi$.

Observe that the stopband cutoff frequency is approximately $\omega_s = \pi/32$, so the transition band extends from approximately $\omega_p = \pi/64$ to $\omega_s = \pi/32$. Recall that the multiplier 2 in Eq. (P12.1) compensates for the cosine modulation so that the bandpass filters also have unity gain in their passbands. Finally, note that the gain of the filter is well below -90 dB for all frequencies above $\omega_s = \pi/32$. This is crucial for minimizing the interaction of the bandpass channels.

The Bandpass Analysis Filters

Since the impulse responses of the bandpass analysis filters are given by Eq. (P12.1), the frequency response of the kth channel is, as determined in part (a), given by Eq. (P12.3), where the center frequencies of the passbands are as in Eq. (P12.2).

(i) Using Eq. (P12.1) write a MATLAB program to compute the frequency responses of the bandpass channels for $k = 0, 1, 2, 3$ over the frequency range $-\pi/64 \le \omega \le \pi/8$. Plot all four curves of $20\log_{10}|H_k(e^{j\omega})|$ on the same axes using different colors or line types. Compare your result to Figure 12.33.

Observe that the transition bands of a given filter overlap with the transition bands of *adjacent* filters. However, only the stopbands (which have gain below -90 dB) of *non-adjacent* filters overlap with the passband and transition bands of a given filter.

Overall Frequency Response of Analysis/Synthesis System

The overall analysis/synthesis system for MPEG audio coding involves two filter banks whose bandpass filters are based upon the same prototype lowpass impulse response, but have different gains and different phase responses. The phase is controlled by the phase factors ϕ_k for the analysis filters and ψ_k for the synthesis filters.

(j) Taking into account the combined effects of filtering, decimation, upsampling, interpolation filtering, and summing, show that the frequency-domain representation of the overall system in Figure P12.2a is

$$\hat{X}(e^{j\omega}) = \sum_{r=0}^{N-1} \tilde{H}_r(e^{j\omega})X(e^{j(\omega-2\pi r/N)})$$

$$= \tilde{H}_0(e^{j\omega})X(e^{j\omega}) + \sum_{r=1}^{N-1} \tilde{H}_r(e^{j\omega})X(e^{j(\omega-2\pi r/N)}), \qquad \text{(P12.9a)}$$

where

$$\tilde{H}_r(e^{j\omega}) = \frac{1}{N}\sum_{k=0}^{N-1} G_k(e^{j\omega})H_k(e^{j(\omega-2\pi r/N)}). \quad \text{(P12.9b)}$$

The two terms on the right in the expanded form of Eq. (P12.9a) show that there are two important considerations. The first term gives the contribution to the output due to the input, $X(e^{j\omega})$, and the second partial sum of the remaining terms gives the contributions of all the aliasing terms introduced by the decimation.

(k) Focusing on a particular channel k and surrounding channels $k-1$ and $k+1$, and invoking the assumption that non-adjacent terms do not interact, show that to a high degree of accuracy, the alias terms completely cancel out and the overall system from input to synthesized output is described by the equation

$$\hat{X}(e^{j\omega}) = e^{-j\omega(L-1)}X(e^{j\omega}). \quad \text{(P12.10)}$$

That is, the overall system has unit gain and a time delay of $L-1 = 2pN = 512$ samples. This delay is comprised of $(L-1)/2 = 256$ samples delay from both the analysis and the synthesis filter bank.

(l) Using the prototype impulse response $h[n]$ given in the MATLAB file `mpeg_window.mat` and Eqs. (P12.1) and (P12.5), compute the bandpass impulse responses $h_k[n]$ and $g_k[n]$ for $n = 0, 1, \ldots, 512$ and use these in the function `freqz( )` to compute the frequency responses of all the bandpass filters, i.e., for $k = 0, 1, \ldots, 31$. Then use Eq. (P12.9b) with $r = 0$ to compute the overall composite frequency response $\tilde{H}_0(e^{j\omega})$ for $0 \leq \omega \leq \pi$. Plot your result. Measure the deviation from the ideal of constant unity gain. Repeat for $r = 4$; i.e., compute and plot the alias frequency response term, $\tilde{H}_4(e^{j\omega})$, from Eq. (P12.9b).

Implementation of the Analysis Filter Bank

The implementation of the MPEG audio analysis filter bank requires the following computations:[20]

$$s_k[n] = \sum_{m=0}^{511} x[n32-m]2h[m]\cos[\omega_k(m-16)], \qquad k = 0, 1, \ldots, 31, \quad \text{(P12.11)}$$

where $\omega_k = 2\pi(2k+1)/128$. In the MPEG audio standard, the implementation of the analysis filter bank is specified by the flow chart in Figure C.4 [156], which shows a sequence of operations that computes one sample of the output of each of the 32 bandpass analysis filters. This flow chart is redrawn here in Figure P12.2c. To see that Figure P12.2c is an implementation of the computations defined by Eq. (P12.11), it is necessary to reorganize the terms of that equation.

[20]Note that even though the implied length of $h[n]$ is 513 samples, the upper limit of the sum in Eq. (P12.11) is 511 because $h[512] = h[0] = 0$.

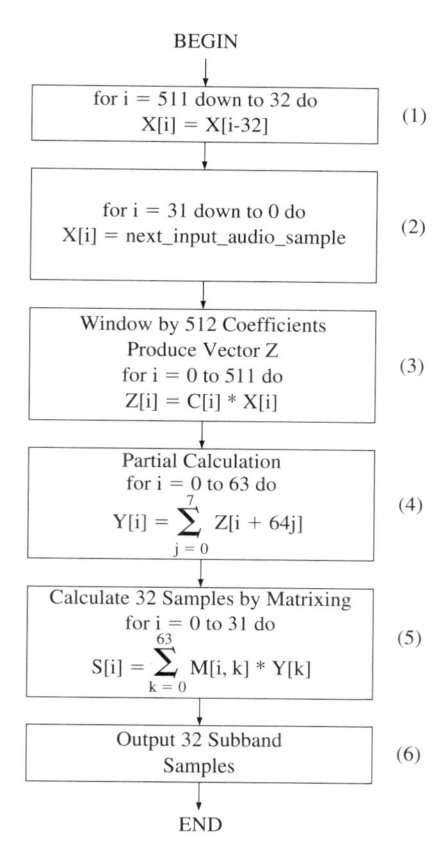

FIGURE P12.2c
Flow chart of MPEG analysis filter bank implementation.

(m) Show that Eq. (P12.11) can be represented as

$$s_k[n] = \sum_{m=0}^{511} w_m[n] \cos\left[\omega_k(m - 16)\right], \qquad k = 0, 1, \ldots, 31, \tag{P12.12}$$

where $w_m[n] = x[n32 - m]2h[m]$ for $m = 0, 1, \ldots, 511$.

In this case, we use a subscript m to index a sequence $w_m[n]$ that is associated with computations at a specific time index n. This form makes it clear that if $w_m[n]$ for $m = 0, 1, 2, \ldots, 511$ is computed once at time n and saved in an array of length 512, then each of the 32 bandpass filter outputs can be computed by modulating with a cosine of frequency ω_k and then summing all terms in the product sequence $w_m[n]$ cos $[\omega_k(m - 16)]$ with m running from 0 to 511. This would save 512 multiplications in 31 out of the 32 filters.

(n) More computations can be saved by using the periodicity and anti-symmetry of the cosine sequences. To see this, Eq. (P12.12) is further transformed by defining the index $m = q + r64$ where q ranges from 0 to 63 and r ranges from 0 to 7. Verify that m would range from 0 to 511 for these values of q and r, and show that with this substitution and some regrouping, Eq. (P12.12) becomes

$$s_k[n] = \sum_{q=0}^{63}\sum_{r=0}^{7} w_{q+r64}[n]\cos\left[\frac{2\pi(2k+1)}{128}(q + r64 - 16)\right], \qquad k = 0, 1, \dots, 31, \tag{P12.13}$$

and since (show this)

$$\cos\left[\frac{2\pi(2k+1)}{128}(q + r64 - 16)\right] = (-1)^r \cos\left[\frac{2\pi(2k+1)}{128}(q - 16)\right], \tag{P12.14}$$

Eq. (P12.13) can finally be written as

$$s_k[n] = \sum_{q=0}^{63}\left(\sum_{r=0}^{7} w_{q+r64}[n](-1)^r\right)\cos\left[\frac{2\pi(2k+1)}{128}(q - 16)\right], \qquad k = 0, 1, \dots, 31. \tag{P12.15}$$

(o) To facilitate even further simplification it is helpful to examine the sequence $w_{q+r64}[n](-1)^r$ in more detail. Expand it to obtain

$$\begin{aligned} z_{q+r64}[n] &= w_{q+r64}[n](-1)^r = x[n32 - q - r64]2h[q + r64](-1)^r \\ &= x[n32 - q - r64]c[q + r64], \end{aligned} \tag{P12.16}$$

where

$$c[q + r64] = 2h[q + r64](-1)^r \qquad \text{for } q = 0, 1, \dots, 63 \text{ and } r = 0, 1, \dots, 7. \tag{P12.17}$$

This is the analysis window plotted in Figure P12.2b(a).

(p) Finally, with this definition of the "analysis window," $c[n]$, as being equal to twice the prototype lowpass filter impulse response with odd-indexed groups of 64 samples changed in sign (see Figure 12.32a), show that the MPEG analysis filter bank equations become

$$s_k[n] = \sum_{q=0}^{63} y_q[n]\cos\left[\frac{2\pi(2k+1)}{128}(q - 16)\right], \qquad k = 0, 1, \dots, 31 \tag{P12.18}$$

where

$$y_q[n] = \sum_{r=0}^{7} z_{q+r64}[n] \tag{P12.19}$$

and

$$z_m[n] = x[n32 - m]c[m], \qquad m = 0, 1, 2, \dots, 511. \tag{P12.20}$$

Equations (P12.18), (P12.19), and (P12.20) are the basis for the algorithm depicted in Figure P12.2c. This figure shows how to compute one sample of each of the 32 bandpass filter outputs.

(q) It is assumed that the input signal is buffered in a sequential array X[ ] of 512 samples. To complete the analysis, we should connect the blocks in the flow chart of Figure P12.2c with the equations that we have derived.

(1) This step shows that the input array is updated by pushing $512 - 32 = 480$ samples down in the input buffer by 32 samples. The last 32 samples (oldest in time) are discarded.

(2) 32 new samples are read into the array. Note that the input array is indexed in reverse time order so that the most recent sample is stored in array position 0 and the oldest sample is in position 511.

(3) The input array X[ ] is multiplied sample-wise by the analysis window C[ ]. What equation does this implement?

(4) A 64 sample array of samples is obtained by summing groups of 8 samples of Z[ ] offset by 64 samples. This results in an array Y[ ] of 64 samples. To what equation does it correspond?

(5) The modulation by cosines for the different bandpass channels is done by this operation. The output samples of the 32 bandpass filters can be thought of as a 32 by 1 column vector, S, obtained by multiplying a 32 by 64 matrix, M, by a 64 by 1 column vector Y

$$\mathrm{S} = \mathrm{MY}.$$

Show that the matrix M is a matrix of cosine values whose elements are defined by

$$\mathrm{M}[\mathrm{i}, \mathrm{k}] = \cos\left[\frac{2\pi}{64}(2\mathrm{i}+1)(\mathrm{k}-16)\right] \quad \begin{cases} \mathrm{i} = 0, 1, \ldots, 31 \\ \mathrm{k} = 0, 1, \ldots, 63. \end{cases} \tag{P12.21}$$

This "matrixing" completes the implementation of Eq. (P12.18).

(6) The 32 by 1 vector S contains one output sample for each of the 32 analysis bandpass filters. Recall that if the sampling rate of the input signal is F_s, then the effective sampling rate of the analysis filter outputs is $F_s/32$.

This completes the derivation and discussion of the implementation of the MPEG analysis filter bank as defined in the standard [156].

12.3. This problem is an extension of Problem 12.2. It explores the use of discrete Fourier transforms for the matrixing operation. It is important to note that while only the MPEG decoder algorithm is standardized, it is really only in the control algorithm for the quantization that there is much flexibility. While different perceptual models can be used for deriving the quantizer parameters, the decoder assumes that the channel signals are obtained from the specified filter bank. Nevertheless, there is still some flexibility in how the filter bank is implemented. Most of variations focus on more efficient ways of implementing step (5) of Figure P12.2c by using fast Fourier transform or fast discrete cosine transform algorithms to implement the multiplications by the cosine terms.

(a) To see how this can be done, show that Eq. (P12.18) in Problem 12.2 can be written as

$$s_k[n] = \mathcal{R}e\left\{\sum_{q=0}^{63} y_q[n]e^{-j2\pi(2k+1)(q-16)/128}\right\}, \qquad k = 0, 1, \dots, 31. \qquad \text{(P12.22)}$$

(b) Show that Eq. (P12.22) can be manipulated into the form

$$s_k[n] = \mathcal{R}e\left\{e^{j\pi(2k+1)/4}\left(\sum_{q=0}^{63}(y_q[n]e^{-j\pi q/64})e^{-j2\pi kq/64}\right)\right\}, \qquad k = 0, 1, \dots, 31. \qquad \text{(P12.23)}$$

(c) Give an algorithm using the FFT for evaluating the term inside the large parenthesis

$$Y_k[n] = \sum_{q=0}^{63}(y_q[n]e^{-j\pi q/64})e^{-j2\pi kq/64}, \qquad k = 0, 1, \dots, 31. \qquad \text{(P12.24)}$$

This approach to saving computation is suggested by Rothweiler [333, 334] and discussed in more detail by Nussbaumer and Vetterli [260], who suggest computing the DFT with an FFT algorithm of Rader and Brenner that uses only real multiplications [316].

Other manipulations of Eq. (P12.18) can lead to other ways to implement the matrixing in step (5) of Figure P12.2c. One example is the method of Konstantinides [198, 199], in which Eq. (P12.18) is manipulated into the form of a 32-point inverse discrete cosine transform (IDCT). If the IDCT is computed with a fast algorithm, this method would generally require less computation than an FFT-based approach.

12.4. The impulse responses of the MPEG analysis filter bank are given by

$$h_k[n] = 2h[n]\cos\left(\frac{\pi(2k+1)}{64}(n-16)\right), \qquad k = 0, 1, \dots, 31, \qquad \text{(P12.25)}$$

Figure P12.4 shows the first four impulse responses plotted as a function of n. Write a MATLAB program to plot all 32 impulse responses using the `strips( )` function.

12.5. (MATLAB Exercise) A complete two-band subband coder is shown in Figure P12.5. In this figure, the filters are all based on the lowpass QMF prototype filter $h_0[n]$; i.e., $h_1[n] = (-1)^n h_0[n]$, $g_0[n] = h_0[n]$, and $g_1[n] = -h_1[n]$. These filters satisfy the condition for alias cancellation, and it can be shown that the overall subband coder frequency response is:

$$H(e^{j\omega}) = \frac{Y(e^{j\omega})}{X(e^{j\omega})} = \frac{1}{2}\left[H_0^2(e^{j\omega}) - H_1^2(e^{j\omega})\right].$$

By choosing $h_0[n]$ to be an even-length, linear phase FIR filter with nominal cutoff frequency at $\omega = \pi/2$ (or equivalently $f = F_s/4$), the overall frequency response can be made nearly flat in magnitude and linear in phase [175]. Under these conditions, and if no quantization is done to the channel signals, $y_0[n]$ and $y_1[n]$, the output of the subband coder is nearly equal to a delayed version of the input to the coder. If the channel signals

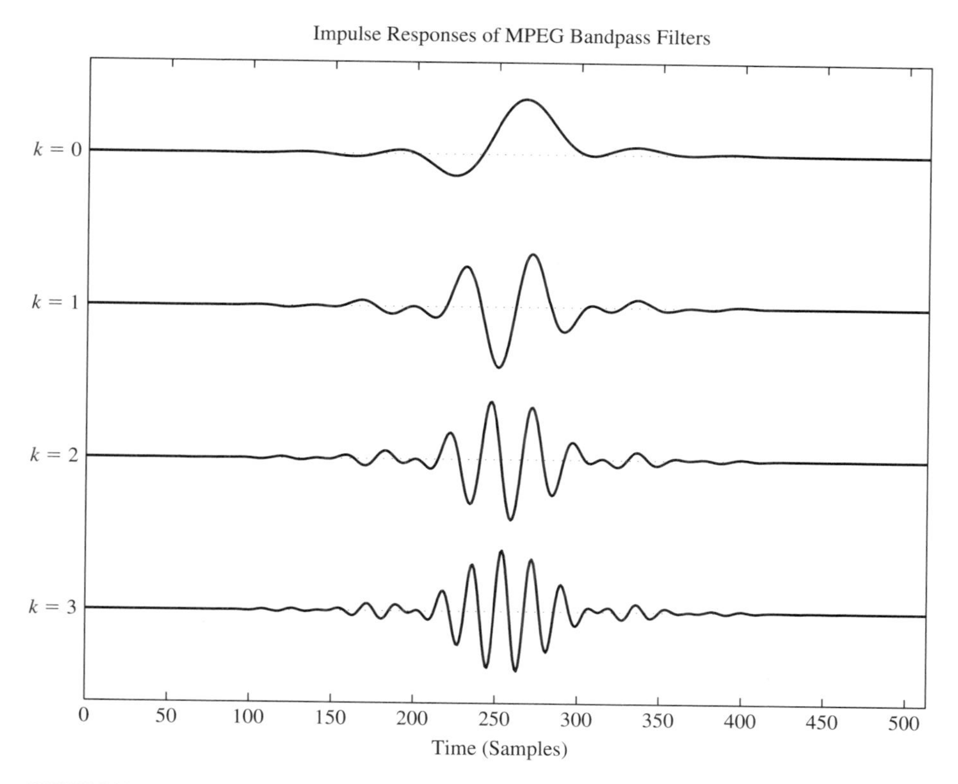

FIGURE P12.4
The impulse responses of the first four MPEG analysis filters.

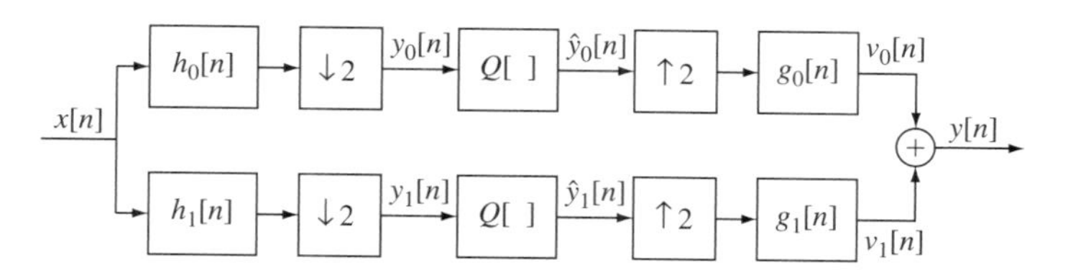

FIGURE P12.5
Two-channel QMF filter bank.

are quantized, it is possible to maintain high perceptual quality while making the total bit rate lower than would be required to quantize the waveform $x[n]$ directly.
Three prototype QMF filters, namely `h24D, h48D, h96D` with filter durations of 24, 48, and 96 samples are provided on the book website. The filter coefficients can be loaded using the MATLAB command `load filters.mat;`

(a) Plot the time and frequency responses of the three QMF filters on a common plot using both linear and log magnitude scales. Assume a sampling frequency of 8000 Hz. How do the three filters compare in terms of sharpness of the cutoff region?

(b) Implement the two-band subband coder as shown in Figure P12.5. Use the function `fxquant( )` (also available on the book website) as a fixed quantizer in each channel. Exploit the fact that the high frequency channel will have lower amplitude than the low frequency channel, so you should be able to use fewer bits for quantization of the high frequency channel signal. You may want to experiment with an adaptive quantizer that adapts the step sizes in blocks to match the signal amplitudes in each channel.

Write your implementation of the two-band subband coder as a MATLAB function with the following input/output structure:

```
function y=subband(x,h),bits)
% Subband decomposition
% y=subband(x,h0,[bits])
% x=input signal vector
% h0=basic QMF filter
% bits=a vector of 2 entries giving the number of bits
% for each channel (optional, is missing do not quantize)
% y=output signal vector
```

(c) Test your system using an impulse as input; i.e., $x[n] = \delta[n]$ to obtain the overall subband coder impulse response $h[n]$. You can then use the DFT to compute the overall frequency response, $H(e^{j\omega})$. Plot the impulse response and frequency response (both the magnitude in dB and the phase in radians) of the overall system using each of the three prototype QMF filters, namely `h24D, h48D, h96D`. Measure the coder time delay and compare it to the theoretical value for the filter length that you are using.

(d) Measure the signal-to-noise ratio (SNR) of the *unquantized* system. You can do this by exciting the system with a speech signal (e.g., `s5.wav`). Use the time delay determined above to align the input and output signals. Also, be sure to avoid end effects where the signals cannot match up. Do this by taking corresponding signal segments of the same length out of the middle of the input and output signal vectors. Do not expect to get infinite SNR since the system is not perfect, i.e., it does not reconstruct the input exactly even with no quantization. It is reasonable to expect an SNR on the order of 70 dB for the long filters, i.e., `h96D`. How does the SNR depend on the filter length? Is the output of the subband coder indistinguishable from the input to the system? (You evaluate this by listening to the speech file processed by the coder).

(e) Using the quantizer provided (`fxquant( )`), measure the SNR for different distributions of the bits. Compare your results to quantizing the signal as a single channel with the same bit rate as the total bit rate for the two half-band signals. Make your comparison on the basis of both SNR and by listening to the input and output signals with quantization.

CHAPTER 13

Text-to-Speech Synthesis Methods

13.1 INTRODUCTION

The quest to build a machine that could speak with the intelligibility and naturalness of a human being has been a subject of great interest for more than 200 years.[1] Flanagan notes that the earliest documented effort at building a talking machine or speech synthesizer was by Kratzenstein in 1779 [105] who built a set of acoustic resonators similar in shape to a human vocal tract. When excited by a vibrating reed, the output of the resonators produced human sounding vowels. Since the time of Kratzenstein, a number of efforts have been directed at creating a voice synthesizer, some with more success than others. In this chapter we will delve broadly (and sometimes deeply) into the research and application area of *text-to-speech synthesis*, or TTS as it is widely known in the field, where the goal of the machine is to convert ordinary text messages into *intelligible* and *natural sounding* synthetic speech so as to transmit information from a machine to a person by voice [355].

A block diagram of a basic TTS system is shown in Figure 13.1. Ideally, the input to the TTS system is arbitrary input text and the output of the TTS system is synthetic speech. There are two fundamental processes that all TTS systems must perform. An analysis of the text must be performed in order to determine the abstract underlying linguistic description of the speech signal. Then the proper sounds corresponding

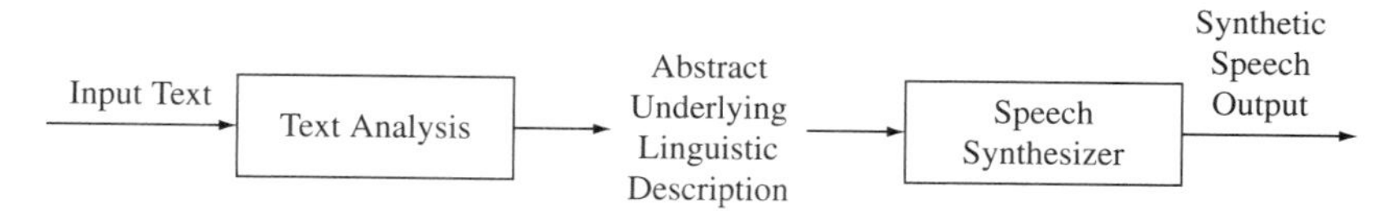

FIGURE 13.1
Block diagram of general TTS system.

[1]Much of the material in this chapter is based on a joint lecture on TTS methods by Dr. Juergen Schroeter of AT&T Labs Research and one of the authors of this book (LRR) and is presented here with the permission of Dr. Schroeter.

to the text input must be synthesized; i.e., a sampled version of the desired spoken output must be created via the speech synthesizer in a form that can be converted to an acoustic signal by a D-to-A converter.

In Section 13.2 we discuss the problem of text analysis. Our goal is to highlight the important linguistic issues and give a necessarily superficial look at how text analysis can produce the requisite phonetic and prosodic data that can be used to control a speech synthesis system. The remaining sections of this chapter focus on various techniques for synthesizing speech from the output of a text analysis system.

13.2 TEXT ANALYSIS

The text analysis module of Figure 13.1 must determine three things from the input text string:

1. **Pronunciation of the text string**: The text analysis process must decide on the set of phonemes that is to be spoken, the degrees of stress at various points in the speech utterance, the intonation of the speech, and the duration of each of the sounds of the utterance.
2. **Syntactic structure of the sentence to be spoken**: The text analysis process must determine where to place pauses, what rate of speaking is most appropriate for the material being spoken, and how much emphasis should be given to individual words and phrases within the final spoken output speech.
3. **Semantic focus and ambiguity resolution**: The text analysis process must resolve homographs (words that are spelled alike but can be pronounced in more than one way), and also must use rules to determine word etymology to decide how best to pronounce words that are not in a conventional dictionary (e.g., names and foreign words and phrases).

Figure 13.2 shows more detail on how text analysis is performed. The input is plain English text and the first stage of processing does some basic text processing operations, including detecting the structure of the document containing the text (e.g., e-mail message versus paragraph of text from an encyclopedia article), normalizing the text (so as to determine how to pronounce words like proper names or homographs with multiple pronunciations), and finally performing a linguistic analysis to determine grammatical information about words and phrases within the text. The basic text processing benefits from an on-line dictionary of word pronunciations along with rules for determining word etymology. The output of the basic text processing step is tagged text, where the tags denote the linguistic properties of the input text string, as determined by the basic text processing block. We now examine each of the three components of the basic text processing block of Figure 13.2 in more detail.

13.2.1 Document Structure Detection

The document structure detection module seeks to determine the location of all punctuation marks in the text, and to decide their significance with regard to the sentence and paragraph structure of the input text. For example, an *end of sentence* marker is

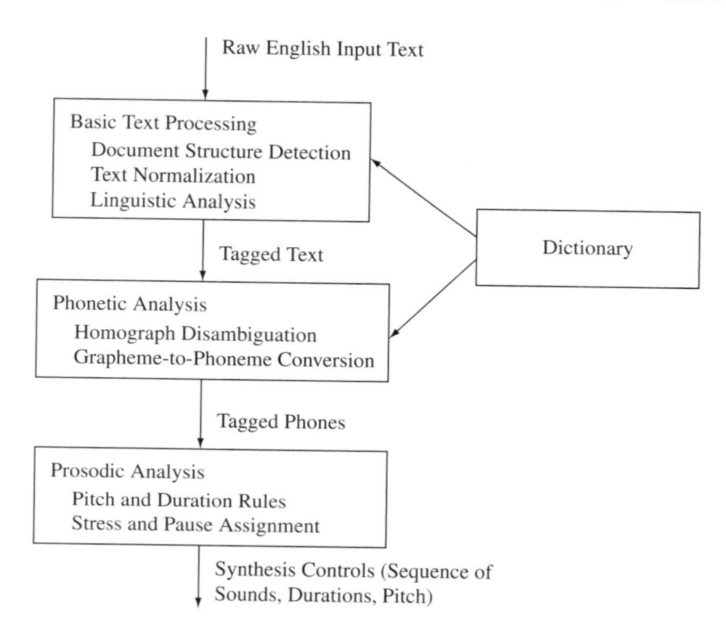

FIGURE 13.2
Components of the text analysis process.

usually a period, ".", a question mark, "?", or an exclamation point, "!". However this is not always the case, as in the sentence,

> "This car is 72.5 in. long"

where there are two periods, neither of which denote the end of the sentence. Another example that illustrates the need to accurately determine document structure is the e-mail message:

> Larry:
> Sure. I'll try to do it before Thursday :-)
> Ed

where the marks at the end of the text message (on the second line above) are a "smiley" message whose pronunciation has little to do with the particular marks in the smiley. Finally, in the case of foreign words, especially those with unusual accent and diacritical marks, special care must be taken to properly handle the text so as not to introduce extraneous and meaningless sounds.

13.2.2 Text Normalization

Text normalization methods handle a range of text problems that occur frequently in real applications of TTS systems. Perhaps the most pervasive problem is the issue of how to handle abbreviations and acronyms. In the following examples, there is ambiguity as to the pronunciation of one or more abbreviations or acronyms:

Example 1: "Dr. Smith lives on Smith Dr."
Example 2: "I live on Bourbon St. in St. Louis"
Example 3: "She worked for DEC in Maynard, MA"

In Example 1, the text "Dr." is pronounced as "Doctor" (when it precedes the name) and as "drive" when the name precedes the text. In Example 2, the text "St." is pronounced as "street" or "Saint," depending on the context of the sentence. Finally, in Example 3, the acronym DEC can be pronounced as either the word "deck" (the spoken acronym) or the name of the company, Digital Equipment Corporation, but it is virtually never pronounced as the letter sequence "D E C."

Similar issues occur with number strings in sentences. For example the number sequence "370-1111" can be pronounced as "three seven zero one one one one" if it is to be interpreted as a raw sequence of digits, or "three seventy dash one thousand eleven" if it is to be interpreted as a so-called "natural number," or even "three seventy model one one one one" if it is an IBM Model 370 computer. Similarly the string "1920" can be pronounced as the year "nineteen twenty" or the number "one thousand nine hundred and twenty."

Finally dates, times, currency, account numbers, ordinals, and cardinal numbers all present their own sets of problems of pronunciation in various contexts. For example, the string "Feb. 15, 1983" probably needs to be converted to the string "February fifteenth, nineteen eighty three," the string "$10.50" should be pronounced as "ten dollars and fifty cents," and the string "Part #10-50" should be pronounced as "part number ten dash fifty" rather than "part pound sign ten to fifty" as might be implied by interpreting the input symbols one character at a time.

Other important text normalization problems include the need to determine how to pronounce proper names, especially those from languages other than English. Thus the name "Rudy Prpch" (a former U.S. Senator from Minnesota) is pronounced as though the name were written "Rudy Perpich." Another important aspect of text normalization is determining the part of speech (POS) of a word, in the context of the sentence, to determine its pronunciation. Thus the word "read" is pronounced differently when used in the past tense (where it is pronounced like the word "red"), than when used in the present tense (where it is pronounced like the word "reed"). Similarly the word "record" is pronounced differently as a noun ("rec-erd") than as a verb ("ri-cord"). Word decomposition is a process of segmenting a complex word into its base form (denoted as a morpheme) along with a set of affixes (both prefixes and suffixes) so as to determine the correct pronunciation. Thus the word "indivisibility" can be broken down into the set of two prefixes, a base form, and a suffix; i.e., "in-di-visible-ity."

Two other problems that text string normalization must handle properly are determining the pronunciation (or even just the meaning) of special symbols in text, e.g., the set of symbols #$%&˜_{}*@, and the problem of resolving character strings such as "10:20," which can be interpreted as a time of day (where it would be pronounced as "twenty after ten") or as a sequence of numbers (where it would be pronounced as "ten to twenty").

13.2.3 Linguistic Analysis

The last step in the basic text processing block of Figure 13.2 is a linguistic analysis of the input text, with the goal of determining, for each word in the printed string, the following linguistic properties:

- the part of speech of the word
- the sense in which each word is being used in the current context
- the location of phrases (or phrase groups) within a sentence (or paragraph), i.e., where a pause in speaking might be appropriate in the ultimate synthesis of the sentence
- the presence of anaphora (i.e., the use of a linguistic unit, such as a pronoun, to refer back to another unit; e.g., the use of her to refer to Anne in the sentence, "Anne asked Edward to pass her the salt.")
- the words on which emphasis are to be placed (for prominence in the sentence)
- the style of speaking, e.g., irate, emotional, relaxed, etc.

A conventional parser could be used as the basis of the linguistic analysis of the printed text, but typically a simple, shallow analysis is performed, since most linguistic parsers are very slow.

13.2.4 Phonetic Analysis

Ultimately, the tagged text obtained from the basic text processing block of any TTS system must be converted to a sequence of tagged phones that describe both the sounds to be produced as well as the manner of speaking, both locally (emphasis) and globally (speaking style). The phonetic analysis block of Figure 13.2 provides the processing that enables the TTS system to perform this conversion, with the help of a pronunciation dictionary. There are two components to the phonetic analysis block, namely a homograph disambiguation step, and a grapheme-to-phoneme step. We discuss the processing of these two blocks next.

13.2.5 Homograph Disambiguation

The homograph disambiguation block must resolve the correct pronunciation of each word in an input string that has more than one pronunciation [424]. Examples of homograph disambiguation include the following:

- "an *absent* boy" versus "do you choose to *absent* yourself?" In the first phrase, the accent is on the first syllable, and in the second phrase, the accent is on the second syllable, for the word "absent."
- "an *overnight* bag" versus "are you staying *overnight*?" Again, in the first phrase, the accent is on the first syllable of "overnight," whereas in the second phrase, the accent moves to the second syllable.
- "he is a *learned* man" versus "he *learned* to play piano." In the first phrase, the accent is on the second syllable of "learned," whereas in the second phrase, the accent is on the first syllable.

13.2.6 Letter-to-Sound Conversion

The second step of phonetic analysis is the process of grapheme-to-phoneme conversion, namely conversion from the text to (marked) speech sounds. Although there are a variety of ways of performing this analysis, perhaps the most straightforward method is to rely on a standard pronunciation dictionary, along with a set of letter-to-sound (LTS) rules for words outside the dictionary.

Figure 13.3 shows the processing for a simple dictionary search for word pronunciation. Each individual word is searched independently. First a "whole word" search is initiated to see if the printed word exists, in its entirety, as an entity of the dictionary. If it exists, the conversion to sounds is straightforward (i.e., copy the sounds directly from the dictionary) and the dictionary search begins on the next word. If not, as is most often the case, the dictionary search attempts to find affixes (both prefixes and suffixes) and strips them from the word (thereby creating what it hopes will be the "root form" of the word), and does another "whole word" search. If the root form is present in the dictionary, an affix reattachment routine is used to modify the word pronunciation to include the effect of each affix on the word pronunciation. If the root form is not present, a set of LTS rules must be used to determine the best pronunciation (perhaps based on the etymology of the word, or the word position, or the word POS) of the root form of the word, again followed by the reattachment of affixes. The use of classification and regression tree (CART) analysis methods to perform the LTS conversion of the root form of the word is one method that has been widely used [44]. More recently, LTS rules have been implemented as finite state machines (FSMs) [246].

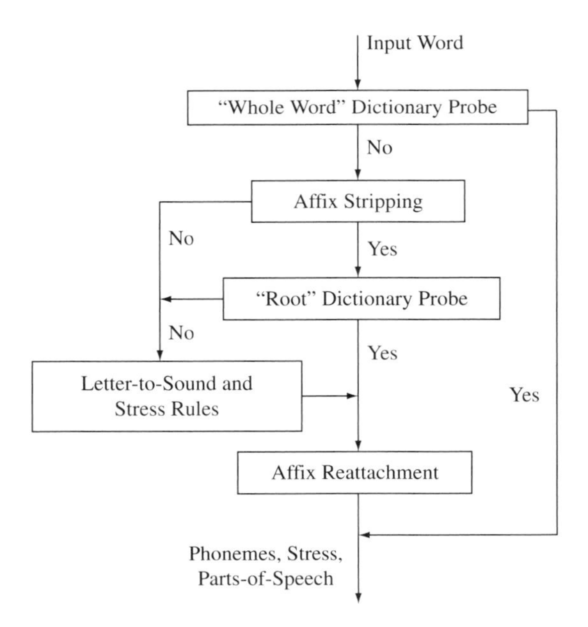

FIGURE 13.3
Flow diagram of dictionary search for proper word pronunciation.

13.2.7 Prosodic Analysis

The last step in the text analysis of Figure 13.2 is prosodic analysis, which provides the speech synthesizer with the complete set of synthesis controls, namely the sequence of speech sounds, their durations, and their pitch. The determination of the sequence of speech sounds is mainly done by the phonetic analysis step, as outlined above. The assignment of duration and pitch to each sound is done by a set of pitch and duration rules, along with a set of rules for assigning stress and determining where appropriate pauses should be inserted so that the local and global speaking rates are as natural as possible.

13.2.8 Prosody Assignment

There are four components to speech prosody, namely pause location and duration assignment (pauses indicate phrasing in speech), pitch assignment (which gives the speech a certain rhythm and liveliness), speech rate determination based on the relative durations of the sounds in the utterance (which determines phoneme durations and timing), and finally sound loudness (which, along with pitch, are cues for emphasis, emotion, and speaking style).

Figure 13.4 shows a flow diagram of the processes used to determine the pitch (or fundamental frequency, F_0) contour of an utterance from the parsed text and phone string output of the phonetic analysis stage of Figure 13.2. As shown in Figure 13.4, all

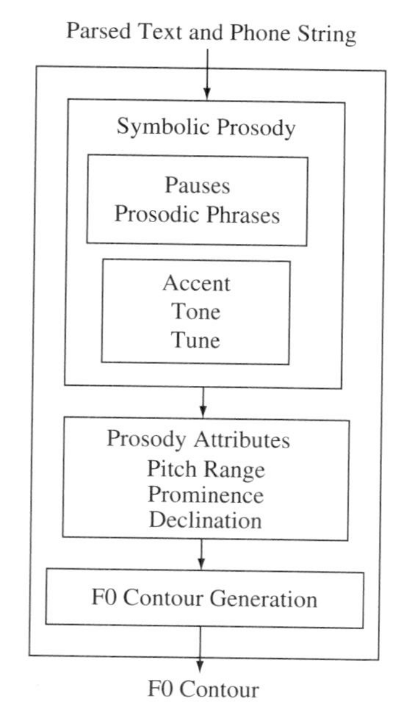

FIGURE 13.4
Flow diagram of general method for determining the pitch contour of a speech utterance.

H*	Peak Accent	
L*	Low Accent	
L*+H	Scooped Accent	
L*+!H	Scooped Downstep Accent	
L + H*	Rising Peak Accent	
!H*	Downstep High Tone	

FIGURE 13.5
Block diagram of general method for determining symbolic and phonetic prosody of a speech utterance. (After Silverman et al. [361].)

the processing steps are modulated (in unspecified ways) by the speaking style of the utterance. The first step in the process is to convert the symbolic prosody contour into a set of pauses and prosody phrases in the utterance so as to create a set of accents, tones, and tunes that describe the general shape and behavior of the pitch contour of the utterance. The use of so-called "ToBI" (tone and break indices) marks, as illustrated in Figure 13.5, provides a description of the general shapes of the pitch contour in local regions, in the form of a set of six distinct pitch accent tones, which can be concatenated to describe the global pitch contour [361, 387]. Depending on the text to be spoken, along with specifications of the pitch range (of both the speaker and the speaking style), the places where prominence or emphasis are to be achieved, and the places where pitch declination occurs, a set of anchor points of the pitch contour are determined, and the final pitch contour of the utterance is obtained by an interpolation process, as illustrated in Figure 13.6. The upper curve shows the set of anchor points obtained from the accent, tone, pitch range, prominence, and declination information determined from the symbolic prosody module, and the lower curve shows the resulting interpolated pitch contour, modulated by the zero-valued pitch frequency during unvoiced portions of the utterance.

With this brief introduction to the main issues in text analysis as context, we can now look at how DSP techniques are employed in synthesizing a signal that is perceived as a spoken utterance of a text message.

13.3 EVOLUTION OF SPEECH SYNTHESIS METHODS

A summary of the progress in speech synthesis, over the period 1962–1997, is given in Figure 13.7. This figure shows that there have been three generations of speech synthesis systems. During the first generation (between 1962 and 1977), formant synthesis of phonemes using a terminal analog synthesizer was the dominant technology, with intense research programs at Bell Labs, the Royal Institute of Technology in Sweden, the Joint Speech Research Unit in the UK, MIT, and Haskins Labs among other places. This method of synthesis used rules that related the phonetic decomposition of the sentence to formant frequency contours.[2] Early on, the synthesis suffered from

[2]First generation synthesis systems had, as input, a string of phonemes and some form of prosody marking obtained by hand; hence such systems were not true TTS systems but merely synthesis systems controlled by human-derived inputs.

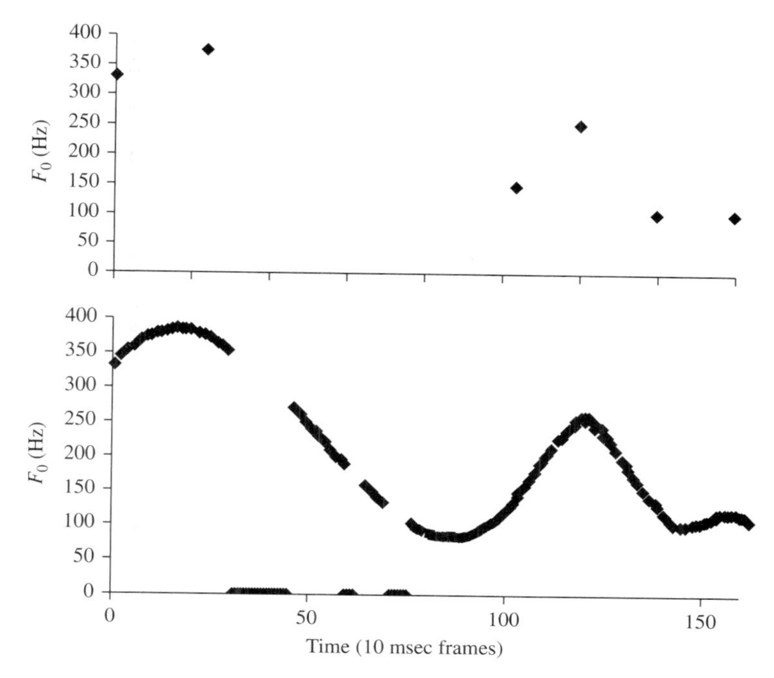

FIGURE 13.6
Illustration of F_0 contour generation from anchor points.

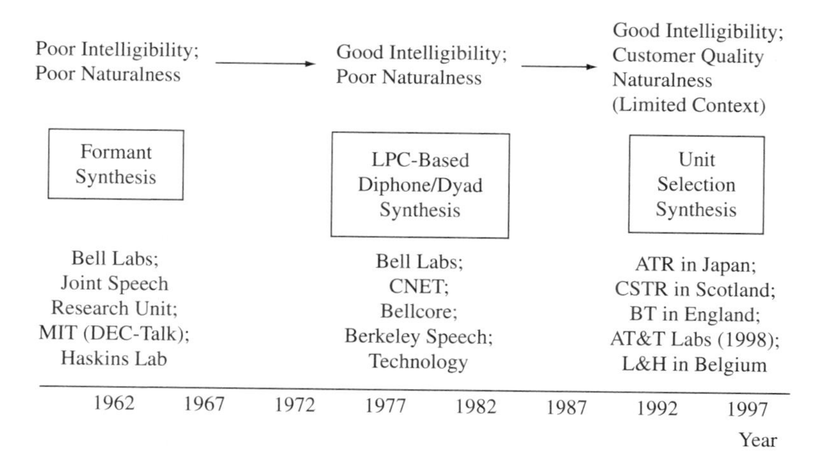

FIGURE 13.7
Timeline of progress in speech synthesis and TTS systems.

both poor intelligibility and poor naturalness as researchers had to learn how best to estimate the control parameters for the synthesizer, and how best to model the speech spectrum to provide intelligible speech output.

The second generation of speech synthesis methods (from 1977 to 1992) followed the period of development of linear predictive coding (LPC) analysis and synthesis methods. During this period, a range of TTS systems started to appear based on using sub-word units such as diphones (units that begin in the middle of the stable state of a phone and end in the middle of the following phone) as the basic unit being modeled, and using a method of concatenating diphones to produce the resulting synthetic speech. The synthesis problem for second-generation synthesis systems was considerably more difficult than in the first generation, since the input usually was printed text, with all the problems of resolving how to properly pronounce the text, as discussed at the beginning of this chapter. The leading players in this second generation of synthesis research were groups at Bell Telephone Labs, Bellcore, MIT, and Berkeley Speech Technology, among others. By carefully modeling and representing diphone units using LPC parameters, it was shown that synthetic speech with good intelligibility could be reliably obtained from text input by concatenating the appropriate diphone units. Although the intelligibility improved dramatically over first generation formant synthesis, the naturalness of the synthetic speech remained low due to the inability of single diphone units to represent all possible combinations of sound using that diphone unit. A detailed survey of progress in TTS conversion up to 1987 was given in a review paper by Klatt [194]. The synthesis examples that accompanied that paper are available for listening at the website http://www.cs-indiana.edu/rhythmsp/ASA/Contents.html.

The quest for a method of speech synthesis that could give good intelligibility along with "customer quality" naturalness served as the motivation for TTS research in the third generation of speech synthesis methods (from 1992 to the present). During this period, the method of unit selection synthesis was introduced and perfected, primarily by Y. Sagisaka at ATR Labs in Kyoto [338]. The resulting synthetic speech from unit selection synthesis had good intelligibility and naturalness that approached that of human generated speech. The use of unit selection synthesis spread rapidly in the 1990s and 2000s, and leaders in this technology included ATR in Japan, CSTR in Scotland, British Telecom in the UK, AT&T Labs in the United States, and L&H in Belgium [97, 153].

13.4 EARLY SPEECH SYNTHESIS APPROACHES

A block diagram of a complete TTS system is shown in Figure 13.8. Once the abstract underlying linguistic description of the text input has been determined via the steps of Figure 13.2, the remaining (major) task of TTS systems is to synthesize a speech waveform whose intelligibility is very high (to make the speech useful as a means of

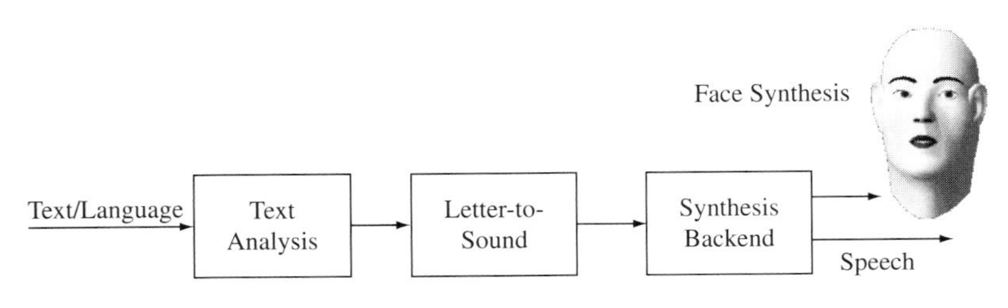

FIGURE 13.8
Block diagram of full TTS system with both speech and visual outputs.

communication between a machine and a human), and whose naturalness is as close to real speech as possible. Both tasks, namely attaining high intelligibility along with reasonable naturalness, are difficult to achieve and depend critically on three issues in the processing of the speech synthesizer "backend" of Figure 13.8, namely:

1. **choice of synthesis units**: including whole words, phones, diphones, dyads, or syllables
2. **choice of synthesis parameters**: including linear predictive features, formants, waveform templates, articulatory parameters, sinusoidal parameters, etc.
3. **method of computation**: including rule-based systems or systems which rely on the concatenation of stored speech units.

We discuss each of these issues in the remaining sections of this chapter. It is interesting to note that traditional TTS systems have only a speech output. As depicted in Figure 13.8, there has been increased interest recently in TTS systems with both a speech output and controls of a facial model so that either an avatar or a human face model can be seen speaking the words of the synthetic utterance. We discuss such "visual TTS" methods at the end of this chapter.

13.4.1 The VODER

Although ideas on how to make a machine speak in a voice that was intelligible and natural had been a quest of speech researchers for more than 100 years, it was only in 1939 that one of the first electronic means for synthesizing speech was demonstrated at the 1939 World's Fair in New York City. The device, known as the VODER (Voice Operated DEmonstratoR), was invented by Homer Dudley and his colleagues at Bell Telephone Laboratories [93]. The VODER was based on a simple model of speech sound production. It was a device that attempted to replicate (or reproduce) the spectrum of a range of speech sounds by the action of a finger keyboard and a set of pedals, as illustrated in Figures 13.9 and 13.10. The excitation source was selected via a wrist bar that selected either periodic pulses or noise, and a foot pedal that controlled the pitch of the periodic pulses for voiced sounds. The 10 fingers on the two hands of the operator controlled a set of 10 contiguous bandpass filters which spanned the speech frequency range and were connected in parallel. There were separate keys for stop consonant sounds, and a means of controlling signal energy was also provided.

Although the training cycle for operators to learn how "to play" the VODER was often on the order of a year, eventually these operators learned how to control the VODER to produce intelligible speech, and a demonstration of the speech quality from the VODER is described in Appendix A. The material in Appendix A is from a radio show where the announcer asks the operator to produce two simple sentences using the VODER. The text of these scenarios is as follows:

- **Announcer**: Will you please say for our Eastern listeners "Good evening radio audience."
- **VODER**: Good evening radio audience.

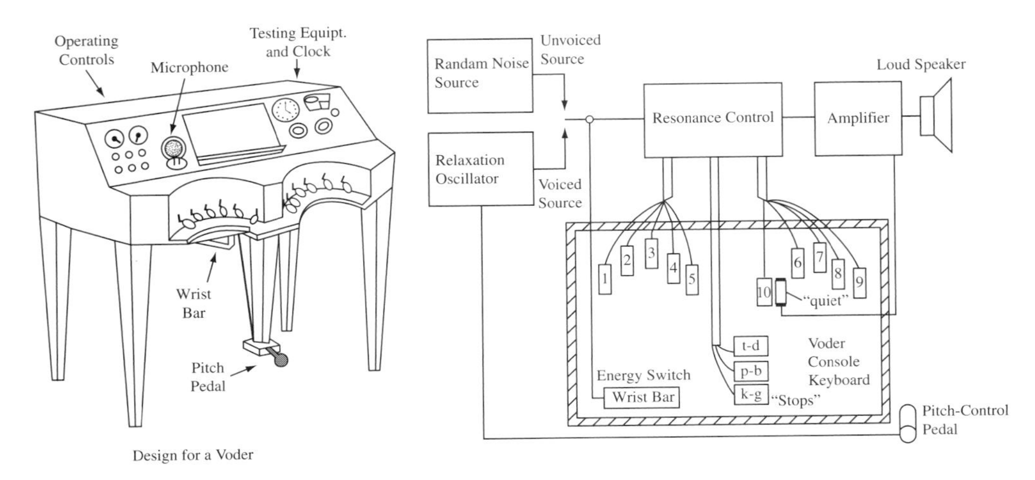

FIGURE 13.9
(Left) Original sketch of the VODER; (Right) VODER block diagram and controls via fingers, wrist, and foot. (After Dudley, Riesz, and Watkins [93]. Reprinted with permission of Alcatel-Lucent USA Inc.)

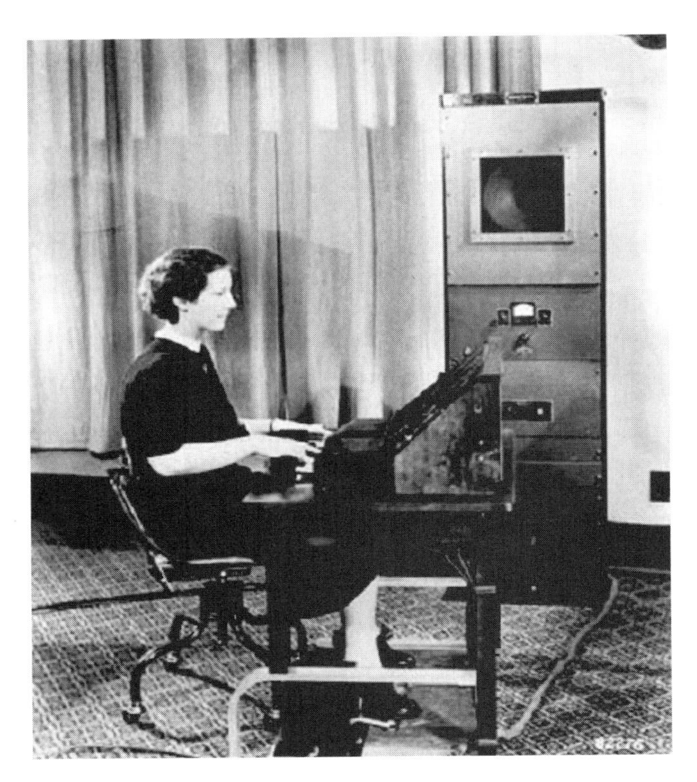

FIGURE 13.10
Illustration of operator at the VODER console. Reprinted with permission of Alcatel-Lucent USA Inc.

- **Announcer**: And now for our Western listeners say "Good afternoon radio audience."
- **VODER**: Good afternoon radio audience.

In the following sections we present overviews of a range of rule-based speech synthesis systems, based on using different synthesis backend methods including a terminal analog (formant, linear predictive) synthesizer and an articulatory synthesizer. We concentrate the discussion on the properties of the synthesizer since it is assumed that the desired speech utterance has been analyzed and parsed into a series of sounds which are then synthesized based on rules for smoothly joining the various speech sound representations.

13.4.2 Terminal Analog Synthesis of Speech

The VODER was an early example of what came to be known as *terminal analog speech synthesis*. The keyboard control made for clever demonstrations, but was not a practical control system. However, the underlying synthesis system was comprised of a continuous-time system that modeled the excitation and vocal tract systems. In terminal analog synthesis, each sound of the language (phoneme) is characterized by a source excitation function and an ideal vocal tract model. Speech is produced by varying the excitation and the vocal tract model control parameters with time at a rate commensurate with the sounds being produced. This synthesis process has been called *terminal analog synthesis* because it is based on a model (an analog) of the human vocal tract production of speech that seeks to produce a signal at its output terminals that is equivalent to the signal produced by a human talker.[3]

The basis for terminal analog synthesis of speech is the source/system model of speech production that we have used many times in this book; namely an excitation source, $e[n]$, and a transfer function of the human vocal tract in the form of a rational system function, i.e.,

$$H(z) = \frac{X(z)}{E(z)} = \frac{B(z)}{A(z)} = \frac{b_0 + \sum_{k=1}^{q} b_k z^{-k}}{1 - \sum_{k=1}^{p} a_k z^{-k}}, \tag{13.1}$$

where $X(z)$ is the z-transform of the output speech signal, $x[n]$, $E(z)$ is the z-transform of the vocal tract excitation signal, $e[n]$, and $\{b_k\} = \{b_0, b_1, b_2, \ldots, b_q\}$ and $\{a_k\} = \{a_1, a_2, \ldots, a_p\}$ are the (time-varying) coefficients of the vocal tract filter.

[3]In the designation "terminal analog synthesis," the terms *analog* and *terminal* result from the historical context of early speech synthesis studies. This can be confusing since today, "analog" implies "not digital" as well as an analogous thing. "Terminal" originally implied the "output terminals" of an electronic analog (not digital) circuit or system that was an analog of the human speech production system.

For most practical speech synthesis systems, both the excitation signal properties (pitch period, N_p, for voiced sounds and voiced–unvoiced classification, V/UV) and the filter coefficients of Eq. (13.1) change periodically so as to synthesize different phonemes.

The vocal tract representation of Eq. (13.1) can be implemented as a speech synthesis system using a direct form implementation. However, it has been shown that it is preferable to factor the numerator and denominator polynomials into either a series of cascade (serial) resonances, or into a parallel combination of resonances. We discussed such details in Chapter 5, where we also showed that an all-pole model ($B(z) = \text{constant}$) is most appropriate for voiced (non-nasal) speech. For unvoiced speech, a simpler model (with one complex pole and one complex zero), implemented via a parallel branch, is adequate. Finally a fixed spectral compensation network can be used to model the combined effects of glottal pulse shape and radiation characteristics from the lips and mouth, based on two real poles in the z-plane. A complete serial terminal analog speech synthesis model, based on the above discussion, is shown in Figure 13.11 [299, 344].

The voiced speech (upper) branch in Figure 13.11 includes an impulse generator (controlled by a time-varying pitch period, N_p), a time-varying voiced signal gain, A_V, and an all-pole discrete-time system that consists of a cascade of three time-varying resonances (the first three formants, F_1, F_2, F_3), and one fixed resonance F_4.

The unvoiced speech (lower) branch in Figure 13.11 includes a white noise generator, a time-varying unvoiced signal gain, A_N, and a resonance/anti-resonance system consisting of a time-varying pole (F_P) and a time-varying zero (F_Z).

The voiced and unvoiced components are added and processed by the fixed spectral compensation network to provide the final synthetic speech output.

The resulting quality of the synthetic speech produced using a terminal analog synthesizer of the type shown in Figure 13.11 is highly variable, with explicit model shortcomings, due to the following:

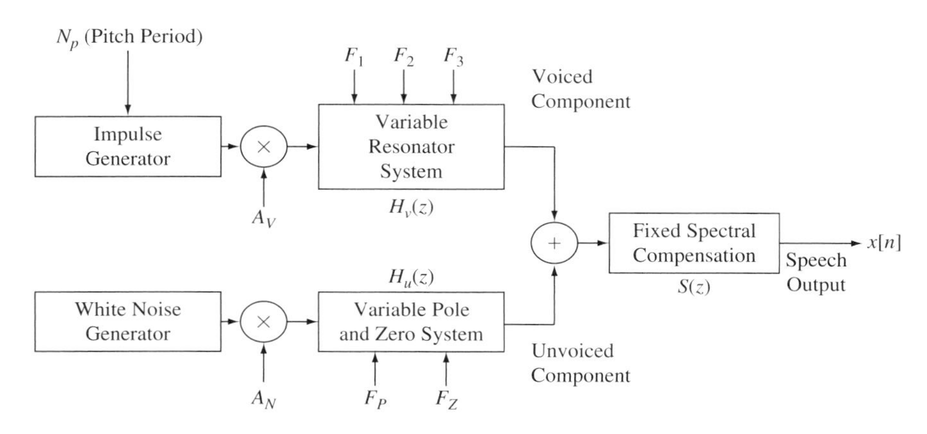

FIGURE 13.11
Speech synthesizer based on a cascade/serial (formant) synthesis model.

- Voiced fricatives are not handled properly since their mixed excitation is not part of the model of Figure 13.11.
- Nasal sounds are not handled properly since nasal zeros are not included in the model.
- Stop consonants are not handled properly since there is no precise timing and control of the complex excitation signal.
- The use of a fixed pitch pulse shape, independent of the pitch period, is inadequate and produces buzzy sounding voiced speech.
- The spectral compensation model is inaccurate and does not work well for unvoiced sounds.

Many of the shortcomings of the model of Figure 13.11 are alleviated by a more complex model proposed by Klatt [193]. However, even with a more sophisticated synthesis model, it remains a very challenging task to compute the synthesis parameters. Nevertheless Klatt's Klattalk system achieved adequate quality by 1983 to justify commercialization by the Digital Equipment Corporation as the DECtalk system. Some DECtalk systems are still in operation today as legacy systems, although the unit selection methods to be discussed later in this chapter now provide superior quality for almost all applications.

13.4.3 Articulatory Methods of Speech Synthesis

Although the earliest electronic means of synthesizing speech was by means of an electrical analog of the vocal tract system, it was widely believed that, at least in theory, more natural sounding speech could be created using realistic motions of the speech articulators (rather than of the formant parameters) by using an articulatory model of speech production. The arguments for trying to build an articulatory model of speech production were the following:

- Known and fairly well-understood physical constraints of articulator movements could be used to create realistic motions of the tongue, jaw, teeth, velum, etc.
- X-ray data (MRI data today) could be used to study the motion of the articulators in the production of individual speech sounds, thereby increasing our understanding of the dynamics of speech production.
- Smooth articulatory parameter motions between sounds could be modeled, either via direct methods [namely solving the wave equation to determine sound pressure at the lips (and nose)], or indirectly by converting articulatory shapes to formants or linear predictive parameters for synthesis on a conventional synthesizer.
- The motions of the articulatory parameters could be constrained so that only natural motions would occur, thereby potentially making the speech more natural sounding.

A schematic diagram of a computational articulatory model used in speech synthesis is shown in Figure 13.12 [63]. The model uses eight independent parameters

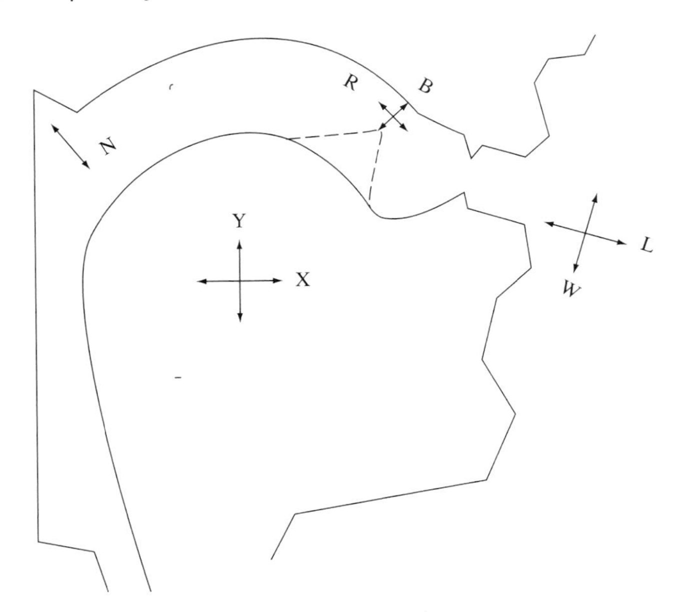

FIGURE 13.12
Schematic diagram of a computational articulatory model used in phoneme synthesis. (After Coker [63]. © [1976] IEEE.)

and one dependent parameter to describe an arbitrary vocal tract model shape. These parameters, as illustrated in Figure 13.12, include the following:

- lip opening—W
- lip protrusion length—L
- tongue body height and length—Y, X
- velar closure—K
- tongue tip height and length—B, R
- jaw raising (dependent parameter)
- velum opening—N.

To show how well the model of Figure 13.12 was able to fit human articulatory data, as measured from X-rays, Figure 13.13 shows plots of the cross-sectional area of the vocal tract along with model fits when the articulatory model parameters have been optimized appropriately, for three vowel sounds (/i/, /a/, and /u/) and for three consonant sounds (/p/, /t/, and /k/). It can be seen that for the vowel sounds, the degree of agreement between the X-ray data and the model is uniformly very good; for the consonant sounds, the degree of agreement is good for most of the length of the vocal tract, with some larger model deviations just prior to the closure for the various stop consonants (especially noticeable for the consonant /k/).

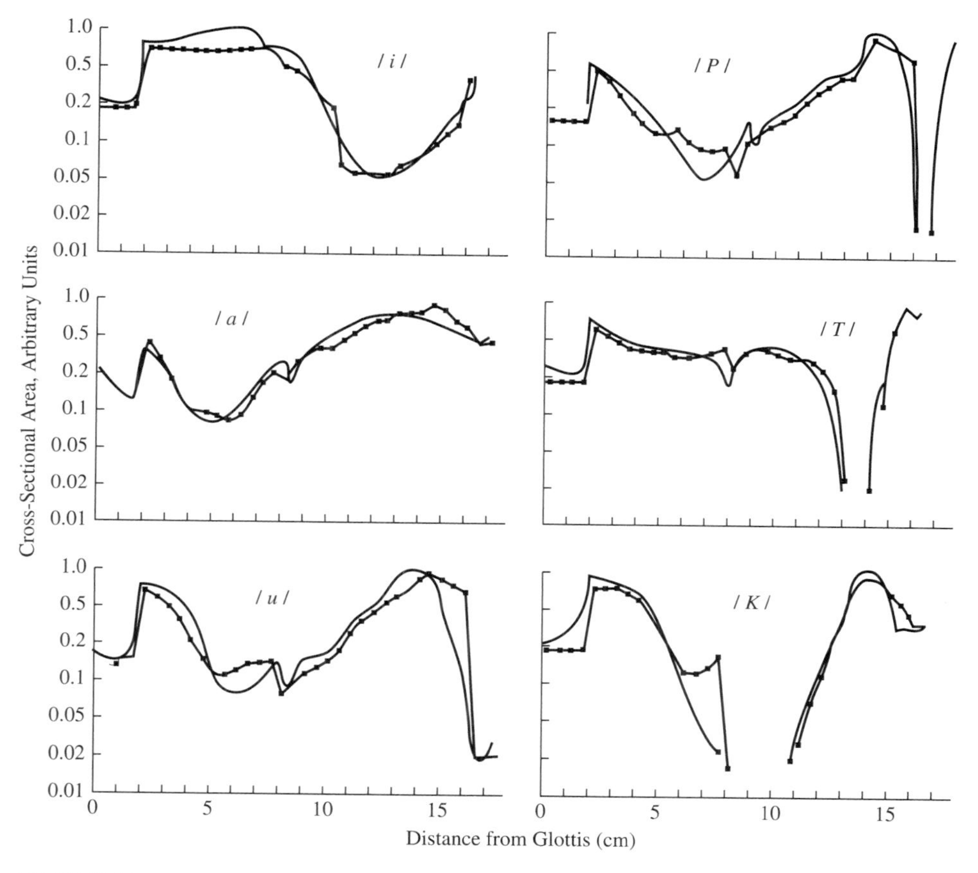

FIGURE 13.13
Examples of the ability of the articulatory model to fit human articulatory data as taken from X-rays. (After Coker [63]. © [1976] IEEE.)

An audio demonstration of articulatory synthesis using a fully automatic closed-loop optimization of the articulatory model parameters, initialized from articulatory codebooks and with the help of neural networks [357], is included on the book website and is described in Appendix A. This demo consists of the sentence "The author lived in a yert in the desert" initially spoken naturally, followed by the articulatory synthesis based on the closed-loop optimization of the articulatory model parameters. The resulting synthetic speech quality is quite reasonable, showing the potential for articulatory models if an automated way of properly controlling the parameters over time could be devised.

What we have learned about articulatory modeling of speech is that it requires a highly accurate model of the glottis and of the vocal tract for the resulting synthetic speech quality to be considered acceptable. It further requires rules for handling the dynamics of the articulator motion in the context of the sounds being produced. Over the years we have yet to create such highly accurate models or learn the correct rules for dynamic control. Hence articulatory speech synthesis methods are interesting, from a theoretical point of view, but are not yet practical for synthesizing speech of acceptable quality.

13.4.4 Word Concatenation Synthesis

Perhaps the simplest approach to creating a speech utterance corresponding to a given text string is to literally splice together the waveforms of pre-recorded words corresponding to the desired utterance. For greatest simplicity, the words can be stored as sampled waveforms and simply concatenated in the correct sequence. This approach generally produces intelligible, but unnatural sounding speech, since it does not take into account the "co-articulation" effects of producing phonemes in continuous speech, and it does not provide either for the adjustment of phoneme durations or the imposition of a desired pitch variation across the utterance. Words spoken in continuous speech sentences are generally much shorter in duration than when spoken in isolation (often up to 50% shorter) as illustrated in Figure 13.14. This figure shows wideband spectrograms for the sentence "This shirt is red" spoken as a sequence of isolated words (with short, distinct pauses between words) as shown at the top of Figure 13.14, and as a continuous utterance, as shown at the bottom of Figure 13.14. It can be seen that, even for this trivial example, the duration of the continuous sentence is on the order of half that of the isolated word version, and further, the formant tracks of the continuous sentence do not look like a set of uniformly compressed formant tracks

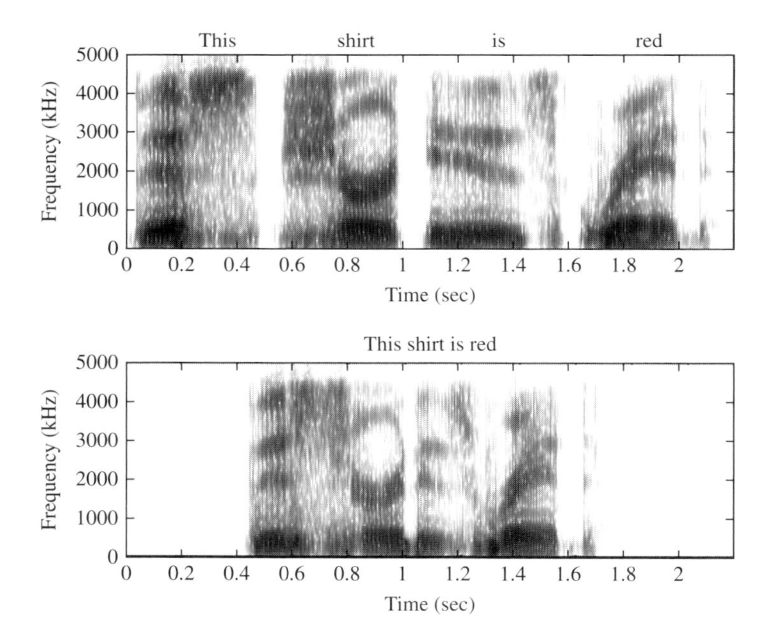

FIGURE 13.14
Wideband spectrograms of a sentence spoken as a sequence of isolated words (top panel) and as a continuous speech utterance (bottom panel).

from the individual words. Furthermore, the boundaries between words in the upper plot are sharply defined, while they are merged in the lower plot.

The word concatenation approach can be made more sophisticated by storing the vocabulary words in a parametric form (formants, LPC parameters, etc.) such as are employed in the speech coders discussed in Chapter 11 [314]. This would result in reduced storage requirement for the words, but the main rationale for this is that the parametric representations, being more closely related to the model for speech production, can be manipulated so as to blend the words together, shorten them, and impose a desired pitch variation. This requires that the control parameters for all the words in the task vocabulary (as obtained from a training set of words) be stored as representations of the words. A special set of word concatenation rules is then used to create the control signals for the synthesizer. Although such a synthesis system would appear to be an attractive alternative for general purpose synthesis of speech, in reality, this type of synthesis is not a practical approach.[4]

There are many reasons for this, but a major problem is that there are far too many words to store in a word catalog for word concatenation synthesis to be practical, except in highly restricted situations. For example, there are about 1.7 million distinct surnames in the United States and each of them would have to be spoken and stored for a general word concatenation synthesis method. Table 13.1 shows the cumulative statistics of the most prominent 200,000 names in the United States, showing that only 93% of surnames of the people in the United States are covered in the top 200,000 names. A second, equally important limitation is that word-length segments of speech are simply the wrong sub-units. As we will see, shorter units such as phonemes or *diphones* are more suitable for synthesis. A third problem with word concatenation synthesis is that words cannot preserve sentence-level stress, rhythm, or intonation patterns, and thus cannot be simply concatenated but instead need to be processed so as to impose external duration and pitch constraints.

A block diagram of a concatenative word synthesis system is shown in Figure 13.15 [314]. The input to the synthesizer is the typed input word string (assuming

TABLE 13.1 Proper name statistics of U.S. surnames.

Number of Names	Cumulative Coverage
10	4.9%
100	16.3%
5,000	59.1%
50,000	83.2%
100,000	88.6%
200,000	93.0%

[4]It should be obvious that whole word concatenation synthesis (from stored waveforms) is also impractical for general purpose synthesis of speech.

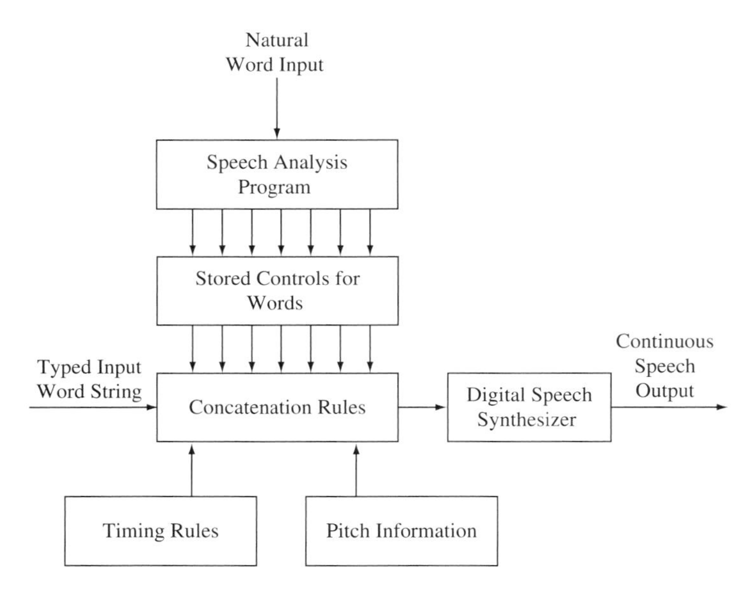

FIGURE 13.15
Block diagram of speech synthesis system using word concatenation.

it is an unambiguous text string) and the synthesizer is assumed to have previously stored the synthesis controls (formant frequencies) for a set of words that constitute the word vocabulary of the system. A set of timing rules determines the modified durations of each of the words in the typed input string, and a pitch contour is superimposed on the utterance, based again on rules appropriate to the text at hand, and the resulting speech is synthesized using a conventional speech synthesizer (of the type to be discussed later in this chapter). Informal evaluations of this concatenative word synthesis output, for very simple tasks such as synthesizing a telephone number in the carrier phrase "The number is XXX-XXXX" led to the conclusion that word-based synthesis methods could not produce acceptable quality speech. In part this was due to the quality of formant synthesis methods, and in part it was due to the more fundamental problems associated with recording and storing exponentially many versions of each word with the range of prosody required for words to match well at the concatenation boundaries.

13.5 UNIT SELECTION METHODS

In principle, the limitations of word concatenation systems are not present in either the terminal analog synthesis approach or the articulatory model approach if we assume an appropriate interface between the output of the text analysis system and the control signals for the synthesizers. However, by the end of the 1980s, it was clear that speech synthesis and TTS technology had reached a roadblock. The state-of-the-art synthesis was highly intelligible but extremely unnatural sounding. Furthermore, a decade of work had not changed the naturalness substantially, although the intelligibility had improved somewhat. A key driving factor in looking at alternative methods for TTS was that computation speed and memory continued to increase at an exponential pace, doubling in capacity and halving in price every 18 months, a trend that continues unabated even today. This Moore's law growth in computational power and memory enabled highly complex concatenative synthesis systems, using multiple tokens of each synthesis unit, to be created, implemented, and highly optimized. As a result, concatenative TTS systems, based on optimal selection of units smaller than words, have proven capable of producing (in some cases) extremely natural sounding synthetic speech, and thus they are the method of choice for almost all modern TTS systems.

The key idea of a concatenative TTS system, using unit selection methods, is to use synthesis segments that are short sections of natural (recorded) speech [35, 36, 97, 153, 215]. It seems plausible that the more segments produced, recorded, annotated, and saved in a database, the better the potential quality of the resulting speech synthesis. Ultimately, if an infinite number of segments were recorded and saved, the resulting synthetic speech would be completely natural for all synthesis tasks. Concatenative speech synthesis systems, based on unit selection methods, are what is conventionally known as "data driven" approaches since their performance tends to get better as more and more data is used for training the system.

In order to design and build a unit selection system for speech synthesis, based on using recorded speech segments, there are several issues that must be resolved, including the following:

1. the speech units used as the basic synthesis building blocks
2. the method of extraction of synthesis units from natural speech utterances
3. the labeling of units for retrieval from a large database of units
4. the signal representation used to represent the units for storage and reproduction purposes
5. signal processing methods for spectrally smoothing the units at unit junctures and for modification of pitch, duration, and amplitude.

We now address each of these issues.

13.5.1 Choice of Concatenation Units

There are a number of possible choices for the units for a concatenative TTS system, including, but not limited to, the following:

- **words**: there are essentially an infinite number of words in English
- **syllables**: there are about 10,000 syllables in English
- **phonemes**: there are only about 45 phonemes in English, but they are highly context dependent
- **demi-syllables**: there are about 2500 demi-syllables (half syllables) in English
- **diphones**: there are about 1500–2500 diphones in English.

Length	Unit	# Units (English)	# Rules, or Unit Mods	Quality
Short ↓	allophone	60–80	Many ↑	Low ↓
	diphone	$< 40^2–65^2$		
	triphone	$< 40^3–65^3$		
	demisyllable	2000		
	syllable	11,000		
	VC*V			
	2-Syllable	$< (11{,}000)^2$		
	Word	100,000–1.5 M		
	Phrase	∞		
Long	Sentence	∞	Few	High

FIGURE 13.16
Comparison of concatenation units in terms of duration, count, context, dependency, and anticipated quality of concatenated units. (After Macchi [215].)

A chart, due to Macchi [215], summarizing the issues involved in the choice of units for concatenation is given in Figure 13.16. This figure shows the following:

- a range of units from allophones (context dependent versions of phonemes) to whole sentences
- an indication of how long the average unit tends to be
- an estimate of the count of each unit in the English language
- an estimate of the number of rules necessary for unit modification due to the context of the preceding and following sounds
- an indication of the anticipated quality of the synthesis based on unit concatenation methods.

It can be seen that as the unit size goes from the shortest (allophones) to the longest (storage of whole words, phrases, or sentences), the effect of context decreases substantially and the quality increases. However the practicality of storing all possible words, phrases, or sentences, even for a limited task domain, is simply unreasonable, and thus a choice must be made among the shorter units.

Figure 13.17, again due to Macchi [215], shows a plot of the coverage of a corpus of 2 million names using several of the unit types from Figure 13.16. Results are shown for the top 50,000 names since the trend is clear by this point on the curve. It can be seen that when using longer units, like words or 2-syllable units, the number of units needed grows almost linearly with the vocabulary size. Using units like syllables and triphones, the growth in units levels off for a vocabulary of 50,000 names, requiring about 5000–10,000 units to cover this database. Using units like diphones or

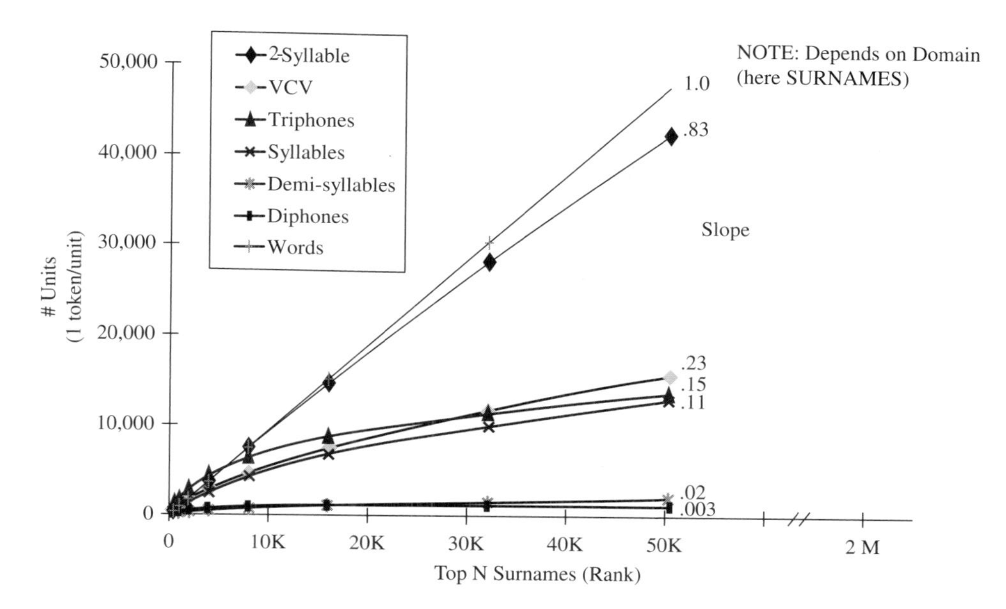

FIGURE 13.17
Unit coverage of vocabulary of top *N* surnames in English by several of the unit types, as a function of *N*. (After Macchi [215].)

demi-syllables, the curve flattens very rapidly at about 150–1000 units for complete coverage of the 50,000 surnames database. The slope of the curve of unit coverage at 50,000 names is indicative of the probability of needing to record an additional unit as a new name is added, and this slope varies from 1 (for words) down to virtually 0 for diphones or demi-syllables. Hence, for a high quality concatenative TTS system, we would expect to use units like diphones, or even possibly syllables, but not words or multi-syllable units.

13.5.2 Selection of Units from Natural Speech

Once the unit type has been selected, the next step is to record a database that is rich in the number of occurrences of each of the units and in the range of phonetic contexts in which each unit appears. Larger recorded databases provide more variations of each of the speech units, and generally this leads to more natural sounding synthetic speech.

The recorded speech database must be labeled according to the basic speech unit for concatenation and segmented into the stream of units using some type of dynamic programming matching procedure, similar to what is conventionally used in most speech recognition systems (see Chapter 14). The recorded speech is generally a set of "over articulated" sentences, read in an almost monotone style so as to make it easier to concatenate units from different training set utterances. The textual material for the database must include several allophonic variations of each of the synthesis units. Typically databases with about 1000 sentences (about 60 minutes of recording) are sufficient for good quality speech synthesis for reasonable-size tasks.

An off-line inventory preparation procedure is used to create the speech database for the synthesis system. Once the location of each of the speech units in each of the recorded sentences is determined (either via automatic or manual means), the speech is coded (with any desired coding method of choice, e.g., cepstral parameters for a feature-based coding or DPCM for waveform coding), and the unit locations (and perhaps also the unit context), for each unit in each sentence, is stored in an inventory that is accessed when a new sentence is to be synthesized [35].

13.5.3 On-Line Unit Selection Synthesis from Text

The on-line synthesis procedure from sub-word units is straightforward, but far from trivial [36]. Following the usual text analysis into phonemes and prosody, the phoneme sequence is first converted to the appropriate sequence of units from the inventory. For example, the phrase "I want" would be converted to diphone units as follows:

text input: I want.
phonemes: /#/ /AY/ /W/ /AA/ /N/ /T/ /#/
diphones: /# AY/ /AY W/ /W AA/ /AA N/ /N T/ /T #/

where the symbol # is used to represent silence (at the end and beginning of each sentence or phrase).

The second step in the on-line synthesis is to select the most appropriate sequence of diphone units from the stored inventory. Since each diphone unit occurs more than once in the stored inventory, the selection of the best sequence of diphone units involves solving a dynamic programming search for the sequence of units that minimizes a specified cost function. The cost function generally is based on diphone matches at each of the boundaries between diphones, where the diphone match is defined in terms of spectral matching characteristics, pitch matching characteristics, and possibly phase matching characteristics. In the next section, we present one possible set of cost functions that can be used for choosing the optimal set of diphone units that best match a given text sentence.

Once the optimal set of diphone units has been chosen, the last step in the on-line unit selection synthesis procedure is to modify the chosen diphones to match more closely in spectrum, pitch, and possibly phase, at each of the diphone juncture points. One method for modifying segments at juncture points is described in Section 13.5.6.

13.5.4 Unit Selection Problem

The "unit selection" problem is basically one of having a given set of *target features* corresponding to the spoken text, and then *automatically* finding the sequence of *units* in the database that most closely match these features. This problem is illustrated in Figure 13.18, which shows a target feature set corresponding to the sequence of sounds (phonemes for this trivial example) /HH/ /EH/ /L/ /OW/ from the word /hello/. Each of the phonemes in this word has multiple representations in the inventory of sounds, having been extracted from different phonemic environments. Hence there are many versions of /HH/ and many versions of /EH/, etc., as illustrated in Figure 13.18. The task of the unit selection module is to choose one of each of the multiple representations of the required sounds, with the goal being to minimize the total perceptual

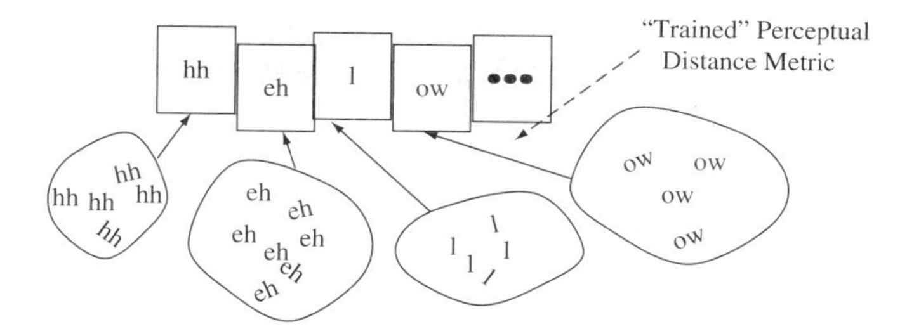

FIGURE 13.18
Illustration of basic process of unit selection. (Figure labeled in lower case ARPAbet symbols.)

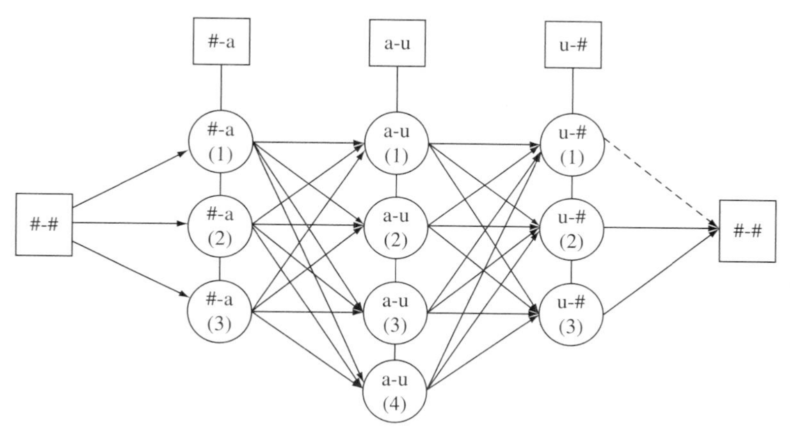

FIGURE 13.19
On-line unit selection based on a Viterbi search through a lattice of alternatives.

distance between segments of the chosen sequence, based on spectral, pitch, and phase differences throughout the sounds and especially at the boundary between sounds. By specifying costs associated with each unit, both globally across the unit, and locally at the unit boundaries with adjacent units, we can find the sequence of units that best "join each other" in the sense of minimizing the accumulated distance across the sequence of units. This optimal sequence (or equivalently the optimal path through the combination of all possible versions of each unit in the string) can be found using a Viterbi search via the method of dynamic programming [308]. This dynamic programming search process is illustrated in Figure 13.19 for a 3-unit search. The Viterbi search computes the cost of every possible path through the lattice and determines the path with the lowest total cost, where the transitional costs (the arcs) reflect the cost of concatenation of a pair of units based on acoustic distances, and the nodes represent the target costs based on the linguistic identity of the unit.

Thus there are two costs associated with the Viterbi search, namely a nodal cost based on the unit segmental distortion (USD), which is defined as the difference

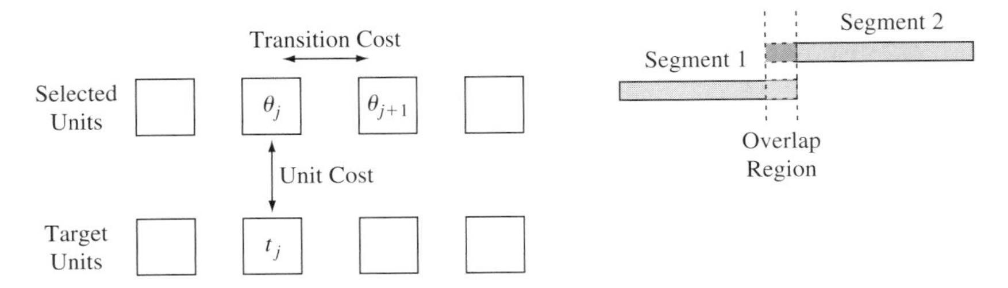

FIGURE 13.20
Illustration of unit selection costs associated with a string of target units and a presumptive string of selected units.

between the desired spectral pattern of the target and that of the candidate unit, throughout the unit, and the unit concatenative distortion (UCD), which is defined as the spectral (and/or pitch and/or phase) discontinuity across the boundaries of the concatenated units. As an example, consider a target context of the word "cart" with target phonemes /K/ /AH/ /R/ /T/, and a source context of the phoneme /AH/ obtained from the source word "want" with source phonemes /W/ /AH/ /N/ /T/. The USD distance would be the cost (specified analytically) between the sound /AH/ in the context /W/ /AH/ /N/ and the desired sound in the context /K/ /AH/ /R/.

Figure 13.20 illustrates concatenative synthesis for a given string of target units (at the bottom of the figure) and a string of selected units from the unit inventory (shown at the top of the figure). We focus our attention on target unit t_j and selected units θ_j and θ_{j+1}. Associated with the match between units t_j and θ_j is a USD unit cost (to be specified later in this section) and associated with the sequence of selected units, θ_j and θ_{j+1}, is a UCD concatenation cost. The total cost of an arbitrary string of N selected units, $\Theta = \{\theta_1, \theta_2, \dots, \theta_N\}$, and the string of N target units, $T = \{t_1, t_2, \dots, t_N\}$, is defined as:

$$d(\Theta, T) = \sum_{j=1}^{N} d_u(\theta_j, t_j) + \sum_{j=1}^{N-1} d_t(\theta_j, \theta_{j+1}), \qquad (13.2)$$

where $d_u(\theta_j, t_j)$ is the USD cost associated with matching target unit t_j with selected unit θ_j and $d_t(\theta_j, \theta_{j+1})$ is the UCD cost associated with concatenating the units θ_j and θ_{j+1}. It should be noted that when selected units θ_j and θ_{j+1} come from the same source sentence and are adjacent units, the UCD cost goes to zero, as this is as natural a concatenation as can exist in the database. Further, it is noted that generally there is a small overlap region between concatenated units. (We will discuss this issue further in the next section.) The optimal path (corresponding to the optimal sequence of units) can be efficiently computed using a standard Viterbi search [116, 308] that scales linearly with both the number of targets and the number of concatenation units.

The Viterbi Search for the Optimal Sequence of Units

Based on the lattice structure of Figure 13.19 and the concepts of unit cost and transition cost of Figure 13.20, a Viterbi search algorithm that computes the optimal sequence of units (in the sense of minimum total accumulated unit and transition costs) is as follows:

- Let j denote the index of the unit position in the J–unit string $(1 \leq j \leq J)$ $(J = 3$ for the example of Figure 13.19).
- Let i denote the index of the unit at each position j, $i = 1, 2, \ldots, I(j)$, where the total number of choices, $I(j)$, is a function of the unit position j. Thus for the simple example of Figure 13.19, there are $J = 3$ units in the desired string (/#-a/, /a-u/, and /u-#/), for unit position $j = 1$, there are $I(1) = 3$ choices, for unit position $j = 2$, there are $I(2) = 4$ choices, and for unit position $j = 3$, there are $I(3) = 3$ choices.
- Using the notation of Figure 13.20, the putative unit at string position j is called θ_j and the "ideal" or target unit at string position j is called t_j.
- The unit cost between units θ_j and t_j is $d_u(\theta_j, t_j)$, and the transition cost between units θ_j and θ_{j+1} is $d_t(\theta_j, \theta_{j+1})$.
- The Viterbi algorithm for finding the optimum sequence of units (i.e., the sequence of units with the minimum accumulated cost) can be stated as follows:
  - Let $\delta_j(i)$ be the optimum (minimum accumulated distance) path to unit $\theta_j(i)$ at index i of unit position j. Also define $\psi_j(i)$ as a two-dimensional array for recording paths through the lattice. We can now solve for $\delta_j(i)$ recursively as:
  - **Initialization** $(j = 1)$

$$\delta_1(i) = d_u(\theta_1(i), t_1), \qquad 1 \leq i \leq I(1),$$
$$\psi_1(i) = 0, \qquad 1 \leq i \leq I(1).$$

  - **Recursion**

$$\delta_j(k) = \min_{1 \leq i \leq I(j-1)} [\delta_{j-1}(i) + d_t(\theta_{j-1}(i), \theta_j(k))] + d_u(\theta_j(k), t_j)$$
$$1 \leq k \leq I(j), \qquad 2 \leq j \leq J,$$
$$\psi_j(k) = \operatorname*{argmin}_{1 \leq i \leq I(j-1)} [\delta_{j-1}(i) + d_t(\theta_{j-1}(i), \theta_j(k))]$$
$$1 \leq k \leq I(j), \qquad 2 \leq j \leq J.$$

  - **Termination**

$$D^* = \min_{1 \leq k \leq I(J)} [\delta_J(k)],$$
$$q_J^* = \operatorname*{argmin}_{1 \leq k \leq I(J)} [\delta_J(k)].$$

  - **Path Backtracking**

$$q_j^* = \psi_{j+1}(q_{j+1}^*), \quad j = J-1, J-2, \ldots, 1.$$

The array $\psi_j(k)$, $1 \leq j \leq J$, $1 \leq k \leq I(j)$, contains the path information, and the optimum sequence of units, q_j^*, $1 \leq j \leq J$, is obtained via path backtracking as shown above.

13.5.5 Transition and Unit Costs

The concatenation cost between two units is essentially the spectral (and/or phase and/or pitch) discontinuity across the boundary between the units, and is defined as:

$$C^c(\theta_j, \theta_{j+1}) = \sum_{k=1}^{p+2} w_k^c C_k^c(\theta_j, \theta_{j+1}), \tag{13.3}$$

where p is the size of the spectral feature vector (typically $p = 12$ mfcc coefficients, often represented as a VQ codebook vector), and the extra two features are log energy and pitch. The weights, w_k^c, are chosen during the unit selection inventory creation phase and are optimized using a trial-and-error procedure. The concatenation cost measures a spectral plus log energy plus pitch difference between the two concatenated units at the boundary frames. Clearly the definition of concatenation cost could be extended to more than a single boundary frame. Also, as stated earlier, the concatenation cost is defined to be zero ($C^c(\theta_j, \theta_{j+1}) = 0$) whenever units θ_j and θ_{j+1} are consecutive (in the database) since, by definition, there is no discontinuity in either spectrum or pitch in this case. Although there are a variety of choices for measuring the spectral/log energy/pitch discontinuity at the boundary, a common cost function is the normalized mean-squared error in the feature parameters, namely:

$$C_k^c(\theta_j, \theta_{j+1}) = \frac{[f_k^{\theta_j}(m) - f_k^{\theta_{j+1}}(1)]^2}{\sigma_k^2}, \tag{13.4}$$

where $f_k^{\theta_j}(l)$ is the kth feature parameter of the lth frame of segment θ_j, m is the (normalized) duration of each segment, and σ_k^2 is the variance of the kth feature vector component.

The USD or target costs are conceptually more difficult to understand, and, in practice, more difficult to instantiate. The USD cost associated with units θ_j and t_j is of the form:

$$d_u(\theta_j, t_j) = \sum_{i=1}^{q} w_i^t \phi_i \left\{ T_i(f_i^{\theta_j}), T_i(f_i^{t_j}) \right\}, \tag{13.5}$$

where q is the number of features that specify the unit θ_j or t_j, w_i^t, $i = 1, 2, \ldots, q$, is a trained set of target weights, and $T_i(\cdot)$ can be either a continuous function (for a set of features, f_i, such as segmental pitch, power, or duration), or a set of integers (in the case of categorical features, f_i, such as unit identity, phonetic class, position in the syllable from which the unit was extracted, etc.). In the latter case, ϕ_i can be looked up in a table of distances. Otherwise, the local distance function can be expressed as a simple quadratic distance of the form:

$$\phi_i \left\{ T_i(f_i^{\theta_j}), T_i(f_i^{t_j}) \right\} = \left[T_i(f_i^{\theta_j}) - T_i(f_i^{t_j}) \right]^2. \tag{13.6}$$

The training of the weights, w_i^t, is done off-line. For each phoneme in each phonetic class in the training speech database (which might be the entire recorded inventory), each exemplar of each unit is treated as a target and all others are treated as candidate units. Using this training set, a least-squares system of linear equations can be derived from which the weight vector can be solved. The details of the weight training methods are described by Schroeter [356].

The final step in the unit selection synthesis, having chosen the optimal sequence of units to match the target sentence, is to smooth/modify the selected units at each of the boundary frames to better match the spectra, pitch, and phase at each unit boundary.

13.5.6 Unit Boundary Smoothing and Modification

The ultimate goal of boundary smoothing and modification is to match the waveforms at the boundaries between pairs of synthesis units so that the discontinuity in pitch, spectrum, and phase is as small as possible, without seriously distorting the speech waveform. There is no simple or clear-cut way of making modifications that preserve pitch, spectrum, and phase, but instead a set of techniques, based on the concepts of time domain harmonic scaling (TDHS) [224], have been developed, enhanced, and applied to the problem of making seamless modifications to the speech waveform to preserve or modify pitch. The simplest such algorithm is the pitch synchronous overlap add (PSOLA) method [54, 249], which is illustrated in Figure 13.21. A pitch analysis is made for the source speech signal and individual periods are marked on the speech waveform. To change the pitch period from its current value, each individual pitch period is either lengthened (by appending zero-valued samples at the end of each period) or shortened (by eliminating samples at the end of each period), to match

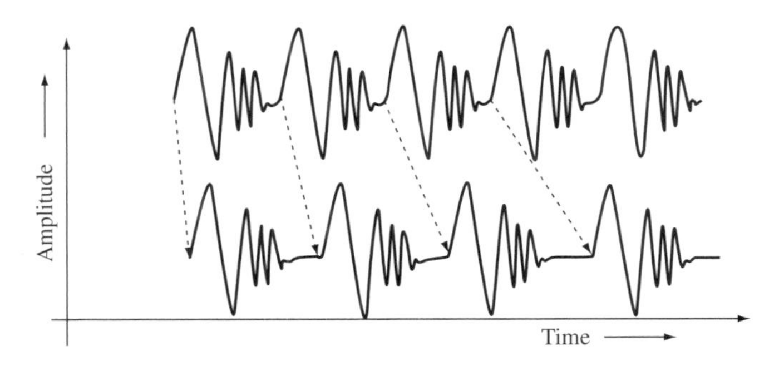

FIGURE 13.21
The PSOLA process used to increase pitch period duration gradually over time. (After Dutoit [97], pp. 251–270.)

the desired pitch as a function of time. The example of Figure 13.21 shows the case of lengthening the pitch period by adding zeros, with increased lengthening over time. For the PSOLA method to work well, both very accurate pitch period estimates (over time) and very accurate estimates of the pitch period phase are required so that the signal can be lengthened or shortened at points of minimum perceptual sensitivity to change.

An important and very practical extension to the PSOLA method is the TD-PSOLA (time-domain PSOLA) method [213], which showed that it was possible to do pitch modification directly on the speech waveform without actually having to do a fine pitch analysis to determine the points of minimal perceptual sensitivity to change. The basic contribution of TD-PSOLA was to recognize the fact that for a periodic signal, $s[n]$, of period, T_0 samples, it is possible to create a modified signal, $\tilde{s}[n]$, with a pitch period of T samples simply by summing windowed and overlapped frames.[5] Let $s_i[n]$ denote the ith period of $s[n]$, i.e.,

$$s_i[n] = s[n]w[n - iT_0] = s[n - iT_0]w[n - iT_0]. \tag{13.7a}$$

The latter form of Eq. (13.7a) follows from the periodicity of $s[n]$. Then, the new periodic signal is defined as

$$\tilde{s}[n] = \sum_{i=-\infty}^{\infty} s_i[n - i(T - T_0)]. \tag{13.7b}$$

Note that this is reminiscent of the overlap-addition (OLA) method of synthesis discussed in Chapter 7 except that the synthesis shift is different from the analysis shift. It can be seen in Eq. (13.7a) that the window is applied pitch-synchronously by moving in multiples of the period T_0. (While our analysis assumes a perfectly periodic signal $s[n]$, in practice we would apply the technique on a short-time basis, so the pitch period must be tracked and individual pitch periods must be marked consistently.) Now substituting Eq. (13.7a) into Eq. (13.7b) results in

$$\begin{aligned}\tilde{s}[n] &= \sum_{i=-\infty}^{\infty} s[n - i(T - T_0) - iT_0]w[n - i(T - T_0) - iT_0] \\ &= \sum_{i=-\infty}^{\infty} s[n - iT]w[n - iT] = \sum_{i=-\infty}^{\infty} s_w[n - iT],\end{aligned} \tag{13.8}$$

where $s_w[n] = s[n]w[n]$. Equation (13.8) shows clearly that $\tilde{s}[n]$ is periodic with period T. Observe that if $T = T_0$, i.e., no pitch change results, and the signal resulting

[5] Ordinarily we would reserve the notation T_0 and T for periods in continuous time. However, to be consistent with the notation in Ref. [97], from which several figures in this section are derived, T_0 and T are integers representing the period in samples.

from Eq. (13.7b) is

$$\tilde{s}[n] = \sum_{i=-\infty}^{\infty} s[n]w[n - iT_0] = s[n]\left(\sum_{i=-\infty}^{\infty} w[n - iT_0]\right). \tag{13.9}$$

Thus, if the window $w[n]$ satisfies the OLA perfect reconstruction conditions discussed in Chapter 7 (e.g., a Bartlett or Hann window), then the signal $s[n]$ is reproduced exactly. This is an important constraint on the window $w[n]$.

In addition to satisfying the perfect reconstruction conditions, it is also important to choose the right length for the window. Using a window that is too small (order of less than a pitch period in duration) leads to a coarse spectrum envelope that is not well defined at the new pitch harmonics. Using a window that is too large (order of several pitch periods) leads to the occurrence of strong spectral lines at the original pitch harmonics, thereby causing a very rough estimate of the spectral envelope at the modified pitch harmonics. A window that is adapted to the pitch period and is of duration of two pitch periods works best in practice and gives good quality pitch-modified speech. With tapering windows like the Hann or Hamming windows of length $2T_0$, this means that if we define $s_w[n] = s[n]w[n]$, the windowing is the time-domain equivalent of a wide-band spectrum analysis where the individual pitch harmonics are not resolved. In other words, the windowing smoothes the spectrum, maintaining the formant resonances, but eliminates the harmonic structure. Thus, if the analysis window is appropriately chosen, the corresponding DTFT, $S_w(e^{j\omega})$, will be a good representation of the spectral envelope at the ith period of the input signal.

The harmonic structure is reintroduced by Eq. (13.8). The DTFT of $\tilde{s}[n]$ is determined by observing that Eq. (13.8) can be written as

$$\tilde{s}[n] = s_w[n] * \sum_{r=-\infty}^{\infty} \delta[n - rT]. \tag{13.10}$$

The DTFT of the periodic impulse train is given by the DTFT pair

$$\sum_{r=-\infty}^{\infty} \delta[n - rT] \Longleftrightarrow \frac{2\pi}{T}\sum_{k=0}^{T-1} \delta(\omega - 2\pi k/T) \quad 0 \leq \omega < 2\pi, \tag{13.11}$$

where $\delta[n]$ represents the unit sample or discrete-time impulse sequence and $\delta(\omega)$ represents the continuous-variable impulse function (Dirac delta function). Therefore, it follows from Eq. (13.11) that

$$\tilde{S}(e^{j\omega}) = \frac{2\pi}{T}\sum_{k=0}^{T-1} S_w(e^{j2\pi k/T})\delta(\omega - 2\pi k/T) \quad 0 \leq \omega < 2\pi. \tag{13.12}$$

The DTFT $\tilde{S}(e^{j\omega})$ is, of course, a periodic function of ω with period 2π. Thus, the spectrum envelope, $S_w(e^{j\omega})$, is resampled by the new harmonic structure at normalized discrete-time frequencies $\omega = (2\pi k/T)$, $k = 0, \ldots, T-1$ over the baseband $0 \leq \omega < 2\pi$. If the window, $w[n]$, of Eq. (13.7a) is chosen as discussed above so that the original harmonic structure is smoothed while leaving an accurate representation of the formant structure, then the processing of Eq. (13.7a) modifies the pitch period of the signal in a simple and a computationally efficient manner, while leaving the spectral envelope (vocal tract frequency response) only slightly modified. The processing of Eq. (13.7a) is what is known as the TD-PSOLA process.

An illustration of the TD-PSOLA method is given in Figure 13.22 [97], which shows an original voiced speech signal with a pitch period of T_0 samples marked in the top left panel, along with a series of windows, of duration L samples, demarcating the individual windowed sections of the speech. The second panel on the left shows the individual windowed speech sections that are each shifted appropriately to match a desired pitch period of $T > T_0$ samples. The overlapped sum of the individual windowed sections of speech (after shifting) is shown in the bottom left panel. The right-hand side of Figure 13.22 shows spectra of the corresponding speech signals shown at the left. The top right hand plot is the DTFT of the entire segment on the left. Thus, the spectrum displays both the spectral envelope and the harmonic structure (additively on the dB scale). The bottom right hand plot likewise shows the DTFT of the entire waveform segment on the left. The middle plot on the right is the DTFT

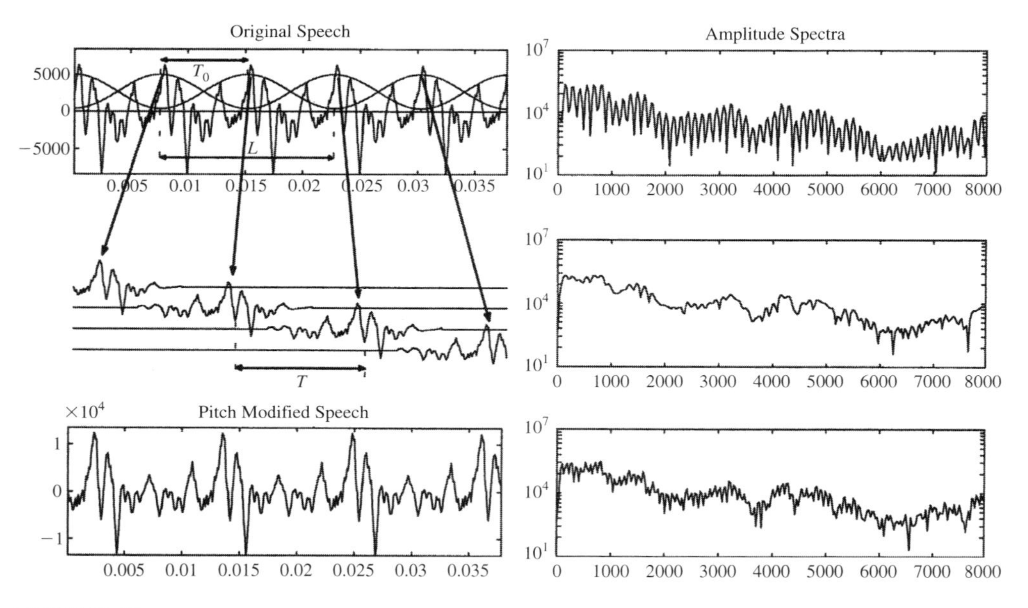

FIGURE 13.22
The TD-PSOLA process used to increase the pitch period of voiced speech from T_0 to T. The upper row, left side, shows the speech waveform comprising approximately five periods with the corresponding narrowband STFT on the right. Also shown on the upper left are several analysis windows. The middle row shows the set of adapted synthesis windows and the resulting wideband STFT of a typical adapted window. The bottom row shows the modified signal with new period T on the left and the corresponding narrowband STFT on the right. (After Dutois [97].)

of one of the windowed segments shown in the middle left hand plot. In effect, this is the spectral envelope extracted from the upper waveform. It can be seen that the modified speech signal essentially has the same spectral envelope as the original signal, as desired, although the pitch harmonics are significantly less well defined for the modified signal as compared to the original speech signal. This is partly due to the fact that the pitch frequency was lowered, resulting in denser harmonics.

The TD-PSOLA method provides a very good way of modifying pitch at the boundary between two segments. We now examine the remaining boundary segment issues. There are three significant problems that can occur at each boundary:

1. **Phase mismatch**: The end of segment 1 does not align in phase (of the pitch period) with the beginning of segment 2, as illustrated in Figure 13.23 [97], where we see that the OLA windows are not centered on the same relative positions within pitch periods. It is almost impossible, algorithmically, to fix this problem due to the difficulty in finding the points of glottal closure so as to mark a consistent point of each pitch period.

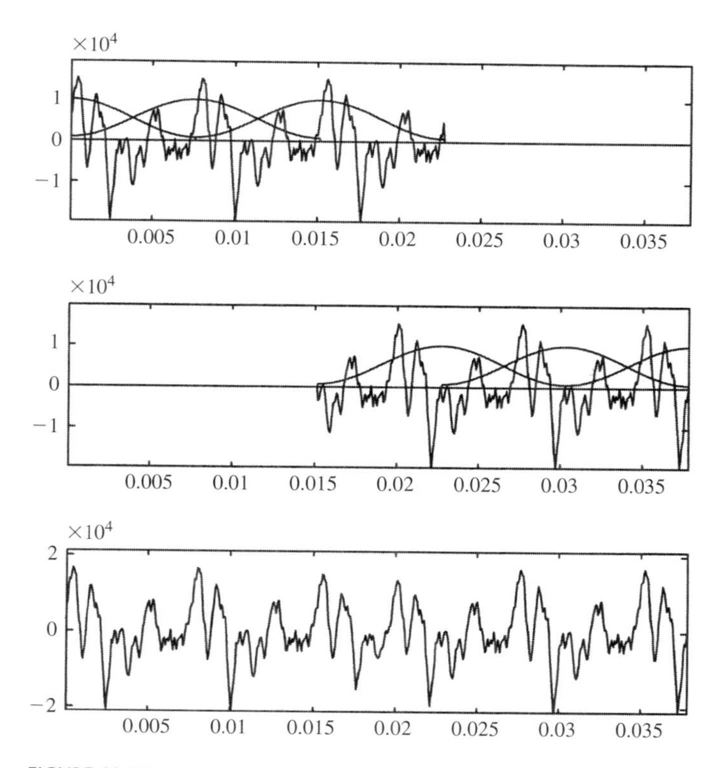

FIGURE 13.23
Illustration of a phase mismatch at the boundary between two segments. Although the pitch periods are essentially the same, as is the waveform shape, the placement of the OLA windows is not centered with regard to the relative positions within periods. (After Dutoit [97].)

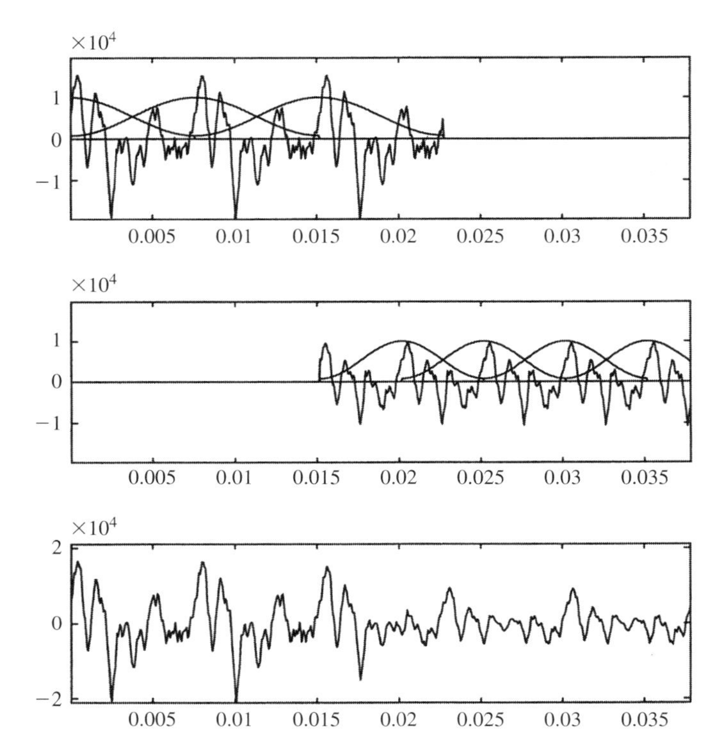

FIGURE 13.24
Illustration of a pitch mismatch at the boundary between two segments. Although the spectral envelopes are essentially the same, the pitch is grossly different between segments. (After Dutoit [97].)

2. **Gross pitch mismatch**: The pitch period at the end of segment 1 is very different from the pitch period at the beginning of segment 2, as illustrated in Figure 13.24 [97]. This situation cannot generally be fixed using the TD-PSOLA method, as this processing works best in the case of small changes in pitch period at the boundary. Fortunately, with large numbers of segments in modern TTS systems, it is generally possible to avoid such gross pitch period mismatch problems in the unit selection process.
3. **Spectral envelope mismatch**: The spectral envelope at the end of segment 1 is very different from the spectral envelope at the beginning of segment 2, as illustrated in Figure 13.25 [97]. Since spectral envelope mismatch is a major cause of quality degradation in concatenative TTS systems, it is essential that this situation be avoided as much as possible. One possible solution is to use representations of the segments such as LPC or the cepstrum, where it is possible to interpolate the spectra at the boundaries between two segments. A second possible solution is to resynthesize all the segments of the database to achieve the ideal situation of all words being pronounced using constant pitch, all phase effects being eliminated by using a coherent positioning of OLA windows, and spectral mismatches being

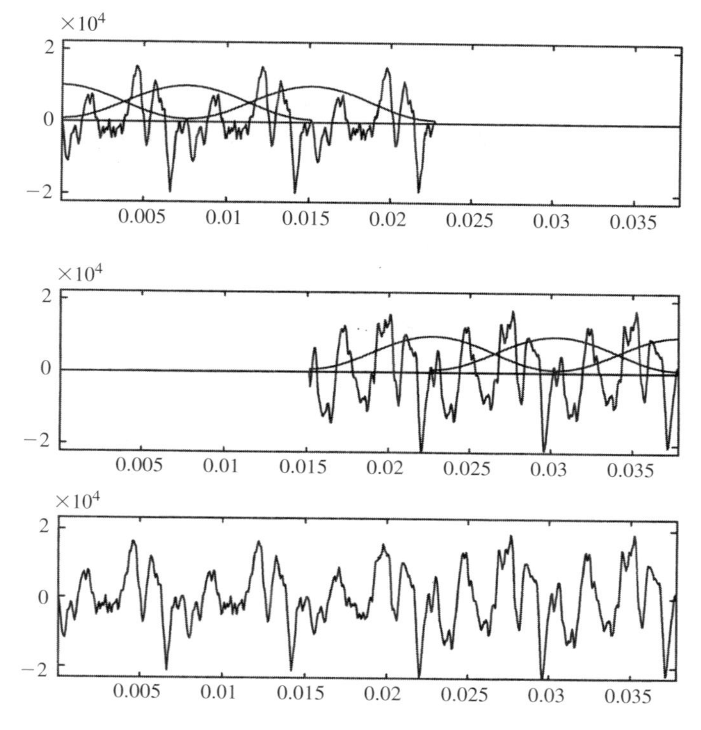

FIGURE 13.25
Illustration of a spectral envelope mismatch between two segments. For this example, the pitch is the same for both segments and the phase is essentially correct. (After Dutoit [97].)

minimized using some type of spectral interpolation process. Dutoit describes this segment resynthesis process in great detail [97].

The techniques of speech waveform modification (especially at the boundaries between concatenated segments) are essential in modern TTS systems since unit selection methods cannot possibly cover all possible combinations of feature variability in a database. This is especially the case for concatenative synthesis using diphone units where there are too many diphones, in too many contexts, to record all possibilities. Since signal processing techniques like TD-PSOLA and LPC spectral interpolation are well understood, easy to use, and are effective for certain pitch and spectral envelope modifications, the use of such techniques is widespread in TTS systems.

Another important use of waveform modification methods is the area of voice alteration; i.e., changing the identity of the speaker via signal processing methods. The goal of voice alteration is to preserve the intelligibility and (as much as possible) the naturalness of the speech, at the same time changing the speaker characteristics so as to match some desired criterion; e.g., from a male talker to a female talker or vice versa. As identified by Yang and Stylianou [423], the major characteristics of the speaker identity reside in the overall shape of the spectral envelope and the locations of the formants of the sounds being produced. To make the speaker identity more female-like, the processing must raise the range of fundamental frequency and shorten the vocal tract. Conversely, to make the speaker identity more male-like, the processing must lower the range of fundamental frequency and lengthen the vocal tract. Using an LPC representation based on pseudo log area ratios (PLARs), the size of the vocal tract could be shortened or lengthened using two controls, namely a front log area ratio modification (for log area ratios at the front of the vocal tract) and a back log area ratio modification (for log area ratios at the back of the vocal tract). By computing a new set of PLARs, based on the original distances from the front of the vocal tract, and suitably modifying the fundamental frequency contour, the speaker identity could gradually become more female- or male-like by adjusting the controls. Yang and Stylianou [423] were able to show that the voice-altered speech was of reasonable quality over a range of voice controls.

13.5.7 Some Experimental Results Using Unit Selection Methods

Based on a series of investigations by Beutnagel et al. [36] on TTS systems based on unit selection methods, it has been found that unit selection methods tend to work better with diphone units (rather than phone units), with no modification of pitch, spectrum, or phase at unit boundaries (assuming that there are enough units in the units database to get reasonably close matches without the need for any modifications), using a simple waveform synthesis method like PSOLA (over simply abutting the diphone waveforms), and using the entire units database for finding the best sequence of units. The system used in these experiments was based on a training set with about 90 minutes of speech.

13.6 TTS FUTURE NEEDS

Modern TTS systems have the capability of producing highly intelligible and surprisingly natural speech utterances, so long as the utterance is not too long, or too detailed, or too technical, etc. The biggest problem with most TTS systems is that they have no idea as to **how** things should be said, but instead rely on the text analysis for emphasis, prosody, phrasing, and all so-called suprasegmental features of the spoken utterance. The more TTS systems learn how to produce context-sensitive pronunciations of words (and phrases), the more natural sounding these systems will become. Hence, by way of example, the utterance "I gave the book to John" has at least three different semantic interpretations, each with different emphasis on words in the utterance; i.e.,

I gave the book to **John**; i.e., not to Mary or Bob.
I gave the **book** to John; i.e., not the photos or the apple.
I gave the book to John; i.e., I did it, not someone else.

The second future need of TTS systems is improvements in the unit selection process so as to better capture the target cost for mismatch between **predicted** unit specification (i.e., phoneme name, duration, pitch, spectral properties) and **actual** features of a candidate recorded unit. Also needed in the unit selection process is a better spectral distance measure that incorporates human perception measures so as to find the best sequence of units for a given utterance.

13.7 VISUAL TTS

Visual TTS (VTTS) is the name given to the process of providing a talking face from textual input, where the speech is obtained from a standard TTS process and the face could be an avatar (of the type shown in Figure 13.26 [233]) or a sampled human head (of the type shown in Figure 13.27 [65]). Visual TTS is predicated on the belief that a personalized friendly agent provides an entertaining and effective user experience. In fact subjective tests have confirmed the belief that visual agents are preferred over text and audio interfaces to information, and that visual agents are more trusted than text and audio interfaces. Interestingly one of the key advantages of using a "talking head" is a user perceived increase in the intelligibility and naturalness of the audio TTS component of the system, even though that component is identical to what is used in an audio-only TTS system [274, 358].

The key proposed application areas for VTTS systems include the following:

- **personal assistant**, as might be used to read your e-mail out loud, or look up information, or manage your calendar or contact list, etc.

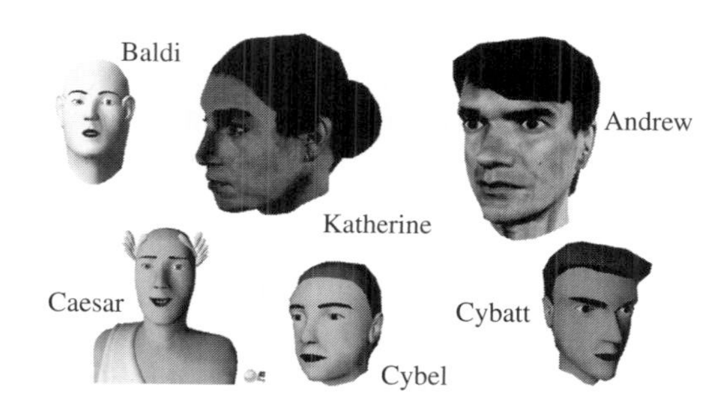

FIGURE 13.26
Examples of family of avatars based on three-dimensional models. (After Masaro [233].)

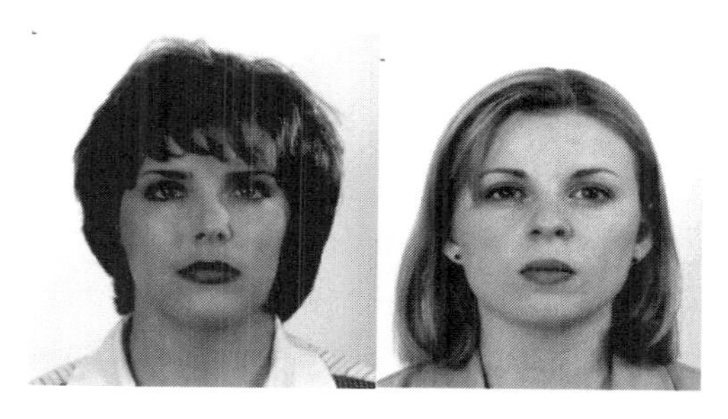

FIGURE 13.27
Examples of sample-based talking heads. (After Cosatto et al. [65]. © [2003] IEEE.)

- **customer service agent**, as might be used in a help or information desk to guide you through a website or enable you to gain access to an appropriate human agent
- **newscaster service**, as might be used to read selected news items, or alert you to breaking stories of special interest; the service "Ananova" is one such agent [5]
- **e-commerce agent**, as might be used to help with on-line purchases
- **game playing companion**, as might be used for both action and strategy games, providing a human-like playmate for entertainment and leisure.

The advantage of using avatars (three-dimensional talking heads) is that they are flexible and can easily be shown in any pose or at any angle, the faces look "cartoon-like," they are easy to modify by adding hats, glasses, etc., and they generally are well liked by most people. On the other hand, the advantage of a sample-based talking head is that it looks like a real person is speaking and users tend to relate to images of real people very well. The disadvantage of sample-based talking heads is that they are highly limited in pose and are not very flexible with regard to changing their appearance.

13.7.1 VTTS Process

Figure 13.28 shows a block diagram of the VTTS process. The input to the system, much like an audio TTS system, is text and the output is the controls needed to render either a three-dimensional model (for the avatar) or a sample-based model, for the human face, onto a display for presentation to a user. The visual rendering model utilizes four types of input to create and animate the talking head display:

1. **audio signal**, which is integrated with the facial expression and face feature movements to create the multimodal user experience of the talking head
2. **speech phonemes and timing information**, often called visemes or visible phonemes, which help the rendering module determine the appropriate mouth shape for each sound in the spoken output (e.g., so that it knows when to close

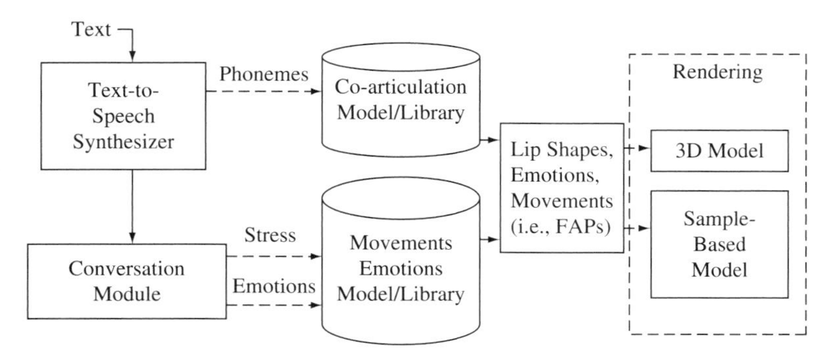

FIGURE 13.28
Block diagram of VTTS process.

TABLE 13.2 Table of 14 static visemes. (After Ostermann et al. [274].)

Viseme #	Phoneme Set	Viseme #	Phoneme Set
1	/p/, /b/, /m/	8	/n/, /l/
2	/f/, /v/	9	/r/
3	/θ/, /ð/	10	/a/
4	/t/, /d/	11	/ε/
5	/k/, /g/	12	/I/
6	/tʃ/, /dʒ/, /ʃ/	13	/a/
7	/s/, /z/	14	/ʊ/

TABLE 13.3 Advantages and disadvantages of the two face rendering methods.

	Advantage	Disadvantage
Synthetic 3D models *parametrized shapes*	Keeps correct appearance under full range of views	Hard to reproduce minute skin details, like wrinkles, that look absolutely natural
Sample-based *parametrized textures*	Reproduces photo realistic appearances; fast	Range of view limited by planar approximation of parts

the lips during a labial stop consonant, e.g., /B/ or /P/); there are 14 static visemes for English and they are listed in Table 13.2, as described in Ref. [274]

3. **face animation parameters (FAPs)**, where the FAPs are the set of parameters that specify the face model and face feature movements (generally non-speech related motions of the eyebrows, nose, ears, head turns, nods, gaze direction, etc., as well as emotions like happiness or fright)
4. **text**, where a separate text analysis is made to determine visual prosody (facial movements that complement the speech sounds) during the production of the speech utterance; such visual prosody cues make the avatar or the sample-based talking head come alive to the user and are essential for high quality VTTS.

A summary of the advantages and disadvantages of the two rendering models for VTTS is given in Table 13.3. It can be seen that neither model is ideal in all circumstances and both have key advantages for a range of applications.

Figure 13.29 illustrates some of the steps in rendering the sample-based face model. Basically, the process steps include the following:

- Snippets of video are concatenated to synthesize the talking head.
- The number of samples which are stored is reduced by decomposing the recorded head into sub-parts (e.g., eyebrow areas, nose and eyes, lips, jaw, etc.).

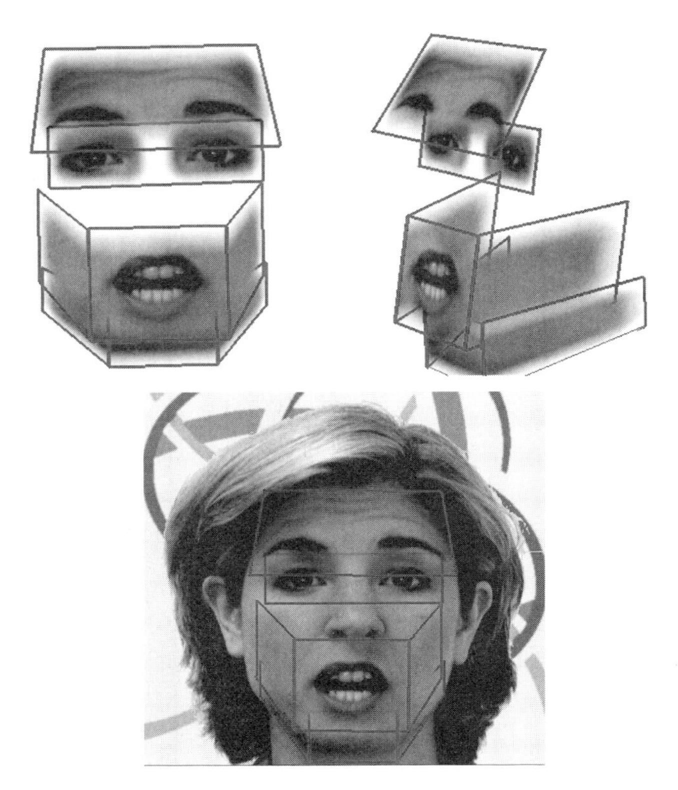

FIGURE 13.29
Illustration of face decomposition for sample-based model.

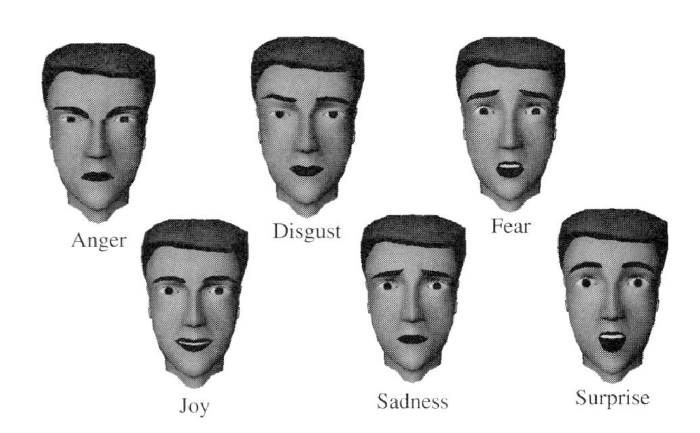

FIGURE 13.30
Examples of facial expressions using avatars. (After Ostermann et al. [274].)

- A background image of the entire head enables the determination of which parts need to be warped to blend in for various visual effects.
- The technique of feathering (providing a transparent gradient at the border between sub-parts) helps provide a smooth blending of facial parts with the head image.
- A smooth transition of each face object (e.g., mouth shape) across the unit boundaries is obtained by using advanced morphing techniques.

An example of a series of facial expressions, shown using the Cybatt avatar of Figure 13.26, is given in Figure 13.30. It is a lot easier expressing the large range of emotions of Figure 13.30 using avatars, where the facial features can be highly articulated (exaggerated), than for the sample-based models.

13.8 SUMMARY

This chapter discussed a range of issues associated with the problem of creating an intelligible and natural sounding voice for a machine, based on a textual input. The first problem in implementing such a system was the definition and realization of a process for converting the input text string to a sequence of labeled sounds (usually phonemes) along with their associated durations. Also generated from the input text was a pitch and intensity contour that provided a means for emphasis of sounds within the resulting speech utterance.

The second problem in the conversion from text to speech was the realization and implementation of the synthesis process that converted the text-based analysis parameter set (i.e., the phonemes along with prosody parameters) to intelligible synthetic speech. Although a variety of synthesis procedures have been investigated, it was shown that a "data-driven" strategy of using a very large speech database and carefully selecting multiple versions of each of a large set of synthesis units was the most successful of synthesis procedures, producing highly intelligible and near natural sounding speech for a variety of tasks. Such unit-selection synthesis methods have dominated the area of text-based speech synthesis for the past decade.

Finally, we ended the discussion of TTS systems with a description of a visual speech synthesis system which integrated either a visual avatar or a sampled human face whose motions (lips, teeth, jaw, brow, head, etc.) were synchronized with the mode of production of the associated speech sounds being produced and spoken. It was argued that such visual TTS systems were highly desirable for a range of human–machine interactions.

PROBLEMS

13.1. (MATLAB Exercise) This MATLAB exercise compares word concatenation synthesis versus naturally spoken utterances for the two sentences:

1. This shirt is red.
2. Are you a good boy or a bad boy.

The source files for the individual words are in the MATLAB directory `chapter_13` on the book website with filenames of the form `word.wav` where word is one of the four unique words in sentence 1 or one of the seven unique words in sentence 2. The source files for the full sentences are also included in the same directory and are of the form `sentence_1.wav` and `sentence_2.wav`.

(a) Concatenate the four words for sentence 1 and play the sequence of concatenated words along with the natural sentence. What is the ratio of durations of the concatenated word sentence and the natural sentence.

(b) Repeat part (a) for the words of sentence 2.

(c) Create a pair of wideband spectrograms of sentence 1 on a single figure with the natural speech spectrogram at the top and the concatenated speech spectrogram at the bottom.

(d) Repeat part (c) for sentence 2.

13.2. (MATLAB Exercise) This MATLAB exercise illustrates the synthesis of a vowel sound using a formant synthesizer of the form:

$$H(z) = \prod_{k=1}^{4} V_k(z),$$

$$V_k(z) = \frac{1 - 2 \cdot e^{-b(k)2\pi/F_s} \cdot \cos(2\pi f(k)/F_s) + e^{-2b(k)2\pi/F_s}}{1 - 2 \cdot e^{-b(k)2\pi/F_s} \cdot \cos(2\pi f(k)/F_s) z^{-1} + e^{-2b(k)2\pi/F_s}},$$

where $b(k)$ are the formant bandwidths (in Hz) and are assumed to be [50, 80, 100, 150] Hz for the first four formants, and $f(k)$ are the formant resonances. The first three formants are associated with the individual vowel sounds and can be obtained from Table 3.4 in Chapter 3, the fourth formant is assumed to be 4000 Hz for all vowel sounds.

(a) Using the values of the first three formants for the vowel /IY/, synthesize 2 seconds of vowel assuming a sampling rate of $F_s = 10{,}000$ Hz, and a constant pitch period, P, of 100 samples. Listen to the vowel; what is the pitch frequency of the synthesized vowel?

(b) Choose a section of 400 samples of the periodic waveform and plot, on a single figure, the windowed waveform and its log magnitude spectrum. Can you see the individual pitch harmonics in the log magnitude spectrum?

(c) Change the pitch period to 200 samples and re-synthesize another 2 seconds of vowel. Listen to the vowel; what is the pitch frequency of the synthesized vowel?

(d) Change the pitch period to 50 samples and re-synthesize another 2 seconds of vowel. Listen to the vowel; what is the pitch frequency of the synthesized vowel?

13.3. (MATLAB Exercise) This MATLAB exercise illustrates the synthesis of a vowel sound with a time-varying pitch period. Assume that the pitch period is now a function of sample index, and is of the form:

$$P[n] = \begin{cases} 50 + \left(\dfrac{150}{10{,}000}\right) \cdot n & 0 \leq n \leq 10{,}000 \\ 200 - \left(\dfrac{150}{10{,}000}\right)(n - 10{,}000) & 10{,}001 \leq n \leq 20{,}000. \end{cases}$$

(a) Plot the pitch period contour for the 20,001 sample interval, $0 \leq n \leq 20{,}000$.

(b) Synthesize 2 seconds of the vowel, /AA/, using the first three formants from Table 3.4 for /AA/, along with a fixed fourth formant at 4000 Hz, and with formant bandwidths of [50, 80, 100, 150] Hz, with a sampling rate of $F_s = 10{,}000$ Hz. Use the pitch period contour above as the pitch during the 2-second interval. Listen to the vowel sound; can you hear the changing pitch frequency?

(c) Re-synthesize the vowel /AA/ with the new pitch period contour:

$$P[n] = \begin{cases} 200 - \left(\dfrac{150}{10{,}000}\right) \cdot n & 0 \leq n \leq 10{,}000 \\ 50 + \left(\dfrac{150}{10{,}000}\right)(n - 10{,}000) & 10{,}001 \leq n \leq 20{,}000. \end{cases}$$

Listen to the new vowel sound. How does the pitch contour differ from that of the previous synthesis of the /AA/ vowel?

CHAPTER 14

Automatic Speech Recognition and Natural Language Understanding

14.1 INTRODUCTION

For more than five decades, speech researchers have worked toward the goal of building a machine that can recognize and understand fluently spoken speech. A major driving factor in the research has been the realization that enabling people to converse with machines has three potential major applications:

1. **Cost Reduction:** By providing a service where humans interact solely with machines, companies eliminate the cost of live agents and thereby significantly reduce the cost of providing the service. Interestingly, as a result, this process often provides humans with a more natural and convenient way of accessing information and services. For example, airlines have been using speech recognition technology for more than a decade, enabling them to inform their customers as to actual flight arrival and flight departure times.
2. **New Revenue Opportunities:** By providing around-the-clock customer helplines and customer care agents, a company can interact remotely with its customers using natural speech input. An example of such a service is the "How May I Help You" (HMIHY©) service of AT&T [135, 136], which automated customer care for AT&T consumer services late in 2000.
3. **Customer Personalization and Retention:** By providing a range of personalized services based on customer preferences, companies can improve the customer experience and raise customer satisfaction with services and information systems. A very simple example would be the personalization of the automotive environment by having the automobile recognize the driver from simple voice commands and then automatically adjust the comfort features of the car to the customer's pre-specified preferences.

Although most of this chapter will be devoted to a discussion of how speech recognition and natural language understanding are implemented, we shall begin by introducing the concept of the *speech dialog circle*, which is depicted in Figure 14.1. This dialog circle is the enabling mechanism for all human-to-machine communications. The speech dialog circle depicts the sequence of events that occur between a spoken utterance (by a human) and the spoken response to that utterance

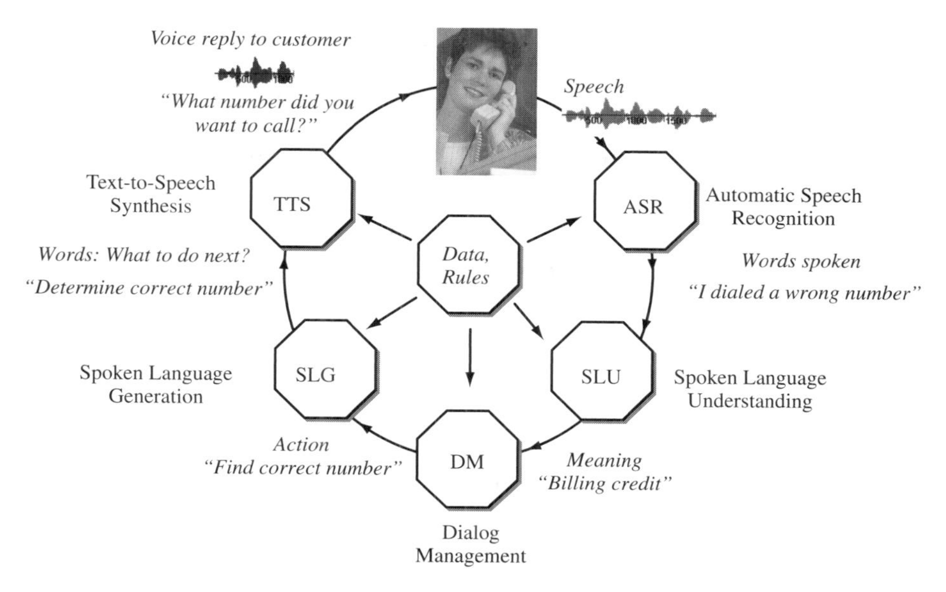

FIGURE 14.1
The speech dialog circle for natural language interactions with a machine.

(which comes from the machine). The input speech is first processed by a module that we denote as an ASR (automatic speech recognition) system whose sole purpose is to convert the spoken input into an orthographic sequence of words which is the best (in a maximum likelihood sense) estimate of the spoken utterance. Next the recognized words are analyzed by a spoken language understanding (SLU) module which attempts to attribute meaning (in the context of the task being performed by the machine) to the spoken words. Once an appropriate meaning has been determined, the dialog management module determines the course of action that would be most appropriate, according to the current state of the dialog and a series of prescribed operation workflow steps for the task at hand. The action taken can be as simple as a request for further information (especially when some confusion or uncertainty exists as to the best action to take) or confirmation of an action that is about to be taken. Once the action is decided upon, the spoken language generation (SLG) module chooses the text that is to be spoken to the human to describe the action being taken, or to request additional instructions as to how to best proceed. Finally the text-to-speech (TTS) module speaks the utterance chosen by the SLG module and the human must decide what to say next so as to continue or complete the desired transaction. The speech dialog circle is traversed once for each spoken input, and is traversed several times during a typical transaction scenario.[1] The more accurate the ASR and

[1] All of the modules in the speech dialog circle can be either data or rule driven in both the learning phases and in the active use phase, as indicated by the central data/rules module in Figure 14.1.

associated SLU modules, the more the spoken dialog circle feels like a natural user interface to the machine, and the easier it is to interact and complete transactions. The speech dialog circle is a powerful concept in modern speech recognition and spoken language understanding systems and is widely used in a range of modern systems.

Chapter 13 discussed text-to-speech synthesis, and in the rest of this chapter, we focus on automatic speech recognition. These are the parts of the dialog circle where the digital signal processing techniques of the earlier chapters are employed. The remaining parts of the dialog circle are no less important, but they are concerned with matters at the text level that are beyond our scope in this book.

14.2 BASIC ASR FORMULATION

The goal of an ASR system is to accurately and efficiently convert a speech signal into a text message transcription of the spoken words, independent of the device used to record the speech (i.e., the transducer or microphone), the speaker's accent, or the acoustic environment in which the speaker is located (e.g., quiet office, noisy room, outdoors). The ultimate goal of ASR, namely recognizing speech as accurately and as reliably as a human listener, has not yet been achieved.

The basis for most modern speech and language understanding systems is the simple (conceptual) model of speech generation and speech recognition shown in Figure 14.2. It is assumed that the speaker intends to express some thought as part of a process of conversing with another human or with a machine. To express that thought, the speaker must compose a linguistically meaningful sentence, W, in the form of a sequence of words (possibly with pauses and other acoustic events such as uh's, um's, er's, etc.). Once the words are chosen, the speaker sends appropriate control signals to the articulatory speech organs, which form a speech utterance whose sounds are those required to speak the desired sentence, resulting in the speech waveform $s[n]$. We call the process of creating the speech waveform from the speaker's intention the *speaker model*, since it reflects the speaker's accent and choice of words used to express a given thought or request. The processing steps of the speech recognizer are shown at the right side of Figure 14.2 and consist of an acoustic processor which analyzes the speech signal and converts it into a set of acoustic (spectral, temporal) features, X, which efficiently characterize the properties of the speech sounds, followed by a linguistic decoding process which makes a best (maximum likelihood) estimate of the words of the spoken sentence, resulting in the recognized sentence $\hat{W}$.

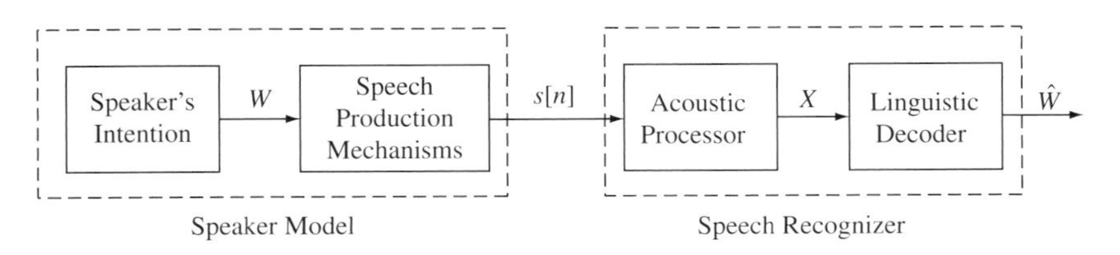

FIGURE 14.2
Conceptual model of speech production (the speaker model) and speech recognition (the speech recognizer) processes.

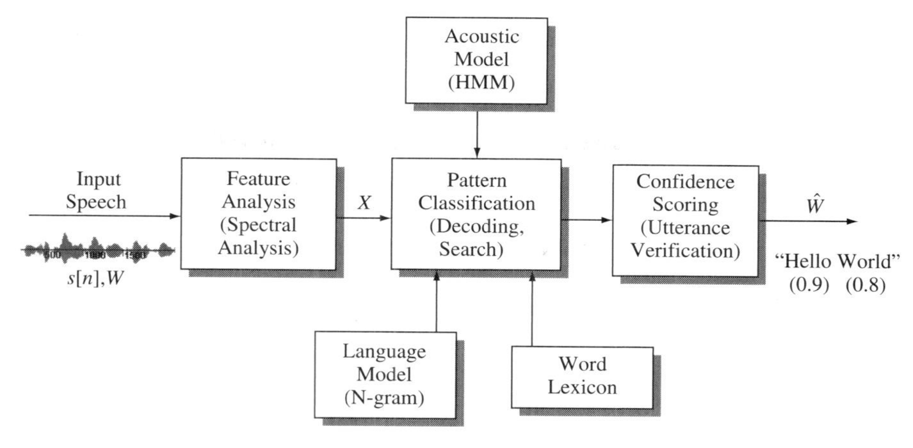

FIGURE 14.3
Block diagram of overall speech recognition system consisting of signal processing (analysis to derive the set of feature vectors that represent the input speech), pattern classification (to best match the set of feature vectors to the optimal set of concatenated word or sound models), and confidence scoring (to get an estimate of the confidence for each word in the recognized output string).

14.3 OVERALL SPEECH RECOGNITION PROCESS

Figure 14.3 shows a block diagram of one possible implementation of the speech recognizer of Figure 14.2. The input speech signal, $s[n]$, is converted to the sequence of feature vectors, $X = \{\mathbf{X}_1, \mathbf{X}_2, \dots, \mathbf{X}_T\}$, by the feature analysis block (also denoted spectral analysis). The feature vectors are computed on a frame-by-frame basis using the techniques discussed in Chapters 6–9. In particular, a sequence of vectors of mel frequency cepstrum coefficients is widely used to represent the short-time spectral characteristics of the speech signal. The pattern classification block (also denoted as the decoding and search block) decodes the sequence of feature vectors into a symbolic representation that is the maximum likelihood string, $\hat{W}$, that could have produced the input sequence of feature vectors (subject to the constraints of the word lexicon and the language model). The pattern recognition system uses a set of acoustic models (represented as hidden Markov models, HMMs) and a word lexicon to provide a set of pronunciations for each word in each possible word string. An N-gram language model is used to compute a language model score for each possible word string. The pattern classification block thereby provides an overall likelihood score for every possible sentence in the task language. The final block in the process is a confidence scoring process (also denoted as an utterance verification block), which is used to provide a confidence score for each individual word in the (maximum likelihood) recognized string. This computation provides a measure of confidence as to the correctness of each content word within the recognized sentence in order to identify possible recognition errors (as well as out-of-vocabulary events) and thereby to potentially improve the performance of the recognition algorithm. To achieve this goal, a word confidence score [278], based on a simple likelihood ratio hypothesis test associated with each recognized word, is performed and the word confidence score is used to determine which, if any, words are likely to be incorrect because of either a recognition error or because it was an *out-of-vocabulary* (OOV) word (that could never be correctly recognized).[2] Each of the blocks in Figure 14.3 requires a range of signal processing operations and, in some cases, extensive digital computation. The remainder of this chapter gives an overview of what is involved in each part of the processing and computation of Figure 14.3.

A simple example of the spoken input to the recognizer consisting of a two-word phrase and the resulting recognized string with confidence scores is given below. This example illustrates the importance and value of confidence scoring.

Spoken Input: credit please
Recognized String: credit fees
Confidence Scores: (0.9) (0.3)

Based on the confidence scores (derived using a likelihood ratio test), the recognition system would realize which word or words (e.g., the word "fees" in the above example) are likely to be in error (or OOV words) and take appropriate steps (in the ensuing dialog) to determine whether or not a recognition error had been made and, if so, to decide how to fix it so that the dialog could move forward to the task goal in an orderly and efficient manner.

14.4 BUILDING A SPEECH RECOGNITION SYSTEM

The four steps in building and evaluating a speech recognition system are the following:

1. Choose the recognition task, including the recognition word vocabulary (the lexicon), the basic speech sounds to represent the vocabulary (the speech units), the task syntax or language model, and the task semantics (if any).
2. Choose the feature set and the associated signal processing for representing the properties of the speech signal over time.
3. Train the set of acoustic and language models.
4. Evaluate the performance of the resulting speech recognition system.

We now discuss each of these four steps in more detail.

14.4.1 The Recognition Task

Recognition tasks vary from simple word and phrase recognition systems to large vocabulary conversational interfaces to machines. For example, for the simple task of recognizing a vocabulary of eleven isolated digits (/zero/ to /nine/ and /oh/) as isolated spoken inputs, we can use whole word models as the basic recognition unit (although

[2]The confidence score for simple *function* words (such as /and/, /the/, /a/, /of/, /it/, etc.) will not generally be high, but usually this will not adversely affect overall sentence understanding since the *content* words are the most important to get correct in order to understand the meaning of the spoken utterance.

it is certainly possible to consider phoneme units, with the digits being composed of sequences of phonemes). Whole word patterns, when viable, are generally the most robust models for recognition since it is reasonable to train simple word models in virtually an unlimited range of acoustic contexts.

Using the digits vocabulary, the task could be recognition of single digits (as in making a choice among a small number of alternatives), or recognition of a string of digits that represents a telephone number, an identification code, or a highly constrained sequence of digits that form a password. Hence a range of task syntaxes is possible, even for such a simple recognition vocabulary.

14.4.2 Recognition Feature Set

There is no "standard" set of features that have been used for speech recognition. Instead, various combinations of acoustic, articulatory, and auditory features have been used in a range of speech recognition systems. The most popular acoustic features have been the mel-frequency cepstral coefficients and their derivatives. Other popular feature vectors have been based on combinations of speech production parameters such as formant frequencies, pitch period, zero-crossing rate, and energy (often in selected frequency bands). The most popular acoustic-phonetic features have been those that describe both manner and place of articulation, often estimated using neural network processing methods [149]. Articulatory features include vocal tract area functions as well as positions of articulators in the production of speech. Feature vectors based on auditory models have also been used. Examples include parameters from representations like the Seneff [360] or Lyon ear models [214], or the Ensemble Interval Histogram method of Ghitza [124], which are discussed in Chapter 4.

A block diagram of the signal processing used in most modern large vocabulary speech recognition systems is shown in Figure 14.4. The analog speech signal is sampled and quantized at rates from 8000 samples/sec up to 20,000 samples/sec. A first-order highpass pre-emphasis network $(1 - \gamma\, z^{-1})$ is often used to compensate for the speech spectral fall-off at higher frequencies. The pre-emphasized signal is next blocked into frames of L samples, with adjacent frames spaced R samples apart. Typical values for L and R correspond to frames of duration 15–40 msec, with frame shifts of 10 msec being most common. A Hamming window is applied to each frame prior to spectral analysis using linear prediction methods. Following (optionally used) simple noise removal methods, the predictor coefficients representing the short-time spectrum, $\alpha'_m[l]$, are normalized and converted to mel-frequency cepstral coefficients (mfcc) $c_m[l]$ via standard analysis methods of the type discussed in Chapter 8 [82].[3] Some type of cepstral bias removal is often used prior to calculation of the first- and second-order cepstral time derivatives. Typically the resulting feature vector is the set of equalized mel-cepstrum cepstral coefficients $c'_m[l]$, and their first- and second-order derivatives $\Delta c'_m[l]$ and $\Delta^2 c'_m[l]$. It is typical to use 13 mfcc coefficients, 13 first-order cepstral derivative coefficients, and 13 second-order derivative cepstral coefficients, resulting in a $D = 39$ component feature vector for each 10 msec frame of speech.

[3] In this chapter, we use the notation $\alpha_m[l]$ to denote the lth prediction coefficient of the mth frame.

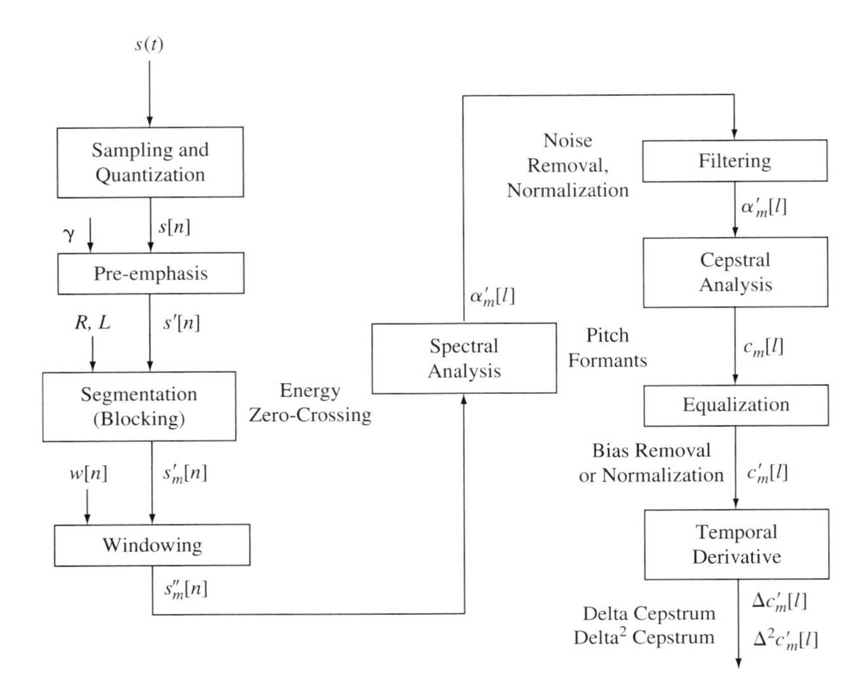

FIGURE 14.4
Block diagram of feature extraction process for feature vector consisting of a set of mfcc coefficients and their first and second derivatives.

A key issue in the choice of acoustic features to represent the speech signal is the lack of robustness of almost every known feature set to noise, background signals, network transmission, and spectral distortions, resulting in a mismatch in the statistical properties of the acoustic feature vectors between the training and testing phases of recognition systems. This lack of robustness of the feature set often leads to significant performance degradation and has led to a wide range of proposed signal processing and information theory fixes to the problems. The range of methods that has been proposed to ameliorate this robustness problem is illustrated in Figure 14.5, which shows the three basic methods of matching, namely at the signal level (via speech enhancement methods), at the feature level (using some type of feature normalization method), and at the model level (using some type of model adaptation method).

14.4.3 Recognition Training

There are two aspects to training models for speech recognition, namely acoustic model training and language model training. Acoustic model training requires recording each of the model units (whole words, phonemes) in as many contexts as possible so that a statistical learning method can create accurate statistical distributions for each of the model states. Acoustic training relies on accurately labeled speech utterances that are segmented according to a corresponding model-unit transcription. Training of the acoustic models first involves segmenting the spoken strings into recognition

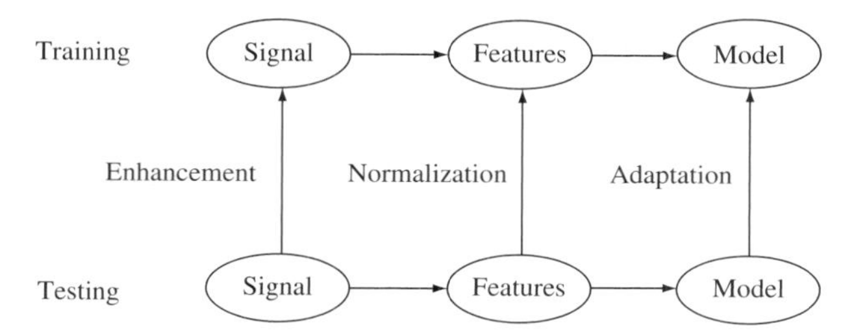

FIGURE 14.5
Overview of methods for making the speech signal, the feature vector, and the set of acoustic models more robust to noise and other distortions via signal enhancement, feature normalization, and model adaptation.

model units (via either a Baum-Welch or Viterbi alignment method as explained later in this chapter), and then using the segmented utterances to simultaneously build statistical model distributions for each state of the vocabulary unit models. As an example, for a simple isolated digit vocabulary, having 100 people speak each of the 11 digits 10 times in isolation forms a powerful training set that provides very accurate and highly reliable whole word digit models for recognition. The resulting statistical models form the basis for the pattern recognition operations that are at the heart of the ASR system.

Language model training requires a large training set of text strings that reflect the syntax of spoken utterances for the task at hand. Generally such text training sets are created by machine, based on a model of grammar for the recognition task. In some cases, text training sets use existing text sources, such as magazine and newspaper articles, or closed caption transcripts of television news broadcasts, etc. Other times, training sets for language models can be created from databases; e.g., valid strings of telephone numbers can be created from existing telephone directories.

14.4.4 Testing and Performance Evaluation

In order to improve the performance of any speech recognition system there must be a reliable and statistically significant way of evaluating recognition system performance based on an independent test set of labeled utterances. Typically we use word error rate and sentence (or task) error rate as measures of recognizer performance. For example, for the digit recognition task, with the task of recognizing valid telephone numbers, we might use a talker population of 25 new talkers, each speaking 10 telephone numbers as sequences of isolated (or connected) digits, and then evaluate the digit error rate and the (constrained) telephone number string error rate.

14.5 THE DECISION PROCESSES IN ASR

The pattern classification and decision operations are at the heart of any ASR system. In this section we give a brief introduction to the Bayesian formulation of the ASR problem.

14.5.1 Bayesian Formulation of the ASR Problem

The problem of ASR is generally treated as a statistical decision problem. Specifically it is formulated as a Bayes maximum *a posteriori* probability (MAP) decision process where we seek to find the word string $\hat{W}$ (in the task language) that maximizes the *a posteriori* probability, $P(W|X)$, of that string, given the sequence of feature vectors, X, i.e.,

$$\hat{W} = \arg\max_{W} P(W|X). \tag{14.1}$$

Using Bayes' rule, we can rewrite Eq. (14.1) in the form:

$$\hat{W} = \arg\max_{W} \frac{P(X|W)P(W)}{P(X)}. \tag{14.2}$$

Equation (14.2) shows that the calculation of the *a posteriori* probability is decomposed into two terms, one that defines the *a priori* probability of the word sequence, W, namely $P(W)$, and the other that defines the likelihood that the word string, W, produced the feature vector, X, namely $P(X|W)$. For all future calculations, we disregard the denominator term, $P(X)$, since it is independent of the word sequence W being optimized. The term $P(X|W)$ is known as the "acoustic model" and is generally denoted as $P_A(X|W)$ to emphasize the acoustic nature of this term. The term $P(W)$ is known as the "language model" and is generally denoted as $P_L(W)$ to emphasize the linguistic nature of this term. The probabilities associated with $P_A(X|W)$ and $P_L(W)$ are estimated or learned from a set of training data that have been labeled by a knowledge source, usually a human expert, where the training sets (both acoustic and linguistic) are as large as reasonably possible. The recognition decoding process of Eq. (14.2) is often written in the form of a 3-step process; i.e.,

$$\hat{W} = \underbrace{\arg\max_{W}}_{\text{Step 3}} \underbrace{P_A(X|W)}_{\text{Step 1}} \underbrace{P_L(W)}_{\text{Step 2}}, \tag{14.3}$$

where Step 1 is the computation of the probability associated with the acoustic model of the speech sounds in the sentence W, Step 2 is the computation of the probability associated with the linguistic model of the words in the sentence, and Step 3 is the computation associated with the search through all valid sentences in the task language for the maximum likelihood sentence.

In order to be more explicit about the signal processing and computations associated with each of the three steps of Eq. (14.3), we need to say a bit more about the feature vector, X, and the word sequence W. We can explicitly represent the feature vector, X, as a sequence of acoustic observations corresponding to each of T frames of the speech, of the form:

$$X = \{\mathbf{X}_1, \mathbf{X}_2, \ldots, \mathbf{X}_T\} \tag{14.4}$$

where the speech signal duration is T frames (i.e., T times the frame shift in msec). Each frame, $\mathbf{X}_t$, $t = 1, 2, \ldots, T$, is an acoustic feature vector of the form,

$$\mathbf{X}_t = (x_{t1}, x_{t2}, \ldots, x_{tD}), \tag{14.5}$$

that characterizes the spectral/temporal properties of the speech signal at time t, and D is the number of acoustic features in each frame (typically $D = 39$ acoustic features are used with 13 mfcc coefficients, 13 delta-mfcc's, and 13 delta-delta mfcc's).

Similarly we can express the optimally decoded word sequence, W, as

$$W = \{W_1, W_2, \ldots, W_M\}, \tag{14.6}$$

where there are assumed to be exactly M words in the decoded string.

Before going into more detail on the implementation and computational aspects of the three steps in the decoding equation of Eq. (14.3), we first describe the statistical model used to characterize the sounds of the recognition vocabulary, i.e., the hidden Markov model.

The Hidden Markov Model

The most widely used method of building acoustic models (for both phonemes and words) is a statistical characterization known as *hidden Markov models* (HMMs) [103, 206, 301, 307]. Figure 14.6 shows a simple $Q = 5$-state HMM for modeling a whole word. Each HMM state is characterized by a mixture density Gaussian distribution that characterizes the statistical behavior of the feature vectors, $\mathbf{X}_t$, across the states of the model [178, 180]. In addition to the statistical feature densities within states, the HMM is also characterized by an explicit set of state transitions, $A = \{a_{ij}, 1 \leq i, j \leq Q\}$, for a Q-state model, which specify the probability of making a transition from state i to state j at each frame, thereby defining the time sequence of the feature vectors over the duration of the word. Usually the state self-transitions, a_{ii}, are large (close to 1.0), and the state jump transitions, a_{12}, a_{23}, a_{34}, a_{45}, in the model, are small (close to 0). The model of Figure 14.6 is called a "left-to-right" model because the state transitions

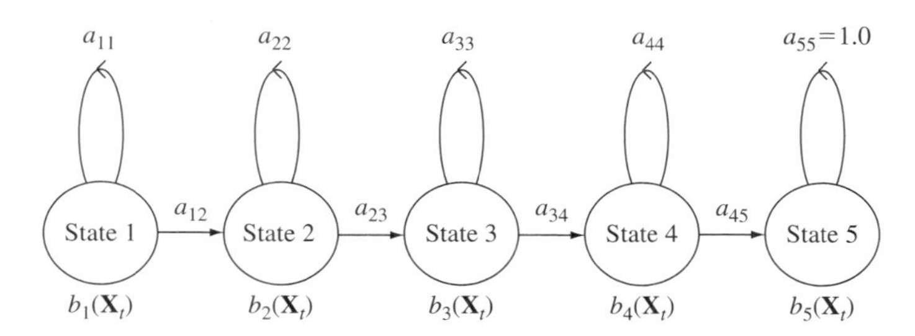

FIGURE 14.6
Word-based HMM with five states and with no possibility of skipping states, i.e., $a_{ij} = 0, j \geq i + 2$. The density function for state i is denoted as $b_i(\mathbf{X}_t)$, $1 \leq i \leq 5$.

satisfy the constraint:

$$a_{ij} = 0, \quad j \geq i+2, 1 \leq i \leq Q; \tag{14.7}$$

i.e., the system proceeds in a sequential manner from state to state, starting at the first state and ending in the last state.

The complete HMM characterization of a Q-state word (or sub-word) model is generally written as $\lambda(A, B, \pi)$, with state transition matrix $A = \{a_{ij}, 1 \leq i, j \leq Q\}$, state observation probability density, $B = \{b_j(\mathbf{X}_t), 1 \leq j \leq Q\}$, and initial state distribution, $\pi = \{\pi_i, 1 \leq i \leq Q\}$, with π_1 set to 1 (and all other π_i set to 0) for the "left-to-right" models of the type shown in Figure 14.6.

In order to train the HMM for each word (or sub-word) unit, a labeled training set of sentences (transcribed into words and sub-word units) is used to guide an efficient training procedure known as the Baum-Welch algorithm [29, 30].[4] This algorithm aligns each of the various word (or sub-word) HMMs with the spoken inputs and then estimates the appropriate means, covariances, and mixture gains for the distributions in each model state, based on the current alignment. The Baum-Welch method is a hill-climbing algorithm and is iterated until a stable alignment of HMM models and speech feature vectors is obtained. The details of the Baum-Welch procedure are beyond the scope of this chapter but can be found in several references on speech recognition methods [151, 308]. The heart of the training procedure for re-estimating HMM model parameters using the Baum-Welch procedure is shown in Figure 14.7. An initial HMM model for each recognition unit (e.g., phoneme, word) is used to begin the training

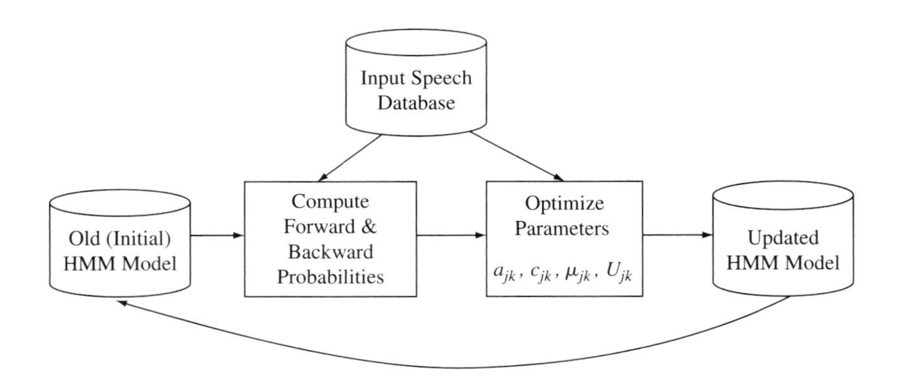

FIGURE 14.7
The Baum-Welch training procedure based on a given training set of utterances. The quantities a_{jk} are the probabilities of transition from state j to state k, and the quantities $\{c_{jk}, \mu_{jk}, U_{jk}\}$ are the parameters for the kth mixture component for the Gaussian mixture model for state j.

[4]The Baum-Welch algorithm is also widely referred to as the forward–backward method.

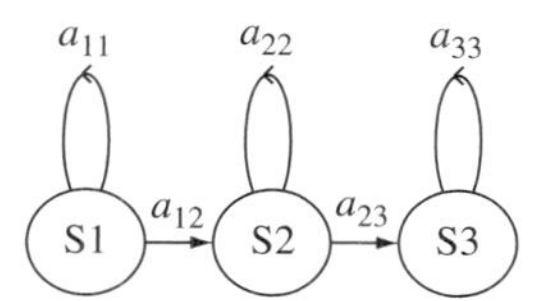

FIGURE 14.8
Sub-word-based HMM with three states, labeled $S1, S2, S3$.

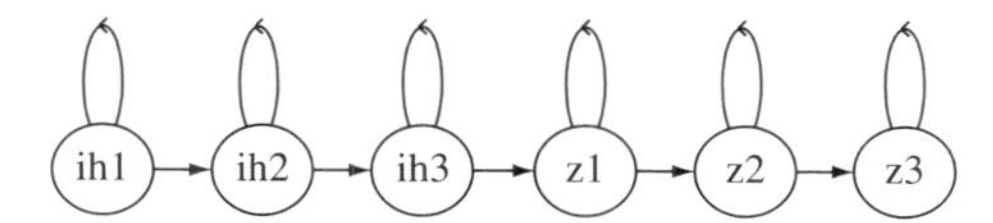

FIGURE 14.9
Word-based HMM for the word /is/ created by concatenating 3-state sub-word models for the sub-word units /ih/ and /z/. (The states of the concatenated model are denoted as ih1, ih2, ih3 for the three states for the sound /ih/, and z1, z2, z3 for the three states for the sound /z/).

process. The initial model can be randomly chosen or selected based on *a priori* knowledge of the model parameters. The iteration loop is a simple updating procedure for computing the forward and backward model probabilities based on an input speech database (the training set of utterances) and then optimizing the model parameters to give a set of updated HMMs, with each iteration increasing the model log likelihood. This process is iterated until no significant increase in log likelihood occurs with each new iteration.

It is a simple matter to go from the left-to-right HMM for a whole word, as shown in Figure 14.6, to an HMM for a sub-word unit (such as a phoneme) as shown in Figure 14.8.[5] This simple 3-state HMM is a basic sub-word unit model with an initial state representing the statistical characteristics at the beginning of a sound, a middle state representing the heart of the sound, and an ending state representing the spectral characteristics at the end of the sound. A word model is realized by concatenating the appropriate sub-word HMMs, as illustrated in Figure 14.9, which concatenates the

[5]We find it convenient to use multiple ways of referring to HMM states of sub-word units, words, and sequences of words or sub-word units. For example, the HMM for a word or sub-word unit model are generally referred to as $q_1, q_2, \ldots, q_Q$, for a Q-state model. Alternatively, we use the notation $S1, S2, S3$ for a 3-state sub-word-based HMM, or equivalently ih1, ih2, ih3 for a 3-state HMM for the particular sub-word unit, /ih/. When we concatenate word or sub-word HMMs to form a sentence HMM, we often index the states as $\{S_1(W_1), S_2(W_1), S_3(W_1), S_1(W_2), S_2(W_2), S_3(W_2), \ldots, S_1(W_M), S_2(W_M), S_3(W_M)\}$ for the concatenation of M words (or sub-word units), $\{W_1, W_2, \ldots, W_M\}$, each with 3-state HMMs.

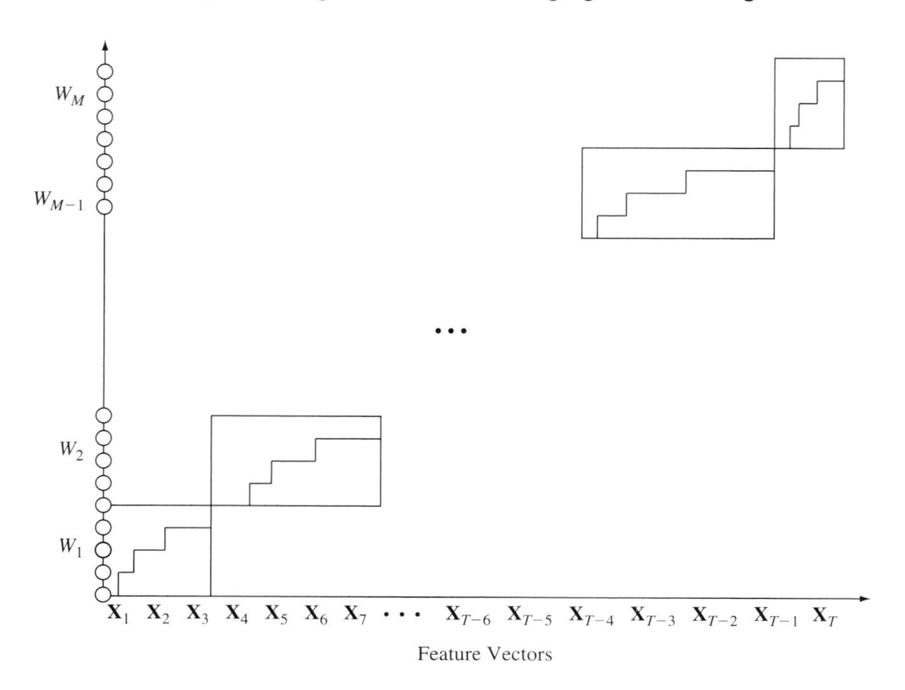

FIGURE 14.10
Alignment of concatenated HMM word models $\{W_1, W_2, \dots, W_{M-1}, W_M\}$ with input speech acoustic feature vectors based on either a Baum-Welch or Viterbi alignment procedure.

3-state HMM for the sound /ih/ with the 3-state HMM for the sound /z/, giving the word model for the word /is/ (pronounced as /ih-z/). In general the composition of a word model (from sub-word unit models) is specified in a word lexicon or dictionary; however, once the word model has been built, it can be used much the same as whole word models for training and for evaluating word strings for maximizing the likelihood as part of the speech recognition process.

We are now ready to complete the picture for aligning a sequence of M word models, $W_1, W_2, \dots, W_M$, with a sequence of feature vectors, $X = \{\mathbf{X}_1, \mathbf{X}_2, \dots, \mathbf{X}_T\}$. The resulting alignment procedure is illustrated in Figure 14.10. The sequence of feature vectors is mapped along the horizontal axis and the concatenated sequence of word states is mapped along the vertical axis. An optimal alignment procedure determines the exact best-matching (in a maximum likelihood sense) sequence between word model states and feature vectors such that the first feature vector, $\mathbf{X}_1$, aligns with the first state in the first word model, and the last feature vector, $\mathbf{X}_T$, aligns with the last state in the Mth word model. (For simplicity we show each word model as a $Q = 5$-state HMM in Figure 14.10, but clearly the alignment procedure works for any size model for any word, subject to the constraint that the total number of feature vectors, T, exceeds the total number of model states, so that every state has at least a single feature vector associated with that state.) The procedure for obtaining the best alignment between feature vectors and model states is based on using either the Baum-Welch statistical alignment procedure (in which we evaluate the probability of every alignment path and add them up to determine the probability of the word string) or a Viterbi alignment procedure [116, 406] for which we determine the single best alignment path and use the probability score along that path as the probability measure for the current word string. The utility of the alignment procedure of Figure 14.10 is based on the ease of evaluating the probability of any alignment path using the Baum-Welch or Viterbi procedures.

14.5.2 The Viterbi Algorithm

The Viterbi algorithm provides a convenient and efficient procedure for determining the best (highest probability) alignment path between the spoken sentence feature vectors, $X = \{\mathbf{X}_1, \mathbf{X}_2, \ldots, \mathbf{X}_t, \ldots, \mathbf{X}_T\}$, and HMM states of the sentence model, $q = \{q_1, q_2, \ldots, q_N\}$, where N is the total number of states in the model, as illustrated in Figure 14.11.[6]

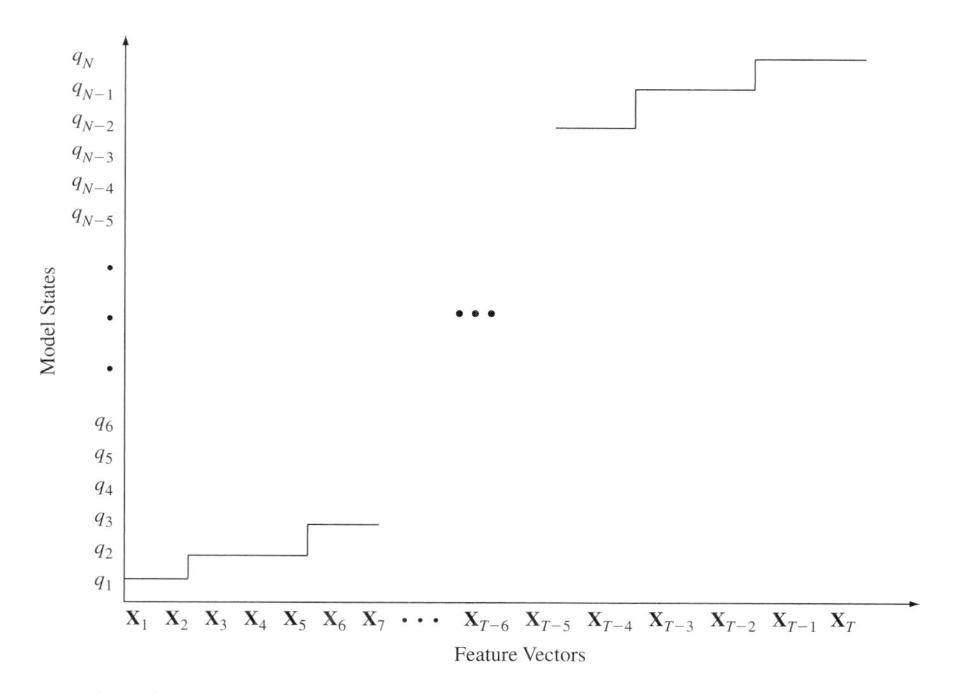

FIGURE 14.11
Alignment of N-state HMM sentence model with the spoken sentence acoustic feature vectors based on a Viterbi alignment procedure.

[6] We have changed notation to an N-state model.

The goal of the Viterbi algorithm is to find the highest probability path that links each speech sentence feature vector to a single model state, consistent with the constraints on how states can be connected (i.e., constraints from the state transition matrix). We define the quantity $\delta_t(i)$ as the probability of the highest probability path, at frame t, that accounts for the first t sentence feature vectors and ends in state i, given the HMM model, λ; i.e.,

$$\delta_t(i) = \max_{q_1, q_2, \ldots, q_{t-1}} P[q_1, q_2, \ldots, q_{t-1}, q_t = i, \mathbf{X}_1, \mathbf{X}_2, \ldots, \mathbf{X}_t | \lambda].$$

The Viterbi algorithm computes the highest probability path accounting for all T speech sentence feature vectors via the following recursion:

Step 1 Initialization

$$\delta_1(i) = \pi_i b_i(\mathbf{X}_1), \qquad 1 \le i \le N,$$
$$\psi_1(i) = 0, \qquad 1 \le i \le N,$$

where π_i is the initial state probability, $b_i(\mathbf{X}_1)$ is the probability of feature vector $\mathbf{X}_1$ being produced in state i, according to the probability density for state 1, namely $b_i(\mathbf{X}_1)$ (usually a mixture Gaussian density on the $D = 39$ feature vector components), and $\psi_1(i)$ is the state back tracking array that keeps track of which state preceded the current state on the optimal path to that state.

Step 2 Recursion

$$\delta_t(j) = \max_{1 \le i \le N} [\delta_{t-1}(i) a_{ij}] b_j(\mathbf{X}_t) \qquad 2 \le t \le T, \qquad 1 \le j \le N,$$
$$\psi_t(j) = \arg \max_{1 \le i \le N} [\delta_{t-1}(i) a_{ij}] \qquad 2 \le t \le T, \qquad 1 \le j \le N.$$

Step 3 Termination

$$P^* = \max_{1 \le i \le N} [\delta_T(i)],$$
$$q_T^* = \arg \max_{1 \le i \le N} [\delta_T(i)].$$

Step 4 State (Path) Backtracking

$$q_t^* = \psi_{t+1}(q_{t+1}^*), \qquad t = T-1, T-2, \ldots, 1.$$

The Viterbi algorithm can be reformulated in terms of logarithmic quantities [i.e., $\tilde{\pi}_i = \log(\pi_i)$, $\tilde{a}_{ij} = \log(a_{ij})$, $\tilde{b}_j(\mathbf{X}_t) = \log(b_j(\mathbf{X}_t))$], leading to a logarithmic Viterbi implementation with all multiplications replaced by additions [308]. Such an implementation generally runs much faster than the standard Viterbi algorithm because addition is generally much faster than multiplication on most modern machines.

We now return to the mathematical formulation of the ASR problem and examine in more detail the three steps in the decoding of Eq. (14.3).

14.5.3 Step 1: Acoustic Modeling

The function of the acoustic modeling step (Step 1) is to assign probabilities to the acoustic realizations of a sequence of words, given the observed acoustic vectors; i.e., we need to compute the probability that the acoustic vector sequence $X = \{\mathbf{X}_1, \mathbf{X}_2, \ldots, \mathbf{X}_T\}$ came from the word sequence $W = \{W_1, W_2, \ldots, W_M\}$ and perform this computation for all possible word sequences. This calculation can be expressed as:

$$P_A(X|W) = P_A(\{\mathbf{X}_1, \mathbf{X}_2, \ldots, \mathbf{X}_T\}|\{W_1, W_2, \ldots, W_M\}). \tag{14.8}$$

If we make the assumption that frame t is aligned with the word model i and HMM model state j via the function w_j^i, and if we assume that each frame is independent of every other frame, we can express Eq. (14.8) as the product

$$P_A(X|W) = \prod_{t=1}^{T} P_A(\mathbf{X}_t|w_j^i), \tag{14.9}$$

where we associate each frame of X with a unique word and state, w_j^i, in the word sequence. Further, we calculate the local probability, $P_A(\mathbf{X}_t|w_j^i)$, given that we know the word and state from which frame t came.

The process of assigning individual speech frames to the appropriate word model in an utterance is based on an optimal alignment process between the concatenated sequence of word models and the sequence of feature vectors of the spoken input utterance being recognized. This alignment process is illustrated in Figure 14.12, which shows the set of T feature vectors (frames) along the horizontal axis, and the set of M word HMMs along the vertical axis. The optimal segmentation of these feature vectors (frames) into the M words is shown by the sequence of boxes, each of which corresponds to one of the words in the utterance and its set of optimally matching feature vectors.

We assume that each word model is further decomposed into a set of states that reflect the changing statistical properties of the feature vectors over time for the duration of the word. We assume that each word is represented by an N-state model, and we denote the states as S_j, $j = 1, 2, \ldots, N$ (where $N = 3$ in Figure 14.12). We further assume that within each state of each word, there is a probability density that characterizes the statistical properties of the feature vectors in that state, and that the probability density of each state, and for each word, is learned during a training phase of the recognizer. Using a mixture of Gaussian normal densities to characterize the statistical distribution of the feature vectors in each state, j, of the word model, which we denote

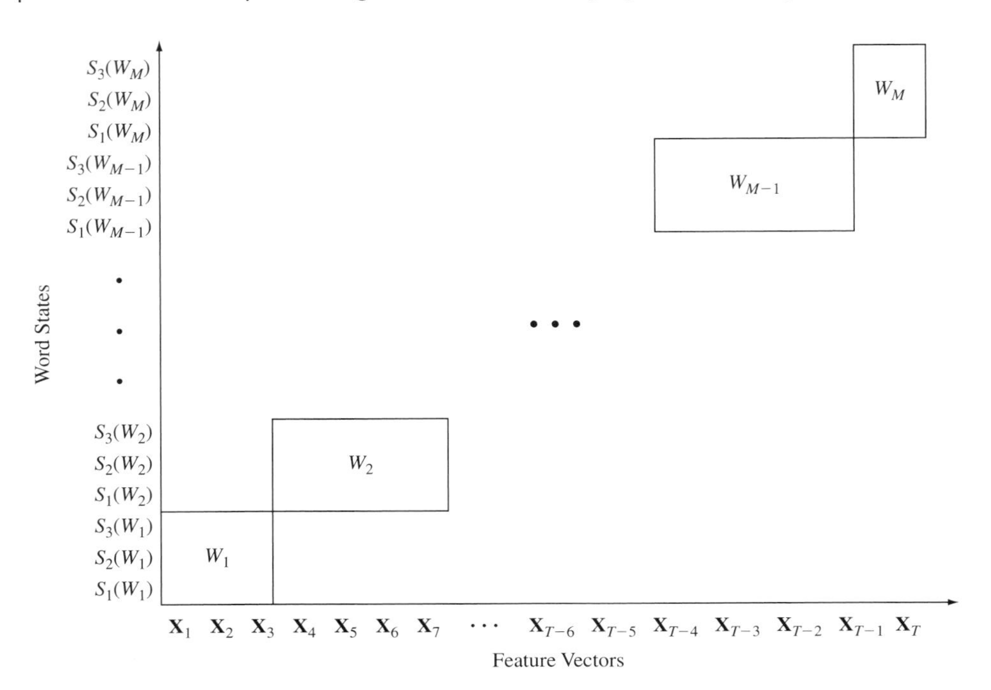

FIGURE 14.12
Illustration of time alignment process between unknown utterance feature vectors ($\mathbf{X}_1, \ldots, \mathbf{X}_T$) and the states of a set of M concatenated word HMM models.

as $b_j(\mathbf{X}_t)$, we get a state-based probability density of the form:

$$b_j(\mathbf{X}_t) = \sum_{k=1}^{K} c_{jk} \mathbb{N}[\mathbf{X}_t, \mu_{jk}, U_{jk}], \tag{14.10}$$

where K is the number of mixture components in the density function, c_{jk} is the weight of the kth mixture component in state j, with the constraint $c_{jk} \geq 0$, and $\mathbb{N}$ is a Gaussian density function with mean vector, μ_{jk}, for mixture k in state j, and covariance matrix, U_{jk}, for mixture k in state j. The density constraints are:

$$\sum_{k=1}^{K} c_{jk} = 1, \quad 1 \leq j \leq N, \tag{14.11}$$

$$\int_{-\infty}^{\infty} b_j(\mathbf{X}_t) d\mathbf{X}_t = 1, \quad 1 \leq j \leq N. \tag{14.12}$$

We now return to the issue of the calculation of the probability of frame t being associated with the jth state of the ith word in the utterance, $P_A(\mathbf{X}_t | w_j^i)$, which is calculated as

$$P_A(\mathbf{X}_t | w_j^i) = b_j^i(\mathbf{X}_t). \tag{14.13}$$

The computation of Eq. (14.13) is incomplete since we have ignored the computation of the probability associated with the links between word states, and we have also not specified how to determine the within-word state, j, in the alignment between a given word and a set of feature vectors corresponding to that word. We come back to these issues later in this section.

The key point is that we assign probabilities to acoustic realizations of a sequence of words by using HMMs of the acoustic feature vectors within words. Using an independent (and orthographically labeled) set of training data, we "train the system" and learn the parameters of the best acoustic models for each word (or more specifically for each sound that composes each word). The parameters, according to the mixture model of Eq. (14.10) are, for each state of the model, the mixture weights, the mean vectors, and the covariance matrix.

Although we have been discussing acoustic models for whole words, it should be clear that for any reasonable size speech recognition task, it is impractical to create a separate acoustic model for every possible word in the vocabulary since each word would have to be spoken in every possible context in order to build a statistically reliable model of the density functions of Eq. (14.10). Even for modest size vocabularies of about 1000 words, the amount of training data required for word models is excessive.

The alternative to word models is to build acoustic-phonetic models for the 40 or so phonemes in the English language and construct the model for a word by concatenating (stringing together sequentially) the models for the constituent phones in the word (as represented in a word dictionary or lexicon). The use of such sub-word acoustic-phonetic models poses no real difficulties in either training or when used to build up word models, and hence it is the most widely used representation for building word models in a speech recognition system. State-of-the-art systems use context-dependent phone models (called tri-phones) as the basic units of recognition [151, 308].

14.5.4 Step 2: The Language Model

The language model assigns probabilities to sequences of words, based on the likelihood of that sequence of words occurring in the context of the task being performed by the speech recognition system. Hence the probability of the text string $W =$ "Call home" for a telephone number identification task is zero since that string makes no sense for the specified task. There are many ways of building language models for specific tasks, including:

1. statistical training from text databases transcribed from task-specific dialogs (a learning procedure)
2. rule-based learning of the formal grammar associated with the task
3. enumerating, by hand, all valid text strings in the language and assigning appropriate probability scores to each string.

The purpose of the language model, or grammar, is to enable the computation of the a priori probability, $P_L(W)$, of a word string, W, consistent with the recognition task [173, 174, 330]. Perhaps the most popular way of constructing the language model is through the use of a statistical N-gram word grammar that is estimated from a large training set of text utterances, either from the task at hand or from a generic database with applicability to a wide range of tasks. To build such a language model, assume we have a large text training set of word-labeled utterances. (Such databases could include millions or even billions of text sentences.) For every sentence in the training set, we have a text file that identifies the words in that sentence. If we make the assumption that the probability of a word in a sentence is conditioned on **only** the previous $N-1$ words, we have the basis for an N-gram language model. Thus we assume we can write the probability of the sentence W, according to an N-gram language model, as:

$$\begin{aligned} P_L(W) &= P_L(W_1, W_2, \dots, W_M) \\ &= \prod_{n=1}^{M} P_L(W_n | W_{n-1}, W_{n-2}, \dots, W_{n-N+1}), \end{aligned} \tag{14.14}$$

where the probability of a word occurring in the sentence only depends on the previous $N-1$ words and we estimate this probability by counting the relative frequencies of N-tuples of words in the training set. Thus, for example, to estimate word "trigram" probabilities [i.e., the probability that a word W_n was preceded by the pair of words (W_{n-1}, W_{n-2})], we compute this quantity as:

$$P(W_n | W_{n-1}, W_{n-2}) = \frac{C(W_{n-2}, W_{n-1}, W_n)}{C(W_{n-2}, W_{n-1})}, \tag{14.15}$$

where $C(W_{n-2}, W_{n-1}, W_n)$ is the frequency count of the word triplet (i.e., the trigram of words) consisting of (W_{n-2}, W_{n-1}, W_n) as it occurs in the text training set, and $C(W_{n-2}, W_{n-1})$ is the frequency count of the word doublet (i.e., bigram of words) (W_{n-2}, W_{n-1}) as it occurs in the text training set.

Although the method of training N-gram word grammars, as described above, generally works quite well, it suffers from the problem that the counts of N-grams are often highly in error due to data sparseness in the training set. Hence for a word vocabulary of several thousand words and even for a text training set of millions of sentences, it has been observed that more than 50% of all possible word trigrams are likely to occur either once or not at all in the training set. This sparseness of data makes estimation of trigram word probabilities highly inaccurate and distorts the probability calculation for a large percentage of sentences if one adheres strictly to the probability rule of Eq. (14.15). Thus when a word trigram does not occur at all in the training set, it is not meaningful to define the trigram probability as 0 [as required by Eq. (14.15)], since this leads to effectively invalidating all strings with that particular trigram from occurring in a real recognition task. Instead, a more statistically meaningful procedure is to employ a smoothing algorithm on all word trigram estimates by interpolating trigram, bigram, and unigram word frequency estimates [20] in the following way:

$$\hat{P}(W_n|W_{n-1}, W_{n-2}) = p_3 \frac{C(W_{n-2}, W_{n-1}, W_n)}{C(W_{n-2}, W_{n-1})} + p_2 \frac{C(W_{n-1}, W_n)}{C(W_{n-1})} + p_1 \frac{C(W_n)}{\sum_n C(W_n)}, \tag{14.16a}$$

$$p_3 + p_2 + p_1 = 1, \tag{14.16b}$$

$$\sum_n C(W_n) = \text{size of text training corpus}, \tag{14.16c}$$

where the smoothing probabilities, p_3, p_2, p_1 are obtained by applying the principle of cross-validation [173, 174].

Language Perplexity

Associated with the language model of a task is a measure of the complexity of the language model, known as "language perplexity" [335]. Language perplexity is the geometric mean of the word-branching factor (the average number of words that follow any given word of the language). Language perplexity, as embodied in the language model, $P_L(W)$, for an M-word sequence, is computed from the entropy [66] as:

$$H(W) = -\frac{1}{M} \log_2 P(W). \tag{14.17}$$

Using a trigram language model, we can write Eq. (14.17) as:

$$H(W) = -\frac{1}{M} \sum_{n=1}^{M} \log_2 P(W_n|W_{n-1}, W_{n-2}), \tag{14.18}$$

where we suitably define the first two probabilities as the unigram and bigram probabilities. As M approaches infinity, the entropy approaches the asymptotic entropy of the source defined by the measure $P_L(W)$. The perplexity of the language is defined as:

$$PP(W) = 2^{H(W)} = P(W_1, W_2, \ldots, W_M)^{-1/M} \quad \text{as } M \to \infty. \tag{14.19}$$

Some examples of language perplexity for specific speech recognition tasks include the following:

- For an 11-digit vocabulary (words /zero/ to /nine/ plus /oh/), where every digit can occur independently of every other digit, the language perplexity is 11.
- For a 2000 word Airline Travel Information System (ATIS) [277], the language perplexity, based on a trigram language model, is 20.
- For a 5000 word *Wall Street Journal* task (reading business articles aloud), the language perplexity (using a bigram language model) is 130 [285].

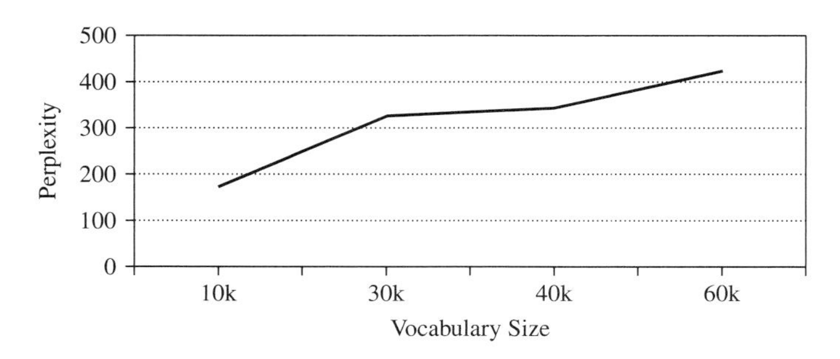

FIGURE 14.13
Bigram language perplexity for the Encarta encyclopedia. (After Huang, Acero, and Hon [151].)

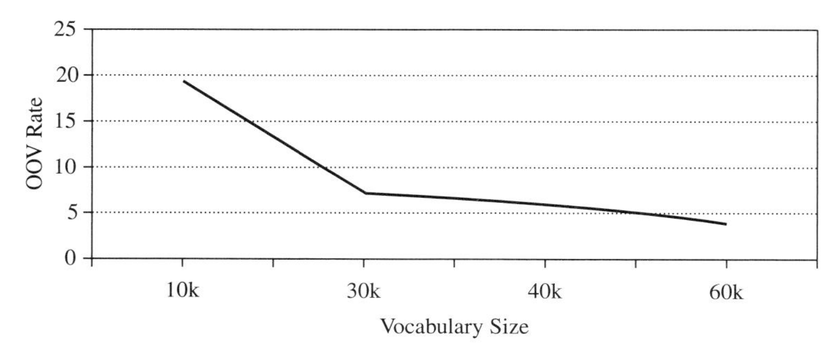

FIGURE 14.14
OOV rate of Encarta encyclopedia as a function of the vocabulary size. (After Huang, Acero, and Hon [151].)

A plot of the bigram perplexity for a training set of 500 million words, evaluated on the Encarta encyclopedia, is shown in Figure 14.13. It can be seen that language perplexity grows slowly with the vocabulary size, reaching a value of about 400 for a 60,000 word vocabulary.

Language Model Coverage

A key issue in the design of a viable language model for speech recognition is the ability of the grammar to represent and include every legal sentence in the task language, while eliminating every non-legal sentence from consideration. Based on this coverage criterion, the language model coverage is ideally 100%, while the language model over-coverage is ideally 0%. Using a bigram or trigram language model, the issue of language model coverage becomes one of determining the rate of "out-of-vocabulary (OOV)" words [187], namely how often a new word appears (for a specific task) that was not seen in a training set size of a specified number of words. To illustrate this concept, Figure 14.14 shows the OOV rate for sentences from the Encarta encyclopedia, again trained on 500 million words of text, as a function of the vocabulary size.

It can be seen that for a vocabulary of the 60,000 most frequently occurring words, the OOV rate is about 4%; i.e., about 4% of the words that are encountered in newly seen text have not been seen previously in the first 500 million words of text, and thus are considered OOV words (which, by definition, cannot be recognized correctly by the recognition system).

14.6 STEP 3: THE SEARCH PROBLEM

The third step in the Bayesian approach to ASR is to search the space of all valid word sequences from the language model, to find the one with the maximum likelihood of having been spoken. The key problem is that the potential size of the search space can be astronomically large (for large vocabularies and high perplexity language models), thereby taking inordinate amounts of computing power to solve by heuristic methods. Fortunately, through the use of methods from the field of finite state automata theory, finite state network (FSN) methods have evolved that reduce the computational burden by orders of magnitude, thereby enabling exact maximum likelihood solutions in computationally feasible times, even for very large speech recognition problems [246].

The basic concept of an FSN transducer is illustrated in Figure 14.15, which shows a word pronunciation network for the word /data/. Each arc in the state diagram corresponds to a phoneme in the word pronunciation network, and the weight is an estimate of the probability that the arc is used in the pronunciation of the word in context. We see that for the word /data/, there are four total pronunciations, namely [along with their (estimated) pronunciation probabilities]:

1. /D/ /EY/ /D/ /AX/—probability of 0.32
2. /D/ /EY/ /T/ /AX/—probability of 0.08
3. /D/ /AE/ /D/ /AX/—probability of 0.48
4. /D/ /AE/ /T/ /AX/—probability of 0.12.

The combined FSN of the four pronunciations is much more efficient than using four separate enumerations of the word, since all the arcs are shared among the four pronunciations and the total computation for the full FSN for the word /data/ is close to 1/4 the computation of the four variants of the same word.

We can continue the process of creating efficient FSNs for each word in the task vocabulary (the speech dictionary or lexicon), and then combine word FSNs into

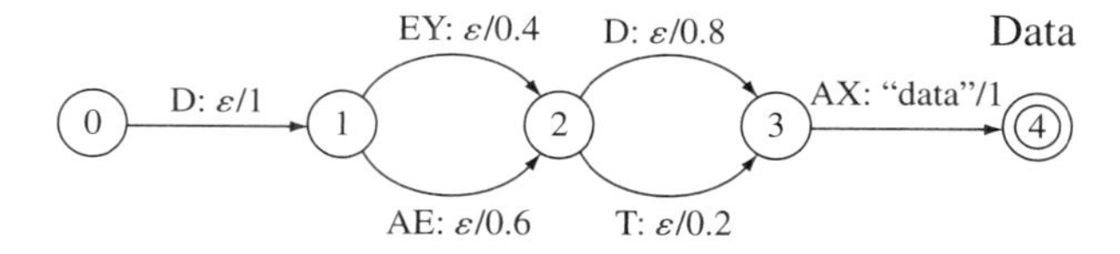

FIGURE 14.15
Word pronunciation transducer for four pronunciations of the word /data/. (After Mohri [246].)

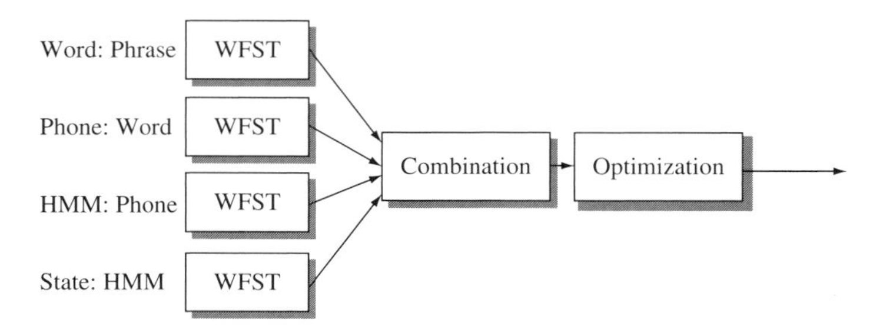

FIGURE 14.16
Use of WFSTs to compile a set of FSNs into a single optimized network to minimize redundancy in the network. (After Mohri [246].)

sentence FSNs using the appropriate language model. Further, we can carry the process down to the level of HMM phones and HMM states, making the process even more efficient. Ultimately we can compile a very large network of model states, model phones, model words, and even model phrases, into a much smaller network via the method of weighted finite state transducers (WFST), which combine the various representations of speech and language and optimize the resulting network to minimize the number of search states (and, equivalently, thereby minimize the amount of duplicate computation). A simple example of such a WFST network optimization is given in Figure 14.16 [246].

Using the techniques of network combination (which includes network composition, determinization, minimization, and weight pushing) and network optimization, the WFST uses a unified mathematical framework to efficiently compile a large network into a minimal representation that is readily searched using standard Viterbi decoding methods [116]. Using these methods, an unoptimized network with 10^{22} states (the result of the cross product of model states, model phones, model words, and model phrases) was able to be compiled down to a mathematically equivalent model with 10^8 states that was readily searched for the optimum word string with no loss of performance or word accuracy [246].

14.7 SIMPLE ASR SYSTEM: ISOLATED DIGIT RECOGNITION

To illustrate the ideas presented in this chapter, consider the implementation of a simple isolated digit recognition system, with a vocabulary of the ten digits (/zero/ to /nine/) plus the alternative /oh/ for /zero/, for a total of 11 words to be recognized. The first step in implementing a digit recognizer is to train a set of acoustic models for the digits. For this simple vocabulary, we use whole word models, so we do not need a word lexicon or a language model since only a single digit is spoken for each recognition trial. For training whole word HMMs for the digits, we need a training set with a sufficient number of occurrences of each spoken digit so as to be able to train reliable and stable acoustic word HMMs. Typically, five tokens of each digit is sufficient for a speaker-trained system (that is a system that works only for the speech of the speaker who trained the system). For a speaker-independent recognizer, a significantly larger

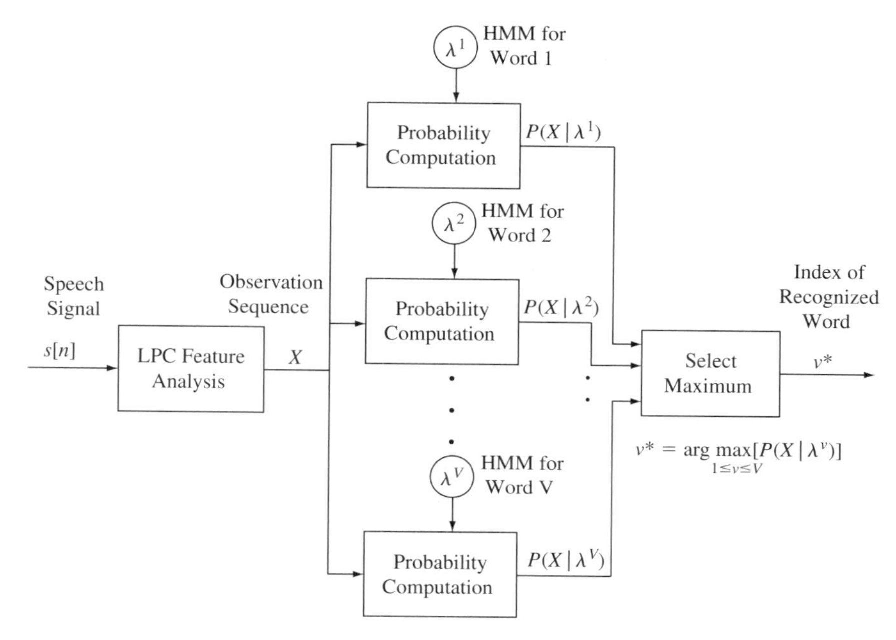

FIGURE 14.17
HMM-based isolated digit recognizer.

number of tokens of each digit is required to completely characterize the variability in accents, speakers, transducers, speaking environments, etc. Typically between 100 and 500 tokens of each digit are used for training reliable word models. Training of word HMMs for each of the digits uses the Baum-Welch method and yields a set of word HMMs, one for each of the 11 digits to be recognized.

Figure 14.17 shows a block diagram of an HMM-based isolated digit recognizer. Not shown explicitly in this diagram is the essential function of isolating the speech signal from its background. Algorithms for doing this were discussed in Chapter 10. Once the speech signal $s[n]$ has been isolated, it is converted to a sequence of mfcc feature vectors, X. Then a probability computation is performed for each of the $V = 11$ digit models, (λ^1) to (λ^V), yielding digit likelihood scores $P(X|\lambda^1)$ to $P(X|\lambda^V)$. Each of the digit likelihood scores uses Viterbi decoding to both optimally align the feature vector, X, with the HMM model states, and to calculate the probability (or equivalently the log likelihood) score for the optimal alignment. The final step in the recognition system of Figure 14.17 is selection of the maximum likelihood score and associating the word index of the maximum likelihood score with the recognized digit.

Figure 14.18 illustrates the Viterbi alignment process of the feature vector frames for the spoken word /six/ and the HMM model for the digit /six/. There are three plots in this figure. The upper plot is the log energy contour of the spoken input for the word /six/ (the phonetic labels for /six/ are /S-IH-K-S/) as a function of the spoken input frame index. The second plot shows the accumulated (negative) log likelihood at each frame of the feature vector. Since all probabilities are less than 1, the accumulated log

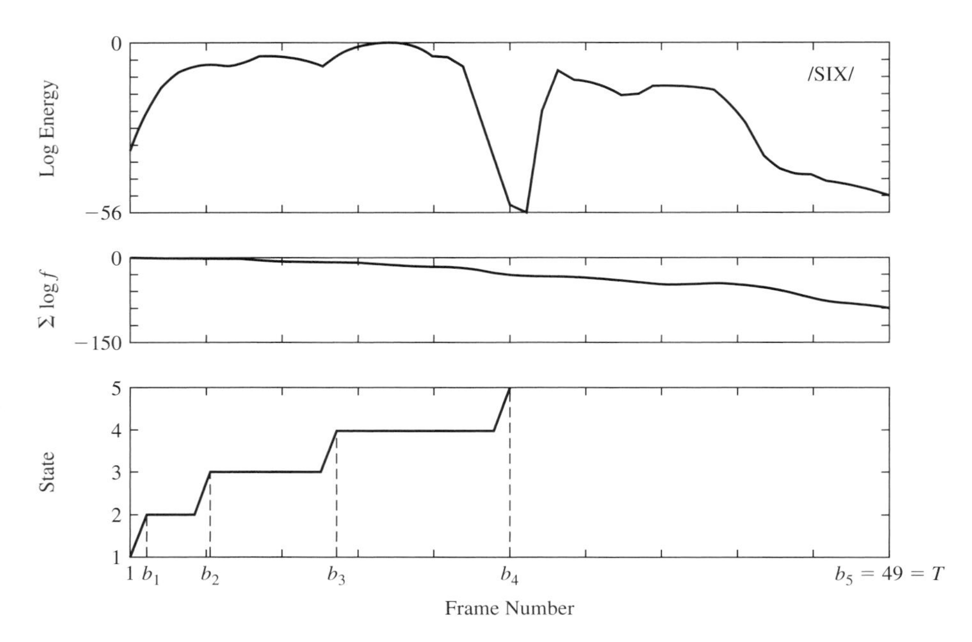

FIGURE 14.18
Optimal alignment of digit model /six/ with set of feature vectors for spoken input (corresponding to the spoken word /six/).

likelihood scores are always negative and decreasing, as shown in Figure 14.18. Finally, the bottom plot shows the alignment between the frames of the spoken input (along the horizontal axis) and the HMM model state for the model corresponding to the spoken word /six/, along the vertical axis. Using a 5-state HMM for each digit, we see that frames 1 to $b_1 - 1$ of the test sequence aligned with model state 1, whereas frames b_1 to $b_2 - 1$ of the test sequence aligned with model state 2, etc. Ultimately we see how each of the five model states aligned with the spoken input frames and we see that we can almost identify the sounds associated with each state. For example, we see that state 5 aligned with the stop gap of the /K/ and the final /S/ sound, whereas states 1–3 aligned primarily with the initial /S/ sound, and finally state 4 aligned with the vowel /IH/ and the initial part of the stop consonant /K/.

Similar plots could be constructed for the test utterance /six/ compared to the HMM models for all the other words in the vocabulary. A correct recognition occurs if the log likelihood score is most negative for reference HMM model for the word /six/ as compared to the test feature vectors for the spoken word /six/. Otherwise, a recognition error occurs.

14.8 PERFORMANCE EVALUATION OF SPEECH RECOGNIZERS

A key issue in speech recognition (and language understanding) system design is evaluation of the system's performance. For simple, isolated word recognition systems, like the digit recognizer of the preceding section, the measure of performance is simply the word error rate of the system. For a more complex system, such as might be used for dictation applications, there are three types of errors that must be taken into account to specify the overall performance of the system. The three types of error that can occur are listed below:

1. **Word insertions,** where more words are recognized than were actually spoken; the most common word insertions are short, function words like /a/, /and/, /to/, /it/, etc., and these most often occur during pauses or hesitations in speaking.
2. **Word substitutions,** where an incorrect word was recognized in place of the actually spoken word; the most common word substitution errors are for similar sounding words (like /fees/ and /please/) but sometimes random word substitution errors occur.
3. **Word deletion,** where a spoken word is just not recognized and the recognition provides no alternative word in its place; word deletions generally occur when two spoken words are recognized as a single (generally longer in duration) word, resulting in both a word substitution error and a word deletion error

Based on the convention of equally weighting all three types of word errors (which often is not the best thing to do since errors on function words are usually significantly less harmful to sentence understanding than errors on content words), the overall word error rate for most speech recognition tasks, WER, is formally defined as:

$$WER = \frac{NI+NS+ND}{|W|}, \tag{14.20}$$

where NI is the number of word insertions, NS is the number of word substitutions, ND is the number of word deletions, and $|W|$ is the number of words in the sentence W being scored. Based on the above definition of word error rate, the performance of a range of speech recognition and natural language understanding systems (a topic we will get to later in this chapter) is shown in Table 14.1.

The word error rates in Table 14.1 were measured at a number of research sites throughout the world over a period of more than a decade, with the earliest results on digit string recognition having been obtained in the late 1980s and early 1990s, and the latest results on natural language conversational material having been measured in the early 2000s [276]. It can be seen that for a vocabulary of 11 digits, the word error rates are very low (0.3% for a very clean recording environment for the TI (Texas Instruments) connected digits database [204]), but when the digit strings are spoken in a noisy shopping mall environment, the word error rate rises to 2.0%, and when embedded within conversational speech (the AT&T HMIHY system), the word error rate increases significantly to 5.0%, showing the lack of robustness of the recognition system to noise and other background disturbances. Table 14.1 also shows the word error rates for a range of DARPA evaluation tasks:

- Resource Management (RM)—read speech consisting of commands and informational requests about a naval ship's database with a 1000 word vocabulary and a word error rate of 2.0%

TABLE 14.1 Word error rates for a range of speech recognition systems.

Corpus	Type of Speech	Vocabulary Size	Word Error Rate
Connected digit strings (TI database)	Spontaneous	11 (0–9, oh)	0.3%
Connected digit strings (AT&T mall recordings)	Spontaneous	11 (0–9, oh)	2.0%
Connected digit strings (AT&T HMIHY©)	Conversational	11 (0–9, oh)	5.0%
Resource Management (RM)	Read speech	1000	2.0%
Airline Travel Information System (ATIS)	Spontaneous	2500	2.5%
North American business (NAB & WSJ)	Read text	64,000	6.6%
Broadcast news	Narrated news	210,000	~15%
Switchboard	Telephone conversation	45,000	~27%
Call-home	Telephone conversation	28,000	~35%

- Air Travel Information System (ATIS) [412]–spontaneous speech input for planning airline travel with a 2500 word vocabulary and a word error rate of 2.5%
- North American Business (NAB)–read text from a range of business magazines and newspapers with a vocabulary of 64,000 words and a word error rate of 6.6%
- Broadcast News–narrated news broadcasts from a range of TV news providers like CNBC with a 210,000 word vocabulary and a word error rate of about 15%
- Switchboard [125]–recorded live telephone conversations between two unrelated individuals with a vocabulary of 45,000 words and a word error rate of about 35%
- Call Home–live telephone conversations between two family members with a vocabulary of 28,000 words and a word error rate of about 35%.

The DARPA research community worked on the above set of tasks for more than a decade (1988–2004) and the progress that was achieved in improving word error rates for several speech recognition tasks, over this period, is plotted in Figure 14.19. It can be seen that for each of the six speech recognition tasks shown in this figure, steady progress was made in reducing the word error rate over time, at almost a steady rate, independent of the specific recognition task being studied.

Figure 14.20 compares machine performance against human performance for different speech recognition system tasks [209]. Human performance is generally very difficult to measure, so the results shown in this figure are more directional than exact. However it is clear that humans outperform machines, for speech recognition, by factors of between 10 and 50. Hence, it should be clear that speech recognition research has a long way to go before machines outperform humans on speech recognition tasks. However, it should be noted that there are many cases when an ASR (and language

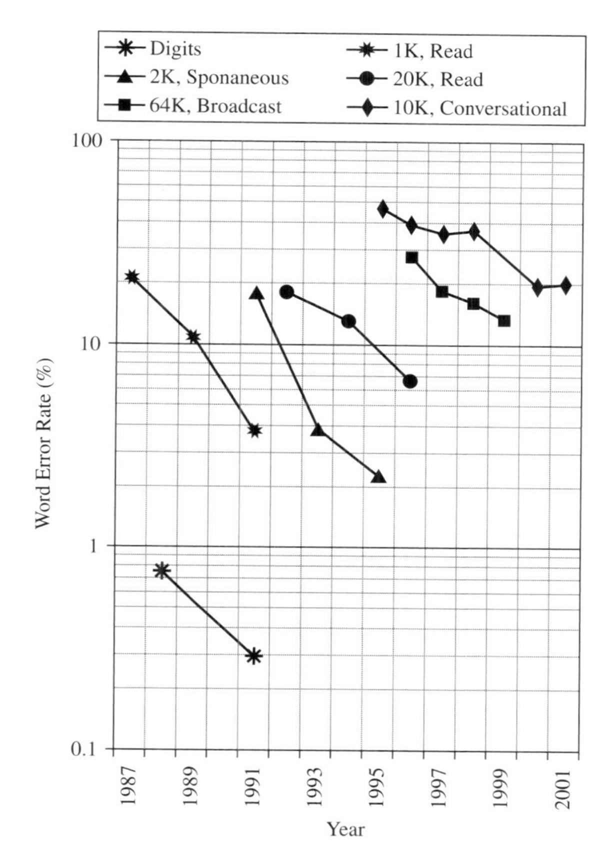

FIGURE 14.19
Word error rate performance for a range of DARPA speech recognition tasks showing the steady reduction in word error rates over time for all tasks. The caption designates the type of recognition task (e.g., digits, read speech, spontaneous speech, broadcast speech, conversational speech) and the size of the vocabulary in thousands of words (K). (After Pallett et al. [277].)

understanding) system can provide a better service than a human. One such example is the recognition of a 16-digit credit card number, spoken as a continuous string of digits. A human listener would have a difficult time keeping track of such a long (seemingly random) string of 16 digits, but a machine would have no trouble keeping track of what was spoken, and utilizing real credit card constraints to provide essentially flawless string recognition accuracy.

14.9 SPOKEN LANGUAGE UNDERSTANDING

Once reliable speech recognition has been achieved, a successful human–machine dialog depends critically on the SLU module of Figure 14.1 to interpret the meaning of the key words and phrases in the recognized speech string, and to map them to actions that

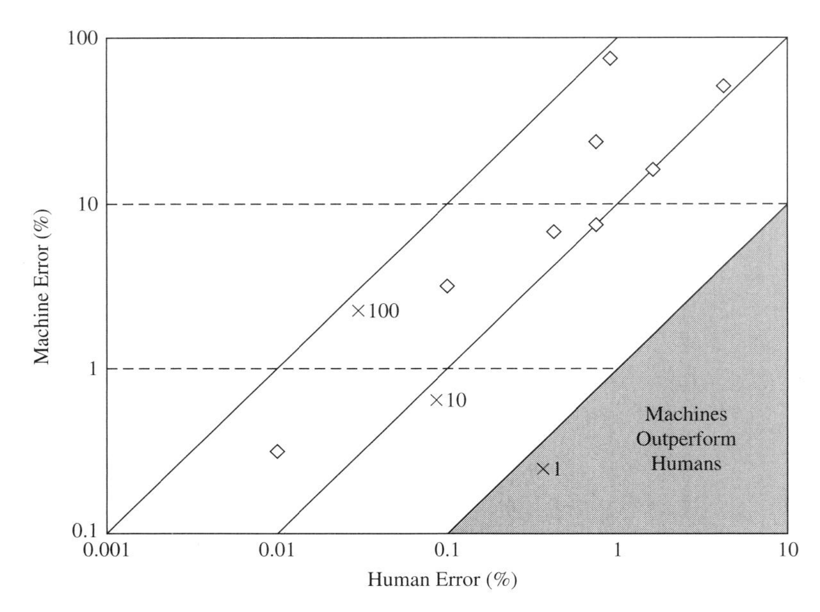

FIGURE 14.20
Comparison of human and machine speech recognition performance for a range of speech recognition tasks. It can be seen that humans outperform machines by factors of between 10 and 50 across the tasks illustrated in this figure. (After Lippmann [209].)

the speech understanding system should take. For accurate speech understanding, it is important to recognize that in domain-specific applications (such as any of the applications shown in Table 14.1), highly accurate speech understanding can be achieved without correctly recognizing every word in the sentence. Hence a speaker can have spoken the sentence, *"I need some help with my computer hard drive,"* and so long as the recognition system correctly recognized the words *"help"* and *"hard drive,"* it understands, for all practical purposes, the context of the speech (needing help) and the object of the context (the hard drive). All of the other words in the sentence can often be mis-recognized (although not so badly that other contextually significant words are recognized in their place) without affecting the understanding of the meaning of the sentence. In this sense, keyword (or key phrase) spotting [418] can be considered a primitive form of speech understanding, without involving sophisticated semantic analysis.

Spoken language understanding makes it possible to offer a range of services where the human can speak naturally without having to learn a specific vocabulary and task syntax in order to complete a transaction and interact with a machine [179]. It performs this task by exploiting the task grammar and task semantics to restrict the range of meanings associated with the recognized word string, and by exploiting a predefined set of "salient" words and phrases that map high information word sequences to this restricted set of meanings. SLU is especially useful when the range of meanings is naturally restricted and easily cataloged so that a Bayesian formulation can be used to optimally determine the meaning of the sentence from the word sequence. This Bayesian approach uses the recognized sequence of words, W, and the underlying meaning, C, to determine the probability of each possible meaning, given the word sequence, and is of the form:

$$P(C|W) = \frac{P(W|C)P(C)}{P(W)}, \tag{14.21}$$

and then finding the best conceptual structure (meaning) using a combination of acoustic, linguistic, and semantic scores, namely:

$$C^* = \arg\max_C P(W|C)P(C). \tag{14.22}$$

This approach makes extensive use of the statistical relationship between the word sequence and the intended meaning.

One of the most successful (commercial) speech understanding systems to date has been the AT&T HMIHY customer care service. For this task, the customer dials into an AT&T 800 number for help on tasks related to his or her long distance or local billing account. The prompt to the customer is simply: "AT&T. How May I Help You?" The customer responds to this prompt with totally unconstrained fluent speech describing the reason for calling the customer care help line. The system tries to recognize every spoken word (but invariably makes a number of word errors), and then uses the Bayesian concept framework to determine the best estimate of the meaning of the speech. Fortunately, the potential meaning of the spoken input is restricted to one of a finite number of outcomes, such as asking about account balances, or new calling plans, or changes in local service, or help for an unrecognized telephone number on a bill, etc. Based on this highly limited set of outcomes, the spoken language component determines which meaning is most appropriate (or else decides not to make a decision but instead to defer the decision to the next cycle of the dialog circle), and appropriately routes the call. A simple overview of the HMIHY system is shown in Figure 14.21.

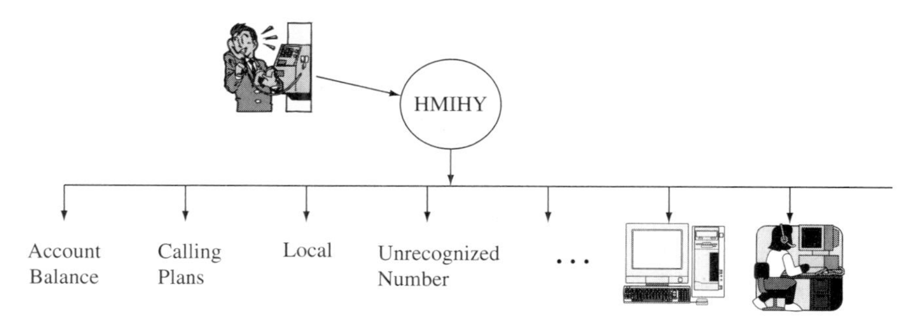

FIGURE 14.21
Conceptual representation of HMIHY© (How May I Help You) natural language understanding system. (After Gorin, Riccardi, and Wright [136].)

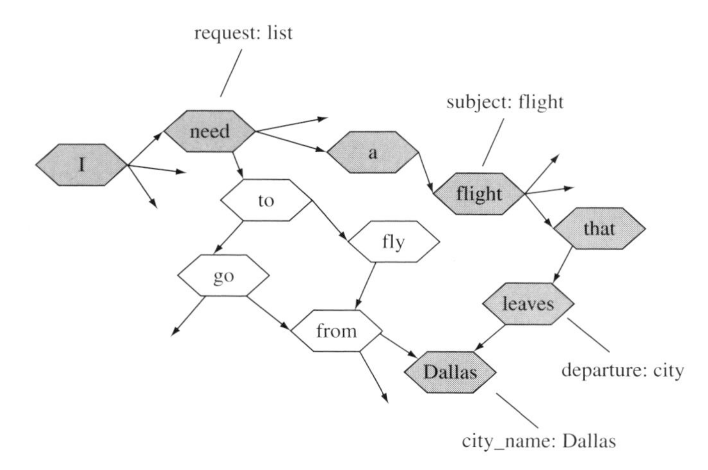

FIGURE 14.22
An example of a word grammar with embedded semantic notions in ATIS.

The major challenge in SLU is to go beyond the simple (flat) classification task of the HMIHY system (where the conceptual meaning is restricted to one of a fixed, often small, set of choices) and to create a true concept and meaning understanding system.

An early attempt at creating a true speech understanding system was employed in the ATIS task by embedding speech recognition in a stylized semantic structure to mimic a natural language interaction between a human and a machine. In this system, the semantic notions encapsulated in the system were rather limited, mostly in terms of originating city and destination city names, fares, airport names, travel times, etc., and could be directly instantiated in a semantic template without much text analysis for understanding. By way of example, a typical semantic template or network is shown in Figure 14.22, where the relevant notions, such as the departing city, can easily be identified and used in dialog management (DM) to create the desired user interaction with the system.

14.10 DIALOG MANAGEMENT AND SPOKEN LANGUAGE GENERATION

The goal of the DM block in Figure 14.1 is to combine the meaning of the current input speech (as determined by the natural language understanding block) with the current state of the system (which is based on the interaction history with the user) in order to decide what the next step in the interaction should be. Once the decision is made, the SLG block translates the action of the dialog manager into a textual representation, and the TTS block converts the textual representation into natural-sounding speech to be played to the user so as to initiate another round of dialog discussion or to end the query. In this manner, the combination of the DM module and the SLG makes viable rather complex services that require multiple exchanges between the system and the human so as to successfully complete a transaction. Such dialog systems can also handle user-initiated topic switching with the domain of the application.

The DM module is one of the most crucial steps in the speech dialog circle for a successful transaction because it enables the human to accomplish the desired task. The way in which the DM module works is by exploiting models of spoken dialog to determine the most appropriate spoken text string to guide the dialog forward toward a clear and well-understood goal or system interaction. The computational models for DM include both structure-based approaches (which models dialog as a pre-defined state transition network that is followed from an initial goal state to a set of final goal states), or plan-based approaches (which consider communication as executing a set of plans that are oriented toward goal achievement).

The key tools of dialog strategy are the following [185]:

- **Confirmation,** used to ascertain correctness of the recognized and understood utterances. Confirmation strategies often merely rephrase the recognized input in the form of a question to confirm key aspects of the dialog before moving forward. A simple confirmation request in response to the recognized utterance "I need a flight to LA tomorrow morning" might be "What time tomorrow morning would you like the flight to Los Angeles to depart?", thereby both confirming the date and destination of the requested flight, but also asking for additional information (i.e., time in the morning for departure) to enable a more focused response at the next iteration of the dialog circle.
- **Error Recovery,** used to get the dialog back on track after a user indicates that the system has misunderstood a spoken request. For example, a user might say the request "I need a flight to St. Louis" and the system might understand it as "I need a flight to San Diego", and respond accordingly. The system could detect, using the word verification strategy mentioned earlier, that the confidence in the destination city San Diego is low, and try to recover by asking the user "Did you want a flight to San Diego or somewhere else." The user would repeat the request for a flight to St. Louis, and, if successfully recognized, the dialog would continue to completion of the requested task.
- **Reprompting,** used when the system expected input but did not receive any input. This tool is clearly needed when the user is confused and does not know what to say next, or when the audio level is too low, etc. A typical reprompting command might be something like "I did not understand your last request. Please repeat it or formulate a new request."
- **Completion,** used to elicit missing input information from the user. A simple example is when a user has requested a flight to a destination city, but didn't specify the departure city and thus the system needs to request additional information to complete the transaction.
- **Constraining,** used to reduce the scope of the request so that a reasonable amount of information is retrieved, presented to the user, or otherwise acted upon. This tool is used when the request is too encompassing, making it difficult to make enough intelligent decisions to complete the transaction properly. Thus if a user requests a flight from Boston to New York, the system needs additional (constraining) information to decide if the destination airport is Kennedy, LaGuardia, or Newark, and therefore must pose a question to the user to constrain the destination airport to one of the three valid possibilities.

- **Relaxation,** used to increase the scope of the request when no information has been retrieved. This tool is used when the recognized and understood utterance does not provide any additional constraining or clarifying information about the request, and therefore the system needs to reduce the scope of the dialog so as to resolve ambiguity and get the dialog back on track. The ultimate form of relaxation is to ask the user simple questions that elicit a single piece of information about the task on each turn of the dialog circle, rather than extracting too much information that is inconsistent or incorrect.
- **Disambiguation,** used to resolve inconsistent input from the user. This tool is used when the system is confident that it recognized what was spoken correctly, but the overall task request is inconsistent or makes no sense in the domain of the current task. In such cases, the best strategy is to attack the problem directly and ask questions that resolve the inconsistent input as simply and as rapidly as possible.
- **Greeting/Closing,** used to maintain social protocol at the beginning and end of an interaction. This tool makes the computer interaction feel more natural and pleasant. A simple greeting of the form "Good morning, Mr. Jones, how can I help you today?" provides a friendly and personalized way of making the human eager and pleased to use the automated dialog system.
- **Mixed Initiative,** allows users to manage the dialog flow. This tool is essential for enabling the user to step in and take over control when things are either going too slowly or in improper directions and the user decides to "take the initiative" away from the system. At other times this tool enables the system to jump in and lead the user by appropriate system responses.

Although most of the tools of dialog strategy are straightforward and the conditions for their use are fairly clear, the mixed initiative tool is perhaps the most interesting one, as it enables a user to manage the dialog and get it back on track. Figure 14.23 shows a simple chart that illustrates the two extremes of mixed initiative for a simple operator services scenario. At the one extreme, where the **system** manages the dialog totally, the system responses are simple declarative requests to elicit information, as exemplified by the system command "Please say collect, calling card, third number." At the other extreme is **user** management of the dialog where the system responses are open ended and the human can freely respond to the system command "How may I help you?"

Figure 14.24 illustrates some simple examples of the use of system initiative, mixed initiative, and user initiative for an airline reservation task. It can be seen that the system initiative leads to long dialogs (due to the limited information retrieval at each query), but the dialogs are relatively easy to design, whereas user initiative leads

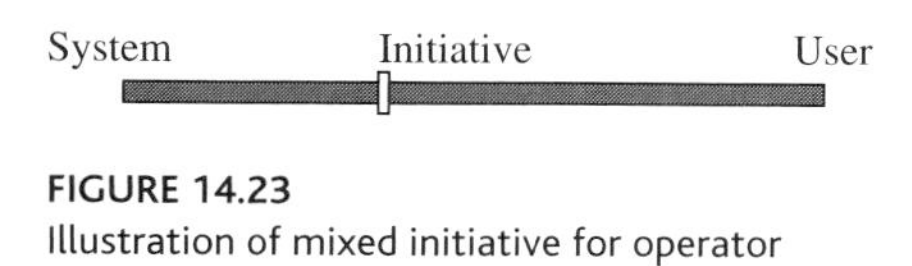

FIGURE 14.23
Illustration of mixed initiative for operator services scenario.

- **System Initiative**

System:	Please say *just* your departure city.
User:	Chicago
System:	Please say *just* your arrival city.
User:	Newark

Long dialogs but easier to design

- **Mixed Initiative**

System:	Please say your departure city
User:	I need to travel from Chicago to Newark tomorrow.

- **User Initiative**

System:	How may I help you?
User:	I need to travel from Chicago to Newark tomorrow.

Shorter dialogs (better user experience) but more difficult to design

FIGURE 14.24
Examples of mixed initiative dialogs.

to shorter dialogs (and hence a better user experience), but the dialogs are more difficult to design. (Most practical natural language systems need to be mixed initiative so as to be able to change initiatives from one extreme to another, depending on the state of the dialog and how successfully things have progressed toward the ultimate understanding goal.)

DM systems are evaluated based on the speed and accuracy of attaining a well-defined task goal, such as booking an airline reservation, renting a car, purchasing a stock, or obtaining help with a service.

14.11 USER INTERFACES

The user interface for a speech communications system is defined by the performance of each of the blocks in the speech dialog circle. A good user interface is essential to the success of any task-oriented system, providing the following capabilities:

- It makes the application "easy-to-use" and robust to the kinds of confusion that arise in human–machine communications by voice.
- It keeps the conversation moving forward, even in periods of great uncertainty for the user or the machine.
- Although it cannot save a system with poor speech recognition or speech understanding performance, it can make or break a system with excellent speech recognition and speech understanding performance.

Although good user interface technology can go a long way toward providing a natural user interface to a machine, it is important that the speech interface system designer carefully match the task to the technology and be realistic about the capabilities of the technology. Thus the most successful speech recognition systems still have non-zero error rates and must have some type of backup strategy to handle errors that inevitably occur. One such strategy is the class of "fail soft" methods, namely those methods which verify key aspects of all recognition decisions and for those content words whose confidence scores are low, rely on either one of the dialog strategies outlined above, or employ something like a multiple recognition utterance strategy where both a recognized (and understood) utterance along with one or more backup alternative meanings are used and the human helps to disambiguate among the possible choices. A second strategy would be to use data dips (into an appropriate database) to resolve ambiguity (e.g., in looking up a credit card number as one of the valid set of issued and working credit cards).

14.12 MULTIMODAL USER INTERFACES

Until this point we have primarily been concerned with speech recognition and understanding interfaces to machines. There are, however, times when a multimodal approach to human–machine communications is both necessary and essential. The potential modalities that can work in concert with speech include gesture and pointing devices (e.g., a mouse, keypad, or stylus). The selection of the most appropriate user interface mode (or combination or modes) depends on the device, the task, the environment, and the user's abilities and preferences. Hence, when trying to identify objects on a map (e.g., restaurants, locations of subway stations, historical sites), the use of a pointing device (to indicate the area of interest) along with speech (to indicate the topic of interest) often is a good user interface, especially for small computing devices like tablet PCs or PDAs. Similarly, when entering PDA-like information (e.g., appointments, reminders, dates, times) onto a small handheld device, the use of a stylus to indicate the appropriate type of information with voice filling in the data field is often the most natural way of entering such information (especially as contrasted with stylus-based text input systems such as graffiti for Palm-like devices). Microsoft's research has shown the efficiency of such a solution with the MiPad (Multimodal Interactive Pad) demonstration, and they claim to have achieved double the throughput for English using the multimodal interface over that achieved with just a pen stylus and the graffiti language.

14.13 SUMMARY

In this chapter we have outlined the major components of a modern speech recognition and natural language understanding system, as used within a voice dialog system. We have shown the role of signal processing in creating a reliable feature set for the recognizer and the role of statistical methods in enabling the recognizer to recognize the words of the spoken input utterance, as well as the meaning associated with the recognized word sequence. We have shown how a dialog manager utilizes the meaning accrued from the current as well as previous spoken inputs to create an appropriate response (as well as potentially taking some appropriate actions) to the human request(s), and finally how the SLG and TTS synthesis parts of the dialog complete the dialog circle by providing feedback to the user as to actions taken and further information that is required to complete the transaction that is requested.

Speech recognition and speech understanding systems have made their way into mainstream applications and services and they have become widely used in society. Among the leading applications of speech recognition and natural language understanding technology include the following:

- **Desktop Applications:** including dictation, command and control of the desktop functionality, control of document properties (fonts, styles, bullets, etc.)
- **Agent Technology:** including simple tasks like stock quotes, traffic reports, weather, access to communications, voice dialing, voice access to directories, voice access to messaging, calendars, appointments, etc.
- **Voice Portals:** including systems capable of converting any web page to a voice-enabled site, where any question that can be answered on-line can be answered via a voice query using protocols like VXML, SALT, SMIL, SOAP, etc.
- **E-Contact Services:** including call centers, customer care centers (such as How May I Help You), and help desks where calls are triaged and answered appropriately using natural language voice dialogs
- **Telematics:** including command and control of automotive features (comfort systems, radio, windows, sunroof)
- **Small Devices:** including cell phones for dialing by name or function, PDA control from voice commands, etc.

There are two broad areas where ASR systems need major improvements before they become ubiquitous in society, namely in system and operational performance. In the system area we need large improvements in accuracy, efficiency, and robustness in order to use the technology for a wide range of tasks, on a wide range of processors, and under a wide range of operating conditions. In the operational area we need better methods of detecting when a person is speaking to a machine and isolating the spoken input from the background. We also need to be able to handle users talking over the voice prompts (so-called barge-in conditions). More reliable and accurate utterance rejection methods are needed so that we can be sure that a word needs to be repeated when poorly recognized the first time. Finally, we need better methods of confidence scoring of words, phrases, and even sentences so as to maintain an intelligent dialog with a human. As speech recognition systems and speech understanding systems become more robust to noise, background disturbances, and transmission characteristics of communication systems, they will find their way into more small devices, thereby providing a natural and intuitive way to control the operation of these devices as well as to ubiquitously and easily access and enter information.

PROBLEMS

14.1. Shown in Figure P14.1 are spectrograms for the set of 11 isolated digits spoken by a male talker. (Arrows are used to visually separate the spectrograms of each of the individual digits.) (Note that the highest frequency in the spectrograms is 8 kHz, indicating a sampling rate of at least 16 kHz.) Using your knowledge of the spectral properties of each of the individual digits (zero, one, two, three, four, five, six, seven, eight, nine, and oh), identify which spectrogram for each row of Figure P14.1 corresponds to each of the digits. Note that each digit occurs only once among the 11 choices.

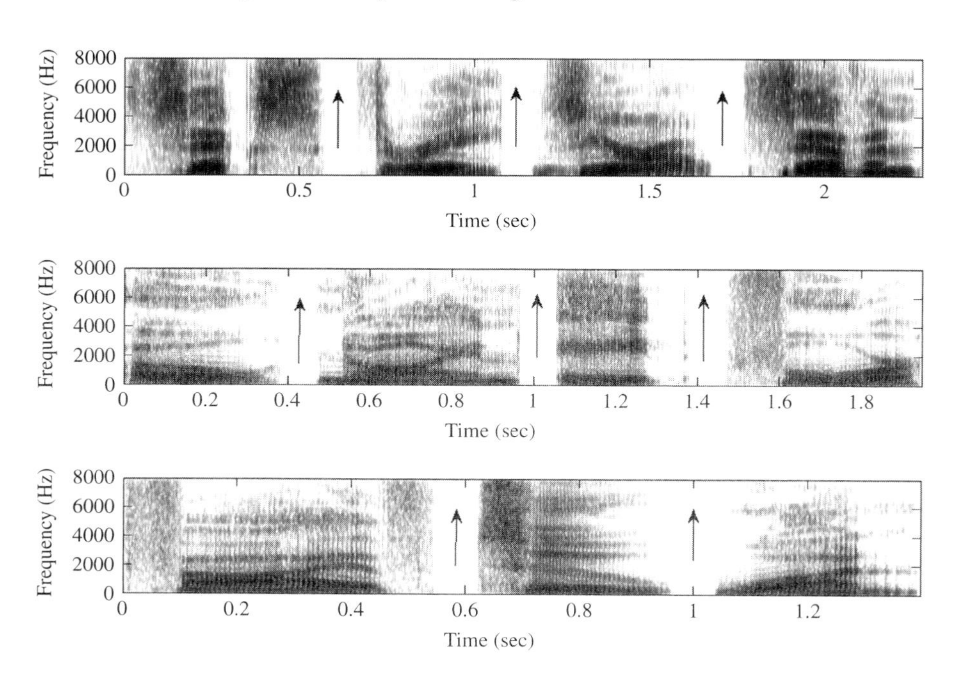

FIGURE P14.1
Digit spectrograms.

14.2. Consider a two-class (class 1 is denoted as ω_1 and class 2 is denoted as ω_2) Bayesian classifier with the class density functions and a priori probabilities of the form:

$$p(x|\omega_1) = N[x, m_1, \sigma_1],$$
$$p(x|\omega_2) = N[x, m_2, \sigma_2],$$
$$P(\omega_1) = \alpha,$$
$$P(\omega_2) = 1 - \alpha,$$

where N is a Gaussian (normal) density function with mean, m, and standard deviation, σ.

(a) Determine the threshold value, x_0, for minimum classification error in terms of $m_1, m_2, \sigma_1, \sigma_2$, and α.

(b) Determine the classification error rate for $\alpha = 0.5$, $m_1 = 3$, $m_2 = -3$, and $\sigma_1 = \sigma_2 = 2$.

14.3. Consider the same two-class Bayesian classifier as in the previous problem but with loss matrix:

$$L = \begin{bmatrix} 0 & \lambda_{12} \\ \lambda_{21} & 0 \end{bmatrix};$$

i.e., instead of minimizing the probability of error, P_e, defined as:

$$P_e = P(x \in R_2|\omega_1)P(\omega_1) + P(x \in R_1|\omega_2)P(\omega_2),$$

where R_1 is the region where $P(\omega_1|x) > P(\omega_2|x)$ and R_2 is the region where $P(\omega_2|x) > P(\omega_1|x)$, we minimize the cost-weighted risk, r, defined as

$$r = \lambda_{12}P(\omega_1)\int_{R_2} p(x|\omega_1)dx + \lambda_{21}P(\omega_2\int_{R_1} p(x|\omega_2)dx.$$

(a) Determine the threshold value, $\hat{x}_0$, for minimum risk in terms of m_1, m_2, σ_1, σ_2, α, λ_{12}, and λ_{21}, assuming that $\sigma_1 = \sigma_2 = 1$.

(b) Determine the minimum risk error rate for $\alpha = 0.5$, $m_1 = 3$, $m_2 = -3$, $\sigma_1 = \sigma_2 = 1$, $\lambda_{12} = 4$, and $\lambda_{21} = 1$.

14.4. Show that the cepstral distortion measure,

$$d(c_1, c_2) = \sum_{n=-\infty}^{\infty} (c_1[n] - c_2[n])^2,$$

obeys the mathematical properties of distance metrics, namely:

1. it is positive definite: $0 < d(c_1, c_2) < \infty$ if $c_1 \neq c_2$ and $d(c_1, c_2) = 0$ only if $c_1 = c_2$;
2. it is symmetric: $d(c_1, c_2) = d(c_2, c_1)$;
3. it satisfies the triangular inequality: $d(c_1, c_2) \leq d(c_1, c_3) + d(c_2, c_3)$.

14.5. Consider the dynamic programming problem of trying to determine the shortest path through a grid. To illustrate the complexities, consider the possible ways of creating a path from one city (call it Seattle) to another city (call it Orlando) as shown in Figure P14.5, where the number on each path line is the path cost between the associated pair of path beginning and path ending node.

The goal of this problem is to determine the shortest distance path from Seattle to Orlando, using dynamic programming methods.

1. The first step in the algorithm is to label each of the nodes in the network sequentially, and then determine the set of predecessor and successor nodes for each node in the network. Label the nodes from 1 to 14, and for each node, make a table of the predecessor nodes (the ones that come into the node) and the successor nodes (the ones that emanate from the node).
2. The second step in the algorithm is to determine the smallest distance path to each node, by computing the distances to all predecessor nodes, and then choosing the predecessor node whose distance, when added to the path cost between the predecessor node and the current node, is smallest. This is the dynamic programming optimization step. You also have to keep track of which predecessor node was used, as this

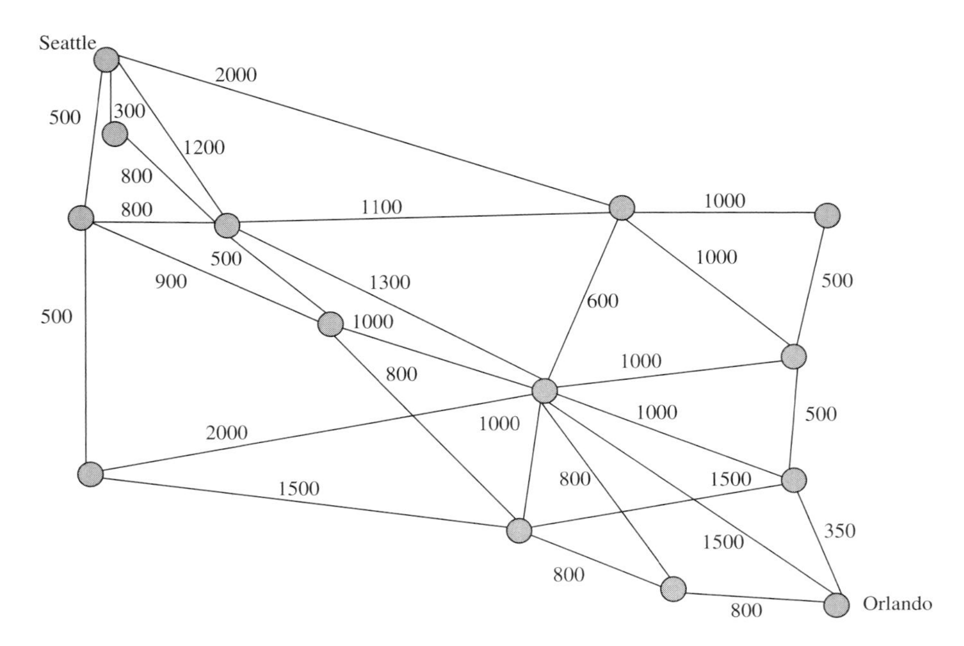

FIGURE P14.5
Shortest path problem.

is essential for backtracking to find the actual shortest path through the network. Perform this second step for all nodes in the network, taking care to choose the order of the nodes so that they are computable (all predecessor nodes have been computed before a successor node is begun).

3. The solution to finding the shortest distance path is the smallest total distance to node 14 (the Orlando node). What is the shortest distance path from Seattle to Orlando?
4. The final step is backtracking from Orlando to Seattle to find the path (or paths in the case of ties for minimum distance) that led to the shortest distance. We do this using the list of predecessor nodes associated with each network node, beginning at the terminal node (Orlando) and working our way backwards to the initial node (Seattle). Determine the actual shortest path (or paths) between Seattle and Orlando.

14.6. (MATLAB Exercise) The purpose of this problem is to write a MATLAB program for dynamic time-warping (DTW) alignment of a test pattern with a reference pattern. The test pattern is available at the book website and is called `test.mat`, and it consists of a set of `ntest` cepstral vectors of size 12 elements each. Similarly the reference pattern is also available at the book website and is called `train.mat`, and it consists of a set of `nref` cepstral vectors of size 12 elements each. Write MATLAB code that optimally time aligns the sets of cepstral vectors in `train.mat` and `test.mat`.

To create the DTW alignment routine, you need to make some assumptions about the alignment path's degrees of freedom. For this problem we make the following assumptions:

1. The initial frames of the reference and test pattern must align; i.e., frame 1 of the reference pattern aligns precisely with frame 1 of the test pattern.

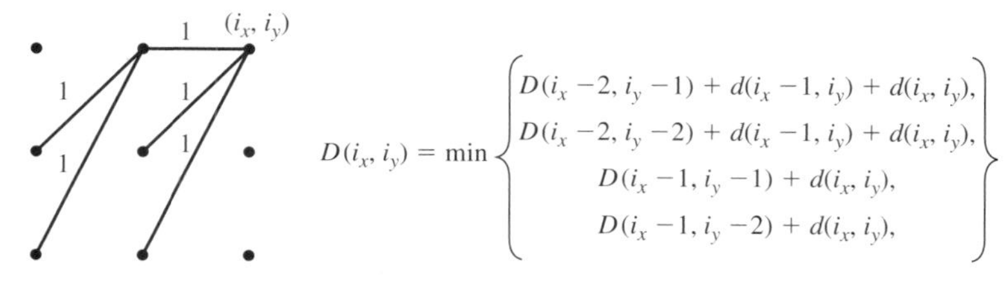

FIGURE P14.6
Shortest path problem.

2. The final frames of the reference and test pattern must align; i.e., frame `nref` of the reference pattern aligns precisely with frame `ntest` of the test pattern.
3. The test pattern maps along the x-axis and the reference pattern maps along the y-axis of the time alignment grid.
4. The test pattern index controls the alignment process; i.e., you index the alignment process by the test pattern frame index, from 1 to `ntest`.
5. The local path constraints are as shown in Figure P14.6. The local path to a node at coordinates (i_x, i_y) is the minimum of the accumulated distance of one of four feeder paths as shown in Figure P14.6, plus the local distance at the node at (i_x, i_y). You must carefully code your DTW algorithm to make sure you keep accurate tabs of which of the four feeder paths actually gave the minimum accumulated distance.
6. The local distance between sets of cepstral vectors is:

$$d(c_{\text{ref}}, c_{\text{ctest}}) = \sum_{n=1}^{12} (c_{\text{ref}}[n] - c_{\text{test}}[n])^2.$$

   Program the DTW algorithm to optimally align the reference and test patterns provided. Determine the minimum average distance through the grid, and plot the alignment path, the accumulated distance through the grid, and the local distance of each node along the optimal path.

Once you have successfully programmed the DTW algorithm and gotten it to provide the optimal alignment of a test and reference pattern, you now can compare the test pattern (corresponding to a spoken isolated digit) to each of the 11 digit reference patterns (one–nine, oh, zero) and determine the "best match" after time alignment. For this exercise the set of 11 reference patterns consist of the sets of cepstral parameters for each digit. These reference patterns are available at the book website and are labeled as `template_isodig_[1-11].mat`. The test pattern is labeled `test.mat` and consists of the cepstral parameter sets for the spoken digit "six." Time align the test pattern with each of the reference patterns, and make a plot of the minimum accumulated distance at each time frame, for each reference pattern—all on a common plot. Which digit has the minimum average distance, and what is the minimum average distance? Which digit has the second lowest average distance and what is the second lowest average distance?

14.7. Consider an ergodic 4-state hidden Markov model (HMM), defined over a discrete set of five symbols, with transition matrix, A, symbol emission matrix, B, and initial state vector,

π, defined as:

$$A = \begin{bmatrix} 0.1 & 0.3 & 0.3 & 0.3 \\ 0.7 & 0.1 & 0.1 & 0.1 \\ 0.25 & 0.25 & 0.25 & 0.25 \\ 0.1 & 0.4 & 0.4 & 0.1 \end{bmatrix},$$

$$B = \begin{bmatrix} 0.6 & 0.1 & 0.1 & 0.1 & 0.1 \\ 0.1 & 0.6 & 0.1 & 0.1 & 0.1 \\ 0.2 & 0.2 & 0.2 & 0.2 & 0.2 \\ 0 & 0 & 0 & 0.5 & 0.5 \end{bmatrix},$$

$$\pi = \begin{bmatrix} 0.25 & 09.25 & 0.25 & 0.25 \end{bmatrix}.$$

(a) What is the probability (according to the HMM), $p(Q)$, of the state sequence: $Q = [1234]$?

(b) What is the probability (according to the model), $p(Q, O)$, of the combined state and observation sequences: $Q = [1234]$, $O = [5321]$?

(c) Using both the forward–backward recursion and the Viterbi recursion, determine $p(O)$, the probability of the observation sequence; what is the optimum state sequence associated with both solutions?

(d) What is the average duration in each of the model states?

14.8. (MATLAB Exercise) Given a training set of end-pointed isolated digit files (one–nine, oh, zero) spoken by a single talker (stored in files that can be accessed from the book website as `digits_train.zip`), create a set of left–right HMMs for each isolated digit, with the following characteristics of the signal processing and the HMM model creation:

1. sampling frequency: 8000 Hz (need to convert sampling frequency for all speech files not initially sampled at 8000 Hz rate using MATLAB function `resample`)
2. speech frame duration, shift: N=320 samples at 8000 Hz (40 msec frames); M=80 samples at 8000 Hz (10 msec shift)
3. Hamming window weighting of speech frame
4. FFT size of 1024 (zero padding using 704 samples) to give log magnitude spectrum
5. convert log magnitude spectrum to real cepstrum and retain c(1) to c(L) (L=12), omitting c(0)
6. number of HMM states in each digit model: 5
7. observations are sets of L=12 cepstral coefficients for each frame
8. observation density is simple diagonal covariance Gaussian with a single mixture component (need to estimate mean and variance vector for each state and for each digit model)
9. disregard the state transition matrix for all calculations (assume it is basically a self-loop of 0.5 probability and a state transition to the succeeding state of 0.5 probability, with the final state just having a self-loop with probability 1.0, so the state transition component is irrelevant).

The process to follow in solving for the individual digit HMM models is as follows:

1. Initially segment all the training files (a list of which can be found in the mat file `files_lrrdig_isodig_train_endpt.mat`, and the actual files can be obtained from `digits_train.zip`; both files are downloadable from the book website) for each digit uniformly into five states; then determine the appropriate set of means and variances for the $L=12$ components of each observation for each state of the model. This creates an initial set of HMM models for each of the 11 isolated digits.
2. Using the models (obtained for each digit) from step 1, use the Viterbi algorithm to segment the training files into states based on the best Viterbi alignment; then re-estimate the digit models (again by determining the new set of means and variances for each model state and for each digit), and iterate this process until convergence (up to five iterations).
3. Score the resulting models (from iteration 5) on the independent test set of end-pointed digits (a list of which can be accessed from the book website as `files_lrrdig_isodig_test_endpt.mat`, with the actual test files obtained from the file `digits_test.zip`) and show the alignment paths between one of the test digits (use the digit /seven/) and the correct HMM model file obtained in step 2 by plotting (on a single plot with four subplots):
   - the log energy of the test digit
   - the Viterbi alignment path of the correct digit
   - the accumulated log probability along the Viterbi path
   - the local distance along the Viterbi path.

Determine the number of correctly recognized digits and the number of digit errors.

Speech and Audio Processing Demonstrations

APPENDIX A

To gain an understanding of the effectiveness of a range of speech and audio processing systems, it is essential to listen to speech and audio utterances that have been processed using systems of the type described throughout this book. To that end, we (with the generous support of numerous colleagues throughout the world) have created a book website with a set of examples of speech and audio signals that have been processed in a variety of ways. Also included on the book website are source speech files and source MATLAB files as needed to solve several of the problems at the end of each chapter.

The website for this book (found at www.pearsonhighered/Rabiner.com) is segmented into directories corresponding to Chapters 1, 2, 7, 11, 12, and 13. These directories contain .wav files that can be played using a wide range of sound programs.

A.1 DEMOS FOR CHAPTER 1

A.1.1 Narrowband and Wideband Speech and Audio Coding

The directory `chapter1_files` of the book website contains examples of narrowband and wideband speech and audio coding at a range of bit rates and includes the following examples:[1]

- filename: `telephone bandwidth speech coders.wav`: coding of narrowband speech (3.2 kHz bandwidth as might be used over ordinary dialed-up telephone lines)
- filename: `wideband speech coding.wav`: coding of wideband speech (7.5 kHz bandwidth as might be used in AM radio), and
- coding of audio signals (20 kHz bandwidth as used in music CDs)

[1] In the top-level directory of the website there is a set of .doc files, one for each chapter in which there are associated demos, that contain information about which file on the website contains which demo.

Narrowband Speech Coding

The narrowband speech coding demo includes the following samples (each sample is separated by an audible beep):

- 64 kbps Pulse Code Modulation (PCM) speech
- 32 kbps Adaptive Differential Pulse Code Modulation (ADPCM) coded speech
- 16 kbps Low Delay Code-Excited Linear Prediction (LDCELP) coded speech
- 8 kbps Code-Excited Linear Prediction (CELP) coded speech
- 4.8 kbps Federal Standard 1016 (FS1016) coded speech
- 2.4 kbps Linear Predictive Coding 10e (LPC10e) coded speech

When listening to these coded narrowband speech utterances, the reader should get a feeling for when the speech quality begins to change (between 8 and 16 kbps) and when the speech quality begins to deteriorate (between 4.8 and 8 kbps). The reader should have an understanding of why this happens based on the discussion of each of these speech coders in Chapter 11.

Wideband Speech Coding

The wideband speech coding demo utilizes both a male talker and a female talker and includes the following samples (again each sample is separated by an audible beep):

- 3.2 kHz bandwidth, uncoded speech sample (narrowband baseline signal)
- 7 kHz bandwidth, uncoded speech sample (wideband baseline signal)
- 7 kHz bandwidth, coded speech sample at 64 kbps
- 7 kHz bandwidth, coded speech sample at 32 kbps
- 7 kHz bandwidth, coded speech sample at 16 kbps

When listening to the samples in the wideband speech coding demo, the reader should immediately notice the quality difference between 3.2 kHz bandwidth speech (narrowband signal) and 7 kHz bandwidth speech (wideband signal) in the form of a dramatic increase in presence and naturalness. Although the intelligibility of speech is adequately preserved in narrowband speech (which is used for both wireless and wireline telephony), the inherent quality of the speech signal is only preserved in wideband speech. Further, by listening to the coded wideband samples at 64, 32, and 16 kbps rates, it is clear that there is no major degradation in quality, even at the 16 kbps rate.

Audio Coding

The third set of demos in this chapter illustrates audio coding of CD quality music using the well known MP3 audio coder. The original CD music was recorded at a sampling rate of 44.1 kHz, in stereo, with 16 bits/sample quantization for an overall bit rate of 1.4 mbps. The MP3-coded material is played at an overall bit rate of 128 kbps; i.e., an 11 to 1 reduction in bit rate over the original audio material in the CD format. Each of four audio and speech files is presented at both the 1.4 mbps CD rate and at the 128 kbps MP3 rate.[2] The reader should listen to each pair of recordings and try to decide whether he or she can determine any auditory difference between the pairs of recordings. (A good test is to play the pairs of recordings in random order and then see if you can determine which is the original CD and which is the MP3 coded version.)

The audio material included on the website consists of pairs of files for the four speech and audio signals, along with an MP3 coded selection from an orchestral piece of music. The specific filenames and their contents are the following:

- filename: `vega_CD.wav`: female vocal at CD rate
- filename: `vega_128kbps`: female vocal at MP3 rate
- filename: `trumpet_CD.wav`: trumpet selection at CD rate
- filename: `trumpet_128kbps.wav`: trumpet selection at MP3 rate
- filename: `baroque_CD.wav`: baroque selection at CD rate
- filename: `baroque_128kbps.wav`: baroque selection at MP3 rate
- filename: `guitar_CD.wav`: guitar selection at CD rate
- filename: `guitar_128kbps.wav`: guitar selection at MP3 rate
- filename: `orchestra_128kbps.wav`: orchestra selection at MP3 rate

Formal listening tests have shown that it is very difficult to determine any consistent differences between original audio and audio files coded at MP3 rates. Discussion of speech and audio coding systems is provided in Chapters 11 and 12.

A.2 DEMOS FOR CHAPTER 2

Included within the website (in the directory `chapter2_files`) are demonstrations of the effects of digitizing speech (and music) as a function of the sampling rate and the number of bits used for quantization. Included in this directory are the following examples:

1. Effects of varying sampling rate on the digitization of a 4-kHz bandwidth analog speech signal

    - filename: `2.1 PCM-effects of sampling frequency.wav`: 5 kHz bandwidth speech at sampling rates of F_s =10, 5, 2.5, and 1.25 kHz rates (notice the aliasing that arises when the sampling rate is below 10 kHz), and quantized at 16 bits/sample

2. Effects of varying the number of bits used for quantization of a 5 kHz bandwidth analog speech signal sampled at $F_s = 10$ kHz

[2]There is a fifth recording of a full orchestra at the MP3 rate for the listening pleasure of the reader.

- filename: `2.2 PCM-effects of number of bits.wav`: number of bits/sample used for quantization varying from 12 to 9 to 4 to 2 and then to 1

3. Effects of varying the number of bits used for quantization of an 8-kHz bandwidth analog music signal sampled at $F_s = 16$ kHz, and initially quantized to 14 bits/sample

   - filename: `2.3 PCM-music quantization.wav`: sequence of the following audio samples
     - 12 bits/sample audio, 2 bits/sample noise (white noise)
     - 10 bits/sample audio, 4 bits/sample noise (white noise)
     - 8 bits/sample audio, 6 bits/sample noise (colored noise)
     - 6 bits/sample audio, 8 bits/sample noise (signal correlated noise)

A.7 DEMOS FOR CHAPTER 7

To better appreciate the state-of-the-art in speeding up and slowing down speech and audio, we have included several audio examples (due to Quatieri [297] and Prof. Dan Ellis' website, Columbia University) on the website. (The set of examples for this chapter can be found in the directory `chapter7_files`.

The first set of examples (due to Quatieri [297]) illustrates variable rates of speed-up and slow-down for a male speaker and consists of the following sequence of samples:

- filename: `tea_party_orig.wav`: original rate of speaking (2.65 seconds duration)
- filename: `tea_party_speeded_15pct.wav`: speeded up by 15% (2.24 seconds duration)
- filename: `tea_party_speeded_37pct.wav`: speeded up by 37.5% (1.66 seconds duration)
- filename: `tea_party_slowed_15pct.wav`: slowed down by 15% (3 seconds duration)
- filename: `tea_party_speeded_35pct.wav`: slowed down by 35% (3.58 seconds duration)

The second set of examples (again due to Quatieri [297]) illustrates variable rates of speed-up and slow-down for a female speaker and consists of the following sequence of samples:

- filename: `swim_orig.wav`: original rate of speaking (2.65 seconds duration)
- filename: `swim_speeded_15pct.wav`: speeded up by 15% (2.24 seconds duration)
- filename: `swim_speeded_37pct.wav`: speeded up by 37.5% (1.66 seconds duration)

- filename: swim_slowed_15pct.wav: slowed down by 15% (3 seconds duration)
- filename: swim_speeded_35pct.wav: slowed down by 35% (3.58 seconds duration)

These two sets of examples illustrate the quality that can be obtained with speed-ups and slow-downs by reasonably small factors.

The next example illustrates a continuous speed change from a very slow to a very fast rate of speaking, with the rate adjusted regularly throughout the duration of the signal. The quality remains high throughout the duration of the utterance.

The continuously variable rate of speaking processed file is found in the following file:

filename: continuous speed change from very slow to very fast.wav

The final set of examples (due to Dan Ellis at Columbia University) illustrates using the phase vocoder to speed-up and slow-down a section of music by factors of 2. The resulting quality of the music is high across the 4-to-1 range of speed variation with very few audible artifacts.

- filename: Maple.wav: the original music section
- filename: FastMaple.wav: the same section of music speeded up by a factor of 2
- filename: SlowMaple.wav: the same section of music slowed down by a factor of 2

A.11 DEMOS FOR CHAPTER 11

On the book website (in the directory chapter11_files) there is included a demonstration of a range of standardized speech coders. The demonstration gives examples of coded speech from six of the speech coders of Table 11.10, in the (decreasing bit rate) order:

- filename: source.wav: original (uncoded) speech file
- filename: g611.wav: G.711 μ−law PCM at 64 kbps
- filename: g726.wav: G.726 ADPCM at 32 kbps
- filename: g.728.wav: G.728 LD-CELP at 16 kbps
- filename: gsm.wav: GSM RPE-LTP at 13 kbps
- filename: lpc10e.wav: lpc10e FS1015 at 2.4 kbps
- filename: melp.wav: NSA MELP at 2.4 kbps

The bit rate falls by a factor of 2 between the first few examples, from 64 kbps down to 16 kbps; then it falls by lesser factors as the quality of the speech degrades, showing the need for finer gradations in the 2.4–13 kbps range.

The narrowband and wideband speech coding examples listed under Chapter 1 and the sampling demos listed under Chapter 2 are also relevant to the material in Chapter 11 of the text.

A.12 DEMOS FOR CHAPTER 12

The audio coding demos listed under Chapter 1 are also relevant to the material in this Chapter 12.

A.13 DEMOS FOR CHAPTER 13

The demos for TTS systems can be found in the directory `chapter13_files` of the website.

A.13.1 Word Concatenation Synthesis

The first demo shows the quality of simple word concatenation synthesis systems. The demo contains a set of seven simple declarative sentences, spoken both as a concatenation of the individual words via the synthesizer of Figure 13.15 and as a continuous sentence. The demo can be found in the directory

`chapter13_files\Word Concatenation Synthesis`

The contents of the file:

- filename: `Voice Response--abutted words and rule-based.wav`

are the following sentences:

1. We were away a year ago.
2. The night is very quiet.
3. You speak with a quiet voice.
4. He and I are not walking.
5. That way is easier.
6. Everyone is trying to rest.
7. Nice people rest when it is night.

It is clear, from listening to these few sentences, that word concatenation output speech is very choppy and difficult to understand.

A.13.2 The VODER

The file `Voder.wav`, which can be found in the directory

`chapter13_files\Voder.wav`

demonstrates the quality of synthesis from the Voder for a simple dialogue. Considering that the Voder was demonstrated at the 1939 World's Fair in New York City, it is amazing that human operators could learn to control the Voder sufficiently accurately to produce intelligible speech.

A.13.3 Articulatory Synthesis of Speech

The file `articulatory_original.wav` which can be found in the directory

`chapter13_files\articulatory synthesis`

was used as the source file for extracting parameters of an articulatory synthesizer. The resulting articulatory synthesis is found in the file:

- filename: `articulatory_synthesis_Sondhi.wav`

The resulting synthetic speech quality is quite intelligible, showing the potential for articulatory models if an automated way of properly controlling the parameters over time could be devised.

A.13.4 Examples of Serial/Cascade Speech Synthesis

Examples of the synthetic speech quality attainable using the serial/cascade formant synthesis model of Figure 13.11 are found in the directory

`chapter13_files\Serial Synthesis`

which includes six sets of synthesis examples, namely:

- filename: `OVE_I.wav`: OVE-1 Synthesis from KTH, Sweden–Gunnar Fant
- filename: `SPASS_MIT.wav`: SPASS Synthesis from MIT, Ray Tomlinson
- filename: `synthesis-by-rule_JSRU.wav`: JSRU Synthesis from British Telecom, John Holmes
- filename: `soliloquy from hamlet--to be or not to be.wav`: Hamlet Soliloquy Synthesis from Bell Labs, John Kelly, and Lou Gerstman
- filename: `We wish you-IBM.wav`: synthesis of singing voice (with one, two, and three voices) from IBM, Rex Dixon
- filename: `daisy-daisy sung.wav`: Singing Synthesis (Daisy-Daisy) from Bell Labs, John Kelly, and Lou Gerstman
- filename: `daisy-daisy with computer accompaniment.wav`: a version of the song Daisy-Daisy with synthesized voice and computer music accompaniment

Most of the examples above were from analysis/synthesis systems, namely systems where someone spoke the desired output and the serial synthesis parameters were estimated from the spoken speech. Clearly the goal of most TTS systems is to provide a model of spoken speech that enables estimation of the synthesizer parameters without having to have anyone speak the specific utterances being synthesized.

A.13.5 Examples of Parallel Speech Synthesis

A demonstration of the quality of synthetic speech that can be obtained using a parallel synthesizer can be found in the directory

`chapter13_files\Parallel Synthesis`

Included in this demo are two examples of speech, namely:

- filename: `holmes_73.wav`: JSRU synthesis of several sentences from British Telecom, UK–John Holmes
- filename: `parallel synthesis_Strong-original then synthetic.wav`: Brigham Young University synthesis of several sentences—Bill Strong

A.13.6 Demonstrations of Complete Speech Synthesis Systems

The set of synthetic speech examples which can be found in the directory

`chapter13_files\evolution of tts systems`

provides a sense of the continuing evolution of speech synthesizers over the three decade period from 1959 to 1987. Almost all of the synthesis systems were rule-based, i.e., the speech was represented as a series of sounds (usually phones) with a parametric representation of each phone, and a set of rules was used to determine appropriate smoothing of the parameters for synthesis by a serial or parallel formant synthesizer, or an LPC synthesizer based on an appropriate set of LPC parameters. One of the systems, the diphone synthesis method of Olive [263], was a concatenative synthesis method based on an LPC representation of the diphone spectra, and was the forerunner of most modern (unit selection) concatenative speech synthesis systems that are widely used today. Included are the following synthesis systems:

- filename: `pattern_playback_1959.wav`: rule-based serial synthesis from Haskins Laboratory, 1959
- filename: `OVE_1962.wav`: rule-based synthesis from KTH Lab, Stockholm, Sweden, 1962
- filename: `coker_umeda_browman_1973.wav`: rule-based synthesis from Bell Labs, 1973
- filename: `MITTALK_1979.wav`: rule-based synthesis from MIT, MI-talk, 1979
- filename: `speak_n_spell_synthesis.wav`: rule-based synthesis from Texas Instruments for Speak-n-Spell toy, 1980
- filename: `Bell Labs_1985.wav`: diphone-based synthesis using diphone concatenation, Bell Labs, 1985
- filename: `klatt_talk.wav`: rule-based synthesis from MIT, Klatt Talk, 1986

- filename: DECTALK_male_1987.wav: rule-based synthesis from DEC, DecTalk, 1987; DecTalk was the first viable commercial synthesis system. Included with DecTalk was a set of voices including:

  1. filename: Klatt_huge_harry.wav
  2. filename: Klatt_kit_the_kid.wav
  3. filename: Klatt_whispering_wendy.wav

By careful listening to the synthesis results from a range of speech research laboratories, over a period of almost three decades, it can be seen that although the intelligibility of the resulting synthetic speech improved greatly over time, the naturalness of the resulting speech consistently fell far below that of naturally spoken speech, giving researchers pause to think about what was required for natural sounding synthesis.

A.13.7 Voice Alteration Methods

Examples of the use of signal processing methods for waveform and voice alterations can be found in the directory

chapter13_files\voice alterations

which includes 3 examples, namely:

1. **prosody modification**, performed using HNM [423] and PSOLA [249] methods; the sequence of files is:

   - filename: source_original.wav: original sentence used to test voice alteration capabilities of synthesizer
   - filename: source_alteration_HNM.wav: prosody alteration using HNM method
   - filename: source_alteration_psola.wav: prosody alteration using the PSOLA (Pitch Synchronous Overlap Add) method

2. **voice alteration**, from female to child's voice; the sequence of files is:

   - filename: source_original.wav: original sentence used to test the ability of the synthesizer to make voice alterations
   - filename: source_alteration_child.wav: spectral modification of speech from the range for an adult male to that of a child

3. **voice alteration**, from child to adult male's voice; the sequence of files is:

   - filename: child_original.wav: original sentence used to test the ability of the synthesizer to make voice alterations

- filename: `source_alteration_male.wav`: spectral modification of speech from the range for a child to that of an adult male

A.13.8 Modern TTS (Unit Selection) System Capabilities

The next set of demonstrations illustrate modern, unit selection, TTS system capabilities. The examples, which can be found in the directory:

`chapter13_files\Natural Voices`

are from the AT&T Natural Voices product and illustrate the range of capability and languages that is available in this unit selection speech synthesis system. The first two examples are synthesized using the voice of a female talker; the next two examples are synthesized using the voice of a male talker; the next three examples are synthesized sentences in Spanish using the voice of a female talker; the final three examples are paragraph-length synthesis examples which illustrate some of the remaining flaws in modern unit selection speech synthesis systems. The specific examples are the following:

- filename: `Crystal_The_set_of_china.wav`: example of the quality of synthetic speech from a female talker
- filename: `Crystal_This_is_a_grand_season.wav`: second example of the quality of synthetic speech from a female talker
- filename: `Mike_The_last_switch_cannot_be_turned_off.wav`: example of the quality of synthetic speech from a male talker
- filename: `Mike_This_is_a_grand_season.wav`: second example of the quality of synthetic speech from a male talker
- filename: `spanish_fem1.wav`: example of the quality of synthetic speech in Spanish using a female talker
- filename: `spanish_fem2.wav`: second example of synthetic speech in Spanish from a female talker
- filename: `spanish_fem3.wav`: third example of synthetic speech in Spanish from a female talker
- filename: `male_US_English_2007.wav`: example of quality of paragraph-length synthetic speech from a male talker
- filename: `female_US_English_2007.wav`: example of quality of paragraph-length synthetic speech from a female talker
- filename: `NV_press-release.wav`: second example of the quality of paragraph-length synthetic speech from a male talker

A.13.9 Synthesis of Widely Known Paragraphs

The material in this section illustrates the effects of pre-knowledge of the spoken material on the resulting quality of the synthetic speech produced by a modern unit selection speech synthesizer. Two well-known paragraphs are synthesized, along with a simple third grade reader paragraph. The directory in which the material for this section can be found is:

`chapter13_files\TTS Paragraphs`

The specific paragraphs are the following:

- filename: `hamlet_2005.wav`: this example contains the well known soliloquy from Hamlet. The listener can evaluate the role of familiarity with the source material in evaluating the quality of the resulting synthetic speech.
- filename: `gettysburg_address_2005.wav`: a second example of widely familiar source material
- filename: `Bob Story_rich_8_2001.wav`: an example of a paragraph from an early grade child's reader, again illustrating the role of simple sentences on the judgment of quality of the resulting synthetic speech.

A.13.10 Foreign Language Synthesis

The material in this section illustrates how the basic principles of unit selection synthesis can be applied to virtually any language. Four examples of synthesis from German, Korean, Spanish, and UK English are included in this session. The directory in which the material for this section can be found is:

`chapter13_files\TTS_Multiple_Languages`

The specific examples of synthetic speech from the above four languages are the following:

- filename: `german_f1.wav`: example from a female German talker
- filename: `korean_f1.wav`: example from a female Korean talker
- filename: `spanish_f1.wav`: example from a female Spanish talker
- filename: `uk_f1.wav`: example from a female UK English talker

A.13.11 Visual TTS

The material in this section gives three examples of visual TTS, namely showing a dynamically changing face (either an avatar or a natural face) along with the resulting speech that accompanies the facial movements and motions. We present three examples of visual TTS; two using an avatar, and one using a natural face. The second avatar example utilizes a singing voice.

The directory in which the material for this section can be found is:

`chapter13_files\Visual TTS`

The specific examples of visual TTS are the following files:

- filename: `jay_messages_avatar.avi`: simple example of visual TTS using an avatar
- filename: `larry_messages_face.avi`: simple example of visual TTS using a natural face
- filename: `au_clair_avatar.avi`: visual TTS using a singing voice

A.13.12 Role of Name Etymology in TTS Name Pronunciation

The example in this section illustrates the role of name etymology in properly pronouncing a set of foreign names. The directory in which the material for this section can be found is:

`chapter13_files\spoken name etymology`

The file in which names are pronounced using the proper rules of name etymology is:

filename: `Name Etymology_Church.wav`

This file gives several examples of name pronunciation using only rules for English names, contrasted with name pronunciation using rules for the language of origin of the spoken name. The improvement in quality of the name pronunciation is remarkable.

APPENDIX

B

Solution of Frequency-Domain Differential Equations

The frequency-domain differential equations of Eqs. (5.28a) and (5.28b) can be solved using the techniques of numerical analysis as discussed for example in Refs. [191] and [295]. The equations to be solved in the closed interval $0 \leq x \leq l$ between the glottis and the lips are

$$\frac{dP}{dx} + ZU = 0, \tag{B.1a}$$

$$\frac{dU}{dx} + (Y + Y_w)P = 0, \tag{B.1b}$$

where $Z = Z(x, \Omega)$ is given by Eq. (5.31a) and $(Y + Y_w) = [Y(x, \Omega) + Y_w(x, \Omega)]$ is given by the sum of Eqs. (5.31b) and (5.29c). For each frequency, Ω, of interest, these equations must be solved subject to the following frequency-domain boundary conditions at the glottis and lips:

$$\text{glottis:} \quad Y_G(\Omega)P(0, \Omega) + U(0, \Omega) = U_G(\Omega), \tag{B.1c}$$

$$\text{lips:} \quad P(l, \Omega) - Z_L(\Omega) \cdot U(l, \Omega) = 0, \tag{B.1d}$$

where $U_G(\Omega)$ is the complex amplitude of the glottal volume velocity source, $Y_G(\Omega)$ is an admittance that may be placed in parallel with the glottal source, and $Z_L(\Omega)$ is the radiation load impedance given by Eq. (5.32b).

To solve Eqs. (B.1a) to (B.1d) numerically, it is necessary to specify the nominal fixed vocal tract area function $A_0(x)$ at a discrete set of $M + 1$ equally spaced points $k\Delta x$ for $k = 0, 1, \ldots, M$, where the spatial sampling interval is $\Delta x = l/M$. An example is shown in Figure B.1. The solid dots represent data measured on sagittal plane X-rays by Fant [101] at 0.5 cm intervals along the length $l = 17$ cm. The points are connected by linear interpolation to give $A_0(x)$ at $\Delta x = 0.5/3$ cm intervals. The interpolated area function thus has $M + 1 = 103$ samples.

A simple approach to numerical solution is to apply trapezoidal integration [191, 289, 295] to the differential equations in Eqs. (B.1a) to (B.1d), which are identical to

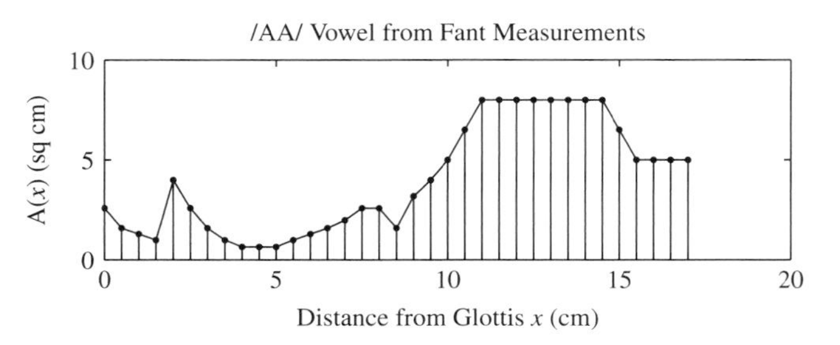

FIGURE B.1
Sampled area function for vowel /AA/ [101].

Eqs. (5.28a) and (5.28b). To do this, it is helpful to define the simpler notation

$$P_k = P(k\Delta x, \Omega), \tag{B.2a}$$

$$U_k = U(k\Delta x, \Omega), \tag{B.2b}$$

for the spatially-sampled complex pressure and volume velocity, and

$$Z_k = \Delta x \cdot Z(k\Delta x, \Omega), \tag{B.2c}$$

$$\tilde{Y}_k = \Delta x \cdot Y(k\Delta x, \Omega) + \Delta x \cdot Y_w(k\Delta x, \Omega), \tag{B.2d}$$

for the spatially sampled complex acoustic impedance and admittance per length Δx. Applying the trapezoidal integration rule to Eqs. (B.1a) and (B.1b) gives the following set of $2M$ equations

$$P_k - P_{k-1} + \frac{1}{2}(Z_k U_k + Z_{k-1} U_{k-1}) = 0, \quad k = 1, 2, \ldots, M, \tag{B.3a}$$

$$U_k - U_{k-1} + \frac{1}{2}(\tilde{Y}_k P_k + \tilde{Y}_{k-1} P_{k-1}) = 0, \quad k = 1, 2, \ldots, M. \tag{B.3b}$$

The boundary conditions at the glottis and lips provide two more equations

$$\text{glottis:} \quad Y_G P_0 + U_0 = U_G, \tag{B.3c}$$

$$\text{lips:} \quad P_M - Z_L U_M = 0, \tag{B.3d}$$

where $U_G = U_G(\Omega)$. Together, Eqs. (B.3a) to (B.3d) comprise $2M + 2$ equations in the $2M + 2$ variables P_k and U_k for $k = 0, 1, \ldots, M$. Defining $\tilde{Y}'_k = \tilde{Y}_k/2$ and $Z'_k = Z_k/2$, the matrix form of these equations is

$$\begin{bmatrix} Y_g & 1 & 0 & 0 & 0 & 0 & \cdot & 0 & 0 & 0 & 0 \\ \tilde{Y}'_0 & -1 & \tilde{Y}'_1 & 1 & 0 & 0 & \cdots & 0 & 0 & 0 & 0 \\ -1 & Z'_0 & 1 & Z'_1 & 0 & 0 & \cdots & 0 & 0 & 0 & 0 \\ 0 & 0 & \tilde{Y}'_1 & -1 & \tilde{Y}'_2 & 1 & \cdots & 0 & 0 & 0 & 0 \\ 0 & 0 & -1 & Z'_1 & 1 & Z'_2 & \cdots & 0 & 0 & 0 & 0 \\ \cdot & \cdot & \cdot & \cdot & \cdot & \cdot & \cdots & \cdot & \cdot & \cdot & \cdot \\ \cdot & \cdot & \cdot & \cdot & \cdot & \cdot & \cdots & \cdot & \cdot & \cdot & \cdot \\ \cdot & \cdot & \cdot & \cdot & \cdot & \cdot & \cdots & \cdot & \cdot & \cdot & \cdot \\ 0 & 0 & 0 & 0 & 0 & 0 & \cdots & \tilde{Y}'_{M-1} & -1 & \tilde{Y}'_M & 1 \\ 0 & 0 & 0 & 0 & 0 & 0 & \cdots & -1 & Z'_{M-1} & 1 & Z'_M \\ 0 & 0 & 0 & 0 & 0 & 0 & \cdots & 0 & 0 & 1 & -Z_L \end{bmatrix} \begin{bmatrix} P_0 \\ U_0 \\ P_1 \\ U_1 \\ P_2 \\ \cdot \\ \cdot \\ \cdot \\ U_{M-1} \\ P_M \\ U_M \end{bmatrix} = \begin{bmatrix} U_G \\ 0 \\ 0 \\ 0 \\ 0 \\ 0 \\ 0 \\ 0 \\ 0 \\ 0 \\ 0 \end{bmatrix}. \tag{B.4}$$

In matrix notation, these equations can be represented as

$$\mathbf{Qr} = \mathbf{s}, \tag{B.5}$$

where the matrix $\mathbf{Q}$ and the source vector $\mathbf{s}$ and response vector $\mathbf{r}$ have obvious definitions when Eq. (B.5) is compared to Eq. (B.4). These equations can be solved by inverting the matrix $\mathbf{Q}$; i.e.,

$$\mathbf{r} = \mathbf{Q}^{-1}\mathbf{s}. \tag{B.6}$$

This gives

$$\mathbf{r}^T = [P_0, U_0, P_1, U_1, \ldots, P_M, U_M], \tag{B.7}$$

the values of pressure and volume velocity at all the sample points within and at the boundaries of the vocal tract tube. To obtain the value of the frequency response between the volume velocity source and the volume velocity at the lips at the analysis frequency Ω, we simply need to take the ratio U_M/U_G. Correspondingly, the value of the frequency response between the glottal volume velocity source and the pressure at the lips is P_M/U_G. For simplicity in computing the frequency response of the acoustic tube alone, we normally set $U_G = 1$. This approach (with $U_G(\Omega) = 1$ for all values of $\Omega = 2\pi F$) was followed in computing the graphs for Figures 5.7, 5.8, 5.11, 5.12, and 5.13–5.16. If we make $U_G = U_G(\Omega)$, i.e., use values of the Fourier transform of a glottal input source, then we can determine the combined effect of the acoustic tube and the glottal source. This was done in creating Figure 5.22.

The inversion of the matrix can be done using standard techniques available in MATLAB or Mathematica, for example. Since it is often of interest to evaluate the frequency response functions at many values of Ω, it is necessary to recompute the matrix $\mathbf{Q}$ and its inverse for each frequency. Thus, it is useful to observe that $\mathbf{Q}$ is a sparse matrix for which efficient inversion algorithms can be derived [289, 295].

Bibliography

1. J. E. Abate, Linear and Adaptive Delta Modulation, *Proceedings of the IEEE*, Vol. 55, pp. 298–308, March 1967.
2. R. B. Adler, L. J. Chu, and R. M. Fano, *Electromagnetic Theory*, John Wiley & Sons, Inc., New York, 1963.
3. J. B. Allen, Short-Term Spectral Analysis and Synthesis and Modification by Discrete Fourier Transform, *IEEE Trans. on Acoustics, Speech and Signal Processing*, Vol. ASSP-25, No. 3, pp. 235–238, June 1977.
4. J. Allen, S. Hunnicutt, and D. Klatt, *From Text to Speech*, Cambridge University Press, Cambridge, UK, 1987.
5. Ananova, http://www.ananova.com/, 2006.
6. J. B. Allen and L. R. Rabiner, A Unified Theory of Short-Time Spectrum Analysis and Synthesis, *Proceedings of the IEEE*, Vol. 65, No. 11, pp. 1558–1564, November 1977.
7. B. S. Atal, Automatic Speaker Recognition Based on Pitch Contours, *J. of Acoustical Society of America*, Vol. 52, pp. 1687–1697, December 1972.
8. B. S. Atal, Towards Determining Articulator Positions from the Speech Signal, *Proc. Speech Comm. Seminar*, Stockholm, Sweden, pp. 1–9, 1974.
9. B. S. Atal, Predictive Coding of Speech at Low Bit Rates, *IEEE Trans. on Communications*, Vol. COM-30, No. 4, pp. 600–614, April 1982.
10. B. S. Atal, *Speech Coding Lecture Notes*, CEI Course, 2006.
11. B. S. Atal, R. V. Cox, and P. Kroon, Spectral Quantization and Interpolation for CELP Coders, *Proc. IEEE Int. Conf. on Acoustics, Speech and Signal Processing*, pp. 69–72, 1989.
12. B. S. Atal and S. L. Hanauer, Speech Analysis and Synthesis by Linear Prediction of the Speech Wave, *J. of Acoustical Society of America*, Vol. 50, No. 2, Part 2, pp. 637–655, August 1971.
13. B. S. Atal and L. R. Rabiner, A Pattern Recognition Approach to Voiced-Unvoiced-Silence Classification with Applications to Speech Recognition, *IEEE Trans. on Acoustics, Speech and Signal Processing*, Vol. ASSP-24, No. 3, pp. 201–212, June 1976.
14. B. S. Atal and J. R. Remde, A New Model of LPC Excitation for Producing Natural-Sounding Speech at Very Low Bit Rates, *Proc. IEEE Int. Conf. on Acoustics, Speech and Signal Processing*, pp. 614–617, 1982.
15. B. S. Atal and M. R. Schroeder, Adaptive Predictive Coding of Speech Signals, *Bell System Technical J.*, Vol. 49, No. 8, pp. 1973–1986, October 1970.
16. B. S. Atal and M. R. Schroeder, Predictive Coding of Speech Signals and Subjective Error Criteria, *IEEE Trans. on Acoustics, Speech and Signal Processing*, Vol. 27, pp. 247–254, 1979.

17. B. S. Atal and M. R. Schroeder, Improved Quantizer for Adaptive Predictive Coding of Speech Signals at Low Bit Rates, *Proc. IEEE Int. Conf. on Acoustics, Speech and Signal Processing*, pp. 535–538, 1980.
18. B. S. Atal, M. R. Schroeder, and V. Stover, Voice-Excited Predictive Coding System for Low Bit-Rate Transmission of Speech, *Proc. Int. Conf. on Communications*, pp. 30–37 to 30–40, 1975.
19. ATSC A/52/10, United States Advanced Television Systems Committee Digital Audio Compression (AC-3) Standard, Doc. A/52/10, December 1995.
20. L. R. Bahl, F. Jelinek, and R. L. Mercer, A Maximum Likelihood Approach to Continuous Speech Recognition, *IEEE Trans. on Pattern Analysis and Machine Intelligence*, Vol. PAMI-5, No. 2, pp. 179–190, 1983.
21. J. M. Baker, A New Time-Domain Analysis of Human Speech and Other Complex Waveforms, *Ph.D. Dissertation*, Carnegie-Mellon Univ., Pittsburgh, PA, 1975.
22. G. L. Baldwin and S. K. Tewksbury, Linear Delta Modulator Integrated Circuit with 17-Mbit/s Sampling Rate, *IEEE Trans. on Communications*, Vol. COM-22, No. 7, pp. 977–985, July 1974.
23. T. P. Barnwell, Objective Measures for Speech Quality Testing, *J. of Acoustical Society of America*, Vol. 6, No. 6, pp. 1658–1663, 1979.
24. T. P. Barnwell, Recursive Windowing for Generating Autocorrelation Coefficients for LPC Analysis, *IEEE Trans. on Acoustics, Speech and Signal Processing*, Vol. 29, No. 5, pp. 1062–1066, October 1981.
25. T. P. Barnwell, J. E. Brown, A. M. Bush, and C. R. Patisaul, Pitch and Voicing in Speech Digitization, *Res. Rept. No. E-21-620-74-B4-1*, Georgia Inst. of Tech., August 1974.
26. T. P. Barnwell, A. M. Bush, J. B. O'Neal, and R. W. Stroh, Adaptive Differential PCM Speech Transmission, *RADC-TR-74-177*, Rome Air Development Center, July 1974.
27. T. P. Barnwell and K. Nayebi, *Speech Coding, A Computer Laboratory Textbook*, John Wiley & Sons, Inc., 1996.
28. S. L. Bates, A Hardware Realization of a PCM-ADPCM Code Converter, *M.S. Thesis*, MIT, Cambridge MA, January 1976.
29. L. E. Baum, An Inequality and Associated Maximization Technique in Statistical Estimation for Probabilistic Functions of Markov Processes, *Inequalities*, Vol. 3, pp. 1–8, 1972.
30. L. E. Baum, T. Petri, G. Soules, and N. Weiss, A Maximization Technique Occurring in the Statistical Analysis of Probabilistic Functions of Markov Chains, *Annals in Mathematical Statistics*, Vol. 41, pp. 164–171, 1970.
31. G. von Bekesy, *Experiments in Hearing*, McGraw-Hill, New York, 1960.
32. J. Benesty, M. M. Sondhi, and Y. Huang (eds.), *Springer Handbook of Speech Processing and Speech Communication*, Springer, 2008.
33. W. R. Bennett, Spectra of Quantized Signals, *Bell System Technical J.*, Vol. 27, No. 3, pp. 446–472, July 1948.
34. L. L. Beranek, *Acoustics*, McGraw-Hill Book Co., New York, 1968.

35. M. Beutnagel and A. Conkie, Interaction of Units in a Unit Selection Database, *Proc. Eurospeech '99*, pp. 1063–1066, Budapest, Hungary, September 1999.
36. M. Beutnagel, A. Conkie, and A. K. Syrdal, Diphone Synthesis Using Unit Selection, *Third Speech Synthesis Workshop*, pp. 185–190, Jenolan Caves, Australia, November 1998.
37. W. A. Blankenship, Note on Computing Autocorrelation, *IEEE Trans. on Acoustics, Speech and Signal Processing*, Vol. ASSP-22, No. 1, pp. 76–77, February 1974.
38. H. W. Bode and C. E. Shannon, A Simplified Derivation of Linear Least-Square Smoothing and Prediction Theory, *Proceedings of the IRE*, Vol. 38, pp. 417–425, 1950.
39. B. Bogert, M. Healy, and J. Tukey, The Quefrency Alanysis of Time Series for Echoes, *Proc. Symp. on Time Series Analysis*, M. Rosenblatt (ed.), Chapter 15, pp. 209–243, John Wiley & Sons, Inc., New York, 1963.
40. R. H. Bolt, F. S. Cooper, E. E. David, Jr., P. B. Denes, J. M. Pickett, and K. N. Stevens, Speaker Identification by Speech Spectrograms, *Science*, Vol. 166, pp. 338–343, 1969.
41. A. M. Bose and K. N. Stevens, *Introductory Network Theory*, Harper and Row, New York, 1965.
42. M. Bosi and R. E. Goldberg, *Introduction to Digital Audio Coding and Standards*, Kluwer Academic Publishers, 2003.
43. H. A. Bourlard and N. Morgan, *Connectionist Speech Recognition—A Hybrid Approach*, Kluwer Academic Publishers, 1994.
44. L. Breiman, J. H. Friedman, R. A. Olshen, and C. J. Stone, *Classification and Regression Trees*, Wadsworth and Brooks, Pacific Grove, CA, 1984.
45. E. Bresch, Y-C. Kim, K. Nayak, D. Byrd, and S. Narayanan, Seeing Speech: Capturing Vocal Tract Shaping Using Real-Time Magnetic Resonance Imaging, *IEEE Signal Processing Magazine*, Vol. 25, No. 3, pp. 123–132, May 2008.
46. J. W. Brown and R. V. Churchill, *Introduction to Complex Variables and Applications*, 8th ed., McGraw-Hill Book Company, New York, 2008.
47. R. V. Bruce, *Bell*, Little Brown and Co., Boston, MA, p. 144, 1973.
48. J. P. Burg, A New Analysis Technique for Time Series Data, *Proc. NATO Advanced Study Institute on Signal Proc.*, Enschede, Netherlands, 1968.
49. C. S. Burrus, R. A. Gopinath, and H. Guo, *Introduction to Wavelets and Wavelet Transforms*, Prentice-Hall, 1998.
50. S. Cain, L. Smrkovski, and M. Wilson, Voiceprint Identification, Expert Article Library, expertpages.com/news/voiceprint_identification.htm.
51. J. P. Campbell, Jr., T. E. Tremain, and V. C. Welch, The Federal Standard 1016 4800 bps CELP Voice Coder, *Digital Signal Processing*, Academic Press, Vol. 1, No. 3, pp. 145–155, 1991.
52. J. P. Carlson, Digitalized Phase Vocoder, *Proc. IEEE Conf. on Speech Communication and Processing*, Boston, MA, November 1967.
53. S. Chandra and W. C. Lin, Experimental Comparison Between Stationary and Non-Stationary Formulations of Linear Prediction Applied to Voiced Speech Analysis, *IEEE Trans. on Acoustics, Speech and Signal Processing*, Vol. ASSP-22, pp. 403–415, 1974.

54. F. Charpentier and M. G. Stella, Diphone Synthesis Using an Overlap-Add Technique for Speech Waveform Concatenation, *Proc. IEEE Int. Conf. on Acoustics, Speech and Signal Processing*, pp. 2015–2018, 1986.
55. J. H. Chen, High Quality 16 Kbps Speech Coding with a One-Way Delay Less than 2 msec, *Proc. IEEE Int. Conf. on Acoustics, Speech and Signal Processing*, pp. 453–456, 1990.
56. J. H. Chen and A. Gersho, Real-Time Vector APC Speech Coding at 4800 bps with Adaptive Postfiltering, *Proc. IEEE Int. Conf. on Acoustics, Speech and Signal Processing*, pp. 2185–2188, April 1987.
57. J. H. Chen and A. Gersho, Adaptive Postfiltering for Quality Enhancement of Coded Speech, *IEEE Trans. on Speech and Audio Processing*, Vol. 3, No. 1, pp. 59–71, January 1995.
58. T. Chiba and M. Kajiyama, *The Vowel, Its Nature and Structure*, Phonetic Society of Japan, 1958.
59. D. G. Childers, *Speech Processing and Synthesis Toolboxes*, John Wiley & Sons, Inc., 1999.
60. N. Chomsky and M. Halle, *The Sound Pattern of English*, Harper and Row, Publishers, New York, 1968.
61. W. C. Chu, *Speech Coding Algorithms*, John Wiley & Sons, Inc., 2003.
62. J. H. Chung and R. W. Schafer, Performance Evaluation of Analysis-by-Synthesis Homomorphic Vocoders, *Proc. IEEE Int. Conf. on Acoustics, Speech and Signal Processing*, Vol. 2, pp. 117–120, March 1992.
63. C. H. Coker, A Model of Articulatory Dynamics and Control, *Proceedings of the IEEE*, Vol. 64, pp. 452–459, 1976.
64. J. W. Cooley and J. W. Tukey, An Algorithm for the Machine Computation of Complex Fourier Series, *Mathematics of Computation*, Vol. 19, pp. 297–381, April 1965.
65. E. Cosatto, J. Ostermann, H. P. Graf, and J. H. Schroeter, Lifelike Talking Faces for Interactive Services, *Proceedings of the IEEE*, Vol. 91, No. 9, pp. 1406–1429, September 2003.
66. T. Cover and J. Thomas, *Elements of Information Theory*, John Wiley & Sons, Inc., 1991.
67. R. V. Cox, Unpublished chart, 2007.
68. R. V. Cox, S. L. Gay, Y. Shoham, S. Quackenbush, N. Seshadri, and N. Jayant, New Directions in Subband Coding, *IEEE J. on Selected Areas in Communications*, Vol. 6, No.2, pp. 391–409, February 1988.
69. R. E. Crochiere and L. R. Rabiner, Optimum FIR Digital Filter Implementation for Decimation, Interpolation and Narrowband Filters, *IEEE Trans. on Acoustics, Speech and Signal Processing*, Vol. ASSP-23, pp. 444–456, October 1975.
70. R. E. Crochiere and L. R. Rabiner, Further Considerations in the Design of Decimators and Interpolators, *IEEE Trans. on Acoustics, Speech and Signal Processing*, Vol. ASSP-24, No. 4, pp. 269–311, August 1976.
71. R. E. Crochiere and L. R. Rabiner, *Multirate Digital Signal Processing*, Prentice-Hall Inc., 1983.
72. R. E. Crochiere, S. A. Webber, and J. L. Flanagan, Digital Coding of Speech in Sub-Bands, *Bell System Technical J.*, Vol. 55, No. 8, pp. 1069–1085, October 1976.

73. A. Croisier, Progress in PCM and Delta Modulation, *Proc. 1974 Zurich Seminar on Digital Communication*, March 1974.
74. A. Croisier, D. Esteban, and C. Galand, Perfect Channel Splitting by Use of Interpolation/Decimation/Tree Decomposition Techniques, *Int. Symp. on Information, Circuits and Systems*, 1976.
75. J. R. Crosmer and T. P. Barnwell, A Low Bit Rate Segment Vocoder Based on Line Spectrum Pairs, *Proc. IEEE Int. Conf. on Acoustics, Speech and Signal Processing*, Vol. 1, pp. 240–243, 1985.
76. P. Cummiskey, Unpublished work, Bell Laboratories.
77. P. Cummiskey, N. S. Jayant, and J. L. Flanagan, Adaptive Quantization in Differential PCM Coding of Speech, *Bell System Technical J.*, Vol. 52, No. 7, pp. 1105–1118, September 1973.
78. C. C. Cutler, Differential Quantization in Communications, U.S. Patent 2,605,361, July 29, 1952.
79. W. Davenport, An Experimental Study of Speech-Wave Probability Distributions, *J. of Acoustical Society of America*, Vol. 24, pp. 390–399, July 1952.
80. W. R. Daumer, Subjective Evaluation of Several Efficient Speech Coders, *IEEE Trans. on Communications*, Vol. COM-30, No. 4, pp. 655–662, April 1982.
81. G. Davidson, Digital Audio Coding: Dolby AC-3, *The Digital Signal Processing Handbook*, V. Madisetti and D. Williams (eds.), CRC Press, pp. 41.1–41.21, 1998.
82. S. B. Davis and P. Mermelstein, Comparison of Parametric Representations for Monosyllabic Word Recognition, *IEEE Trans. on Acoustics, Speech and Signal Processing*, Vol. ASSP-28, No. 4, pp. 357–366, August 1980.
83. A. G. Deczky, Synthesis of Recursive Digital Filters Using the Minimum p-Error Criterion, *IEEE Trans. on Audio and Electroacoustics*, Vol. AU-20, No. 5, pp. 257–263, October 1972.
84. F. E. DeJager, Delta Modulation, A Method of PCM Transmission Using a 1-Unit Code, *Philips Research Report*, pp. 442–466, December 1952.
85. P. C. Delattre, A. M. Liberman, and F. S. Cooper, Acoustic Loci and Transitional Cues for Consonants, *J. of Acoustical Society of America*, Vol. 27, No. 4, pp. 769–773, July 1955.
86. L. Deng and D. O'Shaughnessy, *Speech Processing, A Dynamic and Optimization-Oriented Approach*, Marcel Dekker Inc., 2003.
87. L. Deng, K. Wang, A. Acero, H-W. Hon, J. Droppo, C. Boulis, Y-Y. Wang, D. Jacoby, M. Mahajan, C. Chelba, X. D. Huang, Distributed Speech Processing in MiPad's Multimodal User Interface, *IEEE Trans. on Speech and Audio Processing*, Vol. 10, No. 8, pp. 605–619, November 2002.
88. P. B. Denes and E. N. Pinson, *The Speech Chain*, 2nd ed., W. H. Freeman and Co., 1993.
89. J. Deller, Jr., J. G. Proakis, and J. Hansen, *Discrete-Time Processing of Speech Signals*, Macmillan Publishing, 1993, Wiley-IEEE Press, Classic Reissue, 1999.
90. J. J. Dubnowski, R. W. Schafer, and L. R. Rabiner, Real-Time Digital Hardware Pitch Detector, *IEEE Trans. on Acoustics, Speech and Signal Processing*, Vol. ASSP-24, No. 1, pp. 2–8, February 1976.
91. R. O. Duda, P. E. Hart, and D. G. Stork, *Pattern Classification*, John Wiley & Sons, Inc., 2001.

92. H. Dudley, The Vocoder, *Bell Labs Record*, Vol. 17, pp. 122–126, 1939.
93. H. Dudley, R. R. Riesz, and S. A. Watkins, A Synthetic Speaker, *J. of the Franklin Institute*, Vol. 227, pp. 739–764, 1939.
94. H. K. Dunn, Methods of Measuring Vowel Formant Bandwidths, *J. of Acoustical Society of America*, Vol. 33, pp. 1737–1746, 1961.
95. H. K. Dunn and S. D. White, Statistical Measurements on Conversational Speech, *J. of Acoustical Society of America*, Vol. 11, pp. 278–288, January 1940.
96. S. Dusan, G. J. Gadbois, and J. L. Flanagan, Multimodal Interaction on PDA's Integrating Speech and Pen Inputs, *Eurospeech 2003*, Geneva Switzerland, pp. 2225–2228, 2003.
97. T. Dutoit, *An Introduction to Text-to-Speech Synthesis*, Kluwer Academic Publishers, 1997.
98. T. Dutoit and F. Marques, *Applied Signal Processing: A MATLAB-Based Proof of Concept*, Springer, 2009.
99. L. D. Erman, An Environment and System for Machine Understanding of Connected Speech, *Ph.D. Dissertation*, Carnegie-Mellon Univ., Pittsburgh, PA, 1975.
100. D. Esteban and C. Galand, Application of Quadrature Mirror Filters to Split Band Voice Coding Schemes, *Proc. IEEE Int. Conf. on Acoustics, Speech and Signal Processing*, pp. 191–195, 1977.
101. G. Fant, *Acoustic Theory of Speech Production,* Mouton, The Hague, 1970.
102. D. W. Farnsworth, High-speed Motion Pictures of the Human Vocal Cords, *Bell Labs Record*, Vol. 18, pp. 203–208, 1940.
103. J. D. Ferguson, Hidden Markov Analysis: An Introduction, *Hidden Markov Models for Speech*, Princeton: Institute for Defense Analyses, 1980.
104. J. L. Flanagan, The Design of 'Terminal Analog' Speech Synthesizers, *Jour. Acoustical. Soc. Amer.*, Vol. 29, pp. 306–310, February 1957.
105. J. L. Flanagan, *Speech Analysis, Synthesis and Perception*, 2nd ed., Springer, 1972.
106. J. L. Flanagan, Computers That Talk and Listen: Man-Machine Communication by Voice, *Proceedings of the IEEE*, Vol. 64, No. 4, pp. 416–422, April 1976.
107. J. L. Flanagan and L. Cherry, Excitation of Vocal-Tract Synthesizer, *J. of Acoustical Society of America*, Vol. 45, No. 3, pp. 764–769, March 1969.
108. J. L. Flanagan, C. H. Coker, L. R. Rabiner, R. W. Schafer, and N. Umeda, Synthetic Voices for Computers, *IEEE Spectrum*, Vol. 7, No. 10, pp. 22–45, October 1970.
109. J. L. Flanagan and R. M. Golden, The Phase Vocoder, *Bell System Technical J.*, Vol. 45, pp. 1493–1509, 1966.
110. J. L. Flanagan, K. Ishizaka, and K. L. Shipley, Synthesis of Speech from a Dynamic Model of the Vocal Cords and Vocal Tract, *Bell System Technical J.*, Vol. 54, No. 3, pp. 485–506, March 1975.
111. J. L. Flanagan and L. L. Landgraf, Self Oscillating Source for Vocal-Tract Synthesizers, *IEEE Trans. on Audio and Electroacoustics*, Vol. AU-16, pp. 57–64, March 1968.
112. J. L. Flanagan and M. G. Saslow, Pitch Discrimination for Synthetic Vowels, *J. of Acoustical Society of America*, Vol. 30, No. 5, pp. 435–442, 1958.
113. H. Fletcher, *Speech and Hearing in Communication,* D. Van Nostrand Co., New York, 1953. (Reprinted by Robert E. Krieger Pub. Co. Inc., New York, 1972.)

114. H. Fletcher and W. A. Munson, Loudness, Its Definition, Measurement and Calculation, *J. of Acoustical Society of America*, Vol. 5, pp. 82–108, July 1933.
115. H. Fletcher and W. A. Munson, Relation between Loudness and Masking, *J. of Acoustical Society of America*, Vol. 9, No. 5, pp. 1–10, July 1937.
116. G. D. Forney, The Viterbi Algorithm, *Proceedings of the IEEE*, Vol. 61, pp. 268–278, March 1973.
117. O. Fujimura, Analysis of Nasal Consonants, *J. of Acoustical Society of America*, Vol. 34, No. 12, pp. 1865–1875, December 1962.
118. S. Furui, Cepstral Analysis Technique for Automatic Speaker Verification, *IEEE Trans. on Acoustics, Speech and Signal Processing*, Vol. ASSP-29, No. 2, pp. 254–272, April 1981.
119. S. Furui, Speaker Independent Isolated Word Recognizer Using Dynamic Features of Speech Spectra, *IEEE Trans. on Acoustics, Speech and Signal Processing*, Vol. ASSP-34, pp. 52–59, 1986.
120. S. Furui (ed.), *Digital Speech Processing, Synthesis and Recognition*, 2nd ed., Marcel Dekker Inc., New York, 2001.
121. S. Furui and M. M. Sondhi (eds.), *Advances in Speech Signal Processing*, Marcel Dekker Inc., 1991.
122. C. Galand and D. Esteban, 16 Kbps Real-Time QMF Subband Coding Implementation, *Proc. IEEE Int. Conf. on Acoustics, Speech and Signal Processing*, pp. 332–335, April 1980.
123. A. Gersho and R. M. Gray, *Vector Quantization and Signal Compression*, Kluwer Academic Publishers, 1992.
124. O. Ghitza, Auditory Nerve Representation as a Basis for Speech Processing, *Advances in Speech and Signal Processing*, S. Furui and M. M. Sondhi (eds.), Marcel-Dekker, NY, pp. 453–485, 1991.
125. J. J. Godfrey, E. C. Holliman, and J. McDaniel, SWITCHBOARD: Telephone Speech Corpus for Research and Development, *Proc. IEEE Int. Conf. on Acoustics, Speech and Signal Processing*, pp. 517–520, 1992.
126. B. Gold, Computer Program for Pitch Extraction, *J. of Acoustical Society of America*, Vol. 34, No. 7, pp. 916–921, August 1962.
127. B. Gold and N. Morgan, *Speech and Audio Signal Processing*, John Wiley & Sons, Inc., 2000.
128. B. Gold and L. R. Rabiner, Analysis of Digital and Analog Formant Synthesizers, *IEEE Trans. on Audio and Electroacoustics*, Vol. AU-16, pp. 81–94, March 1968.
129. B. Gold and L. R. Rabiner, Parallel Processing Techniques for Estimating Pitch Periods of Speech in the Time Domain, *J. of Acoustical Society of America*, Vol. 46, No. 2, Part 2, pp. 442–448, August 1969.
130. B. Gold and C. M. Rader, Systems for Compressing the Bandwidth of Speech, *IEEE Trans. on Audio and Electroacoustics*, Vol. AU-15, No. 3, pp. 131–135, September 1967.
131. B. Gold and C. M. Rader, The Channel Vocoder, *IEEE Trans. on Audio and Electroacoustics*, Vol. AU-15, No. 4, pp. 148–160, December 1967.
132. R. Goldberg and L. Riek, *A Practical Handbook of Speech Coders*, CRC Press, 2000.

133. D. J. Goodman, The Application of Delta Modulation to Analog-to-PCM Encoding, *Bell System Technical J.*, Vol. 48, No. 2, pp. 321–343, February 1969.
134. D. J. Goodman, Digital Filters for Code Format Conversion, *Electronics Letters*, Vol. 11, February 1975.
135. A. L. Gorin, B. A. Parker, R. M. Sachs, and J. G. Wilpon, How May I Help You?, *Proc. of the Interactive Voice Technology for Telecommunications Applications (IVTTA)*, pp. 57–60, 1996.
136. A. L. Gorin, G. Riccardi, and J. H. Wright, How May I Help You?, *Speech Communication*, Vol. 23, pp. 113–127, 1997.
137. R. M. Gray, Vector Quantization, *IEEE Signal Processing Magazine*, pp. 4–28, April 1984.
138. R. M. Gray, Quantization Noise Spectra, *IEEE Trans. on Information Theory*, Vol. 36, No. 6, pp. 1220–1244, November 1990.
139. J. A. Greefkes, A Digitally Companded Delta Modulation Modem for Speech Transmission, *Proc IEEE Int. Conf. Communications*, pp. 7-33 to 7-48, June 1970.
140. J. S. Gruber and F. Poza, Voicegram Identification Evidence, 54 Am. Jur. Trials, Lawyers Cooperative Publishing, 1995.
141. J. M. Heinz and K. N. Stevens, On the Properties of Voiceless Fricative Consonants, *J. of Acoustical Society of America*, Vol. 33, No. 5, pp. 589–596, May 1961.
142. H. D. Helms, Fast Fourier Transform Method of Computing Difference Equations and Simulating Filters, *IEEE Trans. on Audio and Electroacoustics*, Vol. 15, No. 2, pp. 85–90, 1967.
143. H. Hermansky, Auditory Modeling in Automatic Recognition of Speech, *Proc. First European Conf. on Signal Analysis and Prediction*, pp. 17–21, Prague, Czech Republic, 1997.
144. W. Hess, *Pitch Determination of Speech Sounds: Algorithms and Devices*, Springer, NY, 1983.
145. E. M. Hoffstetter, An Introduction to the Mathematics of Linear Predictive Filtering as Applied to Speech, *Technical Note 1973-36*, MIT Lincoln Labs, July 1973.
146. A. Holbrook and G. Fairbanks, Diphthong Formants and Their Movements, *J. Speech and Hearing Research*, Vol. 5, No. 1, pp. 38–58, March 1962.
147. H. Hollien, *Forensic Voice Identification*, Academic Press, 2001.
148. J. N. Holmes, Parallel Formant Vocoders, *Proc. IEEE Eascon*, September 1978.
149. J. Hou, On the Use of Frame and Segment-Based Methods for the Detection and Classification of Speech Sounds and Features, *Ph.D. Thesis*, Rutgers University, October 2009.
150. House Ear Institute, www.hei.org, 2007.
151. X. Huang, A. Acero, and H.-W. Hon, *Spoken Language Processing*, Prentice-Hall Inc., Englewood Cliffs, NJ, 2001.
152. C. P. Hughes and A. Nikeghbali, The Zeros of Random Polynomials Cluster Near the Unit Circle, arXiv:math/0406376v3 [math.CV], http://arxiv.org/PS_cache/math/pdf/0406/0406376v3.pdf.

153. A. Hunt and A. Black, Unit Selection in a Concatenative Speech Synthesis System Using a Large Speech Database, *Proc. IEEE Int. Conf. on Acoustics, Speech and Signal Processing*, Atlanta, Vol. 1, pp. 373–376, 1996.
154. http://hyperphysics.phy-astr.edu/hbase/hframe.html.
155. K. Ishizaka and J. L. Flanagan, Synthesis of Voiced Sounds from a Two-Mass Model of the Vocal Cords, *Bell System Technical J.*, Vol. 50, No. 6, pp. 1233–1268, July-August 1972.
156. ISO/IEC JTC1/SC29/WG11 MPEG, IS 11172-3, Information Technology—Coding of Moving Pictures and Associated Audio for Digital Storage Media at Up to About 1.5 Mbits/s – Part 3: Audio, 1992 (MPEG-1).
157. ISO/IEC JTC1/SC29/WG11 MPEG, IS13818-3, Information Technology-Generic Coding of Moving Pictures and Associated Audio, Part 3: Audio, 1994 (MPEG-2).
158. ISO/IEC JTC1/SC29/WG11 MPEG, IS13818-7, Information Technology-Generic Coding of Moving Pictures and Associated Audio, Part 7: Advanced Audio Coding, 1994 (MPEG-2 AAC).
159. ISO/IEC JTC1/SC29/WG11 MPEG, IS14496-3, Coding of Audio-Visual Objects, Part 3: Audio, 1998 (MPEG-4).
160. F. Itakura, Line Spectrum Representation of Linear Prediction Coefficients of Speech Signals, *J. of Acoustical Society of America*, Vol. 57, p. 535, (abstract), 1975.
161. F. I. Itakura and S. Saito, Analysis-Synthesis Telephony Based upon the Maximum Likelihood Method, *Proc. 6th Int. Congress on Acoustics*, pp. C17–20, Tokyo, 1968.
162. F. I. Itakura and S. Saito, A Statistical Method for Estimation of Speech Spectral Density and Formant Frequencies, *Electronics and Communication in Japan*, Vol. 53-A, No, 1, pp. 36–43, 1970.
163. F. Itakura and S. Saito, Digital Filtering Techniques for Speech Analysis and Synthesis, *7th Int. Cong. on Acoustics*, Budapest, Paper 25 C1, 1971.
164. F. Itakura and T. Umezaki, Distance Measure for Speech Recognition Based on the Smoothed Group Delay Spectrum, *Proc. IEEE Int. Conf. on Acoustics, Speech and Signal Processing*, Vol. 12, pp. 1257–1260, April 1987.
165. ITU-T P.800, Methods for Subjective Determination of Transmission Quality, *Int. Telecommunication Unit*, 1996.
166. R. Jakobson, C. G. M. Fant, and M. Halle, *Preliminaries to Speech Analysis: The Distinctive Features and Their Correlates*, MIT Press, Cambridge, MA, 1963.
167. N. S. Jayant, Adaptive Delta Modulation with a One-Bit Memory, *Bell System Technical J.*, pp. 321–342, March 1970.
168. N. S. Jayant, Adaptive Quantization with a One Word Memory, *Bell System Technical J.*, pp. 1119–1144, September 1973.
169. N. S. Jayant, Digital Coding of Speech Waveforms: PCM, DPCM, and DM Quantizers, *Proceedings of the IEEE*, Vol. 62, pp. 611–632, May 1974.
170. N. S. Jayant, Step-Size Transmitting Differential Coders for Mobile Telephony, *Bell System Technical J.*, Vol. 54, No. 9, pp. 1557–1582, November 1975.
171. N. S. Jayant (ed.), *Waveform Quantization and Coding*, IEEE Press, 1976.
172. N. S. Jayant and P. Noll, *Digital Coding of Waveforms*, Prentice-Hall Inc., 1984.
173. F. Jelinek, *Statistical Methods for Speech Recognition*, MIT Press, Cambridge, MA, 1998.

174. F. Jelinek, R. L. Mercer, and S. Roucos, Principles of Lexical Language Modeling for Speech Recognition, *Advances in Speech Signal Processing*, S. Furui and M. M. Sondhi (eds.), Marcel Dekker, pp. 651–699, 1991.
175. J. D. Johnston, A Filter Family Designed for Use in Quadrature Mirror Filter Banks, *Proc. IEEE Int. Conf. on Acoustics, Speech and Signal Processing*, pp. 291–294, April 1980.
176. J. D. Johnston, Transform Coding of Audio Signals Using Perceptual Noise Criteria, *IEEE J. on Selected Areas in Communications*, Vol. 6, No. 2, pp. 314–323, February 1988.
177. M. Johnston, S. Bangalore, and G. Vasireddy, MATCH: Multimodal Access to City Help, *Proc. Automatic Speech Recognition and Understanding Workshop*, Trento, Italy, 2001.
178. B. H. Juang, Maximum Likelihood Estimation for Mixture Multivariate Stochastic Observations of Markov Chains, *AT&T Technology Journal*, Vol. 64, No. 6, pp. 1235–1249, 1985.
179. B. H. Juang and S. Furui, Automatic Recognition and Understanding of Spoken Language—A First Step Towards Natural Human-Machine Communication, *Proceedings of the IEEE*, Vol. 88, No. 8, pp. 1142–1165, 2000.
180. B. H. Juang, S. E. Levinson, and M. M. Sondhi, Maximum Likelihood Estimation for Multivariate Mixture Observations of Markov Chains, *IEEE Trans. on Information Theory*, Vol. 32, No. 2, pp. 307–309, 1986.
181. B. H. Juang, L. R. Rabiner, and J. G. Wilpon, On the Use of Bandpass Liftering in Speech Recognition, *IEEE Trans. on Acoustics, Speech and Signal Processing*, Vol. ASSP-35, No. 7, pp. 947–954, July 1987.
182. B. H. Juang, D. Y. Wong, and A. H. Gray, Jr., Distortion Performance of Vector Quantization for LPC Voice Coding, *IEEE Trans. on Acoustics, Speech and Signal Processing*, Vol. ASSP-30, No. 2, pp. 294–304, April 1982.
183. D. Jurafsky and J. H. Martin, *Speech and Language Processing*, 2nd ed., Prentice-Hall Inc., 2008.
184. J. F. Kaiser, Nonrecursive Digital Filter Design Using the I_0-Sinh Window Function, *Proc. IEEE Int. Symp. on Circuits and Systems*, San Francisco, pp. 20–23, April 1974.
185. C. Kamm, M. Walker, and L. R. Rabiner, The Role of Speech Processing in Human-Computer Intelligent Communication, *Speech Communication*, Vol. 23, pp. 263–278, 1997.
186. H. Kars and K. Brandenburg (eds.), *Applications of Digital Signal Processing to Audio and Acoustics*, Kluwer Academic Publishers, 1998.
187. T. Kawahar and C. H. Lee, Flexible Speech Understanding Based on Combined Key-Phrase Detection and Verification, *IEEE Trans. on Speech and Audio Processing*, Vol. T-SA 6, No. 6, pp. 558–568, 1998.
188. J. L. Kelly, Jr. and C. Lochbaum, Speech Synthesis, *Proc. Stockholm Speech Communications Seminar*, R.I.T., Stockholm, Sweden, September 1962.
189. W. B. Kendall, A New Algorithm for Computing Autocorrelations, *IEEE Trans. on Computers*, Vol. C-23, No. 1, pp. 90–93, January 1974.
190. N. Y. S. Kiang and E. C. Moxon, Tails of Tuning Curves of Auditory Nerve Fibers, *J. of Acoustical Society of America*, Vol. 55, pp. 620–630, 1974.

191. D. Kincaid and W. Cheney, *Numerical Analysis: Mathematics of Scientific Computing*, 3rd ed., American Mathematical Society, 2002.
192. L. E. Kinsler, A. R. Frey, A. B. Coppens, and J. V. Sanders, *Fundamentals of Acoustics*, 4th ed., John Wiley & Sons, Inc., New York, 2000.
193. D. H. Klatt, Software for a Cascade/Parallel Formant Synthesizer, *J. of Acoustical Society of America*, Vol. 67, pp. 971–995, 1980.
194. D. H. Klatt, Review of Text-to-Speech Conversion for English, *J. of Acoustical Society of America*, Vol. 82, pp. 737–793, September 1987.
195. W. B. Kleijn and K. K. Paliwal, *Speech Coding and Synthesis*, Elsevier, 1995.
196. W. Koenig, H. K. Dunn, and L. Y. Lacy, The Sound Spectrograph, *J. of Acoustical Society of America*, Vol. 18, pp. 19–49, July 1946.
197. A. M. Kondoz, *Digital Speech: Coding for Low Bit Rate Communication Systems*, 2nd ed., John Wiley & Sons, Inc., 2004.
198. K. Konstantinides, Fast Subband Filtering in MPEG Audio Coding, *IEEE Signal Processing Letters,* Vol. 1, No. 2, pp. 26–28, February 1994.
199. K. Konstantinides, Fast Subband Filtering in Digital Signal Coding, U.S. Patent 5,508,949, filed December 29, 1993, issued April 16, 1996.
200. P. Ladefoged, *A Course in Phonetics*, 2nd ed., Harcout, Brace, Jovanovich, 1982.
201. S. W. Lang and J. H. McClellan, A Simple Proof of Stability for All-Pole Linear Prediction Models, *Proceedings of the IEEE*, Vol. 67, No. 5, pp. 860–861, May 1979.
202. C. H. Lee, F. K. Soong, and K. K. Paliwal (eds.), *Automatic Speech and Speaker Recognition*, Kluwer Academic Publishers, 1996.
203. I. Lehiste (ed.), *Readings in Acoustic Phonetics*, MIT Press, Cambridge, MA, 1967.
204. R. G. Leonard, A Database for Speaker-Independent Digit Recognition, *Proc. IEEE Int. Conf. on Acoustics, Speech and Signal Processing*, pp. 42.11.1–42.11.4, 1984.
205. S. E. Levinson, *Mathematical Models for Speech Technology*, John Wiley & Sons, Inc., 2005.
206. S. E. Levinson, L. R. Rabiner, and M. M. Sondhi, An Introduction to the Application of the Theory of Probabilistic Functions of a Markov Process to Automatic Speech Recognition, *Bell System Technical J.*, Vol. 62, No. 4, pp. 1035–1074, 1983.
207. H. Levitt, Speech Processing for the Deaf: An Overview, *IEEE Trans. on Audio and Electroacoustics*, Vol. AU-21, pp. 269–273, June 1973.
208. Y. Linde, A. Buzo, and R. M. Gray, An Algorithm for Vector Quantizer Design, *IEEE Trans. on Communications*, Vol. COM-28, pp. 84–95, January 1980.
209. R. P. Lippmann, Speech Recognition by Machines and Humans, *Speech Communication*, Vol. 22, No. 1, pp. 1–15, 1997.
210. S. P. Lloyd, Least Squares Quantization in PCM, *IEEE Trans. on Information Theory*, Vol. IT-28, pp. 127–135, March, 1982.
211. P. Loizou, *Colea: A Matlab Software Tool for Speech Analysis*, http://www.utdallas.edu/~loizou/speech/colea.htm.
212. P. Loizou, *Speech Enhancement Theory and Practice*, CRC Press, 2007.
213. K. Lukaszewicz and M. Karjalainen, Microphonemic Method of Speech Synthesis, *Proc. IEEE Int. Conf. on Acoustics, Speech and Signal Processing*, Dallas, Vol. 3, pp. 1426–1429, 1987.

214. R. F. Lyon, A Computational Model of Filtering, Detection, and Compression in the Cochlea, *Proc. IEEE Int. Conf. on Acoustics, Speech and Signal Processing*, pp. 1282–1285, 1982.
215. M. Macchi, Synthesis by Conatenation, Where We Are, Where We Want to Go, Talk Given to National Science Foundation, http://www.espeech.com/aboutus.htm, August 1998.
216. J. Makhoul, Spectral Analysis of Speech by Linear Prediction, *IEEE Trans. on Audio and Electroacoustics*, Vol. AU-21, No. 3, pp. 140–148, June 1973.
217. J. Makhoul, Linear Prediciton: A Tutorial Review, *Proceedings of the IEEE*, Vol. 63, pp. 561–580, 1975.
218. J. Makhoul, Spectral Linear Prediction: Properties and Applications, *IEEE Trans. on Acoustics, Speech and Signal Processing*, Vol. ASSP-23, No. 3, pp. 283–296, June 1975.
219. J. Makhoul, Stable and Efficient Lattice Methods for Linear Prediction, *IEEE Trans. on Acoustics, Speech and Signal Processing*, Vol. ASSP-25, No. 5, pp. 423–428, October 1977.
220. J. Makhoul and M. Berouti, Adaptive Noise Spectral Shaping and Entropy Coding in Predictive Coding of Speech, *IEEE Trans. on Acoustics, Speech and Signal Processing*, Vol. 27, No. 1, pp. 63–73, February 1979.
221. J. Makhoul, V. Viswanathan, R. Schwarz, and A. W. F. Huggins, A Mixed Source Model for Speech Compression and Synthesis, *J. of Acoustical Society of America*, Vol. 64, pp. 1577–1581, December 1978.
222. J. Makhoul and J. Wolf, Linear Prediction and the Spectral Analysis of Speech, *BBN Report No. 2304*, August 1972.
223. J. N. Maksym, Real-Time Pitch Extraction by Adaptive Prediction of the Speech Waveform, *IEEE Trans. on Audio and Electroacoustics*, Vol. AU-21, No. 3, pp. 149–153, June 1973.
224. D. Malah, Time-Domain Algorithms for Harmonic Bandwidth Reduction and Time-Scaling of Pitch Signals, *IEEE Trans. on Acoustics, Speech and Signal Processing*, Vol. 27, No. 2, pp. 121–133, 1979.
225. C. D. Manning and H. Schutze, *Foundations of Statistical Natural Language Processing*, MIT Press, Cambridge, MA, 1999.
226. J. D. Markel, Digital Inverse Filtering—A New Tool for Formant Trajectory Estimation, *IEEE Trans. on Audio and Electroacoustics*, Vol. AU-20, No. 2, pp. 129–137, June 1972.
227. J. D. Markel, The SIFT Algorithm for Fundamental Frequency Estimation, *IEEE Trans. on Audio and Electroacoustics*, Vol. AU-20, No, 5, pp. 367–377, December 1972.
228. J. D. Markel, Application of a Digital Inverse Filter for Automatic Formant and F_0 Analysis, *IEEE Trans. on Audio and Electroacoustics*, Vol. AU-21, No. 3, pp. 149–153, June 1973.
229. J. D. Markel and A. H. Gray, Jr., On Autocorrelation Equations as Applied to Speech Analysis, *IEEE Trans. on Audio and Electroacoustics*, Vol. AU-21, pp. 69–79, April 1973.
230. J. D. Markel and A. H. Gray, Jr., A Linear Prediction Vocoder Simulation Based upon the Autocorrelation Method, *IEEE Trans. on Acoustics, Speech and Signal Processing*, Vol. ASSP-22, No. 2, pp. 124–134, April 1974.

231. J. D. Markel and A. H. Gray, Jr., Cepstral Distance and the Frequency Domain, *J. of Acoustical Society of America*, Vol. 58, p. S97, 1975.
232. J. D. Markel and A. H. Gray, Jr., *Linear Prediction of Speech*, Springer, 1976.
233. D. Massaro, *Perceiving Talking Faces: From Speech Perception to a Behavioral Principle*, MIT Press, Cambridge, MA, 1998.
234. J. Masson and Z. Picel, Flexible Design of Computationally Efficient Nearly Perfect QMF Filter Banks, *Proc. IEEE Int. Conf. on Acoustics, Speech and Signal Processing*, pp. 541–544, 1985.
235. J. Max, Quantizing for Minimum Distortion, *IRE Trans. on Information Theory*, Vol. IT-6, pp. 7–12, March 1960.
236. R. McAulay, A Low-Rate Vocoder Based on an Adaptive Subband Formant Analysis, *Proc. IEEE Int. Conf. on Acoustics, Speech and Signal Processing*, pp. 28–31, April 1981.
237. S. McCandless, An Algorithm for Automatic Formant Extraction Using Linear Prediction Spectra, *IEEE Trans. on Acoustics, Speech and Signal Processing*, Vol. ASSP-22, No. 2, pp. 135–141, April 1974.
238. J. H. McClellan, T. W. Parks, and L. R. Rabiner, A Computer Program for Designing Optimum FIR Linear Phase Digital Filters, *IEEE Trans. on Audio and Electroacoustics*, Vol. AU-21, pp. 506–526, December 1973.
239. J. H. McClellan, R. W. Schafer, and M. A. Yoder, *Signal Processing First*, Prentice-Hall Inc., Upper Saddle River, NJ, 2003.
240. A. V. McCree and T. P. Barnwell, III, A Mixed Excitation LPC Vocoder Model for Low Bit Rate Speech Coding, *IEEE Trans. on Speech and Audio Processing*, Vol. 3, No. 4, pp. 242–250, July 1995.
241. R. A. McDonald, Signal-to-Noise and Idle Channel Performance of DPCM Systems—Particular Applications to Voice Signals, *Bell System Technical J.*, Vol. 45, No. 7, pp. 1123–1151, September 1966.
242. G. A. Miller, G. A. Heise, and W. Lichten, The Intelligibility of Speech as a Function on the Context of the Test Material, *J. of Experimental Psychology*, Vol. 41, pp. 329–335, 1951.
243. G. A. Miller and P. E. Nicely, An Analysis of Perceptual Confusions among Some English Consonants, *J. of Acoustical Society of America*, Vol. 27, No. 2, pp. 338–352, 1955.
244. F. Mintzer, Filters for Distortion-Free Two-Band Multirate Filter Banks, *IEEE Trans. on Acoustics, Speech and Signal Processing*, Vol. ASSP-33, pp. 626–630, June 1985.
245. S. K. Mitra, *Digital Signal Processing*, 3rd ed., McGraw-Hill, 2006.
246. M. Mohri, Finite-State Transducers in Language and Speech Processing, *Computational Linguistics*, Vol. 23, No. 2, pp. 269–312, 1997.
247. J. A. Moorer, Signal Processing Aspects of Computer Music, *Proceedings of the IEEE*, Vol. 65, No. 8, pp. 1108–1137, August 1977.
248. P. M. Morse and K. U. Ingard, *Theoretical Acoustics*, McGraw-Hill Book Co., New York, 1968.
249. E. Moulines and F. Charpentier, Pitch Synchronous Waveform Processing Techniques for Text-to-Speech Synthesis Using Diphones, *Speech Communication*, Vol. 9, No. 5–6, pp. 453–467, 1990.

250. S. Narayanan and A. Alwan (eds.), *Text to Speech Synthesis: New Paradigms and Advances*, Prentice-Hall Inc., 2004.
251. S. Narayanan, K. S. Nayak, S. Lee, A. Sethy, and D. Byrd, An Approach to Real-Time Magnetic Resonance Imaging for Speech Production, *J. of Acoustical Society of America*, Vol. 115, No. 5, pp. 1771–1776, 2004.
252. A. M. Noll, Cepstrum Pitch Determination, *J. of Acoustical Society of America*, Vol. 41, No. 2, pp. 293–309, February 1967.
253. A. M. Noll, Pitch Determination of Human Speech by the Harmonic Product Spectrum, the Harmonic Sum Spectrum, and a Maximum Likelihood Estimate, *Proc. Symp. Computer Processing in Communication*, pp. 779–798, April 1969.
254. P. Noll, Non-Adaptive and Adaptive DPCM of Speech Signals, *Polytech. Tijdschr. Ed. Elektrotech/Elektron*, (The Netherlands), No. 19, 1972.
255. P. Noll, Adaptive Quantizing in Speech Coding Systems, *Proc. 1974 Zurich Seminar on Digital Communications*, Zurich, March 1974.
256. P. Noll, A Comparative Study of Various Schemes for Speech Encoding, *Bell System Technical J.*, Vol. 54, No. 9, pp. 1597–1614, November 1975.
257. P. Noll, Effect of Channel Errors on the Signal-to-Noise Performance of Speech Encoding Systems, *Bell System Technical J.*, Vol. 54, No. 9, pp. 1615–1636, November 1975.
258. P. Noll, Wideband Speech and Audio Coding, *IEEE Communications Magazine,* pp. 34–44, November 1993.
259. P. Noll, MPEG Digital Audio Coding, *IEEE Signal Processing Magazine,* pp. 59–81, September 1997.
260. H. J. Nussbaumer and M. Vetterli, Computationally Efficient QMF Filter Banks, *Proc. IEEE Int. Conf. on Acoustics, Speech and Signal Processing*, pp. 11.3.1–11.3.4, 1984.
261. H. Nyquist, Certain Topics in Telegraph Transmission Theory, *Trans. of the AIEE*, Vol. 47, pp. 617–644, February 1928.
262. J. P. Olive, Automatic Formant Tracking in a Newton-Raphson Technique, *J. of Acoustical Society of America*, Vol. 50, pp. 661–670, August 1971.
263. J. P. Olive, Rule Synthesis of Speech from Diadic Units, *Proc. IEEE Int. Conf. on Acoustics, Speech and Signal Processing*, pp. 568–570, 1977.
264. J. Olive, A. Greenwood, and J. Coleman, *Acoustics of American English*, Springer, 1993.
265. B. M. Oliver, J. R. Pierce, and C. E. Shannon, The Philosophy of PCM, *Proceedings of the IRE*, Vol. 36, No. 11, pp. 1324–1331, November 1948.
266. A. V. Oppenheim, Superposition in a Class of Nonlinear Systems, *Tech. Report No. 432*, Research Lab of Electronics, MIT, Cambridge, MA, March 1965.
267. A. V. Oppenheim, A Speech Analysis-Synthesis System Based on Homomorphic Filtering, *J. of Acoustical Society of America*, Vol. 45, pp. 458–465, February 1969.
268. A. V. Oppenheim, Sound Spectrograms Using the Fast Fourier Transform, *IEEE Spectrum*, Vol. 7, pp. 57–62, August 1970.
269. A. V. Oppenheim and R. W. Schafer, Homomorphic Analysis of Speech, *IEEE Trans. on Audio and Electroacoustics*, Vol. AU-16, No. 2, pp. 221–226, June 1968.
270. A. V. Oppenheim and R. W. Schafer, *Discrete-Time Signal Processing*, 3rd ed., Prentice-Hall Inc., Upper Saddle River, NJ, 2010.

271. A. V. Oppenheim, R. W. Schafer, and T. G. Stockham, Jr., Nonlinear Filtering of Multiplied and Convolved Signals, *Proceedings of the IEEE*, Vol. 56, No. 8, pp. 1264–1291, August 1968.
272. A. V. Oppenheim, A. S. Willsky, with S. H. Nawab, *Signals and Systems*, 2nd ed., Prentice-Hall Inc., Upper Saddle River, NJ, 1997.
273. D. O'Shaughnessy, *Speech Communication, Human and Machine*, Addison-Wesley, 1987.
274. J. Ostermann, M. Beutnagel, A. Fischer, and Y. Wang, Integration of Talking Heads and Text-to-Speech Synthesizers for Visual TTS, *Proc. ICSLP-98*, Sydney, Australia, November 1998.
275. M. D. Paez and T. H. Glisson, Minimum Mean Squared-Error Quantization in Speech, *IEEE Trans. on Communications*, Vol. COM-20, pp. 225–230, April 1972.
276. D. S. Pallett, A Look at NIST's Benchmark ASR Tests: Past, Present, and Future, *Proc. ASRU'03*, pp. 483–488, 2003.
277. D. S. Pallett, J. G. Fiscus, W. M. Fisher, J. S. Garofol, B. A. Lund, and M. A. Przybocki, 1993 Benchmark Tests for the ARPA Spoken Language Program, *Proc. 1995 ARPA Human Language Technology Workshop*, pp. 5–36, 1995.
278. D. Pallett and J. Fiscus, 1996 Preliminary Broadcast News Benchmark Tests, *Proc. 1995 ARPA Human Language Technology Workshop*, pp. 5–36, 1997.
279. D. Pan, A Tutorial on MPEG/Audio Compression, *IEEE Multimedia,* pp. 60–74, Summer 1995.
280. P. E. Papamichalis, *Practical Approaches to Speech Coding*, Prentice-Hall Inc., 1984.
281. A. Papoulis, *The Fourier Integral and Its Applications*, McGraw-Hill, pp. 47–49, 1962.
282. T. W. Parks and J. H. McClellan, Chebyshev Approximation for Nonrecursive Digital Filter with Linear Phase, *IEEE Trans. on Circuit Theory*, Vol. CT-19, pp. 189–194, March 1972.
283. D. T. Paris and F. K. Hurd, *Basic Electromagnetic Theory*, McGraw-Hill Book Co., New York, 1969.
284. C. R. Patisaul and J. C. Hammett, Time-Frequency Resolution Experiment in Speech Analysis and Synthesis, *J. of Acoustical Society of America*, Vol. 58, No. 6, pp. 1296–1307, December 1975.
285. D. B. Paul and J. M. Baker, The Design for the Wall Street Journal-Based CSR Corpus, *Proc. of the DARPA SLS Workshop*, 1992.
286. J. S. Perkell, *Physiology of Speech Production: Results and Implications of a Quantitative Cineradiographic Study*, MIT Press, Cambridge, MA, 1969.
287. G. E. Peterson and H. L. Barney, Control Methods Used in a Study of the Vowels, *J. of Acoustical Society of America*, Vol. 24, No. 2, pp. 175–184, March 1952.
288. R. K. Potter, G. A. Kopp, and H. C. Green Kopp, *Visible Speech*, D. Van Nostrand Co., New York, 1947. (Republished by Dover Publications, Inc., 1966.)
289. M. R. Portnoff, A Quasi-One-Dimensional Digital Simulation for the Time-Varying Vocal Tract, *M.S. Thesis*, Dept. of Elect. Engr., MIT, Cambridge, MA, 1969.

290. M. R. Portnoff, Implementation of the Digital Phase Vocoder Using the Fast Fourier Transform, *IEEE Trans. on Acoustics, Speech and Signal Processing*, Vol. ASSP-24, No. 3, pp. 243–248, June 1976.
291. M. R. Portnoff and R. W. Schafer, Mathematical Considerations in Digital Simulations of the Vocal Tract, *J. of Acoustical Society of America*, Vol. 53, No. 1 (abstract), pp. 294, January 1973.
292. M. R. Portnoff, V. W. Zue, and A. V. Oppenheim, Some Considerations in the Use of Linear Prediction for Speech Analysis, *MIT QPR No. 106*, Research Lab of Electronics, MIT, Cambridge, MA, July 1972.
293. F. Poza, Voiceprint Identification: Its Forensic Application, *Proc. 1974 Carnahan Crime Countermeasures Conference*, April 1974.
294. Praat Speech Analysis Package, http://www.fon.hum.uva.nl/praat/.
295. W. H. Press, S. A. Teukolsky, W. T. Vetterling, and B. P. Flannery, *Numerical Recipes: The Art of Scientific Computing*, 3rd ed., Cambridge University Press, Cambridge, UK, 2007.
296. S. R. Quackenbush, T. P. Barnwell, III, and M. A. Clements, *Objective Measures of Speech Quality*, Prentice-Hall, New York, 1988.
297. T. F. Quatieri, *Principles of Discrete-Time Speech Processing*, Prentice-Hall Inc., 2002.
298. L. R. Rabiner, Digital Formant Synthesizer for Speech Synthesis Studies, *J. of Acoustical Society of America*, Vol. 43, No. 4, pp. 822–828, April 1968.
299. L. R. Rabiner, A Model for Synthesizing Speech by Rule, *IEEE Trans. on Audio and Electroacoustics*, Vol. AU-17, No. 1, pp. 7–13, March 1969.
300. L. R. Rabiner, On the Use of Autocorrelation Analysis for Pitch Detection, *IEEE Trans. on Acoustics, Speech and Signal Processing*, Vol. ASSP-25, No. 1, pp. 24–33, February 1977.
301. L. R. Rabiner, A Tutorial on Hidden Markov Models and Selected Applications in Speech Recognition, *Proceedings of the IEEE*, Vol. 77, No. 2, pp. 257–286, 1989.
302. L. R. Rabiner, B. S. Atal, and M. R. Sambur, LPC Prediction Error-Analysis of Its Variation with the Position of the Analysis Frame, *IEEE Trans. on Acoustics, Speech and Signal Processing*, Vol. ASSP-25, No. 5, pp. 434–442, October 1977.
303. L. R. Rabiner, M. J. Cheng, A. E. Rosenberg, and C. A. McGonegal, A Comparative Performance Study of Several Pitch Detection Algorithms, *IEEE Trans. on Acoustics, Speech and Signal Processing*, Vol. ASSP-24, No. 5, pp. 399–418, October 1976.
304. L. R. Rabiner and R. E. Crochiere, A Novel Implementation for FIR Digital Filters, *IEEE Trans. on Acoustics, Speech and Signal Processing*, Vol. ASSP-23, pp. 457–464, October 1975.
305. L. R. Rabiner and B. Gold, *Theory and Application of Digital Signal Processing* Prentice-Hall Inc., Englewood Cliffs, NJ, 1975.
306. L. R. Rabiner, B. Gold, and C. A. McGonegal, An Approach to the Approximation Problem for Nonrecursive Digital Filters, *IEEE Trans. on Audio and Electroacoustics*, Vol. 19, No. 3, pp. 200–207, September 1971.
307. L. R. Rabiner and B. H. Juang, An Introduction to Hidden Markov Models, *IEEE Signal Processing Magazine*, 1985.

308. L. R. Rabiner and B. H. Juang, *Fundamentals of Speech Recognition*, Prentice-Hall Inc., 1993.
309. L. R. Rabiner, J. F. Kaiser, O. Herrmann, and M. T. Dolan, Some Comparisons between FIR and IIR Digital Filters, *Bell System Technical J.*, Vol. 53, No. 2, pp. 305–331, February 1974.
310. L. R. Rabiner, J. H. McClellan, and T. W. Parks, FIR Digital Filter Design Techniques Using Weighted Chebyshev Approximation, *Proceedings of the IEEE*, Vol. 63, No. 4, pp. 595–609, April 1975.
311. L. R. Rabiner and M. R. Sambur, An Algorithm for Determining the Endpoints of Isolated Utterances, *Bell System Technical J.*, Vol. 54, No. 2, pp. 297–315, February 1975.
312. L. R. Rabiner, M. R. Sambur, and C. E. Schmidt, Applications of a Non-linear Smoothing Algorithm to Speech Processing, *IEEE Trans. on Acoustics, Speech and Signal Processing*, Vol. ASSP-22, No. 1, pp. 552–557, December 1975.
313. L. R. Rabiner and R. W. Schafer, *Digital Processing of Speech Signals*, Prentice-Hall Inc., 1978.
314. L. R. Rabiner, R. W. Schafer, and J. L. Flanagan, Computer Synthesis of Speech by Concatenation of Formant Coded Words, *Bell System Technical J.*, Vol. 50, No. 5, pp. 1541–1558, May-June 1971.
315. L. R. Rabiner, R. W. Schafer, and C. M. Rader, The Chirp z-Transform Algorithm and Its Application, *Bell System Technical J.*, Vol. 48, pp. 1249–1292, 1969.
316. C. M. Rader and N. Brenner, A New Principle for Fast Fourier Transformation, *IEEE Trans. Acoustics, Speech and Signal Processing*, Vol. ASSP-24, pp. 264–265, 1976.
317. T. Ramstad, Sub-Band Coder with a Simple Adaptive Bit Allocation Algorithm, *Proc. IEEE Int. Conf. on Acoustics, Speech and Signal Processing*, pp. 203–207, 1982.
318. W. S. Rhode, C. D. Geisler, and D. T. Kennedy, Auditory Nerve Fiber Responses to Wide-Band Noise and Tone Combinations, *J. Neurophysiology*, Vol. 41, pp. 692–704, 1978.
319. R. R. Riesz, Description and Demonstration of an Artificial Larynx, *J. of Acoustical Society of America*, Vol. 1, pp. 273–279, 1930.
320. A. W. Rix, J. G. Beerends, M. P. Hollier, and A. P. Hekstra, Perceptual Evaluation of Speech Quality (PESQ)—A New Method for Speech Quality Assessment of Telephone Networks and Codecs, *Proc. IEEE Int. Conf. on Acoustics, Speech and Signal Processing*, Vol. 2, pp. 749–752, May 2001.
321. D. W. Robinson and R. S. Dadson, A Re-determination of the Equal-Loudness Relations for Pure Tones, *British J. of Applied Physics*, Vol. 7, pp. 166–181, 1956.
322. E. A. Robinson, Predictive Decomposition of Time Series with Applications to Seismic Exploration, *Ph.D. Dissertation*, MIT, Cambridge, MA, 1954.
323. E. A. Robinson, Predictive Decomposition of Seismic Traces, *Geophysics*, Vol. 22, pp. 767–778, 1957.
324. P. Rose, *Forensic Speaker Identification*, Taylor & Francis, 2002.
325. R. Rose and T. Barnwell, The Self-Excited Vocoder-Alternative Approach to Toll Quality at 4800 Bigs/Second, *Proc. IEEE Int. Conf. on Acoustics, Speech and Signal Processing*, pp. 453–456, 1986.

326. A. E. Rosenberg, Effect of Glottal Pulse Shape on the Quality of Natural Vowels, *J. of Acoustical Society of America*, Vol. 49, No. 2, pp. 583–590, February 1971.
327. A. E. Rosenberg, Automatic Speaker Verification: A Review, *Proceedings of the IEEE*, Vol. 64, No. 4, pp. 475–487, April 1976.
328. A. E. Rosenberg and M. R. Sambur, New Techniques for Automatic Speaker Verification, *IEEE Trans. on Acoustics, Speech and Signal Processing*, Vol. ASSP-23, pp. 169–176, April 1975.
329. A. E. Rosenberg, R. W. Schafer, and L. R. Rabiner, Effects of Smoothing and Quantizing the Parameters of Formant-Coded Voiced Speech, *J. of Acoustical Society of America*, Vol. 50, No. 6, pp. 1532–1538, December 1971.
330. R. Rosenfeld, Two Decades of Statistical Language Modeling: Where Do We Go from Here?, *Proceedings of the IEEE*, Vol. 88, No. 8, pp. 1270–1278, 2000.
331. M. J. Ross, H. L. Shaffer, A. Cohen, R. Freudberg, and H. J. Manley, Average Magnitude Difference Function Pitch Extractor, *IEEE Trans. on Acoustics, Speech and Signal Processing*, Vol. ASSP-22, pp. 352–362, October 1974.
332. T. D. Rossing, R. F. Moore, and P. A. Wheeler, *The Science of Sound*, 3rd ed., Addison-Wesley, 2002.
333. J. H. Rothweiler, Polyphase Quadrature Filters—A New Subband Coding Technique, *Proc. IEEE Int. Conf. on Acoustics, Speech and Signal Processing*, pp. 1280–1283, 1983.
334. J. H. Rothweiler, System for Digital Multiband Filtering, U.S. Patent 4,691,292, issued September 1, 1987.
335. S. Roukos, Language Representation, *Survey of the State of the Art in Human Language Technology*, G. B. Varile and A. Zampolli (eds.), Cambridge University Press, Cambridge, UK, 1998.
336. M. B. Sachs, C. C. Blackburn, and E. D. Young, Rate-Place and Temporal-Place Representations of Vowels in the Auditory Nerve and Anteroventral Cochlear Nucleus, *J. of Phonetics*, Vol. 16, pp. 37–53, 1988.
337. M. B. Sachs and E. D. Young, Encoding of Steady State Vowels in the Auditory Nerve: Representation in Terms of Discharge Rates, *J. of Acoustical Society of America*, Vol. 66, pp. 470–479, 1979.
338. Y. Sagisaka, Speech Synthesis by Rule Using an Optimal Selection of Non-Uniform Synthesis Units, *Proc. IEEE Int. Conf. on Acoustics, Speech and Signal Processing*, pp. 679–682, 1988.
339. Y. Sagisaka, N. Campbell, and N. Higuchi, *Computing Prosody*, Springer, 1996.
340. M. R. Sambur, An Efficient Linear Prediction Vocoder, *Bell System Technical J.*, Vol. 54, No. 10, pp. 1693–1723, December 1975.
341. M. R. Sambur and L. R. Rabiner, A Speaker Independent Digit Recognition System, *Bell System Technical J.*, Vol. 54, No. 1, pp. 81–102, January 1975.
342. R. W. Schafer, Echo Removal by Discrete Generalized Linear Filtering, *Technical Report No. 466*, Research Lab of Electronics, MIT, Cambridge, MA, February 1969.
343. R. W. Schafer and J. D. Markel (eds.), *Speech Analysis*, IEEE Press Selected Reprint Series, 1979.
344. R. W. Schafer and L. R. Rabiner, System for Automatic Formant Analysis of Voiced Speech, *J. of Acoustical Society of America*, Vol. 47, No. 2, pp. 634–648, February 1970.

345. R. W. Schafer and L. R. Rabiner, Design of Digital Filter Banks for Speech Analysis, *Bell System Technical J.*, Vol. 50, No. 10, pp. 3097–3115, December 1971.
346. R. W. Schafer and L. R. Rabiner, Design and Simulation of a Speech Analysis-Synthesis System Based on Short-Time Fourier Analysis, *IEEE Trans. on Audio and Electroacoustics*, Vol. AU-21, No. 3, pp. 165–174, June 1973.
347. R. W. Schafer and L. R. Rabiner, A Digital Signal Processing Approach to Interpolation, *Proceedings of the IEEE*, Vol. 61, No. 6, pp. 692–702, June 1973.
348. H. R. Schindler, Delta Modulation, *IEEE Spectrum*, Vol. 7, pp. 69–78, October 1970.
349. J. S. Schouten, F. E. DeJager, and J. A. Greefkes, Delta Modulation, A New Modulation System for Telecommunications, *Philips Tech. Report*, pp. 237–245, March 1952.
350. R. Schreier and G. C. Temes, *Understanding Delta-Sigma Data Converters*, IEEE Press and Wiley-Interscience, Piscataway, NJ, 2005.
351. M. R. Schroeder, Vocoders: Analysis and Synthesis of Speech, *Proceedings of the IEEE*, Vol. 54, pp. 720–734, May 1966.
352. M. R. Schroeder, Period Histogram and Product Spectrum: New Methods for Fundamental Frequency Measurement, *J. of Acoustical Society of America*, Vol. 43, No. 4, pp. 829–834, April 1968.
353. M. R. Schroeder and B. S. Atal, Code-Excited Linear Prediction (CELP), *Proc. IEEE Int. Conf. on Acoustics, Speech and Signal Processing*, pp. 937–940, 1985.
354. M. R. Schroeder and E. E. David, A Vocoder for Transmitting 10 kc/s Speech Over a 3.5 kc/s Channel, *Acoustica*, Vol. 10, pp. 35–43, 1960.
355. J. Schroeter, The Fundamentals of Text-to-Speech Synthesis, *VoiceXML Review*, 2001.
356. J. Schroeter, Basic Principles of Speech Synthesis, *Springer Handbook of Speech Processing*, Springer, 2006.
357. J. Schroeter, J. N. Larar, and M. M. Sondhi, Speech Parameter Estimation Using a Vocal Tract/Cord Model, *Proc. IEEE Int. Conf. on Acoustics, Speech and Signal Processing*, pp. 308–311, 1987.
358. J. Schroeter, J. Ostermann, H. P. Graf, M. Beutnagel, E. Cosatto, A. Syrdal, A. Conkie, and Y. Stylianou, Multimodal Speech Synthesis, *Proc. ICSLP-98*, Sydney, Australia, pp. 571–574, November 1998.
359. N. Sedgwick, A Formant Vocoder at 600 Bits Per Second, *IEEE Colloquium on Speech Coding—Techniques and Applications*, pp. 411–416, April 1992.
360. S. Seneff, A Joint Synchrony/Mean-Rate Model of Auditory Speech Processing, *J. of Phonetics*, Vol. 16, pp. 55–76, 1988.
361. K. Silverman, M. Beckman, J. Pitrelli, M. Ostendorf, C. Wightman, P. Price, J. Pierrehumbert, and J. Hirschberg, ToBI: A Standard for Labeling English Prosody, *Proc. ICSLP 1992*, pp. 867–870, Banff, 1992.
362. G.A. Sitton, C. S. Burrus, J. W. Fox, and S. Treitel, Factoring Very-High-Degree Polynomials, *IEEE Signal Processing Magazine*, Vol. 20, No. 6, pp. 27–42, November 2003.
363. M. Slaney, Auditory Toolbox, Ver. 2.0, available at www.auditory.org/postings/1999/21.html, 1999.

364. C. E. Shannon, A Mathematical Theory of Communication, *Bell System Technical J.*, Vol. 27, pp. 623–656, October 1948.

365. H. F. Silverman and N. R. Dixon, A Parametrically Controlled Spectral Analysis System for Speech, *IEEE Trans. on Acoustics, Speech and Signal Processing*, Vol. ASSP-22, No. 5, pp. 362–381, October 1974.

366. B. Smith, Instantaneous Companding of Quantized Signals, *Bell System Technical J.*, Vol. 36, No. 3, pp. 653–709, May 1957.

367. M. J. T. Smith and T. P. Barnwell, III, A Procedure for Designing Exact Reconstruction Filter Banks for Tree Structured Subband Coders, *Proc. IEEE Int. Conf. on Acoustics, Speech and Signal Processing*, pp. 27.1.1–27.1.4, March 1984.

368. M. J. T. Smith and T. P. Barnwell, III, Exact Reconstruction Techniques for Tree Structured Subband Coders, *Proc. IEEE Int. Conf. on Acoustics, Speech and Signal Processing*, Vol. ASSP-34, No. 3, pp. 434–441, June 1986.

369. P. D. Smith, M. Kucic, R. Ellis, P. Hasler, and D. V. Anderson, Mel-Frequency Cepstrum Encoding in Analog Floating-Gate Circuitry, *Proc. Int. Symp. On Circuits and Systems*, Vol. 4, pp. 671–674, May 2002.

370. M. M. Sondhi, New Methods of Pitch Extraction, *IEEE Trans. on Audio and Electroacoustics*, Vol. AU-16, No. 2, pp. 262–266, June 1968.

371. M. M. Sondhi, Determination of Vocal-Tract Shape from Impulse Response at the Lips, *J. of Acoustical Society of America*, Vol. 55, No. 5, pp. 1070–1075, May 1974.

372. M. M. Sondhi and B. Gopinath, Determination of Vocal-Tract Shape from Impulse Response at the Lips, *J. of Acoustical Society of America*, Vol. 49, No. 6, Part 2, pp. 1847–1873, June 1971.

373. F. K. Soong and B. H. Juang, Line Spectrum Pair and Speech Compression, *Proc. IEEE Int. Conf. on Acoustics, Speech and Signal Processing*, Vol. 1, pp. 1.10.1–1.10.4, 1984.

374. A. Spanias, T. Painter, and V. Atti, *Audio Signal Processing and Coding*, John Wiley & Sons, Inc., 2006.

375. R. Steele, *Delta Modulation Systems*, Halsted Press, London, 1975.

376. K. N. Stevens, *Acoustic Phonetics*, MIT Press, Cambridge, MA, 1998.

377. K. N. Stevens, The Perception of Sounds Shaped by Resonant Circuits, *Sc.D. Thesis*, MIT, Cambridge, MA, 1952.

378. S. S. Stevens, J. Volkmann, and E. B. Newman, A Scale for the Measurement of the Psychological Magnitude Pitch, *J. of Acoustical Society of America*. Vol. 8, pp. 1185–1190, 1937.

379. K. Steiglitz and B. Dickinson, Computation of the Complex Cepstrum by Factorization of the z-Transform, *Proc. IEEE Int. Conf. on Acoustics, Speech and Signal Processing*, pp. 723–726, May 1977.

380. T. G. Stockham, Jr., High-Speed Convolution and Correlation, *1966 Spring Joint Computer Conference*, AFIPS Proc., Vol. 28, pp. 229–233, 1966.

381. T. G. Stockham, Jr., T. M. Cannon, and R. B. Ingebretsen, Blind Deconvolution Through Digital Signal Processing, *Proceedings of the IEEE*, Vol. 63, pp. 678–692, April 1975.

382. P. Stoica and R. Moses, *Spectral Analysis of Signals*, Prentice-Hall Inc., Englewood Cliffs, 1997.

383. R. W. Stroh, Optimum and Adaptive Differential PCM, *Ph.D. Dissertation*, Polytechnic Inst. of Brooklyn, Farmingdale, NY, 1970.
384. H. Strube, Determination of the Instant of Glottal Closure from the Speech Wave, *J. of Acoustical Society of America*, Vol. 56, No. 5, pp. 1625–1629, November 1974.
385. N. Sugamura and F. Itakura, Speech Data Compression by LSP Analysis-Synthesis Technique, *Transactions of the Institute of Electronics, Information, and Computer Engineers*, Vol. J64-A, pp. 599–606, 1981.
386. Y. Suzuki, V. Mellert, U. Richter, H. Moller, L. Nielsen, R. Hellman, K. Ashihara, K. Ozawa, and H. Takeshima, Precise and Full-Range Determination of Two-Dimensional Equal Loudness Contours, ISO Document, 1993.
387. A. K. Syrdal, J. Hirschberg, J. McGory, and M. Beckman, Automatic ToBI Prediction and Alignment to Speed Manual Labeling of Prosody, *Speech Communication*, Vol. 33, pp. 135–151, January 2001.
388. P. Taylor, *Text-to-Speech Synthesis*, Cambridge University Press, Cambridge, UK, 2009.
389. E. Terhardt, Calculating Virtual Pitch, *Hearing Research*, Vol. 1, pp. 155–182, 1979.
390. S. Theodoridis and K. Koutroumbas, *Pattern Recognition*, 2nd ed., Chapter 2, Elsevier Academic Press, 2003.
391. Y. Tohkura, A Weighted Cepstral Distance Measure for Speech Recognition, *IEEE Trans. on Acoustics, Speech and Signal Processing*, Vol. ASSP-35, No. 10, pp. 1414–1422, October 1987.
392. J. M. Tribolet, A New Phase Unwrapping Algorithm, *IEEE Trans. on Acoustics, Speech and Signal Processing*, Vol. ASSP-25, No. 2, pp. 170–177, April 1977.
393. J. M. Tribolet and R. E Crochiere, Frequency Domain Coding of Speech, *IEEE Trans. on Acoustics, Speech and Signal Processing*, Vol. ASSP-27, No. 5, pp. 512–530, October 1979.
394. J. M. Tribolet, P. Noll, B. J. McDermott, and R. E. Crochiere, A Study of Complexity and Quality of Speech Waveform Coders, *Proc. IEEE Int. Conf. on Acoustics, Speech and Signal Processing*, pp. 1586–1590, April 1978.
395. J. M. Tribolet, P. Noll, B. J. McDermott, and R. E. Crochiere, A Comparison of the Performance of Four Low Bit Rate Speech Waveform Coders, *Bell System Technical J.*, Vol. 58, pp. 699–712, March 1979.
396. J. W. Tukey, Nonlinear (Non Superpossible) Methods for Smoothing Data, *Congress Record*, 1974 EASCON, p. 673, 1974.
397. F. T. Ulaby, "Fundamentals of Applied Electromagnetics," 5th edition, Prentice-Hall, Inc., Upper Saddle River, NJ, 2007.
398. C. K. Un and D T. Magill, The Residual-Excited Linear Prediction Vocoder with Transmission Rate Below 9.6 kbits/s, *IEEE Trans. on Communications*, Vol. COM-23, No. 12, pp. 1466–1474, December 1975.
399. P. P. Vaidyanathan, *Multirate Systems and Filter Banks*, Prentice-Hall Inc., 1993.
400. J. VanSanten, R. W. Sproat, J. P. Olive, and J. Hirschberg (eds.), *Progress in Speech Synthesis*, Springer, 1996.
401. P. Vary and R. Martin, *Digital Speech Transmission, Enhancement, Coding and Error Concealment*, John Wiley & Sons, Inc., 2006.
402. M. Vetterli and J. Kovacevic, *Wavelets and Subband Coding*, Prentice-Hall Inc., 1995.

403. P. J. Vicens, Aspects of Speech Recognition by Computer, *Ph.D. Thesis*, Stanford Univ., AI Memo No. 85, Comp. Sci. Dept., 1969.
404. R. Viswanathan and J. Makhoul, Quantization Properties of Transmission Parameters in Linear Predictive Systems, *IEEE Trans. on Acoustics, Speech and Signal Processing*, Vol. ASSP-23, No. 3, pp. 309–321, June 1975.
405. R. Viswanathan, W. Russell, and J. Makhoul, Voice-Excited LPC Coders for 9.6 kbps Speech Transmission, *Proc. IEEE Int. Conf. on Acoustics, Speech and Signal Processing*, Vol. 4, pp. 558–561, April 1979.
406. A. J. Viterbi, Error Bounds for Convolutional Codes and an Asymptotically Optimal Decoding Algorithm, *IEEE Trans. on Information Theory*, Vol. IT-13, pp. 260–269, April, 1967.
407. Voicebox, http://www.ee.ic.ac.uk/hp/staff/dmb/voicebox/voicebox.html.
408. W. A. Voiers, W. A. Sharpley, and C. Hehmsoth, Research on Diagnostic Evaluation of Speech Intelligibility, *Air Force Cambridge Research Labs*, MIT, Cambridge, MA, 1975.
409. W. D. Voiers, Diagnostic Acceptability Measure for Speech Communication Systems, *Proc. IEEE Int. Conf. on Acoustics, Speech and Signal Processing*, pp. 204–207, 1977.
410. H. Wakita, Direct Estimation of the Vocal Tract Shape by Inverse Filtering of Acoustic Speech Waveforms, *IEEE Trans. on Audio and Electroacoustics*, Vol. AU-21, No. 5, pp. 417–427, October 1973.
411. H. Wakita, Estimation of Vocal-Tract Shapes From Acoustical Analysis of the Speech Wave: The State of the Art, *IEEE Trans. on Acoustics, Speech and Signal Processing*, Vol. 27, No. 3, pp. 281–285, June 1979.
412. W. Ward, Evaluation of the CMU ATIS System, *Proc. DARPA Speech and Natural Language Workshop*, pp. 101–105, February 1991.
413. http://www.speech.kth.se/wavesurfer/, *WaveSurfer*, Speech, Music and Hearing Dept., KTH University, Stockholm, Sweden, 2005.
414. C. J. Weinstein, A Linear Predictive Vocoder with Voice Excitation, *Proc. Eascon*, September 1975.
415. C. J. Weinstein and A. V. Oppenheim, Predictive Coding in a Homomorphic Vocoder, *IEEE Trans. on Audio and Electroacoustics*, Vol. AU-19, No. 3, pp. 243–248, September 1971.
416. P. D. Welch, The Use of the Fast Fourier Transform for the Estimation of Power Spectra, *IEEE Trans. on Audio and Electroacoustics*, Vol. AU-15, pp. 70–73, June 1970.
417. N. Wiener, *Extrapolation, Interpolation, and Soothing of Stationary Time Series*, MIT Press, Cambridge, MA, 1942; and John Wiley & Sons, Inc., NY, 1949.
418. J. G. Wilpon, L. R. Rabiner, C. H. Lee, and E. Goldman, Automatic Recognition of Keywords in Unconstrained Speech Using Hidden Markov Models, *IEEE Trans. on Acoustics, Speech and Signal Processing*, Vol. 38, No. 11, pp. 1870–1878, 1990.
419. B. Widrow, A Study of Rough Amplitude Quantization by Means of Nyquist Sampling Theory, *IRE Trans. on Circuit Theory*, Vol. 3, No. 4, pp. 266–276, December 1956.
420. B. Widrow and I. Kollár, *Quantization Noise*, Cambridge University Press, Cambridge, UK, 2008.

421. G. Winham and K. Steiglitz, Input Generators for Digital Sound Synthesis, *J. of Acoustical Society of America*, Vol. 47, No. 2, pp. 665–666, February 1970.
422. D. Y. Wong, B. H. Juang, and A. H. Gray, Jr., An 800 Bit/s Vector Quantization LPC Vocoder, *IEEE Trans. on Acoustics, Speech and Signal Processing*, Vol. ASSP-30, pp. 770–780, October 1982.
423. P. F. Yang and Y. Stylianou, Real Time Voice Alteration Based on Linear Prediction, *Proc. ICSLP-98*, Sydney, Australia, November 1998.
424. D. Yarowsky, Homograph Disambiguation in Text-to-Speech Synthesis, Chapter 12 in *Progress in Speech Synthesis*, J. P. Van Santen, R. W. Sproat, J. P. Olive, and J. Hirschberg (eds.), pp. 157–172, Springer, NY, 1996.
425. R. Zelinsky and P. Noll, Adaptive Transform Coding of Speech Signals, *IEEE Trans. on Acoustics, Speech and Signal Processing*, Vol. ASSP-25, pp. 299–309, August 1977.
426. V. Zue, Speech Analysis by Linear Prediction, *MIT QPR No. 105*, Research Lab of Electronics, MIT, Cambridge, MA, April 1972.
427. V. Zue and J. Glass, MIT OCW Course Notes, http://ocw.mit.edu/OcwWeb/Electrical-Engineering-and-Computer-Science/index.htm, 2004.
428. E. Zwicker and H. Fastl, *Psychoacoustics*, 2nd ed., Springer, Berlin, Germany, 1999.

Index

A

I

J

K

L

M

Q

R

S

W

Z

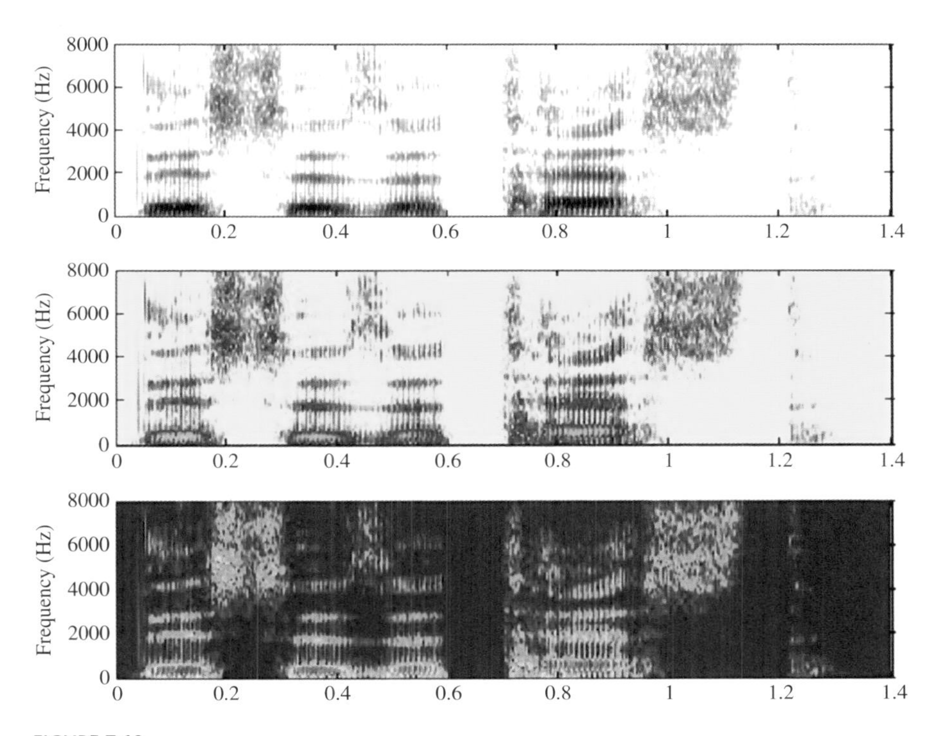

FIGURE 7.19
Comparisons of gray scale wideband spectrogram (top panel) with color wideband spectrograms (middle and bottom panels) of the utterance "This is a test."

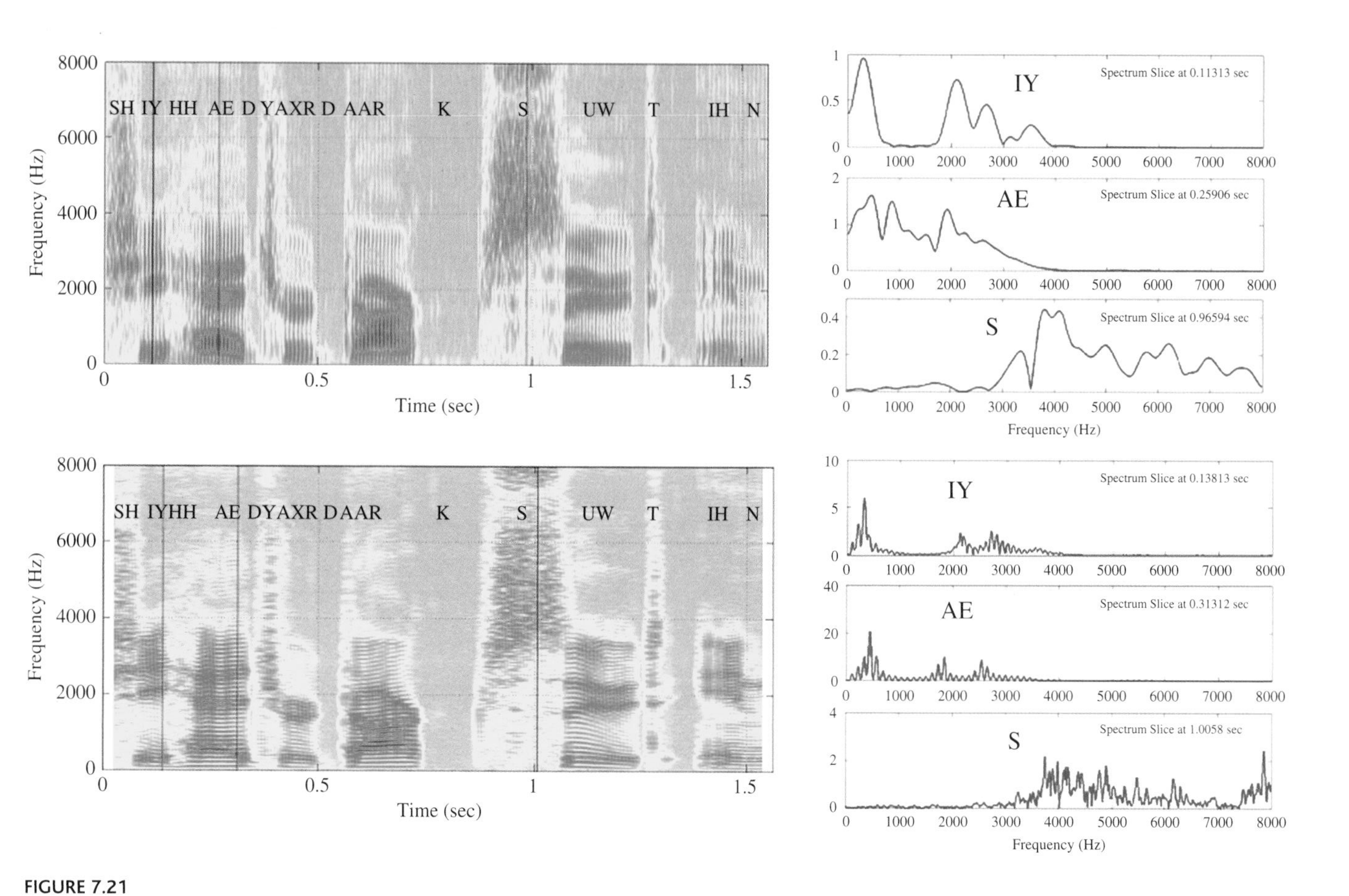

FIGURE 7.21
Spectrograms of a male speaker's utterance of the text "She had your dark suit in." The left column shows wideband and narrowband spectrograms using respectively $L = 80$ sample (5 msec) and $L = 800$ (50 msec) analysis windows, with an 1024-point FFT, and $R = 5$ and $R = 10$ sample shifts between adjacent short-time spectra. The right panel shows wideband and narrowband slices at the three time slots indicated by heavy lines in the sound spectrograms, corresponding to the sounds /IY/, /AE/, and /S/.